AF248652

COSMIC WINDS
AND
THE HELIOSPHERE

COSMIC WINDS
AND
THE HELIOSPHERE

J. R. Jokipii
C. P. Sonett
M. S. Giampapa

Editors

With the editorial assistance of
M. S. Matthews, A. S. Ruskin,
and M. L. Guerrieri

With 52 collaborating authors

THE UNIVERSITY OF ARIZONA PRESS
TUCSON

About the cover:

The front cover is a composite of an optical image of the starburst galaxy NGC 253 and the X-ray emission in the energy range 0.1–0.4 keV, as detected by the ROSAT PSPC instrument (white contour). This extended X-ray halo, signifying hot gas outflow to beyond 10 kpc, is probably not related to the current nuclear starburst—it is most likely the relict signature of an earlier event, which occurred about 10^8 years ago. The total mass ejection is estimated at five million solar masses. Image courtesy of the Max-Planck-Institut für extraterrestrische Physik.

About the back cover:

The back cover is a composite image of the solar corona as captured in white light during a solar eclipse on 30 June 1973, with an H α picture obtained simultaneously. The coronal picture has been processed in order to enhance small-scale intensity gradients. See the Chapter by Esser and Habbal. Figure courtesy of S. Koutchmy.

The University of Arizona Press

© 1997
The Arizona Board of Regents
First printing
All rights seserved
♾ This book is printed on acid-free, archival-quality paper.
Manufactured in the United States of America.
02   01   00   99   98   97   6   5   4   3   2   1
Library of Congress Cataloging-in-Publication Data
Cosmic winds and the heliosphere / J.R. Jokipii, C.P. Sonett, M.S.
    Giampapa, editors ; with the editorial assistance of M.S. Matthews,
    A.S. Ruskin, and M.L. Guerrieri.
        p.   cm. — (Space science series)
    Includes bibliographical references and index.
    ISBN 0-8165-1825-4 (cloth)
    1. Solar wind.   2. Stellar winds.   3. Heliosphere.   I. Jokipii,
J. R. (Jack Randolph), 1939– .   II. Sonett, Charles P.
III. Giampapa, M. S. (Mark S.)   IV. Series.
    QB529.9.C687   1997
    523.5′8—dc21   97-27368
                                                           CIP

British Library Cataloguing-in-Publication Data
A catalogue record for this book is available from the British Library.

CONTENTS

CONTENTS

PART III—THE PHYSICS OF WIND ORIGIN

PART IV—PHYSICAL PHENOMENA IN WINDS

PART V—INTERACTIONS WITH SURROUNDING MEDIA

COLLABORATING AUTHORS

PREFACE

It has been nearly half a century since Biermann first recognized the role of solar plasma in defining the structure of comet tails and argued for the continuous emission of particles from the Sun. Shortly after this, Parker showed that such continuous outflow could be expected theoretically and named the phenomenon the "solar wind." From controversial beginnings, interest in winds has grown exponentially and, as this volume shows, has expanded to become a theme subject in astrophysics. We now know that winds play a major role in a variety of contexts. As in other books in this series, this volume is intended to reflect the present status of astrophysical winds and to provide the advanced student and research worker an up to date summary of key aspects of the subject.

J. R. Jokipii

C. P. Sonett

M. S. Giampapa

PART I
Introduction

MASS EJECTION AND A BRIEF HISTORY OF THE SOLAR WIND CONCEPT

E. N. PARKER
University of Chicago

The universal ejection of gas from gravitating bodies arises from a variety of effects such as explosive heating, rapid spinning of strongly magnetized objects, radiation pressure, and the continuous thermal expansion of the outer portion of a strongly bound quasi-static atmosphere. The solar wind is an example of the last. The development of ideas that led to the concept of the expansion of the solar corona is outlined. An essential step was to recognize the hydrodynamic nature of the thermal expansion, getting away from the older ideas of individual particle acceleration in electromagnetic fields. Once the concept of the solar wind was formulated, the interplanetary magnetic field, the heliosphere, and the gross features of cosmic ray modulation followed directly. One inferred that most stars have similar winds and astrospheres.

I. UNIVERSAL MASS EJECTION

Direct observation and elementary inference combine to suggest that almost all stars and many, if not all, galaxies eject matter into space. The causes of the ejection are diverse, ranging from explosion by sudden production of heat, to the centrifugal ejection from rapidly spinning objects, to the gradual acceleration and departure of the distant portions of a strongly bound thermal atmosphere. The most extreme case of ejection is the high-speed jet of a radio galaxy, representing an energy of the order of 10^{61} erg, equivalent to the annihilation energy of a mass of 5×10^6 $M_\odot$. Such prodiguous energies are available only in self-gravitating objects of 10^8 $M_\odot$ (or more) with radii of the order of the Schwarzschild radius of 2 AU. A spinning black hole with a strong magnetic field is a likely candidate, presumably accreting matter from the space around it (see reviews by Bridle and Perley 1984; Begelman et al. 1984). The idea of centrifugal expulsion as a consequence of relativistic rotation velocities ($r\omega \sim c$) may be considered the extreme relativistic extension of the original scenario of Weber and Davis (1967) on the loss of angular momentum from the Sun through the interaction of the solar wind with the extended magnetic field of the Sun. Novae and supernovae are next in intensity. They suddenly heat a small and large portion, respectively, of their matter to thermal velocities in excess of the gravitational escape velocity, so that the heated matter bursts outward into space (Baade and Zwicky 1934). The matter ejected from supernovae plays an important role in supplying heavy elements to the interstellar medium, from which new stars are formed.

The nonexplosive ejection of matter from a galaxy is referred to as a galactic wind, arising from the upward fountains of hot gas from supernovae and from O and B associations to form a galactic halo of hot gas (10^6–10^7 K) (Spitzer 1956*b*; Lockman 1984; Kulkarni and Heiles 1987; Dickey and Lockman 1990; Parker 1990,1992; Danly et al. 1992; Merrifield 1993; McKee 1993). The galactic wind arises from the thermal expansion of the halo, presumably boosted along by the powerful cosmic ray gas escaping from the disk and halo and perhaps by a broad jet or spray from an active galactic nucleus (Ipavich 1975; Breitschwerdt et al. 1987,1991,1993).

The nonexplosive ejection of matter from stars is universal and involves a variety of contributing effects. Newly formed stellar objects provide a combination of hot and cold outflows, axial jets, and energetic winds presumed to be of thermal origin (Lada 1985). The Wolf Rayet stars present a particularly vigorous and complex display of rotation, mass motions, and mass ejection, evidently combining thermal, magnetic, and rapid rotational effects in their expulsion of matter (Hartmann and MacGregor 1982; Abbott and Conti 1987; Poe et al. 1989; Cassinelli et al. 1989; Lamers et al. 1991). The basic limitation in developing an understanding of such complicated mass motions is the inability to resolve the star in the telescope. One works with the integrated light and infers what hydrodynamic and magnetohydrodynamic effects may account for the complicated time dependent line structure in the spectrum. The O and B stars eject matter through radiation pressure. With such high surface temperatures, and peak intensities well out in the ultraviolet, the strongest resonant absorption cross sections are in such ions as C_{III}, C_{IV}, Ca_{III}, Ca_{IV}, etc., which are among the dominant ions at the surface. Deeper in the star the absorption of radiation is less efficient. In some cases the ultraviolet continuum alone is sufficiently strong as to levitate these ions along with the associated H and He. But it also turns out in some cases that the photospheric emission lines of these ions are of substantially greater intensity, thereby increasing the levitation of each resonant absorption cross section (Lucy and Solomon 1970; Castor et al. 1976; Conti 1978; Cassinelli 1979). Hence the ejection velocity tends to be limited by the width of the photospheric emission line. For as the ions are accelerated upward, their absorption is Doppler shifted toward the blue, taking them off the photospheric emission line as their velocity increases and substantially reducing their levitation as a result.

The radiative ejection of matter from hot stars can be violently unstable, as a consequence of shadowing. That is to say, a local density increase casts a shadow upward so that the gas above is not so strongly lifted, with the result that the gas above is soon swept up by the accelerated density increase, thereby increasing the density maximum. The result can be both upward moving tenuous matter and downward drifting dense matter providing a complicated structure to the general outflow (Brown et al. 1991).

One infers that most cool stars, such as the Sun and later main-sequence stars, as well as many red giants, produce an outflow of matter, presumably of thermal origin (Cassinelli 1979; Dupree 1986). To begin, Adams and Mac-Cormack (1935) and Deutsch (1956,1961) noted the blue shifted absorption lines in the spectra of such red giant stars as α^1 Herculis, which fortunately is an eclipsing binary. The outflow is slow (10 km s^{-1}) and massive (3 $\times$ 10^{-8} M$_\odot$ yr^{-1}), which may be presumed to reduce the mass of the star substantially during its sojourn as a red giant. Then many close binary star systems show evidence of continuing mass loss (Burbidge and Burbidge 1958), probably through the combination of enormous tidal effects that produce extended coronas and the centrifugal expulsion of the orbiting matter. But again one cannot see directly what is happening. It is possible only to construct hypothetical, largely chaotic scenarios and attempt to evaluate their validity by comparison of the theoretical radiative emission with the observed spectrum.

Finally, we come to the tenuous high-speed winds from middle-main-sequence stars like the Sun. Such winds are invisible to the remote observer and their existence is inferred from the fact that the very ordinary star, the Sun, possesses such a wind—the solar wind. For it must be appreciated that even the solar wind provides no detectable electromagnetic emission. The interaction of the solar wind with the planetary magnetospheres and with comets provides a number of detectable effects for the nearby observers, of course, and it was through these indirect effects that the existence of the solar wind was first inferred. The solar wind is simply the outward extension of the quasi-static corona of the Sun. The corona at 1 to 2 $\times$ 10^{6}K, is strongly bound by the gravitational field, with gravitational binding energies of the order of ten times the mean thermal energy 3 kT per ionized hydrogen atom. The solar wind is not an evaporation from an exosphere, as has been sometimes imagined, because the gas presumably begins with a nearly Maxwellian velocity distribution in the low corona and is bound together by collisions, plasma instabilities and the magnetic field into a single fluid. The gas gradually accelerates outward from a few km s^{-1} in the low corona to 10^2 km s^{-1} at $r = 2$ R$_\odot$, to several hundred km s^{-1} at infinity. The corona cannot be static because of its extended temperature, falling off less rapidly than $1/r$. As a result the thermal energy per particle becomes equal to the gravitational energy at some distance $\sim$10 R$_\odot$, and exceeds the gravitational energy beyond. Thus the outer corona expands away into space and the dense near corona, strongly bound and in hydrostatic equilibrium, expands slowly upward to make up for the mass loss of the escaping outer corona. The result is a slow continuous monotonic outward increase of wind velocity. The solar wind exhibits a mass flux of the order of 2 $\times$ 10^8 atoms cm^{-2}s^{-1} (typically 4 atoms cm^{-3} at 400 km s^{-1}), at the orbit of Earth. There is no obvious reason to think that the mass flux is greatly different at high heliocentric latitudes, with the result that the total mass loss is 10^{12} g s^{-1}, amounting to 1.5 $\times$ 10^{-14} M$_\odot$ per year, or 0.7 $\times$ 10^{-4} M$_\odot$ (at the present rate) in the 4.5 $\times$ 10^9 yr age of the Sun.

One infers that the corona of the Sun is a phenomenon common to all middle- and late-main-sequence stars, and therefore all middle-main-sequence stars have fast tenuous stellar winds more or less along the lines of the Sun. Indeed, the quasi-steady thermal expansion of stellar coronas of one form or another is probably a contributing factor in the mass ejection from many other types of stars as well.

It should be emphasized that the heat source that causes the formation of the expanding corona of the Sun has still not been identified, although it has been recognized as a central question in the physics of stars for fifty years.

It has been suggested that the rapid loss of angular momentum from ordinary main sequence stars during their first 10^9 yr is a consequence of strong magnetic fields presumably arising from strong dynamo action, providing strong coronal heating and a massive stellar wind. The strong magnetic fields cause the wind to rotate rigidly out to a substantial distance, of the order of 30 stellar radii or more, so that the loss of a small fraction of the mass may greatly reduce the stellar rotation rate (cf., Weber and Davis 1967; Brandt and Heise 1969a,b; Mestel and Spruit 1987).

II. SOLAR CORPUSCULAR RADIATION

The idea that the Sun emits high-speed particles ($\sim 10^3$ km s^{-1}) goes back over a hundred years to thoughts about the origin of geomagnetic fluctuations and the aurora. There was a growing recognition in the second half of the 19th century that aurorae and geomagnetic activity are related, by virtue of close association in time (Chapman and Bartels 1940). It was also noticed that geomagnetic activity and aurorae were enhanced during the times of sunspot activity on the Sun. The similarity in the appearance of auroral streamers and the cathode ray streamers in an electrical discharge in partial vacuum led Fitzgerald (1892,1900) and Lodge (1900) to suggest that the aurora is the result of charged particles emitted by the Sun and impinging upon Earth. They envisioned Earth as the cathode with the emission caused by a substantial electrostatic potential difference between the Earth and the Sun.

It is curious that Kelvin (1892) pronounced any connection of the geomagnetic fluctuations with the Sun as absurd. He started with the prevailing concept of space being a nearly perfect vacuum. Hence he was free to imagine that the geomagnetic fluctuations could be a consequence only of the fluctuations of an all-pervading solar dipole magnetic field. Thus, a variation ΔB in the geomagnetic field represented a change ΔB in the field of the Sun at the position of Earth. This indicated, at a minimum, switching on or off a field of the order of $B_\odot = (r_E/R_\odot)^3 \Delta B$ at the Sun, where r_E is the radius of the orbit of Earth ($r_E/R_\odot \cong 200$). A variation of $\Delta B \sim 10^{-3}$ Gauss at Earth requires in this way that $B_\odot = 10^4$ Gauss. But the energy $\mathcal{E}$ of such a dipole field at the Sun is of the order of $(1/6)B_\odot{}^2 R_\odot{}^3$ or 5×10^{39} erg. A fluctuation ΔB in the magnetic field of Earth, for instance at the onset of a geomagnetic storm, may occur in a time $\tau = 10^4$s or less, indicating, therefore a power

$\mathcal{E}/\tau \sim 5 \times 10^{35}$ erg s^{-1}, which is 100 times the luminosity of the Sun. Yet the luminosity of the Sun was observed to remain steady at such times. Kelvin declared, therefore, that there could be no connection of geomagnetic activity, and by inference, auroral activity, with the Sun. Any apparent correlations were declared to be pure chance coincidence, however frequently repeated.

The fact is that the contemporary British scientific community proceeded to ignore the facts pointed out by Fitzgerald and Lodge, so that the scientific initiative survived only in Norway where the brilliant researches of Birkeland (1896,1908,1913) showed clearly the likely direct connection of the aurora with solar corpuscular emission. Dessler (1967) provides a brief history of Birkeland's pioneering work. Birkeland constructed a small model Earth in the form of a magnetized sphere placed in a vacuum, which he bombarded with charged particles by introducing an electrostatic potential between the sphere and a hot filament—the terrella experiment. The experiment showed the visible cathode filaments to extend into the sphere along the field lines at high latitudes, in direct analogy to the terrestrial aurora. He determined the electric current patterns around the sphere. He then conducted a geomagnetic data analysis program, using 25 worldwide observing stations, constituting the largest geomagnetic survey up to that time. He coordinated the geomagnetic data with organized auroral observations and pieced together the general picture of an equatorial ring current, together with field aligned current loops completed in the ionosphere, all caused by external magnetic changes. He recognized the implication of long-lived regions on the Sun, providing the 27-day recurring activity through continuing emission of high-speed charged particles. He suggested that comet tails are a consequence of the sputtering of carbonaceous material from the comets. Finally, he noted that the continuing auroral and geomagnetic activity implied continuing solar corpuscular radiation. That is to say, the corpuscular radiation is not an occasional affair, made up of isolated beams of fast particles separated by vacuum, but rather the corpuscular emission pretty well fills interplanetary space. Birkeland's general picture stands up well today, recognizing that it could treat only the gross features of geomagnetic activity and the associated aurora. Oddly, his contributions were largely ignored, and it was not until about 1950 that the scientific study of geomagnetic activity caught up with his work of 1908 and 1913.

Progress in the field of geomagnetic activity in England revived with the generation succeeding Kelvin. Chapman (1918,1919) picked up on solar corpuscular radiation and imagined how it might produce the geomagnetic storm, impacting the geomagnetic field and perhaps penetrating to produce a ring current. Little or no appreciation of Birkeland's work is evident. Subsequent work (Chapman and Ferraro 1931,1932,1933,1940) provided refinement of the ideas, and again made it clear that geomagnetic activity is a consequence of variable solar corpuscular radiation. Magnetic fluctuations at the surface of Earth were expressed in terms of equivalent ionospheric and magnetospheric currents through Ampere's law $4\pi \mathbf{j} = c\nabla \times \mathbf{B}$, and an extensive cumbersome nomenclature was developed. There seemed to be the general view that an equivalent current is an explanation, when in fact the prime mover is the force produced by corpuscular radiation, causing the field fluctuations and thereby inducing currents. It was known that there is at least some minimal magnetic activity and auroral display at high latitudes even on the least active days and nights. But no one seems to have stated the obvious inference that the solar corpuscular radiation, whatever its nature, is always present, therefore filling all of interplanetary space around the Sun.

The big step forward came with Biermann's work on gaseous comet tails. It is an observed fact that comet tails always point in the anti-solar direction regardless of the direction of motion of the comet. The phenomenon was traditionally explained as a consequence of radiation pressure on the individual atoms in the tail. Biermann (1951,1952,1957; Wurm 1943) studied the comet tail phenomenon closely and quantitatively. He showed from observations of the rapid motions of the tiny inhomogeneities in comet tails and from the outward Doppler shift of the emitted lines (Wurm 1943) that the comet tails are accelerated away from the Sun at rates of 20 or more times the sunward gravitational acceleration. That is to say, the radiation pressure would have to exceed gravity by a factor of 20 or more, if it were the primary cause of the anti-solar comet tail. He obtained rough values for the principal resonance absorption lines in the atoms and ions of the comet tail and showed them to be entirely inadequate to provide the observed acceleration. From this he concluded that radiation pressure is not the explanation of the comet tail acceleration. Then he pointed out that, if solar electromagnetic radiation is not the cause of the anti-solar acceleration of comet tails, the only remaining possibility is solar corpuscular radiation.

When I first read Biermann's conclusions, they struck me as interesting but baffling, undoubtedly correct, but what did they imply? Fortunately Biermann visited Chicago for a few days in 1957 at the invitation of J. A. Simpson, so I had a chance to discuss the comet conclusions with him. In the course of our conversation he mentioned that comets pass through all heliocentric latitudes and never fail to show the phenomenon of anti-solar accelerated tails. Therefore, his conclusion could be properly represented by stating that the Sun emits corpuscular radiation in all directions at all times. And that was an interesting statement indeed, because at that time the common feeling was still that solar corpuscular radiation was confined to beams of charged particles, emitted from the Sun by some exotic electromagnetic process. But if the Sun never failed to emit corpuscles in every direction, the acceleration could hardly be considered exotic.

The reader should be aware that there was a growing awareness of solar corpuscular radiation at that time from several different quarters. For instance, the variation of the cosmic ray intensity had been detected fifteen years earlier by S. Forbush, using groundbased ionization chambers that detected the charged secondary particles, mostly μ-mesons, reaching the surface of Earth. The net effect is that the principal signal arises from incoming (primary) cosmic rays (mostly protons and helium nuclei) in the range 20 to 50 GeV, which are responsible for most of the μ-mesons that reach the surface of the Earth. There is a substantial daily variation because the μ-mesons decay in flight, so their intensity at the ground depends upon the height of the atmosphere. But these local variations were recognized and corrected. The inferred primary cosmic ray intensity variations were small but clearly real, and they showed that the cosmic ray intensity diminished slightly when the Sun was particularly active. The largest effect (Forbush 1937,1938) was a decrease over half a day often in coincidence with a geomagnetic storm, called the Forbush decrease in honor of its discoverer. There was also an 11-yr sunspot cycle variation of the cosmic ray intensity (Forbush 1954) with the intensity a maximum at solar minimum. The cause of the cosmic ray variations was not known, but electrostatic fields in interplanetary space were an obvious possibility. The association of the Forbush decrease with geomagnetic activity strongly suggested a connection with solar corpuscular radiation, but the connection was not clear.

Simpson developed the cosmic ray neutron monitor to look at the low-energy cosmic rays where the intensity variations were expected to be much larger, whatever their cause. The neutron monitor detected the neutral component of the secondary cosmic rays, *viz* neutrons, produced by primary particles with energies as low as 1 GeV. The Forbush decreases appeared as abrupt 10 and 20% decreases, and the 11-yr modulation had an amplitude of the order of 50% (Simpson 1951; Meyer and Simpson 1955), and it was clear that the modulation was solar controlled (Simpson 1954). In addition the neutron monitor responded strongly to the bursts of fast particles at $\sim$1 GeV from the occasional large solar flare. The direction and time of arrival of the solar particles at Earth indicated channeling along nonradial interplanetary fields as well as temporary trapping in the inner solar system (Meyer et al. 1956). The cosmic ray energy spectrum up to about 15 GeV was determined by using the geomagnetic field as a magnetic spectrometer, and Simpson (1954) was able to show that the variation of the energy spectrum with time was not the result of an energy change of a fixed amount for each cosmic ray particle, as one might expect from passage of the cosmic ray particles across an interstellar electrostatic potential difference. So the cosmic ray intensity variations indicated the interplanetary presence of magnetic fields manipulated by solar corpuscular radiation in some way. A more detailed review of the development of ideas on cosmic ray modulation can be found in Parker (1963).

Cocconi et al. (1958) suggested that the Forbush decrease might be caused by lobes of magnetic field extended outward from the Sun by bursts of solar corpuscular emission. Any cosmic rays in the field would expand with the expanding field, suffering adiabatic deceleration as well as a decrease in their number density.

Morrison (1956) thought more in terms of plasma clouds ejected from the Sun with internal magnetic fields of tangled topology. The expanding tangled fields within the clouds would have again the double effect of diminished cosmic ray density and adiabatic cooling of the cosmic rays, with access to the relatively empty cloud interior limited by the tangled fields to some form of diffusion. That is to say, both Cocconi et al. and Morrison suggested that the cosmic ray reductions arise from expanding, outward sweeping magnetic fields. The differences were a matter of detail, with Morrison expressing the penetration of cosmic ray particles into the disordered magnetic field of the expanding cloud in diffusion terms. The essential point is that the cosmic ray variations left no alternative to active solar corpuscular radiation in association with interplanetary magnetic fields.

It is interesting to note the estimates of the corpuscular density in interplanetary space at that point in time. Biermann estimated 10^2 cm^{-3} moving at perhaps 10^8 cm s^{-1} to account for the behavior of comet tails in the absence of magnetic fields. Various "detections" were reported in anomalous absorption of electromagnetic radiation and led to estimates of interplanetary particle densities from 10^2 to 10^5 cm^{-3}, with the higher figures associated with times of high solar activity. The polarized component of the zodiacal light followed from 500 electrons cm^{-3} (Chapman 1929; Behr and Siedentopf 1953; Elsasser 1954; van de Hulst 1954). The reader is referred to a summary by Elsasser (1957) detailing the various estimates, which were all large compared to the real solar wind densities of 1 to 20 ions cm^{-3}, indicating their inferential nature.

III. PROPERTIES OF THE SOLAR CORONA

The next crucial step toward a definitive picture of interplanetary activity was made by Chapman, who had become interested in the thermal properties of the solar corona. It should be recalled that the researches of Grotrian, Lyot and Edlen in the 1930s had led to the identification of several coronal emission lines as Fe$_{XII}$, Si$_{XI}$, etc., establishing the astonishing fact that the corona is unquestionably an atmosphere with a temperature of the order of 10^6 K. This explained the observed pressure scale height of 10^{10} cm and the line widths corresponding to thermal velocities of 200 km s^{-1}. Biermann (1946), Alfvén (1950) and Schwarzschild (1948) had independently noted that the heat source responsible for the corona must have its origin in the mechanical work done by the thermal convection beneath the photosphere, suggesting acoustic waves and Alfvén waves as the means for transmitting the energy up into the corona (Schatzman 1949; Piddington 1956; Osterbrock 1961; Uchida 1963; Whitaker 1963).

Chapman had pioneered the calculation of the kinetic properties of gases with the Chapman-Enskog approximation (Chapman 1916,1917; Enskog 1917,1921; Chapman and Cowling 1958). Chapman (1954,1957) calculated the thermal conductivity of the coronal gas (fully ionized hydrogen) to be $\kappa \sim 2 \times 10^{-6} \, T^{5/2}$ erg cm^{-1}s^{-1}K, yielding the enormous coronal value $\kappa \sim 1 \times 10^{10}$ erg cm^{-1}s^{-1}K as a consequence of the mobile electrons (with rms thermal velocities of 10^9 cm s^{-1}) (Spitzer and Harm 1953; Spitzer 1956a;

see corrections for finite temperature gradient in Gray and Kilkenny 1980; Luciani et al. 1983,1985). The point is the large magnitude of κ, essentially independent of the number density $N \sim 10^8$ atoms cm^{-3} of the coronal gas, which is to be contrasted with the dimunitive radiative cooling rate of the order of $10^{-23} N^2$ erg cm^{-3}s^{-1} at coronal temperatures (Weyman 1960), proportional to the square of the density. Thus in the low coronal density of the order of 10^8 atoms cm^{-2} the thermal emission (10^{-7} erg cm^{-3}s^{-1}) is small compared to thermal conduction $\kappa T/R_\odot^2$ (4×10^{-6} erg cm^{-2}s^{-1}) and declines outward from the Sun with the rapid decline of N.

Neglecting thermal emission it follows that radial heat flow with $\kappa \sim T^{5/2}$ provides a temperature $T \sim r^{-2/7}$, falling off very slowly with radial distance from the Sun. This means that the pressure scale height, which has a value of 10^{10} cm close to the Sun for $T = 2 \times 10^6$ K, increases outward in proportion to $r^{12/7}$. Chapman pointed out, then, that the corona extends well beyond the orbit of Earth.

I was fortunate in visiting the High Altitude Observatory in Boulder, Colorado where Chapman was located at the time, and he described to me his conclusions as he obtained them. Static interplanetary coronal models were subsequently worked out in some detail (Chapman 1957,1959; Brandt 1961). Chapman's idea that the solar corona extended beyond the orbit of Earth was inescapable (then and now). But it soon occurred to me that the ideas of Biermann and Chapman were mutually exclusive, because the solar corpuscular radiation could not penetrate through a static corona. Either a transverse magnetic field component or the two-stream plasma instability would lock the two together.

My own work during 1956 and 1957 included various elementary investigations of the dynamics of tenuous plasmas, particularly the limiting case of the collisionless plasma. The subject was new, fascinating, and widely applicable in the astronomical and geophysical universe and in the plasma laboratory. Chew et al. (1956) had developed their formulation of the large-scale dynamics of a collisionless plasma working from the collisionless Boltzmann equations and the transverse and longitudinal invariants $\mu = M w_\perp^2 /2B$ and $v = \ell w_\parallel$ of the individual particles (essentially the guiding center approximation initated by Alfvén 1950) where ℓ is the scale of the system parallel to the field. Their principal results were

$$\frac{\mathrm{d}}{\mathrm{d}t}\left(\frac{p_\parallel B^2}{N^3}\right) = 0$$

$$\frac{\mathrm{d}}{\mathrm{d}t}\left(\frac{p_\perp}{NB}\right) = 0 \tag{1}$$

where

$$p_\parallel = NM\langle w_\parallel^2 \rangle$$

$$p_\perp = \frac{1}{2}NM\langle w_\perp^2 \rangle \tag{2}$$

representing the components of the plasma pressure parallel and perpendicular, respectively, to the magnetic field $\mathbf{B}$. The quantities $\langle w_\parallel^2 \rangle$ and $\langle w_\perp^2 \rangle$ represent the mean square thermal velocities, and the number density of the particles (of mass M) is N, of course.

Watson (1956) and Brueckner and Watson (1956) carried through a different approach, based again on small perturbations of the collisionless Boltzmann equation and the guiding center approximation for particle motions. They noted that the resulting equations for the bulk motion of the plasma in terms of the electric drift velocity

$$\mathbf{u} = c\mathbf{E} \times \mathbf{B}/B^2 \tag{3}$$

were essentially the familiar magnetohydrodynamic induction equation

$$\frac{\partial \mathbf{B}}{\partial t} = \nabla \times (\mathbf{u} \times \mathbf{B}) \tag{4}$$

and the momentum equation for the motion perpendicular to the field, which could be cast in the form

$$\rho\frac{d\mathbf{u}_\perp}{dt} = -\nabla_\perp \left(p_\perp + \frac{B^2}{8\pi} \right) + \frac{[(\mathbf{B} \cdot \nabla)\mathbf{B}]_\perp}{4\pi} \left[1 - \frac{p_\parallel - p_\perp}{B^2/4\pi} \right]. \tag{5}$$

In an isotropic plasma ($p_\parallel = p_\perp$) this reduces to the ordinary magnetohydrodynamic momentum equation with the Lorentz force $(\nabla \times \mathbf{B}) \times \mathbf{B}/4\pi$.

It occurred to me (Parker 1957) that the particle trajectories follow from the guiding center approximation so that the total current density $\mathbf{j}$ can be obtained simply by summing over all the particles passing through unit area. No recourse to the complex mathematics of the collisionless Boltzman equation is needed. The current density $\mathbf{j}$ can then be substituted into Maxwell's equation

$$\frac{\partial \mathbf{E}}{\partial t} = c\nabla \times \mathbf{B} - 4\pi\mathbf{j}. \tag{6}$$

The electric field is related to the electric drift velocity $\mathbf{u}$ with the result that Maxwell's equation provides the momentum equation already obtained by Brueckner and Watson (1957). This simple approach shows directly that the gradient and curvature drifts of the particles, together with the cyclotron motion around the field and the gradient of the plasma density and temperature provide exactly the electric current density $\mathbf{j}$ demanded by Maxwell's equation (or Ampere's law for the steady case) with the result that the bulk motion $\mathbf{u}$ satisfies the ordinary magnetohydrodynamic momentum equation. That is to say, there is no mystery about the cause of $\mathbf{j}$ in the absence of Ohm's law and no mystery about the acceleration of the plasma by the Maxwell stresses of $\mathbf{B}$. The familiar guiding center motions of charged particles both parallel and perpendicular to the magnetic and electric fields provide precisely the electric current demanded by Maxwell's equation. Schlüter (1950,1952) had argued much the same point earlier from a different point of view.

In summary, then the electric drift velocity **u** of a collisionless plasma is described by a conventional hydrodynamic momentum equation and is the velocity appearing in the formal magnetohydrodynamic induction equation. The gradient and curvature drifts of the plasma relative to the electric drift are small, of the order of the rms thermal ion cyclotron radius divided by the characteristic scale ℓ of the plasma and magnetic field. Therefore, in the large-scale plasma the drifts go to zero and **u** becomes the bulk motion of the plasma. Another essential point is that a strong anisotropy ($p_\parallel \neq p_\perp$) of a collisionless plasma cannot long endure in the face of the various internal plasma instabilities which feed on the anisotropy (Parker 1958b). It follows, then, that in the slowly varying large-scale plasma that makes up the solar corona the motion parallel to the magnetic field is also described properly by the hydrodynamic equation. To put the matter differently, the theoretical investigations by myself and others searching for exotic large-scale plasma dynamical effects in collisionless plasmas were a failure. It was hydrodynamics and magnetohydrodynamics as usual.

The other essential point that follows from the dynamical instability of anisotropic thermal motions in a plasma is the mutual exclusion of Biermann's solar corpuscular radiation and Chapman's static extended corona, already mentioned. The two together would comprise a plasma with an extremely anisotropic thermal velocity distribution. The resulting violent plasma oscillations would couple the corona and corpuscular radiation together, so that the one could not pass through the other. So the extended corona precluded solar corpuscular radiation, and the existence of solar corpuscular radiation indicated that there was no static corona. Yet both Chapman and Biermann worked from irrefutable observations and basic physical principles.

IV. THE EXPANDING CORONA

After thinking about the contradiction it became apparent that, however unlikely it seemed, the only possibility was that Biermann and Chapman were talking about the same thing. The static corona at the Sun must somehow become the solar corpuscular radiation in interplanetary space, and it must come about within the framework of hydrodynamics (Parker 1958c).

Examination of the equation for hydrostatic equilibrium of the corona in the gravitational field of the Sun showed a fundamental difficulty with the static coronal model. Write the barometric law

$$\frac{\mathrm{d}p}{\mathrm{d}r} = -\rho \frac{G\mathrm{M}_\odot}{r^2} \tag{7}$$

with $p = 2NKT$ and $\rho = NM$ for fully ionized hydrogen at a temperature $T(r)$ and number density $N(r)$, where M is the mass of the hydrogen atom.

Integrating the hydrostatic equation yields

$$p(r) = p(a)\exp\left[-\frac{w^2 a}{V^2}\int_1^{r/a}\frac{ds}{s^2}\frac{T(a)}{T(s)}\right] \tag{8}$$

where $s \equiv r/a$, and $w \equiv (GM_\odot/a)^{\frac{1}{2}}$ represents the characteristic gravitational velocity ($\sim 4.5 \times 10^7$ cm s^{-1} for $a = 1\,\mathrm{R}_\odot = 7 \times 10^{10}$ cm), and $V = (2kT(a)/M)^{\frac{1}{2}}$ represents the characteristic ion thermal velocity ($\sim 1.3 \times 10^7$ cm s^{-1} for hydrogen at 10^6 K). The pressure $p(a)$ at the base of the corona is 3×10^{-2} dyne cm^{-2} for $N(a) = 10^8$ atom cm^{-3} at 10^6 K. In order that the corona be in static equilibrium, it is necessary that the asymptotic value of $p(r)$ at large r be less than or, at most, equal to the interstellar pressure, of the order of 10^{-12} dyne cm^{-2}, a factor of 3×10^{-11} less than $p(a)$. This requires that the integral reach a value of 26 or more in the limit of large r. With a temperature of the form

$$T(r) = T(a)(a/r)^\chi \tag{9}$$

it follows that (assuming that $\chi < 1$) at large r

$$p(r) \sim p(a)\exp[-w^2/V^2(1-\chi)]. \tag{10}$$

Now $w^2/V^2 \cong 12$ for the values quoted above. Chapman's work indicated $\chi = 2/7$, providing a gas pressure of 0.7×10^{-8} dyne cm^{-2} at infinity, larger by a factor of about 10^4 than of the interstellar pressure. So a simple static corona is not a viable possibility, and, in any case, a static corona is contradicted by the existence of the universal solar corpuscular radiation, as already noted.

If the corona cannot be static, the proper equation would then be the momentum equation for steady radial outflow,

$$\rho v \frac{dv}{dr} + \frac{dp}{dr} = -\rho \frac{GM_\odot}{r^2} \tag{11}$$

and

$$\rho(r)v(r)v^2 = \rho(a)v(a)a^2 \tag{12}$$

for conservation of matter. Treating the simplest case, these equations reduce to

$$\frac{dv}{dr}\left(v - \frac{V^2}{v}\right) = \frac{2V^2}{r} - \frac{w^2 a}{r^2} \tag{13}$$

for an isothermal corona with temperature T. The equation can be integrated immediately to give

$$\frac{v^2}{V^2} - \ln\frac{v^2}{V^2} = 4\ln\frac{r}{a} + \frac{2w^2 a}{V^2 r} + C \tag{14}$$

where C is the arbitrary constant of the integration.

The integration constant C must be chosen to yield a physically acceptable solution. Toward that end note that $2w^2/V^2 \cong 24$ for $T = 10^6$ K, and then consider how to construct a solution that begins at the base of the corona with quasi-static equilibrium ($v^2 << V^2$) and extends to infinity with vanishing pressure, i.e., vanishing density. Hopefully the solution, starting from hydrostatic equilibrium and very small velocity near the Sun, will provide a large velocity, of the order of several hundred km s^{-1} or more at large distance to account for Bierman's solar corpuscular radiation, for otherwise the Chapman-Biermann dilemma cannot be resolved. At $r = a$, where $v^2 \ll V^2$, the solution reduces to

$$\frac{v^2}{V^2} \cong \exp\left(-C - \frac{2w^2}{V^2}\right)\left[1 - \frac{v^2}{V^2} + \frac{1}{2}\frac{v^4}{V^4} + \ldots\right] \tag{15}$$

so that with $2w^2/V^2 \sim 20$ to 30 there is a nearly static solution provided only that C is not large and negative.

In the limit of large r/a, it is evident by inspection of Eq. (14) that v^2/V^2 must either increase without bound, or decline asymptotically to zero because of the unbounded term $4\ln r/a$ on the right-hand side. Consider the possibility that there is a solution at large r/a for which the corona is near static equilibrium, i.e., $v^2 \ll V^2$. Then in the limit of large r/a,

$$\frac{v^2}{V^2} \sim \left(\frac{a}{r}\right)^4 \exp(-C)\left[1 - \frac{v^2}{V^2} + \frac{1}{2}\frac{v^4}{V^4} + \ldots\right] \tag{16}$$

so that

$$N(r) \sim N(a)\exp\left(-\frac{w^2}{V^2}\right). \tag{17}$$

But this is precisely the result for a static corona, providing a pressure at infinity far in excess of the interstellar pressure. If such a solution were viable, the slow outflow at high pressure would eventually overwhelm interstellar space. In any case, a steady solution of this nature exists only in the presence of an enormous interstellar pressure.

The only alternative is the solution for v that is large compared to V. In that case

$$\frac{v^2}{V^2} \sim C + 4\ln\frac{r}{a} + \ln\frac{v^2}{V^2} \tag{18}$$

so that v/V becomes large without bound as r/a increases. The density and pressure fall to zero faster than $1/r^2$ in this case so no interstellar pressure is necessary to maintain the solution.

The question is how the quasi-static solutions at $r = a$, where the corona is strongly bound by gravity, connect to the supersonic solutions ($v > V$) at large r, where the corona expands freely into a vacuum. This requires a dv/dr that is generally positive at all r/a, and Eq. (13) shows the proper choice of C.

The factor $v - V^2/v$ on the left-hand side of Eq. (13) is negative in the near region and positive in the far region. Recalling that the corona is tightly bound by gravity near the Sun $(w^2/V^2 \cong 12)$, it follows that the right-hand side of Eq. (13) is negative in the near region and is positive in the limit of large r/a. So dv/dr can indeed be positive both near and far. The tricky part is to get the solution past the sonic point where the coefficient $v - V^2/v$ vanishes, for at that point dv/dr would be infinite, unless, the right-hand side vanishes at the same point. And that is the requirement that determines C. It follows that $v = V$ at $r = aw^2/2V^2 \equiv r_c$. Thus $r_c \cong 6a$ for 10^6 K. The result is

$$\frac{v^2}{V^2} - \ln\frac{v^2}{V^2} = 4\ln\frac{r}{a} + \frac{2w^2 a}{V^2 r} - 3 - 4\ln\frac{w^2}{2V^2} \tag{19}$$

for the solution extending smoothly and continuously from the base of the corona to vanishing pressure at infinity. There is no other solution (for another choice of C) that extends from $p(a)$ at the Sun to suitably small pressure at infinity. With any other choice of C the solution reaches the zero of the right-hand side of Eq. (13) before v reaches V, so that v declines asymptotically to zero thereafter and provides an overwhelming pressure at infinity, or the solution reaches $v = V$ before the zero of the right-hand side so that dv/dr becomes infinite and the solution heads back toward small r. Steady solutions with the overwhelming pressure at infinity can exist only if there is something at infinity that can push back and match that pressure. This property of the equations is much the same as for the one-dimensional flow in a channel of varying width in which the flow progresses from a high pressure at subsonic speed to supersonic speed at vanishing pressure in a flaring open end of the channel.

The bottom line is simply that the boundary conditions at the Sun and at infinity permit only the solution that passes from quasi-static near the Sun to supersonic at large r/a. This is precisely what is required to reconcile Chapman's static corona with Biermann's universal solar corpuscular radiation. Figure 1 shows $v(r)$ for a variety of coronal temperatures. The *solar wind* seemed to be the appropriate name for the outflow.

There are, of course, obvious ways to avoid the supersonic solution. In particular we do not know (even today) the nature of the heat source that creates the corona (cf., Parker 1991), and one can imagine that the solar corona is so dense that the supersonic expansion of the outer portions requires more energy per particle than thermal conductivity can supply from the observed 10^6 K at the base. Chamberlain (1961) constructed just such a solution, starting with enormous coronal densities, of the order of 10^{10} atoms cm^{-3}, and $T(a) = 2 \times 10^6$ K, so that the thermometric conductivity was small in spite of the large thermal conductivity. The resulting subsonic expansion of about 20 km s^{-1} at $r = 1$ AU follows from conservation of energy, so that $T(r)$ declines asymptotically as $1/r$. This interesting solution was referred to as the solar breeze. It fitted neither the observed density of 10^8 atoms cm^{-3} at

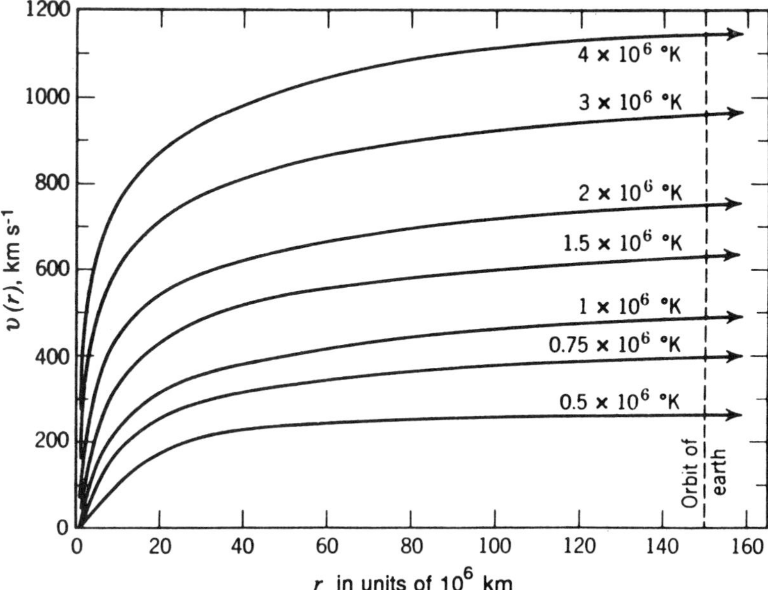

Figure 1.   The radial expansion velocity (km s^{-1}) of an isothermal corona as a function of radial distance in units of 10^6km (1 $R_\odot = 0.7 \times 10^6$km).

the Sun nor did it reconcile solar corpuscular radiation with the static corona observed at the Sun. But, curiously, it was generally accepted because of the widespread aversion to the concept of a static corona going over into a supersonic solar wind at large r (cf., Chapman 1964). Part of the difficulty seemed to be a fixation on the idea of the radial evaporation of particles from a solar exosphere, which provided no more than a tenuous subsonic outflow in its primitive form (Chamberlain 1960,1961; Brandt 1961). The missing ingredient in the simple evaporation models is, of course, the continuing deposition of heat in the departing expanding gas, raising the temperature at large distances so that the temperature falls less rapidly than $1/r^\chi$, where $0<\chi<1$. This provides a total energy per particle that is negative near the Sun but which becomes positive and nonvanishing at infinity, producing a supersonic wind as the temperature falls asymptotically to zero.

As a final comment, the isothermal corona has been used here because of its simplicity as the generic example of the expanding corona. A more general case is the polytropic atmosphere, in which the outward decline of temperature is represented as $T\sim\rho^{\alpha-1}$. The isothermal case corresponds to $\alpha = 1$, and we expect that the value of α appropriate for the corona of the Sun is actually a little larger. It is easy to show (Parker 1960,1963) that any $\alpha<3/2$ provides a supersonic wind at infinity, along the same general lines as detailed for $\alpha = 1$.

V. IMPLICATIONS OF THE SOLAR WIND

The realization that the corona of the Sun transforms with radial distance into the supersonic solar wind opened the way for understanding the previously baffling interplanetary activity that had been inferred from geomagnetic storms, Forbush cosmic ray decreases, the 11-yr cosmic ray modulation, etc. For instance, the simple concept of a wind blowing radially outward from the Sun indicated that the magnetic field in any part of the expanding solar corona is stretched radially outward through space, declining as $1/r^2$ at large r. Thus in spherical polar coordinates (r, θ, φ) one would write

$$B_r(r, \theta, \varphi) \cong B_r(a, \theta, \varphi)(a/r)^2. \tag{20}$$

However, the rotation Ω of the Sun winds the field into an Archimedean spiral, with B_φ given by $(\Omega r \sin\theta/v) \times B_r$. The field lines are given by

$$\varphi = \varphi_0 + \Omega r/v \tag{21}$$

so that

$$B_r(r, \theta, \varphi) = B_r(a, \theta, \varphi - \Omega r/v)(a/r)^2. \tag{22}$$

With $\Omega \cong 2.5 \times 10^{-6}$ radians s^{-1}, a wind velocity of 370 km s^{-1} provides a 45° inclination of the field lines to the radial direction at $r = 1$ AU. Thus, inside the orbit of Earth the field is principally radial and outside the field is primarily azimuthal. The essential point is that the existence of the solar wind unambiguously and quantitatively dictates the basic form of the interplanetary magnetic field (Parker 1958c).

The reduction of the cosmic ray intensity in the inner solar system followed automatically from the outward sweep of the interplanetary magnetic field beyond the orbit of Earth (Parker 1958a). The expected small-scale fluctuations in the large-scale spiral interplanetary field (Parker 1958b) scatter the cosmic ray particles and inhibit their free streaming along the field. The tendency is to transport the particles with the outward moving fluctuations in the field. Thus the magnetic fluctuations play a role in impeding the arrival of galactic cosmic rays, thereby participating in the 11-yr solar cycle modulation of the cosmic ray intensity. They also impede the escape of the occasional enormous bursts of fast particles from large solar flares (Meyer et al. 1956).

Interplanetary blast waves, produced by sudden large outbursts of activity at the Sun, deform the spiral interplanetary magnetic field, producing kinks at the shock surface ahead and at any contact surface or reverse shock behind. The result (Parker 1961a) is an abrupt reduction in cosmic ray density behind the outward advancing front, providing the Forbush decrease phenomenon.

Variations in coronal temperature and density around the Sun provide varying wind speeds from different regions of the corona. The resulting slow and fast streams have essentially the same spiral pattern as the magnetic field at large radial distance, of course, except that the fast streams overtake the slow streams, providing both forward and reverse shocks and the associated spiral interaction regions of compressed plasma and field. Extended magnetic lobes or tongues add to the complexity of the interplanetary magnetic field and a variety of possibilities were sketched out, along with a simple modeling of the associated cosmic ray effects (Parker 1963), generally of the form of a reduction of cosmic ray intensity in the expanding field configurations, including both outward convection of the cosmic rays and adiabatic deceleration.

It was obvious that the solar wind must terminate at some finite distance from the Sun as a consequence of the pressure of the interstellar gas and magnetic field. Lüst and Schlüter (1954) and Davis (1955) had already pointed out the heliocentric cavity in the interstellar medium created by the solar corpuscular radiation. They noted that the radius R of the cavity, now called the *heliosphere*, is determined by the distance at which the ram pressure of the wind falls to a value comparable to the interstellar pressure P, of the order of 10^{-12} dyne cm^{-2}. It follows that if the solar wind density has a value N_1 at the orbit of Earth, $r = r_1 = 1$ AU, then a wind velocity v yields

$$MN_1 \left(\frac{r_1}{R}\right)^2 v^2 \sim P \tag{23}$$

in order of magnitude. The value of R depends upon the mean value of $N_1 v_1^2$. The general result is that R is estimated to be of the order of 10^2 AU or more, far beyond the orbits of the known outer planets.

With the clear hydrodynamic picture provided by the theory of the solar wind, it was possible to calculate simple illustrative models of the cavity in the interstellar wind, including both the 10 to 20 km s^{-1} motion of the interstellar wind blowing by the heliosphere and the galactic field embedded in that wind. Weymann (1960) and Clauser (1960) pointed out that the impact of the supersonic wind against the interstellar medium involves a shock transition in the solar wind. The magnetic field is essentially azimuthal at 10^2 AU so the shock structure is largely magnetic. It was easy to show various forms for the cavity of the shocked solar gas and field and the contact surface (now called the heliopause) with the interstellar wind (Parker 1961b,1963). There is then an interstellar wake, with a diameter of several hundred AU extending away downwind through interstellar space.

The spiral magnetic field in the solar wind is of sufficient strength to influence the dynamics of the coronal expansion and transport angular momentum away from the Sun. This was examined theoretically by Weber and Davis (1967) and by Brandt et al. (1969) followed by an analysis of comet tail inclinations (Brandt and Heise 1969) to determine the extent of the deceleration of solar rotation.

We investigated the complicated interaction of the solar wind with the magnetic field of Earth, treating in a preliminary way the impact of the solar wind ions with the outer boundary of the field (Parker 1958d). The essential feature of the treatment was the recognition that the interaction of the solar wind with the magnetosphere is a magnetohydrodynamic interaction in the large, in which electric fields play only a passive role. That is to say, the electric field $\mathbf{E} = -\mathbf{v} \times \mathbf{B}/c$ is present in the fixed reference frame for the reason that there is no significant large-scale electric field in the frame of reference moving with the plasma. The "actual" electric field depends entirely on the frame of reference in which one chooses to view the action. Oddly, this point is not always appreciated today, and there is still talk of the electric field $\mathbf{E}$ of the solar wind *penetrating* into the magnetosphere as if $\mathbf{E}$ were the prime mover. In fact, the prime mover is the force exerted by the wind on the magnetosphere, and there is an $\mathbf{E}$ produced in the magnetosphere only insofar as those regions are set in motion by the force.

On the small scale there are rapid fluctuations and strong electric fields associated with the structure of the magnetopause and auroral sheets and the neutral sheet in the geomagnetic tail, etc., where the plasma velocity may change by several hundred km s^{-1} across a characteristic distance of the order of the ion cyclotron radius. Thus the structure of the magnetopause lies outside the magnetohydrodynamics that describes the behavior of the plasma throughout the vast volume of space both inside and outside the magnetopause.

Finally, it was obvious that if the Sun, representing a pedestrian, middle-aged, main-sequence type G star, has a solar wind, then we expect that every other main-sequence G star has a similar supersonic stellar wind. Indeed, one may surmise that any star cooler than the middle of class F, and therefore possessing a convective zone, in all probability has a corona of 10^6 K or so, and therefore a supersonic stellar wind. Hence most stars reside in an interstellar cavity with much the same structure as the heliosphere in which we reside.

We made some semiquantitative estimates of the properties of stellar winds from late main sequence stars, based on the dynamics worked out for the solar wind (Parker 1960). First of all, the temperature of the solar corona is of the order of one tenth of the escape temperature at which the mean thermal velocity is equal to the gravitational escape velocity. The temperature cannot be much greater because the mass loss increases so rapidly with increasing temperature that an enormously greater heat input would be needed to sustain the mass loss that goes with any significantly higher temperature. The general result seems to be a supersonic wind with a terminal speed of some substantial fraction of the gravitational escape velocity from the low corona. This is because the wind accelerates only very slowly with radial distance once it has achieved one or two times the ion thermal velocity in the corona. Hence to go faster would require maintaining the coronal temperature for a very large distance out into space.

In summary these many implications of the hydrodynamic theory of the corona and solar wind were collected together and further elaborated in a monograph (Parker 1963). In looking back over that early work today, thirty years later, the oversimplification of many of the illustrative examples is obvious. We know far more about the actual situation today, as a result of three decades of space science. On the other hand, the basic concepts have largely stood the test of time, for the simple reason that the fundamental hydrodynamical principles that describe the solar corona permit only those limited possibilities. And, of course, the theory of the solar wind and its detailed behavior out through interplanetary space has been elaborated and extended over the years by many authors. The lingering outstanding deficiency in the theory of the solar wind is the continuing absence of a clear idea of the heat source that creates the observed temperatures, without which there would be no solar wind. This was a topic of serious concern in the original paper (Parker 1958c) and the question remains unresolved today (Parker 1991). The microflaring in the network fields (supergranule boundaries) is the only theoretically viable hypothesis at present for heating the near corona, and 300 s Alfvén waves are the only viable hypothesis for heating the wind, and perhaps pushing it as well, in the region beyond the critical radius to produce the observed high-velocity wind streams. But neither has been established by observation as adequate for the job.

VI. OBSERVATIONS

The next step, besides some obvious theoretical refinements of the primitive isothermal coronal example, was direct detection of the solar wind by an instrument in space far enough from Earth to be outside the terrestrial magnetosphere (Parker 1958d). As already noted at the end of Sec. II, there had been various claims of groundbased detection of the static interplanetary medium, as well as clouds of solar corpuscular radiation. The results tended to be large, ranging over 10^3 to 10^5 atoms cm^{-3} at the orbit of Earth (Chapman 1929; Behr and Siedentopf 1953; Elsasser 1954,1957; van de Hulst 1954).

Fortunately direct study of the solar wind by plasma detectors carried on spacecraft into the wind was not long delayed. Gringauz et al. (1960,1961,1967) made the first direct detection of the solar wind, with a simple ion trap behind a grid charged to +15 volts. They detected an outward (anti-solar) ion flux of a little more than 10^8 cm^{-2} s^{-1}. Assuming that the ions were mostly protons, it followed that the outward particle velocity was at least 60 km s^{-1}, indicating a particle density of 30 to 50 cm^{-3} or less.

The next detection was with a more sophisticated electrostatic ion velocity analyser developed by the MIT group (Bridge et al. 1961; Bonetti et al. 1963; Scherb 1964) and carried on Explorer 10 (Rossi 1991). They found a quiet-day solar wind from the direction of the Sun with a speed in the range 250 to 400 km s^{-1} (300–800 eV per proton) and a net flux of the order of 6×10^8 atoms cm^{-3}, implying densities of 7 to 20 protons cm^{-3}. The wind switched abruptly on and off at intervals of hours, which was baffling at the time. Subsequently it became apparent that Explorer 10 was moving tailward along a flapping magnetopause. Hence the plasma detector was sometimes in the shocked solar wind and sometimes within the shelter of the magnetosphere.

The JPL plasma instrument carried on Mariner 2, bound for Venus, provided incontrovertible observations (Snyder and Neugebauer 1964; Neugebauer and Snyder 1966,1967). They showed simply that the wind is always present, with velocities ranging from a low of about 300 km s^{-1} to a high near 800 km s^{-1}, and densities ranging from about 2 atoms cm^{-3} to an occasional burst as high as 20 or more. The density varied oppositely from the velocity so that the net particle flux varied less than either. Extensive study of the solar wind since that time has established the perpetual nature of the wind, predicted by the basic hydrodynamics of the solar corona. A particularly interesting aspect of the wind is its locally jittery velocity and density, its distinct and variable ion and electron temperatures, and the sometimes very high velocities going up to 800 km s^{-1} or more. The picture is occasionally punctuated by blast waves from violent active events on the Sun.

The spiral magnetic field configuration was verified by Ness et al. (1964) with magnetometers on the IMP-1 spacecraft. Subsequently the wind and field have been mapped by the Pioneer 10 and 11 and the Voyager 1 and 2 spacecraft to a distance of about 50 AU. at the time of writing (Burlaga et al. 1993, and references therein). The cosmic ray intensity increases slowly out through the solar system with most of the modulation still beyond the farthest spacecraft. This is not the place to review the enormous progress that has been made on all fronts since 1963.

It is important to emphasize again that the precise nature of the coronal heating that provides the solar wind is still not clear. The basic observational difficulty is the small scale of the heating process, which cannot be resolved by groundbased telescopes, but which is directly accessible from a 1 to 1.5 m telescope carried by a balloon to the top of the atmosphere or in an orbiting spacecraft. When such detailed observations of plages, microflares, and magnetic fibril motions are made, we will be much better informed on the cause of the solar wind.

Acknowledgments. The author wishes to express his gratitude to A. J. Dessler for discussion of the early work leading to the concept of solar corpuscular radiation, to K. MacGregor for discussion of recent developments in the theory of winds from hot stars, and to H. Völk for discussion of the present state of the theory of galactic winds. This work was supported in part by a grant from the National Aeronautics and Space Admiministration.

REFERENCES

Abbott, D. C., and Conti, P. S. 1987. Wolf-Rayet stars. *Ann. Rev. Astron. Astrophys.* 25:113–150.

Adams, W. S., and MacCormack, E. 1935. Systematic displacements of lines in the spectra of certain bright stars. *Astrophys. J.* 81:119.

Alfvén, H. 1950. *Cosmical Electrodynamics* (Oxford: Clarendon Press).

Baade, W., and Zwicky, F. 1934. Supernovae and cosmic-rays. *Phys. Rev.* 45:138.

Begelman, M. C., Blandford, R. D., and Rees, M. 1984. Theory of extragalactic radio sources. *Rev. Mod. Phys.* 56:255.

Behr, A., and Siedentopf, H. 1953. Untersuchungen über Zodiakallicht and Gegenschein nach lichtelekfrischen messungen auf den Jungfraujock. *Zeit. Astrophys.* 32:19.

Biermann, L. 1946. Kurze original milteilungen. *Naturwissenschaften* 34:87.

Biermann, L. 1951. KomentenSchweife und solare Korpuskularstrahlung. *Zeit. Astrophys.* 29:274–286.

Biermann, L. 1952. Über den Schweif des kometer Halley im jahre 1910. *Zeit. Naturforsch* 7a:127.

Biermann, L. 1957. Solar corpuscular radiation and the interplanetary gas. *Observatory* 107:109.

Birkeland, Kr. 1896. Sur les rayons cathodiques sens l' action de forces magnetique intenses. *Arch. Sci. Phys. Naturelles* 1:497.

Birkeland, Kr. 1908. On the cause of magnetic storms and the origin of terrestrial magnetism, first section. In *The Norwegian Aurora Polaris Expedition 1902–3*, vol. 1 (Christiania: H. Aschehoug).

Birkeland, Kr. 1913. On the cause of magnetic storms and the origin of terrestrial magnetism, second section. In *The Norwegian Aurora Polaris Expedition 1902–3*, vol. 2 (Christiania: H. Aschehoug).

Bonetti, A., Bridge, H. S., Lazarus, A. J., Lyon, E. F., Rossi, B., and Scherb, F. 1963. Explorer 10 plasma measurements. *J. Geophys. Res.* 68:4017–4063.

Brandt, J. C. 1961. Interplanetary gas. IV. Neutral hydrogen in a model solar corona. *Astrophys. J.* 133:688–700.

Brandt, J. C., and Heise, J. 1969. Interplanetary gas. XV. Nonradial plasma motions from the orientation of ionic comet tails. *Astrophys. J.* 156:1057–1124.

Brandt, J. C., Wolff, C., and Casinelli, J. P. 1969. Interplanetary gas. XVI. A calculation of the angular momentum of the solar wind. *Astrophys. J.* 156:112–156.

Breitschwerdt, D., McKenzie, J. F., and Völk, H. J. 1987. Magnetic field structure of the galaxy and the formation of galactic winds. In *Interstellar Magnetic Fields*, eds. R. Beck and R. Gräve (Berlin: Springer-Verlag), pp. 131–141.

Breitschwerdt, D., McKenzie, J. F., and Völk, H. J. 1991. Galactic winds. I. Cosmic ray and wave-driven winds from the Galaxy. *Astron. Astrophys.* 245:79–98.

Breitschwerdt, D., McKenzie, J. F., and Völk, H. J. 1993. Galactic winds. II. Role of the disk-halo interface in cosmic ray driven galactic winds. *Astron. Astrophys.* 269:54–66.

Bridge, H., Dilworth, C., Lazarus, A. J., Lyon, E. F., Rossi, B., and Scherb, F. 1961. Explorer 10 plasma measurements. *J. Phys. Soc. Japan* 17A–11.

Bridle, A. H., and Perley, R. A. 1984. Extragalactic radio jets. *Ann. Rev. Astron. Astrophys.* 22:319–358.

Brown, J. C., Casinelli, J. P., and Collins, G. W. 1991. Constraints on the physical properties of optical bullets in SS 433. *Astrophys. J.* 378:307–314.

Brueckner, K. A., and Watson, K. M. 1956. Use of the Boltzman equation for the study of ionized gases of low density. II. *Phys. Rev.* 102:19.

Burbidge, E. M., and Burbidge, G. 1958. Stellar evolution. In *Encyclopedia of Physics*, vol. 51, ed. S. Flugge (Berlin: Springer-Verlag), p. 134.

Burlaga, L. F., Perko, J., and Pirraglia, J. 1993. Cosmic-ray modulation, merged interaction regions, and multifractals. *Astrophys. J.* 407:347–358.

Cassinelli, J. P. 1979. Annual review of stellar winds. *Astron. Astrophys.* 17:275.

Cassinelli, J. P., Schulte-Ladbeck, R. E., Poe, C. H., and Abbott, M. 1989. Modelling mass loss from B[e] stars. In *Physics of Luminous Blue Variables*, eds. K.

Davidson, H. J. G. L. M. Lamers and A. F. J. Moffatt (Dordrecht: Kluwer), p. 121.

Castor, J. I., Abbott, D. C., and Klein, R. I. 1976. In *Physique des Mouvements dans les Atmosphères Stellaires*, eds. R. Cayrel and M. Steinberg (Paris: Cont. Natl. Recherche Sci.).

Chamberlain, J. W. 1960. Interplanetary gas II. Expansion of a model solar corona. *Astrophys. J.* 131:47–56.

Chamberlain, J. W. 1961. Interplanetary gas. III. A hydrodynamic model of the corona. *Astrophys. J.* 133:675–687.

Chapman, S. 1916. On the law of distribution of molecular velocities. *Phil. Trans. Roy. Soc. London A* 216:279.

Chapman, S. 1917. On the kinetic theory of a gas. *Phil. Trans. Roy. Soc. London A* 217:115.

Chapman, S. 1918. The energy of magnetic storms. *Mon. Not. Roy. Astron. Soc.* 79:70.

Chapman, S. 1919. An outline of a theory of magnetic storms. *Proc. Roy. Soc. London A* 95:61.

Chapman, S. 1929. Solar streams of corpuscles: Their geometry, absorption of light, and penetrations. *Mon. Not. Roy. Astron. Soc.* 89:456.

Chapman, S. 1954. The viscosity and thermal conductivity of a completely ionized gas. *Astrophys. J.* 120:151–155.

Chapman, S. 1957. Notes on the solar corona and the terrestrial ionosphere. *Smithsonian Contrib. Astrophys.* 2:1–11.

Chapman, S. 1959. Interplanetary space and the Earth's outermost atmosphere. *Proc. Roy. Soc. London A* 253:462–481.

Chapman, S. 1964. *Solar Plasma Geomagnetism and Aurora* (New York: Gordon and Breach), pp. 32–51.

Chapman, S., and Bartels, J. 1940. *Geomagnetism* (Oxford: Oxford Univ. Press).

Chapman, S., and Cowling, T. G. 1958. *The Mathematical Theory of Nonuniform Gases*, 2nd ed. (Cambridge: Cambridge Univ. Press).

Chapman, S., and Ferraro, V. C. A. 1931. A new theory of magnetic storms. *Terrestrial Magn. Atmos. Elect.* 36:77, 171.

Chapman, S., and Ferraro, V. C. A. 1932. A new thoery of magnetic storms. *Terrestrial Magn. Atmos. Elect.* 37:147, 421.

Chapman, S., and Ferraro, V. C. A. 1933. A new theory of magnetic storms. *Terrestrial Magn. Atmos. Elect.* 38:79.

Chapman, S., and Ferraro, V. C. A. 1940. A new theory of magnetic storms. *Terrestrial Magn. Atmos. Elect.* 45:245.

Chew, G. F., Goldberger, M. L., and Low, F. E. 1956. The Boltzmann equation and the one-fluid hydromagnetic equations in the absence of particle collisions. *Proc. Roy. Soc. London A* 236:112.

Clauser, F. 1960. In *4th Symposium on Cosmical Fluid Dynamics*, Varenna, Italy (August), p. 26.

Cocconi, G., Gold, T., Greisen, K., Hayakawa, S., and Morrison, P. 1958. The cosmic ray flare effect. *Nuovo Cimento Suppl. Ser. X* 8(2):161.

Conti, P. 1978. Mass loss in early type stars. *Ann. Rev. Astron. Astrophys.* 16:371–392.

Danly, L., Lockman, F. J., Meade, M. R., and Savage, B. D. 1992. Ultraviolet and radio observations of Milky Way halo gas. *Astrophys. J. Suppl.* 81:125–161.

Davis, L. E., Jr. 1955. Interplanetary magnetic fields and cosmic rays. *Phys. Rev.* 100:1440–1444.

Dessler, A. J. 1967. Solar wind and interplanetary magnetic field. *Rev. Geophys.* 5:1–41.

Deutsch, A. J. 1956. The circumstellar envelope in Alpha Herculis. *Astrophys. J.*

123:210–227.

Deutsch, A. J. 1961. Loss of mass from red giant stars. In *Stellar Atmospheres*, ed. J. L. Greenstein (Chicago: Univ. of Chicago Press), p. 543.

Dickey, J. M., and Lockman, F. J. 1990. H_I in the Galaxy. *Ann. Rev. Astron. Astrophys.* 28:215.

Dupree, A. K. 1986. Mass loss from cool stars. *Ann. Rev. Astron. Astrophys.* 24:377–420.

Elsasser, H. 1954. Die räumliche vertelungder Zodiak allichtmaterie. *Zeit. Astrophys.* 33:274.

Elsasser, H. 1957. Interplanetare Materie. *Mitt. Astr. Gesselsch* II:61.

Enskog, D. 1917. Ph.D. Thesis, Uppsala Univ.

Enskog, D. 1922. Die numerische berechnungder vorgänge in mässig verdünnten gasen. *Svensk, Vet. Akad. Arkiv. mat. Astr. och. Fys.* 16:60.

Fitzgerald, G. F. 1892. Sunspots and magnetic storms. *Electrician* 30:48.

Fitzgerald, G. F. 1900. Sunspots, magnetic storms, comet tails, atmospheric electricity, and aurorae. *Electrician* 46:287.

Forbush, S. E. 1937. On diurnal variation in cosmic-ray intensity. *Terrestrial Magn. Atmos. Elect.* 42:1.

Forbush, S. E. 1938. On cosmic-ray effects associated with magnetic storms. *Terrestrial Magn. Atmos. Elect.* 43:203.

Forbush, S. E. 1954. World-wide cosmic-ray variations, 1937–1952. *J. Geophys. Res.* 59:525–542.

Gray, D. R., and Kilkenny, J. D. 1980. The measurement of ion acoustic turbulence and reduced thermal conductivity caused by a large temperature gradient in a laser heated plasma ionized gas. *Plasma Phys.* 22:81.

Gringauz, K. I., Bezrukikh, V. V., Ozerov, V. D., and Ribchinsky, R. E. 1960. Study of the interplanetary high energy electrons, and solar corpuscular radiation by means of three electrode traps for charged particles on the second Soviet cosmic rocket. *Soviet Phys. Doklady* 53:61.

Gringauz, K. I., Bezrukikh, V. V., Ozerov, V. D., and Ribchinsky, R. E. 1961. Some results of experiments in interplanetary space by means of charged particle traps on Soviet space probes. *Space Res.* 2:539–553.

Gringauz, K. I., Bezrukikh, V. V., and Musatov, L. S. 1967. Solar wind observations with the Venus 3 probe. *Cosmic Res.* 5:216.

Hartmann, L., and MacGregor, K. B. 1982. Protostellar mass and angular momentum loss. *Astrophys. J.* 259:180–192.

Ipavich, F. 1975. Galactic winds driven by cosmic rays. *Astrophys. J.* 196:107–120.

Kelvin, W. T. 1892. Address to the Royal Society at their anniversary meeting, Nov. 30, 1892. *Proc. Roy. Soc. London A* 52:300–317.

Kulkarni, S. R., and Heiles, C. 1987. The atomic component. In *Interstellar Processes*, eds. D. Hollenbach and H. Thronson (Dordrecht: D. Reidel), pp. 87–122.

Lada, C. J. 1985. Cold outflows, energetic winds, and enigmatic jets around young stellar objects. *Ann. Rev. Astron. Astrophys.* 23:267–317.

Lamers, H. J. G. L. M., Maeder, A., Schmutz, W., and Casinelli, J. P. 1991. Wolf-Rayet stars as starting points or as end points of the evolution of massive stars. *Astrophys. J.* 368:538–544.

Lockman, F. J. 1984. The H_I halo in the inner Galaxy. *Astrophys. J.* 283:90–97.

Lodge, O. 1900. Sunspots, magnetic storms, comet tails, atmospheric electricity, and aurorae. *Electrician* 46:249.

Luciani, J. F., Mora, P., and Pellat, R. 1985. Quasi static heat front and delocalized heat flux. *Phys. Fluids* 28:835.

Luciani, J. F., Mora, P., and Virmont, J. 1983. Nonlocal heat transport due to steep temperature gradients. *Phys. Rev. Lett.* 51:1664.

Lucy, L. B., and Solomon, P. M. 1970. Mass loss by hot stars. *Astrophys. J.* 159:879–893.

Lüst, R., and Schlüter, A. 1954. Kraft freie magneticfelder. *Z. Astrophys.* 34:263.

McKee, C. F. 1993. Fountains, bubbles, and hot gas. In *Back to the Galaxy*, eds. S. Holt and F. Verter (New York: American Inst. of Physics), pp. 499–513.

Merrifeld, M. R. 1993. Rotation, scale height and mass. In *Back to the Galaxy*, eds. S. Holt and F. Verter (New York: American Inst. of Physics), pp. 437–446.

Mestel, L., and Spruit, H. C. 1987. On magnetic braking of late-type stars. *Mon. Not. Roy. Astron. Soc.* 226:57–66.

Meyer, P., and Simpson, J. A. 1955. Changes in the low-energy particle cutoff and primary spectrum of cosmic rays. *Phys. Rev.* 99:1517.

Meyer, P., Parker, E. N., and Simpson, J. A. 1956. The solar cosmic rays of February, 1966 and their propagation through interplanetary space. *Phys. Rev.* 104:768.

Morrison, P. 1956. Solar origin of cosmic-ray time variations. *Phys. Rev.* 101:1397–1404.

Ness, N. F., Scearce, C. S., and Seek, J. B. 1964. Initial results of the IMP 1 magnetic field experiment. *J. Geophys. Res.* 69:3531–3569.

Neugebauer, M., and Snyder, C. W. 1966. Mariner 2 observations of the solar wind, 1 Average properties. *J. Geophys. Res.* 71:4469–4484.

Neugebauer, M., and Snyder, C. W. 1967. Mariner 2 observations of the solar wind II. Relation of plasma properties to the magnetic field. *J. Geophys. Res.* 72:1823–1828.

Osterbrock, D. E. 1961. The heating of the solar chromosphere, plages, and corona by magnetohydrodynamic waves. *Astrophys. J.* 134:347–388.

Parker, E. N. 1957. Newtonian development of the dynamical properties of ionized gases of low density. *Phys. Rev.* 107:924.

Parker, E. N. 1958*a*. Cosmic ray modulation by the solar wind. *Phys. Rev.* 110:1445–1447.

Parker, E. N. 1958*b*. Dynamical instability of an anisotropic ionized gas of low density. *Phys. Rev.* 109:1874.

Parker, E. N. 1958*c*. Dynamics of the interplanetary gas and magnetic fields. *Astrophys. J.* 128:664–676.

Parker, E. N. 1958*d*. Interaction of the solar wind with the geomagnetic field. *Phys. Fluids* 1:171.

Parker, E. N. 1960. The hydrodynamic theory of solar corpuscular radiation and stellar winds. *Astrophys. J.* 132:821–866.

Parker, E. N. 1961*a*. Sudden expansion of the corona following a large solar flare and the attendant magnetic field and cosmic ray effects. *Astrophys. J.* 133:1014–1033.

Parker, E. N. 1961*b*. The stellar wind regions. *Astrophys. J.* 134:20–27.

Parker, E. N. 1963. *Interplanetary Dynamical Processes* (New York: Wiley-Interscience).

Parker, E. N. 1990. Spontaneous discontinuities in galactic magnetic fields and the creation of galactic X-ray halos. In *Galactic and Intergalactic Magnetic Fields*, eds. R. Beck, P. P. Kronberg and R. Wielebinski (Dordrecht: Kluwer), pp. 169–175.

Parker, E. N. 1991. Heating solar coronal holes. *Astrophys. J.* 372:719–729.

Parker, E. N. 1992. Fast dynamos, cosmic rays, and the galactic magnetic field. *Astrophys. J.* 401:137–145.

Piddington, H. 1956. Solar atmospheric heating by hydromagnetic waves. *Mon. Not. Roy. Astron. Soc.* 116:314–323.

Poe, C. H., Friend, D. B., and Cassinelli, J. P. 1989. A rotating, magnetic, radiative-driven wind model for Wolf-Rayet stars. *Astrophys. J.* 337:888–902.

Rossi, B. 1991. The interplanetary plasma. *Ann. Rev. Astron. Astrophys.* 29:1–8.

Schatzman, E. 1949. The heating of the solar corona and chromosphere. *Ann. Astrophys.* 12:203–228.

Scherb, F. 1964. Velocity distributions of the interplanetary plasma detected by Explorer 10. *Space Res.* 4:797.

Schlüter, A. 1950. Dynamik des plasma: 1. Grundegleichungen, plasma in gekreuzten feldern. *Z. Naturforsch* 5a:72.

Schlüter, A. 1952. Plasma im magnetfeld. *Ann. Physik* 10:422.

Schwarzschild, M. 1948. On noise arising from the solar granulation. *Astrophys. J.* 107:1–5.

Simpson, J. A. 1951. Neutrons produced in the atmosphere by the cosmic radiations. *Phys. Rev.* 83:1175.

Simpson, J. A. 1954. Cosmic-radiation intensity-time variations and their origin. II the origin of 27-day variations. *Phys. Rev.* 94:426.

Snyder, C. W., and Neugebauer, M. 1964. Interplanetary solar wind measurements by Mariner 2. *Space Res.* 4:89.

Spitzer, L. 1956a. *Physics of Fully Ionized Gases* (New York: Wiley Interscience).

Spitzer, L. 1956b. On a possible interstellar galactic corona. *Astrophys. J.* 124:20–34.

Spitzer, L., and Harm, R. 1953. Transport phenomena in completely ionized gases. *Phys. Rev.* 89:977.

Uchida, Y. 1963. An effect of the magnetic field in shock wave heating theory of the solar corona. *Publ. Astron. Soc. Japan* 15:376.

van de Hulst, H. C. 1954, On the polarization of the zodiacal light. *Mem. Soc. Roy. Sci. Liège* 15:89.

Watson, K. M. 1956. Use of the Boltzmann equation for the study of ionized gases of low density. I. *Phys. Rev.* 102:12.

Weber, E. J., and Davis, L., Jr. 1967. The angular momentum of the solar wind. *Astrophys. J.* 148:217–227.

Weymann, R. 1960. Coronal evaporation as a possible mechanism for mass loss from red giants. *Astrophys. J.* 132:380–403.

Whitaker, W. A. 1963. Heating of the solar corona by gravity waves. *Astrophys. J.* 137:914–930.

Wurm, K. 1943. Die Natur des Kometen. *Mitt. Sternw. Hamburg-Bergendorf* 8(51):57.

PART II
Various Kinds of Winds

THE SOLAR WIND

W. I. AXFORD and J. F. McKENZIE
Max-Planck-Institut für Aeronomie

In order to understand the solar wind it is important to note that it has two forms which may be described as fast and slow. The fast wind is clearly associated with coronal holes and has the characteristics of being relatively uniform and stable. The slow wind is quite variable in terms of temperature, composition, magnetic field strength, etc., and, in contrast to the fast wind, is not in equilibrium with the corona and transition layer at its base. We contend that the fast wind is the normal, equilibrium wind which must first be accounted for in any theory, whereas the slow wind is associated with transiently open magnetic field lines emerging from regions of the corona which are magnetically largely closed and not necessarily associated with coronal holes. The normal (fast) solar wind exhibits microscopic features which indicate that strong wave-particle interactions are prevalent even out to the orbit of Earth. Thus the temperature of the ions is anisotropic with the perpendicular temperature exceeding the parallel temperature. The heavy ion species tend to have the same temperature per atomic mass as the protons and also to have a higher speed, the difference being approximately the Alfvén speed. Within ~ 1 AU the electrons are cooler than the protons but show strong heat flow effects, including a field-aligned beam of higher energy particles apparently arising from the hottest collisionless regions of the corona. In contrast, the slow wind usually has roughly isotropic ion temperature distributions, and the electrons tend to be hotter than the ions. Coulomb collisions appear to play an important role in the slow solar wind, maintaining roughly equal flow speeds and temperatures for all species. The average magnetic field at the base of coronal holes, based on the field observed at large distances, is unipolar and has a strength of 5 to 10 Gauss. Such fields may fluctuate as a result of forcing from below but the associated waves have long periods and are inefficient in transmitting energy into interplanetary space, giving little possibility of accounting for the astonishing amount of energy contained in the wind. This amounts to a factor of about 10 increase in the total energy of an electron-proton pair between the corona and the orbit of the Earth; the energy flux is about the same as that emitted by the transition layer in magnetically closed regions, as one might expect. Because the only source for this energy lies in the magnetic field, one is driven to conclude that the field at the base of the corona is in fact not unipolar and that flare-like eruptions or reconnection events must be prevalent on a small scale. In fact the magnetic field at chromospheric levels in coronal holes is not uniform but tends to be concentrated at the boundaries of supergranules where it is contained in the "network." The latter occupies only about 5% of the surface area of a coronal hole and must therefore contain a magnetic field with strengths in excess of 100 to 200 Gauss at the level where the network is defined by its contrast in brightness with respect to the surrounding chromosphere and photosphere. Because the gas deep within the network is generally heated only by radiative transport and heat conduction from below and at the sides and cools by radiating into space, it must be cooler than the surrounding gas, with a correspondingly smaller scale height and lower density at a given level. As a consequence, the Alfvén speed in the network can be relatively large and reconnection events or flares can be easily produced in complex, stressed fields, provided the field is not strictly unipolar. The energy for such flares may arise partly from below but is probably mainly the result of fresh magnetic flux being pushed into the network from its sides by the large-scale convection associated with the supergranules. Thus, we may regard the energy of the solar wind as being largely derived from the convective motions of the photosphere, coupling through magnetic field amplification and reconnection into nonthermal forms of energy such as waves, shocks and energetic particles. All of these are capable of providing energy to the corona and solar wind without violating any prejudices arising from classical thermodynamics. The consequences of small-scale flare activity as described here are that there should be a continual production of waves, energetic particles and hot, rapidly moving plasma, all of which have useful properties as far as creating the solar wind is concerned. In particular, the wave spectrum may contain a large amount of power at relatively high frequencies. This is ideal for producing the observed properties of the heavier ion components as well as providing the radiation pressure required to push the wind as a whole. Energetic particles may have a similar effect in that they can induce waves which couple them to the plasma and also help drive the wind by their own pressure. Finally, it is important to note that this perspective on the origin of the high speed solar wind forces us to regard it as originating in strong, complex magnetic field regions covering no more that about 1% of the solar surface and involving cool ($\sim$5000 K), relatively low density material. The usually defined "base of the corona" is in no sense a boundary condition for the solar wind: this has implications for such aspects as the composition and state of ionization of the wind and also for the mass flux question, all of which must be considered afresh.

I. INTRODUCTION

The solar wind (then called the "solar corpuscular radiation") was proposed originally by Biermann (1951,1953,1961) to account for the behavior of type 1 (ionized) comet tails. When observed, these point almost directly away from the Sun regardless of the position of the comet. There is usually a small angle between the radial direction and the comet tail, amounting to a few degrees if the latter is assumed to lie in the orbital plane; this can be used to determine the speed of the wind if it is strictly radial. Biermann's estimates for the speed ranged between 400 and 1000 km s^{-1}, which is remarkably good. However, his (less direct) estimates for the density tended to be far too high because he did not take the interplanetary magnetic field into account in considering the interaction between cometary and solar wind plasma and also appeared to be influenced by older zodiacal light observations, which were largely associated with interplanetary dust.

Biermann's hypothesis and the more traditional arguments based on geomagnetic activity (which is more-or-less continuous at high geomagnetic latitudes) stimulated Parker (1958,1961,1963,1965,1969) to consider the solar wind in more detail. It was known at the time that the solar corona was very hot ($\sim$10^6 K) and extended; this led Chapman (1959) to construct a hydrostatic model based on the dominance of heat conduction, ignoring the effects of any wind and the rotation of the Sun. Parker, however, showed that such a model leads to a contradiction in that, with the usual form of the heat conductivity ($K \sim T^{5/2}$), the pressure of the solar atmosphere would overwhelm the pressure of the interstellar medium at large distances and that this should lead to a wind of some sort. In fact, the argument should be modified if the solar magnetic field is included but the result is the same.

Perhaps the most important contribution made by Parker at this stage was to show how the solar wind could be understood, namely as a consequence of the extended heating of the solar corona in association with heat conduction. He used simple hydrodynamic models to illustrate these effects and showed that the expansion of the corona would result in the existence of a critical point in the flow where the Mach number is unity, followed eventually by highly supersonic flow with almost constant speed at large distances. As a corollary he made the important deduction that the interplanetary magnetic field would be weak at large distances from the Sun, with the field lines having the form of Archimedes spirals. He was also able to use these concepts in developing a theory of galactic cosmic ray modulation involving a combination of convection and diffusion of particles in the interplanetary medium.

For reasons that are now difficult to understand, given the comet and geomagnetic observations and the first *in situ* measurements (Gringauz 1961), Chamberlain (1961) challenged the concept of the solar wind as a high-speed flow phenomenon. He suggested instead that there should exist only a feeble "solar breeze" resulting solely from evaporation, which he modeled as heat conduction. The introduction of higher derivatives in theories of this type leads to complexities that can easily obscure one's view. However, if it is realized that heat conduction requires a two-fluid treatment (Sturrock and Hartle 1966) then a very simple theory can be constructed which shows quite clearly that, even relying only on the presence of a hot corona without extended heating, an acceptable if modest solar wind must exist. A similar early attempt to include viscosity in the problem led to completely spurious solutions (Scarf and Noble 1965). These solutions are evident in the treatment of the transonic flow region with both heat conduction and viscosity described by Axford and Newman (1967, Fig. 1).

In this chapter we have not attempted to provide a complete and fully referenced account of the development of the theory of the solar wind but have chosen instead to concentrate on what appear to us to be some of the more important issues that have emerged in recent years. These involve high-speed solar wind streams and their connection to coronal holes and, in turn, their relationship to the chromospheric network and supergranulation. This brings us closer to the problems of the source of the mass and energy flux in the solar wind which, until recently, have been neglected to a large extent in favor of consideration of the behavior of the large-scale flow beyond a few solar radii from the base of the corona. Indeed it is now time for solar and solar wind physicists to get together to try to understand their common problems and perhaps learn a great deal from each other. References to the more general literature which does not usually contain the particular emphasis of the present chapter can be found in the review by Isenberg (1991), the volumes edited by

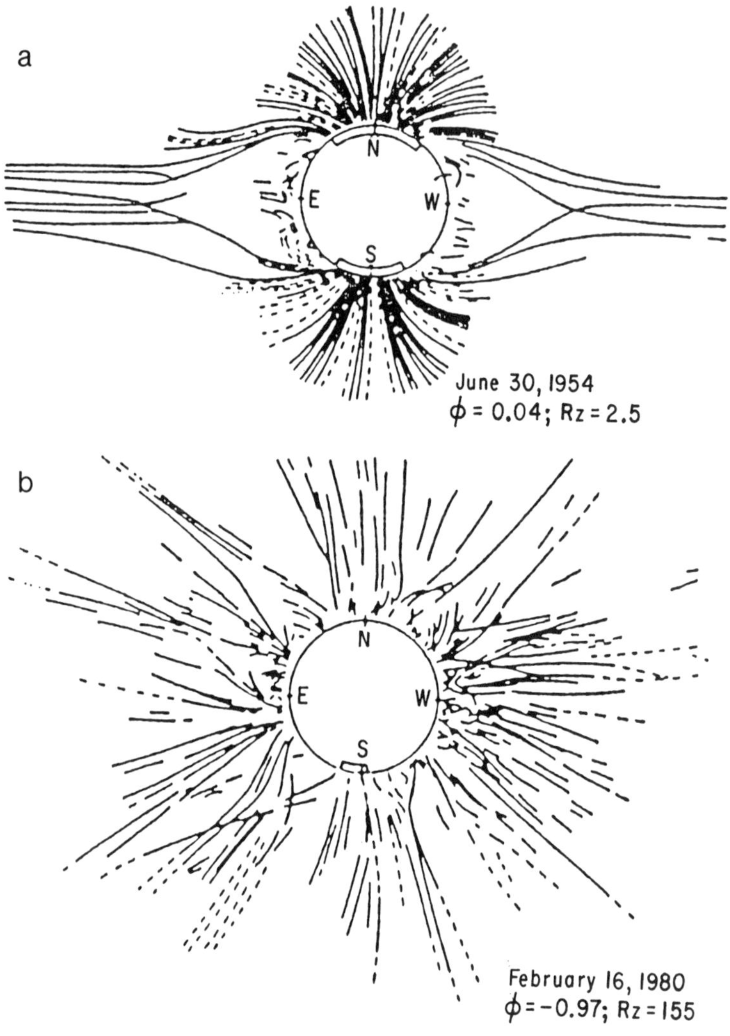

Figure 1.   Drawings of the solar corona made (a) during sunspot minimum (1954) and (b) during sunspot maximum (1980) (figure from Loucif and Koutchmy 1989).

Schwenn and Marsch (1990) and the various Solar Wind Conference volumes (see, e.g., Marsch and Schwenn 1992).

II. CHARACTERISTICS OF THE SOLAR WIND

The solar wind can be regarded as having two distinct modes, namely high speed and low speed (Schwenn 1990). The high-speed wind is clearly associated with coronal holes and as such is especially prominent during sunspot minimum in association with the large polar holes present at such times (see Fig. 1). The low-speed wind appears to originate from the "non-hole" corona where the magnetic field lines are largely closed. This should not be taken as implying that the low-speed wind is simply a superposition of coronal mass ejection events or magnetic bubbles on large and small scales; it is more likely that the region of apparently closed coronal magnetic field associated with the (non-hole) bright EUV-emitting areas contains a substantial component of magnetic flux that is transiently open and disgorges its plasma in the form of a wind with appropriate transient characteristics. As a consequence, the coronal magnetic field may be regarded as having three components, namely long-term open, long-term closed and transiently open and closed (Axford 1977).

TABLE I

Two Modes of Solar Wind Flow

Property (1 AU)	Low Speed	High Speed
Speed (V)	<400 km s^{-1}	700–900 km s^{-1}
Density (n)	~ 10 cm^{-3}	~ 3 cm^{-3}
Flux (nV)	$\sim 3 \times 10^8$ cm^{-2} s^{-1}	$\sim 2 \times 10^8$ cm^{-2} s^{-1}
Magnetic field (B_r)	~ 2.8 nT	~ 2.8 nT
Temperatures	$T_p \sim 4 \times 10^4$ K	$T_p \sim 2 \times 10^5$ K
	$T_e \sim 1.3 \times 10^5 \mathrm{K} > T_p$	$T_e \sim 10^5 \mathrm{K} < T_p$
Coulomb collisions	important	negligible
Anistropies	T_p isotropic	$T_p(\perp) > T_p(\parallel)$
Beams	none	fast ion beams +
		electron "strahl"
Structure	filamentary, highly variable	uniform, slow changes
Composition	He/H$\sim$1–30%	He/H$\sim$5%
Waves	both directions	outwards propagating
Minor species	n_i/n_p variable	$n_i/n_p \sim$constant
	$T_i \sim T_p$	$T_i \sim A T_p$
	$V_i \sim V_p$	$V_i \sim V_p + V_A$
Associated with	streamers,	coronal holes
	transiently open field	
Sunspot minimum	± 15 deg from equator	>30 deg
Sunspot maximum	dominant at most latitudes	less frequent

The average properties of the high-speed and low-speed solar wind modes are summarized in Table I (Schwenn 1990; Mariani and Neubauer 1991). It is important to note that these modes are *qualitatively* different and that it is not therefore possible simply to tune the boundary conditions in any theory in order to account for one or the other at will. The most important qualitative difference is that the slow wind is highly variable; even during periods when the speed may be rather constant, there are fluctuations in density, in magnetic field strength, in temperature and, most noticeably perhaps, in the relative concentrations of helium and minor species (Bame et al. 1977). Thus, the slow wind has a filamentary nature and, because its properties vary from one filament to the next, there is also a lack of equilibrium. This suggests that any given filament is only transiently connected to a corresponding point at the base of the corona and that the point of connection may jump around at intervals that are too short to permit the wind to come into equilibrium with coronal base conditions. This essential time dependence makes it difficult for us to treat the slow wind from a theoretical point of view even though it may be conceptually not difficult to understand. On the other hand, the fast wind, which is inherently stable and has very constant properties, must represent the basic mode of the solar wind and should be treated accordingly as being of first priority for any theoretical consideration.

The significant features of the fast solar wind apart from its speed are (see, e.g., Marsch 1991): (1) the temperature of the protons is generally higher than that of the electrons (within 1 AU) and is anisotropic, with the perpendicular exceeding the parallel temperature; (2) the heavier species are favored relative to the protons in being faster by about the Alfvén speed ($\sim 40/R$ km s^{-1} at R AU) and hotter (having about the same temperature per mass); (3) the electron distribution function exhibits a core/halo structure with a "strahl" of more energetic particles beamed along the magnetic field; and (4) close to the Sun, where there is less interaction with slow streams and there has been less time for nonlinear cascading to occur, the "turbulence" in the wind is comprised largely of outwards-propagating Alfvén waves (see, e.g., Marsch 1990). The density, temperature and speed of the fast wind is such that Coulomb collisions are unimportant except near the base of the corona. In contrast, Coulomb collisions are usually important in the slow wind, bringing the different ion species into equilibrium with the dominant proton component, and the turbulence is more isotropic.

It is evident from the above that wave-particle interactions are very important in high-speed streams. In order to account for the preferential acceleration and heating of heavy species and the heating of the protons perpendicular to the magnetic field, it would appear that gyro-resonant interactions are required. This suggests that the source spectrum should contain waves with high frequencies and comparatively short wavelengths (Axford 1980,1985a; Marsch et al. 1982).

III. CORONAL BASE CONDITIONS FOR THE HIGH-SPEED SOLAR WIND

It is well established that the high-speed solar wind is associated with coronal holes which are well-defined regions of strongly reduced EUV and soft X-ray emission (Withbroe and Noyes 1977). It seems safe to assume that the emission is low relative to that from the surrounding non-hole corona because the temperature and/or the density at the base of the corona are reduced by outflow in the open (unipolar) magnetic field associated with the coronal hole and high speed wind (see, e.g., Gabriel 1976a). However, it is difficult to determine these quantities because measurements must be made at the limb where there is usually foreground and background interference from the denser non-hole corona. It has been suggested that the density in a coronal hole at an altitude of 0.1 $R_\odot$ may be as low as $n \approx 2 \times 10^7$ cm^{-3} (Koutchmy 1977) and the temperature may be significantly less than $T \approx 10^6$ K (Habbal et al. 1993), whereas the corresponding values for the quiet, closed field corona are $\gtrsim 10^8$ cm^{-3} and $\gtrsim 10^6$ K, respectively. We believe that we must be prepared to accept these low values for n and T, noting the consequence that the product $nT \approx 2 \times 10^{13}$ (at the base) is considerably less than has been assumed previously (see, e.g., Hollweg 1992).

There is a clear solar cycle variation in the state of the corona and the solar wind (see, e.g., Kojima and Kakinuma 1987; Rickett and Coles 1991). During sunspot minimum the visible corona is dominated by an equatorial belt of magnetically closed streamers and there are polar coronal holes which emit high speed wind (Fig. 1a). The latter dominates interplanetary space in the range of helio-latitudes $\gtrsim 15$ to 20 degrees (Schwenn et al. 1978). During sunspot maximum the visible corona is much less regular (Fig. 1b), coronal holes are less evident and interplanetary space is dominated by low-speed solar wind with many disturbances associated with coronal mass ejections and flares. Accordingly, we must concentrate on sunspot minimum conditions and high helio-latitudes in comparing observation and theory at this stage. The Ulysses spacecraft is at present sampling such conditions and should provide clear results on the properties of free-streaming, high-speed solar wind, unaffected by stream-stream interactions. Obviously, any solar probe mission, if it is to examine the source region of the solar wind in a useful way, should encounter the Sun during sunspot minimum over the poles so that it traverses coronal hole regions with reasonably well-understood geometries.

The interplanetary magnetic field in high-speed streams is unipolar and is rooted in coronal holes. Thus, from flux conservation it is a simple matter to estimate the average field strength at the base of the corona (Gabriel 1976a,b; Axford 1976). Taking the average radial field strength at 1 AU from Table I, we find that the field at the base of the corona must be between ~ 5 and 10 Gauss; we assume an average value of 8 Gauss. On this basis it is easy to estimate the important parameters defining conditions at the coronal base as given in Table II. There are two columns in this table. The first (which is preferred by us) is based on the relatively low values of plasma density and temperature measured by Koutchmy (1977) and Habbal et al. (1993), respectively ($nT \approx 2 \times 10^{13}$ K cm^{-3}); the second is based on more commonly assumed values ($nT \approx 10^{14}$ K cm^{-3}). In the subsequent discussion quantities given as $X'(X'')$ refer to the first (second) column of this table. It is evident from Table II that the energy flux of the solar wind emanating from coronal holes is approximately the same as that emitted in the form of soft X rays and EUV from the transition zone in magnetically quiet non-hole regions. The plasma in holes must be replenished in a short time (~ 25 (125)

TABLE II

Corona Base Conditions for the High-Speed Solar Wind

Plasma density (n)	2×10^7 cm^{-3}	10^8 cm^{-3}
Magnetic field (B)	8 Gauss	8 Gauss
Alfvén speed (V_A)	2500 km s^{-1}	1200 km s^{-1}
Temperature (T)	10^6 K	10^6 K
Pressure (nT)	2×10^{13} K cm^{-3}	10^{14} K cm^{-3}
Collision time (ν_{ep})	~ 0.6 s	~ 0.2 s
Collision time (ν_{pp})	~ 20 s	6 s
Escape speed (V_e)	620 km s^{-1}	620 km s^{-1}
Plasma beta (β)	$\sim 10^{-3}$	$\sim 10^{-2}$
Sound speed (V_S)	150 km s^{-1}	150 km s^{-1}
Plasma flux (nV)	5×10^{13} cm^{-2} s^{-1}	5×10^{13} cm^{-2} s^{-1}
Plasma speed (V)	25 km s^{-1}	5 km s^{-1}
Thermal scale ($H_T = K/nVk$)	1.5×10^{11} cm	1.5×10^{11} cm
Scale height (H)	50,000 km	50,000 km
Total content (nH)	1×10^{17} cm^{-2}	5×10^{17} cm^{-2}
Time to empty (H/V)	2000 s	10^4 s
Energy flux (F_E)	6×10^5 erg cm^{-2} s^{-1}	6×10^5 erg cm^{-2} s^{-1}
Wave amplitude	75 km s^{-1}	30 km s^{-1}
Area of Sun	$\sim 15\%$	$\sim 15\%$

minutes) in contrast to the several hours required to empty coronal streamers
if they become magnetically open. The very low plasma beta results in the
magnetic field in coronal holes being rigid, current free and almost monopolar
beyond ~ 5 to 10 R$_\odot$ and before rotational effects become important. [This is
shown in the model of Pneumann and Kopp (1971), where a rather low field
strength has been assumed. A simple analytical model can be constructed
using the approach adopted by Gleeson and Axford (1976) for the Jovian
magnetosphere.] If the energy flux appears first in the form of Alfvén waves
emitted from the base of the corona, then the amplitude of the waves must
be ~ 75 (30) km s^{-1} which is larger than usually assumed but still implies
linearity in view of the very high Alfvén speed. It should be noted that
this amplitude refers to coronal base conditions with $n \approx 2 \times 10^7 (10^8)$ cm^{-3}
where the emission from the plasma is relatively weak and may be difficult to
measure. (The equivalent amplitude in closed regions where $n \approx 4 \times 10^8$ cm^3
would be ~ 25 km s^{-1}.)

Table II provides the base conditions for the simplest models of the solar
wind. However, it should be noted that there is a question as to whether
it is proper to take the mass and energy fluxes as primary conditions rather
than the density and temperature because the latter are poorly known and
in any case must somehow be connected to conditions at the top of the
chromosphere. The plasma is collision-dominated for the inner few scale
heights, and heat conduction (downwards from the temperature maximum) is
very important because the nondimensional quantity $VHnk/K$, determining

the relative importance of convection to conduction of thermal energy, is $\gtrsim 0.1$ (here $K \approx 10^{-6}\ T^{5/2}$ cm^2 erg cm^{-1}s^{-1}, is the coefficient of thermal conduction).

It is instructive to consider the energetics of the plasma at the base of the corona and in the distant solar wind. In the corona the gravitational energy of a proton/electron pair is -2170 eV. The enthalpy of the pair is $5\ kT$ which for $T \sim 10^6$ K is about 500 eV and hence the plasma can be regarded as being gravitationally bound. At a distance of 1 AU, the enthalpy and gravitational energy are 30 eV and -10 eV, respectively, and the kinetic energy ($V \approx 800$ km s^{-1}) is 3310 eV. Thus, the total energy required to be given to such a pair of particles is 5480 eV, of which no more than 10% is already apparent as enthalpy at coronal levels. This additional energy must nevertheless pass through the coronal plasma and must therefore be in the form of (hydromagnetic) waves and or suprathermal particles, all of which can dump energy at large distances from the source. If the solar wind were to represent a purely adiabatic expansion of a hot corona, the temperature of the latter would have to be $\sim 11 \times 10^6$ K at the coronal base, which is an order of magnitude higher than usually believed.

IV. SOLAR WIND MODELS

In this section we present a brief review of solar wind models without attempting to provide a complete, fully referenced account. Models of high speed solar wind are usually based on the following conservation equations for steady plasma flow along a radially directed stream/flux tube of cross sectional area $A(r)$ (see, e.g., Alazraki and Couturier 1971; Belcher 1971; Hollweg and Johnson 1988; Tu 1987; Esser et al. 1987; Marsch 1994):

$$\rho V A = \text{const} = J \tag{1}$$

$$B_r A = \text{const} = \phi \tag{2}$$

$$\rho V \frac{dV}{dr} = -\frac{d}{dr}(p_g + p_w) - \rho \frac{GM_\odot}{r^2} \tag{3}$$

$$\frac{1}{A}\frac{d}{dr}\left\{\rho V A\left[\frac{1}{2}V^2 + \frac{\gamma}{(\gamma - 1)}\frac{p_g}{\rho} - \frac{GM_\odot}{r}\right] + A(F_w + F_c)\right\} = 0 \tag{4}$$

in which ρ, V and p_g are, respectively, the plasma density, radial velocity, and pressure, γ is the adiabatic index ($= 5/3$ for hydrogen plasma), $M_\odot$ is the mass of the Sun, F_c is the heat conduction flux ($K\,dT/dr$), F_w is the wave energy flux and Q represents dissipative heating of the plasma by wave damping.

For outwards propagating Alfvén waves the wave energy exchange equation may be written (Hollweg 1973,1974):

$$\frac{1}{A}\frac{d}{dr}(A F_w) = V\frac{dp_w}{dr} - Q$$

where

$$F_w = p_w(3V + 2V_A)$$

and

$$V_A = B_r/\sqrt{4\pi\rho}. \tag{5a}$$

This equation may be cast in the form

$$\frac{1}{A}\frac{\mathrm{d}}{\mathrm{d}r}\left[A\frac{(V + V_A)^2}{V_A}2p_w\right] = -\frac{(V + V_A)}{V_A}Q \tag{5b}$$

which describes the evolution of the wave action. Similarly Eqs. (1) through (5) yield the evolution of the plasma entropy S according to

$$\frac{V}{(\gamma - 1)}\left(\frac{\mathrm{d}p_g}{\mathrm{d}r} - c_g^2\frac{\mathrm{d}\rho}{\mathrm{d}r}\right) = \left(Q - \frac{1}{A}\frac{\mathrm{d}}{\mathrm{d}r}(AF_c)\right) \tag{6a}$$

and

$$c_g^2 = \frac{\gamma p_g}{\rho} \tag{6b}$$

or

$$\frac{\mathrm{d}S}{\mathrm{d}r} = C_v\frac{(\gamma - 1)}{Vp}\left(Q - \frac{1}{A}\frac{\mathrm{d}}{\mathrm{d}r}(AF_c)\right). \tag{6c}$$

If we neglect heat conduction, the above equations enable us to derive the following wind equation for the radial flow speed (see, e.g., Dougherty and McKenzie 1991):

$$(V^2 - V_{ph}^2)\frac{1}{V}\frac{\mathrm{d}V}{\mathrm{d}r} = V_{ph}^2\frac{1}{A}\frac{\mathrm{d}A}{\mathrm{d}r}$$

$$-\frac{Q}{\rho V(V + V_A)}((\gamma - 1)V_A + (\gamma - 3/2)V) - \frac{GM}{r^2} \tag{7a}$$

$$V_{ph}^2 \equiv c_g^2 + c_A^2 \tag{7b}$$

$$c_A^2 \equiv p_w(3V + V_A)/2\rho(V + V_A). \tag{7c}$$

Here V_{ph} is the "long wavelength" compressional wave speed including the effect of the "short wavelength" Alfvén waves through the term c_A^2 in the compound sound speed. Note that the wind equation exhibits critical points where sonic flow ($V = V_{ph}$) occurs simultaneously with the vanishing of the right-hand side of (7a).

At this point it is instructive to consider the case of a static corona ($V = 0$) with both wave and gas pressure determining the density structure. Assuming the electrons to be isothermal, the adiabatic protons and WKB waves ($p_w \propto \rho^{1/2}$), Eq. (3) integrates to yield

$$\frac{3}{8}\ln\bar{\rho} + \frac{15}{16}(\bar{\rho}^{2/3} - 1) + \left(\frac{u_{wo}}{c_{go}}\right)^2(1 - 1/\bar{\rho}^{1/2}) = -\frac{1}{2}\frac{V_e^2}{c_{go}^2}\left(1 - \frac{R_S}{r}\right) \tag{8}$$

where $\bar{\rho} = \rho/\rho_o$, V_e is the escape speed at the coronal base, $r = R_s$, to which the subscript 0 refers in the sound speed (c_{go}) and wave velocity amplitude (u_{wo}). It is straightforward to deduce that the effect of the wave pressure is to increase the effective scale height of the density fall-off. For the conditions given in Table II ($u_{wo}/c_{go} = 1/2(1/3)$) the structure of a static corona on open magnetic field lines is wave dominated beyond a few scale heights and certainly within 0.5 $R_\odot$ (see Fig. 2). The asymptotic solution for large r is $\rho/\rho_o \rightarrow$ constant $\approx 1/4(u_{wo}/V_e)^4$. Note, however, that this is true only provided that the waves are adiabatic and there is no dissipation which would increase the temperature at the expense of the wave pressure.

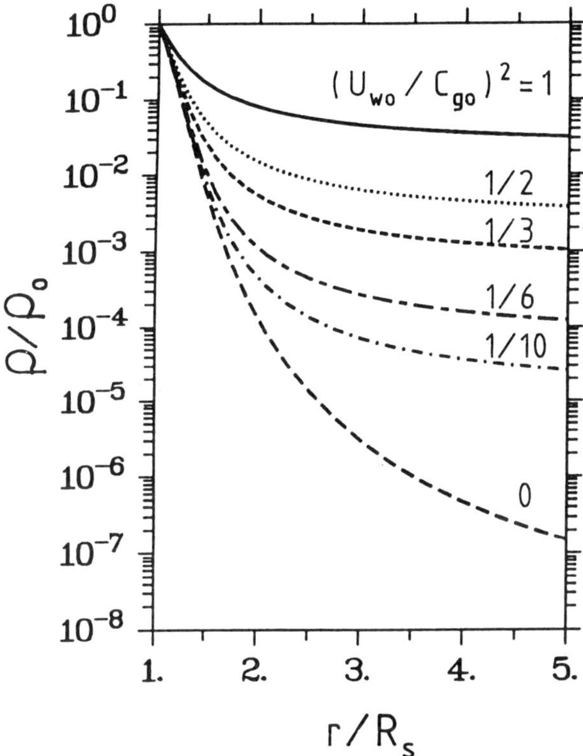

Figure 2.   Density profiles for a hydrostatic corona with isothermal electrons, adiabatic protons and WKB waves. The coronal base temperature is taken to be 10^6K and the different curves are for different ratios of the square of the wave velocity amplitude to the sound speed ($\sim$150 km s^{-1}) at the base.

The Holzer-Axford (1970) model, which may be obtained from Eq. (7a) by neglecting the wave pressure and taking $V_A = 0$, shows that for $\gamma = 5/3$ the presence of extended heating ($Q \neq 0$) enables the flow to go through a critical point ($r = r_c$) thus effecting a smooth subsonic-supersonic transition. (Note that for $\gamma = 5/3$, in the absence of heating, there is no critical point in the wind equation and supersonic solutions begin at the base of the corona with the Mach number $M = V/c_g \geq 1$.) Figure 3 indicates the nature of the solutions in $(M, r/r_c)$ space assuming an exponential form for $Q = Q_o \exp[-(r-r_o)/H]$ with various values of Q_o and H chosen so that all the curves labeled A, B and C have the same asymptotic energy. The corresponding temperature and density profiles are shown in Fig. 4. These calculations confirm that effective coronal temperatures of the order of 10 to 20×10^6 degrees are required to reproduce the characteristics of high-speed streams, which is an order of magnitude higher than observed at the coronal base.

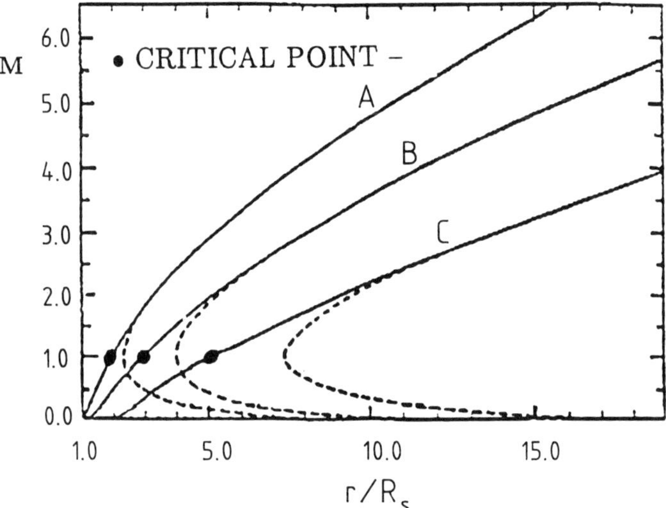

Figure 3.   Curves of Mach number $M = V/c_g$ vs radial distance in units of solar radii for parameters representative of solar conditions, illustrating how distributed heating effects permit smooth subsonic-supersonic critical solutions (full lines). For $Q = 0$ and $\gamma = \frac{5}{3}$ there is no critical point (broken lines). All curves have the same asymptotic energy (see Holzer and Axford 1970).

In a two-fluid (protons and electrons) model (Sturrock and Hartle 1966) the electron temperature profile is approximately determined by assuming that the electron heat conduction flux is constant with the result that $T_e = T_{eo}(r_o/r)^{2/7}$, whereas, with $Q = 0$, the protons essentially cool adiabatically so that $T_p = T_{po}(n/n_o)^{2/3} \sim r^{-4/3}$.

Figure 5 shows temperature profiles for two-fluid models assuming a temperature of $\sim 10^6$ K at the base of the corona (Hartle and Sturrock 1968). These calculations indicate that the corresponding wind speeds are too slow and that the heavy ions are depleted and too cold to provide a satisfactory description of high-speed solar wind. However, the results are not completely inconsistent with the observed properties of the low-speed wind.

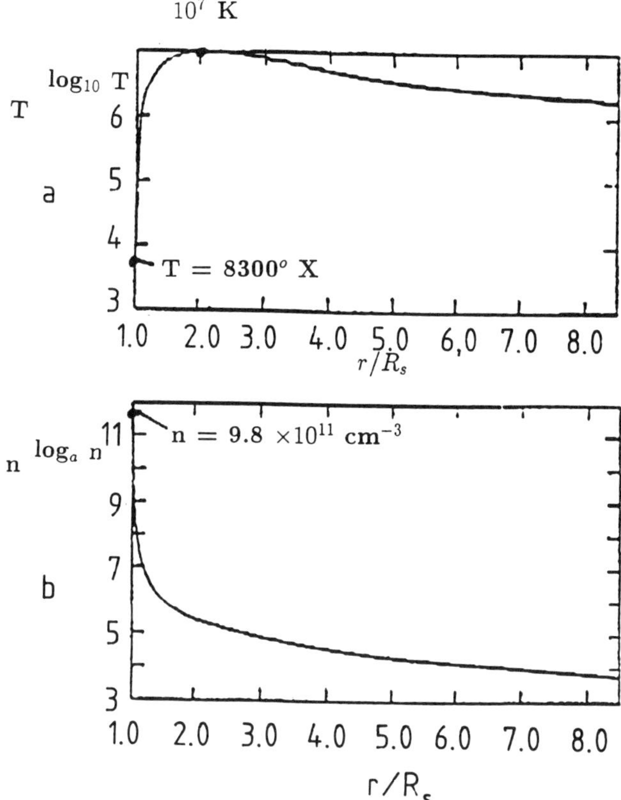

Figure 4. Radial profiles of temperature and density for a critical outflow solution yielding moderately high-speed wind (~500 km s^{-1}). Note that a high temperature is required in the middle corona. There is a very large density gradient in the lower corona which is somewhat artificial as a result of low temperatures at these levels. The parameters assumed correspond to a solution lying between A and B in Fig. 3.

Two-fluid models with anisotropic temperatures and pressures were originally developed by Leer and Axford (1972) using the following equations:

$$\rho V r^2 = J \tag{9}$$

$$\rho V \frac{\mathrm{d}V}{\mathrm{d}r} = -\frac{\mathrm{d}}{\mathrm{d}r}(nk(T_p^{\parallel} + T_e)) - \frac{2}{r}nk(T_p^{\parallel} - T_p^{\perp}) - \rho G \frac{M_{\odot}}{r^2} \tag{10}$$

$$\frac{\mathrm{d}T_p^{\parallel}}{\mathrm{d}r} = -\frac{2T_p^{\parallel}}{V}\frac{\mathrm{d}V}{\mathrm{d}r} + \frac{2}{3Jk}\frac{\mathrm{d}}{\mathrm{d}r}\left(K_p r^2 \frac{\mathrm{d}T_p^{\parallel}}{\mathrm{d}r}\right) + \frac{v_{ep}}{V}(T_e - T_p^{\parallel})$$

$$-2\frac{v_{pp}}{V}(T_p^{\parallel} - T_p^{\perp}) + \frac{2Q_p^{\parallel}}{nVk} \tag{11}$$

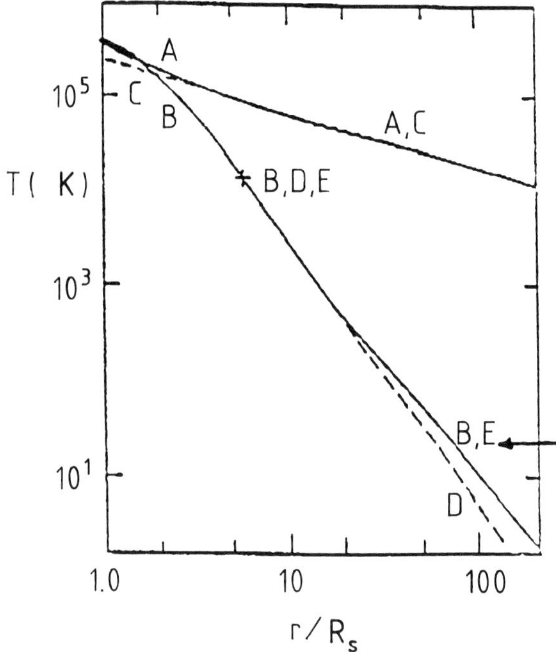

Figure 5.   Temperatures vs radial distance for two-fluid models of the solar wind
(from Holzer and Axford 1970). Curves A and B are the Hartle and Sturrock (1968)
results. Curve C is the electron temperature corresponding to heat conduction alone.
Curve D is the proton temperature appropriate to adiabatic expansion which, beyond
~ 7 R$_\odot$ coincides quite well with the temperature deduced from more accurate
models.

$$\frac{\mathrm{d}T_p^\perp}{\mathrm{d}r} = -\frac{2T_p^\perp}{r} + \frac{\nu_{pe}}{V}(T_e - T_\perp^p) + \frac{\nu_{pp}}{V}(T_p^\parallel - T_p^\perp) + \frac{2Q_p^\perp}{nVk} \qquad (12)$$

$$\frac{\mathrm{d}T_e}{\mathrm{d}r} = \frac{2}{3}\frac{T_e}{n}\frac{\mathrm{d}n}{\mathrm{d}r} + \frac{2}{3}\frac{1}{Jk}\frac{\mathrm{d}}{\mathrm{d}r}\left(K_e r^2 \frac{\mathrm{d}T_e}{\mathrm{d}r}\right) - \frac{\nu_{pe}}{V}(T_e - T_p) + \frac{2}{3}\frac{Q_e}{nVk} \qquad (13)$$

where $T_p = 1/3(T_p^\parallel + 2T_p^\perp)$. It is straightforward to generalize these equations
to a nonspherical geometry and to include waves.

In the absence of collisions and heating one expects $T_p^\parallel > T_p^\perp$ (as $T_p^\perp \propto r^{-2}$
and $T_p^\parallel \propto r^{-4/3}$) which is contrary to what is observed. Hence, a substantial
$Q_p^\perp$ (i.e., $Q_p^\perp >> Q_p^\parallel$), perhaps in the form of wave heating by cyclotron
resonance with protons (and ions), is required to account for this anistropy in
temperature as well as preferential heavy ion heating and acceleration. (Note
that the sample calculations made by Leer and Axford [1972] did not take this
into account and hence indicate that $T_p^\parallel > T_p^\perp$.)

The wave action Eq. (5b) and entropy equation (6c) permit an estimation
of the importance of the thermal pressure relative to wave pressure in driving
the wind. In sub-Alfvénic flow, which is appropriate to the initial acceleration

of the solar wind, the wave action equation approximates to

$$\frac{\mathrm{d}}{\mathrm{d}r}\left(\frac{p_w}{\rho^{1/2}}\right) = -\frac{QA}{2\Phi}.$$ (14)

The plasma entropy, on the other hand, increases according to

$$\frac{\mathrm{d}}{\mathrm{d}r}\left(\frac{p_g}{\rho^{5/3}}\right) = \frac{2}{3}\frac{QA}{\rho^{2/3}J}.$$ (15)

Assuming the length scales, one characterizing the decrease of the wave action and the other the increase of the plasma entropy, are of the same order, we deduce that

$$\frac{p_w}{p_g} \sim \frac{u}{V_A} = \frac{u_o}{V_{Ao}}\left(\frac{\rho_o}{\rho}\right)^{1/2}$$ (16)

whereas in the adiabatic case ($Q = 0$) we have

$$\frac{p_w}{p_g} = \left(\frac{p_{wo}}{p_{go}}\right)\left(\frac{\rho_o}{\rho}\right)^{7/6}.$$ (17)

Therefore, provided the assumption of comparable length scales is valid, dissipation ensures that p_w/p_g remains small until the flow speed attains the Alfvén speed (at, say, between 10 and 20 $R_\odot$); thus, the initial acceleration of the wind is driven primarily by thermal pressure. However, it is important to recognize that, as far as the energetics is concerned, the waves play the dominant role by virtue of the high Alfvén speed and are therefore capable of producing the characteristics of the high-speed wind.

If the waves are predominantly outward propagating so that cascading as described by Tu (1988) is negligible, it is possible to obtain an explicit dissipative heating function $Q_p^\perp$ by postulating that the wave power frequency $P(f, r)$ is absorbed at the local gyrofrequency $f = f_H$. We define

$$P(f, r) = P_1(f)P_2(r),$$
$$f < f_H = 0, \quad f > f_H$$ (18)

where P_1 is only a function of frequency and P_2 is only a function of r. The total wave pressure p_w is then given by

$$p_w(r) = P_2(r)\int_{f_c}^{f_H} P_1(f)\,\mathrm{d}f \equiv P_2(r)\Pi(f_H(r))$$ (19)

where f_c is some low-frequency cut-off.

If we choose $P_2(r)$ to correspond to the WKB (adiabatic) variation, it follows from the wave action Eq. (5b) that the heating rate is given by

$$Q_p^\perp = -\frac{V_A}{(V + V_A)}\frac{1}{A}\left(\frac{A_o(V_o + V_{Ao})^2}{V_{Ao}}2p_{wo}\right)P_1(f_H)\frac{\mathrm{d}f_H}{\mathrm{d}r}$$ (20a)

where the suffix o refers to conditions at the base. Note that for a power spectrum $\propto f^{-1}$ in very sub-Alfvénic flow, this heating rate varies as r^{-3} for spherical geometry. The corresponding variation of the wave pressure is

$$p_w = p_{wo}(\rho/\rho_o)^{1/2}[1 - 2\log(r/r_o)/\log(f_{Ho}/f_c)]. \tag{20b}$$

By using a "WKB-turbulence model" (Tu et al. 1984), Tu (1987) described the solar wind acceleration and heating in a self-consistent fashion which included the evolution of the wave power spectrum through a turbulent cascade process. In its essentials, the model's main features are WKB Alfvén waves and electron heat conduction within 10 $R_\odot$ and beyond this radius, a cascading process is switched on which provides additional plasma heating and acceleration. For a power spectrum varying as f^{-1} at the coronal base, the results are consistent with observational constraints at both the coronal base and in high-speed streams.

As a result of the arguments heading to Eqs. (16) and (20) we see that, provided there is efficient dissipation, it is reasonable to neglect p_w in comparison with p_g and assume that $Q_p^\perp$ is a suitable function of r such that the total heating rates is equal to the total wave energy flux at the base of the corona. This places a simple model such as that of Holzer and Axford (1970) on a sounder basis than might have at first appeared to be the case. It is in turn implied that the (perpendicular) proton temperature should increase to several million degrees beyond a few solar radii from the base of the corona and that this should be regarded as a genuine temperature and not simply transverse motion associated with waves. The problem of electron heating remains; this may be a secondary matter in view of the requirement to heat the ions in order to account for the high speed solar wind, but it is nevertheless important because electron heat conduction contributes to the acceleration of wind and affects the behavior of the plasma at the coronal base where it interfaces the cool chromospheric gas.

The properties of minor ion species in high-speed solar wind, namely that they move faster than the protons by approximately the Alfvén speed, that they are hotter than the protons by approximately their atomic number, and that $T_\perp > T_\parallel$ for each species (see, e.g., Marsch 1991), provides further evidence that preferential acceleration and heating by wave particle interactions through cyclotron resonance is necessary to account for the properties of high speed streams (see, e.g., Isenberg and Hollweg 1983; McKenzie et al. 1979; Axford and McKenzie 1992). As alpha particles in the solar wind are not negligible in terms of their contribution to the overall charge, mass and momentum of the wind, a number of theories in which they are placed on an equal footing with protons have been developed (see, e.g., Yeh 1970; Weber 1973; Bürgi 1992; Leer et al. 1992; McKenzie et al. 1993; McKenzie 1994) but we will not discuss these complications here, except in the following in connection with the so-called mass flux problem.

In-situ observations of the solar wind show that the mass flux is remarkably constant over a wide range of conditions. This appears to a source of concern to solar wind modelers because standard models indicate that the mass flux is sensitive to small variations in coronal conditions. For example, using an isothermal, one-fluid description of the solar wind, Leer and Holzer (1991) point out that an increase of the coronal base temperature from 1×10^6 K to 2×10^6 K results in an increase in the proton mass flux by a factor of about 100. This feature arises because the density falls off exponentially with a scale height which is much smaller than the solar radius $R_\odot$ and the distance to the critical point is a few $R_\odot$. Therefore, in doubling the base temperature, the scale height is increased by a factor of 2 and the distance to the critical point is halved with the result that the mass flux increases by a factor of about $\sqrt{2}e^4 (\approx 76)$. This sensitivity can be reduced somewhat by the presence of Alfvén waves whose pressure acts in such a way as to increase the scale height of the density fall-off. For example, in an Alfvén wave-driven wind the mass flux is proportional to the fourth power of the wave velocity amplitude at the coronal base so that a doubling of the wave "temperature" leads to an increase of the mass flux by only a factor of 4. Furthermore, in models (see, e.g., Holzer and Axford 1970) where a prescribed heating rate raises the temperature to several million degrees very quickly the sensitive dependence of the mass flux on coronal base conditions will be considerably reduced, in a way similar to the effect of waves, because the scale height will increase rapidly and the critical point will approach the Sun. Finally, it has been pointed out by Leer et al. (1992) and Bürgi (1992) that the presence of a non-neglible abundance of helium in the corona may reduce the sensitivity of the mass flux to coronal conditions. However, these models are isothermal containing no heating process, the wave amplitudes are small, and require a large helium (and by implication other heavier ions) abundance in the corona. It is possible that the mass flux in the solar wind is mainly controlled by the rate at which plasma can be supplied from below (see Sec. VI).

V. DIFFICULTIES ASSOCIATED WITH THE SIMPLE MODELS

The various models discussed in the previous section, although not wholly satisfactory, provide some very useful constraints on the nature of the solar wind. First, it is clear that the adiabatic expansion of the hot corona cannot produce the solar wind as the enthalpy available is inadequate by a factor ~ 10. Second, it is clear from the Hartle-Sturrock model that electron heat conduction alone cannot produce the solar wind because the kinetic energies achieved by the ions at large distances are only about 10% of what is needed. However, heat conduction provides a reasonable description of the behavior of the bulk of the electrons and, as we will see later, downwards conduction may be important as a source for the solar wind plasma. Third, (Alfvén) waves appear to offer the best means of producing the solar wind but there are open questions: What is the source of the waves and their frequency spectrum?

What is the relative importance of driving by wave pressure gradients and heating by wave dissipation? What is the cause of wave dissipation? Why are minor species favored? How are the electrons heated?

A major difficulty with the models is that, if one assumes the base of the corona where the solar wind might be considered to have its source, to coincide with the outline of a coronal hole at a level just above the chromosphere, quantities such as the plasma density and temperature are not easy to measure. The electron density is difficult to measure at the limb because of interference from foreground and background closed coronal regions where the density is much higher. Similarly the EUV and soft X-ray emissions from the base of coronal holes facing directly towards the Earth are likely to be dominated by closed loops in the network which are not representative of the more dilute plasma in adjacent open magnetic field lines.

The quantities that are well determined are the integrals of the system of equations, namely the mass, energy and magnetic fluxes, which are equally good as boundary conditions but must lead to values of density, temperature, magnetic field strength and flow speed in the corona that are at least compatible with the observations, however uncertain. Magnetic flux conservation requires that the field strength at the base of the corona, so defined, be ~ 8 Gauss, and there appears to be no difficulty with this result. However, the mass and energy fluxes have implications which do provide grounds for debate.

The energy flux observed in high-speed streams, if it results from Alfvén waves emitted somehow at the base of the corona, requires that the amplitude of the fluctuations at this level be of the order of $75(30)$ km s^{-1}. The observed amplitudes do not usually exceed ~ 30 km s^{-1} (see, e.g., Hassler et al. 1990; Mariska 1992), i.e., comparable to the values we expect to find in the "furnace" described in the next section (see Table III). However, these observations are likely to be dominated by emission from denser, hotter plasma contained in closed field structures and indeed such regions often show red-shifted lines as might be expected from contraction of such field lines. Detailed observations (Dere 1994) indicate that much higher amplitudes and strong blue shifts occur in relatively small explosive events, suggesting that there may be no serious observational constraints on the wave amplitudes in open field lines in coronal holes if the observations are properly interpreted.

The mass flux observed in high-speed streams has given rise to a more serious debate because it is similar to that observed in the low-speed wind and is thought to be a sensitive function of the density and temperature at the base of the corona (see, e.g., Leer and Holzer 1991; Hollweg 1992). Because the two modes of solar wind flow have quite different determining factors it does not seem reasonable to be concerned that they have roughly the same mass fluxes on average. However, the stability of high speed streams themselves may be of more significance. It is argued that for a given Alfvén wave flux (determined by the observed mass and energy fluxes), the requirement that there be a critical point in the flow implies that the density and temperature at the coronal base should be quite tightly constrained in order to produce the observed mass flux. This apparent difficulty is partly associated with the fact that the scale height of the plasma at the base of the corona is very small in comparison with other flow scales, such as the distance to the critical point, and hence small changes in the latter should result in large changes in the local density and therefore the mass flux. However, it should be remembered that in view of the result obtained for a hydrostatic corona in Sec. IV, the scale height determined from the base temperature might not be a useful parameter beyond a few times 0.1 $R_\odot$. On the other hand, it must be emphasized that current models pay no attention at all to the question of the source of the plasma and it is quite possible that this is the controlling factor. That is, the mass flux itself should be regarded as being determined independently, perhaps by the energy flux and by heat conduction downwards and the consequent ionization of neutral chromospheric material. This possibility is addressed in the next section.

Most of the wave observations made in interplanetary space refer to the period range between 30 and 10^5 s and the spectrum is such that the energy flux is dominated by long-period waves (see, e.g., Marsch 1990,1994; Tu 1987,1988). However, if it is true that the effects of the waves are mainly confined to a region well within the Helios orbit (say, within 15 $R_\odot$) and we require that the wave propagation time exceeds, say, 10 periods before damping occurs, then with $V_A \sim 2500(1200)$ km s^{-1}, the important part of the wave spectrum in the region of solar wind acceleration must involve periods $<400(800)$ s. An even stronger constraint is provided by the difficulty for Alfvén waves with periods greater than $\sim 2H/V_A \sim 28(42)$ s to propagate efficiently through the lower corona where the wave speed increases rapidly with altitude (Ferraro and Plumpton 1966).

Given these constraints on the wave spectrum one must consider the possibilities for the generation of the waves in the chromosphere and transition region. It is unlikely that turbulent motions in the chromosphere can produce the desired results because the characteristic periods are of the order of 200 to 300 s and the amplitudes are in any case small ($\gtrsim 1$ km s^{-1}); thus, waves propagating directly from below the base of the corona should be ruled out. A more interesting alternative, which has been proposed by Gold (1964) and Parker (1988) involves stressing the coronal magnetic field as a result of slow twisting motions in the atmosphere at lower levels with the excess energy being released spasmodically in the form of micro- or nano-flares. This is close to what we propose but does not take into account the actual nature of the magnetic field in coronal holes, which is to a large extent open and dominated by the chromospheric network (see Sec. VI).

Small-scale coronal flare activity as described by Gold and Parker is probably very important as a coronal heating mechanism for active regions in the mainly closed magnetic field structures associated with the non-hole corona. However, storage of magnetic energy resulting from twisting motions should not be important in the open magnetic field associated with coronal holes. Because the network beneath the quiet corona is quite similar in coronal holes and non-hole regions and the energy escaping in the form of solar wind and of transition region radiation is approximately the same, we have argued that the source of the solar wind energy and of coronal heating respectively is the same in both cases, namely small-scale flare activity in the chromospheric network (Axford 1985b,1990; McKenzie 1991; Axford and McKenzie 1992). In this case the energy source is the large-scale convection associated with supergranulation that constantly brings small-scale (granular) fields into the network, rather than small-scale chromospheric motions that simply twist the feet of coronal magnetic flux tubes. The spectrum of the waves emitted by such flares bears no relation at all to the time scales of chromospheric motions and hence there is no difficulty in principle in assuming that periods much shorter than 28 s are dominant.

VI. NETWORK FURNACE ORIGIN OF HIGH-SPEED SOLAR WIND

The solar wind models reviewed in Sec. IV have the common characteristic that the base conditions are assumed to be uniform over the solar surface, at least within coronal holes. However, with some thought having been given to the nature and implications of the supergranular convection it became evident some 20 years ago that this assumption is a very poor one (see, e.g., Gabriel 1976a,b). Indeed, conditions at the base of coronal holes are not only difficult to determine because of the low plasma densities that exist there, but the region is quite inhomogeneous as a consequence of the accumulation of magnetic flux at the boundaries of the supergranules, namely in the so called network as shown in Fig. 6. Obviously, the energy and plasma contained in the high-speed solar wind and the accompanying unipolar magnetic flux must originate in a concentrated form from the network and not from the top of the chromosphere as a whole.

The supergranulation, which dominates the chromosphere, is apparent in the form of cells some 30,000 km across in which the hot fluid rises at the center, spreads out and cools and falls at the edges, which constitute the network. The convection speeds are of the order of 1 km s^{-1} and as the typical lifetime of a cell is $\sim 10^5$ s, a cell should properly be regarded as an event rather than a more-or-less steady pattern of flow. Accordingly it must be understood that a steady-state pattern is an idealization. Nevertheless, the magnetic field lines that escape into space because the inwards and outwards fluxes are not equal, can be transported to wherever the network forms within $\sim 2 \times 10^4$ s at most and hence the steady-state picture is adequate for our purposes.

The width of the network varies greatly with height but may be taken as being about 1000 km at mid-chromospheric altitudes where the average magnetic field strength, based on conservation of the coronal flux alone, must exceed ~ 60 Gauss. At photospheric levels the field strength could be as high as ~ 2000 Gauss provided there is a sufficient density contrast inside and outside the network. Such a contrast should be expected because the fluid in the network is that which has undergone most radiative cooling and, because

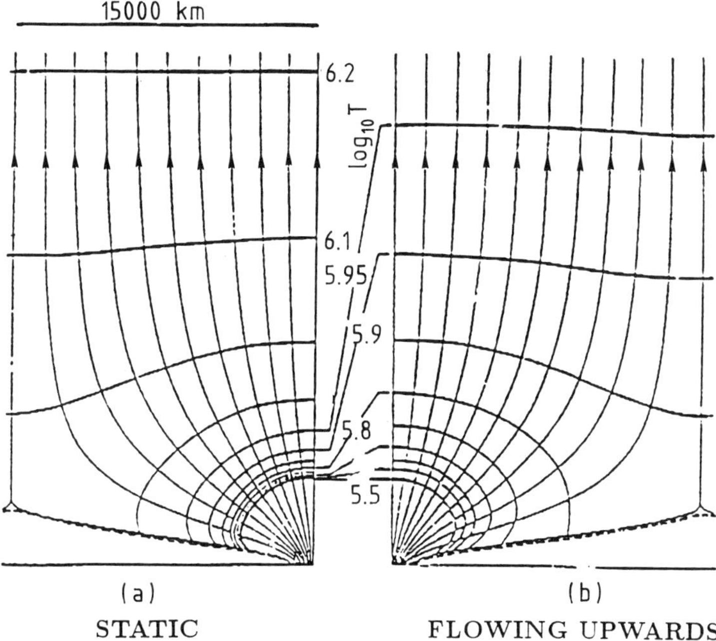

Figure 6. Gabriel's network model comparing (a) a static region without flow and (b) a coronal hole with flow, with the latter possessing a lower coronal temperature and a thicker transition region as a result of the flow.

it has no source of heat other than radiation from the sides, must be expected to drop rapidly under gravity leaving a partial vacuum (Gabriel assumes a factor 10 in density). This should have an important effect on the interaction between the hot corona above the network and the local chromosphere below, which is the source of the coronal and solar wind plasma.

Gabriel (1976a,b) has constructed models of the network based on heat conduction downwards and with and without solar wind. The temperature contours in the lower corona are indicated in Fig. 6 where it can be seen that the upflow of the solar wind leaves the corona cooler than would be expected at the same level in the quiet, closed-field, non-hole corona where there is no such flow. The magnetic structure is that of a potential field, being current free except at the boundaries. The plasma β is small and hence the (coronal) plasma has no role in determining the magnetic field structure. (Similar models have been developed by Elzner and Elwert [1980].)

The Gabriel network model is a useful first-order approximation to the structure of the network and lower corona regardless of the assumption of a steady state. However, there are several important deficiencies. First, it is assumed that the temperature and density of the corona are given at a particular height (30,000 km) with an appropriate downwards heat flux. Second, the calculations are continued only to $\sim 2 \times 10^5$ K and there is no juncture to the cold gas below where the coronal plasma originates. Third, the magnetic field is unipolar and there is no reasonable energy source to produce either the hot corona or the solar wind. The picture must therefore be altered in at least one essential aspect, namely there must be an additional bipolar component of the magnetic field in the network itself so that a source of magnetic energy is available.

The extra magnetic flux must exist in the network side by side with un-cancelled unipolar component and be in the form of closed loops as suggested by Dowdy et al. (1986) (see Fig. 7). Such loops must of course contract downwards as conditions permit but as long as they remain in the network they constitute a source of magnetic energy that can be tapped via magnetic field simplification/reconfiguration resulting from reconnection. It is not clear that all such reconfigurations should appear like flares in the usual sense be-cause the medium is rather different from the lower corona. Nevertheless they must lead to the generation of Alfvén waves and possibly energetic particles (Jokipii and Morfill 1993) which can escape upwards into the corona as re-quired to account for the acceleration of the solar wind. There is evidence for the occurrence of network flares in the form of soft X-ray bright points (see, e.g., Habbal and Harvey 1988), EUV turbulent events, etc. (see, e.g., Dere 1992,1994), but these may represent only a small part of the energy-releasing activity associated with the network.

Dowdy et al. (1986) have described the transition region in the network as a "magnetic junk yard" containing closed loops in which the plasma is relatively dense and cool ($T \lesssim 10^5$ K) and also coronal "funnels" which contain the hotter transition region. We agree in general with this picture but differ in not regarding the field as being static with the closed loops thermally insulated from the funnels and thus requiring an "internal" heating mechanism. Reconnection and flaring should be sufficient to provide heated plasma in the loops which must be regarded as continually cooling and contracting (i.e., producing red-shifted emissions) unless they undergo further reconnection. The loops should contain most of the transition region plasma (10^4 K$\lesssim T \lesssim 10^6$ K) and the plasma in the funnels, although probably hot, is likely to be of low density, escaping upwards and not contributing significantly to the transition region emissions. The effect of reconnection and flaring on the open field lines in the funnels is to produce waves (and possibly high-energy particles) which can escape into the corona and drive the solar wind. The situation is now much more complicated than considered by Gabriel in his first attempts to describe the network and its connection to the corona and solar wind. It is perhaps too complicated to describe more than qualitatively at this point but nevertheless we should pay some attention to the requirements of the solar wind emitted from coronal holes, namely the mass, energy and magnetic fluxes and the generally uniform nature of high-speed streams.

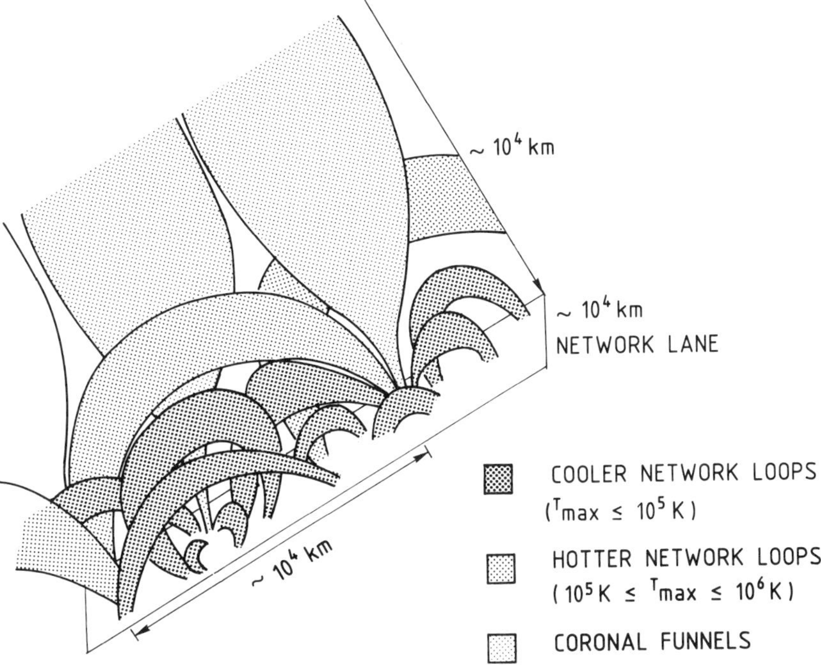

Figure 7.   Closed magnetic flux tubes in the network according to Dowdy et al. (1986). This is a significant modification, based on observations of Gabriel's model (see Fig. 6). Note that only a fraction of the flux tubes are open, connecting to interplanetary field lines (funnels), and that these may contain a much lower plasma density than on closed loops. The latter dominate the transition-region emissions which in general should be red-shifted as a result of contraction of the loops.

The magnetic flux does not provide a problem; however, the field strength implied at the base of the corona (i.e., just above the network) is ~8 Gauss which is somewhat larger than assumed by Gabriel (~1 Gauss). This has an effect on the structure of the magnetic field because the magnetic pressure, which must be balanced by the pressure of the gas in the chromosphere and photosphere, is correspondingly much larger. However, there should be no qualitative difference. The field lines are far from being vertical in the network and indeed are roughly similar to those of a two-dimensional monopole down to about 30° from the horizontal. This is an unduly complicated geometry for our present level of understanding and so for simplicity we take the source region for the solar wind to be a rectangular box having a width of 1000 km, a depth $D = 300$ km (~3 chromospheric scale heights) and an average magnetic field of 100 Gauss. For reasons which will become obvious, we call this the furnace (see Fig. 8). In fact, conditions in the furnace must be time-dependent locally, but for simplicity we consider only time-averaged quantities.

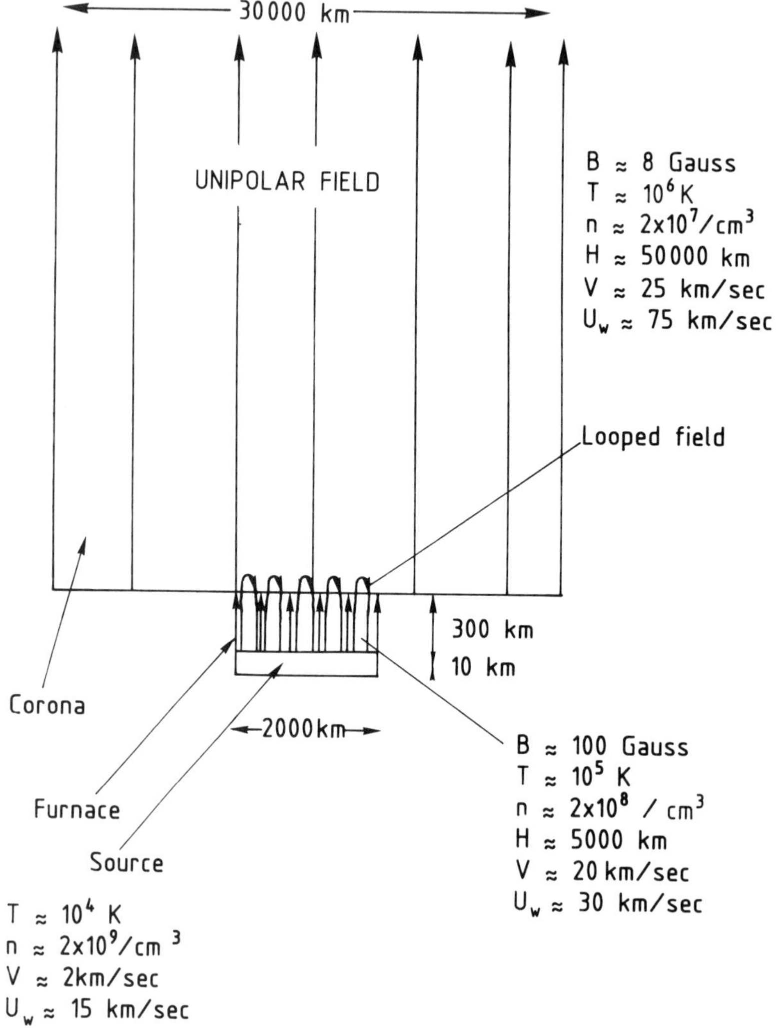

Figure 8.   A schematic representation of the furnace corona model summarized in Table III (first column). It should be understood that the slab configuration is a gross over-simplification of the configuration shown in Figs. 6 and 7.

In constructing Table III, which provides rough estimates of the physical parameters in the furnace, we have used the coronal base conditions given in Table II, increasing the fluxes of mass and energy accordingly and noting that the scale heights for $T \gtrsim 10^5$ K are always larger than the depth D so that the pressure is roughly constant throughout. We anticipate that the temperature structure (defined by a length scale H_T) is determined by a balance between conduction downwards and convection upwards (i.e., $V H_T n k / K \approx 1$) except in the lowest ionizing layer ($T \gtrsim 10^4$ K) where electron collisional ionization may control the thickness of the transition to neutral gas which is the ultimate source of the plasma. Under these conditions, in which nV and nT are both constant, the flow speed is proportional to the temperature. The thickness of the layer in which the temperature is $\sim 10^5$ K is of the order of $H_T = K/nVk \sim 600$ km, which is consistent with the assumed value of D.

TABLE III

Conditions in the Solar Wind Furnace

Magnetic field (B)	~ 100 Gauss	~ 100 Gauss
Temperature (T)	$\sim 10^5$ K	$\sim 10^5$ K
Scale height (H)	~ 5000 km	~ 5000 km
Depth (D)	~ 300 km	~ 300 km
Plasma density (n)	$\sim 2 \times 10^8$ cm^{-3}	$\sim 10^9$ cm^{-3}
Total content (nD)	$\sim 6 \times 10^{15}$ cm^{-2}	$\sim 3 \times 10^{16}$ cm^{-2}
Alfvén speed (V_A)	~ 14000 km s^{-1}	~ 6000 km s^{-1}
Sound speed (V_S)	~ 40 km s^{-1}	~ 40 km s^{-1}
Plasma flux (nV)	$\sim 4 \times 10^{14}$ cm^{-2} s^{-1}	$\sim 4 \times 10^{14}$ cm^{-2} s^{-1}
Thermal scale ($H_T = K/nVk$)	~ 600 km	~ 600 km
Flow speed (V)	~ 20 km s^{-1}	~ 4 km s^{-1}
Emptying time (D/V)	~ 15 s	~ 75 s
Magnetic energy ($W_m = B^2/8\pi$)	~ 400 erg cm^{-3}	~ 400 erg cm^{-3}
Wave energy flux (F_E)	$\sim 7 \times 10^6$ erg cm^{-2} s^{-1}	$\sim 7 \times 10^6$ erg cm^{-2} s^{-1}
Energy time scale ($W_m D/F_E$)	~ 30 min	~ 30 min
Wave amplitude	~ 30 km s^{-1}	~ 12 km s^{-1}
Area of solar surface	$<1\%$	$<1\%$

The plasma density in the ionizing layer, where $T \sim 10^4$ K, is $\sim 2 \times 10^9$ (10^{10}) cm^{-3} and the thickness ~ 10 km. The recombination time in this layer is $\tau_r = 1/\alpha n$ where $\alpha \approx 3 \times 10^{-13}$ cm^{-3} s; hence $\tau_r \approx 1500$ (300) s and, with a flow speed of the order of 2 (0.4) km s^{-1}, we see that the plasma produced in the layer must be transported into the corona rather efficiently and without severe losses. *If this is the case, the mass flux in the solar wind is mainly controlled by the rate at which plasma can be supplied from below.* Recombination losses are important only if $nT \gtrsim 4 \times 10^{14}$ at the base of the corona. On being ionized the plasma must well upwards because the scale height is large compared to the thickness of the furnace and the ionizing layer.

The energy flux requirements of the furnace are quite severe. The mean energy density of the magnetic field assumed is 400 erg cm^{-3} and this must be converted into Alfvén waves and be replaced over a period of ~ 30 minutes in order to maintain the energy flux observed in the high speed solar wind. Because only the looped component of the field in the furnace is available for conversion by reconnection, reconfiguration and relaxation, a large fraction of the field should be in the form of closed loops. Moreover, because the conversion time is much less than the typical lifetime of a supergranule $(\sim 10^5$s), the free magnetic energy in the network must be continually replenished and this can only be done if fresh closed loops, presumably on granular scales ($\lesssim 10^3$ km outside the network), are supplied to the sides of the network by convection. A simple calculation shows that the rate of flux addition is roughly adequate if the average (random) field strength in the supergranules is ~ 20 Gauss and the speed of convection of this flux into the network is ~ 1 km s^{-1}; however, the requirement is a severe one. The source of the energy is of course not the kinetic energy of the convective motions, which are too slow by a large factor, but rather the potential energy of the supergranular material which is capable of generating large field strengths provided the gas carried by the field can cool radiatively and drop under gravity. It is important to note that the unipolar component of the field plays the role of a catalyst in this process in the sense that it cannot be destroyed but provides the possibility for the remaining flux to be "burnt" and the resulting energy to propagate away in the form of waves.

Finally, the relative uniformity of the high speed solar wind as observed beyond ~ 60 R$_\odot$ must be accounted for. The magnetic flux tube associated with a given segment of network passes across an observer at say 1 AU in a period of several hours. During this period the properties of the solar wind remain rather uniform and, if it is the result of a large number of independent reconnection events or flares, there must be some smoothing agent; this could be the coronal reservoir acting as a kind of surge tank. The time scale associated with this reservoir given in Table II is ~ 30 (150) minutes, which must in turn be long compared with time between flares on any given open field line. However, assuming that each flare consumes $(B^2/8\pi)D \sim 1.2 \times 10^{10}$ erg cm^{-2} and noting that the average energy flux must be $\sim 7 \times 10^6$ erg cm^{-2}s, we deduce that the interval between flares is ~ 25 minutes and so the smoothing requirement is just met, which is probably sufficient.

VII. SOME CONSEQUENCES OF THE FURNACE MODEL

Given our knowledge of the solar wind and assuming that the characteristics of the network assumed in constructing Table III are roughly correct, we can derive some consequences of the furnace model.

First, if the flares have horizontal scales L (km) and occupy the full depth D, the requirement that there be a flare every ~ 25 minutes on any given field line leads to a repetition rate such that an observer looking at an area 1 arcsec square in the network would see $250/L^2$ flares s^{-1}. This corresponds to a rate of 1 to 100 s^{-1} for $15 < L < 150$ with the total energy output being $3 \times 10^{22} - 3 \times 10^{24}$ erg. The largest possible events would correspond to a rate of 1 per 30 minutes, have a total energy output of $\sim 10^{26}$ ergs and would fill the field of view ($L \sim 700$). Whether such events can be detected in the UV, EUV or soft X rays is unclear as this depends on the efficiency of energy conversion to hot plasma and this may not be comparable to the energy released upwards in the form of waves. The very large density contrast between the hot coronal funnels and the cooler chromospheric material in the

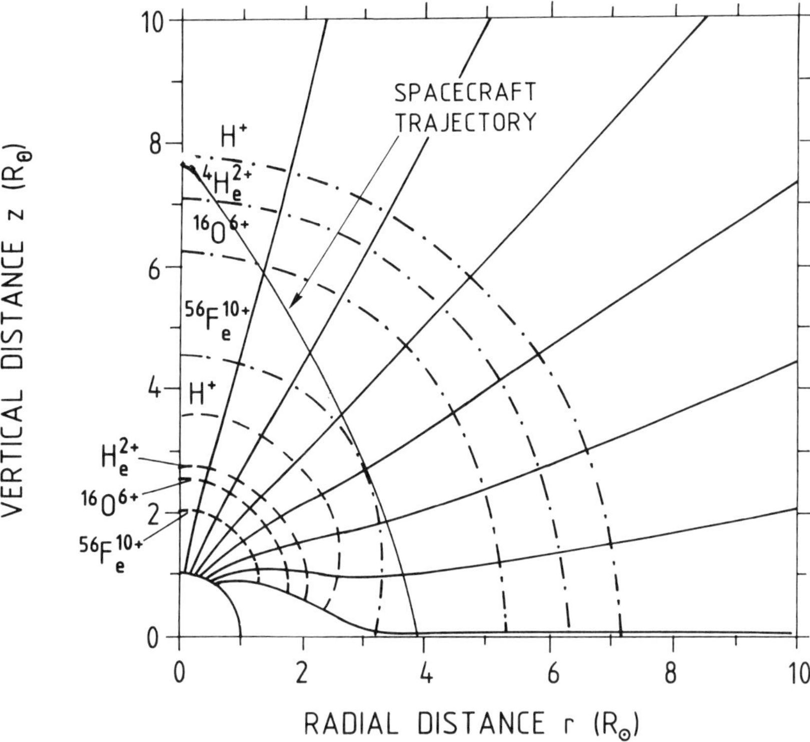

Figure 9.   Magnetic field lines in the corona corresponding to a period around sunspot minimum when coronal holes exist at helio-latitudes $\gtrsim 60°$. Contours of constant magnetic field strength and the corresponding gyrofrequencies (10^2–10^3 He) are shown. Also shown is the tragectory of a possible solar probe.

loops may result in reconnection events having quite different characteristics to those of coronal flares, namely much of the energy released being in the form of escaping Alfvén waves rather than bulk plasma motions and heating. There is, however, ample evidence for the occurrence of such events in or near the network in the form of X-ray bright points and disturbances observed in the far ultraviolet (Habbal and Harvey 1988; Dere 1992,1994). At this stage we should be prepared simply to look for further evidence of events rather than being too concerned about the energetics of the emissions in the particular wavelength chosen. After all we are interested in a furnace in which the fuel is the magnetic field and the output of radiation is Alfvénic, rather than ultraviolet or harder. Given the properties of the solar wind and the nature of the network from which it appears to arise, there is no obvious alternative.

The frequency spectrum of the waves emitting by network flares is of considerable interest given that there are constraints which limit the periods to below ~400 s and to some extent below ~28 s. Magnetic reconnection events have an advantage over the direct production of Alfvén waves by chromospheric turbulence in that the periods of the resulting waves are determined by the Alfvén speed and the geometry of the magnetic configuration involved rather than the periods characteristic of the turbulence. That is, the reconfiguration of the magnetic field associated with a flare converts very low-frequency motions acting over a long time to high-frequency motions generated in a short time. The characteristic period of waves generated by flares in the furnace should be of the order of D/V_A or L/V_A, which, according to Table III, correspond to $1/100(1/10)$ s to as little as $1/2000(1/800)$ s for $L = 15$ km. This suggests that the full spectrum of the waves might cover the range 10^{-4} to 10 s which is quite different from that usually assumed in solar wind models. There is indirect evidence for high-frequency waves in that the "turbulent events" observed in the ultraviolet (Dere 1992,1994) might be expected to involve waves with periods less than the observation time (~ 1 s). (Spatial variations could produce similar effects; however, the size of the field of view places a constraint on the wavelength and hence, for Alfvén waves, also on the period.)

The most interesting aspect of the solar wind source wave spectrum deduced above is that it may contain much higher frequencies than previously considered. This allows linear damping via gyroresonant interactions with the plasma to be a possibility for wave dissipation as suggested by the observations which show that heavy ions are favored relative to protons and that the dissipation converts wave energy into perpendicular thermal energy. In Fig. 9 we show a plot of gyrofrequency for ions with $1 < A/Z < 16$ with an indication of the corresponding distance from the base of the corona. It is important to note that most of the waves contained in the assumed source spectrum will have encountered a resonance within the region where $B \sim 10^{-2}$ Gauss, which corresponds to $\lesssim 10$ $R_\odot$. The possible existence of a linear damping mechanism is important as the waves become nonlinear only at distances $\gtrsim 10\,R_\odot$.

VIII. LOW-SPEED SOLAR WIND

As emphasized in Sec. II, the low-speed solar wind should not be regarded as being in equilibrium with its coronal base (Axford 1977). The filamentary nature of this mode of solar wind flow suggests that each filament is connected to a particular point on the Sun and that neighboring filaments at large distances are not necessarily connected to neighboring points on the Sun. Furthermore, the connections are transient, as a result of continual reconnection in the largely closed field region of the corona, and are not maintained for times long enough to permit equilibrium to be attained along the whole field line to a point of observation at, say, 0.3 to 1 AU. This implies that the characteristic time scales for reconnection (i.e., changing foot points) are short compared with the characteristic times for changes in scalar quantities (temperature, density, composition) to be communicated to 0.3 to 1 A.U. These "communication" times are of the order of 1 to 5 days for convected quantities (i.e., the solar wind travel time) and also, on the basis of a simple dimensional analysis, for thermal conduction.

As a consequence of these arguments, it is clear that any re-arrangement of the field in the "closed" regions of the corona which involves the temporary opening of some field lines to interplanetary space, could create filaments if their time scale is short compared to, say, 1 day. However, it is also necessary that the coronal field lines themselves be sufficiently inhomogeneous in terms of their plasma content for a filamentary structure to become evident in interplanetary space. The time scales for equilibration along closed coronal field lines, including gravitational settling, heat conduction and hydrostatic equilibrium, appear to be of the order of 1 to 3 hr and this is therefore a measure of the time scale for reconnection and filament production. Field lines which remain closed for much longer times tend to come into full hydrostatic and thermal equilibrium with their coronal bases and thereby merge into the more-or-less homogeneous quiet corona. Similarly, field lines which remain open for much longer times will approach a new equilibrium in which the coronal base conditions are those corresponding to a coronal hole. It is therefore necessary that the switching of open and closed field lines should take place on time scales of a few hours at most if filamentary structures are to appear in the slow solar wind region and if these structures are not to be associated with coronal holes.

Some portion of the slow wind is, of course, associated with coronal mass ejection events, at least on a small scale. These events are also associated with the reconfiguration and reconnection of closed coronal field lines but in contrast to the above, result in the formation of magnetic "bottles." They should be identifiable, if the field lines close within the bottles, by the lack of thermal contact with the corona; that is, the electron temperature should be low and there should be no evidence of thermal conduction. Because the slow wind usually exhibits a reasonably high electron temperature, it may be presumed that such events make only a minor contribution to the slow wind as a whole.

Plasma accumulates in the closed field regions of the corona in general over relatively long periods and it is possible for a number of processes to induce changes in the relative abundances of minor species which can become apparent in the slow solar wind if and when the plasma is released into interplanetary space. In addition to gravitational settling and thermal diffusion in fully ionized regions, there are processes occurring at the interface with the chromosphere where the plasma originates, which can lead to abundance variations, especially of those species with small first ionization potential (FIP) (Axford 1985b; von Steiger and Geiss 1989; Ip and Axford 1991). In the source region, the effects of (electron) collisional ionization, photoionization and recombination must be considered, in addition to those mentioned (Marsch et al. 1995). It is important to note that low FIP species are those which can be photo-ionized by Lyman-α radiation, which is, of course, present in abundance in the upper chromosphere and the lower corona (Meyer 1991,1993). As a result, these species (which have small recombination coefficients), are ionized in regions where hydrogen, helium and other high FIP species remain neutral. It is possible to "comb" the low FIP ions out of the mainly neutral regions provided their gyrofrequencies in the ambient magnetic field are large compared with their collision frequencey; in this case, if the magnetic field lines are roughly horizontal and moving upwards, the low FIP ions will be carried preferentially upwards and are thus enriched relative to high FIP ions that are only entrained higher up where they become ionized. Exactly this process occurs in the Earth's ionosphere in the vicinity of the geomagnetic equator where it is observed (Hanson et al. 1972) that Fe^+ is a major ion at altitudes of ~ 1000 km although it could only have originated as a minor species at the ~ 100 km level where it is produced by meteoroid ablation. In the ionosphere, the very large difference between the recombination rates of such metallic ions and molecular ions formed from the major components of the neutral atmosphere is of additional importance but this would not be the case to the same extent for solar material. Horizontal field lines moving upwards are balanced by those moving downwards but the latter are neutral as far as changing the composition is concerned; the net effect is that the relative abundances of low FIP species should become enhanced in the fully ionized regions above.

It should be noted that, because helium is by no means a minor species, a similar effect could occur with respect to hydrogen (which is low FIP relative to the He I 584 radiation background) and helium. That is, in regions of upwards-moving horizontal magnetic field lines, the ionization structure may be such that there is a thin layer containing an abundance of protons and other "low" FIP (<21.1 eV) ions together with the normal low FIP (<10.2 eV) ions, with the helium remaining neutral. By analogy with the previous arguments, it is clear that there is a possibility for helium (and perhaps neon with FIP $= 21.6$ eV) to be depleted relative to hydrogen.

IX. CONCLUSIONS

The approach to understanding the nature of the solar wind given here contains a number of new features and the conclusions are as a consequence, somewhat different from those reached by others. We have emphasized the need to understand the nature of the high-speed solar wind as the basic equilibrium mode of solar wind flow, noting that the low-speed wind is not an equilibrium mode and cannot be understood simply by arbitrarily altering the boundary conditions and assuming steady flow. Most importantly we have paid considerable attention to the source of the solar wind noting that the mechanisms that produce the energy required to account for the wind and the plasma of which it is comprised should not be treated as if they were completely independent matters.

As a consequence of this approach we deduce that, in order to account for the properties of the equilibrium high-speed streams observed by the Helios spacecraft at 60 $R_\odot$ and to provide a sensible match to conditions in the associated coronal holes, it is necessary to consider the wind as originating in the chromospheric network as a consequence of flare-like reconnection events in the strong and complex magnetic fields that exist there (Gabriel 1976a,b; Dowdy et al. 1986). This has led us to the conclusion that the spectrum of the hydromagnetic waves that provide most of the energy requirement is dominated by much higher frequencies than previously assumed and that the dissipation of these waves is relatively rapid and could involve an essentially linear mechanism such as cyclotron damping. In any case, these waves should be absent at 60 $R_\odot$ with those remaining having low frequencies and little relevance to the acceleration of the wind. Energetic particles may also contribute to the energy flux as suggested by Jokipii and Morfill (1993) and these too could be a natural product of the flare-like events proposed.

The question of the mass flux in the solar wind is made complicated by the fact that, whereas the wind itself may be considered as being steady and homogeneous, apart from the presence of hydromagnetic waves, the source of the wind, namely the furnace described here, is strongly heterogeneous and time variations are essential to its role. It is obviously difficult to model such a region although it is essential to our understanding of the solar wind phenomenon. However, the very simple analysis we have made suggests that the ionization process itself could be a controlling factor in determining the mass flux because an admittedly oversimplified model, involving an ionization-conduction front eating into the cool, depleted upper chromosphere in the network, suggests that recombination losses may be unimportant and hence the solar wind mass flux is ultimately determined by the ionization rate. This does not mean that recombination does not occur but rather that it is largely confined to high-density closed loops where the plasma is trapped unless released by a reconnection event. It is also possible that the ionization process may be partly nonthermal in view of the presence of flaring which may give rise to a significant energetic particle population.

As a corollary, the density of the quasi-hydrostatic region which one might wish to describe as the base of the corona above a coronal hole should be regarded as being part of the solution of the overall problem rather that being assumed as a boundary condition. This does not necessarily lead to any conflict with observations because the density obtained from white light coronal measurements is very uncertain in view of the need to eliminate fore- and background contamination and could well be as low as suggested by Koutchmy (1977), for example. Such low densities in the funnels depicted in Fig. 7 would make it almost impossible for useful measurements of the wave amplitude to be made near the heterogeneous coronal base where the much denser and hence more strongly emitting closed coronal regions provide most of the emission. Accordingly, we do not wish to be constrained by observations which suggest that the wave amplitude is as low as 30 km s^{-1}

(see, e.g., Hassler et al. 1990; Mariska 1992) if it is not demonstrated that the measurements refer to coronal funnels rather than closed loops. There is enough evidence that flaring occurs in the network (Dere 1992,1994) with associated broad and Doppler-shifted emission features to suggest that the wave amplitude in funnels could be as large as we have suggested but this must also be contaminated by emission from the brighter closed-field regions.

We are proposing a new approach to solar wind modeling in which the source of the mass and energy involved is brought into consideration, the geometry of the magnetic field in the all-important region near the Sun is more complex than usually assumed and the interpretations of observations traditionally adopted to provide coronal boundary conditions are called into question. We have emphasized that the high-speed solar wind can be regarded as a well-defined time-stationary phenomena with interesting clues concerning its origin contained in the observed plasma properties even at relatively large distances from the Sun. However, we have also made the problem more complex by linking it to the network heterogeneity, flaring and the ionization process(es).

A simplification may be possible if the arguments contained in Sec. IV are correct. For example, if the wave dissipation length is not too large the wave pressure could possibly be neglected in comparison with the gas pressure as a first approximation and the wave dissipation modeled as a heat source for the protons in the form $Q_\perp(r)$ with an appropriate amplitude and length scale. Assuming a two-fluid model with anisotropic proton pressure, and with an assumption concerning the electron heat source, the wind can be modeled on the basis of a combination of the approaches suggested by Sturrock, Hartle, Holzer, Leer and Axford (described in Sec. IV) to give solutions with the required energy and mass fluxes and with the coronal density and electron temperature and (negative) temperature gradient as outputs. If reasonably acceptable, these can in turn be matched to a quasi-hydrostatic model with heat conduction for the region of rapidly diverging magnetic field above the chromospheric network (see Fig. 6) and to a (necessarily) crude model for the furnace below, allowing for ionization and recombination as appropriate. Because the waves have been separated from the solution it is necessary to check *a posteriori* that the wave pressure is indeed small and that the waves propagate and damp in a manner consistent with the assumed $Q_\perp(r)$. If not, the waves must be included explicitly using forms for p_w and $Q_\perp$ such as given in Eqs. (20a,b). It is also necessary to check that the electron heat source, which we have not discussed here, is reasonable. Minor ions can be added at this point and the wave field modified accordingly to take account of their gyroresonances. Without these simplifications the problem is a difficult one indeed.

No model of the solar wind will ever be free of such simplifying assumptions, especially in view of the complexity of the source region. However, in our opinion it should be possible using procedures similar to those outlined above, to achieve satisfactory results which can be checked against observations of the sort that will be made available from the SOHO mission and perhaps ultimately from a solar probe. We believe that the results will in any case show that the density and temperature of the base of the corona will be relatively low as suggested by Koutchmy (1977) and Habbal et al. (1993), and that the wave amplitudes will be higher than usually assumed with a dominance of relatively high frequencies (i.e., periods of 10 s and less).

Acknowledgment. Discussions with E. Marsch, W.-H. Ip, and C.-Y. Tu, and constructive comments from referees are gratefully acknowledged.

REFERENCES

Alazraki, G., and Couturier, P. 1971. Solar wind acceleration caused by the gradient of Alfvén wave pressure. *Astron. Astrophys.* 13:380–389.

Axford, W. I. 1976. Flow of mass and energy in the solar system. In *Physics of Solar Planetary Environments*, ed. D. Williams (San Francisco: American Geophysical Union), pp. 270–285.

Axford, W. I. 1977. The three-dimensional structure of the interplanetary medium. In *Study of Travelling Interplanetary Phenomena*, eds. M. A. Shea, D. F. Smart and S.-T. Wu (Dordrecht: D. Reidel), pp. 145–164.

Axford, W. I. 1980. Very hot plasmas in the solar system. In *Highlights of Astronomy*, vol. 5, ed. P. A. Wayman (Dordrecht: D. Reidel), pp. 351–359.

Axford, W. I. 1985*a*. The solar wind. *Solar Phys.* 100:575–586.

Axford, W. I. 1985*b*. Contribution to the J. Geiss Anniversary Symposium, Univ. of Bern.

Axford, W. I. 1990. Origin of the solar wind. European Geophysical Soc. Meeting, Copenhagen.

Axford, W. I., and McKenzie, J. F. 1992. The origin of high speed solar wind streams. In *Solar Wind Seven*, eds. E. Marsch and R. Schwenn (Oxford: Pergamon Press), pp. 1–5.

Axford, W. I., and Newman, R. C. 1967. Viscous transonic flow in accretion and stellar wind problems. *Astrophys. J.* 147:230–234.

Bame, S. T., Asbridge, J. R., Feldman, W. C., and Gosling, J. T. 1977. Evidence for a structure-free state at high solar wind speeds. *J. Geophys. Res.* 82:1487–1492.

Belcher, J. W. 1971. Alfvénic wave pressure in the solar wind. *Astrophys. J.* 168:509–524.

Biermann, L. 1951. Kometenschwiefe und solare Korpuskularstrahlung. *Zeit. Astrophys.* 29:274–286.

Biermann, L. 1953. Physical processes in comet tails and their relation to solar activity. *Extrait des Memoires Soc. Roy. Sci. Liège Coll.* 4(13):291–302.

Biermann, L. 1961. The solar wind and the interplanetary media. In *Space Astrophysics*, ed. W. Liller (New York: McGraw-Hill), pp. 150–156.

Bürgi, A. 1992. Proton and alpha particle fluxes in the solar wind: Results of a three-fluid model. *J. Geophys. Res.* 97:3137–3150.

Chamberlain, J. W. 1961. Interplanetary gas. III. A hydrodynamic model of the corona. *Astrophys. J.* 133:675–687.

Chapman, S. 1959. Interplanetary space and the earth's outermost atmosphere. *Proc. Roy. Soc.* 253:462–481.

Dere, K. P. 1992. Explosive events and magnetic reconnection in the solar atmosphere. In *Solar Wind Seven*, eds. E. Marsch and R. Schwenn (Oxford: Pergamon Press), pp. 11–20.

Dere, K. P. 1994. Explosive events, magnetic reconection, and coronal heating. *Adv. Space Res.* 14(4):13–22.

Dougherty, M., and McKenzie, J. F. 1991. Compressional instability in the solar wind driven by wave dissipation. *J. Geophys. Res.* 46:179–185.

Dowdy, J. F., Jr., Rabin, D., and Moore, R. L. 1986. On the magnetic structure of the quiet transition region. *Solar Phys.* 105:35–45.

Elzner, L. R., and Elwert, G. 1980. Theoretical network structure of the transition region chromosphere. I and II. *Astron. Astrophys.* 86:181–191.

Esser, R., Holzer, T. E., and Leer, E. 1987. Drawing inferences about solar wind acceleration from coronal minor ion observations. *J. Geophys. Res.* 92:13377–13389.

Ferraro, V. C. A., and Plumpton, C. 1966. *An Introduction to Magneto Fluid Dynamics* (Oxford: Clarendon Press).

Gabriel, A. H. 1976*a*. A magnetic model of the solar transition region. *Phil. Trans. Roy. Soc.* A281:339–352.

Gabriel, A. H. 1976*b*. Structure of the quiet chromosphere and corona. In *The Energy Balance and Hydrodynamics of the Solar Chromosphere and Corona*, ed. R. Bonnet and P. Delache (Clermont-Ferrand: G. de Bussal), pp. 375–418.

Gleeson, L. J., and Axford, W. I. 1976. An analytic model illustrating the effects of rotation on a magnetosphere containing low energy plasma. *J. Geophys. Res.* 81:3403–3406.

Gold, T. 1964. Magnetic energy shedding in the solar atmosphere. In *Proc. AAS-NASA Symp. on The Physics of Solar Flares*, pp. 389–395.

Gringauz, K. I. 1961. Some results of experiments in interplanetary space by means of charged particle traps on Soviet space probes. *Space Res.* 2:539–553.

Habbal, S. R., and Harvey, K. L. 1988. Simultaneous observations of 20 centimeter bright points and He 1 10830 Å dark points in the quiet sun. *Astrophys. J.* 326:966–988.

Habbal, S. R., Esser, R., and Arndt, M. B. 1993. How reliable are coronal hole temperatures deduced from observations? *Astrophys. J.* 413:435–444.

Hanson, W. B., Sterling, D. L., and Woodman, R. F. 1972. Source and identification of heavy ions in the equatorial F-layer. *J. Geophys. Res.* 77:5530–5541.

Hartle, R. E., and Sturrock, P. A. 1968. Two-fluid model of the solar wind. *Astrophys. J.* 151:1155–1170.

Hassler, S. R., Rottman, G. J., Shoub, E. C., and Holzer, T. E. 1990. Line broadening of Mg X 609 and 625 coronal emission lines observed above the solar limb. *Astrophys. J.* 348:77–80.

Hollweg, J. V. 1973. Alfvén waves in a two-fluid model of the solar wind. *Astrophys. J.* 181:547–566.

Hollweg, J. V. 1974. Transverse waves in the solar wind: Arbitrary k, v_o, B_o, and (δB). *J. Geophys. Res.* 79:1539–1541.

Hollweg, J. V. 1990. On WKB expansions for Alfvén waves in the solar wind. *J. Geophys. Res.* 95:14873–14879.

Hollweg, J. V. 1992. Status of solar wind modeling from the transition region outwards. In *Solar Wind Seven*, eds. E. Marsch and R. Schwenn (Oxford: Pergamon Press), pp. 53–60.

Hollweg, J. V., and Johnson, W. 1988. Transition region, corona and solar wind in coronal holes: Some two-fluid models. *J. Geophys. Res.* 93:9547–9554.

Holzer, T. E., and Axford, W. I. 1970. The theory of stellar winds and related flows. *Ann. Rev. Astron. Astrophys.* 8:31–60.

Ip, W.-H., and Axford, W. I. 1991. On the first ionization potential effect of the solar corona. *Adv. Space Res.* 11:247–250.

Isenberg, P. A. 1991. The solar wind. In *Geomagnetism*, vol. 4, ed. J. A. Jacobs (Orlando: Academic Press), pp. 1–85.

Isenberg, P. A., and Hollweg, J. V. 1983. On the preferential acceleration and heating of solar wind heavy ions. *J. Geophys. Res.* 88:3923–3935.

Jokipii, J. R., and Morfill, G. E. 1993. Cosmic rays and the acceleration of the solar wind. *Proc. Cosmic Rays XXIII*, vol. 3, pp. 179–182.

Kojima, M., and Kakinuma, T. 1987. Solar cycle evolution of the solar wind structure between 1973 and 1985 observed with the interplanetary scintillation method. *J. Geophys. Res.* 92:7269–7279.

Koutchmy, S. 1977. Study of the June 30, 1973, trans-polar coronal hole. *Solar Phys.* 51:399–407.

Leer, E., and Axford, W. I. 1972. A two-fluid solar wind model with anisotropic proton temperature. *Solar Phys.* 23:238–250.

Leer, E., and Holzer, T. E. 1991. The solar wind mass flux problem. *Ann. Geophys.* 9:196–201.

Leer, E., Holzer, T. E., and Shoub, E. C. 1992. The solar wind from a corona with a large helium abundance. *J. Geophys. Res.* 97:8183–8201.

Loucif, M. L., and Koutchmy, S. 1989. Solar cycle variations of coronal structures. *Astron. Astrophys. Suppl.* 77:45–66.

Mariani, F., and Neubauer, F. M. 1991. The interplanetary magnetic field. In *Physics of the Inner Heliosphere*, vol. 21, eds. R. Schwenn and E. Marsch (Berlin: Springer-Verlag), pp. 183–206.

Mariska, J. T. 1992. *The Solar Transition Region* (New York: Cambridge Univ. Press).

Marsch, E. 1990. MHD turbulence in the solar wind. In *Physics of the Inner Heliosphere*, vol. 20, eds. R. Schwenn and E. Marsch (Berlin: Springer-Verlag), pp. 159–241.

Marsch, E. 1991. Kinetic physics of the solar wind plasma. In *Physics of the Inner Heliosphere*, vol. 21, eds. R. Schwenn and E. Marsch (Berlin: Springer-Verlag), pp. 45–133.

Marsch, E. 1994. Theoretical models for the solar wind. *Adv. Space Res.* 14(4):103–121.

Marsch, E., and Schwenn, R., eds. 1992. *Solar Wind Seven* (Oxford: Pergamon Press).

Marsch, E., Goertz C. K., and Richter, K. 1982. Wave heating and acceleration of solar wind ions by cyclotron resonance. *J. Geophys. Res.* 87:5030–5044.

Marsch, E., von Steiger, R., and Bochsler, P. 1995. Element fractionation by diffusion in the solar chromosphere. *Astron. Astrophys.*, 301:261–276.

McKenzie, J. F. 1991. Solar corona and wind. *J. Geomag. Geoelec.* 4:45–58.

McKenzie, J. F. 1994. Interaction between Alfvén waves and a multicomponent plasma with differential ion streaming. *J. Geophys. Res.* 99:4193–4200.

McKenzie, J. F., Ip, W.-H., and Axford, W. I. 1979. The acceleration of minor ion species in the solar wind. *Astrophys. Space Sci.* 64:183–211.

McKenzie, J. F., Marsch, E., Baumgärtel, K., and Sauer, K. 1993. Wave and stability properties of multi-ion plasmas with applications to winds and flows. *Annales Geophysicae* 11:341–353.

Meyer, J.-P. 1991. Diagnostic methods for coronal abundances. *Adv. Space Res.* 11(1):269–280.

Meyer, J.-P. 1993. Elemental abundances in active regions, flares and interplanetary medium. *Adv. Space Res.* 13:377–390.

Parker, E. N. 1958. Dynamics of the interplanetary gas and magnetic fields. *Astrophys. J.* 128:664–676.

Parker, E. N. 1961. The solar wind. In *Space Astrophysics*, ed. W. Liller (New York: McGraw-Hill), pp. 157–170.

Parker, E. N. 1963. *Interplanetary Dynamical Processes* (New York: Wiley Interscience).

Parker, E. N. 1965. Dynamical theory of the solar wind. *Space Sci. Rev.* 4:666–708.

Parker, E. N. 1969. Theoretical studies of the solar wind phenomenon. *Space Sci. Rev.* 9:325–360.

Parker, E. N. 1988. Nanoflares and the solar wind X-ray corona. *Astrophys. J.* 330:474–479.

Pneuman, G. W., and Kopp, R. A. 1971. Gas-magnetic field interactions in the solar corona. *Solar Phys.* 18:258–270.

Rickett, B. J., and Coles, W. A. 1991. Evolution of the solar wind structure over a solar cycle: Interplanetary scintillation velocity measurements compared with coronal observations. *J. Geophys. Res.* 96:13849–13859.

Scarf, F. L., and Noble, L. M. 1965. Conductive heating of the solar wind. II. The inner corona. *Astrophys. J.* 141:1479–1491.

Schwenn, R. 1990. Large scale structure of the interplanetary medium. In *Physics of the Inner Heliosphere*, vol. 20, eds. R. Schwenn and E. Marsch (Berlin: Springer-Verlag), pp. 99–181.

Schwenn, R., and Marsch, E., eds. 1990. *Physics of the Inner Heliosphere*, vol. 20 (Berlin: Springer-Verlag), pp. 1–2.

Schwenn, R., Montgomery, M. D., Rosenbauer, H., Miggenieder, H., Mülhäuser, K.-H., Bame, S. J., Feldman, W. C., and Hansen, R. T. 1978. Direct observations of the latitudinal extent of a high-speed stream in the solar wind. *J. Geophys. Res.* 83:1011–1017.

von Steiger, R., and Geiss, J. 1989. Supply of fractionated gases to the corona. *Astron. Astrophys.* 225:222–238.

Sturrock, P. A., and Hartle, R. E. 1966. Two-fluid model of the solar wind. *Phys. Rev. Lett.* 16:628–631.

Tu, C.-Y. 1987. A solar wind model with the power spectrum of Alfvénic fluctuations. *Solar Phys.* 109:149–186.

Tu, C.-Y. 1988. The damping of interplanetary Alfvénic fluctuations and the heating of the solar wind. *J. Geophys. Res.* 93:7–20.

Tu, C.-Y., Pu, Z., and Wei, F. 1984. The power spectrum of interplanetary Alfvńic fluctuations: Derivation of the governing equation and its solution. *J. Geophys. Res.* 89:9695–9702.

Weber, E. J. 1973. Multi-ion plasma in astrophysics. *Astrophys. Space Sci.* 20:391–465.

Withbroe, G., and Noyes, R. W. 1977. Mass and energy flow in the solar wind chromosphere and corona. *Ann. Rev. Astron. Astrophys.* 15:363–387.

Yeh, T. 1970. A three-fluid model of solar winds. *Planet. Space Sci.* 18:199–215.

THE QUEST FOR EVIDENCE OF LONG-PERIOD SOLAR WIND VARIABILITY

C. P. SONETT, G. M. WEBB and A. ZAKHARIAN
University of Arizona

Long-term variability in solar wind parameters inferred from measurements on Earth is the response to a complex mixture of solar and terrestrial forcing (with which may be associated a small interstellar addition), mainly modulation of the galactic cosmic ray flux, which in turn controls the production of cosmogenic nuclides. From the extraterrestrial standpoint, the major issue is the relation of long-period variability to its primarily solar origin and what connection this information might have regarding the structure and dynamics of the Sun. The recent record is augmented by the sunspot index, aurorae, and fluctuations in the geomagnetic field. Radiocarbon (^{14}C), and to a lesser extent ^{10}Be, discloses modulation periods ranging from $\sim$2300 yr downward through the Suess (208 yr), the Gleissberg9 ($\sim$80 yr) and possibly through the 11 yr and beyond. The Suess period is tentatively identified with extension of the Maunder minimum through the entire Holocene. Though evidence for the period of about 2300 yr is present, whether it is extraterrestrially generated is uncertain. That the solar wind is a truly ancient phenomenon is certified by the record in meteorites and lunar regolith. The earliest (Hadean) record, identified with the primeval Sun, is accessed through meteorite metamorphosis and nuclear track detection. The radiocarbon and ^{10}Be variability is conjectured to arise at least partially from eigenmodes of the solar dynamo.

I. PROLOGUE AND BACKGROUND

Although the overall structure of the Sun is generally accepted as being relatively static since the pre-main phase, it is known that the Sun exhibits variability ranging upwards in period from those characterizing the convective turbulence; the decadal, e.g., 11- and 22-yr periods are dominated by the solar dynamo. But all these are thought to be expressions of the dynamics of the convective zone leaving the core in relative quiet—though this view is not entirely universal, e.g., the neutrino flux problem. There is little in the way of direct evidence of longer-period or secular variability, but some information exists from the sunspot record, radioisotopes, geomagnetism, the auroral record, and possibly other relicts preserved in rocks. That evidence is remarkably difficult to unravel as it is based upon a combination of terrestrial and extraterrestrial forcing, and even possibly a small interstellar component leaking through the heliosphere. Though this terrestrially based record can in some instances be traced to the solar wind, the successive steps to the Sun are more conjectural.

The long-term variability of the solar wind is a poorly defined term covering the basic interplanetary plasma parameters (see, e.g., Sonett 1991). By and large, most of the terrestrial proxies tend to be thought of in terms of the interplanetary magnetic field, though not always. The information basic to the issue of long-term variability is divisible into electromagnetic, e.g., aurora, geomagnetic field, sunspots and secondly into that using the production of radioisotopes via the galactic cosmic rays (GCR) and their modulation in the heliosphere as a probe of solar wind behavior (see, e.g., Blinov 1988). (Section IV gives a summary discussion of modulation.) Most of the effects of diffusion and anisotropy present in the 11 to 22 yr range would likely be averaged out for longer periods, as the transit time through the whole heliosphere for the solar wind cannot be more than at the most several years. The terrestrially based parameters are not easily associated with the wind on a one-to-one basis. The series of short commentaries follow this scenario.

It is convenient to define three time categories: (a) that from present epoch high resolution spacecraft instruments, (b) the historical records of aurorae and the geomagnetic field, (c) the long-term prehistorical terrestrial record of cosmogenic isotopes generated by the cosmic ray flux on the top of the atmosphere and whose variations are at least partially attributable to interplanetary modulation, as well as a fourth (d) the extension to the early solar system. (Climate is usually taken as a matter of the radiative output of the Sun and thus independent of the solar wind, though common forcing could be the cause of as yet unrevealed common mode response of climate and the solar wind.) This review bypasses (a) and is centered on (b) and (c), with summary comments on (d). A major problem in understanding the long-term history of the solar wind resides in coordinating data even just from (b) and (c).

Variations of period $\sim$11 to 22 yr in solar wind parameters, density, velocity, turbulence, and magnetic field, are a matter primarily of variability in solar parameters responsible for the genesis of the solar wind. (For brevity we shall refer to these periods simply as 11 and 22 yr except where noted differently.) The 11-yr period of solar activity exhibits a variance of about 2.3 yr; the 22-yr period probably mimics this. Over such long times, changes in the Sun are reflected adiabatically in the solar wind in view of the rapid transit time for the wind through the entire heliosphere.

A division of long-term periodicity can be made by separating periods less and greater than 11 yr. An example is establishing the putative multi-century interruption of the 11-yr period, e.g., Maunder-like minimum in the 11-yr cycle of solar activity (Maunder 1890,1922). Periods of 70 to 90 yr (Gleissberg 1944,1966) and 208 yr (Suess 1968,1970; Houtermans 1971,1973; Sonett 1984) in radiocarbon all exhibit this. Only for the 11-yr period is directly inferred evidence available of a solar wind connection. There is today no broadly accepted recognition that solar parameters vary over a period of 22 yr (the Hale magnetic cycle) or longer, though the cosmogenic evidence seems compelling to these authors. Curiously, the full period of the 22-yr Hale polarity reversal of the solar dynamo is far less obvious than the major 11-yr period.

There is some evidence that side-bands or harmonics of the 11-yr period appear in the sunspot index. These are probably associated with the basic nonlinearity of the dynamo, and thus may not be of intrinsic major importance in themselves. High accuracy radiometry from satellites also discloses a small but significant ($\sim$0.15%) 11-yr solar irradiance variation (Willson et al. 1986; Hudson 1988). And there is some sketchy and uncertain evidence that the neutrino flux varies over 11 yr (see, e.g., Wolfsberg and Kocharov 1991). Standard solar models have not led to insight on mechanisms of multi-year global variability; helioseismology resides at the very high-frequency end of the spectrum of variability and appears to be of no account in the quest for understanding putative solar sources of very long periods. As the 11-yr dynamo period is impressed in a variety of parameters, some of them terrestrial, there is little reason to doubt that variability in the dynamo is propagated to the Earth via the solar wind through the intermediary of the interplanetary magnetic field.

A sensitive and direct indicator of solar wind variability is that of modulation of the GCR. Historically long-period modulation is expressed through the terrestrial, lunar and meteoritic cosmogenic nuclide record (see, e.g., Reedy et al. 1983), while the multi-yearly epoch is covered by neutron monitors from the 1950s on (Simpson 1957). Other kinds of information such as comet tail photography extends the overall record backwards to the turn of the century; more conjectural records are historical terrestrial parameters such as geomagnetic field, aurora, and perhaps relevant but very controversial climate signals.

On into the past, very little is known of what might be inferred to be associated with an underlying solar wind variability beyond at the very most a few million years from meteorite isotope study (see, e.g., Caffee et al. 1991), but so far as to what theory indicates, there is no reason not to suppose that the solar wind has been flowing since some time after the formation of the solar system. There is actually more conjecture regarding the very early solar system from isotope studies and from the thermal history of meteorites and the Moon (Sonett et al. 1970; Herbert et al. 1991).

In the next section the sunspot index (SI) is given a summary revisit. (The major literature on the structure and source of sunspots is bypassed here, being outside the restricted scope of this review.) The association of sunspots with the solar wind remains elusive, and is further complicated by the problem of coronal mass ejections (CMEs); we are reminded that the issue here is long-term solar wind variability and the connection with CMEs is a barely explored matter. This actually applies as well to sunspots vs solar wind variability, although SI does provide a historical context not otherwise available.

As noted earlier the cosmic ray record in meteorites and the lunar regolith are not covered in this chapter. The one exception is the short section on the putative T Tauri phase of the primordial Sun. Finally the variability of the ^{10}Be record from the Antarctic Vostok station is noted as a consequence of geomagnetic variability, neighboring interstellar space, or both including the secular enhancements and the possible consequence of one or more relict supernova-induced interstellar shocks leading in turn to the observed enhanced GCR as suggested from ^{10}Be Vostok antarctic ice core data. Long-term diffusion of GCR into the heliosphere is of course a central fact in the observations of GCR; the problem of pickup ions is important to the outer heliosphere, but is not treated here.

II. THE HISTORICAL AND PROTOHISTORICAL RECORD

A. Sunspots

Of all solar associated parameters, sunspots have by far the central historical role ranging from the ancient oriental record and from the rebirth in the Renaissance and the recognition later by Schwabe of their basically periodic behavior. Among historical records, sunspots extend backwards in time to Chinese antiquity (Bray and Loughhead 1964). But the very early records are fragmentary and not capable of revealing subtleties which turn out to be important. The early oriental observations were heavily biased, for only major spots were observable to the naked eye.

The sunspot index (SI) is determined on a partially qualitative basis from a combination of N_S, the number of sunspots, G the number of spot groups, and a weighting factor W, a figure of merit associated with individual observers and instruments (see, e.g., Gibson 1973) by

$$SI = W(N_S + 10\,G). \tag{1}$$

The modern (telescopic) era began with Galileo and Scheiner, and the invention of the telescope (see Coyne [1991] for a compact and informative history of the Galileo-Scheiner dialogue). A largely unbroken record of increasing reliability exists from 1700 A.D. onward; the record after 1850 is regarded as the most accurate. The prominent 11-yr period is attributed to the transit time for dynamo waves from the solar sunspot zone to the poles where merging results in a residual dipole field (Parker 1955a,b). Key properties of the index are that successive 11-yr periods are of significantly unequal amplitude and of an amplitude modulation envelope with a period of about 75 to 80 yr (Cohen and Lintz 1974; Sonett 1982,1983).

Though establishment of a canonical spectrum for the SI might appear a matter easily resolvable, after many years and countless calculations, it still seems to be an issue not fully settled. A surmise is that the spectrum is complicated, dynamically chaotic (see, e.g., Gudzenko and Chertoprud 1965; Ruzmaikin 1981; Feynman and Gabriel 1990; Morfill et al. 1991), and contains a small but significant quasi-static component (Sonett 1982,1983). It is no help that the SI is everywhere positive; though this corresponds to the obvious statement that sunspots are never negative, this complicates interpretation of the spectrum. Nevertheless, however uncertain, the SI as a measure of convective zone hydromagnetics has retained a considerable degree of popularity, undoubtedly due to its being a window leading towards the dynamics of the Sun's magnetic field source and is today a broadly invoked source of conjecture on oscillatory behavior of the Sun's field.

Figure 1a shows the SI yearly averaged time series. The most obvious property is the envelope of modulation with a period about 8 times that of the 11-yr period. The power spectrum based on the model of Eq. (2) is shown in Fig. 1b, where as expected the 11-yr line dominates and with which are noticeably associated numerous adjacent lines. The previously mentioned envelope emerges at a period of 80 to 90 yr known today as the Gleissberg period; its second harmonic resides at 45 yr. Uncertain evidence of vestigial power at 22 yr is present but at a much lower level than at 11 yr.

It is possible to generate a relatively simple hueristic model for the bulk of this spectrum (including the smallness of the power at 22 yr) by the supposition that the underlying forcing (in amplitude) of the SI is based on the 22-yr Hale cycle but that the forcing energy follows the 11-yr period (following that magnetic energy is the parameter of interest in SI, a reasonable conjecture as the basic Hale period is dominated by the 11-yr cycle). Then

$$ SI\,(t) = \{[1 + \alpha\cos(\omega_m t)][\Delta + \cos(\omega_c t)]\}^2 \qquad (2) $$

where ω_m is the Gleissberg angular frequency, ω_c the angular frequency of the 22-yr Hale period, $0 \le \alpha \le 1$ is the amplitude modulation index, and Δ is included as the aforementioned putative offset from zero of the 11-yr signal (Sonett 1982). The Fourier spectrum of Eq. (2) contains the Gleissberg line, its second harmonic and two quintets about 11 and 22 yr, respectively. See Table I, where Δ can be viewed as a steady field in the solar core (Cowling 1945) extending upwards into the convective zone, thus adding and subtracting from alternate dynamo cycles.

The close association of this model with SI infers that the process leading to the Gleissberg period is nonlinear, resulting in the 45-yr second harmonic. In the absence of the constant term Δ in the model, the quadratic dependence of SI on ω_m would explain the absence of any 22-yr component. That any 22-yr signal is present is a consequence of the nonvanishing contribution of Δ to the dynamo. Though this model remains speculative, it does explain the very weak Hale period and the evidence for the line quintets at 11 and 22 yr. The model is consistent with dynamo nonlinearity and suggests an eigenmode with an $\sim$88-yr period in addition to the Hale cycle.

B. Terrestrial Magnetic Activity and Solar Wind Parameters

The most common functional relation between solar wind parameters and geomagnetic variability is specifically a nonlinear dependence of the latter on

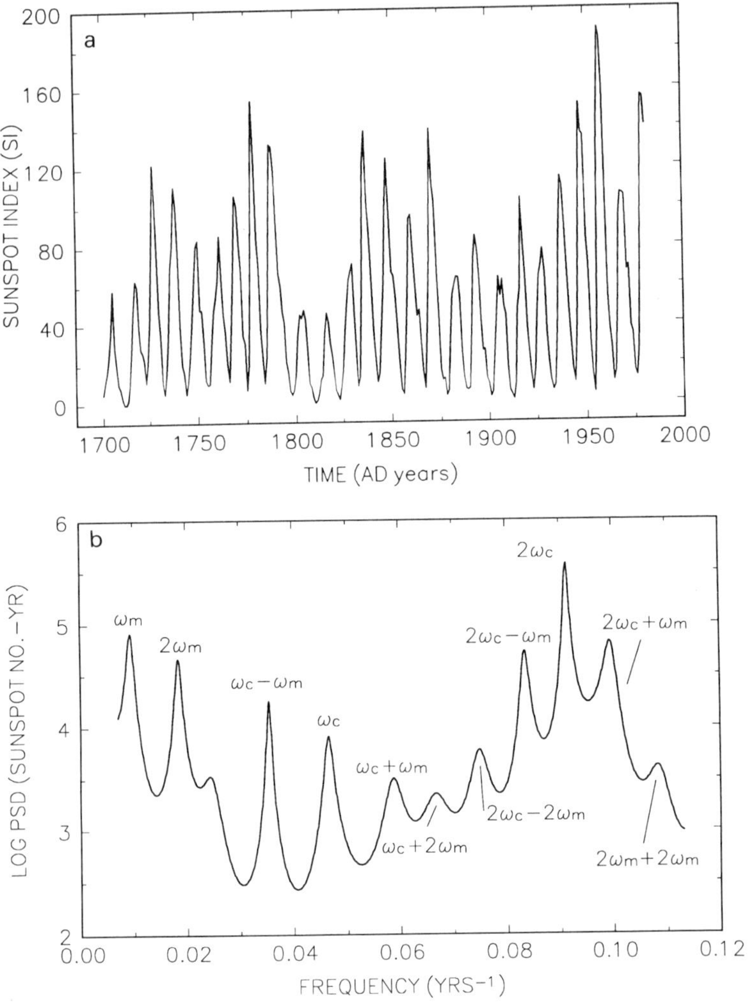

Figure 1.   Sunspot index (SI): (a) yearly averaged time series, (b) maximum entropy power spectrum (order 140). In Table I ω_m designates Gleissberg and its harmonic $2\omega_m$ and ω_c the Wolf period. Spectral line numeral identification numbers correspond to column 1 of Table I.

the south directed field and the product of the solar wind bulk speed having the form (Holzer and Slavin 1982)

$$B_z^m v^n \tag{3}$$

where B_z is the component of the interplanetary magnetic field normal to the ecliptic, v the solar wind bulk speed, and $0.85 \leq m \leq 2$ and $1 \leq n \leq 2$. Holzer and Slavin found that

$$<AL> = 5.6 \times 10^{-4} B_z^{0.85} v^2. \tag{4}$$

TABLE I

Fourier Transform of SI Index for Amplitude Modulation of
Hale Cycle (ω_c) by Gleissberg Frequency (ω_m)[a]

Line Number[b]	Frequency	Line Strength
1	0	$1/2 + \Delta^2(1 + \alpha^2) + \alpha^2/4$
2	ω_m	$\alpha(1 + 2\Delta^2)$
3	$2\omega_m$	$\alpha^2/2(1/2 + \Delta^2)$
4_{-2}	$\omega_c - 2\omega_m$	$\Delta\alpha^2/2$
4_{-1}	$\omega_c - \omega_m$	$2\Delta\alpha$
4_0	ω_c	$\Delta(2 + \alpha^2)$
4_1	$\omega_c + \omega_m$	$2\Delta\alpha$
4_2	$\omega_C + 2\omega_m$	$\Delta\alpha^2/2$
5_{-2}	$2\omega_c - 2\omega_m$	$\alpha^2/8$
5_{-1}	$2\omega_c - \omega_m$	$\alpha/2$
5_0	$2\omega_c$	$1/2(1 + \alpha^2/2)$
5_1	$2\omega_c + \omega_m$	$\alpha/2$
5_2	$2\omega_c + 2\omega_m$	$\alpha^2/8$

[a] $0 < \alpha < 1$ in the table represents amplitude modulation factor.

[b] Line number classification follows the convention, e.g., 4_{-2}, means the lowest frequency member of the 22-yr quintet.

The correlation coefficient between predicted and the actual value of AL was 0.97, decreasing to 0.92 for $B_z v$ and even lower to 0.82 for $B_z^2 v$. This may be compared to Feynman and Crooker (1978) for the computed aa index where

$$aa = 1.3 + 4.7 \times 10^{-5} B_z v^2 \tag{5}$$

for the relation to the aa index (see also Silverman 1992).

All these relations are of the previously defined form $B_z^m v^n$. Overall, the relation is dependent upon several magnetospheric factors; they are important if one aims to use the global magnetosphere as a measure of one or more solar wind parameters. Moreover tests of these parameters are restricted to the past few solar cycles, so strong conclusions are hard to come by. Geomagnetic activity given by aa is shown by Feynman (1982) (Fig. 2) compared to SI. Cross correlation is not shown, but inspection shows that the two are qualitatively correlated. Slavin and Smith disclose a histogram of $\log B$ during both solar maximum and minimum (Fig. 3) showing a change in mean $\log B$ of 25% (78% in amplitude). Such a large change suggests so far unknown spectral features underlying common mechanism of sunspot generation and the intensity of the interplanetary magnetic field.

C. Aurora

A correlation between the solar 11-yr period and the recurrence period of polar aurorae has been reported by Siscoe (1980) and for the past half millennium by Silverman (Fig. 4). As the auroral mechanism is hydromagnetic,

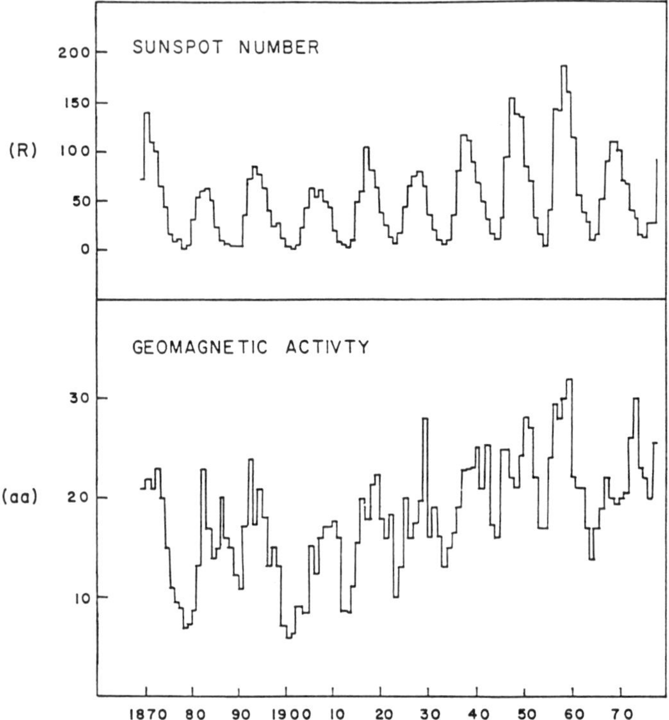

Figure 2.   *aa* index (nanotesla) geomagnetic activity compared to yearly average Wolf sunspot number (figure after Feynman 1982).

this establishes a long-term connection between the Sun and the Earth. Because the auroral mechanism is allied with substorm activity and field line reconnection, it is reasonable to involve field orientation, in short a 11-yr period in the statistics of N–S field orientation. This direct, i.e., nonproxy relation is firm evidence of hydromagnetic coupling of the Sun and Earth. Long-term persistence together with patchy and anecdotal evidence of the Maunder minimum reflecting a hiatus in auroral activity, supports the same conclusion from the sunspot and radiocarbon records. A striking similarity between radiocarbon and aurorae appears when comparing the spectra, the latter showing the distinct Gleissberg period and lines at ∼50 and ∼33 yr.

D. The Maunder Minimum (from the Sunspot Record)

Is the sunspot index actually a proxy for variability of solar wind parameters? In a seminal pair of papers Eddy (1976*a,b*) called attention to an apparent hiatus in solar activity in sunspot number in the period approximately of 1640 to 1730 A.D., generally termed the Maunder minimum (Maunder 1890,1922) and which had attracted slight attention through time. Although the sunspot record is of decreasing reliability with increasing age, there seems

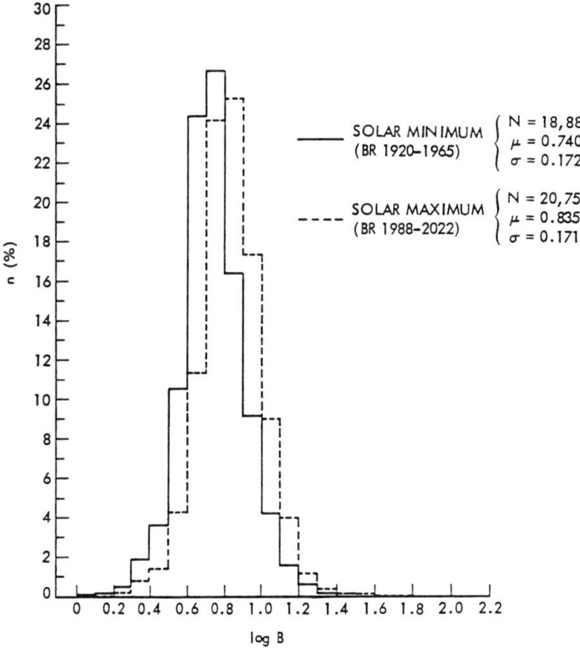

Figure 3. Histogram of log B (interplanetary magnetic field) during solar maximum and minimum disclosing a change of about 25% (figure from Slavin and Smith 1983).

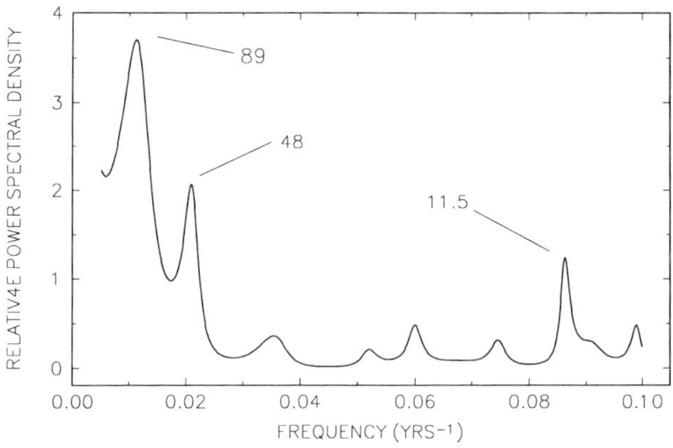

Figure 4. Maximum entropy power spectrum of aurorae. Appearance of the Gleissberg and second and third harmonics (unmarked) is approximately mimicked in the radiocarbon spectrum (Fig. 8 below) inferring an association between the electromagnetic auroral spectrum and the more general $\Delta^{14}C$ spectrum attributable to interplanetary modulation (data courtesy of S. M. Silverman). Callouts are in years.

little doubt that sunspot activity was either very low or absent during the reign of Louis XIV, *le roi Soleil*, as Eddy aptly put it. Eddy made the additional major claim that the ^{14}C record was an indicator of solar activity through GCR modulation and thus provides a measure of the history of decreases in solar activity through corresponding increases in the atmospheric inventory of ^{14}C.

Stuiver (1961) noted early this relation in the radiocarbon record, as have others (see, e.g., Sonett 1984; Suess 1968,1970, 1980; Damon and Sonett 1991; see also Stuiver and Quay 1980,1981). Nevertheless in spite of the identification of numerous other Maunder-like minima in Δ^{14}C, and eventually a full panoply of suggestible $\sim$200 yr-like radiocarbon maxima identifiable in the spectrum, there has always been some uncertainty regarding the Maunder minimum, partially because of the limitation to only part of the last sunspot minimum. The longer Suess period and evidence of minor additional long periods in the radiocarbon record cannot be detected in the SI, if present, because of the restriction of sunspot record length.

S. T. Suess (1979) has suggested a model for the solar wind during the Maunder minimum consisting of a wind of exceptionally low activity in velocity, density, and microturbulence based upon the low frequency of aurorae and the absence of sunspots (a minimum solar wind model). This general hiatus in solar wind related parameters is also extended in his analysis to the putative 22-yr period in climate. The radiocarbon record has an important general role in the development of his model. Recall that the Maunder minimum would have associated with it a maximum in the GCR flux and related production of atmospheric radiocarbon.

Attolini et al. (1988) have shown the presence of the 11-yr Schwabe cycle in both ^{10}Be (1180–1500 A.D.) and the auroral record for pre-Maunder time. ^{10}Be shows an increase by a factor of $\sim$2 over the neighboring record, implying correspondence with the Maunder minimum. Power spectra of the auroral record over 240 yr disclose 10.5 to 11.0 yr lines, but the partitioned time sequence discloses a more complex line system including power at 13.2, 16.5, 20.2, and 21.5 yr which remain unexplained.

The modulation or variability of GCR is ascribed to the changes in the cosmic ray diffusion, drift velocity, and solar wind advection velocity in the heliosphere (see Parker [1965] and Jokipii [1971], and the discussion in Sec. IV).

III. THE NEAR-PREHISTORIC ISOTOPE RECORD

Radiocarbon (^{14}C) and ^{10}Be are the two cosmogenic isotopes which have been intensively studied because of their atmospheric origin and intermediate half-lives (see Lal 1985; Lal and Peters 1967). ^{14}C is the premier isotope for tracing long-period GCR modulation, though a detailed connection with overall heliospheric diffusion is incomplete. Following the existence of a moderated atmospheric neutron sea resulting from the background CR flux,

^{14}C is produced by the reaction

$$^{14}\text{N}\,(n,\,p) \to\,^{14}\text{C} \tag{6}$$

In turn, ^{14}C decays by

$$^{14}\text{C} \to\,^{14}\text{N} + \nu^- + \beta^- \tag{7}$$

where ν^- is the antineutrino and β^- the electron, and ^{14}N is atmospheric. The half life of radiocarbon, $\tau_{1/2}$ = 5730 yr (Lederer et al. 1967). It is neutrons which participate in the $\text{N}(n,\,p)$ reaction yielding ^{14}C (Lingenfelter and Ramaty 1970; O'Brien et al. 1991). The principal distinction between radiocarbon and ^{10}Be is the stronger chemical activity of the former. The production of ^{14}C in the atmosphere is followed directly by its oxidation to CO and CO_2 and rapid chemical interaction with the oceans and ionosphere whereas ^{10}Be is relatively inert; its atmospheric lifetime is more a problem in aerosol physics. It is produced in the atmosphere by spallation reactions of GCR components on the major atmospheric components.

A. Radiocarbon Chronology

The Δ^{14}C record is partially secular (nonstationary) and exhibits a broad spectrum of variability, possibly of mixed origin, e.g., periodic, quasiperiodic, chaotic, and even stochastic. Two continuous decadal records exist, 6000 B.C. to 1945 A.D. (Stuiver and Becker 1993) and 7195 B.C. to 1895 A.D. (see, e.g., Sonett [1984] for sources, and the important annual record of Stuiver and Braziunas [1993]). The Sonett record is a compilation from four sources including some data in common with the Stuiver compilation. They are given in Fig. 5 and Fig. 6, where the upper panels are the raw records and the lower panels show the record reconstructed using singular spectrum analysis (SSA) (see the chapter Appendix) where components 1 and 2 represent the detrend functions shown in dashed lines. Though the two records are qualitatively similar, in the following, primary emphasis is given to the Stuiver record as it is less noisy.

As the decadal record extends over much of the Holocene, periodic and quasiperiodic variability for times as long as 2000 to 3000 yr can be detected; at the high frequency end of the record the signal is strongly damped because of the atmospheric reservoir effect. The variability raises serious issues of dating accuracy for workers in archeology, though it is a boon to the geophysicist. Modulation of the GCR in the atmosphere arising from changes in the geomagnetic field also affects the radiocarbon record strongly through a secular variation in the atmospheric concentration (Elsasser et al. 1956). Bucha (1970) first reported the specific effect of the secular drift of the geomagnetic field on the radiocarbon record.

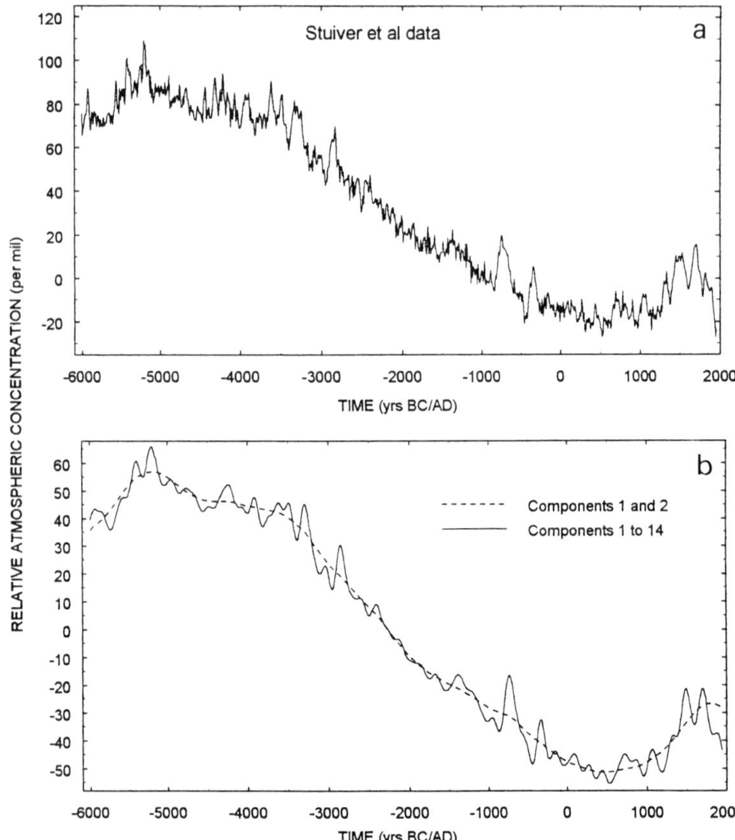

Figure 5.   (a) Decadal radiocarbon (Δ^{14}C) record for the past 8000 yr (Stuiver and Braziunas 1993) (upper panel); panel (b) shows the data reconstructed using singular spectrum analysis and eigenvector components 1 to 14, and reconstruction of detrend function (dashed line) using components 1 and 2.

The Δ^{14}C record is noisy, with several potential sources of noise such as radioactivity and tree-ring counting. For these reasons and the decreasing signal/noise ratio with increasing frequency, considerable effort has gone into analysis of the spectrum which includes singular spectrum analysis (SSA) (Broomhead and King 1986; Vautard and Ghil 1989; Vautard et al. 1992) and Monte Carlo confidence level simulations (chapter Appendix). The upper limit of detection of spectral power should be affected by the Nyquist period of 2 yr in the annual record (resulting simply from the limitation to yearly tree-ring count), and also because of the extreme reservoir attenuation at this period. The corresponding Nyquist period in the decadal record is 20 yr. This is a moot point and the harmonic signal level at this period is undetectable because of reservoir attenuation.

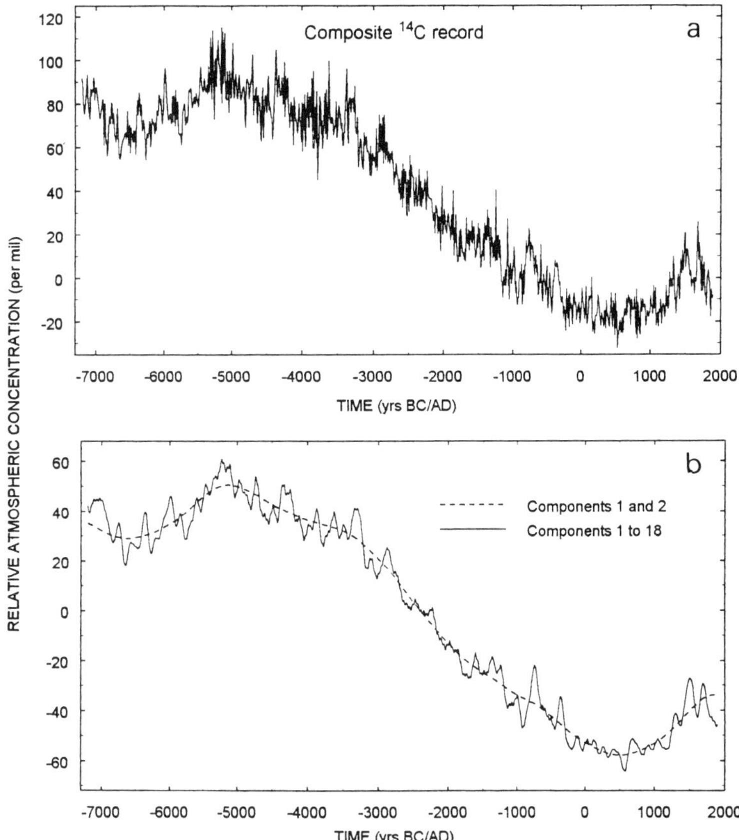

Figure 6.   (a) Decadal radiocarbon record of 9000 yr (Sonett 1992); (b) reconstruction definitions (lower panel) are same as for Fig. 5. Low frequencies are close to those of Fig. 5, but the record is considerably more noisy at the higher frequencies.

The absolute chronology of Δ^{14}C is established by reference to tree-ring counting (Fritts 1976). The difference between tree-ring and the radiocarbon record is called the "delta" radiocarbon, δ^{14}C or Δ^{14}C record. The lower case (δ) refers to the record uncorrected for isotopic fractionation and the upper case (Δ) to the record corrected using ^{13}C. Note that "radiocarbon" is often taken loosely to be synonymous with Δ^{14}C.

The use of radiocarbon to trace solar activity is beset with the complex behavior of the record. Of the eight isotopes of carbon only ^{14}C has a half-life of significant length to be important for archeology and geophysics. Its atmospheric concentration ratio to ^{12}C (the major isotope) is about $1:10^{12}$. The record is divisible into two major parts: (a) that forced by terrestrial activity and for which the global carbon concentration, aside from ^{14}C production, is conserved, and (b) the extraterrestrial production and forcing of ^{14}C which participates in the terrestrial problem as a tracer of total carbon. Thus useful information on terrestrial forcing can be obtained from the ^{14}C record but it also contains the extraterrestrial participation in forcing of ^{14}C.

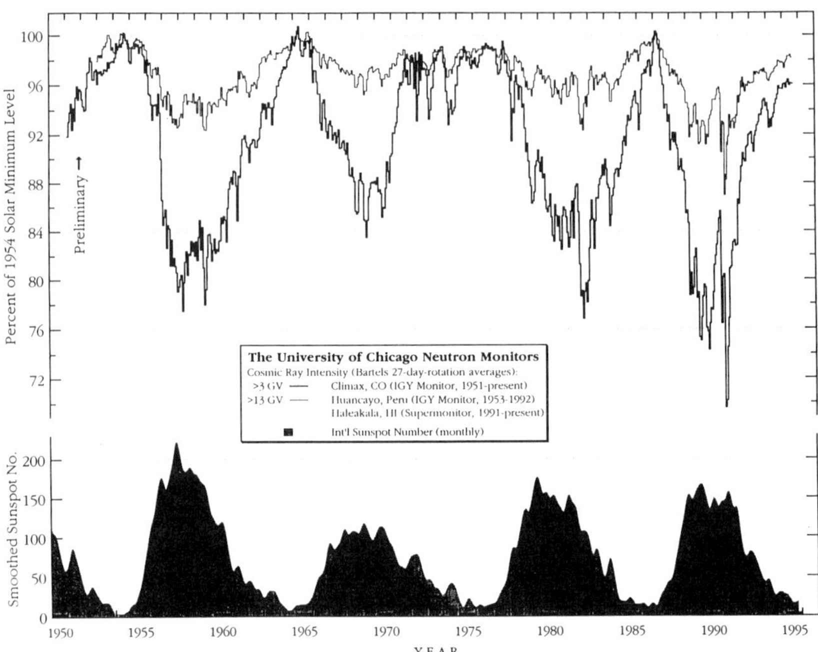

Figure 7.   Climax neutron monitor flux vs time (figure courtesy K. Pyle and J. A. Simpson).

Terrestrial forcing of total carbon is dependent on ocean-atmosphere coupling and transfer (see, e.g., Craig 1957; Siegenthaler et al. 1980; Houtermans et al. 1973). The existence of the relatively large ^{14}C content of the atmosphere has the effect of a low-pass filter with respect to the atmospheric takeup. This reservoir effect can alternatively be regarded as a consequence of the much larger atmospheric pool of ^{14}C than what is created annually by GCR. As a result (depending in detail on the transfer coefficients between ocean and atmosphere) the amplitude of the Δ^{14}C signal at 11 yr is some 100 times smaller than at 1000 yr (see Table II). (What is usually quoted is the ratio of incident signal power to noise.) Thus, although the signal obtained from radiocarbon is extremely small (undetectable in the decadal record), a significant level of modulation of the GCR flux in the atmosphere at 11 yr is confirmed by the neutron monitor record which discloses a peak–peak variation of $\sim$20% at this period and for energy $\sim$3 Gev (Fig. 7).

The periodogram (Fig. 8) of radiocarbon shows a major quasi-period in the neighborhood of 2300 yr which, however, is not equally spaced in the time series. The periodogram of the entire decadal Δ^{14}C record discloses spectral

TABLE II
Ocean–Atmosphere Attenuation Factors[a]

Period (yr)	Attenuation Factor[b]
10,000	0.33−0.2
1000	0.2−0.1
400	0.11−0.08
100	0.06−0.04
10	0.01−0.004

[a] Table from Houtermans et al. 1973.
[b] Amplitude.

features in addition to the long period just discussed, at 180, 150, 120, 80 to 90, 70, and 50 yr. The feature at 80 to 90 yr is generally attributed to the influence of the sunspot Gleissberg period, also noted in the auroral record, but details of how sunspot variability at this period affects the GCR flux has not been explored. A reasonable conjecture is that the 40 to 50-yr period is an expression of the Gleissberg period second harmonic as this is clearly seen in the sunspot spectrum itself (Fig. 1). As for the other features, they are more enigmatic. Speculation suggests their being harmonics of longer periods, but this explanation remains uncertain as harmonic relations are generally difficult to detect unequivocally in the radiocarbon spectrum.

The source of the $\sim$2300-yr forcing is unsettled. Sonett and Finney (1990) have argued for terrestrial or combined solar and terrestrial forcing based upon the deep water circulation period of the global ocean, but by itself this does not infer periodic forcing of the atmospheric concentration. There is also some evidence for a vestigial $\sim$1000-yr period. Figure 9 shows the eigenvalue spectrum of the entire Stuiver decadal record of $\Delta^{14}C$ from singular spectrum analysis (SSA) with error bars defined by the criterion of Ghil and Mo (1991) and 95% confidence levels from 1000 Monte Carlo trials vs an autoregressive AR(1) noise model (discussed in more detail below and in the chapter Appendix). A similar computation vs a white noise floor (dashed lines) is superimposed on the figure. Both calculations are for an autocovariance window $L = 90$ though the results are insensitive in the range $70 \le L \le 130$. This window range corresponds to an upper bound on resolvable periods in the range $700 \le P \le 1300$ yr for the decadal record and $70 \le P \le 130$ years for the ensuing single year record.

In Fig. 9 the first two components provide the input for computing the adaptive detrend function shown in Figs. 5b and 6b for the decadal records. Eigenvalues identified are in four clusters, i.e., (a) λ_{3-8}, (b) λ_{9-14}, (c) λ_{15-22}, and (d) λ_{23-32}. Eigenvectors (EOFs or empirical orthogonal functions) corresponding to $\lambda_{3,4}$, $\lambda_{5,6}$, $\lambda_{7,8}$, λ_{9-12}, $\lambda_{13,14}$ represent periods of 500, 310, 210, 150, and 125 yr.

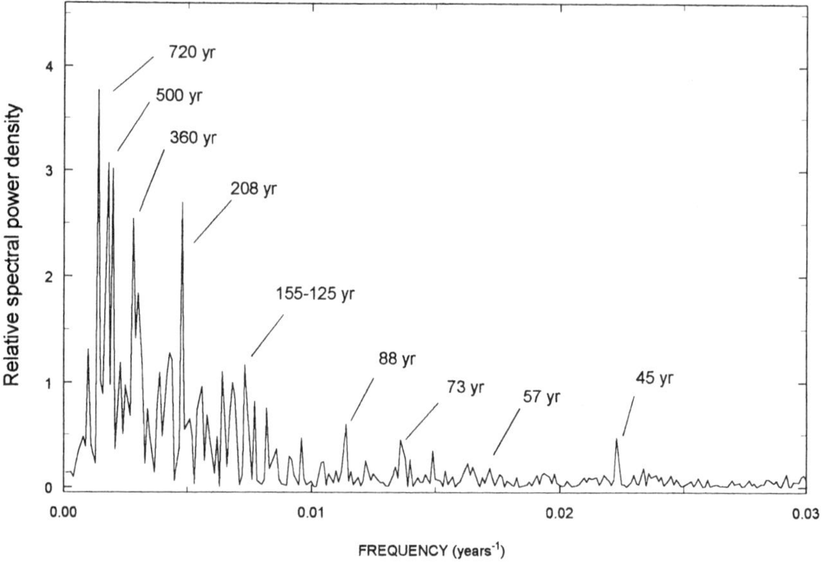

Figure 8.   Power spectral density (periodogram) of Δ^{14}C time sequence for 9000 yr showing major spectral features. The numerical values are line period in (very) approximate years. The nominal Nyquist period is 20 yr. Other computations give a Suess period closer to 210 yr. Because power spectrum is a probability distribution function, variability in the Suess period is reflected in a "most probable" spectral line, not directly reflecting period variability. Amplitudes are subject to ocean–atmosphere low-pass filtering and thus high frequencies are attenuated in the figure. Frequencies are uncertain because of multi-rate sampling.

Figure 10 shows a selected SSA reconstruction of the 500, 200, and 90-yr neighborhoods using $\lambda_{3,4}$ eigenvalues for the 500-yr reconstruction; the center panel uses $\lambda_{7,8}$ which encompasses the 200-yr range and illustrates the remarkable relation of the 200 and 2300-yr periods. ^{14}C has been used to infer a number of Maunder-like minima ever since reports of the approximately 200-yr period in the long-term radiocarbon record (the so called Suess wiggles or de Vries effect; Willis et al. 1960; Houtermans et al. 1971; Suess 1968,1970; Neftel et al. 1981; Damon et al. 1983; Sonett 1984; Damon and Sonett 1991). From the center panel a relation between the neighborhood of $\sim$200 yr and the four $\sim$2300-yr grand maxima is indicated. Side bands are marginally detected in the spectrum suggesting amplitude modulation.

We turn now to the further exploration of these features in the decadal record through Monte Carlo simulation comparisons vs white noise and a first order autoregressive AR(1) "red noise" model (Allen and Smith 1994) $u_{i+1} = \gamma u_i + \alpha w_i$ where u_{i+1} and u_i are the (i+1th) and ith time sequence elements, and where the coefficients γ, α, and w_i are respectively the autocorrelation coefficient, the variance, and the ith noise element. γ and α are fitted to give the synthetic series the same variance and lag-one autocorrelation as those of

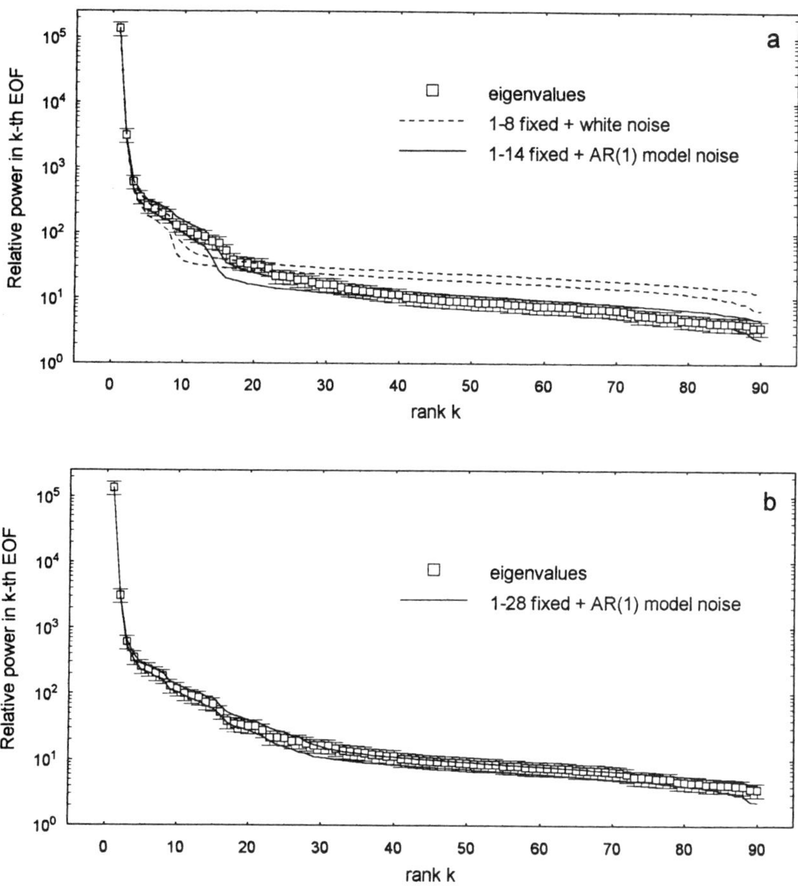

Figure 9.   Eigenvalue spectrum of Stuiver and Braziunas decadal record. EOF components 1 to 8 are held fixed in attempting to isolate the record noise. Error bars for eigenvalues are conservative Ghil and Mo values. The solid lines define the 95% confidence levels for 200 realizations using an autoregressive order 1 [AR(1)] model. The equivalent pure white noise mode yields the dashed lines which show that the higher frequencies (higher rank) eigenvalues are a much better fit to the AR model. This is consistent with the expectation from the atmosphere–ocean reservoir model.

the part of data under consideration. With *only trend components constant*, oscillatory components are identified at significance level of 95% testing against white noise. Eigenvalues of rank up to ~30 are found to be significant against AR(1) and white noise when other low-rank components are fixed. Spectral features described so far (corresponding to low-rank eigenvalues) account for nearly 71% of the total variance of the original detrended time series; the ratio of variance in components 15 to 32 to that in components 1 to 14 is 0.21.

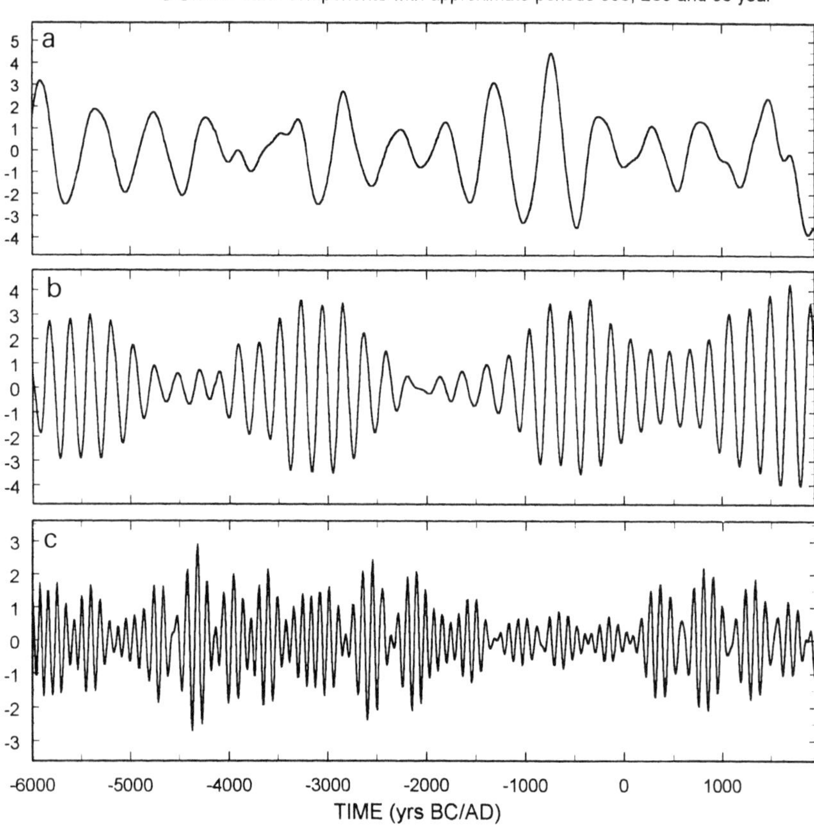

Figure 10. Narrowband reconstructions showing: (a) upper panel using $\lambda_{3,4}$ EOFs to emphasize the 500-yr neighborhood; (b) (center panel) using $\lambda_{7,8}$ corresponding to 200-yr neighborhood, and (c) (lower panel) corresponding to 90-yr neighborhood. Upper panel is very noisy though the envelope (representing signal) appears to be non-Gaussian and thus containing a finite signal. Center panel shows the remarkable beat tone between $\sim$2300- and $\sim$200-yr periods (Sonett 1984). Lower panel appears to have a weak quasi-period envelope of $\sim$500 yr.

The significant period at 208 yr appears to have been first noted by Kruse (see Suess 1980) and confirmed by Houtermans et al. (1971) and Neftel et al. (1981). Sonett and Finney (1990) report a spectrum of 9 major lines and Damon and Sonett report a further extension over the entire span downward from 4000 yr to the Nyquist period (20 yr). Although long discounted, the 208-yr period is now generally accepted and is informally designated as the Suess period. The 208-yr period (seen as an increase in Δ^{14}C in the time series record) has been compared to the partial Maunder minimum noted in the SI record, but the comparison is hindered by the lack of even one complete cycle in the SI.

The Maunder-like minima have been suggested to arise from intermittency of the solar dynamo (Weiss 1994). It is noted first that these Maunder-like cycles appear throughout the entire 9000-yr record of $\Delta^{14}C$ as 34 episodes of increases in radiocarbon, corresponding to decreases in interplanetary modulation but with highly variable duration and spacing. The historical record is too short to show a $\sim$200-yr period in any solar-associated (and presumably electrodynamic) data, so a connection of this period with solar variability remains conjectural; there seems little appeal to the idea that it is some form of terrestrial forcing. The 208-yr period is consistent with a wide spread in actual Maunder-like minima in the $\Delta^{14}C$ record, in that 208 yr represents a value of highest probability in what is really a wide distribution of periods as found by inspection of the time series of $\Delta^{14}C$.

Considering the relation of the 2300- and 208-yr periods, corresponding cyclicity appears to exist in the record of tree-ring growth of Bristlecone pines from timberline in the White Mts. of Eastern California over the shorter period of $\sim$5000 yr (Sonett and Suess 1984), though the tree-ring record has since been corrected and the relationship awaits a new examination. The tree variations are thought to arise from variability of air temperature though a direct dependence of growth upon solar irradiance has not been ruled out (see also Murphy and Palmer 1992).

B. Generalization to Very Long Periods

The periodicity of $\sim$2300 yr is also found in the $\delta^{18}O$ record in ice cores and foraminifera from ocean cores (Pestiaux et al. 1987,1988). Glaciation shows this period as does the Middle Europe oak dendroclimatic record and Dansgaard et al. (1984) report the most prominent period in their Camp Century $\delta^{18}O$ core to be a line at 2550 yr. Early publications of Dansgaard's group discuss periodicities in the $\delta^{18}O$ record for the time interval from 1200 to 2000 A.D. (Johnsen et al. 1970; Dansgaard et al. 1971). Their chronology was based on the assumed average accumulation rate of ice, and their power spectrum for this time interval shows prominent periods at 78 and 181 yr. They associate the 78-yr period with the Gleissberg sunspot periodicity. In a longer section of the core going back to about 10,000 yr, they obtain a period of 350 yr, again using the ice accumulation time scale, and at about 45,000 BP ice accumulation rate years, they find a persistent oscillation with period of $\sim$2000 yr.

The 2300-yr period appears to be free of modulation by the $\sim$10,000-yr secular trend in the geomagnetic field. If forced extraterrestrially, this period should interact nonlinearly with the secular trend from the terrestrial magnetic field, because they are related through $Q\sim B^{-0.52}$ (Elsasser et al. 1956). No such effect has so far been detected. Moreover an extraterrestrial mechanism should have to account for its beat with the 208-yr period. Characteristic deep water residence times for the global oceans range from 1,000 yr for the Atlantic to 2,000 yr for the full Atlantic-Pacific circulation (Broecker and Peng 1970; Broecker and Li 1970). These delay times by themselves cannot explain the global ocean-atmosphere resonance required to drive a 2300-yr total CO_2 cycle (delay times are not periodic), but the times are consistent with such a resonance. It may never be possible to isolate the 2.3-Kyr forcing with confidence. If due to orbit insolation, it may still drive global CO_2 and thus dilution of radiocarbon. If so, a search for an ocean-atmosphere resonance independently of orbit-insolation may be meaningless.

C. The Short-Period Spectrum of $\Delta^{14}C$

Study of short-term variations in the radiocarbon record have had to await the completion of the annual record until now reported by Stuiver and Braziunas (1993) from Northwest Pacific coast trees. Until this the decadal record has proven to be far too noisy and subject to the potential limitations imposed by the Nyquist period of 20 yr. The single year record of $\Delta^{14}C$ from the Pacific Northwest is from 1510 to 1954 A.D. A $\sim$200-yr period is clearly present in the record but is of insufficient length to be included in an accurate spectral density survey.

The annual record is detrended using SSA reconstruction of the long periods ($\sim$200 and 50 yr) captured by the first four EOFs. Figure 11(a) shows the Stuiver and Braziunas annual Pacific Northwest $\Delta^{14}C$ record and (b) gives the reconstruction using the EOFs from 1 to 18 while the dashed line represents the similar reconstruction for 1 to 4. Power spectrum of reconstruction of components 5 to 8 peaks at 26 yr and at $\sim$10 yr for 14 to 18. Tests against white and AR(1) model noise indicate significant variability when trend components are fixed, but no signal significant at 95% confidence level against "red noise" is found when also elements 5 to 9 and 14 to 18 are treated deterministically (Fig. 12). The leading components contribute about 98.5% of the total variance. Figure 13 shows the corresponding MEM spectrum using EOFs 1 to 8 from SSA.

Stuiver and Braziunas report several spectral features including a line at 2.3 yr which they attribute to El Niño. Some form of atmospheric $\Delta^{14}C$ dilution seems to be required as well as a more-or-less global influence of the El Niño. Something of the sort of the latter is suggested by the appearance in sub-Arctic varves of a 27-month period, essentially the period of the quasi-biennial oscillation (QBO) (Sonett et al. 1992).

D. Beryllium

^{10}Be is a companion radioisotope to ^{14}C insofar as they are both produced terrestrially by the incoming GCR flux. But Be does not form a gaseous compound like carbon. Production of ^{10}Be (Raisbeck et al. 1987; Beer et al. 1988) is via atmospheric spallations and is scavenged from the atmosphere in 2 to 3 yr by attachment to aerosols (McHargue and Damon 1991). The production is divided into fractions generated in the stratosphere and the troposphere. The eventual polar deposition on Earth depends upon the transfer rate from stratosphere to troposphere, upon winds aloft which transport the polar and aerosol and dust nucleation centers, and finally upon the very large difference

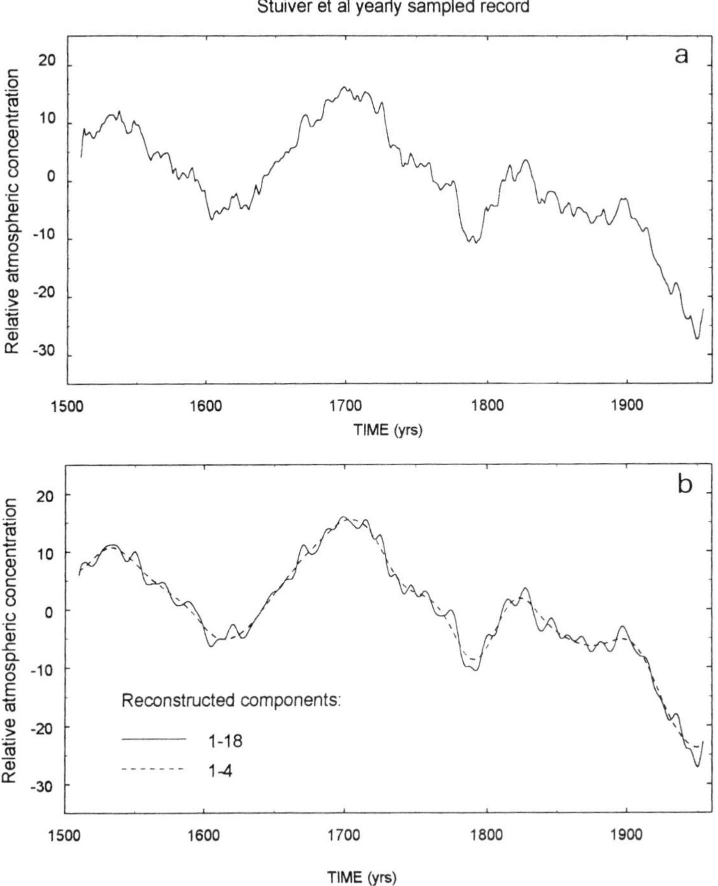

Figure 11.   (a) Stuiver annual radiocarbon record from Pacific Northwest trees; (b) reconstruction of annual radiocarbon record from EOFs 1–18. Also shown is low-frequency record reconstructed for EOFs 1–4 (dashed line).

in production between equatorial and polar regions. The latter affects both the incoming GCR flux and the local energy spectrum (depending on the geomagnetic cosmic ray latitude cutoff), eventually affecting production through the energy dependence of the spallation cross section.

^{10}Be records are available from Antarctica, Greenland, and several ocean bottom sites. Because the snowfall rate in Antarctica is an order less than for Greenland, the time resolution of the latter is correspondingly better, but the record length in Antarctica is usually greater. As the typical concentration of ^{10}Be in polar ice is only 10^4 atoms g^{-1}, counting of radioactive decay is impractical, and it was not until the advent of accelerator mass spectroscopy (AMS) that broad use of this isotope became possible.

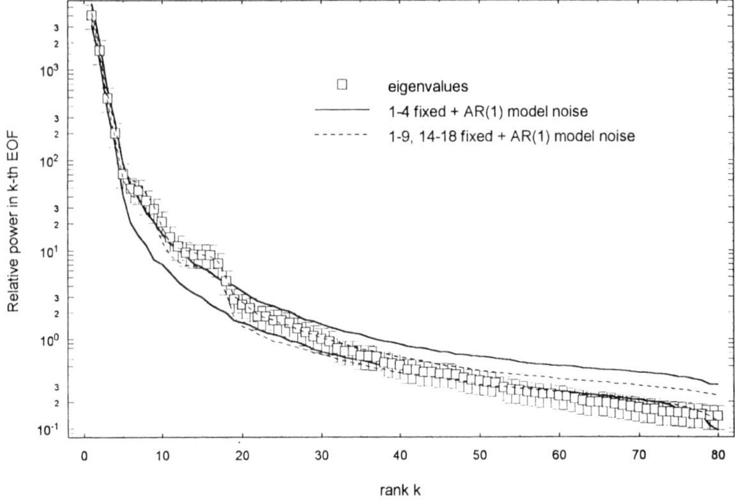

Figure 12. Eigenvalue spectrum of annual Δ^{14}C record with EOFs 1 to 20 retained fixed. Error bars are those defined by Ghil and Mo. Major discontinuities indicate a signal subspace of dimension 32. Calculation is for 1000 realizations and 95% Monte Carlo confidence limits.

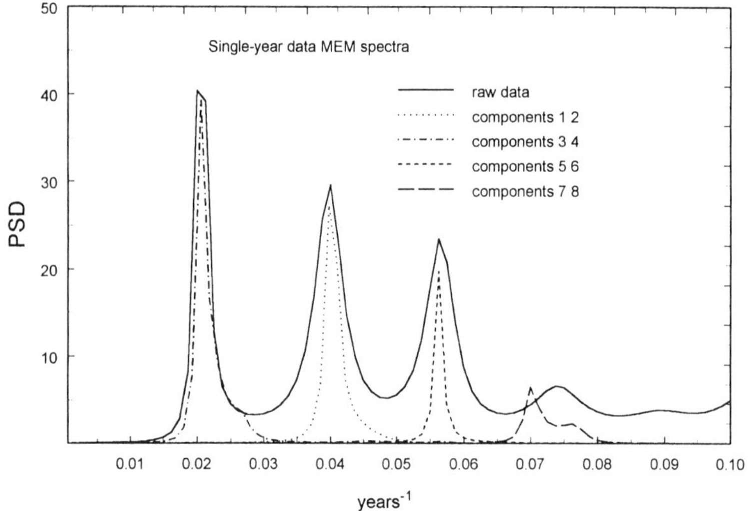

Figure 13. MEM of Stuiver annual Δ^{14}C record detrended using SSA reconstruction.

Major ice coring sites of ^{10}Be in the Antarctic are Vostok, Dome C (Raisbeck et al. 1987), and Byrd Station, Greenland ice at Camp Century, Dye 3, and various ocean sites. The most prominent variation in the Vostok record is from the very steep increases at 35 and 65 Kyr (see Sec. IV.D). For the general concentration found there, the variability, attributed to GCR changes, is likely corrupted by the presence of variations in the intensity of the geomagnetic field marked by changes in the model dipole, though it has been recognized that for the present location of the dipole over Vostok the Störmer cone admits essentially all GCR. Accepting a qualitative value of 25 deg for the nominal half angle of the Störmer cone, if the pole location has been contained within that half angle for the past 150 Kyr (the length of the record), then the possibility arises that fluctuations in the ^{10}Be concentration may be taken as a measure of long period ($>10^3$ yr) GCR variations.

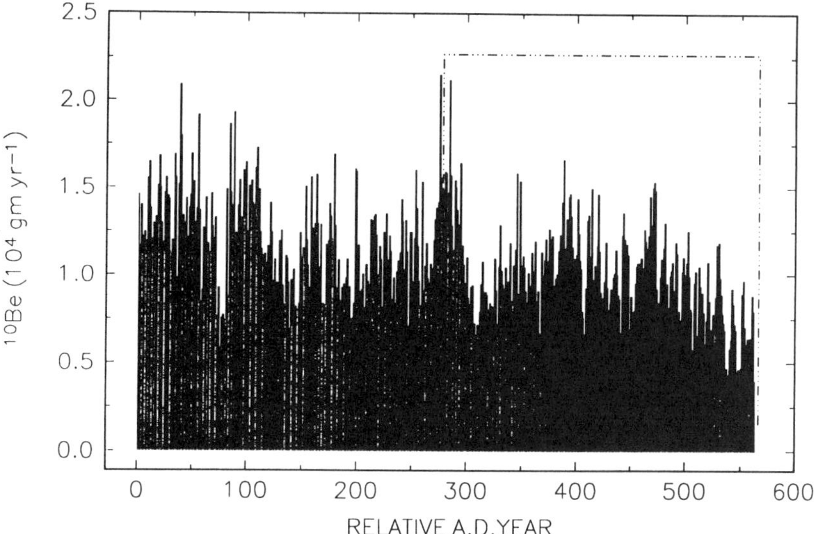

Figure 14. ^{10}Be record from Dye 3 Greenland core. Boxed region contains the 282 years corresponding to the Wolf sunspot record and is used for computing the cross correlation (see Fig. 16 below).

Because Be does not oxidize to a gaseous molecule, and because a dominant atmosphere-ocean reservoir system has no role, shorter periods presumably associated with solar activity should be more easily detectible than for Δ^{14}C. As it turns out the important annual Beer et al. (1990) record of 516 yr (Fig. 14) discloses a significant correlation between ^{10}Be and the SI. In attempting to avoid unnecessary splining to eliminate unequal spacing of data we computed the power spectrum (Fig. 15a) using the Lomb periodogram (Press et al. 1992). In the PSD the dominant lines occur at 188 and 59 yr. Confidence levels in the expanded spectrum (Fig. 15b) are given by the horizontal lines. Some power is present at about 10.9 yr, but it is at a confidence level of only 24%. The corresponding small cross correlation with the Wolf SI index is given in Fig. 16.

An independent attack on the problem of the weak Be response is by reconstruction of the time record using SSA to remove background noise over

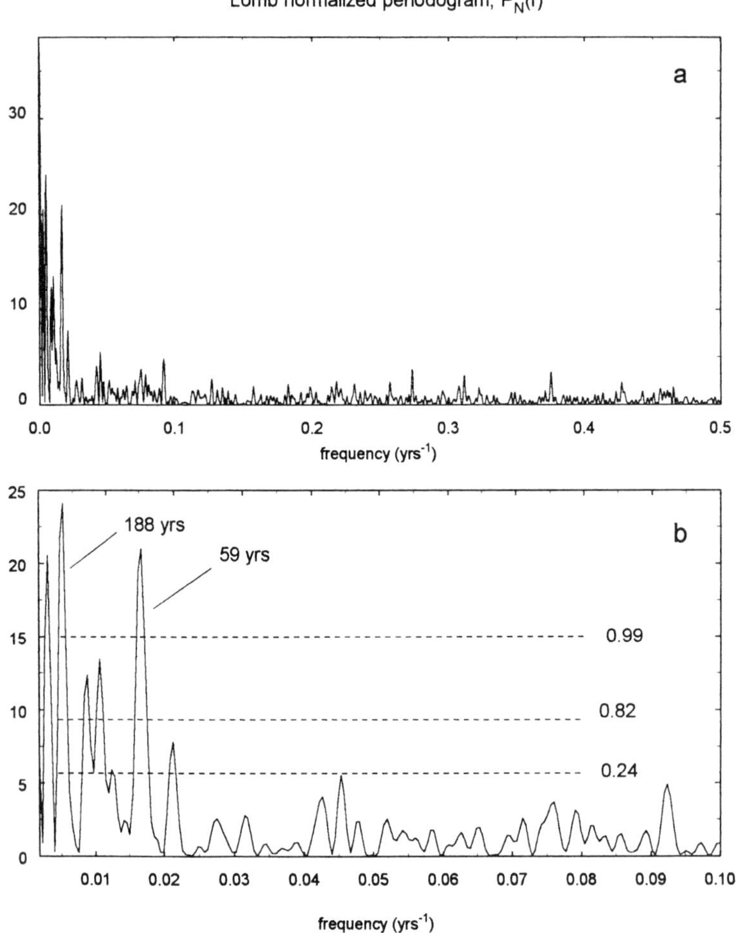

Figure 15.   (a) Lomb periodogram of the ^{10}Be Dye 3 record showing several candidate periods of low confidence level; (b) periodogram expanded to disclose details of long periods. Because of the short record, line at 188 yr may correspond to the Suess line (208 yr in the Δ^{14}C record). Lines between 59 and 188 probably contain Gleissberg variance. Confidence levels are indicated by the dashed horizontal lines. Line of amplitude at right side of figure is probably 11-yr period with confidence level of $\sim$24%.

narrow bands of predefined frequency width. Figure 17a shows the eigen-value spectrum of the Beer record for rank 1 to 95 using an autocovariance lag range of 80 to 100 over which the eigenvalue spectrum is quite stable. The reconstruction uses the eigenvalues of rank listed in Fig. 17b; these eigenvalues have passed the pairing and other tests (Vautard et al. 1992; Vautard and Ghil 1989) and constitute the majority of the total variance of the ^{10}Be record. The break at rank < 10 corresponds to the major lines in the spectrum. Of the two long periods that at 93 yr is probably the Gleissberg line, while the 186-yr line may be the Suess line with both appearing to exhibit corrupted position. Lines of 22 yr and 11 yr are not present in the decadal Δ^{14}C spectrum. That the 22-yr cluster is present is remarkable. The evidence of lines at 22.3 and 13.2-yr periods is uncertain, especially as the data had been splined prior to the computation of spectrum and the position of the 13.2-yr line may really be $\sim$11 yr shifted by the splining.

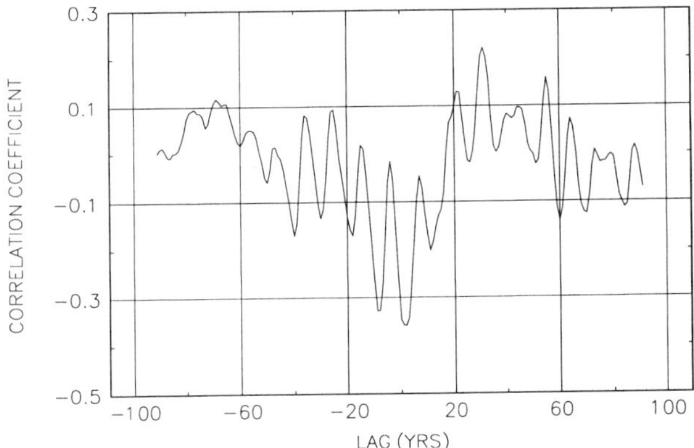

Figure 16.   Cross correlation of splined ^{10}Be Dye 3 record and the annual SI Wolf sunspot record. Section from lag -30 to 0 corresponds closely to -0.4 of Beer et al. (1990) with some differences from lag 0 to 30. Long periods are not present because the Wolf record limit of 282 years restricts the computation.

Neutron monitor data show a peak–peak variation of some 20 to 30% over a solar activity cycle in the production of secondary neutrons in the atmosphere. Assuming that the excitation function for spallation of ^{10}Be tends to an approximately level maximum at an energy of 1 Gev thus it seems inevitable that the ^{10}Be production rate approximates this 20 to 30% (cf. Beer et al. 1990).

The apparent presence of the 11-yr period in all of the several computations of correlations and PSD is inescapable, and though the confidence level from statistics is low, the 11-yr PSD variance is consistent with the cross correlation amplitude. Where the $\sim$50- and/or $\sim$90-year periods are present, it appears that the Gleissberg and/or its second harmonic are making their way through the background, and this appears to be the strongest couple between ^{10}Be and solar variability. The common appearance of the Gleissberg and its second harmonic is seen in Fig. 16 comparing the mid-frequency spectrum of both Δ^{14}C and ^{10}Be (see also Fig. 18). As noted, ^{10}Be shows unexpected and unexplained power at both 22 yr and 13 yr, the latter shared with Δ^{14}C .

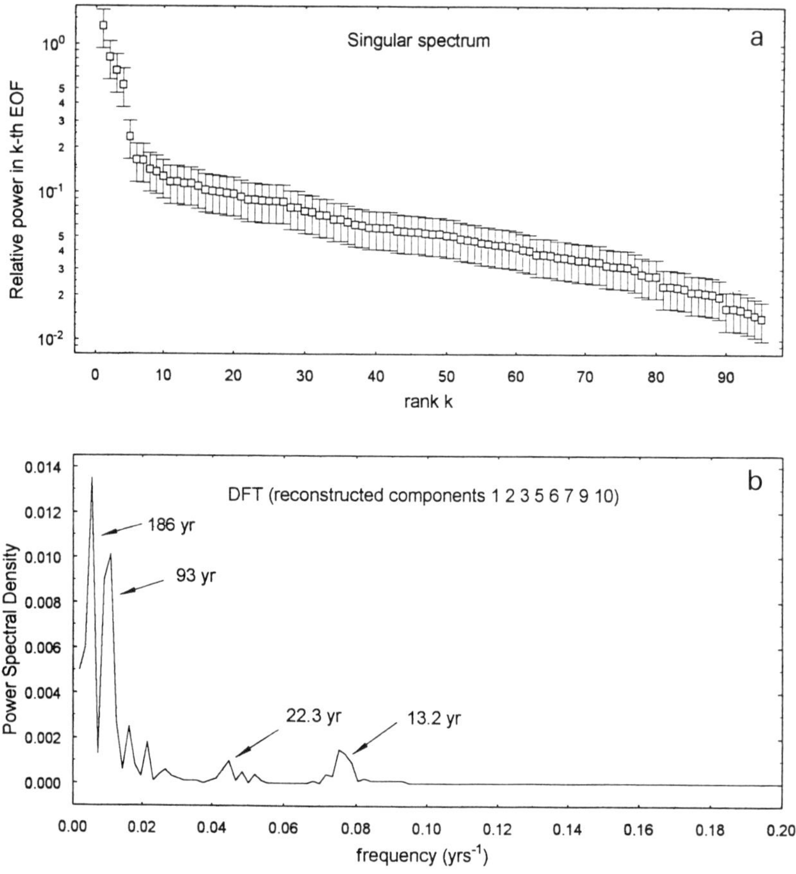

Figure 17. (a) Eigenvalue spectrum of ^{10}Be showing weak structure for rank <10 (b) power spectral density of reconstruction using EOFs 1 to 3, 5 to 7, 9, 10. The long periods correspond closely to the Lomb periodogram. Lines at 22.3 and 13.2 are enigmatic. The former may be associated with the Hale period and the latter a corruption of the 11-yr period.

IV. THE PROBLEM OF INTERPLANETARY MODULATION

A. Curvature and Gradient Drifts, and the 22-yr Modulation

The role of curvature and gradient drifts on the solar modulation of galactic cosmic rays has lead to the prediction of a 22-yr cycle for the variation in the galactic cosmic ray intensity (per unit momentum interval) $j_p = 4\pi p^2 v f_0$, as well as the prediction of a 22-yr cycle in the cosmic ray anisotropy ξ associated with the 11-yr reversal in the polarity of the solar magnetic field (see, e.g., Jokipii et al. 1977). The basic equations determining the solar cycle modulation of galactic cosmic rays consist of the cosmic ray continuity

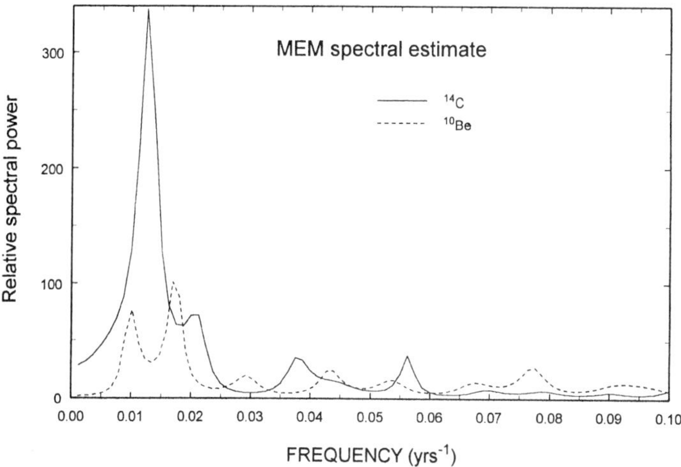

Figure 18.   Maximum entropy spectra of Δ^{14}C and ^{10}Be in the 10 to 100-yr range of period. The coincidence for the Gleissberg is expected; those at $\sim$25 yr may be a compuationally shifted 22-yr (Hale) period, but the power at 13.5 yr is anomalous as before.

equation:

$$\frac{\partial N}{\partial t} + \nabla\cdot\mathbf{S} + \frac{\partial}{\partial p}\left(\frac{p\mathbf{V}}{3}\cdot\nabla N\right) = 0 \tag{8}$$

where

$$N = 4\pi p^2 f_0 \tag{9}$$

is the particle number density per unit momentum interval, and

$$\mathbf{S} = \mathbf{S}_a + \mathbf{S}_d \tag{10}$$

is the particle current. In the above equations p and v denote the cosmic ray particle momentum and speed respectively, and f_0 denotes the average phase space distribution averaged over solid angle in momentum space. The particle current in Eq. 10 consists of an advective component:

$$\mathbf{S}_a = -4\pi p^2 \left(\frac{\mathbf{V}p}{3}\frac{\partial f_0}{\partial p}\right) \tag{11}$$

associated with advection of particles with the solar wind with velocity $\mathbf{V}$, plus a diffusive component:

$$\mathbf{S}_d = -4\pi p^2 \mathbf{K} \cdot \nabla f_0 \tag{12}$$

where $\mathbf{K}$ is the anisotropic diffusion tensor for particle scattering in the random component of the interplanetary magnetic field. The diffusive streaming flux $\mathbf{S}_d$ in turn may be split into the components:

$$\mathbf{S}_d = \mathbf{S}_{d\|} + \mathbf{S}_{d\perp} + \mathbf{S}_{\text{drift}} \tag{13}$$

where

$$\mathbf{S}_{d\parallel} = -4\pi p^2 K_\parallel \frac{\partial f_0}{\partial x_1}\mathbf{e}_B, \quad \mathbf{S}_{d\perp} = -4\pi p^2 K_\perp \left(\frac{\partial f_0}{\partial x_2}\mathbf{e}_2 + \frac{\partial f_0}{\partial x_3}\mathbf{e}_3\right) \quad (14)$$

describe the anisotropic diffusion of cosmic rays parallel ($\parallel$) and perpendicular ($\perp$) to the mean magnetic field, and

$$\mathbf{S}_{\text{drift}} = -4\pi p^2 K_A \left(\nabla f_0 \times \mathbf{e}_B\right) \quad (15)$$

is the particle current associated with curvature and gradient drifts. In the above equations $\mathbf{e}_B$, $\mathbf{e}_2$, and $\mathbf{e}_3$ are orthonormal unit vectors with $\mathbf{e}_B = \mathbf{B}/B$ the unit vector along the mean magnetic field; $K_\parallel$ and $K_\perp$ are the energetic particle diffusion coefficients parallel and perpendicular to $\mathbf{B}$; and K_A is the anti-symmetric component of the diffusion tensor associated with particle drifts. The particle anisotropy in the above development $\xi = 3\mathbf{S}/(vN)$ measures the deviation of the particle distribution function from isotropy.

The only term in the above cosmic ray modulation equations that is naturally associated with a 22-yr cycle is the particle drift current (Eq. 15), which changes direction with the polarity of the magnetic field. An absence of a systematic 22-yr variation in the cosmic ray intensity suggests that the drift current contribution $\nabla\cdot\mathbf{S}_{\text{drift}}$ to the cosmic ray continuity Eq. (8) would be smaller than the advective ($\nabla\cdot\mathbf{S}_a$) and diffusive contributions ($\nabla\cdot\mathbf{S}_d$).

Quasi-linear theory for particle transport in weakly turbulent magnetic fields (Forman et al. 1974) yields the expression

$$K_A = \frac{vr_g}{3}\frac{\omega^2\tau_\perp^2}{1+\omega^2\tau_\perp^2} \quad (16)$$

for the K_A coefficient, where $r_g = pc/(ZeB)$ is the particle gyroradius in the mean field B; $\omega = ZeB/(mc)$ is the particle gyrofrequency, and $\tau_\perp$ is the characteristic time for particle scattering across the field. In the weak turbulence limit $\omega\tau_\perp \gg 1$, and $K_A \simeq vr_g/3$. This value of K_A, is used in most modulation models including the effects of drifts, and maximizes the role of drifts. For the case of hard sphere scattering or BGK Boltzmann scattering models, one obtains a similar formula to Eq. (17), namely

$$K_A = \frac{vr_g}{3}\frac{\omega^2\tau^2}{1+\omega^2\tau^2} \quad (17)$$

where τ is the collision time. If one assumes that the expression (17) for K_A also applies in the limit of strong scattering, with $\omega\tau\sim 1$, then a much lower value of K_A results, with consequent diminution in the role of drifts (note $K_A \sim (vr_g/3)\omega^2\tau^2$ for $\omega\tau \ll 1$). In the strong scattering limit, the particle scatters sufficiently frequently that the ordered particle motion associated with drifts is suppressed. Because interplanetary conditions are more turbulent near solar maximum, drifts should be suppressed at this phase of the solar cycle, and conversely drifts are more important during the quiet phase of the solar cycle at solar minimum. Similarly, the role of drifts should be suppressed in turbulent, merged interaction regions in the outer heliosphere. To make contact with the adiabatic guiding center theory of drifts, substitute the particle current $\mathbf{S}$ described by Eqs. (8) through (15) into the cosmic ray continuity Eq. (8) to obtain the equation:

$$\frac{\partial N}{\partial t} + \nabla\cdot(\mathbf{V}_D N) + \nabla\cdot(\mathbf{S}_a + \mathbf{S}_\| + \mathbf{S}_{d\perp}) + \frac{\partial}{\partial p}\left(\frac{p\mathbf{V}}{3}\cdot\nabla N\right) = 0. \qquad (18)$$

where

$$\mathbf{V}_D = \nabla \times (K_A \mathbf{e}_B) \qquad (19)$$

corresponds in the weak scattering limit to the adiabatic guiding center drift velocity (averaged over a nearly isotropic pitch angle distribution of particles) owing to curvature and gradient drifts, plus the drift parallel to $\mathbf{B}$ associated with field aligned currents. Note that $\nabla\cdot\mathbf{V}_D = 0$, for the drift velocity, Eq. (19). Hence, the effects of drifts on cosmic ray modulation are minimized for small values of $\nabla\cdot(\mathbf{V}_D N)$. A discussion of the role of scattering in modifying drifts may be found in Jokipii (1993).

B. The Double Spiraled Field

An alternative way in which drifts may be suppressed has been explored by Lee and Fisk (1981), in which they assume for the sake of argument, that the interplanetary magnetic field has a special double spiralled configuration that suppresses the role of drifts (even in the weak scattering limit). Parker (1972) has earlier discussed such a field configuration with equilibrium if the field is subject only to simple twisting. As Lee and Fisk note, their configuration has some similarities to the toroidal field confinement design for fusion reactors where cross field diffusion is suppressed in order to avoid plasma cooling by wall contact. A fanciful representation from their paper showing the basic Parker spiral upon which is superimposed a toroidal field (arising presumably from a field aligned current) is in Fig. 19 with the two fields combining to yield net helicity and suppressing diffusion across the Parker spiral field.

Because the field configuration is not necessarily a global self-consistent field structure, it is difficult to assess whether there will be a net suppression of drifts in this model. Some support for such a configuration is suggested by data from Explorer 35 (Lichtenstein and Sonett 1979; Riley et al., work in progress), showing an invariant interplanetary magnetic field vector in minimum variance coordinates moving in arcs with constant angle to the direction of minimum variance (Fig. 20). Such a configuration may arise from field-aligned electric current bunching and must be local. Otherwise fields arising from adjoining current systems would tend to cancel. This can be seen by recognizing that such currents distributed uniformly in longitude (azimuth) in the heliosphere would sum to zero according to Stokes' theorem

Figure 19.   Conjectural double interplanetary spiral field configuration arising from the Parker spiral upon which is superimposed a field aligned current resulting in a total spiraled field (figure from Lee and Fisk 1981).

although this is not true in latitude. Alternatively neighboring flux tubes should contact one another with opposed fields which would be subject to reconnection, making such a configuration unstable. On the other hand, the recent Ulysses data shows that the arc-like motions of the field from time to time do not appear to have a net helicity. Because Feldman has shown evidence for electron beaming, the answer may well lie in currents moving randomly away and towards the Sun depend upon local mirroring.

The division of aligned electron currents into core and halo components (cf., chapter by Feldman and Marsch) for thermal electrons and Anderson et al. (1981,1992) for suprathermal electrons suggests a relation between the data of Lichtenstein and Sonett, and Riley et al., discussed above, and such currents but this matter has not been investigated so far. J. R. Jokipii (personal communication) has shown that under certain constraints on the shape of successive 11-yr periods in the GCR flux, an otherwise detectable 22-yr component in the PSD may vanish, providing another explanation whereby no 22-yr component is present in the GCR spectrum, even though drift modulation may be present.

C. Extraheliospheric Cosmic Ray Variability

Evidence that the cosmic ray flux upon the heliosphere might be variable has in the past relied upon meteoritic data and to some extent Apollo which can provide only an approximate integrated flux. That the radiocarbon spectrum could exhibit time-dependent features in the radiocarbon record assignable to variations in the interstellar cosmic ray flux incident upon the heliosphere was viewed negatively by Sonett (1984). This is reinforced by an argument of G. Morfill (personal communication) as follows. Assume a magnetosound wave

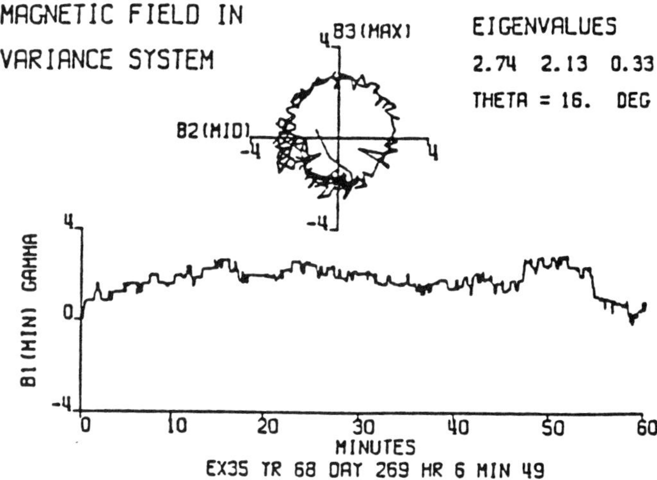

Figure 20.   (a) Interplanetary wave propagation vector **k** locus showing spiral motion about the direction of minimum variance (e_1). Tip of the **k** vector rotates over arcs of random length in the plane of e_2 and e_3 suggesting interrupted field-aligned currents, exhibiting small radial oscillations indicating oscillatory interruptions of aligned current (figure from Explorer 33; Lichtenstein and Sonett 1979).

train subject to periodic adiabatic compression of entrained cosmic rays. Then the diffusion time out of a compressed region $\tau = \lambda^2/\kappa \gg \lambda/\delta v$ where λ is wavelength, δv wave speed, and $\kappa \sim 10^{28} - 10^{30}$ cm^2 s^{-1} is a typical diffusion coefficient ($E\sim 1$ Bev). But for $\delta v \sim 10^6$ to 10^7 cm s^{-1}, $\lambda \gg 10^{26}$ cm $\sim 3 \times 10^4$ kpc. It follows that, treated adiabatically and accepting the value of diffusion coefficient, an interstellar hydromagnetic wave train must be of order galactic dimension to compress effectively its contained cosmic ray inventory, ruling this out as a source of radiocarbon variability on the time scale observed. As the half-life of ^{10}Be is $\sim 1.5 \times 10^6$ yr, sufficient time is available for inward drift against the solar wind as has been suggested for ^{26}Al. Of course the success in penetrating the heliosphere upstream depends upon the charge state. Nevertheless it is possible that some of the ^{10}Be concentration seen in cores could arise extraterrestrially. Fanciful as this might seem, in the next section we briefly regard the Vostok spikes from the standpoint of an interstellar source.

D. Supernovae Shocks or Geomagnetic Field Reversals

In this short section we briefly review the evidence for invasion of the heliosphere by interstellar shock wave-accelerated cosmic rays, the shocks or blast waves originating in ancient local supernovae. Evidence of two anomalies (35 and 60 Kyr ago) in the ^{10}Be concentration in the Vostok and Dome C Antarctic ice cores uncorrelated with corresponding ^{18}O was conjectured by Raisbeck et al. (1987) to be due to superlong Maunder minima; they also made the earlier suggestion of reversals or excursions of the geomagnetic field, during which time the Störmer cutoff would be severely reduced. Sonett et al. (1987) proposed the sweeping of relict supernovae shock waves over the heliosphere as the source of the enhancements. They calculate a 30% chance of heliospheric encounter in 150 Kyr. Castagnoli et al. (1995) have found evidence for the 35-Kyr spike in Mediterranean sea sediments and McHargue et al. (1995) in the Sea of Cortez. L. R. McHargue et al.'s latest results (personal communication) disclose both the 35 Kyr and the lesser 65 Kyr spike in the Blake seamount in the Atlantic Ocean, and Beer et al. (1992) have found evidence for the 35 Kyr spike in Camp Century (Greenland) ice, thus generally confirming the global character of at least the 35 Kyr event.

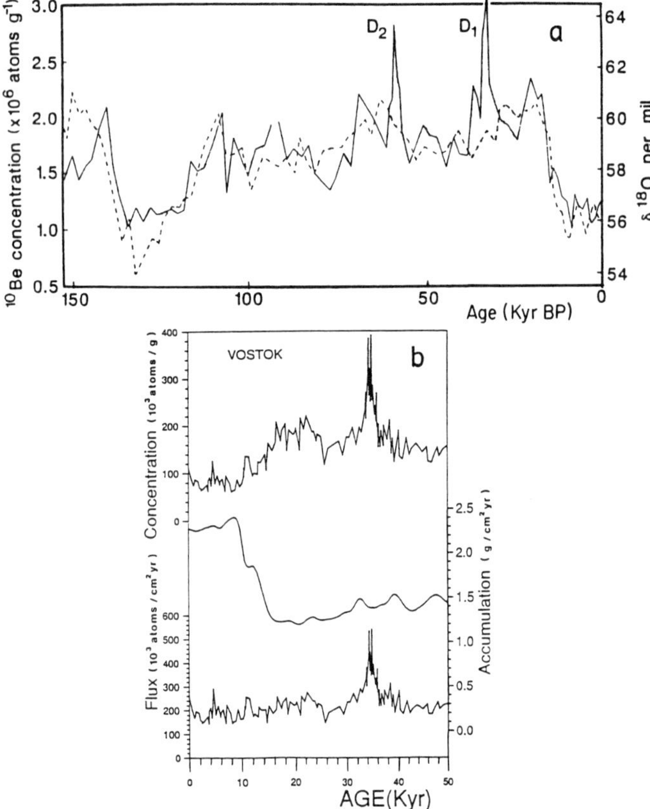

Figure 21.   (a) 150-Kyr record of ^{10}Be and ^{18}O from Vostok ice core. The spike increases in ^{10}Be are putatively attributed to envelopment of the heliosphere by shock accelerated cosmic rays; the alternate model would be based upon geomagnetic field reversals, for which the evidence is presently uncertain (figure after Raisbeck et al. 1987); (b) high-resolution detail (figure from Raisbeck et al. 1992).

Before confirming an interstellar shock model, it is essential to rule out geomagnetic field reversals as the source of the cosmic ray enhancements leading in turn to the [10]Be increases. K. Creer (personal communication) notes no paleomagnetic evidence at 35,000 yr in his data though the question of reversals or excursions in the field appears unresolved[a]. However, G. Kocharov (personal communication) has pointed out that because the atmospheric production peak for [10]Be is at 1 Bev and Vostok is at the south magnetic pole, the Störmer cutoff is unimportant and it follows that field reversals might not be the source of the anomalies (only if large-angle polar wander were not a factor). If an extraheliospheric source is shown to be the initiator of the Vostok spikes, this is mostly of cosmic ray interest, but would have little bearing upon the overall dynamics of the solar wind. Figure 21 shows the beryllium record with the anomalous increases. For an interstellar field of 1 nanotesla and a proton energy of 1 Gev, the Larmor radius is $\sim$0.0002 AU. The 200-yr resolution in the insert of Fig. 21 and a heliospheric speed of 20 km s^{-1} relative to the local system gives a spatial scale of 830 AU, but this calculation depends upon the final time resolution available from core material.

V. PRIMORDIAL SOLAR SYSTEM

Meteorites are classified into some half dozen major species in ascending order of metamorphosis from the very primitive carbonaceous chondrites to the fully differentiated achondrites and metals (see, e.g., Wasson 1974,1985). It is also known that, for those meteorites for which solidification dates are available and do not exhibit resetting due to collisions, the solidification occurred very early in solar system history, likely coeval with the Sun's very early history and even possibly before nuclear onset, though this is conjectural. The enigma of long standing regarding this singular evolutionary event(s) is based upon the intense short-lived heat source required, either short-lived radionuclides and/or electrical induction from an intense solar wind and rapidly spinning Sun as suggested from astronomical observation of early high mass flow stars such as T Tauri objects (Sonett et al. 1970; Herbert et al. 1991).

VI. VARVES AND CLIMATE PROXY

Laminated sediments or rocks (varves) are common on Earth. As such structures are agreed (with a few aeolian exceptions) as having been derived from deposition out of water, many attempts to relate the thickness of such sediments to climate, rainfall, and/or snowfall have been made in the past. This is an extremely difficult subject to work within if the eventual aim is to relate varve thickness to solar irradiance. The difficulties inherent in such research,

---

[a] The terrestrial mantle electromagnetic response to even a step function reversal of the core field is $\sim$10 yr for reasonable mantle conductivity, so the latter has no influence on the [10]Be spike shape if due to field reversals.

aside from the aforementioned, rest also upon the problem of connecting varve thicknesses to the mainstream problem, i.e., solar wind. Sediment deposition relies upon water flow velocity as well as the distribution and shape of particulates. Some of this is alleviated if varve collection is restricted to lake beds, where the yearly record of thickness can be studied aside from other potentially interfering parameters.

For lacustrine deposition (the common source of varves) the dynamics are defined by the internal hydrodynamics of the lake (Anderson and Dean 1988). The relationship between thickness of varves, sedimentation, sediment carrying power of river input, and finally any putative relation between riverine flow, lake thaw, and solar insolation is obviously not simple. To make matter worse we are still faced with the final association to be made (provided the other exist in recognizable form) of solar irradiance and solar wind (see, e.g., Anderson 1991).

This extremely complex matter is further beclouded if the genesis is chemical or biological (stromatolitic). External time benchmarks (such as from radiocarbon) are not available, and time can only be inferred. Nevertheless, the potential importance of this line of inquiry is exceedingly high because over geological time, say from the onset of the Precambrian till recently, they are possibly the only source of information related to ancient solar variability (except for the recent discovery of Miocene tree rings [Kurth et al. 1993]). Thus, whether a rhythmite sequence is tidal or yearly can be difficult to assess. Revision of the late-Precambrian Elatina sediments which had been thought to be solar forced clastic varves but were actually tidal laminae shows how easily a mistaken identity can be made. That is not to say that varves are not seen in a marine environment; see Anderson (1991) for review of various varved laminae sequences, in particular evidence for 11-yr, 90-yr (Gleissberg), 200-yr, and possibly 2500-yr cyclicity. Considering the importance of these results, the complexity of chemical varving, the lack yet of independent confirmation, and the hydro-geological scenario including connection to a larger water body capable of refreshing the varve forming basin makes independent confirmation of these results important.

Even if connections exist between climate and solar irradiance, climate is still a poor proxy for solar wind variability. For isotopes the connection is through modulation, itself not something easily disassembled into various solar wind parameters. Though sparse, some evidence does exist connecting solar irradiance and the radiocarbon record. Changes in irradiance occur over the 11-yr cycle, while the radiocarbon record discloses marginal evidence for a corresponding cyclicity. Actually, because the 11-yr variation of the cosmic ray flux is the largest observed, it ought to produce the largest observed changes in radiocarbon; but what is noted is the response (atmospheric inventory) reduced by a factor $\sim$100 times because of the aforementioned frequency dependent atmospheric attenuation (see, e.g., Siegenthaler et al. 1980). A further difficulty is that nearly all published radiocarbon data is of 10 to 20 year sample intervals, so an 11-yr spectral feature would be indistinguishable from an 11-yr line folded (aliased) downwards from the 1st order (repetition) spectrum.

Though climate lies outside the scope of this review, several aspects of potential interest to the overall problem of long-period solar–terrestrial physics exist due first to the possibility of a long-term analog to the known variable irradiance of the Sun and secondly to the interleaved terrestrial and extraterrestrial forcing of the atmospheric inventory of radiocarbon. There is strong evidence that the Earth's climate has undergone significant changes on the time scale of decades to centuries during the Holocene. Variability of the Sun's total irradiance on these time scales has frequently been suggested as a cause of these climatic variations, notably by Eddy (1976a,b), who pointed out the approximate coincidence of the peak of the Little Ice Age in western Europe with the Maunder minimum in solar activity, and the overall correspondence during the last millennium of warm temperatures with the envelope of solar activity as deduced from the ^{14}C record.

This suggestion was revived and extended by Reid (1987) and Reid and Gage (1988) who noted a strong similarity between the long-term variation of globally averaged sea surface temperatures (SST) (Folland and Kates 1984) and the envelope of the 11-yr sunspot cycle over the past 130 yr. The amplitude of the SST variation was of order 0.5°C, and a calculation based upon a one dimensional model of the ocean's thermal structure showed that variations in irradiance of a few parts per thousand keyed to the solar activity envelope would be sufficient to explain the SST variation. Both the SST and solar activity went through a minimum in the 1900 to 1910 decade and a maximum around 1960. The calculated change in solar total irradiance over the last century was about 0.6%. That the overall temperature variability could fit an approximately 100-yr period (given a few or more cycles) escapes no one's notice nor that this period is approximately the Gleissberg cycle.

That changes in the Sun's magnetic field are more than just a proxy for irradiance changes not only arises from the correlated changes between the sunspot index (as a measure of magnetic activity) and irradiance over the 11-yr solar activity period, but also from some tree-ring growth records (Sonett and Suess 1984) and from study of solar-like stars which show correlated brightness–magnetic activity (Radick 1991; Baliunas 1991).

VII. DISCUSSION

A. The Basis for Solar-Wind Variability

As can be noted from this review the problem of *long-period* solar wind variability is in primitive form. The major difficulty is with the sparse data base and the difficulty with signal/noise ratio. At the present time the theoretical basis for understanding the known solar wind variability, i.e., the Gleissberg and Suess periods is essentially nonexistent, partially because of lack of confidence that the data base is secure. The anecdotal aspects together with the geologic record are extremely complex and mix numerous forcing functions, only some of which are extraterrestrial. Finally even how the solar wind accelerates and presumably discharges a variability eventually seen in the terrestrial record is really a matter of solar physics for which there is presently little understanding of how to force the Sun at long periods (Gough 1988).

Periods ranging upwards to perhaps a century based on appeal to dynamo modes seem plausible; the extension to longer periods must also be accounted for. Though the transit time for acoustic waves in the Sun is only a matter of hours at most, the Alfvén wave transit time can be made to extend much longer. For a field in the core of 1 Gauss, and assuming a core density of 1 g cm^{-3}, admittedly high, the transit time is $\sim$500 yr so it is not impossible to imagine global periods in the range of many years. Gough (1988) has reviewed characteristic times for the Sun but no specifically multi-hundred year periods are identified. Standard stellar models have no place for stellar core variability, yet the possibility is present of changes in the solar radius (Gilliland 1981; Ribes et al. 1991). How this would couple to the solar wind remains unresolved.

B. Dynamos

The 11-yr period appears to be the outgrowth of a regenerative dynamo operating in the Sun's convective zone. There is growing belief that the solar dynamo is a system varying between quasi-periodic and chaotic (see, e.g., Weiss et al. 1984), (though an alternative belief exists that dynamo theory cannot account for the Sun's magnetic field). The sunspot index is a favorite target for attempts to determine dimensionality, usually via the Grassberger-Procaccia (1983a,b) algorithm by which a correlation integral is computed, yielding a dimensionality measure, integral or fractional (and thus fractal). However, interpretation of the dimensionality vs correlation scale is complicated; moreover the computation is subject to risk because of the short length of the customarily analyzed yearly sunspot sequence; Morfill et al. do a correlation analysis on a shorter time scale, thus having a longer though perhaps still insufficient record. Earlier attempts to determine an intrinsic dimensionality ξ for the underlying dynamical system (see, e.g., Gudzenko and Chertoprud 1965; Ruzmaikin 1981) have been mostly European. Also see Feynman and Gabriel (1990) who argue that the solar field exhibits bifurcations, chaos, and subharmonic generation, in particular from 11 yr to 88 yr (approximately the observationally deduced Gleissberg period). A possible connection with the Maunder minimum and its inferred extension throughout the Holocene is a more difficult matter to judge.

If the conventional view is assumed, then the solar dynamo will have to exhibit multiple eigenmodes, a matter which has already been studied by Boyer and Levy (1992). Such a model would have to support at least the Schwabe and Gleissberg period which are separated by a factor of 5 in period. The longer 210-yr Suess period represents a factor of 20 and is perhaps less plausible.

APPENDIX

The following is a summary of the singular spectrum analysis (SSA) algorithm for filtering of a time series by operations on the autocovariance matrix (Vautard and Ghil 1989; Vautard et al. 1992). Consider the linear operator F defined as

$$Fe(t) = \frac{1}{2\tau} \int_{-\tau}^{\tau} C(t-s)e(s)\,ds \qquad (A1)$$

where F acts on functions belonging to $L^2(-\tau, \tau)$ and $C(t)$ is autocorrelation function of time series $x(t)$ at lag t. From Mercer's theorem (Riesz and Sz.-Nagy 1955)

$$C(t-s) = \sum_k \lambda_k e_k(t) e_k(s) \qquad (A2)$$

where λ_k are eigenvalues of operator F,

$$Fe(t) = \lambda e(t). \qquad (A3)$$

For a discreetly sampled time series $x_1, x_2 \ldots x_N$

$$C_{jl} = C(j-l) = \sum_{k=1}^{M} \lambda_k E_j{}^k E_l{}^k \qquad (A4)$$

where $1 \leq j, l \leq M$. Define delayed time vectors as

$$X_i = (x_{i+1}, x_{i+2} \ldots x_{i+M}) \qquad (A5)$$

and

$$(X_i)_j = x_{i+j} = \sum_{k=1}^{M} a_i{}^k E_j{}^k \qquad (A6)$$

where E_i = eigenvectors (EOFs), λ_i = eigenvalues, and

$$a_i{}^k = \sum_{j=1}^{M} x_{i+j} E_j{}^k \qquad (A7)$$

is k-th principal component (projection of the time series on E_k) and

$$\sum_{i=0}^{N-M} \sum_{j=1}^{M} ((Rx)_{i+j} - \sum_{k \in A} a_i{}^k E_j{}^k)^2 = \text{min.} \qquad (A8)$$

Rx_i is reconstruction of components $k \in A$, where $A = 1 \ldots M$. Tests against noise utilize the possibility of reconstruction of a specified part of time series. White noise or a simple AR(1) process

$$u_{t+1} = \gamma u_t + \alpha w_t \qquad (A9)$$

that has a continuous, line-free power spectrum biased towards low frequencies, are used to simulate synthetic time series. Coefficients γ and α are fit to give to the synthetic series the same variance and lag-one coefficient as those of the difference between the original data and reconstructed part.

Acknowledgments. This research was supported in part by NASA Ulysses program via JPL contract and NSF atmospheric sciences grant. We are indebted to J. Beer for providing us with his Be record from Greenland and to S. Silverman for auroral data, and to P. E. Damon for his material comments.

NOTE ADDED IN PROOF

Recent work suggests that both the Suess (200 yr) and Gleissberg (85 yr) periods in radiocarbon are seen in 650 million year old varves from the Dalradian formation in Scotland (Sonett et al., in preparation).

REFERENCES

Allen, M. R., and Smith, L. A. 1994. Investigating the origins and significance of low-frequency modes of climatic variability. *Geophys. Res.* 21:883–886.

Anderson, K. A., McFadden, J. P., and Lin, R. P. 1981. Propagation of low energy solar electrons. *Geophys. Res. Lett* 8:831–834.

Anderson, K. A., Chaizy, P., Lin, R. P., and Sommers, J. 1992. Non-relativistic solar electron events during December 1990: Results from Ulysses. *Geophys. Res. Lett.* 19:1283 (abstract).

Anderson, R. Y. 1991. Changes in solar irradiance recorded in varves and a historical record: A geological perspective. In *The Sun in Time*, eds. C. P. Sonett, M. S. Giampapa and M. S. Matthews (Tucson: Univ. of Arizona Press), pp. 543–561.

Anderson, R. Y., and Dean, W. E. 1988. Lacustrine varve formation through time. *Paleogeog. Paleoclimat. Paleoecol.* 62:215–235.

Attolini, M. R., Cecchini, S., Castagnoli, G. C., Galli, M., and Nanni, T. 1988. On the existence of the 11-yr cycle in solar activity before the Maunder minimum. *J. Geophys. Res.* 93:12729–12734.

Baliunas, S. 1991. The past, present, and future of solar magnetism: Stellar magnetic activity. In *The Sun in Time*, eds. C. P. Sonett, M. S. Giampapa and M. S. Matthews (Tucson: Univ. of Arizona Press), pp. 809–831.

Beer, J., Siegenthaler, U., Bonani, G., Finkel, R. C., Oeschger, H., Suter, M., and Wölfli, W. 1988. Information on past solar activity and geomagnetism. *Nature* 331:675–679.

Beer, J., Blinov, A., Bonani, G., Finkel, R. C., Hofmann, H. J., Lehmann, B., Oeschger, H., Sigg, A., Schwander, J., Staffelback, T., Stauffer, B., Suter, M., and Wölfi, W. 1990. Use of ^{10}Be in polar ice to trace the 11-year cycle of solar activity. *Nature* 347:164–166.

Beer, J., Raisbeck, G. M., and Yiou, F. 1991. Time variations of ^{10}Be and solar activity. In *The Sun in Time*, eds. C. P. Sonett, M. S. Giampapa and M. S. Matthews (Tucson: Univ. of Arizona Press), pp. 343–359.

Beer, J., Johnsen, S. J., Bonani, G., Finkel, R. C., Langway, C. C., Oeschger, H., Stauffer, B., Suter, M., and Wöfli, W. 1992. ^{10}Be peaks as time markers in polar ice cores. In *The Last Deglaciation: Absolute and Radiocarbon Chronologies*, eds. E. Bard and W. S. Broecker (Berlin: Springer-Verlag) pp. 141–153.

Blinov, A. 1988. The dependence of cosmogenic isotope production rate on solar activity and geomagnetic field variations. In *Secular Solar and Geomagnetic Variations in the Last 10,000 Years*, eds. F. R. Stephenson and A. W. Wolfendale (Dordrecht: Kluwer), pp. 329–340.

Boyer, D. W., and Levy, E. H. 1992. Multiple periodicities in the solar magnetic field: possible origin in a multiple-mode solar dynamo. *Astrophys. J.* 396:340–350.

Bray, R. J., and Loughhead, R. E. 1964. *Sunspots* (New York: Dover).

Broecker, W. S., and Li, Y.-H. 1970. Interchange of water between the major oceans. *J. Geophys. Res.* 75:3545–3552.

Broecker, W. S., and Peng, T.-H. 1970. *Tracers in the Sea* (New York: Lamont-Doherty Geological Obs.).

Broomhead, D. S., and King, G. P. 1986. extracting qualitative dynamics from experimental data. *Physica D* 20:217–236.

Bucha, V. 1970. Influence of the earth's magnetic field on radiocarbon dating. In *Radiocarbon Variations and Absolute Chronology* (Stockholm: Almqvist and Wiksell), pp. 501–510.

Caffee, M. W., Hohenberg, C. M., Nichols, R. H., Jr., Olinger, C. T., Wieler, K., Pedroni, A., Signer, P., Swindle, T. D., and Goswami, J. N. 1991. Do meteorites contain irradiation records from exposure to an enhanced-activity Sun? In *The Sun in Time*, eds. C. P. Sonett, M. S. Giamapapa and M. S. Matthews (Tucson: Univ. of Arizona Press), pp. 413–425.

Castagnoli, G. Cini, Albrecht, A., Beer, J., Bonino, G., Shen, Ch., Callegari, E., Taricco, C., Dittrich-Hannen, B., Kubik, P., Suter, M., and Zhu, G. M. 1995. Evidence for enhanced ^{10}Be deposition in Mediterranean sediments 35 Kyr BP. *Geophys. Res. Lett.* 22:707–710.

Cohen, T. J., and Lintz, P. R. 1974. Long term periodicities in the sunspot cycle. *Nature* 250:398–400.

Cowling, T. G. 1945. On the Sun's general magnetic field. *Mon. Not. Roy. Astron. Soc.* 105:166–174.

Coyne, G. V. 1991. Sunspots: The historical background. In *The Sun in Time*, eds. C. P. Sonett, M. S. Giampapa and M. S. Matthews (Tucson: Univ. of Arizona Press), pp. 3–10.

Craig, H. 1957. The natural distribution of radiocarbon and the exchange time of carbon dioxide between atmosphere and sea. *Tellus* 9:1–17.

Damon, P. E., and Sonett, C. P. 1991. Solar and terrestrial components of the atmospheric ^{14}C variation spectrum. In *The Sun in Time*, eds. C. P. Sonett, M. S. Giampapa and M. S. Matthews (Tucson: Univ. of Arizona Press), pp. 360–388.

Damon, P. E., Sternberg, R. S., and Radnell, C. J. 1983. Modeling of atmospheric radiocarbon fluctuations for the past three centuries. *Radiocarbon* 25:249.

Dansgaard, W., Johnsen, S. S., Clausen, H. B., and Langway, C. C. 1971. Climatic record revealed by the Camp Century ice core. In *The Late-Glacial Ages*, ed. K. Turekian (New Haven: Yale Univ. Press), pp. 37–46.

Dansgaard, W., Johnsen, S. J., Clausen, H. B., Dahl-Jensen, D., Gunderstrup, N., Hammer, C., and Oeschger, H. 1984. North Atlantic climate oscillations revealed by deep Greenland ice cores. In *Climate Processes and Climate Sensitivity*, eds. J. E. Hansen and T. Takahashi (Washington, D. C.: American Geophys. Union),

   pp. 288–298.
Eddy, J. A. 1976*a*. The Maunder minimum. *Science* 192:1189–1202.
Eddy, J. A. 1976*b*. The Sun since the bronze age. In *Physics of Solar Planetary Environments*, vol. 2, ed. D.J. Williams (Washington, D. C.: American Geophys. Union), p. 958.
Elsasser, W., Ney, E. P., and Winckler, J. R. 1956. Cosmic ray intensity and geomagnetism. *Nature* 178:1–226.
Feynman, J. 1982. Geomagnetic and solar wind cycles. *J. Geophys. Res.* 87:6153–6162.
Feynman, J., and Crooker, N. U. 1978. The solar wind at the turn of the century. *Nature* 275:626–627.
Feynman, J., and Gabriel, S. B. 1990. Period and phase of the 88 year solar cycle and the Maunder minimum: Evidence for a chaotic Sun. *Solar Phys.* 127:393–403.
Forman, M. A., Jokipii, J.R., and Owens, A. J. 1974 Cosmic-ray streaming perpendicular to the mean magnetic field. *Astrophys. J.* 192:535–540.
Folland, C. K., and Kates, F. 1984. Changes in decadally averaged sea surface temperature over the world 1861–1980. In *Milankovitch and Climate*, part 2, eds. A. L. Berger, J. Imbrie, J. Hays, G. Kukla and B. Saltzman (Dordrecht: D. Reidel), pp. 721–727.
Fritts, H. C. 1976. *Tree Rings and Climate* (New York: Academic Press).
Ghil, M., and Mo, K. 1991. Intraseasonal oscillations in the global atmosphere—Part I. Northern hemisphere and tropics. 1991. *J. Atmos. Sci.* 48:752–790.
Gibson, E. G. 1973. *The Quiet Sun*, NASA SP-303.
Gilliland, R. L. 1981. Solar radius variations over the past 265 years. *Astrophys. J.* 248:1144–1155.
Gleissberg, W. 1944. A table of secular variations of the solar cycle. *Terrest. Magnet. Atmos. Elect.* 49:243–244.
Gleissberg, W. 1966. Ascent and descent in the eighty-year cycles of solar activity. *J. Brit. Astron. Assoc.* 76:265–270.
Gough, D. 1988. Theory of solar variation. In *Solar-Terrestrial Relationships and the Earth Environments in the last Millenia*, ed. G. C. Castagnoli (Amsterdam: North-Holland Press), pp. 90–132.
Grassberger, P., and Procaccia, I. 1983*a*. Characterization of strange attractors. *Phys. Rev. Lett.* 50:346–349.
Grassberger, P., and Procaccia, I. 1983*b*. Measuring the strangeness of strange attractors. *Physica D* 9:189–191.
Gudzenko, L. I., and Chertoprud, V. E. 1965. Some dynamical properties of cyclic solar activity. *Soviet Astron. A.J.* 8:555–561.
Herbert, F., Sonett, C. P., and Gaffey, M. J. 1991. Protoplanetary thermal metamorphism: The hypothesis of electromagentic induction in the protosolar wind. In *The Sun in Time*, eds. C. P. Sonett, M. S. Giampapa and M. S. Matthews (Tucson: Univ. of Arizona Press), pp. 710–739.
Holzer, R. E., and Slavin, J. A. 1982. A quantitative model of geomagnetic activity. *J. Geophys. Res.* 87:9054–9058.
Houtermans, J. C. 1971. Geophysical Interpretation of Bristlecone Pine Radiocarbon Measurements Using a Method of Fourier Analysis of Unequally Spaced Data. Ph.D. Thesis, Univ. of Berne.
Houtermans, J. C., Suess, H. E., and Oeschger, H. 1973. Reservoir models and production rate variations of natural radiocarbon. *J. Geophys. Res.* 78:1897–1908.
Hudson, H. S. 1988. Observed variability of the solar luminosity. *Ann. Rev. Astron. Astrophys.* 26:473–508.
Johnsen, S. J., Dansgaard, W., Clausen, H. B., and Langway, C. C. 1970. Climatic

oscillations 1200–2000 AD. *Nature* 227:482–483.

Jokipii, J. R. 1971. Propagation of cosmic rays in the solar wind. *Rev. Geophys. Space Phys.* 9:27–87.

Jokipii, J. R. 1991. Variations of the cosmic-ray flux with time. In *The Sun in Time*, eds. C. P. Sonett, M. S. Giampapa and M. S. Matthews (Tucson: Univ. of Arizona Press), pp. 205–220.

Jokipii, J. R. 1993. Particle drifts for a finite scattering rate. *Proc. 23rd Intl. Cosmic Ray Conf.* 3:497.

Jokipii, J. R., Levy, E. H., and Hubbard, W. B. 1977. Effects of particle drift on cosmic-ray transport. I. General properties, application to solar modulation. *Astrophys. J.* 213:861–868.

Kurths, J., Spiering, Ch., Müller-Stoll, W., and Striegler, U. 1993. Search for solar periodicities in Miocene tree ring widths. *Terra Nova* 5:359.

Lal, D. 1985. Carbon cycle variations during the past 50,000 years: Atmospheric $^{14}C/^{12}C$ ratios as an isotopic indicator. In *The Carbon Cycle and Atmospheric CO_2: Natural Variations Archean to Present*, eds. E. T. Sundquist and W. S. Broecker (Washington, D. C.: American Geophys. Union), pp. 221–233.

Lal, D., and Peters, B. 1967. Cosmic ray produced radioactivity on the Earth. In *Handbuch der Physik*, vol. 46(2), ed. S. Flügge pp. 551–612.

Lederer, C. M., Hollander, J. M., and Perlman, I. 1967. *Table of Isotopes*, 6th ed. (New York: Wiley).

Lee, M., and Fisk, L. A. 1981. The role of particle drifts in solar modulation. *Astrophys. J.* 248:836–844

Lichtenstein, B. R., and Sonett, C. P. 1979. Dynamic structure of large amplitude Alfvénic variations in the interplanetary magnetic field. *Geophys. Res. Lett.* 7:189–192.

Lingenfelter, R. E., and Ramaty, R. 1970. Astrophysical and geophysical variations in ^{14}C production. In *Radiocarbon Variations and Absolute Chronology*, ed. I. U. Olsson (New York: Wiley), pp. 513–537.

Maunder, E. W. 1890. *Mon. Not. Roy. Astron. Soc.* 1:251.

Maunder, E. W. 1922. The prolonged sunspot minimum of 1645–1715. *J. Brit. Astron. Assoc.* 32:140–145.

McHargue, L. R., and Damon, P. E. 1991. The global beryllium-10 cycle. *Rev. Geophys.* 29:141.

McHargue, L. R., Damon, P. E., and Donahue, D. J. 1995. Enhanced cosmic-ray production of ^{10}Be coincidence with the Mono lake and Laschamp geomagnetic excursions. *Geophys. Res. Lett.* 22:659–662.

Morfill, G. E., Scheingraber, H., Voges, W., and Sonett, C. P. 1991. Sunspot number variations: Stochastic or chaotic. In *The Sun in Time*, eds. C. P. Sonett, M. S. Giampapa and M. S. Matthews (Tucson: Univ. of Arizona Press), pp. 30–58.

Murphy, J. O., and Palmer, J. A. 1992. Ring-width variation in sub-fossil wood samples as an implication of short-term solar variability 2000 yr B.P. *Proc. Astron. Soc. Australia* 10:68.

Neftel, A., Oeschger, H., and Suess, H. E. 1981. Secular non-random variations of cosmogenic carbon-14 in the terrestrial atmosphere. *Earth Planet. Sci. Lett.* 56:127–147.

O'Brien, K., de la Zerda Lerner, A., Shea, M. A., and Smart, D. F. 1991. The production of cosmogenic isotopes in the Earth's atmosphere and their inventories. In *The Sun in Time*, eds. C. P. Sonett, M. S. Giampapa and M. S. Matthews (Tucson: Univ. of Arizona Press), pp. 317–342.

Parker, E. N. 1955*a*. The formation of sunspots from the solar toroidal field. *Astrophys. J.* 121:491–507.

Parker, E. N. 1955*b*. Hydromagnetic dynamo models. *Astrophys. J.* 122:293–314.

Parker, E. N. 1965. The passage of charged particles through interplanetary space. *Planet. Space Sci.* 13(1):9–49.

Parker, E. N. 1972. Topological dissipation and the small-scale fields in turbulent gases. *Astrophys. J.* 174:499–510.

Pestiaux, P., Deplessy, S. C., and Berger, A. 1987. Paleoclimate variability at frequencies from 10^{-4} cycles per year to 10^{-3} cycles per year—evidence for non-linear behavior of the climate system. In *Climate: History, Periodicity, and Predictability*, eds. M. R. Rampino, J. E. Sanders, W. S. Newman and K. L. Königsson (New York: Van Nostrand Reinhold), pp. 285–299.

Pestiaux, P., Duplessy, J. C., van der Mersch, I., and Berger, A. 1988. Paleoclimatic variability at frequencies ranging from 1 cycle per 10000 years to 1 cycle per 1000 years: Evidence for nonlinear behavior of the climate system. *Climate Change* 12:9–37.

Press, W. H., Teukolsky, S. A., Vettering, W. T., and Flannery, B. R. 1992. *Numerical Recipes in Fortran: The Art of Scientific Computing*, 2nd ed. (New York: Cambridge Univ. Press).

Radick, R. R. 1991. The luminosity variability of solar-type stars. In *The Sun in Time*, eds. C. P. Sonett, M. S. Giampapa and M. S. Matthews (Tucson: Univ. of Arizona Press), pp. 787–808.

Raisbeck, G. M., Yiou, F., Bourles, D., Lorius, C., Jouzet, J., and Barkov, N. I. 1987. Evidence for two intervals of enhanced ^{10}Be deposition in Antarctic ice during the last glacial period. *Nature* 326:273–277.

Raisbeck, G. M., Yiou, F., Jouzel, J., Petit, J. R., Barkov, N. I., and Bard, E. 1992. ^{10}Be deposition at Vostok, Antarctica during the last 50,000 years and its relationship to possible cosmogenic production variations during this period. In *The Last Deglaciation: Absolute and Radiocarbon Chronologies*, eds. E. Bard and W. S. Boecker (Berlin: Springer-Verlag), pp. 127–139.

Reedy, R. C., Arnold, J. R., and Lal, D. 1983. Cosmic-ray record in solar system matter. *Ann. Rev. Nucl. Part. Sci.* 33:505–537.

Reid, G. C. 1987. Influence of solar variability on global sea surface temperatures. *Nature* 329:142–143.

Reid, G. C., and Gage, K. S. 1988. The climatic impact of secular variations in solar irradiance. In *Secular, Solar, and Geomagnetic Variations in the Last 10,000 Years*, eds. F. R. Stephenson and A. W. Wolfendale (Dordrecht: Kluwer), pp. 225–243).

Ribes, E., Beardsley, B., Brown, T. M., DeLache, Ph., Laclare, F., Kuhn, J. R., and Leister, N. V. 1991. The variability of the solar diameter. In *The Sun in Time*, eds. C. P. Sonett, M. S. Giampapa and M. S. Matthews (Tucson: Univ. of Arizona Press), pp. 59–97.

Riesz, F., and Sz.-Nagy, B. 1955. *Functional Analysis* (New York F. Ungar).

Ruzmaikin, A. A. 1981. The solar cycle as a strange attractor. *Comments on Astrophys.* 9:85–96.

Siegenthaler, U., Heimann, M., and Öeschger, H. 1980. ^{14}C variations caused by changes in the global carbon cycle. *Radiocarbon* 22:177–191.

Silverman, S. M. 1992. Secular variation of the aurora for the past 500 years. *Rev. Geophys.* 30:333–351.

Simpson, J. A. 1957. Cosmic-radiation neutron intensity monitor. *Ann. IGY*, Vol. IV, Parts IV–VII, pp. 351–373.

Siscoe, G. L. 1980. Evidence in the auroral record for secular solar variations. *Rev. Geophys. Space Phys.* 18:647–658.

Slavin, J. A., and Smith, E. J. 1983. Solar cycle variations in the interplanetary magnetic field. In *Solar Wind Five*, ed. M. Neugebauer, NASA CP-2280, pp. 323–331.

Sonett, C. P. 1982. Sunspot time series: Spectrum from square law modulation of the Hale cycle. *Geophys. Res. Lett.* 9:1313–1316; erratum, *Geophys. Res. Lett.* 10:491 (1983).

Sonett, C. P. 1983. Sunspot index infers a small relict magnetic field in the Sun's core. *Nature* 306:670–673.

Sonett, C. P. 1984. Very long solar periods and the radiocarbon record. *Rev. Geophys. Space Phys.* 22:239–254.

Sonett, C. P. 1991. Long period solar-terrestrial variability. *Rev. Astrophys. Suppl.* 29:909–914.

Sonett, C. P. 1992. The present status of the understanding of the long-period spectrum of radiocarbon. In *Radiocarbon after Four Decades*, eds. R. E. Taylor, A. Long and R. S. Kra (Berlin: Springer-Verlag).

Sonett, C. P., and Finney, S. A. 1990. The spectrum of radiocarbon. *Phil. Trans. Roy. Soc. A* 330:413–426.

Sonett, C. P., and Suess, H. E. 1984. Correlation of Bristlecone pine ring widths with atmospheric carbon-14 variations: A climate-Sun relation. *Nature* 307:141–143.

Sonett, C. P., Morfill, G. E., and Jokipii, J. R. 1987. Interstellar shock waves and ^{10}Be from ice cores. *Nature* 330:458–460.

Sonett, C. P., Colburn, D. S., Schwartz, K., and Kiel, K. 1970. The melting of asteroidal-sized bodies by unipolar dynamo induction from a primordial T Tauri sun. *Astrophys. Space Sci.* 7:446–488.

Sonett, C. P., Williams, C. R., and Nils-Axel, M. 1992. The Fourier spectrum of Swedish riverine varves: Evidence of sub-arctic periodic quasi-biennial (QBO) and Southern oscillations (SO). *Global Planet. Change* 98:57–65.

Stuiver, M. 1961. Variations in radiocarbon concentration and sunspot activity. *J. Geophys. Res.* 66:273–276.

Stuiver, M., and Becker, B. 1993. High-precision decadal calibration of the radiocarbon time scale. *Radiocarbon* 35:35–65.

Stuiver, M., and Braziunas, T. F. 1993. Sun, ocean, climate and atmospheric $^{14}CO_2$: An evaluation of causal and spectral relationships. *The Holocene* 3,4:289–305.

Stuiver, M., and Quay, P. D. 1980. Changes in atmospheric carbon-14 attributed to a variable Sun. *Science* 207:11–19.

Stuiver, M., and Quay, P. D. 1981. Atmospheric ^{14}C changes resulting from fossil fuel CO_2 and cosmic ray variability. *Earth Planet. Sci. Lett.* 53:349–362.

Suess, H. E. 1968. Climate changes, solar activity, and the cosmic-ray production rate of natural radiocarbon. *Meteor. Mon.* 8:146.

Suess, H. E. 1970. The three causes of the secular C14 fluctuations, their amplitudes and time constants. In *Radiocarbon Variations and Absolute Chronology*, ed. I. U. Olsson (New York: Wiley), pp. 595–605.

Suess, S. T. 1979. The solar wind during the Maunder minimum. *Planet. Space Sci.* 27:1001–1013.

Suess, H. E. 1980. The radiocarbon record in tree rings of the last 8000 years. *Radiocarbon* 3:1–4.

Vautard, R., and Ghil, M. 1989. Singular spectrum analysis in nonlinear dynamics with applications to paleoclimatic time series. *Physica D* 35:395–424.

Vautard, R., Yiou, P., and Ghil, M. 1992. Singular-spectrum analysis: A toolkit for short, noisy chaotic signals. *Physica D* 58:95–126.

Wasson, J. T. 1974. *Meteorites: Classification and Properties* (Berlin: Springer-Verlag).

Wasson, J. T. 1985. *Meteorites: Their Record of Early Solar-System History* (New York: W. H. Freeman).

Weiss, N. O. 1994. Chaotic modulation of the solar cycle. *Phil. Trans. Roy. Soc. London A* 348:445.

Weiss, N.O., Cattaneo, F., and Jones, C. A. 1984. Periodic and aperiodic dynamo waves. *Geophys. Astrophys. Fluid Dyn.* 30:305–341.
Willis, E. H., Tauber, H., and Munnich, K. D. 1960. Variations in the atmospheric radiocarbon concentration over the past 1300 years. *Radiocarbon* 3:4.
Willson, R. C., Hudson, H. C., Fröhlich, C., and Brusa, R. W. 1986. Long-term downward trend in total solar irradiance. *Science* 234:1114–1117.
Wolfsberg, K., and Kocharov, G. E. 1991. Solar neutrinos and the history of the Sun. In *The Sun in Time*, eds. C. P. Sonett, M. S. Giampapa and M. S. Matthews (Tucson: Univ. of Arizona Press), pp. 288–313.

THE WINDS OF CATACLYSMIC VARIABLES

CHRISTOPHER W. MAUCHE
Lawrence Livermore National Laboratory

and

JOHN C. RAYMOND
Harvard-Smithsonian Center for Astrophysics

We present an observational and theoretical review of the winds of cataclysmic variables (CVs). Specifically, we consider the related problems of the geometry, ionization state, and mass-loss rate of the winds of CVs. We present evidence of wind variability and discuss the results of studies of eclipsing CVs. Finally, we consider the properties of accretion disk wind and magnetic wind models. We note that some of these models predict substantial angular momentum loss, which could affect both disk structure and binary evolution.

I. INTRODUCTION

Cataclysmic variables (CVs) are a large, diverse class of short-period semi-detached mass-exchanging binaries composed of a white dwarf, a late-type (G, K, or M) main-sequence secondary, and, typically, an accretion disk. The various CV subtypes include novae, which are powered by the thermonuclear burning of material on the surface of the white dwarf, and dwarf novae and nova-like variables, which are powered by the accretion of material transferred from the secondary to the white dwarf. For recent reviews of CVs, refer to Cropper (1990), Livio (1994), and Córdova (1995).

Among the nova-like variables, there exists a class in which the magnetic field of the white dwarf is strong enough to influence the flow of material accreted by the white dwarf. These magnetic nova-like variables are subdivided into polars (AM Her stars), whose white dwarfs have spin periods which are synchronous with the orbital period, and intermediate polars (DQ Her stars), whose spin periods are less than the orbital period. In polars, the strength of the magnetic field of the white dwarf, the small size (short period) of the binary, and the degree of synchronization do not allow an accretion disk to form. In intermediate polars, the lower strength of the magnetic field of the white dwarf, the larger size (longer period) of the binary, and the lack of spin-orbit synchronization allow an accretion disk to form. This disk maintains an essentially Keplerian velocity profile at large radii, but is subsequently disrupted as it moves within the magnetosphere of the white dwarf.

Dwarf novae form a distinctive CV subtype characterized by the frequency, duration, and magnitude of their outbursts. The outbursts of dwarf novae have recurrence times of tens to hundreds of days, durations of between 1 and 10 days, and increases of between 3 and 5 mag in their visual brightness. These properties are thought to result from an instability either in the rate of mass exchange from the secondary to the accretion disk of the white dwarf, or in the rate of mass transfer through the accretion disk itself (Smak 1984; Cannizzo 1993). Nonmagnetic nova-like variables appear to be best described as dwarf novae "stuck" in outburst.

There are two sources of winds in CVs. The first, restricted to nonmagnetic nova-like variables and dwarf novae in outburst, is evidenced by the P Cygni profiles of their ultraviolet resonance lines. In analogy with the winds of early-type stars, the winds of these CVs are generally thought to be driven by radiation pressure, a process which results in high-velocity outflows with mass-loss rates limited by the conservation of momentum to $\dot{M}_{\rm wind} \lesssim L/V_\infty c$, where L is the luminosity of the radiation field and V_∞ is the terminal velocity of the wind. In CVs, these quantities can be written as $L = \zeta\, GM_{\rm wd}\dot{M}/R_{\rm wd}$ and $V_\infty = f\, V_{\rm esc} = f\,(2GM_{\rm wd}/R_{\rm wd})^{1/2}$, respectively, where $M_{\rm wd}$ and $R_{\rm wd}$ are the mass and radius of the white dwarf, respectively, $\dot{M}$ is the mass-accretion rate, and ζ and f are constants of order unity. Hence, the mass-loss rate of the wind is $\dot{M}_{\rm wind} \lesssim \zeta f^{-1}\,(GM_{\rm wd}/2c^2 R_{\rm wd})^{1/2}\,\dot{M} \sim 0.01\,\zeta f^{-1}\,\dot{M}$. The second source of winds in CVs is the solar-type wind of the secondary. Although this wind has no *direct* observational manifestation, it is of great theoretical interest, as it is appealed to to drive the secular evolution of the binary. For earlier reviews of the winds of CVs, refer to Córdova and Howarth (1987) and Drew and Kley (1993).

II. GEOMETRY AND MASS-LOSS RATE

The study of the winds of CVs began with the launch of IUE, the first ultraviolet spectroscopic satellite with sufficient sensitivity and spectral resolution to discover first (Heap et al. 1978), and subsequently to study, the P Cygni profiles of the ultraviolet resonance lines of CVs: namely N V λ 1240, Si IV λ 1400, and C IV λ 1550. The "early" papers reporting IUE observations of "high $\dot{M}$" CVs—classical novae, nova-like variables, and dwarf novae in outburst (see, e.g., Krautter et al. 1981; Klare et al. 1982; Greenstein and Oke 1982; Guinan and Sion 1982; Córdova and Mason 1982,1985; Sion 1985)—all contain a discussion of the P Cygni profiles of these lines.

Observationally, the P Cygni profiles of CVs differ from the profiles of early-type stars in two significant ways, both of which are illustrated by the spectra shown in Fig. 1. First, whereas the deepest part of the blue-shifted absorption component of the line profiles of early-type stars is near the terminal velocity, the deepest part of the line profiles of CVs is near zero velocity. Second, the absorption components of the line profiles of early-

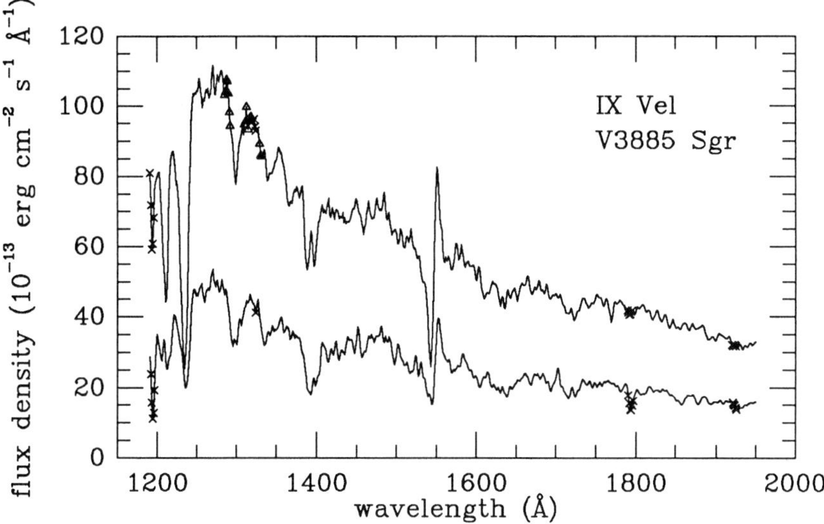

Figure 1.   Typical short-wavelength IUE spectra of the nova-like variables IX Vel
and V3885 Sgr.

type stars often are black, whereas in CVs they are at most roughly half the
depth of the continuum. As judged by the velocity of the blue edge of the
absorption component of the profiles, the terminal velocity of the wind is
roughly 5000 km s^{-1}. This coincidence between the terminal velocity of the
wind and the escape velocity of the white dwarf suggests that the winds of
CVs originate from deep in the accretion disk, the boundary layer between
the disk and the surface of the white dwarf, or the white dwarf itself (Córdova
and Mason 1982).

As in early-type stars, the P Cygni profiles of the ultraviolet resonance
lines of CVs are formed by scattering in their winds (Drew and Verbunt 1985).
Specifically, the line profiles are formed by the scattering of continuum photons of frequency v which come into resonance with the radially accelerating
wind at a point in velocity space given by $V(r) = (v - v_0)c/v_0$, where v_0 is
the rest frequency of the transition. The material in the wind moving toward
the observer in front of the continuum source scatters photons out of the line
of sight, and so produces a blue-shifted absorption feature. From all other
regions of the wind (except the region hidden behind the continuum source),
we receive photons which have been scattered into our line of sight. Because
the projected velocities of these parts of the wind range from positive to negative, the emission feature so produced is roughly symmetric about v_0. In
the Sobolev approximation to the radiation transfer in the expanding wind,
the fraction of scattered photons varies as $1 - \exp[-\tau(r)]$, where the optical
depth $\tau(r) \propto \dot{M}_{\text{wind}}\, \xi(r)\, r^{-2}\, V(r)^{-1}\, [dV(r)/dr]^{-1}$, where $V(r)$ and $\xi(r)$ are
respectively the velocity and ionization laws characterizing the wind.

Possibly different wind geometries aside, the winds of CVs differ from the winds of early-type stars in the source function of the scattered continuum, which is the photosphere of the star in the former case, and the accretion disk in CVs. Because of this differing source of illumination, and because of the inclination dependence introduced by the presence of the accretion disk, the theoretical P Cygni profiles of early-type stars (see, e.g., Olson 1978; Castor and Lamers 1979) cannot be applied to the profiles of CVs. In order to understand why this is so, it is instructive to consider the case of a spherical wind emanating from the center of a luminous accretion disk. If the disk is observed edge-on, no part of the wind lies in front of the disk, and so no absorption results; the P Cygni profile consists simply of an emission component whose full width is given by the terminal velocity of the wind: $\Delta \nu = 2\nu_0 V_\infty / c$. As the inclination from which the disk is observed decreases, the fraction of the wind seen in projection against the disk increases, thereby increasing the strength of the absorption component of the line profile. At the same time, the strength of the emission component decreases due to both the trade-off of emission for absorption regions, and the corresponding increase of the volume of the wind hidden behind the disk. In the limit of a face-on disk, roughly half the wind is hidden behind the disk and roughly half is seen in projection against the disk; the resulting profile has a strong absorption component and little or no emission. This inclination dependence of the P Cygni profiles of CVs is illustrated in Fig. 2.

Given this important difference between the P Cygni profiles of CVs and early-type stars, and the possibility that the geometry of the winds of CVs is more nearly bipolar than spherical, theoretical profiles appropriate to the peculiar conditions of CVs must be constructed in order that well-founded conclusions can be drawn concerning the nature of their winds. In addition to the mass-loss rate, velocity law, and ionization law characterizing the wind, one must specify the source function and geometry of the wind, as well as the source function of the scattered continuum, which is typically taken to be given by the Shakura and Sunyaev (1973) temperature distribution of the accretion disk—itself dependent on the mass-accretion rate and the mass and radius of the white dwarf—and the emissivity of the Planck function: $I_\lambda(r) = B_\lambda[T(r; M_{\rm wd}, R_{\rm wd}, \dot{M})]$. Given this plethora of variables, the first models of the P Cygni profiles of CVs (Drew 1987; Mauche and Raymond 1987) were designed with aerodynamic simplicity: the wind was assumed to originate in the boundary layer or inner disk, the ionization state was assumed to be constant with radius, and the velocity law was assumed to be given by either a power law or linear function of the radius.

In the context of these models, Drew (1987) and Mauche and Raymond (1987) came to the following conclusions regarding the nature of the winds of CVs. First, the wind can originate near the white dwarf only if its acceleration is very slow compared to the winds of early-type stars. Specifically, the wind must not reach a significant fraction of its terminal velocity before reaching a distance from the white dwarf comparable to the distance to the peak of

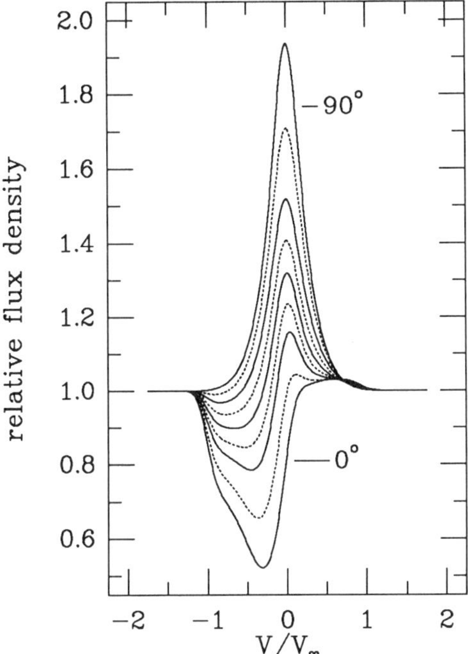

Figure 2. Theoretical P Cygni profiles for a specific mass-loss rate and velocity law for nine values of the inclination (figure from Mauche and Raymond 1987).

the distribution of the resonant continuum in the accretion disk, which for typical parameters is at ~ 10 white dwarf radii. Second, the mass-loss rate of the wind is $\sim$ few $\times\ 10^{-11}\ \xi_{C\,IV}^{-1}\ M_\odot\ yr^{-1}$, where $\xi_{C\,IV}$ is the ionization fraction of C IV. With typical mass-accretion rates of $\dot{M} \sim 10^{-8}\ M_\odot\ yr^{-1}$, the ionization fraction of C IV must then be no less than a few tenths if the wind is driven by radiation pressure. Finally, some type of bipolarity in either the wind density profile [e.g., $n(r, \theta) = n(r, 0)|\cos{}^l\theta|$] or the radiation field [e.g., limb darkening: $I(\theta) = \frac{1}{2}(1 + \frac{3}{2}|\cos\theta|)\,I(0)$] is required to avoid the formation of an absorption component in the line profiles of high-inclination systems.

III. IONIZATION STATE

In order to characterize fully the winds of CVs, we also need to know their ionization state. It is, after all, the absolute concentrations of the ions which ultimately determine the wind's mass-loss rate, the relative concentrations of the ions which determine the relative strengths of the line profiles, and radial variations in the concentrations of the ions which, in collusion with the velocity law, determine the shapes of the profiles. Unfortunately, we have few observational handles on the ionization state of the wind other than those provided by the N V, Si IV, and C IV resonance lines. The Hopkins

Ultraviolet Telescope (HUT) has recently provided information for transitions in the 1200 to 912 Å bandpass, but for only a few systems (see, e.g., Long et al. 1991,1994). Into this near vacuum, theory guides us.

Despite many theoretical attempts to understand the ionization state of the winds of CVs, investigators repeatedly find that if the temperature and luminosity of the radiation generated in the boundary layer is as given by simple theory (see, e.g., Pringle 1977; Patterson and Raymond 1985), and if the wind is smooth and represents a mass-loss rate which is a modest fraction of the accretion rate, the wind will be ionized beyond the observed ionization stages of $N\,V$, $Si\,IV$, and $C\,IV$. Drew and Verbunt (1985) found that a relatively slow velocity law, as well as variations in the temperature and luminosity of the boundary layer, help to increase the ionization fractions of the observed ions, but not by an amount sufficient to explain the strengths of the line profiles. Kallman and Jensen (1985) suggested that the resolution of the related problems of the ionization state of the wind and the apparent discrepancies between observations and theoretical predictions of the luminosity and temperature of the boundary layer radiation was to be found in a wind of sufficient density to absorb photoelectrically the EUV and soft X-ray flux emitted by the boundary layer. However, they found that the required mass-loss rate of such a wind exceeds the mass-accretion rate, a possibility excluded by the conservation of energy, which limits the mass-loss rate to $\dot{M}_{wind} \lesssim \zeta f^{-1} \dot{M}$. A recent ROSAT observation of the dwarf nova SU UMa in outburst shows an increased absorbing column of about 10^{20} cm^{-2} compared with quiescence, a column which does not drastically reduce the observable soft X-rays (Silber et al. 1994). In another attempt to solve this problem, Mauche and Raymond (1987) investigated the consequences of introducing shock compression in the wind in an attempt to reduce its ionization state relative to that of a smooth wind. In order to produce reasonably large $C\,IV$ ionization fractions, they found that mass-loss rates of $\dot{M}_{wind} \sim 0.3\,\dot{M}$ are required if shocks are present with velocities of ~ 100 km s^{-1}. Although this solution allows a wind on energetic grounds, Mauche and Raymond found that such a wind produces much larger ultraviolet line fluxes than are observed and (in their view) is far too efficient in absorbing the soft X-rays generated in the boundary layer.

Since that time, observations (Mauche et al. 1991,1995; van Teeseling et al. 1993) and theory (Hoare and Drew 1991) have (again) called into question the parameters characterizing the boundary layer radiation. Hoare and Drew (1993) have therefore considered wind models with cool, luminous boundary layers and no boundary layer at all. They found that the observed strength of the $C\,IV$ and $N\,V$ lines can be matched by models with luminous, cool ($kT = 5$–9 eV) boundary layers or with disk-only models if the accretion rate is $\dot{M} \sim 2$–4×10^{-8} M$_\odot$ yr^{-1}. The softer radiation fields of these models allow low mass-loss rate ($\dot{M}_{wind} \sim 6 \times 10^{-10}$ M$_\odot$ yr^{-1} $\sim 0.02\,\dot{M}$) winds, capable of being driven by radiation pressure. The presence of blue-shifted absorption in $Si\,IV$ remains a problem, however, as does the presence of a strong $O\,VI$ line (Long et al. 1991,1994). A further difficulty is the origin of the soft X-rays observed in SS Cyg, U Gem, VW Hyi, and OY Car (Córdova et al. 1980,1984; van der Woerd et al. 1986; Naylor et al. 1988). If the wind is driven by radiation pressure, a natural alternative for the origin of the soft X-rays is shocks in the wind (Mauche and Raymond 1987; Naylor et al. 1988). However, this solution fails to explain the production of the quasi-periodic oscillations observed in the soft X-ray flux of SS Cyg, U Gem, and VW Hyi (Córdova et al. 1980,1984; van der Woerd et al. 1987; Mauche 1995), as well as why the X-rays generated in the winds of CVs should be so much cooler than the X-rays produced by the winds of O stars (Chlebowski et al. 1989).

Clearly, more work is needed in this area. Much concerning the ionization state of the wind could be done with the additional information from the far-ultraviolet waveband supplied by HUT and the EUV waveband supplied by the Extreme Ultraviolet Explorer (EUVE). Indeed, EUVE observations of the dwarf nova SS Cyg in outburst (Mauche et al. 1995) have produced results so different from theoretical expectations that they may radically change our understanding of the boundary layers and winds of CVs. First, whereas all models of boundary layer emission (see, e.g., Pringle 1977; Popham and Narayan 1995) predict dramatic changes in the effective temperature with accretion rate, and whereas the brightness of the source rose by a factor of over 100 during the EUVE observations, the *shape* of the EUV spectrum was constant. Second, whereas theory predicts that the luminosities of the boundary and accretion disk should be comparable unless the white dwarf is rotating near breakup, $L_{bl}/L_{disk} \sim 1$, observations imply that this ratio is less than 0.1. Third, contrary to simple theory, the EUV spectrum of SS Cyg is a complex mixture of emission and absorption features dominated by transitions of 5 to 7 times ionized Ne, Mg, and Si. Work to understand these results is ongoing.

IV. VARIABILITY

As is evident from the variability of the P Cygni profiles, the winds of CVs are not time-steady. At one extreme, the winds are observed to develop and subsequently to decay as the outbursts of dwarf novae evolve (see, e.g., Verbunt et al. 1984,1987; la Dous et al. 1985; Hassall et al. 1985). At the other extreme, the winds are observed to vary on time scales of hours down to tens of minutes (see, e.g., Prinja et al. 1992), about as quickly as ultraviolet spectra can be obtained with IUE. Variations on the orbital phase are also observed; the most spectacular being the orbital phase-dependent profile and equivalent width variations observed in the line profiles of the dwarf nova YZ Cnc (Drew and Verbunt 1988; see Fig. 3). The series of observations precipitated by this discovery (see, e.g., Woods et al. 1990,1992) prove, if nothing else, that line profile variations are common on the orbital period. The cause of these variations is less than clear. In general, profile variations can be produced by an azimuthal asymmetry in either the brightness distribution of the accretion disk, or in the wind's geometry, velocity field, or ionization structure. The trick is understanding how these asymmetries are produced, and why they are stable. The answer may tell us something important about the accretion disk or the wind, but little quantitative information exists at the present time to narrow the selection of possibilities.

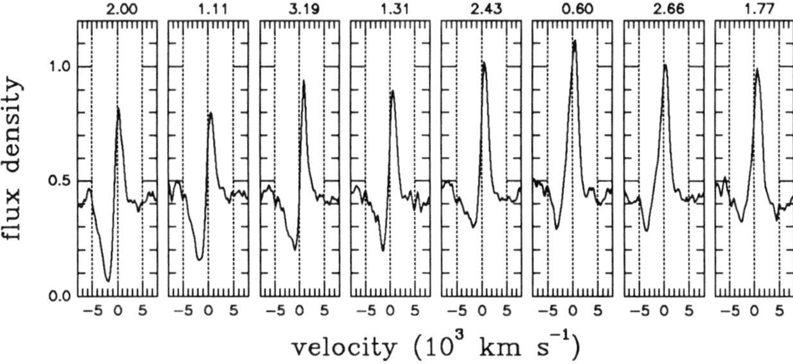

Figure 3.   CIV line profile variations as a function of orbital phase for the dwarf nova YZ Cnc; flux density is measured in units of 10^{-12} erg cm^{-2} s^{-1} Å^{-1} (figure adapted from Drew and Verbunt 1988).

One potentially important, but as yet little-studied, diagnostic observable of the winds of CVs is the radial velocity variations of their P Cygni profiles. Mauche (1991) discovered such variations in high-resolution IUE spectra of the nova-like variable IX Vel. The radial velocity variations of the P Cygni profiles of the Si IV doublet of this system are well fit by $V_{\mathrm{Si\,IV}}(\phi) = -350 - 200\,\sin[2\pi(\phi - 0.1)]$ km s^{-1}. The difference between this solution and the radial velocity solution of the white dwarf and accretion disk is $V_{\mathrm{wind}}(\phi) \approx -400 - 130\,\sin[2\pi(\phi - 0.75)]$; the radial velocity of the wind lags the white dwarf and accretion disk by 90°. One interpretation of this result is that the velocity of the wind in which the Si IV ion dominates varies according to $V_{\mathrm{wind}}(\varphi) \approx 400 - 130\,\cos(2\pi\varphi)$, where φ is the azimuthal angle between the white dwarf and the secondary; the effective velocity of the wind is lowest on the side facing the secondary and highest on the side facing away. This type of variation may not be that uncommon. Drew et al. (1991) observed velocity shifts in the N V and C IV profiles of the dwarf nova DX And; Mauche et al. (1994) found similar radial velocity variations in the N V, Si IV, C IV, and He II emission lines (presumably wind-dominated features) of the eclipsing nova-like variable V347 Pup (see Fig. 4). Given the diagnostic potential of these radial velocity variations, it seems clear that much could be learned from radial velocity studies of other CVs, particularly those for which the orbital ephemeris is well known.

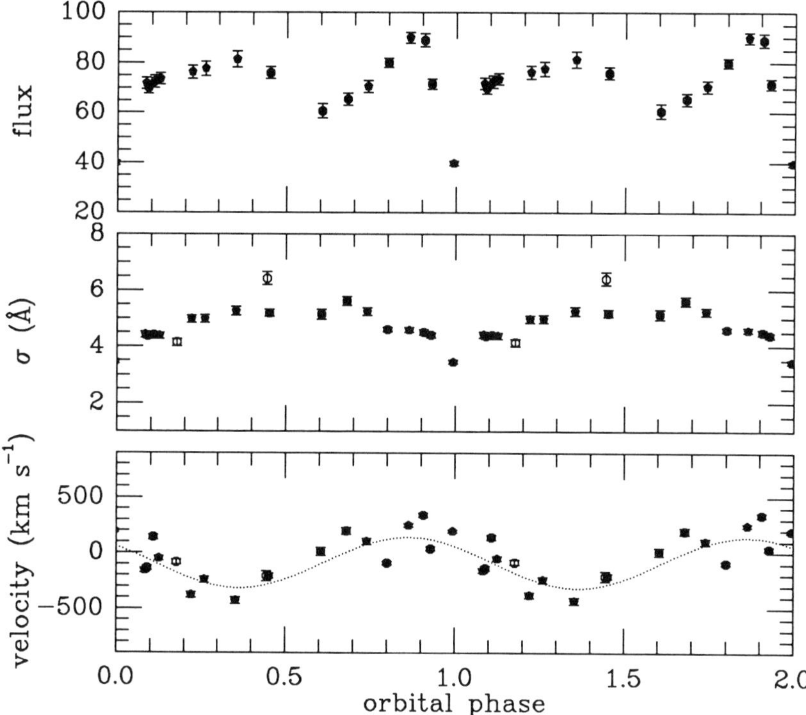

Figure 4.   CIV emission line flux, Gaussian width, and radial velocity as a function
of orbital phase for the nova-like variable V347 Pup. Best sinusoidal fit to radial
velocity variation is overplotted; line flux is measured in units of 10^{-13} erg cm^{-2}s^{-1}
(figure from Mauche et al. 1994).

V. ECLIPSE STUDIES

Unlike T Tauri stars or AGN, CVs present the opportunity of eclipse studies
of the disk and wind. IUE studies of UX UMa, RW Tri, OY Car, Z Cha, and
V347 Pup (Holm et al. 1982; King et al. 1983; Córdova and Mason 1985;
Drew and Verbunt 1985; Naylor et al. 1988; Harlaftis et al. 1992*a,b*; Mauche
et al. 1994) all demonstrate that the ultraviolet emission lines of CVs are not
completely extinguished in eclipse, strongly indicating that they are formed in
a region with dimensions comparable in size to the size of the secondary. Mod-
ulo the possible effects of superposed absorption (Drew 1987), the relative
depths of the eclipses of these lines also give some indication of the relative
dimensions of the regions in which their profiles are formed. V347 Pup is
representative: there, the ratio of the eclipse to mean out-of-eclipse flux of
the N V, Si IV, and C IV emission lines are 35%, 33%, and 53%, respectively
(Mauche et al. 1994; see Fig. 4). The weak eclipse of the C IV profile relative
to the those of N V and Si IV indicates that the region of the wind in which
the C IV ion dominates is more extensive than for these other two ions; a

conclusion which is consistent with the stronger emission component of the P Cygni profile of C IV relative to those of N V and Si IV in lower inclination systems. It is not yet clear whether carbon exists in the form of C IV throughout a larger fraction of the outer wind than N V or Si IV, or whether a wind with constant ionization fractions would simply give $\tau \sim 1$ at a larger radius for the more abundant C IV ion. The former scenario is more consistent with theoretical expectations, because nitrogen and silicon will exist as N IV and Si V in those regions of the wind where C IV dominates (see, e.g., Mauche and Raymond 1987).

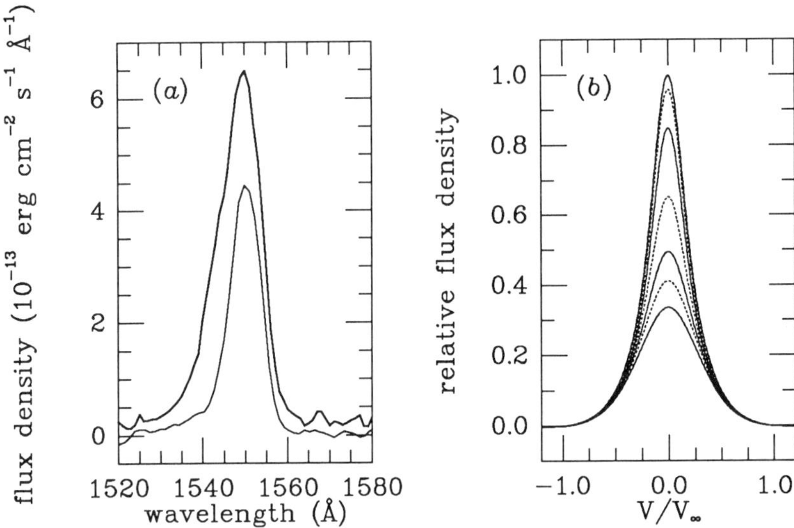

Figure 5. (a) Eclipse (*lower curve*) and mean out-of-eclipse (*upper curve*) CIV line profiles of V347 Pup. (b) Model line profiles at orbital phases ϕ = 0, 0.03, 0.04, 0.05, 0.06, 0.08, and 0.25 (figure from Mauche et al. 1994).

In much the same way that eclipse mapping probes the brightness temperature distribution of the accretion disks of CVs (see, e.g., Horne 1993), eclipse studies also constrain the surface brightness distribution of the winds of CVs, and hence constrain the wind's source function and geometry. Although IUE does not supply the combination of effective area and spectral and temporal resolution required to do true eclipse mapping, some conclusions can be drawn from the existing data. The case of V347 Pup is instructive; as shown in Figs. 4 and 5a, the C IV emission profile decreases both in strength and in width as the wind is eclipsed by the secondary. The same effect is observed in the dwarf nova OY Car (Naylor et al. 1988). In contrast, as shown in Fig. 5b, the profiles of existing wind models (Drew 1987; Mauche et al. 1994) *increase* in width in eclipse. Although this difference between observations and models can be reconciled a number of ways, the results for the optical emission profiles of V347 Pup suggest that the ultraviolet emission profiles may be broadened significantly by rotation. If this is the case, it implies that the wind arises from the surface of the accretion disk rather than from near the white dwarf (Shlosman et al. 1995).

VI. DISK WINDS

Accretion disk winds actually have a number of qualities to recommend them over a wind originating near the white dwarf. First, the gravitational potential at the surface of the accretion disk is much lower than at the surface of the white dwarf. This difference implies a reduction in the energy required to drive the wind of a given mass-loss rate to escape velocity. Second, because a disk wind arises from the same region responsible for the resonant photons which produce the P Cygni profiles, lower mass-loss rates are required to produce a given column density, and hence a given line strength, than the mass-loss rate required of a wind originating near the white dwarf. Third, because a disk wind arises from a spatially extended region, the volume of the wind in which photons are resonantly scattered is likely to be larger than the corresponding volume of a wind originating near the white dwarf. The size of this volume has implications for the depth and width of the eclipse of the line-forming region in systems in which the secondary eclipses the white dwarf. Because the eclipse depths are observed to be fairly small, an extended scattering region is very desirable. Fourth, the $r^{-2}V(r)^{-1}$ increase in the density of the wind associated with a spherically symmetric geometry is softened in a disk wind. This difference could remove the high-density component of the wind overlying the boundary layer and allow soft X-rays to emerge through the wind.

Shlosman and Vitello (1993) and Vitello and Shlosman (1993) have recently explored the properties of disk winds. They assume that the wind originates from an annulus of the accretion disk (see Fig. 6), that it rotates initially with its local Keplerian value and conserves angular momentum as it accelerates outward, and that its terminal velocity scales with the local escape velocity. In total, nine free parameters (plus $M_{\rm wd}$, $R_{\rm wd}$, $\dot{M}$, and $\dot{M}_{\rm wind}$) are required to fully characterize the wind. The model calculates the wind ionization structure self-consistently under the assumptions of constant temperature and local ionization equilibrium and treats the radiation transfer in the Sobolev approximation.

Vitello and Shlosman find that disk winds are capable of providing good fits to the line profiles of RW Sex, RW Tri, and V Sge for reasonable parameters, but note that *unique* fits are not possible, given the many degrees of freedom in the model and the limited number of parameters characterizing the profiles. They find that the introduction of rotation in the wind introduces a radial shear in the velocity which decreases the optical depth, resulting in a reduction in the line center intensity and a broadening of the emission component of the line profile. In agreement with observations, significant

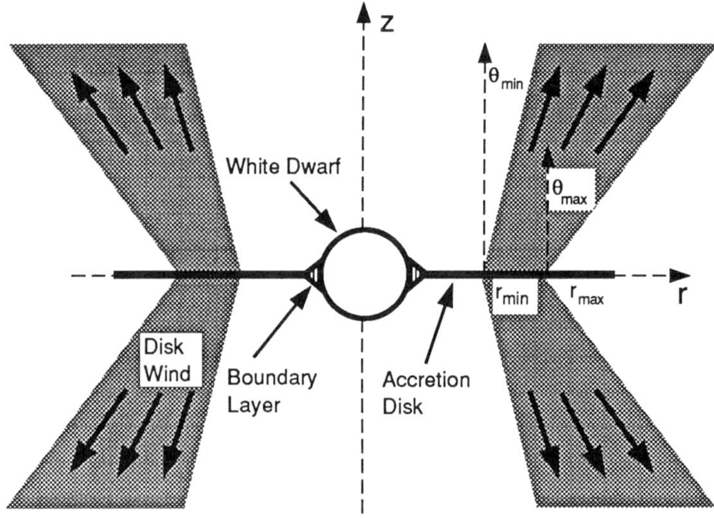

Figure 6.   Geometry for a disk wind (figure from Shlosman and Vitello 1993).

absorption does not occur at high inclination angles because the wind is confined within an angle θ_{max}, which is typically taken to be 65°. As expected,
compared to winds which originate near the white dwarf, disk winds require
lower mass-loss rates to produce a given line strength, and naturally explain
the shallow line eclipse depths of eclipsing CVs. Yet to be demonstrated is
whether a single manifestation of Shlosman and Vitello's kinematic model
can be applied successfully to a wide range of systems with well-constrained
white dwarf masses, accretion rates, and inclination angles. As these authors
point out, what is ultimately required is a self-consistent dynamical model for
the two-dimensional outflow from a disk.

VII. DYNAMICS

All of the models discussed above have assumed various wind geometries
and velocity laws in order to compute P Cygni profiles for diagnostic use. In
order to discuss dynamics, one first needs to specify the driving force. The
general similarity between the P Cygni profiles of CVs and early-type stars
has lulled most researchers into the belief that radiation pressure in spectral
lines drives the wind. However, the only attempt to model such a flow resulted
in velocities as high as those observed, but mass-loss rates which were far
too low, at least in comparison with spherical wind models (Raymond et al.
1988). The reason for the low predicted mass-loss rate is that the critical point
is determined by the behavior of the effective gravity along the streamline,
and it lies far above the disk where the density is low. This problem *might* be
overcome if a rapid change in ionization structure modifies the driving force

enough to lower the critical point to just above the disk surface (see Vitello and Shlosman 1988).

The other dynamical models have employed magnetic fields. While Alfvén waves (see the chapter by MacGregor and Charbonneau) or events similar to coronal mass ejections (see the chapter by Hundhausen) might well be important, the models which have been developed so far rely on centripetal acceleration by a large-scale magnetic field. Blandford and Payne (1982) constructed a self-similarity solution for application to AGN jets which has been applied to CVs by Koen (1986) and Cannizzo and Pudritz (1988). In these models, the field encounters the disk with an angle less than $60°$ to the equatorial plane, so that the rotating field accelerates material along the field lines. This acceleration continues out to the Alfvén radius r_A where the field becomes too weak to enforce corotation. Gas leaving the disk at radius r carries enough angular momentum to drive an accretion rate $\dot{M} = \dot{M}_{\mathrm{wind}}\, r_A^2/r^2$ through the disk, replacing viscous transport of angular momentum through the disk as the fundamental physical process responsible for accretion. The removal of mass and angular momentum from the disk might account for the surprising flatness of the far-ultraviolet spectra of some nova-like variables (la Dous 1991; Rutten et al. 1992; Long et al. 1994). If the mass-loss rate of the wind is large enough, it is possible that only a small fraction of the mass transferred from the secondary reaches the white dwarf, providing at least in principle an explanation for the faintness of the boundary layers in high-$\dot{M}$ CVs. A potential problem with the magnetically driven wind model is that the large-scale magnetic field tends to diffuse outward through the disk, so that some dynamo field generation is probably needed (Van Ballegooijen 1989). A more complete model must also dispense with the assumption that the field lines are straight. The field should be vertical where it passes through the equatorial plane and curve outwards on a scale comparable to the disk thickness. It is the combination of the local field direction with the density fall-off in the disk which actually determines the mass-loss rate (Van Ballegooijen 1989). Van Ballegooijen's model of the field geometry also shows that the field lines are nearly vertical in the inner disk, so that they are bent over enough to drive a wind only from the outer disk, making it difficult to explain the IUE and HUT observations which pertain more to the inner disk.

VIII. EVOLUTION

Winds are crucial to many CVs because these binaries must lose angular momentum to drive mass transfer. Except for systems driven by nuclear evolution (very long-period systems), the companion would recede inside its Roche lobe without angular momentum loss, and mass transfer would cease. Gravitational radiation carries away enough angular momentum to drive mass transfer in the short-period systems ($P \leq 2$ hr), but fails to dispose of angular momentum fast enough to account for the moderate-period systems. In the generally accepted picture, magnetic braking by the wind of the companion, a late-type star spinning synchronously with the orbital period, drives the mass loss and the evolution toward shorter periods. No direct measurements of this wind are available, but scaling from the single-star magnetic braking rates derived from young clusters (Stauffer 1991) leads to good agreement with the observed mass-transfer rates (Verbunt and Zwaan 1981; Rappaport et al. 1983; Patterson 1984). In this picture, magnetic braking ceases when the companion becomes fully convective at a period of 3 hr and mass transfer switches off. The system becomes active again only after gravitational radiation brings the system back into contact as the orbit shrinks and the period reaches 2 hr. Some indirect support for this picture comes from the observation of strong winds in dMe stars (chapter by Mullan). A possible problem is the absence of the expected drop in the X-ray luminosity in field stars at M6, corresponding to the companion in a 3-hr binary. This makes the sudden cessation of magnetic braking due to a drop in magnetic activity less appealing. In any case, the extrapolation of angular momentum loss rate, $\dot{J}$, from young stars to stars in 5-hr binaries is a plausible but very risky undertaking (chapter by Charbonneau et al.).

Magnetic braking by a disk wind is an alternative to magnetic braking by the companion. The connection between the winds observed by IUE and their efficiency in disposing of angular momentum is rather tenuous, however. First, the magnetic field is unknown, and second, the observed winds arise from the inner disk, while most of the angular momentum resides in the outer disk. Cannizzo and Pudritz (1988) estimate the scaling of $\dot{J}$ with period based on simplified assumptions, and they find that braking by a disk wind might well account for the observed mass-transfer rate in systems having periods above 3 hr. In order to shut off mass transfer at a period of 3 hr, they assume that when the companion becomes fully convective its magnetic activity ceases, and there is then no seed field to generate a disk magnetic field. An interesting aspect of this model is its potential feedback. A high mass-transfer rate causes the wind which carries away angular momentum, which in turn shrinks the orbit and causes mass transfer. Livio and Pringle (1994) consider the possibility that this mechanism causes self-excited mass transfer. The period gap occurs when the companion becomes completely covered by starspots, so that the mass transfer is inhibited long enough for the companion to shrink inside its Roche surface, shutting off the feedback loop. This model explains VY Scl "anti-dwarf nova" events as a temporary cessation of mass transfer when a single starspot covers the Lagrangian point.

A possible serious drawback of models which rely on magnetic braking by a disk wind is the likelihood of instability. Lubow et al. (1994) show that accretion through a disk driven by modest angular momentum loss in a wind is unstable. There is also a possibility that if the specific angular momentum of the disk wind is very large and the mass ratio of the system is small, loss of angular momentum from the system could lead to dynamical mass transfer (see, for instance, equation 1 of Melia and Lamb 1987). There is as yet no direct evidence for the large-scale magnetic fields needed to transport angular momentum, or for winds from the outer disk where most of the angular momentum resides, but more detailed studies of CV P Cygni profiles, their eclipses, and their orbital variations may define the wind geometry well enough to discriminate among the competing models.

IX. THE FUTURE

The era of IUE studies of the winds of CVs is coming to a close. Although it remains amazingly productive, IUE is getting old, is in increasingly poor health, and may be put down in the near future for lack of operating funds. However, IUE will not soon be forgotten; active research will continue with the enormous archive that has accumulated over the past 17 plus years. A small subset of this archive has appeared in the form of a 500-page catalogue of low-resolution CV spectra (la Dous 1990), but the real power of the archive lies in the ability to do *quantitative* analysis of thousands of spectra of over one hundred CVs. A small number of high-resolution IUE spectra also reside in the archive (see, e.g., Córdova 1986; Mauche et al. 1988; Mauche 1991; Prinja and Rosen 1995), but these spectra seem to have failed the promise they first appeared to offer to reveal the secrets of the winds of CVs.

For additional observational progress to be made, higher quality (higher signal-to-noise ratio, higher spectral resolution, higher temporal resolution) spectra are required. When it can be wrestled away from the "big boys" working on H_0 and q_0, HST provides the required properties in the near-ultraviolet, but the importance of the far-ultraviolet and the EUV to the study of the winds of CVs cannot be overstated. HST has already observed the eclipsing CVs UX UMa (Mason et al. 1995) and OY Car; HUT has observed UX UMa, Z Cam (Long et al. 1991), and IX Vel (Long et al. 1994); *ORFEUS* recently observed Z Cam and V3885 Sgr; EUVE has observed SS Cyg (Mauche et al. 1995), U Gem, and VW Hyi. With some luck, these and subsequent data sets will prove much of what we understand about the winds of CVs to be wrong.

Acknowledgments. The authors are indebted to F. Córdova, J. Drew, T. Kallman, and F. Verbunt for innumerable discussions relating to the winds of CVs. This work was performed under the auspices of the U. S. Department of Energy under contract to Lawrence Livermore National Laboratory, and under a grant by NASA to the Smithsonian Astrophysical Observatory.

REFERENCES

Blandford, R. D., and Payne, D. G. 1982. Hydromagnetic flows from accretion discs and the production of radio jets. *Mon. Not. Roy. Astron. Soc.* 199:883–903.

Cannizzo, J. K. 1993. The limit cycle instability in dwarf nova accretion disks. In *Accretion Disks in Compact Stellar Systems*, ed. J. C. Wheeler (Singapore: World Scientific), pp. 6–40.

Cannizzo, J. K., and Pudritz, R. E. 1988. A new angular momentum loss mechanism for cataclysmic variables. *Astrophys. J.* 327:840–844.

Castor, J. I., and Lamers, H. J. G. L. M. 1979. An atlas of theoretical P Cygni profiles. *Astrophys. J. Suppl.* 39:481–511.

Chlebowski, T., Harnden, F. R., and Sciortino, S. 1989. The Einstein X-ray observatory catalog of O-type stars. *Astrophys. J.* 341:427–455.

Córdova, F. A. 1986. The promise of high resolution UV spectroscopy for understanding the winds of cataclysmic variables. In *The Physics of Accretion Onto Compact Objects*, eds. K. O. Mason, M. G. Watson and N. E. White (Berlin: Springer-Verlag), pp. 339–355.

Córdova, F. A. 1995. Cataclysmic variable stars. In *X-Ray Binaries*, eds. W. H. G. Lewin, J. van Paradijs and E. P. J. van den Heuvel (Cambridge: Cambridge Univ. Press), pp. 331–388.

Córdova, F. A., and Howarth, I. D. 1987. Accretion onto compact stars in binary systems. In *Exploring the Universe with the IUE Satellite*, ed. Y. Kondo (Dordrecht: D. Reidel), pp. 395–426.

Córdova, F. A., and Mason, K. O. 1982. High-velocity winds from a dwarf nova during outburst. *Astrophys. J.* 260:716–721.

Córdova, F. A., and Mason, K. O. 1985. High-velocity winds in close binaries with accretion disks. II. The view along the plane of the disk. *Astrophys. J.* 290:671–682.

Córdova, F. A., Chester, T. J., Mason, K. O., Kahn, S. M., and Garmire, G. P. 1984. Observations of quasi-coherent soft X-ray oscillations in U Geminorum and SS Cygni. *Astrophys. J.* 278:739–753.

Córdova, F. A., Chester, T. J., Tuohy, I. R., and Garmire, G. P. 1980. Soft X-ray pulsations from SS Cygni. *Astrophys. J.* 235:163–176.

Cropper, M. 1990. The polars. *Space Sci. Rev.* 54:195–295.

Drew, J. E. 1987. Inclination and orbital-phase-dependent resonance line-profile calculations applied to cataclysmic variable winds. *Mon. Not. Roy. Astron. Soc.* 224:595–632.

Drew, J. E., and Kley, W. 1993. Mass loss and the boundary layer. In *Accretion Disks in Compact Stellar Systems*, ed. J. C. Wheeler (Singapore: World Scientific), pp. 212–242.

Drew, J. E., and Verbunt, F. 1985. Investigation of a wind model for cataclysmic variable ultraviolet resonance line emission. *Mon. Not. Roy. Astron. Soc.* 213:191–213.

Drew, J. E., and Verbunt, F. 1988. Regular orbital variations in the ultraviolet resonance lines of YZ Cnc. *Mon. Not. Roy. Astron. Soc.* 234:341–351.

Drew, J. E., Hoare, M. G., and Woods, J. A. 1991. Ultraviolet observations of the long-period dwarf nova DX And in outburst. *Mon. Not. Roy. Astron. Soc.* 250:144–151.

Greenstein, J. L., and Oke, J. B. 1982. RW Sextantis, a disk with a hot, high-velocity wind. *Astrophys. J.* 258:209–216.

Guinan, E. F., and Sion, E. M. 1982. Ultraviolet spectroscopy of the nova-like variable V3885 Sagittarii (=CD −42°14462). *Astrophys. J.* 258:217–223.

Harlaftis, E. T., Hassall, B. J. M., Naylor, T., Charles, P. A., and Sonneborn, G. 1992*a*. UV spectroscopy of Z Chamaeleontis. I. Time dependent dips in superoutburst. *Mon. Not. Roy. Astron. Soc.* 257:607–619.

Harlaftis, E. T., Naylor, T., Hassall, B. J. M., Charles, P. A., Sonneborn, G., and Bailey, J. 1992*b*. UV spectroscopy of Z Chamaeleontis. II. The 1988 January normal outburst. *Mon. Not. Roy. Astron. Soc.* 259:593–603.

Hassall, B. J. M., Pringle, J. E., and Verbunt, F. 1985. Dwarf novae in outburst—Monitoring WX Hydri with IUE. *Mon. Not. Roy. Astron. Soc.* 216:353–363.

Heap, S. R., Boggess, A., Holm, A., Klinglesmith, D. A., Sparks, W., West, D., Wu, C.-C., Boksenberg, A., Willis, A., and Wilson, R. 1978. IUE observations of hot stars—HZ 43, BD +75°325, NGC 6826, SS Cygni, Eta Carinae. *Nature* 275:385–388.

Hoare, M. G., and Drew, J. E. 1991. Boundary-layer temperatures in high accretion rate cataclysmic variables. *Mon. Not. Roy. Astron. Soc.* 249:452–459.

Hoare, M. G., and Drew, J. E. 1993. The ionization state of the winds from cataclysmic variables without classical boundary layers. *Mon. Not. Roy. Astron. Soc.* 260:647–662.

Holm, A. V., Panek, R. J., and Schiffer, F. H., III. 1982. Ultraviolet spectrum variability of UX Ursae Majoris. *Astrophys. J. Lett.* 252:35–37.

Horne, K. 1993. Eclipse mapping of accretion disks. In *Accretion Disks in Compact Stellar Systems*, ed. J. C. Wheeler (Singapore: World Scientific), pp. 117–147.

Kallman, T. R., and Jensen, K. A. 1985. Soft X-rays, winds, and the cataclysmic variable boundary-layer problem. *Astrophys. J.* 299:277–285.

King, A. R., Frank, J., Jameson, R. F., and Sherrington, M. R. 1983. Phase-dependent UV spectra of UX Ursae Majoris. *Mon. Not. Roy. Astron. Soc.* 203:677–683.

Klare, G., Wolf, B., Stahl, O., Krautter, J., Vogt, N., Wargau, W., and Rahe, J. 1982. IUE observations of dwarf novae during active phases. *Astron. Astrophys.* 113:76–84.

Koen, C. 1986. Angular momentum transport in the magnetospheres of cataclysmic variable accretion discs. *Mon. Not. Roy. Astron. Soc.* 223:529–538.

Krautter, J., Vogt, N., Klare, G., Wolf, B., Duerbeck, H. W., Rahe, J., and Wargau, W. 1981. IUE spectroscopy of cataclysmic variables. *Astron. Astrophys.* 102:337–346.

la Dous, C. 1990. A catalogue of low-resolution IUE spectra of dwarf novae and nova-like stars. *Space Sci. Rev.* 52:203–706.

la Dous, C. 1991. New insights from a statistical analysis of IUE spectra of dwarf novae and nova-like stars. I. Inclination effects in lines and continua. *Astron. Astrophys.* 252:100–122.

la Dous, C., Verbunt, F., Pringle, J. E., Schoembs, R., Argyle, R. W., Jones, D. H. P., Schwarzenberg-Czerny, A., Hassall, B. J. M., and Wade, R. A. 1985. Dwarf novae in outburst—Simultaneous ultraviolet and optical observations of RU Pegasi and TZ Persei. *Mon. Not. Roy. Astron. Soc.* 212:231–243.

Livio, M. 1994. Topics in the theory of cataclysmic variables and X-ray binaries. In *Interacting Binaries*, eds. H. Nussbaumer and A. Orr (Berlin: Springer-Verlag), pp. 135–262.

Livio, M., and Pringle, J. E. 1994. Star spots and the period gap in cataclysmic variables. *Astrophys. J.* 427:956–960.

Long, K. S., Blair, W. P., Davidsen, A. F., Bowers, C. W., Dixon, W. V. D., Durrance, S. T., Feldman, P. D., Henry, R. C., Kriss, G. A., and Kimble, R. A. 1991. Spectroscopy of Z Camelopardalis in outburst with the Hopkins Ultraviolet Telescope. *Astrophys. J. Lett.* 381:25–29.

Long, K. S., Wade, R. A., Blair, W. P., Davidson, A. F., and Hubeny, I. 1994. Observations of the bright novalike variable IX Velorum with the Hopkins Ultraviolet Telescope. *Astrophys. J.* 426:704–715.

Lubow, S. H., Papaloizou, J. C. B., and Pringle, J. E. 1994. On the stability of magnetic wind-driven accretion discs. *Mon. Not. Roy. Astron. Soc.* 268:1010–1014.

Mason, K. O., Drew, J. E., Córdova, F. A., Horne, K., Hilditch, R., Knigge, C., Lanz, T., and Meylan, T. 1995. Eclipse observations of an accretion disc wind. *Mon.*

Not. Roy. Astron. Soc. 274:271–286.

Mauche, C. W. 1991. High-resolution IUE spectra of the nova-like variable IX Velorum. *Astrophys. J.* 373:624–632.

Mauche, C. W. 1995. EUVE photometry of SS Cygni: dwarf nova outbursts and oscillations. In *Astrophysics in the Extreme Ultraviolet*, eds. S. Bowyer and B. Haisch (Cambridge: Cambridge Univ. Press), in press.

Mauche, C. W., and Raymond, J. C. 1987. IUE observations of the dwarf nova HL Canis Majoris and the winds of cataclysmic variables. *Astrophys. J.* 323:690–713.

Mauche, C. W., Raymond, J. C., and Córdova, F. A. 1988. Interstellar absorption lines in high-resolution IUE spectra of cataclysmic variables. *Astrophys. J.* 335:829–843.

Mauche, C. W., Raymond, J. C., and Mattei, J. A. 1995. EUVE observations of the anomalous 1993 August outburst of SS Cygni. *Astrophys. J.* 446:842–851.

Mauche, C. W., Wade, R. A., Polidan, R. S., van der Woerd, H., and Paerels, F. B. S. 1991. On the X-ray emitting boundary layer of the dwarf nova VW Hydri. *Astrophys. J.* 372:659–663.

Mauche, C. W., Raymond, J. C., Buckley, D. A. H., Mouchet, M., Bonnell, J., Sullivan, D. J., Bonnet-Bidaud, J.-M., and Bunk, W. H. 1994. Optical, IUE, and ROSAT observations of the eclipsing nova-like variable V347 Puppis (LB 1800). *Astrophys. J.* 424:347–369.

Melia, F., and Lamb, D. Q. 1987. Dynamical mass transfer in cataclysmic variables. *Astrophys. J. Lett.* 321:139–143.

Naylor, T., Bath, G. T., Charles, P. A., Hassall, B. J. M., and Sonneborn, G. 1988. The 1985 May superoutburst of the dwarf nova OY Carinae. II. IUE and EXOSAT observations. *Mon. Not. Roy. Astron. Soc.* 231:237–255.

Olson, G. L. 1978. An analysis of ultraviolet resonance lines and the possible existence of coronae in O stars. *Astrophys. J.* 226:124–137.

Patterson, J. 1984. The evolution of cataclysmic and low-mass X-ray binaries. *Astrophys. J. Suppl.* 54:443–493.

Patterson, J., and Raymond, J. C. 1985. X-ray emission from cataclysmic variables with accretion disks. II. EUV/soft X-ray radiation. *Astrophys. J.* 292:550–558.

Popham, R., and Narayan, R. 1995. Accretion disk boundary layers in cataclysmic variables. I. Optically thick boundary layers. *Astrophys. J.* 442:337–357.

Pringle, J. E. 1977. Soft X-ray emission from dwarf novae. *Mon. Not. Roy. Astron. Soc.* 178:195–202.

Prinja, R. K., and Rosen, S. R. 1995. High-resolution IUE spectroscopy of fast winds from cataclysmic variables. *Mon. Not. Roy. Astron. Soc.* 273:461–474.

Prinja, R. K., Drew, J. E., and Rosen, S. R. 1992. Variability in the UV resonance lines of the cataclysmic variable V795 Herculis (PG 1711+336). *Mon. Not. Roy. Astron. Soc.* 256:219–228.

Rappaport, S., Joss, P. C., and Verbunt, F. 1983. A new technique for calculations of binary stellar evolution, with application to magnetic braking. *Astrophys. J.* 275:713–731.

Raymond, J. C., Van Ballegooijen, A. A., and Mauche, C. W. 1988. Structure and dynamics of cataclysmic variable winds. *Bull. Amer. Astron. Soc.* 20:1020 (abstract).

Rutten, R. G. M., van Paradijs, J., and Tinbergen, J. 1992. Reconstruction of the accretion disk in six cataclysmic variable stars. *Astron. Astrophys.* 260:213–226.

Shakura, N. I., and Sunyaev, R. A. 1973. Black holes in binary systems. Observational appearance. *Astron. Astrophys.* 24:337–355.

Shlosman, I., and Vitello, P. 1993. Winds from accretion disks—Ultraviolet line formation in cataclysmic variables. *Astrophys. J.* 409:372–386.

Shlosman, I., Vitello, P., and Mauche, C. W. 1995. Rotating winds from accretion disks in cataclysmic variables: eclipse modeling of V347 Puppis. *Astrophys. J.* submitted.

Silber, A., Vrtilek, S. D., and Raymond, J. C. 1994. Concurrent X-ray and optical observations of two dwarf novae during eruption. *Astrophys. J.* 425:829–834.

Sion, E. M. 1985. On the nature of the UX Ursa Majoris-type nova-like variables: CPD −48°1577. *Astrophys. J.* 292:601–605.

Smak, J. 1984. Outbursts of dwarf novae. *Publ. Astron. Soc. Pacific* 96:5–18.

Stauffer, J. R. 1991. Rotational velocities of low mass stars in young clusters. In *Angular Momentum Evolution of Young Stars*, eds. S. Catalano and J. R. Stauffer (Dordrecht: Kluwer), pp. 117–134.

Van Ballegooijen, A. A. 1989. Magnetic fields in the accretion disks of cataclysmic variables. In *Accretion Disks and Magnetic Fields in Astrophysics*, ed. G. Belvedere (Dordrecht: Kluwer), pp. 99–106.

van der Woerd, H., Heise, J., and Bateson, F. 1986. The soft X-ray superoutburst of VW Hydri. *Astron. Astrophys.* 156:252–260.

van der Woerd, H., Heise, J., Paerels, F., Beuermann, K., and van der Klis, M. 1987. Discovery of soft X-ray oscillations in VW Hydri. *Astron. Astrophys.* 182:219–228.

van Teeseling, A., Verbunt, F., and Heise, J. 1993. The nature of the X-ray spectrum of VW Hydri. *Astron. Astrophys.* 270:159–164.

Verbunt, F., and Zwaan, C. 1981. Magnetic braking in low-mass X-ray binaries. *Astron. Astrophys.* 100:L7–L9.

Verbunt, F., Pringle, J. E., Wade, R. A., Echevarria, J., Jones, D. H. P., Argyle, R. W., Schwarzenberg-Czerny, A., la Dous, C., and Schoembs, R. 1984. Dwarf novae in outburst—Simultaneous ultraviolet and optical observations of UZ Serpentis, RX Andromedae and AH Herculis. *Mon. Not. Roy. Astron. Soc.* 210:197–221.

Verbunt, F., Hassall, B. J. M., Pringle, J. E., Warner, B., and Marang, F. 1987. Multiwavelength monitoring of the dwarf nova VW Hydri. III. IUE observations. *Mon. Not. Roy. Astron. Soc.* 225:113–130.

Vitello, P., and Shlosman, I. 1988. Line-driven winds from accretion disks. I. Effects of the ionization structure. *Astrophys. J.* 327:680–692.

Vitello, P., and Shlosman, I. 1993. Ultraviolet line diagnostics of accretion disk winds in cataclysmic variables. *Astrophys. J.* 410:815–828.

Woods, J. A., Drew, J. E., and Verbunt, F. 1990. Time dependence of the UV resonance lines in the cataclysmic variables SU UMa, RX And and 0623+71. *Mon. Not. Roy. Astron. Soc.* 245:323–330.

Woods, J. A., Verbunt, F., Collier Cameron, A., Drew, J. E., and Piters, A. 1992. Time dependence of the UV resonance lines in the cataclysmic variables YZ Cnc, IR Gem and V3885 Sgr. *Mon. Not. Roy. Astron. Soc.* 255:237–242.

COLLIMATED OUTFLOWS FROM YOUNG STARS

JOHN BALLY
University of Colorado

and

STEVEN W. STAHLER
University of California at Berkeley

Energetic bipolar outflows are one of the earliest manifestations of the stellar birth process. Why such winds erupt from young stars is not yet known, but their interaction with the surrounding gas and dust is rapidly being elucidated through both observation and theory. We review the characteristics of outflows revealed through the millimeter lines of CO, near-infrared emission from H_2, and forbidden optical emission lines. We argue that the broad outflows observed in CO result from the turbulent entrainment of cloud gas by central, narrow jets. These jets, which have been detected in both near-infrared and optical emission lines, are composed of largely neutral gas ejected from a young star and channeled into a narrow bipolar flow.

Stars are born by gravitational condensation from the cold and diffuse aggregates of interstellar gas known as molecular clouds. For reasons that are still unclear, it appears that all stars at some point emit bipolar winds that eventually disperse surrounding cloud matter. These powerful winds erupt very early, probably during the initial accumulation of stellar mass from molecular cloud cores. Discovered little more than a decade ago, bipolar outflows have already taught us much about the relation of young stars to their environments.

Observed outflows exhibit a broad range of properties. Their temperatures, internal velocities, and sizes vary so widely that a number of different techniques, both in instrumentation and in the wavelengths of choice, have been brought to bear in their study. Most outflow systems have been recognized through their production of millimeter spectral lines in CO, near-infrared radiation from shock-excited H_2, or optical emission lines. Each of these wavelength regimes reveals a different aspect of the basic outflow process. We begin by summarizing the main results obtained with these observational tools (Sec. I). This portion of our review is not meant to be exhaustive; of necessity, we cannot cover other important manifestations of outflows, such as their excitation of maser activity at radio wavelengths. Nor can we give proper credit to the many researchers who have contributed to this active field.

There has been extensive debate concerning both the origin of winds from young stars and the mechanism of their collimation. In the theoretical discussion of Sec. II, we will largely bypass the question the wind origin, and focus instead on the interaction of winds with molecular cloud gas. We will argue for a unified picture in which relatively broad and slow outflows represent cloud matter being entrained by central jets of gas shot out from the central stars. The reader wishing a broader survey of alternative models may consult any of a number of recent review articles (Padman et al. 1991; Bachiller and Gómez-Gonzalez 1992; Cabrit 1993; Fukui et al. 1993).

I. OBSERVATIONAL OVERVIEW

In addition to being spectacular manifestations of stellar birth, outflows may be the crucial agents in the generation of turbulence in molecular clouds, in the self-regulation of the star formation process itself, and may ultimately reveal the recent mass loss and accretion history of their driving sources. Below, we review the general properties of outflows as revealed first by CO observations, second by near infrared observation of shock excited molecular hydrogen, and finally by visual wavelength observations of emission lines.

A. Properties of CO Outflows

Molecular clouds owe their name to the fact that the hydrogen which constitutes the bulk of their mass is in molecular form. Unfortunately, H_2 lacks a permanent electric dipole moment and its first excited rotational level lies at the equivalent temperature of 510 K above the ground state. Thus, the molecule is a poor radiator at typical cloud temperatures of 5 to 50 K. One must turn instead to rarer constituents that do emit more strongly, and can therefore be used to trace the total distribution of outflow matter. The most commonly used tracer is CO, whose low-lying rotational levels are easily excited by collisions with H_2 and He. Over 200 outflows have been discovered through high-velocity wings in CO emission-line profiles. These emission wings are not obscured by quiescent gas along the line of sight. Once the relative abundance of the tracer is known, the observed CO column densities and Doppler shifts can be used to estimate the mass and momentum associated with each outflow.

The spatial extent and flow velocity of outflowing gas can be used to estimate the outflow age. The number of observed outflows, combined with their ages can be used to estimate the formation rate of outflows, which is comparable to the birth rate of stars above a mass of about 0.2 $M_\odot$. Thus, it apears that most stars produce energetic outflows during their birth.

The first outflow mapped in CO was L1551/IRS-5, discovered by Snell et al. (1980) (see Fig. 1). The outflow is revealed by the two well-separated lobes of high-velocity gas. The Doppler shifts of the observed spectral line (in this case, the 2.6 mm $J = 1 \rightarrow 0$ transition of the main CO isotope) show that one lobe is blue-shifted and so is approaching us, while the other is red-shifted and therefore receding. The average internal velocity in the lobes,

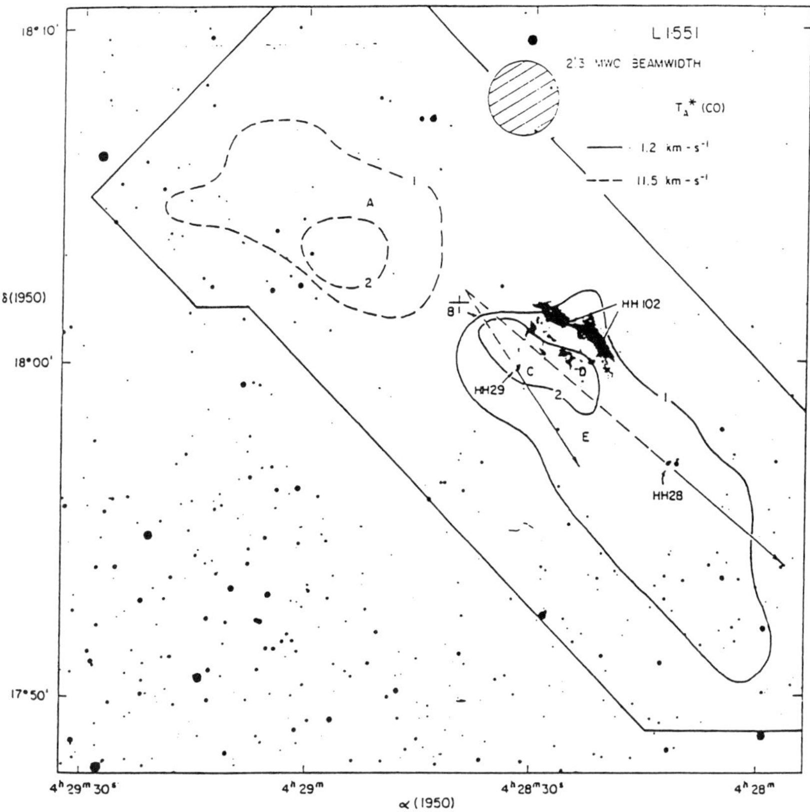

Figure 1.   The molecular outflow in L1551, as traced by the $J = 1 \rightarrow 0$ line of CO (from Snell et al. 1980). The cross near the center of the map marks the position of IRS 5. Note the proper-motion vectors of the HH objects HH 28 and HH 29.

about 5 km s^{-1}, is typical for outflows driven by low-mass stars, as is the maximum linear extent of about 1 pc.

The total mass in each lobe of L1551/IRS-5 has been estimated at $M \approx 2$ M$_\odot$ (Moriarty-Schieven and Snell 1988). Other CO outflow masses range from about 0.01 M$_\odot$ to over 200 M$_\odot$. Measured velocities can be anywhere from a few km s^{-1} to over 200 km s^{-1}, and linear sizes from about 0.1 to 4 pc. All these properties tend to scale with M_*, the inferred mass of the central star. From the total mass and the average velocity $\bar{V}$ within a lobe, a flow kinetic energy $1/2\,M\,\bar{V}^2$ can be estimated. Calculated values range from 10^{43} to nearly 10^{48} erg. Similarly, it is possible to obtain the mechanical luminosity of the flow, $L_{\mathrm{mech}} \equiv 1/2\,M\,\bar{V}^3/R$, where R is the lobe length. Comparing with the stellar luminosity L_*, we find that, for most outflows, $L_{\mathrm{mech}}/L_* \approx 0.001$ to 0.05 (Bally and Lane 1991).

In well-mapped flows, one can estimate dM/dV, the mass associated with

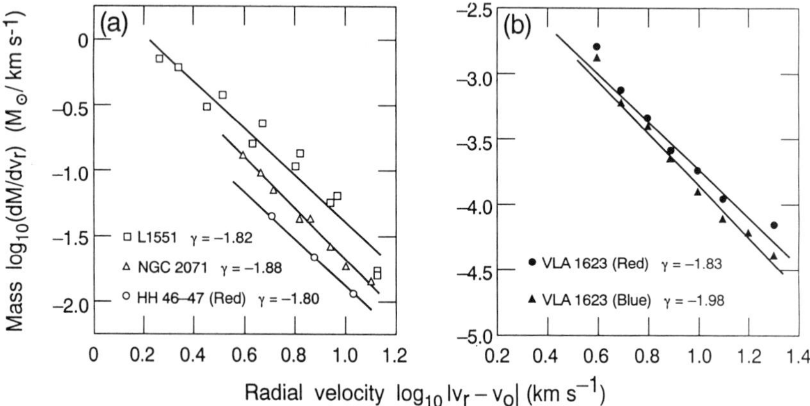

Figure 2.   Observed mass distributions for four outflows. Plotted is the integrated mass per unit radial velocity. The horizontal axis is the absolute value of the radial velocity $v_r - v_o$, the velocity at line center. Figure 2a is taken from Masson & Chernin (1992), who cite the original references. Figure 2b is from André (1993, personal communication). Note that the velocity distributions for L1551-IRS 5 and NGC 2071 are integrated over both outflow lobes, while the other distributions refer to individual lobes, as indicated.

each radial velocity V relative to the quiescent background gas. As shown in Fig. 2, such studies show that $dM/dV \sim V^{-\gamma}$, with $\gamma \lesssim 2$. Thus, most of the CO-bearing gas is moving at relatively low velocity, while very little is at high speed relative to the cloud. This fact implies that much of the matter in outflows could be slowly moving gas that is spectroscopically masked by the intrinsic velocity dispersion of the surrounding molecular gas. Existing mass estimates must therefore be used with due caution.

Similarly, the maximum observed CO velocity is a strong function of receiver sensitivity and source distance. Observations conducted during the early 1980s were usually integrated down to antenna temperatures of about 0.1 K in 0.25 MHz channels. With $1'$ beams, most outflows have full spectral widths between 10 and 30 km s^{-1}, with Orion holding the record at 130 km s^{-1} (Bally and Lada 1983). Increasing sensitivity and smaller beams led to the discovery of much higher CO velocities within previously well-studied outflows such as HH 7–11 (Bachiller and Cernicharo 1990; Masson et al. 1990). This high-velocity material, with total line widths from 100 to 400 km s^{-1}, appears at much lower intensity levels, typically from 1 to 50 mK. Analysis of excitation conditions indicates that, even with $12''$ beams (corresponding to sizes of 0.01 to 0.05 pc) the area filling factor of the beam is less than 1%.

Until the late 1980s, most searches for CO outflows were based on infrared-selected source lists, chosen primarily from the IRAS Point Source Catalog (Fukui 1989; Morgan and Bally 1991; McCutcheon et al. 1991; Fukui et al. 1993). Other searches were based on lists of optically selected sources, mainly T Tauri stars (Edwards and Snell 1983,1984). For the infrared sources, Lada (1987) devised a widely used classification scheme based on the IRAS measured fluxes from 12 to 100 μm. Classes I, II, and III represent a sequence of decreasing infrared excess relative to a normal stellar photosphere. Adams et al. (1987) interpreted the classification as an age sequence. Class I sources are true protostars, i.e., stars still accreting mass through the infall of cloud material; Class II objects are older, pre-main-sequence stars with active, i.e., accreting circumstellar disks; Class III represents reddened main-sequence stars or those with passively heated disks. The detection rate of high-velocity CO wings declines dramatically along this sequence. Over 50% of Class I sources have CO outflows, compared to only a few percent for the Class III stars (Fukui et al. 1993).

These statistics imply that CO outflows are primarily associated with very young stars still deeply embedded in their parent clouds. In some cases, such as IRS-5 in the L1551 outflow, visible light from the star can be seen reflected off adjacent, dusty clumps of gas. Note that there exist stars which are both directly visible optically *and* are associated with CO outflows (Levreault 1988). But these stars, too, are highly reddened, and are evidently being seen through a translucent screen of gas and dust or by scattered light.

Recent observations have found outflows driven by stars so obscured that most of their radiation emerges only at submillimeter wavelengths. Such sources, generally of modest luminosity, are usually not detected in any IRAS band, and therefore cannot be classified within the usual scheme. André and Montmerle (1994) have designated these objects as "Class 0," and have suggested that they, rather than Class I sources, represent true protostars. Thus far, only a handful of examples are known: VLA 1623 (André et al. 1990); L1448 (Bachiller et al. 1990); IRAS 03282+3035 (Bachiller et al. 1991); NGC 2264G (Margulis et al. 1990); NGC 2024B (Richer et al. 1992); and Orion-A South (Schmidt-Burgk et al. 1990). Figure 3 shows the L1448 outflow system, as mapped in the $J = 2 \rightarrow 1$ line of CO.

The outflows driven by Class 0 stars have a number of unusual properties. First, their mechanical luminosities are relatively high, with L_{mech}/L_* ranging from 0.005 to 0.5. While the large majority of CO outflows are rather poorly collimated, with length-to-width ratios between 2 and 5, those from Class 0 sources have ratios from 10 (VLA 1623) to 30 (IRAS 03282+3035). In addition, observations at high angular resolution have revealed small knots of molecular gas with masses of order 10^{-4} $M_\odot$, velocities of 60 km s^{-1}, and internal velocity dispersions of 10 km s^{-1} (see, e.g., Bachiller et al. 1990). Finally, some Class 0 stars exhibit very high abundances of certain molecules, especially SiO, SO, and SO$_2$. In L1448, the SiO abundance is enhanced by 10^6 relative to the surrounding, quiescent cloud. Such extreme chemical enhancements may be an indication of shock processing which may release material from grain mantles or which may drive chemical reactions requiring high temperature environments. The enhancement of SiO in some outflows permits interferometric mapping of the outflow structure with arc-

                J. BALLY AND S. W. STAHLER

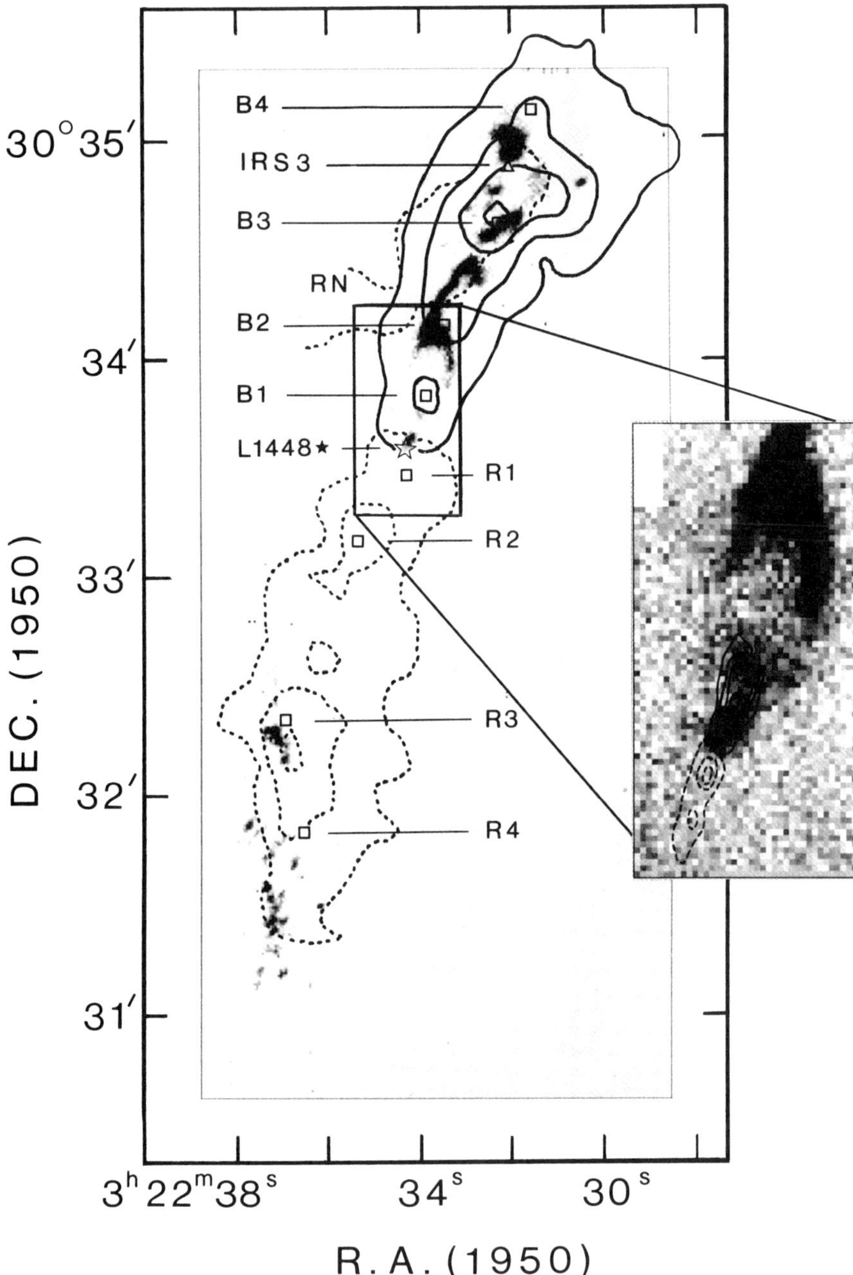

Figure 3.   Combined $J = 2 \rightarrow 1$ CO and 2.12 μm H$_2$ images of the L1448 outflow. The CO map is from Bachiller et al. (1990), while the H$_2$ map is from Bally et al. (1993). The position of the exciting star, L1448*, is indicated.

second resolution. Arc second angular resolution maps of L1448 reveal a compact narrow jet near the driving star with a knotty morphology, a volume density in molecular gas of order 10^6 cm^{-3}, and a radial velocity of ±60 km s^{-1} (Guilloteau et al. 1992).

B. Near-Infrared Observations

Shocks in interstellar gas excite a broad spectrum of emission lines, a result of the large range in temperature and density in the cooling and condensing gas behind the shock. When these shocks occur deep within molecular clouds, the optical post-shock emission is frequently obscured, but near-infrared radiation, which suffers only about 0.1 of the extinction in the visual, may not be. The 2.12 μm $v = 1 \rightarrow 0$ vibrational transition of H$_2$, with an equivalent temperature of 6×10^3 K, is proving to be a powerful tool for studying the most embedded parts of outflows that are still interacting with molecular gas. Although no systematic surveys have been conducted, at least 50 outflows have been detected in this line, usually as part of detailed studies of individual systems already seen in CO. Most of the other CO outflows have not yet been observed with sufficient sensitivity.

The strongest H$_2$ emission tends to be associated with very luminous sources driving CO outflows or with the Class 0 objects. In the first category, Allen and Burton (1993) have presented H$_2$ images of the powerful outflow emerging from the high-luminosity infrared source IRc2 ($L_* \approx 10^5$ L$_\odot$), located in the BN-KL complex of infrared sources in the Orion Nebula. The H$_2$ emission consists of multiple "fingers" extending away from the source toward the blue-shifted CO outflow lobe (see Fig. 4). The fingers are located in the low-density regions between the NH$_3$ filaments recently mapped with arcsecond resolution by Wiseman and Ho (in preparation). At the tip of each finger, knots of intense 1.6 μm emission from [Fe II] are seen. The outflow system contains at least 15 iron knots (dubbed "cannonballs" by Allen and Burton) and associated H$_2$ fingers. Stone et al. (1995) interpret the multiple fingers and cannonballs as hydrodynamical instabilities resulting from the breakout of the outflow into a region of much lower density. Stone et al. show that the dense cooling layers behind a shock can be unstable to Rayleigh-Taylor instabilities if the ambient medium into which the shock is running has a sufficiently steep density gradient, or if the outflow is driven by the right kind of mass loss variability.

Bally and Lane (1990) detected multiple arcs of H$_2$ emission in the Cepheus A outflow. This complex outflow is driven by a 10^4 L$_\odot$ source which, like Orion IRc2, is also associated with high-luminosity and high-velocity H$_2$O and OH masers. The arcs of H$_2$ surround cones of optical [S II] emission with [O III] knots at their tips. The H$_2$ emission may be produced in a system of bow shocks that may have been produced by the hydrodynamic instability of Stone et al. (1995). Near each bow shock apex, the shock may be so strong that H$_2$ is destroyed, while the shock farther downstream could be too weak to excite the molecule. As a result, only an arc-like structure is

Figure 4.   Near-infrared images of the "fingers" of shock-excited [Fe II] in the vicinity of the IRc2 outflow in the Orion Nebula (figure from Allen and Burton 1993). The image displays the 1.64 μm line.

detected in each bow shock.

All the outflows from Class 0 stars exhibit strong H_2 emission. In the L1448 outflow, Bally et al. (1993a) and Davis et al. (1994) have found a narrow H_2 jet near the axis of the blue-shifted CO lobe (see Fig. 3). Each CO knot is surrounded by an arc of H_2 emission. In addition, a continuous, curving filament of H_2 emission connects the eastern edges of these arcs. To within the 15″ angular resolution of the single-dish SiO and CO observations, the knots of both species coincide with the bow shocks seen in H_2. It thus appears that the knots consist of compressed, post-shock gas.

Bally et al. (1993b) have observed shocked H_2 in the IRAS 03282+3035 outflow. This outflow, like others from Class 0 stars, has relatively narrow CO lobes with high velocities (±60 km s^{-1}). Filaments and knots in H_2 are associated with three CO knots located near the end of the blue-shifted CO lobe. From the measured surface brightness, mass and bulk velocity of the H_2 knot farthest from the driving star, it appears that 30% of the knot's kinetic energy is being radiated away in the 2.12 μm line. The CO observations also indicate the existence of more slowly moving outflow gas over a broad region surrounding the high-velocity component.   Such two-component outflows may well be generic, as we discuss in Sec. I.C below.

More evolved young stellar objects (YSOs) tend to be relatively weak emitters of shock excited H_2 emission. However, in these sources (Class I or Class II objects), the outflow has frequently punched out of the parent cloud core and shock waves often become visible at visual wavelengths, forming the so called Herbig-Haro objects.

C. Herbig-Haro Jets

Recent optical observations have shown that many outflows are considerably larger than the extent of their CO emitting contours.   Many YSOs drive parsec scale outflows that have blown out of their parent molecular cloud cores and are interacting with predominantly atomic gas in the surrounding interstellar medium.

Over forty years ago, Herbig (1948,1950) and, independently, Haro (1952) discovered the existence of small patches of optical luminosity in star-forming regions. These Herbig-Haro (HH) objects consist of clusters of knots which vary in luminosity on the time scale of years or decades. The HH spectra are dominated by forbidden emission lines of [O I], [N II], and [S II] and sometimes [O III]. The nature of these curious regions was only gradually revealed. The key breakthrough was made by Schwartz (1975). By comparing the spectra of HH objects with those from supernova remnants, Schwartz was able to show that the optical lines represent the emission from shock-heated gas. It was then natural to conclude that the shocks are asso-ciated with winds from young stars. After unsuccessful searches at optical wavelengths, the stars were located by their infrared emission (Strom et al. 1974; Cohen and Schwartz 1980).

The manner in which the HH shocks are formed has been investigated through numerical modeling.   In some cases, the emission lines indicate a range of shock velocities, from tens to hundreds of km s^{-1}.   Both the excitation conditions and the spectral line shapes in these HH objects were successfully modeled as arising from curved bow shocks (Raga et al. 1986; Hartigan et al. 1987). Such bow shocks arise from stellar winds plowing into either ambient cloud gas, producing terminal working surfaces, or into slower moving stellar ejecta, forming internal working surfaces where faster wind components shock against previously ejected but slower stellar ejecta. Both types of HH objects exhibit large proper motions, indicating spatial velocities ranging from tens of km s^{-1} for terminal working surfaces to several hundred km s^{-1} for internal working surfaces (Cudworth and Herbig 1979; Heathcote and Reipurth 1992; Eisloeffel and Mundt 1992; Reipurth et al. 1992).

In the 1980s, the advent of sensitive CCD imaging led to the discovery of highly collimated jets of optical emission extending from YSOs towards nearby HH objects (Dopita et al. 1982; Mundt and Fried 1983; Reipurth et al. 1986). Following Reipurth (1991), we call these highly collimated outflows

"Herbig-Haro jets." To date, several dozen contiguous jets have been found. Of the more than 300 known clusters of HH objects, most show some degree of alignment, suggesting the existence of underlying jets. An electronic catalog of HH objects is maintained by Reipurth (1995).

Although several Herbig-Haro jets are bipolar, most are one-sided. In most cases, it is the blue-shifted lobe that is visible (HH 32 is a noteworthy counterexample). Class 0 sources that show strong H_2 emission tend to show only faint associated optical HH objects or jets because the obscuration is great (e.g., HH 197 associated with the L1448 outflow). Due to large circumstellar obscuration, it is the retreating, red-shifted lobes of Herbig-Haro jets that are usually hidden from view. In the HH 46/47 system, shown in Fig. 5, a blue-shifted jet lobe is projecting from an infrared star located near the edge of an opaque cloud. A disconnected patch of emission from the red-shifted lobe is visible on the other side of the cloud. While the blue shifted jet is bright, the counter jet is seen only faintly through the nearly opaque cloud.

Some Herbig-Haro jets exhibit chains of internal low excitation knots (cf. HH 34, HH 111). At first, it seemed plausible that they could be the oblique, standing shocks that form the lateral jet-cloud boundary (see also Sec. II.B below). Their large proper motions, however, have excluded this possibility. Low-excitation shocks that travel at the jet speed could arise from slight velocity fluctuations in the gas being injected at the base of the jet. Faster material in the jet overtakes slower gas ahead of it, and their interaction region gradually steepens into a shock. This possibility is being explored through numerical simulations (Stone and Norman 1993).

Some HH objects mark the interface between the jet and surrounding undisturbed ambient gas. Others are excited by internal working surfaces and are propagating into the wakes of bow shock that have passed the region before. Multiple bow shocks tailing each other indicate the presence of internal working surfaces which provide evidence of variations in the ejection velocity of the jet (Reipurth 1989).

Wide-field and narrow-band CCD observations have recently become possible with the introduction of large arrays (2048×2048 pixels and more). Using a CCD camera with a 23' field of view, Bally and Devine (1994) and Bally et al. (1995) have recently found that several HH flows can be traced for more than 1.5 pc from their sources. Some exhibit periodic spacing of HH objects along their flow axes. Figure 5 shows a system exhibiting all these properties, the 3 pc-long outflow emerging from the source HH 34* near the Orion Nebula.

Recent wide-field surveys (Reipurth et al., in preparation) have shown that many Herbig-Haro chains extend well beyond the spatial extent of their associated CO outflows. Over 20 parsec scale chains of HH objects have been found. The outflow traced by the HH 111 jet appears to terminate in the HH objects HH 113 and HH 311 which have a projected angular separation of 1° or 7 pc. In contrast, the associated CO outflow can be traced for a spatial extent of order 0.5 pc, and is confined to the immediate vicinity of

Figure 5.   The HH 46/47 jet in the Gum Nebula (from Reipurth and Heathcote 1991). The photograph is a combined Hα and [S II] image. The blue-shifted side of the jet is projecting toward the left, while a fainter bow shock from the red-shifted side can be seen on the opposite rim of the cloud.

the molecular cloud core containing the exciting source (Reipurth and Olberg 1991). Evidently, the HH 111 jet has blown clear of its parent molecular cloud, and is punching through the tenuous interstellar gas that surrounds it. A similar situation exists in L1551. Recent deep CCD surveys have detected HH objects well beyond the outermost contours of CO emission (Devine et al. 1996). As in the HH 34 system (Fig. 6), most of the parsec-scale chains of HH objects exhibit S-shaped (or point) symmetry about their central sources. One possible explination is that the jets precess or wobble through angles ranging from several to several tens of degrees on time scales of several thousand years.

Since the pioneering study of HH 1–2 by Cohen et al. (1982), it is known that Herbig-Haro jets can produce radio continuum emission at centimeter wavelengths. In principle, such emission could arise from a hot, ionized

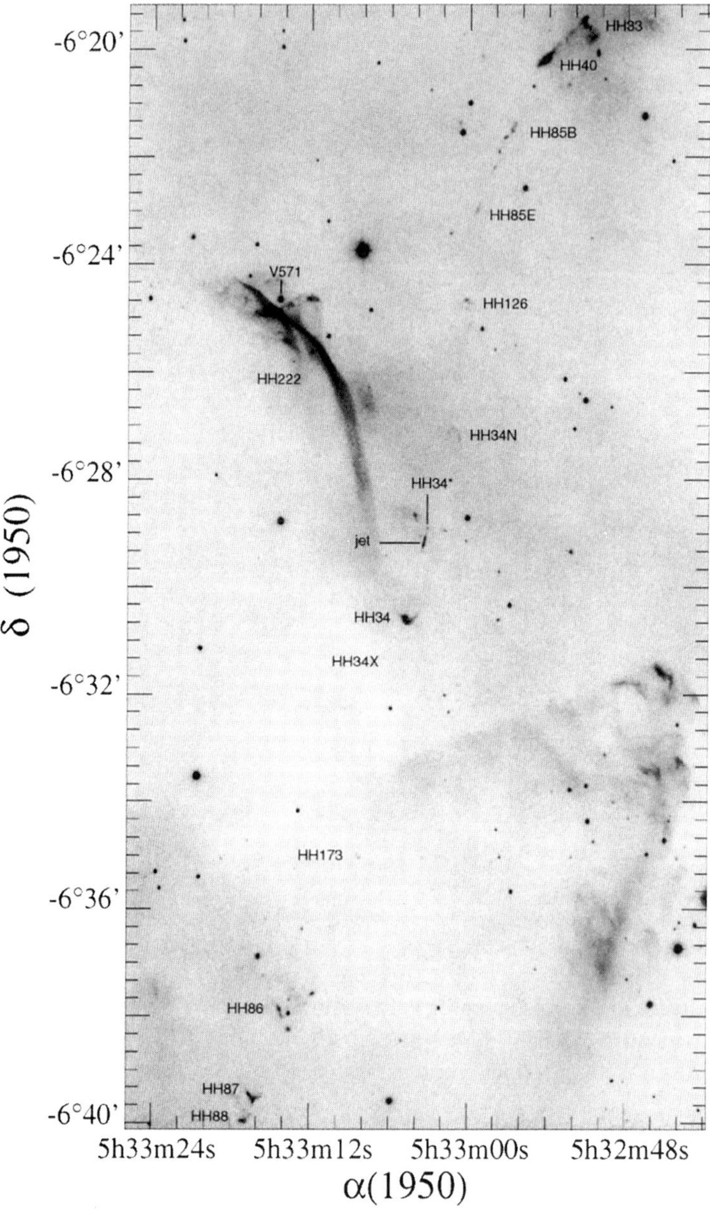

Figure 6.   The HH34 Herbig-Haro jet in Orion. This narrow-band optical image in Hα and [S II] represents a 22 by 10 arcmin field of view centered on the driving source HH 34*. A continuous string of Herbig-Haro objects extends north toward HH 40 and HH 33 near the top of the image. Another string can be traced toward HH 86 in the south. A small linear jet extends toward the south from HH 34* (from Bally and Devine, in preparation).

wind, but the generally low luminosities of the exciting stars render this possibility unlikely. The emission is usually found both near the central star and the strongest HH objects in the outflow, suggesting a shock origin. However, in at least three cases, in Serpens and in Orion, the radio spectra and polarizations indicate nonthermal synchrotron emission (Rodruíguez et al. 1989; Yusef-Zadeh et al. 1990).

The complex shock structures and morphologies present in outflows require that the underlying young stellar objects produce highly collimated jets whose ejection direction and velocity are time variable. Hydrodynamic and cooling instabilities that form in both the internal and terminal working surfaces of these jets produce complex morphologies and result in the fragmentation of the post-shock cooling layers. When these shocks interact with dense molecular gas, or produce sufficiently dense post-shock layers so that molecule formation time scales are short compared to the dynamic time scales of the flow, near infrared emission from the H_2 molecule can be seen in emission. Such shocks may also be sites where ambient CO-bearing gas is entrained into the outflow or where ultra-high-velocity CO is formed in the post-shock cooling layer. We now have evidence that mature outflows blow out of their parent cloud cores, producing visual wavelength radiation from shocks well beyond the extent of associated CO outflows.

II. THEORETICAL PERSPECTIVE

A. Bipolarity and Wind Collimation

The presence of two symmetric (bipolar) lobes is a key feature of CO outflows. Indeed, 80% of these systems are manifestly bipolar (Fukui et al. 1993), though relatively few present as clear a case as L1551/IRS-5 (Fig. 1) or L1448 (Fig. 3). As we have seen, optical outflows are generally monopolar, but give indications of underlying bipolarity. Assuming a wind origin for the outflow phenomenon, the common central axis of the two lobes represents a special direction along which the wind evidently finds it easier to propagate. The existence of such a preferred direction could reflect an intrinsic anisotropy of the wind, or else an environmental effect. Let us briefly consider each alternative.

The possibility of a bipolar stellar wind is difficult to assess without a more detailed understanding of wind formation than is currently available. Most workers agree that some form of centrifugally driven, hydromagnetic flow is most likely (Hartmann and MacGregor 1982; Shu et al. 1988; Wardle and Konigl 1993; Shu et al. 1994a,b,1995; Najita and Shu 1994; Lovelace et al. 1995). In this case, the lightly ionized wind gas is initially constrained to flow out along magnetic field lines rigidly anchored in the surface of the rotating young star. The gas eventually breaks free at the Alfvén surface, where the wind momentum overwhelms the restraining field. If the field has a dipole configuration at the star's surface, the wind sprays out more or less isotropically over a large solid angle (Mestel 1968).

Alternatively, both the wind and field could arise from a circumstellar disk, as a number of workers contend. The twisting of field lines by the disk could then create a pinching force toward the rotation axis, collimating the outflowing gas over large distances (Pudritz and Norman 1986; Newman et al. 1992). Whether the magnetic field can actually maintain a tightly wound spiral pattern is by no means obvious. Investigations in the solar context have shown that the energy in a magnetic flux tube being twisted at one end has a finite upper bound, beyond which the field suffers rapid Ohmic dissipation (see Sturrock 1991, and references therein).

Turning to environmental effects, it has long been appreciated that the molecular cloud cores forming stars have asymmetrical density distributions. From a theoretical viewpoint, flattened configurations arise naturally when a cloud is partially supported against its self-gravity by the interstellar magnetic field (Mouschovias 1976). Under some circumstances, an initially spherical stellar wind propagating into such an anisotropic medium could be distorted into a bipolar configuration.

Such channeling of stellar winds was proposed by Königl (1982), shortly after the discovery of CO outflows. If we consider a truly two-dimensional, slab-like cloud core, its density is naturally highest in the midplane, which will also be the likely site of star formation. Königl envisioned a highly supersonic wind from the central star shocking against cloud material to create an "interstellar bubble" of hot gas. This buoyant bubble expands fastest where the pressure gradient is a maximum, i.e., along the axis perpendicular to the slab plane. Eventually, the bubble erupts into the low-density environment outside the slab, creating the desired bipolar flow.

The model of Königl neglected cooling of the shocked gas, an important effect that calls into question the very existence of the bubble. The theory of interstellar bubbles was originally applied to O and B stars, which emit winds of such high velocity that the shocked wind gas cannot cool over time scales of interest (Castor et al. 1975). However, gas shocked by the slower winds from the more common low-mass stars *can* cool rapidly through the many atomic transitions that would be absent in a hotter, fully ionized plasma. Thus, the wind from a typical protostar or pre-main-sequence star should not drive a bubble at all, but instead produce, upon shocking, a relatively thin, dense, and cool post-shock layer.

With these considerations in mind, let us view environmental collimation in a broader context. Consider again an initially isotropic stellar wind within a cloud of pure molecular hydrogen. Suppose further that we endow the wind with a mass outflow rate ($\dot{M}_w \sim 10^{-7}$ $M_\odot yr^{-1}$) and speed ($V_w \sim 200$ km s^{-1}) characteristic of young, but optically visible, pre-main-sequence stars (Edwards et al. 1993). Such a wind will shock, creating a cavity of average radius R_w, obtained by balancing the wind ram pressure against the external thermal pressure $P_\circ$. The thermal pressure, in turn, is set by $n_\circ$, the total number density of cloud matter, and by $T_\circ$, the cloud temperature. We therefore have:

$$R_\circ = \left(\frac{\dot{M}_w V_w}{4\pi P_\circ} \right)^{1/2} = \left(\frac{\dot{M}_w V_w}{4\pi k_B n_\circ T_\circ} \right)^{1/2}$$
$$= 0.27 \, \text{pc} \, W_{w_1 \, 200}^{1/2} n_4^{-1/2} T_{10}^{-1/2} \tag{1}$$

where $\dot{M}_7$ is the mass loss rate in units of 10^{-7} $M_\odot$ yr^{-1}, $V_{w_1 \, 200}$ is the wind velocity in units of 200 km s^{-1}, n_4 is the ambient density in units of 10^4 cm^{-3}, $T_\odot$ is the ambient temperature in units of 10 K, and k_B is the Boltzmann constant.

The cloud cores in which stars are born have linear sizes of order 0.1 pc (Myers and Benson 1983). Such a length scale is also representative of the *lengths* of most CO outflow lobes, which are naturally truncated by the drop in density at the cloud boundaries. Even in the largest of outflows however, the lobe *width* is considerably smaller than 0.1 pc close to the exciting star. In addition, optical outflows are much narrower along their entire extent.

Could some other source of pressure, such as that associated with interior turbulence, be providing the necessary collimation? The problem is that such pressure pushes outward as well as inward. Molecular cloud cores are known to be roughly in hydrostatic balance, i.e., $P_\circ$ is just sufficient to withstand the compressive force of gravity. Thus, any additional pressure of this sort that could collimate the wind to its observed width would also disperse the cloud core, overwhelming its gravitational binding.

This conclusion holds only for clouds in hydrostatic equilibrium. If, however, it is indeed *protostars* that drive outflows, then at least the innermost regions of cloud cores are *not* static, but are collapsing toward the central stars. The ram pressure of the freely falling gas is not constrained to balance the gravitational force, but can climb to much higher values before the cloud gas impacts the stellar surface. Moreover, if the cloud rotates as it collapses, the ram pressure gradually decreases along the rotation axis (Cassen and Moosman 1981), allowing the wind to break out in a bipolar fashion (Chevalier 1983). It remains to be seen whether such dynamic confinement of protostellar winds yields realistic collimations.

B. Herbig-Haro Jets and Molecular Outflows

Despite the great diversity of observed outflow systems, they tend to fall within two broadly defined categories. In the first are the *Herbig-Haro jets*, outflows with highly collimated lobes and large internal speeds, $V \gtrsim 150$ km s^{-1}. In this same category we may place the narrow structures seen in shocked H_2 and in radio continuum emission.

The second class of *molecular outflows* consists of all the more poorly collimated flows, including the large number detected and mapped in CO. While the optical emission lines in Herbig-Haro jets indicate temperatures near 10^4 K, these broader outflows are much colder, with temperatures no higher than normal molecular cloud values. In addition, their mass-averaged radial velocities are only a few km s^{-1} above those of the background cloud.

Another important distinction between the two classes is the mass contained in the outflow lobes. The mass in Herbig-Haro jets is somewhat problematic, because collisional models for the excitation of the line emission are most sensitive to the electron density only. However, even with the inclusion of a plausibly large neutral component of gas, the jets have orders of magnitude less mass than their driving stars. The masses of molecular outflows can be obtained in a more straightforward manner, through observations with optically thin isotopes of CO. The lobes then turn out to have masses comparable to, or even greater than, those of their central stars. It is generally accepted, therefore, that a molecular outflow does *not* consist of material ejected from the star, but rather represents cloud gas being pushed outward by the stellar wind.

Herbig-Haro jets, however, could well be true stellar gas being shot out at high speed. As we have seen, it is not yet clear how a wind becomes channeled into the observed configurations of the optical outflows. On the other hand, experience with terrestrial jets helps put the problem into perspective. Supersonic jets in the laboratory are collimated by "crossing shocks," which cause the jet to alternately expand and contract as it propagates (Pai 1954). A crossing shock is so oblique that the jet velocity vector is essentially refracted across it with little drop in magnitude. A series of such shocks is created whenever a jet enters a quiescent medium through a relatively narrow nozzle, whose precise shape is unimportant. If stellar jets propagate in this manner, then the central issue becomes the creation of a wind nozzle.

What is the relation between Herbig-Haro jets and molecular outflows? Naturally, a full answer to this question is not yet possible. We can make a start in the right direction, however, by exploring a more modest, purely empirical question, such as, are the two types of outflows actually found together?

Here the answer is a qualified yes. It is rare to see a continuous *optical* jet emanating from a star which is driving a molecular outflow. Optical jets are intrinsically faint except when internal shocks produce bright emission. The clear detection of a molecular outflow implies the existence of a sufficient column density in cloud gas to obscure a jet embedded within. If we ask instead for the fraction of CO outflows that contain HH objects, the answer is over half (Fukui et al. 1993), a fact which strongly suggests the coexistence of the two outflow classes. This impression is further strengthened by the frequent detection of narrow H_2 jets within molecular outflows.

The outflows driven by Class 0 stars fall squarely between our two classes. Their CO emission is reminiscent of molecular outflows, but their narrow lobes are more jet-like. So, too, are their high internal velocities and narrow, clumpy H_2 line emission. Because Class 0 stars are highly embedded, the clear implication is that these hybrid structures represent a very early phase of the outflow process, to be followed by a gradual broadening of the molecular gas surrounding the internal jet. At the other extreme in age, those CO outflows associated with visible stars appear to be unexceptional in terms of their morphology (Levreault 1988). It appears, then, that the broadening process must slow down relatively early in an outflow's life.

C. Cloud Entrainment and Acceleration

Our comments thus far suggest an evolutionary picture which, though largely qualitative at this stage, nevertheless has interesting observational implications. For example, if every CO outflow has evolved from a more jet-like antecedent and still harbors its own jet, then we might expect the presently observed molecular speeds to increase toward the outflow axis. Such "velocity nesting" has indeed been observed in at least three well-mapped systems: L1551/IRS-5 (Moriarty-Schieven and Snell 1988), NGC 2071 (Moriarty-Schieven et al. 1989), and HH 7–11 (Bachiller and Cernicharo 1990). As we mentioned in Sec. I.B, the CO profiles of numerous other outflows point to the presence of relatively small amounts of molecular gas moving at jet-like speeds.

It appears, then, that a molecular outflow lobe consists essentially of nested collars of molecular gas moving at different speeds, all surrounding a central stellar jet. Those collars near the lobe boundary carry the bulk of the outflow's mass, but move at relatively low velocity. The narrower collars near the jet have much higher speeds. Such a distribution in mass and velocity arises naturally if the jet is dragging forward the cloud gas just outside its boundary. This gas, in turn, pulls on its own surrounding matter, which attains a somewhat lower speed. In this manner, a narrow jet containing very little mass is able to impart motion to a far greater quantity of molecular cloud gas. Such an interaction is one aspect of the general fluid dynamical phenomenon of entrainment. De Young (1986) identified two types of entrainment in a jet; "prompt entrainment" which occurs as a terminal shock where a jet sweeps up ambinet gas, and "steady state entrainment" where the turbulent shear along the jet mixes the jet material with the surrounding medium.

For steady-state entrainment to operate, there must be coupling between adjacent collars of gas. In the laboratory, a supersonic jet stirs up a turbulent "mixing layer" at its boundary (see Cantó and Raga 1991 and references therein). This turbulent sheath spreads inward, eventually pinching off the laminar jet. Concurrently, the turbulence also spreads outward into the surrounding medium (for a numerical study in the astrophysical context, see De Young [1986]). A nested velocity pattern is indeed established, with the necessary friction being provided by the eddy viscosity associated with turbulence. Researchers studying CO line profiles in molecular outflows have long noted that their kinematic models require an underlying velocity dispersion of several km s^{-1} (see, e.g., Meyers-Rice and Lada 1991). The physical origin of this dispersion might well be the stochastic motion of a turbulent velocity field that permeates the outflow.

Entrainment is the basic means by which the jet accelerates surrounding matter. In the highly dissipative cloud gas, a fluid element subjected to outside forcing cannot accelerate ballistically. Instead, its velocity (time-averaged, to eliminate the fluctuating component) quickly attains a terminal value in the direction of the jet. Acceleration of the cloud is manifested instead by the increase, away from the star, of the total amount of moving gas. In this form, the acceleration is readily apparent in position-velocity diagrams of outflows with well-separated lobes (see Fig. 7). Their fan-like pattern may be a direct consequence of the nested velocity distribution associated with entrainment (Stahler 1994).

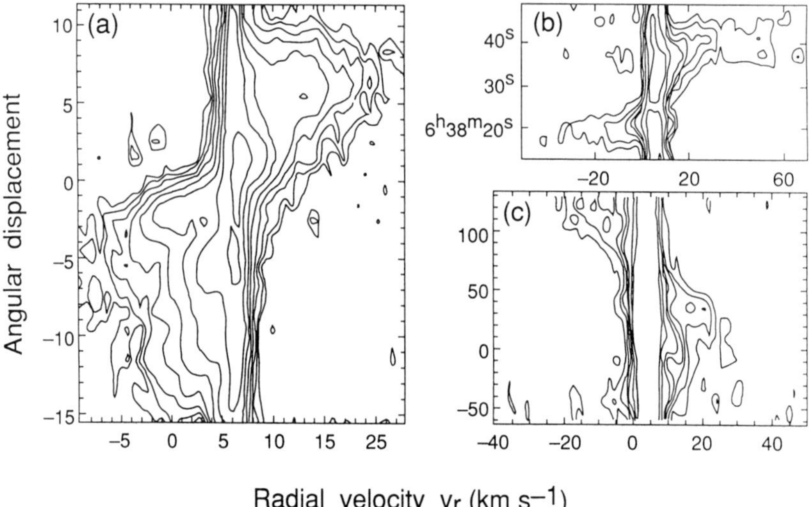

Figure 7.   Position–velocity diagrams along the central outflow axes for: (a) L1551/IRS 5 (Levreault 1988), (b) NGC 2264G (Margulis et al. 1990), and (c) VLA 1623 (André et al. 1990). The units for angular displacement are arcmin in (a), and arcsec for (b) and (c). The L1551/IRS 5 diagram was constructed using the CO $J = 2 \to 1$ line, while the others used the $J = 1 \to 0$ transition.

As we have noted, the velocity within a jet confined by crossing shocks can stay nearly constant for long distances. The fall in the momentum transport rate occurs rather through the decrease in the jet cross section, i.e., through the erosion of the laminar flow by the thickening turbulent mixing layer.

As turbulent cloud gas is dragged forward by the jet, surrounding matter senses the drop in pressure and crowds in to fill the gap. This inward, laminar flow, extending out to large distances from the jet, is the second major aspect of the entrainment process. The external gas approaching the jet is caught up in the turbulent flow and expelled to large distances. The net result is a global decrease in the cloud density. In incompressible fluids, the induced velocities are very small relative to the average speed within the turbulent region (Landau and Lifshitz 1975, ch. 3). If the same is true in our compressible, self-gravitating gas, then the molecular cloud core evolves through a sequence of quasi-static configurations as it is gradually drained of mass. Such a process would explain how a bipolar outflow, though occupying a relatively small solid angle around its star, can nevertheless disperse the enveloping dust and gas to render that star optically visible.

An alternate scenario is that the jet wobbles, and over time, sweeps out a widening and mostly hollow (compared to the surrounding cloud) cone (see, e.g., Masson and Chernin 1993; Raga and Cabrit 1993). Studies of optical emission from the recently discovered parsec-scale chains of HH objects demonstrate that S-shaped (or point) symmetry is common. The most direct interpretation of this result is that the jets precess or wobble on time scales of order several thousand years. Because the visible parts of optical jets are younger than this, such jets will typically appear highly collimated. In contrast, CO dynamic time scales range from 10^4 to 10^5 yr. The poor collimation of CO outflows may result from the sweeping out of a relatively wide cone of ambient gas by a wobbling jet.

Whatever the precise nature of the jet-cloud coupling, the momentum flow within the jet must decline as the molecular gas accelerates. There is now some evidence for a systematic drop in jet speed away from the star. In the HH 34 and HH 111 systems, the HH objects lying farther from the driving source have systematically lower radial velocities and proper motions (Devine et al. 1996; Reipurth et al., in preparation).

Acknowledgments. S. W. S. was funded principally by a grant from the National Science Foundation, and also by the NASA Astrophysics Theory Program through its Center for Star Formation Studies. J. B. acknowledges support for this work from two NASA grants.

REFERENCES

Adams, F. C., Lada, C. J., and Shu, F. H. 1987. Spectral evolution of young stellar objects. *Astrophys. J.* 312:788–806.

Allen, D. A., and Burton, M. G. 1993. Explosive ejection of matter associated with star formation in the Orion nebula. *Nature* 363:54–56.

André, P., and Montmerle, T. 1994. From T Tauri stars to protostars: Circumstellar material and young stellar objects in the ρ Ophiuchi cloud. *Astrophys. J.* 420:837–862.

André, Ph., Martin-Pintado, J., Despois, D., and Montmerle, T. 1990. Discovery of a remarkable bipolar flow and exciting source in the ρ Ophiuchi cloud core. *Astron. Astrophys.* 236:180–192.

Bachiller, R., and Cernicharo, J. 1990. Extremely high-velocity emission from molecular jets in NGC 6334I and NGC 1333 (HH7–11). *Astron. Astrophys.* 239:276–286.

Bachiller, R., and Gómez-González, J. 1992. Bipolar molecular outflows. *Astron. Astrophys. Rev.* 3:257–287.

Bachiller, R., Cernicharo, J., Martin-Pintado, J. Tafalla, M., and Lazareff, B. 1990. High-velocity molecular bullets in a fast bipolar outflow near L1448/IRS3. *Astron. Astrophys.* 231:174–186.

Bachiller, R., Martin-Pintado, J., and Planesas, P. 1991. High-velocity molecular jets and bullets from IRAS 03282+3035. *Astron. Astrophys.* 251:639–648.

Bally, J., and Devine, D. 1994. A parsec-scale "superjet" and quasi-periodic structure in the HH34 outflow? *Astrophys. J. Lett.* 428:65–68.

Bally, J., and Lada, C. J. 1983. The high-velocity molecular flows near young stellar objects. *Astrophys. J.* 265:824–847.

Bally, J., and Lane, A. P. 1990. Shocked molecular hydrogen associated with Herbig-Haro objects and molecular outflows: The Cepheus-A flow. In *Astrophysics with Infrared Arrays*, eds. K. Pilakowski and R. Elston (San Francisco: Astron. Soc. of the Pacific), p. 273.

Bally, J., and Lane, A. P. 1991. In *The Physics of Star Formation and Early Stellar Evolution*, eds. C. Lada and N. D. Kylafis (Dordrecht: Kluwer), pp. 471–495.

Bally, J., Lada, E. A., and Lane, A. P. 1993*a*. The L1448 molecular jet. *Astrophys. J. Lett.* 418:322–327.

Bally, J., Devine, D., Hereld, M., and Rauscher, B. 1993*b*. Molecular hydrogen in the IRAS 03282+3035 stellar jet. *Astrophys. J. Lett.* 418:75–78.

Bally, J., Devine, D., Fesen, R. A., and Lane, A. P. 1995. Twin Herbig-Haro jets and molecular outflows in L1228. *Astrophys. J.* 454:345–360.

Cabrit, S. 1993. In *Stellar Jets and Bipolar Flows*, eds. L. Errico and A. A. Vittone (Dordrecht: Kluwer), p. 1.

Cantó, J., and Raga, A. C. 1991. Mixing layers in stellar outflows. *Astrophys. J.* 372:646–658.

Cassen, P. M., and Moosman, A. 1981. On the formation of protostellar disks. *Icarus* 48:353–376.

Castor, J., McCray, R., and Weaver, R. 1975. Interstellar bubbles. *Astrophys. J. Lett.* 200:107–110.

Chevalier, R. A. 1983. The environments of T Tauri stars. *Astrophys. J.* 268:753–765.

Cohen, M., and Schwartz, R. D. 1980. A search for the exciting stars of Herbig-Haro objects. *Mon. Not. Roy. Astron. Soc.* 191:165–168.

Cohen, M., Beiging, J., and Schwartz, R. D. 1982. VLA observations of mass loss from T Tauri stars. *Astrophys. J.* 253:707–715.

Cudworth, K. M., and Herbig, G. H. 1979. Two large proper-motion Herbig-Haro objects. *Astrophys. J.* 84:548–551.

Davis, C. J., Dent, W. R. F., Matthews, H. G., Aspin, C., and Lightfoot, J. F. 1994. Submillimetre and near-infrared observations of L1448: A curving H_2 jet with multiple bow shocks. *Mon. Not. Roy. Astron. Soc.* 266:933–944.

De Young, D. S. 1986. Mass entrainment in astrophysical jets. *Astrophys. J.* 307:62–72.

Dopita, M. A., Schwartz, R. D., and Evans, I. 1982. Herbig-Haro object 46 and 47: Evidence fo bipolar ejection from a young star. *Astrophys. J. Lett.* 263:73–77.

Edwards, S., and Snell, R. L. 1983. A survey of high-velocity molecular gas in the vicinity of Herbig-Haro objects. I. *Astrophys. J.* 270:605–619.

Edwards, S., and Snell, R. L. 1984. A survey of high-velocity molecular gas near Herbig-Haro objects. II. *Astrophys. J.* 281:237–249.

Edwards, S., Ray, T., and Mundt, R. 1993. Energetic Mass Outflows from Young Stars. In *Protostars and Planets III*, eds. E. H. Levy and J. I. Lunine (Tucson: Univ. of Arizona Press), pp. 567–602.

Eisloeffel, J., and Mundt, R. 1992. Proper motion measurements in HH 34 jet and its associated bow shocks. *Astron. Astrophys.* 263:292–300.

Fukui, Y. 1989. Molecular outflows: Their implication on protostellar evolution.

In *Low Mass Star Formation and Pre-Main-Sequence Objects*, ed. B. Reipurth (Garching: European Southern Obs.), pp. 95–117.

Fukui, Y., Iwata, T., Mizuno, A., Bally, J., and Lane, A. P. 1993. Molecular Outflows. In Protostars and Planets III, eds. E. H. Levy and J. I. Lunine (Tucson: Univ. of Arizona Press), pp. 603–639.

Guilloteau, S., Bachiller, R., Fuente, A., and Lucas, R. 1992. First observations of young bipolar outflows with the IRAM interferometer: 2″ resolution SiO images of the molecular jet in L1448. *Astron. Astrophys.* 265:L49–L52.

Haro, G. 1952. Herbig's nebulous objects near NGC 1999. *Astrophys. J.* 115:572–573.

Hartigan, P., Raymond, J., and Hartmann, L. 1987. Radiative bow shock models of Herbig-Haro objects. *Astrophys. J.* 316:323–348.

Hartmann, L., and MacGregor, K. B. 1982. Protostellar mass and angular momentum loss. *Astrophys. J.* 259:180–192.

Heathcote, S. R., and Reipurth, B. 1992. Kinematics and evolution of the HH 34 complex. *Astron. J.* 104:2193–2212.

Herbig, G. H. 1948. Ph.D. Thesis, Univ. of California, Berkeley.

Herbig, G. H. 1950. The spectrum of the nebulosity surrounding T Tauri. *Astrophys. J.* 111:11–14.

Königl, A. 1982. On the nature of bipolar sources in dense molecular clouds. *Astrophys. J.* 261:115–134.

Lada, C. J. 1987. Star formation: From OB associations to protostars. In *Star Forming Regions*, eds. M. Peimbert and J. Jugaku (Dordrecht: D. Reidel), pp. 1–18.

Landau, L. D., and Lifshitz, E. M. 1975. *Fluid Mechanics* (Oxford: Permagon Press).

Levreault, R. M. 1988. A search for molecular outflows toward pre-main-sequence objects. *Astrophys. J. Suppl.* 67:283–371.

Lovelace, R. V. E., Romanova, M. M., and Bisnovatyi-Kogan, G. S. 1995. Spin-up/spin-down of magnetized stars with accretion discs and outflows. *Mon. Not. Roy. Astron. Soc.* 275:244–254.

Margulis, M. M., Lada, C. J., Hasegawa, T., Hayashi, S. S., Hayashi, M., Kaifu, N., Gatley, I., Greene, T. P., and Young, E. T. 1990. A spectacular molecular outflow in the Monoceros OB1 molecular cloud. *Astrophys. J.* 352:615–624.

Masson, C. R., and Chernin, L. M. 1993. Properties of jet-driven molecular outflows. *Astrophys. J.* 414:230–241.

Masson, C. R., Mundy, L. G., and Keene, J. 1990. The extremely high velocity CO flow in HH 7–11. *Astrophys. J. Lett.* 357:25–28.

McCutcheon, W. H., Dewdney, P. E., Purton, C. R., and Sato, T. 1991. Protostellar candidates in a sample of bright far-infrared IRAS sources. *Astron. J.* 101:1435–1465.

Mestel, L. 1968. Magnetic braking by a stellar wind—I. *Mon. Not. Roy. Astron. Soc.* 138:359–391.

Meyers-Rice, B. A., and Lada, C. J. 1991. The structures and kinematics of bipolar outflows: Observations and models of the Monoceros R2 outflow. *Astrophys. J.* 368:445–462.

Morgan, J. A., and Bally, J. 1991. Molecular outflows in the L1641 region of Orion. *Astrophys. J.* 372:505–517.

Moriarty-Schieven, G. H., and Snell, R. L. 1988. High-resolution images of the L1551 molecular outflow. II. Structure and kinematics. *Astrophys. J.* 332:364–378.

Moriarty-Schieven, G. H., Snell, R. L., and Hughes, V. A. 1989. CO $J = 2 - 1$ observations of the NGC 2071 molecular outflow: A wind driven shell. *Astrophys. J.* 347:358–364.

Mouschovias, T. 1976. Nonhomologous contraction and equilibria of self gravitating, magnetic interstellar clouds embedded in an intercloud medium: Star formation. *Astrophys. J.* 207:141–158.

Mundt, R., and Fried, J. W. 1983. Jets from young stars. *Astrophys. J. Lett.* 274:83–86.

Myers, P. C., and Benson, P. J. 1983. Dense cores in dark clouds: II. NH_3 observations and star formation. *Astrophys. J.* 266:309–320.

Najita, J. R., and Shu, F. H. 1994. Magnetocentrifugally driven flows from young stars and disks. III. Numerical solution of the sub-Alfvénic region. *Astrophys. J.* 429:808–825.

Newman, W. L., Newman, A. L., and Lovelace, R. V. E. 1992. Initiation of bipolar flows by magnetic field twisting in protostellar nebulae. *Astrophys. J.* 392:622–628.

Padman, R., Lasenby, A. N., and Green, D. A. 1991. In *Beams and Jets in Astrophysics*, ed. P. A. Hughes (New York: Cambridge Univ. Press), p. 484.

Pai, S.-I. 1954. *Fluid Dynamics of Jets* (New York: Van Nostrand).

Pudritz, R. E., and Norman, C. A. 1986. Bipolar hydromagnetic winds from disks around protostellar objects. *Astrophys. J.* 301:571–586.

Raga, A. C., and Cabrit, S. 1993. Molecular outflows entrained by jet bow shocks. *Astron. Astrophys.* 278:267–278.

Raga, A. C., Böhm, K. H., and Solf, J. 1986. A new test of bow-shock models of Herbig-Haro objects. *Astron. J.* 92:119–124.

Reipurth, B. 1989. The HH 111 Jet: Multiple outflow episodes from a young star. *Nature* 340:42–44.

Reipurth, B. 1991. In *The Physics of Star Formation and Early Stellar Evolution*, eds. C. Lada and N. D. Kylafis (Dordrecht: Kluwer), p. 479.

Reipurth, B. 1995. *A General Catalogue of Herbig-Haro Objects*, electronically published via anonymous ftp to ftp.hq.eso.org, directory /pub/Catalogs/Herbig-Haro.

Reipurth, B., and Heathcote, S. R. 1991. The jet and energy source of HH 46/47. *Astron. Astrophys.* 246:511–534.

Reipurth, B., and Olberg, M. 1991. Herbig-Haro jets and molecular outflows in L1617. *Astron. Astrophys.* 246:535–550.

Reipurth, B., Bally, J., Graham, J. A., Lane, A. P., and Zealy, W. J. 1986. The jet and energy source of HH34. *Astron. Astrophys.* 164:51–66.

Reipurth, B., Raga, A. C., and Heathcote, S. R. 1992. Structure and kinematics of the HH 111 jet. *Astrophys. J.* 392:145–158.

Richer, J. S., Hills, R. E., and Padman, R. 1992. A fast CO jet in Orion B. *Mon. Not. Roy. Astron. Soc.* 254:525–538.

Rodríguez, L. F., Curiel, S., Moran, J. M., Mirabel, I. F., Roth, M., and Garay, G. 1989. Large proper motions in the remarkable triple radio source in Serpens. *Astrophys. J. Lett.* 346:85–88.

Schmidt-Burgk, J., Güsten, R., Mauersberger, A., Schultz, A., and Wilson, T. L. 1990. A highly collimated outflow in core of OMC-1. *Astrophys. J. Lett.* 362:25–28.

Schwartz, R. D. 1975. T Tauri nebulae and Herbig-Haro nebulae: Evidence for excitation by a strong stellar wind. *Astrophys. J.* 195:631-642.

Shu, F. H., Lizano, S., Ruden, S. P., and Najita, J. 1988. Mass loss from rapidly rotating magnetic protostars. *Astrophys. J. Lett.* 328:19–23.

Shu, F. H., Najita, J., Ostriker, E., Wilkin, F., Ruden, S., and Lizano, S. 1994*a*. Magnetocentrifugally driven flows from young stars and disks. I. A generalized model. *Astrophys. J.* 429:781–796.

Shu, F. H., Najita, J., Ruden, S., and Lizano, S. 1994*b*. Magnetocentrifugally driven flows from young stars and disks. II. Formulation of the dynamical problem. *Astrophys. J.* 429:797–807.

Shu, F. H., Najita, J., Ostriker, E. C., and Shang, H. 1995. Magnetocentrifugally driven flows from young stars and disks. V. Asymptotic collimation into jets. *Astrophys. J. Lett.* 455:155–158.

Snell, R. L., Loren, R. B., and Plambeck, R. L. 1980. Observations of CO in L1551: Evidence for stellar wind driven shocks. *Astrophys. J. Lett.* 239:17–22.

Stahler, S. W. 1994. The kinematics of molecular outflows. *Astrophys. J.* 422:616–620.

Stone, J. M., Xu, J., and Mundy, L. G. 1995. Formation of "bullets" by hydrodynamical instabilities in stellar outflows. *Nature* 377:315–317.

Stone, J. M., and Norman, M. L. 1993. Numerical simulations of protostellar jets with non-equilibrium cooling. I. Method and two-dimensional results. *Astrophys. J.* 413:198–209.

Strom, S. E., Grasdalen, G. L., and Strom, K. M. 1974. Infrared and optical observations of Herbig-Haro objects. *Astrophys. J.* 191:111–142.

Sturrock, P. A. 1991. Maximum energy of semi-infinite magnetic field configurations (Sun). *Astrophys. J.* 380:655–659.

Wardle, M., and Königl, A. 1993. The structure of protostellar accretion disks and the origin of bipolar flows. *Astrophys. J.* 410:218–238.

Yusef-Zadeh, F., Cornwell, T. J., Reipurth, B., and Roth, M. 1990. Detection of synchrotron emission from a unique HH-like object in Orion. *Astrophys. J. Lett.* 348:61–64.

WINDS FROM PULSATING STARS

L. A. WILLSON, C. STRUCK and G. H. BOWEN
Iowa State University

A prerequisite to understanding winds from pulsating stars is to understand the effects of periodic pulsation on the structure of the atmosphere. We review the consequences of periodic shock passage, including both dynamical constraints and thermal balance considerations. The discussion is relevant to all classes of mono-periodic pulsating stars, but particular mention is made of models for long-period variables, Cepheids, and RR Lyrae stars. In the cases considered here, the temperature structure of the pulsating star's atmosphere is determined by the effects of the compression in shocks, expansion between shocks, expansion in a stellar wind, and the exchange of energy between the gas and the radiation field. We argue that under the conditions that apply in or below the region where the wind originates, the appropriate source function is non-LTE, and therefore the cooling rate will be sensitive to the density, with important consequences for the thermal structure of the outer atmosphere and the driving of a wind. In the final section, a variety of published models are reviewed and compared on the basis of differences in assumptions and calculation techniques and the effects of these differences on the results of the calculations.

I. INTRODUCTION

This chapter is primarily concerned with theoretical work on the atmospheric structure of pulsating stars, because, without a thorough understanding of this structure, it is impossible to predict or understand the properties of the wind. In the first sections, we review the analytic theory of the atmospheres and winds of pulsating stars, as this provides the necessary insight to understand the effects of different assumptions on the results of published detailed numerical models, and to avoid deriving incorrect conclusions from observations of, e.g., velocity variations based on Doppler shift observations. In the final section, we summarize basic characteristics of published and pending numerical models for the atmospheres of RR Lyrae stars, Cepheids, and long-period variables including the copiously mass-losing Miras.

It should be noted that there are many papers in the literature that discuss steady-state stellar outflows, i.e., those for which the velocity, density, temperature etc. are independent of time at any given point, following the pioneering work of Parker (1960a,b,1969) for the solar wind. Studies of time-independent flows provide some interesting insights into the physics of winds, including some qualitative features of the limiting flow of the winds from pulsating stars. However, the development of winds in pulsating stars cannot be fully understood if the periodic time variation is ignored; no simple averaging procedure will provide an adequate picture of the genesis of the mass flows. The atmospheric dynamics interact strongly with other atmospheric phenomena and play a central role in the mass loss mechanism. Work on dynamics will therefore be stressed in this review.

We must emphasize the importance of two factors in the models: (1) the dynamical response of the atmosphere to periodic driving; and (2) the essential non-LTE character of the radiative cooling function, and its consequences for the thermal and kinematic structure of the atmosphere. The analytic theory summarized here concerns itself primarily with mono-periodic pulsation, i.e., pulsation in which there is a single dominant period although the variations generally are not purely sinusoidal. The period is perhaps the most important parameter affecting the qualitative character of the results, followed closely by the photospheric isothermal scale height. The period determines the maximum shock amplitude that the atmosphere can sustain without progressive mass loss. Also, the ratio of the period to the atmospheric cutoff period P_{ac} determines whether the atmospheric density gradient is steep enough to reflect small amplitude internal waves, allowing for the buildup of large amplitude pulsation. For $P < P_{ac}$, there is little reflection, and waves propagate rather freely into the atmosphere. For $P > P_{ac}$ some very interesting nonlinear phenomena appear.

In the analytic models, the following dimensionless combinations of physical parameters govern the qualitative behavior, as we shall show in Sec. II:

1. $v_e P / 2r = 38.3 Q(r)$, where $Q(r) = P(M/M_\odot)^{1/2}(R_\star/R_\odot)^{3/2}$ is expressed in days. $Q(R_\star) = P\sqrt{\rho/\rho_\odot}$ is often computed in interior pulsation simulations, and is sometimes referred to as the pulsation constant; it is related to the dynamical time scale of the star as a whole.

2. H_{static}/r, the ratio of the static pressure scale height $H_s = [(1/p)\mathrm{d}p/\mathrm{d}r]^{-1}$ to the radius r. When T is constant and the ideal gas law applies, or for adiabatic exponent $\gamma = 1$, $H_{\text{static}}/r = 2c_T^2/v_{\text{esc}}^2$, where $c_T = \sqrt{P/\rho}$ is the isothermal sound speed.

3. c_T/v_{esc}.

4. P/P_{ac}, where P_{ac} is the acoustic cutoff period $= 4\pi H_{\text{static}}/c_s$, with $c_s = \sqrt{\gamma P/\rho} = \sqrt{\gamma}c_T$.

Because $H_{\text{static}}/r = 2C_T^2/v_{\text{esc}}^2$ and (see Sec. II) $P/P_{ac} \sim Q/(c_s/v_{\text{esc}})$, there are really only two independent parameters to be chosen among (1) through (4) above.

While our analysis concerns the mono-periodic case, some of the discussion will carry over into cases with a broader spectrum of oscillations, whether these be Fourier components of a more complex periodic waveform or multiple incommensurate frequencies excited simultaneously by multiple oscillation modes or by convective turbulence. Thus, for example, one would still expect qualitatively different propagation phenomena for those oscillations in a distribution that have $P > P_{ac}$ than for those with $P < P_{ac}$. One would also expect that the thermal relaxation discussion would be applicable to more general studies of wave propagation in stellar atmospheres. For discussion of detailed nonlinear calculations involving multiple periods or continuous distributions of frequencies, the reader is referred to the chapter by Cuntz and Dorfi and references therein.

For a spectrum of acoustic waves with Mach numbers <1 on a steady background, it is possible to characterize the effects of the waves in terms of a "wave pressure" formalism (Pijpers and Habing 1989; Pijpers and Hearn 1989; Koninx and Pijpers 1993). This can also be used when the background may be considered to be very slowly varying on the scale of the waves. An equivalent condition is that the wavelength must be small compared to the other length scales, i.e., the period must be short. This condition is not met by pulsation with pulsation Q values characteristic of the stars that we will be discussing, for which such averaging does not work. Pijpers (1995) has argued that such short-period waves may be emitted by, be absorbed by, or be amplified by passage through shock waves, so that short-period acoustic waves could prove to be an interesting additional mechanism for the transfer of energy and momentum into and through the atmosphere of a pulsating star. He points out that current numerical models are generally not computed in such a way as to reveal phenomena of much shorter period than the dominant pulsation mode—to see these would require much finer grids, for example, than are used in modeling large-scale pulsation and winds. However, during much of the cycle the gas is expanding, and in these rarefaction zones the sound waves will be damped. Also, waves that are moving at the speed of sound relative to the moving medium are unlikely to play a major role in driving winds from pulsating stars, because such waves are rapidly "absorbed" into the immediate region of the shock fronts (as can be seen by tracing trajectories of material moving at the speed of the gas plus or minus the speed of sound in Fig. 6 or 7 below). Therefore, we expect that more detailed studies will show clearly that short-period sound waves have little or no effect on mass loss from large-amplitude pulsating stars.

II. PERIODIC MOTION IN A STELLAR ATMOSPHERE

In response to periodic driving, a stellar atmosphere may develop a periodic pattern of velocity, temperature, density etc; or, it may respond chaotically (see, e.g., Buchler 1987,1990; Buchler et al. 1987; Buchler and Moskalik 1990). The periodic case is simplest to explore analytically, and is likely to be valid for most stars whose lightcurves are relatively regular and periodic. Numerical calculations of pulsating giants without strongly driven dusty winds can show aperiodicities developing in the upper atmosphere even with period driving (Sec. VI). However, most detailed numerical models for Miras with and without dust, Cepheids, and RR Lyrae stars are well described by the analysis below that assumes simple periodic behavior.

The condition of periodicity, namely that for every physical variable X it must be true that $X(t + P) = X(t)$, limits the shock velocity amplitudes and sets conditions on the thermal structure of the atmosphere. In this section we examine (1) limits on shock velocity amplitudes set by the periodicity condition; and (2) the implications for the thermal structure, assuming simple functions for the dependence of the cooling function for the gas on temperature and density. In Sec. IV we examine constraints on the cooling function in more detail, and argue that density-dependent non-LTE, radiative cooling rates are an essential feature of realistic models.

A. Period and Amplitude

At any point in the atmosphere of a star there is a gravitational force directed inward, there is a pressure gradient force (generally directed outward), and there is a force due to the interaction of the radiation with the matter ("radiation pressure"—also of most interest when it is directed outward). In a static atmosphere these forces balance; in a dynamical atmosphere, acceleration results from differences in their magnitudes. In this section we explore three simple limiting cases, assuming in each that radiation pressure is negligible: (a) when the force of gravity dominates the motion between shocks (free-fall case); (b) when the forces due to pressure gradients are included assuming isothermal or adiabatic expansion between shocks.

1. Free-fall Case. Consider a parcel of gas in the atmosphere of a pulsating star, in the region where the waves have steepened into shocks. The material is hit by a shock once each period, and is given an outward velocity v_{out}. It returns to r_o at time $t_s + P$, when it is hit by another shock (Fig. 1). In the free-fall case, this material is subjected only to the force of gravity between shocks. If v_{out}/v_{esc} is small, then the material will move only a small fraction of a stellar radius over a period P; in this "plane-parallel limit" the material would move in a parabolic trajectory with $r(t) = r_o + v_o t - (1/2)gt^2$, so it would return to its initial position r_o after a time $2t_{ff} = 2v_{out}/g$. For a periodic solution, in the plane-parallel case, $2t_{ff}$ must be equal to the period P.

When v_{out}/v_{esc} is not $\ll 1$, the trajectory passes through a range of g, and the relationship between period and $v_{out}/v_{esc} \equiv \beta$ is given by the more complicated expression:

$$v_{esc} P/2r = 38.3 Q^d = [\beta/(1 - \beta^2)] + (1 - \beta^2)^{-3/2} \arcsin(\beta) \equiv \mathcal{P}_o(\beta) \quad (1)$$

where $Q^d = P\sqrt{\rho/\rho_\odot}$, expressed in days (Hill and Willson 1979; Willson and Hill 1979). Note that Hill and Willson wrote the expression for $\mathcal{P}_o(\beta)$ in a slightly different form including arctangents; it is completely equivalent to Eq. (1). For convenience we define the dimensionless period $\mathcal{P} = v_e P/2r$. Equation (1) then states the condition $\mathcal{P}_o(\beta) = \mathcal{P}$, in the notation used by Hill and Willson (1979) and Willson and Hill (1979). The free-fall trajectory has radial amplitude $r_{max}/r_o = [1/(1 - \beta^2)]$.

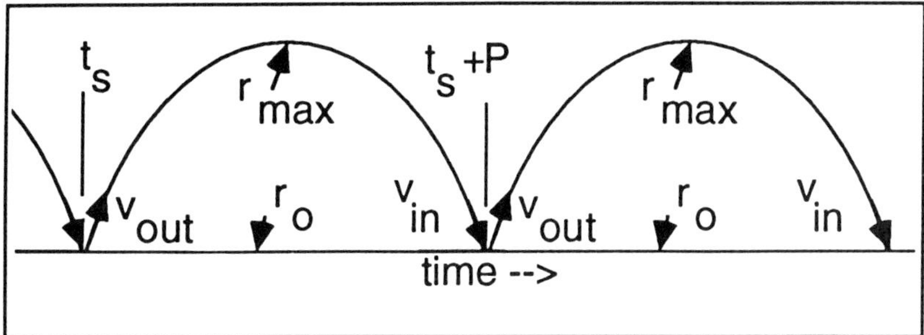

Figure 1.   The time dependence of the radial position of a particle subject to the force of gravity alone between the outward impulse at t_s and the return to r_0 at $t_s + P$. The more general case, where pressure gradients add a force between shocks, will generally have $v_{out} < v_{in}$ and the time of maximum radius will be shifted from $t_s + P/2$ towards $t_s + P$.

2. Condition for Mass Loss Forced by Periodicity. The material escapes from the star when v_{out}/v_{esc} is equal to or greater than that satisfying Eq. (1), so that $P < P_0(\beta)$ and the material does not have enough time to return to r_0 before another shock gives it a new impetus outward. As $P \rightarrow \infty$, $\beta_{critical} \rightarrow 1$ or $v \rightarrow v_{escape} = \sqrt{2GM/r}$ as it must.

3. Limits on the Applicability of the Constant Gravity Solution. The constant gravity solution is $v_{out} = gP/2$, and it applies only if β is small so that $r_{max} - r_0) \ll r_0$). Figure 2 shows that for $Q \sim 0.01$ to 0.1 day, the range of Q at the photosphere for radial modes of most stars, the free-fall β is > 0.2 and *the plane-parallel solution does not apply.* Thus, it is incorrect to set $\Delta v = gP$ and use this to drive g from observed Δv and P. Rather, in the free-fall case $gP = GMP/r^2 = v_e^2 P/2r$ so

$$\Delta v/gP = \Delta v/(v_e^2 P/2r) = 2\beta/P_0(\beta) \qquad (2)$$

which is generally quite small for values of Q appropriate to radial pulsation in low-order modes (Fig. 3).

4. Modifications to the Conditions Given Above when the Effects of Finite Pressure Gradients Between Shocks are Taken into Account. The infall velocity of material returning to r_0 before it is hit by a new shock is v_{in}; in the free-fall case $|v_{in}| = |v_{out}| = \Delta v/2$ where Δv is the shock velocity amplitude. Although in real stars the stellar center-of-mass velocity may be uncertain, leading to large errors in estimating the outward post-shock velocity, it is generally possible to derive limits on the shock velocity amplitude from line doubling at certain phases, or from the strength of emission lines. For the free-fall case, $v_{out} = \Delta v/2$ so $\beta = \Delta v/2v_{esc}$. When the effects of finite pressure gradients are taken into account, typically $|v_{out}| < |v_{in}|$ and $\beta < \Delta v/2v_{esc}$.

The equation of motion for a parcel of gas is

$$v \, dv/dr + GM/r^2 + (1/\rho)dp/dr = 0 \qquad (3)$$

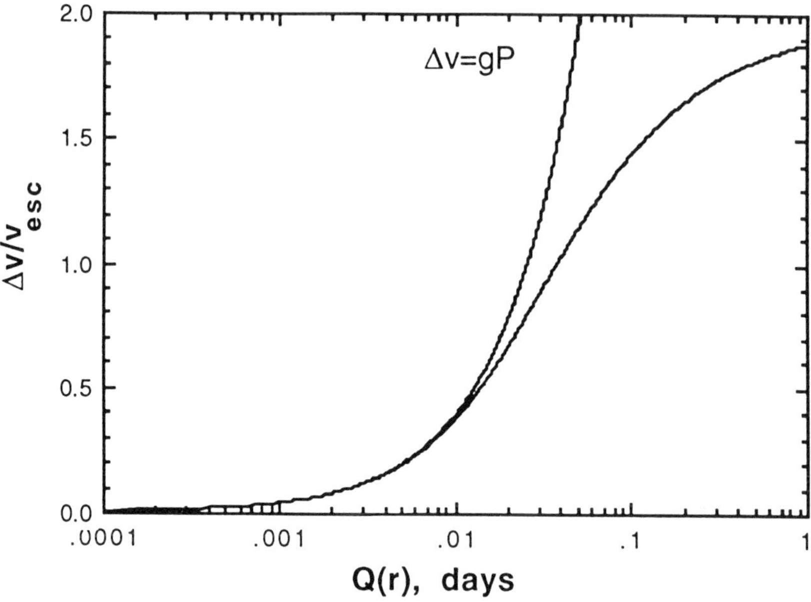

Figure 2. The maximum outward velocity that a parcel of gas may have at time t_s and radius r_o if gravity is to be able to return the parcel to r_o within the pulsation period P.

where p is the pressure and ρ is the density.

While the general case including the pressure gradient requires numerical solution, two interesting limits for the plane parallel case may be found analytically:

(a) Isothermal expansion between shocks. If we use the ideal gas law with T = constant, we can write $p = c_T^2 \rho$ with c_T = the isothermal sound speed = constant between shocks. Integrating Eq. (3) with this simplification and evaluating it for $r = r_o$ ahead of the next shock gives:

$$v_{out}^2 = v_{in}^2 - 2c_T^2 \ln S \qquad (4)$$

where $S \equiv \rho_2/\rho_1$ is the ratio of densities across the shock, the shock compression factor (Willson and Bowen 1985). *(b) Adiabatic expansion with constant ratio of specific heats $\gamma \neq 1$ between shocks.* Again, integrating Eq. (3) using $p \propto \rho^\gamma$ with γ = constant and evaluating the result for $r = r_o$ ahead of the next shock gives:

$$v_{out}^2 = v_{in}^2 - 2(\gamma/\gamma - 1)c_T^2[1 - (1/S)^{\gamma-1}]. \qquad (5)$$

5. Parameterized Description of the Shock Asymmetry. In both cases above, the shock amplitudes are smaller when the pressure term is taken into

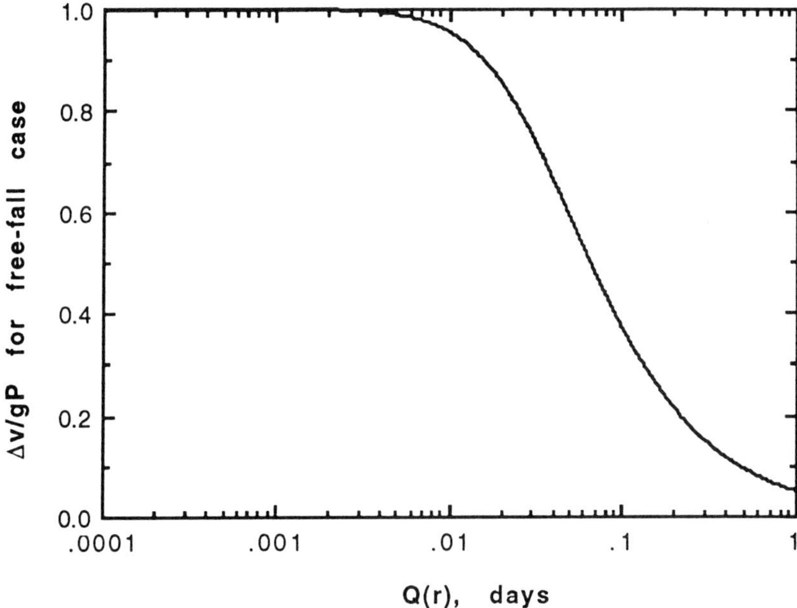

Figure 3.   The maximum (ballistic) shock amplitude divided by gP as a function of $Q = P\sqrt{\rho/\rho_\odot}$. Most radially pulsating stars have photospheric $Q \sim 0.03$ to 0.1 day.

account than they are in the ballistic case. We can write an expression for the asymmetry of the shock in general:

$$(v_{in}^2 - v_{out}^2)/v_{esc}^2 = (\beta_{in}^2 - \beta_{out}^2) = (2c_T^2/v_{esc}^2)f(S) > 0 \qquad (6)$$

where the form of the function $f(S)$ depends on whether the shocks are effectively isothermal, adiabatic, or otherwise specified. The static isothermal scale height $H_s = kT/\mu m_H g = (2c_T^2/v_{esc}^2)r$, so the parameter in the asymmetry condition is naturally related to the degree of extension of the atmosphere, and more compact atmospheres (= atmospheres with smaller H_s/r) should show more symmetric shocks. Note also that in the purely adiabatic case $S \leq 4$ (for $\gamma = 5/3$), while in the isothermal case S may be very large ($S \sim 100$ is seen in some numerical models with isothermal shocks; see, e.g., Hill and Willson [1979]; Willson and Hill [1979]) making larger asymmetry factors more likely in the isothermal case than in the purely adiabatic case. In the hybrid case explored in the analytic models of Struck and Willson (described briefly in Sec. III), an initial phase of very rapid cooling is followed by nearly adiabatic expansion between shocks; in that case, Eq. (5) applies but with $S > 4$.

Observers should note that stars with small ratios of the static scale height to the radius and small Q may have maximum shock amplitudes that come close to satisfying $\Delta v_{max} = gP$. However, stars with relatively large H_s/r at the photosphere and/or larger Q will depart more from the plane-parallel case, and can have a Δv as much as a factor of 3 or 4 less than $(GM_*/r_{\mathrm{phot}}^2)P$. This latter category includes nearly all radially pulsating stars with one single dominant period. Also, in the extended atmosphere, the maximum Δv consistent with a periodic solution depends on $Q(r) = Q_*(r/r_{\mathrm{phot}})^{-3/2}$, so it is important to be confident of the position in the atmosphere where the spectral lines are formed when using Eqs. (1–6) with observed velocities.

In Hill's (1972) RR Lyrae atmosphere calculations, with plane-parallel geometry and isothermal shocks, the shocks rapidly achieved $\Delta v = 0.8 \ gP$ and then stayed at that value throughout the atmosphere (Fig. 4). For his choice of stellar parameters, the photospheric $H/r \sim 0.002$, and the above equations would suggest that the asymmetry should be considerably smaller, hence Δv should be closer to gP. Whether this reflects physical processes overlooked in the above analysis or whether the use of artificial viscosity and rather coarse zoning in these calculations is responsible for the difference is not clear as of this writing.

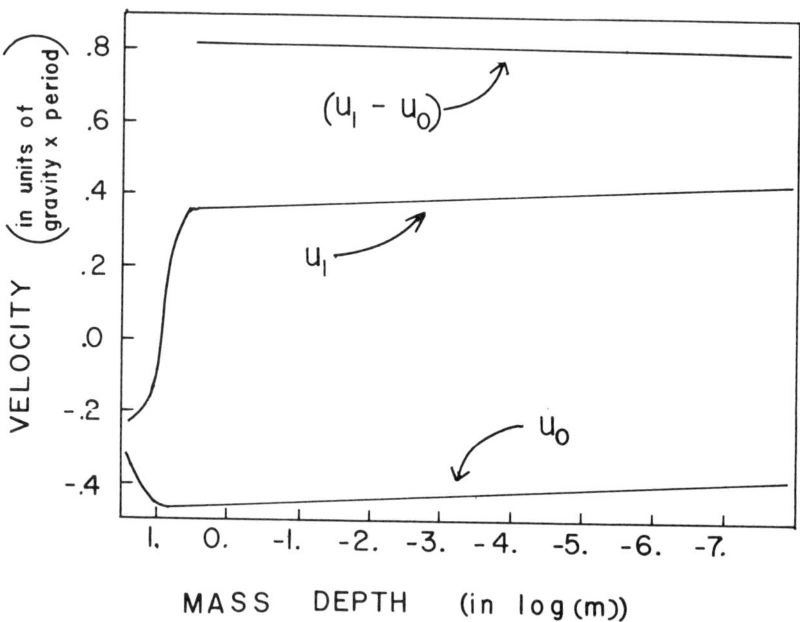

Figure 4.　The velocity discontinuity $(u_1 - u_0)$ pre-shock velocity (u_0) and the post-shock velocity u_1 as a function of mass depth (m in g cm^{-2}) of the shock front as obtained from an isothermal atmosphere calculation for an RR Lyrae star (figure from Hill 1972).

Some detailed numerical models show another phenomenon that gives rise to Δv significantly less than is expected for periodic motion with period P; this is the formation of a secondary shock, formed where infalling material returns to near its initial position well before the full period P has passed. The conditions for the formation of such secondary shocks are considered in Sec. II.B below.

6. Previous Published Work on Analytic Modeling of the Atmospheres of Pulsating Stars. On this topic, the reader is referred to the papers referenced in the above discussion (Hill and Willson 1979; Willson and Hill 1979; Willson and Bowen 1985. A consideration of motions in a pulsating star's atmosphere that pre-dates detailed numerical calculations may be found in Whitney (1956). Bertschinger and Chevalier (1985) have also presented an analytic discussion of this problem; however, they made a number of assumptions that do not agree well with more detailed studies, including: (1) imposition of a scale height $h = u_1{}^2/g$ on the pre-shock densities, where u_1 is the velocity with which the material enters the rising shock front in the frame of the shock; (2) an assumption of isothermal shocks; and (3) an assumption of expansion with an arbitrary adiabatic exponent γ between shocks. Also, their models are truly analytic only in the case of constant gravity, which, as noted above, does not apply to most cases of interest; their solution in the spherically extended case was a numerical patchwork of analytic solutions from the plane-parallel case and an extended free-fall motion, fitted to observed infrared line velocities on the assumption (almost certainly incorrect) that these follow the motion of a single layer of material in the atmosphere. Thus their discussion has little or no applicability to the interpretations of observations and the understanding of motions in real stars.

B. The Acoustic Cutoff Frequency in the Atmosphere

In the classical analysis of stellar pulsation through interior models, the interior of the star is in effect assumed to be an acoustic cavity, and pulsation waves arising from internal driving are assumed to be perfectly reflected at the surface. (In most calculations, this is not assumed explicitly; rather, the condition is used that the pressure goes to zero at the "surface." However, this has the effect of forcing perfect reflection at the surface.) This boundary condition is used to get periods (linear or nonlinear calculations) and amplitudes (nonlinear calculations). The approximation is good if the density gradient in the atmosphere is sufficiently steep, and the pulsation amplitude sufficiently small. Pijpers (1993) has demonstrated (in some idealized cases) that the eigenfrequencies may be substantially different where there is a strongly driven stellar wind that alters the density structure near the photosphere.

The condition for a sufficiently steep density gradient in linear theory is that the wavelength of the pulsation is large compared to the scale height in the atmosphere, or more specifically, that the period is longer than the acoustic cutoff frequency $P_{ac} = 4\pi H/c_s$, where H is the scale height and c_s the sound speed. At the "surface," waves with $P < P_{ac}$ can propagate into the atmosphere, but ones with $P > P_{ac}$ cannot. This condition is based on linearized wave equations, and does not describe what happens when the pulsation amplitudes are large.

Bowen (1990) investigated the effects of varying P/P_{ac} in detailed numerical models of long-period variables. These stars are particularly interesting in this context because: (a) $P_{\text{fundamental}} > P_{ac}$ but $P_{\text{overtones}} < P_{ac}$, and (b) there is considerable difficulty in establishing the modes of pulsation of these stars by classical means, because their radii are observationally ill-defined, and the best discriminant so far is the fitting of dynamical models to observed velocities (see, e.g., Wood 1995, and references therein). Because $P/P_{ac} \propto Q/\sqrt{(H/r)}$ and because Q values for radially pulsating stars generally fall in the range between 0.03 and 0.1 day, radially pulsating stars with photospheric $H/r \ll 0.01$ will generally have $P/P_{ac} > 1$ and the "reflecting" boundary condition should be relatively good for interior models, although for finite amplitudes some "leakage" into the atmosphere will still occur.

In his dynamical (nonlinear) models, Bowen found that with equal driving amplitudes, the average power input at the piston for models with $P < P_{ac}$ (at the photosphere or at the minimum in P_{ac}) is much larger than for $P \gg P_{ac}$ (Fig. 5). For the shorter-period models, waves propagate easily into the atmosphere and carry power out of the standing-wave zone, while for the longer periods, most of the power is reflected from the region of smallest P_{ac}. This is in accordance with what one might expect from the linear analysis. However, for $P > P_{ac}$, large amplitudes can build up, and wave propagation into the atmosphere (with consequent mass loss and some damping of the pulsation amplitude) is a feature of nonlinear calculations only. Typically, there is a minimum in P_{ac} near the photosphere, because $P_{ac} \sim \sqrt{T}/g$, P_{ac} decreases with increasing r at and below the photosphere where T varies with r more steeply than $1/r^2$. Farther out in the atmosphere, the temperature gradient decreases, and g decreases more rapidy than $\sqrt{T}$, so P_{ac} must increase with increasing radius at sufficiently large r/R_*. For modes with periods shorter than the minimum value of P_{ac} in the atmosphere, the nonlinear models show substantial propagation of power into the atmosphere in the form of traveling waves that steepen to become shocks. For periods longer than the minimum P_{ac}, where small amplitude waves are not expected to be able to propagate, the nonlinear models show interesting and complex behavior (Bowen 1990).

The power that was needed to drive pulsations of reasonable amplitude as a function of period for some long-period variable of detailed numerical models is displayed in Fig. 5. Overtone of long-period variable models are difficult to drive (they have large atmospheric damping) because such short-period waves propagate easily into the atmosphere. In this context, it is interesting to note that Wood's (1979) overtone models were driven with a sinusoidal variation of log (pressure) with an amplitude of 1 or 2—i.e., he needed to vary the pressure at the velocity node of the overtone pulsaton by a factor of 10 or more to get large-amplitude oscillations in the atmospheres

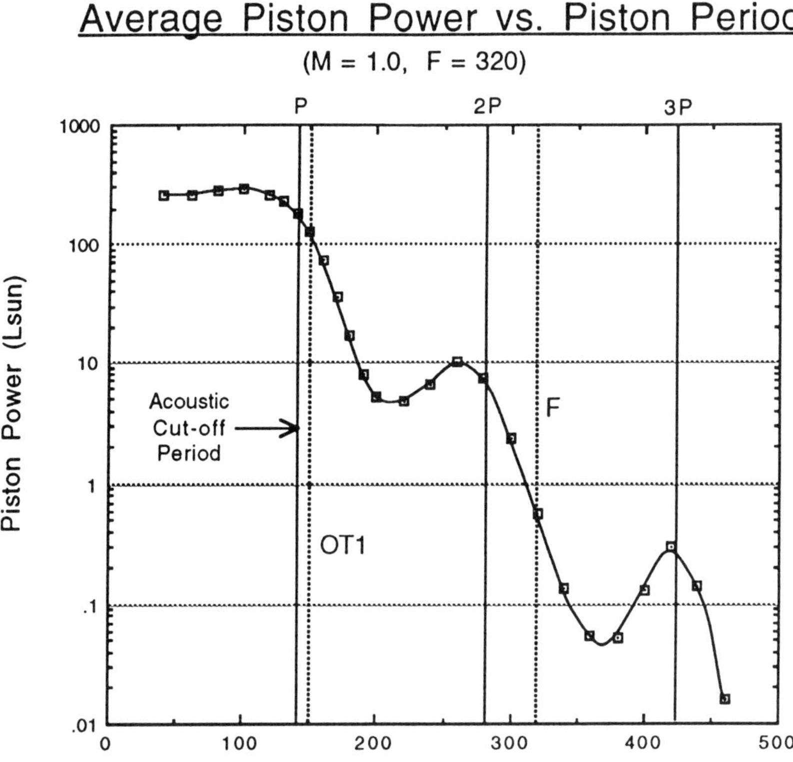

Figure 5.   Average piston power as a function of piston period at a piston velocity amplitude of 1.5 km s^{-1} for a long-period variable star model with $M = 1.0$ M$_\odot$, $R = 240$ R$_\odot$ and $T_{\text{eff}} = 2960$ K (figure from Bowen 1991).

in his models.   Figure 6 shows a long-period variable model atmosphere driven at an overtone period; the structure is strikingly simple and the waves are seen to travel directly from the "piston" inner boundary out into the atmosphere. The "resonant" structure seen in the power plot at $\sim 2P_{ac}$ and $\sim 3P_{ac}$ corresponds to the development in the atmosphere of sub-oscillations and secondary shocks occurring more often than once each pulsation period (see Fig. 7). Observationally, such secondary shocks may produce bumps on the lightcurves, or may be evident as extra components in velocity curves.

Thus, the minimum value of P_{ac} in the atmosphere, approximately equal to P_{ac} at the photosphere, determines whether pulsation at a given period is likely to reach a large amplitude.   For $P < P_{ac}$, atmospheric damping may severely restrict the pulsation amplitude.   For $P > P_{ac}$, large amplitudes are

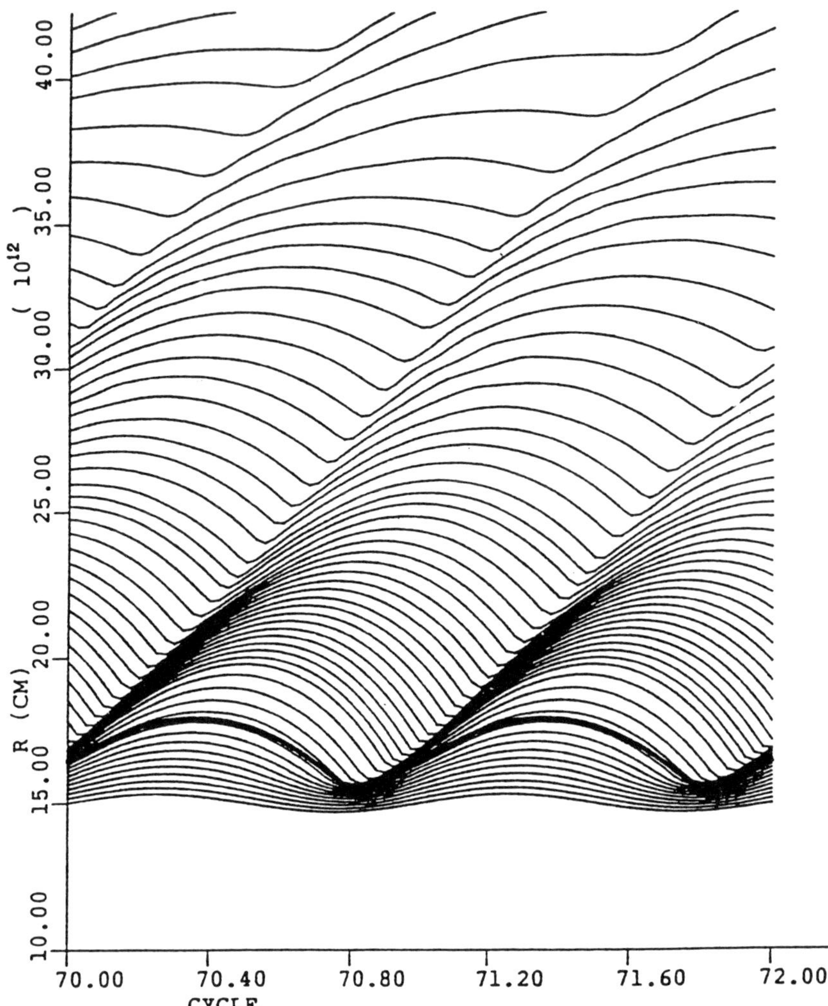

Figure 6.   Positions of representative mass shells in a dynamical model as a function
of time for the same model as in Fig. 5 driven at a period of 152 days corresponding
to first overtone (figure from Bowen 1991).

possible, but dynamical atmosphere calculations demonstrate that in that case
there will still be significant propagation of power into the atmosphere, and
there is likely to be more than one shock per pulsation period.

III. SHOCK-WAVE DEVELOPMENT IN AN ATMOSPHERE WHERE RADIATIVE FORCES ARE NEGLIGIBLE

Features of the detailed numerical models that an analytic theory seeks to
reproduce include $v(r)$, $T(r)$, and $\rho(r)$. Figures 8 and 9 show these variables

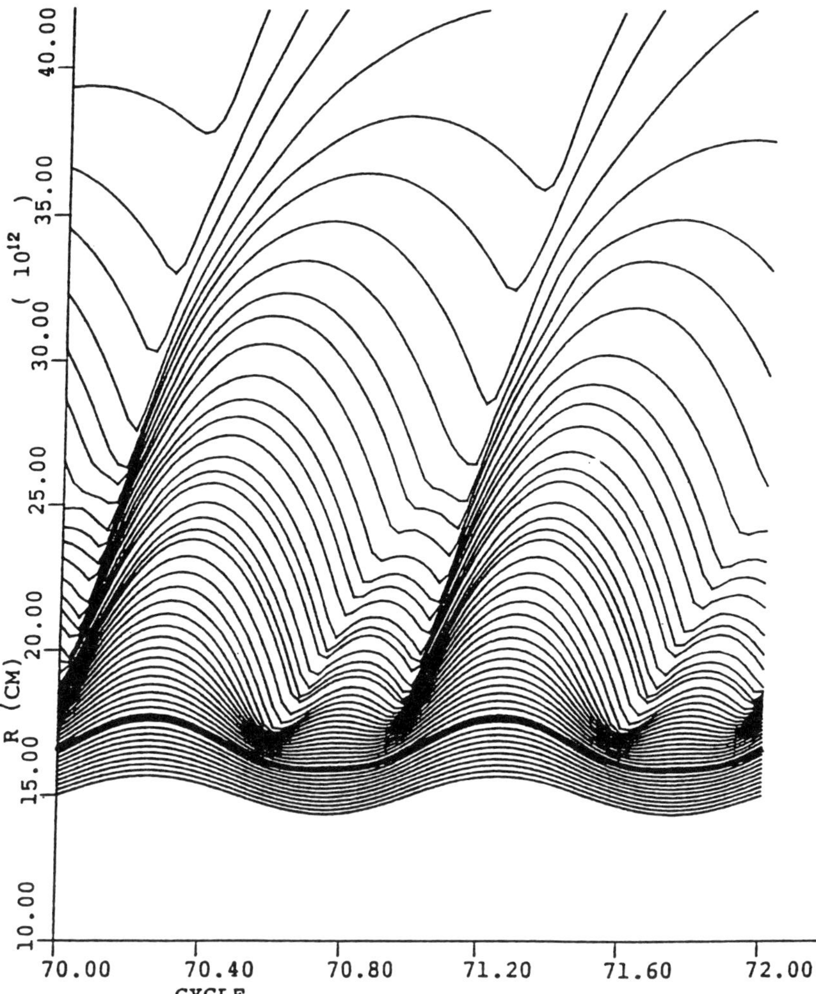

Figure 7.   Positions of representative mass shells in a dynamical model as a function of time for the same model as in Fig. 5 driven with a period of 320 days corresponding to the radial fundamental mode (figure from Bowen 1991).

for a representative calculation by Bowen for parameters appropriate to a long-period variable star but without including factors that are included in his more realistic models for these stars: radiative forces acting on dust and molecules. While these "dust-free Mira" models do not represent any real class of stars, they are useful for their ability to illustrate the importance of factors other than dust. We note the following qualitative features in such models:

1. Shock amplitudes decrease after reaching a maximum amplitude that is consistent with the limits derived in Sec. II.A.

L. A. WILLSON ET AL.

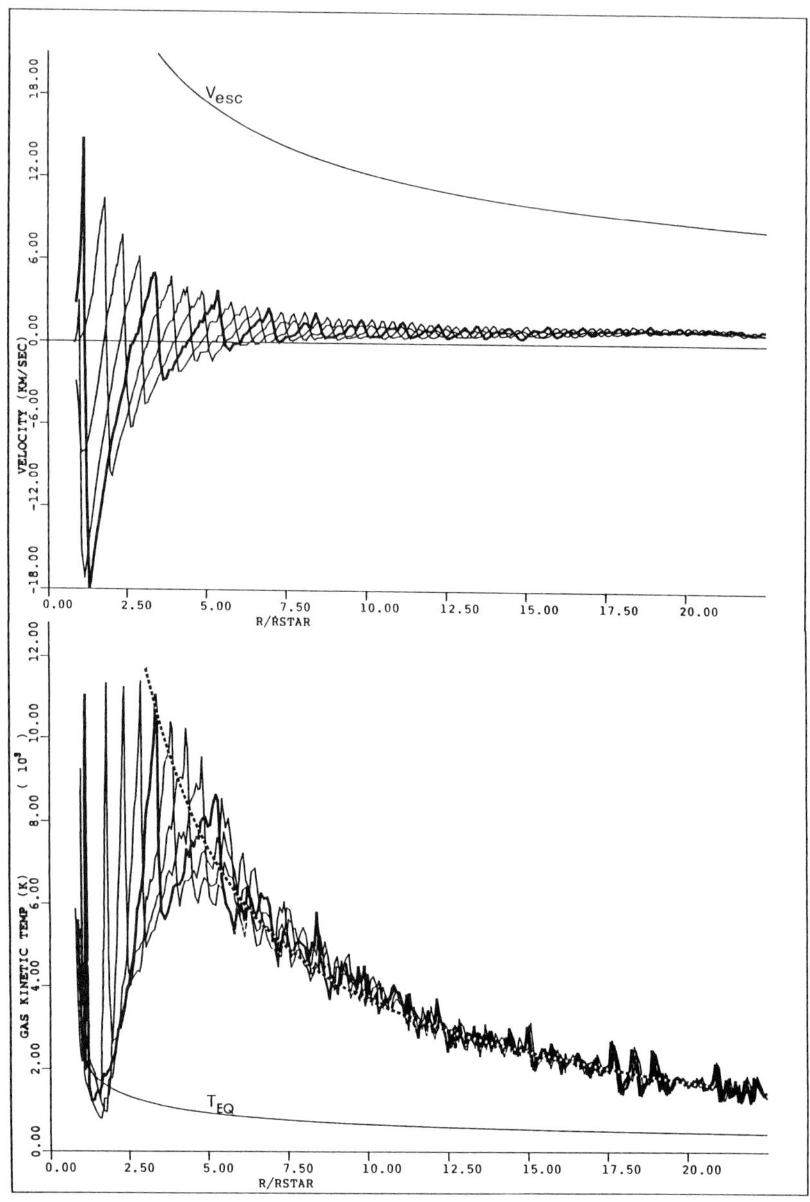

Figure 8. Plots of velocity and temperature as a function of radius for four phases of the oscillation for a model with $M = 1.00\ M_\odot$, $R = 251\ R_\odot$, and $T_{\text{eff}} = 2991$ K. Radiation pressure on dust is not included, even though a normal Mira with these parameters may be expected to form dust; this model is otherwise identical to the more realistic model shown in Fig. 11 below. The dot-dash line in the temperature plot shows the fit obtained from Eq. (28) with $C = 1.6$ and $T_0 = 770$ K at $r_0/R_* = 30$.

2. The temperature shows spikes where shocks pass through, but in the inner atmosphere the post-shock radiative cooling is rapid.
3. There is a region in the inner atmosphere where $T < T_{RE}$ during much of the pulsation cycle.
4. There is a region of elevated temperature, a "calorisphere" (see Willson and Bowen 1986).

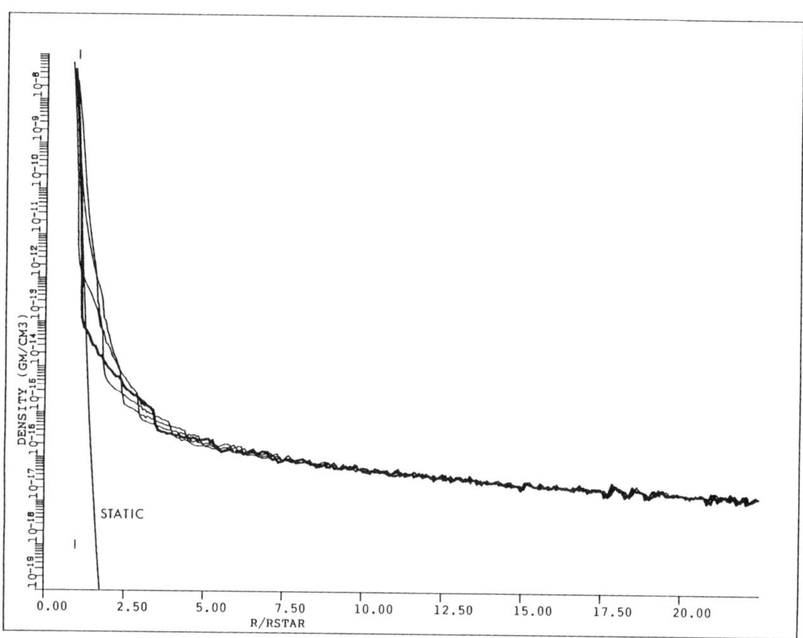

Figure 9.   Density as a function of radius for four phases in the same model as for Fig. 8.

5. The velocity is small (subsonic and $<< v_{\text{escape}}$) and its average is nearly constant through much of the calorisphere.
6. The density is elevated well above that of the corresponding static model, with a scale height that tends to be of the order of the radius in the calorisphere (consistent with expansion at constant v and hence $\rho \sim 1/r^2$).

One goal of an analytic theory of the atmospheres and winds of pulsating stars is to find an expression for the dependence of the mass loss rate on the properties of the star and its pulsation characteristics. The continuity equation tells us that (averaged over a sufficiently long time with respect to the oscillations) we should have

$$\dot{M} = 4\pi \langle r^2 \rho v \rangle = 4\pi r^2 \langle \rho \rangle \langle v \rangle \tag{7}$$

where we must take care to define $\langle \rho \rangle$ and $\langle v \rangle$ to retain this equality. (For example, if we define $\langle v \rangle$ as $[r(t_s + P) - r(t_s)]/P$, where t_s is the time of passage of a shock through the material, then the continuity equation with a given $\dot{M}$ defines $\langle \rho \rangle$.) This equation will be easiest to apply in a region of the atmosphere sufficiently far from the star that the velocity and density variations have achieved small amplitudes. But, in order to apply this far out in the atmosphere, we must determine some reference density ρ_0 (such as the photospheric density) and then determine ρ/ρ_0 and v at the position r/r_0 where we will apply Eq. (7). Thus, a critical step in deriving $\dot{M}$ is to determine the density structure in the atmosphere of the pulsating star.

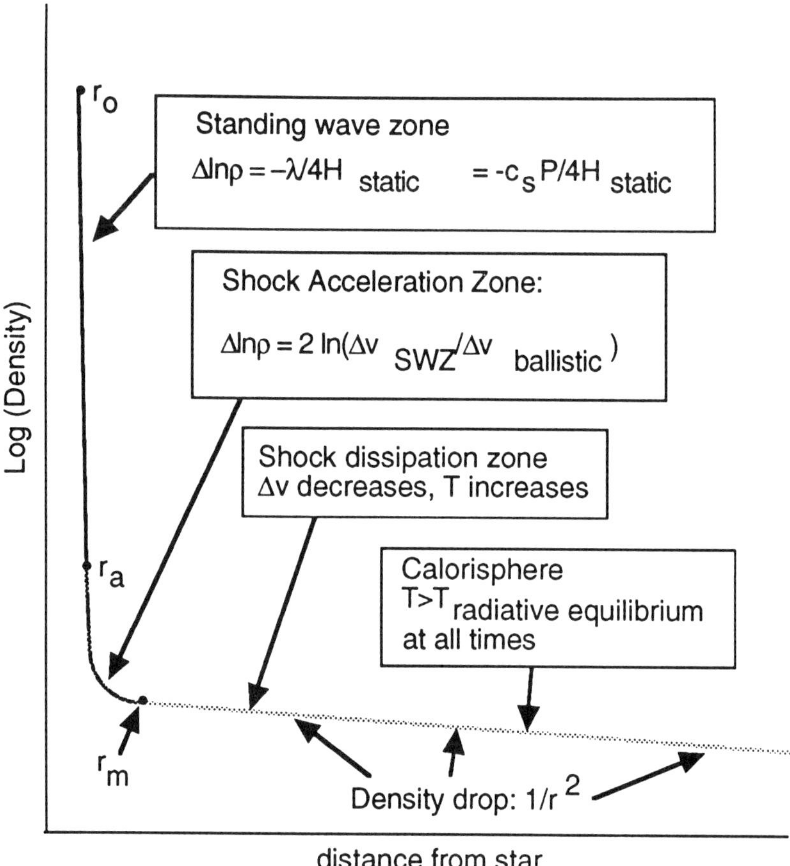

Figure 10.   Analytic model for velocity and density for a pulsating red giant without a radiatively driven (dusty) wind.

In our analytic model (Struck and Willson, in preparation) expressions are found for the density decreases across four regions: (a) the standing-wave zone, (b) the shock-acceleration zone, (c) the dissipation zone, and (d) the calorisphere (Fig. 10).

A. The Standing-Wave Zone

Inside the pulsating star the pulsation is generally well described as a standing wave, with amplitude increasing outwards. For pulsation with $P > P_{ac}$, the standing-wave region is analogous to the classical closed organ pipe, and the extent of this region is about $\lambda/2$, with $\lambda \sim c_s P$. For pulsation with $P < P_{ac}$ (overtone pulsation in long-period variables), the situation is probably closer to that of an open organ pipe with extent $\sim \lambda/4$; however, interior models have derived periods for this case based on the usual reflective surface boundary condition so there is an inconsistency between using published periods and assuming an open organ-pipe configuration. See, for example, the Pijpers (1993) computation of period shifts and strange modes that may occur in a star with a strong outflow or wind. His calculation is analytic and hence serves mainly as a warning of effects that may be overlooked when models are calculated using the traditional boundary conditions.

B. The Shock-Growth Zone

Near the photosphere, an outward-bound traveling wave develops. This traveling wave propagates down a roughly exponential density gradient (equal to the density gradient in the unperturbed atmosphere) and its amplitude grows. This growth is close to that determined from the classical problem of a single shock propagating into a static atmosphere (see, e.g., Zel'dovich and Raizer 1967, pp. 812ff, 859ff). If we take the flux of wave energy to be roughly constant in this region, and assume that the wave is propagating with a speed equal to the speed of sound, then

$$\rho \Delta v^2 c_s = \text{constant} \tag{8a}$$

or

$$\Delta v^2 \propto 1/\rho \tag{8b}$$

if c_s is constant.

Over a few scale heights, this wave will increase in amplitude from $\Delta v \sim c_s$ to the maximum Δv allowed by the periodicity condition (Sec. II). Including finite dissipation will increase the distance needed to build up the shock amplitude to the maximum value; however, this simple estimate comes close to describing the distance over which the shock amplitude grows in the Bowen models, even though the shocks are nearly isothermal (i.e., with rapid radiative cooling) in this region, so it is probably reasonable to neglect the dissipation in deriving an estimate for the extent of this zone.

C. Dissipation Zone

In a spherically extended atmosphere, once the shock reaches the maximum amplitude allowed by the periodicity condition $\mathcal{P}_0(\beta) = 38.3Q$ days, its amplitude must begin to decrease, because $Q(r) \sim r^{-3/2}$. However, without dissipation, Eq. (8) would still apply, and the amplitude would continue to increase. In the dissipation zone, the density and temperature structure adjust so that dissipation allows the shocks to satisfy the periodicity condition on the velocity. First, the density scale height increases, reducing the rate of growth of the shocks, and also the temperature structure adjusts, so that the amplitude can decrease (helped by the effect of increasing $T(r)$ on the pressure gradient and also by the increased rate of cooling at higher T). To visualize how the density adjusts, consider what happens when $\Delta v > \Delta v$ (periodicity), material is given too much outward momentum to be able to return to its initial radius after one period, so it is "levitated" to larger r. This process continues until the density gradient is sufficiently gentle to prevent rapid shock acceleration, and the density (and temperature) are high enough to allow sufficient dissipation of thermal energy after the shock.

D. Calorisphere

As the shock travels to larger r, where the density is smaller, the time that is required for the material compressed and heated by the passage of a shock to cool to the initial pre-shock temperature increases (Sec. IV). Thus, eventually the shock reaches a region where the density is low enough such that the cooling time for gas to reach the radiative equilibrium temperature exceeds the interval between shocks. Then, the minimum temperature reached between shocks adjusts upward until the temperature at a given r satisfies $T(r, t) = T(r, t + P)$. This region is the calorisphere. The cooling function for tenuous gas (chromospheric to interstellar densities) shows very rapid cooling at temperatures above about 8000 K by collisionally excited neutral H, but less rapid cooling below this. In the Bowen models, which incorporate this type of cooling, no gas remains long at temperatures greater than about 8000 K. The minimum temperature reached in the hottest part of the calorisphere tends to settle at $T_{min} < 8000$ K, and T_{min} is lower at larger radii, because the gas cools rapidly to $\sim$8000 K and then continues cooling below that value as adiabatic expansion tends to dominate over radiative exchange during the later parts of the pulsation cycle.

E. Supersonic Wind

Throughout the calorisphere the material is flowing outward at approximately constant speed, but at $v < c_s$. This is *not* the limiting flow that a steady supersonic wind achieves where $v_{esc}^2(r) \ll v^2(r)$: in the calorisphere, $v^2 \ll v_{esc}^2$. At some (large) radius at least one of the following conditions must break down: (a) $v =$ constant; (b) $v < c_T$; (c) $c_T < v_{esc}/2$. In the classical isothermal Parker wind, all three of these conditions are violated where $v = c_T = v_{esc}/2$; beyond that point, the flow is supersonic. However, calculations for the models of Figs. 8 and 9 carried out for enough cycles to show well-relaxed structure out to about 50 stellar radii still do not make a transition to a supersonic wind, but show continued flow with $v \approx$ constant $< c_T \ll v_{esc}$.

Figure 10 displays the analytic scheme described above. Equations for the density decrease across each of these regions will be found in Struck and

Willson (in preparation). The mass loss rate is specified by finding the density at r_m, the inner radius of the calorisphere. Above r_m, v = constant, thus by continuity $\rho \propto 1/r^2$, so knowing r_m, $\rho(r_m)$ and the velocity at any $r > r_m$) gives enough information to find the mass loss rate.

The above arguments suggest that the formation of a calorisphere is an expected consequence of pulsation, with the period and amplitude of pulsation (together with the parameters of the cooling function) determining the maximum temperature and the density of the calorisphere that is formed. One can suppress or eliminate the calorisphere by a general expansion of the atmosphere caused by another mechanism than the pulsation (and resultant $T(r)$) alone, for example, in the presence of a wind driven by radiative forces acting on grains in long-period variables. Or, one can do it by introducing very efficient coupling of the gas to the radiation field at all temperatures, through anomalous composition or very high densities; but such suppositions have readily observable consequences, and these consequences can generally be tested by observation. Thus, if observations establish the absence of a calorisphere in a pulsating star, this fact should be taken as suggesting that there is another effective mass loss mechanism operating in the star's atmosphere.

A false dichotomy is sometimes introduced into the study of chromospheric emission from pulsating stars. It is assumed that pulsation can, at most, provide temporary heating; therefore, if there is steady emission from heated gas, there must be a "chromosphere" heated by stochastic waves with their source in a convection zone, as perhaps describes the heating of the solar chromosphere. For example, Böhm-Vitense and Querci (1987) considered chromospheric emission for Cepheids; they noted that it appears stronger near the red edge of the instability strip, where the convection zone is expected to have more influence on the atmospheric structure, and interpreted this as evidence that the convection zone was the ultimate source of the chromospheric heating. However, the periods, amplitudes, and acoustic cutoff frequencies for these stars are also varying across the strip. Therefore, detailed calculation of expected atmospheric structure should be carried out for these stars before the hypothesis that the observed emission as simply a result of a pulsation-shock heated calorisphere can be eliminated; the observed trend is not sufficient. In fact, the presence of a calorisphere (= material consistently above the radiative equilibrium temperature) is a very general symptom of the presence of nonradiative energy transport and dissipation. Like a fever, the formation of a calorisphere is a very general response to a variety of "diseases."

An inspection of Fig. 10 and the discussion above shows clearly that the mass loss rates for pulsating stars are sensitive to the pulsation properties and, in particular, they are sensitive to the treatment of the thermal relaxation between shocks.

IV. THERMAL RELAXATION PROCESS IN THE ATMOSPHERE AND WIND

The characteristics of the calorisphere of a pulsating star, and the effects of pulsation on the spectrum of the star, depend not only on the character of the internally driven pulsation, but also on the details of the coupling between the radiation field and the matter in the stellar atmosphere. In this section, we will first present arguments justifying a commonly used formula for the thermal relaxation of the gas, relating the rate of change of temperature to radiative exchange and to the $p\mathrm{d}V$ work done on/by the gas. Then, we will examine the validity of the LTE approximation to the source function for these stars, and show that deviations from LTE are likely to be important for conditions appropriate to the calorispheres of pulsating stars.

A. Derivation of the Thermal Relaxation Equation

From the first law of thermodynamics, the rate of change of internal energy per unit volume ($\mathrm{d}U/\mathrm{d}t$) must be equal to the net rate of absorption of energy from the radiation field (Q) minus the rate at which work is done by the gas ($p\mathrm{d}V/\mathrm{d}t$) plus the net rate of acquisition of energy from other sources (e.g., from exchange with the kinetic and thermal energy of grains passing through the gas).

1. Interaction with the Radiation Field. Consider the case most often of interest, namely, well out in the atmosphere, at small optical depths, and with the photosphere subtending less than 2π solid angle. Let R be the photospheric radius, r the distance to the point of interest; then the angular radius of the photosphere as seen from r is $\sin^{-1}(R/r)$. Take the first moment $((1/4\pi)\int_{4\pi}\cdot\mathrm{d}\Omega)$ of the transfer equation in spherical coordinates (see, e.g., Mihalas 1978, p. 244) and integrate:

$$(1/r^2)\mathrm{d}(H_\nu r^2)/\mathrm{d}r = (-J_\nu + S_\nu)\chi_\nu \qquad (9)$$

$$J_\nu = I_\nu{}^{\mathrm{phot}}/2(1-\cos\theta) = I_\nu{}^{\mathrm{phot}}/4\{1-[1-(R/r)^2]\}^{-1/2} \to I_\nu{}^{\mathrm{phot}}/4(R/r)^2 \qquad (10)$$

for $R/r \ll 1$, and

$$H_\nu = I_\nu{}^{\mathrm{phot}}/4(1-\cos^2\theta) = I_\nu{}^{\mathrm{phot}}/4(R/r)^2/[1+(R/r)^2] \to I_\nu{}^{\mathrm{phot}}/4(R/r)^2 \qquad (11)$$

for $R/r \ll 1$.

Well out into the extended atmosphere, $K_\nu = H_\nu = J_\nu$ and the radiation field is strongly forward peaked. Deep in the photosphere, however, the Eddington factor, K/J, is $\sim 1/3$. Thus, one should carry out at least enough of a transfer computation to derive the Eddington factor at various depths. For the sake of the present argument, we consider only the case $H/J = K/J = 1$, as would apply fairly far from the star. Discussions of other cases, and of the use of a variable Eddington factor, may be found in Feuchtinger et al. (1993); a classic reference is Lucy (1971).

Locally, we can consider the transfer through each zone as a simple case of a beam incident on a slab of thickness Δx, optical depth $\tau_\nu = \chi_\nu \Delta x = \kappa_\nu \rho \Delta x$. (Mihalas uses $d\tau_\nu = \chi_\nu dx$ and sets $\chi_\nu = \kappa_\nu + \sigma_\nu$ to distinguish absorption and scattering. We use $\chi_\nu = \kappa_\nu \rho$ so that κ_ν is the opacity per unit mass. Our notation thus departs from Mihalas' at this point.) Assume that we have chosen the zones of the model such that for a single zone, it is appropriate to assign a constant κ_ν and representative values of ρ and T, and to ignore the change in radius across the zone. Each zone is treated locally as a plane-parallel slab; the global r dependence has already been incorporated in the relations between J_ν, H_ν and I_ν. Considering only transfer along r, where J_ν^{in} is J_ν at r and J_ν^{out} is J_ν at $r + \Delta r$, the transfer equation has the solution

$$J_\nu^{\text{out}} = J_\nu^{\text{in}}(e^{-\tau\nu}) + S_\nu(1 - e^{-\tau\nu}) \tag{12a}$$

or

$$J_\nu^{\text{out}} - J_\nu^{\text{in}} = -J_\nu^{\text{in}}(1 - e^{-\tau\nu}) + S_\nu(1 - e^{-\tau\nu}). \tag{12b}$$

But in radiative equilibrium $\int J_\nu^{\text{out}}{}_{RE}\, d\nu = \int J_\nu^{\text{in}}{}_{RE} d\nu$, so $\int J_\nu^{\text{in}}{}_{RE} d\nu = \int S_\nu^{RE} d\nu$ (all integrals taken from $\nu = 0$ to ∞). Therefore, if we use J_ν^{in} to calculate the source function for the gas if it were in radiative equilibrium, assuming that source function to be spherically symmetric, we can write

$$\int J_\nu^{\text{out}} d\nu - \int J_\nu^{\text{in}} d\nu = \int S_\nu^{RE}(1 - e^{-\tau\nu}) d\nu + \int S_\nu(1 - e^{-\tau\nu}) d\nu$$

$$= \int (S_\nu^{RE} - S_\nu)(1 - e^{-\tau\nu}) d\nu \tag{13}$$

and for a thin zone, with small $\tau_\nu = \kappa_\nu \rho \Delta x$ (where Δx is the thickness of the zone),

$$\int (J_\nu^{\text{out}} - J_\nu^{\text{in}}) d\nu = \int (S_\nu^{RE} - S_\nu)\tau_\nu d\nu = \int (S_\nu^{RE} - S_\nu)\kappa_\nu \rho d\nu \Delta x. \tag{14}$$

2. The "Adiabatic Expansion" Term. Assuming a small enough zone so that p, ρ, T may be chosen as constant across the zone, and taking the volume occupied by a mass element m to be ΔV gives $d(p\Delta V)/dt = (\rho/\mu m_H)kT d\Delta V/dt$. The internal energy density $\mathcal{U}$ is related to the temperature through an equation of state; for example, the mass element composed of an ideal gas with $\gamma = 5/3$ has $\mathcal{U} = 3/2NkT = 3/2(\rho/\mu m_H)kT$ and total internal energy $\mathcal{U}\Delta V$. Putting this into the first law of thermodynamics, using the ideal gas law for the equation of state, and considering as our volume the thin shell with $\Delta V = 4\pi r^2 \Delta r$ then gives

$$d(\mathcal{U}\Delta V)/dt = 3/2(\rho/\mu m_H)kdT/dt \Delta V. \tag{15}$$

3. An Expression for the Cooling Incorporating both Radiation and Adiabatic Expansion Terms. The first law of thermodynamics may now be written, making use of Eqs. (14) and (15):

$$3/2(\rho/\mu m_H)k\,\mathrm{d}T/\mathrm{d}t\,\Delta V =$$

$$\int (S_\nu{}^{RE} - S_\nu)\kappa_\nu\rho\,\mathrm{d}\nu\,\Delta V - (\rho/\mu m_H)kT\mathrm{d}\Delta V/\mathrm{d}t. \tag{16}$$

Next, we divide through by $\rho\Delta V$ to get $3/2(k/\mu m_H)\mathrm{d}T/\mathrm{d}t = \int(S_\nu{}^{RE} - S_\nu)\kappa_\nu\mathrm{d}\nu - (k/\mu m_H)T(1/\Delta V)\mathrm{d}\Delta V/\mathrm{d}t$ and use $1/\rho\mathrm{d}\rho/\mathrm{d}r = -(1/\Delta V)\mathrm{d}\Delta V/\mathrm{d}t$ to get

$$3/2(k/\mu m_H)\mathrm{d}T/\mathrm{d}t = \int (S_\nu{}^{RE} - S_\nu)\kappa_\nu\mathrm{d}\nu + (k/\mu m_H)T(1/\rho)\mathrm{d}\rho/\mathrm{d}t. \tag{17}$$

This expression, a similar one for $K/J = 1/3$, or one with variable Eddington factor, are the expressions for cooling that are most often incorporated into dynamical models. There are two very important points to notice about Eq. (17):

Point one: Under many circumstances of interest in the atmospheres of pulsating stars, the second term (representing adiabatic cooling/heating in expansion/compression) will dominate over the first term, particularly at low densities. This term plays an important role whenever there are shocks present; the gas is compressed over a short time interval and then expands over the remainder of the period to satisfy the periodicity condition on the density. Shocks are nearly inevitable in a pulsating star (see Sec. III). When $|\int(S_\nu{}^{RE} - S_\nu)\kappa_\nu\mathrm{d}\nu| \ll |(k/\mu m_H)T(1/\rho)\mathrm{d}\rho/\mathrm{d}t|$, we can say that the process is effectively adiabatic. This applies generally at low densities. In situations where the radiative cooling is rapid (for example, where $S = \sigma T^4$ is used), then as the density decreases during the later part of the intershock expansion the adiabatic cooling term can lead to temperatures well below radiative equilibrium temperatures. This effect is seen in the Bowen models near the innermost shock at some phases of the pulsation, and also shows up dramatically in the RR Lyrae models calculated by Hill (1972) and Fokin (1992); see Sec. VI below.

Point two: In LTE,

$$\kappa(J - S) = \kappa(\sigma T_{RE}{}^4 - \sigma T^4) \tag{18}$$

which is independent of density (except for any density dependence in the flux-averaged opacity per gram, κ).

This LTE relation has been used in many models. It leads to very rapid post-shock cooling from $T \gg T_{RE}$ to $T \sim T_{RE}$, but it fails to apply at low densities, where departures from LTE are expected. In other words, strongly density dependent cooling rates are a phenomenon of non-LTE. The above is well known in the case of the interstellar medium (see, e.g., Osterbrock 1974) but is often overlooked for dynamic stellar atmospheres.

B. Under What Conditions is the Emission per Unit Mass Independent of the Density?

Let us now consider some examples that will allow us to estimate the densities below which the non-LTE details of the thermal relaxation process are likely to be important. The emission per unit volume, η_v in Mihalas' (1978) notation, is given by Eq. (19) (his equation 7-2):

$$\eta_v = (2h v^3 / c^2) \left[\sum_i \sum_{j>i} n_j (gi/gj) a_{ij}(v) + \sum_i n_i{}^* a_{i\kappa}(v) e^{-hv/kT} \right.$$

$$\left. + \sum_\kappa n_e n_\kappa a_{\kappa\kappa}(v, T) e^{-hv/kT} \right]. \tag{19}$$

The first sum in Eq. (19) is over transitions from one level (i) to another (j) within an atom; n_i is the number density in level i, g_i is the degeneracy in level i, and a_{ij} is the probability of a transition from i to j as a function of frequency v. The second sum is over transitions from the continuum κ to levels i, i.e., radiative recombinations. The third sum incorporates the free-free transitions.

The leading term in each sum is the number density of particles in the particular state. The rate of change of the population of level j depends on all the processes leading into and out of it:

$$\mathrm{d}n_i/\mathrm{d}t = \sum_j n_j (C_{ji} + R_{ji}) - n_i \sum_j (C_{ij} + R_{ij}) + \sum_\kappa n_\kappa (C_{\kappa i} + R_{\kappa i}) -$$

$$n_i \sum_\kappa (C_{i\kappa} + R_{i\kappa}) \tag{20}$$

where the first sum represents the rate of collisional (C_{ji}) and radiative (R_{ji}) excitations and de-excitations feeding into state i and the second term represents such leading out of state i; the next two represent ionizations and recombinations similarly. In statistical equilibrium, these populations are determined by the condition that the total number of transitions leading into a given state must be equal to the total of those leading out of it, so that $\mathrm{d}n_i/\mathrm{d}t = 0$. Clearly, solving for the level populations can only be done numerically when the number of levels is large. However, without solving this in the general case, we can determine some general properties of the solution.

At a given temperature $\sum n_j$ is proportional to the density. Thus if we divide η_v by the density, we get an expression for the emissivity per unit mass ϵ_v whose dependence on density comes from the dependence of the relative level populations $n_j / \sum n_j$ on the density.

If the time scale for achieving statistical equilibrium is short compared to other time scales in the problem, then the populations n_i are determined by setting all $\mathrm{d}n_i/\mathrm{d}t = 0$ and solving the resulting set of coupled equations. For the simple but illustrative case of a two-level atom,

$$n_i/n_j = (C_{ji} + R_{ji})/(C_{ij} + R_{ij}) \tag{21}$$

and $C_{ji}/C_{ij} = g_j/g_i e^{-(E_j-E_i)/kT}$ so that the usual Boltzmann distribution is recovered when collisions dominate in both directions.

If collisions dominate both sums in the statistical equilibrium equations, then quite generally (not just for the two-level atom) the populations are in LTE and the source function may be assumed to be the Planck function. But if, for example, the first sum in Eq. (20) is dominated by the collisional excitations and the second sum by radiative de-excitations, then we should expect the source function to depart significantly from the Planck function.

Consider again the general case given by Eqs. (20) and (21). At relatively low temperatures (when H is not excited) the collisional processes will be dominated by transitions with $\Delta E \sim kT \sim 1$ eV. From the explicit expressions for the collisional and radiative rates, we find that there is typically a critical electron density $n_e{}^*$ below which radiative de-excitations dominate over collisional ones for a given pair of levels. (Electron density is the usual critical parameter, because electrons are generally much more effective at exciting transitions than are the other particles in the gas, although at the low temperatures present in the coolest long-period variables, collisions with neutral H may become more important; see, e.g., Dalgarno and McCray 1972; Sutherland and Dopita 1993.)

Explicit approximate expressions for C_{ul} and C_{lu}, R_{ul} and R_{lu} for excitation/de-excitation of an atomic line are given by Jefferies (1968, Sec. 6.2). Equation (21) gives the case where the velocity distribution of the electrons is Maxwellian with $T = T_{\mathrm{kin}}$ and the radiation field is (in the wavelength region of interest) approximately given by $J_\nu = W B_\nu(T_{\mathrm{rad}})$ (W being a geometric dilution factor, equal to the solid angle subtended by the source $/4\pi$) as follows:

$$C_{lu} = 2.16(h_\nu/kT_{\mathrm{kin}})^{-1.68} e^{-(h\nu/kT_{\mathrm{kin}})} T_{\mathrm{kin}}{}^{-1.5}(gf)n_e$$

$$C_{ul} = e^{h\nu/kT_{\mathrm{kin}}} C_{lu} = 2.16(h\nu/kT_{\mathrm{kin}})^{-1.68} T_{\mathrm{kin}}{}^{-1.5}(gf)n_e$$

$$R_{ul} = g_u A_{ul}[1 + W/(e^{h\nu/kT_{\mathrm{rad}}} - 1)]$$

$$R_{lu} = g_u A_{ul}[W/(e^{h\nu/kT_{\mathrm{rad}}} - 1)]. \tag{22}$$

In what follows we neglect stimulated emission and assume that the collisional processes important in cooling the gas will be dominated by transitions for which $h\nu/kT_{\mathrm{kin}}$ is of the order of 1, or λ of the order of λ_{max} in the Planck function for $T = T_{\mathrm{kin}}$. Typically, we need to consider the cooling effects of collisional excitation followed by radiative de-excitation whenever $C_{ul} \ll R_{ul}$ for those individual transitions that are dominant in the cooling term; from the above, this will be true for all such transitions when $n_e < n_e{}^*$, where

$$n_e{}^* = [2.16 T_{\mathrm{kin}}{}^{-1.5}(gf/g_u A_{ul})]^{-1} \tag{23}$$

or, as $gf/gA = 1.5 \times 10^{-8}\lambda_\mu{}^2$ (λ_μ = wavelength in microns),

$$n_e{}^* = [3.3 \times 10^{-8} T_{\mathrm{kin}}{}^{-1.5}\lambda_\mu{}^2]^{-1}. \tag{24}$$

Because we have already assumed $h\nu/kT_{\text{kin}}$ is about 1 for those transitions that will dominate the cooling, we can substitute $hc/kT = (1.44 \times 10^4/T_{\text{kin}})$ microns for λ_μ to get

$$n_e^* \simeq 1.1 T_{\text{kin}}^{3.5}. \tag{25}$$

At 3000 K, n_e^* is about 1.6×10^{12} cm^{-3} and $n_e/n_H \lesssim 10^{-4}$ so nonequilibrium processes (and a density dependent ϵ_ν) are to be expected for number densities below about 1.6×10^{16} cm^{-3} (densities below about 3×10^{-8} g cm^{-3}) and may be important also above that density. At 10,000 K, n_e^* is about 1.1×10^{14} cm^{-3} and $n_e \sim n_H$, so we expect a density-dependent ϵ_ν for densities below about 1.8×10^{-10} g cm^{-3}. Thus, at the densities characteristic of stellar chromospheres and coronae the emissivity per unit mass is expected to be density dependent, with (to first order) $\epsilon_\nu \propto n_e$. A similar conclusion, that non-LTE is the rule rather than the exception in stellar atmospheres, was reached by Mihalas and Mihalas (1984, p. 389).

At very low temperatures the ionization fraction becomes very low, and collisions with neutral hydrogen become more important than collisions with electrons (see, e.g., Dalgarno and McCray 1972). Under those conditions one can derive an n_H^* in a manner fully analogous to the above, and with results also of about the same order of magnitude for the densities below which non-LTE effects become important. For $n_e < n_e^*$ we expect the emission per unit mass to be density dependent because collisional excitations ($\sim n_e n_H$ for the dominant ions of each species) are followed by radiative de-excitations ($\sim n_H$ for the dominant ions), thus removing kinetic energy from the gas and converting it to radiation. According to the arguments presented in Sec. III, a natural consequence of such a density-dependent radiative interaction term is the formation of a calorisphere in the upper atmosphere of a pulsating star, unless the stellar atmosphere is also experiencing rapid global expansion in a radiatively driven wind.

V. APPLICABILITY OF THE PARKER WIND FORMULATION TO THE WINDS OF PULSATING STARS

Many, if not most, studies of stellar winds adopt the steady wind flow assumption, derive the appropriate form of the "wind equation" (incorporating mass conservation, the force equation, and some assumption(s) concerning the energy balance), and then seek to fit a subsonic-to-supersonic solution passing through the critical point(s) of the solution. Can this approach help us to understand the results of the calculations of wind generation for pulsating stars? We first consider the application of the classical thermally driven wind equation to the mean flow in the calorisphere. We then consider the application of a steady-wind equation to the case with substantial radiative driving of a wind. Finally, we consider some examples where theorems derived from analyses of the steady-state wind equation provide some useful qualitative insights.

The classic Parker wind equation comes from the equation of continuity, the momentum equation, and some specification of the thermal properties of the wind (see Parker 1960a,b,1969). One form of this equation is:

$$(1/v)\mathrm{d}v/\mathrm{d}r = \frac{2}{r}\frac{(A(r) - B(r))}{(v^2 - c_T^2)} \qquad (27\mathrm{a})$$

where

$$A(r) = c_T{}^2\left(1 - \frac{1\,\mathrm{d}\ln c_T^2}{2\,\mathrm{d}\ln r}\right) \qquad (27\mathrm{b})$$

for a spherically symmetric wind, and

$$B(r) = \frac{v_{\mathrm{esc}}^2(1 - \Gamma)}{4} \qquad (27\mathrm{c})$$

where Γ = ratio of outward acceleration due to radiation pressure (etc.) to the acceleration of gravity g and the other symbols are as defined in earlier sections (Willson and Bowen 1985).

In this form, this equation is not generally solvable directly, because the temperature distribution (and c_T) will have some dependence on $v(r)$. As Parker (1960a) showed, when $v_{\mathrm{esc}} \gg c_s$, assuming the expansion is adiabatic with $\gamma > 1.5$ leads to a wind equation with no critical point: if $v > c_s$, then $\mathrm{d}v/\mathrm{d}r$ is >0 and if $v < c_s$ then $\mathrm{d}v/\mathrm{d}r < 0$ for all r in that case. Assuming T = constant and $\Gamma = 0$, Eq. (27) gives the original Parker isothermal wind solution, with the velocity increasing through a critical point where $v = c_T = v_{\mathrm{esc}}/2$.

A. Application of the Steady, Thermally Driven Wind Equation to the Mean Flow in the Calorisphere

In the calorisphere of a typical Bowen model, the temperature and density structure adjust in such a way as to give $\langle v \rangle$ = constant $< c_T$. Is this region well described by the equations governing a steady expansion? We argue below that the steady-wind equation derived above is able to reproduce the constant velocity of outflow given the temperature distribution that is found in that part of the model. We interpret this as meaning that the propagating shock waves dissipate their energy in this region so as to produce a temperature gradient that maintains the constant outflow velocity in the wind.

In the calorisphere, $v < c_T$ and $\mathrm{d}\langle v \rangle/\mathrm{d}r = 0$. If the mean flow is described well by the steady-wind equation (Eq. 27), then in this region $A(r) = B(r)$. To see how well this works, set $A = B$ and solve for $T(r)$:

$$T(r)/T(r_{\mathrm{o}}) = (r_{\mathrm{o}}/3H_{\mathrm{o}})(r/r_{\mathrm{o}})^{-1} + [1 - (r_{\mathrm{o}}/3H_{\mathrm{o}})](r/r_{\mathrm{o}})^2 = C/x - (C-1)x^2 \qquad (28)$$

where $x = r/r_{\mathrm{o}}$ and $C = r_{\mathrm{o}}/3H_{\mathrm{o}}$ with H_{o} the static isothermal scale height at the fitting point r_{o}. In Fig. 8 we show $T(r)$ from Eq. (28) with $r_{\mathrm{o}}/R_* = 30$ and $C = 1.6$, and $T_{\mathrm{o}} = 770$ K. At 30 R^*, the model has $\langle T \rangle \sim 770$ and $v_e{}^2/6\langle C_T^2 \rangle \sim 1.3$.

The C that best fits the model's $T(r)$ curve and the C derived from the model's $v_e{}^2/6\langle C_T^2\rangle$ at r_0 approach each other as r_0/R_* are increased in this example, but beyond about 30 R_* the numerical model (even at >300 full amplitude cycles) is not yet fully relaxed, so we cannot push the fitting procedure beyond this without much more extended modeling. Equation (28) does, however, appear to reproduce the general trend in $T(r)$ rather well, suggesting that the wind is reasonably well described as a thermally driven steady subsonic flow at constant velocity.

Through the region of decreasing T and essentially constant v, the contributions to the pressure gradient force from $(T\,d\rho/dr)$ and $(\rho\,dT/dr)$ are comparable, so both the magnitude and the gradient of the temperature are important. Note that $(g + 1/\rho\ dP/dr)\lesssim0.01$ g throughout this part of the atmosphere; thus, the constancy of v reflects a nearly perfect balance between gravity and the pressure gradient forces. Our interpretation of this result is that the shocks in this region maintain the temperature profile, and the temperature gradient ensures that the pressure gradient is able to maintain the constant velocity outflow.

In the inner parts of the calorisphere, near and inside the maximum in $T(r)$, the temperature in the models is not high enough, and the gradient not steep enough, to be the main factor determining the mean outflow that occurs in these models; here, momentum injection by the waves must also be important. The mass loss rate is mostly established in the region where momentum transfer from the waves to the mean flow is still significant, so the steady-state wind equation for a thermally driven wind or breeze is not useful for estimating mass loss rates.

In Bowen's Mira models that include radiative forces on dust, the velocity becomes exclusively positive outward in a region where the shocks have large, supersonic, amplitudes (Fig. 11). Such a complicated velocity field cannot easily be averaged in such a way as to get a steady wind with a well-defined sonic point that will match essential features of the model. Some concepts from the general study of stellar winds do have descriptive, qualitative use in this context, however. Thus, for example, Leer and Holzer (1980; see, also, Holzer and MacGregor 1984; Holzer 1987) have shown that adding energy to the flow below the critical point in a Parker-type steady flow will increase the mass loss rate, while adding it above the flow will increase the limiting velocity without affecting the mass loss rate. In the Bowen models with dust, the time-averaged flow becomes supersonic in about the same region where the dust forms, so much of the energy and momentum added to the gas through its coupling with the dust is incorporated into the flow after this point, altering the wind speed. Also, incorporating radiation pressure on molecules, which is more important deeper in the atmosphere before the mean flow exceeds the speed of sound, tends to increase the scale height in that region, increase the density farther out, and thus have a large effect on the mass loss rate (without much effect on the outflow velocity).

Thus, the classical wind equation has limited applicability—it does not

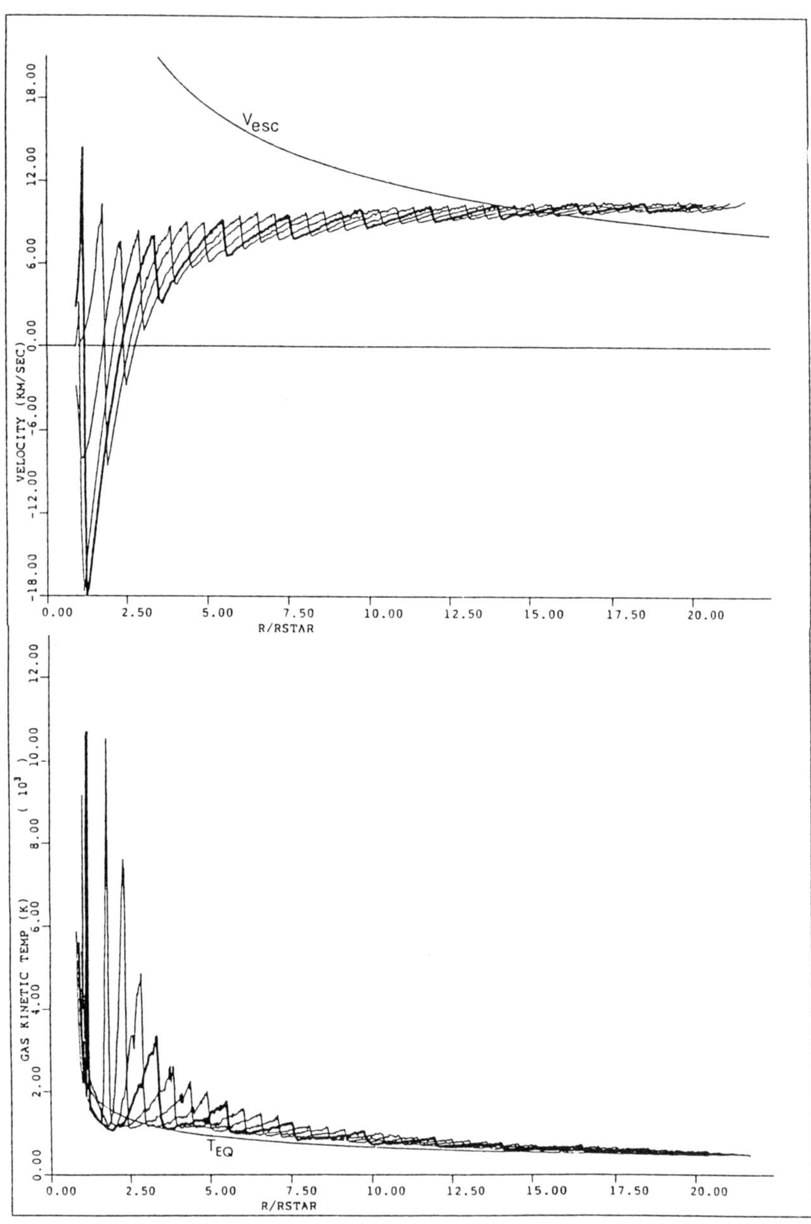

Figure 11.   Plots of velocity and temperature as a function of radius for four phases
of the oscillation for a model with the same parameters as the one in Figs. 8 and 9,
but including also radiation pressure on dust.

provide a useful means of carrying out analytic estimates—in the case of nonlinear periodic oscillations in the regions where the mass loss rates are established in pulsating stars. Still, the theorems of the classical steady-wind theory may provide some assistance in describing qualitative features expected in these atmospheres, as well as a reasonable condition relating the flow to the thermal structure in the outer calorisphere and wind regions. Observers should note that an extended region with the average outflow velocity independent of r may be present throughout much of the atmosphere, giving rise to blue-shifted spectral features but with an outflow velocity less than the speed of sound and much less than the local escape speed.

VI. PUBLISHED AND PENDING SIMULATIONS OF PULSATING STARS

This section consists mainly of a tabulated summary of key properties of numerical models presently in the literature or about to be published. Only models for mono-periodic stars are referenced in this section. Analytic models and multi-periodic calculations are referenced in Sec. II and discussed in more detail in the chapter by Cuntz and Dorfi.

Particularly interesting illustrations of the theoretical considerations presented above include the RR Lyrae models of Hill (1972) and Fokin (1992, 1994); also, the models for long period variables calculated by Wood (1979), Hill and Willson (1979), Willson and Hill (1979), and Bowen (1988).

Radius versus time plots for RR Lyrae models by Hill (1972; plane parallel) and Fokin (1992; spherical) are reproduced in Figs. 12 and 13. Both display aperiodicity, although it is more pronounced in the Fokin models. Both models used $S = \sigma T^4$, i.e., LTE for the radiative cooling term. The temperature structure of the Fokin models (Fig. 14) illustrates the rapid return to $T \sim T_{RE}$ until the adiabatic term becomes greater than the radiative exchange term (in Eq. 18), and the subsequent "refrigeration" phase becomes dramatic.

The series of models by Wood (1979; see Figs. 15, 16 and 17 below), Hill (Hill and Willson 1979; Willson and Hill 1979) and Bowen (1988; his Figs. 8 and 9) for long-period variables *without* radiative acceleration of dust also illustrate quite dramatically the importance of the treatment of thermal relaxation. Wood (see Fig. 15) calculated a model in which the shocks were assumed to be adiabatic throughout the atmosphere; the mass loss rate he estimated from this model was 0.02 $M_\odot$ yr^{-1}. In contrast, his model with isothermal shocks had a net mass loss rate less than about 10^{-12} $M_\odot$ yr^{-1} (Fig. 16). Both of Wood's models were for Q values appropriate for overtone pulsation, and both were driven by large amplitude pressure variations at a velocity node of the internal pulsation as calculated by a linear adiabatic analysis with a zero-pressure surface boundary condition (Fox and Wood 1982). Hill and Willson (1979; Willson and Hill 1979) presented isothermal shock models with similar results ($\dot{M} \sim 10^{-14}$ $M_\odot$ yr^{-1}). The overwhelming mass loss rates found in simulations assuming adiabatic shocks, and the minimal

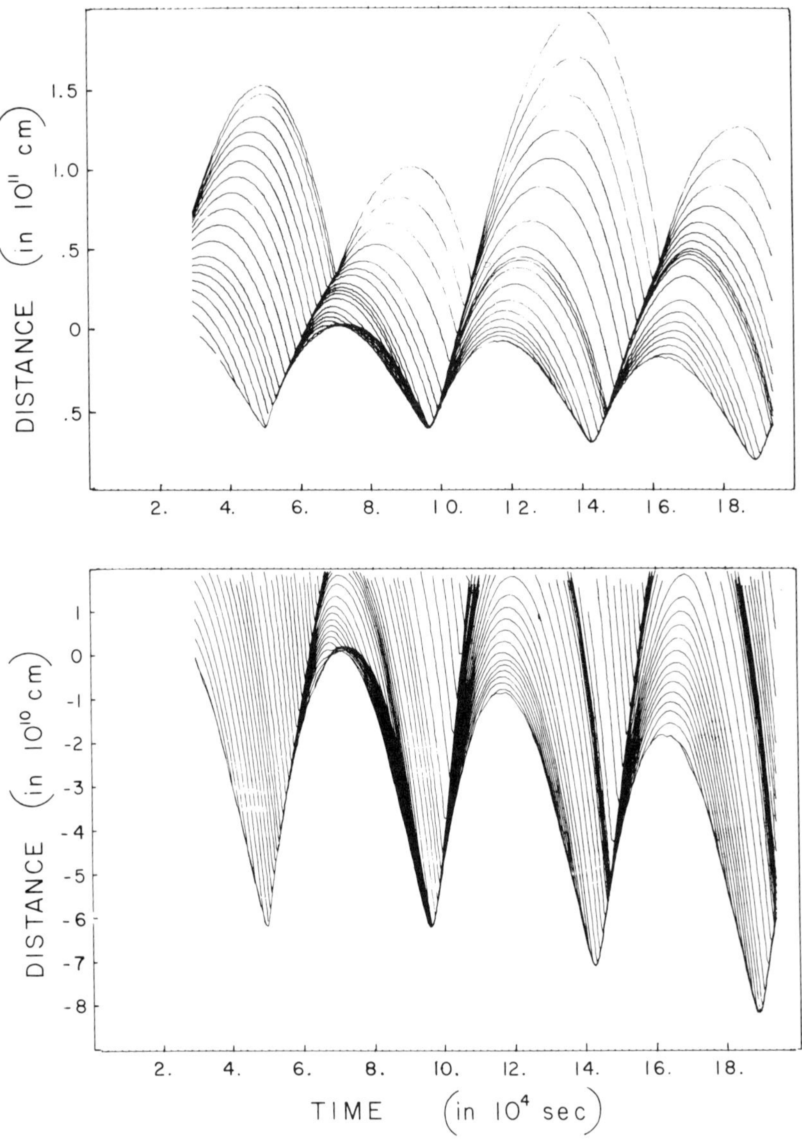

Figure 12.   Radial position of mass shells for an RR Lyrae model with g assumed constant. *Top panel:* Selected mass shells through the range of the entire model. *Bottom panel:* All mass shells nearest the piston and photosphere (figure from Hill 1972).

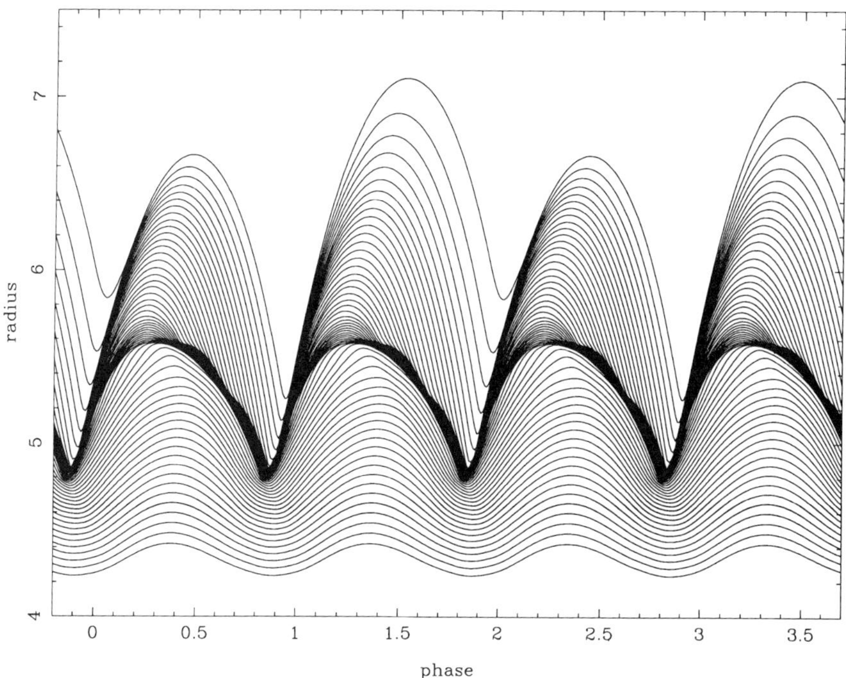

Figure 13.   Radial displacements in solar units of the outer mass zones of an RR Lyrae model in spherical symmetry and with cooling according to $S = \sigma T^4$ (figure from Fokin 1992).

mass loss rates from those assuming isothermal shocks, have been confirmed by Bowen with his code and a range of stellar parameters. Even with a second noncommensurate period added to simulate the effects of irregularities in the pulsation in the hopes of stimulating transients that could eject mass, Hill and Willson found that the mass loss rate did not exceed 10^{-12} $M_\odot$ yr^{-1} in models with isothermal shocks. However, a model in which the shocks were switched from isothermal to adiabatic at the position in the atmosphere where the recombination length $\Delta x = v_{\text{shock}}/\alpha n_e$ (where α = recombination coefficient) became of the same order as H_{static} showing a mass loss rate $\sim 10^{-5}$ $M_\odot$ yr^{-1} from pulsation alone (Willson and Hill 1979)—an early clue that the density dependence of the cooling rate would prove to be of critical importance.

Another feature of isothermal shock models is their tendency to aperiodic (possibly chaotic) behavior. This is evident in the RR Lyrae models of Hill and even more pronounced in the models by Fokin. Wood (1979) displayed the "shock pairing" that occurred in his models, particularly evident when they were very strongly driven (Fig. 17). Such aperiodicities do not require isothermal shock conditions, but are more pronounced when these are applied. Thus, a spectroscopic feature formed high in the atmosphere of a pulsating star may show variability at $2P$ or $3P$. A similar instability

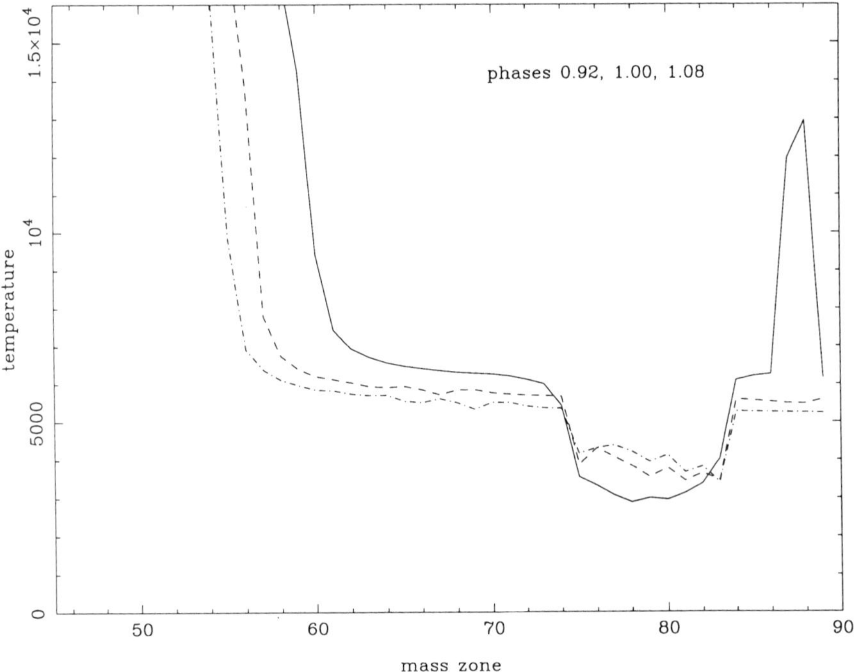

Figure 14.   Temperature versus shell number at three close phases for the RR Lyrae model shown in Fig. 13. The rapid relaxation brings T to T_{RE} rapidly; near shell 80, the adiabatic expansion is dominant over the radiative exchange term where the optical depth is low, giving rise to a dip in the temperature. A spike due to nearly adiabatic compression is also seen at one phase (figure from Fokin 1992).

may cause the characteristic double maxima of the RV Tauri stars and some long-period variables.

When the effects of radiation pressure on dust are included in the models for cool long-period variables, the overall velocity and temperature structure change dramatically. Wood (1979) calculated a model starting from a static flow and found that while shocks developed in the inner atmosphere, at large radii the flow oscillated around a solution similar to the static flow. This model had isothermal shocks and was for an overtone case. Bowen (1988; see, also, Bowen and Willson 1991) has calculated extensive grids of long-period variable models with dust, in most cases assuming fundamental mode pulsation; all of these included density- and temperature-dependent relaxation, and have calculated the radiation pressure on dust as described below. When the values of M, P, R and $T_{\rm eff}$ for the models are chosen to be consistent with the Ostlie and Cox (1986) period-mass-radius relation and the Iben (1984) radius-luminosity-mass relation (with reasonable choices of parameters such as the mixing length/scale height) for evolving asymptotic giant branch (AGB) stars, the mass loss rate for an evolutionary sequence of

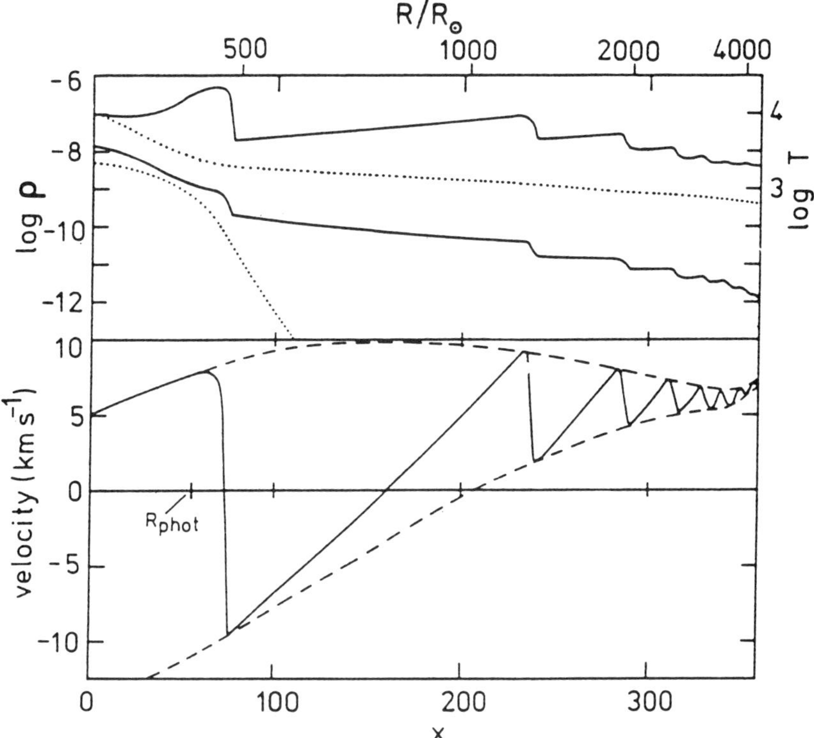

Figure 15.   Temperature, density and velocity profiles at a single phase for a long-period variable star model with adiabatic shocks. Dotted lines show temperature and density distributions in the initial hydrostatic shocks. Dashed lines mark an envelope for the velocity curves (figure from Wood 1979).

models increases as an approximately exponential function of time, from very small values to values so large that mass loss becomes the totally dominant evolutionary process. The outflowing wind, then optically thick with dust and showing all the characteristics expected of a "superwind," rather quickly strips away the stellar envelope and ends the star's life on the AGB (Fig. 18).

A question that naturally arises is: which stellar parameter controls the onset of this superwind? The answer is not simple because all of the parameters are changing in interrelated ways. As the star evolves to higher L, T_{eff} decreases very gradually, R increases, and ultimately the stellar mass M decreases. L/M affects a_{rad}/g from the dust, and for nearly constant T_{eff} the atmospheric scale height $H \sim R^2$, so R, L and M all play important roles in the development of the superwind. Assuming an evolutionary radius-luminosity-mass relation yields a superwind onset luminosity that depends on the mass of the star, and once the superwind starts the decreasing stellar mass helps accelerate the rate at which mass is lost. *Because these factors are*

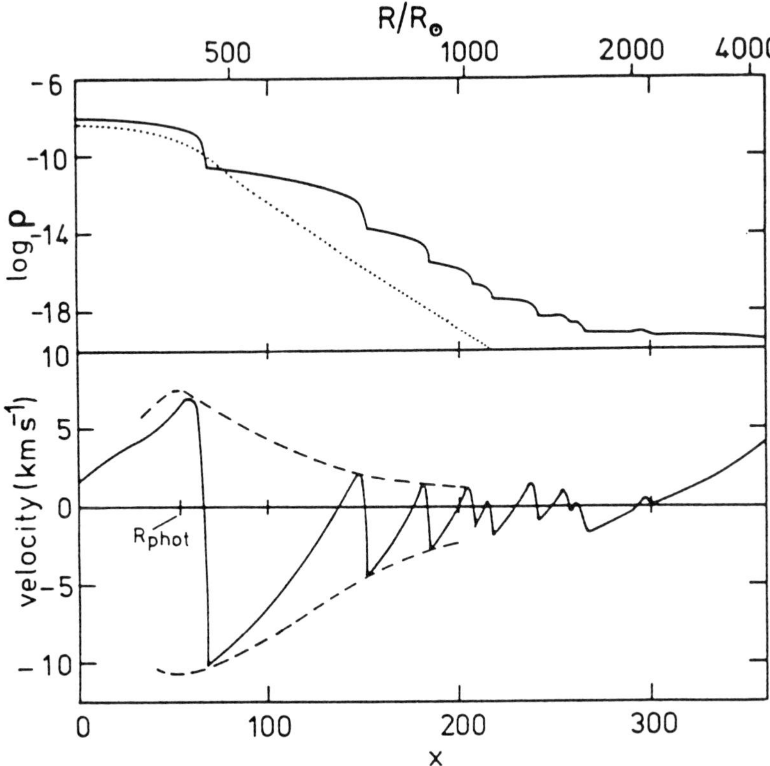

Figure 16.   Temperature, density and velocity profiles at a single phase for the
same stellar parameters as from Fig. 15, but with the shocks here assumed to be
isothermal. The positive velocity at the outer boundary was caused by the passage
of a shock, and thus does not represent the mean mass flow. The mean mass loss
rate was of the ordr of 10^{-12} M$_\odot$ yr^{-1} (figure from Wood 1979).

*so intricately linked, we have refrained from publishing mass loss formulae
based on analytic fits to results published by Bowen (1988), and would caution
the reader against placing much faith in such laws derived from those or any
other published models.* When the roles played by the various factors and
parameters of the models are fully understood, and more extensive grids of
models have been computed, then it will be possible to create a mass loss
formula that can be applied with confidence in evolutionary calculations, as
well as for the interpretation of observations. One lesson that is already
clear, however, is that the mass loss rate increases much more steeply as a
function of stellar parameters near the tip of the AGB than does the widely
used Reimers' relation.

Bowen's (1988; Bowen and Willson 1991) models assume that the cross
section for radiation pressure on dust is a simple function of the local blackbody equilibrium temperature T_{RE}, because the grain temperature at very low

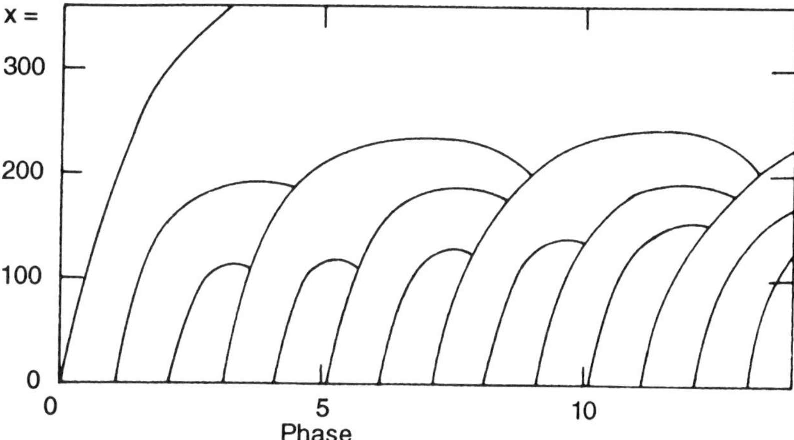

Figure 17.   The "pairing" of shocks in the isothermal model of Fig. 16, from Wood (1979), but with the axes reversed from the original to correspond to the orientation of our Figs. 6 and 7 above.

gas densities is determined primarily by the radiation field, not the gas kinetic temperature. Specifically, the cross section per unit mass in the atmosphere is taken to be

$$\text{dust } \kappa = \kappa_{\max}/\{1 + \exp[(T_{RE} - T_{\text{cond}})/T_{\text{width}}]\} \tag{29}$$

where T_{cond} is the midpoint of the dust grain growth range, typically assumed to be $\sim$1350 K, and T_{width} is a width parameter for this temperature function, typically about 100 K (Bowen 1988). (With this choice of T_{cond}, the dust temperature is expected to be $\lesssim$1000 K and the dust to form at distances from these stars consistent with a variety of recent observational constraints (see, e.g., Bowers 1992; Danchi et al. 1990; Draine 1981; Harvey et al. 1991). Similar prescriptions were used by Wood (1979), and also by Feuchtinger et al. (1993) to construct a test case for comparison with Bowen's models. However, Feuchtinger et al. used $S = \sigma T^4$ and thus their cooling was much more abrupt and their thermal structure quite different from that in Bowen's models.

Sedlmayr, Gail, and their collaborators have been exploring much more detailed dust nucleation schemes, with particular emphasis on the growth of carbon grains in extreme carbon stars (Fleischer et al. 1992; Dominik et al. 1993; also Dorfi and Höfner 1991). Their models show some instabilities that may help explain the episodic outbursts of the R CrB stars (Goeres and Sedlmayr 1992). The factor that "drives" the instability is the density dependence of the dust nucleation that their models include. However, their models do not yet incorporate the essentials of non-LTE cooling; they assume isothermal shocks, giving very large density enhancements and low post-shock temperatures, both of which enhance the rate of dust formation. With these features, their models even include some that showed the development

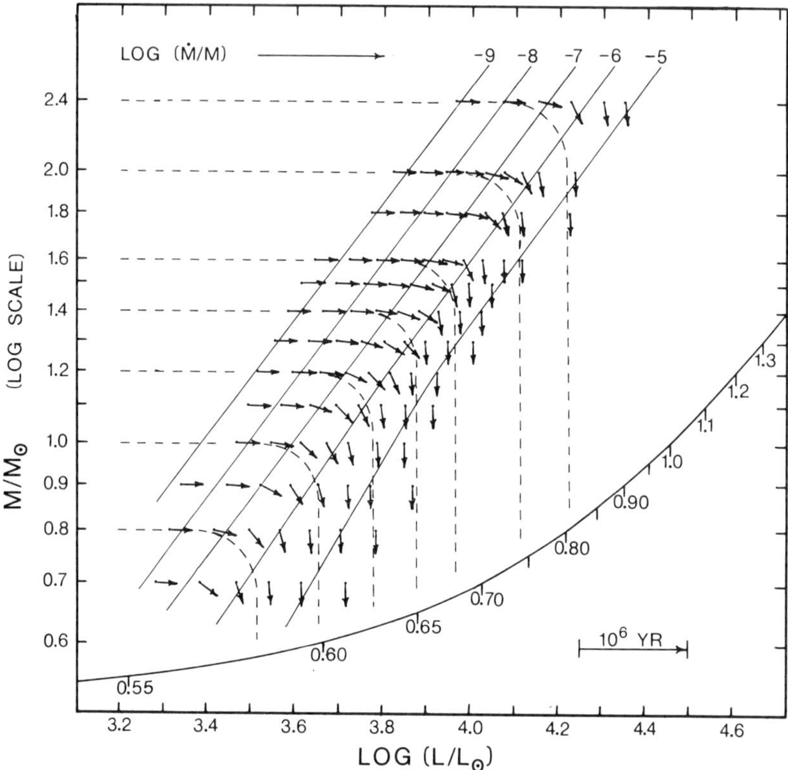

Figure 18.   Mass loss and evolution of stars near the tip of the asymptotic giant branch. Pulsation plus the effects of radiation pressure on dust combine to drive mass loss that increases rapidly as a star evolves up the asymptotic giant branch (AGB), with L increasing, T_{eff} decreasing slightly, and R increasing on time scales of $\sim 10^6$ yr. The solid lines show the approximate loci of lines of models with mass loss rates $= 10^{-9}$ to 10^{-5} $M_\odot$ yr^{-1}. The dashed tracks show approximate evolutionary paths for several representative masses.

of periodic atmospheric shocks without underlying pulsation; the periods selected by the self-exciting models scale as the acoustic cutoff periods for these atmospheres. Note, however, that in addition to the simplified thermal physics, they consider mainly very large C/O (C/O = 1.5 to >3) or a carbon excess (C/O−1)~0.5 to 2, while most observed carbon star C/O have carbon excesses (C/O−1)<0.1. Together, all their assumptions lead to exaggeration of the effects of carbon grains. Thus, their results so far have limited applicability for normal or pulsating carbon stars, and because silicate grains have quite different optical properties, the applicability to the M-type Miras is even more limited.

   In addition to more explicit inclusion of dust formation chemistry in the models of cool pulsating stars, several other promising developments in the modeling of winds from pulsating stars are beginning to appear in print. Feuchtinger et al. (1993) are calculating models using an adaptive grid method; they have made both interior and interior-plus-atmosphere calculations for several categories of stars. The use of an adaptive grid makes the impact of artificial viscosity much less, and this is an advantage in treating the shocks. However, they also still use LTE for the radiative relaxation. Fokin has calculated numerical models for both RR Lyrae stars and W Virginis stars that include the driving zone and are thus "self-excited." Luttermoser (Luttermoser and Bowen 1989,1992) has been carrying out extensive non-LTE radiative transfer calculations using PANDORA and ATLAS to get a better estimate for the cooling function appropriate for use in long-period variables. Sasselov (Sasselov and Lester 1994a,b) has calculated Cepheid models using a method of characteristics for the shocks; this has the advantage that it avoids the use of artificial viscosity for the shocks, but it has the disadvantage that it is not applicable where the shock heating plays a significant role, as it must in the calorispheres of these stars. (Sasselov's published models do not extend to sufficiently large distances from the photosphere, given the levitation of the atmosphere that shocks produce; therefore he underestimates the optical depth and is forced to establish his "chromosphere" where his models suggest that the shocks are still nearly isothermal. We expect that when his models are redone with greater physical extent, the calorisphere will appear naturally at those optical depths where it is needed to reproduce the spectroscopic features.)

VII. CONCLUSIONS AND SUMMARY

Pulsating stars may lose mass due to pulsation alone (although very little); due to pulsation including the effects of shock heating on the atmosphere (substantial mass loss may occur) or due to a combination of the levitation produced by pulsation with the effects of radiation pressure, for example on dust (prodigious mass loss is possible). Pulsation, by levitating the atmosphere, makes mass loss easier and makes small v_∞ more likely.

The pulsation period plays a vital role in determining the maximum allowed shock amplitude for periodic motion, and in most cases of single-mode radial pulsation, $\Delta v_{\text{observed}}$ will be a factor of 2 to 4 less than $(GM^*/r_{\text{phot}}^2)P$. Also, P/P_{ac} determines how easily waves leak into the atmosphere, and therefore plays a role in determining which modes reach large amplitude in pulsating stars. Most radially pulsating stars have $Q > 0.03$ day and $H/r < 0.01$ so they generally have $P > P_{ac}$, and the condition for the trapping of small amplitude waves is met. An interesting exception is a long-period variable of the dimensions of an AGB star or supergiant pulsating in an overtone mode with $Q < 0.05$ day; these will typically have $P < P_{ac}$, and the usual assumption that these waves are reflected at the surface is particularly bad for these "leaky" stars. When $P > P_{ac}$ linear waves may not propagate but nonlinear ones will, with interesting and often complex patterns of atmospheric shocks as a result.

The thermal relaxation process in the atmosphere needs to be treated very carefully, as it has a great impact on the results—temperature structure, velocity structure, and mass loss rates. There are regions in the atmosphere where the adiabatic compression/expansion term will dominate dT/dr; in other regions, the term describing the interaction of radiation and matter dominates. While it has often been assumed that this term may be approximated by using LTE source functions, these probably overestimate the rate of cooling for densities less than about 10^{-9} g cm^{-3} at 3000 K or less than about 10^{-10} g cm^{-3} at 10^4 K, i.e., chromospheric to interstellar densities. Non-LTE effects (collisional excitation followed by radiative de-excitation as dominant processes) lead to very different expressions for the radiative cooling term. Cooling at low densities is much less efficient than in the LTE case, and one consequence is a tendency of pulsating stars to form calorispheres.

Finally, there are complications that remain beyond the reach of present models. These include the possible co-existence and effects of very short-period acoustic waves propagating into the atmosphere and interacting with the shocks; the effects of magnetic fields, if these are present; departures from spherical symmetry due to rotation, large convective cells, interaction with a close companion star or large planet, or chaotic/aperiodic variations in shock propagation; and other phenomena not yet considered. No one model as yet incorporates all the features of the "ideal model" even for the spherically symmetric case: explicit and detailed treatment of the thermal relaxation process, shocks resolved without the need for artificial viscosity; inclusion of the internal driving zone together with the extended atmosphere (out far enough to see the final wind flow); and radiative transfer to allow detailed comparison of the results with spectroscopic and photometric observations, as well as (for the coolest stars) explicit treatment of grain nucleation and growth in the dynamical context. Therefore, predictions concerning spectroscopic features, mass loss rates and wind terminal velocities for specific stars are still very approximate. However, much encouraging progress is being made, with various groups concentrating on specific aspects of this problem.

Acknowledgments. Comments by F. Pijpers and an anonymous referee were very helpful. The authors are grateful to P. Wood, and S. Hill for permission to reproduce figures from their published works, and also to A. Fokin who kindly provided postscript files from a published model for our use here. Finally, we are grateful to the organizers and editors of this volume for the opportunity to participate and to present this material.

REFERENCES

Bertschinger, E., and Chevalier, R. A. 1985. A periodic shock wave model for Mira variable atmospheres. *Astrophys. J.* 299:167–190.

Böhm-Vitense, E., and Querci, M. 1987. Intrinsically variable stars. In *Exploring the Universe with the IUE Satellite*, ed. Y. Kondo (Dordrecht: D. Reidel), pp. 223–257.

Bowen, G. H. 1988. Dynamical modeling of long-period variable star atmospheres. *Astrophys. J.* 329:299–317.

Bowen, G. H. 1990. Dynamical phenomena in pulsating star atmospheres. In *Numerical Modeling of Nonlinear Stellar Pulsations*, ed. J. R. Buchler (Dordrecht: Kluwer), pp. 155–171.

Bowen, G. H. 1991. Instabilities in evolved super- and hypergiants. In *Dynamical Modeling of Cool Supergiant Atmospheres*, pp. 104–110.

Bowen, G. H., and Willson, L. A. 1991. Mass-loss rates for Mira models. *Astrophys. J. Lett.* 375:53–56.

Bowers, P. F. 1992. Location of the dust formation region for Mira variables. *Astrophys. J. Lett.* 390:27–31.

Buchler, J. R. 1987. Chaotic behavior in variable stars. In *Chaotic Phenomena in Astrophysics*, eds. J. R. Buchler and H. Eichhorn (New York: New York Academy of Sciences), pp. 37–54.

Buchler, J. R. 1990. Chaotic pulsations in stellar models. In *Nonlinear Astrophysical Fluid Dynamics*, eds. J. R. Buchler and S. T. Gottesman (New York: New York Academy of Sciences), pp. 17–36.

Buchler, J. R., and Moskalik, P. 1990. Resonances and bifurcations. In *The Numerical Modelling of Nonlinear Stellar Pulsations*, ed. J. R. Buchler (Dordrecht: Kluwer), pp. 143–154.

Buchler, J. R., Goupil, M.-J., and Kovacs, G. 1987. Tangent bifurcations in chaos in stellar pulsations. *Phys. Lett. A* 126(3):177–180.

Dalgarno, A., and McCray, R. A. 1972. Heating and ionization of HI regions. *Ann. Rev. Astron. Astrophys.* 10:375–426.

Danchi, W. C., Bester, M., Degiacomi, C. G., McCullough, P. R., and Townes, C. H. 1990. Location and phase of dust formation in IRC+10216 indicated by 11 micron spatial interferometry. *Astrophys. J. Lett.* 359:59–63.

Dominik, C., Sedlmayr, E., and Gail, H.-P. 1993. Dust formation in stellar winds VI. Moment equations for the formation of heterogeneous and core-mantle grains. *Astron. Astrophys.* 277:578–594.

Dorfi, E. A., and Höfner, S. 1991. Dust formation in winds of long-period variables. I. Equations, method of solution, simple examples. *Astron. Astrophys.* 248:105–114.

Draine, B. T. 1981. Dust formation processes around red giants and supergiants. In *Physical Process in Red Giants*, eds. I. Iben and A. Renzini (Dordrecht: D. Reidel), pp. 317–333.

Feuchtinger, M. U., Dorfi, E. A., and Höfner, S. 1993. Radiation hydrodynamics in atmospheres of long-period variables. *Astron. Astrophys.* 273:513–523.

Fleischer, A. J., Gauger, A., and Sedlmayr, E. 1992. Circumstellar dust shells around long-period variables. I. Dynamical models of C-stars including dust formation, growth and evaporation. *Astron. Astrophys.* 266:321–339.

Fokin, A. B. 1992. Shock waves and Hα profiles in the hydrodynamical model of RR Lyrae. *Mon. Not. Roy. Astron. Soc.* 256:26–36.

Fokin, A. B. 1994. Nonlinear pulsations of the RV Tauri stars. *Astron. Astrophys.* 292:133–151.

Fox, M. W., and Wood, P. R. 1982. Theoretical growth rates, periods, and pulsation constants for long-period variables. *Astrophys. J.* 259:198–212.

Goeres, A., and Sedlmayr, E. 1992. The envelope of R Coronae Borealis stars I. A physical model of the decline events due to dust formation. *Astron. Astrophys.* 265:216–236.

Harvey, P. M., Lester, D. F., Brock, D., and Joy, M. 1991. Dust properties around evolved stars from far-infrared size limits. *Astrophys. J.* 368:558–563.

Hill, S. J. 1972. Hydrodynamic and radiative-transfer effects on an RR Lyrae atmosphere. *Astrophys. J.* 178:793–808.

Hill, S. J., and Willson, L. A. 1979. Theoretical velocity structure of long-period variable star photospheres. *Astrophys. J.* 229:1029–1045.

Holzer, T. E. 1987. Theory of winds from cool stars. In *Circumstellar Matter*, eds. I Appenzeller and C. Jordan (Dordrecht: D. Reidel), pp. 289–306.

Holzer, T. E., and MacGregor, K. B. 1984. Mass loss mechanisms for cool, low-gravity stars. In *Mass Loss from Red Giants*, eds. M. Morris and B. Zuckerman (Dordrecht: D. Reidel), pp. 229–256.

Iben, I., Jr. 1984. On the frequency of planetary nebula nuclei powered by helium burning and on the frequency of white dwarfs with hydrogen-deficient atmospheres. *Astrophys. J.* 277:333–354.

Jefferies, J. T. 1968. *Spectral Line Formation* (London: Blaisdell).

Koninx, J.-P. M., and Pijpers, F. P. 1992. The applicability of the linearized theory of sound-wave driven winds. *Astron. Astrophys.* 265:183–195.

Leer, E., and Holzer, T. E. 1980. Energy addition in the solar wind. *J. Geophys. Res.* 85:4681–4688.

Lucy, L. B. 1971. The formation of resonance lines in extended and expanding atmospheres. *Astrophys. J.* 163:95–110.

Luttermoser, D. G., and Bowen, G. H. 1989. Radiative transfer in the dynamic atmospheres of long period variables. In *Sixth Cambridge Workshop on Cool Stars, Stellar Systems, and the Sun*, ed. G. Wallerstein (San Francisco: Astronomical Soc. of the Pacific), pp. 491–493.

Luttermoser, D. G., and Bowen, G. H. 1992. NLTE synthetic spectra of the Mira-type variable S Car. In *Seventh Cambridge Workshop on Cool Stars, Stellar Systems, and the Sun*, eds. M. S. Giampapa and J. A. Bookbinder (San Francisco: Astronomical Soc. of the Pacific), pp. 558–560.

Mihalas, D. 1978. *Stellar Atmospheres* (San Francisco: W. H. Freeman).

Mihalas, D., and Mihalas, B. W. 1984. *Foundations of Radiation Hydrodynamics* (New York: Oxford Univ. Press).

Osterbrock, D. E. 1974. *Astrophysics of Gaseous Nebulae* (San Francisco: W. H. Freeman).

Ostlie, D. A., and Cox, A. N. 1986. A linear survey of the Mira variable star instability region of the Hertzsprung-Russell diagram. *Astrophys. J.* 311:864–872.

Parker, E. N. 1960a. The hydrohynamic theory of solar corpuscular radiation and stellar winds. *Astrophys. J.* 132:821–866.

Parker, E. N. 1960b. The hydrodynamic treatment of the expanding solar corona. *Astrophys. J.* 132:175–183.

Parker, E. N. 1969. Theoretical studies of the solar wind phenomenon. *Sci. Rev.* 9:325–360.

Pijpers, F. P. 1993. Radial pulsation in variable stars with mass loss. *Astron. Astrophys.* 267:471–489.

Pijpers, F. P. 1995. On the propagation of sound waves in a stellar wind traversed by periodic strong shocks. *Astron. Astrophys.* 295:435–444.

Pijpers, F. P., and Habing, H. J. 1989. Driving the stellar wind of AGB stars by acoustic waves: Exploration of a simple model. *Astron. Astrophys.* 215:334–346.

Pijpers, F. P., and Hearn, A. G. 1989. A model for a stellar wind driven by linear acoustic waves. *Astron. Astrophys.* 209:198–210.

Sasselov, D. D., and Lester, J. B. 1994a. The He I λ 10830 in classical Cepheids. I. Observations. *Astrophys. J.* 423:777–784.

Sasselov, D. D., and Lester, J. B. 1994b. The He I λ10830 in classical Cepheids. II. Mechanisms of formation. *Astrophys. J.* 423:785–794.

Sutherland, R. S., and Dopita, M. A. 1993. Cooling functions for low-density astrophysical plasmas. *Astrophys. J. Suppl.* 88:253–327.

Whitney, C. 1956. Stellar pulsation I. Momemtum transfer by compression waves of finite amplitude. *Ann. Astrophys.* 19:34–43.

Willson, L. A., and Bowen, G. H. 1985. Atmospheric structure and mass loss for pulsating stars. In *Relations Between Chromospheric/Coronal Heating and Mass Loss in Stars*, eds. R. Stalio and J. Zirker (Sunspot, N. M.: National Solar Observatory), pp. 127–176.

Willson, L. A., and Bowen, G. H. 1986. Stellar pulsation, atmospheric structure, and mass loss. In *Cool Stars, Stellar Systems and the Sun: Proc. of the Fourth Cambridge Workshop*, eds. M. Zeilik and D. Gibson (Berlin: Springer-Verlag), pp. 385–396.

Willson, L. A., and Hill, S. J. 1979. Shock wave interpretation of emission lines in long period variable stars. II. Periodicity and mass loss. *Astrophys. J.* 228:854–869.

Wood, P. R. 1979. Pulsation and mass loss in Mira variables. *Astrophys. J.* 227:220–231.

Wood, P. R. 1995. Mira variables: Theory vs. observation. In *Astrophysical Applications of Stellar Pulsation*, eds. R. S. Stoble and P. A. Whitelock (San Francisco: Astronomical Soc. of the Pacific), in press.

Zel'dovich, I. B., and Raizer, Y. P. 1967. *Physics of Shock Waves and High Temperature Hydrodynamic Phenomena* (New York: Academic Press).

WIND PROPERTIES OF EVOLVED STARS: EFFECTS ON PARTICLE ACCELERATION

D. J. MULLAN
University of Delaware

In reviews of winds from evolved stars, the topic which is most commonly discussed is the rate at which each star loses mass $\dot{M}$. Here, however, in the context of interactions between the winds and energetic particles, a somewhat different emphasis is presented. From our knowledge of the solar wind, we are aware that the Sun has carved out a cavity (the heliosphere) for itself in the interstellar medium. The properties of the solar wind are such that energetic particles are accelerated or re-accelerated at various locations in the heliosphere, e.g., at the termination shock of the wind, and also at shocks embedded in corotating structures in the wind. This chapter summarizes the properties of the heliospheres (or "astrospheres") around stars of different classes, and how these may affect energetic particles, as sites of either acceleration or re-acceleration. The problem of re-acceleration is considered from a galactic viewpoint: how far can an energetic particle propagate in the Galaxy before it encounters an astrosphere? Recent estimates of the properties of winds from red dwarf stars are especially intriguing in this regard. Although these winds are not very prominent on a scale of mass loss rates, there are so many red dwarfs in the Galaxy that their astrospheres may play a surprisingly large role in particle-astrosphere interactions.

I. INTRODUCTION

The title of this chapter includes the words "evolved stars." The customary definition of evolved stars includes only giants and supergiants in their post-main-sequence phase of evolution. For present purposes, however, I take the liberty of going beyond this definition and including also cool dwarfs. These stars are, after all, continually evolving (albeit on a long time scale), and although they may not yet have depleted their core hydrogen, they are certainly more evolved at the present time than when they were in the proto-stellar phase. This chapter focuses on winds from cool stars.

A. Potential for Particle Acceleration

The plan of this chapter is to outline what is known about the properties of winds from various cool stars, and then apply this information in a broad-brush way to cosmic ray acceleration. More detailed discussions of the physics of termination shocks and of particle acceleration in winds can be found in the chapters by Suess and Nerney and by Jokipii, respectively.

[197]

In particular, we shall evaluate possibilities for acceleration or re-acceleration. For a discussion of acceleration, the important parameters are the speed of the flow, and the velocity jumps associated with any shocks contained within it. Internal to the wind, there may be shocks associated with corotating interaction regions (CIRs); the potential for such shocks to serve as accelerators of energetic particles was pointed out some time ago for hot star winds (Mullan 1985).

B. Dimensions of the Astrosphere

Of special interest is the termination shock (if any) at the edge of the astrosphere where the wind couples to the interstellar medium (ISM). In a wind with terminal velocity v_t and density $\rho(r)$, the termination shock occurs roughly at the radius r_t where $\rho(r_t)v_t^2 \approx B^2/8\pi$, where B is the ISM magnetic field. We need to adopt values for the field and density in the interstellar medium. In what follows, we shall assume that the ISM magnetic field is 5 ± 3 μG: this is the range in the solar neighborhood (Holzer 1989), but it is broad enough to be also representative of the Galaxy as a whole (Lang 1974). For the number density of ISM gas (neutral plus ionized), we adopt 0.3 ± 0.2 cm^{-3}. This range spans a variety of observational data points; the average neutral hydrogen density at distances of 5 to 10 kpc from the galactic center is 0.5 cm^{-3} (Spitzer 1968), and the total hydrogen density, including both neutral and ionized, in the immediate solar neighborhood is 0.1 to 0.2 cm^{-3} (Cox and Reynolds 1987). The relevant Alfvén speed on the ISM side of the termination shock involves both neutrals and ions (Mullan 1971), and is given by V_A(ISM)$\approx$20 km s^{-1}. (There is a considerable spread in this estimate; using the extremes of fields and densities given above, the values of V_A range from 6 to 55 km s^{-1}.) For the rough estimates which we wish to make here, the condition for the existence of a termination shock can be taken to be $v_t > 20$ km s^{-1}.

Expressing $\rho(r_t)$ in terms of the mass loss rate $\dot{M}$, we find that the radial distance to the termination shock (i.e., the radius of the heliosphere) is r_t(AU)$\approx$140$(\dot{M}_{-13.5}v_{300})^{0.5}$. Here, for convenience, the mass-loss rate and terminal velocity are expressed in units which are typical of the solar wind, $10^{-13.5}$ M$_\odot$yr^{-1}, and 300 km s^{-1}. In order to estimate this typical solar mass loss rate, we use the average particle flux density in the ecliptic at 1 AU given by Zirker (1981), 4.1×10^8 cm^{-2}s^{-1}, and assume spherical symmetry; this actually overestimates the mass loss rate somewhat because mass fluxes in the ecliptic exceed those in the polar regions of the Sun by a factor of 2.3 (Woo and Goldstein 1994). Thus, a more correct estimate of the solar mass loss rate is probably close to 2×10^{-14} M$_\odot$yr^{-1}. Solar wind speeds vary over a range of 300 to 400 km s^{-1}, with occasional streams moving faster than 600 km s^{-1}. Despite the variations in speed, the mass flux of the wind in the ecliptic is remarkably constant (Withbroe 1989); the variations are no more than a factor of 2 throughout the solar cycle.

The above expression places the solar wind termination shock at radial distances of 100 to 200 AU, consistent with recent estimates of radio noise detected by Voyager (chapter by Kurth and Gurnett).

For a discussion of re-acceleration (if this process is important on a galactic scale), we need to discuss the dimensions of the astrosphere surrounding an object of each type. For this purpose, the mass loss rate and the terminal velocity must be known. Once the dimensions of the astrosphere are known, the space density of each class of object must be known in order to determine how likely it is that a cosmic ray will interact with the astrosphere during its motion through the Galaxy.

II. WINDS IN THE COOL HALF OF THE HR DIAGRAM

An overview of the variations of mass-loss rates in stars in the cool half of the HR diagram can be found in Drake (1986). The data discussed by Drake indicate that the winds tend to have the largest mass-loss rates ($\dot{M}$ up to 10^{-4} $M_\odot$ yr^{-1} in certain long-period variables) and the slowest terminal velocities (10–20 km s^{-1}) in the upper right-hand corner of the diagram.

An overview of the combined energy fluxes in the form of mass loss and radiation among cool evolved stars can be found in Judge and Stencel (1991). These authors have found that when the combined energy flux in the forms of radiation and mass loss is expressed as a fraction of the bolometric flux from the star, the ratio tends to remain rather constant across the cool HR diagram ($T_{\text{eff}} \leq 4500$ K). They suggest that the radiative losses from the chromospheres may be powered essentially by short-period acoustic waves, while long-period waves lead to mass loss. Although we will not discuss mass-loss mechanisms in this chapter (this topic is addressed in chapters by Willson et al., by Holzer et al., and by MacGregor and Charbonneau), we will revert briefly to the acoustic suggestion of Judge and Stencel below when we mention population II stars.

A. Gradients Across the HR Diagram

Returning to the HR diagram, there are some well-known gradients across the diagram; e.g., $\dot{M}$ tends to decrease, and the terminal velocities tend to increase, as one approaches the main sequence. This is not meant to imply that the trends are necessarily smooth everywhere as one crosses the HR diagram; certain lines of evidence suggest that there may be a fairly sharp dividing line between red giant winds (cool [$T \leq 10^4$ K] slow [10s of km s^{-1}] flows with relatively heavy mass loss) and solar-like winds (hot [$T \geq 10^6$ K] fast [100's of km s^{-1}] flows with relatively low mass-loss rates) (Stencel and Mullan 1980). Although the trends mentioned above are certainly present as one moves across the HR diagram, the sharpness of the mass-loss "dividing line" has been questioned (Dupree 1986), especially in view of the existence of a class of "hybrid stars"; these are objects which exhibit characteristics of both types of wind simultaneously. It may be that the hot solar-like gas in hybrids is trapped on closed magnetic loops, while a massive cool wind flows from open fields.

However, in the present context, we wish to stress that there are at least two other important gradients in the HR diagram. First, there is what we may call an *information gradient*; data on winds are available for literally hundreds of stars in the upper right-hand corner of the HR diagram (and also incidentally for many stars in the upper left-hand corner), but data on winds from stars of lower luminosity rapidly becomes extremely sparse. In fact, in Drake's (1986) review, there was no discussion of winds from even a single cool dwarf other than the Sun. This lack of information about dwarf star winds is sometimes dismissed as of little consequence; however, in view of some recent data on mass-loss rates from M dwarfs (see below), this prejudice may in fact give rise to serious inadequacies in our view of physical processes in the Galaxy. The reason for this statement can be found in the second significant gradient which exists in the HR diagram: the *population gradient*. Analysis of star counts indicates that there are many more M dwarfs in the galaxy than any other class of star; specifically, according to models of the Galaxy, M dwarfs outnumber the long-period variables by as much as 7 orders of magnitude.

In the following sections, we summarize how information has been obtained by a variety of techniques about winds from cool stars. We also summarize data on numbers of the various kinds of stars in the Galaxy. These data are used to evaluate the potential for particle acceleration.

III. WINDS WITH MASER EMISSION

Certain molecules are subject to population inversion when radiation and/or collisional conditions in a stellar wind are favorable. These molecules then serve as maser sources which can have such high luminosity (and low attenuation) that they can be seen throughout the Galaxy (Bowers 1985).

A. Types of Masers

The most commonly used maser lines are those of OH at 1665 and 1667 MHz (the so-called main lines) and at 1612 MHz, and those of water vapor at 22 GHz, and of SiO at 43 GHz.

The lower level of the SiO maser transition requires collisional excitation of an excited vibrational level, and this is possible only at gas temperatures of order 1700 K or more. Thus, SiO maser emission occurs only in the innermost regions of the wind, essentially in contact with the atmosphere of the star, $r \approx (1 - 2)R_*$. In these regions, the densities are high, $N(H_2) \approx 10^8 \text{cm}^{-3}$. The SiO maser is the first maser to "switch on" when mass loss starts, and this maser switches off when mass loss rates decrease to very low levels (Lewis 1989).

The H_2O maser requires gas temperatures in excess of 650 K to pump the requisite lower level of the transition. This sets an upper limit on the distance from the star that the maser can operate; in Mira variables, this limit is 10 to 100 AU, while in the longest period variables, it is 300 to 700 AU.

(For a classification of the several kinds of variables among cool giants and supergiants, see Querci [1986].) However, there is also a constraint on density; to avoid collisional de-excitation of the upper level, the density must be lower than a critical value ($N(H_2) < 10^7 cm^{-3}$). This constraint means that there is also a lower limit on the radial distance at which the H_2O masers occurs; typically this is between 5 and 10 stellar radii.

The OH masers operate far out in the wind ($r \approx 500-10^4$ AU) where interstellar ultraviolet photons can dissociate the water vapor, and far infrared photons (at $\lambda \approx 35 \mu m$) pump the population inversion. The wind densities in these regions are rather low, $N(H_2) \approx 10^4 cm^{-3}$. To excite the 1612 MHz maser, relatively high mass loss rates are required ($\dot{M} > 10^{-5} M_\odot yr^{-1}$). In such conditions, all the above masers may occur in a single source simultaneously; about 1% of all Miras belong to this group.

B. Information in the Maser Line Profiles and Images

For the maser sources, the emission spectrum appears as a double peak; the peaks are centered at velocities $v_o \pm v_{exp}$, where v_o is the radial velocity of the star relative to the Sun, and v_{exp} is the expansion velocity of the circumstellar shell. An expanding shell model fits the spectral data and also the imaging data very well. Values of v_{exp} can be determined with remarkable precision; typical v_{exp} values are 10 to 20 km s^{-1}, with errors of only ± 0.1 km s^{-1}. Upper limits can be put on rotation (<3 km s^{-1}) and turbulence (<2 km s^{-1}). Phase lags between brightness variations in the redward and blueward emission peaks can be interpreted in terms of a light travel time across the shell. This leads directly to estimates of the radial dimensions of the shell; in sources with all four masers, there is direct confirmation of the radial progression from SiO emission in the innermost wind to H_2O in the intermediate wind, to OH in the outermost wind. Combining the radial dimensions with radio images, distances to the objects can then be obtained with a precision of 10 to 20%.

Thus, a great deal of information about the winds in maser sources is known, and the sampling of the Galaxy is very good. The data suggest that there are $10^{(4-5)}$ Miras in the Galaxy. Note that the terminal speeds are rather low; they are not much greater than what we expect of $V_A(ISM)$. Thus, termination shocks may not exist in these winds. Moreover, within the winds themselves, it seems unlikely that CIR shocks would exist with jumps of larger than 10 to 20 km s^{-1}; although we do not know the magnetic fields in the winds well, it seems likely that Alfvén speeds in the winds will not be small compared to these shock jumps. Thus, acceleration of cosmic rays in these winds does not seem likely.

For the maser sources, mass-loss rates are not easy to extract; the intensity of a maser source depends sensitively on the coherence of velocity fields over certain length scales, and small fluctuations in physical conditions lead to large changes in intensity. Thus, estimates of $\dot{M}$ must come from elsewhere (CO or infrared data; see below). These indicate that for maser sources, $\dot{M}$ ranges from $\approx 10^{-7}$ M$_\odot$yr^{-1} $\approx 10^{-4}$ M$_\odot$yr^{-1}, and the mass-loss rate is found to be linearly proportional to the outer radius of the shell.

C. Astrospheres

Using typical values of, e.g., 3×10^{-7} $M_\odot yr^{-1}$ for $\dot{M}$, and $v_t = 30$ km s^{-1}, we find that the edge of the astrosphere lies at a radius of order 1.4×10^5 AU, i.e., $r_t \approx 0.7$ pc.

The number densities n_* of long-period variables with masses of order 1 $M_\odot$ (these are likely to be maser sources) have been discussed by Jura and Kleinmann (1992) (see also Jura et al. 1993, and references therein). The results suggest that n_* is of order $10^{-6.5}$ pc^{-3}. Combining this with the above estimate of astrospheric radius, we find that the mean free path of a cosmic ray particle between collisions with maser source astrospheres is $\lambda_{MFP} \approx 1/(n_* \pi r_t^2) \approx 2 \times 10^6$ pc. Because the scale height of dust and gas perpendicular to the galactic plane is no more than a few hundred parsecs, a cosmic ray must pass through the disk many times ($\geq 10^4$) before encountering one of these maser source astrospheres.

IV. WINDS WITH THERMAL CO EMISSION

Inter/circumstellar chemistry favors the formation of CO as the most abundant molecule next to H_2. Because the rotational levels of the CO molecule lie relatively low compared to the typical thermal energies which are available, thermal emission from the lowest levels is strong. The $J=1 \to 0$ transition at $\lambda 2.6$ mm is the most commonly studied transition. The spectrum has sharp edges in velocity, and so v_{exp} can again be determined with high precision ($\pm$ a few times 0.1 km s^{-1}).

Radiative transfer in this line can be used to interpret the observed profiles in terms of $\dot{M}$ (Knapp and Morris 1985). To do this, the variation of temperature with radial distance must be known. Although this is not known well, the results are sufficiently insensitive to $T(r)$ that the errors in $\dot{M}$ are not dominated by errors in $T(r)$; the major errors come from uncertainties in the distances to the sources. These may introduce errors of factors of 3 or more in $\dot{M}$.

A recent example of the wealth of information that can be obtained from CO studies can be found in Olofsson et al. (1993). They surveyed 120 carbon stars, and detected every star within 600 pc of the Sun, as well as many beyond that distance; most of them have v_{exp} in the range between 9 and 15 km s^{-1}, although extreme examples range from 4 to 30 km s^{-1}. The values of $\dot{M}$ vary by a factor of at least 10, although most and in the range $\dot{M} \approx (0.9-2.5) \times 10^{-7}$ $M_\odot yr^{-1}$. As a group, the values of $\dot{M}$ for Miras and found to be about 10 times larger than for semi-regular variables. For variables with known periods, the values of $\dot{M}$ are found to increase exponentially rapidly with increasing period (Vassiliadis and Wood 1993): as a result, variables with periods longer than, e.g., 600 days (where mass loss rates are a few times 10^{-4} $M_\odot yr^{-1}$) cannot survive in this state for much longer than 10^4 yr without losing significant fractions of their mass. In fact, much of the overall mass that a star loses during its evolution is almost certainly lost during a relatively small fraction of its overall lifetime; this short phase of "superwind" mass loss may terminate the AGB phase of evolution (Young et al. 1993).

Olofsson et al. (1993) find that the empirical correlation between $\dot{M}$ and v_{exp} is very weak; it appears that the mass loss mechanism (whatever it is) can produce widely differing mass-loss rates with only marginal effects on v_{exp}. Outflows in the CO objects contribute some 0.3 $M_\odot \mathrm{yr}^{-1}$ to the ISM, with roughly equal contributions from each of several decades of $\dot{M}$ (Morris 1985). The outflow velocities of the CO objects are so slow that particle acceleration is not likely in the winds. Termination shocks may not exist.

V. STARS WITH DUST IN THEIR WINDS

Many cool giants have excesses above photospheric emission when they are observed in the far infrared, e.g., in the IRAS bands at 12, 25, and 60 μm. The stars that appear in CO which also have infrared excesses are found to have $\dot{M}$ correlated with the infrared excess. Thus, it appears natural to associate the infrared excesses with the mass-loss process, specifically, with dust in the circumstellar shell.

The shell is in general optically thin at 60 μm. In view of the optical thinness, the flux at 60 μm is related in a straightforward way to the total number of dust particles N_{dust}; each spherical particle of radius a would radiate $B(T, 60\,\mu\mathrm{m})\pi a^2$ erg s^{-1} in our direction if it were a blackbody. To allow for departures from blackbody, we multiply by the radiative efficiency at 60 μm, $Q(60\,\mu\mathrm{m})$, and place the object at distance D. This yields $F(60\,\mu\mathrm{m}) = N_{\mathrm{dust}} B(T, 60\,\mu\mathrm{m}) Q(60\,\mu\mathrm{m})(\pi a^2/4\pi D^2)$ (see, e.g., Hildebrand 1983). Using the value of v_{exp} from the CO data, the dust mass can be converted to a dust mass loss rate $\dot{M}_{\mathrm{dust}}$. Olofsson et al. (1993) find that the ratio of gas to dust mass loss rates is 260 in the most massive outflows and smaller in the less massive outflows.

A recent extensive study of infrared excesses has been reported by Young et al. (1993); these authors are mainly interested in circumstellar shells which can be spatially resolved by IRAS at 60 μm. In their survey of 512 evolved stars and pre-planetary nebulae, they find 76 objects with angular sizes at least 2 arcmin. They model the expansion of a shell plowing into the ISM, and they fit the observed radius to determine an age of the object. Combining this age with the mass-loss rate determined from CO, they obtain a value for the mass of the shell. Ages are found to be a few times 10^4 to a few times 10^5 yr, with carbon stars having lifetimes about twice as long as oxygen-rich stars. The mass of the shell is typically only 0.01 $M_\odot$, whereas the star must lose an amount of mass of order 1 $M_\odot$ in the course of evolution. Apparently, most of the mass loss does not occur in the phase which is being observed by IRAS, but in a short-lived superwind phase. The velocity of expansion of the dust shells is again of order 10 km s^{-1}, probably too small to be of interest in accelerating particles.

A. Limits on Dusty Mass Loss

For cool carbon stars, it seems likely that conditions are favorable for dust to form in the wind. Once the dust forms, the presence of dust can alter the dynamics of the wind in ways which depend on the absorption properties of the dust, and on the interactions between dust grains and gas particles. It is predicted that there should be an upper limit to $\dot{M} (\approx 10^{-4} \, M_\odot yr^{-1})$ because of the wavelength dependence of the opacity of carbon grains. The available CO data are in agreement with this (Netzer and Elitzur 1993). On the other hand, the limit for oxygen-rich stars (if there is one) is at a much higher level.

B. Astrospheres

Oxygen-rich stars may turn out to have the largest astrospheres of any class of cool star (for a given terminal speed). Among many classes of giant stars that have been surveyed by Jura and Kleinmann (1992), it appears that the predominant contributor of mass to the ISM is the class of very dusty oxygen-rich stars; these are losing mass at rates $\dot{M}$ of a few times $10^{-5} \, M_\odot yr^{-1}$. With v_t of order 10 km s^{-1}, the radius of the astrosphere is a few parsecs.

Using data in Jura and Kleinmann (1989), the number densities of these objects can be estimated to be a few times 10^{-8} pc^{-3}. Hence, the mean free path is 6×10^5 pc, somewhat smaller than the result for Miras, but still much larger than galactic dimensions. Thus, the astrospheres of oxygen-rich dusty stars, although the largest of all in linear dimensions, are probably of little importance in the propagation of cosmic rays through the Galaxy.

VI. WINDS FROM WARMER STARS

In stars which are warm enough to have at least partial ionization in the wind, other methods are available to estimate $\dot{M}$. Free-free emission at radio wavelengths from the ionized component of the wind has been used in studies of G-M giants (Drake 1985). Among the warmer stars, 10% of the targets have been detected. No information on v_t is available from this method; another source, such as optical lines or CO data, must be used. Combining the observed free-free flux with the v_t, an estimate of $\dot{M}$ can be obtained. The results indicate ionized mass loss rates of up to a few times $10^{-8} \, M_\odot yr^{-1}$. Comparing with mass-loss rates derived from CO data, it appears that the degree of ionization of the wind is of order 1%. For the cooler stars, the rate of successful detections falls off rapidly; in the later M stars, the degree of ionization in the winds must be very small, less than 0.03%.

A. Partial Ionization in the Wind

In the presence of such low degrees of ionization, ambipolar diffusion effects may become important, especially in shocks where ions and neutrals may drift at several times the thermal speed (Mullan 1971). There have so far been no studies of how these ambipolar effects would affect the structure of the termination shock where such winds encounter the ISM. But the ambipolar effects are already known to be a significant source of energy in some young stellar objects (Safier 1993).

B. The ζ Aurigae Systems

Certain G-K bright giants and supergiants are found in a binary with a hot companion; these are the ζ Aurigae stars. In these systems, the wind from the G-K star can be probed by radiation from the hot star. The hot star acts as a source of continuum which can be absorbed by wind material lying along our line of sight. In this way, mass loss rates and v_t can be determined. Extensive work in this area has been done by Reimers and colleagues (see, e.g., Reimers et al. 1990, and references therein). For other recent work, see also Eaton (1994). Values of $\dot{M}$ of order $10^{-(9-7.5)}$ $M_\odot yr^{-1}$ emerge from these studies. And in these stars, we begin to see values of v_t which are definitely larger than several tens of km s^{-1}, i.e., larger than V_A(ISM). In some of these systems, P Cygni profiles of the lines formed in the wind indicate clearly that v_t may be as large as 200 to 300 km s^{-1} (Eaton 1994). There will certainly be a termination shock at the edge of such winds, and there may be CIR shocks in the winds themselves. The astrospheric radius will be of order 0.1 pc.

To estimate number densities of G and K supergiants and bright giants let us assume that these stars are defined by $M_V \leq -1$. Then referring to Allen (1973), it seems that $n_* \approx 5 \times 10^{-6}$ pc^{-3}. Thus the mean free path for a cosmic ray encountering the astrospheres of these objects will be about 6×10^6 pc.

C. Single G and K Giants and Their Astrospheres

Most G and K giants do not exist in ζ Aurigae systems. Therefore, for most of these stars, we cannot rely on the luxury of having a hot companion shining through the wind of the cool star, and therefore capable of exhibiting absorption features due to the wind. However, for some G and K giants, the star itself may provide a short piece of "quasi-continuum" in the form of broad emission lines. The best examples of this are the Ca II K and Mg II h and k emission cores. In giants, these lines have velocity widths of up to a few hundred km s^{-1}. They can be used to probe the onset of heavy mass loss when the wind speeds are of this order. IUE data have been particularly useful in this regard (see, e.g., Mullan 1984).

The astrospheric radii of these objects are expected to be comparable to those obtained in the ζ Aurigae stars. Regarding the space densities of G and K giants, we may again refer to Allen (1973); assuming that giants have $M_V \leq 0$, we find n_* about 10 times larger than for the bright giants and supergiants. Hence, based on the results for the ζ Aurigae stars mentioned above, we estimate that the mean free path for a cosmic ray encountering the astrospheres of G and K giants will be about 6×10^5 pc, comparable to the result for the stars with dust in the wind.

VII. WINDS FROM POPULATION II STARS

Although the study of mass loss among population II stars is much less developed than among population I stars, various spectral lines have been used to detect wind signatures, including Mg II, Ca II, Hα and He I λ10830 (Dupree et al. 1992,1994). Mass-loss rates derived from these lines are quite uncertain. However, evolutionary considerations indicate that a certain total amount of mass must be lost between the red giant phase and the horizontal branch phase; for globular cluster stars, this amounts typically to a few times 0.1 $M_\odot$ (Iben 1967). But precisely how this amount is lost as a function of time is difficult to establish.

Interestingly, a recent study of the total amount of mass lost in the course of evolution of stars in an open cluster (M67: closer to population I in composition) indicates that the total amount of mass which is lost must be of order 0.6 $M_\odot$, i.e., well in excess of that found for globular cluster stars (Tripicco et al. 1993). This suggests that, whatever the mass loss mechanism is, *it is considerably more effective in population I stars than in population II stars.*

A. Acoustic Heating and Mass Loss

In this regard, we recall the speculation of Judge and Stencel (1991) that acoustic waves might be responsible for mass loss in cool giants. This leads us to ask: is there any reason to believe that population I stars are more efficient at emitting acoustic waves than population II stars? It is known that acoustic waves are created by convective motions, and the acoustic flux is a very sensitive function of the rms convective velocity ($\sim v^8$ according to the Lighthill expression). Now, convective motions are controlled to some extent by the opacity of the medium; it is therefore expected that the acoustic energy flux is different from population I to population II. According to model convection zones, the sense of the difference is that the acoustic flux may be more than 10 times smaller in a metal-poor star than in a metal-rich star. Some evidence in favor of this proposal is found in the tendency for metal-poor stars to rotate faster; the weaker mass loss rates in population II stars provides a less effective "brake" on the angular momentum (other things being equal) (Mullan 1973).

With the weaker mass loss rates in population II stars, and consequently smaller astrospheres, there seems little likelihood that they will play an important role in cosmic ray propagation through the Galaxy.

VIII. SUPERNOVA REMNANTS AND HOT STARS

Although not exactly within the purview of this chapter, it is worthwhile to include the parameters of supernovae remnants (SNRs) in the present discussion of cosmic ray encounters with stellar astrospheres. Because of the overall energetics, acceleration of cosmic rays in the early phases of supernovae (SN) shocks is known to be an efficient process, but here, we are more interested in their potential for long-term contributions, including re-acceleration.

A SNR expands rapidly up to ages of 2×10^4 to 2×10^5 yr, after which radiative losses are sufficiently strong that the remnant enters a much slower "snow-plow" phase (Spitzer 1968). In the latter phase, expansion speeds of only tens of km s^{-1} are expected, and in such a stage, the SNR is not expected to be a significant contributor to cosmic ray acceleration. Thus, lifetimes of SNRs in the present context can be taken to be of order 10^5 yr. With one SN occurring every 30 years in the Galaxy, this means there are roughly 3000 SNRs in the Galaxy which are potentially "cosmic-ray" active at any one time. Thus, their number density is of order 3×10^{-8} pc^{-3}. Entering the snowplow phase, the typical radius is tens of pc (Spitzer 1968). With a radius of 30 pc, i.e., a cross-section of order $1000\,\pi$ pc^2, the mean free path (MFP) between encounters with SNRs is of order 10 kpc. We note that in a discussion of re-acceleration, Ip and Axford (1992) have suggested that cosmic rays may encounter SNRs every 10^3 yr. This implies MFPs of only 300 pc, i.e., 30 times shorter than we have estimated. Ip and Axford do not state how they derived their estimate of λ_{MFP}.

While we are on the subject of supernovae, we must also mention briefly the winds from hot O stars and Wolf Rayet (WR) stars. With $\dot{M}$ up to $\approx 10^{-5}$ M$_\odot$yr^{-1}, and v_t of order 2000 km s^{-1}, we find r_t of order tens of parsecs. To estimate number densities, we note that within 3 kpc of the Sun, some 795 O stars and about one-fourth that number of WR stars are known (Conti 1988); as these stars are confined to a thin galactic disk, the total numbers of both types of stars in the Galaxy is probably of order several thousand. Thus, the parameters for cosmic ray collision are comparable to those for SNRs, and λ_{MFP} is again expected to be of order 10 kpc.

IX. WINDS FROM COOL DWARF STARS

Information on winds from cool dwarfs is difficult to obtain. As a result, the discussion in the present section is inevitably more speculative than the preceding ones. It is only in the recent past that mass loss rates for any cool dwarfs other than the Sun have become available. These results refer to ultraviolet data on a cool dwarf in a detached eclipsing binary (Mullan et al. 1989), and millimeter data on three isolated red dwarfs (Mullan et al. 1992). (Note that the ultraviolet study refers to V 471 Tauri, a sort of miniature ζ Aurigae system, where the hot star is a white dwarf, and the cool star is a K2 dwarf; the principle of the observations is the same, using the eclipses of the small hot star to view the wind from the cool star in absorption.) According to these results, the terminal speeds are comparable to solar v_t, or even larger. Therefore, shock acceleration in CIRs (see Sec. I.A) in the winds of M dwarfs will be at least as efficient as in the solar wind. This means that protons can be accelerated to energies of order MeV or more in the winds (Mullan 1985).

The estimates of mass-loss rates (a few times 10^{-11} to 10^{-10} $M_\odot yr^{-1}$) turn out to be larger than the solar value by up to 3 to 4 orders of magnitude. The millimeter detections are marginal, and it will be important in the future to confirm how typical the results are for red dwarfs in general. Mass-loss rates of the above order would have evolutionary consequences for the stars themselves; a significant fraction of the mass of would be lost in time scales of a few billion years if the rates remained constant over these intervals. And in view of the large numbers of M dwarfs in the Galaxy, the above mass-loss rates suggest that the mass balance of the interstellar medium would be dominated by M dwarfs (Mullan et al. 1992). In view of the importance of the implications of the above mass-loss rates in the above contexts, and also in the context of cosmic ray propagation in the Galaxy, we now summarize four other estimates of mass-loss rates from cool dwarfs.

A. X-ray Emission

Drake (1988) has used X-ray luminosities to put limits on the rate of loss of material which is at temperatures of 10^6 K or higher; assuming a certain velocity law for the wind ($v \sim r^2$), an M dwarf with the lowest measured L_X value ($L_X \sim 10^{26.1}$ erg s^{-1}) is found to have $\dot{M} \approx 5 \times 10^{-14}$ $M_\odot yr^{-1}$, i.e., about twice the solar mass-loss rate. The median L_X for young disk single M dwarfs within 25 pc of the Sun is $10^{28.7}$ erg s^{-1} (Bookbinder et al. 1990). And for an active M dwarf such as AU Mic, $L_X = 10^{29.86}$ erg s^{-1} (Bookbinder et al. 1990). According to Drake's argument, the X rays are ascribed to free-free emission; hence number densities in the wind, as well as $\dot{M}$, scale as $L_X^{0.5}$. Therefore, applying the argument to AU Mic, $\dot{M}$ would be more than 2 orders of magnitude larger than solar, and this refers only to the material at temperatures of 10^6 K or higher.

Actually, there is likely to be significant amounts of cool material also present in the winds from M dwarfs. To see this, note that the radiative cooling time scale t_R of an optically thin gas is given roughly by $10^{3.5}(T_s/N_s)f(T)$ s, where temperature is scaled in units of $10^{7.12}$ K, and number density is scaled in units of $10^{10.84}$ cm^{-3}; $f(T)$ is a factor of order unity which is related to the radiative loss function (Mullan 1976). For gas at temperatures of 10^6 K, $f(T) \approx 0.5$. Using densities of $10^{8-8.5}$ cm^{-3} (typical of the low corona in the polar regions, where most of the solar wind emerges [Newkirk 1967]), we find $t_R(\text{sun}) \sim 10^{4.4-4.9}$ s. The solar wind flow time scale t_f is close to 10^4 s (see, e.g., Mullan 1991). Therefore, $t_f(\text{Sun})$ is shorter than $t_R(\text{Sun})$, and the wind from the Sun escapes before it undergoes significant radiative cooling. However, in M dwarfs, coronal densities are larger than solar by up to 1.5 orders of magnitude (Katsova 1988). As a result $t_R(M)$ may be as small as $10^{2.9-3.4}$ s. On the other hand, stellar masses and radii are nearly proportional to one another on the lower main sequence (Patterson 1984); therefore, the gravitational escape speed of M dwarfs is essentially identical to the solar value, and it is likely that t_f in M dwarf winds will be comparable to the solar value, i.e., $t_f(M) \approx 10^4$ s. This suggests that $t_R(M)$ is shorter than $t_f(M)$. In such a case, significant radiative cooling will occur before the wind has had time to flow very far from the star. Evidence that the wind from the K dwarf in V 471 Tauri is as cool as 2×10^4 K at distances of no more than a few stellar radii from the surface has been presented by Mullan et al. (1989). In view of these estimates of t_f and t_R, we conclude that mass-loss rates based on X-ray emission (Drake 1988) significantly *underestimate* the true values in M dwarfs. Hence, the mass-loss rate from AU Mic may be 3 or more orders of magnitude larger than the solar value.

B. Radio Emission

Free-free radio emission from M dwarf winds is also sensitive only to the ionized component of the wind, and is therefore capable of producing only lower limits on the mass-loss rates. Reports of a 3σ detection of the M dwarf known as Barnard's star at 6 cm correspond to $\dot{M} \approx 2 \times 10^{-11}$ $M_\odot \text{yr}^{-1}$ (S. Drake, personal communication), i.e., the lower limit on mass-loss rate in Barnard's star is almost 3 orders of magnitude larger than the solar value. This refers to an M dwarf which, in terms of magnetic activity, is one of the least active stars known. In such a magnetically quiet star, it is probable that the wind is driven only by the thermal pressure of the corona. In stars with magnetic activity, magnetically driven ejections of mass will add to the thermal wind (see the next sub-section).

C. Discrete Mass Ejections

In very active M dwarfs, flare-related mass ejections may contribute to mass loss at a rate of $\approx 2 \times 10^{-13}$ $M_\odot \text{yr}^{-1}$ (Coleman and Worden 1976). Thus, even if flare ejecta were the only contributors to mass loss from such stars, the mean $\dot{M}$ would exceed the solar value by a factor of 10 or so. However, flares are by no means the only source of mass loss from these stars. Their coronae are hot enough that the mass-loss rate in the form of thermal wind are expected to exceed the solar value by a few orders of magnitude (Mullan et al. 1992). Moreover, flares are not the only source of discrete ejections. It is well known that discrete ejections of mass from the Sun occur not only during flares but also in the form of coronal mass ejections (CMEs); despite a historical prejudice that has often regarded flares as the cause of CMEs, in most cases, CMEs are *not* associated with identifiable flares (Kahler 1992). (In fact, the CMEs may be the dominant phenomenon, causing flares rather than the reverse.) By analogy, in cool dwarfs, mass loss in the form of CMEs which are not associated with flares may dominate the flare-associated events from such stars. Therefore, the estimates of Coleman and Worden (1976) are only lower limits on the mass-loss rate from flare stars in the form of discrete events. And when we add in the effects of a thermal wind, the possibility that mass-loss rates may be at least 2 orders of magnitude larger than solar cannot be discounted.

Regarding contributions of CMEs to mass loss, we note that CMEs contribute at least 16% to the mass flux of the solar wind (Jackson and

Howard 1993); the lower limit would actually be 24% if one were to use a solar mass-loss rate of 2×10^{-14} $M_\odot yr^{-1}$ (see above). In active dwarfs, where magnetic activity levels are much higher than in the Sun, it is expected that magnetically driven CMEs would be an even more important contributor to mass loss than they are in the Sun.

D. Cool Dwarf Wind Accreting onto a Companion

When a cool dwarf is in a binary with a hot white dwarf, accretion of the wind from the cool star may give rise to metallic lines in the white dwarf spectrum. Interpretation of the metallic line strengths leads to an estimate of the accretion rate $\dot{M}_{acc}$ (Sion and Starrfield 1984). Then if one can estimate the fraction ϕ_{WD} of the cool dwarf wind that is actually accreted by the white dwarf, the mass-loss rate $\dot{M}(cool)$ from the cool dwarf follows: $\dot{M}_{acc}/\phi_{WD}$. In the case of Feige 24, $\dot{M}_{acc} \approx 10^{-16}$ $M_\odot yr^{-1}$. A purely hydrodynamic estimate of ϕ_{WD} in this case is of order 10^{-3}, (Sion and Starrfield 1984) and this suggests $\dot{M}(cool) \approx 10^{-13}$ $M_\odot yr^{-1}$. There is no independent method of testing this result in Feige 24.

However, this method of estimating $\dot{M}(cool)$ encounters a serious difficulty. In the presence of magnetic fields and rotation, ϕ_{WD} may become extremely small. A test case occurs in V 471 Tauri, where the mass-loss rate from the cool dwarf can be estimated separately from the accretion rate onto the white dwarf. In that system, the metallic lines in the white dwarf are so weak that the white dwarf must be accreting at a rate which is orders of magnitude smaller than the hydrodynamic predictions; the white dwarf is effectively shielded from the wind of the cool star. The shielding factor ϕ_{sh} is of order $\sim 10^{4.5}$ (Mullan et al. 1991). If we were to try to estimate $\dot{M}(cool)$ solely from what is seen on the white dwarf surface, we would *underestimate* $\dot{M}(cool)$ by factors of ϕ_{sh}. The large value of ϕ_{sh} in V 471 Tauri can be quantitatively accounted for if the magnetic field on the white dwarf has a polar strength of at least a few kilogauss (Mullan et al. 1991). Unfortunately, we do not have enough information to estimate ϕ_{sh} in Feige 24, but if V 471 Tauri is any indication, ϕ_{sh} may be very large. Thus, the Sion and Starrfield (1984) estimate $\dot{M}(cool) \approx 10^{-13}$ $M_\odot yr^{-1}$ (which is already larger than the solar value by a factor of 3) certainly needs to be increased by a factor ϕ_{sh} which may be as large as several orders of magnitude.

E. M Dwarf Astrospheres and Cosmic Rays

Summarizing the discussion up to this point, several lines of evidence point to the conclusion that mass-loss rates in M dwarfs are larger than solar by factors of at least 10 and possibly as much as 10^{3-4}. Let us examine the consequences of this.

The first thing to note is that the radius of the astrosphere of an M dwarf with $\dot{M} \approx 3$ to 4 orders of magnitude larger than solar is of order 0.03 to 0.1 pc. The cross-sectional areas of these astrospheres is therefore 0.003 to 0.03 pc^2.

Now we turn to a crucial property of M dwarfs—their high space density. According to Bahcall's (1986) model of the stellar populations in the Galaxy, the mass of the disk is 6×10^{10} $M_\odot$. Because the mass function rises towards lower masses, most of the mass of the disk is in the form of stars with masses less than 0.6 $M_\odot$. Thus, the total numbers of such stars are at least 10^{11} in the disk population alone. The halo population will only serve to increase this number. With a disk volume of $\leq 3 \times 10^{11}$ pc^3 (i.e., assuming a radius of 10 kpc and a perpendicular scale height of ≤ 1 kpc for low mass stars), the mean density of M dwarfs will be $n_* \geq 0.3$ pc^{-3}. An independent estimate of the space density of M dwarfs which is consistent with this results can be found in Schmitt and Snowden (1990): 0.49 pc^{-3}.

Combining these estimates of space densities of M dwarfs with the cross-sectional area of their astrospheres (0.003–0.03 pc^2), we find that the mean free path for cosmic rays to encounter M dwarf astrospheres is $\lambda_{MFP} \leq 10^{2-3}$ pc. These are by far the shortest MFP among all we have estimated. If the above speculations can be confirmed, it will mean that M dwarfs may play a primary role in cosmic ray re-acceleration in the Galaxy. If the mass-loss rate of M dwarfs exceeds the solar value by a factor of only 2 orders of magnitude, then λ_{MFP} would be comparable to that for SNRs.

The conclusions we have reached here concerning M dwarf astrospheres are speculative; they rely heavily on estimates of mass-loss rates from a few red dwarfs. An important observational program in the coming decade will be to determine how typical the results obtained so far are of red dwarfs in general. The next generation of sub-millimeter detectors are expected to play an important role in addressing this question.

Acknowledgments. This work has been supported by a NASA Grant. Helpful comments by S. Drake and P. Judge are appreciated.

REFERENCES

Allen, C. W. 1976. *Astrophysical Quantities*, 3rd ed. (London: Athlone Press).

Bahcall, J. 1986. Star counts and galactic structure. *Ann. Rev. Astron. Astrophys.* 24:577–611.

Bookbinder, J., Golub, L., and Rosner, R. 1990. Observations of Quiescent X-Ray Emission from K and M Dwarfs. Preprint.

Bowers, P. 1985. Maser emission as a tool to study mass loss from evolved stars. In *Mass Loss from Red Giants*, eds. M. Morris and B. Zuckerman (Dordrecht: D. Reidel), pp. 189–210.

Coleman, G. D., and Worden, S. P. 1976. Mass loss from dwarf M stars through flaring. *Astrophys. J.* 205:475–481.

Conti, P. S. 1988. Overview of O, Of, and Wolf-Rayet populations. In *O Stars and Wolf Rayet Stars*, eds. P. S. Conti and A. Underhill, NASA SP-497, pp. 81–118.

Cox, D. P., and Reynolds, R. J. 1987. The local interstellar medium. *Ann. Rev. Astron. Astrophys.* 25:303–344.

Drake, S. 1985. Radio continuum observations of G-M giants and supergiants and inferred ionized mass loss rates. In *Mass Loss from Red Giants*, eds. M. Morris and B. Zuckerman (Dordrecht: D. Reidel), pp. 187–188.

Drake, S. 1986. Mass loss estimates in cool giants and supergiants. In *Cool Stars, Stellar Systems, and the Sun*, eds. M. Zeilik and D. M. Gibson (New York: Springer-Verlag), pp. 369–384.

Drake, S. 1988. Mass loss from late type stars. Observations. In *Proc. of the 6th Intl. Solar Wind Conf.*, vol. 1, eds. V. J. Pizzo, T. E. Holzer and D. G. Sime, NCAR TN-306 (Boulder: Natl. Center for Atmos. Res.), pp. 129–148.

Dupree, A. K. 1986. Mass loss from cool stars. *Ann. Rev. Astron. Astrophys.* 24:377–420.

Dupree, A. K., Sasselov, D. D., and Lester, J. B. 1992. Discovery of a fast wind from a field Population II giant stars. *Astrophys. J. Lett.* 387:85–88.

Dupree, A. K., Hartmann, L., Smith, G. H., Rodgers, A. W., Roberts, W. H., and Zucker, D. B. 1994. Spectroscopy of chromospheric lines in giants in the globular cluster NGC 6752. *Astrophys. J.* 421:542–549.

Eaton, J. 1994. A deep atmospheric eclipse of AL Velorum. *Astron. J.* 107:729–737.

Hildebrand, R. H. 1983. The determination of cloud masses and dust characteristics from submillimetre thermal emission. *Quart. J. Roy. Astron. Soc.* 24:267–282.

Holzer, T. R. 1989. Interaction between the solar wind and the interstellar medium. *Ann. Rev. Astron. Astrophys.* 27:199–234.

Iben, I. 1967. Stellar evolution within and off the main sequence. *Ann. Rev. Astron. Astrophys.* 5:571–626.

Ip, W.-H., and Axford, W. I. 1992. Particle acceleration up to 10^{20} eV. In *Particle Acceleration in Cosmic Plasmas*, eds. G. Zank and T. Gaisser (New York: American Inst. of Physics), pp. 400–405.

Jackson, B. V., and Howard, R. A. 1993. A CME mass distribution derived from SOLWIND coronagraph observations. *Solar Physics* 148:259–370.

Judge, P., and Stencel, R. 1991. Evolution of the chromospheres and winds of low- and intermediate-mass giant stars. *Astrophys. J.* 371:357–379.

Jura, M., and Kleinmann, S. 1989. Dust enshrouded asymptotic giant branch stars in the solar neighborhood. *Astrophys. J.* 341:359–366.

Jura, M., and Kleinmann, S. 1992. Oxygen-rich semiregular and irregular variables. *Astrophys. J. Suppl.* 83:329–349.

Jura, M., Yamamoto, A., and Kleinmann, S. G. 1993. Long period oxygen rich optical Miras in the solar neighborhood. *Astrophys. J.* 413:298–303.

Kahler, S. 1992. Solar flares and coronal mass ejections. *Ann. Rev. Astron. Astrophys.* 30:113–142.

Katsova, M. 1988. Densities and heating of coronae of the active late-type dwarfs. In *Activity in Cool Star Envelopes*, ed. O. Havnes (Dordrecht: Kluwer), pp. 245–248.

Knapp, G. R., and Morris, M. 1985. Mass loss from evolved stars. III. Mass loss rates for fifty stars from CO $J = 1$-0 transitions. *Astrophys. J.* 292:640–669.

Lang, K. R. 1974. *Astrophysical Formulae* (New York: Springer).

Lewis, B. M. 1989. The chronological sequence of circumstellar masers: Identifying proto-planetary nebulae. *Astrophys. J.* 338:234–243.

Morris, M. 1985. Thermal radio emission from molecules in circumstellar outflows. In *Mass Loss from Red Giants*, eds. M. Morris and B. Zuckerman (Dordrecht: D. Reidel), pp. 129–148.

Mullan, D. J. 1971. The structure of transverse hydromagnetic shocks in regions of low ionization. *Mon. Not. Roy. Astron. Soc.* 153:145–170.

Mullan, D. J. 1973. Fast rotation of metal-poor stars. *Astron. Astrophys.* 27:379–381.

Mullan, D. J. 1976. Thermal X-rays from stellar flares: Re-evaluation of scaling from solar flares. *Astrophys. J.* 207:289–295.

Mullan, D. J. 1984. Asymmetries in stellar Mg II h and k and Ca II H and K profiles: Discrepancies between Mg and Ca asymmetries. *Astrophys. J.* 284:769–773.

Mullan, D. J. 1985. Co-rotating interaction regions in stellar wind: Particle acceleration and non-thermal radio emission in hot stars. In *Radio Stars*, eds. R. Hjellming and D. Gibson (Dordrecht: D. Reidel), pp. 39–42.

Mullan, D. J. 1991. Survival of discrete structures in the solar wind. *Astron. Astrophys.* 248:256–262.

Mullan, D. J., Sion, E. M., Bruhweiler, F. C., and Carpenter, K. 1989. Evidence for a cool wind from the K dwarf in V471 Tauri. *Astrophys. J. Lett.* 339:33–36.

Mullan, D. J., Sion, E. M., Shipman, H. L., and MacDonald, J. 1991. Inefficient accretion by the DA2 white dwarf in V471 Tauri. *Astrophys. J.* 374:707–720.

Mullan, D. J., Doyle, J. G., Redman, R. O., and Mathiondanis, M. 1992. Limits on detectability of mass loss from cool dwarfs. *Astrophys. J.* 397:225–231.

Netzer, N., and Elitzur, M. 1993. The dynamics of stellar outflows dominated by interaction of dust and radiation. *Astrophys. J.* 410:701–713.

Newkirk, G. 1967. Structure of the solar corona. *Ann. Rev. Astron. Astrophys.* 5:213–266.

Olofsson, H., Eriksson, K., Gustafsson, B., and Carlstrom, U. 1993. A study of circumstellar envelopes around bright carbon stars. I. Structure, kinematics, and mass loss rate. *Astrophys. J. Suppl.* 87:267–304.

Patterson, J. 1984. The evolution of cataclysmic and low mass X-ray binaries. *Astrophys. J. Suppl.* 54:443–493.

Querci, F. 1986. Basic properties and photometric variability. In *The M-type Stars*, eds. H. R. Johnson and F. R. Querci, NASA SP-492, pp. 1–112.

Reimers, D., Baade, R., and Schröder, K. P. 1990. Discovery of a highly ionized low mass loss wind and a corona seen in absorption in the G9II star HR 6902. *Astron. Astrophys.* 227:133–140.

Safier, P. N. 1993. Centrifugally driven winds from protostellar disks. I. Wind model and thermal structure. *Astrophys. J.* 408:115–147.

Schmitt, J., and Snowden, S. 1990. Contribution of late type dwarf stars to the soft X-ray diffuse background. *Astrophys. J.* 361:207–214.

Sion, E. M., and Starrfield, S. G. 1984. Feige 24: wind/flare accretion by a hot DAZ1 degenerate. *Astrophys. J.* 286:760–762.

Spitzer, L. 1968. *Diffuse Matter in Space* (New York: Interscience).

Stencel, R. E., and Mullan, D. J. 1980. Detection of mass loss in stellar chromospheres. *Astrophys. J.* 238:221–228.

Tripicco, M. J., Dorman, B., and Bell, R. A. 1993. Mass loss in M67 giants: Evidence from isochrone fitting. *Astron. J.* 106:618–626.

Vassiliadis, E., and Wood, P. R. 1993. Evolution of low- and intermediate-mass stars to the end of the asymptotic giant branch with mass loss. *Astrophys. J.* 413:641–657.

Withbroe, G. L. 1989. The solar wind mass flux. *Astrophys. J. Lett.* 337:49–52.

Woo, R., and Goldstein, R. M. 1994. Latitudinal variation of speed and mass flux in the acceleration region of the solar wind inferred from spectral broadening measurements. *Geophys. Res. Lett.* 21:85–88.

Young, K., Phillips, T. G., and Knapp, G. R. 1993. Circumstellar shells resolved in IRAS survey data. II. Analysis. *Astrophys. J.* 409:725–738.

Zirker, J. B. 1981. The solar corona and the solar wind. In *The Sun as a Star*, ed. S. Jordan, NASA SP-450, pp. 135–162.

WINDS FROM X-RAY BINARIES

T. R. KALLMAN
NASA Goddard Space Flight Center

Winds can play a crucial role in the mass budget of X-ray binaries by providing a source (or sink) of accreting gas. They can also play an important role in determining the observable spectrum of X rays. The interaction between X rays and wind material can be important in determining the wind dynamics. These issues have gained in importance in recent years with the development of improved detectors capable of resolving many of the X-ray spectral and temporal features which may be associated with winds in X-ray binaries. Development of numerical techniques capable of treating the dynamics of winds over a wide range of length scales and incorporating non-spherical geometry have also aided in advancing this field. These advances are discussed in this chapter.

I. INTRODUCTION

X-ray binary systems consist of a collapsed star, such as a neutron star or black hole, in orbit with a normal star. X rays are produced as gas from the normal star is accreted onto the compact object with the conversion of gravitational potential energy into radiation. X-ray binaries divide into two distinct categories: those in which the normal (companion) star has a mass of order 5 to 10 times the mass of the Sun (or greater) and those in which the mass of the normal star is much smaller, roughly comparable to the mass of the Sun. In the former class, known as massive X-ray binaries (MXBs), the normal star emits a strong wind characteristic of single stars of similar type. This affects the X-ray source by fueling the accretion flow, and by absorbing and reprocessing the X rays. In the latter class, known as low-mass X-ray binaries (LMXBs), the accretion flow is likely to be in the form of a dense stream and disk surrounding the X-ray source, and winds can be excited by X-ray heating of this gas or the surface of the normal star.

II. WINDS AND MASSIVE BINARIES

In our Galaxy there are approximately 20 known massive binaries (hereafter referred to as MXBs). Their general properties have been reviewed most recently by White et al. (1993); here we begin by summarizing the properties most relevant to the study of winds. MXBs can be subdivided into systems with supergiant companion stars, and those with Be star companions. The latter are more numerous, while the former are more thoroughly studied. Most of the examples in this chapter will be drawn from the supergiant systems, although the two classes are likely to share many of the physical effects arising from the interaction of wind and compact object.

MXBs typically have X-ray luminosities ranging from $\sim 10^{35}$ erg s^{-1} to $\sim 10^{38}$ erg s^{-1}; in other galaxies the total numbers and luminosities of these objects may be much greater. In the Magellanic clouds, for example, the luminosities of several sources exceed 10^{39} erg s^{-1} (Van Paradijs 1993). This is greater than the Eddington limit for a 1 M$_\odot$ object. In M101, a nearby spiral galaxy, it is likely that the total luminosity of the galaxy in X rays is dominated by such objects, and the same may be true for other galaxies (Fabbiano 1993). Of the MXBs in our Galaxy and the Magellanic clouds, most are likely to contain accreting neutron stars as the source of X rays, and many are observed to pulse in X rays. Pulsations are an indication of the presence of strong magnetic fields near the neutron star surface, although the detailed mechanisms for pulsations are likely to be rather different than for radio pulsars. The orbital separations are typically less than two stellar radii, indicating considerable evolution of the binary orbit. A schematic cartoon of the orbital plane of a massive X-ray binary is shown in Fig. 1. The most widely accepted scenario for producing such an orbit involves the evolution of a binary system consisting initially of two OB stars. The more massive evolves faster, becoming a giant, supernova, and finally a neutron star or black hole. The orbit decays during a common envelope stage prior to the supernova explosion (Vanden Heuvel 1977). Because mass transfer from a more massive to a less massive star is unstable (Frank et al. 1985) the lifetimes of MXBs are expected to be short and to lead to eventual quenching of the X-ray source and possibly coalescence leaving a single or binary neutron star. Many systems also exhibit eclipses of the X-ray source, and the presence of X-ray pulsations and eclipses allows relatively precise determination of the inclination orbital separation, eccentricity and masses of both components.

The theory for the fueling of accretion in MXBs by wind accretion was first worked out by Davidson and Ostriker (1973) based on the Bondi-Hoyle theory of accretion from laminar flow past a compact object (Bondi 1952; Bondi and Hoyle 1944; Hoyle and Littleton 1939). The cross section for gravitational capture of wind material by the compact object is $\sigma_{\mathrm{grav}} = \pi r_{\mathrm{acc}}^2$ where $r_{\mathrm{acc}} = 2GM/v^2$; and M is the mass of the object and v is the wind speed. The accretion rate is then $\dot{M} = 4(GM)^2 n m_H v^{-3} = 1.2 \times 10^{18}$ g s^{-1} $(M/\mathrm{M}_\odot)^2 v_7^{-3} n_{10}$, where n is the density of the wind, n_{10} is the same quantity scaled in units of 10^{10} cm^{-3}, and v_7 is the wind speed scaled in units of 100 km s^{-1}. Assuming an efficiency of 10% for conversion of accreted mass to luminosity, this is sufficient to fuel an X-ray source radiating at 10^{38} erg s^{-1}. Plausible values for the wind density and speed near the X-ray source have been tested against this expression by Conti (1978), and White (1985), who showed that wind accretion can account for the luminosities of approximately half of the known MXBs, primarily those with large mass loss rate in the wind

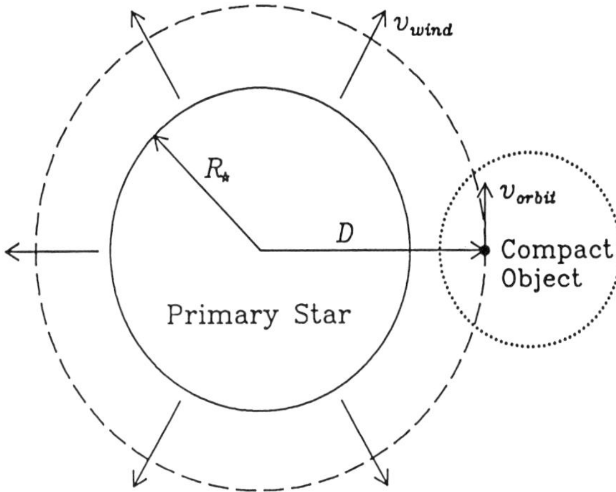

Figure 1.   The assumed geometry of a wind-fed MXRB with a neutron star orbiting counterclockwise around a companion star with velocity $v_{\rm orbit}$. A spherically symmetric wind is driven off the companion and is photoionized within an X-ray Strömgren zone that surrounds the neutron star. The accretion wake and photoionization wake are shown schematically by the shaded regions near the neutron star.

and small or modest X-ray luminosity. This is shown in Fig. 2, in which the ratio $\dot{M}/v_T^4$ required to fuel the X-ray source was derived from the observed X-ray luminosity and the Davidson and Ostriker theory; the expected value of this ratio was derived from mass loss rates and terminal velocities of stars with similar spectral type; the wind speed near the X-ray source was derived assuming a Castor, Abbott, and Klein (1975) velocity law; and the orbital separation measured from the binary parameters. In the objects which fall near the middle of the diagram, there is both a relative plentitude of gas in the wind and a modest accretion requirement. In the remainder of sources, those with weaker winds or larger X-ray luminosity, the mass transfer is presumably aided by the fact that the primary star overflows its Roche lobe. In fact, it is likely that in these sources a combination of the two mechanisms is operating, in which the wind mass loss rate along the line of centers between the two stars is aided by the very weak gravity at the primary's Roche lobe (Friend and Castor 1982). Figure 2 shows that the fueling mechanism in MXBs can be classified according to the ratio of the X-ray luminosity to the wind mass loss rate, $L_X/\dot{M}$; a large value of this ratio places the most severe test upon the wind accretion theory and reveals its breakdown. This ratio also proves useful in distinguishing the importance of other physical effects as is shown below.

MXBs provide the potential for studying the winds from massive stars via the effects on the X-ray source; in addition to fueling the source, the wind can absorb and reprocess the X rays. That is, they may be used to "X-ray" the winds. Among the goals of such a procedure would be to test the validity of the Castor, Abbott, and Klein (1975) theory for the velocity

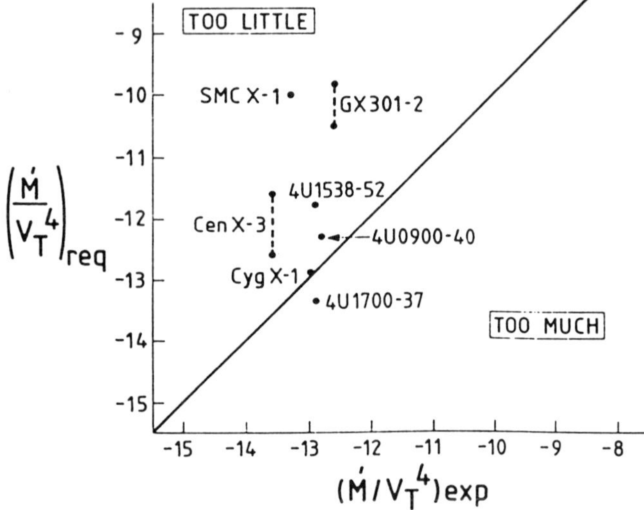

Figure 2.   Comparison of the required and predicted values of the ratio $\dot{M}/v_T^4$ if the X-ray source is fueled by accretion from the stellar wind, where $\dot{M}$ is the wind mass loss rate and v_T is the terminal velocity. The required values are derived from the observed X-ray luminosities and the theory of Davidson and Ostriker (1973), and the predicted values are derived from measured properties of winds from other stars of similar spectral type (figure from White 1985).

structure of the wind, and to search for evidence of shock structures predicted by various authors (Lucy 1984; Lucy and White 1980; Owocki and Rybicki 1984; Owocki et al. 1988). This could be done by measuring the X-ray absorption along various lines of sight to the X-ray source as it orbits the companion star, and then reconstructing the density distribution as a function of radius from the center of the companion star using techniques analagous to those used to reconstruct images from a medical CT scanner. The absorbing column density along the line of sight to the X-ray source can be measured by observations of the X-ray spectrum, because the wind produces photoelectric absorption with a characteristic shape (opacity $\sim v^3$) which is relatively easy to fit to observed spectra.

The first step in X-raying the wind is to check the measured column at various phases of the binary orbit against that expected from a spherical wind with a wind mass loss rate and velocity law which is based on the properties of single OB star winds. Such comparisons show that the column densities are generally consistent with the expectations for systems with the lowest $L_X/\dot{M}$ ratio; these are the extreme objects of the class for which wind accretion is sufficient to fuel the source (White et al. 1983a; Kallman and White 1982). In systems with high $L_X/\dot{M}$ there appears to be little or no photoelectric absorption intrinsic to the wind except near eclipse transitions. The change in absorption near eclipse has been used to infer the density distribution at the base of the wind in these sources by Clark et al. (1988,1990), and reveals what appears to be a much larger scale height for the atmosphere than would be predicted from theory for such atmospheres.

TABLE I

Massive Binaries and Wind Properties[a]

Object	$\log L_x$	$\log \dot{M}$	F_{ecl}/F_0	EW_{Fe}
	erg s^{-1}	M$_\odot$yr^{-1}		ev
Cen X-3	37–38	−6.0	0.1–0.3	228±36
4U1700-37	35.6	−5.0	0.00–0.2	630±90
4U1538-53	36.6	−6.0	0.15–0.20	572±140
SMC X-1	38.7	−6.3	—	≤ 300
Cyg X-1	36.8	−5.6	—	≤ 70
GX301-2	36.3–37.0	−5.7	—	225±76

[a] References: White (1985); White et al. 1983*a,b*.

One shortcoming of the use of column densities based on photoelectric absorption in X-ray spectra is that such absorption is suppressed if the wind is highly ionized. A test which avoids this problem is to measure the ratio of the X-ray flux observed during eclipse to that observed out of eclipse. The X-rays observed during eclipse are likely to be due to electron scattering around the star, and so are unaffected by the ionization of the wind (although this measurement must be carried out at energies greater than, say, 2 KeV in order to be sure to avoid photoelectric absorption). The predicted value of this ratio depends on the column density in the wind, and hence on the ratio $\dot{M}/v_T$. Table I gives measured values of the ratio of eclipsed and uneclipsed fluxes for several MXBs, and Fig. 3a shows a comparison of the predicted and derived ratio $\dot{M}/v_T$. As found previously, in the high $L_X/\dot{M}$ systems the predicted wind properites fail to account for the observed properties, while in low $L_X/\dot{M}$ systems the converse is true. A further test comes from observations of the equivalent width of the K line of iron at 6.4 KeV. This line is produced by reprocessing of X rays from wind material, with an efficiency which is only weakly dependent on the ionization state of the wind. The predicted equivalent width again depends on $\dot{M}/v_T$. Table I gives measured values for this quantity for several systems, and Fig. 3b compares the implied $\dot{M}/v_T$ with that expected from wind theory. In this case, virtually all the MXBs exhibit stronger lines than predicted, with the exceptions of those with very weak lines. This may be due to line formation in the atmosphere of the companion star. More accurate measurements of the spectrum in the vicinity of the line can resolve this ambiguity, because the line from an ionized wind will have a higher energy than the line from the (nearly neutral) stellar atmosphere.

More careful measurements of the photoelectric absorption also reveals significant departures from the behavior expected from a simple spherical wind. These occur at certain phases of the binary orbit for virtually all objects (White 1985; Kallman and White 1982; Watson and Griffiths 1977; Charles

                     T. R. KALLMAN

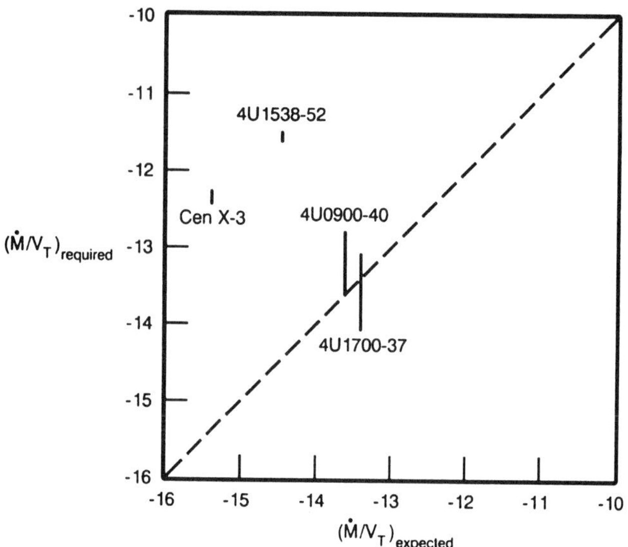

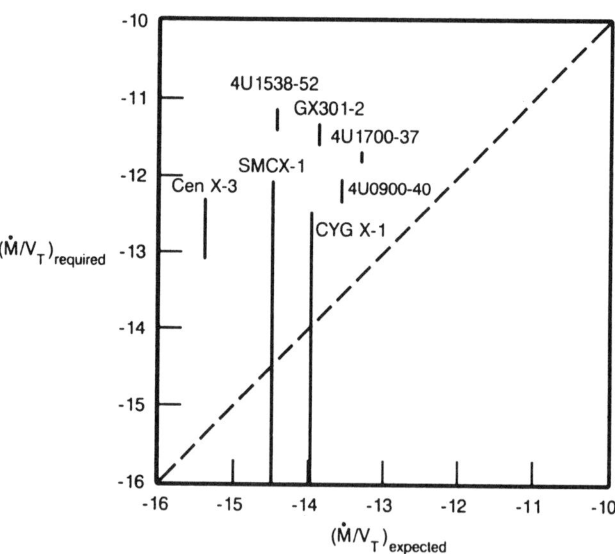

Figure 3.   Comparison of the required and predicted values of the ratio $\dot{M}/v_T$ to explain the eclipsed flux (panel a) and the K line equivalent width (panel b).

et al. 1978; Branduardi et al. 1978). Examples are shown in Fig. 4. Figure 4 shows the X-ray count spectra from two low $L_X/\dot{M}$ objects each at two orbital phases which are symmetric around phase 0.5. Here we adopt the convention that phase 0.5 corresponds to the X-ray source along the line of sight to the companion, and phase 0 is the middle of the X-ray eclipse. These spectra should show identical absorption if the wind were spherically symmetric. Instead, these and other observations show an excess of absorption at later orbital phases, suggesting the presence of excess gas trailing the X-ray source in its orbit. Figure 5 shows the ratio of hard X rays to soft X rays, also a measure of photoelectric absorption, from a high luminosity source (Cen X-3) as a function of orbital phase. In this case, there is excess absorption at phases earlier than 0.5, indicating excess gas leading the X-ray source in its orbit. This spells trouble for the use of MXBs as probes of wind density, because it indicates that the assumption of spherical symmetry for these winds is not adequate, and that the X-ray source plays a role in determining the wind density structure rather than being simply a probe of the wind.

One obvious effect of the X-ray source on the wind is ionization. This was first discussed by Hatchett and McCray (1977) who predicted that the ultraviolet resonance lines formed in the wind would be suppressed by X-ray ionization, and that this effect would be most apparent at phase 0.5 of the binary orbit. This effect was first verified by Dupree et al. (1987) and later observed to be occuring in virtually all of the MXBs which are bright enough to observe in the ultraviolet (see, e.g. Hammerschlag-Hensberge 1980; Hammerschlag-Hensberge et al. 1984,1989). However, in spite of the success in predicting this effect, agreement between model line profiles calculated assuming spherical winds with plausible wind velocity laws and those observed is poor (McCray et al. 1984), providing further evidence for a more complicated interaction between the X-ray source and wind material.

In addition to suppressing ultraviolet lines and X-ray photoelectric opacity, ionization can affect the wind dynamics. This can occur because the ultraviolet resonance lines are responsible for the driving of the wind via radiation pressure from the primary, so ionization will cause the wind to "coast" past the X-ray source. The reduced wind speed can enhance the mass accretion rate, possibly leading to a runaway in accretion, X-ray luminosity, and ionization. This will be limited by the attenuation of X rays in the wind. According to this picture, there would exist two stable states: a high accretion rate high, X-ray luminosity state with a slow ionized wind, and a low X-ray state with an unionized wind (Ho and Arons 1987; MacGregor and Vitello 1982).

The discovery that the absorbing column is asymmetric around phase 0.5 can be understood if the wind is not spherically symmetric. This can occur if the wind dynamics is affected by ionization effects, as discussed in the previous paragraph, or by the gravity of the X-ray source or by X-ray heating. Gravitational focusing of the wind is expected to produce a dense wake which trails the X-ray source in its orbit, thus producing absorption enhancements

T. R. KALLMAN

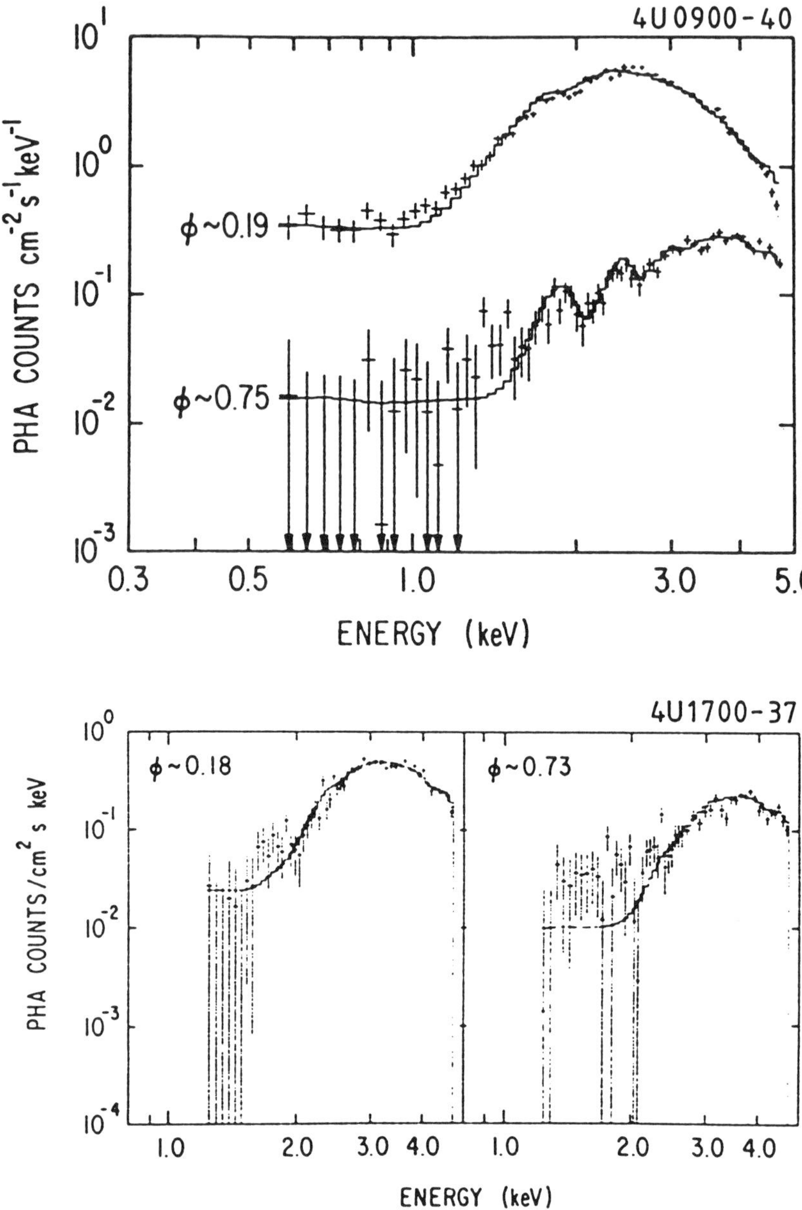

Figure 4. Observations of two MXBs showing evidence for departures from spherical symmetry in the winds. The X-ray count rate spectra were taken by the solid-state spectrometer on the Einstein (HEAO-2) satellite from the low $L_X/\dot{M}$ systems 4U0900-37 (Vela X-1) and 4U1700-37, showing evidence for higher photoelectric absorption near orbital phase 0.8 (eclipse ingress) than phase 0.2 (eclipse egress) (Kallman and White 1982; White et al. 1984).

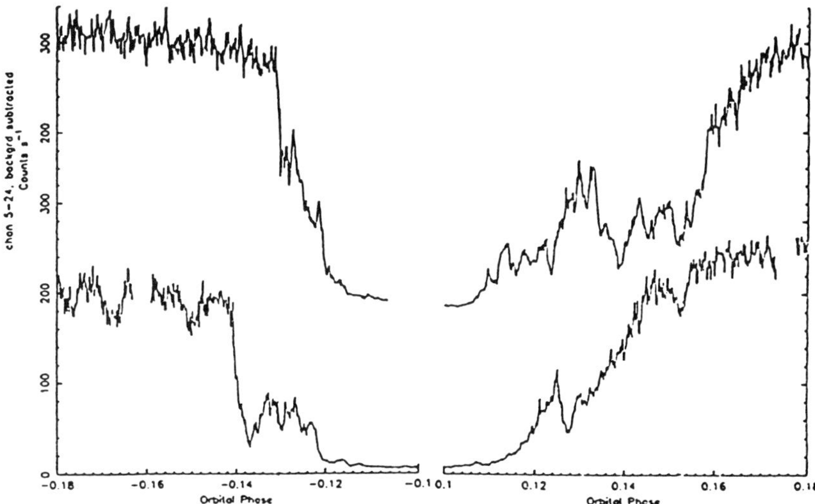

Figure 5.   The hardness ratio observed by the Ginga satellite from the high $L_X/\dot{M}$ system Cen X-3, showing evidence for greater absorption near eclipse egress than ingress (figure from Day and Stevens 1993).

near orbital phase 0.6 (Hunt 1971; Jackson 1975). A bow shock may also be produced, depending on the equation of state of the wind. Figure 6 shows an example of such a wake in an adiabatic wind, taken from two dimensional hydrodynamic calculations of Blondin et al. (1990). X-ray ionization can produce an extended region of shocks trailing the X-ray source, owing to the collision between faster accelerated wind and slower ionized wind in the ionization zone (Fransson and Fabian 1980). This effect is illustrated in Fig. 7, which shows the result of a two-dimensional hydrodynamic simulations of progressively greater $L_X/\dot{M}$ in which the gravity and heating from the X-ray source are neglected, but the ionization effect is not; the photoionization wake is shown to be progressively more dense as the X-ray luminosity is increased (Blondin et al. 1990). In this figure the results are parameterized by the quantity $q = L_X/(n_X a^2)$, (Hatchett and McCray 1977) where n_X is the gas density at the X-ray source and a is the orbital separation. The wake also assumes a very different angle relative to the line of centers of the two stars than does the adiabatic wake. In this case, the wake is nearly parallel to the orbital velocity vector of the neutron star, and can therefore produce excess absorption at a wide range of orbital phases.

Other important observational features also require explanation. X-ray pulsars in massive binary systems exhibit changes in their spin periods. These reflect the torques exerted by the accretion flow on the spinning neutron star, and some objects exhibit stochastic variations in spin periods (Boynton et al. 1984). This may be due to inhomogeneities inherent in the wind driving mechanism itself which are independent of the hydrodynamic effects of the

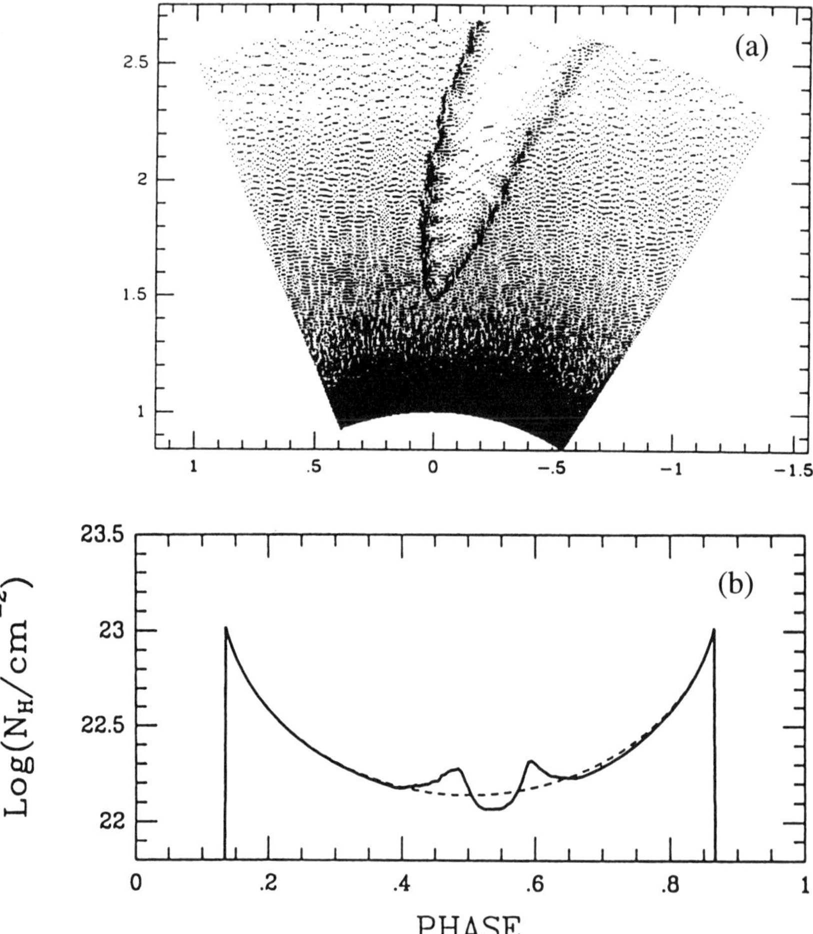

Figure 6.   (a) The accretion wake of a neutron star in an adiabatic, continuously
accelerated stellar wind. The halftone scale displays the density of the wind from
10^9 cm^{-3} (white) to 10^{11} cm^{-3} (black). The scale of this and the following plots
is labeled in units of stellar radii, $R_\star = 2.5 \times 10^{12}$ cm. The accretion radius is
6×10^{10} cm, and thus is not visible on the scale of this plot. (b) The line of sight
column density N_H as a function of orbital phase for the adiabatic wind simulation
shown in panel (a). The dotted line is the column density expected for a spherically
symmetric, undisturbed wind. The solid line is the column density derived from
the simulation results and averaged over several timesteps. We have assumed that
the density of the wind outside of the computational grid is given by that of the
undisturbed wind, and hence have underestimated the deviation from a smooth wind
at phases where the wake extends beyond the grid (figure from Blondin et al. 1990).

neutron star, or it may be due to instabilities introduced by the accretion process or X-ray heating. Simulations of the flow near the neutron star reveal the existence of a "flip-flop" instability which can cause intermittent changes in the sign of the angular momentum of the accreted gas; these have been studied in two dimensions by Fryxell et al. (1987), Fryxell and Taam (1988), Taam and Fryxell (1988,1989), and Matsuda et al. (1987). Recent simulations in three dimensions suggest that the instability is significantly weaker than had been claimed previously based on two dimensional results (Matsuda et al. 1991; Sawada et al. 1989), but there is still some debate about the magnitude of this reduction (Blondin 1993).

Flows farther from the neutron star, on a scale comparable with the binary separation, have been studied by Blondin et al. (1990,1991), who included the effects of X-ray ionization and heating. These papers showed that asymmetric and time-variable structures can be produced by any of the mechanisms proposed earlier, and that which one obtains depends on the parameters of the system. At low $L_X/\dot{M}$ the dominant effects are compressive heating of the wind, leading to relatively steady wake structures resembling the adiabatic wake shown in Fig. 5. At intermediate $L_x/\dot{M}$, X-ray heating leads to very unstable accretion behavior, including dense wake-like structures which are highly variable. The sign and magnitude of the net angular momentum transferred to the neutron star varies on time scales comparable to the wind dynamical time, $\sim 10^3$ s, and the wake structures produce variable absorption which occurs at a wide range of orbital phases approaching the X-ray eclipse (Fig. 8). At high $L_x/\dot{M}$ the ionization effect on the wind dynamics is dominant, leading to a dense, relatively time steady region of shocked gas trailing the neutron star in its orbit.

Further modifications occur if the primary star fills its Roche lobe; mass loss tends to occur from the inner Lagrangian point, producing a dense accretion stream which trails the neutron star in its orbit. Figure 9 shows a simulation of such a flow (Blondin et al. 1991). This suppresses some of the variability seen in non-Roche filling simulations at intermediate $L_x/\dot{M}$, but will still result in excess absorption at late orbital phases. In particular, systems which do not show stochastic changes in their pulse periods may be accreting through structures such as these, because most of the accreted material comes from the stream. It should also be pointed out that the location of the stream depends on whether the companion star corotates with the X-ray source in its orbit; if not, the stream is likely to lag even further behind than shown in Fig. 9.

A complimentary problem arises in the undertanding of the absorption in very high $L_x/\dot{M}$ sources such as Cen X-3 (cf. Figs. 4 and 5), because these show an excess of absorption at early orbital phases. Two possible explanations for this are that it is due to a "self-excited wind" (described in more detail in the following section) in which X-ray heated material is thermally driven from the companion surface (Day and Stevens 1993), or that it is due to a "shadowed wind" (Blondin and Woo 1995). According to this

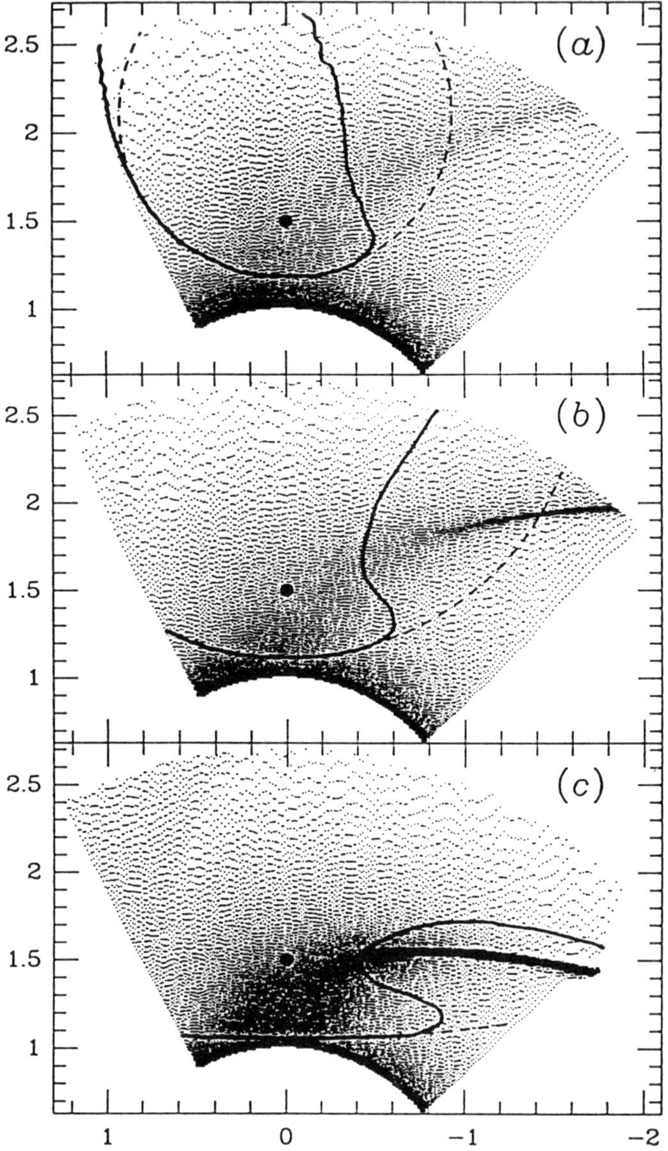

Figure 7.   The density distribution of the photoionization wake as a function of the size of the X-ray Strömgren zone. The values of q are (a) 6.3, (b) 4.0, and (c) 2.0. The density scale varies from 10^9 cm^{-3} to 10^{11} cm^{-3} at saturation (black). The location of the X-ray source is marked by a black dot. The dotted line is the surface of the Strömgren zone in the undisturbed, spherically symmetric wind and the solid line is the self-consistent Strömgren zone, accounting for the changes in the wind density. A very strong photoionization wake begins to show only for relatively small values of q, where the Strömgren zone reaches close to the surface of the primary (figure from Blondin et al. 1990).

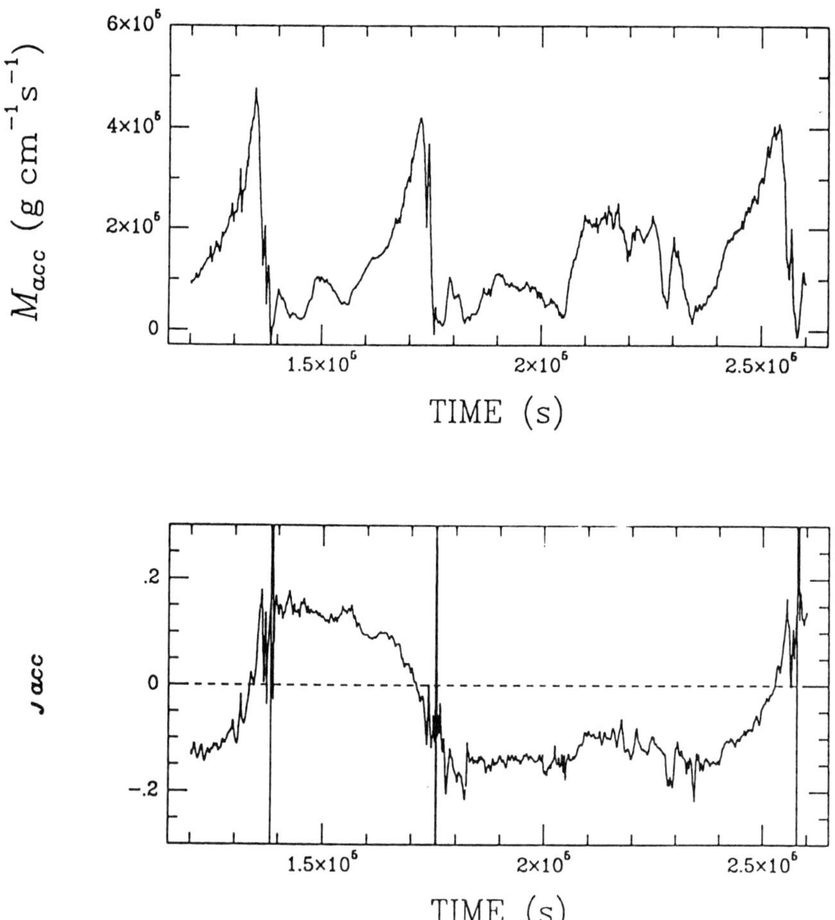

Figure 8.   (a) The mass flux across a circle of radius 2×10^{10} cm for the standard model. The peaks in the mass accretion rate, $\dot{M}_{\mathrm{acc}}$, correspond to episodes of disk oscillation, when the accretion disk changes its sense of rotation. (b) The specific angular momentum of the mass transported across a radius of 2×10^{10} cm in units of $(r_{\mathrm{acc}} v_{\mathrm{rel}})$ for the standard model (figure from Blondin et al. 1990).

latter scenario, the normal radiatively driven wind is suppressed throughout most of the volume of the binary owing to X-ray ionization of the ions responsible for the ultraviolet opacity, and hence the driving of the wind. Only in the shadow of the companion star can a relatively normal wind form. The gas in this wind is swept into the X-ray ionized zone, however, by rotational forces, and produces the observed excess absorption at eclipse egress before it is ionized.

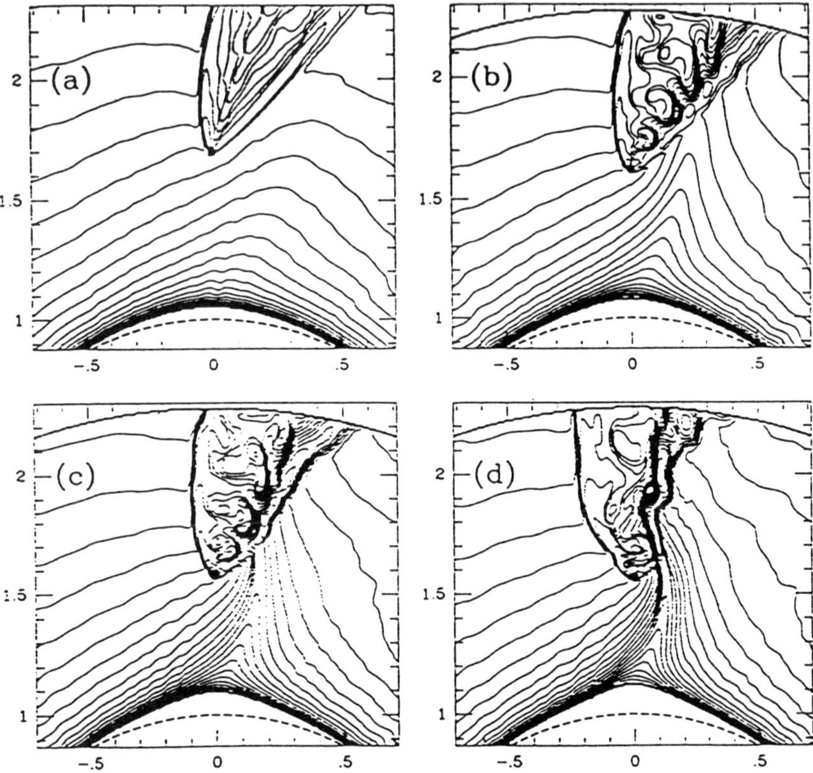

Figure 9.   Formation of a gas stream as the binary separation is decreased in the MXRB models. The density contours are spaced logarithmically in units of $\sqrt{2}$. The four plots are for different binary separations of $D/R_* = 1.70$ (a), 1.62 (b), 1.59 (c), and 1.57 (d). The physical dimensions of these plots are labeled in units of the stellar radius, $R_* = 2.45 \times 10^{12}$ cm, with the coordinate origin positioned at the center of the primary. The dashed line at the bottom of each plot marks the undistorted stellar surface of radius R_*. (figure from Blondin et al. 1992).

X-ray ionization and hydrodynamic effects have a variety of observable consequences, many of which have yet to be explored in detail. In addition to the Hatchett-McCray effect described above, the ultraviolet lines are expected to vary in response to variations in the intrinsic brightness of the X-ray source. In an X-ray pulsar this variation will be periodic at the X-ray period (Kallman et al. 1987) if the pulse period is long compared with the light travel time across the binary orbit, $\sim 10^2$ s. This effect has been observed using HST ultraviolet spectroscopy from one system (Boroson et al. 1995), although at a lower level of modulation than was predicted. Similar variability is expected from non-pulsed sources, in which the X-ray output varies over a wide range of time scales. In this case, the ultraviolet lines will behave as a low pass filter, owing to light travel time smearing of the most rapid variability. X-ray lines are another signature of X-ray heating and ionization which are expected to vary due to changes in both the continuum luminosity and the wind density structure. The iron K fluorescence line is produced with an efficiency which is only weakly dependent on the ionization state and temperature of the wind, so variations in this line are expected to reflect luminosity variations. Other lines, such as the L lines of iron and the K lines of oxygen, silicon and sulfur between 0.6 and 2 KeV, are sensitive to changes in ionization. The fluxes in these lines should reflect variable wake structures associated with the complicated hydrodynamics of the wind.

III. WINDS FROM LOW-MASS BINARIES

The other class of X-ray binaries, those with companion stars with masses less than approximately 1 $M_\odot$, are called low mass X-ray binaries (LMXB). These are more numerous but many aspects of their winds and accretion flows are less well understood than are the more massive systems. There are approximately 100 known systems of this type in our Galaxy, the majority located in the galactic bulge. Orbital periods range from 20 minutes to several days, implying orbital separations 10^{10} to 10^{11} cm. Because the companion star in these systems is not expected to emit a strong wind, the most likely mechanism for mass transfer is Roche lobe overflow. This mechanism requires an additional sink for the orbital angular momentum because, in the absence of such a sink, transfer of matter from a less massive to a more massive star leads to an increase in the binary separation and the loss of the Roche-lobe filling condition. Possible mechanisms include gravitational radiation, or weak winds which are coupled to the star by magnetic fields (Patterson 1984). The gas lost from the companion via Roche lobe overflow initially forms a stream as it leaves the inner Lagrangian point. Closer to the compact object, it settles into a dense disk as it spirals inward and loses its angular momentum. Such accretion disks are also likely to occur in cataclysmic variables and active galactic nuclei; their structure has been extensively explored theoretically (see, e.g. Shakura and Sunyaev 1973), although many uncertainties remain concerning the viscosity mechanism.

Winds, if they exist in LMXBs are most likely to be excited by X-ray heating of the accreting gas or the surface of the companion star. Observational evidence for such heating comes from the photometric changes observed as a function of orbital phase in the companion star in the system Hercules X-1 (Bahcall and Bahcall 1972), and from evidence for a hot corona which scatters X-rays around the companion during X-ray eclipse in several low-mass X-ray binaries (White and Holt 1982). In order to create a wind, X-ray heated gas must reach a temperature comparable to the escape temperature from the local gravitaitional field, $T_{esc} \simeq GM/Rk = 1.6 \times 10^9 \mathrm{K} R_9^{-1}(M/M_\odot)$, where M is the mass, R is the distance from the center of the star or compact object, and R_9 is R scaled in units of 10^9 cm. X rays can heat gas via photoelectric heating, or Compton scattering; the latter dominates when the gas is highly ionized. In the Compton-heated limit the gas cooling is dominated by inverse Compton scattering, and the relative rates for these processes depend only on the shape of the X-ray spectrum. The Compton temperature, which the gas attains in such a situation is typically greater than the temperature attainable by photoelectric heating, and is defined as $T_{IC} = \langle \varepsilon \rangle / 4k \sim 10^7$ K, where $\langle \varepsilon \rangle$ is the mean energy of the continuum photons.

The temperature of X-ray heated gas depends only on the shape of the X-ray spectrum and on the ionization parameter. In situations where the gas pressure is specified, such as in a hydrostatic atmosphere above a star or disk, this quantity is defined as the ratio of the X-ray flux, F (in energy units over the band 1–1000 Ry), to the gas pressure P, according to $\Xi = F/Pc$ (cf. Krolik et al. 1981). For ionization parameter values greater than a critical value, Ξ_c^*, the gas temperature is determined primarily by the effects of Compton scattering. For ionization parameters less than a critical value Ξ_h^*, the gas temperature is determined by the effects of atomic processes; the temperature in this case is $T_{\text{atomic}} \sim 10^4$ K. These critical ionization parameters have values $\Xi_h \simeq 0.3(T_{IC}/10^8 K)^{-3/2}$ and $\Xi_c^* \simeq 10$ (although various effects such as conduction, nonequilibrium effects, strong Compton cooling by the ultraviolet radiation from the disk and other sources of heating may modify these estimates). For intermediate ionization parameters, $\Xi_h^* < \Xi < \Xi_c^*$, there exists a third equilibrium temperature which is unstable according to Field's (1965) criterion.

The existence of an X-ray heated corona depends on the decrease of gravity with distance from the accretion disk midplane. The conditions necessary for the formation of a corona may be expressed in terms of the hydrostatic condition:

$$\frac{dP}{dz} = -\frac{GMm_H}{R^3 kT} Pz = -\frac{Pz}{z_s^2} \leq 0 \tag{1}$$

where P is the gas pressure, z is height above the disk midplane, M is the mass of the central X-ray source, R is the distance from the X-ray source and m_H is the mean mass per particle in the gas. The scale height is $z_s = \sqrt{R^3 kT/(GMm_H)} = 8 \times 10^7 \text{ cm} R_9^{3/2} T_7^{1/2} (M/M_\odot)^{-1/2}$, and the temperature gradient in the disk atmosphere may be written:

$$\frac{dT}{dz} = \frac{dP}{dz} \frac{dT}{d\Xi} \left(\frac{d\Xi}{dP} \right)_F . \tag{2}$$

Because $(d\Xi/dP)_F < 0$, the necessary condition for a corona to form is that $dT/d\Xi > 0$ which is satisfied for $\Xi_h^* \leq \Xi \leq \Xi_c^*$ in the atmospheres of an X-ray illuminated disk or star. The vertical structure in the region where $dT/d\Xi < 0$, for $\Xi_h^* < \Xi < \Xi_c^*$, has not been studied in detail; however, this region is likely to be extremely narrow under the conditions appropriate to most X-ray binaries.

Therefore, we expect a corona at $T = T_{IC}$ to exist wherever $\Xi \geq \Xi_h^*$; it *must* exist for $\Xi \geq \Xi_c^*$. Near the interface between the corona and the disk photosphere must exist a chromosphere and transition region in which X-ray heating is important, but is not sufficiently strong to heat the gas to temperatures comparable to T_{IC}. These regions must dominate the optical and ultraviolet appearance of X-ray illuminated accretion disks. The maximum density of the corona may then be expressed in terms of the incident X-ray flux F according to:

$$n_{\max} = \frac{F}{c\, \Xi_c^* k T_{IC}}$$

$$= 1.9 \times 10^{17}\ \mathrm{cm}^{-3} R_9^{-2} L_{37} f T_{IC7}^{-1} \tag{3}$$

where R_9 is the radius in units of 10^9 cm, L_{37} is the X-ray source luminosity in units of 10^{37} erg^{-1}, and T_{IC7} is the Compton temperature in units of 10^7K. The effects of geometry and the transfer of X rays through the corona are combined in the factor $f = 4\pi R^2 F/L$. $f = 1$ corresponds to unattenuated X-rays normally incident on the disk. The total coronal column is $\sim n_{\max} z_s$ or

$$N_{\mathrm{tot}} = \frac{Lf}{4\pi\, \Xi_c^* c}(RkT_{IC}\mu m_H GM)^{-1/2} E(z_b/z_s)$$

$$= 1.5 \times 10^{25}\ \mathrm{cm}^{-2} L_{37} f R_9^{-1/2} T_{IC7}^{-1/2} E(z_b/z_s) \tag{4}$$

where z_b is the height of the base of the corona, M is the mass of the compact object, μm_H is the mean mass per particle in the gas and E is a slowly varying function related to the error function. It is clear from this equation that the column density of the corona increases with decreasing distance from the source, and that when $R \leq R_{\mathrm{thick}} = 10^9$ cm$L_{37}^2 f^2 T_{IC7}^{-1}$, the Thomson depth through the corona exceeds unity. This buildup must ultimately be limited by the fact that X rays will not penetrate to the base of a corona whose Thomson depth τ_{Th} exceeds some critical value which we define as $\tau_{Th\,\mathrm{crit}} \simeq 1$ to 10. Therefore we may treat the requirement $\tau_{Th} \leq \tau_{Th\,\mathrm{crit}}$ as a second (optically thick) necessary condition for coronal equilibrium. When this condition applies, the minimum ionization parameter in the corona may be shown to be

$$\Xi_{\min} = \frac{Lf\sigma_{Th}}{4\tau_{Th\,\mathrm{crit}}\pi c(RkT_{IC}\mu m_H GM)^{1/2}} E(z/z_s)$$

$$= 10 L_{37} f R_9^{-1/2} T_{IC7}^{-1/2} \tau_{Th\,\mathrm{crit}}^{-1} \frac{M}{\mathrm{M}_\odot} E(z/z_s) \tag{5}$$

where σ_{Th} is the Thomson cross section and z is height at which Thomson depth $\tau_{Th\,\mathrm{crit}}$ occurs ($E(z/z_s)$ is of order unity for our fiducial parameter values and is a very slowly varying function of R). In what follows we assume $\tau_{Th\,\mathrm{crit}} = 1$. As $\Xi_h^* \simeq 0.1$ to 1 (cf. Krolik et al. 1981), this suggests that in the inner region of the disk where the optically thick criterion applies the mean ionization parameter may be much greater than Ξ_c^*, and that it increases with decreasing distance from the source.

At radii $R \geq R_{IC} = 1.6 \times 10^{11}$ cm $T_{IC7}^{-1}(M/M_\odot)$, the corona temperature exceeds the escape temperature and the corona is not bound to the disk. In this region, referred to as the wind region, a crude estimate for the mass loss rate per unit area is (Begelman et al. 1983; Begelman and McKee 1983):

$$\dot{M}/A \simeq n_{\max} m_H v_{\text{wind}} \tag{6}$$

where the density at the base of the wind is $n_{\max} = P_{\min}/(kT_{IC})$ and $P_{\min} = F/(c \Xi_h^*)$, and $v_{\text{wind}} = \sqrt{2kT_{IC}/m_H}$. Therefore

$$\dot{M} \simeq 2 \times 10^{19} \text{ g s}^{-1} F_{15} R_{11}^{-1/2} \Xi^{-1} T_{IC7}^{-1} A_{22} \tag{7}$$

where F_{15} is the flux in units of 10^{15} erg cm^{-2} s^{-1}, R_{11} is radius in units of 10^{11} cm, and A_{22} is the wind emitting area in units of 10^{22} cm^2. This can be compared with the mass accretion rate required to fuel the X-ray source, which is $\dot{M}_{\text{acc}} = L/(\eta c^2) = 10^{18}$ g s$^{-1} L_{38} \eta_{0.1}^{-1}$, where L is the X-ray source luminosity, L_{38} is the corresponding quantity in units of 10^{38} erg s^{-1}, and η is the efficiency with which accreting mass is converted into radiation. Clearly the wind mass loss rate can greatly exceed that required to fuel the X-ray source under suitable conditions. This means that, if the wind is from a heated star, only a small fraction needs to be captured and accreted in order to power the X-ray source. This self-excited wind is thus a possible mechanism for driving mass transfer in LMXBs. If the wind comes from a heated disk, it can quickly evaporate most of the material in the wind region of the disk. This void will propogate inward and eventually reach the compact object, causing the X rays to turn off. The wind driving will then stop, and the disk will reform assuming that the mass supply from the companion star is uninterrupted. Such a disk wind and episodic accretion has been suggested as a possible mechanism for variability in active galaxies. The time scale characterizing this process is the viscous time scale in the disk (see, e.g. Begelman and McKee 1983), $\Delta t \sim \Delta t_{\text{viscous}} \simeq 85s \alpha^{-1}(z_s/R_{IC})^{-2}(M/M_\odot)(R/R_{IC}T_{IC8})^{3/2}$, where α is the disk viscosity parameter.

The self-excited wind has been suggested as a mechanism to explain the evolution of low-mass X-ray binaries, including the loss of orbital angular momentum, mass transfer, and spin-up of the neutron star, leading eventually to evaporation of the companion star and production of a millisecond radio pulsar (see, e.g., Tavani and London 1993). The discovery of PSR 1957+20, a millisecond pulsar with a low mass ($\simeq 0.05$ M$_\odot$) companion, provides support for this picture (see, e.g. Ruderman et al. 1989). Modeling of self excited winds reveals several potential problems with this scenario. First, there are as yet no two- or three-dimensional calculations which calculate what fraction of the wind material is actually captured by the compact object. If too great a fraction is captured, the X-ray source will be smothered, and if the fraction is too small, the observed X rays cannot be accounted for. Second, the mass loss rate depends on the spectrum of the illuminating X rays; a strong flux of soft (≤ 1 KeV) X rays could cool the wind by inverse Compton scattering and produce much less mass loss. Third, there are uncertainties about the importance of emission line cooling in the wind; lines can cool the wind and reduce its mass loss rate (London et al. 1981), but more modeling is required to understand fully the transfer of line photons. Finally, the heating of the star required to produce such a wind must be accompanied by some orbital-phase dependent changes in the photometric properties of the star. Very crudely, the temperature of the illuminated face of the companion star must be $T_{\mathrm{phot}} \simeq (F/\sigma)^{1/4} = 6.4 \times 10^4$ K $F_{15}^{1/4}$, which is much hotter than expected for late spectral types typical of LMXB companions. Observed photometric changes are much smaller than this, suggesting that the companion stars are not strongly heated in the LMXBs we can observe.

In the optically thin coronal region the emission measure of the corona is

$$EM_{\mathrm{thin}} = \left(\frac{fL}{ckT_{IC}\,\Xi_c^*}\right)^2 \frac{1}{4\pi} \left(\frac{kT_{IC}}{GMm_H}\right)^{1/2}$$

$$\times (R_{\mathrm{min}}^{-1/2} - R_{\mathrm{max}}^{-1/2})$$

$$= 4 \times 10^{61} \ \mathrm{cm}^{-3} \frac{L_{38}^2 f^2}{T_7^{3/2} R_{\mathrm{min}9}^{-1/2}}. \tag{8}$$

A wind is expected to emit ~ 1 M$_\odot$ of material over the lifetime of a star. Such large quantities of gas are likely to produce observable spectral features. For plausible line emissivities of $\sim 10^{-25}$ erg cm^3 s^{-1} (e.g., for the O VIII Lα line near 650 eV), such an emission measure would give line luminosities of $\sim 10^{36}$ erg s^{-1} or equivalent widths of ~ 100 eV if even 10% of the gas is capable of line emission. This is likely to be observable with X-ray detectors which will be available in the near future, thus providing a further test for the existence of winds in LMXBs.

REFERENCES

Bahcall, J. N., and Bahcall, N. A. 1972. The peroid and light curve of H2 Herculis. *Astrophys. J. Lett.* 178:1–4.

Begelman, M. C., and McKee, C. F. 1983. Compton heated winds and coronea above accretion disks. II. Radiative transfer and observable consequences. *Astrophys. J.* 271:89–112.

Begelman, M. C., McKee, C. F., and Shields, G. A. 1983. Compton heated winds and coronae above accretion disks. I. Dynamics. *Astrophys. J.* 271:70–88

Blondin, J. 1993. In *Evolution of X-Ray Binaries*, eds. N. White and S. Holt.

Blondin, J., and Woo, J. W. 1995. Wind dynamics in SMC X-1. I. Hydrodynamic simulation. *Astrophys. J.* 445:889.

Blondin, J., Kallman, T., Fryxell, B., and Taam, R. 1990. Hydrodynamic simulations of stellar wind disruption by a compact X-ray source. *Astrophys. J.* 356:591–608.

Blondin, J., Stevens, I. R., and Kallman, T. R. 1991. Enhanced winds and tidal streams in massive X-ray binaries *Astrophys. J.* 371:684.

Bondi, H. 1952. On spherically symmetrical accretion. *Mon. Not. Roy. Astron. Soc.* 112:195–204.

Bondi, H., and Hoyle, F. 1944. On the mechanism of accretion by stars. *Mon. Not. Roy. Astron. Soc.* 104:273–282.

Boroson, B., McCray, R., Kallman, T., and Nagase, F. 1995. *Astrophys. J.*, in press.

Boynton, P. E., Deeter, J. E., Lamb, F. K., Zylstra, G., Pravdo, S. H., White, N. E., Wood, K. S., and Yentis, D. J. 1984. New evidence on the nature of the neutron star and accretion flow in Vela X-1 from pulse timing observations. *Astrophys. J. Lett.* 283:53–56.

Branduardi, G., Mason, K. O., and Sanford, P. W. 1978. Further Copernicus X-ray observations of 3U 1700-37. *Mon. Not. Roy. Astron. Soc.* 185:137–142.

Castor, J. I., Abbott, D. C., and Klein, R. I. 1975. Radiation-driven winds in Of stars. *Astrophys. J.* 195:157–174.

Charles, P. A., Mason, K. O., White, N. E., Culhane, J. L., Sanford, P. W., and Moffat, A. J. F. 1978. X-ray and optical observations of 3U 0900-40 (Vela X-1). *Mon. Not. Roy. Astron. Soc.* 183:813–820.

Clark, G. W., Minato, J. W., and Mi, G. 1988. Discovery of a cyclotron absorption line in the spectrum of the binary X-ray pulsar 4U 1538-52 observed by Gringa. *Astrophys. J.* 324:974–994.

Clark, G., Woo, J., Nagase, F., Makishima, K., and Sakao, T. 1990. *Astrophys. J.* 353:274–280.

Conti, P. S. 1978. *Astron. Astrophys.* 63:225.

Day, C. S. R., and Stevens, I R. 1993. An X-ray excited wind in Centaurus X-3. *Astrophys. J.* 403:322–331.

Davidson, K., and Ostriker, J. P. 1973. Neutron star accretion in a stellar wind: Model for a pulsed X-ray source. *Astrophys. J.* 179:588–598.

Dupree, A. K., et al. 1987. IUE observations of X-ray sources: HD153913 (4U1700-37), HDE226868 (Cyg X-1), HZ Her (Her X-1). *Nature* 275:400–403.

Fabbiano, G. 1993. In *X-Ray Binaries*, eds. W. H. G. Lewin and E. P. J. Vanden Heuvel (Cambridge: Cambridge Univ. Press).

Field, G. B. 1965. Thermal instability. *Astrophys. J.* 142:531–567.

Frank, J., King, A. R., and Raine, D. J. 1985. *Accretion Power in Astrophysics* (Cambridge: Cambridge Univ. Press).

Fransson, C., and Fabian, A. C. 1980. *Astron. Astrophys.* 87:102.

Friend, D. B., and Castor, J. I. 1982. Radiation-driven winds in X-ray binaries. *Astrophys. J.* 261:293–300

Fryxell, B. A., and Taam, R. E. 1988. Numberical simulation of nonaxisymmetric adiabatic accretion flow. *Astrophys. J.* 335:862–880.

Fryxell, B. A., Taam, R. E., and McMillan, S. L. W. 1987. Numerical simulations of adiabatic axisymmetric accretion flow. I. A new mechanism for the formation of jets. *Astrophys. J.* 315:536–554.

Hammerschlag-Hensberge, G. 1980. In *Proc. of the Second European IUE Conference*, ESA SP-157, p. lix.

Hammerschlag-Hensberge, G., Kallman, T. R., and Howarth, I. D. 1984. Ultraviolet high-resolution spectroscopy of the X-ray binary SK 160/SMC X-1. *Astrophys. J.* 283:249.

Hammerschlag-Hensberge, G., Howarth, I. D., and Kallman, T. R. 1989. Orbital variability in the wind of the massive X-ray binary HD153919/4U 1700-37. *Astrophys. J.* 352:698–708.

Hatchett, S. P., and McCray, R. A. 1977. X-ray sources in stellar winds. *Astrophys. J.* 211:552–561.

Ho, C., and Arons, J. 1987. High-luminosity accretion in wind-driven binary X-ray sources. *Astrophys. J.* 316:283–293.

Hoyle, F., and Lyttleton, R. A. 1939 *Proc. Cambridge Phil. Soc.* 35:405.

Hunt, R. 1971. A fluid dynamical study of the accretion process. *Mon. Not. Roy. Astron. Soc.* 154:141–165.

Jackson, J. C. 1975. Parameters of CEN X-$_3$ deduced from observation of its accretion wake. *Mon. Not. Roy. Astron. Soc.* 172:483–492.

Kallman, T. R., and White, N. E. 1982. The anomalous X-ray absorption spectrum of Vela X-1[1]. *Astrophys. J. Lett.* 261:35–39

Kallman, T. R., McCray, R., and Voit, G. M. 1987. Rapid variability of P Cygni lines in massive X-ray binaries. *Astrophys. J.* 317:746–749.

Krolik, J. H., McKee, C. F., and Tarter, C. B. 1981. Two-phase models of quasar emission line regions. *Astrophys. J.* 249:422–442.

London, R., McCray, R., and Auer, L. H. 1981. The structure of X-ray illuminated stellar atmospheres. *Astrophys. J.* 243:970–982.

Lucy, L. B. 1984. Wave amplification in line-driven winds. *Astrophys. J.* 284:351–356.

Lucy, L. B., and White, R. L. 1980. X-ray emissions from the winds of hot stars. *Astrophys. J.* 241:300–305.

MacGregor, K. B., and Vitello, P. A. J. 1982. Stellar winds in binary X-ray systems. *Astrophys. J.* 259:267–281.

Matsuda, T., Inoue, M., and Sawada, K. 1987. Spin-up and spin-down of an accreting compact object. *Mon. Not. Roy. Astron. Soc.* 226:785–811.

Matsuda, T., Sekino, N., Sawanda, K., Shima, E., Livio, M., Anzer, U., and Boerner, G. 1991. *Astron. Astrophys.* 248:301.

McCray, R. A., Kallman, T. R., Castor, J. I., and Olson, G. 1984. Spectral variability in early-type binary X-ray systems. Astrophys. J. 282:245–255.

Owocki, S. P., and Rybicki, G. B. 1984. Instabilities in line-driven stellar winds. I. Dependence on perturbation wavelengths. *Astrophys. J.* 284:337–354.

Owocki, S. P., Castor, J. I., and Rybicki, G. B. 1988. Time-dependent models of radiatively driven stellar winds. I. Nonlinear evolution of instabilities for a pure absorption model. *Astrophys. J.* 335:914.

Patterson, J. 1984. The evolution of cataclysmic and low-mass X-ray binaries. *Astrophys. J. Suppl.* 54:443–493.

Ruderman, M., Shaham, J., and Tavani, M. 1989. Accretion turnoff and rapid evaporation of very light secondaries in low-mass X-ray binaries. *Astrophys. J.* 336:507–518.

Sawada, K., et al. 1989. *Astron. Astrophys.* 221:263.

Shakura, N. I., and Sunyaev, R. A. 1973. Black holes in binary systems: Observational appearance. *Astron. Astrophys.* 24:337–355.

Taam, R. E., and Fryxell, B. A. 1988. On nonsteady accretion in stellar wind fed X-ray sources. *Astrophys. J. Lett.* 327:73–76.

Taam, R. E., and Fryxell, B. A. 1989. Numberical studies of asymmetric adiabatic accretion flow: The effect of velocity gradients. *Astrophys. J.* 339:297–313.

Tavani, M., and London, R. 1993. Hydrodynamic of winds from irradiated companion stars in low-mass X-ray binaries. *Astrophys. J.* 410:281–294.

Vanden Huevel, E. 1977. In *Proc. of the Eighth Texas Symposium on Relativistic Astrophysics* (New York Academy of Sci.).

Van Paradijs, J. 1993. In *X-Ray Binaries*, eds. W. H. G. Lewin and E. P. J. Vanden Heuvel (Cambridge: Cambridge Univ. Press).

Watson, M. G., and Griffiths, R. E. 1977. Ariel V sky survey instrument: Extended observations of 3U 9099-40. *Mon. Not. Roy. Astron. Soc.* 178:513–524.

White, N. E. 1985. In *Interacting Binaries*, eds. P. P. Eggleton and J. E. Pringle (Dordrecht: D. Reidel), pp. 249.

White, N. E., and Holt, S. S. 1982. Accretion disk coronae. *Astrophys. J.* 257:318–337.

White, N. E., Kallman, T. R., and Swank, J. H. 1983*a*. The X-ray absorption spectrum of 4U 1700-37 and its implications for the stellar wind of the compansion HD 153919. *Astrophys. J.* 269:264–272.

White, N. E., Swank, J. H., and Holt, S. S. 1983*b*. Accretion powered X-ray pulsars. *Astrophys. J.* 270:711–734.

White, N. E., Parmar, A., and Nagase, F. 1993. In *X-Ray Binaries*, eds. W. H. G. Lewin and E. P. J. Vanden Heuvel (Cambridge: Cambridge Univ. Press).

PART III
The Physics of Wind Origin

ACCELERATION OF THE SOLAR WIND

T. E. HOLZER
High Altitude Observatory

and

V. H. HANSTEEN and E. LEER
University of Oslo

Parker considered the outflow of coronal plasma from a hot, inner boundary, the coronal base, which can supply any energy and mass flux to the solar wind. This model is well suited to describe the force balance in the corona-solar wind system, and the subsonic-supersonic solar wind. In an effort to study the energy balance of the system, we place the inner boundary in the chromosphere and specify a coronal heating function. It is shown that most of an energy flux deposited over a short distance, in the inner corona, is lost as inward heat conduction (and radiation from the top of the chromosphere), whereas an energy flux deposited in the extended corona, over several density scale heights, goes into driving the solar wind.

I. INTRODUCTION

The concept of a solar wind has been with us for quite some time. The most comprehensive early study leading to the conclusion that an outflow of ionized gas from the Sun must exist (to explain geomagnetic activity) was carried out by Birkeland (1896,1908,1913), although his rather remarkable research was largely neglected by workers in the field until relatively recently. Related ideas of ionized particle streams from the Sun were presented independently by Fitzgerald (1892,1900) and Lodge (1900), but these ideas suffered the same fate of those of Birkeland. The concept of ionized gas outflow from the Sun was sporadically revived for the next half century (see, e.g., Chapman 1919,1929), but it was not until Biermann's (1951,1953,1957) study of comet tails that the concept began to become well established. Making use of Biermann's observational inferences and extensive observational (Grotrian 1933,1939), interpretive (Edlén 1942), and theoretical (Alfvén 1941,1947; Biermann 1946,1948; Chapman 1954,1957) work describing a hot, static, extended solar corona, Parker (1958) developed the first theoretical description of solar wind outflow.

Although our current understanding of solar wind acceleration is based largely on Parker's (1958,1963,1964*a,b*,1965) early theoretical work, his basic ideas were not readily accepted by many in the community who were interested in such phenomena. The reasons for this failure to embrace Parker's ideas (which, in retrospect, are seen to be straightforward, sensible, and clearly explained) appear to be quite diverse, ranging from preconceptions held by many scientists at the time to sociological and psychological factors that seem to be present whenever a breakthrough in understanding is presented by someone who is not firmly situated in the existing scientific establishment. The preconceptions would seem to include the following: the prevailing view that atmospheric escape occurs primarily through single-particle escape, as from an exosphere; the thought that atmospheric temperature profiles are likely to be essentially adiabatic, as in the Earth's troposphere; and the assumption that forces accelerating flows are not dependent on the flow acceleration itself. The sociological and psychological factors might well be described by thoughts such as "it wasn't invented here," "if it were true, I would have noticed it myself," and "who is this guy Parker anyway?"

An illustration of some of these factors that delayed acceptance of Parker's work is provided by two comments of W. H. McCrea, F.R.S., in the Varenna Symposium on Cosmical Gas Dynamics (Thomas 1960), which was an early forum for the discussion of Parker's ideas. The first of McCrea's comments was

> These are indeed exactly the same equations as for symmetrical inflow which were solved first by Bondi. . .Now I spent a lot of time, myself and a pupil, trying to get outflow, and I could not get anything plausible. I got this solution that Parker has talked about. But you have got to give the material the velocity at some level. And that is the whole problem—you can always get the solution if you get the exact initial condition, but we could not see how this could come about in reality.

The last point is particularly intriguing, because it would seem to apply to Bondi's (1952) and McCrea's (1956) own work on accretion, where the velocity would, by McCrea's argument, have to be specified exactly at a location very far from the accreting object. (As we shall soon see, the point is actually applicable neither to Parker's work nor to the accretion work.) The closing comment in the session where Parker's work was discussed was also given by McCrea:

> I make one simple observation, which is not meant to be cynical, although it sounds like it. Fifteen years ago, Hoyle, Bondi, and Lyttleton explained the solar corona by infalling material and got a suitable density gradient. Here, you reverse all the velocities; naturally you get the same density gradient.

Here, McCrea misses the essential point that the coronal density gradients are observationally inferred in the region of subsonic flow, where the atmosphere is essentially hydrostatic, a point which Parker understood quite well. Although McCrea is correct that in a polytropic model the density structure is invariant to the flow direction, the inference McCrea seems to have drawn from this (that somehow Parker's work was not original or was simply trivial) is clearly incorrect.

Opposition from leading astrophysicists like McCrea was not all that Parker faced in explaining his solar wind idea. Perhaps the most strenuous and longest lasting opposition came from Chamberlain (see, e.g., 1961). Chamberlain's "solar breeze" solutions represented a very special type of flow, in which thermal conduction was important close to the Sun, the flow became asymptotically adiabatic at large distances, and (most important of all) the energy per unit mass (E) of the asymptotic adiabatic flow was exactly zero. The interesting characteristic of Chamberlain's singular case ($E \rightarrow 0$) is that asymptotically the flow speed, thermal speed, density, and pressure all vanish (the pressure varying as $r^{-2.5}$) and the Mach number of the flow approaches a constant, whether the flow is subsonic or supersonic (Holzer and Axford 1970). In contrast, for the physically realistic case that Parker considered, where asymptotically $E > 0$, the pressure vanishes at large distances for supersonic flow but approaches a constant for subsonic flow. The introduction of asymptotically vanishing pressure for subsonic flow in Chamberlain's singular case became the essential point in his argument that supersonic solar wind flow is unnecessary. The singular nature of the case Chamberlain presented, however, is the reason that his solutions are physically unacceptable, at least in the absence of an explanation of how a real physical system would be expected to achieve asymptotically a vanishing energy per unit mass in an adiabatic flow regime. Such an explanation has never been provided.

One characteristic of Parker's solar wind research is that physical assumptions made to provide tractable problems are always carefully spelled out. Parker, himself, investigated the effects of removing or modifying various of these assumptions, but the examination of some of the assumptions was left to future researchers. Perhaps, the most significant assumption not thoroughly investigated by Parker involved the use of boundary conditions on the density and temperature at the base of the corona. Assuming that appropriate values of the coronal base density and temperature can be specified, of course, implies that the impact of physical processes below the coronal base on the plasma above the coronal base is adequately represented by these boundary conditions. The impact of this assumption has been partially investigated by Hammer (1982a,b) and Withbroe (1988) and much more thoroughly investigated by Hansteen and Leer (1995). The principal conclusion drawn from these studies is that specification of the coronal base density and temperature as boundary conditions masks important physical effects that tend, among other things, to regulate the solar wind mass flux. Models of the solar wind that place the lower boundary of the system studied in the chromosphere (rather than at the coronal base), and that include a description of energy balance in the upper chromosphere, transition region, and the corona are able to reveal these regulating effects, and the physical implications of these effects will be the focus of the present chapter.

Our approach is to consider a hydrogen atmosphere, which in the corona is a proton-electron gas and in the transition region is a partially ionized hydrogen gas. First we examine the consequences of specifying boundary conditions at the coronal base and illustrate, among other things, the well-known solar wind mass-flux problem. Next we consider a description of mass, momentum, and energy balance in a solar atmosphere whose base is taken to lie in the upper chromosphere, and we contrast the results obtained with those obtained in the former case. It is this latter description that allows us to study the physical effects in the upper chromosphere, transition region, and lower corona that serve to regulate the solar wind mass flux.

II. WIND FROM A CORONA

Let us begin by considering the nature of a solar wind arising from a corona in which the temperature and density of a proton-electron gas are specified at the coronal base. Mass and momentum balance in such a steady, radial, spherically symmetric, thermally driven wind can be described by

$$4\pi \rho u r^2 = \phi = \text{constant} \tag{1}$$

$$u\frac{du}{dr} = -\frac{1}{\rho}\frac{dp}{dr} - \frac{GM}{r^2} \tag{2}$$

where ρ, u, and p are the mass density, flow speed, and thermal pressure, ϕ is the mass flux (i.e., the solar mass loss rate associated with the wind), r is the radial distance from the center of the Sun, and G and M are the gravitation constant and solar mass.

In an ideal gas the pressure is related to the density and temperature ($T = (T_p + T_e)/2$, or the thermal speed, v_t) by

$$p = \rho kT/m = \rho v_t^2 \tag{3}$$

where $m(\approx m_p/2)$ is the mean particle mass. Hence, the thermal pressure gradient force per unit mass in Eq. (2) has the form

$$\frac{1}{\rho}\frac{dp}{dr} = -v_t^2\left(\frac{1}{u}\frac{du}{dr} + \frac{2}{r}\right) + \frac{dv_t^2}{dr} \tag{4}$$

where we have used Eq. (1) to eliminate the density gradient in Eq. (4). We see in Eq. (4) that the pressure gradient force depends on the flow acceleration (first term), the flow geometry (second term), and the temperature gradient (third term).

We can now combine Eqs. (2) and (4) to obtain an equation of motion in the form of a typical stellar wind equation:

$$\frac{1}{u}\frac{du}{dr}(u^2 - v_t^2) = \frac{2v_t^2}{r} - \frac{dv_t^2}{dr} - \frac{GM}{r^2}. \tag{5}$$

In writing Eq. (5), we have split the pressure gradient force into its three parts: the acceleration-dependent part, which appears as the second term on the left side; and the parts depending on the flow geometry and the temperature gradient, which appear as the first two terms on the right side. As discussed in the following paragraph, this form of the equation illustrates the special character of momentum balance in a transonic flow that arises from the acceleration dependence of the pressure gradient force. (Note that it is this special character which seems to have provided one of the major stumbling blocks to the early understanding and acceptance of Parker's work.)

Let us consider, for simplicity, the isothermal form of Eq. (5), in which $v_t = $ constant. In this case, we see first that the solutions to Eq. (5) are independent of flow direction (as noted in Sec. I), and second that there is an elementary singular point (of the saddle type) in the family of solutions of Eq. (5), which occurs at the point where $u = v_t$ and $r = r_c = GM/2v_t^2$, and which is normally referred to as the solar wind critical point. This critical point, which arises from the acceleration-dependent nature of the thermal pressure-gradient force that drives the wind, is the manifestation of the aforementioned special character of solar wind momentum balance, because the wind solution of Eq. (5) passes through this critical point, as does the usual accretion solution (which was mentioned in the preceding section). McCrea's difficulty in understanding the wind solution apparently arose from an assumption that in order to attain this singular solution some process in the corona would have to operate in such a way as to determine precisely the correct speed (corresponding to the singular wind solution) at the coronal base (see Thomas 1960). Instead, of course, as Parker (1963) has pointed out, the requirement for pressure balance between the corona and the interstellar medium (in the presence of an extended coronal temperature) naturally leads to an adjustment in the flow speed so that the critical solution is attained.

The relatively simple description of the solar wind provided by Eq. (5) yields a surprisingly good theory of basic wind dynamics. Parker (1964b, 1965) and many subsequent workers went on to consider energy balance in the wind, including the effects of thermal conduction, heat addition, and energy addition through direct acceleration (viz., momentum addition). Although these many studies have elucidated a variety of important physical processes in the solar wind, they have generally suffered from a limitation that is significant when considering many aspects of wind energy balance. This is the limitation introduced through the specification of the coronal base density and temperature. In order to illustrate why this limitation is important, let us continue a bit further in the application of the basic solar wind theory we have introduced above.

Still focusing on the isothermal form of Eq. (5), we now turn our attention to the solar wind mass flux. The mass flux associated with the ascending critical solution of Eq. (5) (i.e., the wind solution) can be determined by integrating Eq. (5) from the coronal base ($r = r_0$) to the critical point ($r = r_c$)

and by making use of Eq. (1):

$$\phi = 4\pi\rho_0 v_t r_c^2 \exp\left[-\frac{v_{g0}^2}{2v_t^2}\left(1 - \frac{r_0}{r_c}\right) - \frac{1}{2} + \frac{u_0^2}{2v_t^2}\right] \tag{6}$$

where $v_{g0}^2 = 2GM/r_0$ is the square of the gravitational escape speed at the coronal base. Because the solar atmosphere at the coronal base is strongly gravitationally bound ($v_{g0}^2 \gg v_t^2$) and its flow is highly subsonic ($u_0^2 \ll v_t^2$), it is clear from Eq. (6) that variation of the solar wind mass flux is dominated by an exponential dependence on coronal temperature. For example, a factor of 2 increase of coronal temperature (say, from $T = 10^6$ K to $T = 2 \times 10^6$ K) produces a 2 order of magnitude increase of the mass flux. This result holds quite well even when the isothermal assumption is dropped, as long as the isothermal temperature in Eq. (6) is replaced by the average temperature in the region of subsonic flow (see, e.g., Leer and Holzer 1979). Thus, unless there is some mechanism (not included in the basic description of the solar wind presented above) that either very tightly regulates the coronal temperature or regulates the solar wind mass flux in the face of significant coronal temperature variations, one would expect to observe rather large variations of the solar wind mass flux. Such large variations are not, however, observed (see, e.g., Feldman et al. 1977; Schwenn 1983; McComas et al. 1995), and this conflict between observation and basic solar wind theory is the essence of the solar wind mass flux problem.

One mechanism for regulating the proton mass flux in the face of substantial coronal temperature variations involves the possible existence in the lower corona of a significant helium abundance (see, e.g., Bürgi 1992; Leer et al. 1992; Hansteen et al. 1993,1994). This mechanism, while somewhat effective, does not seem to be able to resolve the problem fully. Another mechanism involves the regulation of the coronal temperature and base pressure through thermal conduction from the corona to the chromosphere (Withbroe 1988; see also Hammer 1982a,b). Investigating this mechanism requires consideration of the response of the upper solar chromosphere and the chromosphere-corona transition region to variations in the thermal conduction from the corona to the chromosphere. Specifying boundary conditions at the coronal base on the density and temperature effectively rules out such a consideration, even when a full energy equation is used to describe the coronal plasma. Hence, a self-consistent description of the chromosphere-corona system is required for this type of consideration, and we examine the implications of such a description (developed by Hansteen and Leer [1995]) in the following section.

III. WIND FROM A CHROMOSPHERE-CORONA SYSTEM

In considering a chromosphere-corona system we must add energy balance descriptions for each component of the gas to the mass and momentum balance descriptions provided by Eqs. (1) and (2). Our discussion here is based on the work of Hansteen and Leer (1995), where the complete, time-dependent equations are presented. For our purposes, however, we need only consider steady-state equations, as we did for Eqs. (1) and (2), so that energy balance for the electron-proton gas as a whole and for the electron component of the gas can be represented as follows:

$$\frac{3}{2}\rho u v_t^2 \frac{d \ln T}{dr} = \rho u v_t^2 \frac{d \ln \rho}{dr} - q \frac{d \ln(qr^2)}{dr} + Q - L \tag{7}$$

$$\frac{3}{2}p_e u_e \frac{d \ln T_e}{dr} = p_e u_e \frac{d \ln n_e}{dr} - q_e \frac{d \ln(q_e r^2)}{dr} + Q_e - L_e + C_{ep} \tag{8}$$

with the auxiliary equations for the electron-proton collisional energy exchange rate C_{ep} and the electron-proton collision frequency v_{ep}, taking the form

$$C_{ep} = \frac{3}{2}n_e \kappa v_{ep}\left(T_p - T_e\right) \tag{9}$$

$$v_{ep} = 0.09 n_p T_e^{-3/2}. \tag{10}$$

Here Q is the mechanical heating rate, L is the radiative cooling rate, n is a particle density, the e and p subscripts refer to the electron and proton components of the gas, and the heat flux density q is taken to have the classical form, which for electrons is

$$q_e = -\kappa_e T_e^{5/2} \frac{dT_e}{dr} \tag{11}$$

with $\kappa_e = -7.8 \times 10^{-7}$ erg cm^{-1}s^{-1}K$^{-7/2}$. For the purpose of the following discussion, we shall ignore proton thermal conduction and thus assume that $q = q_e$. This neglect should be reasonable, even in the case where the proton temperature is much higher than the electron temperature, because when this temperature difference leads to the classical proton heat flux becoming comparable to or larger than the classical electron heat flux, a large fraction of the proton heat flux is transferred into flow energy (Olsen and Leer 1995).

Now let us cast Eq. (7), through combination with Eqs. (1) and (2), into two different forms that we make use of in the discussion below. First, Eq. (7) can be rewritten in its conservation form:

$$r^2\left[\rho u\left(\frac{1}{2}u^2 + \frac{5}{2}v_t^2 - \frac{1}{2}v_g^2\right) + q + f_M + f_R\right] = \text{constant} \tag{12}$$

where $v_g^2 = 2GM/r$. f_M and f_R are the mechanical and radiative energy flux densities and are related to their respective heating and cooling rates by $Q = -r^{-2}[d(r^2 f_M)/dr]$ and $L = r^{-2}[d(r^2 f_R)/dr]$. Evidently, Eq. (12) is just a statement that the sum of the advective (associated with bulk flow, enthalpy, and gravitational energy), conductive, mechanical, and radiative energy fluxes is constant. Next, Eq. (7) can be modified slightly to yield an expression for the slope a of the radial profile of the mean temperature:

$$a = \frac{\mathrm{d}\ln T}{\mathrm{d}\ln r} = -\frac{1}{5}\frac{v_g^2}{v_t^2} - \frac{2}{5}\frac{u^2}{v_t^2}\frac{\mathrm{d}\ln u}{\mathrm{d}\ln r}$$
$$- \frac{2}{5}\frac{q}{\rho u v_t^2}\frac{\mathrm{d}\ln(qr^2)}{\mathrm{d}\ln r} + \frac{2}{5}\frac{r}{h_M}\frac{f_M}{\rho u v_t^2} \tag{13}$$

where the mechanical energy flux density f_M is assumed to decline with radial distance owing both to spherical expansion and to exponential damping (over a scale length h_M, beginning at $r = r_1$), so that

$$f_M = f_{M0}\frac{r_0^2}{r^2}e^{-(r-r_1)/h_M}. \tag{14}$$

Note that because the damping begins at $r = r_1$, the last term in Eq. (13) vanishes in the region $r < r_1$. We have set $L = 0$ in this equation, because we intend to use it only in our discussion of coronal energy balance, and radiative cooling is significant only in the chromosphere and lower transition region. From this perspective, we view radiative cooling as an energy sink at the base of our chromosphere-corona system. The terms we have retained on the right side of Eq. (13) correspond to the following physical processes: (1) cooling associated with lifting the flowing gas out of the solar gravitational field through operation of the thermal pressure gradient force; (2) cooling associated with acceleration of the flow by the pressure gradient force; (3) heating (cooling) associated with degradation (enhancement) of the outward heat flux; and (4) heating associated with dissipation of the mechanical energy flux.

Let us now compare the various terms on the right side of Eq. (13) in the region of subsonic flow, and assume that $r_1 = r_0$, so that the heating begins at the coronal base, which we shall take to correspond to $T = 10^6$K. This comparison will give us a feeling for the physical processes that control the temperature in the region of subsonic flow and thus will serve as a basis for understanding the temperature regulation effects that are operative. We use a notation for the terms in Eq. (13) such that the first term (1) on the right (R) side of Eq. (13) is represented by $R_1(13)$. Comparing the gravitational and acceleration terms, we find

$$\left|\frac{R_2(13)}{R_1(13)}\right| = 2\frac{u^2}{v_g^2}\left|\frac{\mathrm{d}\ln u}{\mathrm{d}\ln r}\right|. \tag{15}$$

Near the coronal base, where the flow is highly subsonic, the e-folding length for u may be as small as $0.1r_0$, but the flow speed itself is so much smaller than the gravitational escape speed that $v_g^2 \gg 20u^2$, and the acceleration term ($R_2(13)$) has a negligible effect on the temperature structure throughout most of the subsonic region.

Next we compare the conductive and mechanical heating terms, recalling that for now we are assuming $r_1 = r_0$:

$$\left|\frac{R_3(13)}{R_4(13)}\right| = \left|\frac{h_M}{r}\,\frac{q}{f_M}\,\frac{\mathrm{d}\,\ln(qr^2)}{\mathrm{d}\,\ln r}\right|$$
$$= \left|0.1\frac{aT_{e6}^{7/2}}{f_{M05}}\,\frac{h_M}{r_0}\,e^{(r-r_0)/h_M}\frac{\mathrm{d}\,\ln(qr^2)}{\mathrm{d}\,\ln r}\right| \tag{16}$$

where T_{e6} is the electron temperature in units of 10^6K, and f_{M05} is the mechanical energy flux density at the coronal base in units of 10^5 erg cm^{-2}s^{-1}. Let us first evaluate Eq. (16) at the coronal base. If the mechanical energy flux is of reasonable magnitude (i.e., $f_{M05} > 1$) and the damping length is not too much more than a solar radius, the mechanical heating term will dominate the conductive term unless the electron temperature gradient is quite steep (implying a large magnitude for a). It is possible (Hansteen and Leer 1995), however, to produce such a steep electron temperature gradient if the mechanical heating occurs within a few density scale heights of the coronal base (i.e., if $h_M \ll r_0$). Then the electrons will be heated rapidly, whether or not the mechanical energy deposition is directed primarily into the electron gas, because within a few density scale heights of the coronal base, the proton-electron collisional energy coupling is very strong (i.e., the last term in Eq. (8) is dominant). When this occurs, the mechanical heating and conductive cooling terms in Eq. (13) nearly balance, and most of the outward-propagating mechanical energy flux is eventually returned through inward thermal conduction to the chromosphere, where it is radiated away. (Note that some fraction of the inward heat flux is not radiated away, but rather serves to heat the transition region gas and thus provide the enthalpy flux increase from the chromosphere to the corona. The size of this fraction depends on the magnitude of the solar wind mass flux.) This process which transforms an outward mechanical energy flux into an outward radiative flux through an inward heat flux, represents one aspect of the coronal temperature regulation process referred to at the end of the last section; in this case, the coronal temperature does not increase nearly as much for a given mechanical energy deposition as it would if inward thermal conduction were not explicitly taken into account. In the limit of very small damping length (viz., $h_M < |\mathrm{d}\,\ln \rho/\mathrm{d}r|^{-1}$), this case reduces essentially to the original Parker solar wind description, in which the density and temperature are specified at the coronal base, and the only non-advective energy transport outward from the coronal base is through thermal conduction. Of course, the full range of coronal base temperatures and densities that can be specified arbitrarily (in the original Parker description) is not available here, because of the substantial limitations imposed by the self-consistent description of the chromosphere-corona system.

Returning now to the case first mentioned, in which the mechanical heating term dominates the conductive term, we must consider relative magnitudes of gravitational cooling and mechanical heating, corresponding to the first and fourth terms in Eq. (13):

$$\left|\frac{R_1(13)}{R_4(13)}\right| = \frac{h_M}{r}\frac{\rho u v_g^2}{2f_M}$$

$$= 41\frac{n_{08}T_{06}^{1/2}}{f_{M05}}\frac{u_0}{v_{t0}}\frac{h_M}{r_0}\frac{r_0^2}{r^2}e^{(r-r_0)/h_M}$$

(17)

where n_{08} is the coronal base electron (or proton) density in units of 10^8cm^{-3}. For damping lengths on the order of $h_M = 0.7r_0$, which Hansteen and Leer (1995) chose for their reference model, the radial dependence of Eq. (17) is quite weak for the first one or two solar radii above the coronal base, so let us begin by ignoring this radial dependence. For reasonable values of the coronal base pressure and mechanical energy flux, it follows from Eq. (17) that the ratio of the gravitational cooling term to the mechanical heating term is of the order $100u_0/v_{t0}$, but unless the average temperature in the subsonic region is considerably more than 10^6 K, we know from an analysis like that leading to Eq. (6) that $u_0/v_{t0} \ll 0.01$, so it is clear that the mechanical heating must exceed the gravitational cooling near the coronal base, producing a positive temperature gradient at the base. Of course, beyond a few damping lengths, the gravitational cooling term begins to dominate the mechanical heating term (cf., radial dependence of Eq. [17]), and the coronal temperature reaches its maximum. Beyond this maximum, the flow speed rapidly approaches the thermal speed, and both acceleration cooling (term $R_2(13)$) and conductive heating (term $R_3(13)$) begin to play significant roles in the energy balance.

In this discussion of Eq. (17), we have begun to touch upon two other aspects of the coronal temperature regulation mechanism. Each of these aspects involves an increase of the solar wind mass flux in response to an increase in mechanical heating of the corona. First, for a given coronal temperature structure, the mass flux increases with increasing coronal base density. As noted above, increased mechanical heating leads to an increase in the downward heat flux (even when the heat flux is much smaller than the mechanical energy flux), and this, in turn, increases the coronal base pressure (and density at 10^6K). Note that in this case, it is the increase in coronal density associated with the increased downward heat flux (rather than the removal of energy from the corona by this heat flux) in which we are interested. Second, for a given coronal base density, any increase in temperature associated with an increase in mechanical heating in the subsonic region produces an increase in the flow speed at the coronal base, and thus an increase in the mass flux. (This, of course, is just the effect with which we were dealing when discussing the solar wind mass flux problem.) In both cases, an increase in the solar wind mass flux produces an increase in the gravitational cooling rate, which tends to counter the temperature increase produced by an increase in the mechanical heating rate. Again, we have a tendency toward coronal temperature regulation.

In summary, we have considered three processes that tend to mitigate the coronal temperature increase produced by increased mechanical heating in the corona, and which thus serve as coronal temperature regulation processes. Specifically, increased coronal heating leads to: (1) an increase in the energy loss to the corona by downward heat transport; (2) an increase in the solar wind mass flux and coronal gravitational cooling arising from an increased coronal base density associated with the larger downward heat flux; and (3) an increase in the solar wind mass flux and coronal gravitational cooling arising from an increased coronal base flow speed associated with the higher coronal temperature. The actual effectiveness of these three processes in regulating the coronal temperature is illustrated in the following section, where numerical results obtained by Hansteen and Leer (1995) are presented.

Before leaving this section, though, let us consider one final issue, the question of whether sufficiently large mechanical heating rates can produce a mass flux so large that there is insufficient energy flux to lift it out of the Sun's gravitational field (leading to the absence of a steady state wind). Such a situation is, in fact, encountered in classical solar wind theory (with thermal conduction) when the coronal base density and temperature are sufficiently large (see, e.g., Holzer and Leer 1980). To explore this question, we need to evaluate the relative magnitude of the five terms on the left side of Eq. (12). It is at once obvious that the comparison of these terms is very similar to the comparison that we just carried out for the terms of Eq. (13), particularly if we carry out both comparisons at the coronal base, which we shall do. Indeed, the only significant difference arises from the presence of the enthalpy term in Eq. (12), which is term $R_2(12)$ (note that the radiative term remains irrelevant in this coronal energy balance analysis). It is readily seen that the enthalpy term serves to reduce the effect of the gravitational potential energy term $(R_3(12))$ by some 22% at the coronal base, and it follows that the much smaller bulk flow energy (term $R_1(12)$) is quite negligible. As in the case of Eq. (13), unless the damping length is very small, the magnitude of the heat flux in Eq. (12) (term $R_4(12)$) is less than 10% or 20% of that of the mechanical energy flux and at the coronal base it should be an inward flux and thus serve to reduce the heating effect of the mechanical energy flux by that amount. (We shall return to the case of small damping length in a moment.) Finally, the ratio of the gravitational potential energy flux to the mechanical energy flux $(R_3(12)/R_5(12))$ differs from the equivalent ratio for Eq. (13), which is given by Eq. (17), by the factor r_0/h_M. Thus, for damping lengths on the order of a solar radius or larger, the energy flux given by Eq. (12) should be positive, and there will always be enough energy to lift the wind out of the gravitational field (i.e., a steady state wind solution will exist). This conclusion is consistent with that drawn from the parameter study carried out by Hansteen and Leer (1995). That parameter study did not, however, fully investigate the regime of small damping length, and it appears possible from the preceding analysis that for large mechanical energy fluxes and small damping lengths, a regime in which steady state wind solutions do not exist may be achieved. This would certainly be consistent with the results of classical solar wind studies with thermal conduction, as was mentioned above (see, e.g., Holzer and Leer 1980).

IV. RESULTS FROM NUMERICAL SIMULATIONS

Let us begin by examining the results shown in Fig. 1 from the reference model of Hansteen and Leer (1995), in which $f_{M0} = 1.5 \times 10^5$ erg cm^{-2}s^{-1}, $h_M = 0.7$ R$_\odot$, $r_1 = 1.03$ R$_\odot$, and the heating through dissipation of the mechanical energy flux is distributed by delivering 60% of the heat to the protons and 40% to the electrons. The flow speed and density profiles in panel (a) correspond to those of a relatively typical, moderately high-speed solar wind, with an asymptotic flow speed of more than 600 km s^{-1} and a proton flux density at 1 AU of about 6×10^8 cm^{-2}s^{-1}. The coronal temperature profiles in panel (b) are not, however, what one normally sees in solar wind models, with the proton temperature reaching a maximum of 2.8×10^6 K near $r = 2$ R$_\odot$ and the electron temperature reaching a maximum of 1.5×10^6 K just a few density scale heights above the coronal base. The electron temperature peaks at a lower value than the proton temperature (and nearer the coronal base) for three reasons (cf., Sec. III); more of the mechanical heating is going into the protons; conductive cooling (in the vicinity of the electron temperature peak) is stronger for electrons than for protons; and collisional energy exchange between electrons and protons, which is dominant at the coronal base, ceases to be important a few density scale heights above the coronal base, because of the n^2 dependence of this collisional coupling. Finally, the heat flux profiles in panel (c) differ from those expected on the basis of the discussion in Sec. III only because of the relatively large proton heat flux. As noted earlier, this is an artifact of the classical heat flux description used by Hansteen and Leer (1995), and is not particularly significant in the results shown here. The electron heat flux is seen to transport inward approximately 20% of the outward mechanical energy flux, but only about half of this reaches the chromosphere to be radiated away, and the rest serves to heat the transition region gas, thus producing the outward enthalpy flux associated with the solar wind mass flux. This result contrasts with the small damping length case (not shown here), in which the inward electron heat flux is a much larger fraction of the outward mechanical energy flux, and virtually all of the heat flux is radiated away (cf., Hansteen and Leer 1995).

Now let us turn to the question of regulation of the solar wind mass flux through coronal temperature regulation. Recall from Sec. II that the mass flux varies exponentially with the average coronal temperature in the region of subsonic flow. This means that if the average coronal temperature were to vary linearly with the mechanical energy flux entering the corona, then the mass flux would vary exponentially with the mechanical energy flux. Owing to the effectiveness of the three temperature regulation processes discussed in Sec. III, such an exponential dependence does not occur, as is illustrated

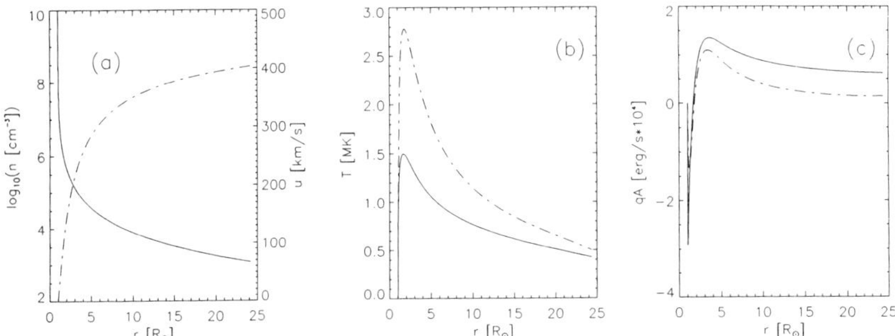

Figure 1.   The reference model of Hansteen and Leer (1995), in which $f_{M0} = 1.5 \times 10^5$ erg cm^{-2}s^{-1}, $h_M = 0.7$ R$_\odot$, $r_1 = 1.03$ R$_\odot$, and the heating through dissipation of the mechanical energy flux is distributed by delivering 60% of the heat to the protons and 40% to the electrons. Panel (a) shows the radial profiles of hydrogen density (solid line) and proton flow speed (dashed line), panel (b) shows the radial profiles of the proton (dashed line) and electron (solid line) temperatures, and panel (c) shows the radial profiles of the proton (dashed line) and electron (solid line) heat fluxes, where A has a value of 1 cm^2 at the coronal base and $A \propto r^2$.

in Fig. 2. Here, the magnitude of the mechanical energy flux in the reference model of Hansteen and Leer (1995) is varied over more than an order of magnitude, and the solar wind mass flux and asymptotic flow speed are calculated. Evidently, the temperature regulation process is quite effective, as the solar wind mass flux is seen to be nearly proportional to the mechanical energy flux entering the corona. For the damping length of $h_M = 0.7$ R$_\odot$ used here, the important coronal temperature regulation processes are the two that increase the mass flux and thus the gravitational cooling rate as the coronal heating rate increases (cf., Sec. III). As it happens these two processes are just about equally effective for the parameters used in this study. Of course, the near proportionality of $\rho_0 u_0$ and f_{M0} means that the asymptotic flow speed will be nearly independent of the magnitude of the mechanical energy flux at the coronal base (Leer and Holzer 1980). This fact is readily deduced from the discussion of Eq. (12) in Sec. III, where it was noted that for this damping length, the dominant terms in Eq. (12) at the coronal base are the gravitational and mechanical energy flux terms. At large distances, the dominant term is clearly the bulk flow term, and the constancy of the total energy flux leads to the result that

$$\frac{u_\infty^2}{v_{g0}^2} \approx \frac{2f_{M0}}{\rho_0 u_0 v_{g0}^2} - 1. \tag{18}$$

In the case at hand, the asymptotic flow speed and the coronal base gravitational escape speed are nearly the same. For larger damping lengths, a larger fraction of the energy will be deposited in the supersonic region, and the ratio u_∞^2/v_{g0}^2 will increase, while the reverse will be true for smaller damping lengths (cf., Leer and Holzer 1980).

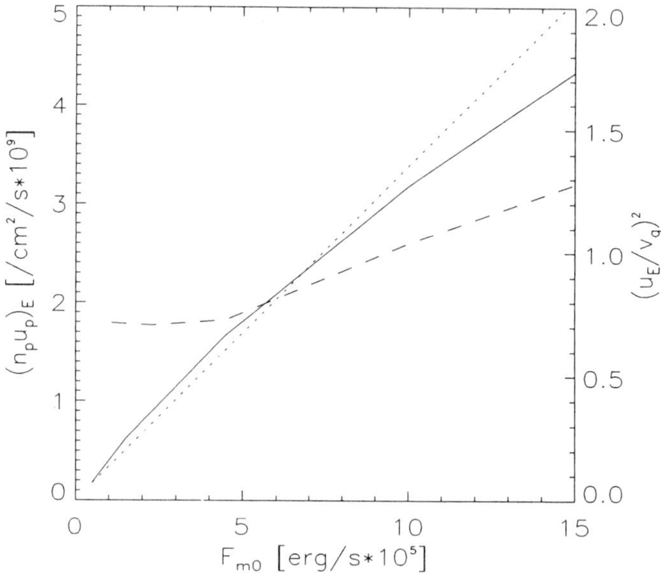

Figure 2. Starting from the reference model of Hansteen and Leer (1995), where $f_{M0} = 1.5 \times 10^5$ erg cm^{-2}s^{-1}, and $h_M = 0.7$ R$_\odot$, the magnitude of the mechanical energy flux is varied over more than an order of magnitude, and the resulting proton flux density at 1 AU (solid line) and square of the ratio of the 1 AU flow speed to the coronal base gravitational escape speed (dashed line) are shown. The proton flux density curve is compared with a dotted curve for which the proton flux density at 1 AU is assumed to be proportional to the mechanical energy flux density at the coronal base.

One of the new and particularly interesting aspects of the Hansteen and Leer (1995) study is the dependence on the damping length of the fraction of the mechanical energy flux going into the wind and the fraction going into chromospheric radiation. This dependence is illustrated in Fig. 3, where it is seen that for very small damping lengths, nearly all the energy is radiated away, while for damping lengths of more than a few density scale heights, about 90% of the energy goes into lifting the solar wind out of the solar gravitational field and accelerating it to its asymptotic flow speed. As noted in Sec. III, the very small damping lengths lead to a situation not dissimilar to the classical solar wind theory of Parker, in which the nonadvective energy transported outward from the coronal base is principally in the form of heat flux. The case studied in detail by Hansteen and Leer (1995) in which the reference model was used with a reduced damping length ($h_M = 0.1$ R$_\odot$) exhibited a very low solar wind flow speed, which is consistent with the studies of conductive solar winds carried out by Holzer and Leer (1980) and others. It is this result that indicates the possibility that there exists a part of the small damping length regime in which there is not enough energy available to lift the solar wind mass flux out of the gravitational field, so that no steady state wind solution

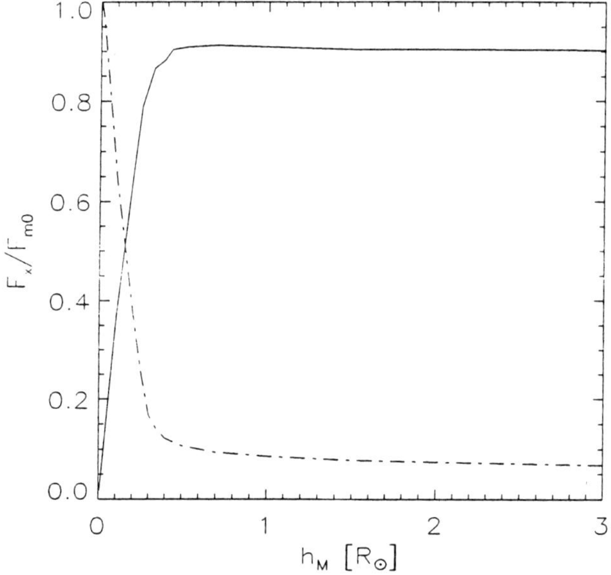

Figure 3. Starting from the reference model of Hansteen and Leer (1995), the damping length is varied from $h_M \ll R_\odot$ to $h_M > R_\odot$, and the fraction of the mechanical energy flux entering the corona that goes into the solar wind (solid line) and the fraction that is radiated away by the chromosphere (dash-dot line) are calculated. For very small damping lengths, nearly all the energy is radiated away, while for damping lengths of more than a few density scale heights, about 90% of the energy goes into lifting the solar wind out of the solar gravitational field and accelerating it to its asymptotic flow speed.

exists. This possibility will be the subject of a future study.

Let us end by considering one last numerical study, in which, for the same mechanical energy input, a static atmosphere (magnetically contained) and an expanding atmosphere (i.e., a wind) are compared. The results of this study are shown in Fig. 4. The principal difference in these two cases arises from the fact that in the static atmosphere, all of the mechanical energy flux deposited in the corona is conducted back to the chromosphere and radiated away, whereas in the expanding atmosphere, virtually all (i.e., some 90%) of the mechanical energy flux deposited in the corona is carried away by the wind. The results are just what we would expect from the preceding discussion. The transition region panels show a much steeper temperature gradient and much higher transition region pressure, which are both reflective of the considerably larger heat flux transported inward to the chromosphere. In the coronal panels, both the temperatures and the density are seen to be higher in the static case, and the electron and proton temperatures are coupled to higher altitudes, owing to the higher density. As a matter of interest, but not of much significance, the heat conducted toward the upper boundary by

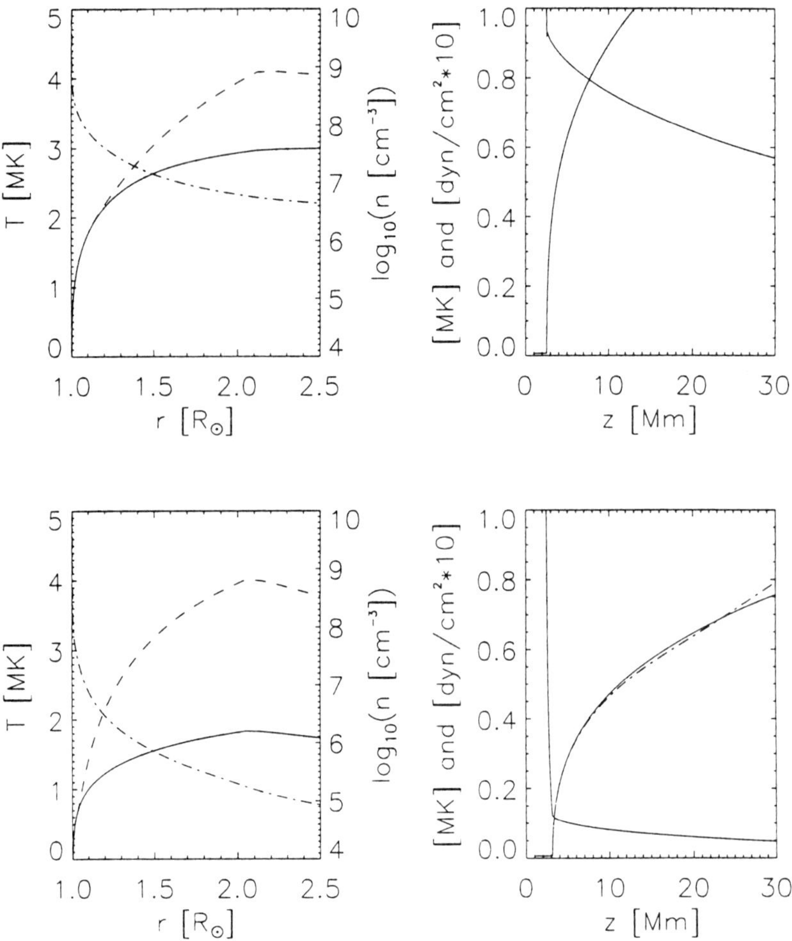

Figure 4. A comparison of static and wind solutions for the same mechanical energy input. In the static case, shown in panels (a) and (b), the flow speed and the temperature gradients are set to zero at $r = 2.5$ R$_\odot$. For both the static and wind solutions, which are shown in panels (c) and (d), the energy input is near $r = 2$ R$_\odot$ (viz., $r_1 = 2.0$ R$_\odot$, $h_M = 0.1$ R$_\odot$, and $f_{M0} = 1.5 \times 10^5$ erg cm^{-2}s^{-1}). The horizontal scale in panels (b) and (d) is in megameters, in order to show the upper chromosphere and transition region, while the horizontal scale in panels (a) and (c) is in solar radii, in order to show the corona. In the coronal panels, the proton density is shown by a dot-dash line and the electron and proton temperatures by solid and dashed lines, respectively. In the transition region panels, the pressure and electron temperature are shown by solid lines, while the proton temperature is shown by a dot-dash line.

the protons in the static case is balanced by a downward conduction of heat away from the upper boundary by the electrons. The energy is collisionally transferred from the protons to the electrons near the upper boundary.

V. CONCLUSIONS

The principal points of this chapter have been to review the early history of solar wind studies and to show how the classical solar wind work of Parker is currently being built upon by considering the effects of removing the assumption that the temperature and density of the corona can be specified as boundary conditions at the coronal base. We have elucidated three physical processes that serve to regulate the coronal temperature when the corona is heated by the spatially dependent dissipation of a mechanical energy flux. These processes involve the following responses to the increase of the mechanical energy flux entering the corona: (1) the increase in the energy loss to the corona by inward heat transport; (2) the increase in the solar wind mass flux and coronal gravitational cooling arising from an increased coronal base density associated with the larger inward heat flux; and (3) the increase in the solar wind mass flux and coronal gravitational cooling arising from an increased coronal base flow speed associated with the higher coronal temperature. We have seen that these processes are effective in mitigating the exponential dependence of solar wind mass flux on coronal temperature, and that they lead to an approximate proportionality between the solar wind mass flux and the mechanical energy flux entering the corona, at least for the parameter ranges considered here. It has also been shown that by approaching very short damping lengths, the results of classical solar wind theory, with coronal base boundary conditions on the density and temperature, begin to be recovered. Thus, the connection between the current work and earlier work is clear, and the limitations of the earlier work are made explicit.

Acknowledgments. We would like to thank Ø. Sandbæk, E. L. Olsen, and Ølie-Svendsen for interesting discussions during the writing of this manuscript, and K. B. MacGregor for comments on the manuscript. This work was supported by the Norwegian Research Council (NFR), and a contract from NASA.

REFERENCES

Alfvén, H. 1941. On the solar corona. *Arkiv för Matematik, Astronomi, o. Fysik* 27A(25):1–23.

Alfvén, H. 1947. Granulation, magneto-hydrodynamic waves, and the heating of the solar corona. *Mon. Not. Roy. Astron. Soc.* 107:211–219.

Biermann, L. 1946. Zur Deutung der chromosphärischen Turbulenz und des Exzesses der UV-Strahlung der Sonne. *Naturwissenschaften* 33:118–119.

Biermann, L. 1948. Über der Ursache der chromosphärischen Turbulenz und des UV-Exzesses der Sonnenstrahlung. *Zs. f. Ap.* 25:161–177.

Biermann, L. 1951. Kometenschweife und solar korpuskularstrahlung. *Zs. f. Ap.* 29:274–286.

Biermann, L. 1953. Physical processes in comet tails and their relation to solar activity. *Mem. Soc. Roy. Sci. Liège Quart. Ser.* 13:291–302.

Biermann, L. 1957. Solar corpuscular radiation and the interplanetary gas. *Observatory*, no. 77.

Birkeland, K. 1896. Sur les rayons cathodiques sons l'action de forces magnetiques intenses. *Arch. Sci. Phys. Naturelles* 1:497–512.

Birkeland, K. 1908. *The Norwegian Aurora Polaris Expedition 1902–3*, vol. 1 (Kristiania: H. Aschehoug and Co).

Birkeland, K. 1913. *The Norwegian Aurora Polaris Expedition 1902–3*, vol. 1 (Kristiania: H. Aschehoug and Co.).

Bondi, H. 1952. On spherically symmetric accretion. *Mon. Not. Roy. Astron. Soc.* 112:195–204.

Bürgi, A. 1992. Proton and alpha particle fluxes in the solar wind: Results of a three-fluid model. *J. Geophys. Res.* 97:3137–3150.

Chamberlain, J. W. 1961. Interplanetary gas. III. A hydrodynamic model of the corona. *Astrophys. J.* 133:675–687.

Chapman, S. 1919. An outline of a theory of magnetic storms. *Proc. Roy. Soc. London A* 95:61–83.

Chapman, S. 1929. Solar streams of corpuscles: Their geometry, absorption of light, and penetration. *Mon. Not. Roy. Astron. Soc.* 89:456–470.

Chapman, S. 1954. The viscosity and thermal conductivity of a completely ionized gas. *Astophys. J.* 120:151–155.

Chapman, S. 1957. Notes on the solar corona and the terrestrial ionosphere. *Smithsonian Contrib. Astrophys.* 2:1.

Edlén, B. 1942. An attempt to identify the emission lines in the spectrum of the solar corona. *Arkiv för Matematik, Astronomi, o. Fysik* 28B(1):1–4.

Feldman, W. C., Asbridge, J. R., Bame, S. J., and Gosling, J. T. 1977. Plasma and magnetic fields from the Sun. In *The Solar Output and Its Variations*, ed. O. R. White (Boulder: Colorado Assoc. Univ. Press), pp. 351–382.

FitzGerald, G. F. 1892. Sunspots and magnetic storms. *Electrician* 30:30.

FitzGerald, G. F. 1900. Sunspots, magnetic storms, comet tails atmospheric electricity, and aurora. *Electrician* 46:287–288.

Grotrian, W. 1933. Über den Intensitätsverhältnis der Koronalinien. *Zs. f. Ap.* 7:26–45.

Grotrian, W. 1939. Zur Frage der Deutung der Linien im Spektrum der Sonnenkorona. *Naturwissenschaften* 27:214.

Hammer, R. 1982*a*. Energy balance of stellar coronae. I. Methods and examples. *Astrophys. J.* 259:767–778.

Hammer, R. 1982*b*. Energy balance of stellar coronae. II. Effect of coronal heating, *Astrophys. J.* 259:779–791.

Hansteen, V. H., and Leer, E. 1995. Coronal heating, densities, and temperatures and solar wind acceleration. *J. Geophys. Res.* 100:21577–21593.

Hansteen, V. H., Holzer, T. E., and Leer, E. 1993. Diffusion effects on the helium abundance of the solar transition region and corona. *Astrophys. J.* 402:334–343.

Hansteen, V. H., Leer, E., and Holzer, T. E. 1994. Coupling of the coronal he abundance to the solar wind. *Astrophys. J.* 428:843–853.

Holzer, T., and Axford, W. 1970. The theory of stellar winds and related flows. *??* 8:31–60.

Holzer, T., and Leer, E. 1980. Conductive solar wind models in rapidly diverging flow geometries. *J. Geophys. Res.* 85:4665–4679.

Leer, E., and Holzer, T. E. 1979. Constraints on the solar coronal temperature in regions of open magnetic field. *Solar Phys.* 63:143–156.

Leer, E., and Holzer, T. E. 1980. Energy addition in the solar wind. *J. Geophys. Res.* 85:4681–4688.

Leer, E., Holzer, T. E., and Shoub, E. C. 1992. Solar wind from a corona with a large helium abundance. *J. Geophys. Res.* 97:8183–8201.

Lodge, O. 1900. Sunspots, magnetic storms, comet tails, atmospheric electricity, and aurora. *Electrician* 46:249–250.

McComas, D. J., Phillips, J. L., Bame, S. J., Gosling, J. T., Goldstein, B. E., and Neugebauer, M. 1995. Ulysses solar wind observations to 56° south. *Space Sci. Rev.* 72:93–98.

McCrea, W. H. 1956. Shock waves in steady radial motion under gravity. *Astrophys. J.* 124:461–468.

Olsen, E. L., and Leer, E. 1995. An 8-moment approximation two-fluid model of the solar wind. *J. Geophys. Res.*, submitted.

Parker, E. N. 1958. Dynamics of the interplanetary gas and magnetic fields. *Astrophys. J.* 128:664–676.

Parker, E. N. 1963. *Interplanetary Dynamical Processes* (New York: Wiley Inter-science).

Parker, E. N. 1964*a*. Dynamical properties of stellar coronas and stellar winds. I. Integration of the momentum equation. *Astrophys. J.* 139:72–92.

Parker, E. N. 1964*b*. Dynamical properties of stellar coronas and stellar winds. II. Integration of the heat-flow equation. *Astrophys. J.* 139:93–122.

Parker, E. N. 1965. Dynamical theory of the solar wind. *Space Sci. Rev.* 4:666–708.

Schwenn, R. 1983. The "average" solar wind in the inner heliosphere: Structures and slow variations. In *Solar Wind Five*, ed. M. Neugebauer, NASA CP-2280, pp. 489–507.

Thomas, R., ed. 1960. *Aerodynamic Phenomena in Stellar Atmospheres*. I.A.U. (New York: Academic Press).

Withbroe, G. L. 1988. The temperature structure, mass, and energy flow in the corona and inner solar wind. *Astrophys. J.* 325:442–467.

CORONAL MASS EJECTIONS

A. J. HUNDHAUSEN
High Altitude Observatory

Coronal mass ejections are the most spectacular disruptions of the solar corona and among the most energetic forms of solar activity. The thousands of examples observed with space-borne and groundbased instruments now provide a fairly thorough statistical description of mass ejection properties and relationships among those properties. Statistical analyses confirm earlier suggestions that mass ejections are disruptions of *large-scale*, magnetically closed regions of the corona. Mass ejections contribute only a minor fraction of the mass and energy in the solar wind. While their physical origins are not clearly established, there is a substantial body of observational evidence discounting solar flares as a common physical cause or "driver" of coronal mass ejections. Although flares seen in soft X-ray emission from the corona are often "associated" with mass ejections (i.e., they occur near the temporal onset and spatial location of the ejection), the intensities or spatial scales of these X-ray flares are poorly or weakly related to the speeds, masses, energies, or spatial scales of the mass ejections.

I. INTRODUCTION

Coronal mass ejections were first clearly identified in observations made with space-borne coronagraphs in the 1970s (see, e.g., Howard et al. 1976; Mac-Queen et al. 1974). Many of the basic questions about coronal mass ejections (or coronal transients, as they have sometimes been called), dealing with their physical nature, their origins, and their effects on the solar wind, were raised at that time (see, e.g., Gosling et al. 1974; Hildner et al. 1976; Munro et al. 1979; and reviews by Gosling 1975; Hildner 1977; and MacQueen 1980); some of these questions remain only partially answered today. For other reviews of this material, see Rust and Hildner (1980), Dryer (1982), Fisher (1984), Hundhausen et al. (1984*a*), Wagner (1984), Hildner (1986), Kahler (1987), Hundhausen (1987,1996), Webb (1992), Gosling (1993*a*), and Dryer (1994).

Coronal mass ejections have now been identified by the thousands in coronagraph data from the Skylab (MacQueen et al. 1974; Gosling et al. 1974; Hildner 1977; MacQueen 1980), Solwind (Michels et al. 1980; Howard et al. 1985,1986; Sheeley et al. 1982,1986), and Solar Maximum Mission (MacQueen et al. 1980; House et al. 1981; Wagner et al. 1981) spacecraft and in coronameter data from the ground (Fisher et al. 1981; Fisher and Poland 1981). These data are our primary source of information on mass ejections in the solar corona and on several of their key properties as they leave the Sun. Conventional "white-light" images produced by these instruments record the photospheric radiation scattered by free electrons in the solar corona. The intensity of scattered radiation is thus a measure of the coronal density (or more precisely, the line of sight integral of that density times a relatively simple scattering function), independent of other physical properties such as temperature or flow speed. Coronagraph data thus reveal the coronal density structure, temporal changes in that structure, and hence bulk motions of the coronal plasma in a very direct manner.

As is common in empirical studies, physical interpretation of these observations has often begun with detailed descriptions of key examples that clearly elucidate some facet of the mass ejection phenomenon and thus guide physical understanding. The large body of observations now available also invites statistical analysis and provides another step in improving our understanding of the phenomenon. Statistical analyses show what is or is not typical and thus test the broad validity of conclusions based on particular examples. They may also reveal general properties and broad relationships that are important clues in developing and extending physical interpretations. In this discussion I will continue to draw upon a few key examples to introduce or illustrate physical conclusions, but will place considerable emphasis on results of recent statistical analyses of the mass ejection characteristics revealed by coronagraph observations as a step toward greater generality. I will deal specifically with two topics central to this conference—the contributions of mass ejections to the solar wind and the relationships of mass ejections to "radiative" manifestations of solar activity (such as solar flares) that might be observable on stars other than the Sun.

II. BACKGROUND

A central concept in much of this discussion will be a simple interpretation of the magnetic nature of mass ejections. This interpretation is based on the chain of reasoning mentioned above, namely as a deduction from a few well-observed, clear examples buttressed by statistical evidence for its broader validity. It is thus worth a brief description here, both for its significance in our discussion and as an example of that chain of reasoning.

The magnetic fields in coronal mass ejections are not directly observed. However, the high electrical conductivity in the ionized coronal plasma ensures that the field is frozen into that plasma. Thus coronagraph observations of a changing coronal density structure imply changes in the geometry of the frozen-in coronal magnetic field. One of the key examples that suggested the magnetic nature of mass ejections occurred on 18 August 1980. It has been described by several authors (see, e.g., Athay and Illing 1986; Illing and Hundhausen 1986; Rompolt 1984) and is illustrated in Figs. 1 and 2. Figure 1 shows the formation of the mass ejection in a time sequence of four coronal

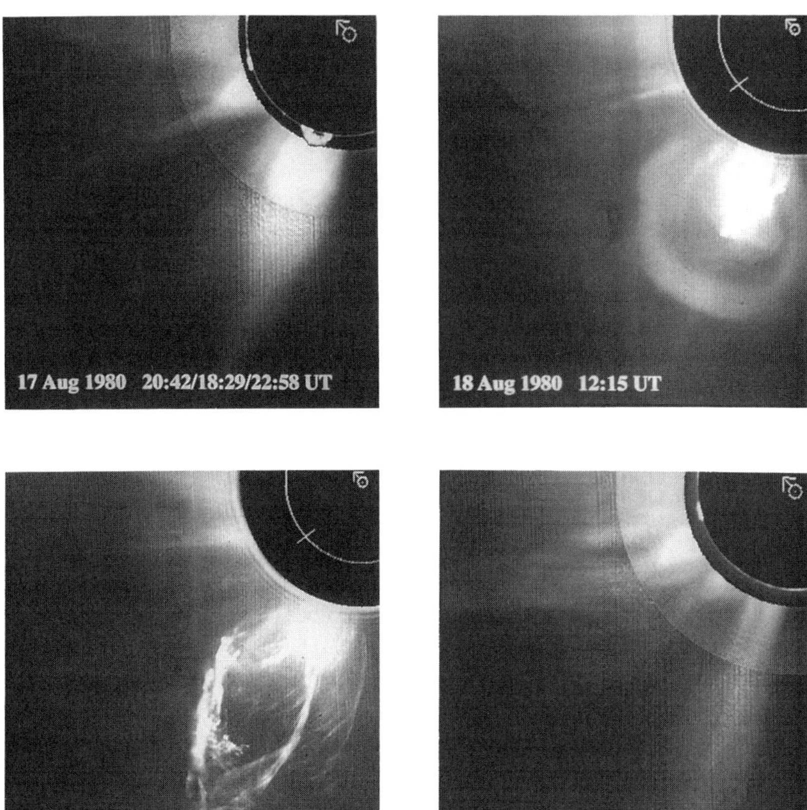

Figure 1.   Four coronal images showing a coronal mass ejection that occurred on 18 August 1980. The first and fourth images are composites of an Hα photograph from Mauna Loa Observatory (rimming the Sun), a coronameter image from Mauna Loa showing the white light corona from 1.15 to 1.8 solar radii from Sun center, and a coronagraph image from SMM showing the white light corona from 1.8 to $\sim$5 solar radii. These images are all from the nearly simultaneous times given on each image. The second and third images are based on Solar Maximum Mission data only.

images. Figure 2 shows sketches of the magnetic field structure suggested by the images under the "frozen-in field" assumption.

The first image of Fig. 1 was constructed from nearly simultaneous coronal images obtained with the Solar Maximum Mission (SMM) coronagraph, showing the outer part of the corona, and the coronameter at the Mauna Loa Observatory (MLO), showing the inner corona. Superposed on these is an image (also from MLO) in the Hα emission line of neutral hydrogen that reveals the presence of any prominences rising above the solar limb. The bright, mound-like coronal feature with a spike-like top is a coronal helmet streamer. As is often observed, the helmet streamer rises above a prominence, seen in Hα, and a dark, "cavity" that surrounds the prominence in white-light images. The mound-like shapes of the cavity and helmet streamer as well as the fine striations seen within them suggest the magnetic structure that would be expected to extend above a magnetic neutral line marked by the prominence; that is, an arcade of "closed" magnetic field lines spanning the neutral line at the base of the corona. The first frame of Fig. 2 shows a cross-sectional sketch of this magnetic structure.

The second and third images in Fig. 1 (based on SMM data only) show the formation of a coronal mass ejection in conjunction with eruption of the prominence and cavity. The second image, from 1215 UT on 18 August shows that the dense coronal plasma in the helmet streamer had been distorted into a bright loop as the dark cavity and embedded bright prominence rose and expanded beneath it. By 1310 UT, when the third image was acquired, the top of the loop had left the field of view of the SMM instrument (with the loop front moving outward at about 600 km s^{-1}); the bright prominence had expanded outward to fill most of this image. The changes in the magnetic field that would have accompanied these observed changes in the plasma density structure are sketched in the second and third frames of Fig. 2. The frozen-in, closed magnetic field lines in the helmet streamer and cavity must have been distorted into broader, closed structures by 1215 UT. The tops of these field lines must then have been stretched out of the instrument field of view along with the bright loop top by 1310 UT.

The fourth image in Fig. 1 is based on SMM and MLO data from close to 1940 UT on 18 August. By this time both the original helmet streamer and prominence seen on 17 August had moved out of the instrument field of view, leaving this region of the corona with a depleted, "blown-out" appearance. This suggests the magnetic structure sketched in the last frame of Fig. 2. The field lines near the Sun have been stretched out to look like a magnetically open region. However, this region is *bipolar*, with a change in the direction of the field above the neutral line that remains rooted in the lower layers of the solar atmosphere. A current sheet must then extend outward where the direction of the magnetic field changes in the midst of the blown-out region.

The coronagraph images of this mass ejection thus suggest disruption of a large coronal region that is initially pervaded by closed magnetic field lines. The mass ejection blows much of the coronal plasma out of the region and distorts this field into a structure that is "open" in the sense that the tops of the closed field lines are carried away from the Sun to become part of the solar wind; the bipolar nature of the region then implies the existence of an extended current sheet within the newly "opened" region. There are many similar examples of mass ejections that disrupt coronal helmet streamers (see, e.g., Figs. 9 and 12 of Hundhausen 1996). Still more mass ejections involve the eruption of prominences, and hence clearly occur over magnetic neutral lines. Several lines of statistical evidence supporting a generalization of this interpretation will emerge below.

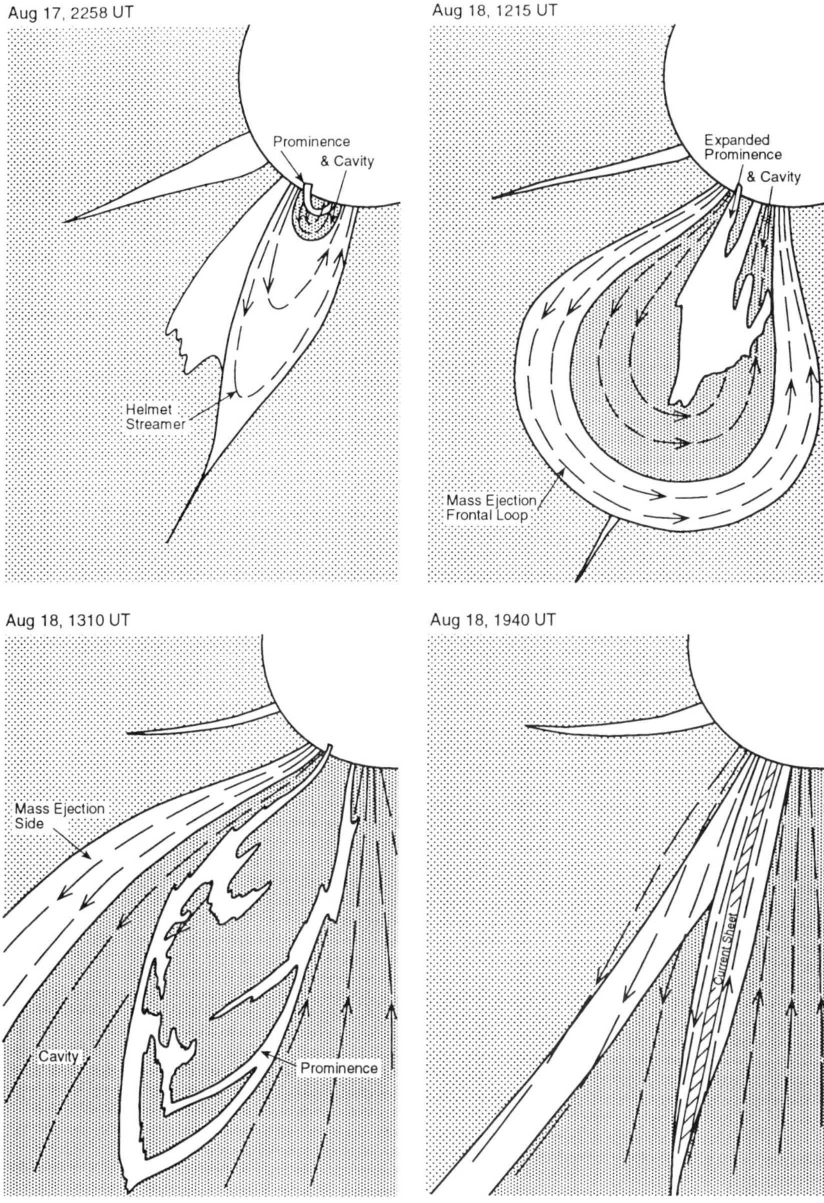

Figure 2.   Sketches of the coronal magnetic geometry at the times of the images shown in Fig. 1.

A few of the implications of this rather simple magnetic interpretation of coronal mass ejections should be stated explicitly here. First, it implies that many of the measured properties of mass ejections, such as their spatial scales and locations, are properties of the magnetic structures that give rise to the ejections and can yield clues as to the physical origins of mass ejections. It also implies that the mass ejection process is basically magnetic in nature. Despite the fact that the phenomenon is observed optically as the ejection of material from the Sun, in a more logical world the acronym CME that is often used for a "coronal mass ejection" would stand for a coronal magnetic ejection. Finally, it raises the intriguing question as to whether mass ejections carry magnetic flux from the Sun, perhaps as a significant "loss mechanism" in the evolution of the field during the solar activity cycle.

III. SOME AVERAGE PROPERTIES OF CORONAL MASS EJECTIONS

Several properties of coronal mass ejections can be easily determined from white-light data like that introduced above. The angular size and the central location (usually expressed as a heliographic latitude) can be measured for any mass ejection feature that is well defined in a single coronal image. A speed of outward motion can be determined for any feature that can be identified in a time sequence of images. These "easy" measurements (none of which depend on the absolute calibration of an instrument) have been made for large numbers of ejections observed with all three of the space-borne coronagraphs mentioned in Sec. I. Integration of the intensity of scattered light permits estimation of the mass contents of ejections (with results that do depend on an absolute calibration). For ejections with both speed and mass determinations, the kinetic, potential (work done in raising the ejected mass out of the solar gravitational field), and total mechanical energy (sum of these) can be estimated. Mass and energy estimates are available for an extensive set of Solwind and SMM ejections.

Figure 3 shows distributions of measured values, based on a subset of the SMM observations, for three important mass ejection properties—angular size in the top panel, speed of outward motion in the middle panel, and the logarithm of the estimated mass in the bottom panel. Each of these distributions shows a distinct peak within a wide range of measured values. For the first two distributions, it is evident from arguments in the published literature that the central portions of these distributions are not seriously distorted by selection effects related to the visibility of mass ejections or measurability of widths or speeds. For example, the angular resolution of the SMM coronagraph is less than $1°$ and the angular sizes of observed mass ejections are determined to within a typical uncertainty of a few degrees (Hundhausen 1993). For a given mass of coronal plasma, a small feature can be seen with greater contrast and is thus easier to detect (as, for example, a mass ejection) than a large feature. It is thus clear (Hundhausen 1996) that the distribution of angular widths does truly fall off at values below the modal width near $45°$, and that this distribution is distinctly different from the distributions that describe sizes of many familiar solar features such as sunspots, active regions, or flares. Detectability of mass ejections at different

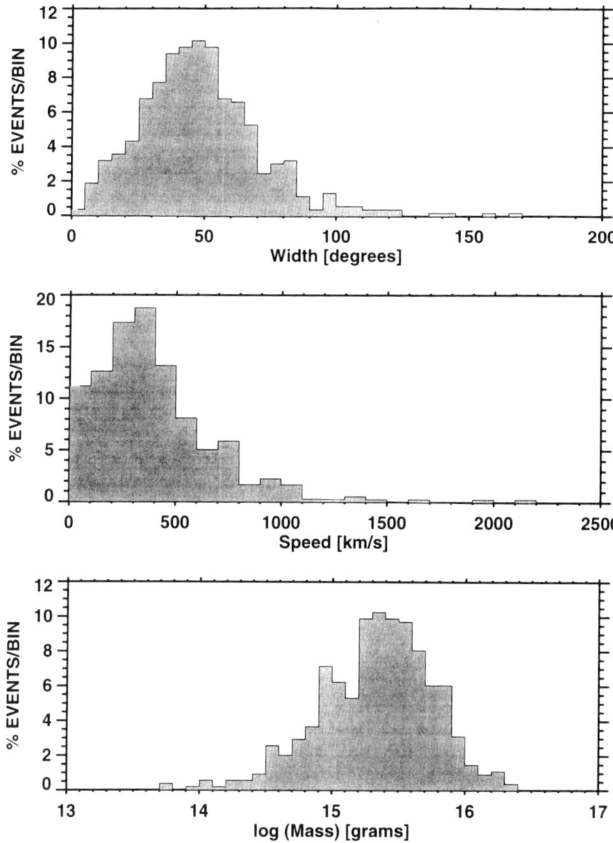

Figure 3.   Distributions of three important properties for a subset of the coronal mass ejections observed with the SMM coronagraph. The mass contents of 546 ejections from 1980 and 1984–1989 have been estimated as in Hundhausen et al. (1994*b*); the distribution of the logarithm of these masses is shown in the bottom panel. The distributions of the 532 angular widths and 357 speeds that could be measured within this same subset of events are shown in the top and middle panels of the figure.

speeds and limitations on the measurability of their motions are discussed in Hundhausen et al. (1984*b*,1994*a*) and, again, are expected to have little effect on the distribution over most of the range of observed speeds. Although such effects have not been as extensively discussed for the third distribution, mass estimates as low as 10^{14} g can be made reliably from SMM images, and it is unlikely that the rapid decrease in frequency of observation below the peak at near 2×10^{15} g is a result of inability to observe and analyze a population of ejections with smaller mass contents.

Average values (when available) for the angular sizes, speeds, masses, and energies deduced from Skylab, Solwind, and SMM data are given in Table I.

The Skylab values are taken from Gosling et al. (1976) and Hundhausen (1993); the Solwind values from Howard et al. (1985,1986); and the SMM values from Hundhausen (1993) and Hundhausen et al. (1994*a,b*). The reader should refer to those sources for details of instrumental and measurement techniques, numbers of measurements, actual ranges of measured values etc. The arguments regarding sampling given above imply that these averages are not seriously compromised by selection effects.

TABLE I

Some Average Characteristics of Coronal Mass Ejections

	Skylab (1973–74)	Solwind (1979–80 and 1984–85)	SMM (1980, 1984–89)
Angular size	42°	43°	47°
Speed	470 km s^{-1}	460 km s^{-1}	350 km s^{-1}
Mass	—	4.0×10^{15} g	2.5×10^{15} g[a]
Kinetic energy	—	3.4×10^{30} erg	3.1×10^{30} erg[a]
Potential energy	—	—	5.4×10^{30} erg[a]
Mechanical energy	—	—	8.5×10^{30} erg[a]

[a] SMM mass and energy analyses completed only through 1988.

Despite some differences in instrumental characteristics, calibrations, and measurement techniques, there is considerable consistency among these averages. It is clear from all these sources that, for example, mass ejections disrupt *large* regions of the solar corona; an average angular size of 40° to 50° implies a spatial scale of the disrupted region in the corona that is an appreciable fraction of a solar radius. This scale is huge compared to the spatial scales of most familiar forms of solar activity, such as sunspots, active regions, or flares seen in radiation at visible wavelengths. Solar features with comparable spatial scales include large prominences (seen as dark "filaments" over the solar disk) and coronal helmet streamers (seen, as in Fig. 1, above the solar limb). The average outward speeds determined from the three data sources are between 350 and 470 km s^{-1}. These values are comparable to the average speed of the distant solar wind near the ecliptic plane. They are less than the coronal Alfvén speed of 500 to 1000 km s^{-1} but greater than the coronal sound speed of 150 to 200 km s^{-1}. The typical mass ejection carries $\sim 3 \times 10^{15}$ g and $\sim 10^{31}$ erg away from the Sun; these values will be placed in the solar wind context in Sec. V.

IV. VARIATIONS IN AND RELATIONSHIPS AMONG THESE CHARACTERISTICS

The Solwind and SMM observations cover almost an entire sunspot cycle and allow a search for temporal variations in mass ejection properties over that cycle. Any such variations could be important clues regarding the relationship of mass ejections to the more familiar forms of solar activity. Examination of the properties tabulated above plus the latitudes and rates of mass ejection occurrence from the same sources reveals several different patterns of variation.

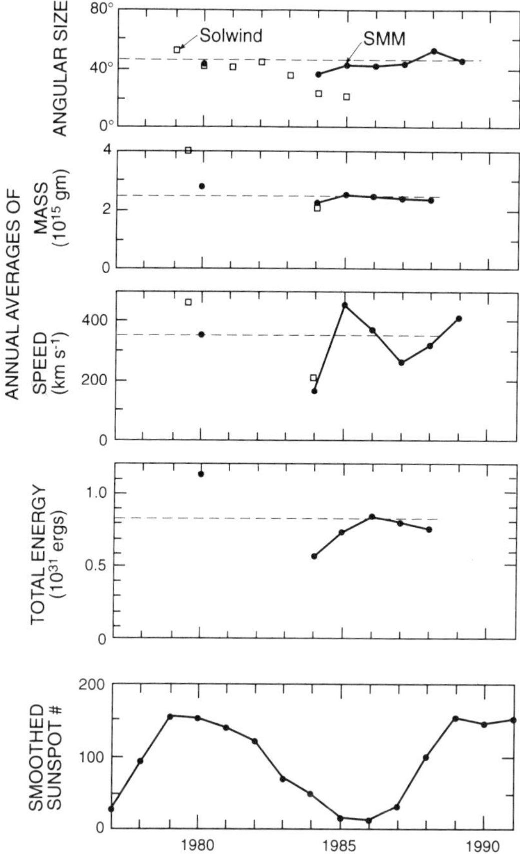

Figure 4.   Annual average values of coronal mass ejection angular size, mass content, speed, and total mechanical energy, derived from Solwind and SMM averages from the epoch 1977–1991. The smoothed sunspot number values in the bottom frame show the evolution of solar activity over the same epoch. The horizontal dashed lines denote the average values of each quantity based on SMM data.

(1) Several well-measured mass ejection properties have shown only small variations during that epoch over which observations are available. For example, Fig. 4 shows annual averages of mass ejection widths and mass contents in the top two panels (Solwind values are from Webb and Howard [1994], SMM values from Hundhausen [1993] and Hundhausen et al. [1994a,b]) and smoothed annual sunspot numbers for the 1977–1991 epoch in the bottom panel. The SMM values reveal little change and the Solwind mass and angular size values suggest changes (related through the Solwind

mass estimation technique) by a factor of $\sim$2 over a complete 11-yr cycle in sunspot number.

(2) Several mass ejection properties have shown larger variations than those described in (1) above, but with *no clear systematic* relationship to the sunspot cycle. For example, annual averages of the mass ejection speed and total mechanical energy are displayed in the third and fourth panels from the top of Fig. 4. Both of these quantities changed by a factor of 2 or 3 over the period 1979–1989. However, these changes do not appear to be *systematic* with respect to the sunspot number. For example, the SMM average speed was as high in the years of minimum sunspot number, 1985 and 1986, as in the years near maximum sunspot number, 1980 and 1989.

(3) There are two properties of coronal mass ejections that changed significantly and with an apparent, systematic relationship to the sunspot number over the epoch of Solwind and SMM observations; these changes may then indicate activity cycle variations in mass ejection characteristics (that should, of course, be verified by further observations over several cycles). The first of these is illustrated in the top frame of Fig. 5, a scatter plot of central latitudes vs time of all mass ejections seen in SMM data from 1980 and 1984– 1989. There appears to be a systematic change in the spread of mass ejections about the heliographic equator; they were confined to a narrow, near-equatorial band at the sunspot minimum epoch but were spread over a much broader band of latitudes at the sunspot maximum epoch. This variation can be quantified in terms of rms average of the observed latitudes (Hundhausen 1993). For purposes of comparison, the second and third frames of Fig. 5 show scatter plots of active region and "optical" flare latitudes for the same period. Both show the well known tendency for this activity to appear at moderate latitudes ($30°$ to $40°$ from the equator) at sunspot minimum and to migrate equatorward as activity increases. There is no poleward spread comparable to that of mass ejections after sunspot minimum (1986–1989 on the figure). The fourth and fifth frames of Fig. 4 show scatter plots for the latitudes of prominences and of bright coronal features (including helmet streamers). These two frames illustrate the well-known poleward drifts (see, e.g., Kiepenheuer 1953) in the maximum latitudes of prominences and coronal helmet streamers during the ascending phase of a solar activity cycle. These drifts or trends are similar to that for mass ejections during the 1986–1989 rise in solar activity. The second systematic variation is illustrated in Fig. 6 where the annual rates of mass ejection occurrence (ejections per day), reported by Webb and Howard (1994) on the basis of Skylab, Solwind, and SMM data, plus interplanetary (Helios) photometer observations are plotted. This rate shows an order of magnitude variation in approximate synchronization with the sunspot cycle.

These different patterns of variation in mass ejection properties have some simple but significant implications concerning the origins of mass ejections and their relationship to other forms of solar activity. The small changes in size and mass suggest that coronal structures of comparable spatial scale and mass content are disrupted by mass ejections at all phases of the activity cycle

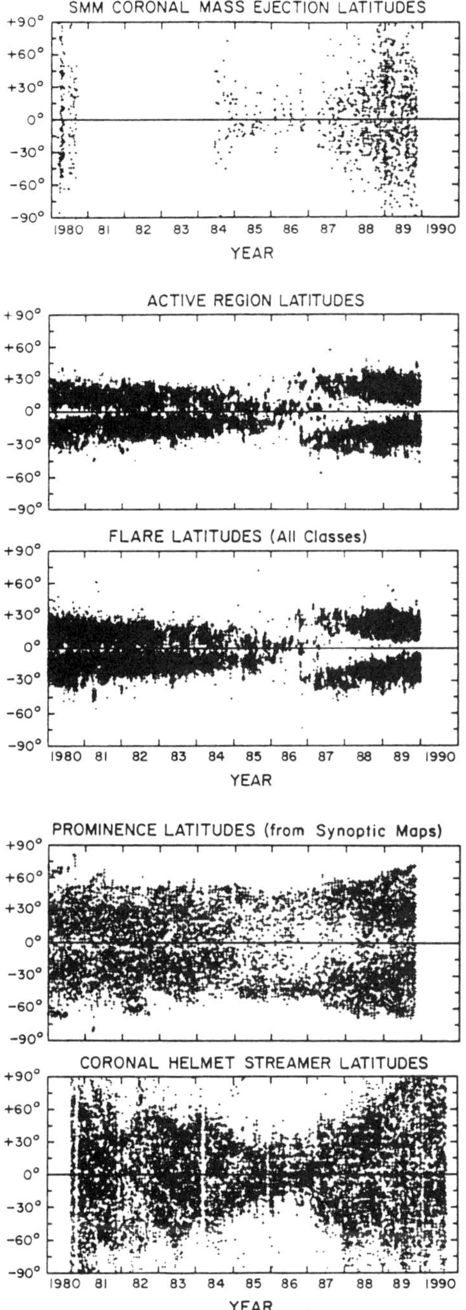

Figure 5. Scatter plots of the latitudes of coronal mass ejections, active regions, optical flares, prominences, and coronal helmet streamers, from 1980–1991.

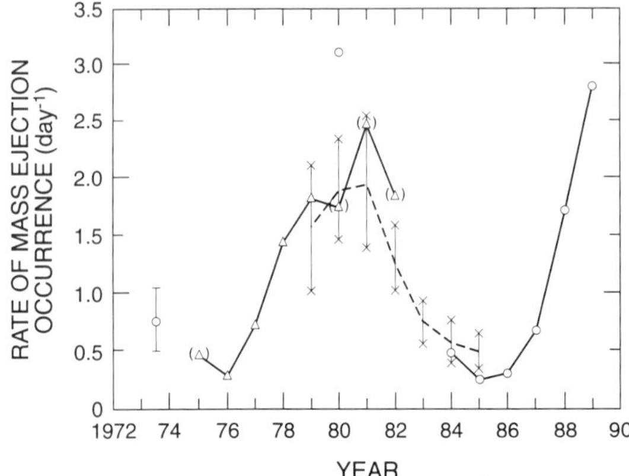

Figure 6. Coronal mass ejection rates (events per day) from Skylab, Solwind, SMM, and Helios interplanetary data (Webb and Howard 1994).

(despite the huge changes in small-scale magnetic features over that cycle). The energy released in the ejections (and the resulting kinetic energy or speed resulting after the mass is lifted out of the solar gravitational field) is more variable but does not directly reflect the overall level of activity. In contrast, the locations of mass ejections do change in a manner similar to those of *large-scale* magnetic features such as prominences and helmet streamers, but not with those of small-scale magnetic features such as sunspots, active regions, or flares. Finally, the rate at which mass ejections occur does undergo a major change in consort with the activity cycle, perhaps in response to the rate of evolution of the coronal magnetic field (Sime 1989).

The observation of mass ejections at a wide range of heliographic latitudes presents another opportunity for statistical examination of relationships to other forms of solar activity. For example, Fig. 7 shows SMM measurements of angular widths, speeds, mass contents, and mechanical energies as functions of distance from the heliographic equator. The visual impression that these are essentially scatter plots can be confirmed in two ways. Table II gives the average values of these quantities in three latitude bands: a 0° to 20° equatorial region, a 20° to 40° active latitude region, and a >40° polar region. The averages in these latitude bands are not greatly different. The linear cross-correlation coefficients of the width, speed, mass, and mechanical energy with angular distance from the heliographic equator are +0.13, −0.03, −0.10, and −0.10, respectively. None is large enough to imply a *significant* systematic variation. It is particularly interesting that these average properties of mass ejections differ so little at active and polar latitudes. Because active latitudes are defined by such small-scale phenomena as sunspots, active regions, or flares, these results again point to the absence of a *direct* relationship of these small-scale magnetic structures with the mass ejection phenomena. The mass ejections seen at high or polar latitudes are, on the average, about as large,

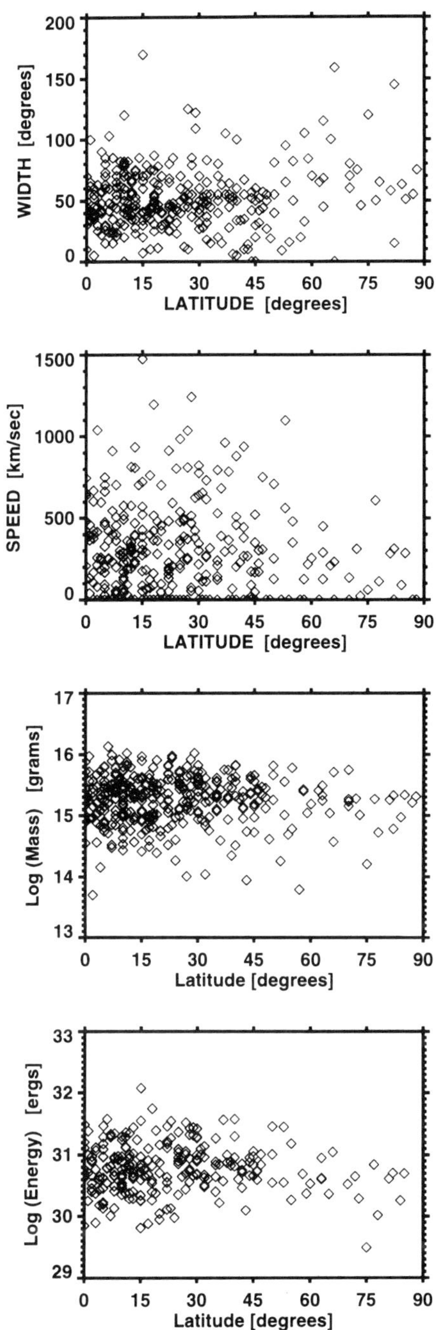

Figure 7.   Scatter plots of coronal mass ejection angular widths, speeds, masses, and total mechanical energies vs angular distance from the heliographic equator.

massive, fast, and energetic as those at active latitudes. This despite the absence at these high or polar latitudes of the strong complex magnetic fields found at active latitudes and regarded by many as the source of energy in the most common forms of solar activity.

TABLE II

SMM Average Characteristics

	Angular Width	Speed (km s^{-1})	Mass (10^{15} g)	Mechanical Energy (10^{-3} erg)
|Latitude| $< 20°$	47°	344	2.5	8.3
$20° \leq$ |Latitude| $< 40°$	45°	359	2.7	10.2
|Latitude| $\geq 40°$	49°	332	1.9	6.2

V. THE CONTRIBUTIONS OF CORONAL MASS EJECTIONS TO THE SOLAR WIND

An obvious, cosmic question regarding coronal mass ejections is that of their contribution to the solar wind. As white-light coronal observations reveal the density of the coronal plasma, they yield a fairly direct estimate of a mass efflux for comparison with the solar wind mass flux inferred from interplanetary observations. An average mass content for an individual ejection of $\sim 3 \times 10^{15}$ g was deduced from the Solwind and SMM observations summarized in Table I and pertains to the plasma ejected in the detectable bright features (such as the bright frontal loop and prominence of Fig. 1) in the observed ejections. The mass ejection occurrence rates given by Webb and Howard (1994) (as in Fig. 5) range from about 0.2 to 3 per day. A typical rate is close to 1 ejection per day; at this rate the mass carried away from the Sun is the 3×10^{15} g average value per day. The solar wind flux density observed in the ecliptic plane at the orbit of Earth is about 3×10^8 protons or 5×10^{-16}g cm^{-2} s^{-1} (Feldman et al. 1977). If this flux density is typical of that over the entire Sun-centered sphere at the orbit of Earth (as indicated by recent Ulysses observations well out of the ecliptic plane; see, e.g., Phillips et al. 1994), the solar wind mass flux is about 1.36×10^{12} g s^{-1} or 1.2×10^{17} g per day. The observed mass ejection efflux of about 3×10^{15} g per day is then only about 2.5% of the average solar wind mass flux. The fraction of the solar wind that could be contributed by detectable motions in observed coronal mass ejections ranges from 0.5% to 7.5% for the actual range of mass ejection occurrence rates mentioned above. At no time during the epoch of coronal mass ejection observations does this simple estimate of the fractional contribution to the solar wind exceed $\sim 10\%$.

Similar results have been obtained many times as mass ejection observations have accumulated since the 1970s (see, e.g., Hildner 1977; MacQueen

1980; Howard et al. 1985). Recently Jackson and Howard (1993) have attempted to account for any population of small, unobserved ejections by extrapolating the approximately exponential mass frequency distribution observed by Solwind downward to small masses; they deduce a contribution to the solar wind (near the ecliptic plane) of 16% for the 1979–1981 epoch of high solar activity (when the rate of occurrence was high [see Fig. 5]). The only exception to the conclusion that the observed efflux in coronal mass ejections are a minor contribution to the solar wind comes from Zirin (1988, p. 260), who concluded that mass ejections "...represent a mass loss as important as the steady solar wind." This disparity is easily understood. Zirin (1988) used a solar wind mass flux of 10^{16} g per day, an order of magnitude too low, and compared it with the upper end of a range of a mass ejection loss rates, "...quoted by the various [unreferenced] authors...", of 10^{15} to 10^{16} g per day, significantly too high. A similar mistake is apparently repeated by Zirin (1994), who invokes a mass ejection occurrence rate of 20 per day at the peak of the solar cycle.

A similar, simple comparison can be made using the mechanical energy (sum of the kinetic energy and the work done in transporting mass out of the solar gravitational field) in mass ejections and in the steady solar wind. The average mechanical energy of 10^{31} erg (Table I) at the typical mass ejection rate of 1 mass ejection per day implies that $\sim 10^{31}$ erg per day is carried away from the Sun by coronal mass ejections. The average ecliptic plane solar wind energy flux density of about 1 erg cm^{-2} s^{-1} (Feldman et al. 1977) implies, again assuming uniformity over a Sun-centered sphere at 1 AU, a steady solar wind energy flux of 2.7×10^{27} erg s^{-1} or 2.3×10^{32} erg per day. Thus observed features in mass ejections can account for about 5% of the energy in the "steady" solar wind.

Our present statistical knowledge of masses and rates confirms earlier conclusions that the coronal disturbances observed in white light and described under the broad rubric of coronal mass ejections provide only a minor part of the mass and energy carried away from the Sun by the solar wind. Further, no significant population of small temporal disturbances, as postulated by Axford (1977) and advocated by others (see the chapter by Axford and McKenzie) as a significant source of solar wind, has been directly observed or detected by extrapolation of observed frequency distributions to small masses or sizes (Jackson and Howard 1993). However, the perturbations of the solar wind produced by coronal mass ejections may be locally, if not globally, important. In fact, the most significant nonrecurrent, temporal disturbances observed in the solar wind, interplanetary shock waves, are clearly a result of the fastest part of the coronal mass ejection population. A major reason for widespread interest in coronal mass ejections is this role in producing solar-activity related disturbances in the solar wind that are transmitted to and have physical effects at the Earth (see, e.g., Hundhausen et al. 1984*a*; Gosling 1993*b*).

Identification of mass ejection features in the solar wind disturbances actually detected in *in-situ* and remote observations of the interplanetary medium is complicated by evolutionary changes as these features propagate outward and interact with the ambient solar wind. The reader is referred to Gosling (1990,1992,1993a) for reviews of this complex topic. However, it may be informative to compare the mass content estimates discussed above with estimates of the mass contents of directly observed solar wind disturbances. Recently Jackson and Webb (1994) have found an average mass slightly above 10^{16} g for some 80 interplanetary density disturbances selected from the 200 such events detected with photometers on the Helios spacecraft between 1975 and 1979. It may be tempting (Webb et al. 1996) to interpret the difference between the Helios interplanetary average and the coronal average given above as evidence for a continued outflow of coronal plasma after passage of the initial bright features seen in coronagraph images. While there is undoubtedly some such outflow, its significance may be greatly overstated by this interpretation. The sampling of 80 events in five years introduces a serious possibility of selection effects. Further, the Helios mass estimates do not distinguish between the material added to the solar wind disturbance by the initial mass ejection and ambient interplanetary material swept into the disturbance as it propagates outward through the solar wind. The significance of this interplanetary contribution to the mass of a solar wind disturbance can be readily estimated. If the typical mass ejection is assumed to occupy a cone with apex at sun center and with half angle of about $22°$ (corresponding to the average angular sizes of $42°$ to $47°$ in Table I), the ejection would subtend about 4% of the solid angle about the Sun. An extension of this cone into interplanetary space is filled (using the solar wind mass flux of 1.2×10^{17} g per day as mentioned above) with about 5×10^{15} g per day, or about 2×10^{16} g of ambient material between the corona and orbit of Earth. The sweeping up of any substantial fraction of this material (a fraction that, of course, depends on the mass ejection speed) would lead to an interplanetary disturbance with significantly larger mass than initially added by the coronal mass ejection.

The separation of initially injected and swept-up ambient material can hypothetically be accomplished by intergration of the mass *flux density* excess (above a normal or ambient level) in solar wind disturbances. Determination of a total mass excess then requires assumption of an overall geometry of the observed disturbances. Application of this technique to 22 interplanetary shock waves observed between 1965 and 1967 (Hundhausen et al. 1970) led, under the assumption that these disturbances subtended about 25% of the solid angle about the Sun, to an average excess mass of 3.5×10^{16} g. Use of the geometry more appropriate to observed mass ejection sizes, as introduced above, would reduce this average value to about 6×10^{15} g. Once again, comparison of either of these values of added mass in shock waves with the lower average value of 3×10^{15} g for coronal mass ejections must be interpreted cautiously. The 22 shock waves in this analysis were clearly selected as the best and largest examples available from this 3-yr period and an *average* value deduced from this sample is clearly an overestimate. Nonetheless, use of the 3×10^{16} g value and the estimated rate of occurrence for interplanetary shock waves for this same 1965–1967 epoch led Hundhausen (1972) to a conclusion remarkably similar to that for mass ejections; less than 10% of the solar wind mass flux could be accounted for by these major solar wind disturbances.

The interpretation of coronal mass ejections as temporary disruptions of large-scale, magnetically closed regions of the corona, introduced by an example in Sec. II and broadened through statistical results on mass ejection sizes, locations, and activity cycle variations in Secs. III and IV, allows us to place this role of mass ejections in a physical context. Since the identification of coronal holes as magnetically-open regions of the corona and the sources of unipolar streams of high-speed solar wind was made in the 1970s (see, e.g., Zirker 1977), a broad view of a global solar wind structure has become widely accepted. In this view, most of the large temporal variations in solar wind properties seen by a stationary interplanetary observer are actually manifestations of a slowly evolving spatial structure that rotates with the Sun; the magnetic sectors and organized patterns of recurrent fast and slow solar wind flows are the results of a quasi-steady expansion of the coronal plasma from different quasi-permanent coronal holes separated by magnetically-closed regions such as coronal helmet streamers. We would now suggest that coronal mass ejections are sporadic additions of coronal material to the solar wind through the disruption of these normally closed regions. The temporary opening of these regions adds, to the wind, new coronal material that was previously frozen onto the closed magnetic field lines and thus not part of the coronal expansion process. These additions occur "between" coronal holes (and the fast solar wind streams that emanate from holes) and can account for the $\sim$10% of the solar wind as deduced above.

VI. CORONAL MASS EJECTIONS AND SOLAR FLARES

Perhaps the most intriguing questions concerning coronal mass ejections are those of their relationship to other forms of solar activity (as touched upon in Sec. IV) and of their origins (in both the phenomenological and physical senses). The relationship of coronal mass ejections to solar flares has historically played a central role in discussions of both questions. This role has been based on well-known associations of flares with phenomena such as geomagnetic storms and interplanetary shock waves, both in turn related to coronal mass ejections. It has probably been reenforced by the emphasis placed on flares in most modern discussions of solar activity.

It may thus be surprising to many who have not been recently involved in the study of coronal mass ejections that there is a large body of observational evidence that casts substantial doubt on the view that mass ejections are driven by flares (through, for example, the release of thermal energy in the flare) and that further indicates a very poor or loose relationship between mass ejection properties (such as size, speed, mass, or energy) and the properties (such as spatial scales or peak intensities) of any associated flares. This evidence has led a number of authors (including this one) to discount flares as a *common* factor in the physical origins of mass ejections. This topic remains controversial; the reader is referred to Gosling (1993*b*), Hudson et al. (1995), and Svestka (1995) for some of the recent contributions to this ongoing argument. This author's views of the observational evidence pertinent to the topic are presented in Hundhausen (1996) and it is probably not appropriate to repeat them here in detail. I will thus only summarize those views, use a few key examples to illustrate them, and then introduce some new results based on recent Yohkoh soft X-ray observations that further clarify the mass ejection-flare relationship.

It is appropriate, however, to mention that the groundwork for these views was laid back in the Skylab era when the first extensive analyses of coronal mass ejections were performed. Two specific results based on Skylab observations have been especially important. First, examination of associations (i.e., near coincidences in time and location) of mass ejections with other forms of solar activity (Munro et al. 1979) showed that eruption of a prominence was the most common form of activity to occur near the onset of a mass ejection; in contrast only a *minority* of mass ejections were associated with *optical* solar flares, as identified by a brightening in the H α emission line of hydrogen. Similar results were found by Webb and Hundhausen (1987) and by St. Cyr and Webb (1991) for mass ejections detected by the SMM coronagraph. Second, a distinctive form of soft X-ray emission accompanied many prominence eruptions and a few mass ejections for which complete observations were available (see, e.g., Sheeley et al. 1975; Webb et al. 1976; Moore et al. 1980). This soft X-ray emission, the thermal radiation from hot coronal plasma, came from loops or arcades of loops that spanned the neutral line beneath the pre-ejection prominence. These loops expanded slowly with time, and the X-ray emission from them decayed slowly (with decay times of tens of minutes to hours) after peak intensity was attained. This type of emission, or X-ray flare, is known as a long-duration event (or LDE). These events were interpreted as the signature of magnetic reconnection proceeding outward along the magnetic neutral line produced (see Sec. II) by a prominence eruption (Hirayama 1974) or coronal mass ejection (Kopp and Pneuman 1976). This interpretation has been subsequently developed in terms of quantitative models (see the reviews by Forbes 1991,1992).

The extensive set of coronal observations made with the coronagraph on the Solar Maximum Mission spacecraft in 1980 and 1984–1989 provided the opportunity of extending and clarifying the concepts developed from Skylab observations. One such important extension has stemmed from comparisons of SMM mass ejection observations and the soft X-ray flux, integrated over an entire Sun, measured by instruments on several GOES spacecraft. These comparisons are a primary source of the conclusion that the relationship between mass ejections and flares is weak or "poor" in a different sense than the Skylab result cited above.

A. Poor Relations

The following pair of ejections serves to illustrate this point. Two major mass ejections seen on 18 March 1989 were comparable in many ways but had significantly different "associated" X-ray emission.

Ejection 1. Figure 8 is one of three SMM coronagraph images from 18 March 1989 that show a large mass ejection over the west limb of the Sun. The bright frontal loop on these images spaned $\sim$60° and was centered above a heliographic latitude of 13°N. The top panel of Fig. 9 shows the heliocentric location of the sharp back edge of the loop in these three images as it moved beyond a heliocentric distance of 4 solar radii between 1959 and 2032 UT. A linear fit to this observed trajectory gives a speed of 330 km s^{-1} in this outer part of the SMM field of view. The estimated mass (Hundhausen et al. 1994b) in the loop is 2×10^{16} g; the kinetic and mechanical energies implied by these values are 1×10^{31} and 5×10^{31} erg, respectively. This mass ejection was somewhat wider and about eight times as massive as the "average" event. Its speed was close to average and the mechanical energy was about five times the average.

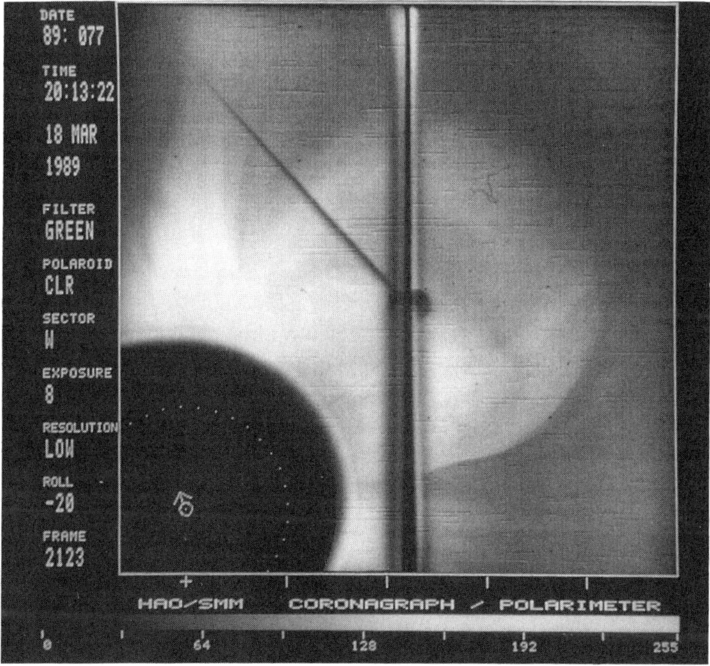

Figure 8.   An SMM coronagraph image from 2013 UT on 18 March 1989 showing a mass ejection over the west limb of the Sun.

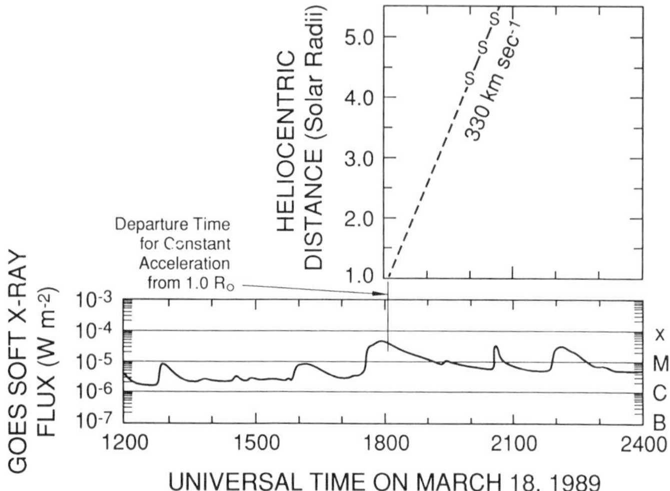

Figure 9.   Trajectory of the top of the dark cavity in the mass ejection shown on Fig. 8, and the integrated soft X-ray flux (at wavelengths from 1 to 8 Å) from the Sun for the last half of 18 March 1989.

The bottom panel of Fig. 9 shows the GOES-7 integrated X-ray (1 to 8 Å) flux from the last twelve hours (UT) of 18 March 1989. The distinct elevation in X-ray flux that began at $\sim$1730 UT has the characteristics of a long duration event or LDE as identified in Skylab studies. For example, the long, smooth decay in intensity after the peak flux (of $\sim$4.4 $\times$ 10^{-5} watts m^{-2}) has an e^{-1} decay time of $\sim$40 min. This X-ray enhancement has been associated with an optical flare at heliographic latitude 36°N and longitude 64°W, or beneath the observed mass ejection. This is a classic example of the mass ejection/soft X-ray association described in the summary of Skylab results given above.

Ejection 2.   Figure 10 is one of four SMM coronagraph images from earlier on 18 March 1989 that show a large mass ejection over the south-west limb of the Sun. The bright frontal loop (extending partly out of the coronagraph field of view) spanned a similar 60° and was centered above a heliographic latitude of 40°S. Within the bright frontal or outer loop was a second, more structured loop-like feature that can be identified with a very large prominence that erupted in conjunction with this mass ejection. The top panel of Fig. 11 shows the heliocentric position of the back edge of the frontal loop as it moved from 2.1 to 4.4 solar radii between 0253 to 0343 UT. A quadratic fit to this trajectory gives a speed of 750 km s^{-1} in the outer part of the field of view of the instrument. The estimated mass in the frontal loop is 1 $\times$ 10^{16} g; this is clearly an underestimate because of the extension of the frontal loop out of the field of view. The kinetic and mechanical energies implied by these values are 2.5 $\times$ 10^{31} and 4.5 $\times$ 10^{31} erg. This mass ejection was slightly wider, twice as fast, and at least four times as massive and nearly 5 times as energetic as the typical event.

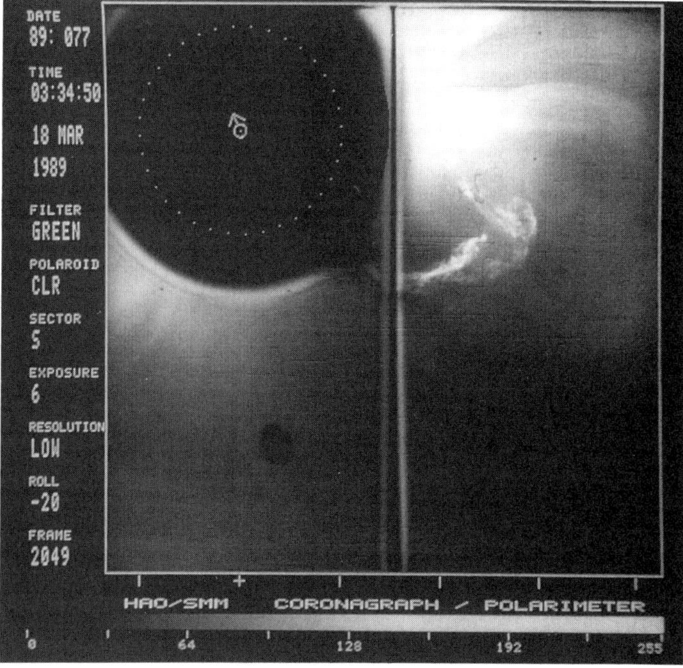

Figure 10.   An SMM image from 0334 UT on 18 March 1989 showing a mass ejection over the southwest limb of the Sun.

The bottom panel of Fig. 11 shows the GOES-7 integrated X-ray (1–8 Å) flux from the first twelve hours (UT) of 18 March 1989. No distinct elevation of the X-ray flux with the characteristics of an LDE can be seen above the slowly varying background flux (of 3 to 4×10^{-6} W m^{-2}) near the time when the mass ejection was observed. The short-lived burst of X-ray emission seen just before the mass ejection was almost certainly not physically related to it. This emission has been associated with a weak optical flare near the center of the solar disk, well away from the mass ejection (and underlying prominence eruption) near the limb. Such moderate intensity, "impulsive" X-ray flares are commonly detected and only rarely associated with large coronal mass ejections (Burkepile et al. 1994).

Consideration of the X-ray fluxes at the times of these two mass ejections illustrates the sense in which the relationship between mass ejections and flares implied by SMM/GOES comparisons can be termed "poor" or "weak." The occurrence of the most intense, long-duration enhancement of the integrated solar X-ray flux at about the time when ejection 1 (shown in Fig. 8) must have formed and been accelerated out of the corona is in accord with the traditional view of an "association" between flares and temporal disturbances in the corona. It does not, however, demonstrate physical cause and effect;

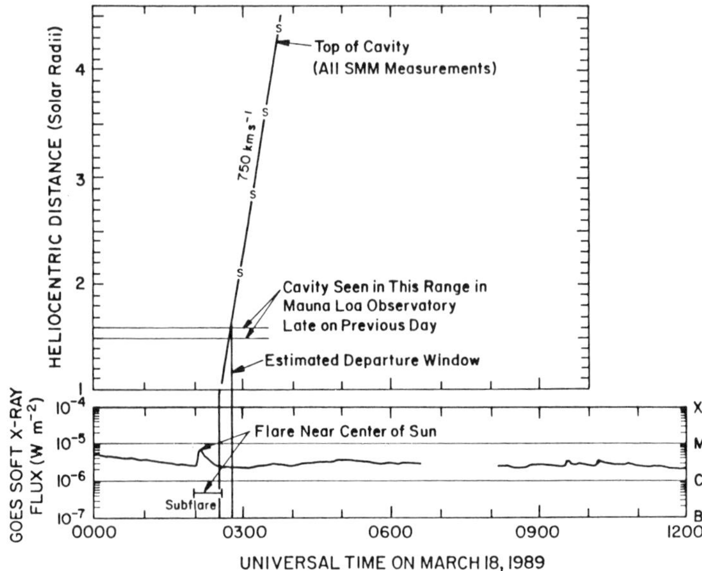

Figure 11.   Trajectory of the dark cavity top in the mass ejection shown on Fig. 10,
and the integrated soft X-ray flux for the first half of 18 March 1989.

the X-ray emission that peaked at about 1800 UT could have produced or
"triggered" the ejection, or it could be the post-ejection emission suggested
in the interpretation of Skylab observations. In contrast, the apparent absence
of any corresponding, long-duration enhancement of the solar X-ray flux at
about the time when ejection 2 (shown in Fig. 10) must have formed and
accelerated is not consistent with such a view. This inconsistency cannot be
avoided by arguing that the X-ray emission was hidden behind the Sun; the
prominence that erupted in conjunction with this mass ejection extended well
onto the visible disk of the Sun and thus part of the coronal region disrupted by
the ejection must have been on the visible "side" of the solar limb. There may
have been enhanced X-ray emission associated with the ejection of Fig. 10, but
at too low an intensity to rise significantly above the background of emission
from the entire visible part of the Sun. Images of the Sun taken in soft X-rays
by an instrument on the Yohkoh spacecraft (discussed below) do show this to
be the case in many mass ejections; we cannot thus rule out the possibility that
there are X-ray flares associated wth most or even all coronal mass ejections.
In this sense our conclusions are different from those drawn in studies of
optical flares cited above and our statement that the relationship between
mass ejections and flares is weak or poor must be understood as follows. The
earlier ejection 2 was as large and left the corona at an appreciably higher
speed than the later ejection 1. While the clearly underestimated mass in the
ejection 2 was half that in the later ejection, the mechanical energies in the two

events are essentially equal. Yet any long-duration X-ray emission related to the earlier event cannot have had a peak intensity higher than about 4×10^{-6} W m^{-2}, or at least a factor of 10 lower than the peak intensity of the emission "associated" with the later event. SMM/GOES comparisons reveal that mass ejections with similar speeds or energies can be associated with X-ray flares of the LDE class with peak intensities that differ by several orders of magnitude.

Many of the known examples of large, fast, massive, and energetic coronal mass ejections with no or very weak associated GOES X-ray flares (several are described in Hundhausen 1996) occurred at high solar latitudes, often in association with eruptions of large, high-latitude solar prominences. These are likely the ejections that appear poleward of the active latitudes on Fig. 5, and contribute to the lack of latitude dependence for observed angular widths, speeds, mass contents, and mechanical energies (Fig. 7). However, a complete statistical analysis of the relationship between the physical properties of mass ejections and the intensities of associated flares remains to be carried out to demonstrate conclusively the validity of these suggestions.

Despite the incomplete nature of these arguments, they have some serious implications regarding the prediction of coronal mass ejections (or even major coronal mass ejections) on the basis of solar X-ray enhancements (or "X-ray flares"), or any analogous assessment of stellar mass ejection occurrences on the basis of stellar X-ray observations. The GOES soft X-ray flux from the entire Sun for 18 March 1989, can be visualized by combining the bottom frames from Fig. 11 (0000 to 1200 UT) and Fig. 9 (1200 to 2400 UT). Eight X-ray flares are listed in the NOAA *Solar-Geophysical Data Report* for this day. Only one of these, the major enhancement near 1800 UT was, as described above, clearly related to an observed mass ejection. Two additional ejections were observed with the SMM coronagraph on this day. One, the major "southwest" event 2 described above, was not related to any significant enhancement of the GOES integrated flux near the time of the mass ejection. The other ejection was seen as a bright loop at 1834 UT, during the major X-ray enhancement mentioned above but at the *opposite* (east) limb of the Sun. Could the number, location, and importance of these observed mass ejections be predicted from the GOES X-ray flux data and any complimentary data on optical flares?

B. The Times Are Out of Joint

A second important extension of the Skylab-era concepts of the mass ejection/solar flare relationship stemmed from SMM studies of the timing of these phenomena. Detailed comparisons of the observed trajectories of mass ejections and the time variations in associated X-ray emissions suggest that most of this emission occurs in the late stages of mass ejections, *after* the "onset or launch" of the ejection. The strongest statement of this result on the basis of SMM observations was given by Harrison et al. (1985) and Harrison (1986):

1. Deduced a "launch time" for mass ejections detected in the SMM field

of view by linear extrapolation of the observed trajectory (based on measurements beyond a heliocentric distance of 1.6 solar radii) to the base of the corona (at 1.0 solar radius).

2. Compared this launch time with the onset of associated X-ray emission detected in a low-energy channel of the Hard X-ray Imaging Spectrometer (HXIS) also on SMM.

For seven events from 1980, for which both instruments observed the same region of the Sun, the X-ray "flare" detected by HXIS *began* 10 to 20 minutes after the mass ejection launch time. Further studies by Harrison et al. (1990) and Harrison (1991,1994) have dealt with a larger number of ejections, often with an integrated X-ray flux (as seen in in Figs. 9 and 11).

An obvious weakness in the analysis by Harrison and colleagues lies in the deduction of the mass ejection launch time. All mass ejections must accelerate from very low speeds as they are formed, and some are observed to undergo continued acceleration well out in the corona. Many ejections are seen to form well above the base of the corona. Thus linear (constant speed) extrapolation of an observed trajectory to the base of the corona may introduce significant errors (of either sign) in determining the true onset of the ejection (see the reviews by Kahler 1992, and Hundhausen 1996). For example, only the very brave (or foolish) would use the launch time indicated on the top panel of Fig. 9, deduced by linear extrapolation from observations beyond 4 solar radii, as a sound basis for comparison with the onset of the X-ray emission in the bottom panel.

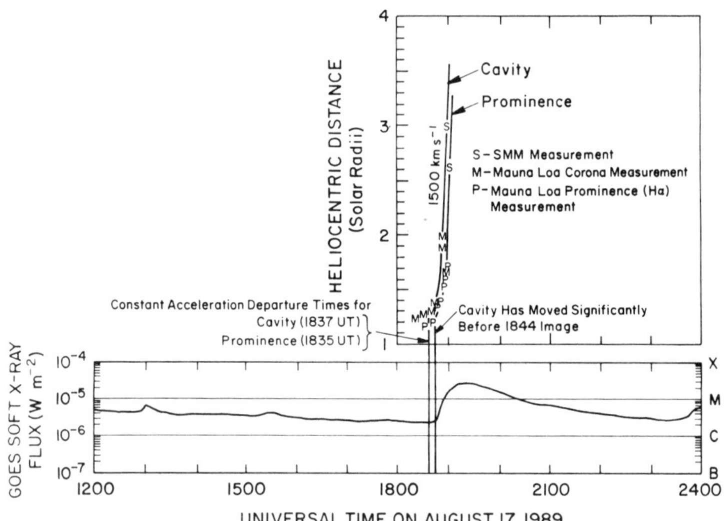

Figure 12.   Trajectories of the prominence and dark cavity in a coronal mass ejection observed with the SMM coronagraph on 17 August 1989, and the integrated soft X-ray flux from the last half of that day.

This difficulty can be circumvented for a minority of mass ejections in which their formation has been directly observed. There are a handful of known mass ejections wherein the trajectory determined by SMM coronagraph observations implies formation within the field of view of that instrument (usually involving disruption of an observable, pre-event coronal feature) or where complimentary data from the lower corona reveals that formation. Hundhausen (1996) has displayed three such events, one of which is summarized in Fig. 12. In this example from 17 August 1989, a dark cavity was seen to rise slowly into the field of view (extending from 1.15 to 2.45 solar radii) of the coronameter operated at the Mauna Loa Observatory; a bright frontal loop formed around the cavity and a prominence also rose within it. At about 1840 UT this entire structure was seen to accelerate outward through the Mauna Loa coronameter field of view and was observed at 1859 UT as a fully developed mass ejection, extending above a heliocentric distance of 3 solar radii, by the SMM coronagraph. The combined Mauna Loa/SMM trajectories for the sharp back of the bright frontal loop (or top of the dark cavity) and the leading edge of the prominence are shown in the top panel of Fig. 12. The later portions of both trajectories can be well fit with constant acceleration curves that imply acceleration from a slow expansion to 1500 km s^{-1} in less than 20 min, with the onset of that acceleration near 1835 UT. Both features had moved significantly in the interval between Mauna Loa images obtained at 1839 and 1844 UT, so that the latter is a latest possible limit for onset of the acceleration. Observations such as these yield an onset time with an accuracy of a few minutes and with no necessity for extrapolation of trajectories to an assumed height of mass ejection formation.

The GOES integrated soft X-ray flux for the second half of 17 August is shown in the bottom panel of Fig. 12. This flux rose above background at 1843 UT, reached a peak intensity an order of magnitude above the whole Sun background at 1924 UT, and displayed the slow decay in intensity of a long-duration event thereafter. In the likely event that this X-ray flare was physically related to the mass ejection/prominence eruption illustrated by the top frame, it began nearly 10 min after the inferred onset of mass ejection acceleration and peaked nearly 50 min thereafter. This result, without the uncertainties implicit in the extrapolations described above, is consistent with the conclusion that this type of X-ray emission is an after-effect of the mass ejection.

The several well-analyzed examples of this nature show that the most significant detectable enhancement of soft X-ray emission (and hence heating of the corona) fits the pattern suggested on the basis of Skylab observations and can be interpreted in terms of magnetic reconnection, as outlined above. However, this argument still lacks the statistical basis necessary to demonstrate widespread validity.

C. The Brave New World of Yohkoh

Another inadequacy in the mass ejection/X-ray flare comparisons shown above stems from the use of an integrated solar X-ray flux. Even as the start or onset time of mass ejections has been more sharply defined and better determined, the onset time of related X-ray emission remains poorly determined when that emission may be masked by a background from the entire Sun. While this problem is minimized in a few examples from the solar minimum epoch when that background was low (see Fig. 32 of Hundhausen 1996), it remains a real limitation in our understanding of the mass ejection/flare relationship. A further problem stemming from the use of an integrated flux is the indetectability of low intensity flares. X-ray emission may indeed occur in mass ejections such as that shown in Figs. 10 and 11 above, but be obscured by the background flux from the entire Sun. The true nature of the relationship between mass ejection properties and X-ray flare intensities may be partially masked by this effect.

These problems have been mitigated since the launch of the Yohkoh spacecraft in late 1991 (albeit in an epoch when no space-borne coronagraphs have been available to give the best mass ejection observations). The Soft X-ray Telescope (SXT) on Yohkoh has provided high-quality *images* of the entire Sun in an 8 to 20 Å wavelength band; this renews the capability of detecting the structure of soft X-ray emission, in a regular, long-term series of observations, for the first time since Skylab. These Yohkoh observations provide the opportunity of alleviating the two difficulties in mass ejection/X-ray flare comparisons mentioned above.

Two examples of soft X-ray emission that was almost surely related to coronal mass ejections have been described by Hiei et al. (1993) and McAllister et al. (1995). Several more examples have been described at scientific meetings. Clearly, Yohkoh studies of the mass ejection/X-ray flare relationship are in their infancy. I will present here one example of a coronal mass ejection that was observed with the coronameter at the Mauna Loa Observatory and for which Yohkoh SXT observations show the pattern of related X-ray emission with a clarity that has seldom been achieved.

Figure 13 displays four coronal images obtained with the Mauna Loa instrument at 1739, 1809, 1831, and 1844 UT on 30 April 1993. A coronal mass ejection was underway over the southwest (lower right) limb of the Sun. The most conspicuous feature of this mass ejection, in the 1.15 to 2.45 solar radii field of view of the first three images, is a bright tongue that can be identified with the eruption of a prominence. The outward motion of this feature is obvious in this sequence. The fourth image was obtained by a second detector system that scans a field of view from 1.45 to 2.75 solar radii. This image shows the presence of a frontal loop in the ejection; formation of the frontal feature ahead of an erupting prominence is a fairly common occurrence in the portion of the corona imaged by the Mauna Loa instrument. The top panel of Fig. 14 shows the trajectories of the sharp top of the central "prominence" and the much fuzzier back of the frontal loop (or top of the dark cavity) based on Mauna Loa images from 1736 to 1913 UT. The prominence top is continually accelerated as it moves from 1.47 to 2.15 solar radii during

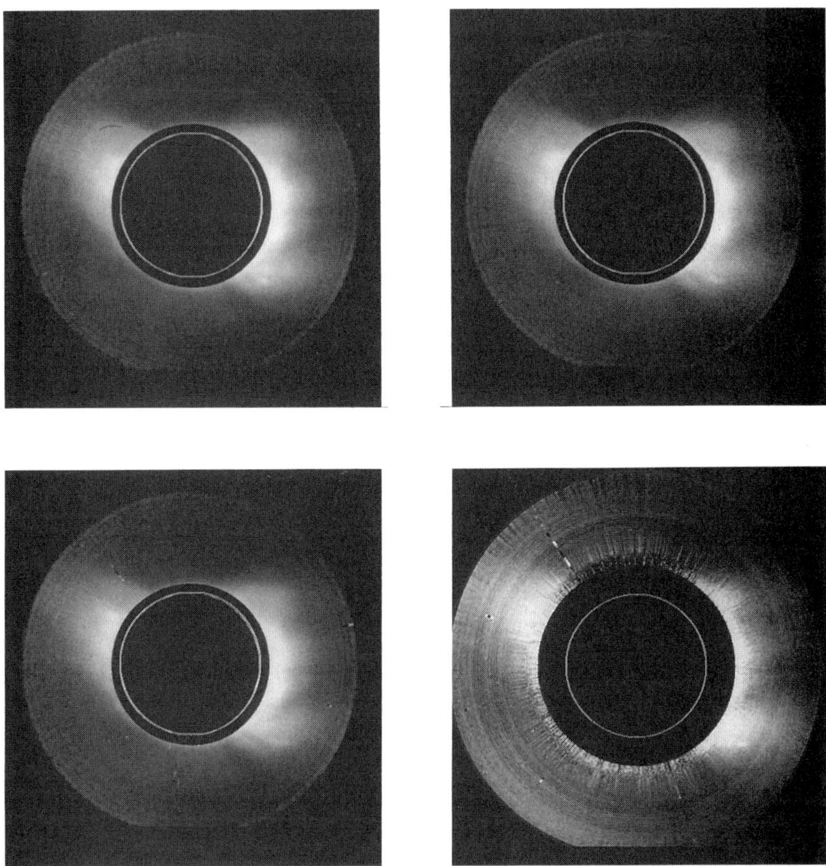

Figure 13.   Four Mauna Loa white-light images of the corona from 30 April 1993. The
bright tongue-like feature is a prominence that erupted on that date. The last image
in the time sequence shows formation of a frontal loop surrounding the prominence.

that time interval; the outward speed rises from 40 to 120 km s^{-1}. Less
accurate measurements of the cavity top also suggest acceleration to a speed
of nearly 200 km s^{-1} at the 1844 time of the last image on Fig. 13.

Figure 15 shows a sequence of Yohkoh SXT images from late on 30
April 1993. The crosses at the lower right indicate the observed (nonradial)
path followed by the prominence top in the mass ejection between 1736 and
1913 UT. The first Yohkoh image from 1647 shows the X-ray corona just
before the start of the mass ejection. The ejection clearly rose from a region,
well poleward of any significant active regions (seen as very bright features
on the image), occupied only by a very faint loop or arcade of X-ray emitting
material. The second image from 1921 shows the corona when the mass
ejection had moved to the end of the indicated path, was fully "formed," and
extended beyond 2 solar radii. No enhanced X-ray emission is discernable

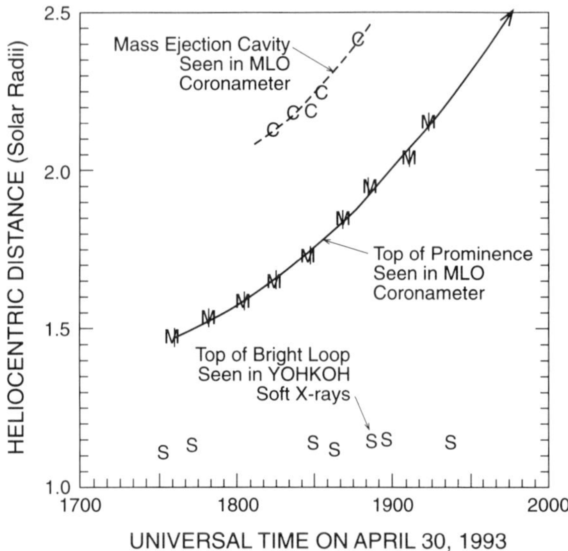

Figure 14.   Trajectories of prominence top (M's) and frontal loops (C's) in the 30
April 1993 mass ejection, and the heliocentric position of the brightest underlying
soft X-ray feature (usually a bright loop) seen in Yohkoh images (as in Fig. 13).

from the mass ejection features themselves, and there were only minor changes
in the faint features beneath the ejection on this image or on other images
obtained between 1647 and 1921. Direct superposition of nearly simultaneous
Mauna Loa and Yohkoh images confirms these conclusions. Most important,
the mass ejection features were not heated to the temperature ($\sim$2 or 3 million
deg) that would render them visible in the Yohkoh SXT images.

The third image of Fig. 15 was acquired at 2228 UT, well after the mass
ejection had left the field of view of the Mauna Loa coronameter. By this
time there is a distinct enhancement in soft X-ray emission from the region
beneath the ejection. This emission came from a system of superposed loops
that was seen to rise slowly over the next several hours. By the time the last
image of Fig. 15 was acquired, 0143 UT on 1 May, the loops in the system
were distinctly higher and wider. The brightest of these loops had a cusp-like
top that was nonradial in the same sense as the path of the prominence. The
heliocentric location of the top of the brightest visible X-ray emitting feature
in the region beneath the mass ejection (including very faint pre-ejection
feature seen on the first two panels of Fig. 15) is also shown on Fig. 14. The
prominence trajectory and height of the brightest X-ray feature are repeated
on a longer time scale in the top panel of Fig. 16; the bottom panel shows the
brightness of the enhanced X-ray emission as an integral of the Yohkoh X-ray
flux over the area ultimately occupied by the bright loops.

It is very clear from Figs. 14 and 16 that the soft X-ray emission detected
by the Yohkoh instrument fits the patterns discussed earlier in this text. The

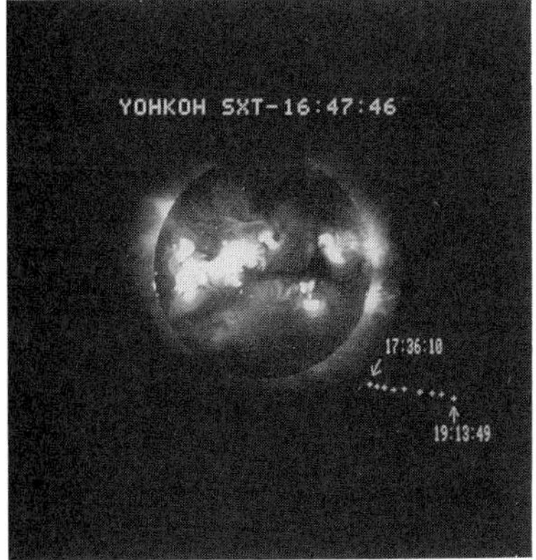

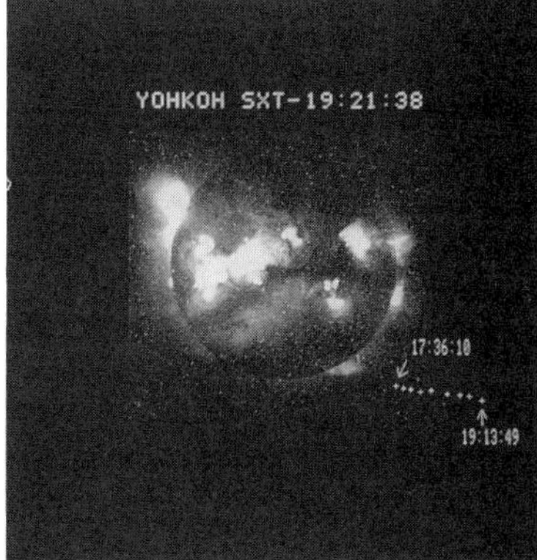

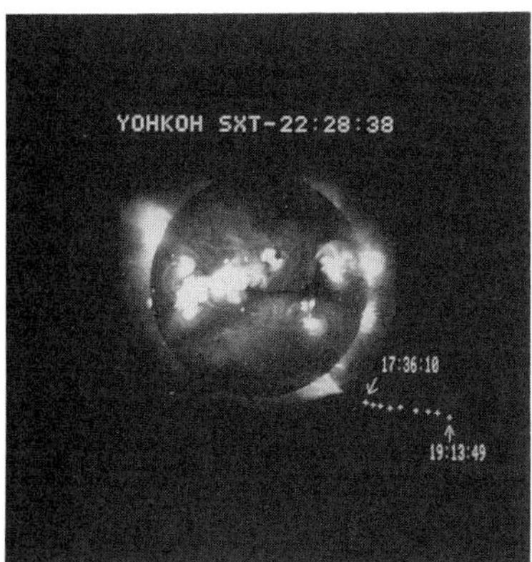

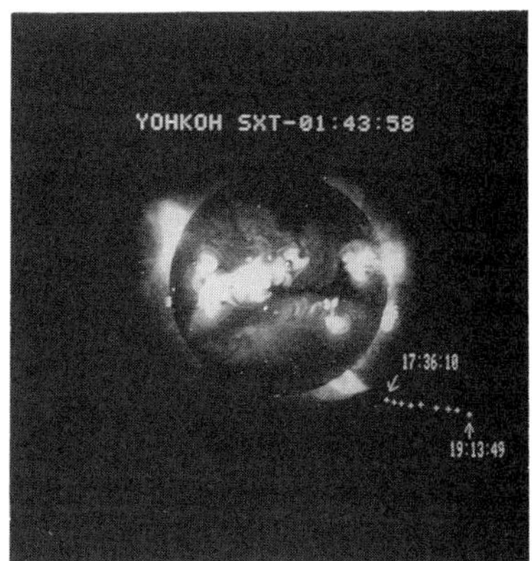

Figure 15. Four Yohkoh soft X-ray (wavelengths from 8 to 20 Å) images from 30 April and 1 May 1993. The crosses show the path followed by the prominence "core" seen in the white-light images of Fig. 13.

long, slow decay in the total X-ray intensity (after $\sim$ 2300 UT) shown in Fig. 16 is that of an LDE (an e^{-1} decay time is almost 24 hr), and most of that emission occurred when the mass ejection was far out in the corona. The top panel of Fig. 14 shows that the emission came from a system of loops that was well beneath the detectable features in the mass ejection. Further, the envelope of X-ray emission rose much more slowly (at a few km s^{-1}) than the observed mass ejection features. The rate of rise of the X-ray emitting features decreased with time, while the mass ejection trajectories showed a clear outward acceleration. This event also fits the *timing* pattern described on the basis of SMM studies; Fig. 16 shows that the onset of *enhanced* X-ray emission began several *hours* after the onset of acceleration of mass ejection features. However, the peak intensity of the X-ray flux remained well below the background from the remainder of the Sun, and this event was *not*

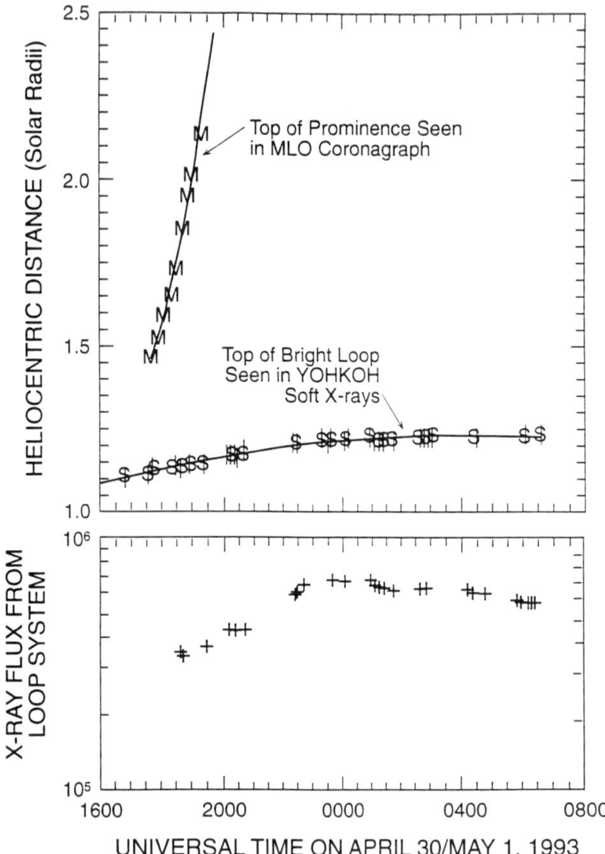

Figure 16.   The prominence trajectory and X-ray loop positions over a longer time
interval, on 30 April and 1 May, plus the integrated Yohkoh soft X-ray flux from
these loops.

detectable in the GOES integrated X-ray flux. It thus is an example of the class
of events where no X-ray flare would have been recorded in an SMM/GOES-
based study (as in Fig. 11). The imaging capability of Yohkoh has allowed
detection of the event, revealed the spatial structure of the emitting regions
and placed the event in the timing context set forth above.

This same pattern has been seen in other examples described in Hiei et al.
(1993) and Sime et al. (1994). Three of these events occurred at relatively high
solar latitudes, well poleward of any bright active regions. The 24 January
1992 example of Hiei et al. exhibited the most striking X-ray emission of
the three. The bright X-ray emitting loops seen in that case continued to rise
and spread until they could be identified as a new helmet streamer (replacing
the streamer blown out in the mass ejection) in the MLO images. Again,
the onset of enhanced X-ray emission followed by several hours the eruption

of a prominence that was part of the mass ejection. A natural explanation for the emission was proposed by Hiei et al. in terms of the same magnetic reconnection process, proceeding along the current sheet produced by the mass ejection. The clear reformation of a closed helmet streamer in this event strongly suggests a return to a "pre-ejection" state on the ~ 1 day time scale of the X-ray emission process. The sporadic addition of "new" coronal material to the solar wind (as in Sec. V) must have diminished on a similar time scale.

Figure 17.   A composite of Mauna Loa white-light and Yohkoh soft X-ray images from 15 and 16 March 1993. The Mauna Loa data showed a mass ejection that blew out the coronal region indicated by the arc at the right side of the white-light image from 2118 UT on 15 March. Intense X-ray emission came from the small loops shown on the later (0009 UT on 16 March) Yohkoh image.

While these conclusions and interpretations seem clear in these examples of high-latitude ejections, one must be careful in generalizing them, especially to mass ejections from active latitudes. Figure 17 is a superposition of MLO and SXT images from an active latitude event that was observed on 15 March 1993. MLO observations began at 2115 UT and showed a mass ejection well underway; it blew out a region of the corona with the typical angular width of $\sim 40°$, as indicated by the arc at the right edge of the figure. SXT images show a flare in progress after 2032 UT, with emission from a loop-like feature that rose and spread slowly. In this example, the X-ray flux from the loop was sufficiently high to be detectable (as a long-duration event) in the GOES integrated flux. However, the spatial scale of the loop remained much smaller than that of the mass ejection itself; the loop faded from visibility before attaining a horizontal scale larger than $\sim 1/10$ that of the mass ejection. This is clearly post-ejection emission (at the time of the observations in Fig. 17) as was often seen in the GOES integrated soft X-ray flux to follow mass ejections (as in Figs. 9 or 12) identified in the SMM data (see, e.g., Hundhausen 1996). The SXT images show that this emission comes from a spatial structure similar to that in the high-latitude examples, but with significant differences in intensity and spatial scale.

There are hints that some of these differences may be systematic. As illustrated by the examples above, the post-ejection emission related to high latitude mass ejections often spreads to large spatial scales but is at a low intensity; the post-ejection emission related to active latitude ejections is often on smaller scales but is at a higher intensity. These differences could result from magnetic reconnection in simple, large-scale, weak magnetic fields at the high (near-polar) latitudes, but in complex, small-scale, strong magnetic fields in the active latitudes. Yet the statistical analysis of SMM ejections has suggested that there is little systematic difference between high latitude and active latitude mass ejections. This may be additional evidence that the properties of mass ejections are poorly related to the intensity of any associated soft X-ray emission (and the implied heating of the corona), as suggested in Hundhausen (1996). Only a thorough statistical analysis of these Yohkoh observations will confirm or disprove the validity of this suggestion, as well as many other impressions gained in a preliminary examination of the Yohkoh SXT data set.

VII. CONCLUSIONS

Observations of the solar corona in "white light" (i.e., the photospheric radiation scattered by free electrons in the highly ionized coronal plasma) are the traditional basis for identifying and describing the phenomenon of coronal mass ejections. Twenty years of these observations have yielded an extensive description of mass ejections, with numerous examples that are both beautiful and suggestive of the physical conditions and processes in the phenomenon. Statistical analysis of the thousands of coronal mass ejections identified in space-borne observations made with the Skylab, Solwind, and SMM coronagraphs and the groundbased Mauna Loa coronameter give a detailed description of several basic physical properties of mass ejections, relationships among these properties, and connections to other manifestations of solar activity seen lower in the solar atmosphere with other observational techniques. Since their discovery little more than twenty years ago, coronal mass ejections have become an extensively observed and studied form of solar activity.

The physical interpretation of this large body of observations has led to considerable agreement and some controversy. There should be little remaining doubt that the *typical* coronal mass ejection is a disruption of a large-scale region of the solar corona wherein the magnetic field is initially closed; the disruption often involves eruption of prominences nested within the same field structure. There is a large body of evidence that *many* mass ejections occur well away from small-scale regions of photospheric or chromospheric activity. The prominences involved in these ejections are very large, relatively simple quiescent structures.

The insensitivity of white-light coronal observations to all coronal properties other than density dictates the use of other observations in the physical study of the connections between coronal mass ejections and most other forms of solar activity. The most useful data in such studies has been the soft X-ray emission from the corona as a fairly direct indication of the thermal state of the coronal plasma. Skylab images taken in soft X-rays were the basis of a general picture (and subsequent quantitative models) of heating of the corona by magnetic reconnection in a magnetic field structure produced by mass ejections. The extensive Solwind and SMM observations of coronal mass ejections were made in an era when X-ray images of the Sun were not routinely available. However, comparison of the SMM observations with integrated solar X-ray fluxes (measured from several GOES spacecraft) produced evidence that (1) the major X-ray emission (and hence coronal heating) associated with many mass ejections followed the ejections in a manner consistent with this magnetic reconnection picture, and (2) the intensity of this X-ray emission was weakly or poorly related to such mass ejection properties as speed or energy. The availability of X-ray images from the Yohkoh soft X-ray telescope has restored our ability to pursue these relationships in greater detail, a process that is presently in a very early state.

This author's interpretation of this entire body of observations leads to the view that mass ejections are the result of the long-term evolution of the large-scale components of the solar magnetic field that extend well out into the corona to form the slowly evolving pattern of open magnetic regions (coronal holes) that are the everyday sources of most solar wind and intervening, closed magnetic regions (such as coronal helmet streamers) within which the coronal plasma is contained on closed field lines. The slow evolution of the magnetic field at the base of the corona (that may or may not be closely related to smaller-scale features beneath it) stresses the closed magnetic structures so that they sporadically are "blown" open, releasing new material into the solar wind flow. Some or all of the blown-open regions return to a less stressed, closed configuration, probably through a magnetic reconnection process that heats newly reformed magnetic loops and produces the soft X-ray emission seen in Skylab, GOES, Yohkoh, and related data sets. Smaller-scale regions beneath the ejection may also be disturbed and undergo a similar reconnection/flaring process. The energy released in the mass ejection itself would come from the initial "stressing" of the closed field region and is largely converted into the mechanical energy of the material that leaves the Sun. Re-closure of the blown-open magnetic fields releases thermal (or internal) energy on the reformed, closed magnetic loops to produce the radiation identified as a flare (as in Fig. 36 of Hundhausen 1996). There need then be no close relationship between the energy, speed, mass, or other properties of the ejected material and the energy or intensity of associated flare emission. For physical arguments regarding this view the reader is directed to Low (1990,1993,1994).

Acknowledgments. The author wishes to thank L. Boyd for producing this manuscript, and J. Burkepile, K. MacGregor, and A. McAllister for comments on the manuscript. The comparisons of Mauna Loa and Yohkoh observations in Sec. VI are part of a collaborative effort with E. Hiei and D. G. Sime; I thank them for permission to describe them here. The High Altitude Observatory is part of the National Center for Atmospheric Research which is sponsored by the National Science Foundation under the management of the University Corporation for Atmospheric Research.

REFERENCES

Athay, R. G., and Illing, R. M. E. 1986. Analysis of the prominence associated with the mass ejection of Aug. 18, 1980. *J. Geophys. Res.* 91:10961–10973.

Axford, W. I. 1977. The three-dimensional structure of the interplanetary medium. In *Study of Travelling Interplanetary Phenomena/1977*, eds. M. A. Shea et al. (Dordrecht: D. Reidel), pp. 145–164.

Burkepile, J. T., Hundhausen, A. J., and Seiden, J. A. 1994. A study of GOES X-ray events associated with coronal mass ejections. In *Proc. of the Third SOHO Workshop—Solar Dynamic Phenomena and Solar Wind Consequences*, ESA SP-373, pp. 57–60.

Dryer, M. 1982. Coronal transient phenomena. *Space Sci. Rev.* 33:233–275.

Dryer, M. 1994. Interplanetary studies: Propagation of disturbances between the Sun and the magnetosphere. *Space Sci. Rev.* 67:363–419.

Feldman, W. C., Asbridge, J. R., Bame, S. J., and Gosling, J. T. 1977. Plasma and magnetic fields from the Sun, in *The Solar Output and Its Variation*, ed. O. R. White (Boulder: Colorado Univ. Press), pp. 351–381.

Fisher, R. R. 1984. Coronal mass ejection events. *Adv. Space Res.* 4:163–174.

Fisher, R. R., and Poland, A. I. 1981. Coronal activity below 2 $R_\odot$: 1980 February 15–17. *Astrophys. J.* 246:1004–1009.

Fisher, R. R., Garcia, C. J., and Seagraves, P. 1981. On the coronal transient-eruptive prominence of 1980 August 5. *Astrophys. J. Lett.* 246:161–164.

Forbes, T. G. 1991. Magnetic reconnection in solar flares. *Geophys. Astrophys. Fluid Dyn.* 62:15–36.

Forbes, T. G. 1992. Field opening and reconnection. In *Eruptive Solar Flares*, eds. Z. Svestka, B. V. Jackson and M. E. Machado (New York: Springer-Verlag), pp. 79–88.

Gosling, J. T. 1975. Large-scale inhomogeneities in the solar wind of solar origin. *Rev. Geophys. Space Sci.* 13:1053–1058.

Gosling, J. T. 1990. Coronal mass ejections and magnetic flux ropes in interplanetary space. In *Physics of Magnetic Flux Ropes*, eds. C. T. Russell, E. R. Priest and L. C. Lee (Washington, D. C.: American Geophys. Union), pp. 343–364.

Gosling, J. T. 1992. In situ observations of coronal mass ejections in interplanetary space. In *Eruptive Solar Flares*, eds. Z. Svestka, B. V. Jackson and M. E. Machado (New York: Springer-Verlag), pp. 258–267.

Gosling, J. T. 1993*a*. Coronal mass ejections: The link between solar and geomagnetic activity. *Phys. Fluids B* 5(7):2638.

Gosling, J. T. 1993*b*. The solar flare myth. *J. Geophys. Res.* 98:18937–18949.

Gosling, J. T., Hildner, E., MacQueen, R. M., Munro, R. H., Poland, A. I., and Ross, C. L. 1974. Mass ejections from the Sun: A view from Skylab. *J. Geophys. Res.* 79:4581–4587.

Gosling, J. T., Hildner, E., MacQueen, R. M., Munro, R. H., Poland, A. I., and Ross, C. L. 1976. The speeds of coronal mass ejection events. *Solar Phys.* 48:389–397.

Harrison, R. A. 1986. Solar coronal mass ejections and flares. *Astron. Astrophys.* 162:283–296.

Harrison, R. A. 1991. Coronal transients and their relation to solar flares, *Adv. Space Res.* 11:25–26.

Harrison, R. A. 1994. A statistical study of the coronal mass ejection phenomenon. *Adv. Space Res.* 14:23–28.

Harrison, R. A., Waggett, P. W., Bentley, R. D., Phillips, K. J. H., Brunner, M., Dryer, M., and Simnett, G. M. 1985. The X-ray signature of solar coronal mass ejections. *Solar Phys.* 97:387–400.

Harrison, R. A., Hildner, E., Hundhausen, A. J., and Sime, D. G. 1990. The launch of solar coronal mass ejections: Results from the coronal mass ejection onset program. *J. Geophys. Res.* 95:917–937.

Hiei, E., Hundhausen, A. J., and Sime, D. G. 1993. Reformation of a coronal helmet streamer by magnetic reconnection after a coronal mass ejection. *J. Geophys. Res.* 20:2785–2788.

Hildner, E. 1977. Mass ejections from the corona into interplanetary space. In *Study of Travelling Interplanetary Phenomena*, eds. M. A. Shea et al. (Dordrecht: D. Reidel), pp. 3–21.

Hildner, E. 1986. Do we understand coronal mass ejections yet? *Adv. Space Res.* 6:297–306.

Hildner, E., Gosling, J. T., MacQueen, R. M., Munro, R. H., Poland, A. I., and Ross, C. L. 1976. Frequency of coronal transients and solar activity. *Solar Phys.* 48:127–135.

Hirayama, T. 1974. Theoretical models of flares and prominences, I: Evaporating flare model. *Solar Phys.* 34:323.

House, L. L., Wagner, W. J., Hildner, E., Sawyer, C., and Schmidt, H. U. 1981. Studies of the solar corona with the Solar Maximum Mission coronagraph/polarimeter. *Astrophys. J. Lett.* 224:117–121.

Hoawrd, R. A., Koomen, M. J., Michels, D. J., Tunsey, R., Detwiler, C. R., Roberts, D. E., Seal, R. T., and Whitney, J. D. 1976. *NOAA World Data Center A for Solar-Terrestrial Physics Rept. UAG-48A.*

Howard, R. A., Sheeley, N. R., Jr., Koomen, M. J., and Michels, D. J. 1985. Coronal mass ejections, 1979–1981. *J. Geophys. Res.* 90:8, 173.

Howard, R. A., Sheeley, N. R., Jr., Koomen, M. J., and Michels, D. J. 1986. The solar cycle dependence of coronal mass ejections. In *The Sun and the Heliosphere in Three Dimensions*, ed. R. G. Marsden (Dordrecht: D. Reidel), p. 107.

Hudson, H., Haisch, B., and Strong, K. T. 1995. Comments on the solar flare myth. *J.*

Geophys. Res. 100:3473–3477.

Hundhausen A. J. 1972. *Coronal Expansion and Solar Wind* (New York: Springer-Verlag), pp. 206–207.

Hundhausen, A. J. 1987. The origin and propagation of coronal mass ejections. In *Proc. of the Sixth International Solar Wind Conference*, eds. V. J. Pizzo, T. E. Holzer and D. G. Sime (Boulder: National Center for Atmospheric Research), pp. 181–214.

Hundhausen, A. J. 1993. The sizes and location of coronal mass ejections: SMM observations from 1980 and 1984–1989. *J. Geophys. Res.* 98:13177–13200.

Hundhausen, A. J. 1996. Coronal mass ejections: A summary of SMM observations from 1980 and 1984–1989. In *The Many Faces of the Sun; Scientific Highlights of the Solar Maximum Mission*, eds. K. T. Strong et al. (Berlin: Springer-Verlag), in press.

Hundhausen, A. J., Bame, S. J., and Montgomery, M. D. 1970. Large-scale characteristics of flare-associated solar wind disturbances. *J. Geophys. Res.* 75:4631–4642.

Hundhausen, A. J., et al. 1984*a*. Coronal transients and their interplanetary effects. In *Solar Terrestrial Physics: Present and Future*, eds. D. M. Butler and K. Papadopoulos, NASA Reference Publ. 1120, pp. 6–1 to 6–32.

Hundhausen, A. J., Sawyer, C. B., House, L., Illing, R. M. E., and Wagner, W. J. 1984*b*. Coronal mass ejections observed during the Solar Maximum Mission: Latitude distribution and rate of occurrence. *J. Geophys. Res.* 89:2639–2646.

Hundhausen, A. J., Burkepile, J. T., and St. Cyr, O. C. 1994*a*. The speeds of coronal mass ejections: SMM observations from 1980 and 1984–1989. *J. Geophys. Res.* 99:6543–6552.

Hundhausen, A. J., Stanger, A. L., and Serbicki, S. A. 1994*b*. Mass and energy contents of coronal mass ejections: SMM results from 1980 and 1984–1988. In *Proc. of the Third SOHO Workshop—Solar Dynamical Phenomena and Solar Wind Consequences*, ESA SP-373, pp. 409–412.

Illing, R. M. E., and Hundhausen, A. J. 1986. Disruption of a coronal streamer by an eruptive prominence and a coronal mass ejection. *J. Geophys. Res.* 91:10951–10960.

Jackson, B., and Howard, R. A. 1993. A CME mass distribution derived from Solwind coronagraph observations. *Solar Phys.* 148:359–370.

Kahler, S. 1987. Observations of coronal mass ejections near the Sun. In *Proc. of the Sixth International Solar Wind Conference*, eds. V. J. Pizzo, T. E. Holzer and D. G. Sime (Boulder: National Center for Atmospheric Research), pp. 181–214

Kahler, S. 1992. Solar flares and coronal mass ejections. *Ann. Rev. Astron. Astrophys.* 30:113–141.

Kiepenheuer, K. O. 1953. Solar activity. In *The Sun*, ed. G. P. Kuiper (Chicago: Univ. Chicago Press), chpt. 6.

Kopp, R. A., and Pneuman, G. W. 1976. Magnetic reconnection in the corona and the loop prominence phenomenon. *Solar Phys.* 50:85–98.

Low, B. C. 1990. Equilibrium and dynamics of coronal magnetic fields. *Ann. Rev. Astron. Astrophys.* 28:491–524.

Low, B. C. 1993. Mass acceleration processes: The case of the coronal mass ejection. *Adv. Space Res.* 13:63–69.

Low, B. C. 1994. Magnetohydrodynamic processes in the solar corona: Flares, coronal mass ejections, and magnetic helicity. *Phys. Plasmas* 1:1684–1690.

MacQueen, R. M. 1980. Coronal transients: A summary. *Phil. Trans. Roy. Soc. London A* 297:605–620.

MacQueen, R. M., Eddy, J. A., Gosling, J. T., Hildner, E., Munro, R. H., Newkirk, G. A., Jr., Poland, A. I., and Ross, C. L. 1974. The outer corona as observed from Skylab: Preliminary results. *Astrophys. J. Lett.* 187:85–88.

MacQueen, R. M., Csoeke-Poecke, A., Hildner, E., Reynolds, R., Stanger, A., Te Poel, H., and Wagner, W. 1980. The High Altitude Observatory coronagraph/polarimeter on the Solar Maximum Mission. *Solar Phys.* 65:91–107.

McAllister, A. H., Dryer, M., McIntosh, P., Singer, H., and Weiss, L. 1995. A large polar crown CME and a "problem" geomagnetic storm: April 14–23, 1994. *J. Geophys. Res.*, submitted.

Michels, D. J., Howard, R. A., Koomen, M. J., and Sheeley, N. R., Jr. 1980. Satellite observations of the outer corona near sunspot maximum. In *Radio Physics of the Sun*, eds. M. R. Kundu and T. Gergely (Hingham, Mass.: D. Reidel), pp. 439–442.

Moore, R. L. 1980. The thermal X-ray plasma. In *Solar Flares*, ed. P. A. Sturrock (Boulder: Colorado Assoc. Univ. Press), pp. 341–409.

Munro, R. H., Gosling, J. T., Hildner, E., MacQueen, R. M., Poland, A. I., and Ross, C. L. 1979. The association of coronal mass ejection transients with other forms of solar activity. *Solar Phys.* 61:201–215.

Phillips, J. L., Balough, A., Bame, S. J., Goldstein, B. E., Gosling, J. T., Hoeksema, J. T., McComas, D. J., Neugebauer, M., Sheeley, N. R., Jr., and Wang, Y. M. 1994. Ulysses at 50° south: Constant immersion in the high-speed solar wind. *Geophys. Res. Lett.* 21:1105–1108.

Rompolt, B. 1984. Eruption of huge magnetic systems from the Sun. *Adv. Space Res.* 7:357–361.

Rust, D. M., et al. 1980. Mass ejections. In *Solar Flares*, ed. P. A. Sturrock (Boulder: Colorado Assoc. Univ. Press), pp. 273–339.

Sheeley, N. R., Jr., Bohlin, J. D., Bruekner, G. E., Purcell, J. D., Scherrer, V. E., Tousey, R., Smith, J. B., Speich, D. M., Tandberg-Hanssen, E., Wilson, R. M., De Loach, A. C., Hoover, R. B., and McGuire, J. P. 1975. Coronal changes associated with a disappearing filament. *Solar Phys.* 45:377–392.

Sheeley, N. R., Jr., Howard, R. A., Koomen, M. J., Michels, D. J., Harvey, K., and Harvey, J. 1982. Observations of coronal structure during sunspot maximum. *Space Sci. Rev.* 33:219–231.

Sheeley, N. R., Jr., Howard, R. A., Koomen, M. J., and Michels, D. J. 1986. Solwind observations of coronal mass ejections during 1979–1985. In *Solar Flares and Coronal Physics using P/OF as a Research Tool*, eds. E. Tandberg-Hanssen, R. M. Wilson and H. S. Hudson, NASA CP-2421, pp. 241–245.

Sime, D. G. 1989. Coronal mass ejection rate and the evolution of the large-scale K-coronal density distribution. *J. Geophys. Res.* 94:151–158.

Sime, D. G., Hiei, E., and Hundhausen, A. J. 1994. Coronal eruptive events on April 4 and May 4, 1992. In *X-ray Solar Physics from Yohkoh*, eds. Uchida et al. (Tokyo: Universal Academy Press), pp. 197–200.

St. Cyr, O. C., and Webb, D. F. 1991. Activity assocaited with coronal ejections at solar minimum: SMM observations from 1984 to 1986. *Solar Phys.* 136:379.

Svestka, Z. 1995. On "the solar flare myth" postulated by Gosling. *Solar Phys.* 160:153–156.

Wagner, W. J. 1984. Coronal mass ejections. *Ann. Rev. Astron. Astrophys.* 22:267–289.

Wagner, W. J., Hildner, E., House, L. L., Sawyer, C. B., Sheridan, K. V., and Dulk, G. A. 1981. Radio and visible light observations of matter ejected from the Sun. *Astrophys. J. Lett.* 244:123–126.

Webb, D. F. 1992. The solar sources of coronal mass ejections. In *Eruptive Solar Flares*, eds. Z. Svestka, B. V. Jackson and M. Machado (New York: Springer-Verlag), pp. 234–247.

Webb, D. F., and Howard, R. A. 1994. The solar cycle variations of the occurrence rate of coronal mass ejections and the solar wind mass flux. *J. Geophys. Res.*

99:4201–4220.

Webb, D. F., and Hundhausen, A. J. 1987. Activity associated with the solar origin of coronal mass ejections. *Solar Phys.* 108:383.

Webb, D. F., Krieger, A. S., and Rust, D. M. 1976. Coronal X-ray enhancements associated with the Hα filament disappearances. *Solar Phys.* 48:159–186.

Webb, D. F., Howard, R. A., and Jackson, B. V. 1996. Comparison of CME masses and kinetic energies near the Sun and in the inner heliosphere. In *Proc. of Solar Wind 8*, in press.

Zirin, H. 1988. *Astrophysics of the Sun* (New York: Cambridge Univ. Press).

Zirin, H. 1994. Solar storminess. *Sky & Teles.* 88:9.

Zirker, J. B., ed. 1977. *Coronal Holes and High Speed Wind Streams* (Boulder: Colorado Assoc. Univ. Press).

CORONAL HOLES AND THE SOLAR WIND

R. ESSER and S. R. HABBAL
Harvard-Smithsonian Center for Astrophysics

The observational characteristics of coronal holes and the solar wind streams that they produce are summarized. The accuracy to which these characteristics can be derived from remote measurements is investigated. Emphasis is placed on the uncertainties in remote observations due to the convolution of plasmas with sometimes completely different properties along the line of sight. We discuss whether the observational limits derived for plasma parameters in the inner corona are sufficiently precise to place constraints on solar wind models, and to distinguish between the predictions resulting from different models. The observational techniques that are chosen as a demonstration are also relevant for the interpretation of measurements from future SOHO instruments.

I. INTRODUCTION

The basic theoretical concept of the solar wind was established by Parker in 1958 (see the chapter by Parker). The first observations of solar wind streams were carried out from the Mariner 2 spacecraft in 1962. Since then it has been known that the flow speed of the solar wind particles is a few hundred km s^{-1} at 1 AU, but can reach speeds of more than 600 to 700 km s^{-1} at times. During the Skylab period it was established that these high-speed streams originate from coronal holes which are regions of low densities and temperatures with predominantly open magnetic fields (see, e.g., review by Zirker 1977). The question where the low-speed solar wind has its origin seems to be much more difficult to settle. It might arise from the edges of large coronal holes, from small coronal holes (see, e.g., Hundhausen 1977, and references therein), or it might come from regions overlaying the closed magnetic field structures (see the chapter by Axford and McKenzie). To avoid any ambiguity we will consider only the high-speed wind in this chapter.

The thermally driven solar wind model suggested by Parker (1958) describes the overall behavior of the low-speed electron-proton solar wind rather well (see the chapter by Holzer et al.), but for reasonable temperatures and densities at the base of the corona it yields flow speeds at the orbit of the Earth which are far lower than the values typically observed in high-speed streams. Parker (1965) suggested that magnetohydrodynamic waves traveling outward from the Sun could transfer energy to the solar wind and thus play a role in accelerating the electron-proton plasma to the speeds characteristic of high-speed streams. Since Parker's early studies, a great number of different theoretical approaches have been discussed in the literature. In most of these models the solar wind is driven by thermal pressure and Alfvén waves (see the chapter by MacGregor and Charbonneau), and is treated separately from the transition region by assuming an already hot corona as inner boundary condition (for a different approach see recent review by Hollweg [1992]).

Other mechanisms that have been suggested as additional energy sources, instead of Alfvén waves, are the ejection of diamagnetic plasmoids from the Sun (Pneuman 1986), and suprathermal tails on the electron velocity distribution (chapter by Scudder). The details in these models are, however, much less worked out than in the Alfvén wave assisted models.

The results of any solar wind model have to match observations at 1 AU, and the parameters derived from observations describing the plasma close to the coronal base. While the outer boundary conditions are known in great detail from *in-situ* observations (chapter by Feldman and Marsch), much less is known about the inner boundary conditions. On the other hand, solar wind models are known to be extremely sensitive to the input parameters at the coronal base, such as electron and proton temperatures, electron density, helium abundance, wave velocity amplitude and magnetic field. It can, for example, be shown that the mass flux calculated from these models increases almost exponentially with the coronal base temperature, unless a mechanism is invoked that decreases this sensitivity, such as a significant helium abundance at the coronal base (chapter by Holzer et al.).

In this chapter we will concentrate on the manifestation of coronal holes using different observational techniques, and discuss the major problems that have to be considered when plasma parameters describing the inner corona are derived from observations. Emphasis is placed on temperature and density inferences. In Sec. IV we address the implications of this inquiry for solar wind modeling and future observations.

II. OBSERVATIONAL CHARACTERISTICS OF CORONAL HOLES

Krieger et al. (1973) were the first to identify coronal holes as the source of high-speed solar wind streams following several earlier predictions (see, e.g., Snyder and Neugebauer 1966; Wilcox 1968). When observed in coronal lines formed at temperatures around or greater than 10^6 K, coronal holes emerge as regions of depressed emission (see Fig. 1). Although they persist on average over five solar rotations (Bohlin 1977), coronal holes can often be temporarily and totally obscured by the quiet Sun emission in the foreground or background along the line of sight. As shown in the example of Fig. 1, the south polar coronal hole seems absent; yet, it was present in X-ray observations made a few days earlier and half a solar rotation later (see, for example, the time sequence in Zombeck et al. [1978]). Despite their seemingly long lifetime, their boundaries as defined by the arch-like structures in the quiet regions bordering them, undergo significant changes over time (see, for example, Fig. 1 in Withbroe 1983).

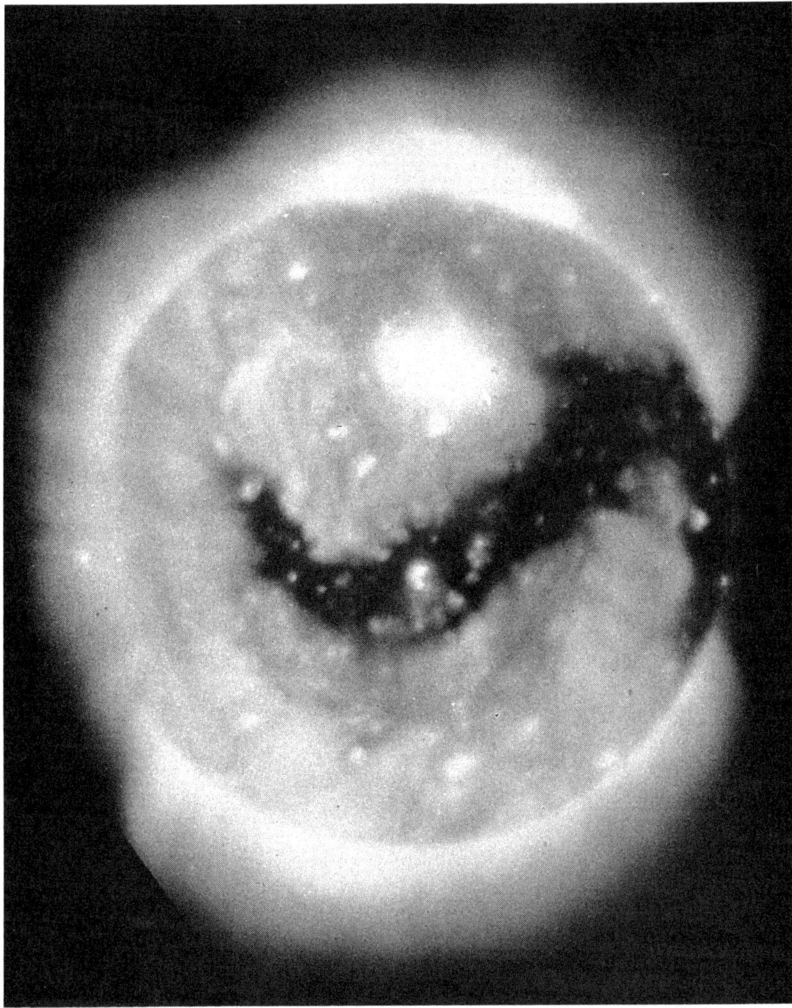

Figure 1.   Skylab X-ray photograph of the corona taken on 21 August 1973 (figure courtesy American Science and Engineering and Harvard College Observatory).

We review in this section the observational characteristics of coronal holes, the different spatial structures observed within them, and their implications for our understanding of the solar wind flow.

A. Polar Plumes

White-light eclipse photographs provided the first evidence that magnetic field

                R. ESSER AND S. R. HABBAL

lines in coronal holes extend away from the Sun (see, e.g., van de Hulst 1950*a*; Koutchmy 1977). Such an *open* magnetic structure is particularly evident from the filamentary nature of polar plumes seen in polar coronal holes (see Fig. 2). One of the earliest quantitative studies connecting polar plumes to structures in interplanetary space was made by Newkirk and Harvey (1968) who envisioned polar plumes to be rooted at the intersections of supergranular cells and to expand into interplanetary space. The connection between the spatial variations in the solar wind observed near the Earth and the solar granulation pattern was also proposed by Michel (1967). More recently, analyses of interplanetary space observations showed evidence for spatial structure in the solar wind with a typical size matching that of supergranular cells (Thieme et al. 1989).

TABLE I

Temperature of Maximum Line Intensity

Ion	Wavelength (Å)	Temperature (K)
Mg IX	369	1.0×10^6
Mg X	625	1.1×10^6
C III	977	7.4×10^4
Ne VII	465	5.0×10^5
O VI	1032	3.0×10^5
Fe X	6374	1.0×10^6
Fe XIV	5303	1.8×10^6

From the analysis of polar plumes observed at EUV wavelengths, Ahmad and Withbroe (1977) estimated that polar plumes contain about 15% of the mass in a typical polar coronal hole. Using a hydrodynamic model based on the analysis of a polar plume observed in X rays, Ahmad and Webb (1978) concluded that the mass flux in polar plumes is consistent with that measured in high-speed solar wind streams. The connection between polar plumes in the extended corona and magnetic structures at the solar surface is best seen at XUV and EUV wavelengths characteristic of coronal emission. As shown in Fig. 3, in XUV observations made in Mg IX 368 Å (formed around 10^6 K; see Table I), polar plumes (labeled a, b, c, d and e in Fig. 3) appear as searchlights overlying a thicker compact emission on the disk identified as coronal bright points, which are miniature active regions. These observations also show that plumes have an average lifetime of hours or even days (cf., Bohlin et al. 1975*a*). The connection between polar plumes and fast solar wind streams remains unresolved. At present there is not even sufficient observational evidence to establish clearly whether the polar plumes are embedded in the central parts of the coronal holes or at their edges. It therefore remains an open question whether the solar wind originates from the "ambient" regions forming the bulk volume of coronal holes, from polar plumes, or from both.

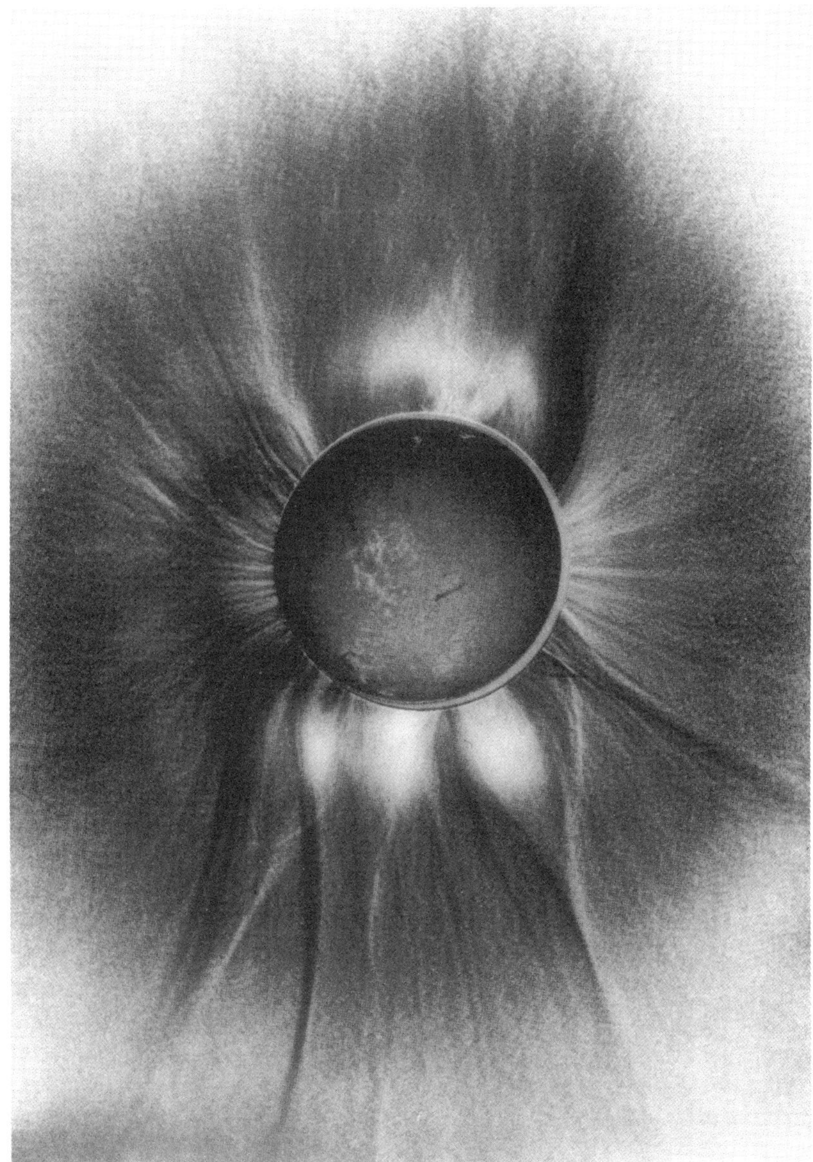

Figure 2. Composite image of the solar corona as captured in white light during a solar eclipse on 30 June 1973, with an Hα picture obtained simultaneously. The coronal picture has been processed in order to enhance small-scale intensity gradients (figure from Koutchmy 1977).

B. Macrospicules

Observed only in polar coronal holes, macrospicules refer to jets of chromospheric material, reminiscent of giant spicules, extending anywhere from 10 to 60 arcsec above the limb, with lifetimes ranging from 8 to 45 min (see, e.g., Waldmeier 1955; Tousey 1972; Bohlin et al. 1975b; Withbroe et al. 1976; Dere et al. 1989a). Most prominent in EUV spectral lines formed around 5×10^4 K and in Hα, they are invisible in lines formed above 3×10^5 K. In addition to their distinct morphological structure, as shown in the time sequence given in Fig. 4, they have also been observed to undergo occasional ballooning and pinching off at the top of their magnetic structure (Withbroe et al. 1976; Habbal and Gonzalez 1991).

Macrospicules and polar plumes define the fine-scale structure of coronal holes, as observed off the limb, yet, they have different magnetic structures. They are prominent at different temperature ranges and have different lifetimes. While polar plumes might extend into interplanetary space as proposed by Thieme et al. (1989), the extent of macrospicules into the outer corona rarely exceeds 1 arcmin. The contribution of macrospicules to the solar wind flow, however, as a consequence of changes in their magnetic topology, might not be negligible (see, e.g., Mullan and Ahmad 1982).

C. The Small-Scale Structure at the Base of Coronal Holes

The small-scale structure of coronal holes on the solar surface, on scales of 10 to 40 arcsec, is dominated by bright points. Several observational studies have shown that the emission from bright points is characterized by a very pronounced variability on a time scale of minutes (see, e.g., Sheeley and Golub 1979; Habbal and Withbroe 1981; Habbal et al. 1990). At coronal temperatures, bright points observed in a coronal hole are characterized by their enhanced emission against the coronal hole background (shown encircled in the left panel of Fig. 5a in Mg X 625 Å). In cooler lines; such as C III 977 Å (as shown by the circles in the right panel of Fig. 5a), bright points are readily seen to coincide with the intersection of supergranular cells, but are not visually distinct from the rest of the enhanced network emission. (Usually identified in coronal lines, bright points have an average intensity that is significantly larger than the average network emission in cooler lines, see Habbal and Grace [1991] for details.) The temporal changes of the emission in their archlike structures indicate the existence of substructures that evolve differently in time (see Fig. 5b and c). Furthermore, comparison of structures observed simultaneously in different temperature lines, such as in the examples of Fig. 5b and c (top panels are for Mg X and bottom panels for C III), shows that substructures can be heated to different temperatures. The changes in time of the emission are not necessarily correlated spatially at these different temperatures. The temporal variability of the emission from coronal bright points led Parker (1991) to propose that they could provide the energy input into coronal holes through their incessant flaring activity. Habbal and Grace (1991), however, showed that within the observational limits of data

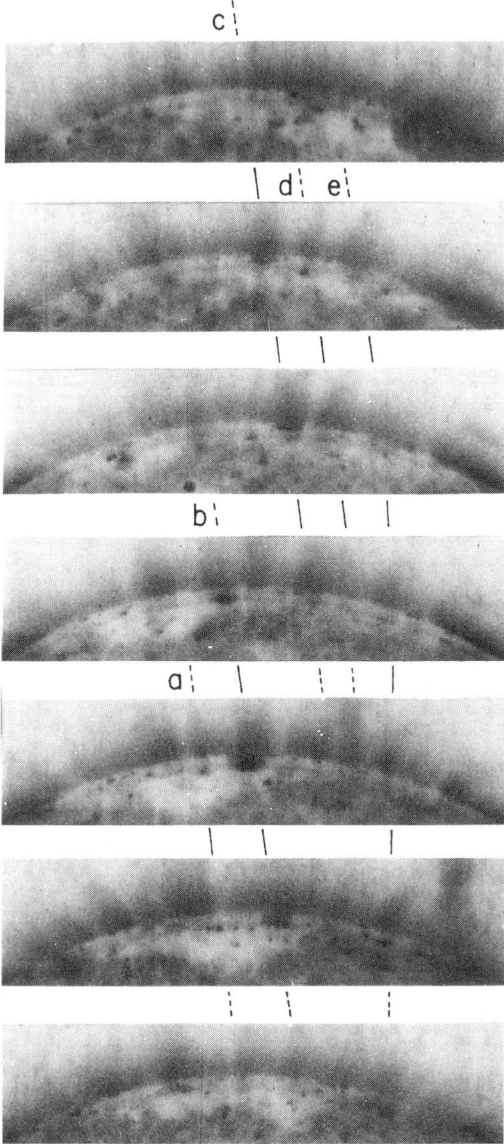

Figure 3.   A time sequence of observations of a north polar coronal hole taken in Mg IX 368 Å, between 30 December 1973 and 5 January 1974. In this negative print, the most intense emission is shown as black and the weakest as white. The time intervals between consecutive frames is approximately one day. The dashed and solid lines, drawn with labels a, b, c, d, and e, between consecutive frames, point to the evolution of some polar plumes. The spatial extent of the emission above the limb is about 60 arcsec (figure courtesy Naval Research Laboratory).

currently available, the variable component of the energy output from bright points, in the form of radiation, cannot account for the energy requirements in coronal holes (see also Habbal [1992] for a more detailed discussion of this topic).

D. Comparison Between Small-Scale Structures in Coronal Holes and Quiet Regions

Although the large-scale magnetic field bestows a distinct character to coronal holes and quiet regions, with an open configuration in the former as opposed to a closed, arch-like structure in the latter, the properties of the small-scale closed magnetic structures are not observationally different in these two regions. For example, bright points emerge indiscriminately on the Sun, and their empirical properties in coronal holes are remarkably similar to those in quiet regions (Habbal et al. 1990; Habbal and Grace 1991). Dere et al. (1989b) also showed that the nonthermal broadening of spectral lines formed around 10^5 K is only marginally different in coronal holes from quiet regions. They derived an average rms nonthermal velocity of 29 km s^{-1} in coronal holes as compared to 28 km s^{-1} in quiet regions. If these amplitudes are a reflection of the energy available to heat the corona and accelerate the solar wind, then this energy is the same in a coronal hole and a quiet region.

At temperatures below 800,000 K, coronal holes are no longer readily distinguishable from the surrounding quiet regions, as seen in the example of Fig. 5a. This is also seen in Fig. 6 which shows a composite sequence of scans in a south polar coronal hole taken simultaneously in lines typical of coronal emission (Mg X 625 Å) and transition region temperatures (Ne VII 465 Å and O VI 1032 Å) (see Table I). The observational results described so far show that coronal holes and quiet regions cannot be differentiated on the basis of the characteristics of the small-scale structures that pervade them. The differences in the characteristics between these two regions might then be determined by the funneling of the energy through the overlying large-scale magnetic field.

In summary, while magnetic structures of different spatial scales abound in coronal holes, their role in channeling the energy into the solar wind remains an open question. Furthermore, the exact nature of polar plumes and their source region within coronal holes is still unknown. Although polar plumes seem to be anchored in bright points at the intersection of supergranular cells, it is not clear if they originate within coronal holes or at their boundaries, and whether EUV and X-ray polar plumes are the same as white light polar plumes. Consequently, their mapping into interplanetary space, and their role in carrying a fraction or the bulk of the solar wind flow have yet to be resolved.

III. PLASMA PARAMETERS IN THE INNER CORONA

The observations described in the previous section give a reasonable description of the overall behavior of coronal holes and the structures inside them.

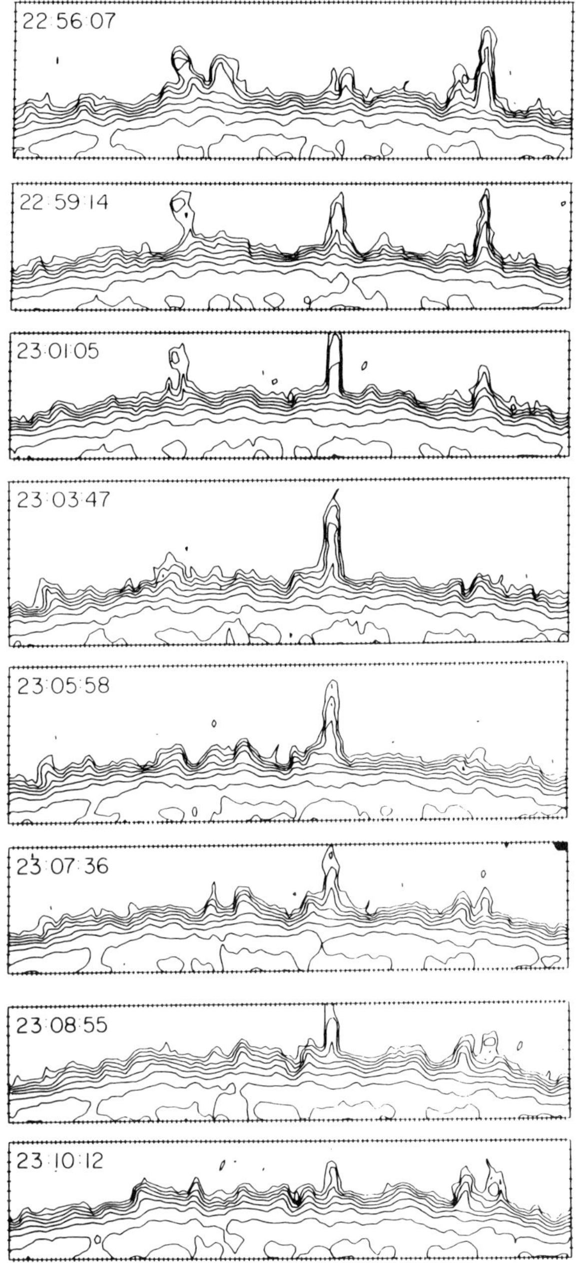

Figure 4.   A time sequence of intensity contours of the C III 977 Å emission from macrospicules observed in a polar coronal hole on 11 December 1973, with the Skylab EUV spectroheliometer. The distance between tick marks is 5 arcsec in the vertical direction and 2.5 arcsec in the horizontal direction (figure from Withbroe et al. 1976).

To model the emerging solar wind, however, an overall description is not sufficient. As boundary conditions to constrain solar wind models we need to place limits on the plasma parameters in the inner corona. Not only are the values of these parameters at the coronal base extremely important, but so is their behavior as a function of height above the solar limb. Such parameters cannot be directly measured, but are derived from observations using models and model assumptions. Measurements from which the electron density and temperature can be inferred with sufficient accuracy have not been carried out on a regular basis over time periods covering one or more solar rotations. Our current knowledge of electron temperature and density in the inner corona must therefore be considered rather sporadic. Other important parameters such as the helium abundance which, in principle, can be determined from intensity measurements of helium lines in the inner corona, have so far never been derived from observations. Although attempts have been carried out to measure such lines, to date these measurements have not been successful (see, e.g., Leer et al. 1992, and references therein). Recently Habbal and Esser (1994) have suggested a different approach to place limits on the helium abundance using empirical inferences of electron density, temperature and their respective gradients in this region.

In the following we will discuss some of the major problems involved in deriving plasma parameters in the inner solar wind region, using the electron density and temperature as examples. In Sec. III.E we address some of the other important plasma parameters, such as wave velocity amplitude, magnetic field strength and helium abundance. We consider plasma parameters describing the large-scale structures of coronal holes only. Deriving parameters for the small-scale structures is exceedingly complicated; in addition the relevance for the solar wind of small-scale structures such as macrospicules and bright points is not well understood.

In Table II, taken from Habbal et al. (1993), we have summarized the temperatures and densities previously inferred from a variety of measurements. We selected measurements that were clearly carried out in or above coronal holes. Table II, although by no means complete, contains several examples of the different techniques most commonly used to infer coronal temperatures and densities. The values quoted in the table cover the range of temperatures and densities found in the literature. The temperature derived in most techniques is the electron temperature, except in the case of the resonantly scattered Lyman-α line width measurements which represent the sum of the thermal and kinetic temperature of the scattering particles (see Sec. III.F). The temperatures and densities from Table II are plotted as a function of height in Fig. 7a and b, respectively. Disk observations of the temperature are plotted at 1 R_S (open circles), those derived from charge state measurements are plotted at 1.6 R_S (triangles) because no distance can be attributed to these measurements without modeling. Limb observations are plotted at their respective heights (dots), values derived from white light are marked with Xs. The arrows indicate the temperature range, radial span, or uncertainties, in

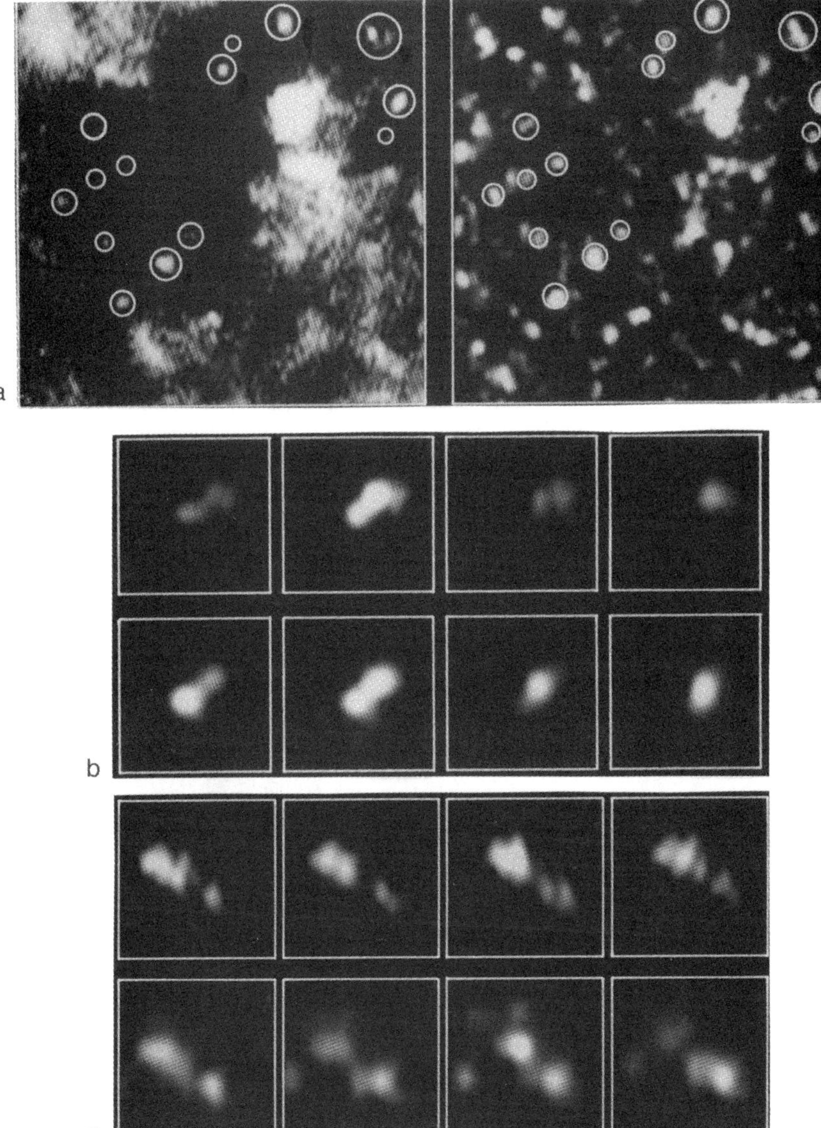

Figure 5. (a) EUV spectroheliograms taken on 21 August 1973, with the Skylab EUV spectroheliometer, of a $5' \times 5'$ area of a coronal hole at Sun center (namely the central meridian part of Fig. 1, around the equator) in Mg X 625 Å, formed around 1.1×10^6 K and C III 977 Å formed at 7.4×10^4 K. The coronal hole, distinguishable by the substantially reduced Mg X emission, is surrounded by quiet regions with two small active regions in the west. Only bright points in the coronal hole are circled. Although Mg X emission is absent from some of the circled bright points, it was present in earlier or later scans. (b) A time sequence showing changes in the emission from two different coronal hole bright points. The top panels correspond to Mg X emission and the lower to C III. The scale is magnified over that shown in (a) above. The frames are 5.5 min apart (figure from Habbal et al. 1990).

TABLE II

Temperatures and Densities in Coronal Holes[a]

Data Type	Radial Span (R_S)	T (10^6 K)	n_e (10^8 cm^{-3})	References
Limb				
White light	1.04–1.4	1.15	2 (at 1.2 R_S)	van de Hulst 1950*a,b*;
	1.1–1.6	1.2	2.3 (at 1.1 R_S)	Saito 1965.
X ray	1–1.4	1.3	27	Krieger et al. 1973;
	1.04–1.1	0.77–0.81	7–8 (at 1.04 R_S)	Ahmad and Webb 1978.
XUV	1.025–1.35	0.7–1.2	2 (at 1.1 R_S)	Bohlin et al. 1975*a*.
EUV	1.0–1.1	1.1 (at 1.06 R_S)	5 (at 1.03 R_S)	Ahmad and Withbroe 1977.
	1.0–1.15	0.76 (at 1.03 R_S)	1.7 (at 1.03 R_S)	Mariska 1978.
		1.1 (at 1.1 R_S)	0.41 (at 1.1 R_S)	
UV	1.01–1.02	<1.0		Doschek and Feldman 1977;
Ly-α	1.5–3.5	1.22±0.1 (at 1.5 R_S)	0.07 (at 1.5 R_S)	Withbroe et al. 1985;
	1.5–2.0	1.7±0.3 (at 1.8 R_S)	0.5 (at 1.5 R_S)	Withbroe et al. 1986.
Disk				
EUV		1.05 (at CH center)	3 (at CH center)	Munro and Withbroe 1972;
		1.1–1.3 (in network)		Raymond and Doyle 1981.
		1.2–1.6 (in cell)		
In situ				
Charge state	<1.5>	1.3±0.2 (CNO ions)		Galvin et al. 1984;
	<1.5>	1.3±0.3 (CNO ions)		Ipavich et al. 1986.
	<3.0>	1.4±0.2 (Fe ions)		

[a] Table taken from Habbal et al. 1993.

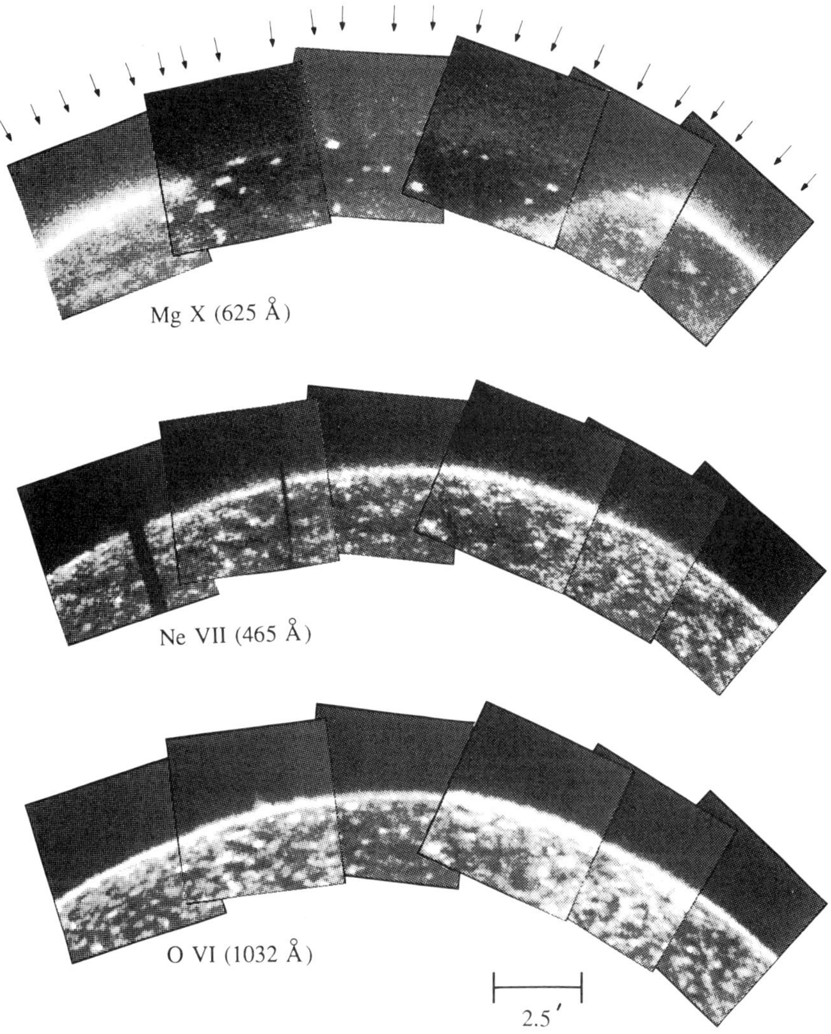

Figure 6.   South polar region observed through a consecutive set of scans obtained with the Skylab EUV spectroheliometer on 11 December 1973. The extent of the coronal hole is identified by the bordering quiet regions. Observations in Mg X 625 Å and O VI 1032 Å were made simultaneously, while the Ne VII 765 Å observations were made with a different grating position at the end of each Mg X/O VI scan, hence 5.5 min later. Arrows point to the radial directions where line ratios were used to infer temperatures (see Sec. III.D) (figure from Habbal et al. 1993).

the measurements as provided by the authors. For the density we have chosen the example given by Saito (1958,1965) who measured the polarization brightness as a function of height in plumes (dashed line) and the thinner regions between the plumes (solid lines) separately, and the lowest values for the densities from Table I (Mariska 1978) (dotted line). We have also added the lowest density profile given by Lallement et al. (1986) (dash-dotted line). The whole range of inferred temperatures and densities in the inner corona is rather large, spanning from about 0.7 to 1.6×10^6 K for the temperatures, and about 2 orders of magnitude for the densities below 1.6 R_S. This large variation could have several reasons: it could be due, at least partly, to the use of different observational techniques (note, for example, that the highest temperatures are derived using charge state measurements at 1 AU), it could reflect a true variation of temperatures and densities across the coronal hole, or from one hole to another, or, in case of the temperature, it could be due to an increase with distance from the coronal base. Another important source for the observed range of the temperature and density could be the varying contribution from surrounding regions due to the integration along the line of sight. To discriminate between these different causes, observations of the same solar wind region using several techniques simultaneously would have to be carried out, both as a function of height from the coronal base, and over time periods long compared to the rotational period of the Sun. Such measurements are, however, not available. In the following, we give a short presentation of the observational techniques, their underlying assumptions, and the estimated errors that could be inherent in these assumptions. We then give an example of the expected variations of the temperature with height and across the coronal hole, and we discuss the effect of the contribution from surrounding regions using measurements of the red Fe X 6374 Å and green Fe XIV 5303 Å spectral line intensities.

A. Polarization Brightness Measurements

Eclipse white light observations are primarily used to derive the radial density profile in the inner corona from the intensity of the polarized brightness of the white light corona. Assumptions about the contribution from the dust scattered component (F corona), and about the density distribution along the line of sight have to be made. In the mathematical derivations originally worked out by Baumbach (1939), and subsequently by van de Hulst (1950*a,b*) and Saito (1965), the density distribution is assumed to be cylindrically symmetric and homogeneous along the line of sight. The coronal temperature of the electrons, T_e, is then derived from the inferred density gradient, i.e., from the solution of the force balance equation between pressure gradient and gravity which can be written as:

$$\frac{dp}{dr} = -\rho \frac{GM}{r^2}. \tag{1}$$

Here $p = n_e k T_e + n_p k T_p + n_\alpha k T_\alpha$ is the total gas pressure, with the subscripts e, p and α referring to electrons, protons and fully ionized helium, respectively. G is the gravitational constant, M is the solar mass, r is the heliocentric distance, k is the Boltzmann constant, and $\rho = n_e m_e + n_p m_p + n_\alpha m_\alpha$ is the mass density. Equation (1) is then solved with the assumptions $T_e = T_p = T_\alpha = $ const, and $n_\alpha / n_p = \alpha = $ const. The ratio n_α / n_p is usually taken to

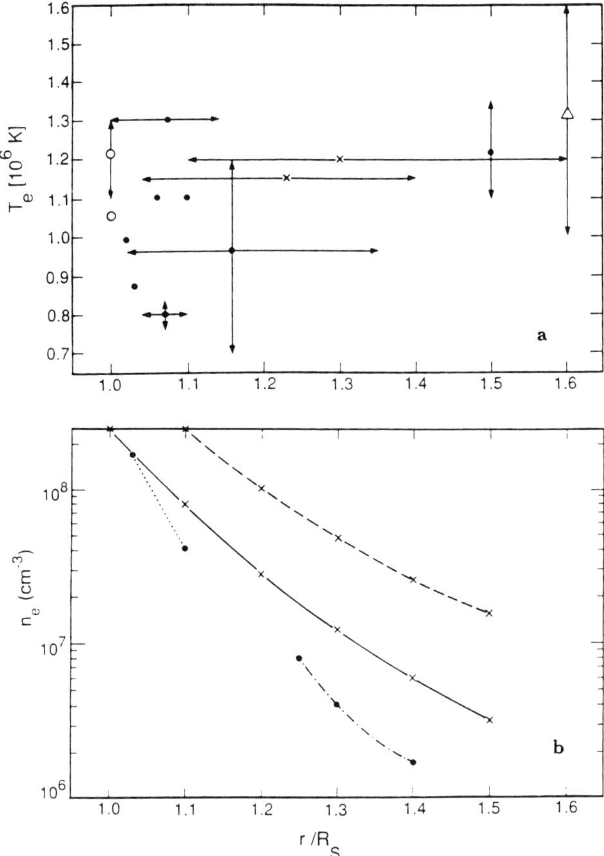

Figure 7.   (a) Temperatures in the inner corona given in Table I. Disk observations (open circles) are plotted at 1 R_S, temperatures drived from *in situ* charge state measurements (triangles) at 1.6 R_S. Ultraviolet limb observations (dots) and white light observations (crosses) are plotted at their respective heights. Vertical arrows indicate the temperature range at a given heliocentric distance, while the horizontal arrows show the radial extent of a given inferred temperature, whenever provided by the corresponding authors. (b) Densities derived from white light observations given by Saito (1965) for polar plumes (dashed line) and the regions in between the plumes (solid line). A density profile derived by Lallement et al. (1986) is also shown (dash-dotted line). The values given by Mariska (1978) (dotted line) were derived from EUV observations.

be 10%. Inherent in Eq. (1) are also the assumptions that flow speeds, any additional force (e.g., wave pressure forces), as well as the effects of heavy ions, other than fully ionized helium, are negligible in the inner corona. To date, there is no study to investigate the effect of all of these assumptions on the accuracy of the derived parameters. The uncertainty of the helium abundance alone could be of the order of 20% or more if the helium is overabundant at the coronal base (see Sec. III.F).

B. In-situ Charge State Measurements

The equation describing the relative changes in the i-th ionization state of a given element in a steady radial flow is given by:

$$\frac{\mathrm{d}}{\mathrm{d}r}(n_i v_i r^2) = n_e[n_{i-1}C_{n-1} - n_i(C_i + R_i) + n_{i+1}R_{i+i}] \tag{2}$$

where C_i and R_i describe the ionization and recombination coefficients. If the electron density n_e is below a certain limit ($n_e \leq 10^9$ cm^{-3}; Jordan [1969]), and the velocity distribution of the electrons is Maxwellian, then the rate coefficients depend only on the electron temperature. Close to the coronal base the ion velocity v_i is small, the divergence of the particle flux is negligible, and therefore the recombination and ionization processes balance locally. In this region we can write for the simple example of two ions (see, e.g., Hundhausen 1972):

$$\frac{n_i}{n_{i+1}} = \frac{R_{i+1}}{C_i}. \tag{3}$$

At larger distances from the coronal base where the electron density becomes sufficiently small and the velocity becomes large, the time scale of the ionization and recombination processes becomes small compared to the expansion time scale. In that case we have for any charge state:

$$\frac{1}{r^2}\frac{\mathrm{d}}{\mathrm{d}r}(n_i v_i r^2) = 0 \tag{4}$$

and for two neighboring states we can write:

$$\frac{n_i v_i}{n_{i+1} v_{i+1}} = \text{const.} \tag{5}$$

The relative fluxes are then independent of the electron temperature, and are said to be frozen into the solar wind. Under the assumption that all charge states have the same flow speed (n_i/n_{i+1} = const) one can determine the electron temperature at the freezing-in distance by comparing the *in-situ* measured charge state ratio n_i/n_{i+1} to the calculated ratio R_{i+1}/C_i from Eq. (3).

Using oxygen as an example, Esser and Leer (1990) showed that different charge states of the same ion can easily flow with different speeds close to the coronal base (below 3 R_S). The freezing-in density for oxygen is about 2×10^7 cm^{-3}, while it is about 10^6 cm^{-3} for iron. In both cases this might be very close to the coronal base in low density coronal holes. Owocki (1983) showed that the assumption $v_i = v_{i+1}$ can lead to an overestimation of the temperature ΔT at the freezing-in distance of 0.5×10^6 K, if the actual flow speed ratios are $v_{i+1}/v_i = 5$ for oxygen, and $v_{i+1}/v_i = 2$ for iron ions.

The effect of non-Maxwellian velocity distributions on the ratio of the recombination to ionization coefficients (R_{i+1}/C_i) has been investigated by Owocki and Scudder (1983) and Bürgi (1987). Owocki and Scudder found that the effect on the freezing-in temperatures of iron is small, while for oxygen (O^{+6}/O^{+7}) the temperature could be overestimated by as much as $\Delta T = 0.75 \times 10^6$ K, depending on how much the distribution function deviates from a Maxwellian.

C. Line Ratio Technique

Another technique commonly used to derive electron temperatures in the inner corona is the line ratio technique. The ratio of measured intensities from different ions, or different charge states of the same ion, are compared to the ratios calculated theoretically. The intensity of an optically thin spectral line can be approximated by (see, e.g., Dere and Mason 1981):

$$I = \int f(\lambda, T_e) n_e^\gamma \, ds. \tag{6}$$

The function $f(\lambda, T_e)$ depends on the atomic data, the elemental abundance, the ionization balance, as well as on the electron temperature T_e. For allowed lines the function $f(\lambda, T_e)$ is given by Pottasch (1964). The exponent describing the density dependence is $\gamma = 2$ for allowed lines that are collisionally excited from the ground state, $1 < \gamma < 2$ for forbidden lines coming from metastable levels, and $2 < \gamma < 3$ for allowed lines from metastable levels. The integration in Eq. (6) is along the line of sight.

Figure 8 shows two examples of the calculated emissivity curves. In Fig. 8a we have chosen the allowed lines Mg X 625 Å and O VI 1032 Å; the numerical code and the atomic data sources are described in Raymond and Doyle (1981), Cox and Raymond (1985) and Doyle and Keenan (1992). Figure 8b was calculated for the forbidden Fe XIV 5303 Å and Fe X 6374 Å lines using a model by Brickhouse et al. (1995). The emissivity curves show the temperature range over which the spectral lines are formed. When comparing the intensity ratio calculated from the emissivity curves to the measured intensity ratios it is assumed that the temperature and density are constant along the line of sight, e.g., that the lines are emitted from the same spatial element. If there is a distribution of different temperatures and densities present along the line of sight this assumption fails, particularly if the spectral lines peak at widely different temperatures as in the cases shown in Fig. 8. Using the line ratio technique, in this case, is misleading. This will be discussed in more detail in Sec. III.E.

Not all atomic data inherent in the above equation are equally well known and there is to date no detailed study about how the uncertainties in the atomic data could affect the derived temperatures (see, e.g., Brickhouse et al. 1995). Also the unknown coronal abundances of the different elements pose a problem if spectral lines from different elements are used. This problem

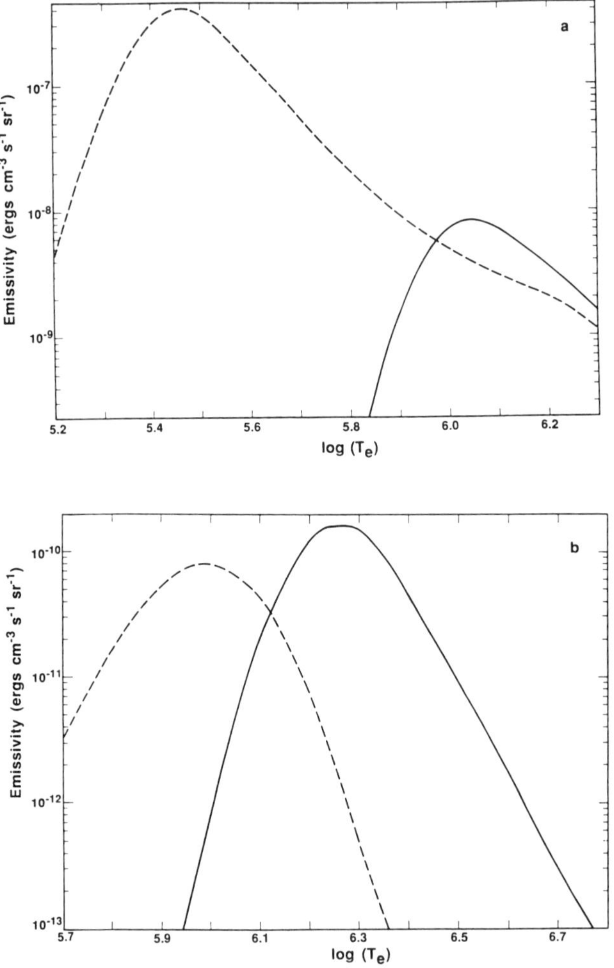

Figure 8.   (a) Emissivity of the O VI 1032 Å (dashed line) and Mg X 625 Å (solid line) lines vs temperature calculated from a computer code provided by J. Raymond. (b) Emissivity vs temperature for a forbidden Fe X 6374 Å (dashed line) and Fe XIV 5305 Å (solid line) calculated from a computer code provided by N. Brickhouse and J. Raymond (Brickhouse et al. 1995).

can be overcome partly by measuring lines from many different elements. The temperatures deduced from the different line ratios will then only be consistent for a certain range of coronal abundances, thus placing limits on both the temperature and the abundance (see, e.g., Habbal et al. 1993). The best approach, however, is to find suitable spectral line ratios from the same element.

D. Temperature Variation Inside Coronal Holes

The Skylab measurements described in Sec. II were used to study the variation of the temperature across a coronal hole and into the bordering quiet regions (Habbal et al. 1993). The intensity ratios between the three lines shown in Fig. 6 were calculated from the observed intensities for each column 5″ wide in the radial directions indicated by the arrows. For each radial direction the ratios were calculated as a function of height between 1.02 and 1.07 R_S in steps of 5 arcsec. This leads to a range of temperatures in each radial direction indicated by the vertical bars in Fig. 9. Also included in these ranges are uncertainties due to the elemental abundance. Uncertainties due to atomic data (unknown) or statistical errors (estimated to be of the order of 30%) were not included. Judging from Fig. 9 it seems that the temperatures inside the coronal hole are relatively constant. A small increase in the temperature of 20 to 30% can be seen at the edges of the hole into the quiet regions. A similar ratio between the temperature inside the hole and the temperature in the bordering quiet regions was also found by Munro and Withbroe (1972) for a coronal hole in the center of the disk. Whether small temperature variations inside coronal holes are a characteristic for all holes remains to be seen in the future when more data become available.

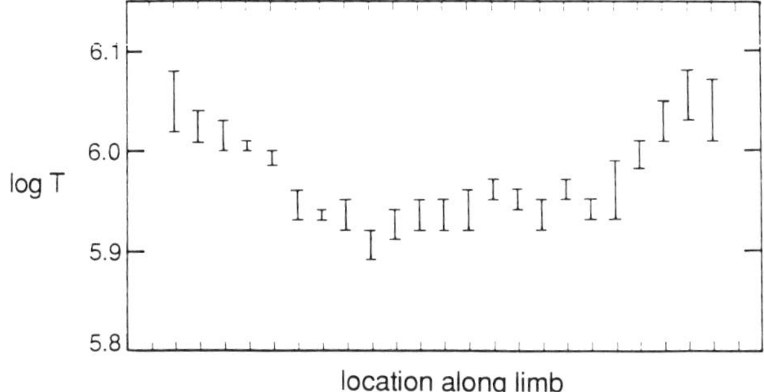

Figure 9.   The logarithm of the temperature as derived from the Skylab observations shown in Fig. 7. The location along the limb corresponds to the arrows in Fig. 7. The temperatures were calculated for different heights above the limb, giving a range of temperatures at each location. This range is indicated by the vertical bars.

The overall trend in the behavior of the temperature as a function of distance in Fig. 7, seems to indicate an increase of the temperature with distance. This trend can also be seen in observations as a function of height above the limb. In Fig. 10a we have used the same data as in Fig. 6, but have only chosen one radial direction in the hole, and have plotted this as a function of height. The temperature very close to the base is about 8×10^5 K and increases slightly to less than 10^6 K, which is in agreement with many of the observations given in Table II (see, e.g., Mariska 1978; Bohlin et al. 1975a). This does not of course necessarily mean that the temperature maximum is less than 10^6 K, the temperature maximum could simply be situated further away from the base. To date there are no measurements giving the electron temperature at distances from the limb out to 3 R_S, for example. We have therefore taken the temperature derived from the width of the Lyman-α line (Withbroe et al. 1985) and included them in Fig. 10b. The Lyman-α line width reflects the total (thermal plus kinetic; see Sec. III.F) temperature of the protons at these distances (Olsen et al. 1994). On the other hand, the electron temperature could very well be different from the proton temperature in the inner corona (Habbal et al. 1995). Comparison of Fig. 10a and b indicates that the temperature maximum might be below 2 R_S, and that it does not exceed 1 to 1.2×10^6 K. In the other example given in Table II (Withbroe et al. 1986) also taken above a polar coronal hole, there is evidence that the temperature inferred was much more influenced by surrounding regions (see next section) and might not necessarily reflect the temperature in the coronal hole. Comparing the temperature profile in Fig. 10a with the freezing-in temperatures derived from the charge state measurements for iron and oxygen (see Table II or Fig. 7a), instead of with Fig. 10b, seems to indicate a larger increase of temperature with distance from the coronal base. However, this discrepancy might be explained by differential flow speeds and/or non-Maxwellian velocity distributions (Sec. III.B).

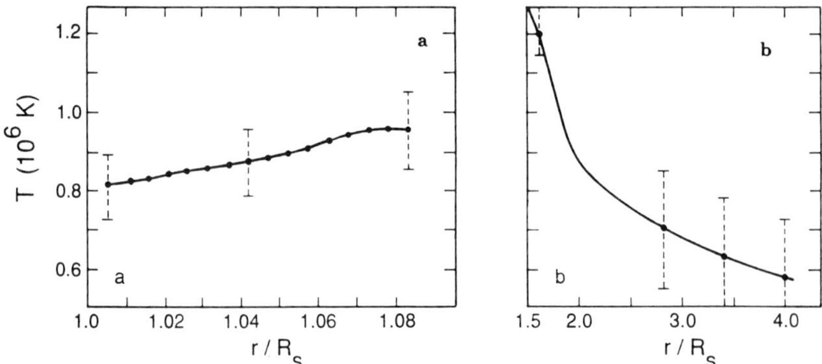

Figure 10.	(a) Temperatures drived from the Skylab observations in Fig. 6 as a function of height for the central part of the coronal hole. The error bars take into account only the statistical error which was estimated to be of the order of 30%. (b) Kinetic temperatures as derived from the resonantly scattered Lyman-α line (Withbroe et al. 1985) as a function of heliocentric distance.

E. Line-of-Sight Effect

Due to their low temperature and density, coronal holes can be completely obscured by foreground and/or background emission (see, for example, Fig. 1). Even in cases where the coronal hole is rather large on the solar disk, there can be a significant contribution to the emission from surrounding regions.

This can be seen in Fig. 11 which shows solar X-ray images from the Yohkoh mission from 27 March and 10 April 1993. For a description of the SXT instrument see Tsuneta et al. (1991). In Fig. 11a the coronal hole looks small on the disk; on the other hand, there is relatively little contribution from the foreground or background above the limb. This can be seen by comparing this figure to Fig. 11b which was taken about 1/2 solar rotation later. In Fig. 11b the coronal hole, although large and clearly visible on the disk, has a significant contribution from the background above the limb. To investigate the contribution of the surrounding regions and its effects on derived plasma parameters, we have plotted in Fig. 12a and b the emission of the forbidden Fe X 6374 Å (dashed lines) and Fe XIV 5303 Å (solid lines) spectral lines measured at $0.15\ R_S$ above the solar limb (Esser et al. 1995). These observations were carried out at the National Solar Observatory at Sacramento Peak on 27 March (Fig. 12a) and 10 April, 1993 (Fig. 12b). The southern coronal hole is clearly visible on both days, its center is located at about $180°$ from the north. However, the green Fe XIV line should not be visible in coronal holes at all because its emission at coronal hole temperatures (of about 10^6 K) is negligible (see Fig. 8b). The presence of the green line in the center of the coronal hole in Fig. 12b is therefore an indicator that the emission from the surroundings is rather high on that day, while it is much lower in Fig. 12a. Because the emission on 10 April is obviously coming from regions other than the coronal hole itself, using the line ratio technique described in Sec. III.C (or other remote sensing techniques) can yield misleading results. To get an indication for the magnitude of this line of sight effect, we have plotted the electron temperatures derived from the line ratio technique using the two iron lines (lower panels). We find that the minimum temperature of the hole varies by more than 0.5×10^6 K. Because the temperatures derived inside coronal holes on the solar disk do not show such large temperature variations, the temperature variations seen in Fig. 12 are most likely due to contributions from the background. This can also be seen in Fig. 13 which shows the minimum temperatures derived from the red and green line measurements at each day over about four solar rotations. We see that the derived temperatures vary from day to day, but in addition the minima (maxima) also show a certain periodicity of about 25 to 27 days which is the rotational period of the Sun as seen from the Earth. Overall it seems that the uncertainty due to foreground and/or background contributions are of the order of 0.5×10^6 K.

F. Additional Crucial Plasma Parameters

So far we have only discussed the essential information that can be found in the literature regarding electron temperature and density. In addition to these two parameters the wave velocity amplitude is probably most important for models assuming that the solar wind is driven by thermal pressure and Alfvén waves. It can be shown that velocity fluctuations associated with hydrodynamic waves will produce a significant broadening of spectral lines observed above the limb if the velocity amplitude of the waves is more than

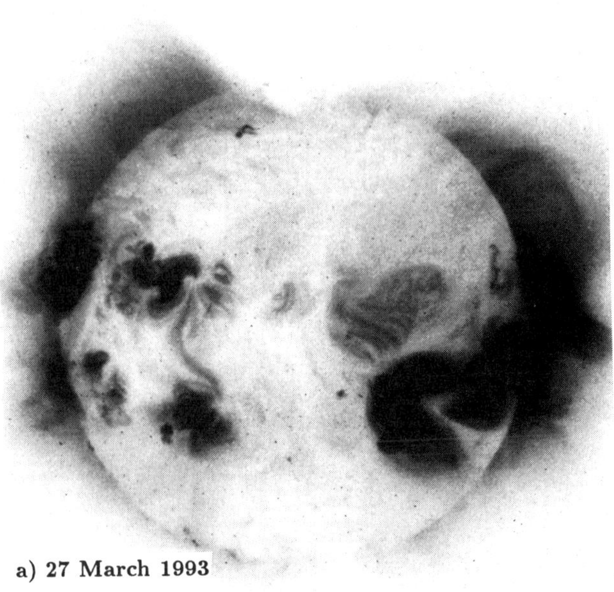

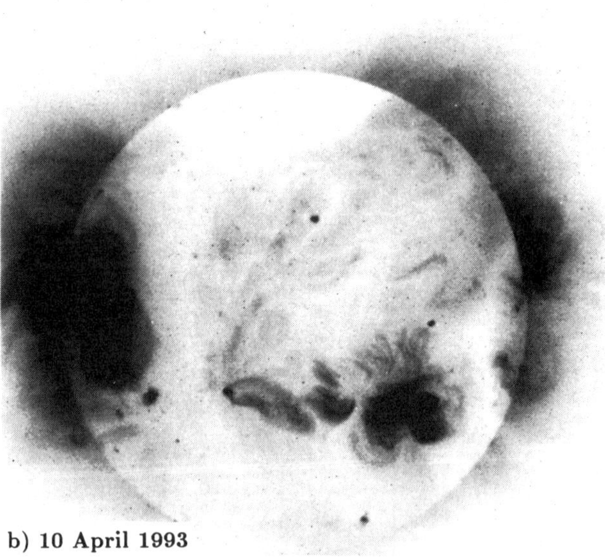

Figure 11. Soft X-ray images of the Sun provided by the Yohkoh team (figure courtesy of H. S. Hudson).

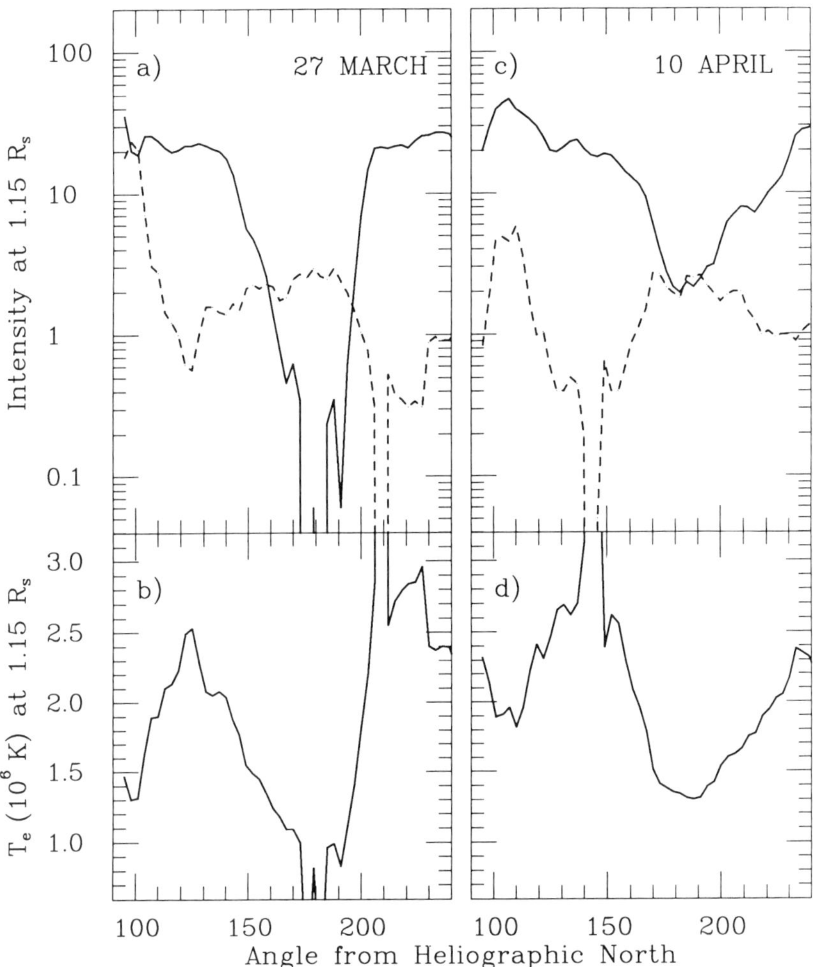

Figure 12.   Intensities of the red Fe X 6374 Å (dashed lines) and green Fe XIV 5303 Å (solid) observed at the National Solar Observatory at Sacramento Peak in 1993 (observations courtesy of R. C. Altrock). (a) 27 March and (b) 10 April. The minima in the green line correspond to the southern coronal hole visible in the Yohkoh images in Fig. 11. Also plotted are the temperatures (solid lines) derived from the intensity ratio using the emissivity given in Fig. 8b.

10 km s^{-1} (see, e.g., Hollweg 1973; Withbroe et al. 1982; Esser et al. 1987). Due to the line-of-sight integration and the integration in time inherent in spectral line observations, the particle motion associated with the waves will most likely appear random, thus contributing to the broadening of the line. The total line broadening associated with the thermal and wave motion can

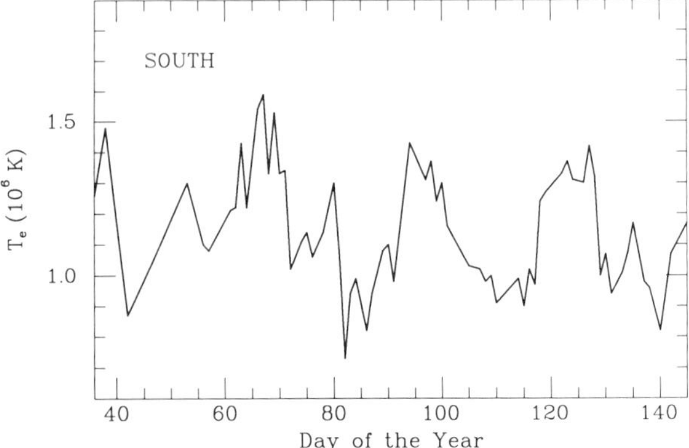

Figure 13.   Minimum temperatures in the southern coronal hole derived from the Sacramento Peak Fe X and Fe XIV intensity measurements for each day when data were available.

then be characterized by an effective temperature:

$$T_{\text{eff}} = T_i + \frac{m_i < \delta v^2 >}{3k}. \tag{7}$$

Here T_i is the temperature of the particles, k the Boltzman constant, and $<\delta v^2>^{1/2}$ is the velocity amplitude of the waves. The widths of spectral lines observed above the limb in coronal holes have been found to be larger than expected from thermal broadening alone. This increased broadening is consistent with the broadening described in Eq. (7) (see, e.g., Withbroe et al. 1985,1986). In addition to spectral line observations there are other observations that are consistent with the predictions made by Alfvén wave assisted models, such as the rapid decrease of the density anisotropy with distance from the coronal base as observed using radio interferometers (Coles and Esser 1992). There is, however, to date no direct proof of the importance of Alfvén waves for the acceleration of the solar wind. A somewhat stronger evidence for the presence and importance of MHD waves in the inner corona could be provided if several spectral lines from ions of different masses were measured as a function of height in conjunction with white light observations of high accuracy. The effect of the nonthermal line broadening is expected to increase with the mass of the line-producing ion (cf., Eq. 7) if all ions have the same thermal temperature (see, e.g., Esser et al. 1987), and for Alfvén waves propagating radially outward the rms wave velocity amplitude is inversely proportional to the quadratic root of the density if the waves are undamped. Given a set of spectral line measurements, one could try to establish a connection between the line broadening due to thermal and nonthermal motions, the mass of the ions, and the density gradient. As long as such observations are not available, the values derived for the wave velocity amplitudes in the inner corona from the different observational techniques remain suggestive at best.

Because the energy flux of the Alfvén waves increases monotonically with the magnetic field (chapter by MacGregor and Charbonneau), the flow velocity in the Alfvén wave assisted models also increases with the magnetic field. The field values used in the models are the values extrapolated inwards from *in-situ* measurements. The magnetic field values available from photospheric measurements are strongly model dependent, and are believed to underestimate the true coronal field (see, e.g., Wang and Sheeley 1988).

The third parameter that we want to discuss briefly is the helium abundance at the coronal base as it might be of importance for regulating the mass flux of the solar wind (see, e.g., the chapter by Holzer et al.). Model calculations show that the small values of the helium abundance at 1 AU of 4 to 5 % do not necessarily imply the same low helium abundance at the coronal base (see, e.g., Joselyn and Holzer 1978; Leer et al. 1992; Bürgi 1992). It has previously been suggested to use the intensity of the H II 304 Å line together with the H I Lyman-α 1216 Å line to determine the ratio of the helium to neutral hydrogen densities in the inner corona (Ahmad 1977). However, both lines consist of a resonantly scattered component which is proportional to the density, and a collisionally excited component which depends on the density squared. The abundance can therefore only be inferred from these lines via model calculations (Leer et al. 1992). Another method using inferences of the electron density and temperature from ultraviolet spectral line measurements for example, has recently been proposed by Habbal and Esser (1994). Replacing the mass density and pressure by their respective expressions, and assuming the same temperature for all species, Eq. (1) can be rewritten as:

$$\frac{1}{n_\alpha}\frac{\mathrm{d}n_\alpha}{\mathrm{d}r} = \frac{1+4\alpha}{\alpha}\frac{m_p G M_S}{k\,r^2}\frac{1}{T} + \frac{2+3\alpha}{\alpha}\frac{1}{T}\frac{\mathrm{d}T}{\mathrm{d}r} + \frac{2(1+2\alpha)}{\alpha}\frac{1}{n_e}\frac{\mathrm{d}n_e}{\mathrm{d}r}. \quad (8)$$

The variables and gradients are substituted by their empirical values for a given heliocentric distance. Using the condition that $\mathrm{d}n_\alpha/\mathrm{d}r < 0$, Eq. (8) can be solved for all values of α for which the right-hand side of the equation becomes negative. Applying this technique to the observations described in Sec. II.D, Fig. 6 and Sec. III.D, Fig. 10a, Habbal and Esser (1994) found a minimum value for the α abundance for these particular observations of about 25% at the coronal base. This minimum value decreases sharply as a function of distance. The assumptions that are inherent in the derivation of the α abundance are the same as the ones that enter into the derivation of the temperatures and densities described mainly in Secs. III.A and III.C. In spite of the inaccuracies inherent in these techniques, the above described procedure indicates that the helium abundance at the coronal base could indeed be large. A better constraint on the helium abundance could be placed with this technique if coordinated observations of white light and spectral lines were available.

IV. SUMMARY AND DISCUSSION

The observational characteristics of coronal holes show that they are far from being uniform regions of open magnetic field lines from which the solar wind streams into interplanetary space. Although the fine-scale magnetic structures in coronal holes have been observed for quite some time, we have ample knowledge at present to start investigating theoretical models of the relationship between these structures and the properties of the solar wind. For example, the influence of the prevalent activity in the underlying small-scale structure on the solar wind flow has not been addressed. Also, the magnetic topology of polar plumes and macrospicules has not been explored, nor has their importance for the acceleration and heating of the solar wind been investigated.

Most of the properties of the solar wind emanating from coronal holes as observed at 1 AU, e.g., flow speed, $500 \leq v_E \leq 900$ km s^{-1}, and mass flux $1.5 \leq n_E v_E \leq 5.8$ cm^{-2} s^{-1} (see, e.g., the chapter by Feldman and Marsch), can very well be reproduced using MHD models in which the solar wind is driven by thermal pressure and Alfvén waves (see, e.g., the chapters by MacGregor and Charbonneau, and by Holzer et al.). Given the almost exponential increase of the mass flux with temperature at the coronal base in these models (see, e.g., Leer et al. 1982), reasonable values for the mass flux can be attained only for a limited range of temperatures and densities in the inner corona. The three-fluid models which include fully ionized helium (chapter by Holzer et al.) are less sensitive to temperature changes at the coronal base. Even for those models, however, a 30% increase in temperature, which is below the accuracy of inferred temperatures at present, increases the mass flux by about a factor of 2.

The main goal for future observations must therefore be to place tighter constraints on the temperature in the inner corona, and on other plasma parameters such as density and helium abundance.

Some of the uncertainties present in currently available data and the corresponding analysis techniques should improve with the new space missions, such as SPARTAN, SOHO and ULYSSES. For example, the line of sight effect can be estimated much better given the possibility of long-term observations with SOHO. Ultraviolet observations in the source region of the solar wind, and charge state measurements at large distances from the Sun will be attainable simultaneously for the first time. Given the accuracy promised by the investigators associated with these instruments it should be possible to estimate the differential flow speeds of different ions in the inner solar wind region (see Sec. III.B). Knowledge of the differential flow speed would in turn help to model the behavior of the minor ions in the solar wind. With the instruments on board SOHO it should also be feasible to trace polar plumes out to larger distances, and place constraints on plasma parameters both inside the plumes and the regions between them. The advent of these space missions could provide the basis for an improved interpretation of the data in the inner corona, if coordinated observations are used in conjunction with careful theoretical studies.

Acknowledgments. We are greatly indebted to J. V. Hollweg for his very constructive remarks on an earlier version of the manuscript and for bringing the reference to Michel (1967) to our attention. We are also thankful to R. C. Altrock for providing the daily coronal data from the Sacramento Peak Observatory of the National Solar Observatory. The NSO/Sacramento Peak data used here are produced cooperatively by USAF/AFMC/PL/ GPSS and NSF/NOAO/NSO/SP. We are also greatful to H. S. Hudson for supplying the solar X-ray images from the Yohkoh mission of ISAS, Japan. The X-ray telescope was prepared by the Lockheed Palo Alto Research Laboratory, the National Astronomical Observatory of Japan, and the University of Tokyo with the support of NASA and ISAS. This work was supported by two NASA grants and a grant from the United States Air Force.

REFERENCES

Ahmad, I. A. 1977. Coronal He^+ λ304 radiation. *Solar Phys.* 53:409–415.

Ahmad, I. A., and Withbroe, G. L. 1977. EUV analysis of polar plumes. *Solar Phys.* 53:397–408.

Ahmad, I. A., and Webb, D. F. 1978. X-ray analysis of a polar plume. *Solar Phys.* 58:323–336.

Baumbach, S. 1939. Die Polarisation der Sonnenkorona. *Astronomische Nachrichten* 267:273–296.

Bohlin, J. D. 1977. Extreme-ultraviolet observations of coronal holes, I: Locations, sizes and evolution of coronal holes, June 1973–January 1974. *Solar Phys.* 51:377–398.

Bohlin, J. D., Sheeley, N. R., Jr., and Tousey, R. 1975*a*. Structure of the sun's polar cap at wavelengths 240–600 Å. In *Space Research*, ed. M. J. Rycroft (Berlin: Akademie-Verlag), pp. 651–656.

Bohlin, J. D., Vogel, S. N., Purcell, J. D., Sheeley, N. R., Jr., Tousey, R., and Van-Hoosier, M. E., 1975*b*. A newly observed solar feature: Macrospicules in He II 304 Å. *Astrophys. J. Lett.* 197:133–135.

Brickhouse, N. S., Raymond, J. C., and Smith, B. W. 1995. New model of iron spectra in the extreme ultraviolet and application to SERTS and EUVE observations: A solar active region and *Capella. Astrophys. J. Suppl.* 97:531–570.

Bürgi, A. 1987. Effects of non-Maxwellian velocity distribution functions and non-spherical geometry on the minor ions in the solar wind. *J. Geophys. Res.* 92:1057–1066.

Bürgi, A. 1992. Proton and alpha particle fluxes in the solar wind: Results of a three-fluid model. *J. Geophys. Res.* 97:3137–3150.

Coles, W. A., and Esser., R. 1992. An observational limit on the amplitude of Alfvén waves in the solar wind and comparison with an acceleration model. *J. Geophys. Res.* 97:19139–19148.

Cox, D. P., and Raymond, J. C. 1985. Preionization-dependent families of radiative shock waves. *Astrophys. J.* 298:651–659.

Dere, K. P., and Mason, H. E. 1981. Spectroscopic diagnostics of the active region: Transition zone and corona. In *Solar Active Regions*, ed. F. Q. Orall (Colorado: Colorado Associated Press), pp. 129–164.

Dere, K. P., Bartoe, J.-D. F., Brueckner, G. E., Cook, J. W., Socker, D. G., and Ewing, J. W., 1989*a*. UV observations of macrospicules at the solar limb. *Solar Phys.* 229:55–63.

Dere, K. P., Bartoe, J.-D. F., Brueckner, G. E., and Recely, F. 1989*b*. Transition zone flows observed in a coronal hole on the solar disk. *Astrophys. J. Lett.* 345:95–97.

Doyle, J. G., and Keenan, F. P. 1992. A recalculation of the line emissivities for the stronger UV and extreme-UV lines in the 400–2800 Å wave length range. *Astron. Astrophys.* 264:173–176.

Doschek, G. A., and Feldman, U. 1977. The coronal temperature and nonthermal motions in a coronal hole compared with other solar regions. *Astrophys. J. Lett.* 212:143–146.

Esser, R., Holzer, T. E., and Leer, E. 1987. Drawing inferences about solar wind acceleration from coronal minor ion observations. *J. Geophys. Res.* 92:13377–13389.

Esser, R., and Leer, E. 1990. Flow of oxygen ions in the solar wind acceleration region. *J. Geophys. Res.* 95:10269–10272.

Esser, R., Brickhouse, N. S., Habbal, S. R., Altrock, R. C., and Hudson, H. S. 1995. Using Fe X 6374 Å and Fe XIV 5303 Å spectral line intensities to study the effect of line of sight intergration on coronal temperature inferences. *J. Geophys. Res.*, in press.

Galvin, A. B., Ipavich, F. M., Gloeckler, G., Hovestadt, D., Klecker, B., and Scholer, M. 1984. Solar wind ionization temperatures inferred from the charge state composition of diffuse particle events. *J. Geophys. Res.* 89:2655–2671.

Habbal, S. R. 1992. Coronal energy distribution and X-ray activity in the small scale magnetic field of the quiet sun. *Ann. Geophys.* 10:34–46.

Habbal, S. R., and Esser, R. 1994. On the derivation of empirical limits on the helium abundance in coronal holes below 1.5 R_S. *Astrophys. J. Lett.* 421:59–62.

Habbal, S. R., and Gonzalez, R. D. 1991. First observations of macrospicules at 4.8 GHz at the solar limb in polar coronal holes. *Astrophys. J. Lett.* 326:25–27.

Habbal, S. R., and Grace, E. 1991. The connection between coronal bright points and the variability of the quiet sun EUV emission. *Astrophys. J.* 382:667–676.

Habbal, S. R., and Withbroe, G. L. 1981. Spatial and temporal variations of EUV coronal bright points. *Solar Phys.* 69:77–97.

Habbal, S. R., Dowdy, J. F., Jr., and Withbroe, G. L. 1990. A comparison between bright points in a coronal hole and a quiet sun region. *Astrophys. J.* 352:333–342.

Habbal, S. R., Esser, R., and Arndt, M. B. 1993. How reliable are coronal hole temperatures deduced from observations? *Astrophys. J.* 413:435–444.

Habbal, S. R., Esser, R., Gulathakurta, M., and Fisher, R. 1995. Flow properties of the solar wind derived from a two-fluid model with contraints from white light and in situ interplanetary observations. *J. Geophys. Res.*, in press.

Hollweg, J. V. 1973. Alfvén waves in a two-fluid model of the solar wind. *Astrophys. J.* 181:547–566.

Hollweg, J. V. 1992. Status of solar wind modeling from the transition region outwards. In *Solar Wind Seven*, eds. E. Marsch and R. Schwenn (Oxford: Pergamon Press), pp. 53–60.

Hundhausen, A. J. 1972. *Coronal Expansion and Solar Wind* (Berlin: Springer-Verlag), pp. 114–119.

Hundhausen, A. J. 1977. An interplanetary view of coronal holes. In *Coronal Holes and High Speed Streams*, ed. J. B. Zirker (Colorado: Colorado Associated Univ. Press), pp. 225–329.

Ipavich, F. M., Galvin, A. B., Gloeckler, G., Hovestadt, D., Bame, S. J., Klecker, B., Scholer, M., Fisk, L. A., and Fan, C. Y. 1986. Solar wind Fe and CNO measurements in high-speed flows. *J. Geophys. Res.* 91:4133–4141.

Jordan, C. 1969. The ionization equilibrium of elements between carbon and nickel. *Mon. Not. Roy. Astron. Soc.* 142:501–521.

Joselyn, J.-A., and Holzer, T. E. 1978. A steady three-fluid expansion for nonspherical geometries. *J. Geophys. Res.* 83:1019–1026.

Koutchmy, S. 1977. Study of the June 30, 1973 trans-polar coronal hole. *Solar Phys.* 51:399–407.

Krieger, A. S., Timothy, A. F., and Roelof, E. C. 1973. A coronal hole and its identification as the source of a high velocity solar wind stream. *Solar Phys.* 29:505–525.

Lallement, R., Holzer, T. E., and Munro, R. H. 1986. Solar wind expansion in a polar coronal hole: Inferences from coronal white light and interplanetary Lyman alpha. *J. Geophys. Res.* 91:6751–6759.

Leer, E., Holzer, T. E., and Flå, T. 1982. Acceleration of the solar wind. *Space. Sci. Rev.* 33:161–200.

Leer, E., Holzer, T. E., and Shoub, E. C. 1992. Solar wind from a corona with a large helium abundance. *J. Geophys. Res.* 97:8183–8201.

Mariska, J. T. 1978. Analysis of extreme ultraviolet observations of a polar coronal hole. *Astrophys. J.* 225:252–258.

Michel, F. C. 1967. Model of solar wind structure. *J. Geophys. Res.* 72:1917–1932.

Mullan, D. J., and Ahmad, I. A. 1982. Coronal holes: Mass loss driven by magnetic reconnection. *Solar Phys.* 75:347–350.

Munro, R. H., and Withbroe, G. L. 1972. Properties of a coronal "hole" derived from extreme-ultraviolet observations. *Astrophys. J.* 176:511–520.

Newkirk, G., Jr., and Harvey, J. 1968. Coronal polar plumes. *Solar Phys.* 3:321–343.

Olsen, L. O., Leer, E., and Holzer, T. E. 1994. Neutral hydrogen in the solar wind acceleration region. *Astrophys. J.* 420:913–925.

Owocki, S. P. 1983. Interpreting the solar wind ionization state. In *Solar Wind Five*, ed. M. Neugebauer, NASA CP-2280, pp. 623–641.

Owocki, S. P., and Scudder, J. D. 1983. The effect of a non-Maxwellian velocity distribution on oxygen and iron ionization balances in the solar corona. *Astrophys. J.* 270:758–768.

Parker, E. N. 1958. Dynamics of the interplanetary gas and magnetic fields. *Astrophys. J.* 128:664–676.

Parker, E. N. 1965. Dynamical theory of the solar wind. *Space Sci. Rev.* 4:666–708.

Parker, E. N. 1991. Heating coronal holes. *Astrophys. J.* 372:719–729.

Pneuman, G. W. 1986. Driving mechanisms for the solar wind. *Space Sci. Rev.* 43:105–138.

Pottasch, S. R. 1964. On the interpretation of the solar ultraviolet emission line spectrum. *Space Sci. Rev.* 3:816–855.

Raymond, J. C., and Doyle, J. G. 1981. Emissivities of strong ultraviolet lines. *Astrophys. J.* 245:1141–1144.

Saito, K. 1958. Polar rays of the solar corona. *Publ. Astron. Soc. Japan* 10:49–78.

Saito, K. 1965. Polar rays of the solar corona. II. *Publ. Astron. Soc. Japan* 17:1–26.

Sheeley, N. R., Jr., and Golub, L. 1979. Rapid changes in the fine structure of a coronal bright point and a small active region. *Solar Phys.* 63:119–126.

Snyder, C. W., and Neugebauer, M. 1966. The relation of Mariner-2 plasma data for solar phenomena. In *The Solar Wind*, ed. R. J. Mackin (Oxford: Pergamon Press) pp. 25–34.

Thieme, K. M., Schwenn, R., and Marsch, E. 1989. Are structures in high speed streams signatures of coronal fine structures? *Adv. Space Res.* 9:127–130.

Tousey, R. 1972. High angular resolution observations from rockets: Solar XUV observations. In *Space Research* vol. 12, eds. S. A. Bowhill, L. D. Jaffe and M. J. Rycroft (Berlin: Akademie-Verlag), pp. 1719–1737.

Tsuneta, S., Acton, L., Bruner, M., Lemen, J., Brown, W., Caravalho, R., Catura, R., Freeland, S., Jurcevich, B., Morrison, M., Ogawara, Y., Hirayama, T., and Owens J. 1991. The soft x-ray telescope for the Solar-A mission. *Solar Phys.* 136:37–68.

van de Hulst, H. C. 1950*a*. On the polar rays of the corona. *Bull. Astron. Soc. Netherlands* 11:150–160.

van de Hulst, H. C. 1950*b*. The electron density of the solar corona. *Bull. Astro. Soc. Netherlands* 11:135–150.

Waldmeier, M. 1955. Ergebnisse und Probleme der Sonnenforschung (Leipzig: Akademische Verlagsgesellschaft).

Wang, Y.-M., and Sheeley, N. R. 1988. The solar origin of long-term variations of the interplanetary magnetic field strength. *J. Geophys. Res.* 93:11227–11236.

Wilcox, J. M. 1968. The interplanetary magnetic field. Solar origin and terrestrial effects. *Space Sci. Rev.* 8:258-328.

Withbroe, G. L. 1983. Solar wind and coronal structure. *ESA Journal* 7(4):341–356.

Withbroe, G. L., Jaffe, D. T., Foukal, P. V., Huber, M. C. E., Noyes, R. W., Reeves, E. M., Schmahl, E. J., Timothy, J. G., and Vernazza, J. E. 1976. EUV transients observed at the solar pole. *Astrophys. J.* 203:528–532.

Withbroe, G. L., Kohl, J. L., Weiser, H., and Munro, R. H. 1982. Probing the solar wind acceleration region using spectroscopic techniques. *Space Sci. Rev.* 33:17–52.

Withbroe, G. L., Kohl, J. L., Weiser, H., and Munro, R. H. 1985. Coronal temperatures, heating, and energy flow in a polar region of the sun at solar maximum. *Astrophys. J.* 297:324–337.

Withbroe, G. L., Kohl, J. L., and Weiser, H. 1986. Analysis of coronal H I Lyman-alpha measurements in a polar region of the sun observed in 1979. *Astrophys. J.* 307:381–388.

Zirker, J. B. 1977. Coronal holes—an overview. In *Coronal Holes and High Speed Wind Streams*, ed. J. B. Zirker (Colorado: Colorado Associated Univ. Press), pp. 1–26.

Zombeck, M. V., Vaiana, G. S., Haggerty, R., Krieger, A. S., Silk, J. S., and Timothy, A. 1978. An atlas of soft X-ray images of the solar corona from SKYLAB. *Astrophys. J. Suppl.* 38:69–85.

ALFVÉN WAVE-DRIVEN WINDS

K. B. MacGREGOR and P. CHARBONNEAU
High Altitude Observatory

We examine how the dynamical properties of stationary, wind-type outflows from stars are modified by the interaction between small-amplitude, propagating Alfvén waves and the expanding stellar atmosphere. We first establish a theoretical and mathematical framework within which both the spatial dependences of the wave amplitudes and the force they exert on the flow can be derived. We then consider the effect of short-wavelength (i.e., WKB), outwardly traveling disturbances on winds which would otherwise be thermally driven, as well as on winds in which the thermal pressure gradient force plays a negligible role. We subsequently relax the requirement that the perturbation wavelength be short in comparison with the characteristic scale length of the background medium. By so doing, we provide for the process of Alfvén wave reflection, making it possible to treat flows containing finite wavelength (i.e., non-WKB) fluctuations that propagate in both the inward and outward directions.

I. INTRODUCTION

In the early 1960s, the first generation of satellite-borne experiments detected the presence of a continuous, supersonic outflow of plasma from the Sun into interplanetary space, thereby confirming the predictions of Parker (1958) concerning the existence and properties of the solar wind. While many of the observed characteristics of the flow appeared to conform with Parker's description of stationary coronal expansion driven by the thermal pressure gradient force, it soon became evident that the solar wind was by no means steady, spherically symmetric, and structureless. In particular, *in-situ* measurements of the physical properties of the solar wind by instruments in Earth orbit showed the flow to be variable (both spatially and temporally) on a variety of scales. After more than 30 years of nearly continuous observations, it is now recognized that much of this variability is due to the presence of magnetohydrodynamic (MHD) waves in the interplanetary medium (Coleman 1968; Belcher and Davis 1971; Denskat and Neubauer 1983; Roberts 1989; Marsch and Tu 1990; Roberts and Goldstein 1991; Marsch 1991). Because vector fluctuations in the plasma velocity are frequently observed to be well correlated with those in the magnetic field, Alfvén waves have been identified as one of the primary constituents of the microscale structure of the solar wind at 1 AU. Most of these waves presumably originate in the solar coronal source regions of the outflow, and undergo considerable modification of their physical properties as a consequence of the interaction between them and large-scale structures in the wind. The interested reader is referred to the previously cited papers for detailed discussion of MHD turbulence and waves in the solar wind.

It was also Parker (1965) who, in that year, showed that the presence of traveling Alfvén waves in the solar wind could have implications for the dynamics of coronal expansion. He considered a stationary, spherical outflow containing waves with (i) propagation vectors parallel to a radially directed background magnetic field, and (ii), ϕ-directed magnetic and velocity fluctuations. Parker solved for the spatial evolution of the amplitudes of perturbations having suitably short wavelengths, and noted that both the time-averaged energy density and energy flux density of such waves decreased with increasing distance from the Sun. By examining conservation of energy for the system composed of the waves and the flow, he concluded that the reason for this behavior was because the waves performed work on the wind. That is, outwardly propagating Alfvén waves in a spherical, inhomogeneous, expanding atmosphere exert a force on the gas through which they travel, providing an acceleration above and beyond that which is available from the thermal pressure gradient alone. It is this property of Alfvénic disturbances under stellar atmospheric conditions that is the primary focus of this chapter.

In the case of the Sun, the force associated with propagating Alfvén waves has been invoked in an effort to account for the requisite acceleration of recurrent high-speed streams in the solar wind. In the context of outflows from stars other than the Sun, wave-driven wind models have been applied to objects for which the inferred thermal acceleration appears to be insufficient to induce the observed expansion of the stellar atmosphere. Specifically, Alfvén wave acceleration has been utilized in the construction of models for both the massive, low-velocity outflows from cool, evolved stars and the winds of low-mass, pre-main-sequence stars. On yet larger scales, it has been suggested that Alfvén waves produced by cosmic rays streaming through the magnetized interstellar plasma might be capable of initiating and sustaining a galactic wind.

In this chapter, we conduct a detailed examination of MHD flows whose dynamical properties derive primarily from their interaction with propagating Alfvén waves. The approach we adopt is to study simplified wind models in an effort to understand the basic physics of the wave acceleration process. Toward this goal, we initially consider the ways in which wind dynamics are affected by the presence of traveling fluctuations having wavelengths that are short in comparison with the length scale over which the wave propagation speed varies in the flow. We assess the influence of such WKB disturbances on wind properties for thermodynamic conditions corresponding to two rather different atmospheric dynamical states in the absence of waves. In the first case, the assumed gas temperature is high enough to ensure the existence of a thermally driven wind without waves, while in the second, thermal acceleration alone is incapable of producing significant mass loss. We then reformulate the basic wind model without resorting to the WKB approximation to describe the evolution of wave amplitudes, and consider effects due to linear Alfvénic perturbations of arbitrary wavelength. For finite-wavelength fluctuations, wave reflection can take place throughout the flow, and wind properties are further modified as a consequence of the simultaneous presence of both inward and outward propagating waves. We conclude by noting a number of deficiencies in the theoretical framework discussed herein, and by enumerating several extensions and improvements for incorporation in the next generation of wave-driven wind models.

II. WIND MODEL CONSTRUCTION

In accordance with the objectives stated in the preceding section, we focus our attention on a stationary, spherically symmetric outflow emanating from the outer atmospheric layers of a nonrotating star of mass M_* and radius R_*. We assume that the wind consists of a fully ionized, perfectly conducting gas, threaded by a stellar magnetic field whose only component (in the absence of waves) lies along the radial direction of a spherical polar coordinate system with origin at the center of the star. At a reference level $r = r_0$ in the stellar atmosphere are introduced small-amplitude, monochromatic, Alfvénic (i.e., noncompressive) perturbations having velocity amplitudes δu and magnetic amplitudes δB in the ϕ-direction, and propagation vectors $\mathbf{k}$ parallel to the background magnetic field.

Throughout the domain $r_0 \leq r < \infty$, we seek wind solutions that can be represented as the sum of a stationary (i.e., time-averaged) component and a fluctuating, wave-like component. Specifically, we write

$$\mathbf{u}(r, t) = u(r)\mathbf{e}_r + \delta u(r, t)\mathbf{e}_\phi \tag{1a}$$

and

$$\mathbf{B}(r, t) = B(r)\mathbf{e}_r + \delta B(r, t)\mathbf{e}_\phi \tag{1b}$$

where $u(r)$ is the flow speed and $B(r)$ the background, stellar field. The quantities δu and δB have time averages $\langle \delta u \rangle = \langle \delta B \rangle = 0$, and (as noted above) orientations such that $\mathbf{k} \cdot \delta \mathbf{u} = \mathbf{k} \cdot \delta \mathbf{B} = 0$. For flow and magnetic fields having the forms given above, application of the laws for mass and magnetic flux conservation yield

$$\rho(r) = \rho_0 \left(\frac{r_0}{r}\right)^2 \left(\frac{u_0}{u}\right) \tag{2a}$$

and

$$B(r) = B_0 \left(\frac{r_0}{r}\right)^2 \tag{2b}$$

where ρ is the mass density, and the subscript 0 denotes evaluation at the reference radius r_0.

The equation of motion for a stationary wind is derived by substituting Eqs. (1a) and (1b) for $\mathbf{u}$ and $\mathbf{B}$ into the equation expressing conservation of momentum for the gas. In order to include the dynamical effect of the waves on the background flow, it is necessary to retain terms of second order in the wave amplitudes while performing this substitution. By time-averaging the radial component of the equation that results from this procedure, we obtain

$$u \frac{du}{dr} = -\frac{1}{\rho} \frac{dp}{dr} - \frac{GM_*}{r^2} + \frac{\langle f_w \rangle}{\rho} \tag{3}$$

where, for a gas with temperature T and mean mass per particle μ,

$$p = \frac{\rho k_B T}{\mu} \tag{4}$$

is the pressure. Ordinarily, T would be derived by supplementing Eq. (3) with an explicit energy equation for the wind. However, because our primary intent is the elucidation of the dynamical interaction between the waves and the flow, we henceforth omit detailed consideration of processes affecting the thermal energy balance of the wind and instead assume that the outflow is isothermal with $T = T_0$ throughout.

The quantity $\langle f_w \rangle$ appearing on the right-hand side of Eq. (3) is the time-averaged, radial component of the force (per unit volume) exerted on the wind by propagating Alfvénic fluctuations. Insight into the physical origin of this acceleration can be gained by noting that the wave force density is given by

$$f_w = \left[\rho \left(\delta \mathbf{u} \cdot \nabla \right) \delta \mathbf{u} + \frac{1}{4\pi} \left(\nabla \times \delta \mathbf{B} \right) \times \delta \mathbf{B} \right]_r \tag{5}$$

so that for $\delta \mathbf{u}$ and $\delta \mathbf{B}$ as assumed in Eqs. (1),

$$\langle f_w \rangle = \frac{\rho \langle \delta u^2 \rangle}{r} - \frac{\langle \delta B^2 \rangle}{4\pi r} - \frac{d}{dr} \left(\frac{\langle \delta B^2 \rangle}{8\pi} \right). \tag{6}$$

Written in this form, it is evident that one contribution to the acceleration is the centrifugal force arising from the oscillatory, azimuthal motion of the gas in the wave. A second contribution results from the Lorentz force $(\delta \mathbf{j} \times \delta \mathbf{B})/c$ produced by the interaction between the magnetic amplitude $\delta \mathbf{B}$ and the fluctuating current density $\delta \mathbf{j}$ associated with it. As discussed by Hollweg (1978), $\langle f_w \rangle$ is nonzero because the wind is spherical and inhomogeneous. For waves in a homogeneous, planar medium, the first term in Eq. (5) is identically equal to zero, while the second vanishes upon time-averaging.

The wind model is completed by providing the means for determining the wave amplitudes δu and δB as functions of radius and time. For this purpose, we utilize the azimuthal component of the gas momentum equation

$$\frac{\partial}{\partial t} \delta u + \frac{u}{r} \frac{\partial}{\partial r} (r \delta u) = \frac{B}{4\pi \rho r} \frac{\partial}{\partial r} (r \delta B) \tag{7}$$

together with the induction equation

$$\frac{\partial}{\partial t}\delta B = \frac{1}{r}\frac{\partial}{\partial r}[r(B\delta u - u\delta B)]. \tag{8}$$

Equations (3), (7), and (8), along with relations (2a), (2b), (4), and (6) constitute a complete set which, when supplemented by suitable boundary and initial conditions, can be solved to obtain the values of u, δu, and δB throughout the domain. In succeeding sections of this chapter, we consider the physical properties of wind solutions obtained using two different assumptions regarding the relative sizes of the perturbation wavelength λ and the scale length ℓ for variations in ρ, u, and B.

III. WKB WAVE AMPLITUDES AND FORCE

In the limit $\lambda/\ell \ll 1$, wave propagation and evolution can be treated using the so-called WKB approximation. The short wavelength assumption makes wave reflection from the Alfvén speed gradient in the wind impossible, so that only outwardly propagating disturbances are present in the flow. To obtain the WKB solutions for the wave amplitudes, we first suppose that both δu and δB vary with time according to $e^{-i\omega t}$. Defining the small parameter $\epsilon \equiv \lambda/\ell = (2\pi/k\ell)$, we then assume that the spatial dependence of the velocity and magnetic perturbations can be written in the form (see, e.g., Belcher 1971; Hollweg 1973b)

$$\delta u(r, t) = [\delta u_1(r) + \epsilon\delta u_2(r) + \epsilon^2\delta u_3(r) + \cdots]e^{i[\psi(r)-\omega t]} \tag{9}$$

with a similar expansion for δB. The length scales of the quantities δu_1, δu_2,....appearing in Eq. (9) are also presumed to be of order ℓ, and

$$k(r) = \frac{\mathrm{d}\psi(r)}{\mathrm{d}r} \tag{10}$$

with ω constant. The convergence and consistency of expansions of the form given in Eq. (9) have been discussed by Hollweg (1990) and Barnes (1992).

The wave properties are derived by inserting the expansions for δu and δB into Eqs. (7) and (8), performing the indicated differentiations, and equating terms of equal order in ϵ. To lowest (i.e., zeroth) order, this procedure yields

$$\omega = k(u + u_A) \tag{11}$$

and

$$\delta u_1 = \mp\frac{\delta B_1}{\sqrt{4\pi\rho}} \tag{12}$$

where $u_A \equiv B/\sqrt{4\pi\rho}$ is the Alfvén speed and the $-$ $(+)$ sign in Eq. (12) applies to outward (inward) propagating waves. Substitution of these results in the first order equations obtained from Eqs. (7) and (8) leads to a differential equation for the function δB_1 (cf. Belcher 1971). After some manipulation, this equation can be integrated analytically, with the result

$$\delta B(r) = \delta B_0 \left(\frac{M_{A0}}{M_A}\right)^{1/2} \left(\frac{1 + M_{A0}}{1 + M_A}\right) \tag{13}$$

where

$$M_A \equiv (u/u_A) = M_{A0}(\rho_0/\rho)^{1/2} \tag{14}$$

is the Alfvénic Mach number and we have dropped the subscript on δB_1 for clarity. The value of δu follows directly from using this expression for δB in Eq. (12).

An alternative and, in many respects, simpler approach to the determination of the wave amplitudes in the WKB approximation is through use of the conservation law for the wave action density (Dewar 1970; Bretherton 1970; Jacques 1977). As will be discussed in greater detail in a subsequent section, the wave energy density

$$\langle \varepsilon_w \rangle = \frac{1}{2}\rho\langle \delta u^2 \rangle + \frac{\langle \delta B^2 \rangle}{8\pi} \tag{15}$$

is not conserved because the waves do work accelerating the wind. Instead, for a steady flow in the absence of dissipation, it can be shown that the wave action density

$$S = \frac{\langle \varepsilon_w \rangle}{\omega - \mathbf{k} \cdot \mathbf{u}} \tag{16}$$

obeys a conservation law of the form

$$\nabla \cdot [(\mathbf{u} + \mathbf{u}_A)S] = 0 \tag{17}$$

(see, e.g., Jacques 1977; Barnes 1992). For a spherically symmetric wind with $\mathbf{u}$, $\mathbf{B}$, and $\mathbf{k}$ all in the radial direction, Eq. (17) is readily integrated to obtain

$$\frac{r^2(u + u_A)\langle \varepsilon_w \rangle}{\omega u_A} = \text{constant} \tag{18}$$

from which it follows that

$$\langle \varepsilon_w \rangle = \frac{\langle \delta B^2 \rangle}{4\pi} = \frac{\langle \delta B_0^2 \rangle}{4\pi} \left(\frac{M_{A0}}{M_A}\right) \left(\frac{1 + M_{A0}}{1 + M_A}\right)^2 \tag{19}$$

where we have used Eqs. (12) and (15), and evaluated the constant in Eq. (18) at r_0. The result given in Eq. (19) is completely equivalent to expression (13) for the wave magnetic amplitude.

For later reference, we also give the WKB expressions for the wave energy flux density $\langle \Phi_w \rangle$ and the force per unit volume $\langle f_w \rangle$. As a vector, the energy flux density is radially directed, and is the sum of: (i) the advected kinetic energy density of the wave $\rho\langle\delta u^2\rangle\mathbf{u}/2$; and (ii), the electromagnetic Poynting flux $(c/4\pi)\langle\delta\mathbf{E}\times\delta\mathbf{B}\rangle$, where $\delta\mathbf{E}$ is the fluctuating electric field associated with the wave. For the assumed flow geometry and wave polarization, $\langle\Phi_w\rangle$ can be written as

$$\langle\Phi_w\rangle = \rho u\left(\frac{1}{2}\langle\delta u^2\rangle + \frac{\langle\delta B^2\rangle}{4\pi\rho} - \frac{\langle\delta u\delta B\rangle}{M_A\sqrt{4\pi\rho}}\right) \tag{20}$$

a result which, through use of the relation (12), is expressible in the form

$$\langle\Phi_w\rangle = \langle\varepsilon_w\rangle\left(\frac{3}{2}u + u_A\right) \tag{21}$$

where $\langle\varepsilon_w\rangle$ is given by Eq. (19). To derive an expression for the wave force density, we first use Eq. (12) to eliminate $\langle\delta u^2\rangle$ in favor of $\langle\delta B^2\rangle$ in Eq. (6). As a result of this substitution, the centrifugal and magnetic tension components of the wave force cancel, and $\langle f_w\rangle$ becomes

$$\langle f_w\rangle = -\frac{d}{dr}\left(\frac{\langle\delta B^2\rangle}{8\pi}\right) = \frac{\langle\varepsilon_w\rangle}{4}\left(\frac{1+3M_A}{1+M_A}\right)\left(\frac{2}{r} + \frac{1}{u}\frac{du}{dr}\right) \tag{22}$$

where $\langle\delta B^2\rangle$ has been evaluated from Eq. (19). The expression (22) suggests that for WKB Alfvén waves, the force can be regarded as arising from the gradient of a time-averaged wave pressure, $\langle p_w\rangle = \langle\varepsilon_w\rangle/2$ (cf. Jacques 1977,1978).

IV. WIND SOLUTIONS WITH WKB ALFVÉN WAVES

Upon substituting Eq. (22) for the wave force in Eq. (3) and using Eqs. (4) and (2a) to re-express p in terms of r and u, the wind equation of motion assumes the form

$$\left[u^2 - a^2 - \frac{\langle\varepsilon_w\rangle}{4\rho}\left(\frac{1+3M_A}{1+M_A}\right)\right]\frac{r}{u}\frac{du}{dr} =$$

$$\left[2a^2 - \frac{GM_*}{r} + \frac{\langle\varepsilon_w\rangle}{2\rho}\left(\frac{1+3M_A}{1+M_A}\right)\right] \tag{23}$$

where $a = (k_B T/\mu)^{1/2}$ is the isothermal sound speed. A complete wind solution $u(r)$ is obtained by integrating Eq. (23) over the interval $r_0 \leq r < \infty$. In order to accomplish this task, values must be assigned to the quantities $M_*, r_0, \mu, N_0 (= \rho_0/\mu), T, B_0$, and δB_0, and Eq. (23) must be supplemented by an appropriate boundary condition. For this latter purpose, we choose to impose the requirement that the gas pressure in the flow tend toward zero asymptotically. As is the case for spherical, thermally driven winds (see, e.g., Parker 1963), Eq. (23) contains a critical point (r_c, u_c) at which location the bracketed quantities on the left- and right-hand sides of the equality simultaneously vanish. Holzer et al. (1983) have shown that this critical point occurs where the expansion velocity equals the propagation speed of a radially traveling sound wave in the presence of short wavelength Alfvén waves. The single solution to Eq. (23) that satisfies the stipulated boundary condition and attains super-Alfvénic flow speeds at large distances from the star is the one that passes smoothly through the critical point. We select this solution by varying the reference level wind velocity u_0 and integrating Eq. (23) until a profile $u(r)$ that behaves regularly at (r_c, u_c) is obtained.

A. Modifications to Thermally Driven Flow

In Figs. 1 through 4, we show results pertaining to wind models with WKB Alfvén waves. The solutions depicted were obtained for the following values of the stellar and reference level physical properties: $M_* = M_\odot$, $r_0 = 1.25$ $R_\odot$, $\mu = m_p/2$, $N_0 = 2 \times 10^7$ cm^{-3}, $T = 10^6$ K, and $B_0 = 1$ Gauss. Detailed, numerical models for the steady expansion of the solar corona including the acceleration produced by outwardly propagating Alfvén waves have been constructed by (among others) Belcher (1971), Alazraki and Couturier (1971), Hollweg (1973a,1978,1986), Jacques (1978), Leer et al. (1982), Tu (1987), and Jatenco-Pereira and Opher (1989c). For the adopted parameter values, a thermally driven wind is present even in the limit of vanishing Alfvén wave amplitudes. This flow has $u_0 = 1.19$ km s^{-1}, and attains a speed $u = 444.3$ km s^{-1} at the distance $r = 100\,r_0$ chosen as the outer boundary of the computational domain. The magnitude of the magnetic perturbation δB_0 is specified by assigning a value to the quantity

$$\alpha \equiv (\delta B_0/B_0)^2 \tag{24}$$

which is the ratio of the wave magnetic energy density to the energy density of the background field at r_0. The results shown in Figs. 1 through 4 correspond to solutions having $\alpha = 0.01, 0.03, 0.10$.

In Fig. 1, we show profiles of the wind flow speed $u(r)$ and Alfvén speed $u_A(r)$ for $r_0 \leq r \leq 100\,r_0$ and the three α values given above. The dashed curve represents the velocity profile of the thermally driven wind that exists in the absence of waves. As is evident from the results depicted, the expansion speed of a wind containing waves is everywhere greater than that of a wind accelerated solely by the thermal pressure gradient force. In addition, note that the velocity at the coronal base of the flow is an increasing function of the wave amplitude δB_0, having the value $u_0 = 4.63$ km s^{-1} for $\alpha = 0.01$ and growing to $u_0 = 37.35$ km s^{-1} for $\alpha = 0.10$. Because the coronal number density is the same for each of the three solutions shown, the mass flux density is likewise an increasing function of the parameter α. Comparison of the profiles of u and u_A indicates that the flows become super-Alfvénic at the locations $r_A/r_0 = 13.53, 8.54$, and 4.85 for $\alpha = 0.01, 0.03,$

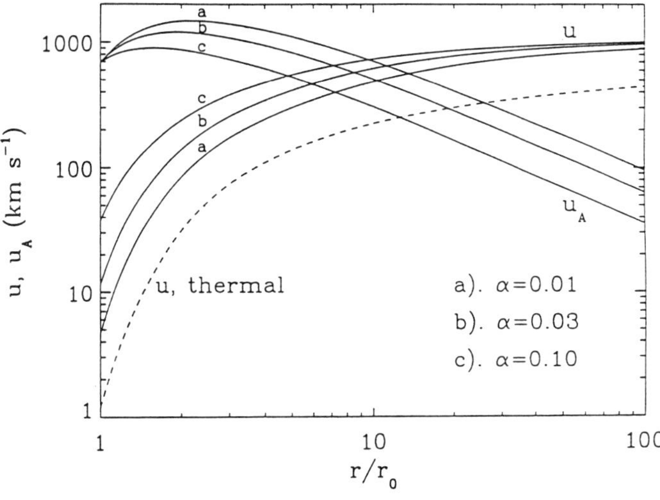

Figure 1.   Profiles of the flow and Alfvén speeds for wave-driven wind models in the WKB approximation (see Sec. IV.A). The solutions depicted were obtained for three different values of the quantity $\alpha \equiv (\delta B_0/B_0)^2$. The dashed curve is the velocity profile of the thermally driven wind that exists for $\alpha = 0$.

and 0.10, respectively, and, as a consequence of continued acceleration by Alfvén waves, attain the velocities 888.5, 974.9, and 1005 km s^{-1} at $r = 100\, r_0$. Because the density distribution in the subsonic region at the base of the wind closely resembles that of an isothermal, hydrostatic atmosphere, the decrease in ρ with increasing height is sufficiently rapid in comparison to that of B that the Alfvén speed initially increases with r. The maximum value of u_A is lower for larger α values, a manifestation of the higher densities that prevail throughout wind solutions having larger wave amplitudes at the reference radius. At larger distances where the atmospheric expansion is both supersonic and super-Alfvénic, the rate of decrease of ρ becomes slower than that of B, and u_A is a decreasing function of r.

In Fig. 2, we show the time-averaged amplitudes $\langle \delta u^2 \rangle^{1/2}$ and $\langle \delta B^2 \rangle^{1/2}$ as functions of r for the three solutions of Fig. 1. The profiles of these quantities have been normalized to their respective values at r_0 which, for future reference, are $\langle \delta B_0^2 \rangle^{1/2} = 0.100, 0.173$, and 0.316 Gauss, and $\langle \delta u_0^2 \rangle^{1/2} = 69.0, 119.5$, and 218.1 km s^{-1} for $\alpha = 0.01, 0.03$, and 0.10. As is easily deduced from Eqs. (12), (13), and (14), in the sub-Alfvénic ($M_A \ll 1$) portion of the flow, the magnetic and velocity perturbations vary according to $\langle \delta B^2 \rangle^{1/2} \sim \rho^{1/4}$ and $\langle \delta u^2 \rangle^{1/2} \sim \rho^{-1/4}$. The same formulae indicate that when the wind attains a nearly constant, super-Alfvénic ($M_A \gg 1$) speed at large distances, the dependence of the wave amplitudes on r is given by $\langle \delta B^2 \rangle^{1/2} \sim r^{-3/2}$ and $\langle \delta u^2 \rangle^{1/2} \sim r^{-1/2}$.

As discussed in Sec. III, among the quantities that are expressible in terms

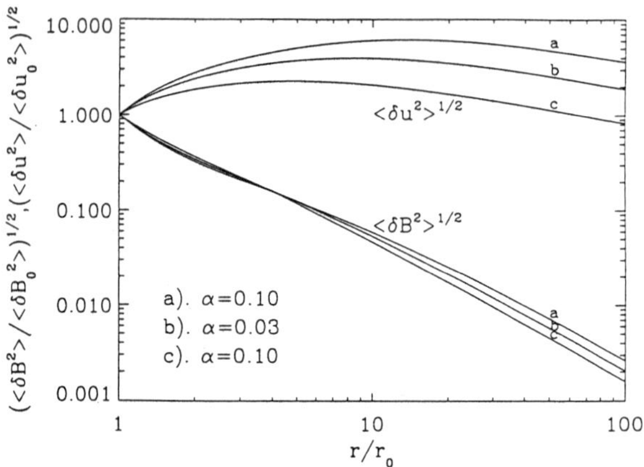

Figure 2. The time-averaged magnetic and velocity amplitudes, $\langle \delta B^2 \rangle^{1/2}$ and $\langle \delta u^2 \rangle^{1/2}$, as functions of r/r_0 for the three wind solutions of Fig. 1. The perturbation quantities have been normalized to their respective values at r_0, as described in the text.

of the magnetic and velocity fluctuations $\langle \delta B^2 \rangle^{1/2}$ and $\langle \delta u^2 \rangle^{1/2}$ are the wave energy density $\langle \varepsilon_w \rangle$ and the wave energy flux density $\langle \Phi_w \rangle$. The profiles of $\langle \varepsilon_w \rangle$ and $\langle \Phi_w \rangle$ corresponding the three wind solutions of Fig. 1 are given in Fig. 3. From the results derived above, it is readily seen that $\langle \varepsilon_w \rangle$ varies as $\rho^{1/2}$ for $M_A \ll 1$, and as r^{-3} for $M_A \gg 1$. The wave energy density decreases with increasing distance from r_0 because the waves do work accelerating the flow against the opposing gravitational force. Indeed, as was noted in connection with the derivation of Eq. (22), the force exerted on the wind material by short wavelength Alfvén waves is proportional to the radial gradient of $\langle \varepsilon_w \rangle$. Additional understanding of the relation between the wave energy and force densities can be gained by examining profiles of the time-averaged energy flux density of the waves, depicted in Fig. 3b. Consideration of Eq. (21) indicates that in the sub-Alfvénic portion of the flow, the first of the two terms that constitute $\langle \Phi_w \rangle$ varies as $r^{-1} u^{1/2}$, while the the second varies as r^{-2}. At large distances where $M_A \gg 1$ and u tends toward a constant value, $\langle \Phi_w \rangle$ has the same dependence on radius as $\langle \varepsilon_w \rangle$, $\langle \Phi_w \rangle \sim r^{-3}$. A quantitative expression of energy conservation for Alfvén waves in a stationary outflow can be derived by explicitly calculating the divergence of the energy flux density. Using Eqs. (19) and (21) to evaluate $\langle \Phi_w \rangle$, it follows that

$$\nabla \cdot \langle \Phi_w \rangle = -u \, \langle f_w \rangle \tag{25}$$

where the time-averaged force $\langle f_w \rangle$ is given by Eq. (22). Equation (25) indicates that the nonzero divergence of the energy flux density is precisely equal to rate at which work is done on the flow by the force associated with the waves. Hence, the decrease in $\langle \varepsilon_w \rangle$ observed in Fig. 3a is indicative of the conversion of a portion of the wave energy into increased kinetic energy of the flow.

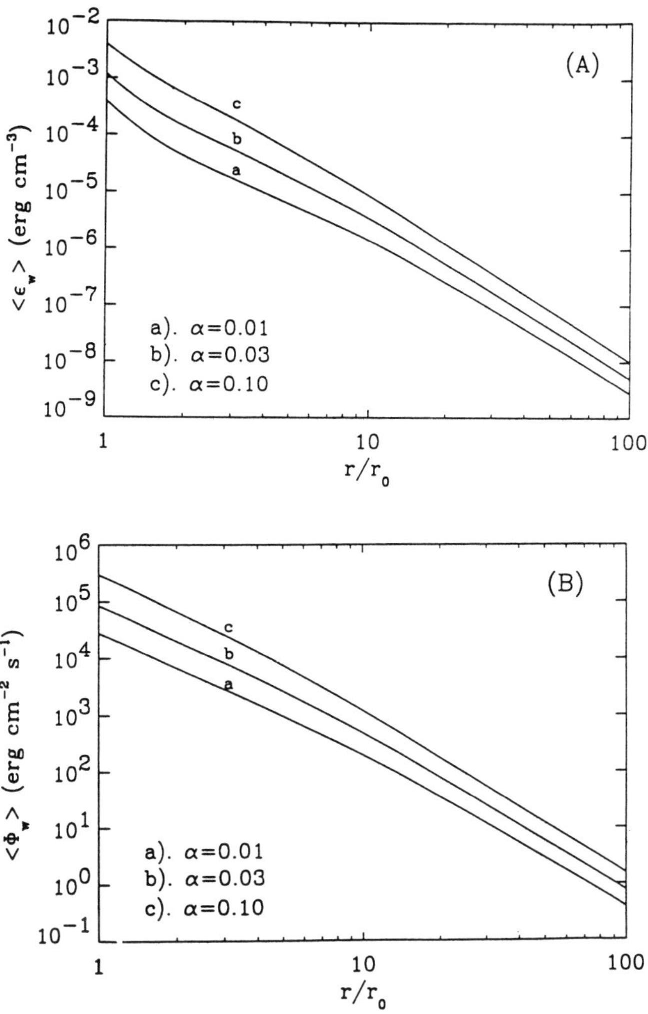

Figure 3. The time-averaged wave energy density $\langle \varepsilon_w \rangle$ (panel A) and energy flux density $\langle \Phi_w \rangle$ (panel B) as functions of r/r_0 for the WKB Alfvén wave-drive wind models of Fig. 1.

The ways in which the dynamical structure of an otherwise thermally driven wind is modified by the inclusion of the force due to propagating Alfvén waves is illustrated in Fig. 4, which shows the balance of forces throughout the solutions with $\alpha = 0.01$ and 0.10. In each panel, we show profiles of the accelerations produced by the thermal pressure gradient force $(-a^2 d \ln \rho / dr)$, the gravitational force (GM_*/r^2), and the wave force $(\langle f_w \rangle / \rho)$, as well as the net outward acceleration arising from the wave and pressure gradient forces combined. The subsonic portion of the model having $\alpha = 0.01$ (see Fig. 4a)

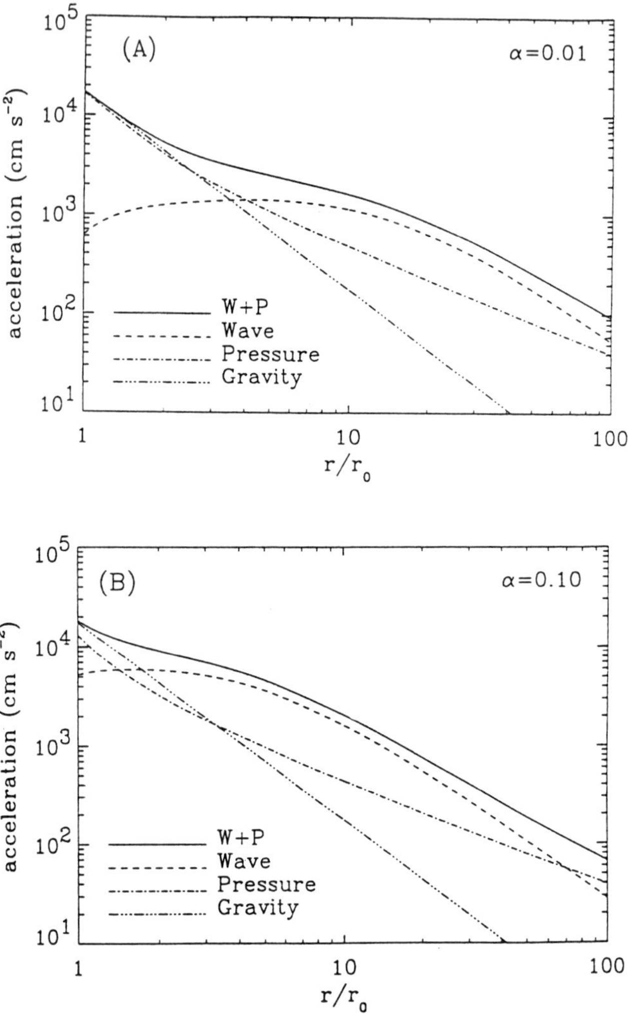

Figure 4.   Profiles of the accelerations produced by the indicated forces for two of the
wind solutions of Fig. 1. Panel A contains results for the model having $\alpha = 0.01$,
while the results in panel B pertain to the model with $\alpha = 0.10$. The solid curve
denoted W+P in each panel represents the sum of the accelerations due to the wave
and thermal pressure gradient forces.

is characterized by an approximate balance between the gravitational and
thermal pressure gradient forces, with the wave force more than an order of
magnitude smaller for r near r_0. Note, however, that within the sub-Alfvénic
region $r \leq r_A$ ($\approx 13.5\, r_0$ for the solution under consideration)

$$\langle f_w \rangle / \rho \approx -\frac{\langle \varepsilon_w \rangle}{4\rho} \frac{\mathrm{d} \ln \rho}{\mathrm{d}r} \tag{26}$$

(cf. Eq. [22]), so that the ratio of the acceleration provided by waves to that provided by the thermal pressure gradient varies as

$$\frac{\langle f_w \rangle}{(-a^2 \mathrm{d} \ln \rho / \mathrm{d}r)} \approx \frac{\langle \varepsilon_w \rangle}{4\rho a^2} \sim \rho^{-1/2}. \tag{27}$$

The ratio Eq. (27) therefore increases throughout much of the flow interior to r_A, and the wave force becomes the dominant source of outward acceleration within a few reference radii of the base of the wind. Asymptotically, because $\langle \varepsilon_w \rangle \sim r^{-3}$ for $M_A \gg 1$, $\langle f_w \rangle \sim r^{-4}$ and the wave acceleration $\langle f_w \rangle / \rho$ declines as r^{-2}. When the initial magnetic amplitude of the wave is increased to $\alpha = 0.10$ (see Fig. 4b), the Alfvén wave acceleration is significantly enhanced for $r \approx r_0$ (cf. Eq. [26]), with the result that u_0 is correspondingly larger than the value that obtains in the case $\alpha = 0.01$. Note that although the wave acceleration for the solution having $\alpha = 0.10$ remains greater than that of the solution with $\alpha = 0.01$ throughout the interval $r_0 \leq r \lesssim 10\, r_0$, at larger distances, the former becomes smaller than the latter. This behavior is a consequence of the fact that the mass density is greater in the outer portions of models with larger α values, thereby leading to a reduction in the magnitude of the wave acceleration, $\langle f_w \rangle / \rho$.

B. Wave-Dominated Flows

As noted in Sec. I, the interaction between propagating hydromagnetic waves and an expanding, spherical atmosphere has been utilized to account for the existence of winds from stars with physical properties that differ considerably from those of the Sun. For example, Alfvén wave-driven wind models have been constructed for the massive, low-velocity outflows from late-type giants and supergiants (see, e.g., Hartmann and MacGregor 1980,1982; Hartmann and Avrett 1984; Holzer et al. 1983; Jatenco-Pereira and Opher 1989a), as well as for mass loss from low-mass, pre-main-sequence, T Tauri stars (DeCampli 1981; Hartmann et al. 1982; Jatenco-Pereira and Opher 1989b) These particular applications have in common the fact that the atmospheric temperatures inferred from observations of such objects are too low to make acceleration by the thermal pressure gradient a viable mass loss mechanism. We use the term "wave-dominated" to describe outflows driven primarily by the wave force, with little (if any) of the atmospheric expansion deriving from thermal sources. The properties of stellar winds of this type have been systematically studied by Belcher and Olbert (1975). In addition, the conditions of temperature and density believed to exist in the atmospheres of some pre- and post-main-sequence stars imply that linear Alfvén waves, if present, could sustain appreciable damping as a result of the friction between neutral and ionized components of the gas (see, e.g., Holzer et al. 1983, and references therein).

In view of these considerations, we now amend the previously developed decription of WKB Alfvén waves to provide for the possibility of wave damping. To do this, we follow Jacques and characterize the dissipation of Alfvén waves in an expanding stellar atmosphere by a damping length L. As noted by Jacques (1977), for waves propagating through a moving, dissipative medium, the action density S (cf. Eq. 16) is no longer conserved. Instead, with the inclusion of wave damping the conservation law (Eq. 17) for the action density is modified to become

$$\nabla \cdot [(\mathbf{u} + \mathbf{u}_A)S] = -(u + u_A)S/L. \tag{28}$$

Because the primary focus of the present chapter is the elucidation of dynamical effects associated with the force due to Alfvén waves, we adopt the simplest treatment of wave damping by assuming (for the purpose of illustration) that the length scale L is constant throughout the wind. Moreover, we neglect any modifications to the thermal structure of the flow which might arise as a result of dissipative heating, and retain the assumption of isothermality. The interested reader is referred to the paper of Holzer et al. (1983) for discussion of wind models with specific processes for wave damping, including effects on the thermal and ionization state of the flow.

These suppositions, together with the assumed flow geometry, make possible the direct integration of Eq. (28), from which it follows that

$$\langle \varepsilon_w \rangle = \langle \varepsilon_{w0} \rangle \left(\frac{M_{A0}}{M_A} \right) \left(\frac{1 + M_{A0}}{1 + M_A} \right)^2 \exp \left[-\frac{(r - r_0)}{L} \right] \tag{29}$$

where $\langle \varepsilon_{w0} \rangle = \langle \delta B_0^2 \rangle / 4\pi$. Calculation of the force exerted on the flow by damped, WKB Alfvén waves then proceeds as in the derivation of expression (22), yielding

$$\langle f_w \rangle = \frac{\langle \varepsilon_w \rangle}{4} \left(\frac{1 + 3M_A}{1 + M_A} \right) \left(\frac{2}{r} + \frac{1}{u} \frac{du}{dr} \right) + \frac{\langle \varepsilon_w \rangle}{2L} \tag{30}$$

with $\langle \varepsilon_w \rangle$ given by Eq. (29). Substitution of the force (Eq. 30) into the time-averaged radial momentum Eq. (3) leads to an equation of motion for the wind which, apart from the addition of the L-dependent term ($\langle \epsilon_w \rangle r / 2\rho L$) inside the brackets on the right-hand side of the equality, is otherwise identical in form to Eq. (23). Determination of a solution $u(r)$ throughout the computational domain requires that we specify the values of M_*, r_0, μ, N_0, T, B_0, α, and L. As in the case of Eq. (23), the single flow solution that begins subsonically at r_0, and accelerates continuously to super-Alfvénic speeds with vanishing gas pressure at large distances is the solution that passes smoothly through a critical point in the wind equation of motion. In the present problem, an additional complication arises from the fact that when wave dissipation is included, two distinct critical solutions exist for sufficiently small values of the damping length L (see, e.g., Hartmann and MacGregor 1980; Holzer et al. 1983). In this instance, the solution containing the innermost of the two

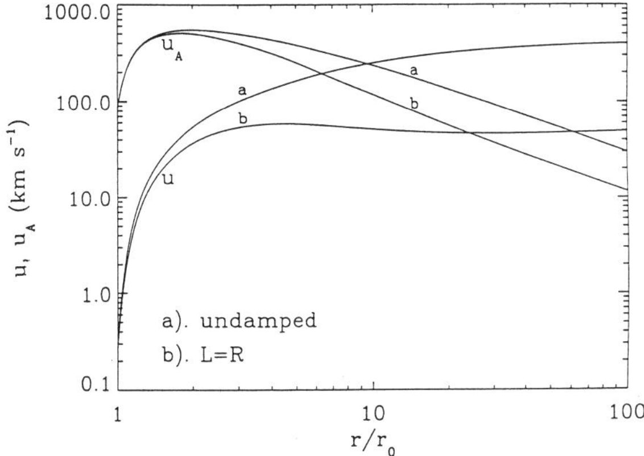

Figure 5. Profiles of the flow and Alfvén speeds for the two WKB wave-dominated wind models discussed in Sec. IV.B. For these solutions, significant thermally driven mass loss does not exist in the absence of waves. The curves labeled *a* refer to a model which does not include wave dissipation ($L = \infty$), while the curves labeled *b* refer to a model in which wave damping occurs with the constant length scale $L = R_*$ throughout the flow.

critical points is the one appropriate for describing the wave-dominated flows currently under consideration.

In Figs. 5 through 8, we show some of the properties of wave-dominated wind solutions, computed both with and without damping. The results depicted were obtained by assigning values characteristic of a cool, evolved, supergiant star to the requisite input quantities: $M_* = 16\ M_\odot$, $r_0 = R_* = 400\ R_\odot$, $\mu = 0.667\ m_H$, $N_0 = 10^{11}\ \mathrm{cm}^{-3}$, and $T = 10^4$ K. We furthermore adopted $B_0 = 10$ Gauss and $\alpha = 0.10$, so that the magnetic and velocity perturbations at r_0 have the values $\delta B_0 = 3.16$ Gauss and $\delta u_0 = 26.7$ km s^{-1} for these solutions. In Fig. 5, we show profiles of the flow and Alfvén speeds for wind models containing undamped ($L = \infty$) and damped ($L = R_*$) waves. In the case of the former solution, the outflow begins with speed $u_0 = 0.32$ km s^{-1} at the reference radius, becomes super-Alfvénic at the location $r_A = 9.56\ r_0$, and attains a speed ≈ 400 km s^{-1} at $r = 100\ r_0$. When wave dissipation with $L = R_*$ is included, the properties of the flow for $r \approx r_0$ are not significantly modified; for example, the initial expansion speed of the wind driven by damped Alfvén waves is only slightly reduced, having the value $u_0 = 0.26$ km s^{-1}. However, the asymptotic behavior of the flow is markedly different from that exhibited by the solution having $L = \infty$. In particular, the wind speed at the distance $r = 100\ r_0$ is 49 km s^{-1}, a diminution of almost an order of magnitude. Moreover, inspection of the velocity profile corresponding to flow with wave damping reveals the presence of a region at intermediate distances ($4.5\ r_0 \lesssim r \lesssim 31\ r_0$) within which the wind is decelerated (i.e., $du/dr < 0$) before being re-accelerated ($du/dr > 0$) farther from the star.

In Figs. 6 and 7, we depict the wave amplitudes $\langle \delta u^2 \rangle^{1/2}$, $\langle \delta B^2 \rangle^{1/2}$, the

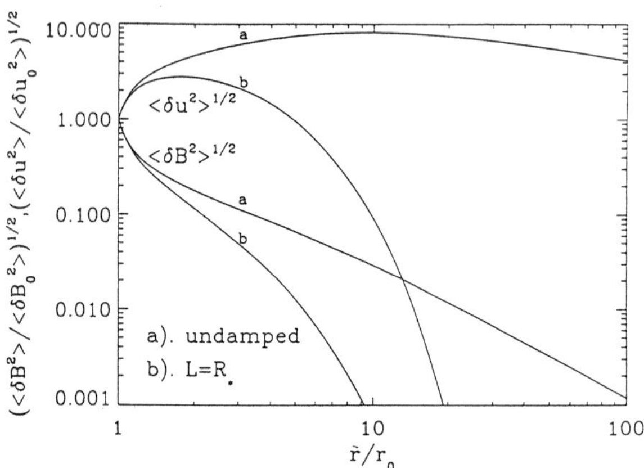

Figure 6. Profiles of the wave magnetic and velocity amplitudes for the two wave-
dominated wind models of Fig. 5. As in Fig. 2, the perturbation quantities are
normalized to their respective values at the base of the flow.

wave energy density $\langle \varepsilon_w \rangle$, and the wave energy flux density $\langle \Phi_w \rangle$ as functions
of r/r_0 for the two wind models of Fig. 5. For the model computed without
wave damping, it is readily verified that the dependences of each of these
quantities on flow properties in the sub- and super-Alfvénic portions of the
wind are the same as those derived in Sec. IV.A. For the model which includes
the effect of damping, the dependences of wave-related quantities are likewise
unchanged in that part of the flow having $M_A \ll 1$ and $(r - r_0) \lesssim L$. However,
at distances $\gtrsim L$ from the reference radius, both the magnetic and velocity
amplitudes suffer significant attenuation (cf. Fig. 6), a consequence of the
adopted prescription for wave dissipation. Corresponding damping-induced
reductions in $\langle \varepsilon_w \rangle$ and $\langle \Phi_w \rangle$ are also evident in Figs. 7a and 7b. Because the
wave force depends directly on $\langle \varepsilon_w \rangle$ (cf. Eq. 30), the rapid depletion apparent
in Fig. 7a must in turn lead to a decrease in $\langle f_w \rangle$. As was seen in Fig. 5, such
a substantial lessening of the outward acceleration causes profound changes
in the wind velocity profile since, for the solutions depicted, the outflow is
driven almost exclusively by the wave force.

The dynamical structure of a wave-dominated wind is further considered
in Fig. 8, wherein we show the balance of forces throughout each of the
solutions discussed above. As in Fig. 4, the panels depict profiles of the individual accelerations produced by the thermal pressure gradient, gravitational,
and wave forces, together with the profile of the sum of the thermal and wave
accelerations. For both of the solutions shown, the acceleration $\langle f_w \rangle / \rho$ at the
base of the flow is well-approximated by Eq. (26), and is smaller than the acceleration due to the gradient in thermal pressure. The relative importance of the
wave force increases with increasing distance from r_0 (cf. Eq. 27), however, so
that acceleration by Alfvén waves quickly becomes the principal mechanism
for maintaining the steady, radial expansion of the atmosphere. In the case

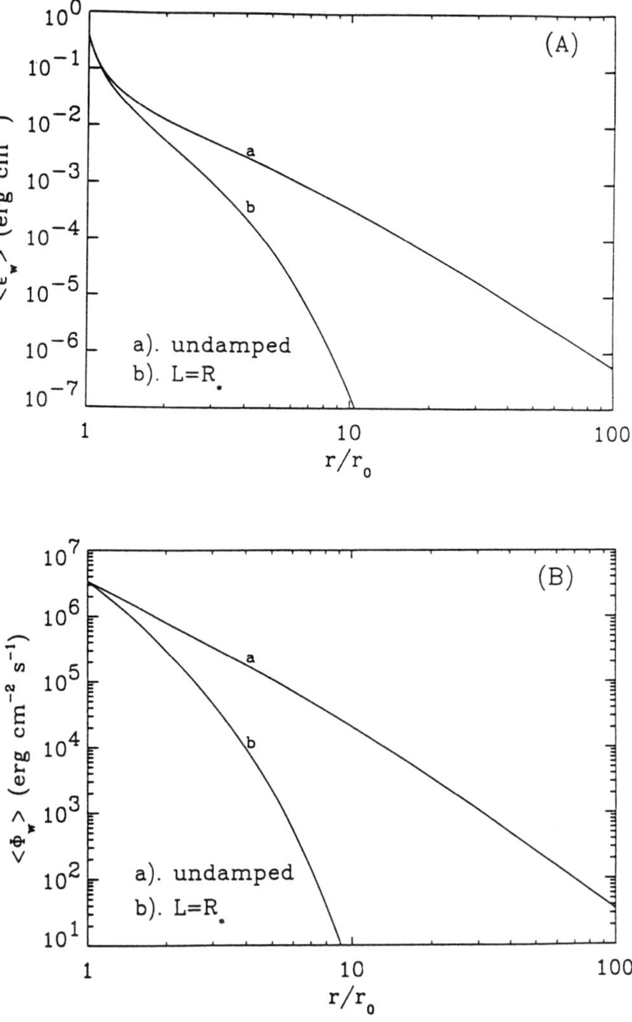

Figure 7.   The time-averaged wave energy density (panel A) and energy flux density (panel B) as functions of r/r_0 for the wave-dominated wind models of Fig. 5.

of flow with undamped Alfvén waves, $\langle f_w \rangle / \rho$ declines like r^{-2} at distances such that $M_A \gg 1$, but remains significantly larger than either the thermal or gravitational acceleration to the outer boundary of the computational domain. The addition of momentum to the supersonic region of the flow through the agency of the wave force shown in Fig. 8a contributes to the high streaming velocities ($\gtrsim 400$ km s^{-1}) that the wind achieves at large distances from the star (see, e.g., Leer and Holzer 1980). In the case of flow with damped Alfvén waves (cf. Fig. 8b), the wave acceleration becomes just slightly greater than gravity before decreasing as a result of the exponential decay of $\langle \varepsilon_w \rangle$. As

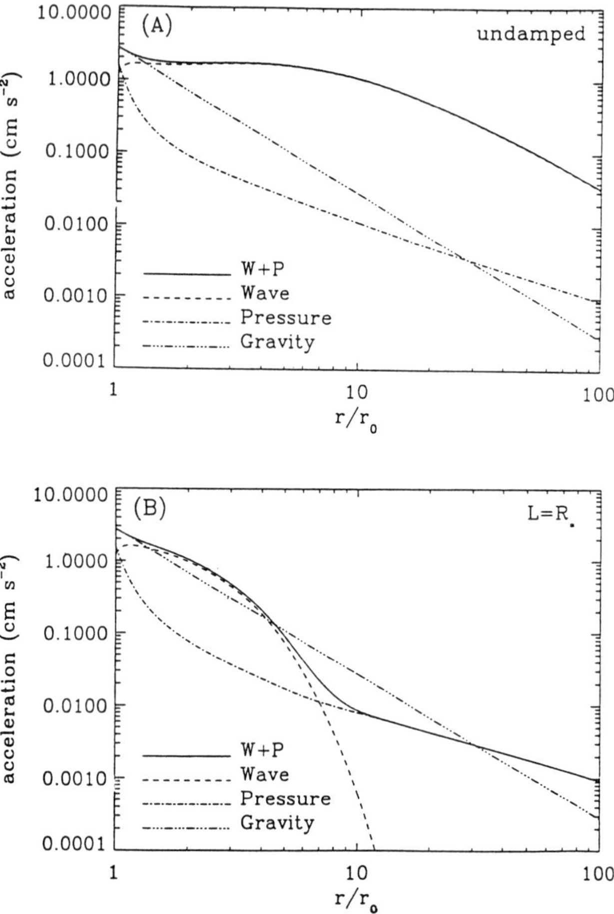

Figure 8. The balance of forces throughout the wave-dominated wind models of Fig. 5. The curves representing the accelerations produced by individual forces acting on the flows are coded identically to those in Fig. 4.

can be seen in the figure, at the location $r \approx 4.5\ r_0$ noted above, the outward acceleration produced by waves and thermal pressure collectively becomes less than the inward gravitational acceleration. This location marks the inner boundary of the deceleration region seen in the velocity profile of Fig. 5. For distances $r \gtrsim 10\ r_0$, virtually all of the outward acceleration of the wind is provided by the thermal pressure force which, although initially smaller than gravity, decreases outward more slowly (i.e., approximately as r^{-1}). At the location $r \approx 31\ r_0$, the thermal ($\approx 2a^2/r$) and gravitational (GM_*/r^2) accelerations become equal. Because the velocity gradient is quite small in the outer portions of the wind, this position is nearly coincident with the location of the sonic critical point appropriate to thermally driven flow in the absence of waves. Beyond this radius, $(2a^2/r) > (GM_*/r^2)$, and the gas is re-accelerated by the thermal pressure gradient force.

We conclude this section by noting that an approximate, analytic solution to the wind equation of motion can be derived for the limit in which the force due to undamped waves is much greater than that due to the thermal pressure gradient (i.e., wave-dominated flows). By a straightforward series of manipulations (see, e.g., Leer et al. 1982; MacGregor 1983), expressions for the critical point location r_c, the initial wind velocity u_0, and the asymptotic flow speed u_∞ can be obtained; these are

$$r_c/r_0 \approx \frac{7}{4(1 + \beta/2)} \tag{31a}$$

$$u_0/u_{\mathrm{esc}} \approx \frac{1}{8}\beta^2 (r_c/r_0)^{7/2} \tag{31b}$$

and

$$u_\infty/u_{\mathrm{esc}} \approx (\beta/M_{A0} - 1)^{1/2} \tag{31c}$$

where $u_{\mathrm{esc}} = (2GM_*/r_0)^{1/2}$ is the reference level escape speed and $\beta \equiv \langle \varepsilon_{w0}\rangle/(\rho_0 u_{\mathrm{esc}}^2/2)$. Equation (31c) is valid under the conditions that both β and u_0/u_{esc} are $\ll 1$.

V. WIND SOLUTIONS WITH NON-WKB ALFVÉN WAVES

Use of the WKB approximation to determine the amplitudes of propagating Alfvén waves in the outflow is justified only if the quantity λ/ℓ (cf. Sec. III) remains $\ll 1$ throughout $r_0 \leq r \leq 100\, r_0$. Given the physical properties of the wind solutions described in Sec. IV above, we can now ascertain whether or not this condition is fulfilled for waves of various frequencies. For this purpose, we chose the solution with $\alpha = 0.01$ from Sec. IV.A, and compare the length scale $\ell = |\mathrm{d}\ln u_A/\mathrm{d}r|^{-1}$ with the wavelength $\lambda = 2\pi(u + u_A)/\omega$ of an Alfvénic disturbance having frequency ω. For this particular model, we find that all waves with frequencies in the range $10^{-5} \leq \omega \leq 10^{-2}$ s^{-1} (i.e., waves with periods between about 10 minutes and 10 days) violate the criterion for applicability of the WKB approximation at the base of the flow where the gradient in u_A is substantial (cf. Fig. 1). Moreover, for waves with frequencies $\omega \lesssim 10^{-4}$ s^{-1}, $\lambda/\ell > 1$ everywhere within the computational domain. These results suggest that the theoretical framework developed in Sec. III should be modified to accommodate physical effects associated with finite wave frequencies and wavelengths. While many investigators have treated the propagation of non-WKB Alfvén waves in stellar atmospheres and winds with specified properties (see, e.g., Heinemann and Olbert 1980; Leer et al. 1982; An et al. 1989,1990; Rosner et al. 1991; Barkhudarov 1991; Velli 1993; Lou and Rosner 1993; Krogulec et al. 1994), few dynamically consistent models have been constructed to date.

A. Non-WKB Wave Amplitudes and Force

We adopt the model presented at the outset of Sec. II almost without change, the single revision being to restrict the domain of interest to the equatorial plane ($\theta = \pi/2$) of the star. Because finite frequency waves at other latitudes can produce meridional acceleration, this restriction is necessary in order that the wind velocity have only a radial component. To obtain an expression for the force due to non-WKB Alfvén waves, we return to the starting point for determining the wave amplitudes as functions of radius and time, namely, the azimuthal momentum and induction Eqs. (7) and (8). After some manipulation, these equations can be cast in the form (see, e.g., Heinemann and Olbert 1980; Leer et al. 1982; MacGregor and Charbonneau 1994)

$$\left[\frac{\partial}{\partial t} + (u - u_A)\frac{\partial}{\partial r} \right] f = \frac{1}{2}(u - u_A)g \frac{d \ln u_A}{dr} \tag{32a}$$

and

$$\left[\frac{\partial}{\partial t} + (u + u_A)\frac{\partial}{\partial r} \right] g = \frac{1}{2}(u + u_A)f \frac{d \ln u_A}{dr} \tag{32b}$$

where the functions $f(r, t)$ and $g(r, t)$ are defined in terms of the perturbations δu and δB according to

$$\delta u + \frac{\delta B}{\sqrt{4\pi\rho}} = \frac{\sqrt{M_A}}{M_A - 1} f \tag{33a}$$

and

$$\delta u - \frac{\delta B}{\sqrt{4\pi\rho}} = \frac{\sqrt{M_A}}{M_A + 1} g. \tag{33b}$$

The left-hand sides of Eqs. (32a) and (32b) describe the propagation of the wave-like quantities f and g along the inward- and outward-pointing characteristics, $dr/dt = (u \pm u_A)$. The right-hand sides of these equations express the fact that waves traveling in the outward ($+\mathbf{e}_r$) and inward ($-\mathbf{e}_r$) directions are coupled through reflection from the gradient in u_A. That is, partial reflection of the outwardly propagating g waves constitutes a source of inwardly propagating f waves, and vice versa. It is readily verified that the WKB wave amplitudes derived in Sec. III are recovered from Eqs. (32) and (33) when the coupling terms are omitted and f is set equal to zero.

To obtain the equation of motion for wind flow in the presence of non-WKB Alfvén waves, we first prescribe a harmonic time dependence for the perturbation quantities in Eqs. (33), $(f, g) = [F(r), G(r)]e^{-i\omega t}$. In terms of the complex functions F and G, the wave velocity and magnetic amplitudes are given by

$$\delta u = \frac{\sqrt{M_A}}{2}\left(\frac{F}{M_A - 1} + \frac{G}{M_A + 1} \right) e^{-i\omega t} \tag{34a}$$

and

$$\delta B = \frac{\sqrt{M_A}}{2} \left(\frac{F}{M_A - 1} - \frac{G}{M_A + 1} \right) e^{-i\omega t}. \tag{34b}$$

The time-averaged wave acceleration $\langle f_w \rangle / \rho$ can then be calculated by using Eqs. (34) to compute $\langle \delta u^2 \rangle$ and $\langle \delta B^2 \rangle$, substituting the results in Eq. (6), and performing the indicated differentiation. The differential equation for $u(r)$ analogous to Eq. (23) follows upon using the wave force so-derived in the radial momentum Eq. (3). Explicit expressions for the wave force and other amplitude-dependent quantities in terms of the functions F and G are given in the paper by MacGregor and Charbonneau (1994). As is evident from inspection of Eqs. (34), for non-WKB waves

$$|\delta u| \neq \frac{|\delta B|}{\sqrt{4\pi\rho}} \tag{35}$$

(see, e.g., Eq. 12). Hence, unlike the case of WKB waves, the centrifugal and magnetic tension components of the wave force (Eq. 6) do not cancel.

B. Modifications to Thermally Driven Flow

In practice, wind solutions are computed by first using the properties of a known WKB solution to specify the coefficients in Eqs. (32a) and (32b), and numerically integrating to obtain F and G everywhere within $r_0 \leq r \leq 100\, r_0$. The wave acceleration is then determined as a function of r, as described in the preceding paragraph. With this result, Eq. (3) is easily integrated to derive updated profiles for $u(r)$ and $u_A(r)$. The steps in this procedure are then repeated until convergence is attained. The boundary conditions applied to the solutions of the three differential equations that form the model are: (i) that $F = 0$ at $r = r_A$; (ii) that the value of G at $r = r_0$ be such that the amplitude $|\delta u_0|$ is equal to the corresponding WKB value there; and (iii), that the value of u at r_0 be such that the flow passes smoothly through the sonic critical point that is contained in the wind equation of motion. A complete description of the method, together with detailed discussion of the boundary conditions imposed on the functions F, G, and u is presented in the previously cited paper by MacGregor and Charbonneau (1994).

In Figs. 9 through 12, we show results pertaining to wind models with non-WKB Alfvén waves. The iterative solution procedure was initiated using the $\alpha = 0.01$, WKB wind model of Sec. IV.A, so that the values of M_*, r_0, μ, N_0, T, and B_0 are identical for the two sets of results. Moreover, as was true for the solutions considered in Sec. IV.A, in the present case, a thermally driven outflow exists even if no waves are present in the wind. In Fig. 9, we display the flow and Alfvén speeds as functions of r/r_0 for models containing waves with frequencies $\omega = 10^{-3}$, 10^{-4}, and 10^{-5} s^{-1}. For comparison, the velocity profile of the WKB wind model used to start the solution process is also included in the figure. The results depicted indicate that the flow speeds of winds with non-WKB Alfvén waves are generally smaller than

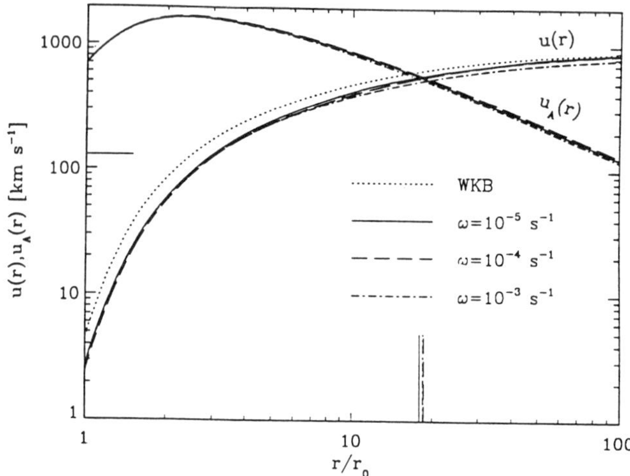

Figure 9. Profiles of the flow and Alfvén speeds for the non-WKB wind solutions discussed in Sec. V.B. The dotted curves in the figure correspond to the WKB solution that has the same value of δu_0 as the three finite frequency solutions. The vertical line segments on the abscissa denote the location of the Alfvén point for each solution, while the horizontal tic mark on the ordinate indicates the value of the isothermal sound speed.

those of the corresponding WKB model, particularly in the sub-Alfvénic region $r \lesssim 20\ r_0$. This reduction is a result of the fact that finite-frequency waves are continuously reflected as they propagate through a medium with an Alfvén speed profile of the type shown in the figure. One consequence of the interaction between the outward and inward traveling waves that are produced in this way is that the dependences of the amplitudes $\langle \delta B^2 \rangle^{1/2}$ and $\langle \delta u^2 \rangle^{1/2}$ on flow properties and position are changed from those appropriate to WKB waves. As will be seen below, because of this the time-averaged wave force for $r \lesssim r_A$ is smaller than that due to high-frequency, short-wavelength waves, and flow speeds are therefore diminished. For the non-WKB solutions shown in Fig. 9, the values of u_0 are reduced relative to that of the WKB solution by about a factor of $1/2$.

In Fig. 10, we show profiles of the normalized wave amplitudes (cf. Fig. 2) for the WKB and non-WKB wind solutions of Fig. 9. The reference level values of these quantities for the latter solutions are $\langle \delta u_0^2 \rangle^{1/2} = 69.0$ km s^{-1} as before, and $\langle \delta B_0^2 \rangle^{1/2} = 4.72 \times 10^{-2}$, 5.60×10^{-2}, and 1.17×10^{-1} Gauss for $\omega = 10^{-3}$, 10^{-4}, and 10^{-5} s^{-1}, respectively. For each of the three finite-frequency models shown, the departures of $\langle \delta B^2 \rangle^{1/2}$ and $\langle \delta u^2 \rangle^{1/2}$ from WKB-like behavior are most noticeable in the innermost portion of the flow where u_A varies most rapidly with r. In the super-Alfvénic region ($r \gtrsim 20\ r_0$), the solutions with $\omega = 10^{-3}$ and 10^{-4} s^{-1} exhibit dependences on radius that are similar to that of the WKB solution. For $\omega = 10^{-5}$ s^{-1}, the magnetic and velocity perturbations display non-WKB behavior even at large distances, an

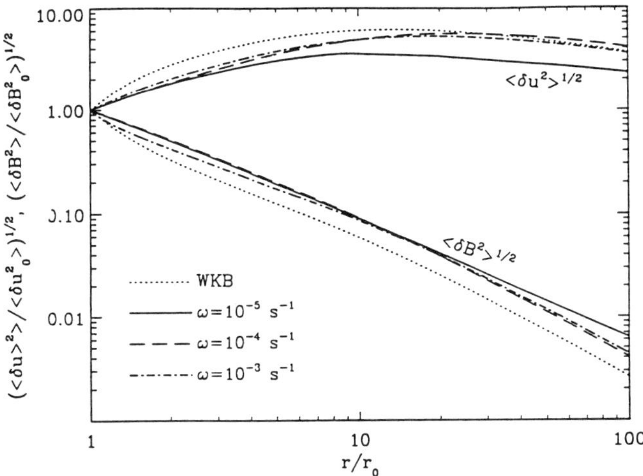

Figure 10. The time-averaged wave magnetic and velocity amplitudes as functions of r/r_0 for the three non-WKB wind models of Fig. 9. As in that figure, the dotted curves represent the amplitudes of waves in the corresponding WKB solution.

indication that such long wavelength disturbances are reflected throughout the wind. Note that in the limit $\omega \equiv 0$, Eqs. (32a) and (32b) can be solved analytically, thereby enabling us to obtain the approximate dependences of the amplitudes on r for very low frequency waves. When $M_A \ll 1$, we find that $\langle \delta B^2 \rangle^{1/2} \sim r^{-1}$ and $\langle \delta u^2 \rangle^{1/2} \sim r$, while for $M_A \gg 1$, both time-averaged amplitudes vary like r^{-1}.

Similar behavior is observed in profiles of the wave energy density $\langle \varepsilon_w \rangle$ (Fig. 11a) and energy flux density $\langle \Phi_w \rangle$ (Fig. 11b). For the range of wave frequencies considered, deviations from the WKB results for both of these quantities are most pronounced near the base of the flow where the length scale ℓ (defined at the beginning of this section) is shortest. For the lowest frequency waves, reflection also takes place in the super-Alfvénic portion of the wind, causing $\langle \varepsilon_w \rangle$ and $\langle \Phi_w \rangle$ to decline outward more slowly at large distances ($\sim r^{-2}$) than in the case of very high frequency disturbances ($\sim r^{-3}$; cf. Fig. 3 and the discussion thereof).

As noted in the preceding paragraphs, for wind models with non-WKB Alfvén waves, the dependences of wave-related quantities on flow properties can differ considerably from those characteristic of wind models containing WKB waves. The dynamical consequences of these modifications are illustrated in Fig. 12, which shows profiles of the accelerations produced by the forces operative throughout the three finite-frequency solutions of Fig. 9. Also shown in each panel are profiles of (i) the WKB wave acceleration for $\alpha = 0.01$, and (ii), the sum of this acceleration and that due to the thermal pressure gradient force. For all of the non-WKB solutions depicted, the time-averaged wave acceleration $\langle f_w \rangle / \rho$ in the region of the flow interior to r_A is smaller

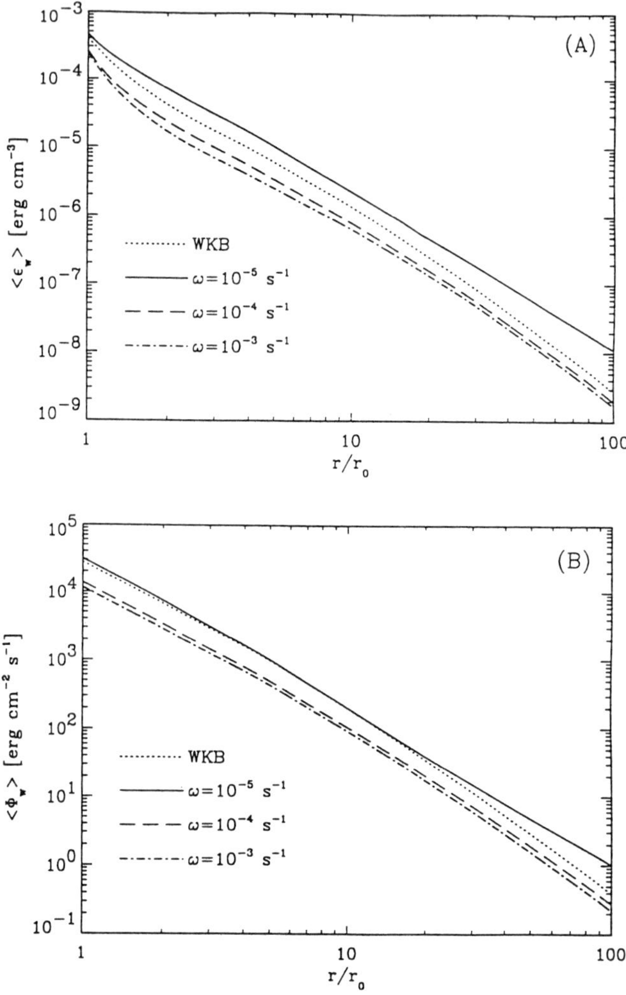

Figure 11. Profiles of the time-averaged wave energy density (panel A) and energy flux density (panel B) for the three non-WKB wind solutions shown in Figs. 9 and 10.

than the the wave acceleration in the corresponding portion of the WKB solution. For high-frequency solutions ($\omega = 10^{-3}$ s^{-1}), non-WKB effects are limited to the sub-Alfvénic flow, wherein the centrifugal ($\langle \delta u^2 \rangle / r$), magnetic tension ($-\langle \delta B^2 \rangle / 4\pi\rho r$), and wave pressure ($\langle \delta B^2 \rangle / 8\pi$) gradient forces all make distinct contributions to the total wave acceleration. At distances $\gtrsim 3\, r_0$ in this solution, the first two of these components become equal in magnitude, so that virtually all of the wave force arises from the gradient in wave pressure, as is the case for WKB waves (cf. Eq. 22). For solutions with very low frequency waves ($\omega = 10^{-5}$ s^{-1}), non-WKB effects are apparent in both the

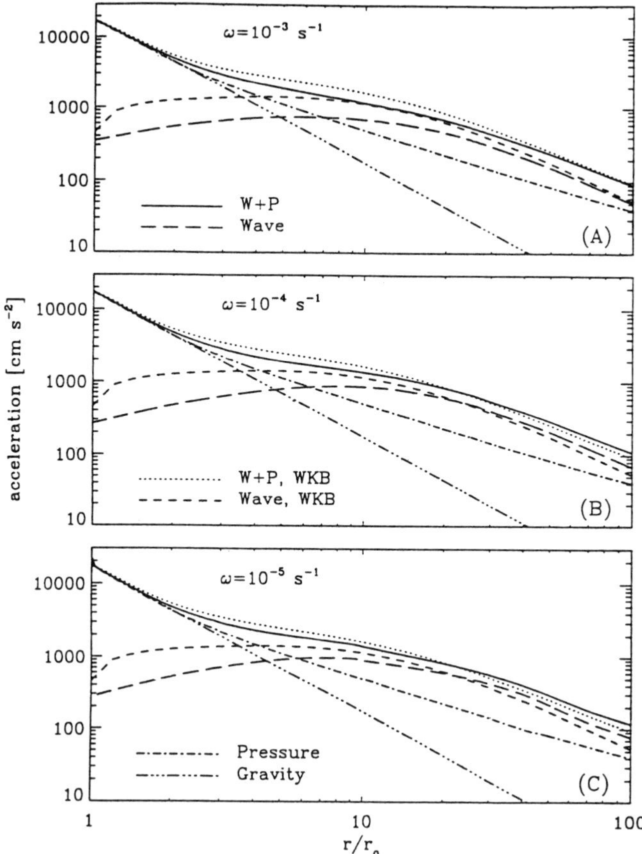

Figure 12. Profiles of the accelerations produced by the thermal pressure gradient, gravitational, and Alfvén wave forces for the WKB and non-WKB wind models of Fig. 9.

sub- and super-Alfvénic portions of the wind. Because $\langle \delta B^2 \rangle \sim r^{-2}$ where $M_A \ll 1$, the magnetic tension and wave pressure contributions to $\langle f_w \rangle$ cancel in this region, and most of the wave acceleration is produced centrifugally. At larger distances where $M_A \gg 1$, the magnetic tension and wave pressure components of the wave acceleration no longer have equal magnitudes, and are each individually larger than the centrifugal contribution. Because, as was noted in connection with the discussion of Fig. 10, the amplitudes of low-frequency waves decrease more slowly with r than do those of very high-frequency waves, the acceleration associated with these non-WKB perturbations becomes larger asymptotically than that obtained in the WKB limit. Such behavior can be seen for $r \gtrsim r_A$ in the profiles for the $\omega = 10^{-4}$ and 10^{-5} s^{-1} solutions.

C. Wave-Dominated Flows

An obvious extension to the work described in Sec. V.B is the application of the model to stars having atmospheric thermodynamic conditions for which significant thermally driven mass loss is not possible. As an example, in the present section, we briefly consider the nature of the interaction between un-damped, non-WKB Alfvén waves and the expanding circumstellar envelope of a cool, evolved star. For this purpose, we adopt the values of M_*, r_0, μ, N_0, B_0, and α used to construct the $L = \infty$ wind model discussed in Sec. IV.B, but (for reasons which will subsequently become clear) increase the assumed temperature of the outflow to $T = 3 \times 10^4$ K. With this higher value of T, the flow driven by WKB Alfvén waves has $u_0 = 0.67$ km s^{-1}, $r_A/r_0 \approx 8$, and $u(100 \, r_0) = 282$ km s^{-1}.

A sample of the results obtained from calculations performed assuming $\omega = 10^{-7}$ s^{-1} and using the WKB solution given above to start the iteration procedure is displayed in Fig. 13. In panel (a), we show profiles of u, u_A, and ρ/ρ_0 for both the WKB and non-WKB solutions, while in panel (b), we depict the corresponding acceleration profiles, as in Fig. 12. As was the case for the solutions considered in Sec. V.B, the reflection of non-WKB Alfvén waves at the base of the wind leads to a reduction (relative to WKB values) in the magnitudes of both the time-averaged wave acceleration and flow velocity for $r \lesssim 5 \, r_0$. For $\omega = 10^{-7}$ s^{-1}, the flow speed at r_0 is $u_0 = 0.028$ km s^{-1}, a value nearly 24 times smaller than the initial speed of the WKB solution shown in the figure. Note also that the asymptotic speed of the wind containing non-WKB waves is considerably higher than that of the WKB solution; for the former flow, $u(100 \, r_0) = 577$ km s^{-1}, a better than two-fold increase from the WKB result. Figure 13b indicates that the origin of this enhancement is the large value of the acceleration produced by Alfvén waves at distances $r \gtrsim 10$ r_0. As noted in Sec. V.B, a portion of the increase in $\langle f_w \rangle/\rho$ at large distances stems from the modified dependences of non-WKB wave amplitudes on the asymptotic properties of the flow. In addition, in the low frequency solution presently under consideration, the wave acceleration at large radii is further augmented by the lower densities that prevail in the outer portions of the wind (see, e.g., Fig. 13a).

From the illustrative computational results presented herein, it does not appear that models with undamped, non-WKB Alfvén waves are well suited to the problem of accounting for the observed properties of winds from late-type giant and supergiant stars. The massive, low-temperature outflows from these stars have measured asymptotic flow speeds that are generally smaller than the gravitational escape speed from the stellar photosphere (see, e.g., Dupree 1986). As the solutions discussed in the present and preceding sections indi-cate, for decreasing Alfvén wave frequencies, the changing character of the wave acceleration leads to winds with smaller mass fluxes and larger stream-ing velocities at great distances from the star. Moreover, for gas temperatures just slightly lower than the value $T = 3 \times 10^4$ K appropriate to the models of

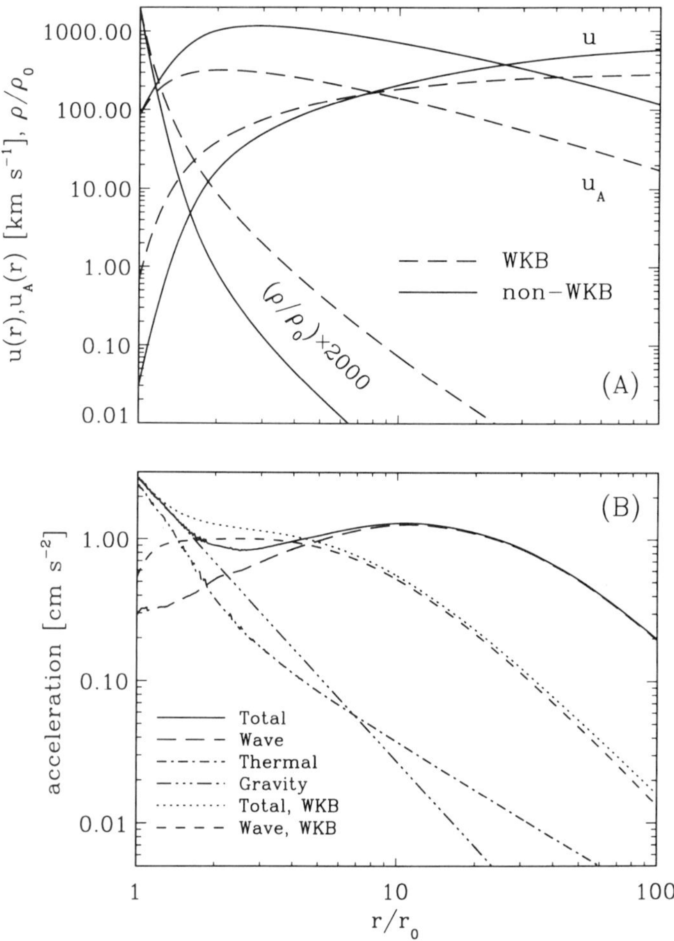

Figure 13. Velocity and force balance profiles for a wave-dominated flow containing undamped, non-WKB, Alfvénic fluctuations (see Sec. V.C). In panel A, the quantities u, u_A, and ρ/ρ_0 are shown as functions of r/r_0. The density distribution has been renormalized to have its starting value equal to the upper limit of the scale for u and u_A. The dashed curves in the panel depict the results obtained using the WKB approximation, as described in the text. In panel B, the accelerations produced by the forces acting on the wind are shown as functions of r/r_0. The coding of the curves in the panel is identical to that used in Fig. 12.

Fig. 13, the decline in flow speed near the base of the wind is so precipitous that the solution scheme outlined in Sec. V.B fails to converge. Additional details concerning the application of non-WKB Alfvén wave-driven wind models to mass loss from cool, evolved stars are available in the paper by Charbonneau and MacGregor (1995).

VI. CONCLUSION

Our purpose throughout the foregoing analysis has not been to exhaustively review all extant research on the acceleration of solar/stellar wind-type outflows by hydromagnetic waves. Instead, we have used simplified models to examine the interaction between small amplitude, Alfvénic disturbances and a spherical, expanding stellar atmosphere. In particular, we have investigated both the physical origin of the force produced by MHD waves propagating in such an environment, and the consequences of the attendant acceleration for the dynamical state of the atmosphere. For high-frequency, short-wavelength (i.e., WKB) waves, only outward traveling perturbations are present in the flow, and the force they exert on the wind material derives from the gradient of a time-averaged wave pressure, $\langle p_w \rangle = \langle \delta B^2 \rangle / 8\pi$ (see, e.g., Sec. III). For low-frequency, long-wavelength (i.e., non-WKB) waves, reflection can occur, so that outward and inward traveling waves are simultaneously present in the flow. Under these conditions, the time-averaged wave acceleration is comprised of contributions from both the fluctuating centrifugal and Lorentz forces (see, e.g., Sec. V.A). In either case, when atmospheric temperatures are high enough to ensure the existence of a thermally driven flow, the inclusion of waves can lead to significant changes in many computed wind properties. Likewise, when atmospheric temperatures are low enough that little, if any, thermally driven mass loss can take place, the acceleration provided by propagating Alfvén waves can, by itself, initiate and maintain a steady outflow.

As noted in the preceding paragraph, the illustrative calculations described in the present chapter were performed using simplified models for the system consisting of waves and background flow. These simplifications are not inherent to the models, but are instead introduced through a variety of assumptions, many of which merit reconsideration in later work. Among the improvements and/or extensions which would eliminate several of the limitations of the models developed herein are: (i) the inclusion of a spectrum of waves rather than a single, monochromatic disturbance; (ii) the incorporation of effects associated with rotation of the star; (iii) a more realistic treatment of processes affecting both the thermal energy balance in the flow and wave interactions and damping; (iv) development of the capability for treating two dimensional, axisymmetric flow in the presence of more complicated (and realistic) poloidal field geometries; and (v), removal of the (dubious) presumption that very low-frequency, non-WKB waves can be time-averaged in favor of a fully time-dependent treatment of the waves and wind. It is to be hoped that a number of these enhancements can be implemented in future models for Alfvén wave-driven winds.

Acknowledgments. The authors acknowledge numerous discussions on wave-driven winds with E. Leer, and are grateful to P. Judge for carefully reading and commenting on the manuscript.

REFERENCES

Alazraki, G., and Couturier, P. 1971. Solar wind acceleration caused by the gradient of Alfvén wave pressure. *Astron. Astrophys.* 13:380–389.

An, C. H., Musielak, Z. E., Moore, R. L., and Suess, S. T. 1989. Reflection and trapping of transient Alfvén waves propagating in an isothermal atmosphere with constant gravity and uniform magnetic field. *Astrophys. J.* 345:597–605.

An, C. H., Suess, S. T., Moore, R. L., and Musielak, Z. E. 1990. Reflection and trapping of Alfvén waves in a spherically symmetric stellar atmosphere. *Astrophys. J.* 350:309–323.

Barkhudarov, M. R. 1991. Alfvén waves in stellar winds. *Solar Phys.* 135:131–161.

Barnes, A. 1992. Theory of magnetohydrodynamic waves: The WKB approximation revisited. *J. Geophys. Res.* 97:12105–12112.

Belcher, J. W. 1971. Alfvénic wave pressures and the solar wind. *Astrophys. J.* 168:509–524.

Belcher, J. W., and Davis, L., Jr. 1971. Large-amplitude Alfvén waves in the interplanetary medium, 2. *J. Geophys. Res.* 76:3534–3563.

Belcher, J. W., and Olbert, S. 1975. Stellar winds driven by Alfvén waves. *Astrophys. J.* 200:369–382.

Bretherton, F. P. 1970. The general linearized theory of wave propagation. In *Mathematical Problems in the Geophysical Sciences*, ed. W. H. Reid (Providence, R. I.: American Mathematical Soc.), pp. 61–102.

Charbonneau, P., and MacGregor, K. B. 1995. Stellar winds with non-WKB Alfvén waves. II. Wind models for cool, evolved stars. *Astrophys. J.*, in press.

Coleman, P. J. 1968. Turbulence, viscosity, and dissipation in the solar wind plasma. *Astrophys. J.* 153:371–388.

DeCampli, W. M. 1981. T Tauri winds. *Astrophys. J.* 244:124–146.

Denskat, K. U., and Neubauer, F. M. 1983. Observations of hydromagnetic turbulence in the solar wind. In *Solar Wind Five*, ed. M. Neugebauer, NASA CP-2280, pp. 81–91.

Dewar, R. L. 1970. Interaction between hydromagnetic waves and a time-dependent, inhomogeneous medium. *Phys. Fluids* 13:2710–2720.

Dupree, A. K. 1986. Mass loss from cool stars. *Ann. Rev. Astron. Astrophys.* 24:377–420.

Hartmann, L., and Avrett, E. H. 1984. On the extended chromosphere of α Orionis. *Astrophys. J.* 284:238–249.

Hartmann, L., and MacGregor, K. B. 1980. Momentum and energy deposition in late-type stellar atmospheres and winds. *Astrophys. J.* 242:260–282.

Hartmann, L., and MacGregor, K. B. 1982. Wave-driven winds from cool stars. I. Some effects of magnetic field geometry. *Astrophys. J.* 257:264–268.

Hartmann, L., Edwards, S., and Avrett, E. 1982. Wave-driven winds from cool stars. II. Models for T Tauri stars. *Astrophys. J.* 261:279–292.

Heinemann, M., and Olbert, S. 1980. Non-WKB Alfvén waves in the solar wind. *J. Geophys. Res.* 85:1311–1327.

Hollweg, J. V. 1973a. Alfvén waves in a two-fluid model of the solar wind. *Astrophys. J.* 181:547–566.

Hollweg, J. V. 1973b. Alfvén waves in the solar wind: Wave pressure, Poynting flux, and angular momentum. *J. Geophys. Res.* 78:3643–3652.

Hollweg, J. V. 1978. Some physical processes in the solar wind. *Rev. Geophys. Space Phys.* 16:689–720.

Hollweg, J. V. 1986. Transition region, corona, and solar wind in coronal holes. *J. Geophys. Res.* 91:4111–4125.

Hollweg, J. V. 1990. On WKB expansions for Alfvén waves in the solar wind. *J. Geophys. Res.* 95:14873–14879.

Holzer, T. E., Flå, T., and Leer, E. 1983. Alfvén waves in stellar winds. *Astrophys. J.* 275:808–835.

Jacques, S. A. 1977. Momentum and energy transport by waves in the solar atmosphere and solar wind. *Astrophys. J.* 215:942–951.

Jacques, S. A. 1978. Solar wind models with Alfvén waves. *Astrophys. J.* 226:632–649.

Jatenco-Pereira, V., and Opher, R. 1989*a*. Effect of diverging magnetic fields on mass loss in late-type giant stars. *Astron. Astrophys.* 209:327–336.

Jatenco-Pereira, V., and Opher, R. 1989*b*. Alfvén-driven protostellar winds. *Mon. Not. Roy. Astron. Soc.* 236:1–20.

Jatenco-Pereira, V., and Opher, R. 1989*c*. Observational limits on the coronal hole flow geometry in an Alfvén wave-driven solar wind. *Astrophys. J.* 344:513–521.

Krogulec, M., Musielak, Z. E., Suess, S. T., Nerney, S. F., and Moore, R. L. 1994. Reflection of Alfvén waves in the solar wind. *J. Geophys. Res.* 99:23489–23501.

Leer, E., and Holzer, T. E. 1980. Energy addition in the solar wind. *J. Geophys. Res.* 85:4681–4688.

Leer, E., Holzer, T. E., and Flå, T. 1982. Acceleration of the solar wind. *Space Sci. Rev.* 33:161–200.

Lou, Y. Q., and Rosner, R. 1993. Reflection of Alfvén waves in stellar atmospheres: The case of open magnetic fields. *Astrophys. J.* 424:429–435.

MacGregor, K. B. 1983. Theory of winds in late-type evolved and pre-main sequence stars. In *Solar Wind Five*, ed. M. Neugebauer, NASA CP-2280, pp. 241–261.

MacGregor, K. B., and Charbonneau, P. 1994. Stellar winds with non-WKB Alfvén waves. I. Wind models for solar coronal conditions. *Astrophys. J.* 430:387–398.

Marsch, E. 1991. MHD turbulence in the solar wind. In *Physics of the Inner Heliosphere*, vol. II, eds. R. Schwenn and E. Marsch (Heidelberg: Springer-Verlag), pp. 159–241.

Marsch, E., and Tu, C. Y. 1990. On the radial evolution of MHD turbulence in the inner heliosphere. *J. Geophys. Res.* 95:8211–8229.

Parker, E. N. 1958. Dynamics of the interplanetary gas and magnetic fields. *Astrophys. J.* 128:664–675.

Parker, E. N. 1963. *Interplanetary Dynamical Processes* (New York: Interscience).

Parker, E. N. 1965. Dynamical theory of the solar wind. *Space Sci. Rev.* 4:666–708.

Roberts, D. A. 1989. Interplanetary observational constraints on Alfvén wave acceleration of the solar wind. *J. Geophys. Res.* 94:6899–6905.

Roberts, D. A., and Goldstein, M. L. 1991. Turbulence and waves in the solar wind. *Rev. Geophys. Suppl.* 29:932–943.

Rosner, R., An, C. H., Musielak, Z. E., Moore, R. L., and Suess, S. T. 1991. Magnetic confinement, Alfvén wave reflection, and the origins of X-ray and mass loss "dividing lines" for late-type giants and supergiants. *Astrophys. J. Lett.* 372:91–94.

Tu, C. Y. 1987. A solar wind model with the power spectrum of Alfvénic fluctuations. *Solar Phys.* 109:149–186.

Velli, M. 1993. On the propagation of ideal, linear Alfvén waves in radially stratified stellar atmospheres and winds. *Astron. Astrophys.* 270:304–314.

PART IV
Physical Phenomena in Winds

OVERVIEW OF HELIOSPHERIC SHOCKS

A. BALOGH
Imperial College

and

P. RILEY
University of Arizona

In-situ plasma and magnetic field measurements of the solar wind have demonstrated the existence of collisionless interplanetary shocks for over thirty years. In this review, we summarize these observations and emphasize the most recent results, namely, shocks observed by the Ulysses spacecraft at high latitudes. Although observed interplanetary shocks are almost always fast-mode, slow-mode and possibly intermediate-mode shock waves are also present. Fast shocks, in the ecliptic plane are generated principally in two ways; these, together with a new high-latitude generation model, are discussed. The association of transient shocks with coronal mass ejections (as opposed to solar flares) is also reviewed.

I. INTRODUCTION AND HISTORICAL PERSPECTIVE

Shock waves are among the most impressive phenomena in the solar wind, and there is no doubt that the presence of interplanetary shocks plays a major role in the physics of the heliosphere. Ranging from the modulation of cosmic rays to magnetic storms, interplanetary shock waves have far reaching effects.

Some shocks arise from disturbances in the solar atmosphere, while others are a consequence of interactions in the streaming solar wind. Additionally, a large-scale shock wave is thought to exist ahead of the heliopause—the boundary separating the solar wind and interstellar medium. The classification is completed by stationary shocks, formed in the solar wind upstream of planets and other bodies.

Extensive research has covered all aspects of heliospheric shock waves. Such research is in fact a quintessential example of the cross-fertilization of theory and the interpretation of observations. From relatively modest beginnings, numerical simulation techniques have been developed, which have provided important insight into the detailed physical processes which control the collisionless shock transition process. Furthermore, much that has been learned from the most comprehensively studied shock wave, the Earth's bow shock, it has been used to understand better those shock waves in the heliosphere which, by necessity, are only observed once. Despite the generally different parameter regimes from those encountered in heliospheric shocks, the richness of the observations and theoretical interpretation of phenomena observed at the bow shock provides a significant source of understanding, particularly of the phenomenological details of the shock transition itself.

There is now no doubt that shock waves are recognizable structures in the solar wind, and that the concept originally inherited from collision-dominated media has been successfully extrapolated to collisionless plasmas in the heliosphere both qualitatively and quantitatively. There is, nevertheless, an inherent contradiction at the heart of the concept of collisionless shock waves. On the one hand, a discontinuous transition between two states of the plasma is assumed, represented by nominally well-defined parameters which characterize unambiguously the upstream and downstream states, and which are linked by continuity equations, augmented as necessary by additional equations. On the other hand, the necessary dissipation in the shock transition itself generates phenomena which can propagate information, primarily in the form of particle populations which have interacted with the shock, away from the shock front, upstream as well as downstream in general, and which, in turn, modify in a complex manner the states of the plasma and the temporal and spatial functions of its constituents. This is illustrated schematically in Fig. 1. As a result, the collisionless shock possibly never propagates into a region which has not already been modified, to a greater or lesser extent, by phenomena which originate at the shock and which therefore affect, through a set of complex feedback processes, the shock transition itself. This picture is fully borne out by numerically simulated shocks, in which the initial conditions, or "states" of the upstream and downstream plasma are fully defined at the outset; but once the shock is generated, the characteristic parameters of the medium evolve as a result of particles propagating away from the shock, generating, through a set of (usually nonlinear) wave particle interactions and a range of instabilities, significantly more complex "states" than was assumed at the start of the simulation.

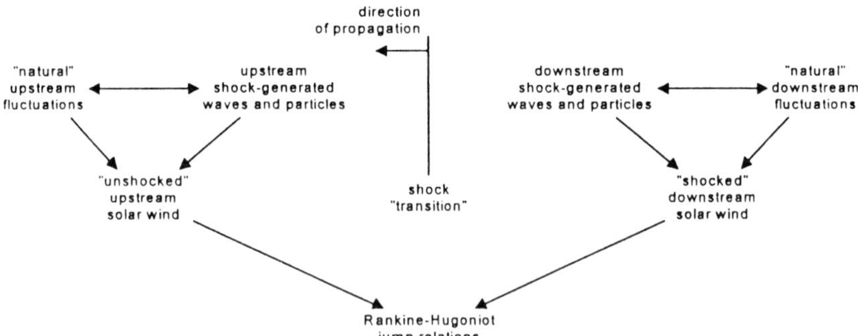

Figure 1.   A schematic indicating the relationship between various states upstream and downstream of a shock transition. See text for details.

Shock observations in space, often restricted to a single spatial point, also bear out the conclusion that collisionless shock waves cannot be studied or understood without considering carefully the wider neighborhood, upstream and downstream of the shock, and even, whenever possible, its extent transverse to the direction of propagation in the frame of the "unshocked" plasma. This fundamental characteristic of collisionless shock waves, i.e., their intimate relationship with a much wider region than would be the case either for a hydrodynamic shock or for a collision-dominated MHD shock, indicates that concepts inherited from shock waves in these more fluid-like media need to be used with great care.

Although shock waves have been studied for over a century, the collisionless shock is a relatively new feature. It was not until 1950 that de Hoffman and Teller (1950) extended the theory of fluid shocks to conducting fluids, possibly as an outgrowth of research on nuclear explosions in World War II.

It is hard to imagine a time where belief in the existence of collisionless shocks was less than overwhelming but such was the case until the early 1960s when direct observation in space (Sonett et al. 1964) and in the laboratory (Paul et al. 1967; Alikhanov et al. 1968; Pugh and Patrick 1967) showed their reality and the need for a noncollisional dissipation. The setting for the evolution of such shock waves as a real and important aspect of astrophysical plasmas appears to have begun with a seminal observation by Gold (1959). The then widely known sudden commencement (SC), a five minute or so world-wide rise in the geomagnetic field, was known to follow the occurrence of a flare in the solar atmosphere by several days. Gold proposed that unless a steep wave-like disturbance carried the flare information to the Earth, velocity dispersion would rule out the SC. Moreover a medium, the solar wind, as proposed by Parker (1958), was required for the "event" to propagate in. Today it is recognized that collisionless shock waves propagating in interplanetary space do not, necessarily, arise from flares on the Sun, but are caused either by coronal mass ejections (CMEs) or evolve in space from the interaction between fast and slow solar wind streams.

Heliospheric research has tended to cluster around interplanetary space-craft missions: Explorer, Pioneer, Voyager, Helios, and Galileo. We are currently in the latest campaign—that of the Ulysses mission. For the first time, we are affording ourselves the opportunity to literally take a bird's-eye view of the heliosphere. Launched in October 1990, Ulysses encountered Jupiter in February 1992, after which it began a southward trajectory, reaching S80° in September 1994. Ulysses crossed the Sun's equator in February 1995, and completed its primary mission, passing over the north polar regions of the Sun in June–September 1995.

The scope of this overview is to provide a summary of the current status on heliospheric shocks, with a reference list which is illustrative rather than comprehensive. Many excellent reviews and collections of reviews have appeared over the past dozen years or so on the different aspects of collisionless shock research. This review follows the traditional MHD classification of shock waves, and concentrates on shocks propagating in the solar wind as distinct from shock waves created by planets, comets, and the interaction between the heliosphere and the local interstellar medium. The termination shock is discussed in the chapter by Suess and Nerney. Although this review emphasizes the observations, we have tried, wherever possible, to illustrate the close connection of the interpretation of these observations with theory and numerical simulations.

The microscale structure of shock waves is an important topic that could easily fill an entire volume. It is only briefly touched upon in this review. The interested reader is referred to several recent reviews (see, e.g., Mellot 1985; Gurnett 1985). Particle acceleration at shocks is covered in detail in the chapter by Jokipii.

II. THE COLLISIONLESS SHOCK

A. Basic Properties

Shock waves arise as a consequence of compressibility effects. In the simplest terms, if we assume the medium can be described by an adiabatic law (such that the sound speed, $c_s = (\gamma P/\rho)^{1/2} \propto \rho^{(\gamma-1)/2}$), then a propagating finite-amplitude compressive wave necessarily steepens because c_s is greatest at the peak of the wave. A shock wave forms when a balance is reached between energy dissipation and wave steepening.

Plasma shocks are more complex than fluid dynamical shocks, partly because several characteristic speeds replace the one sound speed of the fluid shock. Collisionless shocks, in turn, are more complex still because of the range of wave-particle interactions needed for providing the necessary dissipation involved in the shock transition. In a collisional medium, this dissipation is provided by binary collisions. In a collisionless medium, it is still not clear exactly what mechanisms operate and what contribution is made by each.

The conservation equations of magnetohydrodynamics are the MHD extension of the Navier-Stokes equations of fluid dynamics. They express the conservation laws for flux, momentum and energy. The general set of these partial differential conservation equations are governed by parameters for a macroscopic, one particle MHD fluid with isotropic pressure and bulk electrical resistivity (but anisotropic viscosity) are given by Landau and Lifshitz (1960). The conservation laws and Maxwell's equations are combined to provide the Rankine-Hugoniot (R-H) conditions (see, e.g., Tidman and Krall 1971) which need to be satisfied by the "discontinuous" jump in plasma parameters across the shock front.

In ordinary fluids, pressure is isotropic, and there is a single characteristic sound speed with which information on disturbances is propagated. In MHD fluids, the magnetic field introduces an anisotropy in the plasma parameters. There are three wave modes: the fast, intermediate and slow. Of these, the fast and slow mode waves are compressive wave modes, whereas the intermediate wave is purely transverse. The propagation velocities of the three modes depend on the angle between the wave vector and the local magnetic field vector.

B. Useful Shock Parameters

When discussing collisionless shock waves it is convenient to define the following parameters. These can be used, either independently or together, to describe various shock regimes.

θ_{Bn}: The angle between the magnetic field direction ($\mathbf{B}/B$) and the unit normal to the shock ($\mathbf{n}$). More commonly (and for the remainder of this chapter), this refers to the angle made between the unit normal and the upstream magnetic field direction ($\mathbf{B}_1$).

Magnetosonic Mach Number: The ratio of the solar wind speed to the speed of magnetosonic waves.

Plasma-β : The ratio of particle thermal pressure to the magnetic field pressure.

It is further useful to subdivide shocks into parallel ($\theta_{Bn} = 0°$) and perpendicular ($\theta_{Bn} = 90°$). More realistically though, one talks of quasi-parallel ($0° < \theta_{Bn} < 45°$) and quasi-perpendicular ($45° < \theta_{Bn} < 90°$) shocks. Oblique shocks ($30° < \theta_{Bn} < 60°$) form an overlapping category.

C. MHD Shock Solutions

A convenient way to categorize the MHD R-H shock solutions is to define four "states":

state 1: $v_{\text{flow}} > v_{\text{fast}}$
state 2: $v_{\text{fast}} > v_{\text{flow}} > v_{\text{int}}$
state 3: $v_{\text{int}} > v_{\text{flow}} > v_{\text{slow}}$
state 4: $v_{\text{flow}} < v_{\text{slow}}$

where v_{flow} is the flow speed in the shock frame, and $v_{\text{fast,int,slow}}$ refer to the fast-mode, intermediate, and slow-mode propagation speeds respectively. Entropy considerations restrict the possible transitions between these states to: $1\rightarrow2$, $1\rightarrow3$, $1\rightarrow4$, $2\rightarrow3$, $2\rightarrow4$, and $3\rightarrow4$ (Wu and Kennel 1992). $1\rightarrow2$ represents a fast-mode shock, $3\rightarrow4$ a slow-mode shock, and the remaining four transitions are intermediate shocks.

1. Slow Shocks. In contrast to fast (-mode) shocks (see below), very few slow (-mode) shocks have been identified in the solar wind. Chao and Olbert (1970) first observed two forward slow shocks using Mariner 5 data during its flight to Venus. Subsequently, Burlaga and Chao (1971) presented two examples of a forward and reverse slow shock from a scan of the Pioneer 6 data set.

There is strong evidence from both theory and MHD simulations (Richter et al. 1985, and references therein) that the probability of a slow mode wave steepening into a shock decreases with increasing heliocentric distance. Thus it appears unlikely that slow shocks would be found at distances much greater than 1 AU. While the observed slow shocks have typically been associated with the leading edges of high-speed streams (Burlaga 1974), Hundhausen et al. (1987) have suggested a mechanism for their formation and evolution via CMEs (coronal mass ejections). Based primarily on the fact that the observed speed of CMEs is frequently below the Alfvén speed (but above the sound speed), they argued that a slow shock could form above the CME, as it sweeps through the corona. Furthermore, the geometry would be fundamentally different from the draped fast shock-CME configuration; with the shock front being concave upward.

More recently Hu and Habbal (1993), using a one-dimensional isentropic MHD flow model, have shown that an initial velocity enhancement disturbance can evolve into a double shock pair in the solar wind. The system (in order of distance from the Sun) consists of a reverse fast shock, a reverse slow shock, a forward slow shock, and a forward fast shock. They concluded that such a system (which has yet to be confirmed by observations) would most likely be found in the inner heliosphere, where conditions favor the velocity disturbances that are required to generate the system of shocks.

2. Intermediate Shocks. A thorough description of intermediate shocks is beyond the scope of this review; however, we make a few comments and reference the most recent literature. In particular we emphasize that the existence of such shocks is still somewhat questionable (Hada 1994). It was argued by Kantrowitz and Petschek (1966) that intermediate shocks cannot exist in real physical systems, based on evolutionary arguments. However, one-dimensional MHD Navier-Stokes (see, e.g., Lyu 1991), two-fluid, and hybrid simulations (Wu and Hada 1991) have inferred that such intermediate shocks can in fact exist. Hada (1994) has argued that the inclusion of dissipation modifies the evolutionary conditions (via the presence of dissipative wave modes) to the extent that intermediate shocks can be stable (i.e., evolutionary).

Chao et al. (1993) claimed to have identified an intermediate shock in interplanetary space. Using Voyager 1 data (bulk velocity, number density, temperature, and magnetic field components) they identified an event for which: (i) the R-H relations for an intermediate shock were satisfied; (ii) the slow Mach number and normal Alfvén Mach number were greater than unity in the pre-shock region, but less than unity in the post-shock region; (iii) the fast Mach number was less than unity in the pre-shock state; (iv) the tangential magnetic field reversed sign through the shock; and (v) the magnetic field magnitude decreases across the shock. Thus the shock-candidate is of a type $2 \rightarrow 4$. These results, however, do not preclude the event being either a tangential or rotational discontinuity. Comparing the high-resolution magnetometer data with hybrid simulations, they argued that the structure of the field is not consistent with a tangential discontinuity. Finally, Walen's relation

($\Delta \mathbf{v} \propto \Delta \mathbf{B}$) is not satisfied, suggesting the event is not a rotational discontinuity.

3. Fast Shocks. Fast shocks represent the bulk of interplanetary shock observations. For the remainder of this review we focus on their properties.

D. Dissipative MHD Shocks

Far away from the transition itself, the upstream and downstream states of the plasma are related by the R-H equations. The physical processes at the shock transition itself are highly complex; the dissipation of the kinetic energy into heat, implied by the irreversible nature of the shock transition and the consequent increase in entropy require a range of kinetic and wave-particle interaction processes which depend on the speed of the shock, the upstream plasma parameters and the angle of propagation between the shock wave and the upstream magnetic field. In particular, a critical Mach number has been identified (M_{crit}) for the case where the normal component of the downstream flow velocity equals the downstream sound speed. The theoretical importance of this is that for shocks defined by $M_s > M_{\text{crit}}$, resistivity alone can no longer provide the necessary dissipation. In the classical MHD fluid formulation, the excess dissipation is provided by viscosity; in collisionless shocks, the conversion of flow momentum into heat is provided by the complex interactions of reflected, incident and transmitted particles, microinstabilities, wave generation and wave-particle interactions.

E. Shock Classifications

1. Classification by θ_{Bn}. The dissipation process which nominally takes place at the shock front in fact takes place over several characteristic scale lengths ordered by the direction of the magnetic field with respect to the shock front. Because of the anisotropy of the kinetic and dynamic processes involved in the shock transition, the angle θ_{Bn} is an important qualitative and quantitative indication for the nature of the shock transition, and shock waves can therefore be classified according to it. At the extremes, the consideration of parallel and perpendicular shocks is useful in providing theoretical limits for the derivation of asymptotic solutions for the shock transition, even if in reality such shocks may not strictly exist. However, the related classification of shocks into quasi-parallel and quasi-perpendicular categories is a very important one. These categories are qualitatively and quantitatively different, and their main features can be recognized from the respective asymptotic cases. Oblique shocks represent not so much an intermediate category, but are often quoted to indicate the variability of the limiting value of θ_{Bn} between the two major categories. Such shocks in fact show characteristic features which belong to one or other of the two major categories, dependent on upstream plasma pandarameterp.

Quasi-Perpendicular Shocks. Quasi-perpendicular shocks, at Mach numbers less than the critical value, have a relatively simple, laminar structure; the plasma parameters and the magnetic field show a step-like increase at the shock front. In a reference system in which the shock is stationary, the upstream plasma flows into the shock, where a charge separation occurs as the electrons, due to their much smaller gyroradius, drift across the magnetic field, while the plasma ions pass through the thin shock transition, essentially unaffected. This electron-ion separation leads to instabilities and electrostatic or electromagnetic fluctuations which affect the particles in a way which can be interpreted as a form of resistivity, leading to the heating and dissipation implied in the shock transition. For supercritical quasi-perpendicular shocks, the magnetic field magnitude has a characteristic, relatively complex profile consisting of a small precursor increase, usually called the foot of the shock wave, followed by a steep ramp at the shock transition, an overshoot and an undershoot just past the transition, before settling down to the downstream value. The excess dissipation needed in supercritical shocks is provided by ion reflection which is nearly specular; at quasi-perpendicular shocks, the ions' guiding center motion leads them through to the downstream region where they contribute significantly to the thermalization process. It is this initially reflected ion population which appears to control both the foot and the overshoot in supercritical, quasi-perpendicular shocks.

Quasi-Parallel Shocks. In the case of quasi-parallel shocks, at least a part of the ion population interacting with the shock is able to carry information away from the shock into the upstream region, and to modify the upstream plasma into which the shock is propagating, through a range of wave-particle interaction phenomena, as well as through the ion beam instability between incident and reflected ions. In particular, long wavelength whistler waves may contribute significantly to local ion heating, thus providing some of the necessary dissipation (Omidi et al. 1990). Although much of the research effort on quasi-parallel shocks (as for most other types of shocks) has concentrated on the Earth's bow shock, interplanetary shocks should, in principle provide a better context for their study. The scale lengths involved in quasi-parallel shocks are in general much greater than the transverse dimensions of the bow shock. A very extensive foreshock region is normally observed, in which shock-associated processes significantly modify the upstream plasma.

The increasing complexities encountered in the MHD view of shock waves have led to the widespread use of numerical simulation techniques. It is impossible, in the context of this overview, to do justice to the extensive results which have been published over the past dozen years or so. The ever-increasing computing power available to handle more complex simulation algorithms has ensured that parametric studies of shock waves have identified many of the underlying physical processes of both quasi-perpendicular and quasi-parallel shocks. In the case of quasi-parallel shocks, simulation results (carried out for parameter regimes applicable to the Earth's bow shock) have uncovered an important process whereby the shock front can propagate by re-formation, as the waves ahead of the shock steepen in turn and generate a new shock front, ahead and spatially distinct from the previous location of the shock (Burgess 1989; Schwartz and Burgess, 1991). This possible spatial differentiation leads to a view of quasi-parallel shock waves in which significant structuring can occur in the direction transverse to the shock front on scales that are comparable to, or smaller than, the shock thickness; thus highlighting the need to treat with considerable care all shock parameters derived from plasma and field observations averaged over distance scales significantly larger than the shock thickness.

2. Classification by Mach Number.   The critical Mach number, separating the so-called subcritical and supercritical shock regimes, has been calculated by Edminston and Kennel (1984) as a function of the upstream plasma parameters. A more detailed analysis is in fact necessary, as shown by Kennel (1987,1988), to take into account all aspects of the shock dissipation process. By considering the relative contributions of resistivity, thermal conduction and viscosity in the MHD formulation of the shock transition, other critical Mach numbers have been identified to characterize the shock-generated dissipation of flow energy into heat. However, even these detailed MHD calculations can only be used as qualitative indicators of the detailed shock transition process. Kinetic effects by the different particle species, electrons, protons and even heavier ions introduce a large number of possible dissipative scales which combine in real shock waves to produce the usually very complex shock profiles observed in practice. In particular, it is now recognized that accurate shock modeling must take into account the contribution of α-particles, the second most abundant ion species in the solar wind.

3. Other Classifications. The literature abounds with other classifications and categories of shock waves. Of these we mention just two: switch-on/off and forward/reverse shocks. The switching refers to the tangential component of the magnetic field ($\mathbf{B}_t$). So that for switch–off shocks, $\mathbf{B}_t \rightarrow 0$ across the shock. This can occur, only in the limit that the upstream flow speed (v_1) equals the upstream Alfvén speed (v_{A1}). For switch-on shocks, $v_1 > v_{A1}$, and $\mathbf{B}_t$ turns on. Thus, we see that the switch-off/on shock represents a limiting case of a slow/fast shock. In fact, this forms one of the basic properties of shocks, namely that for a fast/slow shock the magnetic field is refracted away/towards the shock normal. Because the normal component of the field is conserved, this further implies that the field strength increases/decreases across a fast/slow shock. Recently, Farris et al. (1994) presented evidence for an "almost switch-on shock" at the Earth's bow shock using ISEE data. So far, no switch-on/off shocks have been identified in interplanetary space. Forward/reverse shocks are a classification specific to the solar wind, and are categorized based solely on the direction of propagation of the shock in the solar wind rest frame (i.e., with the correct Galilean transformation—the two are indistinguishable). Specifically, forward shocks propagate outwards, away from the Sun, and reverse shocks towards the Sun. Note however that because the solar wind velocity is much higher than any shock velocity, even the reverse shock propagates outwards, with respect to a stationary observer.

F. Further Complexities

1. Heavy ions. Borrini et al. (1982) carried out a superposed epoch analysis of approximately 100 fast shocks at near 1 AU and found that by separating the shocks into those that show a He enhancement within two days after the shock passage and those that do not, the former showed significantly larger jumps in: (1) flow speed; (2) proton temperature; (3) magnetic field strength; and (4) total pressure. Ogilvie et al. (1982) noted that directly behind the shock, v_p can exceed v_α by more than 50 to 100 km s^{-1}. Richter et al. (1985) have argued that this and other observational evidence strongly suggests that one cannot neglect the role of α–particles in the MHD structure of interplanetary shocks. In particular, they argued that when one applies mass-flux conservation to the shock to determine the shock velocity (once the unit normal direction is determined), the inclusion of α particles could alter the mass flux behind the shock by up to 20%. Thus the shock parameters (θ_{Bn}, v_s) might be significantly altered.

2. Anisotropic Plasmas. Interplanetary MHD shock analysis has restricted itself, for the most part, to isotropic plasmas. A notable exception was that of Lepping and Argentiero (1971) who used a subset of the R-H relations to investigate the effects of thermal anisotropy on shocks observed by Explorers 33 and 35. Their main conclusion was that, even for liberal values of thermal anisotropy, the resultant normal did not vary by more than 2°. However, by using the reduced set of R-H relations, the thermal anisotropy was relegated to just one equation (conservation of the tangential momentum flux). In the full equations the anisotropies present themselves in two further equations; conservation of the normal momentum flux and energy flux. Furthermore, in the presence of plasma anisotropies, the usual rules for delineating between rotational discontinuities and oblique shocks are no longer sufficient (Hudson 1970). In particular, rotational discontinuities do not generally conserve density and magnetic field strength, leading to the possibility of misinterpreting these as either shocks or tangential discontinuities.

III. COLLISIONLESS SHOCKS IN THE HELIOSPHERE

The solar wind is a complex, structured flow, composed primarily of ions and electrons, originating in the solar atmosphere, and moving approximately radially at a highly variable speed, ranging from about 3×10^5 m s^{-1} to more that 10^6 m s^{-1}. The major constituents are protons and electrons. The next most abundant constituent is He^{++}; its abundance relative to protons is variable, from a few percent to more than 20%. Other atomic species, fully or partly ionized, are also present in the solar wind. The relative abundances of the different species, as well as the ionization state of the heavier elements, is an important indicator of the region of the corona in which particular streams originate. Proton densities at 1 AU range from $n_p \sim 3$ cm^{-3} to $n_p \sim 20$ cm^{-3} or more. The temperatures of the dominant species are also variable; for protons it is between 4×10^4 K and 2×10^5 K.

Away from the immediate proximity of the Sun, the mean free path of particles making up the solar wind plasma is much larger than the scale lengths of plasma processes. For example, at 1 AU the mean free path is approximately 1 AU. Thus it is reasonable to conclude that coulomb collisions cannot contribute any significant dissipation on scales less than ~ 0.1 AU.

Shock waves in the low-latitude solar wind are now known to be generated by two principal mechanisms. At heliocentric distances less than 1 to 3 AU, most if not all shock waves are caused by a transient injection of mass and momentum from the solar corona, creating an outward propagating compression wave which can steepen into a shock wave in the ambient solar wind (see, e.g., Richter et al. 1985). At larger heliocentric distances, the compression of a slow solar wind stream by a following fast stream can also steepen to generate a shock wave (see, e.g., Smith 1985). In both cases, once a shock wave has formed, there is a discontinuous transition between the upstream and downstream regimes. The transition is nominally confined to a thin surface separating two states of the plasma. The meaning given to the thinness of the shock transition hides the wide range of complex phenomena involved in the transition between the two flow regimes; the shock transition depends critically on the mechanisms which enable the plasma to dissipate the energy involved in the transition.

The overwhelming bulk of the observations of collisionless shocks come from single spacecraft measurements. This means that inferences have to be made concerning the context of the observations, based on upstream and downstream measurements, which lead to the usual ambiguities between spatial and temporal variations. However, statistical studies of the Earth's bow shock over almost 30 yr have provided a great deal of evidence for the characteristic features of this shock, over a wide range of solar wind and interplanetary magnetic field regimes, so that single spacecraft observations can be extrapolated with greater confidence. Furthermore, a number of multi-spacecraft observations have also been available for the bow shock. The applicability of bow shock studies to interplanetary shocks, however, is limited, partly because of the much larger scale of interplanetary shocks, and partly because the upstream-downstream flow regimes are usually very different from those prevailing at the bow shock. Nevertheless, at least qualitative inferences can be drawn, when care is exercised, on the nature of the shock transition process, in particular on the dissipation processes as a function of flow regimes and upstream magnetic field geometry.

IV. SHOCKS GENERATED BY SOLAR TRANSIENTS

The first intimation of the existence of collisionless shock waves in interplanetary space arose from the association of major solar flares with storm sudden commencements (SCs) in the Earth's magnetic field (Gold 1955). As it was noted that SCs often happened within a few days of large flares on the Sun, it seemed likely that the eruption of solar material into interplanetary space following the flare generated a compressive shock wave, traveling at some 800 to 1000 km s^{-1} (to match the observed delay between the flare and the SC), and the impact of this shock wave resulted in the compression of the Earth's magnetic field, compatible with the SC signature.

A. The Solar Flare Myth

"The solar flare myth," a phrase coined by Gosling (1993) succinctly summarizes the dethroning of solar flares as the cause of transient shock waves, and their replacement by CMEs. Central to the solar flare myth (also referred to as "the big flare syndrome" by Kahler [1982]) is the idea of cause and effect, and that while there have been many apparent correlations of transient shocks with parent flares, such correlation may be only fortuitous.

Early interplanetary shock studies followed the original suggestion that shock waves originated near the Sun by the energy explosively released in solar flares. Fast-forward shocks observed by spacecraft in Earth orbit show a very good correlation with storm SCs (see, e.g., Chao and Lepping 1974; Cane 1985; Smith et al. 1986). This close association between shocks and SCs is well understood. The sudden increase in solar wind ram pressure caused by shocks acts as a step function input to the magnetosphere, and causes a compression that is observed by groundbased magnetometers. Until at least the mid-1970s, the temporal association between SCs and solar flares strongly supported a causal association of shock waves with flaring activity on the Sun. Solar flares are easily monitored phenomena on the solar disk and have a long history of observations. Very good statistics have been available covering a very wide range of the electromagnetic spectrum, from X rays through the visible spectrum to radio waves. The total energy involved in the flare process is considerable and until CMEs, also called coronal transients, were identified in the 1970s (see, e.g., MacQueen 1980), flares could be identified with some justification as the sources of major transient interplanetary disturbances.

From the 1970s, however, it has been recognized that: (a) the frequency of CMEs shows a temporal dependence on the solar cycle similar to that of solar flares; (b) CMEs represent an energy and mass release from the Sun commensurate with or greater than those involved in solar flares; and (c) at least for the more energetic category of CMEs, there is often a temporal association with solar flares. While the coronal processes at the origin of CMEs are not fully understood, it has been shown that they originate in large-scale, magnetically closed coronal regions (see, e.g., Hundhausen 1993), over a much wider range of heliolatitudes than is the case for solar flares. There is a tendency for CMEs to originate in or close to the coronal streamer belt which follows the magnetic equator, while flares occur preferentially in the high activity belt around the solar equator.

There is now a strong presumption, yet to be fully demonstrated and modeled, that coronal instabilities which give rise to CMEs may also act as a trigger to the solar flare phenomenon. Therefore a temporal association between solar flares and CMEs can occur which, if not fortuitous, is likely to be indirect (Kahler 1992; Gosling 1993).

The careful study of the association of interplanetary observations with CMEs leaves little doubt now that CMEs are the dominant sources of transient interplanetary disturbances. There is a problem, however, in that it is difficult or even impossible to observe CMEs against the solar disk, and therefore only those which appear in coronal observations above the limb can be visually identified. The association of shock waves with CMEs rather than flares was strongly supported by a series of observations by the Helios 1 space probe which, during the four years from 1979 to 1982, spent alternate six month periods approximately over the east and west limbs of the Sun. Combining the shock waves detected at Helios 1 with coronagraph observations from the Earth-orbiting P78-1 satellite, a positive association could be made between the majority of shock waves and the CMEs (Sheeley et al. 1983).

On the visible disk of the Sun, disappearing filaments, in the absence of any flaring activity, have been identified with shock waves observed at 1 AU (see, e.g. Sanahuja et al. 1983). It now seems that disappearing filaments constitute a signature of at least one class of CMEs on the solar disk. Further evidence concerning the origin and characteristics of CMEs, as related to the generation of transient interplanetary shocks, can be found in the study of radio emission from the corona (Sheeley et al. 1984). Although it is generally accepted that Type II metric radio bursts in the corona are emitted by energetic electrons accelerated by a propagating shock wave (see, e.g., Wild and Smerd 1972), the true physical relationship between the shock and burst is not completely clear (Bougeret 1985).

At this stage, the conclusion concerning the association of CMEs with heliospheric shocks is that some CMEs, if their velocity and total mass (and therefore energy) are sufficient, will generate a shock wave in the solar wind which can propagate to several AU and can therefore be observed by interplanetary probes. The answer to the question as to what is meant by "sufficient" may also depend on the velocity and density characteristics of the solar wind regime in which the CME is propagating. The complex coronal processes that give rise to such highly energetic CMEs are more likely to be accompanied by energetic flares which had been previously identified as the "parents" of the interplanetary shocks.

The association between fast CMEs and interplanetary shocks remains an observational challenge. Space-borne instrumentation provides evidence for the passage of a transient disturbance in interplanetary space, but independent observations are required to detect the launch of the CME from the Sun that was responsible for the observed disturbance. Earth orbiting spacecraft are not well placed to be supported by such independent and direct confirmation of coronal transients. Observations by heliocentric probes, such as Helios, ISEE-3 and Ulysses which have spent some time above the solar limb, as seen from the Earth, have been used successfully to associate transient shock waves with CMEs launched in their direction, as observed by Earth-based or Earth-orbiting coronagraphs. However, the manifestation of CMEs in interplanetary space can also be recognized in most cases by *in-situ* measurements. A set of signatures, some of which are more important and consistent than others, has been identified for recognizing in the solar wind plasma the remnants of CMEs (Zwickl et al. 1983; Gosling 1990). Because of the predominantly closed geometry of the magnetic field lines in CMEs, suprathermal electrons as well as energetic protons (Marsden et al. 1987) are trapped and show strong bi-directional streaming along the field lines. The magnetic field in CMEs is usually stronger and has a smaller variability of its strength than in the normal solar wind flow; it often shows a consistent rotation of long duration representative of a magnetic flux rope or magnetic cloud (Klein and Burlaga 1982; Burlaga 1991). The plasma itself is of low plasma-β, has a higher concentration of helium than the average solar wind but a lower temperature. The heavier ions often show unusual charge states associated with the lower temperature (Bame et al. 1979).

B. The Properties of Transient Shock Waves

The statistical study of shock waves generated by solar transients, observed by the two Helios spacecraft in heliocentric orbits between 0.3 and 1 AU, was carried out by Volkmer and Neubauer (1985). The positive correlation between solar activity (as represented by the sunspot number) and the frequency of shock waves was found to be, as expected, quite clear. At solar minimum, there were very few shock waves, but as soon as the activity began to increase, several shock waves (between 4 and 8) were observed during each solar rotation. Looking more closely, however, the frequency is seen to fluctuate, with the highest values during the years preceding and following solar maximum. Shock waves observed at the orbit of Venus (0.72 AU) have shown a similar solar cycle dependence (Lindsey et al. 1994). The same study showed that the plasma characteristics associated with the majority (80%) of the shocks were accompanied by the recognized interplanetary signatures of CMEs. The observed characteristics of CME-associated shock waves at 1 AU (which cover virtually all shocks at 1 AU) vary according to the position of the observer with respect to the large-scale geometry of the shock front and to the relative helio-longitude of the CME. In a study covering 10^3 shock waves, Borrini et al. (1982) found that about half were followed by clear CME signatures, principally in the form of a high alpha to proton ratio, but also showing an increased magnetic field strength ahead of the CME which is now recognized to be due to the draping of the magnetic field lines around the leading edge of the CME. Based on the statistical study of these shocks, he proposed a simplified sketch of the large scale geometry of transient shocks driven by CMEs, as shown in Fig. 2. While it is now recognized that the magnetic geometry of CMEs is quite complex (Gosling 1990), the sketch still indicates the way in which the CME driver generates a much larger shock front as it propagates from the Sun to 1 AU and beyond. The detailed structure of the driver is also more complex than shown in the figure. There is considerable temporal variability in the different CME signatures. In particular, low plasma temperatures, bi-directional suprathermal electron fluxes and the α to proton ratio do not show a consistent spatial distribution behind a CME-driven shock wave (Zwickl et al. 1983). The low variance of the magnetic field and of the solar wind velocity, together with low plasma temperatures, appear to signal the main magnetic structure of the CME driver; however, the α-to-proton ratio, while associated with the CME, may be a less direct signature of the driver itself. The spatial variability of these parameters is likely to be representative of the complex geometry and dynamics of the CME itself, from its origin in the corona through its propagation in the ambient solar wind.

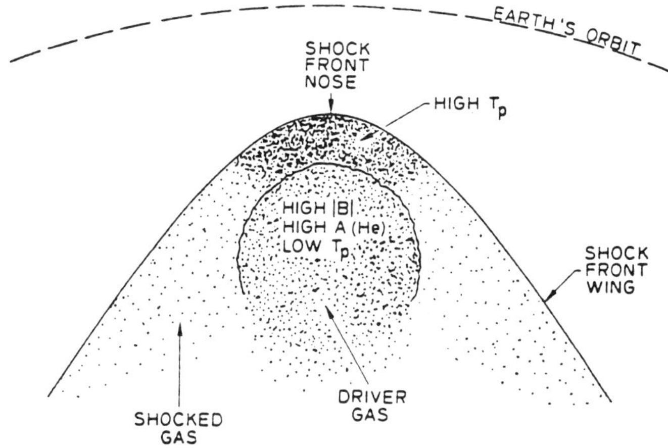

Figure 2.   A simplified sketch of the large-scale geometry of transient shock disturbances driven by CMEs (figure from Borrini et al. 1982).

Generally speaking, transient shocks at 1 AU have low magnetosonic Mach numbers when compared to that of the Earth's bow shock. Shocks have been observed down to a detectability limit of about $M_c = 1.1$ and up to about $M_c = 4$, whereas the bow shock routinely has a Mach number $M_c = 8$ or more. Nevertheless, given that the critical Mach number (above which resistivity can no longer provide the necessary dissipation) strongly depends on the upstream solar wind conditions and on the angle between the shock normal and the upstream magnetic field, a surprisingly high proportion of interplanetary shocks are supercritical. In a study of 34 fast forward shocks, Bavassano-Cattaneo et al. (1986) showed that 19 were supercritical and 7 had a Mach number close to the critical value. In particular, most quasi-parallel shocks appear to be supercritical. Interplanetary shocks therefore also demonstrate the features associated with dissipation mechanisms associated with supercritical shocks, namely a foot (a small increase) in the magnitude of the magnetic field preceding the shock transition, and an overshoot at the transition itself. Furthermore, a range of upstream waves detectable at most shocks (Tsurutani et al. 1983) also indicate the presence of shock-processed suprathermal particles, reflected from the shock front or due to leakage from the downstream region.

A simple, one-dimensional numerical method for evaluating the effect of a sudden release of energy from the Sun, in the form of a pressure pulse, or in the form of a sudden but more persistent increase in plasma pressure on the interplanetary medium has been available since the late 1960s (Hundhausen and Gentry 1969). The objective of this, and more detailed models has been to gain a first order understanding of the interplanetary evolution shock waves generated by solar transient events. This technique was thoroughly reviewed by Hundhausen (1985).

V. SHOCKS GENERATED BY INTERPLANETARY PROCESSES

Parker's original derivation of the origin of the solar wind in the hot solar corona assumed spherical symmetry, i.e., a constant temperature, density and magnetic field at some assumed source surface around the Sun (cf. Parker 1963). This assumption naturally yielded a solar wind flow at constant velocity. From what was already known of the corona, from eclipse photographs as well as from spectroscopic measurements of density and temperature distributions, this was clearly an oversimplification. Nevertheless, it was a very useful one, and sufficient to derive the existence of a supersonic solar wind. Sarabhai (1963) and Parker (1963) recognized the likelihood of nonuniform flow velocities from different solar longitudes, with the immediate consequence that, as the Sun rotates, and as faster flow is emitted behind a slow flow, this will result in the compression of the slow flow in interplanetary space. This interaction between fast and slow solar wind streams is now recognized as one of the global dynamic processes which shape the structure of the interplanetary medium, at least at low to medium helio-latitudes.

The relative stability of the regions on the Sun from which fast and slow solar wind streams originate (if we exclude the periods near solar maximum activity) results in the approximate recurrence, from one solar rotation to the next, of a succession of slow and fast solar wind streams. This is the case for an observer near the Earth, and, in practice, for any interplanetary probe, at least near the ecliptic plane and at heliocentric distances out to 10 or so AU. The large-scale structure created by the interaction of the solar wind streams thus appears to corotate with the Sun, hence the name corotating interaction region (CIR) given to the regions of the solar wind flow that are affected by the compression process operating at the interface between a fast stream following a slow stream.

Slow and fast solar wind streams are generated apparently in different regions of the corona. The association of high solar wind speeds with large coronal holes was established in the 1970s, when, during the solar minimum years of 1973–1975, such coronal holes were first observed by Skylab (Krieger et al. 1973; Nolte et al. 1976). The origin of the slow solar wind is still under debate: it has not been established whether it originates, in a manner still unknown, above magnetically closed, hot and active parts of the corona, or from small and scattered coronal holes. The properties of the two kinds of solar wind are different; fast streams, with velocities in excess of 550 km s^{-1}, have lower densities and temperatures, and a lower proton to α ratio than the slow wind, which has typical velocities below 400 km s^{-1}.

A. The Formation of Shock Pairs

Observations of the way the slower solar wind is compressed by the velocity gradient due to the following faster flowing solar wind were first made at 1 AU. It was noted that an interaction region already formed between the two streams, resulting in a pile-up of the magnetic field, and a compressed, turbulent plasma in the region of interaction. The interface between the two streams, assumed to be a tangential discontinuity, was identified in some of these regions. Early modeling of this process (Carovillano and Siscoe 1969; Hundhausen 1972; Hundhausen 1973*a,b*; Hundhausen and Gosling 1976) formulated many of the key questions and provided first, approximate answers concerning the evolution of CIRs, as a function of heliocentric distance. In particular, the formation of shock waves at the leading and trailing edges of the CIRs was predicted, based on the steepening of the compression region with distance from the Sun.

In parallel with the modeling, the study of numerous such interaction regions at 1 AU (see, e.g., Gosling et al. 1972) confirmed the likelihood of continued dynamical evolution of interaction regions, as they traveled out from 1 AU. The relatively simple MHD simulations showed that as the forward shock moved anti-sunward, overtaking the upstream slow stream, and that the reverse shock moved sunward (in the frame of the flowing plasma), the region would widen as it traveled outwards. The heliocentric distance at which the steepening of the CIR edges would develop into shock pairs was expected to be 2 to 4 AU. In fact, a restricted number of CIRs were found to have developed shock pairs even at 1 AU, under special conditions (Chao et al. 1972). The first observations of fully developed forward-reverse shock pairs were made by Pioneer 10 at about the orbit of Mars, at 1.4 AU (Smith and Wolfe 1976).

The question as to the heliospheric distance at which CIRs develop forward and reverse shocks appears to have only a first-order answer, in that most observations and models agree that shocks are more likely formed at distances greater than 1 AU, at least near the solar equatorial plane. The observations of corotating shocks at distances less than 1 AU (Richter et al. 1980), at 1 AU (see, e.g., Burlaga et al. 1985) and at larger distances (Smith 1985) indicate that both forward and reverse shocks can develop at all distances from about 0.5 AU. Observations and modeling also show that the width of interaction regions at distances greater than 5 or 10 AU is such that merging between them is highly likely, and that therefore beyond about 10 AU, a more complex regime of merged interaction regions is likely to dominate the heliosphere.

While the general picture (in a largely one-dimensional sense) of the development of fast forward-reverse shock pairs at the leading and trailing edges of CIRs is well understood and documented, significant complexities arise from the three-dimensional nature of the interaction regions, as well as from nonuniformities in the original flow velocities in both slow and fast streams. As mentioned in Sec. II, Hu and Habbal (1993) recently suggested that a velocity disturbance near the Sun can develop a double shock pair, at least at distances less than 1 AU, containing a forward-reverse pair of slow shocks between the leading fast-forward shock and the trailing fast reverse shock. In this context, observations of CIRs and even some corotating shock waves by the Helios spacecraft between 0.3 and 1 AU (Richter et al. 1980) represent an extreme case of stream dynamics, likely to be a function of the solar cycle (because of the different configuration of the source regions of the high-speed streams with respect to the streamer belt) and not just of heliocentric distance.

B. The Evolution of CIRs

A special feature of the evolution of the CIRs as a function of heliocentric distance is the likelihood that the forward shock leading a CIR will catch up with the reverse shock of the preceding CIR, resulting in the merging of the two interaction regions. Using a relatively simple one dimensional MHD code (Dryer and Steinolfson 1976), the interaction between the forward shock wave associated with a CIR was shown to interact with the reverse shock wave of the preceding CIR. This aspect of the evolution of CIRs in the outer heliosphere was further studied in two-dimensional MHD simulations by Pizzo (1983), and, using a more accurate modeling technique in which the solar wind flow is taken to be radial and the magnetic field is normal to the flow, but which also includes the R-H relations, by Whang and Burlaga (1985,1988). As reviewed by Whang (1991), a series of studies based on this latter technique have shown that given a CIR velocity profile at, say, 1 AU, the development of shock waves and their interaction can be successfully modeled at larger heliocentric distances.

Two questions, relevant to determining the large-scale structure of CIRs in the outer heliosphere, concern, first, the distance at which interaction between shock waves occurs to generate merged interaction regions, and, second, the three-dimensional evolution of CIRs. These questions are relevant for determining the way in which nonuniform pressure waves shape the outer structure of the heliosphere for the propagation of cosmic rays and for the interaction of the solar wind with the heliospheric termination shock.

C. Solar Cycle Effects

During solar minimum, the recurrent pattern of alternating slow and fast solar wind streams (observed near the solar equatorial plane) promotes the relatively steady occurrence of CIRs (and their associated shock pairs) from one solar rotation to the next. This pattern is not present near solar maximum and CIRs do not develop in the same clear way. The dynamics of the heliosphere are dominated by CMEs and other solar transient phenomena which generate the shock waves seen even at very large heliocentric distances.

During solar maximum conditions, a different kind of interaction region has been identified. There are no identifiable large coronal holes to generate fast solar wind streams; instead, there are many transient fast flows presumably associated with CMEs. The slow solar wind streams, originating possibly in the smaller coronal holes that can be identified on the Sun, are compressed by the fast flows associated with the CMEs. In an analogous manner to the formation of CIRs observed near solar minimum, interaction regions created by the fast transient flow compressing the "steady" slower flow also develop a characteristic forward-reverse shock pair. Such regions were designated as transient interaction regions (TIRs) (Smith 1985) and were observed at distances up to 10 AU. At distances beyond 20 AU global merged interaction regions (GMIRs) have been observed by the Pioneer and Voyager spacecraft (see the chapter by McDonald and Burlaga).

The Ulysses spacecraft, traveling in the ecliptic from 1 to 5 AU in 1991, just after solar maximum, only started observing the clear forward-reverse shock pairs characteristic of fully developed CIRs from about 3 AU (Balogh et al. 1993). Near the ecliptic plane, the most stable patterns of CIRs have been observed prior to solar minima, probably associated with the low latitude extensions of the developing polar coronal holes (Thieme et al. 1990). The number of CIRs observed during any 27-day period is a function of the solar activity cycle, as manifested by the development state and equatorial extension of coronal holes from the two polar regions of the Sun. At least during the declining phase of the cycle, when CIRs appear to be the dominant large-scale, dynamic features in the solar wind at 1 AU and beyond, two such regions have usually been observed in each solar rotation at 1 AU and out to moderate distances.

VI. CURRENT RESEARCH: INTERPLANETARY SHOCKS AT HIGH LATITUDES

Until the last year or so, observations of interplanetary shock waves (as of other phenomena in the solar wind) had been restricted to near the ecliptic plane. Since the beginning of 1992, the Ulysses spacecraft has followed an interplanetary trajectory which took it over the southern solar pole during June–November 1994 and the northern pole during June–September 1995, thus providing the first exploration of interplanetary phenomena, including the presence and characteristics of shock waves, as a function of solar latitude. Ulysses has, for the first time, allowed us to explore fully the three-dimensionality of interplanetary space. Of the many aspects of heliospheric research being expanded by new discoveries, interplanetary shocks are no exception.

Shock waves are naturally three-dimensional phenomena; it is therefore important to determine their propagation characteristics in the high heliolatitude flow regimes of the solar wind. It was expected that the spiral magnetic field lines are less tightly wound at high solar latitudes, yielding a large-scale geometry different from that near the ecliptic plane. The more parallel geometry between the solar wind flow vector and the average interplanetary magnetic field vector was expected to affect the propagation of shock waves of solar origin, as well as the formation and large-scale geometry of shock waves generated through stream-stream interactions.

Balogh et al. (1995) provided a comprehensive list of interplanetary shocks observed by Ulysses from the start of the mission (October 1990) until the end of 1993 comprising of 160 shocks. Also tabulated were several fundamental shock parameters: θ_{Bn}, $B_{\text{downstream}}/B_{\text{upstream}}$, M_s, and whether the shock was forward or reverse. The trends in this list will be explored below.

A. The Three-Dimensional Nature of Shock Propagation

Between July 1992 and April 1994 (S12°→S59°) Ulysses observed approximately 50 shocks (Gosling et al. 1995). These are shown in Fig. 3, where the shock strength (as inferred from the ratio of the upstream to downstream density minus one) is plotted against heliographic latitude. The two shock pairs at S32.5° and S54° have been associated with the over-expansion of a CME (see below), and most of the lone forward shocks equatorward of S26° appear to be driven by CMEs. Of particular interest is that at ~S26° the forward shocks seem almost to disappear. Not coincidentally, this latitude is also roughly the same as the tilt of the coronal streamer belt with respect to the solar equatorial plane (~29°). Gosling et al. (1993) reported several case studies of mid-latitude CIRs bounded by forward/reverse shocks (or waves). Figure 4 shows one such event. We focus on the last two panels showing the meridional and azimuthal flow components. θ is positive for northward flow, and ϕ is positive into the direction of planetary motion (westward). The forward shock and reverse wave bounding the CIR are marked. Thus the flow downstream of the forward shock turned northward and westward, while the flow downstream of the reverse wave turned southward and eastward.

These observations can be explained as a natural consequence of large-scale pressure gradients associated with the stream interface (i.e., the transition layer separating the fast and slow flows), and are a direct result of the inherent three-dimensionality of the heliosphere. Following speculations by McNutt (1988), Pizzo (1991,1994) carried out three-dimensional MHD numerical simulations of the inner heliosphere to explain these flow deflections. We present only a brief heuristic argument to outline the essential details. Figure 5 shows a sketch of the tilted geometry of the Sun, where α is the angle between the dipole and rotation axes. Consider a parcel of plasma moving radially out from the western limb of the Sun (1). Some time later parcel 2, which was ejected from a predominantly faster region, will catch up to parcel 1. Because the first part of parcel 2 to hit the slow region is the north/west corner, and south/east will be last, one can see that a shearing motion will take place such that the normal direction to the interaction region is rotated away from the

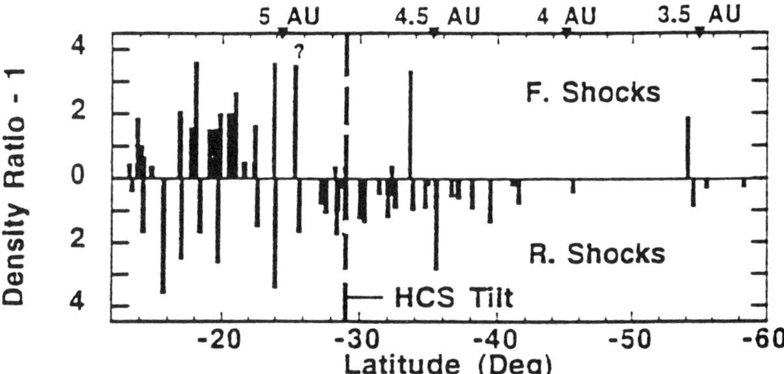

Figure 3. Shock strength (as inferred from the ratio of the upstream to downstream density minus one) is plotted against heliographic latitude. Forward/reverse shocks are above/below the horizontal line (figure from Gosling et al. 1993).

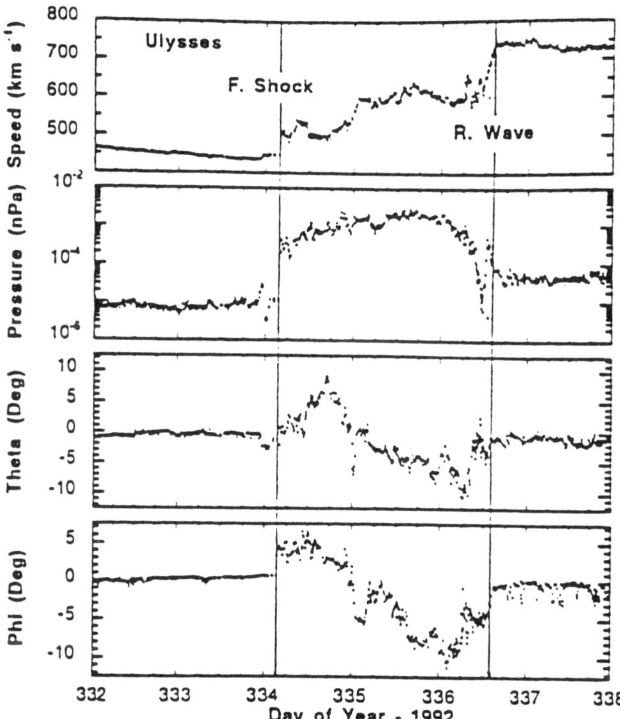

Figure 4. CIR observed by the Ulysses spacecraft. The panels show (from top to bottom): solar wind speed, proton thermal pressure, meridional flow, and azimuthal flow (figure from Gosling et al. 1993).

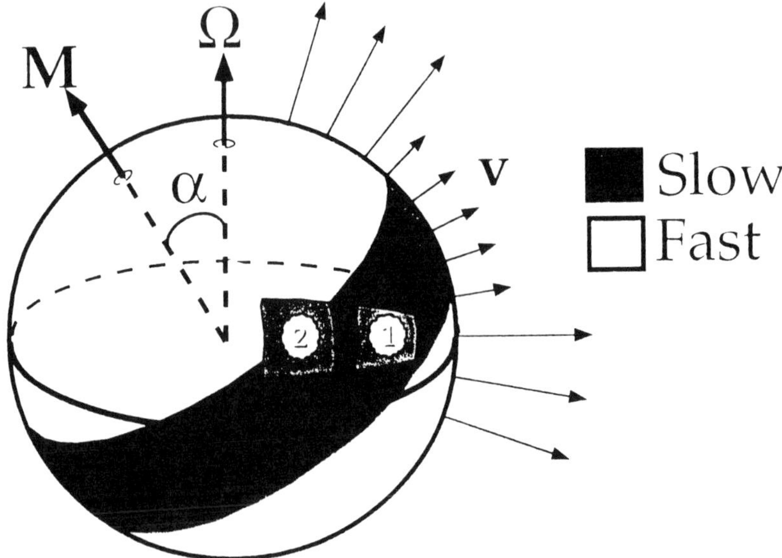

Figure 5.   Tilted-dipole geometry of the Sun. The angle α represents the tilt of the dipole axis with respect to the rotation axis. The meridional and azimuthal deflections observed in stream interfaces can be understood in simple terms by considering two parcels of plasma ejected at positions 1 and 2. See text for details.

radial direction, to the southeast. Figure 6 demonstrates how two interfaces (each emanating from a different hemisphere) rotate away from the radial direction. Adding fast forward and reverse shocks to these interfaces now allows us to explain the results of Fig. 6; namely that the forward shocks tend to propagate equatorward, and the reverse shocks tend to propagate poleward. One can readily verify that the observed flow deflections are consistent with this picture (see, e.g., Siscoe 1972). This "third dimension" of CIRs is discussed in more detail by Gosling (1995).

B. Expansion Shock Pairs at High Latitudes

At low latitudes there are two primary mechanisms for the generation of interplanetary shocks, as discussed above. At higher latitudes, Ulysses observations have led to the possibility of a new class of shock formation (Gosling et al. 1994a,b). These structures are high-latitude ($>$S30$°$) CMEs bounded by forward and reverse shock waves. The fundamental difference between these and low-latitude CME structures is that the CMEs are observed to be traveling at roughly the same speed as the ambient solar wind. Hence it is improbable that the shock pairs could form via the fast/slow interaction discussed above. Although only three such events have so far been observed, it appears that these shock pairs must be formed by an over-expansion of the CME (Gosling et al. 1994a). Figure 7 depicts the evolution of a high-latitude CME. The essential requirement is that the internal pressure is sufficiently high to drive

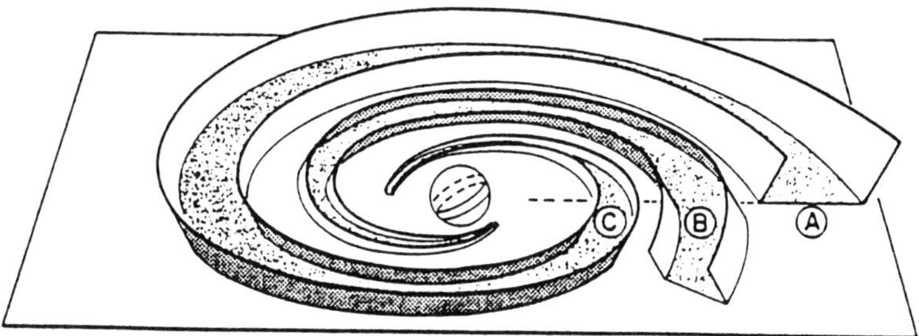

Figure 6.   The three-dimensional nature of stream interfaces is illustrated here, where interplanetary shock surfaces are shown. Notice that regardless of the hemisphere from which the disturbance is connected the inferred forward shocks propagate equatorward, and the referse shocks propagate poleward (figure adapted from Pizzo 1994).

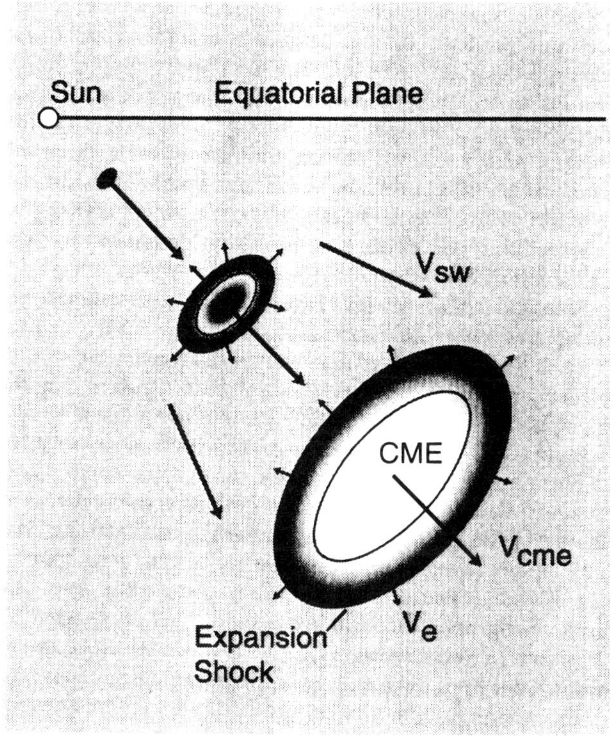

Figure 7.   The conjectured evolution of a high-latitude CME. Shading within the CME indicates the distribution of pressure (figure from Gosling et al. 1994a).

the over-expansion. As the CME moves out (such that $v_{CME} \leq v_{SW}$) the high pressure causes the CME to expand, leading to an expansion-shock wrapped around the CME. Thus, the observed forward/reverse-shock pair is in fact the same expansion shock sampled twice by the spacecraft as it traverses the structure. The pressure enhancement localizes in the region immediately downstream of the shock, and the central region of the CME shows a pressure deficit.

Such shock pairs may occur more frequently, at lower latitudes as well, but conditions for observing them may be severely constrained by the geometry of the reverse shock front being restricted to near the center-line of the expanding CME. The fact that the Ulysses observations of these over-expanded CMEs are restricted to high helio-latitudes, where the solar wind flow from the solar polar coronal hole is relatively uniform, may also indicate that dynamic effects in the equatorial regions of the Sun will prevent their unambiguous identification.

VII. SUMMARY

In the thirty years since the first positive identification of collisionless shocks in the interplanetary medium, we have confirmed, identified and discovered many new and interesting heliospheric shock-related phenomena. In this review we have highlighted just some of these results.

For the most part, we chose to restrict ourselves to an essentially macroscopic description, but in closing, we cannot help remarking on the importance of the microstructure of collisionless shocks. Its importance can be appreciated by the often-quoted adage that, "Collisionless shocks are the simplest configuration in which a macroscopic flow is regulated by microscopic dissipation" (see, e.g., Stone and Tsurutani 1985); and therein lie some of the greatest puzzles yet to be solved.

Currently, with the Ulysses mission, we are embarking on a new era of discoveries, and are for the first time probing the true three-dimensionality of the heliosphere. If the initial results from the Ulysses mission are indicative of what is to come, we can expect the next thirty years to be as least as exciting.

Acknowledgments. We are very grateful to C. P. Sonett for many useful discussions, as well as comments and suggestions on the manuscript. The work at the University of Arizona was supported by a contract with the Jet Propulsion Laboratory.

REFERENCES

Alikhanov, S. G., Alinovskii, N. I., Dolgov-Savelev, G. G., Eselevich, B. G., Kurtmalaev, R. Kh., Nesterikhin, V. K., Yu, E., Pilskii, V. I., Sagdeev, R. Z., and Semenov, V. N. 1968. *Development of a Collisionless Shock Wave Programme* (Novosibirsk: Culham Translation Office).
Balogh, A., Forsyth, R. J., Ahuja, A., Southwood, D. J., Smith, E. J., and Tsurutani, B. T. 1993. The interplanetary magnetic field from 1 to 5 AU: Ulysses observations. *Adv. Space Res.* 13:15–24.

Balogh, A., Gonzales-Esparza, J. A., Forsyth, R. J., Burton, M. E., Goldstein, B. E., Smith, E. J., and Bame, S. J. 1995. Interplanetary shock waves: Ulysses observations in and out of the ecliptic plane. In *The High Latitude Heliosphere*, ed. R. G. Marsden (Dordrecht: Kluwer), pp. 171–180.

Bame, S. J., Asbridge, J. R., Feldman, W. C., Fenimore, E. E., and Gosling, J. T. 1979. Solar wind heavy ions from flare-heated coronal plasma. *Solar Phys.* 62:179–201.

Bavassano-Cattaneo, M. B., Tsurutani, B. T., Smith, E. J., and Lin, R. P. 1986. Subcritical and supercritical interplanetary shocks: Magnetic field and energetic particle observations. *J. Geophys. Res.* 91:11929–11935.

Borrini, G., Gosling, J. T., Bame, S. J., and Feldman, W. C. 1982. An analysis of shock wave disturbances observed at 1 AU from 1971 through 1978. *J. Geophys. Res.* 87:4365–4373.

Burgess, D. 1989. Cyclic behavior at quasi-parallel collisionless shocks. *Geophys. Res. Lett.* 16:435–348.

Bougeret, J.-L. 1985. Observations of shock formation and evolution in the solar atmosphere. In *Collisionless Shocks in the Heliosphere: Reviews of Current Research*, eds. B. T. Tsurutani and R. G. Stone (Washington: American Geophysical Union), pp. 13-32.

Burlaga, L. F. 1974. Interplanetary stream interfaces. *J. Geophys. Res.* 79:3717–3725.

Burlaga, L. F. 1991. Magnetic clouds. In *Physics of the Inner Heliosphere II*, eds. R. Schwenn and E. Marsch (New York: Springer-Verlag), pp. 1–22.

Burlaga, L. F., and Chao, J. K. 1971. Reverse and forward slow shocks in the solar wind. *J. Geophys. Res.* 76:7516–7521.

Burlaga, L. F., Pizzo, V. J., Lazarus, A. J., and Gazis, P. 1985. Stream dynamics between 1 AU and 2 AU: A comparison of observations and theory. *J. Geophys. Res.* 90:7377–7388.

Cane, H. V. 1985. The evolution of interplanetary shocks. *J. Geophys. Res.* 90:191–197.

Carovillano, R. L., and Siscoe, G. L. 1969. Corotating structure in the solar wind. *Solar Phys.* 8:401–414.

Chao, J. K., and Lepping, R. P. 1974. A correlative study of ssc s, interplanetary shocks and solar activity. *J. Geophys. Res.* 79:1799–1807.

Chao, J. K., and Olbert, S. 1970. Observation of slow shocks in interplanetary space. *J. Geophys. Res.* 75:6394–6397.

Chao, J. K., Formisano, V., and Hedgecock, P. C. 1972. Shock pair observation. In *Solar Wind*, eds. C. P. Sonett, P. J. Coleman, Jr., and J. M. Wilcox, NASA SP-308, pp. 435–443.

Chao, J. K., Lyu, L. H., Wu, B. H., Lazarus, A. J., Chang, T. S., and Lepping, R. P. 1993. Observations of an intermediate shock in interplanetary space. *J. Geophys. Res.* 98:17443–17450.

de Hoffman, F., and Teller, E. 1950. Magnetohydrodynamic shocks. *Phys. Rev.* 80:692–703.

Dryer, M., and Steinolfson, R. S. 1976. MHD solution of interplanetary disturbances generated by simulated velocity perturbations *J. Geophys. Res.* 81:5413–5419.

Edminston, J. P., and Kennel, C. F. 1984. A parametric survey of the first critical Mach number for a fast MHD shock. *J. Plasma Phys.* 32:429–441.

Farris, M. H., Russell, C. T., Fitzenreiter, R. J., and Ogilvie, K. W. 1994. The subcritical, quasi-parallel, switch-on shock. *Geophys. Res. Lett.* 21:837–840.

Gold, T. 1955. Discussion of shock waves and rarified gases. In *Gas Dynamics of Cosmic Clouds*, eds. J. C. van de Hulst and J. M. Burgers (New York: North Holland), pp. 103–105.

Gold, T. 1959. Plasma and magnetic fields in the solar system. *J. Geophys. Res.*

64:1665–1674.

Gosling, J. T. 1990. Coronal mass ejections and magnetic flux ropes in interplanetary space. In *Physics of Magnetic Flux Ropes*, eds. C. T. Russell, E. R. Priest and L. C. Lee (Washington D. C.: American Geophysical Union), pp. 343–364.

Gosling, J. T. 1993. The solar flare myth. *J. Geophys. Res.* 98:18937–18949.

Gosling, J. T. 1995. Solar wind corotating interaction regions: The third dimension. *Rev. Geophys. Suppl.* (Washington, D. C.: U. S. National Report to the International Union of Geodesy and Geophysics 1991–1994), pp. 597–601.

Gosling, J. T., Hundhausen, A. J., Pizzo, V. J., and Asbridge, J. R. 1972. Compression and rarefaction in the solar wind: Vela 3. *J. Geophys. Res.* 77:5442–5454.

Gosling, J. T., Bame, S. J., McComas, D. J., Phillips, J. L., Pizzo, V. J., Goldstein, B. E., and Neugebauer, M. 1993. Latitudinal variation of solar wind corotating stream interaction regions: Ulysses. *Geophys. Res. Lett.* 20:2789–2792.

Gosling, J. T., Bame, S. J., McComas, D. J., Phillips, J. L., Scime, E. E., Pizzo, V. J., Goldstein, B. E., and Balogh, A. 1994*a*. A forward-reverse shock pair in the solar wind driven by over-expansion of a Coronal Mass Ejection: Ulysses observations. *Geophys. Res. Lett.* 21:237–240.

Gosling, J. T., McComas, D. J., Phillips, J. L., Weiss, L. A., Pizzo, V. J., Goldstein, B. E., and Forsyth, R. J. 1994*b*. A new class of forward-reverse shock pairs in the solar wind. *Geophys. Res. Lett.* 21:2271–2274.

Gosling, J. T., Bame, S. J., McComas, D. J., Phillips, J. L., Pizzo, V. J., Goldstein, B. E., and Neugebauer, M. 1995. Solar wind corotating stream interaction regions out of the ecliptic plane: Ulysses. In *The High Latitude Heliosphere*, ed. R. G. Marsden (Dordrecht: Kluwer), pp. 99–104.

Gurnett, D. A. 1985. Plasma waves and instabilities. In *Collisionless Shocks in the Heliosphere: Reviews of Current Research*, eds. B. T. Tsurutani and R. G. Stone (Washington: American Geophysical Union), pp. 207–224.

Hada, T. 1994. Evolutionary conditions in the dissipative MHD system: Stability of intermediate MHD shock waves. *Geophys. Res. Lett.* 21:2275–2278.

Hu, Y. Q., and Habbal, S. R. 1993. Double shock pairs in the solar wind. *J. Geophys. Res.* 98:3551–3561.

Hudson, P. D. 1970. Discontinuities in an anisotropic plasma and their identification in the solar wind. *Planet. Space Sci.* 18:1611–1622.

Hundhausen, A. J. 1972. *Coronal Expansion and the Solar Wind* (New York: Springer-Verlag).

Hundhausen, A. J. 1973*a*. Nonlinear model of high-speed solar wind streams. *J. Geophys. Res.* 78:1528–1542.

Hundhausen, A. J. 1973*b*. Evolution of large scale solar wind structures beyond 1 AU. *J. Geophys. Res.* 78:2035–2042.

Hundhausen, A. J. 1985. Some macroscopic properties of shock waves in the heliosphere. In *Collisionless Shocks in the Heliosphere: A Tutorial Review*, eds. R. G. Stone and B. T. Tsurutani (Washington: American Geophysical Union), pp. 37–58.

Hundhausen, A. J. 1993. The size and locations of coronal mass ejections: SMM observations from 1980 and 1984–1989. *J. Geophys. Res.* 98:13177–13200.

Hundhausen, A. J., and Gentry, R. A. 1969. Numerical simulation of flare-generated disturbances in the solar wind. *J. Geophys. Res.* 74:2908–2918.

Hundhausen, A. J., and Gosling, J. T. 1976. Solar wind structures at large heliocentric distances: An interpretation of Pioneer 10 observations. *J. Geophys. Res.* 81:1436–1440.

Hundhausen, A. J., Holzer, T. E., and Low, B. C. 1987. Do slow shocks precede some coronal mass ejections? *J. Geophys. Res.* 92:11173–11178.

Kahler, S. W. 1982. The role of the big flare syndrome in correlations of solar energetic proton fluxes and associated microwave burst parameters. *J. Geophys. Res.* 87:3439–3448.

Kahler, S. W. 1992. Solar flares and coronal mass ejections. *Ann. Rev. Astron. Astrophys.* 30:113–141.

Kantrowitz, A. R., and Petschek, H. E. 1966. MHD characteristics and shock waves. In *Plasma Physics in Theory and Application*, ed. W. B. Kunkel (New York: McGraw-Hill), p. 148.

Kennel, C. F. 1987. Critical Mach numbers in classical magnetohydrodynamics. *J. Geophys. Res.* 92:13427–13437.

Kennel, C. F. 1988. Shock structure in classical magnetohydrodynamics. *J. Geophys. Res.* 93:8545–8557.

Klein, L. W., and Burlaga, L. F. 1982. Interplanetary magnetic clouds at 1 AU. *J. Geophys. Res.* 87:613–624.

Krieger, A. S., Timothy, A. F., and Roelof, E. C. 1973. A coronal hole and its identification as the source of a high velocity solar wind stream. *Solar Phys.* 29:505–525.

Landau, L. D., and Lifshitz, E. M. 1960. *Electrodynamics of Continuous Media* (Oxford: Pergamon Press).

Lepping, R. P., and Argentiero, P. D. 1971. Single spacecraft method of estimating shock normals. *J. Geophys. Res.* 76:4349–4359.

Lindsey, G. M., Russell, C. T., Luhmann, J. G., and Gazis, P. 1994. On the sources of interplanetary shocks at 0.72 AU. *J. Geophys. Res.* 99:11–17.

Lyu, L. H. 1991. A Study of 1–D Nonlinear Hydromagnetic Waves and Collisionless Shocks. Ph.D. Thesis, Univ. of Alaska, Fairbanks.

MacQueen, R. M. 1980. Coronal transients: A summary. *Phil. Trans. Roy. Soc. London A* 297:605–620.

Marsden, R. G., Sanderson, T. R., Tranquille, C., and Wenzel, K.-P. 1987. ISEE-3 observations of low-energy proton bi-directional events and their relation to isolated interplanetary magnetic structures. *J. Geophys. Res.* 92:11009–11019.

Mellot, M. M. 1985. Subcritical collisionless shock waves. In *Collisionless Shocks in the Heliosphere: Reviews of Current Research*, eds. B.T. Tsurutani and R. G. Stone (Washington, D. C.: American Geophysical Union), pp. 131–140.

McNutt, R. L., Jr. 1988. Possible explanations of north-south plasma flow in the outer heliosphere and meridional transport of magnetic flux. *Geophys. Res. Lett.* 15:1523–1526.

Nolte, J. T., Krieger, A. S., Timothy, A. F., Gold, R. E., Roelof, E. C., Vaiana, G., Lazarus, A. J., Sullivan, J. D. and McIntosh, P. S. 1976. Coronal holes as sources of solar wind. *Solar Phys.* 46:303–322.

Oglivie, K. W., Copland, M. A., and Zwickl, R. D. 1982. Helium, hydrogen, and oxygen velocities observed on ISEE 3. *J. Geophys. Res.* 87:7363–7369.

Omidi, N., Quest, K. B., and Winske, D. 1990. Low Mach number parallel and quasi-parallel shocks. *J. Geophys. Res.* 95:20717-20730.

Parker, E. N. 1958. Dynamics of the interplanetary gas and magnetic fields *Astrophys. J.* 128:664–676.

Parker, E. N. 1963. *Interplanetary Dynamical Processes* (New York: Interscience).

Paul, J. W. M., Goldenbaum, G. C., Hyoshi, A., Holmes, L. S., and Hardcastle, R. A. 1967. Measurements of electron temperatures produced by collisionless shocks in magnetized plasma. *Nature* 216:363–364.

Pizzo, V. J. 1983. Quasi-steady solar wind dynamics. In *Solar Wind Five* ed. M. Neugebauer, NASA CP-2280, pp. 675–691.

Pizzo, V. J. 1991. The evolution of corotating stream fronts near the ecliptic plane in the inner solar system. Three-dimensional tilted dipole fronts. *J. Geophys. Res.* 96:5405–5420.

Pizzo, V. J. 1994. Global, quasi-steady dynamics of the distant solar wind. 1. Origin of North-south flows in the heliosphere. *J. Geophys. Res.* 99:4173–4183.

Pugh, E., and Patrick, R. 1967. Plasma tunnel studies of collision-free flows and shocks. *Phys. Fluids* 10:2579–2585.

Richter, A. K., Schwenn, R., and Neubauer, F. M. 1980. Nature and Origin of Corotating Shock Waves Within 1 AU. Report No. MPAE-W-79-80-38 (Katlenburg-Lindau: Max-Planck-Institut für Aeronomie).

Richter, A. K., Hsieh, K. C., Luttrell, A. H., Marsch, E., and Schwenn, R. 1985. Review of interplanetary shock phenomena near and within 1 AU. In *Collisionless Shocks in the Heliosphere: Reviews of Current Research*, eds. B. T. Tsurutani and R. G. Stone (Washington: American Geophysical Union), pp. 33–50.

Sanahuja, B., Domingo, V., Wenzel, K. -P., Joselyn, J. A., and Keppler, E. 1983. A large proton event associated with solar filament activity. *Solar Phys.* 84:321–337.

Sarabhai, V. 1963. Some consequences of non-uniformity of solar wind velocity. *J. Geophys. Res.* 68:1555–1557.

Schwartz, S. J., and Burgess, D. 1991. Quasi-parallel shocks: A patchwork of three-dimensional structures. *Geophys. Res. Lett.* 18:373–376.

Sheeley, N. R., Jr., Howard, R. A., Koomen, M. J., Michels, D. J., Schwenn, R., Muhlhauser, K. H., and Rosenbauer, H. 1983. Associations between coronal mass ejections and interplanetary shocks. In *Solar Wind 5*, ed. M. Neugebauer, NASA CP-2280, pp. 693–702.

Sheeley, N. R., Jr., Stewart, R. T., Robinson, R. D., Howard, R. A., Koomen, M. J., and Michels, D. J. 1984. Association between coronal mass ejections and metric type II bursts. *Astrophys. J.* 279:839–847.

Siscoe, G. L. 1972. Structure and orientation of solar wind interaction fronts: Pioneer 6. *J. Geophys. Res.* 77:27–34.

Smith, E. J. 1985. Interplanetary shock phenomena beyond 1 AU. In *Collisionless Shocks in the Heliosphere: Reviews of Current Research*, eds. B. T. Tsurutani and R. G. Stone (Washington: American Geophysical Union), pp. 69–83.

Smith, E. J., and Wolfe, J. H. 1976. Observations of interaction regions and corotating shocks between one and five AU: Pioneer 10 and 11. *Geophys. Res. Lett.* 3:137–140.

Smith, E. J., Slavin, J. A., Zwickl, R. D., and Bame, S. J. 1986. Shocks and sudden storm commencements. In *Solar Wind Magnetosphere Coupling* eds. Y. Kamide and J. A. Slavin (Tokyo: Terra Scientific), pp. 345–365.

Sonett, C. P., Colburn, D. S., Davis, L., Jr., Smith, E. J., Coleman, P. J., Jr. 1964. Evidence for a collison-free magnetohydrodynamic shock in interplanetary space. *Phys. Rev. Lett.* 13:153–156.

Stone, R. G., and Tsurutani, B. T. 1985. Introduction. In *Collisionless Shocks in the Heliosphere: Reviews of Current Research*, eds. B. T. Tsurutani and R. G. Stone (Washington, D. C.: American Geophysical Union), pp. vii–viii.

Thieme, K. M., Marsch, E., and Schwenn, R. 1990. Spatial structures in high-speed streams as signatures of fine structures in coronal holes. *Ann. Geophysicae* 8:713–724.

Tidman, D. A., and Krall, N. A. 1971. *Shock Waves in Collisionless Plasmas* (New York: Wiley-Interscience).

Tsurutani, B. T., Smith, E. J., and Jones, D. E. 1983. Waves observed upstream of interplanetary shocks. *J. Geophys. Res.* 88:5645–5656.

Volkmer, P. M., and Neubauer, F. M. 1985. Statistical properties of fast magnetoacoustic shock waves in the solar wind between 0.3 and 1 AU: Helios-1, 2 observations. *Ann. Geophysicae* 3:1–12.

Whang, Y. C. 1991. Shock interactions in the outer heliosphere. *Space Sci. Rev.*

57:339–388.

Whang, Y. C., and Burlaga, L. F. 1985. Evolution and interaction of interplanetary shocks. *J. Geophys. Res.* 90:10765–10778.

Whang, Y. C., and Burlaga, L. F. 1988. Evolution of recurrent solar wind structures between 14 AU and the termination shock. *J. Geophys. Res.* 93:5446–5460.

Wild, J. P., and Smerd, S. F. 1972. Radio bursts from the solar corona. *Ann. Rev. Astron. Astrophys.* 10:159–196.

Wu, C. C., and Hada, T. 1991. Formation of intermediate shocks in both two-fluid and hybrid models. *J. Geophys. Res.* 96:3769–3778.

Wu, C. C., and Kennel, C. F. 1992. Structural relations for time-dependent intermediate shocks. *Geophys. Res. Lett.* 19:2087–2090.

Zwickl, R. D., Asbridge, J. R., Bame, S. J., Feldman, W. C., Gosling, J. T., and Smith, E. J. 1983. Plasma properties of driver gas following interplanetary shocks observed by ISEE-3. In *Solar Wind 5*, ed. M. Neugebauer, NASA CP-2280, pp. 711–717.

GLOBAL MERGED INTERACTION REGIONS IN THE OUTER HELIOSPHERE

F. B. McDONALD
University of Maryland

and

L. F. BURLAGA
Goddard Space Flight Center

At heliocentric distances beyond some 20 AU, multiple interplanetary disturbances generated by coronal mass ejections and corotating fast streams coalesce to form a new, long-lived, large-scale structure known as a global merged interaction region (GMIR). These GMIRs produce the discrete step-decreases in the galactic cosmic rays that are the dominant factor in producing the long-term 11-yr modulation cycle and are also the source of a new type of energetic particle event observed in the outer heliosphere. The properties of these GMIRs are discussed in terms of the principal interplanetary disturbances observed in the inner heliosphere and in the context of corotating and local merged interaction regions in the outer heliosphere. Detailed observations of three GMIRs in 1980, 1989 and 1991 are presented using selected sets of interplanetary magnetic field, solar wind and energetic particle data from the Pioneer and Voyager spacecraft in the outer heliosphere. The larger GMIRs, such as the 1991 event, appear to be the most plausible "trigger" and energy source for the low-frequency, long-duration radio emission events observed by the Voyagers 1 and 2 plasma wave experiment.

I. INTRODUCTION

Our heliosphere is a great bubble carved out of the local interstellar medium by the pressure of the outflowing solar wind. The interface formed by this interaction is complex with an inner boundary, the termination shock, located at a heliocentric distance, estimated to be on the order of 100 AU. The vast interplanetary region extending out to the termination shock is under the control of the Sun and reflects the continuing flow of the solar wind as well as transient solar events and the long-term (11-yr) changes in solar activity.

At 1 AU in the inner solar system, the major disturbances observed in the interplanetary medium are shock waves associated with coronal mass ejections (CMEs) and fast corotating streams that originate from coronal holes on the Sun. The effects of the passage of a shock wave are measured on time scales of the order of a day, while corotating streams may persist over a period of 5 to 10 days. The rate of occurrence of both of these phenomena varies markedly over the 11-yr solar cycle. CMEs follow the general level of solar activity, while recurrent high-speed solar wind streams are most prevalent during the period of declining solar activity. Energetic particle events are frequently associated with both types of these interplanetary disturbances. In the solar corona, CME-driven shocks are believed to be the source of most major solar energetic particle events. The resulting interplanetary shock can be a continuing source of MeV ions. A corotating stream generally forms forward and reverse shocks beyond 1 AU that accelerate solar wind ions to MeV energies and produce decreases in the intensity of galactic cosmic rays. There is also a 11-yr variation of the galactic cosmic ray intensity where the intensity level varies approximately inversely with the level of solar activity.

As these disturbances move beyond 1 AU toward the outer heliosphere, there is a superposition of solar related phenomena leading to the creation of a new type of large-scale disturbance in the interplanetary medium known as a global merged interaction region (GMIR). Originally identified by Burlaga et al. (1984b,1985,1993a), GMIRs form and evolve with increasing heliocentric distance through the coalescence of interplanetary shocks and interaction regions produced by CMEs and corotating streams. These systems are a major element in producing the long-term 11-yr cosmic ray modulation (Burlaga et al. 1991,1993b; Perko and Burlaga 1992; le Roux and Potgieter 1993; Potgieter 1993; McDonald et al. 1994). They also accelerate and transport MeV ions that provide a useful diagnostic tool for studying these systems. GMIRs along with the heliospheric neutral current sheet are two of the major dynamic features of the outer heliosphere.

The combination of Pioneers 10 and 11, Voyagers 1 and 2 at large heliocentric distances (Fig. 1), with Ulysses at high heliographic latitudes, and IMP 8 at 1 AU, constitute a unique network for studying the large-scale disturbances in the interplanetary medium and their effects on the three major energetic particle populations, anomalous, galactic cosmic rays and low-energy solar/interplanetary particles. In mid-1993, Pioneer 10 (57.3 AU) and Voyager 2 (40.4 AU) were near the plane of the ecliptic but separated by $153°$ in heliolongitude, while Voyager 1 (52.6 AU) at a heliolatitude of $\lambda = 33°$N and Ulysses in the inner heliosphere extend a three-dimensional perspective to these studies.

In Sec. II, the evidence from galactic cosmic ray and solar/interplanetary energetic particle studies for the existence of GMIRs is presented along with some of their properties that are prescribed by these observations. In Sec. III, the basic characteristics of interplanetary shocks and high-speed solar wind streams are reviewed and the process by which they merge to form GMIRs is outlined. In Sec. IV, examples of three GMIRs are given using selected sets of the available interplanetary magnetic field, solar wind and energetic particle data. Section V summarizes our conclusions, discusses theoretical studies of the modulation effects of GMIRs, the relation of GMIRs to low-frequency heliospheric radio emission, and suggests new areas for future studies.

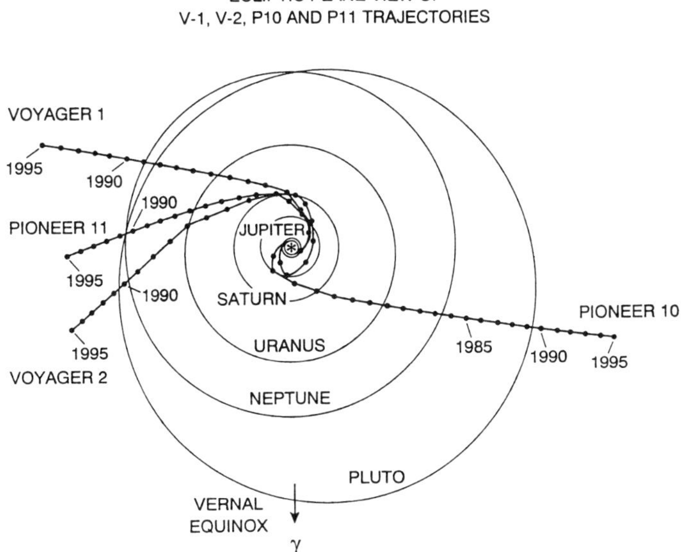

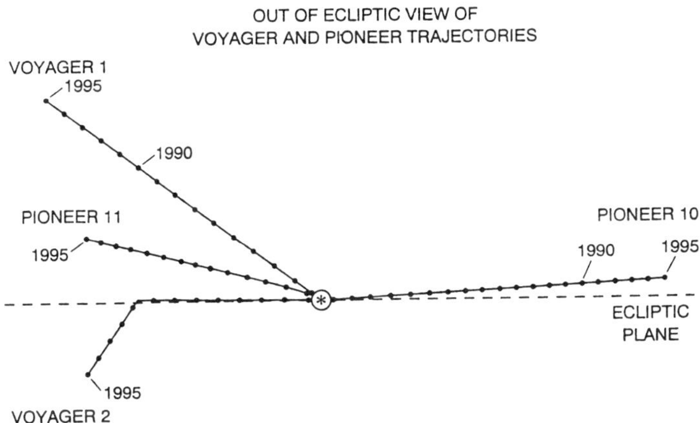

Figure 1.   Ecliptic and out-of-the-ecliptic plane projections of the Voyagers 1 and 2 and Pioneers 10 and 11 trajectories from launch through the end of 1994.

These heliospheric phenomena should have broader astrophysical application, because GMIRs are thought to be present around other stars with stellar winds and transient coronal disturbances.

II. THE EVIDENCE FOR GLOBAL MERGED INTERACTION REGIONS FROM ENERGETIC PARTICLE STUDIES

A. Studies of Cosmic Ray Modulation

The long-term (11-yr) modulation of galactic cosmic rays generally begins soon after the onset of solar activity in the new cycle and the initial recovery phase starts about 0.5 yr after the maximum of solar activity. For the limited period of observations ($\sim$4 cycles), there is a strong indication that alternate cycles have different time histories over the solar minimum period suggesting the existence of a 22-yr cycle. This long-term modulation is most probably a combination of diffusion, convection and adiabatic energy loss processes, as well as the effects of large-scale gradient and curvature drifts in the interplanetary magnetic field. Establishing the relative role of these different processes and determining the nature of the coupling between solar activity and the interplanetary medium that governs them, are vital to understanding cosmic ray modulation.

Jokipii and his coworkers (Jokipii et al. 1977; Jokipii and Thomas 1981; Kota and Jokipii 1983) have emphasized that drift effects should play a major role in determining the flow pattern of cosmic rays in the heliosphere, and the cosmic ray intensity should vary inversely with the angle between the neutral current sheet and the solar equator. With the reversal of the solar magnetic field near solar maximum, there is a reversal of the drift-produced flow patterns of cosmic rays in the heliosphere. When the Sun's magnetic field in the northern hemisphere is outwardly directed ($qA>0$), positively charged particles flow in over the solar poles and out along the neutral current sheet. This condition prevailed from the maximum of cycle 20 (1969/1970) to the maximum of cycle 21 (1980). When the direction of the solar field reversed ($qA<0$) (1980–1990), the flow pattern for positively charged particles is inward along the neutral current sheet and outward over the solar poles.

With the availability of cosmic ray data for several different particle types over a broad range of rigidities from multiple spacecraft that spanned significant portions of a solar cycle and extended to heliocentric distances beyond 15 AU, it was established (McDonald et al. 1981) that the long-term modulation for cycle 21 (Fig. 2) occurred in a series of well-defined "step decreases" that propagated radially outward from 1 AU to $\sim$25 AU with the solar wind speed on the order of 450 km s^{-1} with essentially the same percentage decrease at these large heliocentric distances as at 1 AU. The galactic cosmic ray data in Fig. 2 have been averaged over 26-day periods ($\sim$1 solar rotation) which acts as a filter for reducing the effects of shorter-term transient decreases. The three intensity decreases over the 1978–80 time period are approximately exponential with time and extend over a period of

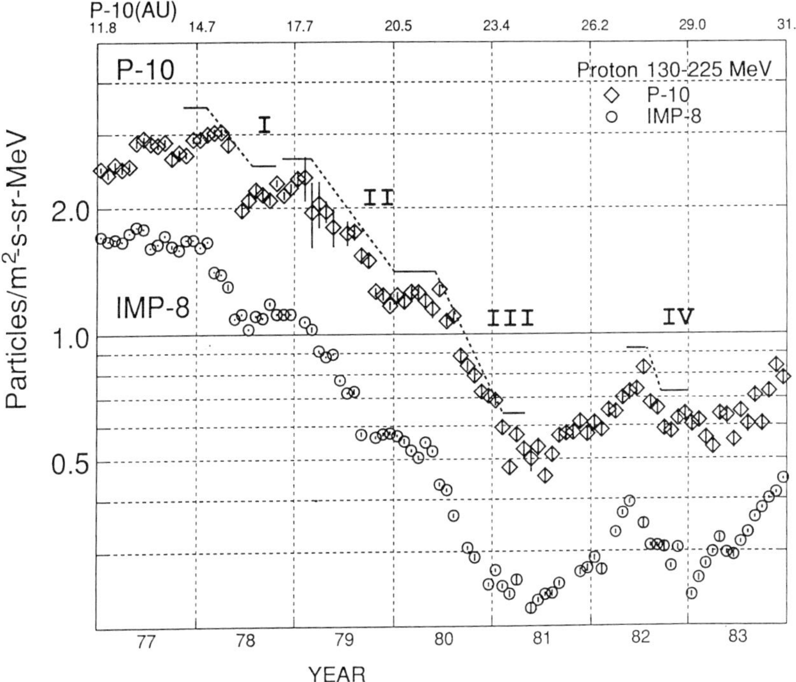

Figure 2.   Time histories (26-day averages) of 130 to 225 MeV galactic cosmic ray hydrogen from IMP 8, and Pioneer 10 for 1977 through the end of 1983. The Pioneer 10 heliocentric distances are indicated along the abscissa at the top of the plot. The plateau and step decreases discussed in the text are identified in the plot (figure after Fujii and McDonald 1995).

4 to 6 months. They occur simultaneously for ions and electrons over an extended rigidity range (0.2–3 GV for ions). At the lower rigidities (<1 GV), the intensity decreases are followed by plateau periods, while at higher rigidities (~2 GV) there is a small rate of recovery. These same decreases are present at neutron monitor energies but the plateau regions are distorted by the larger, short-term partial recovery of these higher-rigidity particles (McDonald et al. 1994). Some 22 years earlier in the same phase of cycle 19, Lockwood (1960) concluded that the intensity decrease from 1955–59 occurred in a series of several sudden drops followed by periods of partial recovery.

Measurements of the high-energy electron intensity over the same time interval provides a different perspective on the nature of these step-decreases, because the drift-produced flow pattern of the negatively charged particles is the inverse of that experienced by the positively charged ions with the same value of βR (where β is the ratio of the particle velocity to the velocity of light and R is the particle rigidity). The combined time history of 1.24-GeV electrons ($\beta R = 1.24$) (27-day averages) from the University of Chicago

ISEE/ICE electron experiment (Tuska 1990) and of 175- to 450-MeV/nucleon He^{++} ($\beta R = 1.1$) (see Fig. 3) shows that steps II and III (Fig. 2) occur simultaneously for ions and electrons as previously reported by Evenson and Meyer (1984). Furthermore, the relative intensity changes in steps II and III are essentially the same for these positively and negatively charged particles. There is a small positron contribution to the electron intensity measurements ($\leq$15–20% at 1 GeV) but it is not large enough to distort the basic similarity of the electron and helium response to step-decreases. The more rapid recovery of electrons (Fig. 3) following the 1980 reversal of the solar magnetic fields (Hoeksema 1989) suggests a significant influence of drifts on particle transport during the recovery period.

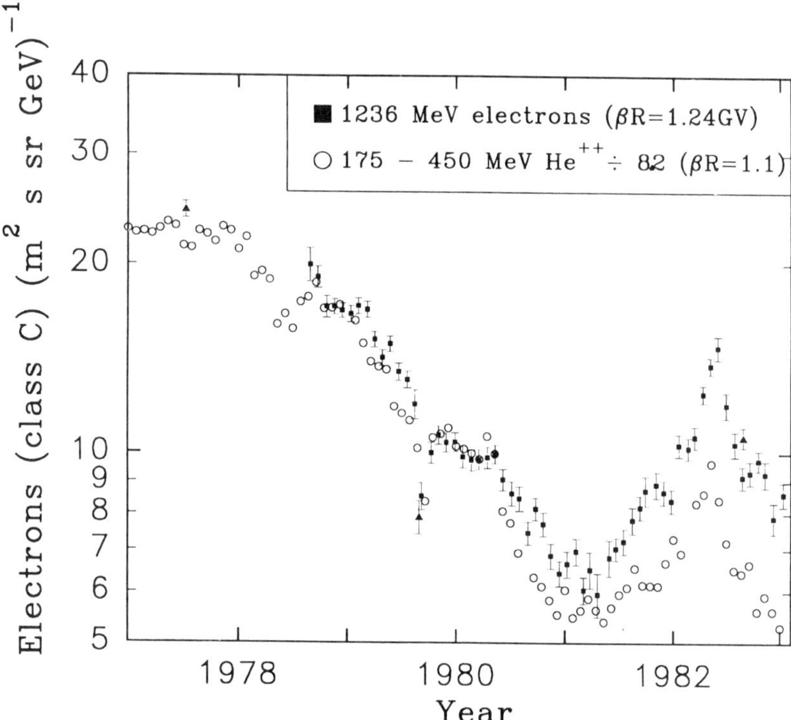

Figure 3. Balloon (solid triangles) and ISEE-3 measurements (solid squares, University of Chicago electron experiment, 27-day averages) of 1.24 GeV electrons ($\beta R = 1.24$ GV) (Tuska 1990) and of 1175 to 450 MeV/n He^{++} ($\beta R = 1.1$ GV) (IMP 8). The He^{++} measurements have been normalized to the electron data over the time period 1978.78 to 1979.0 (figure after McDonald et al. 1994).

In cycle 22 the Voyager 1 data (Fig. 4) between 29 and 42 AU at a heliolatitude of 32°N show step-decreases that are very similar to those observed in the previous cycle. Near the plane of the ecliptic, Pioneer 10, Voyager 2 and IMP 8 observed similar step-decreases, but there were also intensity changes produced by the increasing inclination of the current sheet at these lower values of heliolatitude (McDonald et al. 1994). Again the magnitude of the step decreases for the comparable energy interval of a given cosmic ray species is essentially the same at 48 AU as at 1 AU (McDonald et al. 1994).

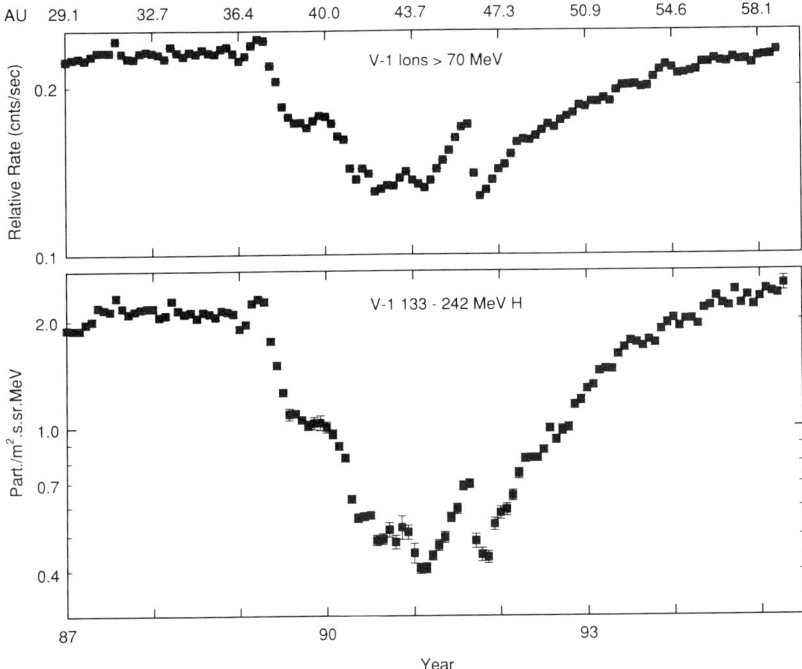

Figure 4.   Voyager 1 time histories of 133 to 242 MeV galactic cosmic ray hydrogen and the integral intensity of galactic cosmic ray ions >70 MeV/n from 1987 through 1993.6

These observations of the intensity changes of galactic cosmic ray nuclei and electrons establish that most of the long-term modulation occurs through a relatively small number ($\sim$4–5) of step-decreases that, as will be shown in Secs. III and IV, are produced by GMIRs. However, there is a significant modulation role for the current-sheet-related drift effects, particularly in the $qA<0$ periods.

The role of GMIRs in producing the long-term modulation imposes severe requirements:

1. The modulation effectiveness of a given GMIR must persist over an extended period of time ($\sim$1.5–1.8 yr) in order to explain the lack of any significant recovery at rigidities <3 GV until after the period of maximum solar activity.

2. The presence of a plateau region following the step-decreases in the $qA>0$

cycle suggest that GMIRs must extend to high latitudes and essentially form closed structures.

The geometrical picture that emerges from these requirements is that of a closed, thick shell that is convected out to the modulation boundary by the solar wind. It is not known whether the modulation effects of the GMIR extend beyond the termination shock. Processes for producing modulation in the boundary layer of the heliopause have been discussed by Potgieter and le Roux (1987) and Quenby et al. (1990).

B. Low-Energy Ions of Solar/Interplanetary Origin

In the outer heliosphere the MeV ions offer another means for studying large-scale interplanetary disturbances. Beyond some 25 AU a different type of energetic particle event has been observed that represents the evolution and superposition of energetic particle events from the inner heliosphere and the continuing acceleration of these ions by large-scale interplanetary disturbances (Van Allen and Mihalov 1990; McDonald and Selesnick 1991; Decker et al. 1991; Dröge et al. 1992). In the later sections these disturbances will be identified with GMIRs.

Five examples of these events are seen in the time history of MeV protons (Fig. 5) at heliocentric distances between 25 and 53 AU for the 1989–1992 cycle 22 period of high solar activity. Van Allen and Mihalov (1990) showed that event II (Fig. 5) was associated with a large decrease in the cosmic ray intensity. McDonald et al. (1994) showed that the three step-decreases observed at Pioneer 10 over the 1989–1990 period had the same temporal boundaries as the 5 MeV proton events.

The study of GMIRs and their effects on the various energetic particle populations in the heliosphere is important (1) as a vital element in the long-term modulation of galactic and anomalous cosmic rays, and (2) for understanding particle acceleration and transport in the heliosphere.

A key question that has yet to be answered is whether the MeV ions and electrons observed in the outer heliosphere 4 to 5 months after an outburst of solar activity are the solar energetic particles accelerated in the inner heliosphere and continually re-energized by the shocks within the GMIR or whether they are more recently accelerated by these shocks from a seed population of superthermal solar wind ions. Indeed, there may be contributions from both processes. Establishing the relative contributions will require more detailed studies of the radial evolution of these events as well as detailed modeling of the processes involved.

III. MERGED INTERACTION REGIONS

A. Interaction Regions at 1 AU

The earliest measurements of the solar wind showed that the density N, proton temperature T, and magnetic field strength B are relatively high ahead of fast

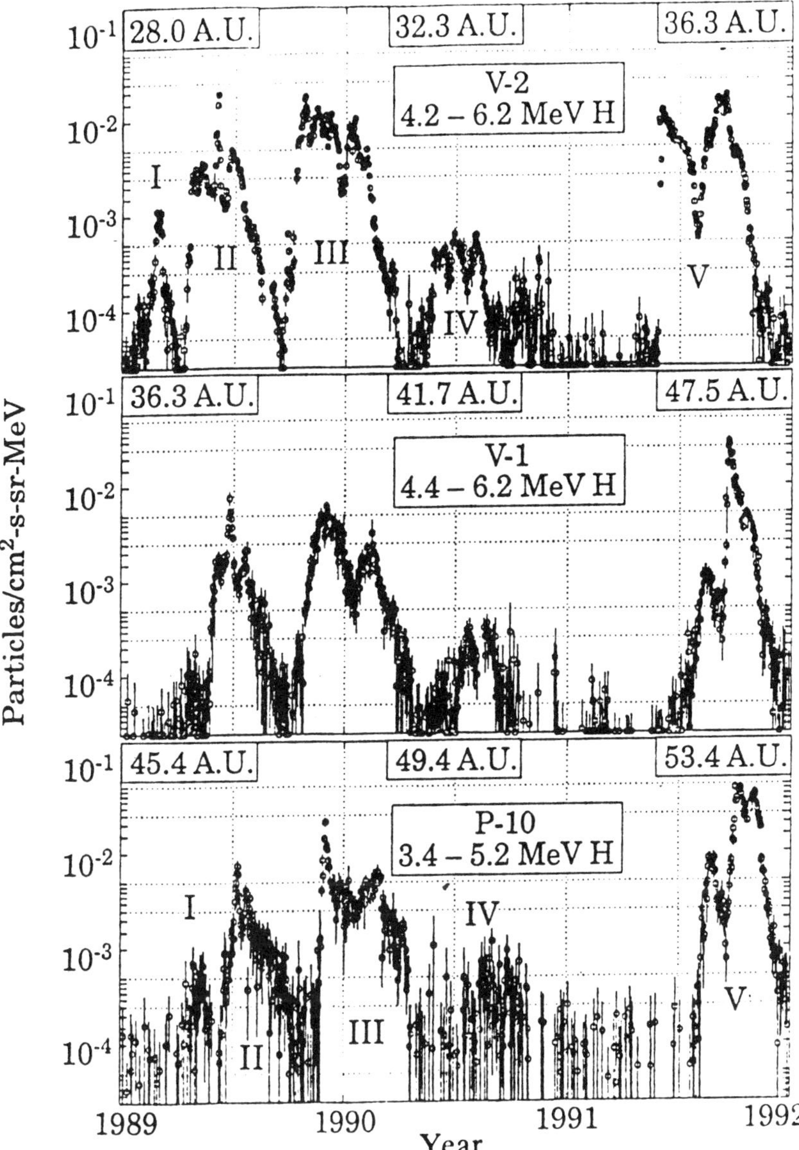

Figure 5.   Time histories (24-hr averages) of low-energy protons from the CRS experiment on Voyagers 1 and 2 and the Pioneer 10 CRT experiment for the period of maximum solar activity of cycle 22 1989–1992. The five major energetic particle events of solar/interplanetary origin observed over this period are identified by roman numerals.

streams (see, e.g., Neugebauer and Snyder 1966) implying that the pressure $P = Nk(T_p + T_e + B^2/8\pi)$ is enhanced in this region. Burlaga and Ogilvie (1970) pointed out that these high-pressure regions play a fundamental role in the dynamics of the solar wind, and they introduced the term "interaction regions" for such regions. At 1 AU interaction regions are generally of two types: corotating interaction regions (CIRs) ahead of large-scale corotating streams extending over at least a few AU, and local interaction regions (LIRs) ahead of more localized transient ejecta and behind transient shocks. CIRs are large-scale, long-lasting, quasi-stationary features with a spiral geometry similar to that of spiral magnetic field lines. LIRs are transient features that tend to be concentric with the Sun but are limited in longitudinal, latitudinal and radial extent.

The term "corotating interaction region" (CIR) was introduced by Smith and Wolfe (1976) to describe the recurrent interaction regions observed by Pioneer just beyond the orbit of Earth, but CIRs are also observed at 1 AU. An example of a corotating stream at 1 AU is shown in Fig. 6 from Burlaga et al. (1984b). Corotating streams have low densities (because they originate in coronal holes) and high temperatures. The density ahead of a corotating stream is high, both because it typically advances into a high-density region (the heliospheric plasma sheet, which is presumably an extension of a coronal streamer into the heliosphere as described by Borrini et al. [1981]) and because the speed gradient compresses the plasma kinematically (see, e.g., Parker 1963). The magnetic field strength is always high ahead of a corotating stream at 1 AU because of kinematic compression of the ambient field. The front of a stream is generally bounded by a thin stream interface (Burlaga 1974) across which the density drops abruptly, the temperature and speed increase abruptly, and the magnetic field strength reaches a maximum. The stream interface is marked by a vertical line in Fig. 6. The maximum pressure in an interaction region generally occurs at the stream interface. Corotating streams and CIRs at 1 AU generally produce no net change in the cosmic ray intensity, as evident in Fig. 6, although they usually produce a transient decrease in the intensity that happens to be absent in this figure (see, e.g., Burlaga 1984a; Kota and Jokipii 1983; Iucci et al. 1984a).

Local interaction regions near 1 AU occur ahead of ejecta and behind transient shocks. When an ejection moves fast enough to drive a shock, the LIR is most intense between the shock and the ejection, but it extends beyond the ejection following the shock. Examples of two ejecta-driving shocks are shown in Fig. 7 (Burlaga et al. 1984b). In the ejecta, the speeds are only moderately high, the densities are variable but generally low, and the temperatures are typically highly variable and either low or near average solar wind values. The magnetic field strengths in an ejection are typically greater than or equal to the average solar wind value. The magnetic field is irregular in most ejecta (Hundhausen 1972), but it is highly ordered in the subset of ejecta called magnetic clouds (Burlaga 1991). Ahead of an ejection, the density, proton temperature and magnetic field strength are higher than

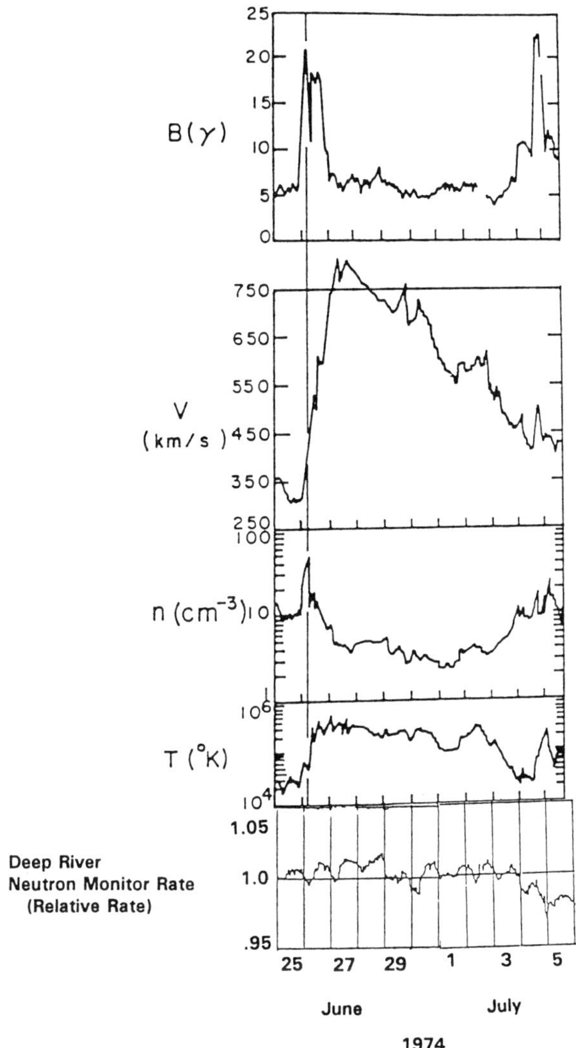

Figure 6.   Corotating stream and interaction region. The very fast stream is a corotating stream, identified by its low density n, high proton temperature T, and the presence of a stream interface indicated by the vertical solid line. The magnetic field strength B is strong at the leading edge of the stream, and marks the corotating interaction region (CIR). This corotating stream had no net effect on the cosmic ray intensity measured at Deep River (bottom panel), although, in general, there is a depression but no net decrease in cosmic ray intensity associated with the passage of a corotating stream.

average, owing both to kinematic compression by the advancing ejection and to compression by the shock. The pressure is obviously high ahead of such an ejection, giving a localized transient interaction region, i.e., a LIR.

Local interaction regions at 1 AU generally produce decreases in the cosmic ray intensity lasting a few days ("Forbush decreases"). The cosmic ray intensity decrease is primarily related to the enhanced magnetic field strength (Barouch and Burlaga 1975; Iucci et al. 1984b), as one can see in Fig. 7. In the case where a fast magnetic cloud drives a strong shock, the sheath between the shock and the magnetic cloud is very turbulent and the magnetic fields in the magnetic cloud are relatively undisturbed. In such a case, the cosmic ray intensity decrease is produced primarily by the turbulent sheath and only to a lesser extent by the smooth fields of the magnetic cloud (Zhang and Burlaga 1988). However, slow-moving magnetic clouds and ejections with strong magnetic fields and no shocks can also produce decreases in the cosmic ray intensity (see, e.g., Cane et al. 1993). The largest cosmic ray intensity decreases at 1 AU are produced by local interaction regions associated with fast ejecta and shocks. The longest lasting cosmic ray intensity decreases (long-lasting Forbush decreases; Lockwood 1971) are caused by a series of closely spaced interaction regions including LIRs (Barouch and Burlaga 1975). The magnetic field strengths in LIRs associated with transient ejecta and Forbush decreases tend to be larger than those associated with CIRs (Burlaga and King 1979).

B. Merged Interaction Regions

Beyond 1 AU interaction regions tend to coalesce (Burlaga 1983; Burlaga et al. 1983) and form merged interaction regions (MIRs) (Burlaga et al. 1984b,1985). The subject of MIRs has been reviewed by Burlaga (1988), and the mechanisms that can produce MIRs are discussed in more detail in Burlaga (1984) as well as in the forthcoming book by Burlaga (1995). Only some of the essential characteristics of MIRs are reviewed here.

Evidence for the formation of MIRs with increasing distance from the Sun and the corresponding restructuring of the heliosphere was first presented by Burlaga (1983; Burlaga et al. 1983,1984b). The radial variation of the magnetic field strength is shown in Fig. 8, where 10-hr averages of the magnetic field strength (normalized by a nominal value of the Parker magnetic field strength B_p) are shown as a function of time for three 170-day intervals. Measurements during the first interval were obtained in the range from 1.0 to 2.6 AU, those in the second interval from 4.0 AU to 5.2 AU, and those in the third interval from 6.9 AU to 8.2 AU. Near 1 AU, there are several narrow interaction regions per solar rotation and there is no clear pattern among them. At intermediate distances, fewer, broader and more intense interaction regions appear, approximately two per solar rotation. Between 6.9 AU and 8.2 AU only a single, broad intense MIR per solar rotation period is observed. That this evolution is a radial effect rather than a temporal effect was demonstrated by Burlaga and Mish (1987) using simultaneous observations from ISEE-3

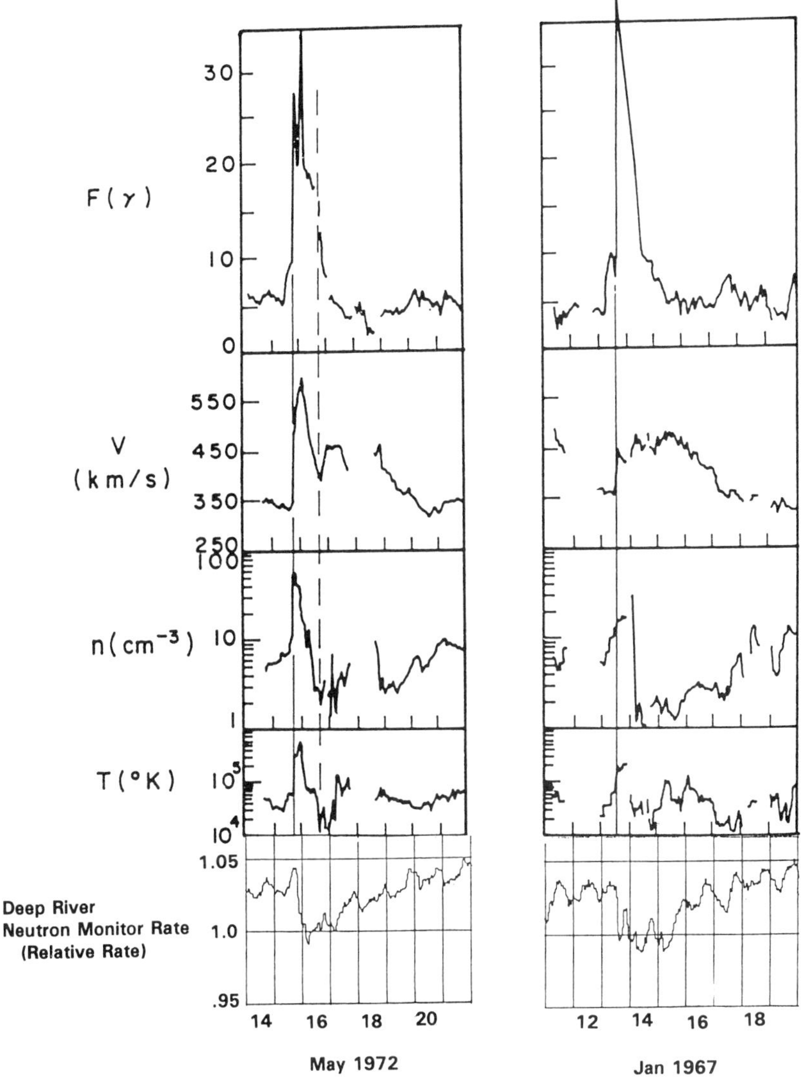

Figure 7.   Transient ejecta, local interaction regions (LIRs) and Forbush decreases. The ejecta are streams with relatively low density preceeded by shocks (the vertical solid lines) and local interaction regions with very intense magnetic fields F. Each ejection is associated with a Forbush decrease shown by the decrease in the cosmic ray intensity measured at Deep River (bottom panel). The Forbush decrease occurs primarily in the turbulent sheath between the shock and the ejection.

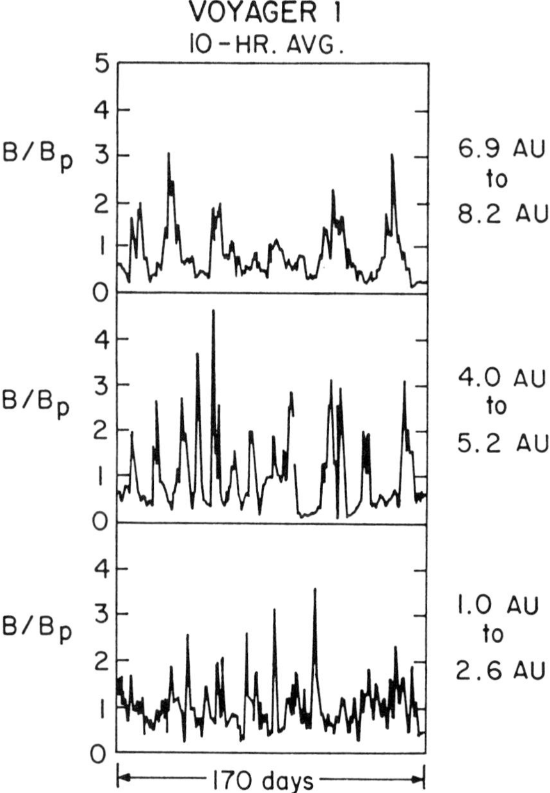

Figure 8. Formation of merged interaction regions with increasing distance from the Sun. The magnetic field strength, normalized with respect to a nominal value B_p from Parker's spiral field model, are shown for three different distance intervals (figure from Burlaga 1984).

and Voyager 1. With increasing distance from the Sun, small-scale structures merge to form larger-scale structures, and the interplanetary medium becomes increasingly ordered. The process is irreversible, i.e., one cannot reconstruct the conditions at 1 AU from the observations at 8 AU, so that memory of the source conditions is lost as a result of the formation of MIRs (Burlaga 1983).

One mechanism by which MIRs can form is illustrated in Fig. 9, which shows speed measurements made by Helios 1 between 0.3 AU and 1 AU (middle panel) and Voyager 1 between 8.0 AU and 8.5 AU (top panel). Several streams were observed by Helios 1, including one near day 180 in which the speed exceeded 800 km s^{-1}. A kinematic projection of the speed profile observed by Helios 1 to the position of Voyager 1 is shown in the bottom panel of Fig. 9. One cannot use kinematic models to describe dynamical processes in the outer heliosphere, because the effects of pressure gradients are very important there. Nevertheless, the kinematic mapping shows intuitively how

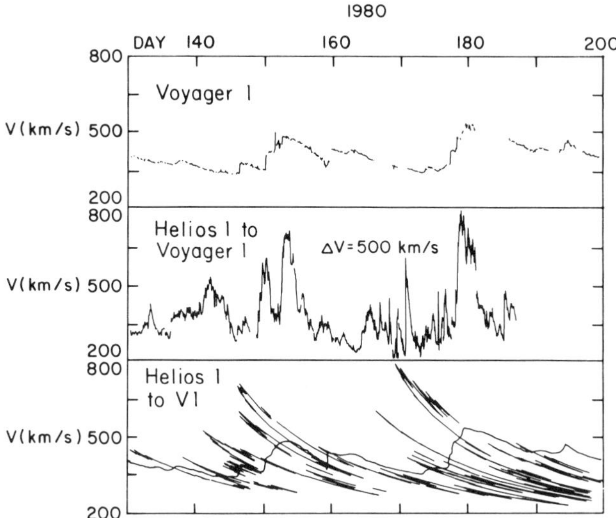

Figure 9. *Middle panel:* Observations of a series of streams at Helios 1 near 0.3 AU, shifted in time to allow for the propagation time to Voyager 1. *Top panel:* Observations of the speed at Voyager 1 demonstrating the damping and merging of the streams observed by Helios 1. *Bottom panel:* A kinematic, constant-speed projection of the speeds observed by Helios 1 to the position of Voyager 1, together with a smoothed connected version of the Voyager 1 observations.

the complex speed profile at Helios 1 tended to evolve to the simpler speed profile at Voyager 1. The three fast streams observed by Helios 1 between day 140 and day 160 interacted with one another to produce a very different speed profile at Voyager 1, with only vestiges of the original streams evident. The three interaction regions (not shown in Fig. 9) associated with the three streams at Helios 1 coalesced to form a single MIR at Voyager 1. Similarly, the two fast streams and other irregular speed variations observed by Helios 1 between days 165 and 185 interacted to produce a single compound stream at Voyager 1, and the interaction regions at Helios 1 coalesced to form a single MIR at Voyager 1. Beyond the orbit of Voyager 1 the two MIRs described above presumably coalesced to form a single, more complex MIR.

MIRs are essential to understanding cosmic ray modulation, because the decreases in the cosmic ray intensity are related to the strong magnetic fields in MIRs (Burlaga et al. 1985). One can compute the local cosmic ray intensity profile from local measurements over a period of a year or more by solving a simple first-order differential equation for the rate of change of cosmic ray intensity with the magnetic field profile observed at a spacecraft as an input. However, a complete understanding of the effect of MIRs on cosmic rays requires consideration of the geometry of the MIRs.

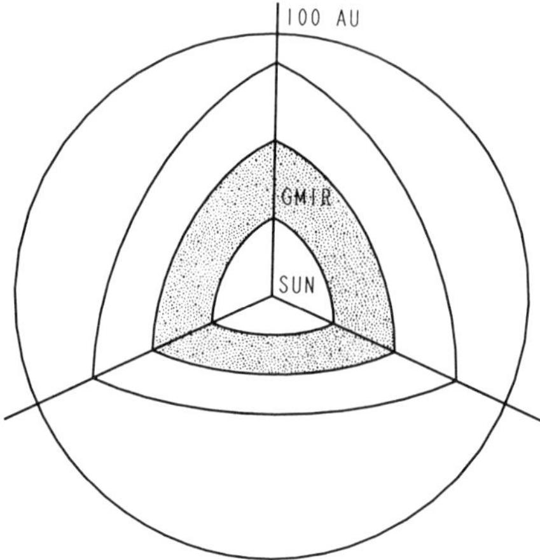

Figure 10.   A global merged interaction region (GMIR). The GMIR has the topology
of a closed shell containing intense irregular magnetic fields. The actual geometry
of a GMIR can be much more complex than the spherical geometry shown here.
The essential characteristic is a closed shell-like structure that propagates away from
the Sun at the solar wind speed.

C. Classification of Merged Interaction Regions

There are many configurations of MIRs in the outer heliosphere. However,
it is convenient and meaningful to identify three broad categories of MIRs
based on their geometry: corotating MIRs (CMIRs), local MIRs (LMIRs)
and global MIRS (GMIRs) (Burlaga et al. 1993).

Corotating MIRs are MIRs with spiral forms extending many AU in
the radial direction; they are produced by the coalescence of two or more
corotating interaction regions. The corotating MIRs and rarefaction regions
can produce successive decreases and increases in the cosmic ray intensity,
typically quasi-periodically over a period of several months, but the CMIRs
generally produce little or no net modulation (Burlaga et al. 1985,1987,1991).

The local MIRs have a limited radial, latitudinal and longitudinal extent.
They do not extend around the Sun, and they do not necessarily extend to
high latitudes. LMIRs are formed in many ways by interactions among
transient ejecta and shocks and possibly also with CIRs. Because the analysis
of such transient structures requires simultaneous observations from several
spacecraft with separations on the order of 1 AU, little is known about their
geometries. LMIRs can have a significant effect on the cosmic rays in some
localized volume in the outer heliosphere, just as LIRs can produce large
Forbush decreases at 1 AU. However, it is essential to understand that LMIRs
are different from LIRs, being larger structures formed from the irreversible
coalescence of one or more LIRs, shocks and CIRs. Thus, the decrease in
cosmic ray intensity associated with a LMIR is not a Forbush decrease, which
is the result of a single LIR related to a single shock, sheath and ejection.

Global MIRs are quasi-spherical, shell-like MIRs extending 360° around the Sun in the ecliptic, reaching at least 30° in heliographic latitude, and probably closing at the poles (Fig. 10). GMIRs correspond to the structures identified by Burlaga et al. (1984*b*) as the shells that caused the propagating step-decreases in the cosmic ray intensity identified by McDonald et al. (1981) during the last solar cycle. GMIRs are associated with systems of transient flows, although not every system of transient flows produces a GMIR. When the Sun becomes episodically active, transient ejecta and shocks can interact with one another and with corotating streams to form a large-scale MIR. This MIR persists over several solar rotations and moves outward through the heliosphere at the solar wind speed, acting as a barrier to cosmic rays. Cliver et al. (1987) showed that a system of transient flows can be related to multiple solar flares and coronal mass ejections.

The magnetic fields in a GMIR are strong over a large volume of space; their variation is extremely complex, possibly multifractal in some cases (Burlaga et al. 1993*b*). The internal structure or a GMIR probably varies considerably from one point to another, depending on the particular ejecta, shocks, LIRs, and CMIRs that interact in various radial sections of the GMIR. The essential features of a GMIR are its strong magnetic fields and its three-dimensional, shell-like topology, which forms a barrier to cosmic rays entering the heliosphere from all directions. An illustration of the shell-like topology of a GMIR is shown in Fig. 10, but one must remember that the actual boundaries and internal structure can have a very complex, inhomogeneous and variable form. A GMIR might be preceded by a strong shock (Burlaga 1993) just like a LIR ahead of a fast ejection, but the GMIR shock might also have been produced by the coalescence of two or more shocks, and there might be multiple shocks associated with a GMIR.

GMIRS and the associated step-decreases in the cosmic ray intensity are fundamentally different from the LIRs associated with ejecta, shocks and Forbush decreases. As this review demonstrates, the passage of a GMIR is related to a long-lasting step-decrease in the cosmic ray intensity. In particular, the decrease in the cosmic ray intensity is related to an increase in the intensity of the magnetic fields in the GMIR (Burlaga et al. 1985). The step-decrease in cosmic ray intensity is observed over a large range of latitudes and longitudes, by definition; decreases in cosmic ray intensity of almost any sort are related to the enhanced magnetic fields in MIRs (Burlaga et al. 1985). Thus, the presence of a GMIR can be inferred from the cosmic ray data alone, provided that the data are from spacecraft widely separated in longitude and latitude (McDonald et al. 1994).

IV. EXAMPLES OF GLOBAL MERGED INTERACTION REGIONS AND EVENTS LEADING TO THEIR FORMATION

A. The Step-Decrease of 1980

Figure 11 shows that a major step-decrease in the cosmic ray intensity occurred during 1980, when Voyager 2 was near 8.5 AU and Pioneer 10 was near 25 AU. The region near 8.5 AU is a transition region, where streams are still present but are damping out and where interaction regions are beginning to merge to form MIRs. Thus, these Voyager data are not optimal for studying the relation between MIRs and the cosmic ray intensity changes, because the velocity affects the cosmic ray intensity profile as well as the magnetic field strength. There is no simple relation between the changes in cosmic ray intensity and the magnetic field strength in this region relatively close to the Sun. However, the Voyager 2 data for 1980 do provide valuable insights concerning how the transition may occur from the isolated flows and interaction regions in the region within 8 AU to the compound streams and MIRs significantly beyond 8 AU.

The decrease in the cosmic ray intensity observed by Voyager 2 during 1980 is shown in Fig. 11 with the solar wind speed, the magnetic field strength and the 4.2 to 6 MeV hydrogen intensities. While the step-decrease is actually very broad, it cannot be related to any single GMIR in the data in this figure, because GMIRs are not formed at this distance. A corresponding step-decrease at 1 AU was related to a system of transient flows (Burlaga et al. 1984b), which led them to propose that these broad, step-decreases are related to shells of intense disturbed magnetic fields related to systems of transient flows. The 1980 decrease in the cosmic ray intensity observed by Voyager 2 can be resolved into three components, which we shall call I, II and III. Component I of the decrease in the cosmic ray intensity observed by Voyager 2 during 1980 is a large, abrupt decrease at about 1980.47. This component is associated with an interaction region containing very strong magnetic fields. This is shown in more detail in Fig. 12 where the interaction region is marked "B" (Burlaga et al. 1984b, Fig. 10). The interaction region B that produces the abrupt decrease in cosmic ray intensity occurs ahead of a moderately fast stream (see also Fig. 11), which is probably a transient ejection, because it is associated with two forward fast shocks and no reverse shock. A peak in the low-energy protons in Fig. 11 is probably related to the shocks driven by the stream. The interaction regions ahead of component A were corotating interaction regions associated with forward-reverse shock pairs. They produced temporary decreases in the cosmic ray intensity, but no net modulation.

Component II of the cosmic ray intensity decrease observed by Voyager 2 during 1980 is the slowly varying decrease from 1980.5 to 1987.8 (Fig. 11). During this interval, the interaction regions tend to be periodic with moderate magnetic field strengths, and there are no broad streams with speeds greater than 500 km s^{-1}. Thus, many, if not most, of the interaction regions during this interval are probably corotating interaction regions, which produce small, localized, recurrent decreases in the cosmic ray intensity but no large net change. There are probably some shocks and transient flows during this period, because the speed fluctuations are not periodic and there is a large flux of low energy particles throughout most of the period, suggesting the presence of both transient and corotating shocks. Thus, there is probably a mixture of

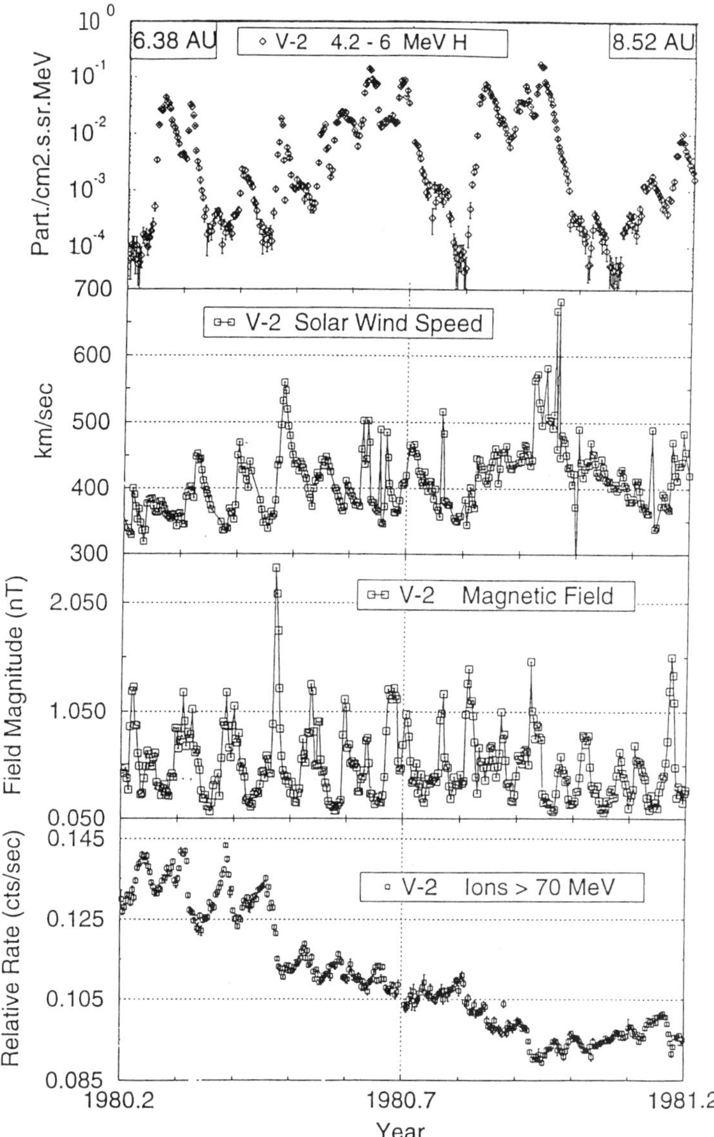

Figure 11.   Time history of Voyager 2 (24-hr averages) 4.2 to 6.2 MeV H along with the daily averages of the solar wind speed (from the Voyager 2 solar wind experiment), the magnitude of the interplanetary magnetic field and the counting rate of galactic cosmic rays with energies >70MeV from the CRS experiment for 1980.2 to 1981.2 time period.

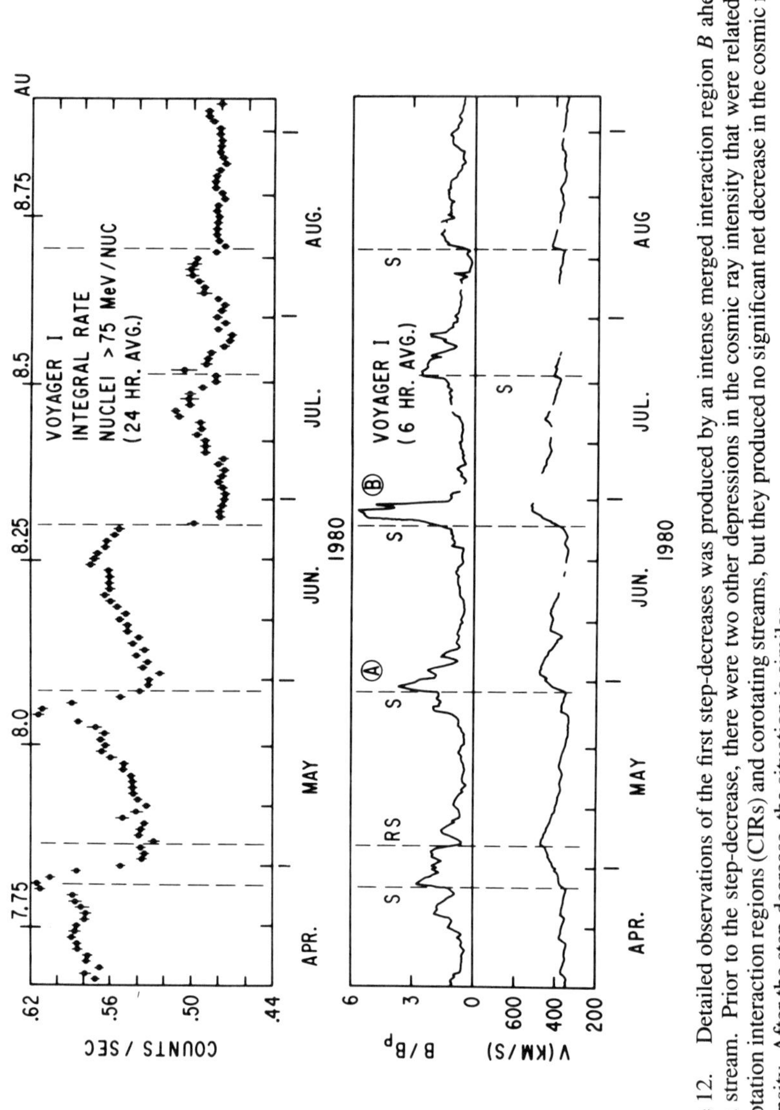

Figure 12.   Detailed observations of the first step-decreases was produced by an intense merged interaction region B ahead of a stream. Prior to the step-decrease, there were two other depressions in the cosmic ray intensity that were related to corotation interaction regions (CIRs) and corotating streams, but they produced no significant net decrease in the cosmic ray intensity. After the step-decrease, the situation is similar.

transient and corotating flows, the former being primarily responsible for the small net decrease in the cosmic ray intensity during the interval.

Component III of the cosmic ray intensity decrease observed by Voyager 2 during 1980 is the relatively large and abrupt change from 1980.8 to 1980.92 (Fig. 11). This decrease is associated with the leading edge of a broad, fast compound stream. The front of the stream is accompanied by intense low-energy particle activity, suggesting the presence of strong shocks, hence, the possibility that at least one or two components of the transient stream are transient ejecta. The front of the stream is also associated with two relatively strong interaction regions and only a short interval with weak magnetic fields. Thus, we are probably observing the formation of a MIR ahead of the compound stream.

In summary, the broad step-decrease observed by Voyager 2 during 1980 is associated with a complex system of flows and interaction regions. There are two fast streams that probably contain transient ejecta and shocks, which are organizing the formation of MIRs, but the MIRs are not fully formed at this distance.

The corresponding cosmic ray and solar wind speed observations at Pioneer 10 near 24.5 AU are shown in Fig. 13. Again, the decrease is rather broad, extending from 1980.5 to 1981.1. Again, there are three components: I, a large relatively abrupt decrease from 1980.4 to approximately 1980.7; II, a plateau from 1980.4 to 1980.95; and III, a large abrupt decrease from 1980.95 to 1990.1. Unfortunately, there are no magnetic field data for this event. The similarity between the cosmic ray intensity at Pioneer 10 with that at Voyager 2, on the other side of the Sun, suggests that there is an underlying global flow configuration. This is manifested by the two fast compound streams seen in the Pioneer 10 data of Fig. 13, which correspond to the two fast flows observed by Voyager 2. The two large, step-like cosmic ray intensity decreases observed by Pioneer 10 occur at the leading edges of the two fast-flow systems, and we can be confident that there were related regions of intense magnetic fields there, probably MIRs. The plateau in the cosmic ray intensity, II, is associated with relatively slow and recurrent speed fluctuations, which probably represent the remnants of corotating streams.

The evolution of the flow configuration seen by Pioneer 10 beyond 24.5 AU has not been computed. We would anticipate that the flow might evolve as follows. The second, fast-flow system (region III) will advance into the region II containing slow corotating flows, making the plateau region narrower, and possibly eventually eliminating it. Then, the cosmic ray intensity will decrease almost continually throughout the interval. The corotating interaction regions in region II will probably coalesce with the MIR ahead of the compound stream in region III to form a broader MIR. Eventually, this MIR might coalesce with that ahead of the stream in region I, forming a single GMIR. The net result might be similar to that observed by Voyager 1 at 39.9 AU during 1989, as described below.

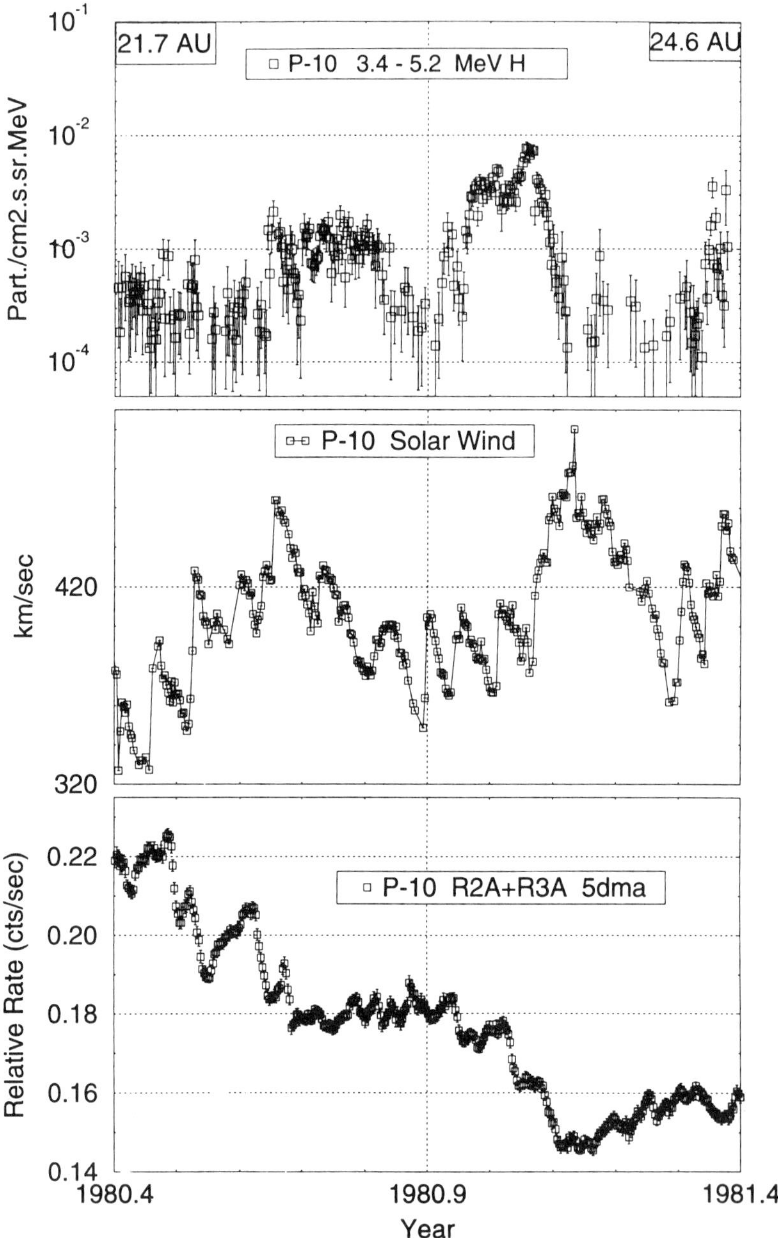

Figure 13.   Time history of Pioneer 10 (24-hr averages) 3.4 to 5.2 MeV H, solar wind speed (from the Pioneer 10 Ames plasma analyzer) and the counting rate of galactic cosmic rays >56 MeV/n for the 1980.4 to 1981.4 time period.

B. The Step-Decrease of 1989 and the Associated Global Merged Interaction Region

In solar cycle 22 a major step-decrease in the cosmic ray intensity was observed by Voyager 1 from approximately 1989.35 to approximately 1989.55, as shown at the bottom of Fig. 14. This figure also shows that the decrease in cosmic ray intensity was associated with an extended region of intense magnetic fields, a GMIR, which has been discussed in detail by Burlaga et al. (1993*a*). They showed that the cosmic ray intensity observed by Voyager 1 can be computed from the magnetic field strength observed in the GMIR using the CR-B relation of Burlaga et al. (1985). The magnetic field strength in the GMIR is filamentary, and the two principal decreases occur in association with the two filaments with the strongest fields. A broad region of intense low-energy particles was associated with the GMIR and the cosmic ray intensity decrease, but the low-energy particles arrived after the GMIR and after the cosmic ray intensity began to decrease.

A similar step-decrease in cosmic ray intensity was observed by Voyager 2 near the ecliptic, approximately 30° below Voyager 1. This decrease in the cosmic ray intensity was also associated with an extensive region of very intense magnetic fields, a GMIR (Burlaga et al. 1993*a*). The magnetic field strength profile at Voyager 2 differed in detail from that at Voyager 1, but in both cases the magnetic field strength was stronger than average for an extended period of time. Again, the cosmic ray intensity observed by Voyager 2 can be computed from the magnetic field strength observed in the GMIR, using the CR-B relation of Burlaga et al. (1985).

Step-decreases in the cosmic ray intensity were also observed by Pioneer 10 and Pioneer 11, and one can assume that each was accompanied by intense magnetic fields. Thus, the step-decreases in cosmic ray intensity during 1989 indicate the passage of a single, shell-like region extending around the Sun and to high latitudes, containing intense magnetic fields, a GMIR.

C. The Global Merged Interaction Region Associated with the Intense Solar Activity of March/June 1991

In cycle 22, some 20 months after the time of maximum solar activity, there began a new increase in activity that was especially strong in the southern solar hemisphere in March 1991 and in the northern solar hemisphere in June 1991. These solar events were some of the largest of that cycle and the Forbush decrease in June was the biggest ever recorded by neutron monitors (Webber and Lockwood 1993). The isolated occurrence in time of these major episodes of solar activity, their north-south asymmetry, the large increase of low energy ions and decreases in the galactic cosmic ray intensity in the outer heliosphere offer a unique opportunity to study the generation of a GMIR, as well as its relation to events on the Sun, to the acceleration and to transport of MeV ions, and to cosmic ray modulation (McDonald et al. 1994).

By early 1991 the recovery of the galactic cosmic rays from their minimum value after solar maximum was well underway both at 1 AU and at

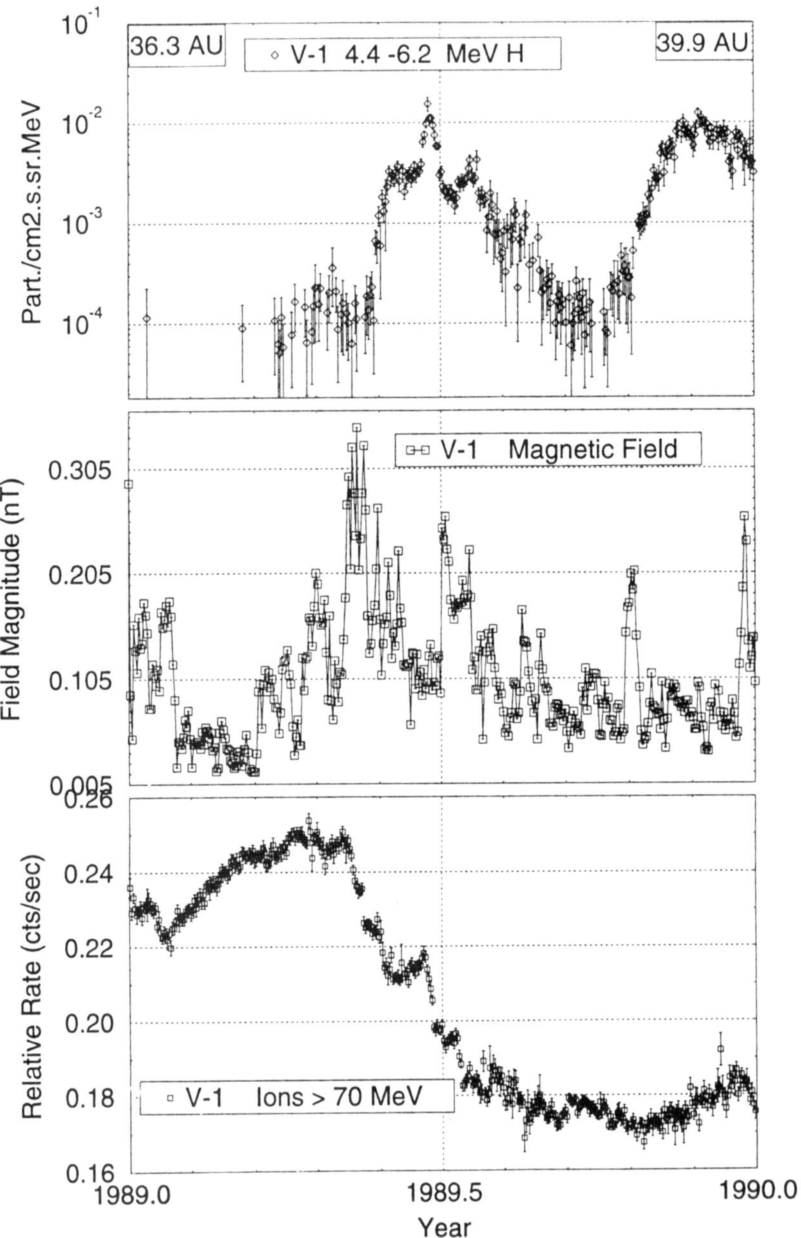

Figure 14.   Time history of Voyager 1 (24-hr averages) 4.4 to 6.2 MeV H, the magnitude of the interplanetary magnetic field and the counting rate of galactic cosmic rays >70 MeV/n for the year 1989.

the four spacecraft in the outer heliosphere. At Voyager 2, this recovery is interrupted by the sharp increase in the interplanetary magnetic field and solar wind velocity at 1991.4 (Fig. 15). Belcher et al. (1993) interpreted the plasma data in terms of an interplanetary shock. The net effect of this disturbance is to create a cosmic ray plateau that extends to 1991.68. At this time there are large increases in B and V that produce a 21% decrease in the galactic cosmic ray intensity.

This combined plot of V, B and the intensity of low-energy ions (Fig. 15) clearly show the relation of the interplanetary structures to the MeV particles. The increase in 5 MeV ions occurs after the arrival of the initial shock associated with the GMIR and the three most significant peaks in V and B (indicated by the dash lines extending across all four data sets of Fig. 15) produce peaks in the particle intensity.

At Pioneer 10, near 53 AU, and separated from Voyager 2 by 153° in heliolongitude, the time history and the changes (20.5%) in the galactic cosmic ray intensity (Fig. 16) are very similar to those at Voyager 2 except that in the latter case the plateau region occurs earlier and extends over a longer period of time. The initial increase in the solar wind velocity at Pioneer 10 is more modest while the two following velocity peaks are comparable. The low-energy ion data reflect these differences and provide a more extended view of the structure of the GMIR.

The changes in the cosmic ray intensity at Pioneer 11 (17°N) are similar to those of Voyager 2 at nearly the same heliocentric distance (Van Allen and Fillius 1992). However, at 32°N the Voyager 1 galactic cosmic ray time history is very different; there is an absence of the initial plateau region and the sharp decrease of 28.3±2% (Webber and Lockwood 1993). McDonald et al. (1993a) argued that when the effects of the on-going recovery are taken into account, the net changes at the four spacecraft are essentially the same and the observed temporal differences are a reflection of the north-south asymmetry in solar activity.

These major interplanetary events in 1991 must represent the superimposed effect of the numerous large CMEs that, at the Sun, were centered first about the March period and later about the first half of June 1991. The two epochs of solar activity are clearly visible in the double peaked structure of the low-energy particle increases and in the magnetic field peaks, as well as the solar wind data. Nevertheless the individual solar events in each of these periods have merged their identity to create this large-scale, long-lived interplanetary disturbance whose composite structure acts as a global barrier to galactic cosmic rays.

V. DISCUSSION

The observational data clearly establish a relationship between large-scale interplanetary disturbances in the outer heliosphere (GMIRs) and the step-decreases in the galactic cosmic ray intensity that constitute the major part

F. B. McDONALD AND L. F. BURLAGA

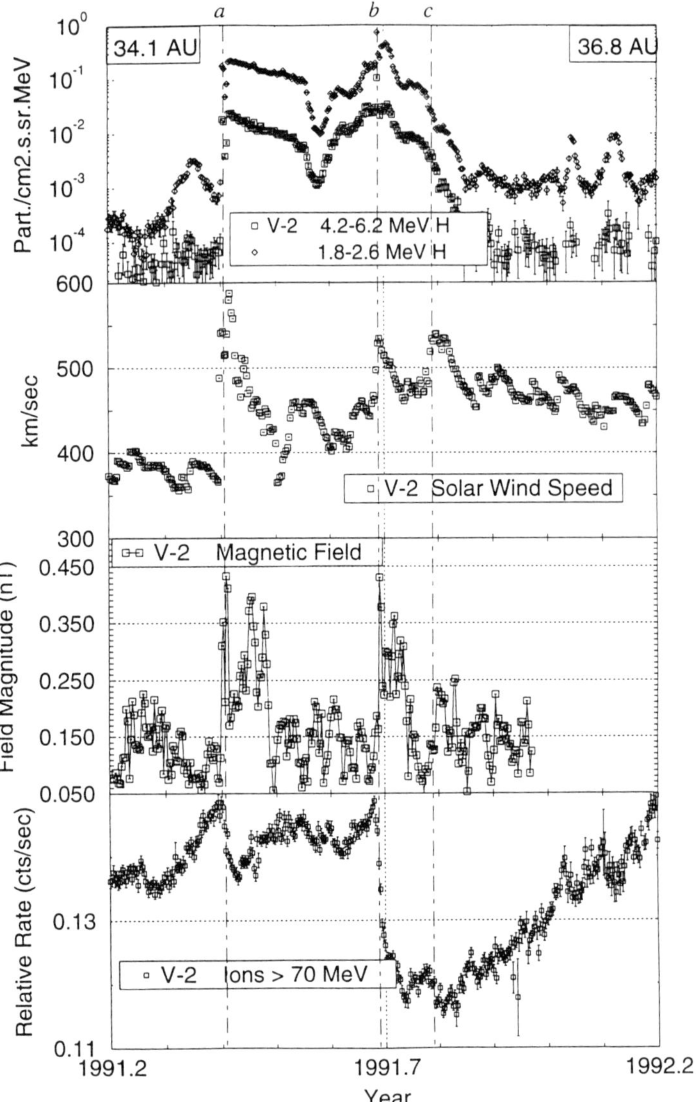

Figure 15.   Time history of Voyager 2 (24-hr averages) 1.8 to 2.6 MeV and 4.2 to 6.2 MeV H along with the daily averages of the solar wind speed (from the Voyager 2 solar wind experiment), the magnitude of the interplanetary magnetic field and the counting rate of galactic cosmic rays with energies >70 MeV/n from the CRS experiment. The three lines mark the passage of the three major peaks in the solar wind over this period and provide fiducial marks for the analysis of this event (figure from McDonald et al. 1994).

of the long-term (11-yr) cosmic ray modulation. Burlaga et al. (1993b) and Perko (1993) incorporated the observations of the magnetic fields in GMIRs into a one-dimensional, solar modulation diffusion equation. These studies used the actual interplanetary magnetic field intensity for the determination of the energetic-particle diffusion coefficients which are then propagated out to a modulation boundary at 120 AU. This model reproduces the time-intensity profiles at 1 AU for both Voyager 2 and Pioneer 10 over a complete solar cycle, as well as that of Voyager 1 at the higher latitude. Le Roux and Potgieter (1993) used a two-dimensional, time-dependent drift model with global merged interaction regions. In their simulations, GMIRs were modeled as idealized spherical regions of enhanced interplanetary magnetic field (relative to the nominal field) which propagated radially outward with a velocity of 400 km s^{-1}. As in the Perko model, the diffusion tensor is defined to be inversely proportional to the time-dependent interplanetary magnetic field so that the GMIRs become partial barriers against the diffusing and drifting cosmic rays.

The form of the field change assumed by le Roux and Potgieter for two GMIRs is shown in the top panel of Fig. 17. The corresponding particle-diffusion coefficient is shown in the lower half of the same panel. The long-term effects of these GMIRs for $qA>0$ and $qA<0$ are shown in the middle and lower panels of this figure. In each of these two panels, the dashed curves are for the case where GMIRs were neglected and the inclination of the neutral current sheet changed from 10° to 75° over the first 2 years.

From these simulations, le Roux and Potgieter (1993) concluded the following:

1. The changing heliospheric neutral sheet is not a sufficient modulation parameter to explain the increase in cosmic ray modulation for $qA>0$ cycles (e.g., 1977–1979). On the other hand, this lack of response of cosmic rays to the changes in the heliospheric current sheet (HCS) is most suitable for explaining the plateau-like period in the cosmic ray intensity from 1972 to 1976.

2. The combination of the HCS and two successive GMIRs improved the simulation of the decreasing phase of the $qA>0$ cycle from 1977 to 1979 significantly, but only after the features of the GMIRs were modified as follows: they were assumed to be spherical, expanding radially inward and outward with the Alfvén velocity and increasing their strength as they propagated outward; and a strong solar polar field modification (Jokipii and Kota 1989) was utilized. All these measures were necessary to counteract the very effective diffusion, and especially the drifts from the polar regions of the heliosphere during $qA>0$ periods. The inclusion of GMIRs, nevertheless, produced a very natural and convincing explanation for the observed large step-decreases on long-term modulation (McDonald et al. 1981; Burlaga et al. 1993a).

3. In contrast to the $qA>0$ simulations, the combination of two GMIRs and a changing HCS proved to be very effective in generating long-term

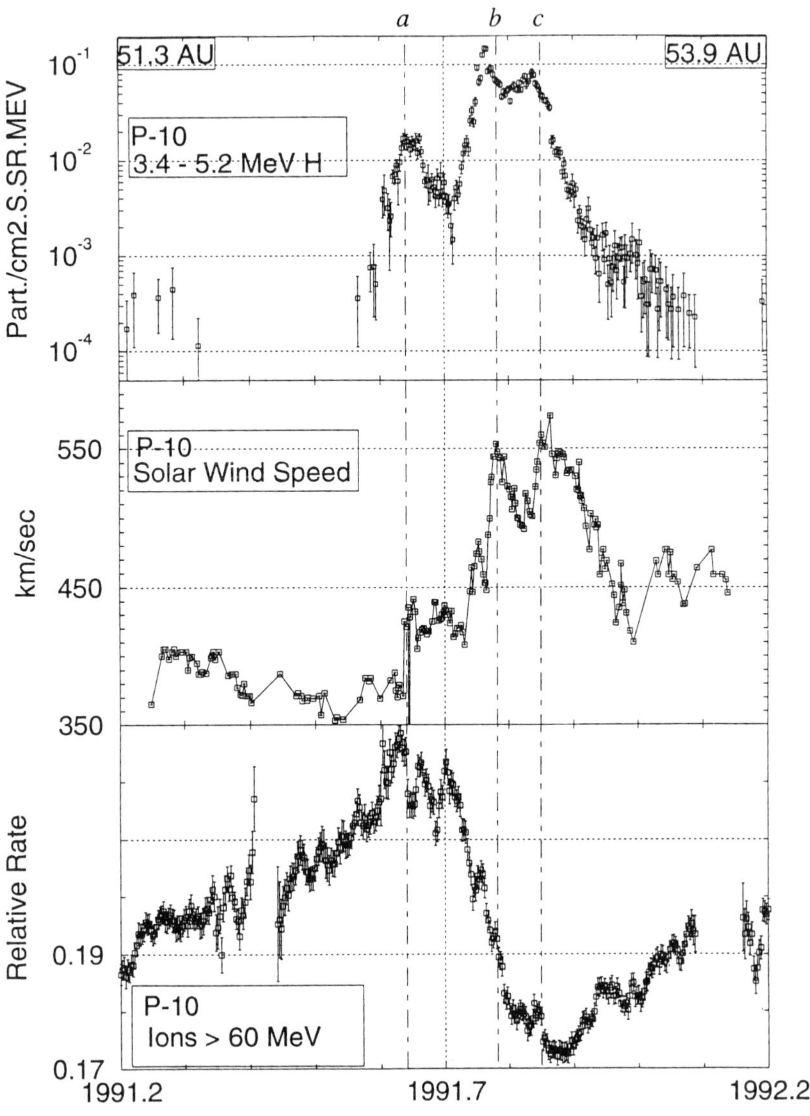

Figure 16.   Time history of Pioneer 10 (24-hr averages) 3.5 to 5.2 MeV H, along with the solar wind speed (from the Pioneer 10 Ames plasma analyzer) and the integral intensity of galactic cosmic rays >56 MeV/n. The choice of the three peaks in the solar wind data (indicated by lines) is based on their apparent correspondence to the Voyager 2 data. These Voyager 2 and Pioneer 10 peaks probably reflect the same intervals of solar activity rather than specific events (figure from McDonald et al. 1994).

modulation when $qA<0$, because of the higher sensitivity of cosmic ray protons to changes in the HCS, and especially to the successive GMIRs which, under these circumstances, became more effective diffusion/drift barriers than when $A>0$.

4. A study of the radial and energy dependence of the declining intensity phase of the proton modulation cycle revealed that two successive GMIRs produced a significant part of the total long-term modulation. In fact, GMIRs are expected to dominate this process relative to the HCS at energies above $\sim$5 GeV and radial distances larger $\sim$40 AU.

B. Relation Between Global Merged Interaction Regions and the Voyager Observations of Long-Duration, Low-Frequency Radio Emission

Starting in 1992.57 the Voyager 1/2 plasma wave experiment (PWS) observed the onset of a long-duration, low-frequency radio event (Gurnett et al. 1993). The two primary spectral components of this event were well above the local plasma frequency with a main emission band near 2.0 kHz and a series of narrowband emissions that drifted upward in frequency at a rate of $\sim$3.0 kHz to a peak frequency of 3.6 kHz. Approximate radio-direction finding was possible during the monthly spacecraft roll maneuvers (for magnetometer calibration purposes) and indicated a source location toward either the nose or tail of the heliosphere. The Voyager experiment has observed several such events in the past after the spacecraft had moved beyond some 12 AU but the 1992 emission was the strongest recorded up to the present time.

The most plausible trigger for this event is the GMIR produced by the solar activity of March/June 1991 (Figs. 15,16). The time history of this radio emission (Fig. 18) shows a remarkable resemblance to that of the low-energy ions at Voyager 2 when a time delay of 1.2 yr is introduced. This delay was adopted to line up the a particle peak at Voyager 2 with the initial plateau of the 3.11 kHz radio emission. While this choice of delay times is not unique, it is clear that a time shift of this order will bring the two data sets into general alignment. While the GMIR structure will continue to evolve over such a lengthy period, it is interesting to note that features corresponding to the b and c peaks can be found in the radio data. The concept of a solar wind "trigger" for these low-frequency rapid emissions was first advanced by McNutt (1988) to explain the 1983/84 PWS observations (Kurth et al. 1984). The concept of a GMIR as the driving force for these events represents an extension of McNutt's original hypothesis.

The PWS investigators have proposed that a possible source for this emission is the interaction of shock waves with the enhanced plasma density in the region of the nose of the heliopause beyond the termination shock (Gurnett et al. 1993). The extended temporal duration of the GMIR event in the outer heliosphere and the energy available from the superposition of multiple CMEs make this largest GMIR of the last two solar cycles an attractive candidate for producing the strongest radio emission yet observed in the outer heliosphere by the Voyager PWS (McDonald et al. 1994).

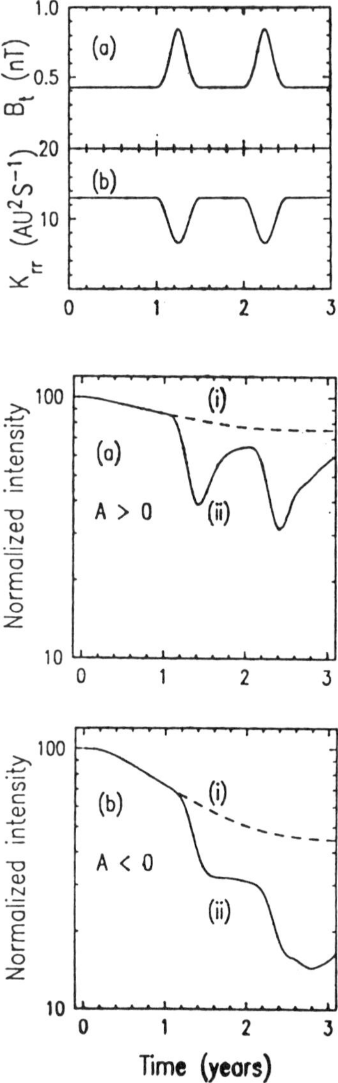

Figure 17.   *Top panel:* The data in the upper part of this panel is the simulated time variation of the interplanetary magnetic field at 20 AU in the equatorial plane. The two peaks in this plot represent an idealistic simulation of two large GMIRs, while the *b* part of this panel shows the expected change in the radial diffusion coefficient. *Center panel:* This panel is the calculated normalized intensity with time for 1 GeV hydrogen at 20 AU in the ecliptic plane for $qA > 0$ as the current sheet inclination increases from $10°$ to $75°$ and the two GMIRs propagate through the heliosphere. The dashed line shows the change expected due only to the effects of the changing inclination of the current sheet. *Bottom panel:* This panel is the same as the middle one except that $qA < 0$ (le Roux and Potgieter 1993).

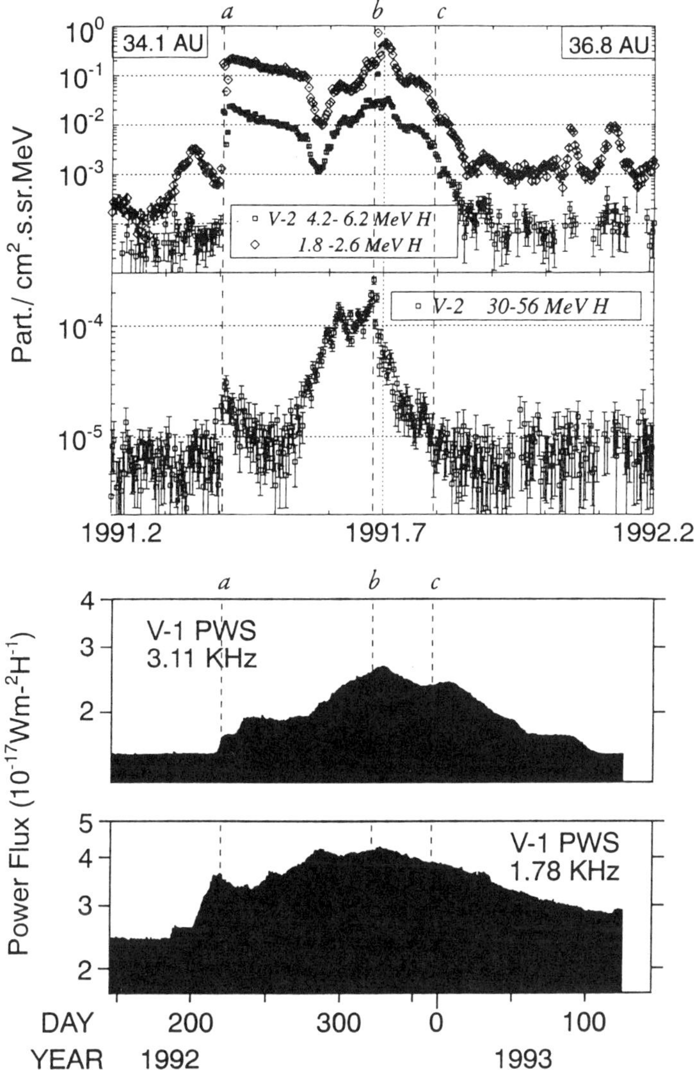

Figure 18.   A comparison of the time history of low-and medium-energy S/IP ions at Voyager 2 (Fig. 11) and the low-frequency radio emission detected at Voyagers 1 and 2 starting on day 209, 1992. A time delay of 1.2 yr is used to align the *a* peak with the onset of the first plateau in the 3.11 KHz emission (figure from McDonald et al. 1994).

C. Future Directions

There are several important studies that need to be undertaken to extend our knowledge of the properties of GMIRs. Detailed modeling of the evolution of these events with heliocentric distance that examines (1) the initial merging of CMEs and high speed streams producing the GMIRs, and (2) the further changes as these disturbances move toward the termination shock and the modifications that occur as they approach the heliopause. The cosmic ray data and the detailed modeling of le Roux and Potgieter indicate that GMIRs must be closed shells that become progressively stronger in the polar region. Jokipii and Kota (1989) have shown that polar drift effects must be significantly reduced in order to reproduce the observed cosmic ray gradients near the ecliptic plane. Although Ulysses is only at a few AU when it crosses this polar region, its interplanetary observations may provide further insight into the dynamics of the disturbances in this region.

At the Sun, there is continual activity that is frequently episodic and is modulated by the 11-yr solar cycle. The occurrence of GMIRs and cosmic ray step-decreases are well defined, discrete happenings that only occur 6 to 7 times over an 11-yr solar cycle. Studies are needed to identify the convergence of solar events that leads to the generation of a GMIR in the outer heliosphere.

REFERENCES

Barouch, E., and Burlaga, L. F. 1975. Causes of Forbush decreases and other cosmic ray variations. *J. Geophys. Res.* 80:449–456.

Belcher, J. W., Lazarus, A. J., and McNutt, R. L. 1993. Large scale density structures in the outer heliosphere. *Adv. Space Res.* 13(6):41–48.

Borrini, G., Gosling, J. T., Bame, S. J., Feldman, W. C. and Wilcox, J. M. 1981. Solar wind helium and hydrogen structure near the heliospheric current sheet—A signal of coronal streamers at 1 AU. *J. Geophys. Res.* 86:4565–4576.

Burlaga, L. F. 1974. Interplanetary stream interfaces. *J. Geophys. Res.* 79:3717–3735.

Burlaga, L. F. 1983. Corotating pressure waves without fast streams. *J. Geophys. Res* 88:6085–6094.

Burlaga, L. F. 1984. MHD processes in the outer heliosphere. *Space Sci. Rev.* 39:255–316.

Burlaga, L. F. 1988. Interaction regions in the distant heliosphere. In *Proc. of the Sixth Solar Wind Conf.*, eds. V. J. Pizzo, T. E. Holzer and D. G. Sime, NCAR TN-306 (Boulder: Natl. Center for Atmos. Res.), p. 562.

Burlaga, L. F. 1991. Magnetic clouds. In *Physics of the Inner Heliosphere II*, eds. R. Schwenn and E. Marsch (Berlin: Springer-Verlag), pp. 1–19.

Burlaga, L. 1993. Shocks in the outer heliosphere: Voyager 2 observations from 18.9 AU to 30.2 AU (1986–1989). *J. Geophys. Res.* 99:4161–4171.

Burlaga, L. F. 1995. *Interplanetary Magnetohydrodnamics* (New York: Oxford Univ. Press).

Burlaga, L. F., and King, J. H. 1979. Intense interplanetary magnetic fields observed by geocentric spacecraft during 1963–1975. In *J. Geophys. Res.* 84:6633–6640.

Burlaga, L. F., and Mish, W. H. 1987. Large-scale fluctuations in the interplanetary medium. *J. Geophys. Res.* 92:1261–1266.

Burlaga, L. F., and Ogilvie, K. W. 1970. Magnetic and thermal pressures in the solar wind. *Solar Phys.* 15:61–71.

Burlaga, L. F., Schwenn, R., and Rosenbauer, H. 1983. Dynamical evolution of interplanetary magnetic fields and flows between 0.3 AU and 8.5 AU: Entrainment. *Geophys. Res. Lett.* 10:413–416.

Burlaga, L. F., Klein, L, Lepping, R. P., and Behannon, K. W. 1984*a*. Large-scale interplanetary magnetic fields: Voyager 1 and 2 observations between 1 AU and 9.5 AU. *J. Geophys. Res.* 89:10659–10668.

Burlaga, L. F., McDonald, F. B., Ness, N. F., Schwenn, R., Lazarus, A. J., and Mariani, F. 1984*b*. Interplanetary flow systems associated with cosmic ray modulation in 1977–1978. *J. Geophys. Res.* 89:6579–6587.

Burlaga, L. F., McDonald, F. B., Goldstein, M. L., and Lazarus, A. J. 1985. Cosmic ray modulation and turbulent interaction regions near 11 AU. *J. Geophys. Res.* 90:12127–12139.

Burlaga, L. F., Ness, N. F., and McDonald, F. B. 1987. Large-scale fluctuations in B between 13 AU and 22 AU and their effects on cosmic rays. *J. Geophys. Res.* 92:13647–13652.

Burlaga, L., F., McDonald, F. B., Ness, N. F., and Lazarus, A. J. 1991. Cosmic ray modulations: Voyager 2 observations: 1987–1988. *J. Geophys. Res.* 96:3789–3799.

Burlaga, L. F., McDonald, F. B., and Ness, N. F. 1993*a*. Cosmic ray modulation and the distant heliospheric magnetic field: Voyager 1 and 2 observations from 1986 through 1989. *J. Geophys. Res.* 98:1–11.

Burlaga, L. F., Perko, J., and Pirraglia, J. 1993*b*. Cosmic ray modulation, merged interaction regions and multifractals. *Astrophys. J.* 407:347–358.

Cane, H.V., Richardson, I. G., and von Rosenvinge, T. T. 1993. Cosmic ray decreases and particle acceleration in 1978–1982 and the associated solar wind structures. *J. Geophys. Res.* 98:13295–13339.

Cliver, E. W., Mihalov, J. D. , Sheeley, N. R. Jr., Howard, R. A., Koomen, M. J. and Schwenn, R. 1987. Solar activity and heliosphere-wide cosmic ray modulation in mid-1982. *J. Geophys. Res.* 92:8487–8501.

Decker, R. B., Gold, R. E., and Krimigis, S. M. 1991. Distribution of 30–4000 KeV ions associated with an interplanetary shock at Voyager 2 (30 AU) and Voyager 1 (38 AU) in 1989. *Proc. 22nd Intl. Cosmic Ray Conf.* 3:296–299.

Dröge, W., Müeller-Mellin, R., and Cliver, E. W. 1992. Superevents: their origin and propagation through the heliosphere from 0.3–35 AU. *Astrophys. J. Lett.* 387:97–100.

Evenson, P., and Meyer, P. 1984. Solar modulation of cosmic ray electrons 1978–1983. *J. Geophys. Res.* 89:2647–2654.

Fujii, Z., and McDonald, F. B. 1995. Study of the properties of the step decreases in galactic and anomalous cosmic rays over solar cycle 21. *J. Geophys. Res.* 100:17043–17052.

Gurnett, D. A., Kurth, W. S., Allendorf, S. C., and Poynter, R. L. 1993. Radio emission from the heliopause triggered by an interplanetary shock science. *Science* 262:199–203.

Hoeksema, J. T. 1989. Extending the Sun's magnetic field through the three-dimensional heliosphere. *Adv. Space Res.* 9(4):141–142.

Hundhausen, A. J. 1972. *Coronal Expansion and the Solar Wind* (New York: Springer-Verlag).

Iucci, N. Parisi, M., Storini, M., and Villorese, G. 1984*a*. Interplanetary disturbances during Forbush decreases. *Nuovo Cimento* 7C:467–488.

Iucci, N., Parisi, M., Storini, M., and Villorese, G. 1984*b*. Some properties, of flare-not-associated Forbush decreases. *Nuovo Cimento* 7C:722–731.

Jokipii, J. R., and Kota, J. 1989, The polar heliospheric magnetic field, 1989. *Geophys. Res. Lett.* 16:1–4.

Jokipii, J. R., and Thomas, E. T. 1981. Effects of drifts on the transport of cosmic rays, IV, Modulation by a wavy interplanetary current sheet. *Astrophys. J.* 243:1115–1122.

Jokipii, J. R., Levy, E. H., and Hubbard, W. B. 1977. Effects of particle drifts on cosmic ray transport, I, General properties, applications to solar modulation. *Astrophys. J.* 213:861–868.

Kota, J., and Jokipii, J. R. 1983. Effects of drift on the transport of cosmic rays, VI, A three-dimensional model including diffusion. *Astrophys. J.* 265:573–581.

Kurth, W. S., Gurnett, D. A., Scarf, F. L., and Poynter, R. L. 1984. Detection of a radio emission at 3 Khz in the outerheliosphere. *Nature* 312:27–31.

le Roux, J. A., and Potgieter, M. S. 1993. The simulation of merged interaction regions in the outer heliosphere. *Adv. Space Res.* 13(6):251–256.

le Roux, J. A., and Potgieter, M. S. 1995. The simulation of complete 11 and 22 year modulation cycles for cosmic rays in the heliosphere using a drift model with global merged interaction regions. *Astrophys. J.* 442:847–851.

Lockwood, J. A. 1960. On the long-term variation in the cosmic radiation. *J. Geophys. Res.* 65:19–26.

Lockwood, J. A. 1971. Forbush decreases in the cosmic radiation. *Space Sci. Rev.* 12:658–715.

McDonald, F. B., and Selesnick, R. S. 1991. Solar interplanetary energetic particles in the outer heliosphere. *Proc. 22nd Intl. Cosmic Ray Conf.* 3:189–192.

McDonald, F. B., Lal, N., Trainor, J. H., Van Hollebeke, M. A. I., and Webber, W. R. 1981. The solar modulation of galactic cosmic rays in the outer heliosphere. *Astrophys. J. Lett.* 249:71.

McDonald, F. B., Lal, N., and McGuire, R. E. 1993. Role of drifts and merged interaction regions in the long-term modulation of cosmic rays. *J. Geophys. Res.* 98:1243–1256.

McDonald, F. B., Barnes, A., Burlaga, L. F., Gazis, P., Mihalov J., and Selesnick, R. S. 1994. The effects of the intense solar activity of March/June 1991 observed in the outer heliosphere. *J. Geophys. Res.* 99:14705–14715.

McNutt, R. L., Jr. 1988. A solar wind "trigger" for the outer heliosphere radio emissions and the distance to the terminal shock. *Geophys. Res. Lett.* 15:1307–1310.

Neugebauer, M., and Snyder, C. W. 1966. Mariner 2 observations of the solar wind 2. Average properties. *J. Geophys. Res.* 71:4469–4484.

Parker, E. N. 1963. *Interplanetary Dynamical Processes* (New York: Wiley-Interscience).

Perko, J. S., and Burlaga, L. F. 1992. Intensity variations in the interplanetary magnetic field measured by Voyager 2 and the 11-year solar cycle modulation gathered by galactic cosmic rays. *J. Geophys. Res.* 97:4305–4308.

Potgieter, M. S. 1993. Time dependent modulation: Role of drifts and interaction regions. *Adv. Space Res.* 13:239–249

Potgieter, M. S., and le Roux, J. A. 1987. On a possible modulation barrier in the outer heliosphere. *Proc. 22nd Intl. Cosmic Ray Conf.* 3:291–249.

Quenby, J. J., Lockwood, J. A., and Webber, W. R. 1990. Cosmic-ray gradient measurements and modulation beyond the inner solar wind termination shock. *Astrophys. J.* 365:365–371.

Smith, E. J., and Wolfe, J. H. 1976. Observations of interaction regions and corotating shocks. *Geophys. Res. Lett.* 3:137–140.

Tuska, E. B. 1990. Charge-Sign Dependent Solar Modulation of 1–10 GV Cosmic Rays. Ph.D. Thesis, Univ. of Delaware.

Van Allen, J. A., and Fillius, R. W. 1992. Propagaion of a large Forbush decrease in cosmic-ray intensity past the Earth, Pioneer 11 at 34 AU, and Pioneer 10 at

53 AU. *Geophys. Res. Lett.* 19:1423–1426.

Van Allen, J. A., and Mihalov, J. D. 1990. Forbush decreases and particle acceleration in the outer heliosphere. *Geophys. Res. Lett.* 17:761–764.

Webber, W. R., and Lockwood, J. A. 1993. Giant transient decreases of cosmic rays in the outer heliosphere in September 1991. *J. Geophys. Res.* 98:7821–7832.

Zhang, G., and Burlaga, L. R. 1988. Magnetic clouds, geomagnetic disturbances, and cosmic ray decreases. J. Geophys. Res. 93:2511–2518.

SOLAR WIND MAGNETIC FIELDS

EDWARD J. SMITH
Jet Propulsion Laboratory

The magnetic fields originate as coronal fields that are convected into space by the supersonic, infinitely conducting, solar wind. On average, the Sun's rotation causes the field to wind up and form an Archimedes spiral. However, the field direction changes almost continuously on a variety of scales and the irregular nature of these changes is often interpreted as evidence that the solar wind flow is turbulent. To a surprising extent, the large-scale field near the ecliptic (solar equator), where observations have been made (between 0.3 and 35 AU), appears to be dominated by dipole-like fields from the Sun's polar corona. The (open) field lines, however, are attached to the Sun at only one end, the other end being carried off into the outer heliosphere by the solar wind. The oppositely directed fields in the northern and southern hemispheres are separated by a thin, wavy current sheet that encloses the Sun and extends throughout the heliosphere. A complication to this otherwise straightforward description is the presence of solar wind structure associated with high-speed streams that corotate with the Sun. Another cause of structure is the (frequent) presence of coronal mass ejections whose origins, magnetic topology, internal dynamics and interaction with the pre-existing solar wind are still being actively studied. Recent interest has also been directed to understanding the topology of the field between the shock that is thought to terminate the supersonic solar wind flow and the heliopause. The three-dimensional field properties at high latitudes and in the vicinity of the polar heliosphere have attracted recent interest as a result of the Ulysses Mission. A brief description of the recent results obtained when this spacecraft reached the Sun's south pole is included.

I. INTRODUCTION

The heliospheric magnetic field (HMF) was called the interplanetary magnetic field (IMF) for many years until observations became available beyond the outermost planets of the solar system and significantly far above and below their orbit planes. Our understanding of the large-scale field and the processes which shape it are ultimately derived from Parker's solar wind model. The model provides the baseline against which all the observations have been compared and has been modified as necessary to accommodate departures from the simple set of assumptions that underlie it.

In presenting his model, Parker anticipated significant departures of the solar wind from simple radial solar wind flow at constant speed. Thus, solar activity was expected to generate shock waves that would propagate rapidly outward through the pre-existing solar wind. At great distances, the solar wind was anticipated to interact with the magnetized interstellar plasma, forcing it out of the solar system and thereby creating the heliosphere. A termination shock was visualized inside the heliosphere at which the supersonic solar flow would become subsonic before it reaches the boundary of the heliosphere called the heliopause. The above are only a few examples of the large-scale solar wind structure which have been the object of study ever since the first extensive space measurements became available.

This review addresses both the average properties and the many forms of large-scale structure in the solar wind magnetic field. Properties of this pervasive magnetic field based on observation and theory will be described from its origin in the solar corona all the way to the heliopause as well as from the equator to the pole.

II. THE SPIRAL ANGLE

The basic elements of the solar wind field in the solar equator are shown in Fig. 1 (Parker 1963). The field lines are seen to all originate at the Sun as coronal fields that are carried off by the radially flowing supersonic solar wind represented by the radial arrows. The most basic property of the fields is the spiral form which results from their connection at one end to the rotating Sun. The analogy is often made to a rotating lawn sprinkler where the water droplets although moving radially outward lie on a spiraled locus. Alternatively, one can imagine the solar wind plasma to be sliding along a rotating field line while simultaneously moving radially outward (consider the pickup arm of a phonograph). The dashed circle represents the distance to the Earth's orbit (1 AU).

The original Parker figure has been modified to include information regarding the field polarity. Arrows have been added to distinguish outward (positive) fields from sunward (negative) fields. In addition, appropriate plus and minus signs have been added just inside the circle to denote polarity. This representation explains the origin of the term, magnetic sector, to describe the polarity. The circle is thereby divided into pie-shaped sectors (two sectors as shown here but occasionally the circle is divided into four or even six sectors).

The expression for the spiral angle ψ in the usual spherical coordinates (r, θ, ϕ) with the Sun's rotation axis as the pole, is:

$$\tan \psi = -r\, d\phi/dr = B_\phi/B_r = -\Omega r \sin \theta / V_r. \tag{1}$$

Parker assumed a steady state with a radial solar wind velocity V_r. The angular rotation rate of the Sun is $\Omega = 2\pi/T$ where T is usually assumed to be the sidereal equatorial period of 25.4 days. A typical value for V_r is 420 km s^{-1} so that, at the orbit of Earth, $\psi = 45°$ (rather than 55° as shown in Fig. 1).

Given B_r and B_r, V_ϕ will automatically adjust to yield the spiral angle, Eq. (1). The sign of B_ϕ depends on the sign of B_r. It might be supposed that a B_θ component would be possible and it is commonly stated that Parker simply assumed it was zero. However, both the spiral angle and $B_\theta = 0$ are

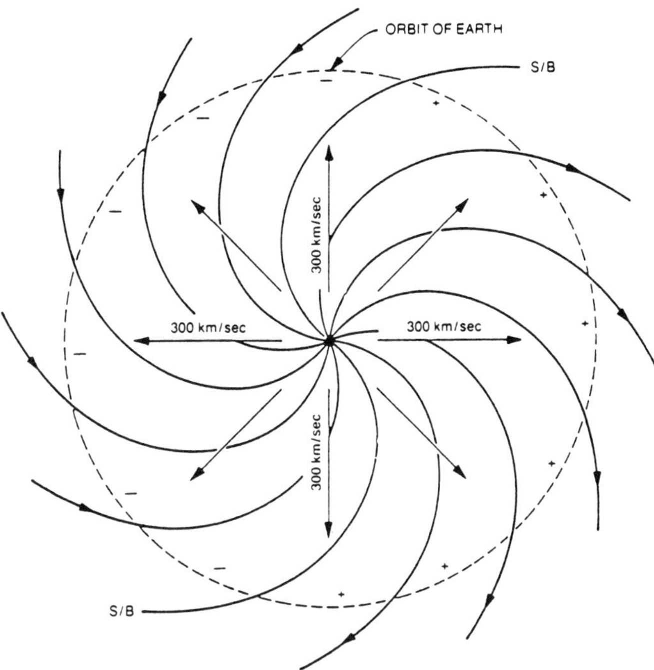

Figure 1. The Parker spiral. Spiraling magnetic field lines are shown emanating from the Sun and being carried out to 1 AU (the dashed circle). The spiral angle between the radial direction and the field is larger than the more typical angle of 45° because a relatively low velocity (300 km s^{-1}) was assumed in deriving the figure. The radial lines represent the solar wind velocity. Arrows have been added showing the field polarity and the corresponding plus signs and minus signs give rise to the so-called magnetic sectors (two in this instance). A pair of opppositely directed field lines coincide with the sector boundaries (S/B).

a consequence of the electric field vanishing in the solar wind frame (along with the assumptions of a steady state and vector $V = V_r, 0, 0$). If an electric field was present in the infinitely conducting, collisionless plasma, infinitely large currents would develop. The mathematical details involving the electric and magnetic fields and currents associated with Parker's model have been relegated to the appendix, at the end of this chapter.

Confirmation that the field lies in the $r\phi$ plane is presented in Fig. 2. This histogram was obtained from Pioneer 10 and 11 measurements of the field latitude angle δ_B over the radial range from 1 to 8.5 AU (Thomas and Smith 1980). A very large number of hourly averages (43,484) are involved. The north-south component shows considerable variability with a width at half maximum of $\approx\pm30°$. Nevertheless, the statistically significant average of δ_B is zero.

A histogram of the observed longitude angle of the field, ϕ_B is shown in

                    E. J. SMITH

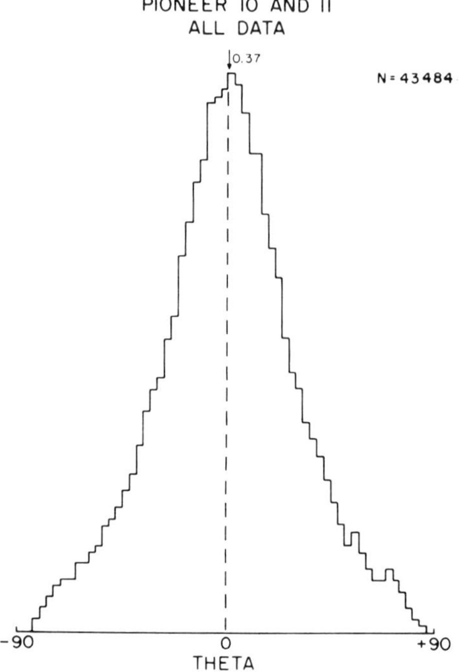

Figure 2.   Latitudinal angle of the magnetic field. The instantaneous latitudes of $\vec{B}$ (called θ here) relative to the radial direction have been assembled to form this histogram. Pioneer 10 and 11 data acquired between 1 and 8.5 AU are represented. The prediction of the Parker model that there would be no nonzero average latitude angle or, equivalently, no average north-south component is borne out by the observations.

Fig. 3. The hourly averages represented here come from the same data set as in Fig. 2. In order to accommodate the changing spiral angle with distance, the measured fields were transformed into coordinates with one axis along the Parker Spiral assuming $\psi = 45°$ at 1 AU ($\tan \psi = r$). Two peaks are seen corresponding to outward ($0°$) and inward ($180°$) spiral fields. The two peaks agree with the Parker Spiral angle with a high level of certainty.

Significant variability in the angles is evident in Figs. 2 and 3. What are the causes? Differentiation of the basic equation for the spiral angle leads to

$$\delta(\tan \psi) = \frac{\Omega r}{V_r} \sin \theta \left(\frac{\delta V_r}{V_r} \right) - \frac{r \sin \theta}{V_r} (\delta \Omega). \qquad (2)$$

Two causes are seen: variations in the radial solar wind speed and departures of the magnetic field from strict corotation at the solar wind source. The latter are called stochastic variations by Jokipii and Parker (1970) because they are expected to accompany irregular motions (random walk) of the field lines in the photosphere and chromosphere.

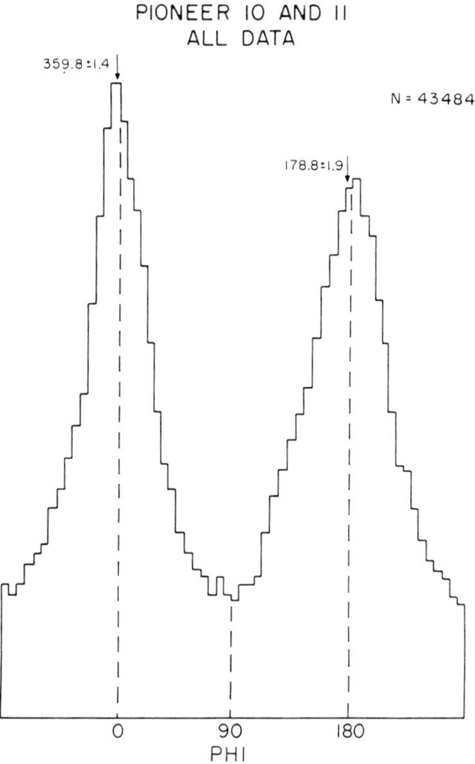

Figure 3.   Longitude angle of the magnetic field. In order to test the prediction that the longitude angle (ϕ) coincides on the average with the Archimedes spiral, the observed field longitude angles were transformed into a coordinate system with one axis along the spiral direction. The Pioneer spacecraft were traveling from 1 to 8.5 AU during the interval represented by these data and this choice of coordinates automatically took into account the radial dependence of ϕ. The two peaks correspond to a spiral angle of 45° at 1 AU and show the presence of the two sector polarities.

The relative effectiveness of these two sources of variability can easily be estimated numerically. What solar wind speeds would cause $\delta(\tan \psi) = \pm 0.5$ or ψ to vary between 27° and 56°? The low and high value of V_r are 260 and 700 km s^{-1}, rather extreme values observationally. This calculation shows that it is difficult to attribute the large variations observed over time scales of minutes to hours to variations in the solar wind speed.

The second term in Eq. (2) implies that, for the same variation, $\delta\Omega = (0.5)$ V_r/r. At the Sun (one solar radius, r_S), the equivalent transverse velocity would be $\delta V = r_s \delta\Omega = (0.5) V_r r_s/r \cong 1$ km s^{-1}. Thus, modest motions on the Sun, perhaps associated with supergranules, can cause large deviations in the field direction.

Another major source of variability of the spiral angle is the presence of waves or turbulence. The Alfvén relation between the perturbations in the velocity and magnetic field associated with the waves is $\delta\vec{V}/\delta\vec{B} = C_A/B$, where C_A is the Alfvén speed. Observationally, short-period ($\leq$3 hr) fluctuations in the field can cause deflections in direction of $\pm 45°$ corresponding to $\delta B/B\approx 1$. It then follows that $\delta V\approx C_A$, typically 60 km s^{-1} at 1 AU. Thus, relatively small transverse velocity variations, $\cong V_r/10$, are associated with large angular field deflections.

The fact that $\delta V/C_A\approx 1$ is noteworthy as a ratio of the velocity perturbations to the wave speed of approximately one is generally taken to be a condition of strong turbulence. Because the solar wind is highly supersonic, it has often been supposed that the flow is turbulent and the language of turbulence carried over from fluid dynamics has often been applied in studies of the ever-present field fluctuations.

III. SECTOR STRUCTURE

That the sector structure is a large-scale, persistent feature of the heliospheric field is borne out by Fig. 4 (Smith et al. 1986). This figure represents a standard display of the sectors with each horizontal line, consisting of plus signs, minus signs and gaps (for missing data or indeterminate polarity), equivalent to a solar rotation of 27 days. The solar rotations are identified by their Bartels rotation number, designated BR, running between numbers 1910 to 2039. Eight years of data, a major fraction of a solar cycle, are shown extending from 1974 to 1982. The right half of the figure contains the polarities observed by Pioneer 11 as it traveled outward from 1.0 to beyond 10 AU. Because the heliographic latitude of the spacecraft is changing throughout this interval, the latitudes are listed in the extreme right-hand column. The corresponding sector structure at 1 AU as observed by the two spacecraft, ISEE-3 and IMP-8, is shown in the left half of the figure for comparison. The general correspondence in the polarities between 1 AU and the outer heliosphere is evident.

An alternate representation of the stability of the sector structure is that in Fig. 5, another histogram of the field longitude angle. The data shown are from more recent observations by Pioneer 11 in 1991 while the spacecraft was located at 35 AU. The striking feature of this histogram is the narrowness of the two peaks corresponding to the inward ($\approx 90°$) and outward ($\approx 270°$) Parker spiral. The solar heliospheric (SH) coordinate system in which the angles are measured has the x (or R) axis radially outward from the Sun, the z (or N) axis is northward in the plane of x and the Sun's rotation axis H while the y (or T) axis completes the orthogonal set. Spacecraft typically do not lie in the solar equatorial plane so that R is not perpendicular to H. The field vector in curvilinear coordinates in this and the coordinate system introduced in the discussion of the spiral angle are simply related: $B_R = B_r$, $B_T = B_\phi$, $B_N = -B_\theta$.

Figure 4.   The sector structure at 1 AU and in the outer heliosphere. The column on the left shows the sector structure each solar rotation as determined from IMP and/or ISEE-3 data at 1 AU. In the right column are the corresponding sectors observed by Pioneer 11 as it progressed from 1 to 12 AU. The 9-yr interval includes solar minimum (1976), maximum (1979) and the descent toward the next minimum. There is substantial agreement in the sector structure in spite of the increasing radial distance of Pioneer 11.

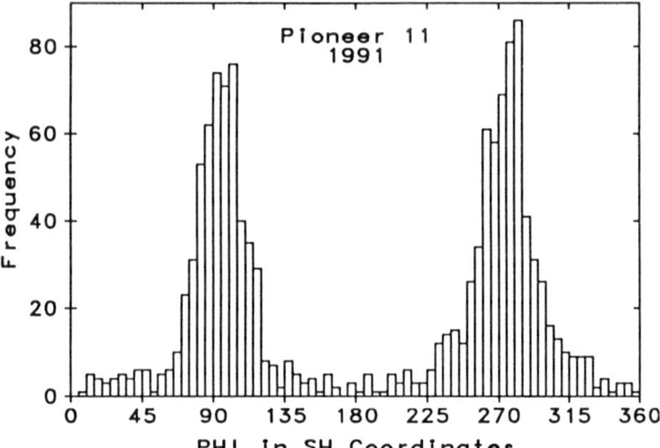

Figure 5.   Field longitude angle observed at 35 AU. The field longitude angle is shown based on hourly averages acquired by Pioneer 11 in 1991 near 35 AU. The spiral angle and the inward and outward sectors ($\approx \pm 90°$ here corresponding to the wrapping-up of the field with distance) are clearly identified. There is a relative absence of other directions (note the depth of the minimina near $0°$ and $180°$) such as might be anticipated if the current sheet was becoming "tattered" or tending to disappear.

Commonly occurring tracking gaps in these kinds of data often make the assignment of daily polarities a hazardous undertaking because of the large implicit uncertainties. This representation, however, shows the extent to which the observations agree with the spiral angle and reveal the extent to which the field coincides with one of the polarities. There has been considerable speculation that the oppositely directed fields would reconnect across the current sheet and that this would disrupt the sector structure at large distances. In spite of such prognostications, there is little, if any, evidence of a "filling-in" of the region between the two dominant polarities such as might be anticipated if reconnection was on-going.

IV. THE SOLAR ORIGIN OF THE MAGNETIC FIELD

Many of the observed properties are consistent with the field originating at high solar latitudes. A commonly held view of the relation between solar and solar wind magnetic fields is shown schematically in Fig. 6. This sketch is a qualitative three-dimensional representation of the field topology derived rigorously by Pneuman and Kopp (1971). Basically, the "open" fields with one end at high latitude and the other end carried off by the solar wind, overlie "closed" transequatorial field lines which have both ends rooted in the Sun. The open field lines can be identified with dark coronal holes. The closed fields contain trapped electrons which scatter visible radiation from the photosphere and are generally seen in coronagraph images in which they have the appearance of a bright arcade. The field lines at the top of the loop-like lower-lying fields tend to form a cusp or streamer so that the entire structure, which typically extends around the Sun in the form of a thick disk, is called the coronal streamer belt (CSB).

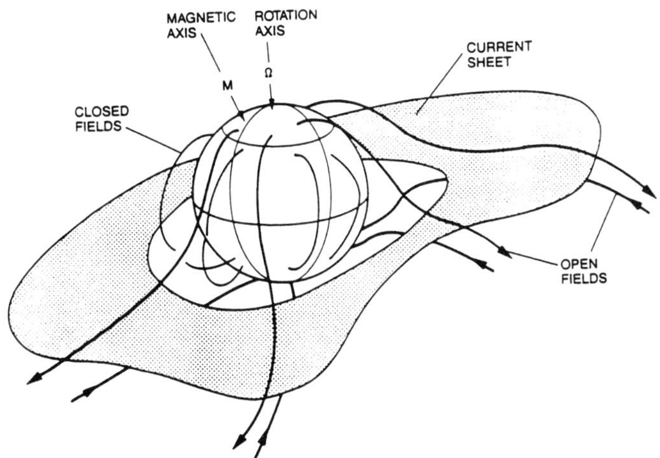

Figure 6.   Schematic presentation of the solar wind magnetic field lines near the Sun. The closed fields (loops) begin and end on the Sun. The open field lines have only one end on the Sun, the other "end" being carried off by the solar wind. The dotted surface is the heliospheric current sheet. It is shown as warped because the equivalent magnetic symmetry axis M does not coincide with the Sun's rotation axis.

It is seen that the heliospheric current sheet (HCS), the dotted surface, maps back to the Sun along the streamer belt (Hundhausen 1977; Smith et al. 1978; Borini et al. 1982). The figure shows that the symmetry axis of the HCS/CSB, represented here by the equivalent magnetic axis M does not generally coincide with the Sun's rotation axis. This tilt angle causes the current sheet to develop the form of a wavy ballerina skirt or hat brim as the Sun rotates. Various representations of this wavy current sheet on a variety of heliospheric scales and for various tilt angles have appeared in the literature (see, e.g., Jokipii and Thomas 1981).

There are several reasons why knowledge of the current sheet location is important to studies of the solar wind. Although deformed, the HCS serves as the magnetic equator of the heliosphere and organizes many properties of the heliospheric particles. Figure 7 (Zhao and Hundhausen 1981) shows the average dependence of the solar wind proton density and speed on the latitude above and below the current sheet (called heliomagnetic latitude). The density tends to be a maximum and the speed a minimum in the vicinity of the current sheet, general characteristics which have been widely observed in a large number of other studies.

The HCS has also been found to influence the intensities of high-energy

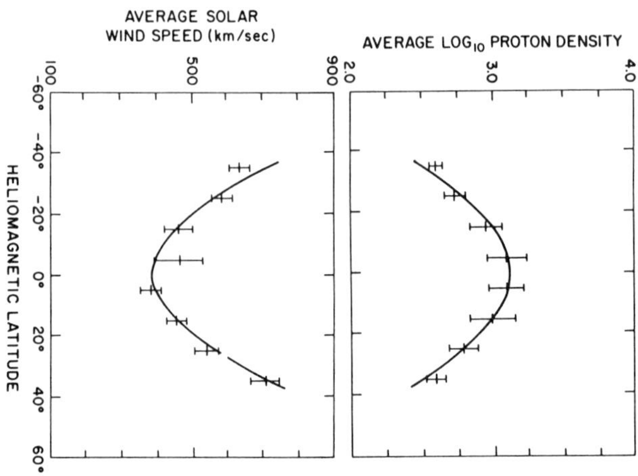

Figure 7.   Variation of solar wind density and velocity with latitude above and below
the heliospheric current sheet. The heliomagnetic latitude is the spacecraft latitude
relative to the magnetic equator or the current sheet rather than to the Sun's rotation
equator. The speed is low and the density is high near the current sheet.

galactic cosmic rays (GCR) (Jokipii et al. 1977) and the anomalous cosmic
ray (ACR) component (Lockwood et al. 1988) (which originates inside the
heliosphere as a consequence of the pick up of freshly ionized interstellar
neutrals followed by acceleration to high energies ≥ 10 MeV). The identification of fast or slow solar wind as originating in the northern or southern solar
hemisphere can also be established by their location relative to the current
sheet, i.e., their magnetic polarity.

Detailed comparisons of the HCS with actual solar magnetic fields are
possible by extrapolation of observed photospheric fields to a "solar wind
source surface" (Hoeksema et al. 1983). Line-of-sight magnetic fields derived
from synoptic observations by groundbased magnetographs are extrapolated
upward from the photosphere into a spherical shell using a potential field
model. In addition to the fields being prescribed at the inner boundary, the
fields at the outer spherical boundary, the source surface, are required to be
radial and the resulting solution is computed as a sum of spherical harmonics.

On the source surface, there is a wavy neutral line along which $B =
B_r = 0$ that separates oppositely directed large-scale fields. This neutral
line is identified with the heliospheric current sheet. Figure 8 provides an
example derived in October 1991 (Hoeksema 1992). The upper half of the
figure contains the stronger, smaller-scale photospheric fields while the lower

half of the figure shows the weaker, larger-scale fields after extrapolation into the "corona." Notice that the low-order, dipole term is clearly dominant at the source surface.

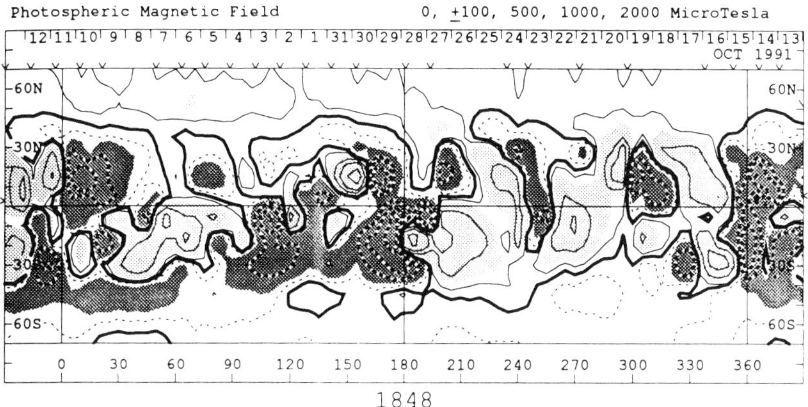

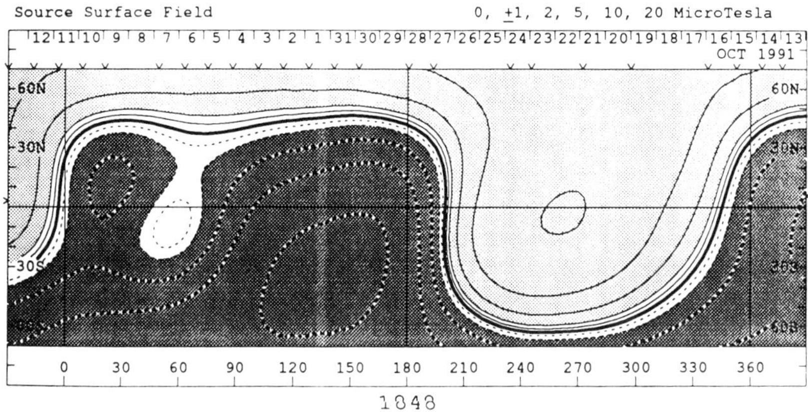

Figure 8.   Magnetic fields at the photosphere and on the solar wind source surface. The upper figure is a synoptic representation of the photospheric field strength and polarity along the line-of-sight to an Earth-based magnetograph. A single solar Carrington rotation (1848) is shown. The lower figure shows the corresponding smoother contours after the observed fields are extrapolated to a spherical surface at 2.5 AU and subjected to a boundary condition that the field be radial there.

Application of this procedure to many successive solar rotations has revealed how the solar field changes with the solar cycle. Figure 9 shows the secular variation in the dipole term between 1976 and 1991, i.e., over the most recent sunspot cycle. The upper panel contains the north polar field strength which shows a maximum of ≈ 1 Gauss (10^{-4} Tesla) near solar minimum (1976), vanishes at solar maximum (1979–1980) and then reverses polarity.

The field strength increases to -2 Gauss at the succeeding minimum and the field is seen again to reverse polarity in mid-1990.

Solar Cycle Variations

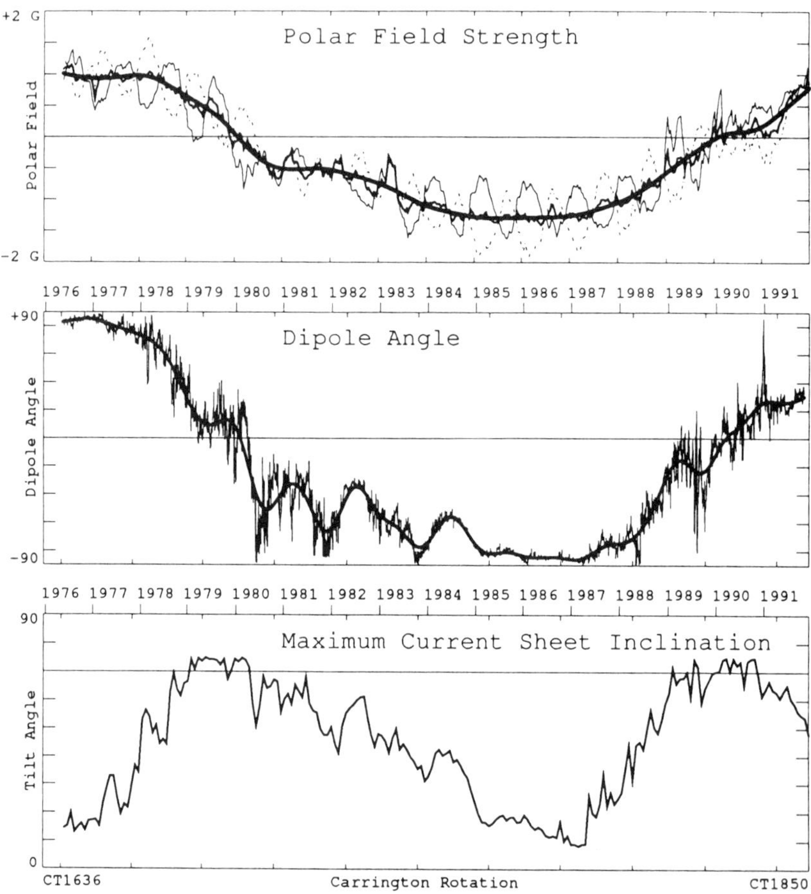

Figure 9. Solar cycle variations in the Sun's magnetic field. Inferred properties of the solar field are shown during the recent sunspot cycle (1976–1991). The upper curve of the polar field strength shows polarity reversals in 1979–1980 and 1990. The dark line is the smoothed average of the annual variations in the north (light solid line) and south (dashed line) polar fields. The middle panel shows the dipole angle which systematically rotated from being aligned with the rotation axis (90°) during the 1976 minimum to being equatorial (0°) at the maxima in 1979 and 1990. The bottom curve shows the maximum current sheet inclination (tilt angle) on the source surface and follows the middle panel.

SECTOR STRUCTURE – PIONEER 11

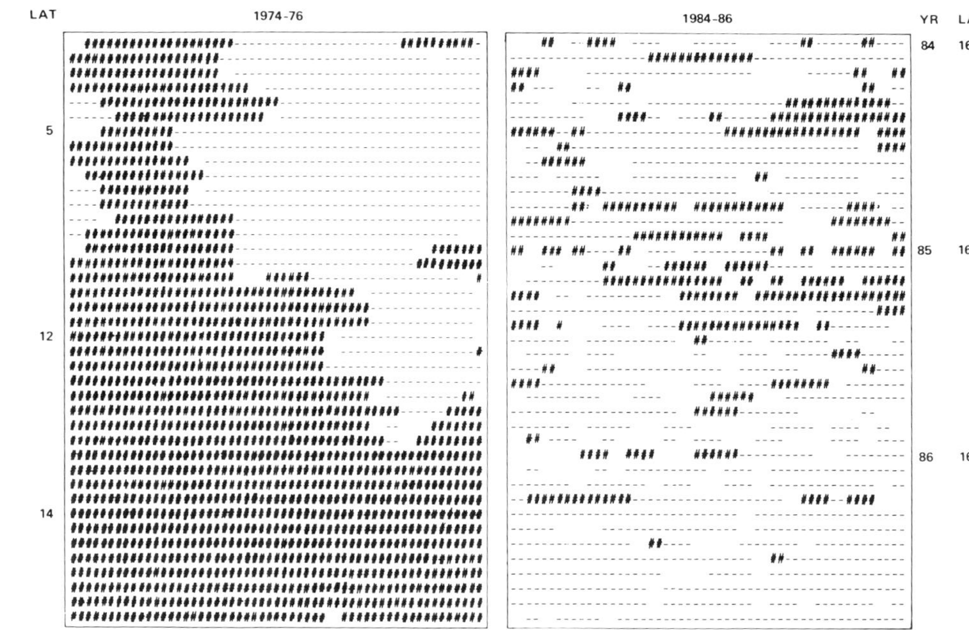

Figure 10.   Pioneer 11 observations of the disappearance of the sector structure at 15° latitude. Data from two successive minima in 1976 and 1986 are shown. In the left figure, fields corresponding to positive polarity or outward fields are seen to be continuously present at northern latitudes because the current sheet inclination is less than the latitude of the spacecraft. In the right figure, ten years later, the opposite situation prevails. Negative polarity or inward-directed fields are seen continuously, again corresponding with the fields in the Sun's northern hemisphere.

In addition to these changes with field strength or moment, the dipole changes orientation in a systematic fashion as shown in the middle panel. The latitude angle of the dipole is $\approx\pm90$ degrees near solar minimum (i.e., aligned with the Sun's rotation axis) and decreases to $\approx0°$ when the field reverses sign. The bottom panel addresses the inclination of the source surface neutral sheet (derived from its maximum variation in latitude during a given solar rotation). This parameter is closely related to the dipole "tilt" angle and is consistent with a low inclination of the neutral sheet at minimum and a high inclination at maximum.

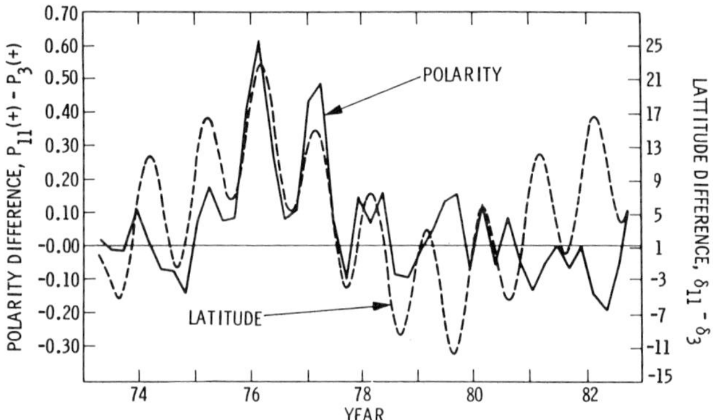

Figure 11.   Correlation between the differences in magnetic polarity at two spacecraft and their differences in latitude. The observations were obtained simultaneously at ISEE-3 and Pioneer 11 which were located at significantly different heliographic latitudes as shown. The variations are caused by the annual excursion of ISEE-3 as it orbits the Sun in the ecliptic plane at an inclination of 7.25° to the rotation axis. At each spacecraft, the fraction of each solar rotation that a positive polarity was observed, $P(+)$, was found and the differences are shown. During 1976, when Pioneer 11 was above ISEE-3, an excess positive polarity was seen. The phase between the two curves reversed in 1979–1980 when the Sun's polar caps reversed polarity.

The effect of the changing current sheet inclination and the reversal in polarity can both be observed in the sector structure. Figure 10, which is in basically the same format as Fig. 4, shows observations during two successive solar minima in 1976 and 1986 (Smith 1989). The polarities were observed by Pioneer 11 between 1 and 20 AU at the north heliographic latitudes indicated in the columns denoted LAT. Note the disappearance of the typical two sectors in the bottom half of both panels with the field pointing continuously outward in 1976 and inward in 1986. The obvious interpretation is that the current sheet inclination has decreased in both instances to less than $\sim15°$ so that the spacecraft is continuously above the current sheet and so records only a single polarity.

A third disappearance of the sector structure at the approach to solar minimum has been observed recently by the Ulysses spacecraft at 30°S latitude

(Smith et al. 1993). Thus, by virtue of having spacecraft positioned above or below the solar equator, the low inclination has been seen in three successive solar cycles.

The reversal in the Sun's polarity was first observed by Rosenberg and Coleman (1969) using spacecraft data and subsequently by Svalgaard and Wilcox (1974) who based their analyses on interplanetary field polarities inferred from groundbased magnetic observations at high latitude. Both studies exploited the annual excursion of the Earth in heliographic latitude of $\pm 7.25°$ which leads to a yearly variation in $P(+)$, the relative fraction of the time a positive polarity is observed during each solar rotation.

Figure 11 contains the results of a similar type of analysis but one which is capable of higher time resolution. The solid curve is the difference in $P(+)$ at two spacecraft, Pioneer 11 which was enroute to 12 AU and ISEE-3 which was stationed in a halo orbit about the first Sun-Earth libration point L_1 just inside 1 AU. The dashed curve is the difference in the instantaneous latitudes δ of the two spacecraft. The principal features of note are the overall correlation between the differences in P and δ and the 180° change in phase that occurred in 1979–1980 just at the time that the polar caps of the Sun reversed polarity.

V. SOLAR WIND STRUCTURE: EFFECT ON THE HELIOSPHERIC MAGNETIC FIELD

The solar wind exhibits considerable structure which is visible each solar rotation. It is caused by the wind originating from specific solar regions, such as coronal holes, and by the rotation of the Sun which allows fast wind to overtake and interact with slow wind from a different longitude.

The characteristic stream-stream interaction is shown schematically in Fig. 12 (Belcher and Davis 1971). The upper half of the figure characterizes the interaction in terms of (solid) streamlines emanating from four corotating solar sources which lead to a slow-fast-slow-fast configuration. The stream lines are spiraled because the coordinate system is corotating with the Sun. A spacecraft will appear to proceed along the circular arc in a clockwise sense. The dotted curves are lines of force of the solar wind magnetic field. The bottom half of the figure shows the characteristic variation observed in the solar wind speed V_W and density N, the field strength B, the thermal speed (equivalent to temperature) V_T, the standard deviation in the magnetic field fluctuations σ_S, and the azimuthal velocity component V_ϕ.

The essential physics of the interaction leads to the development of compression regions (corotating interaction regions or CIRs) alternating with rarefaction regions. The compression region consists of slow wind (S) that has been accelerated (S') and fast wind (F) that has been decelerated (F') separated by an interface at which the total solar wind internal pressure $(NkT + B^2/8\pi)$ reaches a peak. The pressure gradient supplies the forces

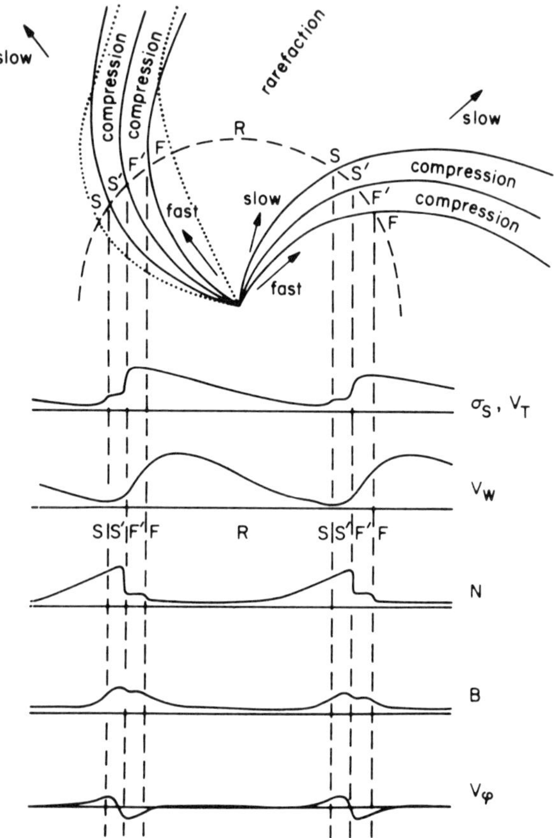

Figure 12.   Schematic representation of corotating interaction regions. Magnetic fields and solar wind streamlines are shown in the upper figure in a coordinate system rotating with the Sun. This schematic and the variations in the field and solar wind parameters shown in the lower half of the figure are appropriate to conditions near 1 AU. At larger distances the edges of the interaction/compression region steepen into shocks.

which alternately speed up and slow down the solar wind and cause the interaction region to widen continuously with distance.

The intersection of the magnetic field lines with the streamlines correspond to their crossing through a collisionless shock into the CIR. A forward shock (propagating outward from the Sun) develops along the leading edge of the CIR and a reverse shock (propagating sunward but convected outward by the more rapidly moving supersonic solar wind) forms at the trailing edge. The consequences of the forward-reverse shock pair for the magnetic field and solar wind speed profiles are shown in Fig. 13 (Smith and Wolfe 1977). The upper two panels contain measurements over several days made at 4.3 AU by Pioneer 10. The middle panel and the schematic below it show that the gradual increase in speed typical of observations at 1 AU has been replaced by two steps at the shocks.

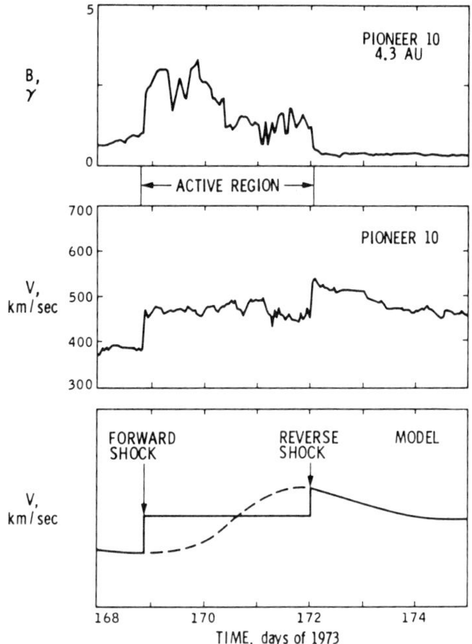

Figure 13.   A pair of forward and reverse shocks bounding a corotating interaction region (CIR). The field magnitude and speed are shown at 4.3 AU as observed by Pioneer 10 over an interval of several days. The abrupt increase in *B* and *V* corresponds to a forward shock as indicated in the diagram at the bottom. The increase in *V* with a simultaneous decrease in *B* coincides with a reverse shock.

The solar wind structure associated with the alternating compression and rarefaction regions becomes even more striking with increasing distance. As the interaction regions widen, they eventually begin to encounter one another and merge into even wider regions. Beyond 5 to 10 AU, a single large CIR is customarily observed each solar rotation. Evidently, the fastest-moving structure per rotation has overtaken and compressed the slower-moving structure ahead of it.

A good example of merging and its effect on solar wind structure can be seen in Fig. 14 (Smith 1985). The field magnitudes are shown at two different locations in 1984, at 16 AU by Pioneer 11 and as measured at 1 AU by ISEE-3 (lower panel). At Pioneer, the field is seen to increase and decrease by a factor ≤ 5 at periodic intervals corresponding to a solar rotation. The lower figure shows many more smaller increases associated with the passage by the spacecraft of several streams and CIRs each rotation.

The merging process is a dominant feature of the outer heliosphere and

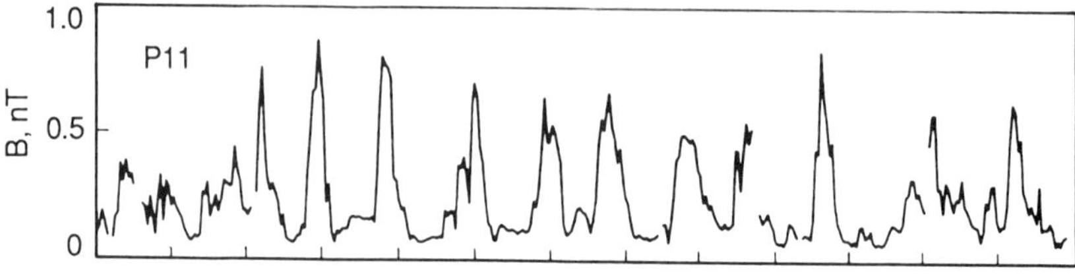

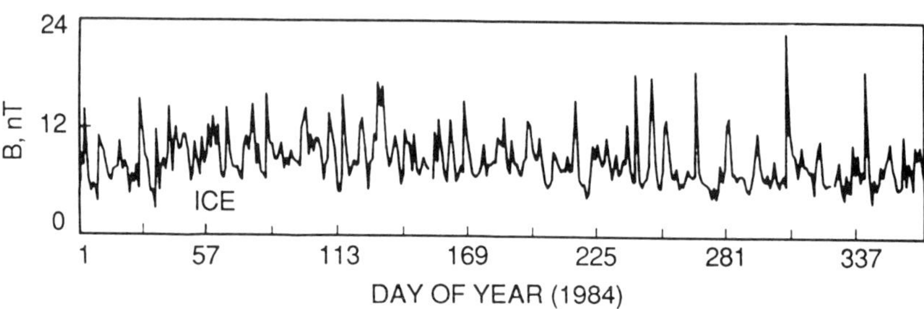

Figure 14.   Magnetic field variations at large distance compared with those at 1 AU. The upper panel contains daily averages of B at Pioneer 11 which was located at 16 AU. A single large increase tends to be seen each solar rotation. The lower panel contains comparable observations made at 1 AU by the International Cometary Explorer/ICE (ISEE-3). Many increases in B are seen each rotation which must have merged by the time they reached 16 AU.

continues to operate to distances ≥ 50 AU. By the time the solar wind reaches such distances, the successive CIRs evident in Fig. 14 have also merged so that the solar wind structure over a year is typically dominated by a few very large merged interaction regions (MIRs) in which enhancements in the field, density and temperature extend over several solar rotations (see, e.g., the chapter by McDonald and Burlaga, and references therein). During solar maximum, this merging is especially effective when it is driven by very energetic coronal mass ejections (CMEs) associated with major outbursts of solar activity (e.g., in March–June 1991).

VI. VARIATIONS IN FIELD MAGNITUDE WITH TIME AND DISTANCE

The magnetic field magnitude varies with the solar cycle in a systematic manner. In Fig. 15 (Winterhalter et al. 1990), annual variations of B at 1 AU are shown by the open squares (ISEE-3) and circles (omnibus values assembled by the National Space Science Data Center/NSSDC). The field has minima near solar minimum in 1976 and 1986–87. The field maxima occur near or just following solar maximum (1982, 1990). It is interesting that B is largest near the time that the Sun's polar cap fields reverse (rather than becoming small or vanishing). The magnetic fields equivalent to the highly inclined HCS or to the magnetic axis having been rotated through $\approx 90°$ now presumably originate in the Sun's equatorial regions. The filled dots in Fig. 15 represent measurements obtained by Pioneer 11 in the outer heliosphere (at the distances indicated along the top) which have been extrapolated back to 1 AU using Parker's solar wind model and assuming $\psi = \pi/4$ at 1 AU (i.e., without correcting for variations in solar wind speed). Although the extrapolated field reproduces the solar cycle variation observed at 1 AU, there is a systematic difference with B being lower than expected at large distances.

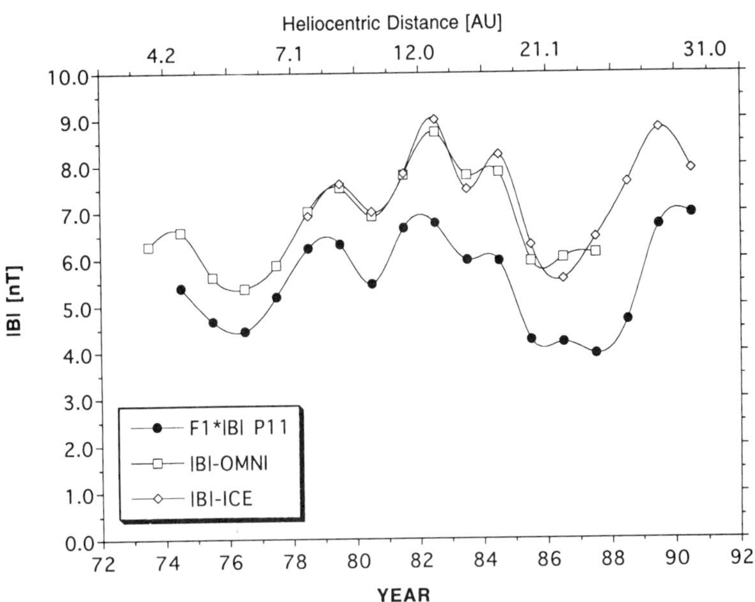

Figure 15.   A comparison of field magnitudes at 1 AU and in the outer heliosphere. The upper curve (open symbols) consists of annual averages of B at 1 AU as measured by several spacecraft between 1973 and 1992. The lower curve (solid dots) are annual averages of the field magnitude measured at Pioneer 11 but extrapolated back to 1 AU using the Parker solar wind model. During this 20-yr interval, Pioneer traveled out to beyond 30 AU. The comparison shows the solar cycle variation in B at both locations but with lower fields at the larger distances than would be expected on the basis of the model. The difference, called the magnetic flux deficit, has generated a considerable controversy in the recent scientific literature.

This difference has been described as a magnetic "flux deficit." The solar magnetic flux is given by $\Phi = \int B_r dA$. Because B_r is proportional to r^{-2}, it becomes negligibly small compared to B, which is $\approx B_\phi$ at large r and is difficult to determine in the presence of ever-present large transverse fluctuations. However, Eq. (1) implies $\Phi = \int r V_r B_\phi d\theta dt$ because $d\phi = \Omega dt$. Thus, the time average of $r V_r B_\phi \approx r V_r B$ is at large r is also a measure of Φ. In the Parker model, ϕ is constant and $r V_r B_\phi$ is an invariant.

The flux deficit seen in the Pioneer data has not been confirmed by Voyager observations over the same range of radial distances (Burlaga and Ness 1993). Various possibilities have been proposed to explain the discrepancy, in particular, the relatively "high latitude" of Pioneer 11 and the influence of changes in V_r on the extrapolated field magnitude. However, the latitude of Pioneer has varied significantly between -5 and $15°$ over the life of the mission and the inclusion of the measured speed at Pioneer in the extrapolation has failed to account for the flux deficit (Winterhalter et al. 1990). At present, the reason for the deficit appearing in the Pioneer data but not in the Voyager data is unknown.

Apart from the observations, theories have been developed that anticipate a departure from the Parker model. The original suggestion was that the enhanced spiraling of the field near the solar equator leads to increased pressure there which then deflects the solar wind and magnetic field slightly poleward (Suess and Nerney 1975; Suess et al. 1985). In a more recent model, the tilted magnetic dipole and the characteristic heliomagnetic latitude dependence of the solar wind speed give rise to excessive equatorial pressure associated with stream-stream interactions (Pizzo and Goldstein 1987). Either model can account for a deficit of the magnitude observed by Pioneer. This issue and its successful resolution in terms of the physical processes involved are not only of heliospheric interest. Astrophysicists have found processes such as that proposed by Suess and Nerney of interest in accounting for the formation and poleward displacement of jets.

VII. CORONAL MASS EJECTIONS

Thus far, the discussion of solar wind structure has concentrated on the effect of fast and slow streams. Not all of the solar wind, however, issues more or less steadily from stable sources such as coronal holes. The issue of corotating versus transient sources of solar plasma has a long history. Before direct solar wind observations became available, a major point of controversy was the open spiraled field topology characteristic of Parker's model in contrast to magnetic "tongues" or "bottles," a model advocated by Gold (1962). In retrospect, as often happens, both models were valid.

The Parker model was confirmed first beginning in 1962 (Snyder and Neugebauer 1964; Neugebauer and Snyder 1966). Another ten years passed before coronal mass ejections were conclusively identified, first in Skylab white light coronagraph images (see, e.g., Zirker 1977) and subsequently by HELIOS and other spacecraft. Therein lies a tale; it is difficult to distinguish the two types of solar wind on the basis of *in-situ* observations.

Although, as Hundhausen (see his chapter) has shown, CMEs make a contribution to the total solar wind mass of only 1 to 10%, they exert a profound influence on solar wind structure on a global scale. The reason is related to the enormous expansion which CMEs undergo as they travel outward into the heliosphere. Near the Sun, as the coronagraph images show, their characteristic scale is a solar radius or diameter. However, by the time they reach 1 AU, they have grown to scales that are a significant fraction of an astronomical unit, often 0.1 to 0.2 AU in radius with a comparable transverse dimension. Two complementary reasons for this expansion have been suggested. A speed gradient between the leading and trailing edges of 50 to 100 km s^{-1} could be responsible. Alternatively, the expansion could be driven by excess internal pressure rather than by a free expansion.

The expansion is presumably responsible in part for the difficulty in identifying CME plasma and fields by direct observation. Another factor is the nature of the interaction between the CME and the preceding solar wind which bears a resemblance to fast-slow stream interactions. CMEs are typically accompanied by a forward shock, a region of high pressure and irregular fields and, occasionally, by a reverse shock. Identification depends on a number of features being present more or less simultaneously such as lower than usual proton temperatures and densities, stronger and quieter than usual field strengths, the enhancement of helium relative to hydrogen, the presence of bi-directional electron and/or ion streams, etc.

A subset of CMEs that have a distinctive magnetic topology which assists identification are so-called magnetic "clouds" (Burlaga 1991). An example is shown in Fig. 16 based on HELIOS data. Notable features, used to define clouds, are the enhanced field strength and a characteristic north-south field deflection which, in combination with the x component, indicates a large-scale rotation of the field. The variations in the basic solar wind parameters are also fairly typical, namely, a decreasing speed from front to back, and regions of low density and temperature.

A schematic representation of two CME field topologies is presented in Fig. 17 (Gosling 1990). The upper figure represents magnetic loops (tongues or bottles) drawn out from the Sun by the plasma. The lower figure represents the fields as helical, a configuration that is often thought to be associated with force-free fields ($\mathbf{j} \times \mathbf{B} = 0$ implying a balance between magnetic pressure and magnetic tension forces). A number of comparisons between observed fields inside magnetic clouds and best-fit force-free fields based, for example, on models such as those of Burlaga (1988), have yielded a close correspondence. However, as a final warning on the difficulty of generalizing the properties of CMEs, it has been estimated that only 10% or less of CMEs fit the definition of magnetic clouds (Gosling 1990). Figure 17 actually presupposes (solely for convenience) the answer to a long-standing question. Do both ends of magnetic field lines inside the CME remain attached to the Sun or do they close on themselves to form detached loops? This global question has been difficult to answer with available *in-situ* measurements. Perhaps, both configurations occur at various times.

Near solar maximum, the Sun produces numerous very energetic CMEs, which lead to a restructuring of the global properties of the heliosphere. Active regions have the capacity to generate several ejections during a solar rotation which then propagate into the outer heliosphere over a range of longitudes.

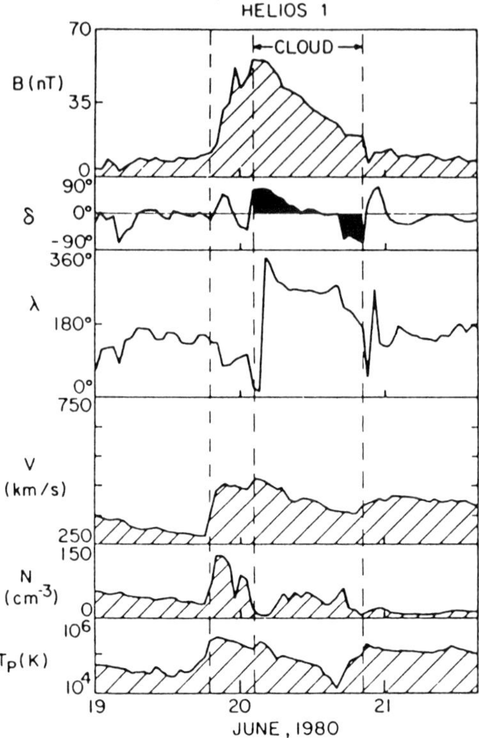

Figure 16.   Example of a "magnetic cloud." The cloud shown here is based on Helios 1 data. The magnitude and two angles (latitude, longitude) of the magnetic field and the solar wind velocity, density and temperature are shown. "Magnetic clouds" are a subset of CMEs identified solely by their magnetic properties of which this is a representative example.

They can compress the slower-moving solar wind which preceded them to form global or large-scale merged interaction regions (MIRs) that occupy a large solid angle about the Sun.

A kinematic simulation of this process is shown in Fig. 18 (Akasofu and Hakamada 1983). A series of six CMEs are shown after they have left the Sun and traveled into the outer heliosphere. This "snapshot" shows the formation of several merged regions represented by compressions of the magnetic field (the dark densely packed azimuthal field lines) which alternate with very large rarefaction regions (the light nearly radial field lines). The contrast with spiral structures of the kind represented in Fig. 12 is evident.

VIII. THE MAGNETIC FIELD NEAR THE HELIOPAUSE

Although the Pioneers and Voyagers have penetrated deeply into the outer heliosphere, they are still apparently a long distance from the bounding surface

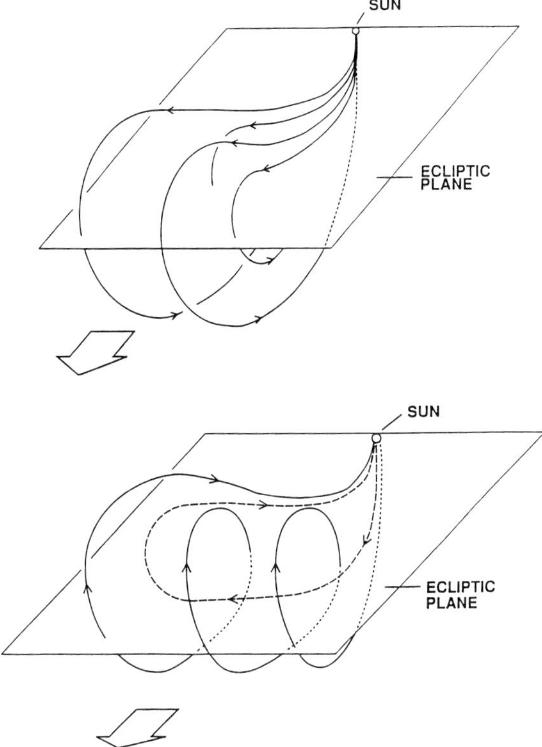

Figure 17.   Schematic of the CME as a flux rope. The possible magnetic topology of a CME is shown. Both examples show the CME as a magnetic "tongue" or "bottle" with the field lines attached to the Sun (closed). In the lower half of the figure, the field lines form a helix which could correspond to a force-free field configuration.

(the heliopause) between the solar and interstellar media. The scale of the heliosphere is uncertain at present but indirect evidence suggests a distance to the nose (closest approach) of ≈ 100 AU.

Before reaching the heliopause, the spacecraft are expected to pass through an inner shock at which the supersonic flow of the solar wind terminates, changes abruptly to a subsonic flow (hence, a termination shock) and is deflected away from or along the outer boundary to flow into the heliotail (Parker 1963; Axford 1972; Holzer 1989). The crossing of the termination shock is being awaited anxiously (Pesses et al. 1993) and when observed should establish more definitively the distance to the heliopause. The shocked solar wind lies in the region between the termination shock and the heliopause called the heliosheath.

There is considerable research interest in several related questions. What is the nature of the shock? In particular, will its structure be modified fundamentally by energetic particles (anomalous cosmic rays which are thought to

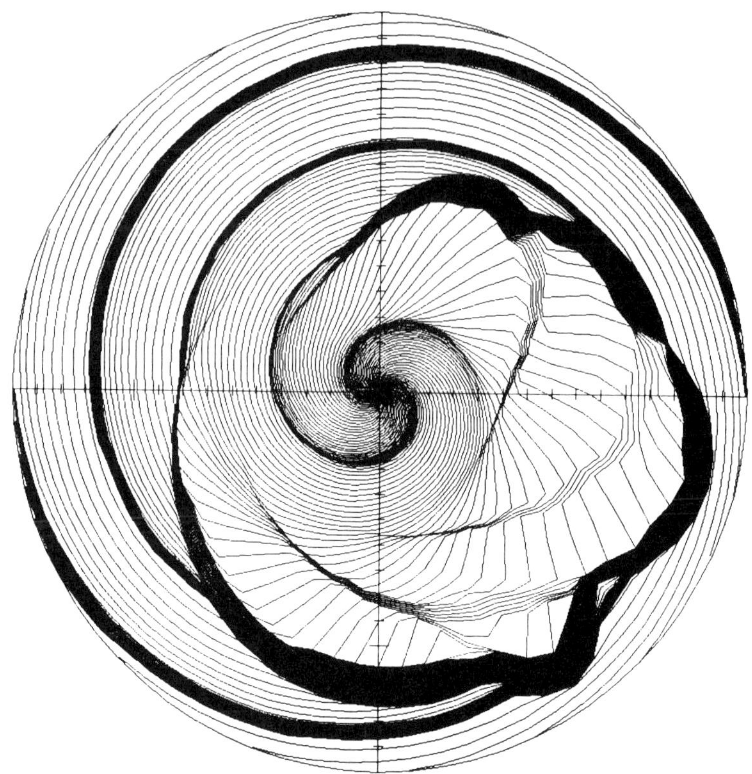

Figure 18.	Simulation of a series of CMEs as they expand into the outer heliosphere. The expansions of a series of six successive CMEs from the same solar longitude are shown between the Sun and 15 AU. The dark regions are magnetic fields that are compressed by the on-coming CME. The lighter regions are expansion regions within the CME proper. The combined effect of the several CMEs could be to form a barrier (a global merged interaction region) that opposes the entry of cosmic rays.

be accelerated at the shock and whose energy density may equal or exceed the solar wind energy density)? What is the character of the flow in the heliosheath (compressive or noncompressive)? Is merging or reconnection of the heliospheric and interstellar magnetic fields important?

The character of the magnetic field in the sheath and the influence of sector structure on magnetic merging are topics which have attracted attention recently. The results of a recent model are shown in Fig. 19 based on potential flow (Nerney et al. 1993). The sphere at the center of the figure represents the inner termination shock. Streamlines of the flow are shown which appear to emanate from the substagnation point to form a cylinder whose axis lies in the direction of the approaching interstellar wind. Field lines are shown along the side of the cylinder and are parallel to the flow in the plane corresponding to the solar equator and transverse to the flow in the plane containing the Sun's

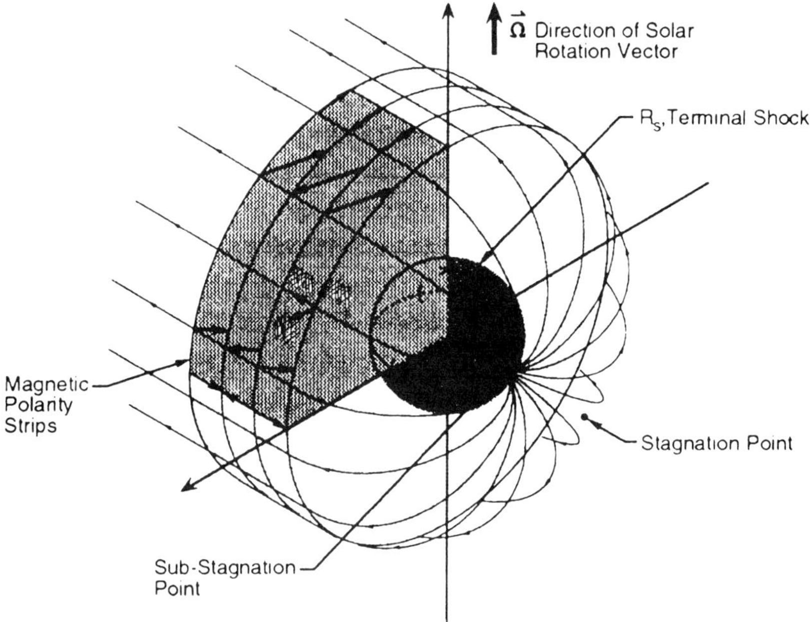

Figure 19.   Magnetic field just inside the heliopause. In this cylindrical coordinate system, the streamlines are shown as originating at the stagnation point between the heliopause and the interstellar plasma wind. They then turn and flow parallel to the heliopause along the inside. Representative magnetic field lines are also shown. They are parallel to the flow along the equator (the plane perpendicular to the solar rotation axis) but perpendicular to the flow and significantly greater in magnitude over the solar poles.

pole or rotation axis. The shaded sector coincides with three magnetic sectors whose polarities alternate as shown by the arrow heads.

Important features of the model are: (1) the origin of the field lines lying on or near the heliopause, which all originate from or near the stagnation region; (2) a strong asymmetry in the field orientation and strength between the equator and the pole, with the field being significantly stronger in the polar region; and (3) the varying polarity, which could imply a sequence of regions or "stripes" in which magnetic merging may take place. Certainly, it will be interesting to determine ultimately to what extent provocative modes such as this are correct.

IX. THE THREE-DIMENSIONAL MAGNETIC FIELD: RECENT ULYSSES RESULTS

Thus far, the discussion has emphasized the region of the heliosphere that we know best, namely, the fan-shaped region near the ecliptic plane in which all previous measurements have been made. The Parker model is basically three dimensional and the field topology is specified at all latitudes. Figure 20 shows three open field lines, one originating at the equator, the second at mid-latitude and the third in the polar region. The field lines are basically helices lying on the surfaces of cones whose axes are the Sun's rotation axis and with half-angles equal to the co-latitude θ. As the figure shows, the fields are strongly spiraled near the equator, having the appearance of a coiled spring, and nearly radial in the polar regions.

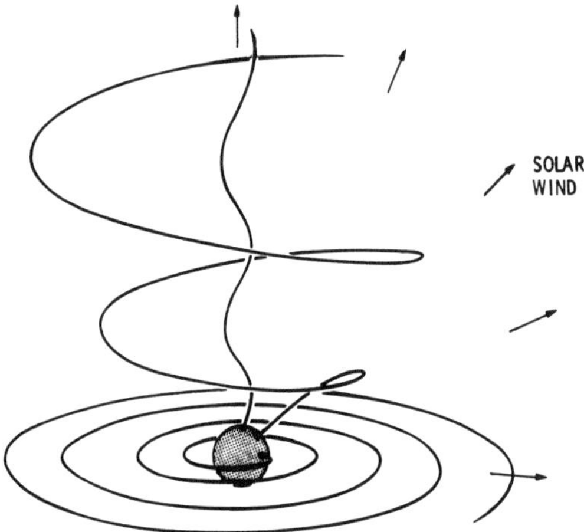

Figure 20.   The heliospheric magnetic field in three dimensions. Open magnetic field lines originating at three latitudes are shown in the form of helices. In the equator, the field lines are tightly wound, take the form of a coiled spring and are transverse to the flow. At higher latitudes the field lines lie on the surface of a cone and are less tightly wound, especially near the Sun's pole where they are nearly radial and parallel to the solar wind (represented by the arrows).

A comparison of measured fields at ISEE-3 and Pioneer Venus Orbiter over several years has exposed a latitude dependence in which the field strength is asymmetric about the equator (Luhmann et al. 1988). The field is weaker in the north in some years and stronger in the north in others. A comparison of Voyager 1 and 2 data at significantly different latitudes ($\sim$30°) also showed years in which the source field extrapolated back to 1 AU was either stronger or weaker at high latitude (Burlaga and Ness 1993).

There are reasons to expect deviations from a simple dependence of B on $\sin\theta$, because the photospheric/coronal field is inhomogeneous and the solar wind speed and angular rotation rate of the Sun vary with latitude. Much new information, a testing of our current ideas and some unexpected surprises are eagerly anticipated as a result of the Ulysses Mission.

The Ulysses Mission profile appears in Fig. 21 (Smith et al. 1991; Wenzel et al. 1992). It shows that the spacecraft was launched in October 1990, traveled outward to Jupiter, arriving in February 1992, and used Jupiter's gravity

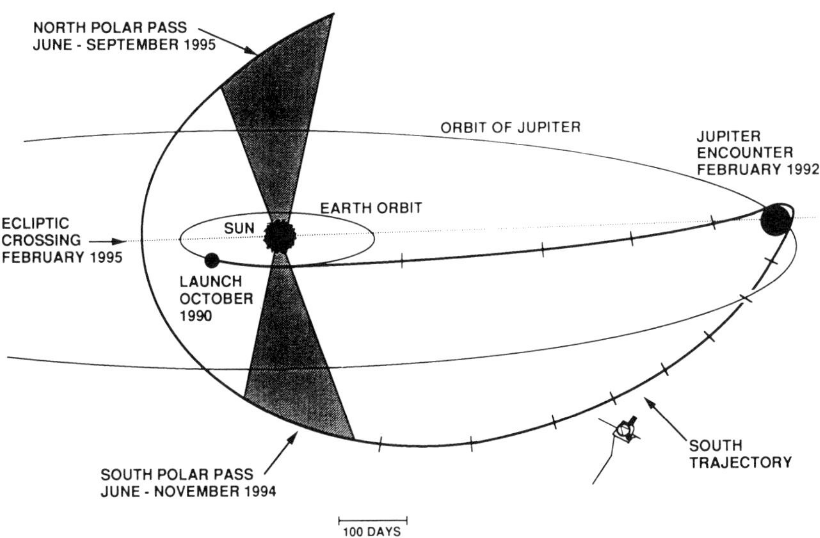

Figure 21.   Profile of the Ulysses Mission. Ulysses was launched in October 1990 into a low inclination elliptical orbit with aphelion near the orbit of Jupiter. The spacecraft encountered Jupiter in February 1992 and was gravitationally redirected into a high inclination orbit. The resulting trajectory is a large ellipse inclined 80° to the solar equator with a perihelion of 1.1 AU, an aphelion of 5.2 AU and a period of 6.3 yr. It reaches maximum latitudes in September 1994 (south) and August 1995 (north).

to rotate the inclination of the elliptical orbit to traverse the Sun's south polar region in 1994. After leaving the equatorial plane in 1992, Ulysses observations of the solar wind and magnetic field came to be dominated by a series of large corotating high-speed streams with peak speeds ≥ 700 km s^{-1} (Phillips et al. 1994). The presence of these streams was a temporal effect, rather than a latitude dependence, similar streams being observed simultaneously by spacecraft in or near the ecliptic. When Ulysses reached $-30°$ in May 1993, the sector structure disappeared showing that the spacecraft had risen above the maximum latitudinal extent of the HCS (Smith et al. 1993). Above this latitude, only inward-directed fields corresponding to the magnetic polarity of the Sun's south pole were seen. Above about $-50°$, the large streams and other solar wind structure essentially disappeared and Ulysses was continuously immersed in solar wind from the south polar coronal hole with a steady speed of ≈ 750 km s^{-1} (Phillips et al. 1995). On a few occasions, the tops of a CME were detected. The magnetic field was found to be highly variable in direction with $\Delta B/B \approx 1$. Correlation of the field changes with simultaneous changes in the velocity demonstrated that the variations were

452   E. J. SMITH

caused by outward-propagating Alfvén waves having large amplitudes and
periods extending up to tens of hours.

   One of the most striking results involved the latitude gradient in the
radial field component which is most easily related to the solar magnetic field.
Many scientists, including those responsible for the source surface models
described above, expected the field strength to increase with latitude owing to
the influence of the Sun's magnetic dipole (Wang 1995; Zhao and Hoeksema
1995). However, the measurements revealed the absence of a gradient with
B_R being essentially uniform and independent of latitude (Balogh et al. 1995).
Figure 22 shows $B_R r^2$ at Ulysses, averaged over successive latitude intervals
of $5°$, as a function of time and latitude. Measurements are also shown that
were obtained simultaneously in the ecliptic by the spacecraft IMP-8. The
two data sets fail to show a significant difference as Ulysses traveled from the
equator to the pole.

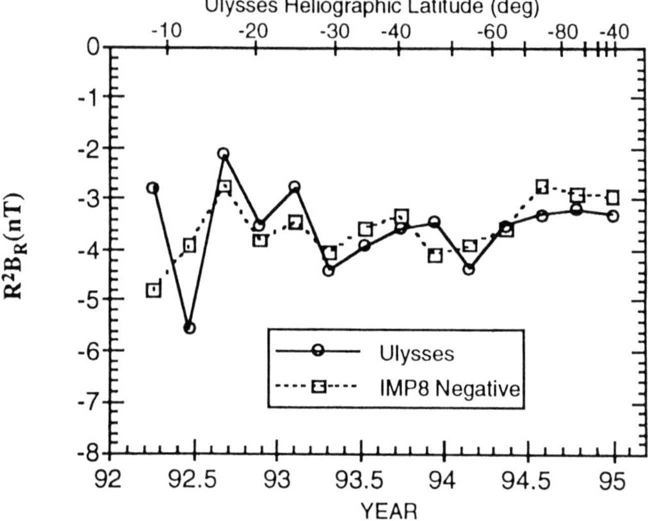

Figure 22.   The radial magnetic field at Ulysses and in the ecliptic. Averages of
   $B_R r^2$ at Ulysses over $5°$ increments in latitude (top) are shown as a function of time
   (bottom). Also shown are averages of B_R over the same time intervals measured by
   the IMP-8 spacecraft inside negative magnetic sectors. The correspondence of the
   time variations at both locations and the absence of a significant latitude gradient
   are evident.

   This result has important implications. Because the global photospheric
field is much stronger in the polar caps, the absence of a gradient in the solar
wind implies that magnetic flux is being transported equatorward. There must
be a corresponding divergence of the solar wind flow from the polar coronal
hole. The likely explanation is that the magnetic field is exerting a stress in the
solar wind acceleration region that causes this divergence. The stronger polar

cap fields near the Sun represent a gradient in magnetic pressure $(B^2/8\pi)$ tending to push the plasma to lower latitudes. The observed uniformity of B_R at larger distances would be the end-result of these stresses having finally relaxed to a stable configuration. Models of coronal magnetic fields which include such stresses and the resulting solar wind divergence (Pneuman and Kopp 1971; Suess et al. 1977) are confirmed by these Ulysses observations.

The spiral angle has also been studied as a function of latitude. The average angle has been computed from the averages of the components (indicated by brackets): $\langle\phi_B\rangle = \langle\tan^{-1}(\langle B_T\rangle/\langle B_R\rangle)\rangle$. It has been compared with the Parker spiral angle, $\phi_P = \tan^{-1}(\Omega r \cos\delta/V_r)$, resulting in the histogram of the difference shown in Fig. 23 which includes observations above $60°$. The average differs from 0 with $\langle\phi_B\rangle - \phi_P > 0$ which, for inward-directed fields, implies a departure toward a more radial direction (describable as an "underwinding" of the Parker spiral). This result agrees qualitatively with similar analyses carried out near the ecliptic using multi-spacecraft measurements.

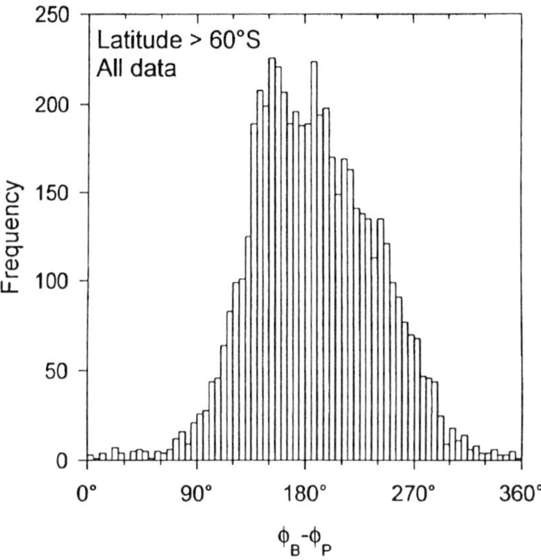

Figure 23.   The spiral angle observed by Ulysses at high latitudes. Hourly averages of ϕ_B obtained above $60°$ latitude were used to prepare this histogram. The value of the Parker spiral, obtained from the hourly average of the solar wind speed, was subtracted from ϕ_B so that the histogram represents deviations from the Parker spiral. The most probable value is negative implying the field is too tightly wound. However, because of the asymmetry in the distribution, the overall average is positive corresponding to the field being underwound.

However, the observed distribution function shows a large departure from a normal or Gaussian distribution. A marked asymmetry is present which favors large values of $\phi_B - \phi_P$ and which causes a significant difference between the average and most probable value. Up to about $-50°$, the latter agreed with ϕ_P, but at higher latitudes neither measure agrees with the Parker model. Presumably, ϕ_B is being strongly affected by the large amplitude Alfvén waves and, until their effect is eliminated somehow, it is difficult to demonstrate the extent of the agreement with the model or to use the spiral angle to infer the solar rotation rate at high latitude.

APPENDIX

The spiral angle originates from basic physical considerations. An electric field E in inertial space will transform into the frame of reference moving with the solar wind as $E_W = E + \mathbf{V} \times \mathbf{B}$, where $\mathbf{V}$ is the solar wind velocity and $\mathbf{B}$ is the magnetic field. Because the solar wind is basically a collisionless plasma with infinite electrical conductivity, $E = J/\sigma$ must vanish for finite currents J implying $E = -\mathbf{V} \times \mathbf{B}$. Alternatively, the latter condition can be described conceptually by saying the magnetic field is frozen-into and moves with the plasma so that there is no relative motion to generate an E_W.

The E field in inertial space is the result of the Sun's B field rotating past the point of observation and is given by $\mathbf{E} = E_\Theta = -\Omega r \sin\theta B_r$, where Ω is the angular velocity of the Sun (or, more strictly, of the end of the field line) and θ is the colatitude in the usual spherical coordinates. For the case of a radial velocity, i.e., $\mathbf{V} = V_r$, $\mathbf{V} \times \mathbf{B} = V_r B_\phi$. Hence, $E_\Theta = -\Omega r \sin\theta B_r = V_r B_\phi$ or

$$B_\phi/B_r = \tan\psi = -\Omega r \sin\theta / V_r. \tag{3}$$

According to this equation, given V_r and B_r (a property of the solar field and generated by currents at the Sun), B_ϕ will adjust to yield the appropriate spiral angle.

It might be supposed that a B_θ component would be possible but it is excluded from Parker's model. It is all too often claimed that Parker simply assumed $B_\theta = 0$ and that misunderstanding has led to confusion. For that reason, a derivation of $B_\theta = 0$ now follows. Consider the electric field that would be associated with B_θ, i.e., $E_\phi = -V_r B_\theta$. Then, the voltage along a circle about the Sun would be $V = \oint E_\phi \cdot \mathrm{d}s = -2\pi r V_r B_\theta$. However, $V = -\mathrm{d}\phi/\mathrm{d}t = 0$, so that $B_\theta = 0$.

The model also prescribes the currents that are responsible for B_ϕ. The current associated with B_ϕ can be computed from $\nabla \times \mathbf{B} = \mu_o \mathbf{J}$ and, as expected, is transverse to B_ϕ, i.e., $j_r = -2(B_{ro}r_o)\sin\delta/\mu_o r^2$ where the radial field component is B_{ro} at $r_o = 1$ AU and δ is heliographic latitude. Thus, for an *outward* field in the northern hemisphere ($\delta > 0$), the radial current is inward, and is 0 on the equator and maximum at the pole. Because the Sun cannot be realistically represented as a magnetic monopole, a more physical model would have *inward* radial fields in the southern hemisphere. Because both B_{ro} and δ are negative, the associated radial currents are also inward. The inward currents north and south are compensated by an outward surface current at the equator such that the total current is 0 as required (Smith et al. 1978).

The electromagnetic boundary condition applied at the equatorial interface between the two oppositely directed fields leads to a linear surface current density $\mathbf{K}$:

$$K_r = 2B_{ro}r_o/\mu_o r, \quad K_\phi = 2B_{ro}r_o^2/\mu_o r^2 \tag{4}$$

(conservation of magnetic flux requires that $B_r = B_{ro}r_o^2/r^2$).

It can readily be shown that the net current out of the Sun is

$$I_r = 2\int_o^{\frac{\pi}{2}}\int_o^{2\pi} j_r \cos\delta r^2 \mathrm{d}\phi \mathrm{d}\delta - \int_o^{2\pi} K_r r\mathrm{d}\phi = 0. \tag{5}$$

It is also seen that the field lines form Archimedes spirals (as is well known), because

$$-r\mathrm{d}\phi/\mathrm{d}r = B_\phi/B_r = -\Omega r \sin\theta/V_r \ \text{ or } \ r = (V_r/\Omega\sin\theta)\phi. \tag{6}$$

The current streamlines on the equator form hyperbolic or reciprocal spirals. As above,

$$r\mathrm{d}\phi/\mathrm{d}r = K_\phi/K_r = r_o/r, \ \text{ so that } \ \phi = \phi_o + 1 - r_o/r. \tag{7}$$

Acknowledgments. Portions of this manuscript present work done at the Jet Propulsion Laboratory of the California Institute of Technology for the National Aeronautics and Space Administration. The summary of recent Ulysses results was not possible at the time of the conference and was not presented there. However, it has been added at the request of one of the referees to update the review with more recent information.

REFERENCES

Akasofu, S.-I., and Hakamada, K. 1983. Solar wind disturbances in the outer heliosphere, caused by six successive solar flares from the same active region. *Geophys. Res. Lett.* 10:577–579.

Axford, W. I. 1972. The interaction of the solar wind with the interstellar medium. In *Solar Wind*, NASA SP-308, pp. 609–660.

Balogh, A., Smith, E. J., Tsurutani, B. T., Southwood, D. J., Forsyth, R. J., and Horbury, T. S. 1995. Structure and characteristics of the heliospheric magnetic field over the south polar region of the Sun. *Science* 268:1007–1010.

Belcher, J. W., and Davis, L., Jr. 1971. Large-amplitude Alfvén waves in the interplanetary medium, 2. *J. Geophys. Res.* 76:3534–3563.

Borini, G., Gosling, J. T., Bame, S. J., and Feldman, W. C. 1982. Helium abundance enhancements in the solar wind. *J. Geophys. Res.* 87:7370–7378.

Burlaga, L. F. 1988. Magnetic clouds and force-free fields with constant alpha. *J. Geophys. Res.* 93:7217–7224.

Burlaga, L. F. 1991. Magnetic clouds. In *Physics of the Inner Heliosphere II*, eds. R. Schwenn and E. Marsch (New York: Springer-Verlag), pp. 1–22.

Burlaga, L. F., and Ness, N. F. 1993. Radial and latitudinal variations of the magnetic field strength in the outer heliosphere. *J. Geophys. Res.* 98:3539–3549.

Gold, T. 1962. Magnetic storms. *Space Sci. Rev.* 1:100–114.

Gosling, J. T. 1990. Coronal mass ejections and magnetic flux ropes in interplanetary space. In *Physics of Magnetic Flux Ropes*, eds. C. T. Russell, E. R. Priest and L. C. Lee (Washington, D. C.: American Geophysical Union), pp. 343–364.

Hoeksema, J. T. 1992. Large scale structure of the heliospheric magnetic field: 1976–1991. In *Solar Wind Seven*, eds. E. Marsch and R. Schwenn (New York: Pergamon Press), pp. 191–196.

Hoeksma, J. T., Wilcox, J. M., and Scherrer, P. H. 1983. The structure of the heliospheric current sheet: 1978–1982. *J. Geophys. Res.* 88:9910–9918.

Holzer, T. E. 1989. Interaction between the solar wind and the interstellar medium. *Ann. Rev. Astron. Astrophys.* 27:199–234.

Hundhausen, A. J. 1977. An interplanetary view of coronal holes. In *Coronal Holes and High-Speed Wind Streams*, ed. J. B. Zirker (Boulder: Colorado Associated Univ. Press), pp. 225–329.

Jokipii, J. R., and Parker, E. N. 1970. On the convection, diffusion and adiabatic deceleration of cosmic rays in the solar wind. *Astrophys. J.* 160:735–744.

Jokipii, J. R., and Thomas, B. T. 1981. Effects of drifts on the transport of cosmic rays, IV, Modulation by a wavy interplanetary current sheet. *Astrophys. J.* 243:1115–1122.

Jokipii, J. R., Levy, E. H., and Hubbard, W. B. 1977. Effects of particle drift on cosmic ray transport, I, General properties, application to solar modulation. *Astrophys. J.* 213:861–868.

Lockwood, J. A., Webber, W. R., and Hoeksema, J. T. 1988. Interplanetary cosmic ray latitudinal gradient in 1984 to 1987 using IMP8 and Voyager data. *J. Geophys. Res.* 93:7521–7526.

Luhmann, J. G., Russell, C. T., and Smith, E. J. 1988. Asymmetries of the interplanetary field inferred from observations. In *Proc. of the Sixth Intl. Solar Wind Conf.*, vol. 1, eds. V. J. Pizzo, T. E. Holzer and D. G. Sime, NCAR TN-306 (Boulder: Natl. Center for Atmos. Res.), pp. 323–327.

Nerney, S., Suess, S. T., and Schmahl, E. J. 1993. Flow downstream of the heliospheric termination shock. *J. Geophys. Res.* 98:15169–15176.

Neugebauer, M., and Snyder, C. W. 1966. Mariner 2 observations of the solar wind, 1, average properties. *J. Geophys. Res.* 71:4469–4484.

Parker, E. N. 1963. *Interplanetary Dynamical Processes* (New York: Wiley).

Pesses, M. E., Jones, W. V., and Forman, M. 1993. Voyager and Pioneer missions to the boundaries of the heliosphere. *J. Geophys. Res.* 98:15123–15127.

Phillips, J. L., Balogh, A., Bame, S. J., Goldstein, B. E., Gosling, J. T., Hoeksema, J. T., McComas, D. J., Neugebauer, N., Sheeley, Jr., N. R., and Wang, Y.-M. 1994. Ulysses at 50 degrees South: Constant immersion in the high speed solar wind. *Geophys. Res. Lett.* 21:1105–1108.

Phillips, J. L., Bame, S. J., Feldman, W. C., Goldstein, B. E., Gosling, J. T., Hammond, C. M., McComas, D. J., Neugebauer, M., Scime, E. E., and Suess, S. T. 1995. Ulysses solar wind plasma observations at high southerly latitudes. *Science* 268:1030–1033.

Pizzo, V. J., and Goldstein, B. E. 1987. Meridional transport of magnetic flux in the solar wind between 1 and 10 AU: A theoretical analysis. *J. Geophys. Res.* 92:7241–7253.

Pneuman, G. W., and Kopp, R. A. 1971. Gas-magnetic field interactions in the solar corona. *Solar Phys.* 18:258–270.

Rosenberg, R. L., and Coleman, P. J., Jr. 1969. Heliographic latitude dependence of the dominant polarity of the interplanetary magnetic field. *J. Geophys. Res.* 74:5611–5622.

Smith, E. J. 1985. Interplanetary shock phenomena beyond 1 AU. In *Collisionless Shocks in the Heliosphere: Reviews of Current Research*, eds. B. T. Tsurutani and R. G. Stone (Washington, D. C.: American Geophysical Union), pp. 69–83.

Smith, E. J. 1989. Interplanetary magnetic field over two solar cycles and out to 20 AU. *Adv. Space Res.* 9:159–169.

Smith, E. J., and Wolfe, J. H. 1977. Pioneer 10, 11 observations of evolving solar wind streams and shocks beyond 1 AU. In *Study of Traveling Interplanetary Phenomena*, eds. M. A. Shea, D. F. Smart and S. T. Wu (Dordrecht: D. Reidel), pp. 227–257.

Smith, E. J., Tsurutani, B. T., and Rosenberg, R. L. 1978. Observations of the interplanetary sector structure to heliographic latitudes of 16°: Pioneer 11. *J. Geophys. Res.* 83:717–724.

Smith, E. J., Slavin, J. A., and Thomas, B. T. 1986. The heliospheric current sheet: 3 dimensional structure and solar cycle changes. In *Proc. of the ESLAB Symposium on the Three Dimensional Structure of the Heliosphere* (Dordrecht: D. Reidel), pp. 267–274.

Smith, E. J., Page, D. E., and Wenzel, K.-P. 1991. Ulysses: A journey above the Sun's poles. *Eos: Trans. AGU* 72:241.

Smith, E. J., Neugebauer, M., Balogh, A., Bame, S., J., Erdös, G., Forsyth, R. J., Goldstein, B. E., Phillips, J. L., and Tsurutani, B. T. 1993. Disappearance of the heliospheric sector structure at Ulysses. *Geophys. Res. Lett.* 20:2327–2330.

Snyder, C. W., and Neugebauer, M. 1964. Interplanetary solar wind measurements by Mariner 2. *Space Res.* 4:89–113.

Suess, S. T., and Nerney, S. F. 1975. The global solar wind and predictions for Pioneers 10 and 11. *Geophys. Res. Lett.* 2:75–77.

Suess, S. T., Richter, A. K., Winge, C. R., and Nerney, S. F. 1977. South polar coronal hole—A mathematical simulation. *Astrophys. J.* 217:296–305.

Suess, S. T., Thomas, B. T., and Nerney, S. F. 1985. Theoretical interpretation of the observed interplanetary magnetic field radial variation in the outer solar system. *J. Geophys. Res.* 90:4378–4382.

Svalgaard, L., and Wilcox, J. M. 1974. The spiral interplanetary magnetic field: A polarity and sunspot cycle variation. *Science* 186:51–53.

Thomas, B. T., and Smith, E. J. 1980. The Parker spiral configuration of the interplanetary magnetic field between 1 AU and 8.5 AU. *J. Geophys. Res.* 85:6861–6867.

Wang, Y.-M. 1995. Latitude and solar cycle dependence of radial IMF intensity. *Space Sci. Rev.* 72:193–196.

Wenzel, K.-P., Marsden, R. G., Page, D. E., and Smith, E. J. 1992. The Ulysses mission. *Astron. Astrophys. Suppl.* 92:207–219.

Winterhalter, D., Smith, E. J., Wolfe, J. H., and Slavin, J. A. 1990. Spatial gradients in the heliospheric magnetic field: Pioneer 11 observations between 1 AU and 24 AU, and over solar cycle 21. *J. Geophys. Res.* 95:1–11.

Zhao, X.-P., and Hoeksema, J. T. 1995. Modeling the out-of-ecliptic interplanetary magnetic field in the declining phase of sunspot cycle 22. *Space Sci. Rev.* 72:189–192.

Zhao, X.-P., and Hundhausen, A. J. 1981. Organization of solar wind plasma properties in a tilted, heliomagnetic coordinate system. *J. Geophys. Res.* 86:5423–5430.

Zirker, J. B. 1977. Coronal holes—An overview. In *Coronal Holes and High Speed Wind Streams*, ed. J. B. Zirker (Boulder: Colorado Associated Univ. Press), pp. 1–26.

NEAR-SUN MAGNETIC FIELDS AND THE SOLAR WIND

N. R. SHEELEY, Jr., and Y.-M. WANG
Naval Research Laboratory

and

J. L. PHILLIPS
Los Alamos National Laboratory

Intensive research during the Skylab era of the 1970s led to the paradigm that fast wind comes from coronal holes and slow wind comes from other parts of the Sun. However, the accumulation of complementary solar and interplanetary observations since that time have led to other concepts, including the idea that all of the nontransient wind comes from open-field regions whose coronal flux-tube divergence rates determine the speed of the wind. These ideas are currently being put to detailed theoretical scrutiny and the observational test of Ulysses out-of-ecliptic measurements.

I. INTRODUCTION

Solar wind origin was the subject of intense study during the Skylab era of the early 1970s. It was the declining phase of sunspot cycle 20 when new solar instruments were beginning to track the evolution of large, low-latitude coronal holes and new in-ecliptic spacecraft were recording "monstrous" solar wind streams with speeds of 700 to 800 km s^{-1} (Neupert and Pizzo 1974; Bell and Noci 1976; Nolte et al. 1976; Sheeley et al. 1976; Wagner 1976; Hundhausen 1977; Feldman et al. 1978). The persistent correlation between the central-meridian-passage times of the holes and the arrival times of the streams at Earth (three days later) left little doubt that the holes were the sources of the streams (see Zirker 1977, and references therein). A two-component view emerged, in which fast, hot, low-density plasma came from holes and slow, cool, high-density plasma came from closed-field regions.

But now nearly 20 years have passed since the Skylab era, and coronal holes, solar magnetic fields, and in-ecliptic solar wind have been tracked through all phases of two sunspot cycles (see Marsch and Schwenn 1992, and references therein). In addition, radio interplanetary scintillation (IPS) measurements have provided inferences about the high-latitude wind (Coles and Rickett 1976; Sime and Rickett 1978; Kojima and Kakinuma 1987,1990; Watanabe 1989; Rickett and Coles 1991), and measurements from some spacecraft in the outer heliosphere have provided direct information out of, but close to, the ecliptic plane (Gazis et al. 1988; Gazis et al. 1989; Barnes et al. 1992).

As the observations accumulated, puzzles and exceptions were found, causing refinements and rationalizations. Sometimes, as in 1986, recurrent patterns of high speed occurred at Earth without coronal holes in the $\pm 40°$ range of solar latitudes, suggesting that the fast wind came from the polar holes. Sometimes, as in 1979-1981, low-latitude holes occurred without high-speed streams, suggesting that the holes were "too small" or "too far from the ecliptic" to affect Earth. Sometimes, as in 1977, 1985, and 1990, recurrent streams in excess of 700 km s^{-1} were clearly associated with "weak holes" whose contrast on X-ray and helium images was unusually low (Nolte et al. 1977; Sheeley and Harvey 1978; Sheeley and Wang 1991).

The growing data base not only provided exceptions, but also provided a large statistical sample for deriving a new rule. In fact, the data supported an old idea (Levine et al. 1977) that had not caught on during the Skylab era: all of the nontransient wind comes from open-field regions (holes), but the speed of the wind on a given flux tube is determined by the coronal expansion of that tube, low expansion corresponding to high speed and vice versa. Thus, individual flux tubes that fan out greatly in passing through the corona (like those at the edge of a polar hole at sunspot minimum) would have slow wind at Earth. Conversely, tubes that fan out little (like those at the center of a large polar hole) would have fast asymptotic wind.

The advantage of the expansion-factor concept is that with a coronal model it can be used to derive solar wind speed empirically from observed maps of photospheric field. Thus, the concept can be tested quantitatively against observations of solar wind speed. The disadvantage is that there is as yet no magnetic model simple enough to permit routine calculations from observed fields, but realistic enough to include the dynamic effects of the real Sun.

To analyze real solar observations, it has been necessary to use approximations such as the source-surface method in which the field is assumed to be current free between the photosphere and an idealized spherical source surface at 2.5 R$_\odot$ where the nonradial components are constrained to vanish. The surviving open field lines then extend radially outward with the help of assumed volume currents in the heliosphere (Schatten et al. 1969; Altschuler and Newkirk 1969; Hoeksema 1984; Wang and Sheeley 1992). Despite its overly simplistic character, this source-surface method has led to considerable progress and insight into the nontransient properties of coronal holes, sector boundaries, and solar wind streams (Hoeksema 1992; Wang et al. 1990; Hakamada et al. 1991; Wang and Sheeley 1990*b*,1993; Sheeley and Wang 1991), including the correlation between coronal expansion factor and in-ecliptic solar wind speed at Earth (Levine et al. 1977; Wang and Sheeley 1990*a*). Evidently, the properties of these large-scale structures are relatively insensitive to the details of the coronal model, or the lower corona is approximately current free, or both.

Not only does the present declining phase of the sunspot cycle provide us with a renewed opportunity to test the wind-speed/expansion-factor relation with high-speed streams in the ecliptic, but also the Ulysses spacecraft is providing an unprecedented opportunity to test it at increasingly high solar latitudes (Balogh et al. 1993*a,b*; Bame et al. 1992*a,b*,1993; Smith et al. 1993). As we shall see below, these high-latitude observations are helping to resolve the issue of solar wind origin and raising new questions about the acceleration mechanism.

II. THE EXPANSION-FACTOR CONCEPT AND IMPLICATIONS

Nearly four decades of solar magnetic observations have familiarized us with the characteristic patterns of magnetic fields on the Sun: relatively compact bipolar regions of erupting flux in the sunspot belts, elongated regions of alternating polarity being stretched eastward and poleward by photospheric transport processes, and unipolar regions at the Sun's poles. As individual field lines extend upward from these photospheric locations, some bend sideways in the lower corona and return to their opposite-polarity counterparts in the photosphere. These "closed" field lines have no effect in the outer corona and beyond. Other loops are caught up in the coronal expansion and extend far from the Sun, leaving "open-field" legs rooted in coronal holes at the Sun's surface.

We may think of an open field line as an infinitesimal flux tube extending upward through the corona into the solar wind, and define its expansion factor at the radial distance r by the solid angle its cross section subtends at the center of the Sun:

$$f(r) = \frac{d\Omega(r)}{d\Omega(R)} = \left(\frac{R}{r}\right)^2 \frac{dA(r)}{dA(R)} = \left(\frac{R}{r}\right)^2 \frac{B(R)}{B(r)} \tag{1}$$

where R is the solar radius. Thus, a radial field whose strength falls off as $1/r^2$ will have an expansion factor of 1, independent of distance from the Sun. A more divergent dipole-like field whose strength falls off as $1/r^3$ will have an expansion factor of r/R, larger than 1 and increasing with distance from the Sun. Finally, a convergent field whose strength has the slower $1/r$ fall off will have an expansion factor of R/r, smaller than 1 and decreasing as r increases.

Next, we consider the observational consequences of an inverse relation between solar wind speed and coronal flux-tube expansion, leaving the physical basis for such a relation to Sec. V of this chapter. We will begin by considering the kinds of topologies that ought to produce fast wind, and then use the current-free model to derive coronal expansion factors from the observed fields at representative phases of the sunspot cycle. Most of these results have been described previously (Wang and Sheeley 1990*b*,1994; Sheeley and Wang 1991; Sheeley 1992).

If each solar hemisphere were dominated by a single large polar hole whose field is peaked at the center, then the flux tubes would diverge least at the pole where the field lines are nearly radial and most at the edge of the hole where the field lines curve out to the magnetic equator. This is the situation at sunspot minimum, when we would therefore expect fast polar wind and slow equatorial wind.

However, if the latitude distribution of the field should change so that the field has a local minimum at the pole, then the field lines from the surrounding flux would converge over the pole as they begin their way out through the corona. The expansion factor would become less than 1, and an even faster wind would originate at the pole. This is the situation shortly after the polar fields reverse—weak polar fields surrounded by strong incipient trailing-polarity flux migrating poleward from the sunspot belts.

Actually, the poleward flux migration occurs in episodic bursts, with alternations of leading and following polarity over a few years before the following polarity finally wins out (Howard and LaBonte 1981; Wang et al. 1989). Thus, one would expect the high-latitude wind speed to fluctuate for a year or two before settling down to a steady value associated with the peaked field distribution.

If the central field were so weak that it actually reversed sign, then there would be a gap in the polar coronal hole and low-expansion field lines from its facing edges would converge over the pole before heading out through the corona. In effect, there would be an azimuthal coronal hole surrounding the pole, and its poleward edge ought to be the source of fast polar wind.

Similarly, when a polar hole is accompanied by a low-latitude hole of like polarity, the adjacent edges of these two holes would be the source of lower flux-tube expansion factors and higher wind speeds than occur over the center of the polar hole. This is the configuration that typically occurs during the declining phase of the sunspot cycle.

Plates 1 through 4 in the Color Section illustrate these concepts at four representative phases of the sunspot cycle. Sunspot minimum is represented by Plate 1, showing the observed photospheric field (top panel), together with derived maps of open-field at the photosphere (second from top), coronal ("source-surface") field at 2.5 $R_\odot$ (third from top), and source-surface expansion factor (bottom panel), all for Carrington rotation 1775 (May 3–30, 1986). Open-field regions are color-coded by their coronal expansion factors, with the colors red, white, yellow, green, blue being separated by the factors 3, 4.8, 8, 18. The configuration is nearly axisymmetric with expansion factors in the range 3 to 8 at the polar coronal holes, and much greater factors, well in excess of 18, at the magnetic equator.

The most interesting aspect of Plate 1 is the red region of very low expansion factors toward the left end of the map. They originate in a low-latitude spur of the polar coronal hole, evidently produced by the eruption of the bipolar magnetic region nearby. Thus, by breaking the overall symmetry, the erupting bipolar magnetic region produced a lower expansion factor and presumably a faster ecliptic wind than one would ideally obtain by tipping the Sun sideways and aiming the north pole at Earth.

This interaction between polar and low-latitude holes is most striking during the declining phase of the cycle, as Plate 2 illustrates. Here in Carrington rotation 1741 (October 18–November 15, 1983), there are two low-latitude spurs in each hemisphere, corresponding to a four-sector global magnetic configuration. The lowest expansion factors (<3) tend to originate at the facing edges of the polar holes and their low-latitude extensions, not at the poles themselves, where the expansion factors are mostly in the range between 3 and 8. Whether or not these low-expansion regions will actually affect Earth depends on Earth's heliocentric latitude (3–6°N at this time of year) and the details of the magnetic field model (stronger polar fields and the neglected current sheets would push flux to lower latitudes).

The configuration at sunspot maximum is illustrated in Plate 3, which shows Carrington rotation 1813 (March 4–April 1, 1989) about one year before polar-field reversal. Here, only vestiges of the polar holes remain, and the source surface is dominated by expansion factors greater than 18. The only regions of very low expansion occur in a small spur of the south polar hole and in an equatorial region toward the left end of the map. The latter region coincided with a "weak" helium coronal hole and with a recurrent wind stream whose speed at Earth exceeded 750 km s^{-1} on March 31 and 650 km s^{-1} when it reappeared in April and May (Sheeley and Wang 1991).

The configuration soon after polar-field reversal is illustrated in Plate 4, which shows Carrington rotation 1847 (September 17–October 15, 1991). Here a considerable amount of red appears at high latitudes, and the magnetic polarities have the new-cycle configuration in each hemisphere (positive in the north and negative in the south). Comparing Plates 1 and 4, we see that expansion factors less than 3 were much more common at the poles in 1991 when the polar fields were weak than in 1986 when the polar fields were strong.

We can see this effect more clearly in Plate 5, which shows solar wind speed as a function of latitude and time during the sunspot cycle. This speed is derived from the source-surface expansion factor using an in-ecliptic calibration. Each column is a stack of latitude strips obtained from source-surface maps like those in the bottom panels of Plates 1 through 4 over the interval 1977–1993. Red regions of low expansion (<3) are intermittently visible at high northern latitudes during 1980–1982 and at high southern latitudes during 1981–1982. They migrate equatorward as the cycle advances, almost reaching the equator in 1983–1986 during the declining phase of the cycle. The remnants recede poleward after sunspot minimum, leaving regions of high expansion (>18) behind them at all latitudes in 1990 at the next polar-field reversal. Soon afterward, the low-expansion regions reform near the poles and begin their next cycle of equatorward migration.

Figure 1 summarizes the evolution of source-surface polarity in the same stackplot format as Plate 5. The polarity reversals at high latitude are clearly visible around 1980 and 1990. During the intervening time, the sector structure first contracts toward the equator (reaching it around sunspot minimum in 1986) and then expands back toward the poles with the onset of the new sunspot cycle. The sector patterns are nearly identical over a wide range of latitudes, reflecting the rigid rotation rate of the current-free coronal magnetic field (Hoeksema and Scherrer 1987; Wang et al. 1988; Wang and Sheeley 1994). Even at the equator, these interlocking patterns match the shapes and sizes of the regions of very low flux-tube expansion lying on either side of the equator (cf., Plate 5). This suggests that in-ecliptic observations of sector polarity can be used to infer the properties of the very fast wind streams that migrate toward the equator, but do not quite reach it.

The overall trend for wind speed to be fast around sunspot minimum and slow at maximum has been known for some time, not only from *in-situ* measurements in and near the ecliptic plane, but also from radio IPS measurements at all latitudes (see the reviews by Kojima and Kakinuma [1990] and Rickett and Coles [1991] and references contained therein). Figure 2 compares annual averages of IPS speed with annual averages of speeds derived from source-surface expansion factors in the ecliptic, at $30°$N, and at $60°$N during 1972–1989 (Sheeley et al. 1991). The magnetically derived speeds shown here were obtained with a different calibration than those of the previous figures (factors of 3.5, 9, 18 now corresponding to speeds of 650, 550, 450 km s^{-1}), and have somewhat lower peak values. Nevertheless, the out-of-ecliptic agreement is about as good as that obtained between the derived speed and the *in-situ* speed in the ecliptic plane. In particular, both the IPS and magnetically derived speeds show the same cyclic behavior and have slow wind at high latitude only during a brief interval near sunspot maximum.

III. IN-ECLIPTIC COMPARISONS

Carrington maps like those of Plates 1 through 4 can also be used to generate stackplots of source-surface expansion factor and magnetic polarity at the heliographic latitude of Earth (which ranges from $7.25°$S to $7.25°$N during the year) for comparison with in-ecliptic solar wind measurements. Indeed, stackplot comparisons of source-surface expansion factor and *in-situ* wind speed provided the statistical basis for the inverse correlation in the first place (Wang and Sheeley 1990*a*).

Plate 6 shows such a comparison based on observations from the Wilcox Solar Observatory during 1976–1994. Here, the solar observations have been transformed to the Earth and placed in Bartels format (time running left to right in 27-day rows) for comparison with in-ecliptic measurements of wind speed and magnetic polarity. The color-coding of expansion factor is the same as in Plate 5, so that values of 3, 4.8, 8, and 18 separate the colors red, white, yellow, green, and blue.

The overall agreement of the complicated magnetic patterns leaves little doubt that they are related. On the other hand, the patterns of low coronal

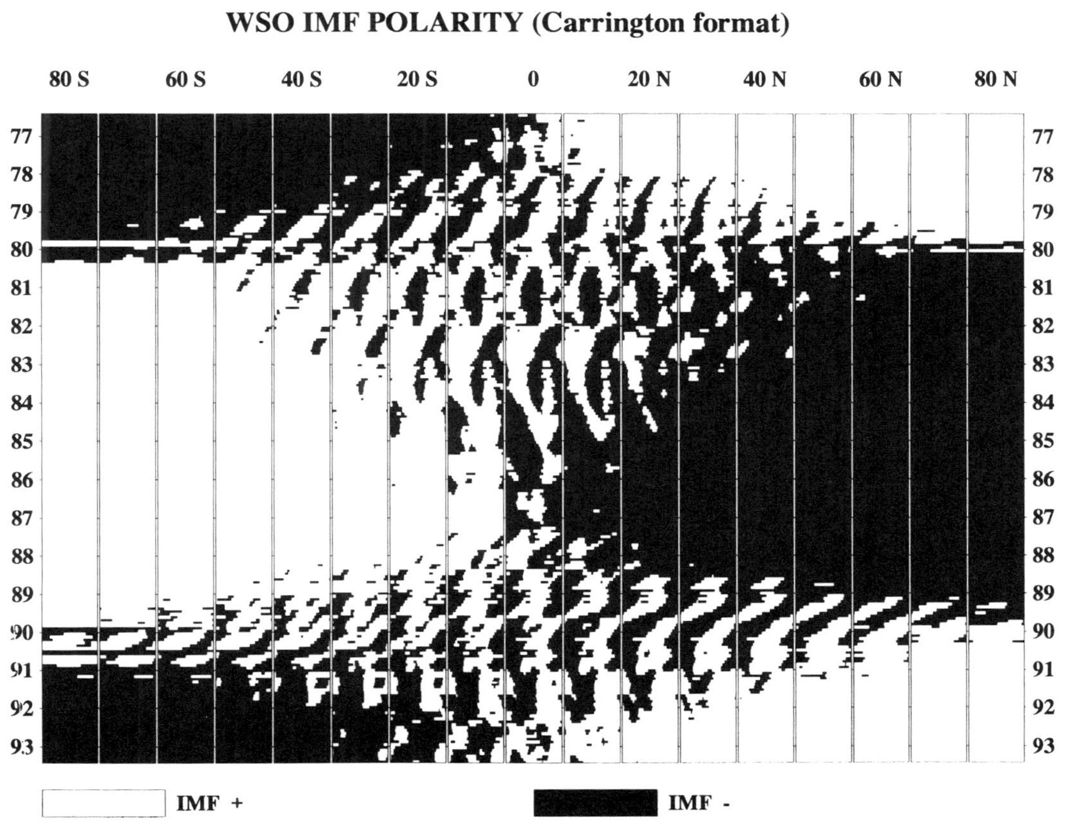

WSO IMF POLARITY (Carrington format)

Figure 1.   A display of source-surface magnetic polarity similar to color Plate 5 (Wang and Sheeley 1994).

expansion and high wind speed seem less well correlated. It is true that there is a rough correspondence with most of the enhanced regions occurring in 1983–1986 during the declining phase of cycle 21. It is also true that there are some detailed agreements, not only during 1983–1986, but also in 1976 at the previous minimum and in 1991 near the maximum of cycle 22. However, there are also disagreements such as the ones in 1986 and 1992 when some regions of low expansion were not accompanied by fast wind and vice versa. Some of these disagreements seem to occur where the magnetic polarities also disagree. Possible sources of these discrepancies would include measurement errors in the photospheric field (especially at high latitude in the hemisphere tipped away from Earth) and the neglect of currents in the coronal model (especially around sunspot minimum when the neutral line is near the ecliptic).

Figures 3a and b illustrate the effect of adding an equatorial current sheet to an axisymmetric field similar to the observed field at sunspot minimum (Wang and Sheeley 1990a). Comparing the current-free (dashed) field lines in Fig. 3a with their current-sheet (solid) counterparts, we see that the current sheet redistributes polar flux closer to the equator. Comparing the corresponding expansion factors in Fig. 3b, we see that there are qualitatively different behaviors above and below 2.5 $R_\odot$, which is the location of both the source surface and the inner edge of the current sheet in this model.

Below 2.5 $R_\odot$, both models show a monotonic increase of expansion factor with colatitude (polar angle), eventually becoming infinite at the edge of the hole (28° colatitude for the current-sheet model and 32° for the source-surface model). Thus, by pulling polar flux to lower latitudes, a current sheet would produce greater divergences and larger expansion factors at all latitudes, but it would not change the inverse sense of the correlation with asymptotic solar wind speed. However, above 2.5 $R_\odot$, the current sheet reverses the dependence of expansion factor on latitude, so that flux tubes that expand least in passing through the lower corona now expand most in going through the outer corona into the heliosphere. As will be discussed in Sec. V, this refocusing of flux tubes may provide an explanation for the observed constancy of solar wind mass flux at 1 AU (see Wang 1993).

It is relatively straightforward to modify the above example so that each polar hole is accompanied by a large axisymmetric like polarity hole in its hemisphere. With the source-surface model, the expansion factor at 2.5 $R_\odot$ would still be about 4 over the pole, but then decrease to a smaller value at mid latitudes (where field lines from the facing edges of the polar and low-latitude holes converge) before increasing without limit as the equatorial neutral line is approached (Sheeley and Wang 1991). Our preliminary calculations indicate that if a current sheet were added, the trend would be the same, but the location of the mid-latitude minimum would shift toward the equator (by 10–20° depending on whether the expansion factor is evaluated at 2.5 $R_\odot$ or at greater distances where the field lines become radial). Thus, the inclusion of a current sheet might improve the correspondence between coronal expansion

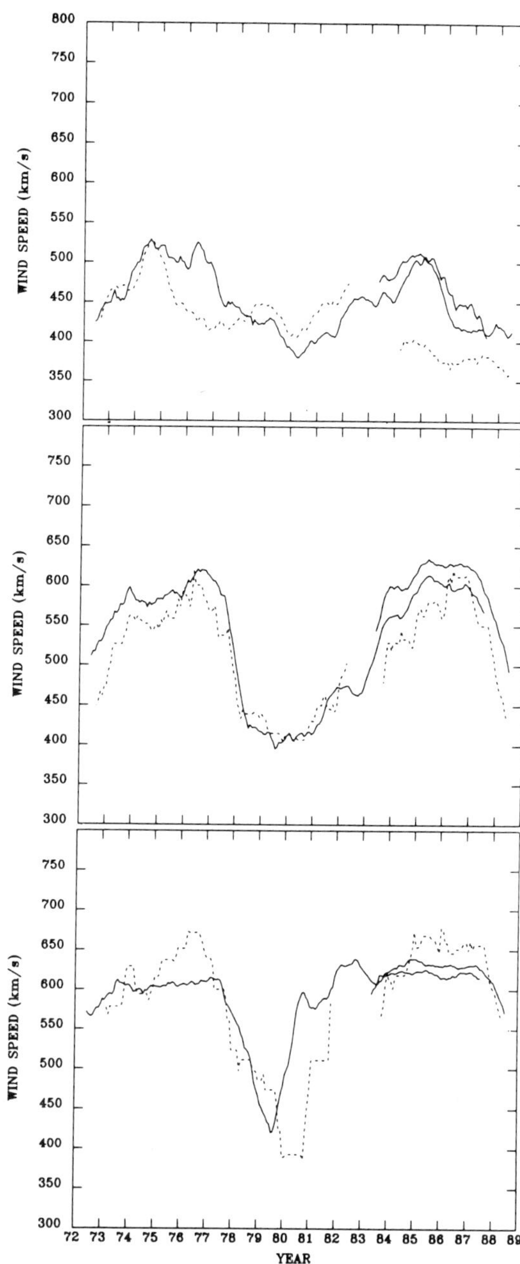

Figure 2.   Thirteen-rotation running means of interplanetary scintillation (IPS) speed (dashed lines) and magnetically derived speed (solid lines) for the combined interval 1972–1988, illustrating the variation during the sunspot cycle: *Top:* In ecliptic-speeds. *Middle:* Speeds at 30°N. *Bottom:* Speeds at 60°N. IPS observations are from the University of California at San Diego and Nagoya University, and solar magnetic field measurements are from the Mt. Wilson Observatory during 1972–1987 and the Wilcox Solar Observatory during 1983–1988 (Sheeley et al. 1991).

factor and solar wind speed at low latitudes, but it would probably not improve the agreement at high latitudes where Ulysses has been observing faster speeds than the model predicts (see Sec. IV below).

IV. THE SOLAR WIND AT ULYSSES

The Ulysses spacecraft was launched on October 6, 1990 on a unique trajectory which will carry it over the poles of the Sun. Ulysses initially proceeded near the ecliptic plane to Jupiter for a gravity assist, then turned southward after the Jupiter encounter in February 1992 at 5.4 AU. The spacecraft reached a maximum southerly latitude of 80.2° in September 1994, at 2.3 AU. Data described here extend through mid-January 1994, at which time the spacecraft was at 3.7 AU and 50°S heliographic latitude.

The Ulysses solar wind plasma experiment (see, e.g., Bame et al. 1992) comprises two separate analyzers, one for electrons and one for positively charged ions. The ion analyzer incorporates an active tracking scheme and commandable measurement matrices which together enable accurate and telemetry-efficient determination of the solar wind proton and helium three-dimensional velocity vectors, as well as other ion parameters. Observations shown here (Fig. 4) are mainly the solar wind proton bulk parameters; supporting ion and electron measurements were used for interpretation of transient and corotating solar wind features.

Figure 4 shows a time series of daily-averaged solar wind speed from instrument turn-on in November 1990 through December 1993. Its general features have been described by Bame et al. (1993) and Phillips et al. (1994) and include: (1) an irregular series of small solar wind streams during late 1990 and throughout 1991, with occasional transient disturbances such as the two major ones marked CME (for their origin in coronal mass ejections); (2) strongly damped corotating and transient streams during the first half of 1992; (3) repeated occurrence of a pronounced high-speed stream during the remainder of 1992 and the first half of 1993; (4) disappearance of low-speed, high-density flows beginning in April 1993 as the spacecraft climbed to the 29°S location of the heliomagnetic streamer belt; and (5) a further increase in minimum wind speed to nearly 700 km s^{-1} in late July 1993.

Figure 5 shows the trend of speed and density subsequent to the Jupiter encounter. Here, six-hour averaged mean density, minimum speed, and maximum speed are binned by Carrington rotation for solar latitudes from 7°S to 50°S (March 1992 through mid-January 1994). Thus, the maximum speeds roughly represent the peaks in Fig. 4, the minimum speeds represent the valleys, and the mean densities tend to be dominated by the high-density plasma regions observed in the vicinity of the heliospheric current sheet. After the onset of the well-defined solar wind streams near 12°S latitude (June 1992), there is a weak trend for increasing peak speed with increasing latitude until 33°S, after which the trend is irregular or declining. Note the jump in minimum speed and the marked decrease in mean density near 29°S. This is

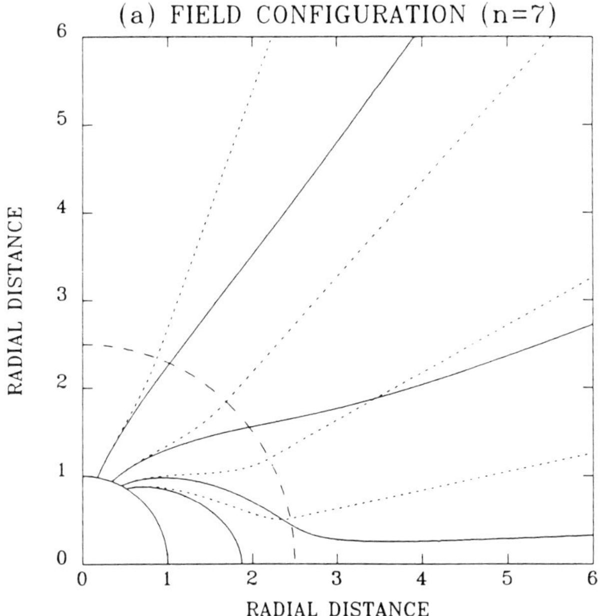

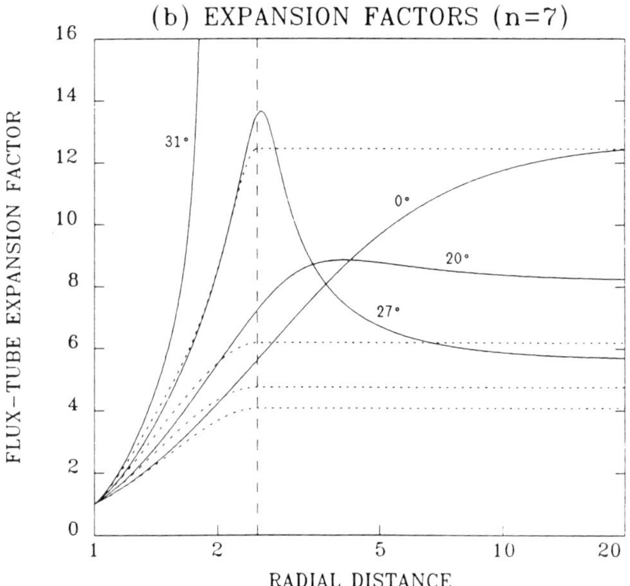

Figure 3.   Magnetic flux-tube geometry calculated for an axisymmetric photospheric field $B_r = \cos^7\theta$, using both a current-sheet (solid lines) and current-free (dashed lines) model. (a) Field lines with footpoint colatitudes 10, 20, 27, and 31°. (b) The radial dependence of the corresponding flux-tube expansion factors (Wang and Sheeley 1990a).

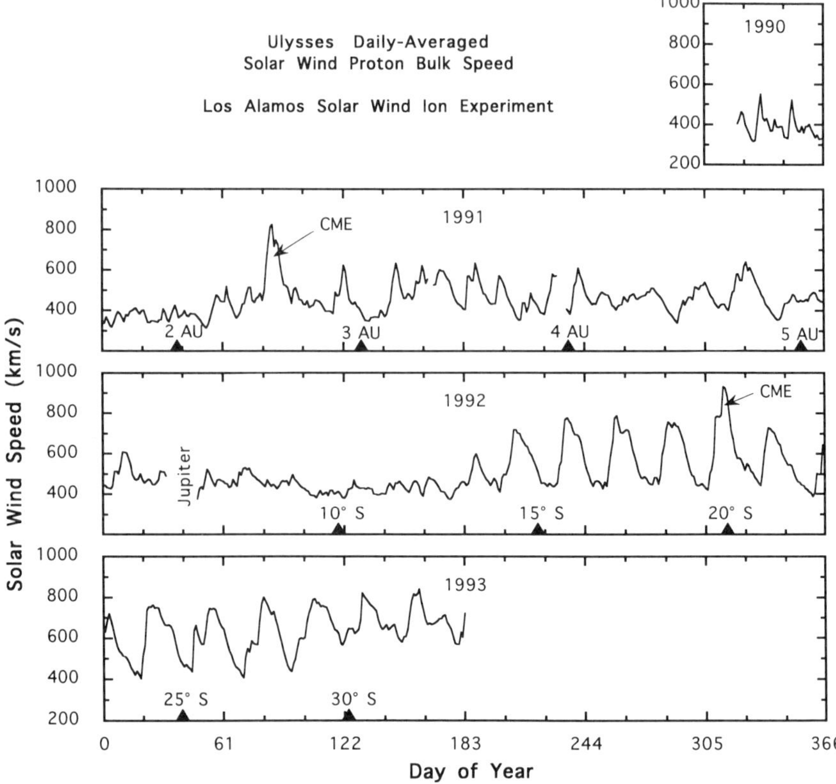

Figure 4.   Daily averaged solar wind proton bulk speed from turn-on on the solar wind ion experiment in November 1990, through December 1993. Heliocentric distance is marked by heavy triangles through the Jupiter encounter in February 1992; after the encounter the triangles show heliographic latitude in 5° increments.

due to the fact that the heliospheric current sheet was tilted ∼29° from the heliographic equatorial plane, such that Ulysses no longer crossed the current sheet after its latitude exceeded 29° (Smith et al. 1993; Bame et al. 1993). A further jump in minimum speed occurred at ∼37°S; poleward of that latitude, Ulysses encountered only fast wind (Phillips et al. 1994).

Although this chapter is concerned mainly with the steady solar wind and its origin back at the Sun, it is important to recognize that the steady wind is modified by dynamical propagation effects and by the effects of solar transients. Next, we shall describe these effects briefly to gain some insight on how they may influence the comparison between Ulysses wind speed and coronal flux-tube expansion factor.

The propagation effects are particularly important at heliocentric distances beyond 1 AU where compressions and rarefactions often steepen into forward-propagating and reverse-propagating magnetosonic shock waves

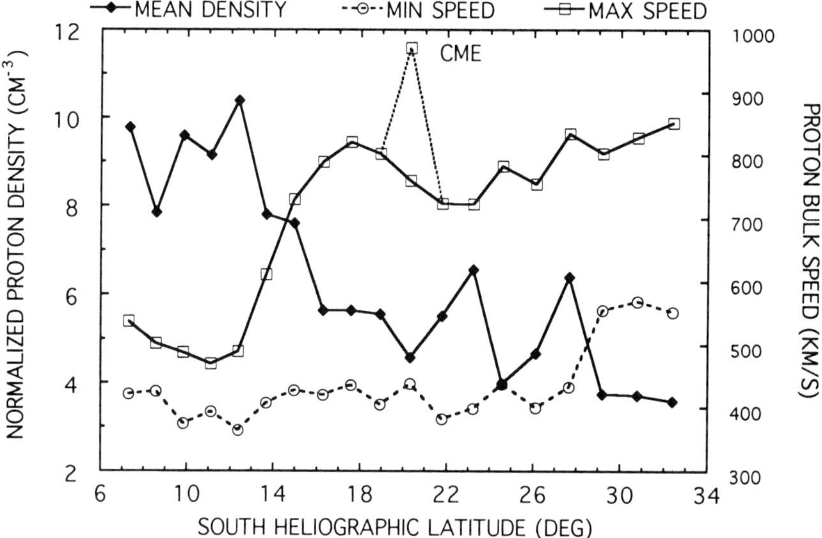

Figure 5.   Mean density, minimum speed, and maximum speed, for Ulysses six-hour averaged solar wind ion measurements, binned by Carrington rotation, from March 1992 through mid-December 1993, plotted versus heliographic latitude. A high-speed coronal mass ejection (CME) interval in November 1992 (the same interval as shown in Fig. 7) is plotted separately (Bame et al. 1993).

(Gosling et al. 1976; Smith and Wolfe 1976). Such propagation effects are clearly visible in Fig. 6, which shows the speed, density, and temperature profiles of three high-speed streams in 1992 when Ulysses was above 15°S and beyond 5 AU. The first high-speed stream is associated with a relatively simple corotating interactive region (CIR), formed when fast wind overtakes slower wind ahead of it. The forward and reverse shocks (indicated by vertical solid and dashed lines, respectively) on days 226 and 234 bound a compressed (high-density) region which is expanding into the lower-density regions which precede and follow it. The third stream has a similar profile, with a forward-reverse shock pair on days 279 and 285. The second stream, however, is associated with a much more complex interaction region with multiple shock waves. These effects of momentum transfer illustrate the difficulties of mapping even these relatively stable wind streams back to the Sun.

In addition to solar wind dynamical processes driven by solar rotation and the interaction of recurrent streams, solar transient events also play an important role in modifying the solar wind structure. The most dramatic of these events are coronal mass ejections (CMEs), in which discrete plasma structures are ejected from the Sun, driving shock waves ahead of them as they move out through the heliosphere (Sheeley et al. 1985). Figure 7 illustrates the effect of one such shock at Ulysses during November 1992. The sequence

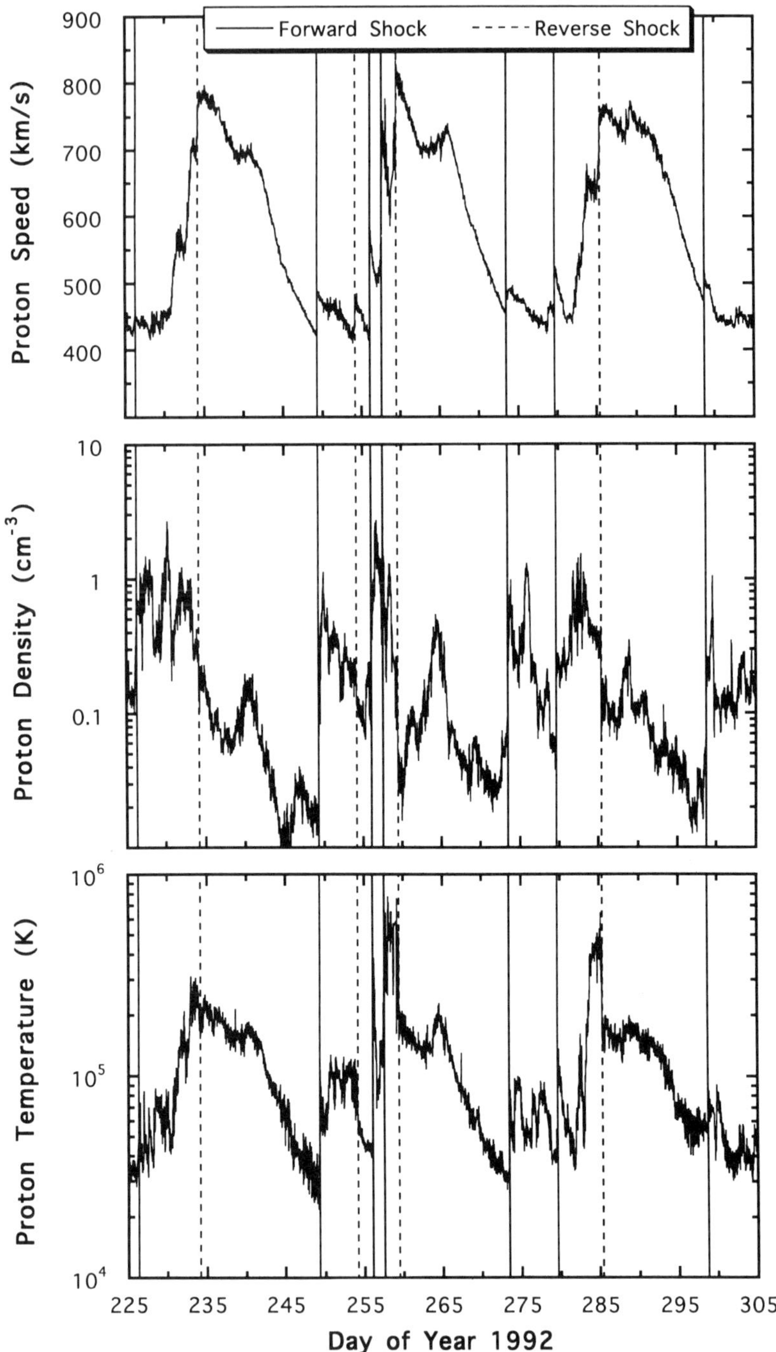

Figure 6.   Solar wind proton density, bulk speed, and temperature, for an 80-day (roughly three rotations) interval in the second half of 1992. Forward and reverse shocks (indicated by sudden jumps of all three parameters, or an increase in speed accompanied by decreases of both density and temperature, respectively) are marked by vertical solid and dashed traces, respectively.

begins with a reverse shock wave on the trailing edge of a CIR, after which the solar wind speed is roughly 800 km s^{-1}. Three days later, a forward shock is observed, causing the already high ambient speed to increase by a few 100 km s^{-1} more, reaching speeds as high as 1000 km s^{-1}. To date, we have not allowed for such transients in our comparison between Ulysses wind speed and coronal magnetic structure.

Despite the above caveats, if we map the Ulysses speeds back to the Sun and display the results in Carrington strips similar to those of Plate 5, we obtain the result shown in the right column of Plate 7 for rotations 1835–1878. In fact, the results are plotted three times for comparison with source-surface magnetic polarity (left column) and source-surface expansion factor (middle column) derived from observations at the Wilcox Solar Observatory, the Mt. Wilson Observatory, and the National Solar Observatory, all plotted at Ulysses' latitude. The color coding is the same as used in Plates 1 through 5.

The vertical (27.3-day) patterns of positive and negative polarity during 1991 and early 1992 contain low-expansion features that are matched by somewhat blurred-out speed enhancements at Ulysses. We expect that at least part of this blurring is due to the momentum transfer involved in the stream interactions described above. Although the transformation back to the Sun probably introduces additional blurring, we have attempted to minimize this artificial contribution by using a running 27-day average speed rather than the instantaneous speed in the transformation. This is analogous to using a constant 3-day shift to compare *in-situ* measurements at Earth with coronal holes at the Sun.

Also, in Carrington rotation 1840 (March 11–April 7, 1991), the Ulysses panel has a short segment of red pixels indicating very fast speed while its counterpart in the middle panels lies in the blue corridor of high expansion factors separating the two recurrence patterns. This is the March 1991 CME event indicated in Fig. 4.

In mid-1992 (Carrington rotation 1856–1857), a slanted (28–29-day) pattern of even lower flux-tube expansion (faster wind speed) is matched by a very fast wind stream at Ulysses. The adjacent speed minima also agree in location, meandering from upper right to lower left during the first half of 1993 and then becoming vertical again during the second half of 1993.

By the end of 1993, the observed speed minima have become much more filled in than the derived minima, and the Ulysses speeds are systematically faster than those of the model (cf., Phillips et al. 1994). We could understand this discrepancy if the source-surface neutral line were located too far to the south, but this is not the case. Its location near 30°S is in good agreement with the value of 29°S obtained with the Ulysses magnetometer (Balogh et al. 1993; Smith et al. 1993; Phillips et al. 1994).

In the model, the region of very low source-surface expansion factor was rooted in a lobe of the south polar hole. This is indicated in Plate 8, whose middle panels compare the derived holes (third panel from top) with those observed (as light areas) in the He I 10830 Å map (second from top) during

N. R. SHEELEY ET AL.

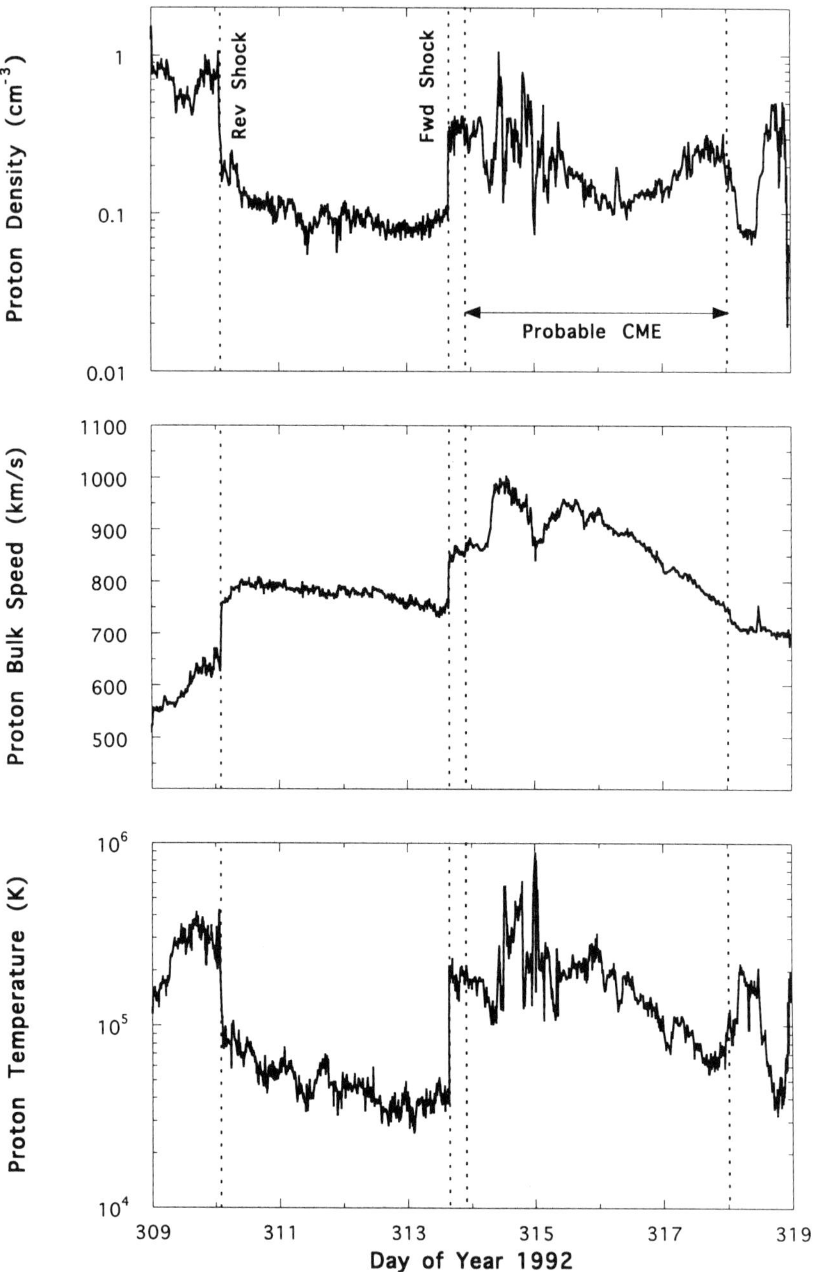

Figure 7. Solar wind proton density, bulk speed, and temperature, for a 10-day interval in late 1992. Vertical dashed traces mark a reverse shock, a forward shock, and the boundaries of a coronal mass ejection.

Carrington rotation 1862 (October 31–November 27, 1992) when Ulysses was seeing the stream for the 6th time. Plate 9 shows that, except for the eastward drift of the hole and stream, the situation changed little by Carrington rotation 1869 (May 10–June 6, 1993) when Ulysses was encountering some of the fastest speeds to date. This suggests that the increase in speed reflects Ulysses movement to higher southern latitudes, rather than a change in solar conditions.

On the other hand, the configuration did change in mid-1992 when the high-speed stream was first observed. A detailed examination of the global stackplot of coronal expansion factor (Plate 5) shows that the vertical (27.3-day) recurrence pattern during 1990–1991 stopped in mid-1992, and a slanted ($\sim$28–29-day) pattern formed at high south latitude. This effect is more visible in Fig. 8, which is similar to Fig. 1, except that the axisymmetric component of the field has been removed. Here, one can clearly see the abrupt change in mid-1992 from a vertical pattern at low latitude to a slanted pattern at high south latitude. Indeed, Fig. 8 shows that such changes are characteristic of the nonaxisymmetric field, which evolves in a sequence of long-lived recurrence patterns separated by sudden transitions, much like eons of geologic time.

We can inspect the mid-1992 boundary in detail using Plates 10 and 11, which show the derived coronal holes and source-surface expansion factors during Carrington rotations 1855–1858 (April 23–August 10, 1992). At the beginning of this interval, there was a low-latitude extension of the south polar hole near 90° longitude, but by the end of the interval, this hole had died and a new one had formed near 270°, although not extending to such low latitudes. Similarly, in Plate 11, the pattern of "red wind" near 90° died, and was replaced by a new pattern near 270°.

The expansion-factor concept predicts that Ulysses will eventually pass through a mid-latitude region of very low expansion factor and very fast wind, and arrive at high latitudes where the expansion factor will be slightly higher and wind speed slightly lower. This seems to have happened during 1993, but the magnitude of the effect was relatively small: peak speeds around 850 km s^{-1} decreased to only about 800 km s^{-1} as Ulysses moved from 30 to 50°S (Phillips et al. 1994).

V. FLUX-TUBE EXPANSION, CORONAL HEATING, AND SOLAR WIND ACCELERATION

The wind-speed expansion-factor correlation has important implications for the energy sources of the solar wind and for the physical properties of the source regions (Wang and Sheeley 1991; Wang 1993). The most significant implication is that the energy flux that accelerates the solar wind is roughly invariant within open field regions. This result is readily apparent in the idealized case where the coronal temperature is assumed to be a constant independent of the expansion factor (Wang and Sheeley 1991). In that case, the mass flux at the sonic point is roughly invariant, being mainly a function

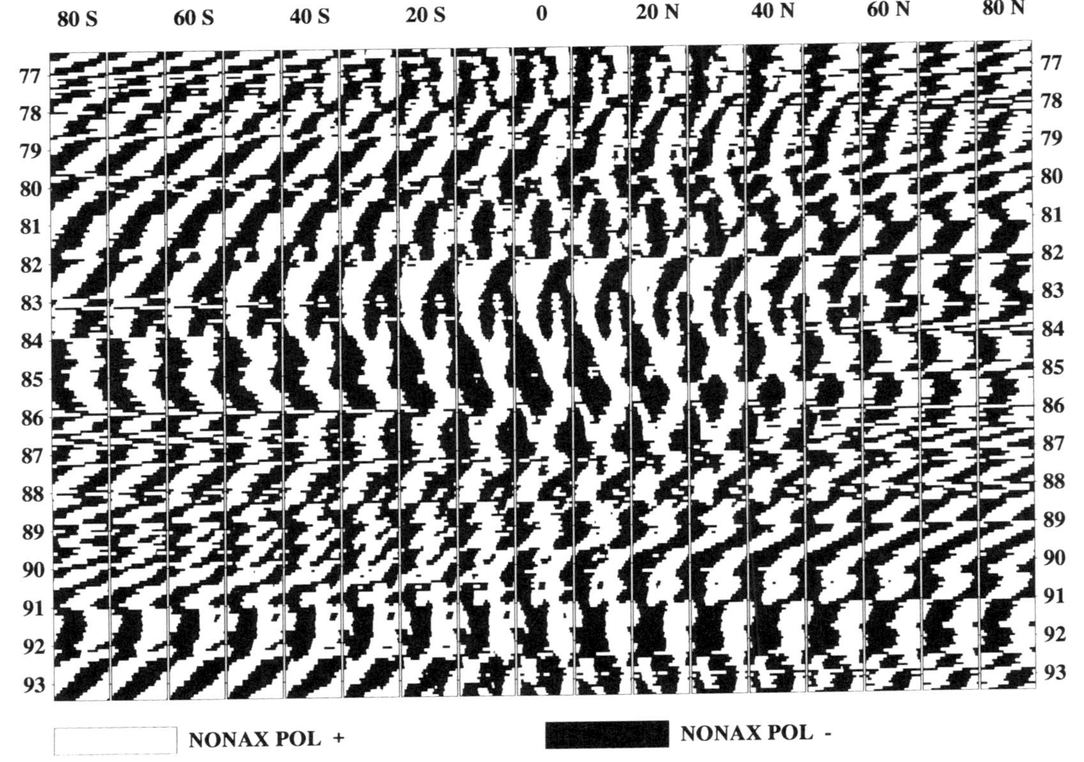

Figure 8.   Stackplots of the nonaxisymmetric component of source-surface field, showing the episodic evolution of the field (same format as Fig. 1) (Wang and Sheeley 1994).

of the coronal temperature or gravitational scale height. By continuity along a flux tube, the mass flux at the coronal base then increases almost linearly with the areal expansion factor. Therefore, if the wave or mechanical energy flux is constant at the coronal base, the wave energy available per particle and the resulting final wind speed must decrease as the expansion factor increases.

As emphasized by Leer et al. (1982), the solar wind mass flux can be extremely sensitive to the assumed coronal temperature, so that a full understanding of the role of the expansion factor requires a self-consistent determination of the temperature. Wind models that incorporate the energy balance requirements of the corona and transition region have been developed by Hammer (1982), Hollweg (1986), Withbroe (1988), and others. This type of model was used by Wang (1993) to calculate the coronal temperatures, densities, and asymptotic wind speed as a function of the expansion factor near the Sun.

In that study, a set of single-fluid equations was solved numerically for the flow along a radially oriented flux tube; the flow was driven by thermal and Alfvén-wave pressure gradients. The energy equation included the effects of heat conduction, mechanical energy dissipation, Alfvén-wave propagation, and radiative cooling. The mechanical energy flux was assumed to be damped over a characteristic length scale which was adjusted to satisfy the imposed boundary conditions. As in the models of Hammer (1982) and Withbroe (1988), the downward heat flux at the coronal base (defined to be at a temperature 5×10^5 K) was required to balance the total radiative and enthalpy losses from the transition zone, which was otherwise not explicitly included in the model. This boundary condition essentially provided a physical constraint on the plasma density at the coronal base.

For a constant flux of mechanical and Alfvén-wave energy at the coronal base, the maximum coronal temperature was found to decrease as the expansion factor increased (because the same energy input then had to heat a greater volume of plasma), whereas the mass flux at the Sun gradually increased despite the lower temperature. Because of the decreased thermal acceleration and the decreased wave energy per particle, the net result of increasing the expansion factor was to decrease the asymptotic wind speed, in agreement with the empirical studies.

An important characteristic of the energy balance models is that the mass flux at the coronal base is remarkably insensitive to variations of the wave energy input and of the expansion factor (Withbroe 1989; Wang 1993). The fact that the observed mass flux at Earth is also remarkably constant (see, e.g., Schwenn 1983) suggests that the net expansion undergone by flux tubes between the Sun and Earth is roughly invariant, despite the large differences in their divergence rates near the Sun. Thus, flux tubes that expand rapidly near the Sun must later "reconverge," whereas flux tubes that expand slowly near the Sun must later diverge more rapidly. Physically, this behavior follows from the requirement of transverse force balance in the region where the plasma β approaches unity (Withbroe 1989); the net effect is that the thermal pressure becomes more or less uniform at large distances from the Sun. As discussed in Sec. III, the nonmonotonic variation of the expansion factor along a flux tube is consistent with simple current-sheet models for the interplanetary field (see Fig. 3).

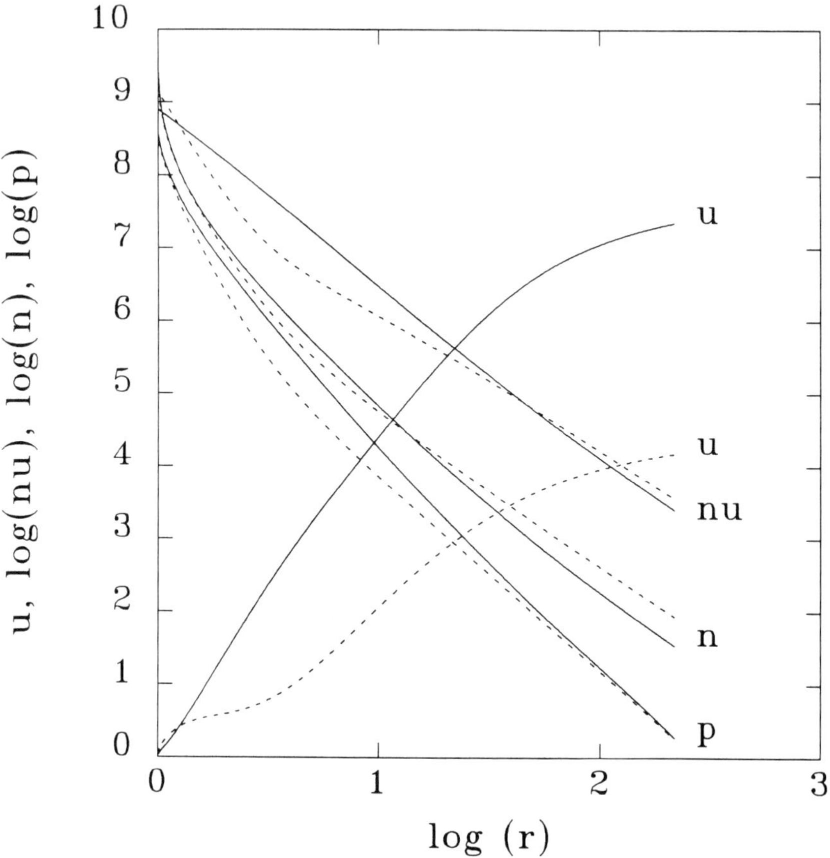

Figure 9.   Radial variation of wind speed u, proton flux nu, proton density n, and thermal pressure p for a fast wind model (solid lines) and a slow wind model (dotted lines). Horizontal axis gives $\log_{10}(r/R)$; vertical axis gives $u/(100\,\mathrm{km\,s^{-1}})$, $\log_{10}(nu/10^5\,\mathrm{cm^{-2}s^{-1}})$, $\log_{10}(n/0.1\,\mathrm{cm^{-3}})$, and $\log_{10}(p/10^{-10}\mathrm{dyn\,cm^{-2}})$. In the fast wind model, the magnetic field declines as r^{-2} near the Sun, as r^{-3} at intermediate distances, and as r^{-2} again far from the Sun. In the slow wind model, the field declines as r^{-5} near the Sun, as r^{-1} at intermediate distances, and as r^{-2} far from the Sun. Both models assume a mechanical energy flux of 5×10^5 ergs $\mathrm{cm^{-2}s^{-1}}$ and an Alfvén-wave flux of 2×10^5 ergs $\mathrm{cm^{-2}s^{-1}}$ at the coronal base (Wang 1993).

Figure 9 shows two model wind solutions that allow for "post-expansion" and "reconvergence" of the flux tubes. In the fast wind solution (solid lines), the magnetic field falls off as r^{-2} near the Sun, as r^{-3} at intermediate distances, and as r^{-2} again far from the Sun. In the slow wind solution (dotted lines), the magnetic field falls off as r^{-5} near the Sun, as r^{-1} at intermediate distances, and as r^{-2} far from the Sun. Both models assume the same mechanical and Alfvén-wave energy fluxes at the coronal base. Near the Sun, the thermal pressures and mass fluxes in the slow wind model fall well below those in the fast wind model, but far from the Sun the values become nearly equal again. This is because the wind speed at Earth is determined mainly by the flux-tube divergence rate near the Sun, whereas the mass flux at Earth depends on the global behavior of the flux tube all the way from Sun to Earth.

VI. SUMMARY, CONCLUSIONS, AND FUTURE DIRECTIONS

Our observational knowledge of the Sun and solar wind has increased considerably since the Skylab era of the 1970s, and it continues to increase during the 1990s as Ulysses maps uncharted territory away from the ecliptic plane. The old concept that coronal holes are the source of high-speed streams has provided a framework for understanding some properties of solar wind speed, including its increase with latitude around sunspot minimum. However, subsequent observations have provided some apparent contradictions, including the much steeper speed-latitude relation near sunspot minimum than near maximum, and the emission of faster wind from some weak or fragmentary holes than from much larger well-defined holes.

The expansion-factor concept provided a way of understanding these new observations. Weak holes turned out to be regions of open magnetic field with low coronal expansion. Also, the in-ecliptic comparison of wind speed and expansion factor provided a good inverse correlation throughout the sunspot cycle without the need for *ad hoc* modifications at sunspot maximum and minimum.

The expansion-factor concept also led to a prediction of how the wind speed ought to vary with time and latitude over the sunspot cycle. During the declining phase of the cycle, the speed ought to be faster at mid latitudes (where field lines from the polar and low-latitude holes "collide" to produce very low expansion factors) than at the centers of the polar holes. Ulysses observations through mid-January 1994 (50°S) support this prediction, but not by a wide margin. Whereas the peak speeds encountered in the 30 to 40°S range lie around 850 km s^{-1}, the peak speeds over the range between 40 and 50°S have been about 800 km s^{-1} (Phillips et al. 1994). Based on in-ecliptic calibrations, the flux-tube model predicts high-latitude speeds of about 650 to 750 km s^{-1} toward sunspot minimum.

At present, it is unclear how to resolve this quantitative discrepancy. During its in-ecliptic passage, Ulysses observed minimum speeds of about 400 km s^{-1} which are systematically larger than the minimum speeds of 300 km s^{-1} typically found near Earth. Thus, part of the discrepancy may be related to Ulysses greater distance from the Sun. In principle, part of the discrepancy might also lie in the source-surface model, which does not include the effects of heliospheric current sheets. An equatorial current sheet would pull high-latitude flux to lower latitudes and thereby shift fast mid-latitude wind to lower latitude. However, such a rearrangement would not explain the systematically faster wind that Ulysses sees in the 40 to 50°S region. Another source of discrepancy might be the groundbased measurements of high-latitude photospheric fields, which are subject to large uncertainties. Overestimates of polar field strength would cause the coronal expansion factors to be too large at high latitude and the derived wind speeds to be too small.

Despite its underestimate of wind speed, the model has provided a good account of the speed modulation (i.e., the locations of fast and slow wind). The high-speed streams clearly originate in regions of low coronal expansion. Their birth, evolution (including recurrence rate), and ultimate dissolution are all well matched to corresponding properties of the Sun's large-scale field. Thus, although some additional process may be necessary to provide detailed quantitative agreement with the observed speed at high latitude, there is little doubt that coronal flux-tube expansion must play a major role in determining the global structure of the wind.

Finally, we note two additional predictions of the expansion-factor model that can be tested by future Ulysses measurements. First, at high latitudes, the wind speed ought to decrease during the present approach to sunspot minimum as the polar fields become stronger and more divergent. This suggests that Ulysses may detect slightly slower wind at the north pole in July 1995 than at the south pole in September 1994. Second, patterns of very fast wind ought to form at the poles soon after the next sunspot maximum when trailing-polarity flux arrives there and changes the strongly divergent configuration to a convergent or weakly divergent one. Ulysses may sense this transition by finding slow wind over the south pole in the year 2000 and a recurrent stream of very fast wind over the north pole in 2001.

Acknowledgments. We would like to acknowledge our colleagues S. J. Bame and B. E. Goldstein for their contributions to the Ulysses observations, and J. T. Gosling for helpful discussions. We also are grateful to D. M. Hurley for programming help in the initial phase of our Sun-Ulysses comparison. We thank J. T. Hoeksema and P. H. Scherrer (Wilcox Solar Observatory) and J. E. Boyden and R. K. Ulrich (Mt. Wilson Observatory/UCLA) for providing the solar magnetic observations used in this study. The NSO/Kitt Peak helium images were produced cooperatively by NSF/NOAO, NASA/GSFC, and NOAA/SEL and made available courtesy of J. W. Harvey. We appreciate the efforts of J. T. Gosling (LANL), J. H. King (NASA/GSFC), A. J. Lazarus (MIT), E. J. Smith (JPL), and R. D. Zwickl (NOAA), who provided the interplanetary observations used in Plate 6. Work at Los Alamos was performed under the auspices of the U. S. Department of Energy. Work at JPL was supported in part by the Naval Reserve Science and Technology Program. Financial support at NRL was provided by the Office of Naval Research and the Solar Physics Branch of the NASA Space Physics Division.

REFERENCES

Altschuler, M. D., and Newkirk, G. 1969. Magnetic fields and the structure of the solar corona. I. Methods in calculating coronal fields. *Solar Phys.* 9:131–149.

Balogh, A., Forsyth, R. J., Ahuja, A., Southwood, D. J., Smith, E. J., and Tsurutani, B. T. 1993*a*. The interplanetary magnetic field from 1 to 5 AU: Ulysses observations. *Adv. Space Res.* 13(6):15–24.

Balogh, A., Erdös, G., Forsyth, R. J., and Smith, E. J. 1993*b*. The evolution of the interplanetary sector structure in 1992. *Geophys. Res. Lett.* 20:2331–2334.

Bame, S. J., Phillips, J. L., McComas, D. J., Gosling, J. T., and Goldstein, B. E. 1992*a*. The Ulysses solar wind plasma investigation: Experiment description and initial in-ecliptic results. In *Solar Wind Seven*, eds. E. Marsch and R. Schwenn (Oxford: Pergamon Press), pp. 139–142.

Bame, S. J., McComas, D. J., Berraclough, B. L., Phillips, J. L., Sofaly, K. J., Chavez, J. C., Goldstein, B. E., and Sakurai, R. K. 1992*b*. The Ulysses solar wind plasma experiment. *Astron. Astrophys. Suppl.* 92:237–265.

Bame, S. J., Goldstein, B. E., Gosling, J. T., Harvey, J. W., McComas, D. J., Neugebauer, M., and Phillips, J. L. 1993. Ulysses observations of a recurrent high speed solar wind stream and the heliomagnetic streamer belt. *Geophys. Res. Lett.* 20:2323–2326.

Barnes, A., Mihalov, J. D., Gazis, P. R., Lazarus, A. J., Belcher, J. W., Gordon, G. S., Jr., and McNutt, R. L., Jr. 1992. Global properties of the plasma in the outer heliosphere I. Large-scale structure and evolution. In *Solar Wind Seven*, eds. E. Marsch and R. Schwenn (Oxford: Pergamon Press), pp. 143–146.

Bell, B., and Noci, G. 1976. Intensity of the Fe XV emission line corona, the level of geomagnetic activity, and the velocity of the solar wind. *J. Geophys. Res.* 81:4508–4516.

Coles, W. A., and Rickett, B. J. 1976. IPS observations of the solar wind speed out of the ecliptic. *J. Geophys. Res.* 81:4797–4799.

Feldman, W. C., Asbridge, J. R., Bame, S. J., and Gosling, J. T. 1978. Long term variations of solar wind properties. *J. Geophys. Res.* 83:2177–2189.

Gazis, P. R., Barnes, A., and Lazarus, A. J. 1988. In *Solar Wind Six*, eds. V. J. Pizzo, T. E. Holzer and D. G. Sime, NCAR TN-306 (Boulder: National Center for Atmospheric Res.), p. 563.

Gazis, P. R., Mihalov, J. D., Barnes, A., Lazarus, A. J., and Smith, E. J. 1989. Pioneer and Voyager observations of the solar wind at large heliocentric distances and latitudes. *Geophys. Res. Lett.* 16:233–226.

Gosling, J. T., Hundhausen, A. J., and Bame, S. J. 1976. Solar wind stram evolution at large heliocentric distances: Experimental demonstration and the test of a model. *J. Geophys. Res.* 81:2111–2122.

Hakamada, K., Kojima, M., and Kakinuma, T. 1991. Solar wind speed and He I (1083 nm) absorption line intensity. *J. Geophys. Res.* 96:5397–5403.

Hammer, R. 1982. Energy balance of stellar coronae. I. Methods and examples. *Astrophys. J.* 259:767–778.

Hoeksema, J. T. 1984. Ph.D Thesis, Stanford University.

Hoeksema, J. T. 1992. Large-scale structure of the heliospheric magnetic field: 1976–1991. In *Solar Wind Seven*, eds. E. Marsch and R. Schwenn (Oxford: Pergamon Press), pp. 191–196.

Hoeksema, J. T., and Scherrer, P. H. 1987. Rotation of the coronal magnetic field. *Astrophys. J.* 318:428–436.

Hollweg, J. V. 1986. Transition region, corona, and solar wind in coronal holes. *J. Geophys. Res.* 91:4111–4125.

Howard, R. A., and LaBonte, B. J. 1981. Surface magnetic fields during the solar activity cycle. *Solar Phys.* 74:131–145.

Hundhausen, A. J. 1977. An interplanetary view of coronal holes. In *Coronal Holes and High Speed Wind Streams*, ed. J.B. Zirker (Boulder: Colorado Associated Univ. Press), pp. 225–329.

Kojima, M., and Kakinuma, T. 1987. Solar cycle evolution of solar wind speed structure between 1973 and 1985 observed with the interplanetary scintillation method. *J. Geophys. Res.* 92:7269–7279.

Kojima, M., and Kakinuma, T. 1990. Solar cycle dependence of global distribution of solar wind speed. *Space Sci. Rev.* 53:173–222.

Leer, R., Holzer, T. E., and Fla, T. 1982. Acceleration of the solar wind. *Space Sci. Rev.* 33:161–200.

Levine, R. H., Altschuler, M. D., and Harvey, J. W. 1977. Solar sources of the interplanetary magnetic field and solar wind. *J. Geophys. Res.* 82:1061–1065.

Marsch, E., and Schwenn, R., eds. 1992. *Solar Wind Seven* (Oxford: Pergamon Press).

Neupert, W. M., and Pizzo, V. 1974. Solar coronal holes as sources of recurrent geomagnetic distubrances. *J. Geophys. Res.* 79:3701–3709.

Nolte, J. T., Krieger, A. S., Timothy, A. F., Vaiana, G. S., and Zombeck, M. V. 1976. A comparison of solar wind streams and coronal structure near solar minimum. *Solar Phys.* 46:291–294.

Nolte, J. T., Davis, J. M., Gerassimenko, M., Lazarus, A. J., and Sullivan, J. D. 1977. A comparison of solar wind streams and coronal structures near solar minimum. *Geophys. Res. Lett.* 4:291–294.

Phillips, J. L., Balogh, A., Bame, S. J., Goldstein, B. E., Gosling, J. T., Hoeksema, J. T., McComas, D. J., Neugebauer, M., Sheeley, N. R., Jr., and Wang, Y.-M. 1994. Ulysses at 50 south: Constant immersion in the high-speed solar wind. *Geophys. Res. Lett.* 21:1105–1108.

Rickett, B. J., and Coles, W. A. 1991. Evolution of the solar wind structure over a solar cycle: IPS velocity measurements compared with coronal observations. *J. Geophys. Res.* 96:1717–1736.

Schatten, K. H., Wilcox, J. M., and Ness, N. F. 1969. A model of the interplanetary and coronal magnetic fields. *Solar Phys.* 6:442–455.

Schwenn, R. 1983. The "average" solar wind in the inner heliosphere: Structures and solar variations. In *Solar Wind Five*, ed. M. Neugebauer, NASA CP-2280, pp. 489–507.

Sheeley, N. R., Jr. 1992. Coronal holes and solar wind streams during the sunspot cycle. In *Solar Wind Seven*, eds. E. Marsch and R. Schwenn (Oxford: Pergamon Press), pp. 263–271.

Sheeley, N. R., Jr. and Harvey, J. W. 1978. Coronal holes, solar wind streams, and geomagnetic activity during the new sunspot cycle. *Solar Phys.* 59:159–173.

Sheeley, N. R., Jr., and Wang, Y.-M. 1991. Magnetic field configurations associated with fast solar wind. *Solar Phys.* 131:165–186.

Sheeley, N. R., Jr., Harvey, J. W., and Feldman, W. C. 1976. Coronal holes, solar wind streams, and recurrent geomagnetic disturbances: 1973–1976. *Solar Phys.* 49:271–278.

Sheeley, N. R., Jr., Howard, R. A., Koomen, M. J., Michels, D. J., Schwenn, R., Muhlhauser, K. H., and Rosenbauer, H. 1985. Coronal mass ejections and interplanetary shocks. *J. Geophys. Res.* 90:163–175.

Sheeley, N. R., Jr., Swanson, E. T., and Wang, Y.-M. 1991. Out-of-ecliptic tests of the inverse correlation between solar wind speed and coronal expansion factor. *J. Geophys. Res.* 96:13861–13868.

Sime, D. G., and Rickett, B. J. 1978. The latitude and longitude structure of the solar wind speed from IPS observations. *J. Geophys. Res.* 83:5757–5762.

Smith, E. J., and Wolfe, J. H. 1976. Observations of interaction regions and corotating shocks between one and five AU: Pioneer 10 and 11. *Geophys. Res. Lett.* 3:137–143.

Smith, E. J., Neugebauer, M., Balogh, A., Bame, S. J., Erdös, G., Forsyth, R. J., Goldstein, B. E., Phillips, J. L., and Tsurutani, B. T. 1993. Disappearance of the heliospheric sector structure at Ulysses. *Geophys. Res. Lett.* 20:2327–2330.

Wagner, W. J. 1976. Coronal holes observed by OSO-7 and interplanetary magnetic sector structure. *Astrophys. J.* 206:583–588.

Wang, Y.-M. 1993. Flux-tube divergence, coronal heating, and the solar wind. *Astrophys. J. Lett.* 410:123–126.

Wang, Y.-M., and Sheeley, N. R., Jr. 1990a. Solar wind speed and coronal flux-tube expansion. *Astrophys. J.* 355:726–732.

Wang, Y.-M., and Sheeley, N. R., Jr. 1990b. Magnetic flux transport and the sunspot-cycle evolution of coronal holes and their wind streams. *Astrophys. J.* 365:372–386.

Wang, Y.-M., and Sheeley, N. R., Jr. 1991. Why fast solar wind originates from slowly expanding coronal flux tubes. *Astrophys. J. Lett.* 372:45–48.

Wang, Y.-M., and Sheeley, N. R., Jr. 1992. On potential-field models of the solar corona. *Astrophys. J.* 392:310–319.

Wang, Y.-M., and Sheeley, N. R., Jr. 1993. Understanding the rotation of coronal hole Understanding the rotation of coronal holes. *Astrophys. J.* 414:916–927.

Wang, Y.-M., and Sheeley, N. R., Jr. 1994. Global evolution of interplanetary sector structure, coronal holes, and solar wind streams during 1976–1993: Stackplot displays based on solar magnetic observations. *J. Geophys. Res.* 99:6597–6608.

Wang, Y.-M., Sheeley, N. R., Jr., and Nash, A. G. 1988. The quasi-rigid rotation of coronal magnetic fields. *Astrophys. J.* 327:427–450.

Wang, Y.-M., Nash, A. G., and Sheeley, N. R., Jr. 1989. Evolution of the Sun's polar fields during sunspot cycle 21: Poleward surges and long-term behavior. *Astrophys. J.* 347:529–539.

Wang, Y.-M., Sheeley, N. R., Jr., and Nash, A. G. 1990. Latitudinal distribution of solar-wind speed from magnetic observations of the Sun. *Nature* 347:439–444.

Watanabe, T. 1989. Solar wind latitude/longitude variations derived from interplanetary scintillations. *Adv. Space Res.* 9:99.

Withbroe, G. L. 1988. The temperature structure, mass, and energy flow in the corona and inner solar wind. *Astrophys. J.* 325:442–467.

Withbroe, G. L. 1989. The solar wind mass flux. *Astrophys. J. Lett.* 337:49–52.

Zirker, J. B., ed. 1977. *Coronal Holes and High Speed Wind Streams* (Boulder: Colorado Associated Univ. Press).

STOCHASTIC PROCESSES AND THE ORIGIN OF STELLAR WINDS

M. CUNTZ
High Altitude Observatory, National Center for Atmospheric Research

and

E. A. DORFI
Institute for Astronomy, University of Vienna

In this chapter, we present results which provide insight into the origin of stellar winds in five selected stars: a typical RR Lyrae star (A7-F5 III), δ Cephei (F5-G2 Ib), α Bootis (K1.5 III), α Orionis (M2 Iab), and a representative AGB star. These stars have been chosen because up-to-date *ab initio* models exist which attempt to explain stellar mass loss behavior. The stars exhibit different ranges of wave modes including acoustic modes and radial pulsation modes, which have been incorporated in recent models of nonlinear pulsation. For α Boo and α Ori, wave models based on Alfvén waves have also been discussed. The generation of wave modes within the stellar bodies operates as an inner boundary condition for the atmospheres which often leads to the generation of mass loss. Important processes also occur in the atmospheres and winds themselves. These processes include shock-shock interaction and the "sudden" onset of molecule and dust formation which usually occurs in a highly stochastic manner. Both processes can result in the generation of episodic mass loss events.

I. INTRODUCTION

It is a well-known fact that most types of stars have significant mass loss, which often impacts the course of stellar evolution. The mass loss of stars depends crucially on the position of the star in the HR diagram. It is controlled by the evolutionary status of the star and is related to a relatively broad set of physical parameters, including "basic parameters" like effective temperature, gravity and stellar rotation rate and other parameters like the magnetic field strength on the stellar surface and the presence of radial and nonradial oscillation modes. In some cases a direct dependence on the metal abundances also exists. In this chapter, it is our task to discuss the origin of stellar winds considering that stochastic processes occur. We argue that two distinct types of stochastic processes are relevant: processes that occur in the atmospheric layers and processes associated with the stellar body. We consider processes as stochastic, when accurate predictions of the relevant physical variables fail. This can occur due to the nature of the atmospheric boundary conditions or due to nonlinear processes operating in the atmospheric computational domain. Examples attributable to the behavior of the atmospheric boundary condition include the excitation of wave modes in an irregular manner. Examples for stochastic processes operating in atmospheric layers include episodic molecule and dust formation due to small changes in the atmospheric thermodynamic conditions. These effects can promote complicated hydrodynamic effects including the generation of episodic outflow events. Processes associated with the stellar body include stellar pulsation. Stellar pulsation modes as well as acoustic modes with extremely long periods (if existent; see below) and certain types of MHD modes tend to increase drastically the scale height of the atmospheric pressure and density, which is an important condition in order that the mass loss generated becomes appreciable. These modes also generate atmospheric velocity fields which impact many physical processes.

Over the last two decades or so, various mass loss mechanisms have been identified (see, e.g., Holzer and MacGregor 1985). These mechanisms include the thermal pressure gradient, momentum transfer by acoustic waves, Alfvén waves, stellar pulsation modes, and radiation force in atomic and molecular lines, and on dust grains. As discussed in the literature, it seems that all of these mechanisms are relevant in certain parts of the HR diagram, but for most stars one or a few of these mechanisms dominate. In some cases, the relevance of a proposed mechanism can be evaluated without performing model computations in detail. For instance, mass loss rates based on the thermal pressure gradient are significant only when the star possesses coronal-type structures for solar gravity. In noncoronal stars, on the other hand, the mass loss rates remain extremely low. Assuming an outer atmospheric temperature of 20,000 K as inferred from IUE data, Haisch et al. (1980) have explored the possibility of a thermally driven wind in α Boo (K1.5 III). They found a very low mass loss rate of $\sim 10^{-16}$ M$_\odot$ yr^{-1}. In most cases, mass loss cannot occur via acoustic energy dissipation. Acoustic waves in the context of this chapter are nonmagnetic waves of relatively short periods compared to the acoustic cutoff period and small or moderately large velocity amplitudes compared to the sound speed at the inner atmospheric boundary. It is found that acoustic waves tend to dissipate most of their energy immediately beyond the stellar photospheres and therefore fail to produce significant stellar winds. For cases in which the energy requirement of the wind is dominated by the potential energy term of the stellar wind momentum equation, it has been shown that the length scale of the dissipated mechanical energy flux must be on the order of or larger than a stellar radius (see, e.g., Holzer and MacGregor 1985). Acoustic waves usually fail to meet this criterion. For α Boo, Cuntz (1990) investigated the mass loss rate produced by propagating shock waves. He found a mass loss rate between 10^{-14} and 10^{-16} M$_\odot$ yr^{-1} depending on the adopted model parameters, which is well below the values suggested by observations (see Judge and Stencel 1991, and references therein).

In this chapter we discuss the origin of stellar winds in five selected stars: δ Cephei (F5-G2 Ib), α Bootis (K1.5 III), and α Orionis (M2 Iab), as well as a typical RR Lyrae, and an AGB star. These stars have been chosen because up-to-date *ab initio* models exist that attempt to explain the stellar mass loss behavior. Note that we have not selected o Ceti (M5-9 IIIe), which is considered in the chapter by Willson et al. The star o Cet is characterized by photospheric fundamental mode pulsation with a period of 332 days caused by the hydrogen ionization instability of the stellar envelope. Gillet et al. (1985) and others have presented studies of shock-excited emission lines such as $H\alpha$ and have shown that line profile variations follow the stellar pulsation cycle in a very characteristic manner. Early calculations of mass loss from AGB stars (Wood 1979; Willson and Hill 1979) using the two limiting cases of isothermal or adiabatic shock waves, lead either to too low or too high mass loss rates. Bowen (1988) has performed numerical calculations of so called piston models for Mira-type stars based on a number of simplifying assumptions concerning the radiation transport and the formation of dust. In these computations the stellar pulsation is initiated by a moving inner boundary. More recent results have been given by Bowen (1990,1992). Fleischer et al. (1992) have extended such calculations by including the detailed treatment of time-dependent dust formation and evaporation, and used the Lucy approximation for the radiation field (Lucy 1971,1976). These results show the development of a number of dust driven wind shells. Feuchtinger et al. (1993) have done similar calculations, solving the full radiation hydrodynamical equations which reveal the importance of the location of the pulsation zone for the resulting lightcurves. Höfner et al. (1995) have performed the first numerical simulations containing the full coupled system of gas, radiation and time-dependent dust formation. Some results of these computations will be presented in Sec. III.D.

Now we shall discuss some basic features of our target stars: δ Cephei is the prototype of the Cepheid variables, which are highly regular pulsators. The pulsation period of δ Cephei itself is 5.3663 days. We note that because of their period-luminosity relation, Cepheid stars have received a great deal of attention as "standard candles" for galactic and extragalactic distance determinations. A recent discussion has been provided by Sasselov and Karovska (1994), who also considered the effect of pulsational shocks on the δ Cephei brightness distribution. The study of the δ Cephei mass-luminosity relationship was particularly relevant to clarifying stellar evolution scenarios and was an important motivation for computing up-to-date opacities for stellar interiors. For more information see Buchler et al. (1990), Moskalik et al. (1992), and references therein. Buchler and his group have studied stellar oscillation modes in Cepheids (and other types of variable stars) using the so called amplitude equation formalism. In this method, photospheric oscillation modes are deduced based on a detailed stability analysis of the stellar interior. An updated version of this method which now also includes stochastic driving has been given by Buchler et al. (1993). It is important to note, however, that this type of method treats the stellar atmosphere as a "perfect filter" and therefore ignores feed-back mechanisms whenever they exist.

The red supergiant α Ori (M2 Iab), on the other hand, is a semi-regular pulsator on various different time scales. Dupree et al. (1990) have presented results from a six-year monitoring program and found unequivocal evidence for variabilities in visible and ultraviolet continua as well as in the Mg II h and k lines. Dupree et al. (1987,1990) identified a major $\sim$1.15-yr period which was also found by Smith et al. (1989) in photospheric low-amplitude radial velocity variations. We note that the atmosphere of α Ori is also shaped by time-dependent stochastic processes. Not long ago, Carpenter et al. (1994a) presented novel results from a HST-GHRS study of flow velocities based on emission and absorption features of various metallic lines. These lines indicate a broad range of velocities, ranging from (supersonic) inflows of 8 km s^{-1} to (supersonic) outflows of 13 km s^{-1}, indicative of stochastic velocity structures. A further feature in the outer atmosphere of α Ori is the occurrence of hydrochemical processes leading to episodic formation and destruction of molecules and dust. Molecules and dust are relevant for generating mass loss as these species are able to capture photon momentum. Up-to-date models which include a detailed description of nonequilibrium chemical reaction rates as well as the influence of the chromospheric or interstellar ultraviolet radiation field are given by Glassgold and Huggins (1986), Mamon et al. (1987), and Beck et al. (1992). Evidence that radiation pressure in molecular lines can also support mass loss was described by Maciel (1976,1977), Elitzur et al. (1989), and Jørgensen and Johnson (1992). Hydrochemical processes have also received special attention as they have been identified as the probable cause of phenomena revealed by "nonstandard" observations such as the occurrence of sudden dips in the radio continuum flux at 2, 3.6, and probably 6 cm (Drake et al. 1992). These observations may indicate radiative cooling instabilities which lead to the onset of molecule formation at different positions in the outer stellar atmosphere (see, e.g., Muchmore et al. 1987; Cuntz and Muchmore 1994). All these processes are potentially relevant to the origin of mass loss in this star. Last but not least, we included α Boo (K1.5 III) in our sample, a star, which shows little activity at all.

We now discuss aspects of the origin of mass loss in more detail. Philosophically, mass loss is generated in the stellar body itself. Stars have distinct internal structures which give rise to certain types of wave modes present on the stellar surface. A further basic feature in stars is the level of magnetic activity produced by stellar rotation (Noyes et al. 1984) or by the decay of an active dynamo which existed at an earlier stage of stellar evolution (chapter by Charbonneau et al.). Magnetic activity can also lead to mass loss. Stellar winds produced by Alfvén waves are important examples of this phenomenon. Nevertheless, the physical processes operating on the stellar surface are only a part of the physical picture of mass loss generation. It is also crucial to focus on physical processes that operate in stellar envelopes. Stellar winds start in stellar photospheres, but in many cases the velocity fields present there are very small when compared to the thermal velocity or the sound speed. At greater atmospheric heights, the velocity increases and in many types of stars (for instance, the Sun) it exceeds the sound speed. In all cases, however, the stellar wind needs to exceed the local escape speed as otherwise the matter would remain bound to the star. In this respect we point out that in many stars, detailed observations indicate the presence of outer atmospheric flow patterns which in some cases even consist of supersonic motions. Nevertheless, it is clearly a misconception to infer the presence of mass loss from this type of observation or even to present determinations of "observed" mass loss rates as long as estimates of the local escape speed are not at hand. Important exceptions include supergiant stars, that are primary stars of ζ Aurigae binary systems. For such stars, information about the geometrical resolution of the wind is available allowing the approximate determination of the local escape speed, and improved estimates of the mass loss rates can be given (see, e.g., Schröder 1985).

In the last two decades or so, much work has been done to explore the physical nature of processes relevant to the generation of mass loss. A "homogeneous" approach, however, does not exist. The most important reason is that most relevant processes are usually strongly coupled leading to very complicated physical scenarios. This prevents us from considering all relevant physical processes simultaneously. This limitation makes it impossible to gain a full picture of all relevant processes occurring, even when considering the full set of the observations available. *Ab initio* models are therefore never able to explain the full set of observations that exist. As a consequence, we have decided to explore the most important theoretical and observational features in a limited number of well-studied objects and to consider existing *ab initio* models whenever possible. The chapter is structured as follows: in Sec. II we describe pivotal observations for our selected stars; in Sec. III we present results from theoretical models and conclusions are given in Sec. IV.

II. OBSERVATIONAL RESULTS

A. RR Lyr Stars (A7-F5 III)

RR Lyrae stars are low mass $M \simeq 0.5$ $M_\odot$ variable stars with a relatively low metal content as they are Population II stars. These stars exhibit a narrow range of visual magnitudes and their periods range typically from about 0.2 to 1.2 days. The longest known period is that of UX Nor which is $P = 2.4$ days (Diethelm 1981). A large sample containing 90 RR Lyrae lightcurves in 6 colors has been given by Lub (1977). Depending on the amplitude, period and the shape of the lightcurve, several different classes of pulsations have been introduced. The catalog of variable stars given by Kukarin et al. (1976) lists about 5800 RR Lyrae stars. The RRab subclass is characterized by a steep ascending branch with generally asymmetric lightcurves, whereas the RRc subclass exhibits almost sinusoidal lightcurves. The spectral types of RR Lyrae stars fall between A7 and F5. Note also that the boundaries of this range depend on the ΔS-factor which has been introduced by Preston (1959) to describe the spectra and the metal abundances of the stars. Note that the spectral range deduced from the Ca II lines can differ from the spectral range obtained from the hydrogen lines. Since early observations, it has become evident that a number of RR Lyrae stars exhibit hydrogen emission lines that are split into two components during the ascending part of the lightcurve (Struve 1947). Without having a detailed theory of this emission phenomenon available, it is generally believed that shock waves run through the outer stellar atmospheres. The typical range of luminosities peaks at about $60\,L_\odot$ which makes these stars less suitable to estimate the distances of external galaxies. Hence, observational studies are restricted to our Galaxy. Furthermore, it is well known that these stars undergo nonlinear pulsation in the fundamental mode, in the first overtone or in both modes simultaneously. A large number of numerical computations have been carried out by various authors (see, e.g., Stellingwerf 1975; Kovács and Buchler 1988) to explore the stability of the RR Lyrae models.

B. δ Cep (F5-G2 Ib)

Cepheids are the most important building block for calibrating the cosmic distance scale for nearby galaxies (see, e.g., Feast and Walker 1987). Extensive information concerning observations and theory of Cepheids can be found in the book edited by Madore (1985). The understanding of Cepheid pulsations is a long-standing problem in astrophysics with relevance to several other topics. It is known that a number of discrepancies are encountered when theoretical models are compared with observations (see, e.g., the reviews of Cox 1980 and Simon 1987). In particular, mass determinations of Cepheids differ depending on the method used, i.e., there are mass determinations from stellar evolution calculations, from the Baade-Wesselink method, from the pulsation theory and from special features of the lightcurves, usually referred to as bump and beat masses. As stated by Moskalik et al. (1992), the discrepancies are decreasing (but not disappearing) when recent theoretical and observational improvements are taken into account. The most significant improvement in this respect is due to more accurate stellar opacities (Rogers and Iglesias 1992; Seaton et al. 1994). The most recent calculations favor a smaller mass, radius and luminosity for δ Cep. All results presented in this chapter are obtained using these up-to-date opacity tables. Specifically, we show some very recent results for the nonlinear pulsation of δ Cep.

In order to relate the observations to a theoretical model of δ Cep we must discuss the values of fundamental stellar parameters of δ Cep like mass, radius, effective temperature and luminosity given in the literature. Due to the discrepancies mentioned above, the parameter estimates for δ Cep vary between 4.7 $M_\odot$ and 6.6 $M_\odot$, 44 $R_\odot$ and 51 $R_\odot$, 3130 $L_\odot$ and 2460 $L_\odot$, respectively (see, e.g., Cox 1980). The effective temperature is about $T_{\mathrm{eff}} = 6100$ K. More recent measurements of the properties of δ Cep (as well as T Vul and X Cyg) using infrared photometry lead to a mean radius of $R_{\mathrm{ph}} = 37.28\ R_\odot$, an evolutionary mass of $M_{\mathrm{E}} = 5.7\ M_\odot$, a Baade-Wesselink mass of $M_{\mathrm{W}} = 2.51\ M_\odot$, a luminosity of $L = 1723\ L_\odot$ and an effective temperature of T_{eff} = 6056 K (Fernley et al. 1989). Estimates of the radius of δ Cep are summarized in Turner (1988, Table I). These estimates range from 37 $R_\odot$ to 45 $R_\odot$. Turner gets a mean radius of 42.7 $R_\odot$ based on the Baade-Wesselink method. Some recent spectral observations of δ Cep have been performed by Breitfellner and Gillet (1993) indicating that shock waves induce atmospheric motions which modify certain spectral features related to turbulence during the pulsational cycles. They found that the amplitude of the pulsation modes at the photospheric level is about 11% to 13% of the photospheric radius R_{ph} depending upon which value for R_{ph} is adopted. The values which are used include $R_{\text{ph}} \simeq 43$ $R_\odot$ (Turner 1988) and $R_{\text{ph}} \simeq 37$ $R_\odot$ (Fernley et al. 1989). We furthermore note that the value of the observed period of $P = 5.3663$ days of δ Cep is slowly decreasing by 7.6×10^{-9} days per cycle or 0.04 s per year (Szabados 1980).

C. α Boo (K1.5 III)

The star α Boo (K1.5 III) is the prototype of an inactive, noncoronal, yellow giant star. Ayres and Linsky (1975), Ayres et al. (1982,1986,1995) and Judge (1986) have studied observational and semi-empirical constraints of the outer atmosphere, which point to chromospheric structures with temperatures up to about 20,000 K, perhaps accompanied by tiny layers of somewhat higher temperatures. A search for detectable X-ray fluxes failed (Ayres et al. 1981,1991; Maggio et al. 1990), which is a strong indication for the lack of magnetic activity. The mass loss rate of α Boo for the ionized component of the wind has been estimated as 6.9×10^{-11} $M_\odot$ yr^{-1} based on radio continuum observations at 6 cm (Drake and Linsky 1986), suggesting a total mass loss rate of about 2×10^{-10} $M_\odot$ yr^{-1} (see, e.g., Judge and Stencel 1991). Schrijver (1987) and Rutten et al. (1991) noted that the Ca II and Mg II emission line fluxes in α Boo coincide with the deduced chromospheric "basal" flux limits, which is evidence that the chromosphere is heated by acoustic shocks. Stencel and Mullan (1980) and Stencel et al. (1980) analyzed the Mg II h and k line emission cores which indicate an outward directed chromospheric wind flow. A high level of atmospheric activity can be seen in the He I λ 10,830 line. O'Brien and Lambert (1979) reported that the He I λ 10,830 line shows profiles which vary between a P Cygni shape, a pure absorption feature, and no feature at all. Lambert (1987) showed a sequence spanning about 6 weeks in which the profile varied smoothly from a P Cygni profile (very weak emission) to a broad, shallow, blue-shifted absorption line. Another observational property of α Boo is the presence of low-amplitude radial velocity variations. Smith et al. (1987) reported the detection of variations with a 1.842±0.005-day (or 2.183±0.005-day) period and an amplitude of 160±10 m s^{-1}. Further detections were reported by Cochran (1988) and Irwin et al. (1989). Belmonte et al. (1990) presented new evidence for radial velocity variations in Arcturus in a relatively broad range of periods with the highest amplitude at $\sim$2.7 days. The authors also presented weak evidence for an $\sim$8.3-day period, which could possibly be interpreted as the fundamental pulsation mode. Hatzes and

Cochran (1994) have meanwhile re-analyzed the presence of photospheric low-amplitude oscillations in α Boo. They found periods of $\sim$2.46, $\sim$4.03, and $\sim$8.52 days, periods that are similar to those previously given by Belmonte et al.

D. α Ori (M2 Iab)

The star α Ori (M2 Iab) is the visually apparent brightest red supergiant and has been the most thoroughly studied. It is a variable star with respect to many different observational features. Dupree et al. (1990) have presented detailed results from a six-year monitoring program and found unequivocal evidence for variabilities in the $\lambda\lambda$ 2950–3050 ultraviolet continuum and the Mg II h and k lines, which are reflected in variations of the B magnitude. Dupree et al. (1987,1990) identified a major $\sim$1.15-yr period, which was also found in photospheric low-amplitude oscillations (Smith et al. 1989). α Ori is a non-coronal star, as indicated by the absence of detectable transition layer emission lines (Linsky and Haisch 1979; Haisch et al. 1990). Judge and Stencel (1991) analyzed the global energy and momentum requirements of the chromosphere and wind of this star and found that the chromospheric emission is relatively weak and is consistent with the extrapolated chromospheric basal flux limits. This result provides evidence that the chromosphere of α Ori is dominated by non-magnetic heating. α Ori shows no detectable X-ray flux, indicative of coronal activity, to a low upper limit. Radio emission measurements suggest a spherically symmetric, partially ionized, chromospheric region extending from 1 to 4 R_* (Newell and Hjellming 1982). Evidence for episodic chromospheric activity has been given by Toussaint and Reimers (1989). They reported a 35% increase of the flux in the chromospheric Ca II K flux from February 1988 to February 1989 implying a localized increase in the chromospheric nonradiative heating rate. This result coincides with the appearance of a bright spot on the stellar surface at wavelengths 633, 700, and 710 nm (Buscher et al. 1990). The mass loss rate of α Ori is between 2 and 4 $\times$ 10^{-6} $M_\odot$ yr^{-1} inferred from circumstellar K I emission line profiles as observed 5 and 7.5 arcsec from the star (Mauron 1990). α Ori is surrounded by a huge shell of dust and gas extending from a few R_* above the photosphere out to $\sim$10^3 R_* or more (see, e.g., Bernat et al. 1978). Different components moving at different radial velocities are seen, but the direction of flow apparently alternates between infalling and outfalling motion, possibly as a result of drastic events such as sporadic mass ejections or simply a consequence of stellar pulsation or waves. Irregular short-period fluctuations of the photospheric energy flux also occur (see, e.g., Goldberg 1984). Schwarzschild (1975) and Antia et al. (1984) suggested that the surface of a low-gravity star like α Ori might be covered by a relatively small number of extremely large convective cells, which are expected to impact largely the photospheric and chromospheric dynamics and initiate convective overshooting.

Querci and Querci (1986) have summarized observations of the blue and ultraviolet Fe II emission core velocities which have been described in detail by Boesgaard and Magnan (1975), Boesgaard (1979), and Carpenter (1984). These observations were obtained between 1970 and 1978. They indicate complex velocity structures in the atmosphere of this star, including changes from infall to outfall and from outfall to infall. The changes are significant and are unpredictable. Due to the lack of data, however, it is impossible to determine the relevant time scales precisely. They possibly range between 10^6 and 10^8 s. An important problem for interpreting the data correctly is that changes in overlying absorption features can easily imitate velocity fields in the line formation region which do not exist. Moreover, the observations described were not performed at the same wavelengths implying that they may refer to different atmospheric layers. A further result which demonstrates the existence of complicated dynamic features in the atmosphere of α Ori is the presence of multiple absorption lines in Na I, K I, and CO (see, e.g., Goldberg et al. 1975; Bernat et al. 1979; Bernat 1981). Jura (1984) interpreted these results as evidence for multiple circumstellar shells produced by the action of radiative forces on dust grains formed during distinct epochs of mass ejection. These observations again suggest that different atmospheric components moving at different radial velocities are present, and that the direction of the flow varies greatly with time and atmospheric height.

Drake et al. (1992) reported results from a radio continuum observation monitoring program for α Ori. These authors found variations in the form of dips, most pronounced at 2 cm, evident also at 3.7 cm, and possibly present at 6 cm. The variations behave stochastically and the changes at different wavelengths do not appear to be correlated. The amplitude of the variations is about 25% which shows that a large portion of the outer atmosphere must be involved. The authors also found that the variations can occur on a time scale of one month, which is unequivocal evidence that the physical cause is not related to rotation. A tempting explanation of these observations is that radiative instabilities in the wind region occur as suggested by Muchmore et al. (1987) and Cuntz and Muchmore (1994). Radiative instabilities potentially lead to rapid molecule formation and to a drastic decrease in the degree of hydrogen ionization. Similar results for the solar chromosphere regarding CO have been given by Ayres (1981) and Kneer (1983).

Further evidence for the presence of stochastic stellar wind flows in α Ori was given by Carpenter et al. (1994a). They presented results from a HST-GHRS study of flow velocities based on emission and absorption features from multiple ions indicative of dynamic chromospheric structures. The results obtained are extremely intriguing. The ultraviolet chromospheric emission lines indicate a broad range of velocities, ranging from (supersonic) inflows of 8 km s^{-1} to (supersonic) outflows of 13 km s^{-1}. The majority of the lines indicate a (subsonic) outflow of about 3 km s^{-1}. For the circumstellar shell probed by Fe I and Mn I, an outflow velocity of about 10 km s^{-1} was found which is commensurate with earlier determinations of Knapp et al. (1980) and others. The most dramatic difference in the radial velocities and velocity bisectors was found for the Mg II h and k lines. The Mg II h line indicates an accelerating *outflow* from 10 to 21 km s^{-1}, whereas the Mg II k line indicates an *inflow* of 8 km s^{-1}. Almost all of these results were completely unexpected and are still unexplained. They definitely indicate complicated subsonic and supersonic velocity structures at different chromospheric heights which can probably be attributed to stochastic energy and momentum transfer. Carpenter et al. (1994b) have meanwhile investigated the formation of O I, C I, Fe II, and S I lines in the HST-GHRS spectra and have constructed two semi-empirical chromospheric models based on these lines. The models seem to indicate the presence of some dust within the chromosphere near the temperature minimum. This result is a clear indication of gross chromospheric inhomogeneities—a likely consequence of radiative instabilities in α Ori.

III. STELLAR WIND CALCULATIONS

A. RR Lyr and δ Cep Stars: Examples for Radial Pulsators

The state of the art of theoretical modeling of radial stellar pulsations up to the year 1989 is contained in a book edited by Buchler (1990). Since then a number of new developments based on the method of adaptive grids (Dorfi and Feuchtinger 1991; Gehmeyr 1992a,b) have led to better numerical techniques for simulating a pulsating star. The most recent models of Feuchtinger and Dorfi (1994) are able to treat the pulsating object simultaneously with a gray stellar atmosphere. In such cases, nonlinear waves can be followed running through the outer atmospheric layers. Nevertheless, mass loss rates estimated from such theoretical pulsation calculations are not yet available as the models do not extend to atmospheric layers in which the flow velocities generated by the shock waves surpass the local escape velocities. However, recent improvements in nonlinear stellar pulsation calculations may provide theoretical mass loss rates for a number of pulsating stars in the near future. Because RR Lyrae stars are extensively studied objects observationally as well as by theoretical modeling, any result based on a new development can be compared to a number of results available in the literature. In Fig. 1 we present the nonlinear full-amplitude pulsations of a radiative RR Lyrae star with the following stellar parameters: $M = 0.578$ M$_\odot$, $L = 64.3$ L$_\odot$, and $T_{\text{eff}} = 6500$ K. These computations adopt recent opacity tables from Rogers and Iglesias (1992) and Seaton et al. (1994) for a chemical composition of $X = 0.7$, $Y = 0.299$ and $Z = 0.001$ and an appropriate equation of state (Wuchterl 1991) for the same mixture. Fokin (1992) has performed calculations for RR Lyrae variables using an explicit numerical scheme. An advantage of such an approach consists in an easier implementation of frequency-dependent radiative transfer and its coupling to the gas dynamics. However, this explicit nature makes it very hard to detect the limiting cycle of the pulsation even when running the calculations over a long time period.

The variations in the stellar luminosity are given in panel A of Fig. 1, the gas velocity at the photosphere is given in panel B, and the changes in the radius of the photosphere are presented in panel C, where the mean

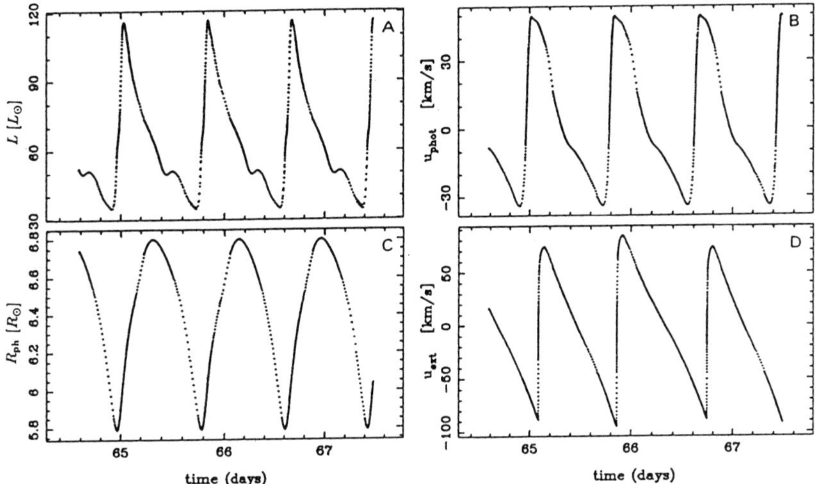

Figure 1.   Pulsating model for a typical RR Lyrae star. The stellar parameters are $M = 0.578$ M$_\odot$, $L = 64.3$ L$_\odot$, $T_{\mathrm{eff}} = 6500$ K, the chemical composition of $X = 0.7$, $Y = 0.299$ and $Z = 0.001$. The adopted period is $P = 0.828$ days; (A) the stellar luminosity at the photospheric radius $R_{\mathrm{ph}} = 6.29$ R$_\odot$; (B) the gas velocity at R_{ph}; (C) the location of the photospheric radius in units of R$_\odot$; (D) the negative gas velocity at the radius of $R_{\mathrm{ext}} = 6.464$ R$_\odot$.

photospheric radius is given as $R_{\mathrm{ph}} = 6.29$ R$_\odot$. The period of the pulsation is $P = 0.828$ days. Every grid point denotes a computational timestep showing that the temporal evolution of the pulsation is also well represented by the adaptive radiation hydrodynamical code. The gas velocity at the external computational boundary at $R_{\mathrm{ext}} = 6.464$ R$_\odot$ can be seen in panel D where the pulsations trigger shock waves running through the atmosphere. At this radius R_{ext}, a Lagrangian boundary condition is formulated and hence R_{ext} represents only the mean value during the pulsation cycle. Clearly at these low densities the velocity variations are much more pronounced compared to the photospheric velocities, which have an amplitude of about ± 100 km s^{-1} at the radius of our outer boundary R_{ext}. Note that these large velocities are still below the local escape velocity, which is $v_{\mathrm{esc}} = 185$ km s^{-1}. Although the pulsation is very regular in photospheric layers, we note that the velocity variations above the photosphere also depend on the nonlinear response of the stellar atmosphere. This effect leads to a less regular velocity pattern (Fig. 1B vs D) which can be accompanied by episodic mass loss behavior. Note that the gas velocity shows a steepening due to the outgoing nonlinear waves. However, a more quantitative estimate of mass loss from RR Lyrae stars must await further computations based on adaptive grids, that are able to resolve the structure of a pulsating star from the driving zones up to the atmospheric layers that escape from the star.

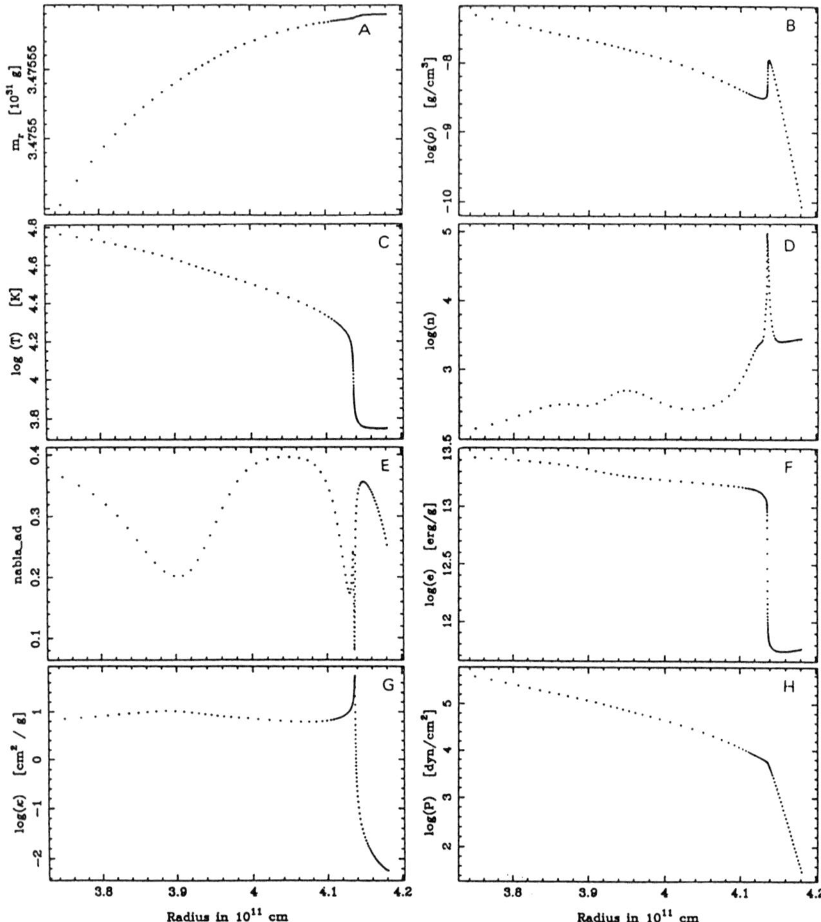

Figure 2. Radial structure of an initial hydrostatic RR Lyrae model between 3.8×10^{11} cm and $R_{\text{ext}} = 4.127 \times 10^{11}$ cm. The total mass is given by $M = 0.578 \, \text{M}_\odot$, the luminosity by $L = 64.3 \, \text{L}_\odot$, the effective temperature $T_{\text{eff}} = 6710$ K. All relevant physical variables are resolved on the adaptive grid using 300 grid points; (A) the integrated mass; (B) the gas density; (C) the temperature; (D) the point concentration showing the resolution achieved at the opacity peak; (E) the adiabatic gradient $\nabla_{\text{ad}} = \partial \ln T / \partial \ln P$; (F) the specific internal gas energy; (G) the opacity; (H) the gas pressure.

In contrast to all earlier pulsation calculations these models also include the stellar atmosphere. The radial structure of the hydrostatic initial model is plotted in Fig. 2. Every grid point is shown individually to illustrate the good resolution of the physical quantities. In this case, the stellar parameters are given by $M = 0.578 \, \text{M}_\odot$, $L = 64.3 \, \text{L}_\odot$, and $T_{\text{eff}} = 6700$ K. Focusing on the outer parts of the model, we can identify the stellar atmosphere by the rapid spatial decrease of the density (panel B), temperature (panel C), gas energy (panel F), and pressure (panel H). Within this structure we see a density inversion due to the opacity (panel G). In this layer, it is important to resolve the very steep gradients of the physical quantities. The thermodynamic quantity $\nabla_{ad} = \partial \ln T / \partial \ln P$ mirrors the stellar driving zones (panel E) and varies within the ionization zones (see, e.g., Kippenhahn and Weigert 1990). It is therefore essential to distribute enough grid points in these regions to obtain an accurate description of the stellar pulsation. At a radius of about 3.9×10^{11} cm, we find the He II ionization zone. At 4.14×10^{11} cm, the H ionization zone is superimposed on the broader He I ionization zone. The quantity in panel D displays the point concentration and reflects the clustering of the grid points near steep physical features.

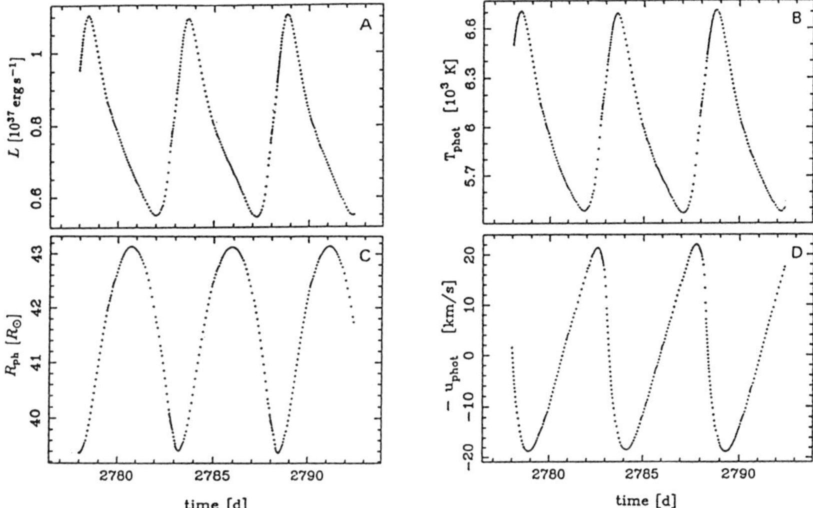

Figure 3.   Pulsating model of δ Cephei with the adopted stellar parameters of $M = 5\ M_\odot$, $L = 2000\ L_\odot$, $T_{eff} = 6000$ K. The adopted period is $P = 5.25$ days; (A) the stellar luminosity; (B) the photospheric temperature; (C) the photospheric radius; (D) the negative photospheric gas velocity.

Figure 3 shows the pulsation modes obtained for δ Cep computed using the new adaptive radiation hydrodynamics computer code (Dorfi and Feuchtinger 1991; Feuchtinger and Dorfi 1994). Every computational time-step is again plotted by a single dot. At this stage of numerical modeling we did not try a detailed fit to the large number of observations available for δ Cep. We present, however, a pulsation calculation performed with the following stellar parameters: $M = 5\ M_\odot$, $L = 2000\ L_\odot$, $T_{eff} = 6000$ K, OP-opacities (Seaton et al. 1994), and an equation of state with solar composition

(Wuchterl 1991) yielding a pulsational period of $P = 5.25$ days. Panel A displays the stellar luminosity in units of 10^{37} erg s^{-1}. The effective temperature in panel B varies between 6720 K and 5490 K, which is at the maximum found to be very close to the observed values of 6800 K but shows a large difference at the observed minimum of 5760 K. Two points are important for this comparison. First, as already mentioned, pulsational models were only calculated for a single set of stellar parameters. Considering the uncertainties discussed above, a better agreement between theoretical results and observations can easily be achieved. Second, it is not always straight forward to identify the location within the pulsating atmosphere of observationally deduced temperatures. The variations of the photospheric radius are plotted in panel C in units of the solar radius $R_\odot$ changing between 43.1 $R_\odot$ and 39.6 $R_\odot$. The negative photospheric gas velocity is shown in panel D and plotted in this way to allow an easier comparison with observations. An expanding velocity (positive values as the radius of the star increases) is directed towards the observer and hence mostly displayed by negative values. Note that these figures represent the full amplitude nonlinear pulsation where the kinetic energy (not shown here) of the star remains constant indicating that the limit cycle of the pulsation cycle has been reached.

B. α Boo (K1.5 III): A Difficult Candidate

For α Boo a broad variety of observations exists, but the origin of its stellar wind is far from understood. The wind model which received most attention is that of Hartmann and MacGregor (1980). They investigated the response of an outer atmosphere model appropriate to α Boo due to momentum transfer by Alfvén waves. Important features of these models include the following: (1) the propagation of the Alfvén waves is treated in a time-independent manner leading to stationary outflows; (2) the treatment of the Alfvén waves is based upon the WKB approximation; (3) the degree of hydrogen ionization is assumed to be constant; and (4) the dissipation length of the waves is also assumed to be constant, either 1 $R_\star$ or infinity. In order to be relevant, these models must face three significant tests: first, they must reproduce (or must be consistent with) the observed thermal structures. These structures can in principle be inferred from optical and IUE data (see, e.g., Ayres and Linsky 1975). Second, the models have to predict the mass loss rate somewhat correctly. For the single inactive K giant Arcturus, it seems to be most appropriate to rely on the interpretation of the 6 cm radio continuum flux measurements. A third criterion regards the final flow speed of the wind. In Arcturus it is believed that the final flow speed of the wind is 40 km s^{-1} (Drake 1985). This last criterion is most crucial when constraining the computed models. It was found that when a dissipation length of 1 $R_\star$ was assumed, the models computed were most useful. Hartmann and MacGregor found that the dissipation of Alfvén waves with an energy flux of 3×10^6 erg cm^{-2} s^{-1} and a magnetic field strength of $\sim$10 Gauss lead to a mass loss rate of $\sim 10^{-9}$ M$_\odot$ yr^{-1}. The final flow speed of the wind is $\sim$60 km s^{-1} and the adopted temperature of the atmosphere is 5000 K or 10,000 K, depending on the model. These results have been considered a big success for the model proposed. Nevertheless, a detailed comparison between computed and observed spectral features was never made. Equally important is the fact that these models suffer from serious technical restrictions, including the choice of the damping length of the waves as a free parameter, the treatment of the atmospheric thermodynamics, and the so called WKB approximation for the waves. Wave models with variable damping length have meanwhile been computed by Holzer et al. (1983). Rosner et al. (1991) have presented results, which also include Alfvén wave reflection. MacGregor and Charbonneau (see their chapter) have presented a preliminary study of non-WKB waves in an atmosphere somewhat appropriate to that of α Ori (see below). The results of this study are relevant when the wave periods are long enough that non-WKB effects dominate. They found that the mass loss rate is reduced and the final flow speed of the wind increases in comparison with the WKB results. Both tendencies, however, make the agreement between theoretical and observed values worse. Recent improvements in the treatment of Alfvén wave propagation including the consequences on the associated magnetic field topology have been given by Rosner et al. (1995).

Cuntz (1987) studied the response of the outer atmosphere of α Boo to acoustic shock waves, which are an ultimate consequence of stellar convection. These models thus consider solely short-period waves as predicted by the models of acoustic energy generation by Bohn (1981,1984). The models of Cuntz show strong time-dependent episodes of momentum and energy deposition which give rise to substantial chromospheric heating. Note, however, that the short-period shock-wave models fail completely in producing significant mass loss. Cuntz (1990) re-investigated the mass loss in α Boo produced by propagating shock waves by exploring the impact of a large range of wave periods. It was found again that wave modes associated with stellar convection are able to produce chromospheric temperatures, but the mass loss rates remain extremely low (i.e., between 10^{-14} and 10^{-16} $M_\odot$ yr^{-1} depending on the model parameters). These models confirm that mass loss occurs only when the dissipation length of the mechanical energy flux of the shocks is on the order of or larger than a stellar radius, because the global energy requirement of the wind is completely dominated by the potential energy term (see, e.g., Holzer and MacGregor 1985; Hammer 1988). Cuntz (1990) also found that *adiabatic* shock waves with periods larger than 5.6×10^5 s (6.5 days) lead to mass loss rates between 10^{-10} and 10^{-11} $M_\odot$ yr^{-1}. Unfortunately, the wave periods and amplitudes considered have no observational support.

This topic was meanwhile revisited by Sutmann and Cuntz (1995). They have studied the possibility that the radial velocity variations in α Boo have significant influence on the outer atmospheric dynamics of this star, including the generation of mass loss. The models calculated were based on the observational data from Belmonte et al. (1990) and Hatzes and Cochran (1994). Sutmann and Cuntz found that this is not the case as most of the wave modes remain evanescent ("mode trapping") as the wave periods are well above the acoustic cutoff. The propagation of energy into higher atmospheric layers remains therefore marginal. Nonlinear effects are found to be negligible because of the small wave amplitudes. They also checked to see if low-amplitude radial velocity variations can support mass loss. They found that this is not the case as the energy required to lift the wind out of the gravitational potential of the star remains many orders of magnitude below the wind energy flux constrained by observations, which have been discussed by Judge and Stencel (1991) and others. They argued that the energy contained in the observed oscillation modes is about 20 times greater than the energy required to drive the wind. Unfortunately, the authors have not considered the phase difference between the velocity and density, which drastically reduces the energy flux associated with these modes. In case that the modes are assumed to be associated with waves having sawtooth profiles, the energy still remains insufficient for mass loss generation (Sutmann and Cuntz 1995). Considering the fact that acoustic wave models also fail to produce significant *time-averaged* mass loss rates (Cuntz 1990), it is safe to say that mass loss in noncoronal yellow giants like α Boo cannot be initiated by a nonmagnetic mechanism. This result is particularly astonishing, when taking into account that chromospheric heating in this type of stars can probably be explained fully by acoustic energy dissipation (Schrijver 1987; Rutten et al. 1991). Cuntz and Luttermoser (1990) also argued that the appearance and time-dependent behavior of the He I λ 10,830 line in this star might also be attributable to the propagation of stochastic acoustic waves.

C. α Ori (M2 Iab): Ab-initio Calculations Based on Shock Waves, Alfvén Waves, Molecules and Dust

For α Ori several different models exist which attempt to explain the onset of the stellar wind. Unfortunately, each model is able to explain only a small subset of the existing observations. The model which is cited most is that of Hartmann and Avrett (1984) which is based on momentum transfer by Alfvén waves. They found that Alfvén waves are able to drive a steady outflow having a realistic mass loss rate, and the integrated line fluxes of the Mg II k and C II λ 2325 lines match the observations to within a factor of 3. They also found that the dissipation of Alfvén waves with an energy flux of 1×10^5 erg cm^{-2} s^{-1} and a magnetic field strength of 2 Gauss leads to a mass loss rate of 1.4×10^{-6} $M_\odot$ yr^{-1}. The final flow speed of the wind was found to be 25 km s^{-1} and the maximum of the chromospheric temperatures 8050 K. Hartmann and Avrett have computed a further model in which an energy flux of 9×10^5 erg cm^{-2} s^{-1} and a magnetic field strength of 5 Gauss has been used. In this case, the mass loss rate increases by more than a factor of 10. The final flow speed of the wind is now 20 km s^{-1} and the maximal chromospheric temperature is 7100 K.

We note that most dynamical properties of this wind model seem to be consistent with values expected for a "typical" inactive cool supergiant. On the other hand, the models of Hartmann and Avrett are not very successful in reproducing observed chromospheric emission line features inferred from IUE data. They computed fluxes and profiles for various chromospheric emission lines utilizing the PANDORA radiative transfer code and found that the line fluxes agree within a factor of 3 with observations, except for Ca II K, where the disagreement reaches a factor of 16. Gross discrepancies, however, were found for the line profiles. These discrepancies suggest that the α Ori chromosphere is much more turbulent than indicated by the Alfvén wave models computed and that episodic inflow and outflow events play a pivotal role in the energy and momentum balance of the outer α Ori atmosphere. Moreover, in the view of recent HST-GHRS spectra (Carpenter et al. 1994a), which provide unequivocal evidence for stochastic chromospheric velocity fields (see Sec. II.D), it seems that the models of Hartmann and Avrett have only limited relevance for most atmospheric layers. A further important criticism inherent to all Alfvén wave-driven wind models of cool giants and supergiants is the lack of reliable magnetic field measurements in the photospheres. Marcy and Bruning (1984) have shown that the magnetic fields on the surfaces of cool giants are below a relatively high upper limit (i.e., several hundred Gauss). The magnetic field strengths adopted in the models of Hartmann and MacGregor (1980) and Hartmann and Avrett (1984) range from 1 to 10 Gauss. We have already noted that the Alfvén wave-driven wind models computed also observe several serious limitations. One of these limitations is the adoption of the so called WKB approximation. MacGregor and Charbonneau (see their chapter) have presented a preliminary study of non-WKB waves in an atmosphere somewhat appropriate to that of α Ori. They have concluded that in models in which non-WKB waves dominate, the mass loss rate is reduced and the final flow speed of the wind increases. Both tendencies make the agreement between theoretical and observed values worse. On the other hand, non-WKB wave modes may be able to reproduce some of the observed time-dependent behavior of the atmosphere but detailed model computations must still be performed. Also note that Alfvén waves now deserve some special attention because they are possibly capable of reproducing the coronal dividing line in the HR diagram (Rosner et al. 1991,1995), which is still a major unsolved problem in modern stellar astrophysics.

A further study aimed at investigating the onset of the stellar wind in α Ori addresses the propagation and interaction of short-period acoustic waves in chromospheric layers (Cuntz 1992). Cuntz has studied the response of an α Ori atmosphere model to the propagation of stochastic wave modes, which have been selected according to results from traditional acoustic convection zone models. Figure 4 shows a panel of 8 snapshots of a time-dependent stochastic wave calculation. The snapshot series has been started from a monochromatic wave model computed for a fixed wave period of 2×10^6 s and a fixed initial wave amplitude of 0.10 Mach. Then shock waves with stochastically changing wave periods are introduced into the atmosphere. The panel demonstrate the generation of an episodic inflow and outflow event caused by efficient shock-

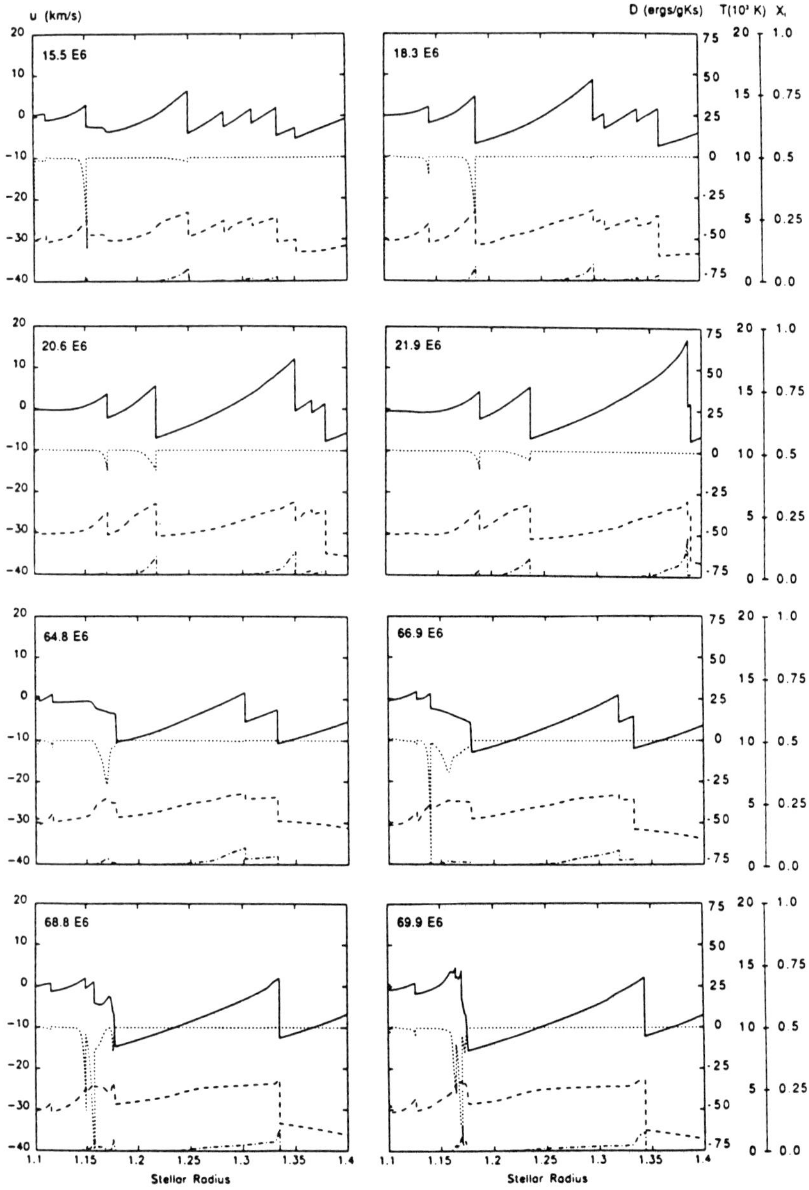

Figure 4.   Snapshots of a stochastic acoustic wave calculation for chromospheric layers in α Ori (M2 Iab) given by Cuntz (1992). The stellar effective temperature is $T_{\text{eff}} = 3900$ K, the radius is $R = 860\,R_\odot$, and the gravity is $\log g_* = -0.4$ (Tsuji 1989). The flow speed U (solid line), the radiation damping function D (dotted line), the temperature T (dashed line), and the hydrogen ionization degree X_i (dashed-dotted line) are shown as functions of height. The numbers in the upper left corners denote the time after switching on the acoustic frequency spectrum that was employed.

shock interaction. The numbers in the upper left corners denote the time after allowing the wave period to change stochastically. The time scale for typical changes in the wave propagation is 1×10^6 s. The time between the episodic outflow and inflow event is about 5×10^7 s and ranges from 2×10^7 to 2×10^8 s in further models, which is consistent with constraints from observations (see, e.g., Querci and Querci 1986, and references therein). The atmospheric models are characterized by a complicated hydrodynamic structure containing a nonuniform distribution of shocks. The shock strengths and the shock speeds differ substantially and change nonmonotonically with height. This result is a clue to the basic physics which is going on; after allowing the wave period to change stochastically, shocks with different strengths are introduced into the atmosphere. Different shock strengths cause different shock speeds, which lead to interacting, overtaking and merging of shocks (= *shock cannibalism*). As the strength of an overtaking shock combines with the shocks it engulfs, its speed increases, so it overtakes more and more shocks in front of it attaining an even greater strength. Consequently, the amount of momentum and energy deposition that occurs in the atmospheric layer varies drastically. The direction of the flow alternates between infalling and outflowing motions depending on the strengths of the shocks and the hydrodynamic history of the flow. The radiation losses which predominantly occur behind shocks decrease with increasing atmospheric height showing that the waves behave more and more adiabatically when propagating outward. The temperatures behind the shocks are somewhat similar to those of the semi-empirical chromosphere model of Basri et al. (1981), who computed a semi-empirical chromosphere model for α Ori based upon the Ca II K and Mg II h and k emission lines from IUE spectra. Note, however, that a detailed comparison between observed and computed line spectra is still unavailable. Basri et al. found a steady increase of the atmospheric temperature starting from 2820 K in the temperature minimum layer up to 7000 K at a column mass density of 1.0×10^{-6} g cm^{-2}.

Figure 5 shows the minimal and maximal flow speed in a time-dependent monochromatic and time-dependent stochastic wave model as functions of the atmospheric height. The monochromatic model is based upon a wave period of 5×10^6 s. For the computation of the stochastic model, 16 wave periods have been introduced into the atmosphere covering a timespan of 6.9×10^7 s. The minimum wave period introduced is 1.2×10^6 s, and the maximum is 1.6×10^7s. The results are quite impressive: in the stochastic wave model the maximal outflow velocity is 24.1 km s^{-1} and the maximal inflow velocity is -17.4 km s^{-1}. The corresponding Mach numbers are 3.10 and 3.33, respectively, demonstrating that in stochastic wave models supersonic flow patterns can easily be generated without *ad-hoc* assumptions. In monochromatic wave models which do not include shock-shock interaction, it was found that the flow speeds are substantially lower and never exceed the sound speed. The fact that in the stochastic wave model relatively high inflow and outflow velocities are found within chromospheric layers provides some

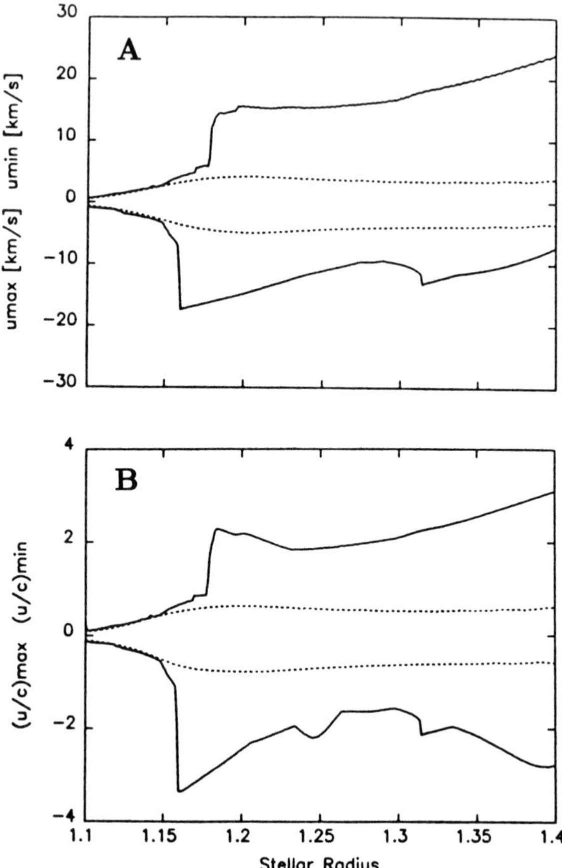

Figure 5.   Chromospheric flow speeds predicted by time-dependent wave models;
    (A) minimal and maximal flow speed in a time-dependent monochromatic (dotted
    line) and stochastic wave calculation (solid line); (B) minimal and maximal ratios
    of flow speed to sound speed (i.e., Mach number).

confidence that this model can possibly be used to interpret observational
results based on HST-GHRS data (Carpenter et al. 1994a). Chromospheric
flow velocities derived from emission and absorption features from a broad
variety of ions show that sub- and supersonic inflow and outflow events
occur (see Sec. II.D). Note, however, that the wave models of Cuntz fail
to produce any realistic time-averaged mass loss rates. This is because the
models rely on short-period wave modes generated by stellar convection,
which dissipate most of their energy immediately beyond the photosphere.
This result confirms again that mass loss occurs only when the dissipation
length of the mechanical energy flux of the shocks is on order of or larger than
a stellar radius as the global energy requirement of the wind is dominated by

the potential energy term (see, e.g., Holzer and MacGregor 1985). Significant mass loss is expected only when waves with long periods are used. For α Ori it might be important also to consider the 1.15-yr period caused by pulsation (Dupree et al. 1987,1990; Smith et al. 1989) as well as other periods (Goldberg 1984).

A further type of model aimed to explain the onset of mass loss in α Ori considers radiation pressure on dust grains. Beck et al. (1992) have studied the nonequilibrium chemistry in inner regions of an oxygen-rich circumstellar shell appropriate to that of α Ori. They considered a total of 38 species and a system of 342 reactions. They assumed a steady (i.e., time-independent) outflow with a prescribed mass loss rate and temperature structure chosen in accordance to the model of Glassgold and Huggins (1986). As a crucial feature of the models, the authors also included the effect of the ultraviolet chromospheric radiation field. The chromospheric ultraviolet radiation was also included in the models of Clegg et al. (1983) and Glassgold and Huggins (1986). These authors, however, have only treated steady-state concentrations for the species which considerably simplified the models. Most importantly, Beck et al. showed that the ultraviolet radiation field largely impacts the atmospheric chemistry. Figure 6 shows the number densities for various species at a fixed atmospheric height as function of a changing chromospheric ultraviolet radiation field; this chromospheric ultraviolet radiation plays a pivotal role in the molecular chemistry. Whenever the ultraviolet radiation flux becomes appreciable, photoions and radicals determine the physical structure of large portions of the circumstellar shell. In this case, the main modes of molecule formation involve charge-exchange, exothermal ion-radical, radiative dissociation, and radiative association reactions. It was found that the ultraviolet flux level attributed to α Ori is very close to the borderline at which dust formation is possible. If the ultraviolet flux in α Ori would be slightly stronger, the ultraviolet photons would completely suppress any nucleation process leading to the formation of dust.

We note, however, that the models of Beck et al. (1992) do not properly treat the physical structures of the stellar photosphere and chromosphere. The models start at $2\ R_*$, where the density has a prescribed value. This approach implies that stars like α Ori must have a further mechanism, which carries a sufficient amount of matter to the dust formation radius. Based on results from observational studies, it seems that stellar pulsation modes are the most promising candidate mechanism (besides Alfvén waves) to provide this missing link. The models of Beck et al. (1992) also provide important clues as to why outer atmospheric structures in stars like α Ori behave stochastically; as the wave modes traveling from underneath are changing the density and temperature of the wind in an irregular manner, it is unavoidable that the mass loss rate enhanced by radiation pressure on dust grains also varies. In addition, variability in the formation of dust and by implication in the resulting mass loss rate can also be introduced by fluctuation of the ultraviolet radiation field. As discussed in Sec. II.D, it is known that in α Ori (and probably also in similar

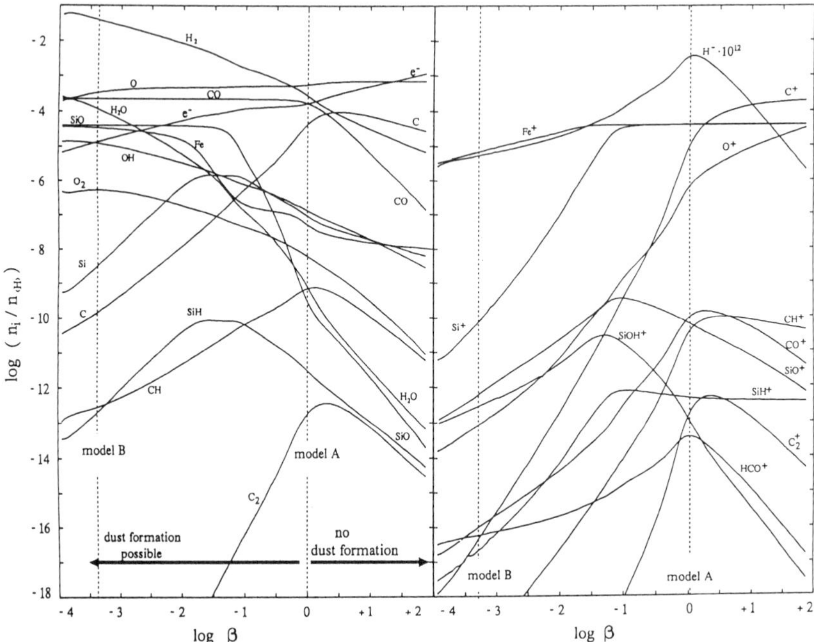

Figure 6.   Dust chemistry in the outer atmosphere of α Ori as given by Beck et al. (1992). The figures show the number densities for various species at $r = 5\ R_*$ as functions of the adopted ultraviolet flux which is parameterized by β. $\beta = 1$ denotes the standard ultraviolet flux as given by observations.

stars) that the emerging chromospheric radiation is not constant, but shows a distinct level of periodic, quasi-periodic, and stochastic variability produced by chromospheric heating process(es) (see, e.g., Dupree et al. 1987,1990; Toussaint and Reimers 1989).

There is a further complication in stars like α Ori; it is found that cool dusty layers and hot chromospheric-type layers most likely co-exist at various atmospheric heights, which also has important dynamic consequences for the onset of the stellar wind. This behavior is probably due to radiative instabilities caused by CO and SiO molecules and distinct hydrodynamic effects such as quasi-adiabatic cooling (see, e.g., Cuntz and Muchmore 1994). In this respect, it is intriguing to realize that the work of Carpenter et al. (1994b) provided important theoretical evidence for dust existing inside the α Ori chromosphere. Further studies are needed to obtain insight into the underlying physical and chemical atmospheric processes.

D. AGB Stars: Models Based on Pulsation and Self-Excited Stochastic Dust Formation

This section deals with mass loss due to nonstationary processes occurring at late stages of stellar evolution. It is centered on theoretical models of mass loss in AGB stars. We concentrate our description on low mass stars (up to an initial main-sequence mass less than about 8 $M_\odot$) with cool extended atmospheres as mass loss in more massive and consequently in hotter stars occurring via radiation force in atomic lines is covered elsewhere. Note that the winds in these stars can also exhibit a number of fluctuations which are attributable to an instability in the expanding wind (see, e.g., review by Owocki 1990). A similar phenomenon occurring in dust-forming atmospheres will be illustrated in this section. The stellar evolution properties of low mass stars have been reviewed by Iben and Renzini (1983).

Most of these late-type stars have extended convection zones in their outer hydrogen-rich envelopes and are pulsationally unstable. As summarized by Wood (1990), the theoretical models do not match the observed properties in a satisfactory way. We are therefore faced with the problem that no pulsation theory of these long-period variables can explain their temporal variations. Because mass loss plays a fundamental role in these stars, Pijpers (1993) suggested that modifications of the outer boundary conditions may be necessary to get an adequate description for pulsating stars. Again, it is clear that only a fully hydrodynamical calculation ranging from the stellar interior up to the wind zone can give a quantitative answer to this problem.

From a number of observations of cool extended stars it is well documented that these atmospheres are sites of heavy dust production (see, e.g., the review of Lafon and Berruyer 1991). The radiation pressure on the newly condensed dust grains provides an effective force to drive a slow but massive stellar outflow with mass loss rates up to $\dot{M} \simeq 10^{-5}\,M_\odot\,\mathrm{yr}^{-1}$. These winds provide the most important source of dust particles for the interstellar medium (see, e.g., Gail 1990). Self-consistent models for stationary outflows have been constructed by Dominik et al. (1989). These models show that certain combinations of stellar parameters (e.g., low photospheric temperatures and high luminosities) can produce such stellar winds. Nevertheless, the observed correlation between the infrared excess, which is interpreted as mass loss, and the pulsation period (de Gioia-Eastwood et al. 1981; Jura 1986; Anandarao et al. 1993) indicates that mass loss is a time-dependent phenomenon and can be characterized by a two-step process (see, e.g., Jones et al. 1981). The pulsations seem to be necessary to levitate the stellar material into the region where dust forms. If enough material condenses into dust particles, a stellar outflow develops, which is clearly modulated by the stellar pulsations as pulsation lifts the material up to the condensation zone. Based on these assumptions, a number of models with various levels of simplified physical assumptions have been constructed. These models simulate the stellar pulsation by an oscillating inner boundary condition (Wood 1979; Bowen 1988; Fleischer et al. 1992; Feuchtinger et al. 1993; also see the chapter by Willson et al. for more details).

The process of dust formation is linked to a relaxation time scale τ_{rel} towards equilibrium values for given thermodynamic conditions (Höfner and Dorfi 1992). Taking typical values for the gas density and the temperature, this time scale can be on the order of the pulsation period of AGB stars leading to a rather subtle interplay between stellar pulsations and dust formation. A shock wave running through a zone of dust formation will generally increase the density and temperature. However, due to the interaction with the radiation field, it is *a priori* not clear at which location in the downstream region optimal conditions for nucleation and growth of dust particles are produced. Hence, in one case it might be possible to create dust particles immediately behind the shock transition, whereas in other cases the gas must first cool down and only in far downstream regions effective dust formations may take place (Höfner et al. 1995). In other words, we expect that dust-driven winds of pulsating stars in certain stellar parameter regimes should be a rather irregular and time-dependent process. In this case, the stochasticity of the mass loss behavior is not directly initiated by the boundary conditions, but due to the nonlinearity of the processes operating in the computational domain. (For the time-dependent α Ori chromosphere models previously described, the stochasticity of the flow occurred due to the randomly changing wave frequencies which have been specified at the inner atmospheric boundary.) Figures 7 and 8 illustrate the irregular behavior of a dust driven mass loss from an AGB star.

The results presented are obtained by solving the full system of radiation hydrodynamical equations coupled to the equations which describe the time-dependent formation, growth, and evaporation of dust particles. These equations consist of a system of 12 nonlinear partial differential equations (see Dorfi and Feuchtinger [1991] for the description of the numerical method, and Höfner et al. [1995] for a discussion of the physical equations). The initial conditions are stated for an atmosphere of a carbon-rich AGB star, $M = 1 \, M_\odot$, $T_{\mathrm{eff}} = 2600$ K, $L = 10^4 \, L_\odot$ and a ratio of carbon to oxygen as $\varepsilon_C/\varepsilon_O = 2.5$. (i.e., a cool carbon star). These parameters lead to a stellar radius of $R_* = 3.43 \times 10^{13}$ cm corresponding to 493 $R_\odot$. The computational domain extends from 1 R_* to 30 R_*. The simulations run on an adaptive grid with 500 radial grid points. Figure 7 shows the temporal evolution of the velocity in km s^{-1}, the gas temperature in K, and the mass loss rate in $M_\odot$ yr^{-1} between 40 and 120 yr at a radius of 30 R_* after the computation has been started from an initially dust-free hydrostatic configuration. It is easy to trace the different shock waves reaching this radius of 30 R_* based on the increase of the velocity, the mass loss rate and the gas temperature which are followed by radiating cooling zones. A corresponding radial structure is plotted in Fig. 8.

We note that this model does not rely on stellar pulsation modes employed at the inner boundary of the atmosphere, but on a self-excited instability associated with the formation of dust. The combination used of stellar parameters with a large value of the carbon abundance ε_C shows that the previously mentioned effect that the dust formation itself influences the radiation field can result in an unstable situation provoking two entirely different solutions (Winters et al. 1994; Höfner et al. 1995). Because the radiation is blocked by the formation of dust particles, the gas inside the dust layer is heated up to a temperature which inhibits further dust formation. This dust-free zone is

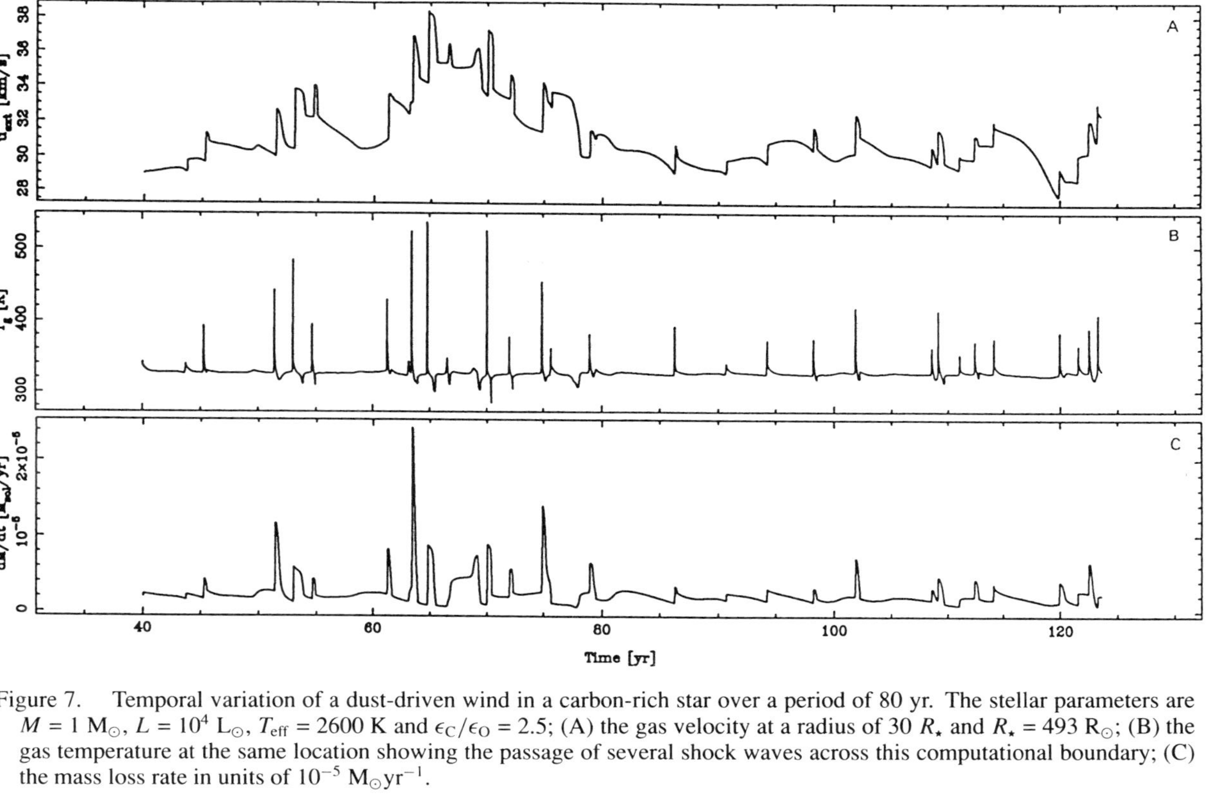

Figure 7. Temporal variation of a dust-driven wind in a carbon-rich star over a period of 80 yr. The stellar parameters are $M = 1\ M_\odot$, $L = 10^4\ L_\odot$, $T_{\mathrm{eff}} = 2600$ K and $\epsilon_C/\epsilon_O = 2.5$; (A) the gas velocity at a radius of 30 $R_\star$ and $R_\star = 493\ R_\odot$; (B) the gas temperature at the same location showing the passage of several shock waves across this computational boundary; (C) the mass loss rate in units of $10^{-5}\ M_\odot\mathrm{yr}^{-1}$.

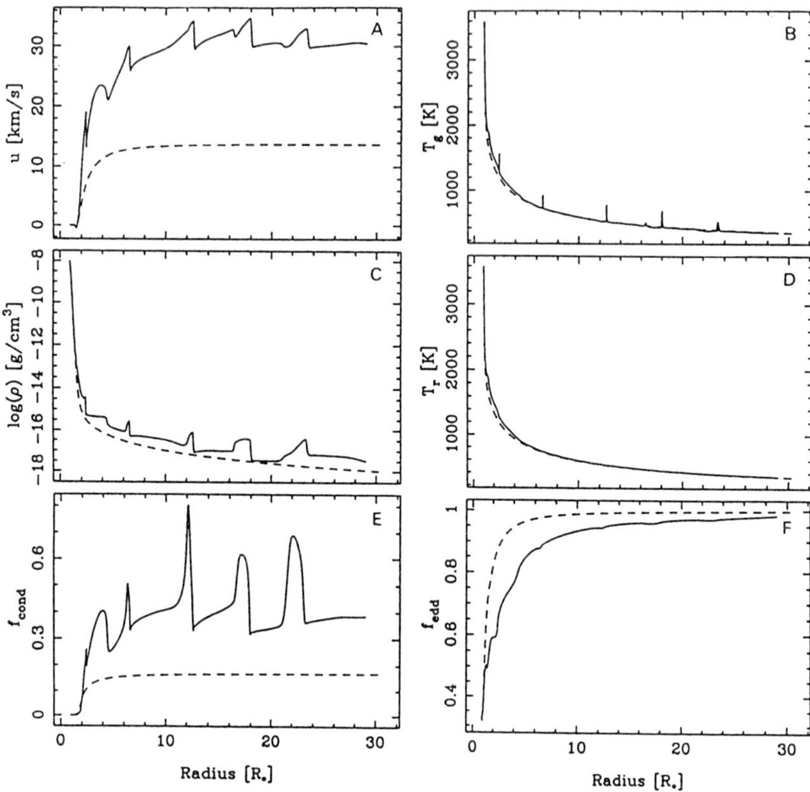

Figure 8. A typical radial structure of the dust driven wind of Fig. 7 between 1 R_* and 30 R_*. The dashed lines correspond to a stationary solution with $\epsilon_C/\epsilon_O = 2.3$ and all other parameters are unchanged; (A) the gas velocity in km s^{-1}; (B) the gas temperature in K exhibits 6 shocks in this radius interval; (C) the gas density; (D) the radiation temperature which runs smoothly across the shock waves; (E) the fraction of carbon condensed into grains; (F) the variable Eddington-factor f_{edd}.

able to fall back towards the star whereas the dust shell is being accelerated outwards decreasing the blocking of the strong radiation field. The gas cools again giving rise to the next cycle of dust formation and the whole process repeats. In order to prove the physical nature of this instability, the authors have performed a number of computations showing the dependence of these solutions on the four parameters M, T_{eff}, L and on the carbon to oxygen ratio $\varepsilon_C/\varepsilon_O$. It is possible to produce stationary as well as periodic solutions with the same numerical method by changing the basic stellar parameter (Höfner et al. 1995).

Figure 8 depicts with the solid line a typical radial structure of the physical variables also presented in Fig. 7 between 1 R_* and 30 R_* revealing a number of nonlinear waves running through the extended stellar atmosphere. The dashed line represents a stationary solution obtained with the same numerical code and this solutions differs by only one parameter from the irregular mass loss of Fig. 7, with the lower carbon to oxygen ratio, i.e., $\varepsilon_C/\varepsilon_O = 2.3$. The figure displays the gas velocity u, the gas temperature T_g, the gas density ρ, the radiation temperature T_r, the fraction of carbon f_{cond} condensed into grains and the Eddington factor f_{edd}. Comparing the stationary solutions, we observe a number of significant differences. First, the average velocity is about a factor of 3 larger, which is about 30 km s^{-1}. This is due to the lower amount of carbon condensed into grains which reduces the radiative acceleration for stationary winds. Second, the gas temperature follows nicely the stationary structure except in the vicinity of shock waves where we find shock-heated gas cooling down towards the equilibrium temperature. Third, the temperature increases inside the innermost shock waves caused by the effective blocking of the radiation field as seen, e.g., in the behavior of the radiation temperature. This behavior is also illustrated in the deviation of the Eddington factor f_{edd} from the stationary value. These lower values of f_{edd} are typical for a more isotropic radiation field produced by the dense dust shells. Although this particular example exhibits an irregular solution for a dust-driven wind, we point out that such deviations from the stationary values also occur in regular dust driven winds (see, e.g., Höfner et al. 1995). Again, we emphasize that the AGB star provides constant inner boundary conditions in these computations and that the wind modulations obtained are solely initiated by an atmospheric dust-induced κ-mechanism. Similar results which point to the same physical mechanism have meanwhile been given by Fleischer et al. (1995).

We should point out here that ultraviolet observations of optically bright carbon stars demonstrate that many of these stars have also chromospheric emission (Querci and Querci 1985; Johnson and Luttermoser 1987). Whether this emission comes from a chromospheric-type layer, as in warmer stars, or from outward propagating individual shocks, as shown in the models above, remains open to debate. Luttermoser et al. (1989) have modeled the ultraviolet emission lines seen in carbon stars and have demonstrated that this emission must come from circumstellar regions at temperatures in excess of 5000 K. A radio survey of N-type (i.e, cool) carbon stars at 6 cm have further demonstrated that this chromospheric-type region cannot have temperatures in excess of 10,000 K (Luttermoser and Brown 1992). As can be seen in the models presented above, the weak shocks generated have insufficient temperatures to give rise to the ultraviolet emission seen in some carbon stars. This demonstrates that long-period pulsational-driven waves may be important in these stars as well as for the AGB Mira-type variables.

IV. CONCLUSIONS

It has been the purpose of this chapter to discuss the origin of stellar winds considering the impact of stochastic processes. Stochastic processes can be initiated by the stellar body by providing distinct boundary conditions for the stellar atmosphere. They can also occur in the atmospheres themselves.

Shock-shock interaction and the "sudden" onset of molecule and dust formation due to marginal changes in the atmospheric thermodynamical conditions are key examples for this behavior. In fact, it can be said that a broad variety of different physical processes is usually involved when mass loss behavior occurs. Most significantly, it is found that in late-type stars, the energy requirement of the wind is dominated by the potential energy term of the stellar wind momentum equation (see, e.g., Holzer and MacGregor 1985), implying that the length scale of the dissipated mechanical energy flux must be on the order of or larger than a stellar radius. Acoustic waves usually fail to meet this criterion as they dissipate the lion's share of their mechanical energy flux immediately beyond the stellar photospheres (Hartmann and MacGregor 1980; Cuntz 1990; Sutmann and Cuntz 1995). On the other hand, as found by Cuntz (1987,1992), acoustic waves are very efficient in producing episodic outflow events, largely due to shock-shock interaction in the stochastic wave field, particularly in stars of low gravity. In addition, they enhance the atmospheric turbulence which is an important feature in most stellar photospheres and chromospheres. Stellar pulsation is very efficient in producing significant continuous outflows. Momentum transfer due to strong shocks can expand the atmospheres enormously and can drive the atmospheric matter to velocities beyond the escape speed. Model calculations for o Ceti (M5-9 IIIe) given by Bowen (1988) and Willson et al. (see their chapter), among others are key examples. Another important mechanism for producing continuous outflows in late-type stars might be momentum transfer by Alfvén waves. In this case, however, significant discrepancies between theoretical models and observational results still remain, indicating the need for further studies. In all these cases, it is very important to consider results both from observations and theory to ensure an appropriate assessment of the mechanism. Because of the complexity of the phenomena involved we have focused on a list of distinct stars: δ Cep (F5-G2 Ib), α Boo (K1.5 III), α Ori (M2 Iab) and a representative RR Lyrae star and an AGB star. These stars differ vastly in their stellar parameters as well as types of atmospheric activities, thus providing an interesting set selected for tutorial purposes.

For RR Lyr and δ Cep stars, improved models in conjunction with accurate stellar opacities have meanwhile become available. These models include the simulation of the driving zones together with the stellar atmosphere by applying an adaptive grid to resolve the running waves in the atmospheres. Current simulations on RR Lyrae variables are found to be in better agreement with the observations and can eventually also provide theoretical mass loss rates. This task requires that the stellar atmospheres be extended to lower densities where the material reaches escape velocities. The stellar material is accelerated by the shock waves. It is expected that the resulting mass loss may not be regular due to the nonlinear response of the outer layers on the pulsational modes. Most recent efforts in modeling pulsating δ Cepheids also open up the possibility of considering photospheric pulsation on the stellar photospheric brightness distribution. This is extremely relevant when

δ Cephei stars are utilized for extragalactic distance determinations (Sasselov and Karovska 1994). A difficult candidate with respect to the generation of mass loss is the noncoronal yellow giant α Boo. Acoustic waves and global photospheric oscillation are both present and lead to a variety of stochastic effects in the stellar atmosphere. Nevertheless, they are found to be unimportant for the origin of the stellar wind in this type of star (Sutmann and Cuntz 1995). This result points to the importance of a magnetic field-related mechanism or mechanisms, as discussed by Hartmann and MacGregor (1980). More studies are needed to clarify the basic physics of mass generation in this type of star.

For α Ori, several different models exist which attempt to explain the onset of the stellar wind. Unfortunately, each model is able to explain only a small subset of the existing observations. It seems that Alfvén wave-driven wind models (Hartmann and Avrett 1984) may be relevant for explaining some of the observed properties, despite the fact that major discrepancies between the results from the theoretical models and the observations remain. It is also noteworthy that the photospheric magnetic field strength which is a key quantity in this type of model is unconstrained by observations as it is in the model for α Boo; this reduces the significance of the models substantially. Another important feature in this star is the effect of the $\sim$1.15-yr pulsation period and of other periods found in extensive monitoring programs. Unfortunately detailed model calculations are still missing. In addition, flow patterns associated with radiation pressure on molecules and dust are expected to play a significant role in the outer atmospheric dynamics in this star. Beck et al. (1992) as well as Glassgold and Huggins (1986) and Mamon et al. (1987) have demonstrated that radiation pressure on dust grains can be an important mass loss mechanism. In addition, the chromospheric ultraviolet radiation field is found to be pivotal for the molecule and dust chemistry in α Ori-type stars as the ions and radicals formed change the efficiency of most molecular reaction rates. Hydrodynamic refrigeration due to quasi-adiabatic cooling by strong shocks formed in the stochastic wave field besides thermal instabilities due to newly formed molecules and dust may also play an important role for the mass loss. The complexity of these different mechanisms and effects ensures that mass loss in stars like α Ori should behave stochastically which is in agreement with observations. Nevertheless, further theoretical studies which employ the action of stellar oscillations (and perhaps also momentum transfer by magnetic field-related mechanisms) as well as molecular and dust chemistry are needed to clarify the origin of the stellar wind in this star. Höfner and Dorfi (1992) have already focused on the generation of dust-driven winds in AGB stars, caused by the interaction of different effects operating on different time scales. They focused particularly on episodic dust formation due to self-excited thermal instability. In this case, episodic mass loss is expected to occur even in the absence of the propagating shocks that exist. Höfner and Dorfi found that backwards reactions due to backwarming are important for explaining stochastic dust formation and can furthermore change the stellar pulsation period. The development of shell-like dust-driven structures in this type of stars impinges on the stellar radiation field which necessitates calculations of the Eddington factor. The different features involved in the mass loss of AGB stars demand a fully consistent treatment of the coupled system of dust, radiation and gas dynamics which is a basic issue of current research.

Acknowledgments. We are thankful to K. B. MacGregor, L. A. Willson and D. G. Luttermoser for comments on an earlier version of the chapter. M. C. is pleased to acknowledge support by the German Research Foundation, by NASA, and by the Advanced Study Program at the National Center for Atmospheric Research under sponsorship of the National Science Foundation. E. A. D. would like to thank M. Feuchtinger and S. Höfner for providing some recent computational data and for their continuous support to maintain theoretical astrophysical research in Austria. This work was also supported by the Austrian Fonds zur Förderung der wissenschaftlichen Forschung.

REFERENCES

Anandarao, B. G., Pottasch, S. R., and Vaidya, D. B. 1993. Circumstellar dust in Mira variables and the mass loss mechanisms. *Astron. Astrophys.* 273:570–574.

Antia, H. M., Chitre, S. M., and Narasimha, D. 1984. Convection in the envelopes of red giants. *Astrophys. J.* 282:574–583.

Ayres, T. R. 1981. Thermal bifurcation in the solar outer atmosphere. *Astrophys. J.* 244:1064–1071.

Ayres, T. R., and Linsky, J. L. 1975. Stellar model chromospheres, III. *Astrophys. J.* 200:660–674.

Ayres, T. R., Linsky, J. L., Vaiana, G. S., Golub, L., and Rosner, R. 1981. The cool half of the H-R diagram in soft X-rays. *Astrophys. J.* 250:293–299.

Ayres, T. R., Simon, T., and Linsky, J. L. 1982. Evolution of chromospheres and coronae in solar mass stars: A far-ultraviolet and soft X-ray comparison of Arcturus (K2 III) and Alpha Centauri A (G2 V) *Astrophys. J.* 263:791–802.

Ayres, T. R., Judge, P. G., Jordan, C., Brown, A., and Linsky, J. L. 1986. High-dispersion observations of α Bootis (K1 III) with the International Ultraviolet Explorer. *Astrophys. J.* 311:947–959.

Ayres, T. R., Fleming, T. A., and Schmitt, J. H. M. M. 1991. Digging in the coronal graveyard: A ROSAT observation of the red giant Arcturus. *Astrophys. J. Lett.* 376:45–48.

Ayres, T. R., Fleming, T. A., Simon, T., Haisch, B. M., Brown, A., Lenz, D., Wamsteker, W., De Martino, D., Gonzalez, C., Bonnell, J., Mas-Hesse, J. M., Rosso, C., Schmitt, J. H. M. M., Trümper, J., Voges, W., Pye, J., Dempsey, R. C., Linsky, J. L., Guinan, E. F., Harper, G. M., Jordan, C., Montesinos, B. M., Pagano, I., and Rodonó, M. 1995. The RIASS coronathon: Joint X-ray and ultraviolet observations of normal F-K stars. *Astrophys. J. Suppl.* 96:223–260.

Basri, G. S., Linsky, J. L., and Eriksson, K. 1981. Outer atmospheres of cool stars, VIII. *Astrophys. J.* 251:162–180.

Beck, H. K. B., Gail, H.-P., Henkel, R., and Sedlmayr, E. 1992. Chemistry in circumstellar shells, I. *Astron. Astrophys.* 265:626–642.

Belmonte, J. A., Jones, A. R., Pallé, P. L., and Cortés, T. R. 1990. Global acoustic oscillations on α Bootis. *Astrophys. J.* 358:595–609.

Bernat, A. P. 1981. Observations of circumstellar carbon monoxide and evidence for multiple ejections in red giants. *Astrophys. J.* 246:184–192.

Bernat, A. P., Honeycutt, R. K., Kephart, J. E., Gow, C. E., Sandford, M. T., II, and Lambert D. L. 1978. The remarkable extent of the circumstellar gas shell surrounding Betelgeuse. *Astrophys. J.* 219:532–537.

Bernat, A. P., Hall, D. N. B., Hinkle, K. H., and Ridgway, S. T. 1979. Observations of CO circumstellar absorption in the 4.6 micron spectrum of Alpha Orionis. *Astrophys. J. Lett.* 233:135–139.

Boesgaard, A. M. 1979. Velocity fields in the shell of α Orionis. *Astrophys. J.* 232:485–495.

Boesgaard, A. M., and Magnan C. 1975. The circumstellar shell of Alpha Orionis from a study of the Fe II emission lines. *Astrophys. J.* 198:369–378.

Bohn, H. U. 1981. On the Generation of Acoustic Energy in the Convection Zones of Late Type Stars. Ph.D. Thesis, Univ. of Würzburg (in German).

Bohn, H. U. 1984. Generation of acoustic energy from convection zones of late type stars. *Astron. Astrophys.* 136:338–350.

Bowen, G. H. 1988. Dynamical modeling of long-period variable star atmospheres. *Astrophys. J.* 329:299–317.

Bowen, G. H. 1990. Dynamical phenomena in pulsating star atmospheres. In *The Numerical Modelling of Nonlinear Stellar Pulsations: Problems and Prospects*, ed. J. R. Buchler (Dordrecht: Kluwer), pp. 155–171.

Bowen, G. H. 1992. Dynamical modeling of cool supergiant atmospheres. In *Instabilities in Evolved Super- and Hypergiants*, eds. C. de Jager and H. Nieuwenhuijzen (Amsterdam: North-Holland), pp. 104–110.

Breitfellner, M. G., and Gillet, D. 1993. Atmospheric motions in classical Cepheid stars, I. *Astron. Astrophys.* 277:524–540.

Buchler, J. R. 1990. *The Numerical Modelling of Nonlinear Stellar Pulsations: Problems and Prospects* (Dordrecht: Kluwer).

Buchler, J. R., Moskalik, P., and Kovács, G. 1990. A survey of bump Cepheid model pulsations. *Astrophys. J.* 351:617–631.

Buchler, J. R., Goupil, M.-J., and Kovács, G. 1993. Stellar pulsation with stochastic driving. *Astron. Astrophys.* 280:157–168.

Buscher, D. F., Haniff, C. A., Baldwin, J. E., and Warner, P. J. 1990. Detection of a bright feature on the surface of Betelgeuse. *Mon. Not. Roy. Astron. Soc.* 245:7p–11p.

Carpenter, K. G. 1984. Characteristics of the Fe II and C II emission in high-resolution IUE spectra (2300–3000 Å) of Alpha Orionis. *Astrophys. J.* 285:181–189.

Carpenter, K. G., Robinson, R. D., Wahlgren, G. M., Linsky, J. L., and Brown, A. 1994*a*. GHRS observations of cool, low-gravity stars, I. *Astrophys. J.* 428:329–344.

Carpenter, K. G., Robinson, R. D., Judge, P. G., Ebbets, D. C., and Brandt, J.C. 1994*b*. GHRS observations and analysis of the O I and C I resonance lines in the UV spectrum of α Ori. In *Cool Stars, Stellar Systems, and the Sun*, ed. J.-P. Caillault (San Francisco: Astronomical Soc. of the Pacific), pp. 56–58.

Clegg, R. E. S., van Ijzendoorn, L. J., and Allamandola, L. J. 1983. Circumstellar silicon chemistry and the SiO maser. *Mon. Not. Roy. Astron. Soc.* 203:125–146.

Cochran, W. D. 1988. Confirmation of radial velocity variations in Arcturus. *Astrophys. J.* 334:349–356.

Cox, A. N. 1980. The masses of Cepheids. *Ann. Rev. Astron. Astrophys.* 18:15–41.

Cuntz, M. 1987. Episodic mass loss in late-type stars due to acoustic wave packets. *Astron. Astrophys.* 188:L5–L8.

Cuntz, M. 1990. On the generation of mass loss in cool giant stars due to propagating shock waves. *Astrophys. J.* 353:255–264.

Cuntz, M. 1992. Stochastic stellar wind models: Evolved stars. In *Cool Stars, Stellar Systems, and the Sun*, eds. M. S. Giampapa and J. A. Bookbinder (San Francisco: Astronomical Soc. of the Pacific), pp. 383–393.

Cuntz, M., and Luttermoser, D. G. 1990. Stochastic shock waves as a candidate mechanism for the formation of the He I λ10830 line in cool giant stars. *Astrophys. J. Lett.* 353:39–43.

Cuntz, M., and Muchmore, D. O. 1994. The CO/SiO radiative instability in cool star atmospheres revisited. *Astrophys. J.* 433:303–312.

de Gioia-Eastwood, K., Hackwell, J. A., Grasdalen, G. L., and Gehrz, R. D. 1981. A correlation between infrared excess and period for Mira variables. *Astrophys. J. Lett.* 245:75–78.

Diethelm, R. 1981. Photometric classification of pulsating variables with periods between one and three days. *ESO Messenger* 25:29–31

Dominik, C., Gail, H.-P., and Sedlmayr, E. 1989. The size distribution of dust particles in a dust-driven wind. *Astron. Astrophys.* 223:227–236.

Dorfi, E. A., and Feuchtinger, M. U. 1991. Nonlinear stellar pulsations, I. *Astron. Astrophys.* 249:417–427.

Drake, S. A. 1985. Modeling lines formed in the expanding chromospheres of red giants. In *Progress in Stellar Spectral Line Formation Theory*, eds. J. E. Beckman and L. Crivellari (Dordrecht: D. Reidel), pp. 351–357.

Drake, S. A., and Linsky, J. L. 1986. Radio continuum emission from winds, chromospheres, and coronae of cool giants and supergiants. *Astron. J.* 91:602–620.

Drake, S. A., Bookbinder, J. A., Florkowski, D. R., Linsky, J. L., Simon, T., and Stencel R. E. 1992. Four years of monitoring α Orionis with the VLA: Where have all the flares gone? In *Cool Stars, Stellar Systems, and the Sun*, eds. M. S. Giampapa and J. A. Bookbinder (San Francisco: Astronomical Soc. of the Pacific), pp. 455–457.

Dupree, A. K., Baliunas, S. L., Guinan, E. F., Hartmann, L., Nassiopoulos, G. E., and Sonneborn, G. 1987. Periodic photospheric and chromospheric modulation in Alpha Orionis (Betelgeuse). *Astrophys. J. Lett.* 317:85–89.

Dupree, A. K., Baliunas, S. L., Guinan, E. F., Hartmann, L., and Sonneborn, G. 1990. Alpha Ori: Evidence for pulsation. In *Confrontation Between Stellar Pulsation and Evolution*, eds. C. Cacciari and G. Clementini (San Francisco: Astronomical Soc. of the Pacific), pp. 468–471.

Elitzur, M., Brown, J. A., and Johnson H. R. 1989. On the onset of mass loss in late-type stars. *Astrophys. J. Lett.* 341:95–98.

Feast, M. W., and Walker, A. R. 1987. Cepheids as distance indicators. *Ann. Rev. Astron. Astrophys.* 25:345–375.

Fernley, J. A., Skillen, I., and Jameson, R. F. 1989. Cepheid radii and effective temperatures. *Mon. Not. Roy. Astron. Soc.* 237:947–971.

Feuchtinger, M. U., Dorfi, E. A., and Höfner, S. 1993. Radiation hydrodynamics in atmospheres of long-period variables. *Astron. Astrophys.* 273:513–523.

Feuchtinger, M. U., and Dorfi, E. A. 1994. Nonlinear stellar pulsation, II. *Astron. Astrophys.* 291:209–225.

Fleischer, A. J., Gauger, A., and Sedlmayr, E. 1992. Circumstellar dust shells around long-period variables, I. *Astron. Astrophys.* 266:321–339.

Fleischer, A. J., Gauger, A., and Sedlmayr, E. 1995. Circumstellar dust shells around long-period variables, III. *Astron. Astrophys.* 297:543–555.

Fokin, A. B. 1992. Shock waves and Hα profiles in the hydrodynamical model for RR Lyrae. *Mon. Not. Roy. Astron. Soc.* 256:26–36.

Gail, H.-P. 1990. Winds of late type stars. *Rev. Mod. Astron.* 3:156–173.

Gehmeyr, M. 1992*a*. On nonlinear radial oscillations in convective RR Lyrae stars, I. *Astrophys. J.* 399:265–271.

Gehmeyr, M. 1992*b*. On nonlinear radial oscillations in convective RR Lyrae stars, II. *Astrophys. J.* 399:272–283.

Gillet, D., Ferlet, R., Maurice, E., and Bouchet, P. 1985. The shock-induced variability of the Hα emission profile in Mira, II. *Astron. Astrophys.* 150:89–96.

Glassgold, A. E., and Huggins, P. J. 1986. The ionization structure of the circumstellar envelope of Alpha Orionis. *Astrophys. J.* 306:605–617.

Goldberg, L. 1984. The variability of Alpha Orionis. *Publ. Astron. Soc. Pacific* 96:366–371.

Goldberg, L., Ramsey, R. W., Testerman, L., and Carbon, D. 1975. High-resolution profiles of sodium and potassium lines in Alpha Orionis. *Astrophys. J.* 199:427–431.

Haisch, B. M., Linsky, J. L., and Basri, G. S. 1980. Outer atmospheres of cool stars, IV. *Astrophys. J.* 235:519–533.

Haisch, B. M., Bookbinder, J. A., Maggio, A., Vaiana, G. S., and Bennett J.O. 1990. IUE and Einstein survey of late-type giant and supergiant stars and the dividing line *Astrophys. J.* 361:570–589.

Hammer, R. 1988. The winds from cool stars including the sun. In *Pulsation and Mass Loss in Stars*, eds. R. Stalio and L. A. Willson (Dordrecht: Kluwer), pp. 51–72.

Hartmann, L., and Avrett, E. H. 1984. On the extended chromosphere of α Orionis. *Astrophys. J.* 284:238–249.

Hartmann, L., and MacGregor, K. B. 1980. Momentum and energy deposition in late-type stellar atmospheres and winds. *Astrophys. J.* 242:260–282.

Hatzes, A. P., and Cochran, W. D. 1994. Short-period radial velocity variations of α Bootis: Evidence for radial pulsations. *Astrophys. J.* 422:366–373.

Höfner, S., and Dorfi, E. A. 1992. Dust formation in winds of long-period variables. *Astron. Astrophys.* 265:207–215.

Höfner, S., Feuchtinger, M. U., and Dorfi, E. A. 1995. Dust formation in winds of long period variables, III. *Astron. Astrophys.* 297:815–827.

Holzer, T. E., and MacGregor, K. B. 1985. Mass loss mechanisms for cool, low-gravity stars. In *Mass Loss from Red Giants*, eds. M. Morris and B. Zuckerman (Dordrecht: D. Reidel), pp. 229–255.

Holzer, T. E., Flå, T., and Leer, E. 1983. Alfvén waves in stellar winds. *Astrophys. J.* 275:808–835.

Iben, I., and Renzini, A. 1983. Asymptotic giant branch evolution and beyond. *Ann. Rev. Astron. Astrophys.* 21:271–342.

Irwin, A. W., Campbell, B., Morbey, C. L., Walker, G. A. H., and Yang, S. 1989. Long-period radial-velocity variations of Arcturus. *Publ. Astron. Soc. Pacific* 101:147–159.

Johnson, H. R., and Luttermoser, D. G. 1987. Ultraviolet spectra and chromospheres of cool carbon stars. *Astrophys. J.* 314:329–340.

Jones, T. W., Ney, E. P., and Stein, W. A. 1981. Pulsations, grain condensation, and mass loss in long-period variable stars. *Astrophys. J.* 250:324–326.

Jørgensen, U. G., and Johnson, H. R. 1992. Radiative force on molecules and its possible role for mass loss in evolved AGB stars. *Astron. Astrophys.* 265:168–176.

Judge, P. G. 1986. Constraints on the outer atmospheric structure of late-type giant stars with IUE: methods and application to Arcturus (α Boo K2 III). *Mon. Not. Roy. Astron. Soc.* 221:119–153.

Judge, P. G., and Stencel, R. E. 1991. Evolution of the chromospheres and winds of low- and intermediate-mass giant stars. *Astrophys. J.* 371:357–379.

Jura, M. 1984. Multiple circumstellar shells and radiation pressure on grains in the outflows from late-type giants. *Astrophys. J.* 282:200–205.

Jura, M. 1986. Mass loss from carbon stars. *Astrophys. J.* 303:327–332.

Kippenhahn, R., and Weigert, A. 1990. *Stellar Structure and Evolution* (Berlin: Springer-Verlag).

Knapp, G. R., Phillips, T. G., and Huggins, P. J. 1980. Detection of CO emission at 1.3 millimeters from the Betelgeuse circumstellar shell. *Astrophys. J. Lett.* 242:25–28.

Kneer, F. 1983. A possible explanation of the Wilson-Bappu relation and the chromospheric temperature rise in late-type stars. *Astron. Astrophys.* 128:311–317.

Kovács, G., and Buchler, J. R. 1988. A survey of RR Lyrae models and search for steady multimode pulsations. *Astrophys. J.* 324:1026–1041.

Kukarin, B. V., Kholopov, P. N., Efremov, N. B., Kukarina, N. P., Kurochkin, N. E., Medvedeva, G. I., Perova, N. B., Pskovsky, Yu. P., Fedorovich, V. P., and Frolov, M. S. 1976. *General Catalog of Variable Stars*, Suppl. 3 to the 3rd ed. (Moscow: Shternberg Astronomical Inst.).

Lafon, J.-P. J., and Berruyer, N. 1991. Mass loss mechanisms in evolved stars. *Astron. Astrophys. Rev.* 2:249–289.

Lambert, D. L. 1987. Further observations of the He I 10830 Å chromospheric line in stars. *Astrophys. J. Suppl.* 65:255–271.

Linsky, J. L., and Haisch, B. M. 1979. Outer atmospheres of cool stars, I. *Astrophys. J. Lett.* 229:27–32.

Lub, J. 1977. An atlas of light and colour curves of field RR Lyrae stars. *Astron. Astrophys. Suppl.* 29:345–378.

Lucy, L. B. 1971. The formation of resonance lines in extended and expanding atmospheres. *Astrophys. J.* 163:95–110.

Lucy, L. B. 1976. Mass loss by cool carbon stars. *Astrophys. J.* 205:482–491.

Luttermoser, D. G., and Brown, A. 1992. A VLA 3.6 centimeter survey of N-type carbon stars. *Astrophys. J.* 384:634–639.

Luttermoser, D. G., Johnson, H. R., Avrett, E. H., and Loeser, R. 1989. Chromospheric structure of cool carbon stars. *Astrophys. J.* 345:543–553.

Maciel, W. J. 1976. Mass loss from Mira variables by the action of radiation pressure on molecules. *Astron. Astrophys.* 48:27–31.

Maciel, W. J. 1977. Mass loss from Mira variables, II. *Astron. Astrophys.* 57:273–277.

Madore, B. F. 1985. *Cepheids: Theory and Observations* (Cambridge: Cambridge Univ. Press).

Maggio, A., Vaiana, G. S., Haisch, B. M., Stern, R. A., Bookbinder, J., Harnden, F. R., Jr., and Rosner, R. 1990. Einstein Observatory magnitude-limited X-ray survey of late-type giant and supergiant stars. *Astrophys. J.* 348:253–278.

Mamon, G. A., Glassgold, A. E., and Omont, A. 1987. Photochemistry and molecular ions in oxygen-rich circumstellar envelopes. *Astrophys. J.* 323:306–315.

Marcy, G. W., and Bruning, D. H. 1984. Magnetic field observations of evolved stars. *Astrophys. J.* 281:286–291.

Mauron, N. 1990. New observations of atomic line scattering around Betelgeuse. *Astron. Astrophys.* 227:141–146.

Moskalik, P., Buchler, J. R., and Marom, A. 1992. Toward a resolution of the bump and beat cepheid mass discrepancies. *Astrophys. J.* 385:685–693.

Muchmore, D. O., Nuth, J. A., III, and Stencel, R. E. 1987. SiO cooling instability in the envelopes of cool giant stars. *Astrophys. J. Lett.* 315:141–146.

Newell, R. T., and Hjellming, R. M. 1982. Radio emission from the extended chromosphere of Alpha Orionis. *Astrophys. J. Lett.* 263:85–87.

Noyes, R. W., Hartmann, L. W., Baliunas, S. L., Duncan, D. K., and Vaughan, A. H. 1984. Rotation, convection, and magnetic activity in lower main-sequence stars.

Astrophys. J. 279:763–777.

O'Brien, G. T., and Lambert, D. L. 1979. He I 10830 Å emission from α Bootis and α^1 Herculis. *Astrophys. J. Lett.* 229:33–37.

Owocki, S. P. 1990. Winds from hot stars. *Rev. Mod. Astron.* 3:98–123.

Pijpers, F. P. 1993. Radial pulsation in variable stars with mass loss. *Astron. Astrophys.* 267:471–489.

Preston, G. W. 1959. A spectroscopic study of the RR Lyrae stars. *Astrophys. J.* 130:507–538.

Querci, M., and Querci, F. 1985. Temporal variations in UV spectra of the red giant C star, TW Hor. *Astron. Astrophys.* 147:121–126.

Querci, M., and Querci, F. 1986. Dynamical chromospheric structure of α Orionis. In *Cool Stars, Stellar Systems, and the Sun*, eds. M. Zeilik and D. M. Gibson (Berlin: Springer-Verlag), pp. 492–495.

Rogers, F. J., and Iglesias, C. A. 1992. Radiative atomic Rosseland mean opacity tables. *Astrophys. J. Suppl.* 79:507–568.

Rosner, R., An, C.-H., Musielak, Z. E., Moore, R. L., and Suess, S. T. 1991. Magnetic confinement, Alfvén wave reflection, and the origin of X-ray and mass-loss "dividing lines" for late-type giants and supergiants. *Astrophys. J. Lett.* 372:91–94.

Rosner, R., Musielak, Z. E., Cattaneo, F., Moore, R. L., and Suess, S. T. 1995. On the origin of "dividing lines" for late-type giants and supergiants. *Astrophys. J. Lett.* 442:25–28.

Rutten, R. G. M., Schrijver, C. J., Lemmens, A. F. P., and Zwaan, C. 1991. Magnetic structure in cool stars, XVII. *Astron. Astrophys.* 252:203–219.

Sasselov, D. D., and Karovska, M. 1994. On Cepheid diameter and distance measurement. *Astrophys. J.* 432:367–372.

Schrijver, C. J. 1987. Magnetic structure in cool stars, XI. *Astron. Astrophys.* 172:111–123.

Schröder, K.-P. 1985. A study of ultraviolet spectra of ζ Aurigae VV Cephei systems, VII. *Astron. Astrophys.* 147:103–110.

Schwarzschild, M. 1975. On the scale of photospheric convection in red giants and supergiants. *Astrophys. J.* 195:137–144.

Seaton, M. J., Yu Yan, Mihalas, D., and Pradhan, A. K. 1994. Opacities for stellar atmospheres. *Mon. Not. Roy. Astron. Soc.* 266:805–828.

Simon, N. R. 1987. Cepheids: Problems and possibilities. In *Stellar Pulsation*, eds. A. N. Cox, W. M. Sparks and S. G. Starrfield (Berlin: Springer-Verlag), pp. 148–158.

Smith, M. A., Patten, B. M., and Goldberg, L. 1989. Radial-velocity variations in α Ori, α Sco, and α Her. *Astron. J.* 98:2233–2248.

Smith, P. H., McMillan, R. S., and Merline, W. J. 1987. Evidence for periodic radial velocity variations in Arcturus. *Astrophys. J. Lett.* 317:79–84.

Stellingwerf, R. F. 1975. Nonlinear effects in double-mode Cepheids. *Astrophys. J.* 199:705–709.

Stencel, R. E., and Mullan, D. J. 1980. Detection of mass loss in stellar chromospheres. *Astrophys. J.* 238:221–228; addendum: 240:718.

Stencel, R. E., Mullan, D. J., Linsky, J. L., Basri, G. S., and Worden, S. P. 1980. The outer atmospheres of cool stars, VII. *Astrophys. J. Suppl.* 44:383–402.

Struve, O. 1947. Peculiar hydrogen lines in the spectrum of RR Lyrae. *Publ. Astron. Soc. Pacific* 59:192–194.

Sutmann, G., and Cuntz, M. 1995. Generation of mass loss in K giants: The failure of global oscillation modes and possible implications. *Astrophys. J. Lett.* 442:61–64.

Szabados, L. 1980, In *Photoelectric UBV Photometry of Northern Cepheids, II*, vol.

76 (Budapest: Mitteilungen der Sternwarte der ungarischen Akademie der Wissenschaften), pp. 3–119.

Toussaint, F., and Reimers, D. 1989. Variations in the chromospheric Ca II lines of α Orionis. *Astron. Astrophys.* 226:L17–L19.

Tsuji, T. 1989. Testing the models against observations. In *The Evolution of Peculiar Red Giant Stars*, eds. H. R. Johnson and B. Zuckerman (Cambridge: Cambridge Univ. Press), pp. 81–100.

Turner, D. G. 1988. A new pulsational radius for delta Cephei. *Astron. J.* 96:1565–1569.

Willson, L. A., and Hill, S. J. 1979. Shock wave interpretation of emission lines in long period variable stars, II. *Astrophys. J.* 228:854–869.

Winters, J. M., Fleischer, A. J., Gauger, A., and Sedlmayr, E. 1994. Circumstellar dust shells around long-period variables, II. *Astron. Astrophys.* 290:623–633.

Wood, R. P. 1979. Pulsation and mass loss in Mira variables. *Astrophys. J.* 227:220–231.

Wood, P. R. 1990. Pulsation and evolution of Mira variables. In *From Miras to Planetary Nebulae: Which Path from Stellar Evolution?*, eds. M. O. Mennessier and A. Omont (Gif-sur-Yvette: Editions Frontières), pp. 67–84.

Wuchterl, G. 1991. Hydrodynamics of giant planet formation, II. *Icarus* 91:39–52.

MAGNETOHYDRODYNAMIC TURBULENCE IN COSMIC WINDS

M. L. GOLDSTEIN and D. A. ROBERTS
NASA Goddard Space Flight Center

and

W. H. MATTHAEUS
Bartol Research Institute

We review research on the nonlinear, turbulent evolution of the solar wind plasma, regarded as an expanding, compressible magnetofluid. Direct spacecraft observations from 0.3 to over 20 AU, remote sensing observations of plasmas near the Sun, numerical simulations, and turbulence modeling provide a complementary set of methods that have convincingly shown that the fluctuations in the wind parameters undergo significant dynamical evolution. The roles of nearly spherical expansion, compression, and shear—the major important ingredients of the picture—in producing and controlling turbulent cascades have begun to be elucidated. The solar wind provides an excellent "laboratory" for answering the many basic open questions about turbulent cosmic plasmas; the lessons learned here should be applicable in many other regions, from solar flares to extragalactic jets.

I. INTRODUCTION

The term "turbulence" conjures up visions of violent, chaotic motions, and indeed such motions are typically turbulent in a more technical sense. There is little doubt that the complex plasma flows associated with supernova remnants, hinted at in photographs, involve the nonlinear coupling of motions on a wide range of scales, driven by a massive energy release at the center, and dissipated in thin shocks and other small-scale processes. Extragalactic jets must also be in this class of flows as shown by their knotty contortions and relativistic particles contained in the strikingly ordered large-scale motion. However, such violence is not necessary, and a summer's breeze or the flow in a stream will exhibit generically similar features although without the shocks and other small structures that are accompanied by energetic particle acceleration when a plasma and magnetic field are present. The plasma flow from the Sun is typically somewhere between these extremes, both in size and in some rough measure of its "violence," although energetically the solar wind, like many other astrophysical flows, is not substantial enough to rustle a leaf except perhaps near its source region. Although generally turbulent, under specific conditions the solar wind may generate relatively stable structures that also depend for their existence on small dissipation. As an example, observations at solar minimum in the outer heliosphere indicate the possible existence of a stable vortex street generated by the strong velocity shears existing on either side of the heliospheric current sheet.

Despite its tenuousness, the solar wind can be readily measured in great detail by *in-situ* spacecraft, and we now have considerable knowledge of its electromagnetic fields and particle distributions. The spatial scales sampled range from tens of AU down to a few km; the corresponding times are from years to fractions of a second. It is rare to find a system so broadly accessible, and consequently the heliosphere provides an excellent laboratory for the study of turbulence in a magnetofluid. In time the lessons learned here will be helpful in understanding processes in many other places, from the arches and plumes near the solar surface to the supernova remnants and jets mentioned above. We are presently in a very good position to make progress on these issues due to a wealth of observations coupled with simulations on increasingly powerful computers and a growing sophistication in analytical approaches. Here we will review these aspects of the problem of heliospheric turbulence with an emphasis on open questions and issues of general interest.

The underlying question to which turbulent processes provide at least part of the answer is what is the origin, nature, and dynamics of the deviations from the Parker spiral magnetic field with uniform flow. In principle this question encompasses everything from the scale of streams, which may be regarded as large-scale fluctuations, down to the scale of the ion Larmor radius. In practice the streams are often taken as an inhomogeneous background in which smaller fluctuations evolve, and the gyromotion of ions (and electrons) is described by kinetic techniques that are generally too unwieldy to use on the intermediate-scale fluctuations on which we concentrate here. In general terms, the origin of these deviations is clear: the Sun's surface and inner atmosphere are highly time-variable and inhomogeneous. In addition to the ever-changing coronal hole structure, there are smaller-scale striated arches, plumes, helmets, and other structures continuously appearing and fading on scales of seconds to days or months. Exactly how this activity gives rise to the fluctuating fields observed by spacecraft has proven to be one of the thorniest questions in heliospheric physics, and remains open despite many efforts. Farther from the Sun, the wealth of observations by both *in-situ* spacecraft and radio-frequency remote sensing have given us a more detailed picture, and thus this region shall be the focus of most of our discussion.

Ignoring for the moment the problem of making high-speed streams, the smaller-scale solar variability will in general give rise to both compressive and shear variations in the flow, corresponding to velocity fluctuations that are, respectively, either irrotational or solenoidal. Much of the compressive component will be rapidly damped, but some of the "fast mode" waves and, probably more importantly, most of the nonpropagating, nearly pressure-balanced compressive structures will survive the trip through the corona into the solar wind. Fluctuations with wave vectors perpendicular to the local mean field may be of the compressive type (transverse pressure balances) or dynamical and incompressive (two-dimensional incompressible turbulence). Both these types of nearly two-dimensional structures may have shear motions associated with them (microstreams) that can stir the flow in its later evolution, and may also have other important effects that are discussed in later sections. The propagating incompressive component of the fluctuations (the Alfvénic fluctuations) are very difficult to damp, and thus are expected to be a major component of what will be observed by spacecraft. The wind flow becomes faster than the Alfvén wave speed ($B/\sqrt{4\pi\rho}$, where B is the magnetic field strength and ρ is the mass density) at a critical point somewhere between 10 and 20 $R_\odot$. A standard view of the critical point is that it is spherically symmetric and static. In this case, any Alfvén waves propagating inward in the solar wind frame observed beyond this point are convected outward and thus must have originated beyond the critical point. This implies that until the wind has had a chance to generate waves *in situ*, the predominant direction of propagation will be outward. Combined with the frozen-in flux relation that is expected over a wide range of scales for the high Reynolds number of the solar wind plasma, the outward propagation implies a definite relationship between the magnetic and velocity fluctuations of the Alfvénic component, and this has provided a powerful diagnostic for the analysis of the solar wind dynamics. A more precise treatment of the critical point may include the possibility that inward type waves coexist with outward waves at or near the critical point, forming a standing or nearly standing wave packet. Some observations suggest that there may be enhanced fluctuations near the critical point which may be associated with interactions of the standing and outward traveling waves.

What arrives at a spacecraft at 0.3 AU or beyond, then, is a mix of Alfvénic fluctuations, convected structures, microstreams, and some propagating compressive structures that have interacted with each other and been influenced by the general nearly spherical expansion of the flow. The expansion is clearly the dominant influence on the amplitude of the fluctuations, and, because the different fields may be affected differently, it may also affect the apparent nature of the mix. Certainly the largest-scale Alfvénic fluctuations will tend to be reflected off the large-scale gradients of the Alfvén speed, and this may be why these scales do not appear to be outward propagating by 1 AU. However, there are also more subtle influences of the expansion such that, for example, the transverse velocity fluctuations associated with convected structures decay much faster than the magnetic fluctuations associated with them even for high wave number structures.

Apart from the expansion, shears between and within streams and the inhomogeneities associated with convected structures interact nonlinearly with the Alfvénic population and with dynamically active two-dimensional fluctuations. Interactions among the Alfvénic fluctuations can produce at least three kinds of couplings. First, they can excite compressive fluctuations due to the variations in magnetic field strength inevitably associated with a superposition of different wave numbers. Second, the Alfvén waves will also interact with each other to the extent that they are propagating in opposite directions; this interaction will get stronger and self-sustaining once some inward waves are generated by reflection or other processes. Third, the Alfvénic fluctuations and the two-dimensional incompressive fluctuations can interact to produce spectral anisotropy in the incompressible component of the flow, and, especially, it can transfer to higher wave numbers perpendicular to the large-scale magnetic field.

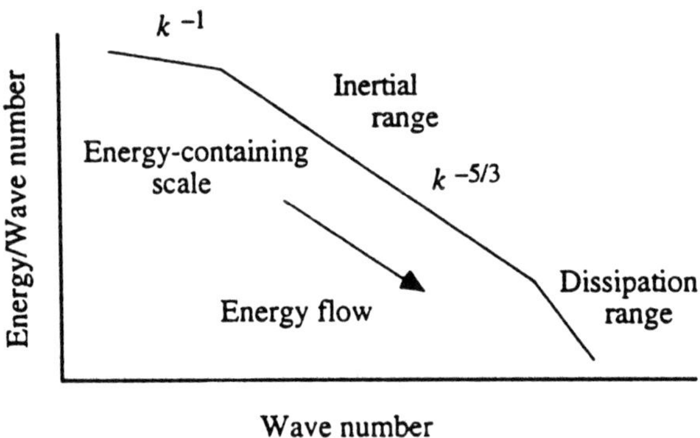

Figure 1.   A schematic representation of a power spectrum of either magnetic fluctuations or fluctuations of the total energy of solar wind fields.

The nonlinear interactions produce a flow of energy in wave number space that is predominantly from large to small scales. This can be described by interactions between neighboring wave number fluctuations, but in configuration space it often is manifested as compressive or shear-driven steepening. The resulting small-scale structures are sometimes localized and coherent (such as electric current and vorticity structures near reconnection zones seen in MHD simulations), and may give rise to intermittency in solar wind turbulence. As indicated schematically in Fig. 1, the spectrum of fluctuations is often, although somewhat artificially, divided into three ranges: the largest, "energy containing" scales, so called because the amplitudes at low wave numbers are large, that provide the reservoir tapped by the cascade; the intermediate "inertial range," characterized by power law spectra, in which the nonlinear, inertial, term in the equations of motion dominates over the dissipation; and the small-scale "dissipation range" in which the fluctuations are finally converted to thermal energy. The flow of energy to different scales is also associated with changes in the degree to which the fluctuations at a particular scale are purely Alfvénic and with changes in a variety of correlations. That dissipation leads to heating of the wind, and turbulence may also play a role in coronal heating. Similarly, the fluctuations may help to accelerate winds; these areas will be touched on briefly below. The solar wind provides a wide range of parameter regimes in which to study turbulent processes, and it is by comparison of solar wind observations with simulations and theoretical models that we have been able to make substantial progress in understanding.

The next two sections provide overviews of the observations and theories of solar wind turbulence. Given this general framework, we then deal with specific issues of current study, including the origin of the fluctuations near the Sun, and the relative roles of expansion, shear, and compression. We also deal briefly with the problem of the dissipation of the turbulent energy, and in this context discuss the issues that must be resolved to determine the role of turbulence in coronal heating and solar wind acceleration. The final section gives a (biased) view of where we stand. We have tried to note all points of view, although the emphasis is at times on work by us or our collaborators; this should be taken as a measure of our limitations, not of relative quality. The presentation focuses on the physical understanding of the processes involved, with outlines of the mathematical details. The references, while not complete, are intended to give the reader a sufficient number of points of contact to trace the history and details of any particular area easily.

II. BRIEF HISTORY AND OVERVIEW OF OBSERVATIONS

A. Early Observations

Even the earliest observations of the solar wind showed that no quantity measured with sufficient accuracy was steady. Not only were there large changes in speed, density, and temperature associated with high-speed "streams," but at all scales there was persistent variability (see Figs. 2 and 3, which will be referred to throughout the chapter). In the late 1960s the formalism of MHD modes began to be applied to these fluctuations, with considerable success. Both propagating MHD modes (Coleman 1966) and convected, nonpropagating structures (Burlaga and Ogilvie 1970) were identified. The Alfvén mode in particular was found to provide a dominant contribution to the fluctuations for substantial periods; this was studied observationally in a paper by Belcher and Davis (1971) (also see Unti and Neugebauer 1968), that is still well worth reading, and it confirmed the ideas of Barnes (1966) who had concluded that the magnetoacoustic modes were generally damped strongly by kinetic effects. A most striking aspect of the observations was that the dominant Alfvén mode was nearly always propagating outward from the Sun. (Note that although the above statement is conventional, a more precise statement would be that the sense of correlation between velocity and magnetic fluctuations is that associated with outward propagation. The nonpropagating two-dimensional structures may also contribute to the observed correlation.)

When power spectra of the magnetic fluctuations were constructed, Coleman (1968) was struck by the similarity of the spectral slopes to those of velocity fluctuations seen often in isotropic, homogeneous fluid turbulence (Kolmogoroff 1941). Similar spectral behavior is predicted for isotropic magnetofluid turbulence (Kraichnan 1965), although it is not obvious how such

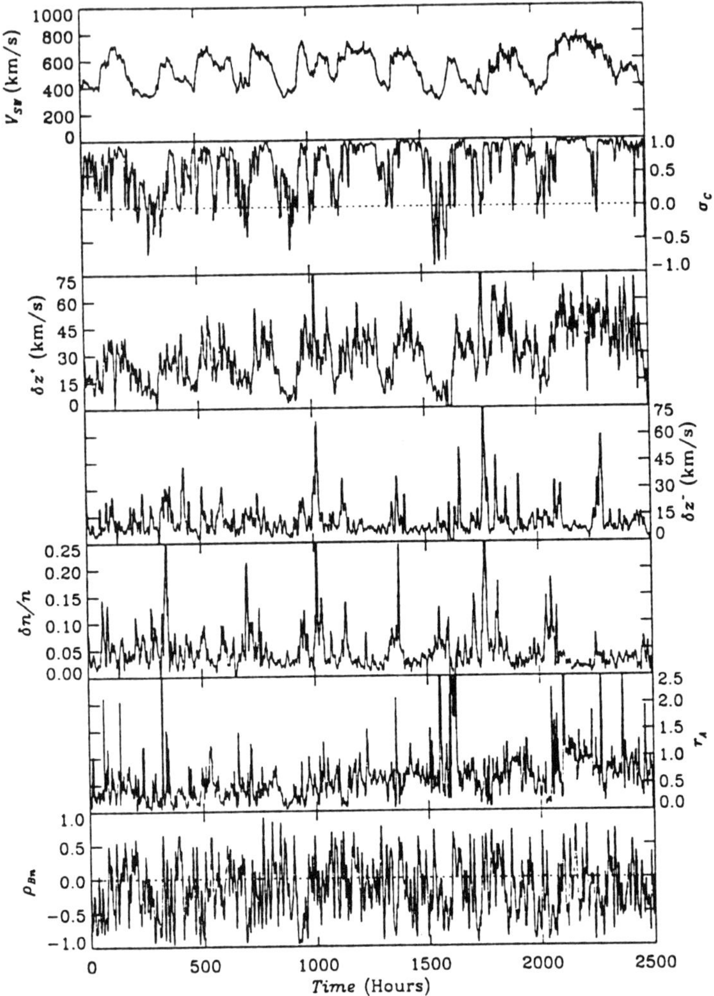

Figure 2.   (a) Large-scale features of about 100 days of hour-averaged data from the primary mission of Helios 2 (early 1976). From top to bottom the panels are: (1) the speed of the solar wind, showing recurrent high-speed streams; (2) the magnitude of the magnetic field; (3) the distance of the spacecraft from the Sun; (4) the radial component of the magnetic field divided by the field strength, showing how the relative fluctuations increase with increasing distance; (5) the density of the wind; (6) the Alfvén ratio of kinetic to magnetic energy at a ten-day scale, showing the dominance of the radial velocity changes; (7) the correlation between density and magnetic field strength.

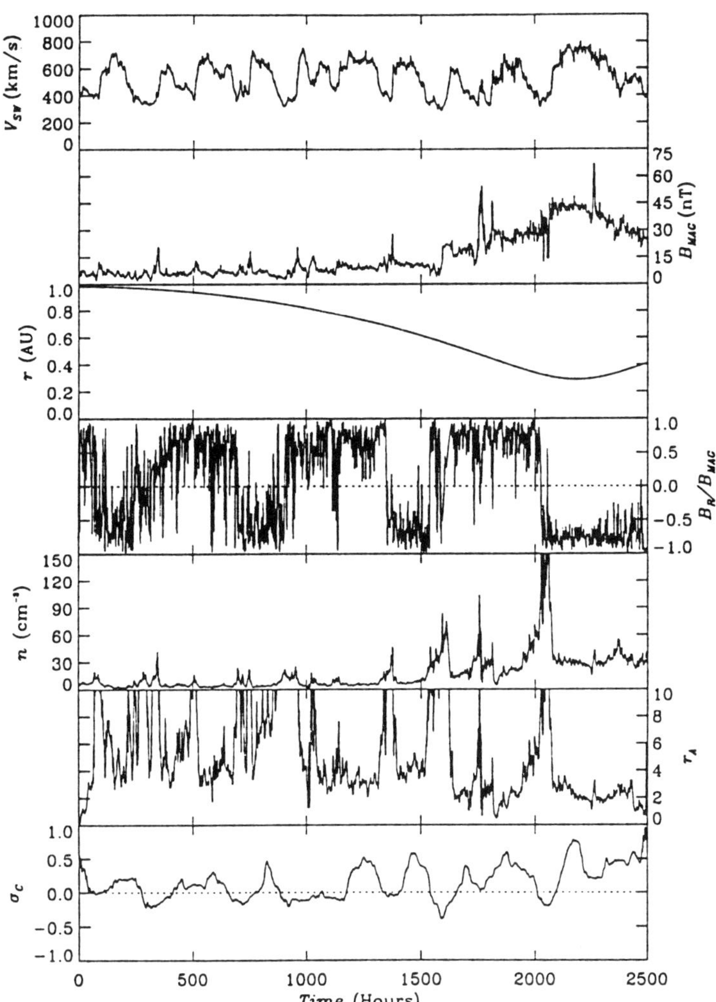

Figure 2.   (b) Small-scale features constructed from the same Helios data set. All fluctuating quantities are at the 3-hr scale, smoothed with a 7-hr window. From top to bottom: (1) the speed of the solar wind (repeated for clarity from Fig. 2a); (2) the normalized cross helicity; (3) the (declining) amplitude of the δz^+ fluctuations; (4) the (relatively constant) amplitude of the δz^- fluctuations; (5) the relative density fluctuations; (6) the Alfvén ratio; (7) the correlation between density and magnetic field strength.

theories, which are based on the assumption of isotropy, are applicable to a magnetofluid with a preferred direction induced by the local mean magnetic field. Coleman suggested that the solar wind was a turbulent and dynamically evolving medium driven by stream-shear instabilities. He argued further that turbulent dissipation at high wave numbers could account for the anomalously high proton temperatures observed in the solar wind at 1 AU. This picture of the solar wind as a turbulent magnetofluid suffered initially from several shortcomings: first, as argued by Belcher and Davis (1971), it was difficult to reconcile turbulent generation by stream-shear with the observed outward propagation of the Alfvén waves; second, in an incompressible and ideal magnetofluid (i.e., one with no dissipation) pure Alfvén waves are a solution of the MHD equations and therefore no nonlinear interactions could arise. Furthermore, it was thought that the presence of the mean magnetic field of the solar wind would damp the shear-driven Kelvin-Helmholtz instability (Parker 1964), thus further suppressing nonlinear, turbulent, interactions. In addition, Bavassano et al. (1978) argued that even if excited, shear instabilities could not produce fluctuations on the large scales observed.

Other mechanisms for generating turbulence *in situ*, including, for example, kinetic instabilities, have also been discarded because they fail to generate outward propagating waves and because they tend to produce waves at higher frequencies and wave numbers than is required to account for the power law inertial range of the fluctuations that extends from about $k = 10^{-12}$ cm^{-1} to 10^{-7} cm^{-1} (Barnes 1979*b*; Bavassano et al. 1978; Hollweg 1975). In light of these difficulties, the turbulence viewpoint espoused by Coleman was not generally accepted and the major source of interplanetary fluctuations was assumed to lie below the Alfvénic critical point in the solar corona as Belcher and Davis had suggested because that naturally accounted for outward traveling waves. Subsequent analysis of the role of the Kelvin-Helmholtz instability in the solar wind by, for example, Korzhov et al. (1984) showed that Parker's analysis, because it ignored oblique wave vectors, was too pessimistic. Also, as conjectured by Southwood (1968), instability might always exist in compressible magnetofluids (see Miura and Pritchett [1982] for a more recent review and discussion). Similarly the case for complete suppression of nonlinearities due to the high Alfvénic correlation in the solar wind also weakens under closer examination; the fluctuations have a high, but not perfect degree of correlation, and MHD simulations (see, e.g., Matthaeus and Montgomery 1984) indicate that initial data with similar levels ($\approx 90\%$) of correlation can still generate broadband power law spectra rapidly.

B. Waves vs Turbulence: More Recent Observations and a Proposed Resolution

It is not possible to distinguish between solar or *in-situ* origins and evolution of interplanetary fluctuations from observations obtained at a single point in the heliosphere. One needs to know how the plasma properties of the solar wind, including proton temperature, evolve with heliocentric distance.

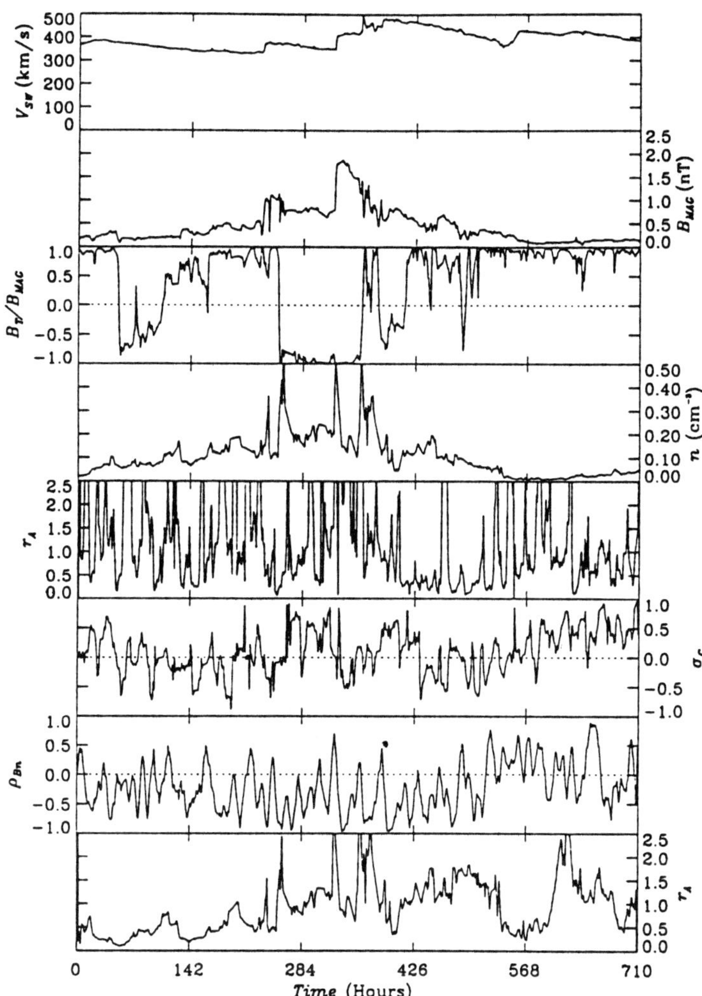

Figure 3.   Hour-averaged Voyager 1 data gathered near 8 AU in mid-1980, along with various derived quantities. From top to bottom: (1) the solar wind speed; (2) the magnitude of the magnetic field; (3) the tangential component of the magnetic field compared to the field magnitude; (4) the density, showing large compressions and a rarefaction toward the end of the interval (see also $|B|$); (5) Alfvén ratio at the 3-hr scale, smoothed with a 7-hr window; (6) the normalized cross helicity at the 3-hr scale, smoothed with a 7-hr window; (7) the correlation between magnetic field and density at the 3-hr scale, smoothed with a 7-hr window; (8) the Alfvén ratio at the 10-day scale, showing the loss of stream dominance in the outer heliosphere.

Fortunately, over the past three decades the launch of several spacecraft has made such studies possible. The excellent particle and fields experiments on the Voyager and Helios spacecraft in particular have permitted studies spanning heliocentric distances from 0.3 to beyond 30 AU.

1. Determining the Global Invariants of MHD: Spectral Studies. Matthaeus and Goldstein (1982*a*) began a systematic study of the extent to which the turbulence model of the solar wind was consistent with observations. Their stated goal was to understand both the magnetic and velocity spectra and their relationship to macroscopic structure. Because compressive effects in the inertial range appeared relatively unimportant, they used the language of incompressible, single-fluid MHD in their initial studies. They emphasized the role of the "rugged" integral invariants of these equations. These invariants are not changed by the truncation of the Fourier representation of the fields. The three known invariants in the absence of a mean magnetic field are the energy (per unit mass)

$$E = \frac{1}{2} \int d^3x (v^2 + b^2) \tag{1}$$

the cross helicity

$$H_c = \frac{1}{2} \int d^3x \, \mathbf{v} \cdot \mathbf{b} \tag{2}$$

and the magnetic helicity

$$H_m = \int d^3x \, \mathbf{A} \cdot \mathbf{B} \tag{3}$$

where $\mathbf{A}$ is the magnetic vector potential defined so that $\mathbf{B} = \nabla \times \mathbf{A}$, $\mathbf{v}$ is the velocity, and $\mathbf{b}$ is the magnetic field in Alfvén speed units, $\mathbf{B}/(4\pi\rho)^{1/2}$. When a mean field is present, the magnetic helicity is modified (Matthaeus and Goldstein 1982*a*). Of particular use in analysis and interpretation of solar wind data is the normalized measure of the reduced cross helicity, $\sigma_c \equiv 2H_c^r/E^r$, and its spectrum $\sigma_c(f) \equiv 2H_c^r(f)/E^r(f)$, where the term "reduced" refers to integrating the spectral tensor over the two directions transverse to the direction of the solar wind velocity so that the spectra are functions of wave number parallel to $\mathbf{V}_{SW}$ only. In this chapter, as in many of the references cited, interpretation of solar wind data is simplified considerably by defining the cross helicity so that positive values always indicate outward propagation from the Sun regardless of the direction of the mean magnetic field, and we shall adopt that convention unless otherwise specified.

Matthaeus et al. (1982) were the first to show how to determine the magnetic helicity and its reduced spectrum from single point measurements of the magnetic field, using a method that depended on the superAlfvénic nature of the solar wind so that the frozen-in-flow assumption held (Taylor 1938). From a single spacecraft, however, it is not possible to obtain a full three-dimensional spectrum of the invariants (Matthaeus and Goldstein 1982b), thus limiting analyses to reduced spectra (Batchelor 1970).

The power spectra of magnetic fluctuations that Coleman constructed could be described, over a substantial range of scales, as a power law with spectral index α somewhere between -1 and -2. One of the goals of early determinations of the power spectrum was to quantify the value of α better; a value of -1 seemed very low because in the absence of dissipation it implied infinite energy content; conversely, $\alpha = -2$ possibly implied a medium dominated by shocks and other discontinuities. Fully developed homogeneous and isotropic fluid turbulence was known to have a value of $\alpha = -5/3$ (Kolmogoroff 1941) while Kraichnan's generalization to MHD (Kraichnan 1965) suggested a value of $\alpha = -3/2$. Matthaeus et al. (1982), using several intervals of Voyager data when the spacecraft was between 1 to 5 AU, were able to show that, at least at times, $\alpha = -1.7\pm0.1$. A sample power spectrum constructed from Mariner 10 magnetometer data from 20 March 1974 when the spacecraft was near 0.5 AU can be seen in Fig. 4. The spectral index of this example is significantly closer to $\alpha = -5/3$ than it is to $-3/2$. Note also that this spectrum shows a break to a steeper dissipation range at about 0.8 Hz, a frequency that can be directly related to the local ion cyclotron frequency and that is thus indicative of ion cyclotron damping. Subsequent analyses by others have tended to verify the $\alpha = -5/3$ result for the inertial range (see, e.g., Marsch 1991).

The (reduced) power spectrum of the cross helicity that Matthaeus and Goldstein (1982a) reported, generally indicated propagation of Alfvénic fluctuations outward from the Sun, consistent with the Belcher and Davis (1971) result, but they also noted that "mixed" cross helicity periods could be found beyond 1 AU. In contrast, the spectrum of the reduced magnetic helicity, which was also a power law with index $\alpha = -7/3$, indicated random fluctuations between positive and negative values throughout the inertial range of the spectrum. This has been confirmed by Goldstein et al. (1994) who found, using many high-resolution intervals, that only the dissipation range showed any tendency for a systematic value of helicity. The latter effect may be associated with ion cyclotron damping.

2. The Alfvén Effect. Kraichnan (1965) had predicted that the magnetic and kinetic energy (per unit wave number) should be equal on average in the inertial range of magnetofluid turbulence. The physical reasoning behind this conjecture can be understood by noting that Alfvénic solutions ($\delta\mathbf{v} = \pm\delta\mathbf{b}$) of arbitrary amplitude are solutions of the incompressible ideal (i.e., dissipationless) MHD equations. A perturbation of $\delta\mathbf{v} \neq 0$ (with $\delta\mathbf{b} = 0$) can therefore be described as an initial condition for two Alfvén wave packets with magnetic fluctuations $+\delta\mathbf{b}$ and $-\delta\mathbf{b}$, respectively. Because the packets are expected to propagate away from each other with speeds $\pm B_0/(4\pi\rho)^{1/2}$, where $\mathbf{B} = \mathbf{B}_0 + \delta\mathbf{B}$ ($\mathbf{B}_0 =< \mathbf{B} >$), approximate equipartition should result. A measure of this is the "Alfvén ratio" $r_A = E_V(k)/E_B(k)$, where $E_V(k)$ and $E_B(k)$ are the reduced spectra of the kinetic energy and magnetic energy,

Mariner 10 1974:79:12 (0.12 sec)

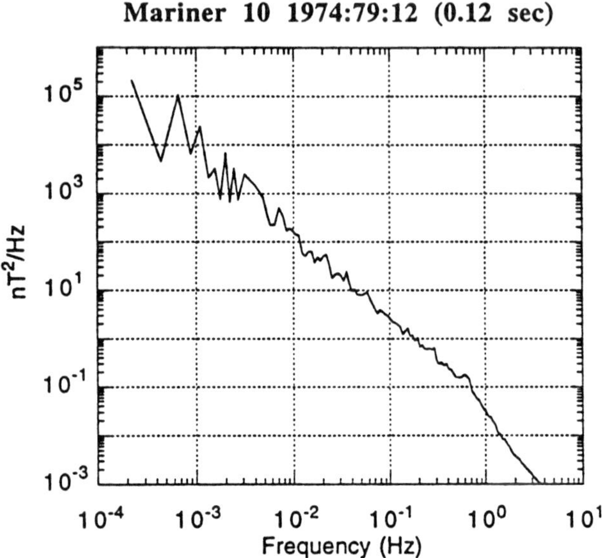

Figure 4.   The trace of the power spectral matrix of the magnetic field fluctuations from a 74-min interval of 0.12-s average Mariner 10 magnetometer data obtained on 20 March 1974. Note that the inertial range has a spectral slope very close to $-5/3$ and a clearly defined dissipation range is apparent at the highest frequencies. This time series is too short to see the energy containing scales.

respectively. Although some evidence for such an equipartition was reported in two-dimensional simulations (Fyfe and Montgomery 1976), solar wind observations show that average values of r_A computed for fluctuations in the inertial range of the spectrum are generally between 0.4 and 1.0 throughout the heliosphere (Matthaeus and Goldstein 1982a; Roberts 1992; Roberts et al. 1987a,b,1989,1990,1992). The behavior of r_A in the heliosphere is illustrated in Figs. 2 and 3. At the large $\sim$10-day scale, r_A is dominated by the large fluctuations in the radial component of the solar wind velocity (Fig. 2a), even within streams, while at the smaller scales illustrated in Fig. 2b at $\sim$0.3 AU, $r_A \approx 1$, but by 1 AU it decreases to $\sim$0.5. In the outer heliosphere, as seen by Voyager 1 at 8 AU (Fig. 3), despite brief spikes (some of which are associated with data gaps), r_A still has a most probable value $\sim$0.5.

   3. Elsässer Variables.   We have used "Alfvénicity" to refer to the degree to which the interplanetary fluctuations resembled outward propagating Alfvén waves, i.e., a medium with high cross helicity. This is somewhat confusing because a fluid with Alfvén waves propagating both parallel and anti-parallel to a mean magnetic field would have a small cross helicity and would not be "Alfvénic" in the sense used above. To better describe the behavior of inward and outward propagating fluctuations, the Elsässer variables (Elsässer 1950,1956) are often used. They are defined by $\mathbf{z}^{\pm} = \delta\mathbf{v} \pm \delta\mathbf{b}$, where z^+ and z^- refer to waves propagating "outward" and "inward" with respect to $\mathbf{B}_0$. Note that in an expanding medium such as the solar wind, outward propagating modes in fact have a minor contribution from z^-, but we will neglect this subtlety. Elsässer variables are most useful for describing incompressible MHD because in the ideal, nonexpanding, limit either z^+ or $z^- = 0$ is an exact solution. When $z^\pm$ are nonzero they are coupled nonlinearly. Kraichnan (1965) constructed his arguments for equipartition and the $\alpha = -3/2$ spectrum in terms of these variables. Dobrowolny et al. (1980a,b) used Elsässer variables to argue that solar wind fluctuations will have increasing cross helicity if they are predominantly of one sign initially (dynamic alignment). Later, Goldstein et al. (1986) showed how the variables could be used to correct a systematic bias toward outward propagation in the interpretation of solar wind data in low Mach number flows. The Elsässer formalism was subsequently generalized for the compressible MHD equations by Marsch and Mangeney (1987), and the variables have proven useful in trying to ascertain the relative roles of velocity shear and compressibility in driving interplanetary turbulence (Marsch 1991; Roberts et al. 1991; Roberts and Goldstein 1991).

An example of the general behavior and evolution of the fluctuations in $z^\pm$ is shown in Fig. 2b. The quantities δz^+ and δz^- are measures of the relative behavior of the Elsässer variables obtained by subtracting 3-hr averages of the vector components of the $z^\pm$ from the total vectors, computing the magnitude of the resulting variations, and then plotting the 7-hr average of the result. What is noteworthy in Fig. 2b is that in the inner heliosphere δz^+ is large, but decreases as one approaches 1 AU. In contrast, δz^- is relatively small and constant from 0.3 to 1 AU. This tendency of δz^+ to evolve rapidly in the inner heliosphere, approaching z^- by 1 AU is also reflected in the spectra of these quantities (Fig. 5). In highly Alfvénic regions near the Sun, z^+ (outward) has a flat spectrum at low frequencies and $\alpha \approx -5/3$ at high frequencies while z^- shows nearly the reverse behavior (Fig. 6). This tendency may be qualitatively understood in terms of the "minority species effect" developed by Matthaeus and Montgomery (1984) to explain dynamic alignment, and is also predicted in closure calculations (see, e.g., Grappin et al. 1983). Farther out, the spectra of both δz^+ and δz^- tend to be nearly parallel except at low frequencies where they merge, as illustrated in Fig. 5.

4. The Spectrum of Density Fluctuations. One of the puzzles of MHD turbulence in general and solar wind observations in particular is trying to understand the observed power spectrum of the density. Whether computed from *in-situ* solar wind observations (Goldstein and Siscoe 1972) or using radio scintillation techniques to probe the interplanetary (Armstrong et al. 1990; Scott et al. 1983; Woo and Schwenn 1991; Woo and Armstrong 1979) or interstellar medium (Armstrong et al. 1981; Higdon 1984), the power spectrum of density fluctuations has the same "Kolmogoroff" $\alpha = -5/3$ spectral slope as do components of the magnetic field and velocity, although the spectra of density, temperature, z^-, and magnetic field magnitude all tend to flatten at

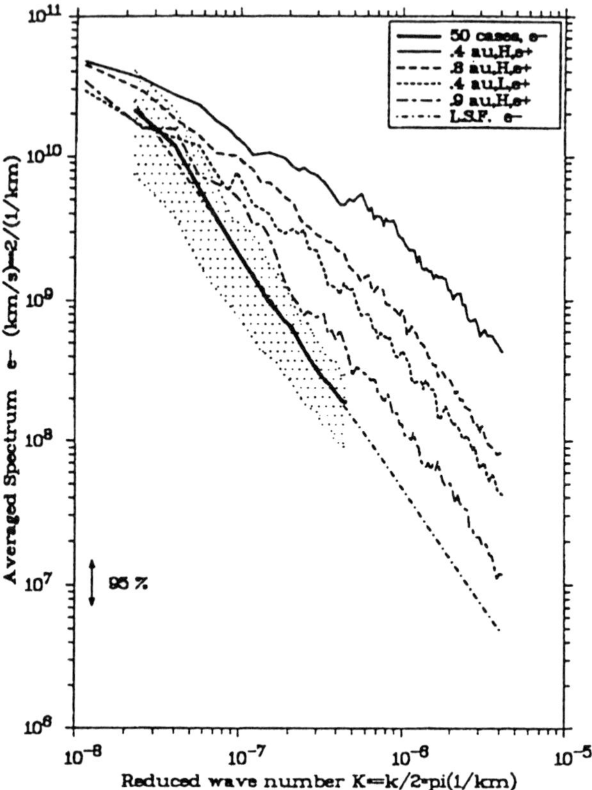

Figure 5. The evolution of the Elsässer variable spectrum of z^+ from 0.4 to 1 AU plotted along with a least-squares fit to 50 z^- spectra (bottom dash-dot line) and the spectral range of the 50 spectra indicated by the dappled band (figure from Tu and Marsch 1990a).

high wave numbers near the Sun in Alfvénic streams.

5. *Stationarity: The Validity of Spectral Analysis.* The construction of a power spectrum assumes that the medium being studied is either stationary or spatially homogeneous (or both). For example, the theory of pitch-angle scattering of energetic particles in magnetic turbulence (see, e.g., the review by Fisk [1979] and theories of field line random walk [Jokipii 1971]) utilize ensemble averages of the magnetic field. Implicit in these descriptions of particle propagation is the assumption that the ergodic theorem holds, i.e., averaging individual representations of the medium from an ensemble is equivalent to averaging over stationary time series. Power spectra constructed from time series mimic the ensemble averaging procedure by smoothing over a large number of spectral estimates.

That the solar wind is often stationary was shown by Matthaeus and Goldstein (1982b) who demonstrated that time intervals ranging from days to

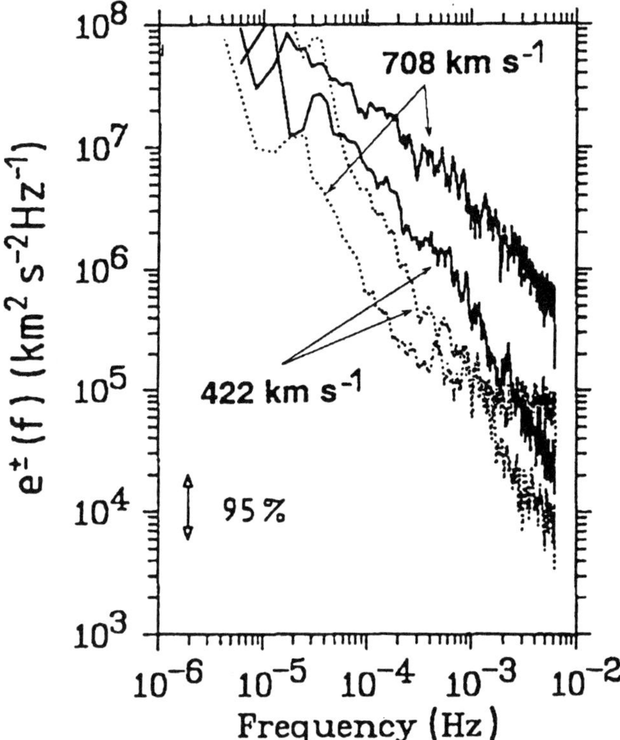

Figure 6.   The Elsässer variable spectra for high (solid line) and low (dotted line) speed streams from the Helios data described in Fig. 2 (figure adapted from Marsch and Tu 1990*a*).

years did indeed satisfy the conditions of "weak" stationarity when the effects of solar rotation were included, thus justifying the use of power spectra to describe interplanetary phenomena. Using their technique for selecting stationary intervals, Matthaeus et al. (1986) constructed an interplanetary field ensemble from data collected near 1 AU and used the ensemble to determine various statistical properties of the interplanetary magnetic field, including the correlation time and length of the magnetic field ($\approx 4.9 \times 10^{11}$ cm), the average direction of the field, etc., along with the variances of those quantities. One consequence of this type of analysis is that, for the kind of spectra observed in the solar wind, quantities that depend sensitively on low-frequency power, such as $\delta B/B$ or the correlation scale, are not well determined and depend upon the duration of the averaging period.

6. *The Evolution of Cross Helicity.* Bruno et al. (1985) using Helios magnetometer and plasma data and Roberts et al. (1987*a,b*), using both Helios and Voyager data, examined the global heliospheric evolution of the normalized reduced cross helicity. Both Bruno et al. and Roberts et al. showed that there was a significant reduction in the Alfvénicity of long period ($\gtrsim 10$ hr)

fluctuations between 0.3 and 0.87 AU. This is illustrated in Fig. 2a which shows a clear decrease in the large-scale large values of $\sigma_c > 0$ from large positive values to nearly random small values as Helios moved from 0.3 to near 1 AU. (The data set comprise about 100 days of hour-averaged data from the primary mission of Helios 2 in early 1976.) A similar evolution is apparent in the smaller-scale σ_c values (Fig. 2b) which are strongly positive near 0.3 AU, but show a gradual radial decline as well as low values in "structured" regions at all distances. Note that at neither scale is this decrease in Alfvénicity accompanied by any significant increase in negative values of σ_c as might be expected if sufficient time had elapsed for the faster evolving small amount of z^- fluctuations ("minority species" effect) to become important (Matthaeus and Montgomery; Grappin et al. 1983). Similarly, the highly Alfvénic character of many intervals near 0.3 AU down to scales longer than the transit time to the spacecraft indicate that no reservoir of large-scale z^- fluctuations is present as suggested by Matthaeus et al. (1984). Roberts et al. (1987a,b) concluded that the solar atmosphere had to be the main source of the outward wave flux at 0.3 AU.

The same data set provides some evidence that regions of strong shear accompanied a rapid decrease in σ_c from a nearly pure Alfvénic state to a more mixed state (see Bavassano and Bruno 1989). This is especially clear in Fig. 2b where comparison of the top two panels (V_{SW} and σ_c) shows that rapid decreases of σ_c are associated with sharp changes in V_{SW}. Regions of nearly pure outward propagating Alfvén waves are associated with small velocity gradients and not necessarily with the trailing edges of high-speed streams. That this is also true in the outer heliosphere can be seen in Fig. 3, which shows that near the end of the interval, in a rarefaction region which has small velocity gradients and has been relatively undisturbed in its journey to 8 AU (Whang and Burlaga 1985), σ_c is still quite large and positive.

Roberts et al. (1987a,b) also argued that the degree of Alfvénicity was nearly independent of the degree of compression in the wind, being only slightly less in compression regions than in stream rarefaction regions (cf. Fig. 2), although, as shown in Fig. 2b the relative density fluctuations do show some correlation with z^- fluctuations. It is also clear from Figs. 2 and 3 that interaction of fast and slow streams to form compression regions is not associated with an increase in negative values of σ_c.

7. *Additional Analyses Supporting the Turbulence Viewpoint.* Pressure balance structures are another feature of the solar wind that vary markedly with heliocentric distance. The strongest indicator of the evolution is that the percentage of the time that the correlation between $|\mathbf{B}|$ and $\delta\rho$ is less than -0.8 increases with distance (cf. Figs. 2 and 3), indicating that the compressive component of the fluctuations becomes dominated by pressure balance structures in the outer heliosphere (Roberts 1992). Recent studies of the evolution of large-amplitude Alfvén waves (Agim et al. 1995; Ghosh and Goldstein 1994a,b; Roberts and Wiltberger 1995) suggest that the formation of pressure balance structures plays an important role in the long-time evo-

lution of large-amplitude Alfvén waves such as are found in the solar wind. Strong anti-correlations exist in the filamentary, structured regions in the inner heliosphere (see, e.g., Klein et al. 1993a; Fig. 2b). Anticorrelations of the same type as are expected in pressure balanced structures also appear in nearly incompressible solutions to the MHD equations (Matthaeus et al. 1991).

In the absence of any sources of heating, the solar wind temperature should decrease adiabatically due to the expansion of the wind. Coleman (1968) noted that in turbulent MHD flows, the energy supplied at large scales dissipates at small scales (see, e.g., Marsch 1990; Zhou and Matthaeus 1990a). The details of the decrease in temperature T with distance r depend on the type of flow, but generally the decrease in temperature is slower than expected from adiabatic cooling alone (Freeman 1988; Freeman et al. 1992; Lopez and Freeman 1982; Schwenn 1983), suggesting that heat must be added to the fluid either from turbulent processes or some other source such as heating by shock waves. Adiabatic cooling always dominates because the density decreases as r^{-2}. Therefore, if the polytropic index of the fluid is approximately 5/3, the adiabatic temperature variation is $T(r) \propto r^{-4/3}$, which is too strong to be overcome by either turbulent or shock heating. The observed power law index for the temperature in the inner heliosphere ranges between -0.78 and -1.33. The most recent studies indicate that there is little difference between the behavior of the temperature with distance between fast and slow streams (Marsch and Richter 1987; Freeman et al. 1992). Gazis (1984) studied the temperature evolution in the outer heliosphere using Voyager 1 data and found that various velocity streams had approximately similar temperature gradients, although the faster ones were hotter. He concluded that $T \propto r^{-0.7}$ for all streams (see, e.g., Tu 1987). All of these studies suggest that additional heating sources must be present in the solar wind. However, as will be discussed in more detail below, it is not yet known whether heating arising from turbulent dissipation is sufficient to account for the observations.

A comprehensive account of solar wind phenomenology should also explain the observed tendency for the direction of minimum variance of magnetic field fluctuations to lie along the direction of the mean magnetic field. This property of the interplanetary field was first reported by Belcher and Davis (1971) using data obtained near 1 AU, and confirmed more recently using data from the outer heliosphere (Klein et al. 1991; Parker 1980). A clear trend is present as one moves outward in the heliosphere: fluctuations in velocity, which initially are more isotropic than the magnetic fluctuations, show little evolution with distance, while magnetic fluctuations become less anisotropic and by 2 AU have the same 3:1 anisotropy (the ratio in power transverse to the minimum variance direction to that along it) as do the velocity fluctuations. Most curious, however, is that while the minimum variance direction of the magnetic fluctuations lies along the average field direction, the minimum variance direction of the velocity fluctuations is more nearly aligned with the radial direction (Klein et al. 1991). There has been considerable controversy about the interpretation of these results (Denskat and

Burlaga 1977; Solodyna and Belcher 1976) because the observations are at variance with the predictions of WKB theory (Hollweg 1975; Völk and Alpers 1973). There have also been more recent studies using inner heliosphere data, and the reader should see the work of Bavassano et al. (1982) and Klein et al. (1993b).

Although we have known since the early work of Belcher and Davis (1971) that the solar wind fluctuations are not isotropic, the assumption of isotropy underlies some ways of viewing solar wind turbulence. For example, Kolmogoroff's dimensional argument leading to the 5/3 power law was made originally for isotropic fluid turbulence. To determine the symmetry properties of the fluctuations, we must specify the three-dimensional form of the correlation tensor, or spectral tensor, of the magnetic or velocity fluctuations. This, however, has not been done and models of such phenomena as energetic particle propagation in the interplanetary medium have had to assume a form for the spectral tensor. Three models are commonly used: the "slab," in which all nonzero spectral amplitudes are associated with wave vectors $\mathbf{k}$ parallel to an ambient mean magnetic field; the isotropic model, in which wave vectors with the same magnitude all have equal spectral amplitudes; and the quasi-two-dimensional model in which wave vectors are nearly perpendicular to the large-scale magnetic field $\mathbf{B}$ while the magnetic fluctuations are orthogonal to both $\mathbf{k}$ and $\mathbf{B}$.

Evidence that quasi-two-dimensional turbulence exists was found in several laboratory experiments (Robinson and Rusbridge 1971; Robinson et al. 1968; Rusbridge 1969; Zweben et al. 1979; Zweben and Taylor 1981). In turn, the experiments motivated development of a theory of turbulence (Montgomery 1982) based on the Strauss equations (Strauss 1976), which describe an incompressible model of MHD in the presence of a strong mean magnetic field. The tendency of magnetic fluctuations to become highly anisotropic when a strong magnetic field is present has been confirmed in a series of numerical simulations (Shebalin et al. 1983; Goldstein et al. 1987,1989; Matthaeus and Lamkin 1985; Roberts et al. 1992; Oughton et al. 1994).

The occurrence of strong anisotropies in laboratory magnetofluids and in simulations suggests looking for similar anisotropies in the solar wind, which is difficult to do using data from a single spacecraft (but see Sari and Valley 1976). Some indirect information of the symmetry of the turbulence has been obtained from radio scattering observations (Armstrong et al. 1990; Woo and Armstrong 1979) which indicate that density spectra near the Sun are highly anisotropic with the irregularities stretched out approximately along the radial direction. Matthaeus et al. (1990) took long stationary data intervals, constructed an ensemble, and used it to estimate the two- and three-dimensional correlation function. Their analysis used 5-min averages of 16 months of nearly continuous ISEE 3 data, but even the two-dimensional correlation function was too sparse to transform into a spectrum. Nonetheless, they presented fairly strong evidence that the ensemble of data was neither "slab-like" Alfvén waves nor isotropic. Rather, the correlation function consisted of two populations: fluctuations with large correlation lengths transverse to $\bar{\mathbf{B}}$ (Alfvénic) and fluctuations with large correlation lengths parallel to $\bar{\mathbf{B}}$ (quasi-two-dimensional). ($\bar{\mathbf{B}}$ is the direction of the average magnetic field in the data interval used for the correlation calculation.) At small separations, $r \leq 5 \times 10^{10}$ cm, Alfvénic, or slab, symmetry appeared dominant, whereas contributions from a quasi-two-dimensional component became noticeable for $r \geq 15 \times 10^{10}$ cm.

These results indicate that the solar wind at 1 AU contains a population of Alfvénic fluctuations, originating in the solar corona, along with a quasi-two-dimensional component that could be evidence of turbulent evolution between the Sun and 1 AU. Because anisotropy appears to be a feature of turbulent evolution, Matthaeus et al. predicted that quasi-two-dimensionality should become more important in the outer heliosphere—a prediction that has yet to be tested. In quasi-two-dimensional turbulence, pitch-angle scattering by resonant wave-particle interactions is suppressed, as observed (see, e.g., Kunow et al. 1991). Bieber et al. (1994) calculated cosmic ray mean free paths based on quasi-linear theory using a dynamic turbulence model and a mixture of quasi-two-dimensional and slab magnetic fluctuations. Their results indicate reasonable agreement with observed particle mean free paths using such a model. In addition, in quasi-two-dimensional turbulence, the direction of minimum variance of magnetic fluctuations is parallel to $\mathbf{B}_0$, as is also observed.

C. Coherent Structures Observed at 20 to 25 AU

Although the concept of turbulence usually brings to mind a chaotic or disordered medium, there are phenomena that can arise in high Reynolds number flows that are both highly ordered and long lived. An example from fluids is the "Karman vortex street" (Karman 1911), seen in the wake of flow past a solid body, such as an island. Abernathy and Kronauer (1962) showed that under certain conditions two neighboring vortex sheets could create a vortex street without the need for a solid body. The relevance of this phenomenon to solar wind observations is the report by Lazarus et al. (1988) and McNutt (1986) that between 20 and 25 AU Voyager 2 at solar minimum observed an extended interval of quasi-periodic meridional flow having a period of ≈ 25.5 days (one solar rotation). The flow deflected the solar wind by $\approx 6°$. A distinguishing feature was the presence of two maxima in the bulk speed per solar rotation, but only a single pressure maximum at the beginning of the bulk velocity two-peaked structure. Veselovsky (1990), Veselovsky and Triskova (1990), and Burlaga (1990) proposed that the observations were consistent with a vortex street produced by the two vortex sheets lying on either side of the solar minimum heliospheric current sheet. In the inner heliosphere, the mean magnetic field has a sufficiently large component along the direction of the solar wind velocity vector to inhibit the formation of a vortex street, but at 20 AU the mean magnetic field, now transverse to the solar wind flow, makes excitation of a vortex street possible. Recently, Siregar et al. (1992,1993,1994,1995) investigated the conditions under which a vortex street can arise in compressible MHD in two and three dimensions and concluded that excitation of such a street appears possible in the outer heliosphere. Pizzo (1994a), however, has suggested that the observations may be understood without formation of a vortex street; we defer further discussion to a later section (IV.E).

III. THEORETICAL AND SIMULATION MODELS OF TURBULENCE

A. Applicability of MHD for Description of Solar Wind Phenomena

Solutions of the MHD equations have proven invaluable in understanding many aspects of the turbulent evolution of fluctuations in the solar wind. In the following sections we discuss problems that have been tackled using these tools, including radial evolution of fluctuation amplitudes and temperature, the role played by velocity shear in driving the evolution of magnetohydrodynamic turbulence in the solar wind, the role of compressibility and compressive structures, and the role of decay-like processes. Incompressible MHD is, perhaps, the simplest mathematical model that may be capable of describing important aspects of solar wind phenomenology (Montgomery 1983). Although a surprising amount of progress in understanding solar wind evolution has been made using this model, it is clear that a complete description of solar wind evolution and dynamics must include compressive effects.

Kinetic effects are clearly important in the solar wind—the heat flux, for example, arises from the long collision mean free path of electrons in the solar corona and thus cannot be described readily by a local set of fluid equations. Our present theoretical and computational capabilities are inadequate to describe the inertial scale or energy-containing scales of solar wind fluctuations using kinetic theory, and the questions we are addressing here, namely to account for the physical processes that drive the evolution of the solar wind fluctuations, require a more coarse-grained fluid approach. Even that approach is too ambitious, however, because to model accurately the overall global aspects of the solar wind with the complete set of equations that describes all the relevant physical properties of the system and its boundaries would exceed available computer resources. Such an approach could, of necessity, only be approximated by limiting the resolution and accuracy of solutions and by parameterizing subgrid-scale phenomena, forfeiting realistic description of small-scale fluctuations.

The approach most frequently taken is to approximate the complete set of physical equations with a simplified set of model equations—here, MHD—which contain essential elements of the physics. These equations permit accurate solution, but the physical dissipation at small scales is still described using simplified parameterizations. Using MHD permits modeling of phenomena larger than the kinetic scale of wave-particle interactions. The magnetofluid equations themselves are nonlinear and are too complicated to permit analytical solution; nonetheless, current numerical methods and computers permit highly accurate solutions of turbulence.

B. MHD Equations

The MHD equations can be derived by taking moments of the Boltzmann or Vlasov equations, noting that there are no contributions to the moment equations from the Coulomb collision term (Montgomery and Tidman 1964), and combining the result with Maxwell's equations assuming quasi-neutrality. One further assumes that the relevant time scales are longer than the proton Larmor period and that the relevant length scales are longer than the proton Larmor radius. The moment equations themselves form an infinite hierarchy that must be closed by additional assumptions that imply either particle-particle or particle-wave collisions. We achieve closure either by using an equation of state after the second moment or an energy equation. The equations are then characterized by eight fields: the density ρ, velocity $\mathbf{v}$, temperature T, and magnetic field $\mathbf{b}$. In dimensionless units the resulting (compressible) MHD equations are continuity

$$\frac{\partial \rho}{\partial t} = -\nabla \cdot (\rho \mathbf{v}) \tag{4}$$

momentum

$$\frac{\partial \mathbf{v}}{\partial t} = -\mathbf{v} \cdot \nabla \mathbf{v} + \frac{1}{\rho} \left(\mathbf{j} \times \mathbf{B} - \nabla p + \nabla \cdot \underline{\phi} \right) \tag{5}$$

Maxwell's equations

$$\frac{\partial \mathbf{B}}{\partial t} = -\nabla \times \mathbf{E}$$

$$\mathbf{j} = \nabla \times \mathbf{B} \tag{6}$$

$$\nabla \cdot \mathbf{B} = 0$$

together with Ohm's Law

$$\mathbf{E} = -\mathbf{v} \times \mathbf{B} + \eta \mathbf{j} + \frac{\varepsilon}{\rho} (\mathbf{j} \times \mathbf{B}) \tag{7}$$

and either a polytropic equation of the form $p = \rho^{\gamma}$, or an energy equation with the ideal gas law, which yield an equation for the temperature evolution. In Eqs. (4) through (7), η is the resistivity, ε is the Hall parameter, important as frequencies approach the ion cyclotron frequency, and ν is the viscosity. The tensor $\underline{\phi} = \nu[\nabla \mathbf{v} + (\nabla \mathbf{v})^{T}] - \frac{2}{3}\nu(\nabla \cdot \mathbf{v})\underline{\mathbf{I}}$ is the hydrodynamic viscous stress tensor where $\underline{\mathbf{I}}$ is the unit tensor and ν is the dimensionless kinematic fluid viscosity. We ignore anisotropies in the dissipation and the bulk viscosity. In the results discussed here, we used a polytropic closure. In the incompressible limit (ρ = const), and neglecting the Hall term, Eqs. (4) through (7) become:

$$\frac{\partial \mathbf{v}}{\partial t} + \mathbf{v} \cdot \nabla \mathbf{v} = \mathbf{b} \cdot \nabla \mathbf{b} - \tfrac{1}{2}\nabla b^2 + \nu \nabla^2 \mathbf{v} - \nabla p$$

$$\frac{\partial \mathbf{b}}{\partial t} + \mathbf{v} \cdot \nabla \mathbf{b} = \mathbf{b} \cdot \nabla \mathbf{v} + \eta \nabla^2 \mathbf{b} \qquad (8)$$

$$\nabla \cdot \mathbf{v} = \nabla \cdot \mathbf{b} = 0.$$

The unit of time is the "eddy turnover time" for a unit velocity field of unit length scale.

As illustrated below, often much of the physical behavior of the system can be obtained by ignoring compressibility and/or by reducing the number of dimensions to one or two.

C. MHD Turbulence: Definitions and Comparison to Other Viewpoints

The central issue separating the "wave" and "turbulence" viewpoints is the role of nonlinear interactions. The linearized, compressible MHD equations yield three traveling modes (see Barnes [1979a], for a good review):

1. The slow or acoustic mode that is almost a pure sound wave when propagating along the mean magnetic field;
2. The Alfvén, or intermediate, or shear Alfvén mode that always moves with a constant group velocity along the magnetic field and is noncompressive;
3. The fast or "compressional Alfvén" mode that combines magnetic and thermal pressure in phase to produce a higher propagation speed.

In addition, the equations admit solutions for nonpropagating structures in which the magnetic and thermal pressures balance.

In the limit of thin structures these are called tangential discontinuities. The slow and fast modes are strongly damped except for some special directions of propagation of the fast mode (Barnes 1966); this is presumably what leads to the dominance of the Alfvén mode in the observed fluctuations.

Compressive nonlinear interactions couple the Alfvén mode with the compressive traveling modes such that the beats between Alfvén waves of different wave number and/or propagation direction give rise to $|\mathbf{B}|$ fluctuations that drive density fluctuations and, through the equation of state or an energy equation, couple to thermal pressure variations. For the perturbation of a large amplitude, circularly polarized Alfvén wave, this coupling can be described by a linear "parametric" instability (Viñas and Goldstein 1991a,b, and references therein). The damping of these acoustic modes leads to an indirect dissipation of the Alfvén wave known as nonlinear Landau damping. We can also think of the parametric process as scattering of the original Alfvén wave by small-scale density fluctuations to produce backward traveling Alfvén waves; this process may be a source of "reflected" waves (Marsch and Tu 1993). Any mechanism that produces backward propagating Alfvén waves is of potential importance in the generation of turbulence because it is the interaction of Alfvén waves propagating in opposite directions that forms the nonlinear cascade in incompressive MHD (see below). For broadband initial spectra, it is the beats between pre-existing Alfvén waves that are the dominant source of density fluctuations (Agim et al. 1995; Montgomery et al. 1987; see also Umeki and Terasawa 1992). Thus the specific linear parametric growth mechanism may not be of particular importance in the generation of solar wind turbulence.

Most of MHD turbulence theory has dealt with the incompressive interaction of Alfvénic fluctuations propagating in opposite directions. The momentum and induction equations in terms of the Elsässer variables are

$$\frac{\partial \mathbf{z}^{\pm}}{\partial t} = -\mathbf{z}^{\mp} \cdot \nabla \mathbf{z}^{\pm} \pm \mathbf{V}_A \cdot \nabla \mathbf{z}^{\pm} - \frac{1}{\rho} \nabla p . \tag{9}$$

Note that if either $\mathbf{z}^{+}$ or $\mathbf{z}^{-}$ is zero, there are no nonlinear interactions. The nonlinear terms serve to couple different scales or wave vectors.

In fluid turbulence, neighboring wave numbers couple most strongly, leading to a conservative transport of energy in wave number space that is only stopped by the formation and dissipation of very small structures (see Fig. 1). The Reynolds number, $R_e = UL/\nu$, measures the relative strength of the nonlinear and viscous stresses at the scale L of the correlation length of the fluctuations. As R_e increases, the scale for dissipation decreases such that the rate of energy cascade through the inertial range (where the nonlinear terms are dominant) is on average equal to the small-scale dissipation rate in steady state. The cascade rate is assumed to be independent of R_e, and this assumption yields the well-verified Kolmogoroff spectrum. In particular, assuming the dissipation rate ϵ is equal to the cascade rate and that the latter only depends on the energy per wave number E_k and the scalar wave number $k = |\mathbf{k}|$ for an isotropic shell in the k-space, then dimensional analysis gives

$$E_k = C_K \epsilon^{2/3} k^{-5/3} \tag{10}$$

where C_K is a universal "Kolmogoroff constant." To complete this picture, the large, energy-containing scales may either be stirred at a rate that determines the level of the fluctuations by equating the inertial range flux to the energy input rate, or they may be freely decaying. In the latter case a quasi-equilibrium can be set up with the input into the inertial range given by, for example, the Kolmogoroff rate of the decay of the large-scale eddies, U^3/L (Batchelor 1970; Hollweg 1986). The application of quasi-equilibrium theory to the solar wind has recently been developed by Matthaeus et al. (1994).

The situation for a magnetized plasma is more complex than the fluid case, with both the mean magnetic field and the cross helicity having possible strong effects on the cascade rate (Grappin et al. 1983; Matthaeus and Zhou 1989; Zhou and Matthaeus 1990a; Tu and Marsch 1990b). The strength of the background magnetic field controls the propagation of the waves and may also produce spectral anisotropies, which we ignore in the present simplified discussion. Strong fields lower the interaction time of the eddies (Kraichnan 1965), as can be seen as follows: If τ_{NL} $(= 1/kv_k = 1/(k^3 E)^{1/2})$ denotes the characteristic time taken for the nonlinear term to change the velocity substantially in the absence of magnetic fields, and if one assumes that the field is strong, then Alfvénic propagation limits the interaction time to τ_A. It would take τ_{NL}/τ_A coherent interactions each of time τ_A to produce the same change as τ_{NL} without the field, but if we suppose that each interaction is independent of the others, it will take $(\tau_{NL}/\tau_A)^2$ interactions. This means the time for spectral transfer will be increased from τ_{NL} to $\tau_s \equiv (\tau_A)(\tau_{NL}/\tau_A)^2 = \tau_{NL}^2/\tau_A$. A more general case can be obtained by including a nonlinear time for each of $z^{\pm}$, and by replacing τ_A in the above argument with τ_3, the triple decorrelation time arrived at by taking the sum of the rates for nonlinear and Alfvénic processes (Kraichnan 1965; Matthaeus and Zhou 1989; Pouquet et al. 1976); thus,

$$(\tau_3^{\pm})^{-1} = \tau_A^{-1} + (\tau_{NL}^{\pm})^{-1} \tag{11}$$

and

$$\tau_s = \tau_{NL}^2/\tau_3 . \tag{12}$$

Now $\tau_A = 1/(kV_A)$ and $\tau_{NL}^{\pm} = 1/(k^3 E^{\mp})^{1/2}$ because it is z^+ that convects z^- and visa versa. Combining the above gives the rate of spectral transfer

$$\frac{1}{\tau_s^{\pm}} = \frac{k^2 E^{\mp}}{V_A + (kE^{\mp})^{1/2}} \tag{13}$$

and dissipation rates

$$\epsilon^{\pm} \equiv \frac{(z_k^{\pm})^2}{\tau_s^{\pm}} = \frac{k^3 E_k^+ E_k^-}{V_A + (kE^{\mp})^{1/2}} . \tag{14}$$

In this phenomenology, cascades are slowed both by a strong external field and high cross helicity, and the cascade rates of z^+ and z^- are not equal as they are if the cross helicity is ignored. The latter case was considered by Kraichnan (1965), and gives a $k^{-3/2}$ energy spectrum.

These considerations apply to steady state turbulence, whereas the previous sections have documented that significant spectral evolution occurs in the solar wind. The simplest model for near steady state cases, initially applied by Tu et al. (1984) and Tu (1988), is one in which the cascade rate, ϵ, above is taken to give the local flux F_k out of a shell of radius k in wave vector space. Then with $F_k = \epsilon(k)$,

$$\left(\frac{\partial E_k}{\partial t}\right)_{NL} = -\frac{\partial F_k}{\partial k} . \tag{15}$$

A more complex but in some ways preferable approximation is to treat the flow of energy in k-space as diffusive (Zhou and Matthaeus 1990a; Leith

1967). This approach is used for numerical solutions of two-scale separated equations by Oughton (1993).

D. Simulations: Overview of Spectral Codes and Algorithms

To examine the nature of MHD turbulence and its relationship to solar wind evolution and structure, numerical algorithms must be designed to describe accurately the inertial range of turbulence. Modeling the nearly three decades of scale sizes in this range is not possible in three dimensions (a little more than two decades is the present practical limit), but one must still use high Reynolds numbers and accurate codes to compute inertial range spectra accurately. Current spectral methods contain no numerical dissipation, but have the disadvantage of using highly idealized boundary conditions. The incompressible MHD simulations we describe have a uniform density; more recently, compressive effects have been included. Fourier method codes use periodic boundary conditions while use of Chebyshev transforms permits nonperiodic boundary conditions. Most of the results discussed here did not have sufficient resolution (with up to 256^2 Fourier modes) to resolve a clearly defined inertial range; to do that requires resolution of order 512^2 together with hyperresistivity and hyperviscosity (Passot and Pouquet 1988; Biskamp and Welter 1989, Verma 1994; Siregar et al. 1995).

As discussed above, spectral MHD codes involve a drastic over simplification of the problem, in terms of boundary conditions, dimensionality, and composition. This single-fluid formulation ignores altogether the influence of the substantial ($\approx 5\%$ by number) α-particle density in the wind. Even simulations which include an energy equation will typically not account for the observed anisotropies in the proton temperature. As a further simplification, consistent with neglect of kinetic effects, typical codes use scalar dissipation coefficients and equal magnitudes for the values of the resistivity and viscosity. While the assumption of periodic boundaries clearly limits the domain of applicability of the simulations, the assumption of incompressibility has some theoretical justification (Matthaeus and Brown 1988; Matthaeus et al. 1991; Montgomery et al. 1987; Zank and Matthaeus 1990,1991,1992,1993; Zank et al. 1990). Results from simulations of the incompressible MHD equations provide a basis for comparison both with results of compressive simulations (Roberts et al. 1991) and incompressive multi-length-scale MHD turbulence modeling (Marsch and Tu 1989; Zhou and Matthaeus 1989,1990a,b,c; Zhou et al. 1990). MHD simulation models have also included some of the effects of the expansion of the medium with increasing heliocentric distance (Grappin et al. 1993a,b).

Two-and-a-half-dimensional simulations (in which all three components of the fluctuating magnetic and velocity fields are included), or three-dimensional simulations, are necessary to study the evolution of the magnetic helicity (in two dimensions, the analogous invariant is $< a^2 >$, the mean-squared vector potential; see Fyfe and Montgomery 1976). Preliminary results on the evolution of the magnetic helicity were discussed first by Stribling et al.

(1994,1995,1996), Stribling and Matthaeus (1995) and also by Ghosh et al. (1995). In three dimensions, because of the increased number of degrees of freedom available for nonlinear coupling, perturbations orthogonal to the two-dimensional plane may excite new nonlinear couplings exhibiting both fast growth and new physics. However, there is some evidence that spectral transfer in three dimensions will occur analogously to, and at least as quickly as, it does in two dimensions (Stribling et al. 1996). In a sense, two-dimensional simulations provide a lower limit on the rate of evolution of turbulence.

E. Turbulence Modeling

Efforts to understand the evolution of Alfvén waves propagating in the (inhomogeneous) solar wind began with Coleman (1966,1967), Unti and Neugebauer (1968), and Belcher and Davis (1971). WKB theory was used in an effort to account for the relatively weak inhomogeneities in the free-flowing wind (see, e.g., Hollweg 1973a,b,1974). The observed fact that $\delta B/B \lesssim 1$ supported the expectation that WKB perturbation theory could describe correctly wave propagation in the solar wind. As mentioned above, WKB theory, however, fails in many important respects. First, WKB predicts the occurrence of only outward propagating fluctuations. Second, WKB analysis predicts that the direction of minimum variance of magnetic fluctuations should be radial, which is also inconsistent with observations. Third, WKB predicts that $r_A \approx 1$, while $r_A \approx 1/2$ is more commonly observed. Heinemann and Olbert (1980) generalized WKB to include non-WKB (finite wavelength) effects. The basic structure of their equations is

$$\frac{\partial f^{\pm}}{\partial t} + (U \pm V_A)\frac{\partial f^{\pm}}{\partial s} = f^{\mp}(U \pm V_A)\frac{d\psi}{ds} \tag{16}$$

where spherical geometry and incompressive fluctuations were assumed. The variable $f^{\pm}$ is $z^{\pm}$ normalized to take into account the WKB variation in amplitudes, s is the arc length along a field line, and ψ is a normalized solar wind density. The gradient terms on the right-hand side serve as "drivers" according to which inward (outward) waves are generated from outward (inward) waves; a process described as "reflection," "scattering," or "mixing." When the wave numbers along the direction of the magnetic field are high, the non-WKB terms have rapidly changing phases with respect to the driven terms so that inward and outward waves pass through each other without resonant interaction or coupling. Thus for small-scale parallel propagating waves, an initially outward Alfvénic disturbance will remain Alfvénic, its amplitude decay following WKB ($\delta B \propto r^{-3/2}$), and velocity fluctuations will remain equipartitioned and aligned with the magnetic fluctuations.

A variety of subsequent models have been developed to account for the observations and to correct inadequacies in the WKB approach. The most ambitious ones try to unify the wave propagation and turbulence evolution paradigms (Marsch and Tu 1989,1990a,b; Tu 1988; Tu and Marsch 1990a,b,1992; Tu et al. 1984; Zhou and Matthaeus 1989,1990a,b,c). The basic approach taken is to begin with the ideal single fluid MHD equations. Dissipation is dropped and the assumption is made that the average large-scale fields vary on the scale of R, the local heliocentric distance, whereas the plasma fluctuations, which have a correlation length $\lesssim 1/10R$, vary only on short scales.

One proceeds by writing the velocity, magnetic field, density, and pressure as ensemble-average slowly varying quantities plus rapidly varying small-scale perturbations. The original MHD equations are ensemble averaged and the results are subtracted from the original equations to yield a set of equations for the fluctuating quantities. From these, equations are derived for the evolution of a variety of correlation functions. The equations are then in a form amenable to further approximation. The simplifications begin by treating the small-scale quantities as incompressible so that $\delta\rho = \nabla\cdot\mathbf{v} = 0$, the justification being both observational and theoretical (see Sec. III.F, below).

Although in principle the ensemble-averaged magnetic field, solar wind velocity, and density can be influenced by the fluctuating fields, one assumes that they are known: $\langle\mathbf{V}\rangle = $ const., $\rho \sim 1/R^2$, and $\angle\mathbf{B}\rangle$ is given by the Archimedian spiral Parker field. Because evolution of the magnetic helicity is not important for describing effects of inhomogeneity, the plasma is assumed to be mirror symmetric. A further approximation, less well justified (Carbone et al. 1995; Matthaeus et al. 1990), but needed to make the problem tractable, is to assume a form for the three-dimensional spectral tensor. The common forms used are "slab," isotropic, two-dimensional, or axisymmetric. As an illustration, we outline the isotropic case. The equations are Fourier transformed with respect to the small-scale spatial variable $\mathbf{r}$ to obtain equations describing the evolution of inward and outward propagating fluctuations.

After considerable simplification and algebra, one arrives at the following system of equations whose solutions can be compared with observations (Oughton 1993):

$$\left[\frac{\partial P^{\pm}}{\partial t} + (U \mp V_{Ar})\frac{\partial P^{\pm}}{\partial R}\right] + \left(\frac{U \pm V_{Ar}}{R}\right)P^{\pm} + M^{\pm}F = NL^{\pm}$$

$$\left(\frac{\partial F}{\partial t} + U\frac{\partial F}{\partial R} + \frac{U}{R}F\right) + 2\left(P^{-}M^{+} + P^{+}M^{-}\right) = NL^{F}. \tag{17}$$

The first terms on the left-hand side describe the linear evolution of $P^{\pm}$ (the power in $z^{\pm}$) and F (essentially difference between the kinetic and magnetic energy spectra), the second term in the first of the equations is the WKB approximation and the last term on the left-hand side contains the terms describing mixing of the inward and outward propagating fluctuations. The right-hand side of Eq. (17) contains the small-scale nonlinear terms that include the diffusive flow of energy in Fourier space. The mixing terms $M^{\pm}$ are given, in a simple case, by

$$M^{\pm} = \frac{\alpha}{\alpha + 2}\frac{1}{2R}\left[U \mp \left(\frac{4}{\alpha} - 1\right)V_{Ar}\right] \tag{18}$$

where α is the index of the power spectrum of the total energy.

The nonlinear terms on the right-hand side of Eq. (18) must be modeled before solving for the coupling of turbulence, WKB effects, and linear convection. The terms $NL^{\pm}$ and NL^{F} describe complex turbulent phenomena that are not well understood in inhomogeneous media. Assuming that the separation of scales is valid, it is possible to treat the turbulence locally, using homogeneous theory and phenomenological models (Marsch and Tu 1993; Verma 1994; Zhou and Matthaeus 1990a,c). Because we are treating the inertial range as an incompressible magnetofluid, we can follow, for example, Kolmogoroff's approach for $NL^{\pm}$ as outlined in the previous section. Modeling NL^{F} is more difficult because less is known about it (e.g., F is not even a conserved quantity). Oughton (1993) treats this term phenomenologically by dividing it into two terms, one of which describes the tendency for r_A to become less than one, and the other of which tends to enforce equipartition and drives F toward zero. The model is completed by energy input from the low-k end of the inertial range using Kolmogoroff's hypothesis for the decay of the energy-containing eddies (Bachelor 1970; Matthaeus et al. 1994). The resulting set of equations now forms a closed system which can be integrated numerically to give the evolution of the power spectra with heliocentric distance. Consequences of this approach will be discussed in Sec. IV.B.

F. Nearly Incompressible Turbulence

Application of the standard Kolmogoroff arguments to weakly compressible hydrodynamics predicts an omnidirectional pressure spectrum with power law index $\alpha = -7/3$ (Batchelor 1951; George et al. 1984). A linear (perturbative) relationship between pressure and density would imply for that case a $-7/3$ spectral index also for the density, which is at odds with solar wind observations. A treatment of density spectra for MHD was proposed by Montgomery et al. (1987), who showed that under certain circumstances a spectral index of $-5/3$ for the density could be obtained when the magnetic (component) fluctuation spectrum index has this value. This tracking of the magnetic spectrum by the density spectrum emerges at high wave numbers, and is essentially a statistical form of pressure balance between thermal and magnetic pressures (see Shebalin and Montgomery 1988). The theory requires low turbulent Mach number (and, as was found later by Matthaeus and Brown [1988], bounds on acoustic activity in the initial data), but does not require the special geometry of velocity fluctuations that are needed for exact (static) pressure balance.

Further elaboration of the ideas of Montgomery et al. (1987) has led to the development of a theory for "nearly-incompressible MHD" (Matthaeus and Brown 1988; Zank and Matthaeus 1990,1993), which describes the nature of compressible MHD solutions in the limit of low turbulent acoustic Mach number, $M_s = \delta v/c_s$. Conditions for these flows to approach solutions to the incompressible equations are obtained by carrying out a systematic two-time-scale, two-length-scale expansion of the compressible equations, using the turbulent Mach number as a small parameter. This theory has important antecedents in the hydrodynamic theory of acoustic wave generation by vortical flows (Lighthill 1952) and in the rigorous connections that have been established between low Mach number and incompressible hydrodynamics (Klainerman and Majda 1982).

For conditions in which near-incompressibility is obtained, the lowest-order solution is identical to the corresponding solution of the incompressible MHD equations, with higher-order terms providing the compressive modifications. At low M_s the slow time scale is that for the incompressive fields, and the fast scale is associated with acoustic modes. When acoustic modes are of sufficiently small amplitude in the initial data the density fluctuations are enslaved to the incompressive fields. It is possible that the range of applicability of nearly incompressible theory might be greater for cases in which the acoustic modes are damped (as is likely to be the case in the solar wind) or when the turbulence as a whole is decaying (because the small parameter decreases in time). Simulations have shed some light on the duration of applicability of nearly incompressible theory for hydrodynamics (Ghosh and Matthaeus 1992), but the analogous study for MHD with its greater parameter space has yet to be accomplished.

Discussion of solar wind observations in terms of the earlier (high β; β is the ratio of thermal to magnetic pressures) nearly incompressible theory can be found in Matthaeus et al. (1991), who focused on the "pseudosound" character of density fluctuations presented by Montgomery et al. (1987). Tests of the expected scaling of the density fluctuation amplitude with Mach number ($\delta n/n \propto M_s^2$) for a polytropic equation of state have had limited success, although at very low M_s in the outer heliosphere the scaling seems to work well (Matthaeus et al. 1991). In the inner heliosphere, even at solar minimum, the situation is more complex (Grappin et al. 1991; Klein et al. 1993a) with $\delta n/n \propto M_s$ between the high-speed streams and both this and the M_s^2 scaling in the high-speed wind regions. This is to be expected because the turbulence is in general less developed and the density "fluctuations" in the striated regions are probably due largely to initial conditions rather than dynamical evolution. Nearly incompressible theory also allows for $\delta n/n \propto M_s$ in flows for which the density and temperature are anticorrelated (Zank et al. 1990), but this typically occurs only in the striated regions.

The nearly incompressible formulation of MHD has somewhat distinctive properties in the cases of high and low plasma β (Zank and Matthaeus 1993). For example, one interesting implication is that nonpropagating density fluctuations are expected to be anisotropic (quasi-two dimensional) for $\beta \approx 1$ with the propagating density fluctuations isotropically distributed. In contrast, for low $\beta \ll 1$ the propagating acoustic waves also become anisotropic. The latter case would be more appropriate in the lower corona, where some indication of elongated density fluctuations has been seen (see, e.g., Coles et al. 1991). A number of other implications of this theory for solar wind observations have been discussed by Zank and Matthaeus (1992).

IV. CENTRAL ISSUES AND PROBLEMS IN SOLAR WIND TURBULENCE

A. Origin Near the Sun: Photospheric Motions, Reconnection, or Shear?

What region of the solar atmosphere gives birth to the initial population of solar wind fluctuations? What free-energy source do they tap? How do the initial fluctuations evolve up to the Alfvénic critical point that, in effect, selects the outward propagating Alfvén waves? These are among the most fundamental questions regarding solar wind turbulence, but they are all unresolved. The correct solution to these problems involves both understanding how the free energy source generates fluctuations in the fields and how those fluctuations propagate from the source region to interplanetary space. Considerable work has been done on the wave propagation problem in the lower corona from the standpoint of linear theory; this was reviewed and extended recently by Velli et al. (1991), Velli (1993), and Krogulec et al. (1994). A recent review of observational evidence pertaining to sources of small-scale structure has been given by Mullan (1990).

The propagation theories predict a frequency dependent reflection coefficient with possible resonances and even trapping of the waves if the circumstances are correct. One might expect this to lead to features in the spectrum observed by spacecraft. However, no such features are seen (see, e.g., Tu et al. 1990), and the observed spectrum at 0.3 AU puts strong constraints on the generation mechanism. In large, relatively uniform regions removed from sector boundaries and stream interfaces, the observed spectrum from the scale of days to minutes consists of a power law with spectral index $\alpha \approx -1$ and no other features. The scale of a day is that of the transit time to the spacecraft, and thus part of this regime is more properly temporal and linked to boundary conditions than spatial and linked to turbulent processes. Various ingenious ideas have been advanced for producing an f^{-1} spectrum in an expanding, turbulent medium (see, e.g., Velli et al. 1989), but these mechanisms all meet the same objection that the plasma will have much less than one eddy-turnover time to evolve, and the structures involved are too big to fit in the "box" between the Sun and the point of observation (Zhou et al. 1990). Moreover, cascade mechanisms generally produce steeper spectra ($\alpha \simeq 5/3$).

The above argument implies that structure is required in the boundary conditions of the initial population of interplanetary fluctuations. Such structure clearly exists: filaments, polar plumes, and X-ray bright points, to name a few of particular significance, all indicate a highly irregular surface and corona. Perhaps the simplest model is one in which motions in the photosphere wiggle the ends of the field lines like rubber bands. This model is probably ruled out by the f^{-1} spectrum in that motions to produce such a spectrum appear chaotic, in contrast to the rather smooth motions seen in time series of solar images. It is unlikely that transmission of fluctuations through the corona flattens the spectrum, and moreover such an evolution would decrease the already low power available from the footpoint motion. As noted above, it is also difficult to see how nonlinear coupling effects could modify a photospheric flux to yield the flat spectral index for the scales involved.

A more probable region for the origin of the fluctuations is the lower corona. Here there are two potential sources of energy, namely, the magnetic and the velocity fields. Reconnection will tap the magnetic energy source, leading both directly to field fluctuations and to jets of plasma that can distort field lines leading to further fluctuations. Matthaeus and Goldstein (1986) argued that a superposition of power law distributed fluctuations from uncorrelated sources can lead to an f^{-1} spectrum in the solar wind, provided that the correlation scales of the differing sources are log-normally distributed. The particular model presented by Matthaeus and Goldstein is based upon a scale invariant distribution of structures produced by magnetic reconnection in the lower corona. There is no direct evidence for the existence of such a reconnection cascade; however, the basic idea of scale invariant superposition might be applied to other dynamical processes as well. Feldman et al. (1993) suggested that the observation of ion jets in the solar wind is consistent with a reconnection source (described by others as "microflares" and "nanoflares"; see Parker 1990, and references therein) near the transition region in the solar atmosphere. This nanoflare energy flux could possibly contribute to both the acceleration of the high-speed streams and to the production of fluctuations. Even without reconnection, the corona may be as filamentary in its velocity structure ("microstreams") as it is observed to be in its density structure (Woo et al. 1995). The shears between stream tubes could then lead to fluctuations and a concomitant reduction in the shear gradients. It is clear that very structured flows in the inner heliosphere are associated with highly developed turbulence spectra (Klein et al. 1993b, and references therein). The f^{-1} spectrum could be imposed by the affect of the acceleration and expansion of the wind (Schmidt and Marsch 1995; Schmidt 1995).

Both high-resolution observations near the Sun and simulations of appropriately structured flows will be needed to test the above ideas on the generation of the fluctuations. Ideally the latter will be done in an expanding flow with a structured atmosphere, but many insights will be possible using simpler cases.

B. The Role of Expansion in the Evolution of Fluctuations

From the standpoint of the heliosphere as a whole, the Sun is nearly a point source of plasma, and thus the flow expands roughly spherically. This expansion dominates all other effects in determining the mean (Parker) field magnitude and direction, the amplitude of the fluctuations, and the temperature of the plasma. The effect on the fluctuations should be quite different depending on the relative size of the disturbance and the scale for gradients in plasma properties. Gradients will have little effect on disturbances whose scale lengths parallel to the large-scale gradients are small. The gradient scale varies as the distance from the Sun (e.g., $(1/U)\nabla\cdot\mathbf{U} = 2/R$, where $\mathbf{U}$ is a constant radial expansion speed), and thus the expansion will affect a wider range of scales closer to the Sun. Two limiting cases are of particular interest: parallel propagating Alfvén waves (i.e., "slab" geometry), and structures or turbulence with wave vectors perpendicular to the Parker field (i.e., "two-dimensional" geometry). Exactly perpendicular wave vectors are nonpropagating convecting, structures that are a combination of quasi-two-dimensional turbulence and field-aligned filamentary structures originating near the solar surface; these appear in coronal photographs as polar plumes or as strong density contrasts in helmet streamers (Koutchmy 1988).

There are many approaches to finding the evolution of the fluctuations with distance; the WKB and approach of Heinemann and Olbert (1980), mentioned above, are two examples both of which have proven inadequate. The implications of the recent work by Marsch and Tu (1993), Zhou and Matthaeus (1990a,b,c), and Oughton (1993), which combine evolution of waves in the large-scale flow with modeling of turbulence in the inertial range, are now being developed and tested against observations. The driving term in Eq. (16) or the mixing term in Eq. (17) can become large in several distinct circumstances (Zhou and Matthaeus 1990b). Essentially, it is when the phase interference between the outward and inward propagating waves, proportional to $\mathbf{k}\cdot\mathbf{V}_A$, becomes small that reflection or mixing become important. This happens if the fluctuations are strongly turbulent and thus no dispersion relation exists, as may well be the case for solar wind fluctuations.

There are also, however, three additional possibilities within the context of linearized waves when the driving terms can be large; namely, when k becomes small (large wavelengths for any propagation direction with respect to the magnetic field), when V_A becomes small (as it must over the poles of the Sun; see Jokipii and Kóta [1989]), or when the wave vector becomes nearly perpendicular to the mean magnetic field (the convected filamentary structures mentioned above). As pointed out by Jokipii and Kóta, for the low field case, the main effect of a small value of $\mathbf{k}\cdot\mathbf{V}_A$ is that the tension in the background magnetic field lines can be ignored. In this case, neglecting compression, velocity and magnetic fluctuations transverse to the magnetic field evolve independently according to their own conservation laws. As shown by Tu and Marsch (1993), the conservation of transverse magnetic flux and vorticity implies that both the velocity and magnetic fluctuations have $1/r$ dependences on heliocentric distance, and thus, when measured in Alfvén speed units, magnetic field fluctuations are constant with distance, the Alfvén ratio is proportional to r^{-2} and

$$\sigma_c \propto \frac{r^{-1}}{r^{-2} + \text{const}}. \tag{19}$$

Behavior of this type for the radial evolution of the cross helicity, the Alfvén ratio and the kinetic and magnetic energies can be obtained from scale-separated spectral transport theory (Zhou and Matthaeus 1990c; Oughton 1993; Tu and Marsch 1993) under a variety of approximations. Simple analytic solutions emerge only when nonlinear coupling effects are ignored. For a number of different cases the normalized cross helicity behaves as

$$\sigma_c \propto \frac{r^{-a}}{r^{-2a} + \text{const}} \tag{20}$$

with the parameter a generally in the range of about $1/6$ to 1, depending upon the specific approximations made. For example, Zhou and Matthaeus ($1990c$) showed that $a = 1$ for a slab model (corresponding to the model of Heinemann and Olbert [1980]) in the weak magnetic field ($V_A \rightarrow 0$) limit. For isotropic fluctuations, also neglecting large-scale magnetic field effects, Zhou and Matthaeus ($1990c$) found $a = 5/11$ for a Kolmogoroff $-5/3$ spectral index. Generally speaking, in the linear model (Oughton 1993), the $k = 0$ slab model is identical to the two-dimensional fluctuation model, even when large-scale propagation effects are not ignored as done by Tu and Marsch (1993).

Although analytical solutions have not been presented for the latter case, the numerical results for two-dimensional fluctuations including large-scale propagation effects (Oughton 1993) show a radial decrease of σ_c only slightly slower than the above case with $a = 1$. All of the above indicate, within the context of the linear model, that when the fluctuations are very large-scale, parallel-propagating Alfvén waves or two-dimensional fluctuations with wave vectors nearly perpendicular to the magnetic field, the magnetic energy should quickly become dominant and the cross helicity should disappear with increasing radial distance. The isotropic assumption gives a qualitatively similar effect when the large-scale magnetic field is ignored, but when it is included (Oughton 1993) the reduction of cross helicity is greatly suppressed. In effect, isotropic fluctuations act very much like shorter-wavelength slab fluctuations, which are accurately described by WKB theory.

The appearance of very different radial evolution for small-scale disturbances with wave vectors parallel to the magnetic field and fluctuations at any scale perpendicular to the magnetic field suggests that much of evolution of the solar wind fluctuations arises from expansion alone (Oughton and Matthaeus 1992; Tu and Marsch 1993; Zhou and Matthaeus $1990c$; Schmidt and Marsch 1995; Schmidt 1995). If we suppose that both components were superposed near the Alfvénic critical point, then the right mix of them could lead to cross helicity evolution similar to that observed as well as to correlation functions peaked parallel and perpendicular to the average magnetic field, as reported by Matthaeus et al. (1990). The magnetic fluctuations of the convected filamentary structures decay more slowly than do those of parallel waves, and thus the structures should become more dominant with increasing distance from the Sun. Even if the structures start out Alfvénic, they rapidly become less so and their dominance will lead to an overall lower cross helicity, as observed. There are no nonlinear couplings in this model (except, importantly, to maintain the orientation of the wave vectors with respect to the field), implying that expansion is responsible for most of the observed behavior. Supposing that the structures have a well-developed turbulence spectrum, it might be possible to make this model fit the observed spectral evolution as well; the increasing dominance of the structures would lead to increasingly Kolmogoroff-like spectra.

However, further comparison of observation with the predictions of this linear superposition model shows that the model at the very least neglects important effects. The observations indicate that r_A does not approach the value of zero that would be expected for pure two-dimensional fluctuations, but rather never goes below 0.5, on average (Roberts et al. 1990). The solar wind fluctuations never stray very far from kinetic/magnetic energy equipartition. Oughton (1993) has shown that the transport models can be extended to include this tendency toward equipartition. He found that even for pure two-dimensional fluctuations the mixing effect is nullified to a large extent. For specific forms of the model, σ_c decreases only to a level of about 0.8 by 2 AU. Thus it appears doubtful that the observed reduction of σ_c can be fully accounted for by an approach in which the mixing of inward and outward sense of cross helicity is produced solely by expansion.

Both Tu and Marsch (1993) and Oughton (1993) also point out that a superposition model predicts that the observations should show a dependence on the angle between the solar wind velocity and the mean magnetic field. For example, near 1 AU r_A should be small when the field is nearly tangential and near 1 when it is radial because the spacecraft only samples wave vectors along the wind direction. However, it is clear from Helios 2 primary mission data obtained between 0.75 and 1 AU that no correlation exists between r_A and direction of **B**. The cross helicity in this model should be uniformly low in the outer heliosphere, but there are regions, such as the example observed near 8 AU seen in Fig. 3 in the refraction region toward the end of the interval (see also Roberts et al. 1987; Fig. 7, bottom plot) that are highly Alfvénic with $r_A \approx 1$, although the field is in the tangential direction (unpublished observations).

These considerations imply that while there should be a tendency for $k_\perp$ structures to produce reductions in cross helicity, other processes dominate the linear effects. This also seems to be the case for large-scale parallel Alfvénic fluctuations in that $r_A \approx 1$ for the transverse fluctuations at large scales in radial mean magnetic fields in the inner heliosphere. It is true that the cross helicity of the such fluctuations decays rapidly, more or less according to what might be expected from Eq. (19) or Eq. (20); perhaps this is due to phase shifts coming from the gradients rather than the decay of the velocity field. The fluctuation amplitude evolution was recently studied from a very general viewpoint by Verma and Roberts (1993). They found that any fluctuations about time-averaged fields would evolve at nearly the WKB rate if r_A were held fixed in a range about the observed values and the band of fluctuations in k-space had no net dissipation. This is consistent with the observed evolution of the amplitudes in the inner heliosphere at large scales (Roberts 1989) and again implies that the large scales (as long as a day or so in the spacecraft frame) do

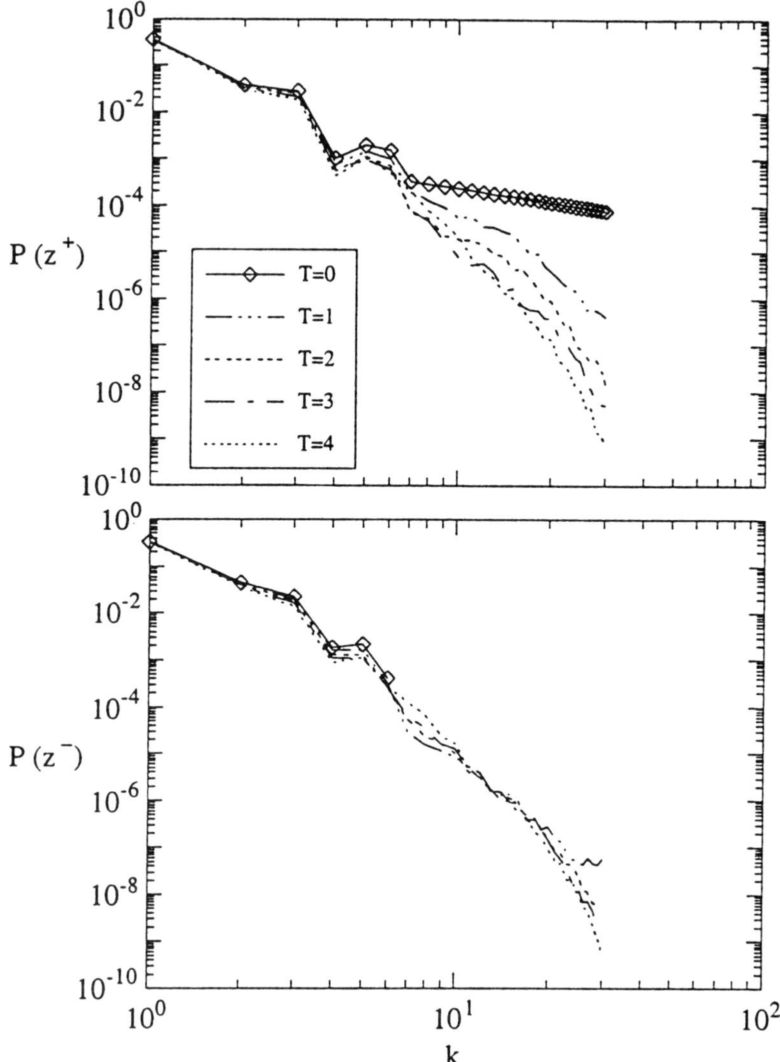

Figure 7.   Time evolution of the power spectra of the Elsässer variable z^+ and z^- for the incompressible MHD simulation that modeled the heliospheric current sheet (figure from Roberts et al. 1991).

undergo some nonlinear interactions that maintain near equipartition of the magnetic and velocity fields.

Further study is needed to determine the full range of effects possible in a scale-separated transport model. However, it seems fairly certain at this point that the expansion effects contained in the linear transport models cannot completely explain the observed radial evolution of solar wind fluctuations.

C. Shear: General Cascade Driver or Localized Mixer?

That velocity shear plays an important role in the evolution of solar wind fluctuations appears clear from the observations discussed above. What is not clear, however, is whether velocity gradients produce local effects only, or if they globally control the cross helicity, Alfvén ratio and other physical parameters (Bavassano and Bruno 1989; Klein et al. 1993b). Simulations should provide a means for determining the relative importance of shear compared to compression and expansion. For example, as Alfvénic fluctuations scatter from or interact with convecting pressure balance structures, will the cross helicity be destroyed, and, if so, how does the rate of change in cross helicity compare with that driven by velocity shear? As discussed above, some models suggest that the global expansion of the solar wind will itself cause a decrease in the Alfvénicity of solar wind fluctuations. No systematic comparisons have been made of the relative importance of the global expansion and local velocity gradients in controlling the cross helicity and Alfvén ratio evolution (see Grappin et al. [1993a,b] for an initial study), and for the most part each piece has been dealt with separately. In this section we concentrate on the role of shear, particularly as revealed through incompressible, MHD simulations of a nonexpanding flow; here shear is the sole agent of change, and its effects can be isolated.

Various simulation studies intended to model the solar wind using MHD have been carried out in one to three dimensions, with and without compressibility, and in one case, with expansion effects included. A set of studies by us and our colleagues has been specifically focused on the problems raised by observations, in the context of the discussion in Sec. II, and has attempted to answer the criticisms of the Coleman viewpoint that shear is the primary driver of the solar wind dynamical evolution. Some of the findings of these periodic, incompressible MHD studies, are discussed in more detail below. Nonlinear shear interactions can produce turbulent cascades even when the linear Kelvin-Helmholtz modes are stable. Shear interactions produce a cascade flux with a cross helicity that reflects the large scales, and thus the observed interplanetary evolution requires some other mechanism, such as expansion, to produce low cross helicity at these scales. The effects of shear are maximized near strong shears and where the magnetic field is along the plane of the shear, such as near the heliospheric current sheet. Coherent shear layers are not required to produce cross helicity evolution similar to that observed, but only small cross helicity at large scales and a mean field that is not too strong. Shear simulations can reproduce approximately the right distribution of values of r_A. Finally, the above conclusions, mostly obtained in two-dimensions, remain valid when three-dimensional variations are included. The rest of this section will give more details of these studies, and place them in the context of the work of others.

Our discussion of the numerical simulations constitutes a progression from a simple two-dimensional velocity shear layer to a three-dimensional situation that includes both velocity and magnetic shear intended to model solar wind conditions; additional details can be found in Roberts et al. (1991,1992). The simplest case contained no energy above the low wave number coherent modes used to define a weakly varying magnetic field and two shear layers. Random noise (1% of the energy) initiated the nonlinear evolution. The strength of the mean field, in dimensionless units, equaled the Alfvén speed so that the energy in the mean field nearly equaled the fluctuating energy. The value of σ_c was initially only 0.12. Although the strong mean field and the relatively large width of the shear layer made only the longest wavelength Kelvin-Helmholtz unstable, the wave number spectra of both magnetic energy and kinetic energy evolved rapidly. Within a single eddy-turnover time a significant amount of energy had cascaded to high-k modes. The physical time scale of this evolution was estimated roughly by dividing the correlation length of fluctuations at 1 AU of about 5×10^6 km (Matthaeus and Goldstein 1982a) by a typical rms speed fluctuation of ≈ 100 km s^{-1} for scales that include streams to find a characteristic time of about a day, and thus within a few eddy-turnover times the wind has convected the plasma to 1 AU. After some 9 eddy-turnover times, a fairly well-developed power law spectrum formed. The initially small global σ_c remained small and the inertial range σ_c fluctuated about zero, indicating that the cascade of energy from large scales favored neither "outward" nor "inward" propagation (Bavassano et al. 1978). The Alfvén ratio r_A was close to one. Thus, while the velocity shear instability can generate a nearly equipartitioned spectrum of fluctuations, it cannot produce the outward traveling Alfvén waves characteristic of the inner heliosphere, in contrast to the idea presented by Matthaeus et al. (1984).

To study the effect of velocity shear on initially highly Alfvénic fluctuations, an isotropic high wave-number power law distribution of Alfvénic modes with maximal cross helicity was added to the above shear modes. In this case, σ_c systematically decreased in time at high wave numbers, reminiscent of the observed decrease reported by Roberts et al. (1989a,b). The Alfvén ratio r_A remained nearly unity throughout the run. Thus velocity shear appears capable of decreasing the cross helicity of the small-scale fluctuations. As the magnitude of the background magnetic field increased, these effects became weaker; the cross helicity evolution essentially stopped for sufficiently large B_0 as the interactions became more nearly linear (see Grappin et al. 1982; Matthaeus and Zhou 1989). For strong background magnetic fields, r_A approached 1, somewhat larger than the typical solar wind values of 0.5. The dominance of the magnetic energy when B_0 is small may arise from small-scale reconnection (see, e.g., Grappin 1985; Matthaeus and Lamkin 1986; Weisshaar 1988).

The cross helicity evolution seen in the above simulations is closely associated with strong shear layers, but the observed decrease is nearly ubiquitous. Although shear certainly exists within streams, it is smaller and less organized than the shear at stream boundaries. To address this limitation, Roberts et al. (1992) performed freely decaying and driven two-dimensional incompressible simulations that did not contain any organized shear layers. In the decay case, the large-scale cross helicity was zero. For the driven case, the effect of the larger scales on the initially Alfvénic population was modeled as a forcing in a low k band of wave vectors by adding, at each time step, uncorrelated (thus zero cross helicity) $\delta \mathbf{v}$ and $\delta \mathbf{B}$ increments to low-k modes. In both sets of simulations, the observed changes in σ_c resembled those seen in the earlier cases containing large-scale shear; namely, the initially large high-k cross helicity gradually decreased as the run proceeded, indicating that even small, but nontrivial, velocity shear affects initially highly Alfvénic flows.

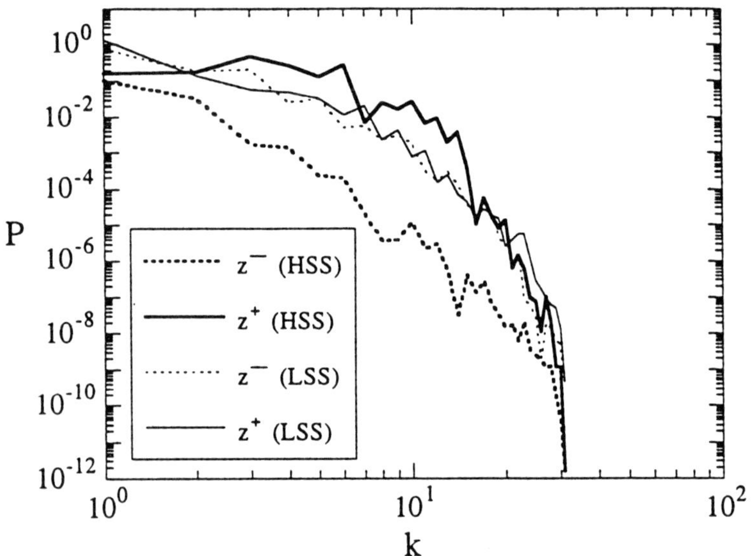

Figure 8.   Elsässer spectra for high- and low-speed streams taken from cuts along x in the simulation box (figure from Roberts et al. 1991).

In Roberts et al. (1991), these studies were further generalized by including a fairly narrow "low-speed stream" surrounded on either side by high-speed streams with a current sheet embedded in the middle of the slow wind, thereby mimicking solar wind conditions at solar minimum. The initial population of purely Alfvénic fluctuations at high k was given a flat spectrum (modal spectral index -1), again in accordance with observations assuming that both slow and fast wind were initially Alfvénic and outward propagating. Figure 7 shows one-dimensional energy spectra in terms of Elsässer variables for the solution. In the top panel the dominant "outward propagating" (z^+) fluctuations relax rapidly toward a steep spectrum, and then evolve slowly, similar to the observed spectra (Marsch and Tu 1990a) shown in Fig. 5 (see also the evolution of δz^+ and δz^- in Fig. 2b). Shear driven turbulence again is producing nearly equal large-scale input of both z^+ and z^- to a cascade; the z^- energy attains a steady state matched by dissipation, whereas z^+ dissipates due to the flat input spectrum of z^+ power at high k. In the outer heliosphere the spectra gradually evolve toward each other, presumably with a slow cascade matched by slow dissipation (Roberts et al. 1987a,b), and this quasi-steady state leads to an evolution of the amplitudes that is nearly that of dissipationless Alfvénic turbulence (Roberts et al. 1990; Verma and Roberts 1993). In Fig. 8 we show Elsässer spectra constructed from the simulation for a typical cut in x in the middle of the "high-speed stream," and for a cut near the current sheet—the "low-speed stream." We see the characteristic "breathing" (see Fig. 6) seen in the observations as described by Grappin et al. (1990) and Marsch and Tu (1990a), although the z^- increases somewhat more here than is observed. Spectra for a cut near the current sheet at the edge of the box, in the middle of the high-speed stream look very much like the low-speed stream spectra, indicating that proximity to the current sheet is the critical factor and not the speed of the wind or even the proximity to shear layers.

The spatial structure of the fluctuations can be analyzed by filtering out the low wave number power. At $T = 1$ for the "solar minimum" case, σ_c calculated by local correlations of these "high-pass" fields deviates significantly from $+1$ at the location of the current sheets rather than where the vorticity is large. By contrast, without current sheets the strong shear layers are the dominant source of the evolution. Figure 9 shows the spatial evolution of B_x, V_x, σ_c, and ω for the incompressible simulation with current sheets described in Roberts et al. (1991). The top panel of the figure shows that the field and vorticity structure have been maintained on average, but that σ_c at high k (originally $+1$ everywhere) has been significantly decreased in the region of the current sheets near the edge and the middle of the box and especially so when the current sheet is near the strong velocity shear layers. This effect is observed in the solar wind (cf. Fig. 2b, second panel from the top). By $T = 4$ the cross helicity is low everywhere, on average, although both positive and negative regions contribute. There is no significant spreading of the current sheet region, considered as bounded by the vorticity layers, as would be required in a "spreading turbulent wake" picture (Grappin et al. 1991), although note that expansion effects have been neglected. The distribution of r_A values, while not exactly the same as deduced from Voyager observations, does show a preponderance of values near 0.5, as observed (see the figure 4 and discussion in Roberts 1992).

Incompressible two-dimensional simulations may be too idealized. We deal with the question of compressibility in the next section, and consider here the possibility of three-dimensional effects in the incompressible regime. One concern is that in modeling the heliospheric current sheet in two dimensions the magnitude of the magnetic field goes to zero at the center of the sheet.

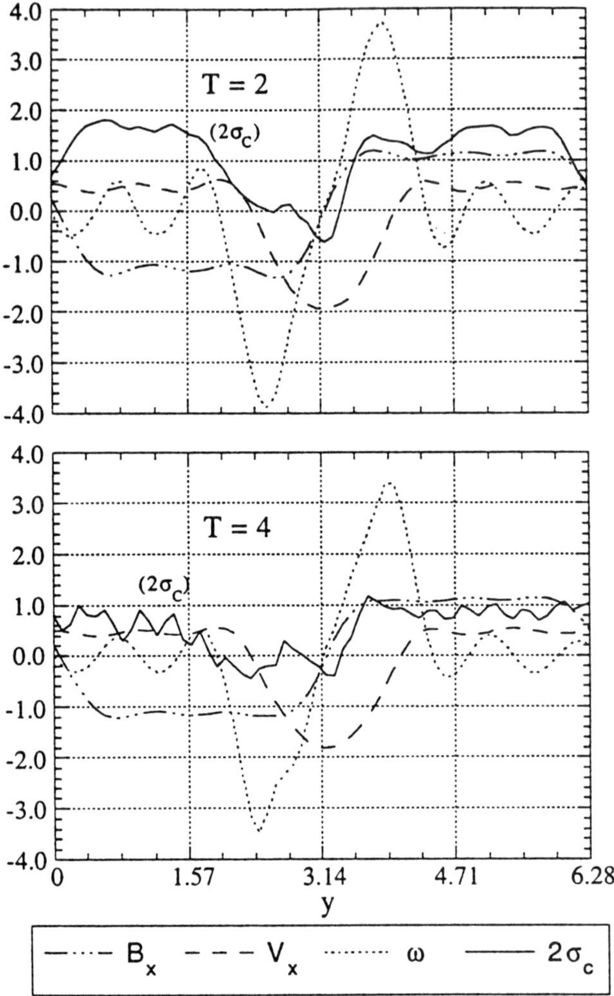

Figure 9.   Quantities averaged over x at $T = 2$ and $T = 4$ for the incompressible run with current sheets (figure from Roberts et al. 1991).

Observations of the current sheet fields show that the field magnitude rarely decreases by more than 20% or so. Its typical behavior is to rotate out of the meridional plane as it changes direction from, say, outward to inward. Thus it must be determined whether or not the decrease in cross helicity in the simulations is caused by the near-zero value of B or, as we have argued, is a result of the diminished ability of the magnetic field to stabilize velocity shear instabilities when it is transverse to the flow direction. An answer to that question has been given by Stribling et al. (1996) who performed a series of three-dimensional (incompressible) simulations of the solar minimum current sheet with shear layers at resolutions reaching 128^3 and Reynolds numbers up to 1000. The runs went to 5 eddy-turnover times. The large-scale fields were perturbed by adding isotropic circularly polarized Alfvén waves containing 10% of the energy of the large-scale fields. The evolution of the cross helicity was essentially identical to that seen in two dimensions. As before, σ_c is depleted within the regions of velocity and magnetic shear, while outside the shear layers the flow remains highly Alfvénic for a longer time. Consequently, the assumption that it is the change in direction of the magnetic field in the center of the current sheet, not its magnitude, that is controlling the evolution of the cross helicity appears to be justified. A similar evolution of cross helicity near the current sheet is seen in the simulations of Grappin et al. (1993a,b), who included the transverse expansion of the flow, and thus this result seems quite robust.

D. Compressive Structures: Scattering Centers or Along for the Ride?

Differences in velocities imply a divergence in the flow yield compressions of the plasma; these are common features in the solar wind and suggest that any realistic description of the evolution of solar wind fluctuations must attempt to ascertain their importance. The effects of compressive flows on large-scale fields are obvious in the changes in the plasma time series with increasing heliocentric distance. The fast streams catch up to the slow streams even when the flow is largely latitudinally stratified at solar minimum because of the tilt of the solar rotation axis with respect to the magnetically determined flow plane. The "interaction regions" where the fast and slow flows converge are thus compressed regions with increased density, temperature, and magnetic field strength. Beyond a few AU from the Sun (and earlier in the case of fast transient flows) these regions become shocked and irreversibly heated, but the compression is isentropic until then. Interaction regions are also the locus of increased shear (Roberts et al. 1987b; Pizzo 1994a,b,1989). Although the shear is not directly observable, the compressions, velocity jumps, and deflections predicted from this simple viewpoint are all readily observed.

In the early picture of Belcher and Davis (1971), the interaction regions were thought to be where waves were amplified or generated, and indeed wave amplitudes are higher there. Belcher and Davis also argued that the Alfvénicity should be lower in these regions due to the generation of both inward propagating Alfvén waves and compressive fluctuations. They observed the expected correlation, but more extensive observations by Roberts et al. (1987a) showed that in general there is only a small additional depletion of cross helicity in interaction regions in the outer heliosphere. Another study (Roberts et al. 1990) showed that the increased amplitudes of the fluctuations were most strongly correlated with the compression of the average magnetic field, and thus it is more appropriate to view the amplitude increase as resulting from the enhancement of pre-existing fluctuations, rather than as due to local generation. There is no quantitative study of this increase, but simple flux conservation yields an increase in the transverse field components, thus making such increases plausible. The phenomenon is not one of shock generation, for although there is enhanced turbulence near interplanetary shocks, the amplitude increases occur long before shocks form and in a much broader region of space than the shock layer.

While the large-scale compressions due to streams may not provide the answer to the problem of the evolution of the turbulent fluctuations, it is nonetheless true that compressive effects may be significant in other ways. In particular, the level of the small-scale z^- fluctuations is clearly correlated with relative density enhancements, and especially at solar minimum the regions with highly structured density and other plasma properties show both highly developed turbulence spectra and strong cross helicity depletions (Bavassano and Bruno 1989; Grappin et al. 1991; Marsch and Tu 1989,1990a,b). Moreover, the spectra of density, temperature, magnetic field strength, and z^- all show similar flattening at the high wave-number end of the inertial range, seemingly implying links between them. Note that an enhancement at high wave numbers of the minority z^- was predicted by Grappin et al. (1983) for incompressible turbulence, and thus it is possible that this is the primary effect that drives the others. Perhaps most puzzling is the observation that z^+ is somewhat better correlated with temperature than with any other indicator of macro-scale structure (Grappin et al. 1990).

These observations have led to various ideas for the importance of compressive effects, including the "nondynamic" suggestion that the structures perpendicular to the field represent non-Alfvénic disturbances that are convected with the wind and that are seen readily when the field is not radial (see the remarks on the linear superposition model in Sec. IV.B), and that the Alfvénic fluctuations interact dynamically with the structures to produce the observed cross helicity evolution (see, e.g., Klein et al. 1993a). It is also possible that the regions observed to be structured in "compressive" quantities were also highly striated in velocity as well near to the Sun. Evidence for this is provided by radio occultation measurements that at least hint at regions of very strong fast and slow flow differences within what is nominally a single stream (Lotova 1988; Woo 1995). Some indication for striated flows has been reported in fast streams by Thieme et al. (1989,1990).

A counter current within the community maintains that the small-scale compressive effects are results rather than causes. The earliest relevant observations showed that frequently the compressive structures are nearly pressure balanced (Burlaga and Ogilvie 1970), and later observations have shown that this is often true to very fine scales (Roberts et al. 1987a). Moreover, the tendency for pressure balance becomes greater with increasing heliocentric distance (Roberts 1990; Vellante and Lazarus 1987). Both interstellar and interplanetary density spectra tend to have $-5/3$ spectral indices, and the observed values of the relative density fluctuations is usually small (about 0.1) outside of the highly structured regions near the Sun.

The nearly incompressive viewpoint is supported to some extent by simulations. For example, a compressible-MHD simulation (Roberts et al. 1991),

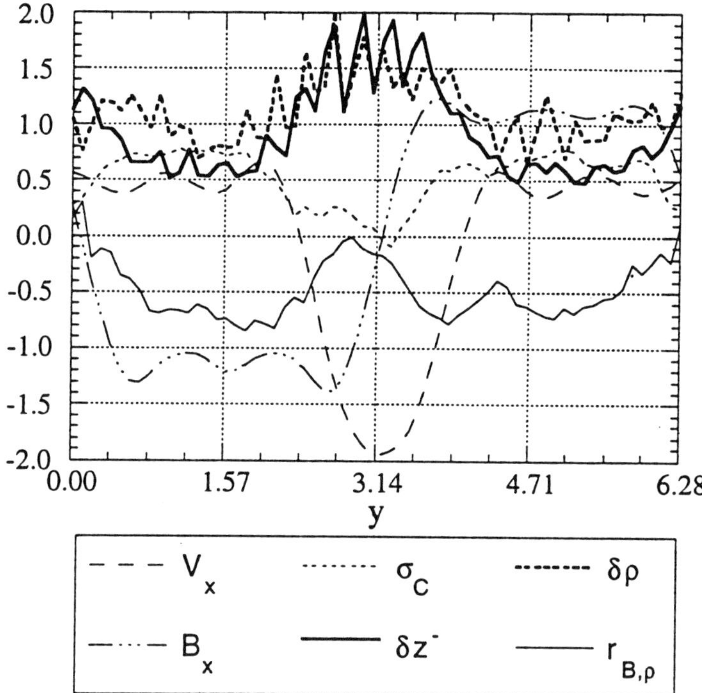

Figure 10.   Quantities averaged over x at $T = 2$ for the compressible run with current sheets. The density and z^- fluctuations are normalized to twice their maximum values (0.042 and 0.075, respectively) (figure from Roberts et al. 1991).

similar to that analyzed above, with current sheet, an initial turbulent sonic Mach number of 0.3 (typical of solar wind values), and $\gamma = 5/3$, yields nearly identical evolution of the "incompressive" quantities σ_c, v_x, and b_x as was found in the incompressible case (see Fig. 10). The new features were a persistent anti-correlation between the magnetic field magnitude and the density ($r_{B,\rho}$), suggestive of nearly pressure balanced "pseudosound" (see Matthaeus et al. 1991), and a density fluctuation that correlated well with the fluctuations in z^-. This correlation has been pointed out previously (Bavassano and Bruno 1989; Bruno and Bavassano 1991; Grappin et al. 1990) associated with arguments that the appearance of higher compression (measured, for example, by higher $\delta\rho/\rho$) indicated the essential role of compression in understanding the plasma evolution. The present simulations indicate that, on the contrary, these density fluctuations arise naturally from the incompressive evolution (Matthaeus et al. 1991; Montgomery et al. 1987).

The simulations also show essentially the correct behavior of r_A. The values of r_A, found by calculating ratios of fluctuating energy in small boxes in the simulation domain, remain near 1, and, although the mean value is in fact very close to 1 at $T = 2$ (somewhat higher than the observed 0.5), the distribution of r_A values is strongly peaked near 0.5. There are regions of lower r_A associated with low cross-helicity regions such as were recently reported by Tu and Marsch (1991). While there is no simple association between r_A and σ_c in either the simulations or observations, both quantities tend to have low values in regions that are more strongly developed, and both tend to be near +1 in more purely Alfvénic regions, as expected.

A major failing of the above simulation is that the anti-correlation of density and magnetic field strength is observed to be most pronounced in the region near the current sheet, just the opposite of what is observed here. It is likely that the structures responsible for this anti-correlation are convected from the Sun, and thus should be in the initial conditions, whereas in fact the density was initially constant. The observations indicate that the initial predominance of the pressure balanced structures near the current sheet is eroded, but that farther out the $r_{B,\rho}$ anti-correlation returns as the plasma relaxes into a state consistent with dynamically enforced pressure balance. Further simulation efforts will be needed to decide if this scenario is correct.

A second important problem with the simulations is that the shear regions are much too broad compared to observations. As remarked by Bavassano and Bruno (1989), this leads to difficulties because waves will not propagate fast enough away from the observed narrow stream interfaces to produce cross helicity depletions at the required distances from the interfaces. For the shear picture to be valid, either small shears must affect the plasma or larger shears must have existed closer to the Sun. The former possibility holds at least to some extent, in that the simulations show that shear well below the threshold for Kelvin-Helmholtz instability produces significant nonlinear effects. It may be, however, that for the turbulence to become as well developed as observed in the plasma sheet region, microstreams (see Sec. IV.A) must pervade the striated regions in the corona. On the other hand, the explanation for the well-developed spectra may lie in initial conditions, in the interaction of the Alfvénic population with the convected, striated structures, or in compressive effects. Simulations now in progress will shed light on these issues.

E. Coherent Structures in the Outer Heliosphere: Vortex Streets and Other Possibilities

We have concentrated above on trying to account for the observed turbulent evolution of interplanetary fluctuations, but as noted above, even high Reynolds number MHD flows can, under special circumstances, produce long-lived stable and coherent structures. The same strong velocity gradients that produce rapid evolution of the turbulence in the inner heliosphere may give rise to stable vortex streets in the outer heliosphere at solar minimum. It is no surprise that if such a phenomenon is going to occur that it be found at solar minimum and at large distances. At solar minimum the large-scale structure of the heliospheric fields is stable, unperturbed by flare-driven shocks, coronal mass ejections, and other transients that destroy the heliospheric plasma sheet. Furthermore, it is only at large distances, where the mean Parker field is wound into a tight spiral that the magnetic field cannot inhibit the fluid's tendency to produce vortex rolls characteristic of the nonlinear state of the Kelvin-Helmholtz instability. However, in contrast to the simple Kelvin-Helmholtz geometry, the situation in the outer heliosphere is characterized by two high-speed flows sandwiching in the middle a thin slow flow region containing the heliospheric current sheet. Theoretically, it is known from fluid studies that no-slip boundary conditions will create vorticity in the boundary layer. If an initial perturbation forms a staggered vorticity distribution, then a stable vortex street can be formed. These streets are usually associated with a solid body; e.g., a cylinder or an island, but there is theoretical evidence that they can also be formed even without a solid body present. The question for solar wind observations is whether such a street can be generated in a magnetofluid.

This question was answered affirmatively by Siregar et al. (1992,1993, 1994) who found, initially using two-dimensional MHD simulations, that a specific class of perturbations could be added to two shear layers or vortex streets that would produce stable vortex structures. They further argued that conditions in the outer heliosphere at solar minimum could reasonably be expected to produce the needed perturbations. They also were able to show that vortex streets once generated would remain stable until damped by viscous dissipation. The important parameters controlling the excitation and evolution of the vortex street are the ratio S of the initial distance between the vortex streets to the perturbation wavelength; the Mach number M_a; and the (viscous) Reynolds number R_e. S must be less than some critical value for vortex streets to form; as S approaches that value, the time scale for formation of the street increases. When the magnetic field has a strong component in the direction of the solar wind flow, vortex street development is inhibited, consistent with the absence of any observed signatures of a vortex street in the inner heliosphere. For the parameters of their simulations, Siregar et al. found that the vortex street develops and becomes fully developed by about 37 AU, which is slightly farther out than the observations. The first signs of the double maxima profiles for the solar wind speed, however, occur during the early stages in the development of the street.

While the simple vortex model (Burlaga 1990; Veselovsky 1990; Veselovsky and Triskova 1990) also yields two maxima in the velocity per period, they are of equal intensity, which is not observed. The MHD simulations produce two qualitatively unequal velocity maxima per rotation. The model also shows a substantial deflection of the flow from the radial direction with one period per solar rotation as observed. Strong correlation of pressure/density maxima with increase in total velocity are also seen in the simulations, although the density fluctuations are smaller than observed—a difference that is perhaps due to neglecting the warp of the current street which in the solar wind causes interaction regions to form with associated large density enhancements.

The net effect of the presence of the vortex street in the equatorial region is a transport of plasma and magnetic flux out of that region. The local relationship between plasma density and magnetic energy is not a simple one, even in the frozen-in approximation (Siregar et al. 1993). The magnetic field decreases even after vortices start to decay under the influence of dissipation. During this time the density, which initially was decreasing, begins to rise. Qualitatively, these results suggest that flux can be transported to higher and lower latitudes by the presence of a central vortex street, and this may be related to the observation of a "flux deficit" in the outer heliosphere found in Voyager magnetometer data during the period of the quasi-periodic flows (Burlaga and Ness 1993).

The role of the magnetic field in controlling properties of the vortex street cannot be examined fully in two dimensions. Siregar et al. (1995) have investigated this issue in a series of three-dimensional Fourier-Fourier solutions of the incompressible MHD equations; more recent, unpublished, work has extended the investigation to three-dimensional compressible MHD as well (Siregar et al. 1993). When the sheared magnetic field in the plane perpendicular to the velocity shear plane is weak, three-dimensional modes are excited and the vortex tubes kink. In contrast, strong magnetic fields prevent the vortex breakdown and the evolution of the street is essentially two-dimensional.

F. Dissipation and Heating: The Problem of Solar Wind Acceleration

Coleman (1968) first suggested that the solar wind was heated by turbulent evolution. His idea has been made more explicit by Tu et al. (1984) and Tu (1988) who embedded either Kolmogoroff or Kraichnan-type nonlinear inertial range phenomenology in the spatial transport framework of WKB theory. Tu (1987) also utilized the same basic approach to attempt a calculation of solar wind heating including feedback on the large-scale fields. Another approach to modeling the effects of turbulent dissipation and heating on the acceleration of the wind was made by Hollweg (1986) and Hollweg and Johnson (1988), who employed nonlinear phenomenology appropriate to the energy containing range of hydrodynamic turbulence, along with a WKB transport equation. Both of these approaches showed, at least roughly, that turbulent heating was consistent with the observed spectral evolution in the inner heliosphere, taking into account the dissipation of the fluctuations by a cascade and the general decay of the fluctuations due to expansion. However, these solar wind models fall short of providing a quantitative explanation of heating and acceleration, as pointed out, for example, by Isenberg (1990), and by Matthaeus et al. (1994).

There are two areas that have been suggested in which the models by Tu and Hollweg might be improved. One is the extension of the spatial transport formalism, from WKB theory to a full multiple scales treatment. Various possibilities that might emerge from this generalization, including the influence of fluctuation geometry and the role of mixing of inward and outward fluctuations, were discussed earlier in this review (see Secs. III.E and IV.B, above).

Another aspect of heating models that appears to require further development is the phenomenological treatment of the nonlinear effects. Currently, there are several efforts in progress to evaluate and possibly improve upon the simple nonlinear models that have been used by Tu et al. (1984), Tu (1987,1988), Zhou and Matthaeus (1990*b*) and Matthaeus et al. (1994). Generally speaking, the development of these models falls into two classes: models for the energy containing range of fluctuations (Hollweg 1986; Hollweg and Johnson 1988; Matthaeus et al. 1994), and models for the inertial range of the fluctuations (Tu et al. 1984; Tu 1987,1988; Zhou and Matthaeus 1990*b*).

More recently, Verma (1994) and Verma et al. (1995) used the phenomenologies discussed in Sec. III.C to evaluate the role of turbulent dissipation in the inner and outer heliosphere. He found that there was no simple, consistent set of cascade constants that would fit all the data for both Alfvénic and non-Alfvénic streams, but in many cases it seemed that the predicted heating had to be suppressed compared to that found with a nominal Kolmogoroff cascade to account for the observed evolution, and that power laws for temperature versus distance comparable to those observed were not difficult to generate. The lack of a completely self-consistent model is troublesome, but Verma also found that the predicted cascade rates do not agree in detail with simulations, especially at high cross helicity, and thus there is reason to believe the nonlinear terms have yet to be modeled properly.

Other recent preliminary studies (Matthaeus et al. 1993; Pontius et al. 1993) have focused on various models of nonlinearities that affect the larger-scale energy-containing eddies in the solar wind, which likely provide the energy source that drives the inertial range cascade. By comparison with decaying three-dimensional MHD simulations, Pontius et al. (1993) concluded that phenomenological models work reasonably well in predicting, to a reasonable accuracy and for some parameter ranges, the overall decay of Elsässer energies, and the energy difference (kinetic minus magnetic). However, the greatest difficulties in the energy containing models occurs when the cross helicity is fractionally large compared with the fluctuation energy. This is consistent with the similar deficiency found by Verma and Roberts (1993) for inertial range cascade models.

Unfortunately, the high cross helicity parameter regime is precisely that which is most relevant to solar wind evolution in the inner heliosphere and the problem of acceleration. Therefore, further development appears to be needed in these areas. Given the uncertainties involved, turbulent heating remains a likely source of significant heating well into the outer heliosphere where it competes with shock and ion-pickup heating. However, during 1986–1987, few shocks were observed between 19 and 30 AU (Burlaga 1994). Thus, turbulent heating and interstellar ion-pickup heating appear to be the most attractive options for heating the solar wind in the outer heliosphere. There are a number of additional studies that have placed some constraints of models of acceleration by waves or turbulence, which we review briefly here.

It is commonly held that the high-speed solar wind streams are largely accelerated by Alfvén waves generated near the Sun. The waves are invoked because standard thermal models of the acceleration fail to account for the constraints on mass flux, density, and temperature at the coronal base and 1 AU. A fundamental problem with these models is that the wave amplitudes at 1 AU, while a substantial fraction of the background Parker field, are not nearly large enough near the Sun to accelerate the wind if they have not been dissipated on their way to the Earth (Hollweg 1975,1978; Roberts 1989). (Consideration of finite wavelength non-WKB effects does not change this conclusion [see Krogulec et al. 1994].) Early models, as well as many more recent ones, invoked "saturation" to circumvent this difficulty. In this view, the amplitude of the transverse wave fluctuations increases compared to the background magnetic field, as predicted by the WKB approximation, until $\delta B/B$ is order unity when "nonlinear effects" lead to dissipation and a subsequent tying of the fluctuation amplitude to the mean field strength. One specific mechanism for this is that the large amplitude waves induce density fluctuations via a parametric process or more directly through intrinsic $|\mathbf{B}|$ fluctuations, and the density fluctuations are then damped by the usual Landau mechanism ("nonlinear Landau damping") leading to a capping of the fluctuation amplitude at the magnitude of the mean field. Some recent models (Hollweg and Johnson 1988; Isenberg 1990) invoke turbulence to dissipate the waves, although with limited success in achieving a self-consistent picture when attempting to account for both coronal heating and wind acceleration. It is possible to choose a correlation scale that gives a reasonable dissipation with a Kolmogoroff phenomenology for the cascade, thereby replacing the nonlinear Landau damping hypothesis with one involving a different as yet not fully understood mechanism. Clearly at this point, both observational constraints and further simulations are needed to sort out the issues.

There has been considerable evidence produced for the presence of the waves required to make high-speed streams. Both Doppler (Withbroe 1988) and Faraday rotation (Hollweg et al. 1982) measurements imply wave levels at least marginally consistent with acceleration. The validity of the interpretation of the anomalous Doppler widths has recently been questioned by Scudder (1992) who argued that the data could be explained by nonthermal particle distributions rather than turbulent bulk motion. The Faraday rotation measurements have been refined recently by Sakuri and Spangler (1994) with the result that the amplitudes are not as high as originally estimated. Recently, Coles and Esser (1992) tried to model the flows observed with radio remote sensing measurements using a wave acceleration model, and they found that no one model was adequate for all the cases.

In addition to the thermal constraints on wind acceleration provided by spacecraft, there are also important constraints provided by the measurements of the turbulence. For example, in the Belcher and Davis picture we see Alfvénic fluctuations where they were needed to accelerate the wind, namely in the high-speed streams. However, there are large regions of slow wind that have equal Alfvénic fluctuations (Marsch et al. 1981; Roberts et al. 1987b), so it must be explained why the wind was slow in these instances. Moreover, a saturation mechanism must be able to act continuously between a few tens of solar radii and the Earth's orbit without destroying the Alfvénic correlations. If nonlinear Landau damping were active, we would expect the fluctuations to be non-Alfvénic due to the wave scattering implicit in the interaction of the original waves with the compressive fluctuations built up (see, e.g., Viñas and Goldstein 1991a,b; Agim et al. 1995). The turbulent cascade mechanism must not only keep the fluctuations Alfvénic, but also act in a region in which a strong mean magnetic field and large cross helicity act to slow the cascade. The problems with the standard cascade phenomenologies must be addressed to clarify this issue. The relevant parameter space must be mapped out in simulations to determine a proper way of extrapolating experience in observed regions to regions near the Sun. Also, before firm conclusions are reached in this area, the effects of expansion must be accounted for, because Grappin et al. (1993b) have shown that expansion may inhibit cascades.

The saturation picture can also be tested directly, both observationally and in simulations. Roberts (1989) showed that, at scales of a few hours or more in the solar wind frame, the amplitude of the fluctuations in the inner heliosphere continues to increase compared to the mean field strength such that the WKB prediction holds quite well. This leaves smaller scales to account for the dissipation, presumably through a turbulence mechanism, but unless there are large features in the spectrum at high wave numbers, there cannot be enough energy available in the spectrum to do this. Although the evidence is indirect, remote sensing observations of density spectra show no evidence for large unusual spectral features at high k (Coles et al. 1991; Roberts 1989). Recent compressible MHD simulations (Agim et al. 1995; Ghosh and Goldstein 1994a,b; Roberts 1995) suggest that the occurrence of large values of the wave amplitude have no drastic consequences for the evolution, and thus that saturation does not occur.

V. SUMMARY AND SUGGESTIONS FOR FURTHER DEVELOPMENT

The last decade or so of research has finally established that turbulence is relevant to the evolution of the interplanetary fluctuations. The free energy due to the inhomogeneity and nonstationarity of the solar atmosphere manifests itself in both large-scale coherent structures such as interaction regions (and possibly at times as vortex streets), but also in a general cascade of energy from large to small scales accompanied by heating. It has not been emphasized here, but the tangled fields that result also affect high-energy particle motion, and turbulent processes that we begin to understand in the heliosphere may help us to fathom the puzzles of the origin of cosmic rays and other high-energy populations. More generally, the solar wind continues to serve as an excellent laboratory for the study of nonlinear processes.

A (biased) scenario of the turbulent dynamics in the heliosphere is as follows. Either structured magnetic fields or strong shears near the Sun produce a quite flat spectrum of fluctuations that become largely incompressive through damping of acoustic modes and become largely Alfvénic due to the "selection effect" associated with the Alfvénic critical point. The shear in the velocity field both between and within streams leads to approximately equal injection of power in z^+ ("outward") and z^- ("inward") at large scales. The low cross helicity at large scales required for this picture is due to non-Alfvénic large-scale structures and, perhaps more importantly, to the effects of the nearly spherical expansion of the flow. Because z^- is initially small at high wave numbers past the Alfvénic critical point, the low-k injection leads to a spectrum that grows until it is balanced by dissipation; both injection and dissipation then remain at a relatively low, nearly constant, level. The z^- spectrum thus is well developed, or "old," very near the Sun. Regions where the z^- fluctuation levels are high lead to stronger compressive effects, and apart from compressive structures built in by the initial conditions at the Sun, the compressive fluctuations are largely a response to the incompressive evolution.

The injection of z^+ is inadequate to balance the rapid dissipation of the large high-k fluctuations, and thus the z^+ spectrum initially decreases rapidly. When the z^+ level nearly equals the z^- level, the two spectra continue to evolve slowly toward each other with their overall amplitudes decreasing at nearly the rate they would if no spectral transfer were occurring. Overall, the expansion may slow the evolution due to nonlinear interactions. However, the turbulent evolution is accelerated in regions, such as the heliospheric current sheet at solar minimum, where the local mean field has only a small component parallel to a nearby strong shear layer. In these regions, the flat z^+ spectrum has already dissipated by 0.3 AU and thus the turbulence is more fully developed. These regions also tend to be highly structured, with sharp changes in density and other quantities, and this may contribute to the evolution both directly and possibly indirectly through the existence of strong shears near the Sun. The combined effects of dissipation and shear-driven cascades produce the evolution of spectral slopes and levels observed in the inner heliosphere. Broad slow-speed regions that do not contain a current sheet have properties similar to high-speed streams at solar minimum and the essential difference between such slow flows and high-speed flows is that the low-speed wind has had more time to evolve. Without the current sheet, the wind loses its Alfvénic character slowly due to a variety of causes including the nonlinear effects of small velocity shears and the (weaker) mixing effect associated with the expansion.

Where do we go from here? The above picture has many aspects that are speculative or that are based on a piecemeal approach to the problem. MHD turbulence modeling that attempts to produce a unified picture is only in its infancy, and only a few of the simplest cases have been studied in detail.

Further efforts along these lines must go hand in hand with simulations, for the understanding of such things as the role of the residual energy, the importance of compressive effects, and the proper closure for the nonlinear flux functions are currently intractable by analytical methods. Simulations with adequate resolution (say, 1024^3) in three dimensions to answer many of these questions should be within reach within the foreseeable future, and these can be used both to model specific solar wind situations and to develop better nonlinear closures. Modeling efforts must also continually be checked against the observations, as shown by the crucial role played by the limiting of the residual energy to give the observed value of $r_A \approx 1$; imposition of this constraint drastically changes the qualitative solution of the simple turbulent expansion model, greatly limiting the effect of mixing.

Observationally, the Ulysses spacecraft measurements of fields and particles over the solar poles will provide extremely useful information from regions of uniform high-speed wind, perhaps with a relatively smaller mean field. SOHO will provide new constraints on the existence and role of fluctuations near the Sun, as will continuing radio remote sensing investigations. The spacecraft that would probably do the most for understanding the issues raised here is Solar Probe that would make *in-situ* measurements to within four solar radii; this mission may have to wait for sunnier economic times, but it would resolve many outstanding issues. Finally, the Cluster mission will provide a new look at small scales, allowing the separation of space and time variables to an unprecedented extent, and this will help to resolve some of the uncertainty about dissipative processes. Taken together, the new sets of observations coordinated with simulations and theory should bring the study of solar wind turbulence to a state of maturity where the ideas can be applied with some confidence to other solar and astrophysical problems.

REFERENCES

Abernathy, F. H., and Kronauer, R. E. 1962. The formation of vortex streets. *J. Fluid Mech.* 13:1–20.

Agim, Y. Z., Viñas, A. F., and Goldstein, M. L. 1996. Magnetohydrodynamic and hybrid simulations of broadband fluctuations near interplanetary shocks. *J. Geophys. Res.* 100:17081–17106.

Armstrong, J. W., Cordes, J. M., and Rickett, B. J. 1981. Density power spectrum in the local interstellar medium. *Nature* 291:561–564.

Armstrong, J. W., Coles, W. A., Kojima, M., and Rickett, B. J. 1990. Observations of field-aligned density fluctuations in the inner solar wind. *Astrophys. J.* 358:685–692.

Barnes, A. 1966. Collisionless damping of hydromagnetic waves. *Phys. Fluids* 9:1483.

Barnes, A. 1979a. Hydromagnetic waves and turbulence in the solar wind. In *Solar System Plasma Physics*, eds. E. N. Parker, C. F. Kennel and L. J. Lanzerotti

(Amsterdam: North-Holland), pp. 249.

Barnes, A. 1979*b*. Physics of the solar wind. *Rev. Geophys.* 17:596.

Batchelor, G. K. 1951. Pressure fluctuations in isotropic turbulence. *Proc. Cambridge Phil. Soc.* 47:359.

Batchelor, G. K. 1970. *Theory of Homogeneous Turbulence* (New York: Cambridge Univ. Press).

Bavassano, B., and Bruno, R. 1989. Evidence of local generation of Alfvénic turbulence in the solar wind. *J. Geophys. Res.* 94:11977–11982.

Bavassano, B., Dobrowolny, M., and Moreno, G., 1978. Local instabilities of Alfvén waves in high-speed streams. *Solar Phys.* 57:44.

Bavassano, M., Dobrowolny, M., Fanfoni, G., Mariani, F. and Ness, N. F., 1982. Statistical properties of MHD fluctuations associated with high-speed streams from Helios-2 observations. *Solar Phys* 78:373.

Belcher, J. W., and Davis, L. 1971. Large-amplitude Alfvén waves in the interplanetary medium, 2. *J. Geophys. Res.* 76:3534–3563.

Bieber, J. W., Matthaeus, W. H., Smith, C. W., Wanner, W., Kallenrode, M.-B., and Wibberenz, G. 1994. Proton and electron mean free paths: The Palmer consensus revisited. *Astrophys. J.* 420:294–306.

Biskamp, D., and Welter, H. 1989. Dynamics of decaying two-dimensional magneto-hydrodynamic turbulence. *Phys. Fluids B* 1:1964.

Bruno, R., and Bavassano, B. 1991. Solar wind fluctuations at large scale: A comparison between low and high solar activity conditions. *J. Geophys. Res.* 96:7841–7851.

Bruno, R., Bavassano, B., and Villante, U. 1985. Evidence for long period Alfvén waves in the inner solar system. *J. Geophys. Res.* 90:4373–4377.

Burlaga, L. F. 1990. A heliospheric vortex street? *J. Geophys. Res.* 95:4333–4336.

Burlaga, L. R. 1994. Shocks in the outer heliosphere: Voyager 2 observations from 18.0–30.2 AU (1986–1989). *J. Geophys. Res.* 99:4161–4171.

Burlaga, L. F., and Ness, N. F. 1993. Radial and latitudinal variations of the magnetic field strength in the outer heliosphere. *J. Geophys. Res.* 98:3539–3550.

Burlaga, L. F., and Ogilvie, K. W. 1970. Magnetic and thermal pressures in the solar wind. *Solar Phys.* 15:61.

Carbone, V., Malara, F., and Veltri, P. 1995. A model for the 3-D magnetic field correlation spectra of low frequency solar wind fluctuations during Alfvénic periods. *J. Geophys. Res.* 100:1763–1778.

Coleman, P. J. 1966. Hydromagnetic waves in the interplanetary medium. *Phys. Rev. Lett.* 17:207.

Coleman, P. J. 1967. Wave-like phenomena in the interplanetary plasma: Mariner 2. *Planet. Space Sci.* 15:953–973.

Coleman, P. J. 1968. Turbulence, viscosity, and dissipation in the solar wind plasma. *Astrophys. J.* 153:371–388.

Coles, W. A., and Esser, R. 1992. An observational limit to the amplitude of Alfvén waves in the solar wind and comparison with an acceleration model. *J. Geophys. Res.* 97:19139–19148.

Coles, W. A., Liu, W., Harmon, J. K., and Martin, C. L. 1991. The solar wind density spectrum near the Sun: Results from Voyager radio measurements. *J. Geophys. Res.* 96:1745–1755.

Denskat, K. U., and Burlaga, L. F. 1977. Multispacecraft observations of microscale fluctuations in the solar wind. *J. Geophys. Res.* 82:2693–2704.

Dobrowolny, M., Mangeney, A., and Veltri, P. 1980*a*. Fully developed anisotropic turbulence in interplanetary space. *Phys. Rev. Lett.* 45:144.

Dobrowolny, M., Mangeney, A., and Veltri, P. 1980*b*. Properties of mhd turbulence in the solar wind. *Astron. Astrophys.* 83:26.

Elsässer, W. M. 1950. The hydromagnetic equations. *Phys. Rev.* 79:183.

Elsässer, W. M. 1956. Hydromagnetic dynamo theory. *Rev. Mod. Phys.* 18:135.

Feldman, W. C., Gosling, J. T., McComas, D. J., and Phillips, J. L. 1993. Evidence for ion jets in the high speed solar wind. *J. Geophys. Res.* 98:5593–5605.

Fisk, L. A. 1979. The interactions of energetic particles with the solar wind. In *Solar System Plasma Physics*, eds. E. N. Parker, C. F. Kennel and L. J. Lanzerotti (Amsterdam: North-Holland), p. 177.

Freeman, J. W. 1988. Estimates of solar wind heating inside 0.3 AU. *Geophys. Res. Lett.* 15:88–91.

Freeman, J. W., Totten, T., and Arya, S. 1992. A determination of the polytropic index of the free streaming solar wind using improved temperature and density radial power-law indices. *Eos: Trans. AGU* 73:238.

Fyfe, D., and Montgomery, D. 1976. High beta turbulence in two-dimensional magnetohydrodynamics. *J. Plasma Phys.* 29:181.

Gazis, P. R. 1984. Observation of plasma bulk parameters and the energy balance of the solar wind between 1 and 10 AU. *J. Geophys. Res.* 89:775–785.

George, W. K., Buether, P. D. and Arndt, R. E. A. 1984. Pressure spectra in turbulent free shear flows. *J. Fluid Mech.* 148:155–191.

Ghosh, S., and Goldstein, M. L. 1994*a*. Nonlinear evolution of a large-amplitude circularly polarized Alfvén wave: High beta. *J. Geophys. Res.* 99:19289–19300.

Ghosh, S., and Goldstein, M. L. 1994*b*. Nonlinear evolution of a large amplitude circularly polarized Alfvén wave: Low beta. *J. Geophys. Res.* 99:13351–13562.

Ghosh, S., and Matthaeus, W. H. 1992. Low Mach number two-dimensional hydrodynamic turbulence: Energy budgets and density fluctuations in a polytropic fluid. *Phys. Fluids A* 4:148.

Ghosh, S., Stribling, T., Goldstein, M. L., and Matthaeus, W. H. 1995. Evolution of magnetic helicity in compressible magnetohydrodynamics with a mean magnetic field. In *Space Plasmas: Coupling Between Small and Medium Scale Processes*, eds. M. Ashour-Abdalla and T. Chang (Washington, D. C.: American Geophysical Union), p. 1.

Goldstein, B., and Siscoe, G. L. 1972. Spectra and cross spectra of solar wind parameters from Mariner 5. In *Solar Wind II*, eds. C. P. Sonnett, P. J. Coleman and J. M. Wilcox, NASA SP-506.

Goldstein, M. L., Roberts, D. A., and Matthaeus, W. H. 1986. Systematic errors in determining the propagation direction of interplanetary Alfvénic fluctuations. *J. Geophys. Res.* 91:13357–13365.

Goldstein, M. L., Roberts, D. A., Ghosh, S., and Matthaeus, W. H. 1987. Numerical simulation of solar wind and magnetospheric phenomena. In *Proc. 21st ESLAB Symposium*, eds. B. Battrick and E. J. Rolfe ESA SP-275, p. 115.

Goldstein, M. L., Roberts, D. A., and Matthaeus, W. H. 1989. Numerical simulation of interplanetary and magnetospheric phenomena: The Kelvin-Helmholtz instability. In *Solar System Plasma Physics*, eds. J. H. Waite, J. L. Burch and R. L. Moore (Washington, D. C.: American Geophysical Union), p. 113.

Goldstein, M. L., Roberts, D. A., and Fitch, C. A. 1994. Properties of the fluctuating magnetic helicity in the inertial and dissipation ranges of solar wind turbulence. *J. Geophys. Res.* 99:11519–11538.

Grappin, R. 1985. Onset and decay of two-dimensional MHD turbulence with velocity-magnetic field correlation. In *Quelques aspects de la turbulence MHD dévelopée*, Ph.D. Thesis, Univ. of Paris.

Grappin, R., Frisch, U., Léorat, J., and Pouquet, A. 1982. Alfvénic fluctuations as asymptotic states of MHD turbulence. *Astron. Astrophys.* 105:6.

Grappin, R., Pouquet, A., and Léorat, J. 1983. Dependence of MHD turbulence spectra on the velocity field-magnetic field correlation. *Astron. Astrophys.* 126:51.

Grappin, R., Mangeney, A., and Marsch, E. 1990. On the origin of solar wind MHD turbulence: Helios data revisited. *J. Geophys. Res.* 95:8197–8209.

Grappin, R., Velli, M., and Mangeney, A. 1991. "Alfvénic" versus "standard" turbulence in the solar wind. *Ann. Geophys.* 9:416–426.

Grappin, R., Velli, M., and Mangeney, A. 1993a. MHD simulations of solar wind turbulence in comobile coordinates. In *Spatio-Temporal Analysis for Resolving Plasma Turbulence (START)*, ESA WPP-325 (Aussois, France: European Space Agency), p. 325.

Grappin, R., Velli, M., and Mangeney, A. 1993b. Nonlinear wave evolution in the expanding solar wind. *Phys. Rev. Lett.* 70:2190–2193.

Heinemann, M., and Olbert, S. 1980. Non-WKB Alfvén waves in the solar wind. *J. Geophys. Res.* 85:1311–1327.

Higdon, J. C. 1984. Density fluctuations in the interstellar medium: Evidence for anisotropic magnetogasdynamic turbulence. I. Model and astrophysical sites. *Astrophys. J.* 285:109–123.

Hollweg, J. V. 1973a. Alfvén waves in a two-fluid model of the solar wind. *Astrophys. J.* 181:547–566.

Hollweg, J. V. 1973b. Transverse Alfvén waves in the solar wind: Wave pressure, Poynting flux, and angular momentum. *J. Geophys. Res.* 78:3643–3652.

Hollweg, J. V. 1974. Transverse Alfvén waves in the solar wind: Arbitrary $\mathbf{k}$, v_o, B_o, and $\delta\mathbf{b}$. *J. Geophys. Res.* 79:1539–1541.

Hollweg, J. V. 1975. Waves and instabilities in the solar wind. *Rev. Geophys.* 13:263–289.

Hollweg, J. V. 1978. Some physical processes in the solar wind. *Rev. Geophys.* 16:689–720.

Hollweg, J. V. 1986. Transition region, corona, and solar wind in coronal holes. *J. Geophys. Res.* 91:4111–4125.

Hollweg, J. V., and Johnson, W. 1988. Transition region, corona, and solar wind in the coronal holes: Some two fluid models. *J. Geophys. Res.* 93:9547–9554.

Hollweg, J. V., Bird, M. K., Volland, H., Edenhofer, P., Stelzried, C. T., and Seidel, B. L., 1982. Possible evidence for coronal Alfvén waves. *J. Geophys. Res.* 87:1–8.

Isenberg, P. A. 1990. Investigations of a turbulence-driven solar wind model. *J. Geophys. Res.* 95:6437–6442.

Jokipii, J. R. 1971. Propagation of cosmic rays in the solar wind. *Rev. Geophys. Space Phys.* 9:27.

Jokipii, J. R., and Kóta, J. 1989. The polar heliospheric magnetic field. *Geophys. Res. Lett.* 16:1–4.

Karman, T. V. 1911. Uber den mechanismus des widerstandes, den ein bewegter kopper in einer flussigkeit erfahrt. *Nachrichten Math.-Phys.* K1:509.

Klainerman, S., and Majda, A. 1982. Compressible and incompressible fluids. *Commun. Pure Appl. Math.* 35:629.

Klein, L., Bruno, R., and Bavassano, B. 1993a. Scaling of density-fluctuations with Mach number and density-temperature anticorrelations in the inner heliosphere. *J. Geophys. Res.* 98:7837–7841.

Klein, L., Bruno, R., Bavassano, B., and Rosenbauer, H. 1993b. Anisotropy and minimum-variance of magnetohydrodynamic fluctuations in the inner heliosphere. *J. Geophys. Res.* 98:17461–17466.

Klein, L. W., Roberts, D. A., and Goldstein, M. L. 1991. Anisotropy and minimum variance directions of solar wind fluctuations in the outer heliosphere. *J. Geophys. Res.* 96:3779–3788.

Kolmogoroff, A. N. 1941. The local structure of turbulence in incompressible viscous fluid for very large Reynolds numbers. *C. R. Acad. Sci. URSS* 30:301.

Koutchmy, S. 1988. Space-borne coronagraphy. *Space Sci. Rev.* 47:91.

Korzhov, N. P., Mishin, V. V., and Tomozov, V. M. 1984. On the role of plasma parameters and the Kelvin-Helmholtz instability in a viscous interaction of solar wind streams. *Planet. Space Sci.* 32(9):1169–1178.

Kraichnan, R. H. 1965. Inertial range of hydromagnetic turbulence. *Phys. Fluids* 8:1385.

Krogulec, M., Musielak, Z. E., Suess, S. T., Nerney, S. F., and Moore, R. L. 1994. Reflection of Alfvén waves in the solar wind. *J. Geophys. Res.* 99:23489–23501.

Kunow, H., Wibberenz, G., Green, G., Müller-Mellin, R., and Kallenrode, M.-B. 1991. Energetic particles in the inner solar system. In *Physics of the Inner Heliosphere*, eds. R. Schwenn and E. Marsch (Heidelberg: Springer-Verlag), p. 243.

Lazarus, A. J., Yedida, B., Villanueva, L., and McNutt, R. L. 1988. Meridional plasma flows in the outer heliosphere. *Geophys. Res. Lett.* 15:1519–1522.

Leith, C. E. 1967. Diffusion approximation to inertial energy transfer in isotropic turbulence. *Phys. Fluids* 10:1409.

Lighthill, M. J. 1952. On sound generated aerodynamically. *Proc. Roy. Soc. London* A 211:564.

Lopez, R. E., and Freeman, J. W. 1982. Solar wind proton temperature-velocity relationship. *J. Geophys. Res.* 87:6011–6028.

Lotova, N. A. 1988. The solar wind transonic region. *Solar Phys.* 117:399.

Marsch, E. 1990. Turbulence in the solar wind. In *Reviews in Modern Astronomy*, ed. G. Klare (Heidelberg: Springer-Verlag).

Marsch, E. 1991. MHD turbulence in the solar wind. In *Physics of the Inner Heliosphere*, eds. R. Schwenn and E. Marsch (Heidelberg: Springer-Verlag), p. 159.

Marsch, E., and Mangeney, A. 1987. Ideal MHD equations in terms of compressive Elsässer variables. *J. Geophys. Res.* 92:7363–7377.

Marsch, E., and Richter, A. K. 1987. On the equation of state and collision time for a multicomponent, anisotropic solar wind. *Ann. Geophys.* 5A:71.

Marsch, E., and Tu, C.-Y. 1989. Dynamics of correlation functions with Elsässer variables for inhomogeneous MHD turbulence. *J. Plasma Phys.* 41:479.

Marsch, E., and Tu, C.-Y. 1990a. On the radial evolution of MHD turbulence in the inner heliosphere. *J. Geophys. Res.* 95:8211–8229.

Marsch, E., and Tu, C.-Y. 1990b. Spectral and spatial evolution of compressive turbulence in the inner solar wind. *J. Geophys. Res.* 95:11945–11956.

Marsch, E., and Tu, C.-Y. 1993. Modeling results on spatial transport and spectral transfer of solar wind Alfvénic turbulence. *J. Geophys. Res.* 98:21045–21060.

Marsch, E., Mühlhäuser, K.-H., Rosenbauer, H., Schwenn, R., and Denskat, K. U. 1981. Pronounced proton core temperature anisotropy, ion differential speed, and simultaneous Alfvén wave activity in slow solar wind at 0.3 AU. *J. Geophys. Res.* 86:9199–9203.

Matthaeus, W. H., and Brown, M. R. 1988. Nearly incompressible magnetohydrodynamics at low Mach number. *Phys. Fluids* 31:3634.

Matthaeus, W. H., and Goldstein, M. L. 1982a. Measurement of the rugged invariants of magnetohydrodynamic turbulence. *J. Geophys. Res.* 87:6011–6028.

Matthaeus, W. H., and Goldstein, M. L. 1982b. Stationarity of magnetohydrodynamic fluctuations in the solar wind. *J. Geophys. Res.* 87:10347–10354.

Matthaeus, W. H., and Goldstein, M. L. 1986. Low-frequency 1/f noise in the interplanetary magnetic field. *Phys. Rev. Lett.* 57:495.

Matthaeus, W. H., and Lamkin, S. L. 1985. Rapid magnetic reconnection caused by finite amplitude fluctuations. *Phys. Fluids* 28:303.

Matthaeus, W. H., and Lamkin, S. L. 1986. Turbulent magnetic reconnection. *Phys. Fluids* 8:2513.

Matthaeus, W. H., and Montgomery, D. 1984. Dynamic alignment and selective decay in magnetohydrodynamics. In *Statistical Physics and Chaos in Fusion Plasmas*, eds. C. W. Horton and L. E. Reichl (New York: J. Wiley), p. 285.

Matthaeus, W. H., and Zhou, Y. 1989. Extended inertial range phenomenology of magnetohydrodynamic turbulence. *Phys. Fluids* B1:1929.

Matthaeus, W. H., Goldstein, M. L., and Smith, C. 1982. Evaluation of magnetic helicity in homogeneous turbulence. *Phys. Rev. Lett.* 58:1256.

Matthaeus, W. H., Goldstein, M. L., and Montgomery, D. C. 1984. Turbulent generation of outward traveling interplanetary Alfvénic fluctuations. *Phys. Rev. Lett.* 51:1484.

Matthaeus, W. H., Goldstein, M. L., and King, J. H. 1986. An interplanetary field ensemble at 1 AU. *J. Geophys. Res.* 91:59–69.

Matthaeus, W. H., Goldstein, M. L., and Roberts, D. A. 1990. Evidence for the presence of quasi-two-dimensional, nearly incompressible fluctuations in the solar wind. *J. Geophys. Res.* 95:20673–20683.

Matthaeus, W. H., Klein, L. W., Ghosh, S., and Brown, M. R. 1991. Nearly incompressible magnetohydrodynamics, pseudosound, and solar wind fluctuations. *J. Geophys. Res.* 96:5421–5435.

Matthaeus, W., Oughton, S., Pontius, D., and Zhou, Y. 1993. Model dynamical solutions for solar wind energy containing eddies. *Eos: Trans. AGU* 74:237.

Matthaeus, W. H., Oughton, S., Pontius, D., and Zhou, Y. 1994. Evolution of energy-containing turbulent eddies in the solar wind. *J. Geophys. Res.* 99:19267–19287.

McNutt, R. 1986. Possible explanations of north-south plasma flow in the outer heliosphere and meridional transport of magnetic flux. *Geophys. Res. Lett.* 15:1523–1526.

Miura, A., and Pritchett, P. L. 1982. Nonlocal stability analysis of the MHD Kelvin-Helmholtz instability in a compressible plasma. *J. Geophys. Res.* 87:7431–7444.

Montgomery, D. 1982. Major disruptions, inverse cascades, and the Strauss equations. *Physica Scripta* T2/1:83.

Montgomery, D. 1983. Theory of hydromagnetic turbulence. In *Solar Wind Five*, ed. M. Neugebauer, NASA CP-2280, p. 107.

Montgomery, D. C., and Tidman, D. A. 1964. *Plasma Kinetic Theory* (New York: McGraw-Hill).

Montgomery, D., Brown, M., and Matthaeus, W. H. 1987. Density fluctuation spectra in magnetohydrodynamic turbulence. *J. Geophys. Res.* 92:282–284.

Mullan, D. J. 1990. Sources of the solar wind: What are the smallest-scale structures? *Astron. Astrophys.* 232:520.

Oughton, S. 1993. Transport of solar wind fluctuations: A Turbulence Approach. Ph.D. Thesis, Univ. of Delaware.

Oughton, S., and Matthaeus, W. H. 1992. Evolution of solar wind fluctuations and the influence of turbulent "mixing." In *Solar Wind Seven*, eds. E. Marsch and R. Schwenn (Oxford: Pergamon Press), pp. 523–526.

Oughton, S., Priest, E., and Matthaeus, W. H. 1994. The influence of a mean magnetic field on three-dimensional magnetohydrodynamic turbulence. *J. Fluid Mech.* 380:95–117.

Parker, E. N. 1964. Dynamical properties of solar and stellar winds. III. *Astrophys. J.* 139:690–709.

Parker, E. N. 1990. Intrinsic magnetic discontinuities and solar X-ray emission. *Geophys. Res. Lett.* 17:2055–2058.

Parker, G. D. 1980. Propagation directions of hydromagnetic waves in interplanetary space: Pioneer 10 and 11. *J. Geophys. Res.* 85:4283–4287.

Passot, T., and Pouquet, A. 1988. Hyperviscosity for compressible flows using spectral methods. *J. Comp. Phys* 75:300.

Pizzo, V. J. 1989. The evolution of corotating stream fronts near the ecliptic plane in the inner solar system 1. Two-dimensional fronts. *J. Geophys. Res.* 94:8673–8684.

Pizzo, V. J. 1994*a*. Global, quasi-steady dynamics of the distant solar wind. 1. Origin of north-south flows in the outer heliosphere. *J. Geophys. Res.* 99:4173–4183.

Pizzo, V. J. 1994*b*. Global, quasi-steady dynamics of the distant solar wind. 2. Deformation of the heliospheric current sheet. *J. Geophys. Res.* 99:4185–4191.

Pontius, D., Gray, P., Hossain, M., and Matthaeus, W. 1993. Phenomenological modeling of homogeneous MHD turbulence and comparisons with numerical simulations. *Eos: Trans. AGU* 74:477.

Pouquet, A., Frisch, U., and Léorat, J. 1976. Strong MHD helical turbulence and the nonlinear dynamo effect. *J. Fluid Mech.* 77:321–354.

Roberts, D. A. 1989. Interplanetary observational constraints on Alfvén wave acceleration of the solar wind. *J. Geophys. Res.* 94:6899–6905.

Roberts, D. A. 1990. Heliocentric distance and temporal dependence of the interplanetary density-magnetic field magnitude correlation. *J. Geophys. Res.* 95:1087–1090.

Roberts, D. A. 1992. Observation and simulation of the radial evolution and stream structure of solar wind turbulence. In *Solar Wind Seven*, eds. E. Marsch and R. Schwenn (Oxford: Pergamon Press), pp. 533–538.

Roberts, D. A., and Goldstein, M. L. 1991. Turbulence and waves in the solar wind. In *U. S. National Report to International Union of Geodesy and Geophysics*, ed. M. A. Shea (Washington, D. C.: American Geophysical Union), pp. 932–943.

Roberts, D. A., and Wiltberger, M. 1995. Nonequilibrium, large-amplitude MHD fluctuations in the solar wind. *J. Geophys. Res.* 100:3405–3415.

Roberts, D. A., Klein, L. W., Goldstein, M. L., and Matthaeus, W. H. 1987*a*. The nature and evolution of magnetohydrodynamic fluctuations in the solar wind: Voyager observations. *J. Geophys. Res.* 92:11021–11040.

Roberts, D. A., Goldstein, M. L., Klein, L. W., and Matthaeus, W. H. 1987*b*. Origin and evolution of fluctuations in the solar wind: Helios observations and Helios-Voyager comparisons. *J. Geophys. Res.* 92:12023–12035.

Roberts, D. A., Goldstein, M. L., Matthaeus, W. H., and Klein, L. W. 1989. Observation and simulation of MHD turbulence in the solar wind. In *Turbulence and Nonlinear Dynamics in MHD Flows*, eds. M. Meneguzzi, A. Pouquet and P. L. Sulem (New York: Elsevier Science), p. 87.

Roberts, D. A., Goldstein, M. L., and Klein, L. W. 1990. The amplitudes of interplanetary fluctuations: Stream structure, heliocentric distance, and frequency dependence. *J. Geophys. Res.* 95:4203–4216.

Roberts, D. A., Ghosh, S., Goldstein, M. L., and Matthaeus, W. H. 1991. Magnetohydrodynamic simulation of the radial evolution and stream structure of solar wind turbulence. *Phys. Rev. Lett.* 67:3741.

Roberts, D. A., Goldstein, M. L., Matthaeus, W. H., and Ghosh, S. 1992. Velocity shear generation of solar wind turbulence. *J. Geophys. Res.* 97:17115–17130.

Robinson, D. C., and Rusbridge, M. G. 1971. Structure of turbulence in the Zeta plasma. *Phys. Fluids* 14:2499.

Robinson, D. C., Rusbridge, M. G., and Saunders, P. A. H. 1968. Partition of energy in a turbulent plasma. *J. Plasma Phys.* 10:1005.

Rusbridge, M. G. 1969. Energy flow in the Zeta discharge. *J. Plasma Phys.* 11:35.

Sakurai, T., and Spangler, S. R. 1994. The study of coronal plasma structures and fluctuations with Faraday rotation measurements. *Astrophys. J.* 434:773–785.

Sari, J. W., and Valley, G. C. 1976. Interplanetary magnetic field power spectra: Mean field radial or perpendicular to radial. *J. Geophys. Res.* 81:5489–5499.

Schmidt, J. M., and Marsch, E. 1995. Spatial transport and spectral transfer of solar wind turbulence composed of Alfvén waves and convective structures I: The theoretical model. *Ann. Geophys.*, in press.

Schmidt, J. M. 1995. Spatial transport and spectral transfer of solar wind turbulence composed of Alfvén waves and convective structures II: Numerical results. *Ann. Geophys.*, in press.

Schwenn, R. 1983. The "average" solar wind in the inner heliosphere: Structure and slow variations. In *Solar Wind Five*, ed. M. Neugebauer, NASA CP-2280, p. 485.

Scott, S. L., Coles, W. A., and Bourgois, G. 1983. *Astron. Astrophys.* 123:207.

Scudder, J. D. 1992. Why stars should possess temperature inversions. *Astrophys. J.* 398:319–349.

Shebalin, J. V., and Montgomery, D. 1988. Turbulent magnetohydrodynamic density fluctuations. *J. Plasma Phys.* 39:339.

Shebalin, J. V., Matthaeus, W. H., and Montgomery, D. 1983. Anisotropy in MHD turbulence due to a mean magnetic field. *J. Plasma Phys.* 29:525.

Siregar, E., Ghosh, S., and Goldstein, M. L. 1993. Compressible effects in a three-dimensional plasma vortex street. Application to the solar wind. *Eos: Trans. AGU* 74:477.

Siregar, E., Ghosh, S., and Goldstein, M. L. 1995. Nonlinear entropy production operators for magnetohydrodynamic plasmas. *Phys. Plasmas* 2(5):1480.

Siregar, E., Roberts, D. A., and Goldstein, M. L. 1992. An evolving MHD vortex street model for quasi-periodic solar wind fluctuations. *Geophys. Res. Lett.* 19:1427–1430.

Siregar, E., Roberts, D. A., and Goldstein, M. L. 1993. Quasi-periodic transverse plasma flow associated with an evolving MHD vortex street in the outer heliosphere. *J. Geophys. Res.* 98:13233–13246.

Siregar, E., Roberts, D. A., and Goldstein, M. L. 1995. Magnetohydrodynamics of interacting vortex sheets. In *Space Plasmas: Coupling Between Small and Medium Scale Processes*, eds. M. Ashour-Abdalla and T. Chang (Washington, D. C.: American Geophysical Union), p. 49.

Siregar, E., Stribling, W. T., and Goldstein, M. L. 1994. On the dynamics of a plasma vortex street and its topological signatures. *Phys. Plasmas* 1:2125.

Solodyna, C. V., and Belcher, J. W. 1976. On the minimum variance direction of magnetic field fluctuations in the azimuthal velocity structure of the solar wind. *Geophys. Res. Lett.* 3:565–568.

Southwood, D. J. 1968. The hydromagnetic stability of the magnetospheric boundary. *Planet. Space Sci.* 16(5):587–605.

Strauss, H. R. 1976. Nonlinear, three-dimensional magnetohydrodynamics of noncircular tokamaks. *Phys. Fluids* 19:134.

Stribling, T., and Matthaeus, W. H. 1995. Decay of magnetic helicity in ideal magnetohydrodynamics with a DC magnetic field. In *Space Plasmas, Coupling Between Small and Medium Scale Processes*, eds. M. Ashour-Abdalla and T. Chang (Washington, D. C.: American Geophysical Union), p. 55.

Stribling, T., Matthaeus, W. H., and Ghosh, S. 1994. Nonlinear decay of magnetic helicity in magnetohydrodynamic turbulence with a mean magnetic field. *J. Geophys. Res.* 99:2567–2576.

Stribling, T., Matthaeus, W. H., and Oughton, S. 1995. Magnetic helicity in magnetohydrodynamic turbulence with a mean magnetic field. *Phys. Plasmas*, 2:1437–1452.

Stribling, T., Roberts, D. A., and Goldstein, M. L. 1996. A three-dimensional magnetohydrodynamic model of the inner heliosphere. *J. Geophys. Res.*, in press.

Taylor, G. I. 1938. The spectrum of turbulence. *Proc. Roy. Soc. London A* 164:476.

Thieme, K. M., Schwenn, R., and Marsch, E. 1989. Are structures in high-speed streams signatures of coronal fine structures? *Adv. Space Res.* 9:127.

Thieme, K. M., Marsch, E., and Schwenn, R. 1990. Spatial structures in high-speed streams as signatures of fine structures in coronal holes. *Ann. Geophys.* 8:713.

Tu, C.-Y. 1987. A solar wind model with the power spectrum of Alfvénic fluctuations. *Solar Phys.* 109:149–186.

Tu, C.-Y. 1988. The damping of interplanetary Alfvénic fluctuations and the heating of the solar wind. *J. Geophys. Res.* 93:7–20.

Tu, C.-Y., and Marsch, E. 1990a. Evidence for a "background" spectrum of solar wind turbulence in the inner heliosphere. *J. Geophys. Res.* 95:4337–4341.

Tu, C.-Y., and Marsch, E. 1990b. Transfer equations for spectral densities of inhomogeneous MHD turbulence. *J. Plasma Phys.* 44:103.

Tu, C.-Y., and Marsch, E. 1991. A case study of very low cross-helicity fluctuations in the solar wind. *Ann. Geophys.* 9:319.

Tu, C.-Y., and Marsch, E. 1992. The evolution of MHD turbulence in the solar wind. In *Solar Wind Seven*, eds. E. Marsch and R. Schwenn (Oxford: Pergamon Press), pp. 549–554.

Tu, C.-Y., and Marsch, E. 1993. A model of solar wind fluctuations with two components: Alfvén waves and convective structures. *J. Geophys. Res.* 98:1257–1276.

Tu, C.-Y., Marsch, E., and Rosenbauer, H. 1990. The dependence of MHD turbulence spectra on the inner solar wind stream structure near solar minimum. *Geophys. Res. Lett.* 17:283–286.

Tu, C.-Y., Pu, Z.-Y., and Wei, F.-S. 1984. The power spectrum of interplanetary Alfvénic fluctuations: Derivation of the governing equation and its solution. *J. Geophys. Res.* 89:9695–9702.

Umeki, H., and Terasawa, T. 1992. Decay instability of incoherent Alfvén waves in the solar wind. *J. Geophys. Res.* 97:3113–3119.

Unti, T. W., and Neugebauer, M. 1968. Alfvén waves in the solar wind. *Phys. Fluids* 11:563.

Vellante, M., and Lazarus, A. J. 1987. An analysis of solar wind fluctuations between 1 and 10 AU. *J. Geophys. Res.* 92:9893–9900.

Velli, M. 1993. On the propagation of ideal, linear Alfvén waves in radially stratified stellar atmospheres and winds. *Astron. Astrophys.* 270:304.

Velli, M., Grappin, R., and Mangeney, A. 1989. Turbulent cascade of incompressible unidirectional Alfvén waves in the interplanetary medium. *Phys. Rev. Lett.* 63:1807.

Velli, M., Grappin, R., and Mangeney, A. 1991. Waves from the Sun? *Geophys. Astrophys. Fluid Dyn.* 62:101.

Verma, M. K. 1994. Magnetohydrodynamic Turbulence Models of Solar Wind Evolution. Ph.D. Thesis, Univ. of Maryland.

Verma, M., and Roberts, D. A. 1993. The radial evolution of the amplitudes of "dissipationless" turbulent solar wind fluctuations. *J. Geophys. Res.* 98:5625–5630.

Verma, M. K., Roberts, D. A., and Goldstein, M. L. 1995. Turbulent heating and temperature evolution of the solar wind plasma. *J. Geophys. Res.* 100:19839–19850.

Veselovsky, I. S. 1990. Solar wind vortex flow in the outer heliosphere. In *Physics of the Outer Heliosphere*, eds. S. Grzedzielski and D. E. Page (New York: Pergamon Press), p. 277.

Veselovsky, I. S., and Triskova, L. 1990. MHD spiral vortex tubes of the solar wind in the outer heliosphere. *Studies Geophys. Geod.* 34:362.

Viñas, A. F., and Goldstein, M. L. 1991a. Parametric instabilities of circularly polarized, large amplitude, dispersive Alfvén waves: Excitation of parallel propagating electromagnetic daughter waves. *J. Plasma Phys.* 46:107.

Viñas, A. F., and Goldstein, M. L. 1991b. Parametric instabilities of circularly polarized, large amplitude, dispersive Alfvén waves: Excitation of obliquely propagating daughter and sideband waves. *J. Plasma Phys.* 46:129.

Völk, H. J., and Alpers, W. 1973. The propagation of Alfvén waves and their direction anisotropy in the solar wind. *Astrophys. Space Sci.* 20:267.

Weisshaar, E. 1988. Nonlinear Alfvénic fluctuations in uniform magnetic background fields. *Geophys. Astrophys. Fluid Dyn.* 41:141.

Whang, Y. C., and Burlaga, L. F. 1985. Evolution and interaction of interplanetary shocks. *J. Geophys. Res.* 90:10765–107788.

Withbroe, G. L. 1988. The temperature structure, mass, and energy flow in the corona and inner solar wind. *Astrophys. J.* 325:442–467.

Woo, R. 1995. Solar wind speed structure near the Sun at 3–12 R_S. *Geophys. Res. Lett.*, submitted.

Woo, R. W., and Armstrong, J. W. 1979. Spacecraft radio scattering observations of the power spectrum of electron density fluctuations in the solar wind. *J. Geophys. Res.* 84:7288–7296.

Woo, R., and Schwenn, R. 1991. Comparison of doppler scintillation and in situ spacecraft plasma measurements of interplanetary disturbances. *J. Geophys. Res.* 96:21227–21244.

Woo, R., Armstrong, J. W., Bird, M., and Pätzold, M. 1995. Variation of fractional electron density fluctuations near 0.1 AU from the Sun observed by Ulysses dual-frequency ranging measurements. *Geophys. Res. Lett.*, submitted.

Zank, G. P., and Matthaeus, W. H. 1990. Nearly incompressible hydrodynamics and heat conduction. *Phys. Rev. Lett.* 64:1243.

Zank, G. P., and Matthaeus, W. H. 1991. The equations of nearly incompressible fluids. *Phys. Fluids A* 3:69.

Zank, G. P., and Matthaeus, W. H. 1992. Waves and turbulence in the solar wind. *J. Geophys. Res.* 97:17189–17194.

Zank, G. P., and Matthaeus, W. H. 1993. Nearly incompressible fluids, II, Magneto-hydrodynamics, turbulence and waves. *Phys. Fluids A* 5:257–273.

Zank, G. P., Matthaeus, W. H., and Klein, L. W. 1990. Temperature and density anti-correlations in solar wind fluctuations. *Geophys. Res. Lett.* 17:1239–1242.

Zhou, Y., and Matthaeus, W. H. 1989. Non-WKB evolution of solar wind fluctuations: A turbulence modeling approach. *Geophys. Res. Lett.* 16:755–758.

Zhou, Y., and Matthaeus, W. H. 1990*a*. Models of inertial range spectra of interplanetary magnetohydrodynamic turbulence. *J. Geophys. Res.* 95:14881–14892.

Zhou, Y., and Matthaeus, W. H. 1990*b*. Remarks on transport theories of interplanetary fluctuations. *J. Geophys. Res.* 95:14863–14871.

Zhou, Y., and Matthaeus, W. H. 1990*c*. Transport and turbulence modeling of solar wind fluctuations. *J. Geophys. Res.* 95:10291–10311.

Zhou, Y., Matthaeus, W. H., Roberts, D. A., and Goldstein, M. L. 1990. Physical consistency in modeling interplanetary magnetohydrodynamic fluctuations. *Phys. Rev. Lett.* 64:2591.

Zweben, S. J., Menyuk, C. R., and Taylor, R. J. 1979. Small-scale magnetic fluctuations inside the Macrotor tokamak. *Phys. Rev. Lett.* 42:1270.

Zweben, S. J., and Taylor, R. J. 1981. Phenomenological comparison of magnetic and electrostatic fluctuations in the Macrotor tokamak. *Nucl. Fusion* 21:193.

COMPOSITION OF THE SOLAR WIND

R. von STEIGER
University of Bern

J. GEISS
University of Bern

and

G. GLOECKLER
University of Maryland

The abundances of at least ten elements in the solar wind have been determined by space-borne energy and mass spectrometers or by the foil collection technique, namely H, He, C, (N), O, Ne, Mg, Si, S, Ar, and Fe. Indirect measurements in lunar regolith inclusions enlarge this list by Kr and Xe. We review the published values for the abundances of these elements. Particularly, we discuss the variations of these abundances as a function of solar wind properties. Such variations, which possibly hold a key to theoretical models for solar wind origin and expansion, have been observed notably in the He/H and the Mg/O abundance ratios. We then analyze the temperatures obtained from different charge state ratios, which are frozen-in in the corona. These ratios can be used to discriminate between different solar wind flow types, because they carry information about the source region into interplanetary space. Finally, we turn to the first ionization potential (FIP) fractionation. Low-FIP (<10 eV) elements are enriched by a factor of $\sim$4 to 5 relative to the photosphere in the predominantly slow, in-ecliptic solar wind. On the other hand, this enrichment is found to be considerably less in the high-speed streams from coronal holes. We review different models which have been proposed to explain this FIP fractionation effect, and compare the model predictions to the observed data.

I. INTRODUCTION

A. Motivation

Solar wind composition studies are motivated by open questions in several major fields: solar physics, heliospheric and planetary physics, and astrophysics and cosmology. In this chapter, we discuss the current status of knowledge about the solar wind elemental and charge state composition mainly in relation to the first of these fields, and touch the other two only very briefly.

After more than three decades of theoretical and experimental research on the solar wind, there still remain some fundamental open questions such as: how is the mass transported from the photosphere through the chromosphere into the corona, and how is its composition affected thereby? How is the corona heated to a plasma temperature 200 to 400 times higher than that of the photosphere? How is the mass from the corona, where it is basically at rest, accelerated to form the supersonic, super-Alfvénic solar wind? In order to address these questions, observational data on the solar wind elemental and isotopic composition, of as many elements as possible, are very important and useful.

On its way from the photosphere into the corona and wind, the solar material first encounters a temperature minimum of $\sim$4400 K, low enough for many elements to exist as neutral atoms. Later, in the chromospheric temperature plateau of $\sim 10^4$ K, these atoms are re-ionized, mainly by solar EUV radiation. Atoms with a low first ionization potential (FIP) are ionized more quickly than high-FIP atoms. Apparently, high-FIP elements are fed preferentially to the corona, leading to an overabundance by a factor of $\sim$4 to 5 of these relative to the low-FIPs in the slow, in-ecliptic solar wind, as well as in the solar energetic particle population. While this fact is observationally well established by now, the mechanism or mechanisms leading to such an atom-ion separation in the chromosphere are not fully identified, and are subject to some debate. We turn to this question in Sec. III.B.

In the lower corona, the elements are further ionized to high charge states, approaching collisional equilibrium with the local electron temperature of $\sim$1 to 2 $\times$ 10^6 K. As they are dragged out of the corona, by Coulomb collisions with the protons and/or by momentum transfer from waves, the ionization/recombination rates rapidly decrease with decreasing density, and the heavy ion charge states are frozen-in at a few solar radii. Consequently, heavy ion charge state and abundance measurements in interplanetary space provide information about the conditions in the source region of the solar wind.

Solar wind material may be accelerated in interplanetary space by interaction with various shocks, e.g., the forward and reverse shocks bounding a corotating interaction region (CIR), or the shocks driven by coronal transients such as coronal mass ejections (CMEs). These ions, which can be detected by energetic particle instruments, provide information on both the solar wind and coronal composition, and on the acceleration mechanism at these shocks. Moreover, the average composition of the solar energetic particles (SEPs) can be used to infer the coronal composition, thus complementing the solar wind measurements.

For the history and evolution of our planetary system, composition data of the planetary bodies and atmospheres, comets, and meteorites are crucial. For interpretation, these data need to be compared to a baseline composition. Because the Sun contains $\sim$99.9% of matter in the solar system, its surface composition closely reflects that of the protosolar nebula (with the exception of ^{4}He, which may be slightly depleted, of D, which has been destroyed, and of ^{3}He, which in turn is enhanced, see below). The solar wind carries this material to 1 AU, where it can be analyzed *in situ*; a good understanding of the FIP fractionation is therefore also important in order to derive the photospheric baseline composition.

Finally, the solar wind ^{3}He abundance has been used to derive the proto-solar D/H ratio (Geiss and Reeves 1972). Deuterium has been burnt to ^{3}He in the outer regions of the Sun, due to its low Coulomb barrier. Allowing for the original ^{3}He, known from meteorites, D/H may be obtained from solar wind ^{3}He. The primordial value of this ratio is of great importance in standard big bang cosmology, because it is a measure for ρ_0, the density of the universe today. [More precisely, D/H measures $\eta(\rho_0, H_0, T_0)$, the entropy per baryon, but with the microwave background temperature $T_0 = 2.74$ K well known from COBE, and the Hubble constant $H_0 \simeq 50$ km s^{-1} Mpc, this amounts to a density measurement.] Since the solar wind measurements, D/H has also been measured directly in interstellar clouds, using absorption lines in stellar spectra (cf. Boesgaard and Steigman 1985), and recently even in intergalactic clouds, using quasar spectra (Songaila et al. 1994). In accord with abundance measurements of the other primordially produced elements, ^{3}He, ^{4}He, and ^{7}Li, the deuterium abundance indicates an undercritical ρ_0, i.e., an open universe (except for the possible existence of nonbaryonic dark matter).

B. Solar Wind Measurement Methods

In the past three decades, solar wind composition instruments of increasing complexity, resolution, and dynamic range have been flown on many spacecraft. Here we briefly discuss the four main types of sensors. The solar wind abundance data, presented in Sec. II, is based on the more recently flown of those instruments. Thereby, the earlier missions are underrepresented, even though they have been equally important to our understanding of the solar wind.

The first solar wind instruments were electrostatic analyzers, either Faraday cups combined with retarding potential grids, or curved-plate analyzers, selecting the energy per charge E/q of the incident ions. By stepping the E/q analyzer through a series of (usually log-spaced) steps, the distribution function (phase space density as a function of energy) of the ion beam was obtained. With these sensors, helium could be readily resolved from hydrogen (Neugebauer and Snyder 1966) at times when the kinetic temperature was sufficiently low for the two ion peaks to be separated. This was possible only because both species flow at nearly the same speed, and because the flow speed is highly supersonic, allowing to convert the E/q scale into a m/q scale. Thus, the individual H$^+$ and He^{++} distribution functions were well separated by a factor of nearly 2 in E/q. Still, the interpretation of these data "required faith in the assumption that the primary peak is due to H$^+$ and that the secondary peak is due to He^{++} ions" (Neugebauer 1981). With equal faith, this assumption was later extended self-consistently and successfully to the heavy ions. More advanced E/q sensors were able to record the low charge states of O, Si, and Fe (Bame et al. 1975) at times when the solar wind kinetic temperature was sufficiently low. However, at high kinetic temperatures the individual peaks could no longer be resolved. Moreover, the high charge states of C and O always remained hidden behind the large He peak.

The next generation of sensors had a velocity selector, e.g., a Wien filter, added after the E/q analyzer. This technique was first used by Ogilvie and Wilkerson (1969) on Explorer 34. By stepping both the analyzer and the selector voltages, it was possible to obtain v and m/q of the incident ions independently. Of course, this results in a longer stepping sequence, reducing the time resolution of such instruments. With these sensors, of which ICI on ISEE-3 is the only example flown in interplanetary space (Coplan et al. 1978), it became possible to verify the equal velocity hypothesis, and to find deviations therefrom. Rare cases were identified where the familiar double peak is not caused by H^+ and He^{++} at nearly equal velocity, but by a double proton beam, accidentally separated by a factor of $\sqrt{2}$ in velocity (Rosenbauer et al. 1977). Also, it could be shown that the He and heavier ions indeed all flow at very nearly the same speed, and that they are faster than the protons. The differential velocity amounts to about the Alfvén speed in high-speed streams, but is less in the slow solar wind, and almost vanishes at instances where the density is unusually high, e.g., at the interface of a high-speed stream running into the preceding slow wind (Grünwaldt and Rosenbauer 1978; Neugebauer 1981). Also, it was shown that the differential velocity increases with decreasing heliocentric distance in proportion with the Alfvén speed by Helios between 0.3 AU and 0.83 AU (Marsch et al. 1981). Because of the isotachic property of the solar wind heavy ions, E/q spectra are almost equivalent to m/q spectra, but at times of high kinetic temperature, they are difficult to interpret due to significant overlap of the ion peaks. With the true m/q sensor ICI/ISEE-3 (as opposed to E/q analyzers), it was possible to determine the abundances and the kinetic properties of ^{3}He, ^{4}He, O, Si, and Fe for all solar wind conditions. Furthermore, ICI additionally measured Ne (Kunz et al. 1983), but other important elements such as C and Mg still remained hidden due to m/q ambiguities with more abundant species. [Ne was already known from the lunar foil experiment (see below), but with ICI it could be directly related to O.]

A different type of second-generation instrument consisted of an E/q analyzer with an array of solid-state energy detectors (SSD) added, such as ULECA on ISEE-3 (Hovestadt et al. 1978). It was thus possible to determine the charge states of the incoming ions, but measurements were limited to high-speed solar wind due to the inefficiency of the SSD at low energy ($\lesssim 4$ keV/e). Still, this was no disadvantage, because ordinary E/q spectra were difficult to interpret in the fast solar wind due to the associated high kinetic temperature. Using ULECA, Ipavich et al. (1986) reported the first measurements of Fe charge states in high-speed streams, along with those of the CNO group.

The experimental situation was further improved by the third generation of instruments, which are using time-of-flight (Gloeckler 1990). These sensors, of which SWICS/Ulysses is the first one to be flown in the solar wind in interplanetary space (Gloeckler et al. 1992), combine a classical E/q analyzer with a time-of-flight measurement, yielding m/q, and a total energy measurement in a solid state detector, yielding m. The three measurements then allow to calculate energy, mass, and charge of each incoming ion separately. The main advantages of such an instrument are the following:

1. Mass resolution, resolving important m/q ambiguities such as C^{6+}–He^{2+}, Mg^{10+}–C^{5+}, and the high charge states of Si and Fe.
2. Low background, due to the coincidence technique used in registering the start and stop pulse of the time-of-flight path and the energy measurement. These triple coincidence counts are virtually background free, and also the double coincidence counts (no energy measurement) contain very little background.
3. Because solid state detectors are inefficient at solar wind energies (~ 1 keV/amu), a post-acceleration of >20 keV is needed to boost the ions. As a side effect, this guarantees that solar wind of very different speeds is measured under nearly the same conditions inside the sensor. Comparisons of high-speed and low-speed solar wind composition can thus reliably be made.

With SWICS/Ulysses, the solar wind abundances of C, (N), Mg, and S can be measured for the first time, and O, Si, and Fe are measured more accurately, because their high charge states are also directly observed. These data confirm and extend earlier findings by CHEM on AMPTE/CCE (Gloeckler et al. 1989), which sampled the shocked solar wind plasma in the Earth's magnetosheath during short times of enhanced solar wind pressure. While it is unlikely that the plasma composition is altered by the passage of the bow shock, abundance determinations are still difficult in the magnetosheath. Unlike in the solar wind, where the full distribution function is sampled, only the portion above a constant, instrument dependent minimum $(E/q)_0$ of the thermalized distribution function is seen behind the shock, and the missing low-energy part must be accounted for appropriately.

A completely different type of solar wind sensor was flown on the Apollo missions: the SWC foil collection experiment (Geiss et al. 1970b). Thin Al and Pt foils were exposed to the solar wind on the lunar surface. A large, well known fraction of the solar wind was trapped in the foils, which were returned and could later be analyzed by laboratory mass spectrometers. Noble gas abundances and highly accurate isotopic compositions could be obtained by this technique. For Ar, these still are the only solar wind data available (except for values obtained from trapped gases in the lunar soil, which are not fully reviewed in this work).

Finally, solar wind ions are also implanted into the lunar regolith, and the composition may be obtained therefrom. Recently, Wieler et al. (1993) have carefully analyzed the strictly surface-correlated component of the noble gases, including Kr and Xe, in a recently ($<10^8$ yr) exposed regolith sample, and obtained solar wind abundance ratios (relative to Ar) of them.

II. SOLAR WIND COMPOSITION DATA

A. Average Abundances of Elements in the Solar Wind

We now turn to an element-by-element review of the most reliable published values of the known solar wind abundances. The main emphasis is on the heavy ($A>4$) constituents, and the values will be given relative to oxygen. The total (hydrogen) density of the solar wind is highly variable, but most element ratios are fairly constant, at least in specific solar wind types. If possible, we cite an average value for fast streams ($v>600$ km s^{-1}) and another for slow, interstream solar wind ($v<600$ km s^{-1}). (Note that the term 'fast streams' strictly refers to coronal hole-associated solar wind, not to transient, flare associated phenomena, which may also produce very high solar wind speeds, but have otherwise very different properties.) Variations of these average values will be discussed in the subsequent section. The major constituents, H and He, are discussed only briefly, and we refer to the extensive literature on these elements (see, e.g., Neugebauer 1981; Schwenn 1990). All values are collected in Table I, along with the references.

Hydrogen. H is the main solar wind gas by mass under almost all conditions. With 30 years of *in-situ* observations (and a record of comet tail observations throughout human history), it is now firmly established that the solar wind is an ubiquitous and continuous phenomenon. Even so, its properties at 1 AU vary considerably. Average proton densities are $n_p = 8.3$ cm^{-3} in the slow wind, and $n_p = 2.73$ cm^{-3} in the fast streams (Feldman et al. 1977; Schwenn 1990). H is most accurately measured relative to He, and the average values are He/H = 0.038 in the slow wind, and He/H = 0.048 in fast streams (Schwenn 1990), but in this chapter, we use O as the reference element (see the paragraph on O below). Relative to O, Bame et al. (1975) give H/O = 1900±400 for the slow, in-ecliptic solar wind, indicating a slight enrichment of the H/O ratio with respect to the photosphere (H/O = 1175). Recently, Wimmer Schweingruber et al. (1994) have analyzed the period between July 1992 and April 1993, using data from SWICS/Ulysses. The solar wind speed varied between 400 and 800 km s^{-1} once per solar rotation (see Fig. 1), making the period ideal for a comparison of the two solar wind types. They find H/O = 1890 ± 600 in the slow wind (in accord with Bame et al. 1975), and H/O = 1590 ± 500 in fast streams. If the latter value is confirmed in the polar coronal hole, these new measurements indicate that H does not behave much differently than the other, heavy high-FIP elements (due to its being the major component, or to its low mass). We return to the FIP fractionation in Sec. III.

Helium. He was detected in the solar wind by Neugebauer and Snyder (1966), and has since been measured extensively (see Neugebauer 1981; Schwenn 1990). It normally amounts to $\sim$4% by number of the total solar wind, or to about 15% by mass. In the fast solar wind, the He/H ratio at 1 AU is a remarkably constant 4.8%, as compared to an average of 3.8% in the slow wind (Feldman et al. 1977; Schwenn 1990). However, the latter value is highly variable, as discussed in Sec. II.B below. Relative to O, Bochsler et al. (1986) report He/O $= 75 \pm 20$ from the 46-month ICI/ISEE-3 data set, taken around solar maximum. These data are representative for the in-ecliptic solar wind around solar maximum, which is dominated by low speeds ($v \lesssim 400$ km s^{-1}), because in-ecliptic high-speed streams are not common during that portion of the solar cycle. This value has recently been confirmed by Villanueva et al. (1994), who find an average He/O $= 69 \pm 11$, using data of the Plasma Science (PLS) sensor on both Voyager spaceprobes, obtained during the pre-Jupiter phase of the missions in 1977–79. Wimmer Schweingruber et al. (1994), using data from SWICS/Ulysses, find He/O $= 93 \pm 31$ in the slow wind, in agreement with Bochsler et al. (1986), and He/O $= 83 \pm 25$ in nine recurrences of a high-speed stream at mid-latitudes. Further SWICS measurements over the polar coronal holes, traversed in 1994 and 1995, will show whether this value is compatible with, or significantly different from, the interstream value.

Carbon. C is a very important element for the FIP effect, because its 11.26 eV ionization potential places it near the step between low-FIP and high-FIP elements (see Sec. III). It can only be measured by $m - m/q$ sensors because the main charge state C^{6+} is subject to an ambiguity with the more abundant He^{2+}, as is C^{5+} with Mg^{10+}. In a comprehensive survey of all AMPTE/CCE magnetosheath periods, von Steiger et al. (1992a) find C/O $= 0.55 \pm 0.08$ in high-speed streams, and C/O $= 0.77 \pm 0.12$ in high-density, flare associated events. In Fig. 1 (third panel), we give daily averages of the C/O ratio measured by SWICS/Ulysses during the second half of 1992. During this period, the solar wind started to become structured in a simple way, with one high-speed stream appearing per solar rotation (see Figs. 1 and 7, top panels). From the whole Ulysses mission data up to day 325, 1993, we deduce an average C/O $= 0.72 \pm 0.10$, with no significant difference detected so far between slow wind and fast streams. This is in reasonably good agreement with the AMPTE data to reinforce the assumption that the magnetosheath composition is unfractionated relative to the solar wind. However, the C/O variation found by AMPTE appears to be missing. This may be due to a data selection effect; we will discuss the abundance variations (or lack thereof) in Sec. III.A.

Nitrogen. N is difficult to observe even with $m - m/q$ sensors because its main charge states are close to the 5 to 10 times more abundant C and O. In the Earth's magnetosheath, Gloeckler et al. (1986,1989) report N/O $= 0.129 \pm 0.008$ in flare associated events, and N/O $= 0.145 \pm 0.011$ in fast streams. The two values agree within their error bars.

Oxygen. O is the most abundant of the heavy trace components of the solar wind under almost all conditions. It thus usually serves as the reference element, as in this work. However, this convention discriminates against elements which can be measured with greater accuracy relative to some other reference, such as the noble gases, and have to be renormalized. Such values are marked in Table I, and the original data from which they were derived are

TABLE I

Average Solar Wind Abundances Relative to Oxygen in the Slow, Interstream Solar Wind and in High-Speed Streams from Coronal Holes ($T_O < 1.3 \times 10^6$ K)*

	FIP	Solar Wind in Ecliptic	Solar Wind from Coronal Holes	SEP-based Corona	Photosphere
H	13.60	1900 ± 400[a] 1890 ± 600[s]	1590 ± 500[s]	1170 ± 89[m]	$1175°$
He	24.59	75 ± 20[b] 93 ± 31[s]	83 ± 25[s]	55 ± 3[m]	$115°$
C	11.26	0.72 ± 0.10[c]	0.70 ± 0.10[c]	0.428 ± 0.043[n]	0.468[p]
N	14.53	0.129 ± 0.008[d,1]	0.145 ± 0.011[d,1]	0.123 ± 0.009[n]	0.117[q]
O	13.62	$\equiv 1$	$\equiv 1$	$\equiv 1$	$\equiv 1$
Ne	21.56	0.17 ± 0.02[b] 0.14 ± 0.02[e,2] 0.10 ± 0.03[t]	0.136 ± 0.011[d,1] 0.12 ± 0.04[t]	0.142 ± 0.014[n]	$0.138°$

Mg	7.65	0.16 ± 0.03[c]	0.083 ± 0.02[c]	0.193 ± 0.011[n]	0.0447[c]irc
Si	8.15	0.19 ± 0.04[f]	0.054 ± 0.009[d,1]	0.164 ± 0.0099[n]	0.0417[c]irc
		0.18 ± 0.02[g]			
S	10.36	0.038 ± 0.009[d,1]	0.019 ± 0.003[d,1]	0.0377 ± 0.0016[n]	0.0191[c]irc
		0.05 ± 0.02[h,2]	0.022 ± 0.008[h,2]		
Ar	15.76	0.004 ± 0.001[i,2]		0.0037 ± 0.0006[n]	0.00447[c]irc
Fe	7.87	$0.19\pm^{0.10}_{0.07}$[j,2]	0.057 ± 0.007[d,1]	0.172 ± 0.023[n]	0.0355[r]
		0.12 ± 0.03[k]			
Kr	14.00	3.4 ± 0.4[l,2,3]			1.89[o,3]
Xe	12.13	0.88 ± 0.12[l,2,3]			0.197[o,3]

* Data not originally referred to O are marked, and original values are given in Table II. Coronal abundances as inferred from solar energetic particles measurements, and photospheric abundances are given for comparison.

[1] Values determined in the Earth's magnetosheath.

[2] Original values not referred to O; cf. Table II.

[3] Values $\times 10^6$.

[a] Bame et al. 1975; [b]Bochsler et al. 1986; [c]this chapter; [d]Gloeckler et al. 1989; [e]Geiss et al. 1970b; [f]Bochsler 1989; [g]Galvin et al. 1993; [h]Shafer et al. 1993; [i]Cerutti 1974; [j]Schmid et al. 1988; [k]Ipavich et al. 1992; [l]Wieler et al. 1993; [m]Reames 1994; [n]Garrard and Stone 1993; [o]Anders and Grevesse 1989; [p]Grevesse et al. 1992; [q]Grevesse and Anders 1991; [r]Hannaford et al. 1992; [s]Wimmer Schweingruber et al. 1994; [t]Geiss et al. 1994a.

TABLE II

Solar Wind Original Abundance Ratio from Measurements
Not Referred to Oxygen[a]

He/Ne	530 ± 70[b]
S/Si (IS)	0.30 ± 0.12[c]
S/Si (CH)	0.40 ± 0.15[c]
Ne/Ar	47 ± 7[d]
He/Fe	$400\pm^{200}_{130}$[e]
Kr/Ar	0.00086 ± 0.00010[f]
Xe/Ar	0.00022 ± 0.00003[f]

[a] These values have been renormalized to O and appear in Table I; [b]Geiss et al. 1970b; [c]Shafer et al. 1993; [d]Cerutti 1974; [e]Schmid et al. 1988; [f]Wieler et al. 1993.

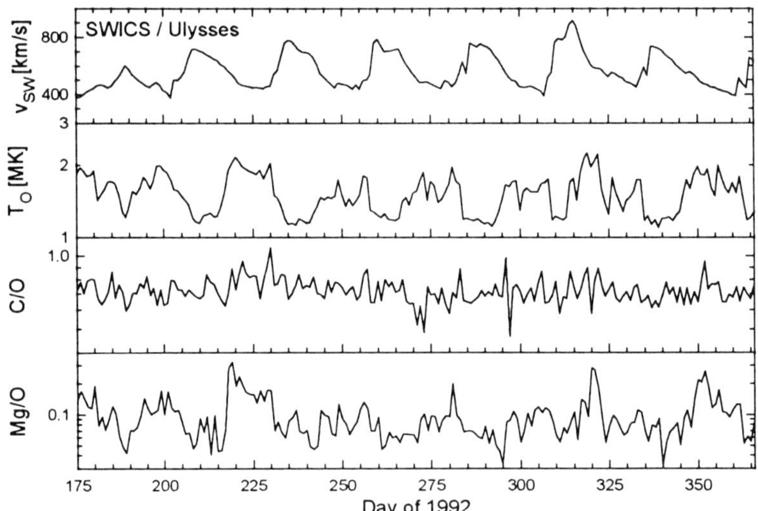

Figure 1.   Solar wind He^{++} velocity, oxygen freezing-in temperature, C/O, and Mg/O abundance ratio (top to bottom panel) from SWICS/Ulysses during the second half of 1992. A distinct stream structure developed, with a high-speed stream appearing once per solar rotation ($\sim$26 day sidereal rotation period). Ulysses was between 5.4 AU and 5.0 AU heliocentric distance and between $-12°$ and $-23°$ heliographic latitude.

given in Table II.

Neon. Neon was first observed with the Apollo SWC foil experiment by Geiss et al. (1970b), giving He/Ne $= 530 \pm 70$. This includes a 7% contribution from ^{22}Ne, and is normalized to He, because O is unavailable from the foil. Later, Ne was also observed, and directly related to O, by the ICI/ISEE-3 instrument, clearly manifesting itself between C^{5+} and O^{6+} in the m/q spectra (Kunz et al. 1983). Bochsler et al. (1986) give Ne/O $= 0.17 \pm 0.02$, again for the 1978–1982 period around solar maximum. This agrees well with the SWC measurements, as shown in Fig. 2. In a number of fast streams, observed in the magnetosheath by CHEM, Gloeckler et al. (1989) find $\text{Ne/O} = 0.136 \pm 0.011$. Results from SWICS/Ulysses, measured during a 100-day period starting on day 210 of 1992, confirm this or even a somewhat lower value, both in fast streams at low T_O, $\text{Ne/O} = 0.101 \pm 0.028$, and in the interstream solar wind at high T_O, $\text{Ne/O} = 0.116 \pm 0.039$ (Geiss et al. 1994a). The values were obtained from the Ne^{8+} ion exclusively, but because it is the most abundant Ne charge state by more than an order of magnitude over the full range of freezing-in temperatures encountered, there is little uncertainty introduced by disregarding the other charge states. They also include the 7% contribution from ^{22}Ne.

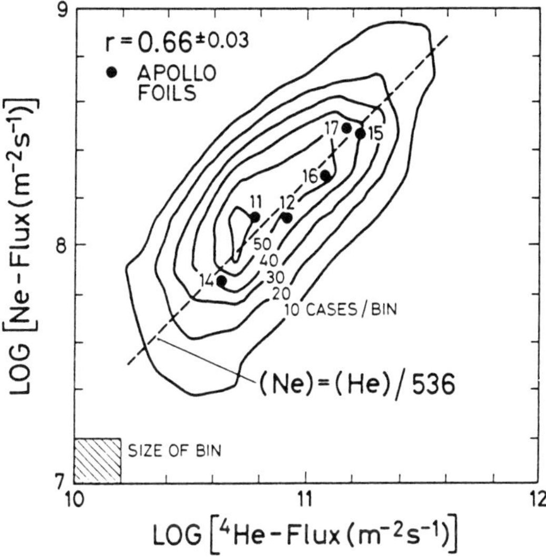

Figure 2. Solar wind Ne/He flux ratios as determined by ICI/ISEE-3 (contours), and from the SWC/Apollo experiment (dots) (figure from Bochsler et al. 1986).

Magnesium. Mg, like C, can only be observed with $m - m/q$ type sensors. In Fig. 1 (bottom panel), we give daily averages of the Mg/O ratio measured by SWICS/Ulysses. These were calculated from the main charge state, Mg^{10+}, using the oxygen freezing-in temperature T_O (second panel, cf. Sec. II.C.), and the tables of Arnaud and Rothenflug (1985) to include the contribution from other charge states. Furthermore, we have added a 21% correction to account for the heavy isotopes, ^{25}Mg and ^{26}Mg. It is obvious from Fig. 1 that the Mg/O ratio is much more variable than, say, the C/O ratio, as will be discussed below. Here, we only report average values for high-speed streams, $\text{Mg/O} = 0.083 \pm 0.02$, and for interstream solar wind, $\text{Mg/O} = 0.16 \pm 0.03$. These values are based on the whole Ulysses mission up to day 325, 1993; they extend earlier reportings based on smaller data sets (von Steiger et al. 1992*b*; von Steiger and Geiss 1993). We note that the fast-stream value agrees within the error bars with the one given by Gloeckler et al. (1989), which was determined in the Earth's magnetosheath.

Silicon. The Si abundance was determined by Bochsler (1989) as $Si/O = 0.19 \pm 0.04$ from the ICI/ISEE-3 data between Aug. 1978 and June 1982, which is representative for the in-ecliptic solar wind around solar maximum (see the section on He above). More recently, Galvin et al. (1993) found $Si/O = 0.18 \pm 0.02$, both inside and after a coronal mass ejection (CME) observed on days 88–97, 1991, by SWICS/Ulysses. In high-speed streams observed in the Earth's magnetosheath by AMPTE, Gloeckler et al. (1989) give $Si/O = 0.054 \pm 0.009$. Preliminary results from SWICS confirm that Si/O, much like Mg/O, is significantly lower in fast streams than in the interstream, if not quite as low as the AMPTE value.

Sulfur. S, like C, is an important element, because it is near the division between low- and high-FIP elements. Shafer et al. (1993) have determined the S/Si abundance ratios in a driver plasma after a CME, and in a fast stream, both observed in 1991 by SWICS/Ulysses. Renormalized to O, using the Si/O ratio above, the values are $S/O = 0.05 \pm 0.02$ in the driver plasma, which we take as typical for the interstream value, and $S/O = 0.022 \pm 0.008$ in the fast stream. They agree well with the ones found earlier in the Earth's magnetosheath by CHEM/AMPTE.

Argon. Solar wind Ar still is exclusively available from the SWC lunar foil experiment (except for gas inclusions in the lunar soil, which are not reviewed in this chapter), from which the Ar/He flux ratio can be determined (Geiss et al. 1972). From the Ar/He value of Cerutti (1974), which we renormalize to O using the He/O ratio of Bochsler et al. (1986) in Table I, we obtain $Ar/O = 0.004 \pm 0.001$. We take this value as representative for the slow, in-ecliptic solar wind. The Ar/O abundance ratio in fast streams is unknown, but from its FIP (cf. Sec. III.A.) we believe that it is not significantly different from the interstream value.

Iron. Fe has a long history of solar wind abundance determinations, because it can be observed by m/q or E/q type sensors, thanks to the fact that its low charge states, Fe^{7+} to Fe^{12+}, are not masked by other, more abundant elements in their spectra. Bame et al. (1979) found an average of $Fe/He = 1.9 \times 10^{-3}$, with variations of a factor of 4 in both directions. From the ICI/ISEE-3 data set, Schmid et al. (1988) obtain $Fe/He = (2.5^{+1.2}_{-0.8}) \times 10^{-3}$, which is compatible with the earlier value. As in the case of He and Si, we adopt this value (renormalized to O, $Fe/O = 0.19^{+0.10}_{-0.07}$) as representative for the interstream solar wind in our table. More recently, Ipavich et al. (1992) have determined a lower value of $Fe/O = 0.12 \pm 0.03$ from a three-month survey in 1990–1991, using SWICS/Ulysses. With this instrument, the high charge states, Fe^{13+} to Fe^{16+}, can be observed directly, with little contamination from Si (or S). This new, lower value is in accord with an earlier reporting by Bame et al. (1975) of $Fe/O = 0.1$, and a more comprehensive survey is needed to decide whether the SWICS value is significantly lower in the interstream than the one from ICI. Fe in high-speed streams was observed for the first time by Mitchell and Roelof (1980), using IMP 7/8, and later by Ipavich et al. (1986), using the ULECA/ISEE-3 instrument. In two fast streams, the latter authors find Fe/O = 0.063 (renormalized to O). (The same authors also quote a Fe abundance observed in the driver plasma behind a shock, which agrees better with the high ICI value rather than the low SWICS value.) This value is significantly lower than in the interstream solar wind, as it is also the case for Mg/O and Si/O (see above). It was confirmed by Gloeckler et al. (1989), who find Fe/O = 0.057±0.007 in fast streams observed in Earth's magnetosheath.

Krypton and Xenon. The very low abundances of the heavy noble gases so far made them unaccessible to both space-borne mass spectrometers and to the foil collection technique. However, due to the longer integration time by orders of magnitude, they may be detected as implants in the lunar regolith. Still, there are two problems with the lunar data: (1) to identify the true solar wind component; and (2) to make sure that the elemental composition is not shifted by diffusive losses. Wieler et al. (1993) have most rigorously solved this problem with their on-line etching technique and derived reliable Ar/Kr/Xe ratios. They used for this work the mineral ilmenite, which has an exceptionally high retentivity for noble gases, enabling the authors to exclude changes in element composition by diffusive losses. From figure 2 of Wieler et al. (1993), we read off (and again renormalize to O) Kr/O = $(3.4\pm0.4) \times 10^{-6}$, and Xe/O = $(0.88 \pm 0.12) \times 10^{-6}$ for recently ($\lesssim 10^8$ yr) irradiated lunar soil. We take these values as representative for the predominantly slow, in-ecliptic solar wind.

B. Variations in Solar Wind Abundances

It is obvious from Table I that significant differences exist between the abundance ratios in the slow, in-ecliptic solar wind, and in the fast, coronal hole associated streams for some element ratios (e.g., Mg/O), while others (e.g., Ne/O) show no such behavior. This can be viewed by comparing the two panels of Color Plate 12, where we simply have accumulated all pulse height data of SWICS/Ulysses between February 1992 and July 1993. The upper panel contains all counts from spectra where the solar wind bulk speed (of α particles) was higher than 600 km s^{-1} and thus represents the coronal hole composition, while the lower panel was accumulated from the low-speed ($v_\alpha < 600$ km s^{-1}) spectra. The total accumulation time was roughly equal for the two panels. Even though these are raw data, the relative abundances and particularly the charge states of the heavy ($A>4$) elements can readily be compared between the panels. In fast streams, all heavy elements exist in significantly lower charge states, indicating a lower temperature at the source of these streams. Also, the reduced Mg/O ratio in high-speed streams is quite clearly visible (the comparison between the adjacent Mg^{10+} and Ne^{8+} makes this clear, because the Ne/O ratio is the same in both panels).

Additionally, transient solar wind events (e.g., flare associated mass ejections) may also show strong abundance anomalies. We now briefly discuss the He/H abundance variations, which have been found ever since this ratio can be routinely monitored. Then we turn to the variations in the Mg/O and C/O ratios in some more detail. Finally, we discuss the abundance anomalies found in the coronal Ne/O and Ne/Mg ratios and compare them to the neon abundance variations in the solar wind.

He/H. The He/H abundance ratio is observed to be variable on a wide range of time scales. Long-term variations became evident as soon as the *in-situ* observations covered a full solar cycle; in recurring high-speed streams, He/H is higher ($\sim$5%) during several days than in the slow solar wind (average $\sim$4%). The variations about the latter average are much larger (by a factor of $\sim$3) than in the high-speed streams, leading to the conclusion that these streams represent a structure-free state of the solar wind, while the slow wind is filamentary in structure (Bame et al. 1977).

Much larger variations on short time scales of the order of hours are produced by transient phenomena. After the passage of a flare associated interplanetary shock, the He/H ratio may be observed to increase by almost an order of magnitude in the driver gas behind the shock (Hirshberg et al. 1972; Borrini et al. 1982). Similar enrichments in the driver gas have also been observed for Fe by Ipavich et al. (1986) and Schmid et al. (1988). This may be interpreted as evidence for compositional stratification in the low corona and the subsequent ejection of a large body of unfractionated coronal material in a flare associated event (see, e.g., Hundhausen 1972). Such a dynamical accumulation of heavy ions in the low corona is indeed predicted by solar wind expansion models (Geiss et al. 1970*a*; Bürgi and Geiss 1986), who find that the ions heavier than H are dragged through the corona at lower speeds than H and thus are dynamically accumulated there. This effect could be particularly strong for He, because it has the least favorable drag factor (Geiss et al. 1970*a*)

$$\Gamma = Q^2/(2A - Q - 1) \qquad (1)$$

for collisions with protons, where Q is the charge and A is the mass number of the ion.

Another type of short-term He/H variations is observed at magnetic sector boundaries, which occur near the leading edge of high-speed streams. By superimposing 74 well-defined sector boundary crossings, which are marked by a magnetic field reversal, a density peak, and a minimum in both the flow speed and the proton temperature, Borrini et al. (1981) found a significant depletion of He/H to $\sim$3% on the average, and at some crossings virtually no He was present. The very slow solar wind at sector boundaries is generally believed to emerge from coronal streamers, but it is currently unclear how the material may leave these magnetically closed structures. Two possibilities are:

1. Flow from the low corona around the streamer along open field lines, or
2. Quasistationary reconnection at the top of the streamer.

In both cases, a depletion of He/H may be expected, in case (1) by the unfavorable Coulomb drag of He in H, as discussed above, and in case (2) by static, gravitational stratification of the material inside the streamer (Geiss 1985). It is therefore crucial to repeat the analysis of Borrini et al. (1981) for other heavy elements, such as O/H, in order to discriminate between these possibilities. While in case (2) an even stronger depletion of O/H is expected (Bürgi 1992), due to the higher mass of O, no strong depletion of O/H at sector boundaries is expected in case (1), because the Coulomb drag factor is larger due to the higher charge state of O, normally O^{6+}. Results from SWICS/Ulysses indicate that the O/H ratio does not drop significantly at sector boundary crossings with a reduced He/H ratio (Wimmer Schweingruber 1994; von Steiger et al. 1995), and thus favor case (1).

C/O. From Fig. 1, it is apparent that the C/O abundance ratio is not strongly varying; specifically, there is no systematic difference between C/O in high-speed streams as compared to the interstream solar wind (see Table I). This is further illustrated in Fig. 3. We have accumulated daily C/O averages using the data from SWICS/Ulysses for the whole Ulysses mission up to day 325, 1993. These are plotted vs the freezing-in temperature T_O obtained from the O^{7+}/O^{6+} charge state ratios, indicative for the solar wind type (see Sec. II.C.). Two populations may be easily distinguished in Fig. 3. The data obtained in fast streams from coronal holes fall below $T_O \lesssim 1.3 \times 10^6$K, and they are clearly separated from the interstream population above $T_O \gtrsim 1.3 \times 10^6$K. Both populations yield the same average C/O within the error bars (see Table I); it should also be noted that both populations are enriched in C relative to the photosphere by a factor of $\lesssim 2$, and virtually no data points fall below the photospheric value. We will return to this point in Sec. III.

No significant trend of C/O with T_O is discernible in Fig. 3, and only a slightly more positive correlation is obtained if the interstream data alone are considered. This result is somewhat at variance with that obtained by von Steiger et al. (1992*a*), who found a moderately strong positive correlation of C/O with T_C (see Fig. 6 below for the T_O to T_C correlation), using data of CHEM/AMPTE, taken in the Earth's magnetosheath. However, their data set was much smaller, and it differs in one important aspect from the SWICS data: Only high-pressure solar wind could be observed, because only at such instances the magnetopause was sufficiently compressed to allow the apogee of AMPTE/CCE to reach the magnetosheath. Such high pressures can only be produced either by fast streams or by transient high density events, usually associated with solar flares, but not by the slow, interstream solar wind. The CHEM data are thus strongly biased towards high pressure, particularly for $T_O \gtrsim 1.3 \times 10^6$ K. By mixing the subpopulation of high-pressure events with the ordinary interstream solar wind, as in the SWICS data presented here, the positive correlation of C/O with T_O in the former is swamped by the latter, much more frequent cases, and only a marginal correlation survives. It remains, of course, to be shown that the trend found by CHEM is recovered from the SWICS data if they are restricted to high-pressure solar wind.

Mg/O. From Fig. 1, a clear positive correlation of the Mg/O abun-

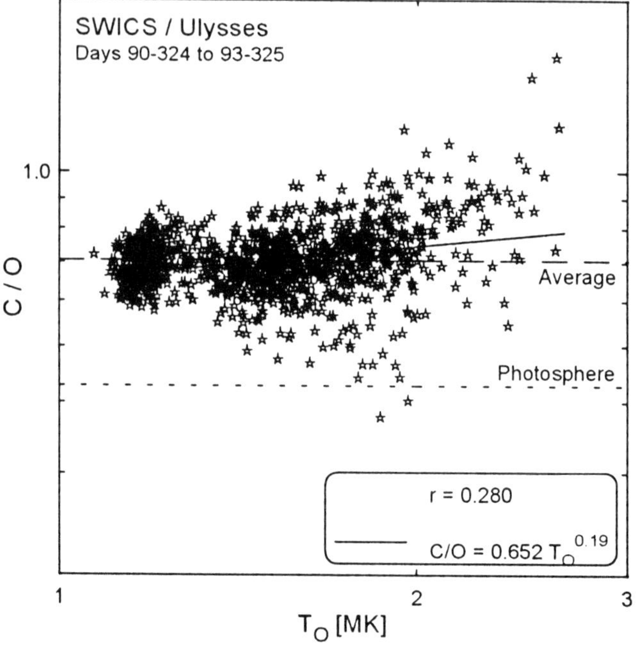

Figure 3.   Solar wind C/O abundance ratio vs oxygen freezing-in temperature from SWICS/Ulysses. The data covers the whole mission up to day 325, 1993, and each symbol corresponds to a daily average. Two populations may be readily distinguished: fast streams at $T_O < 1.3 \times 10^6$ K, and interstream solar wind at $T_O > 1.3 \times 10^6$ K. Both yield the same average C/O, and no significant trend with T_O is discernible.

dance ratio with the coronal freezing-in temperature T_O is evident. In Fig. 4, we have plotted daily averages of the Mg/O abundance ratio obtained by SWICS/Ulysses vs T_O. Mg was derived from the Mg^{10+} charge state density. The contribution from the other charge states (mainly Mg^{9+}) was added, assuming the freezing-in temperatures of Mg and O to be the same, and using the tables of Arnaud and Rothenflug (1985). This contribution ranges from $\sim 5\%$ at high T_O ($\sim 2 \times 10^6$ K) to $\sim 50\%$ at low T_O ($\sim 1.2 \times 10^6$ K). This procedure is necessary at present because of the difficulty to extract the low, less abundant charge states of Mg from the $m - m/q$ –matrix (Color Plate 12), particularly Mg^{9+} due to the interference with O^{6+}. Preliminary results on Mg^{8+} indicate that the freezing-in temperature calculated from Mg^{10+}/Mg^{8+} indeed agrees with T_O sufficiently well, and thus justify the correction applied to Mg^{10+} to first order. [However, it is possible that the temperature $T(Mg^{10+}/Mg^{8+})$ is slightly, but systematically lower than T_O, which implies that our correction to Mg^{10+} may be underestimated, and hence that the total Mg density may be higher than given in Table I, both in the interstream and in coronal hole flows.] As in the case of C/O, in Fig. 4 the same two populations are clearly visible and well separated from each other. High-speed stream data at $T_O \lesssim 1.3 \times 10^6$ K, and interstream solar wind at $T_O \gtrsim 1.3 \times 10^6$ K. However, unlike in the case of C/O, a significant correlation of Mg/O with T_O exists, with a correlation coefficient of $r > 0.8$. The trend is most clearly seen in the interstream data at high T_O, where the correction discussed above for charge states other than Mg^{10+} is small, and it is therefore not produced by said correction. Furthermore, contributions from other elements to the region of Mg^{10+} in the $m - m/q$ –matrix (Color Plate 12), notably by Si^{12+} were carefully subtracted, thus avoiding a spurious correlation,

From this comprehensive survey, we may firmly conclude that the Mg/O abundance ratio is significantly lower (and also less variable) in fast streams than in the interstream solar wind, a fact which was first pointed out by Gloeckler et al. (1989), using data obtained in the Earth's magnetosheath. If we take Mg as a representative for all low-FIP elements, this has immediate consequences for the strength of the FIP bias (solar wind abundance ratio divided by solar photospheric abundance ratio). It appears to be significantly lower (only a factor of ~ 2) in fast streams than in the slow solar wind, where it amounts to a factor of ~ 4.

In order to justify the assumption that Mg indeed is a fiducial low-FIP element, it has yet to be shown that other low-FIP elements show a similar behavior. So far, the published evidence is somewhat ambiguous. Galvin et al. (1992) find a positive correlation of Si/O with freezing-in temperatures obtained from Si charge states during a two-month survey of SWICS data in late 1991, even though it is somewhat smaller and less significant than what we find for $Mg/O(T_O)$. On the other hand, Ipavich et al. (1992) find only a very marginal correlation of Fe/O with T_O during a three-month survey in late 1990 and early 1991, also using SWICS data. We believe that this data set is too small to conclude that there is no $Fe/O(T_O)$ correlation, particularly because it contains no coronal hole data. Preliminary work by Gloeckler (1993, personal communication) now finds such a $Fe/O(T_O)$ correlation in a larger survey in 1992. Larger surveys on both Si/O and Fe/O are needed, and will be undertaken in the future, to address the question of the strength of the FIP bias in coronal holes as compared to interstream solar wind. We discuss the significance of this parameter in Sec. III.

Ne/O. Ne and O both are high-FIP elements, and according to the standard FIP pattern (Sec. III.A), the Ne/O ratio should be unbiased in the corona and solar wind when compared to the photospheric ratio. However, Widing and Feldman (1992), and McKenzie and Feldman (1992), using EUV and soft X-ray data from SKYLAB, report variations in the coronal Ne/O abundance ratio of up to a factor of 2. High chromospheric Ne/C and Ne/O ratios were found by Reames et al. (1988), deduced from γ-ray spectroscopy in the solar atmosphere, and from ^{3}He-rich flares. These may be caused by localized (spatial and temporal) preferential ionization of Ne by soft X rays and a subsequent enhancement of Ne by charge-dependent upward mass transport (Shemi 1991). Based on such observations, it has been suggested that there

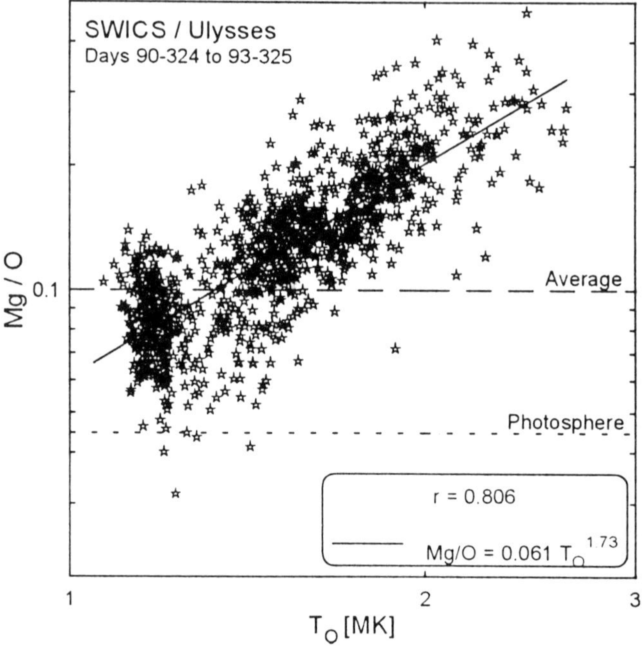

Figure 4.   Solar wind Mg/O abundance ratio vs oxygen freezing-in temperature from SWICS/Ulysses. The data covers the whole mission up to day 325, 1993, and each symbol corresponds to a daily average. The same two populations as in C/O (T_O) (see Fig. 3) may be readily distinguished. However, unlike in C/O (T_O), there is a clear positive correlation of the Mg/O abundance ration with T_O.

exists a region where the FIP step function is inverted to account for this observation (see Antiochos, as cited in Schmelz [1993]). A recent FIP model also predicts the existence of such an inversion layer (Marsch et al. 1994) (see Sec. III.B). On the other hand, *in-situ* observations by SWICS/Ulysses show quite a constant Ne/O ratio, where variations by a factor of 2 are the exception rather than the rule. It follows that the regions of enhanced Ne/O observed in the corona are either too small to make an observable imprint on the solar wind Ne/O ratio because of dilution with the ambient material, or that they are isolated from the part of the solar atmosphere contributing to the solar wind.

More puzzling is the finding of a strongly varying Ne/Mg abundance ratio, which is correlated with the magnetic field morphology of the region where it is measured (Widing and Feldman 1992). In diverging field regions, as well as in polar plumes, the observed Mg/Ne is *higher* by almost an order of magnitude than in active regions and flare loops, where it is again higher by a factor of $\sim$4 with respect to the photosphere. While the latter factor is clearly interpreted as the familiar FIP bias, the increase of Mg/Ne in open field regions is in contradiction to the *in-situ* observations of Mg/O (and Mg/Ne) ratio discussed above, which hold that the Mg/Ne ratio is *reduced* in fast streams, known to emanate from these regions (cf. Sec. II.C.). Again, these regions with an overenhancement of Mg/Ne must be isolated from the solar wind source region, but how this is achieved is difficult to envisage in open field regions. It is hoped that new data from Ulysses, taken directly above the polar coronal holes, will shed more light on this question.

C. Charge States of Solar Wind Ions

The charge states of heavy elements observed in the solar wind constitute a powerful diagnostic of the corona. While the solar material is streaming up through the corona, the abundance ratio of any two charge states follows its local reaction equilibrium value with the electrons (Arnaud and Rothenflug 1985) up to the point where the expansion time scale becomes smaller than the ionization/recombination time scale. Around this point, which is situated between $\sim$1.5 and $\sim$3.5 $R_\odot$ depending on the particular pair considered and on the flow state of the nascent solar wind, the charge state ratio freezes in and remains virtually unaltered out to 1 AU (Bame et al. 1974; Owocki et al. 1983). Thus, solar wind charge state ratios may be used as coronal thermometers (see Fig. 5). Because different ratios freeze in at different coronal altitudes, a coronal temperature profile may be obtained from them. In Fig. 6, we compare the coronal temperatures obtained from the O^{7+}/O^{6+} and the C^{6+}/C^{5+} ratios by SWICS/Ulysses. An excellent correlation between the two is observed, but, on the average, the oxygen temperature is 23% higher than the carbon temperature. This may be interpreted as O freezing in nearer to the coronal temperature maximum than C, but to derive a temperature gradient, accurate knowledge of the freezing in altitudes would be required.

Bürgi and Geiss (1986) have developed theoretical models for the coronal expansion of a main gas (p-e or p-α-e) with an admixture of heavy minor ions as trace gases (C, N, O, Ne, Mg, and Si). They followed numerically the charge state ratios from the base of the corona out to 10 $R_\odot$; freezing in occurred between 1.5 and 2 $R_\odot$ for all heavy ions considered. However, the freezing-in temperatures obtained in the simpler p-e model proved to be lower than normally observed in the slow solar wind. The principal result of Bürgi and Geiss (1986) was that this problem could be solved by taking the finite abundance of the α particles into account, in a p-α-e model. Thereby, the scale height in the corona is altered by a dynamical accumulation of α particles, increasing the temperature maximum and thus the freezing-in temperatures. Agreement with charge state ratios measured *in situ*, particularly for the well-determined pair O^{7+}/O^{6+}, could be obtained both for slow solar wind (their model 4) and for a wave-driven coronal hole model (their model 5). The model was extended to considering non-Maxwellian electron distribution functions in the corona (Owocki and Scudder 1983; Bürgi 1987), and to

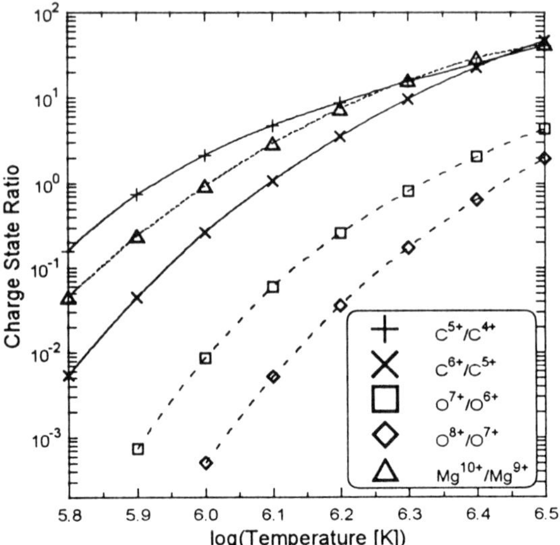

Figure 5.   Charge state ratios as coronal hole thermometers. In collisional equilibrium, each pair of charge states assumes a ratio indicative of the local electron
temperature (data from Arnaud and Rothenflug 1985).

differential speeds between the individual charge states of an element (Bürgi
1989, unpublished).

By probing the conditions in the solar wind source region, the heavy
ion charge states and abundances are also an important diagnostic for the
solar wind type itself. It has been known for a long time that coronal holes
are the sources of recurrent high-speed streams (Bame et al. 1977; Zirker
1977); however, only equatorial extensions of the coronal holes could be
observed by the in-ecliptic missions then available. Evidence for high speeds
in the solar wind from the polar coronal holes also came from observations of
interplanetary scintillations (Coles et al. 1980). Recently, this fact has been
impressively confirmed by *in-situ* measurements of the Ulysses spaceprobe,
which is currently on the transition from the ecliptic to high solar latitudes
(Smith et al. 1993). After Jupiter encounter in February 1992, which deflected
Ulysses towards high heliographic latitudes on an orbit with an inclination of
80°, a recurrent high-speed stream started to emerge with the sidereal rate at
mid latitudes of ~1/26 days (Fig. 7). At first, the speed maxima increased to
~800 km s⁻¹, and later the interstream minima also increased, until Ulysses
was completely immersed in the south polar coronal hole stream after mid-
1993. The key observation is that the coronal freezing-in temperature is
decreasing in every recurrence of the stream and stays low in the polar stream,
unambiguously identifying its coronal hole origin.

High solar wind speeds are not unique to fast streams from coronal holes,
but may also be produced by flare-associated transient events. A second

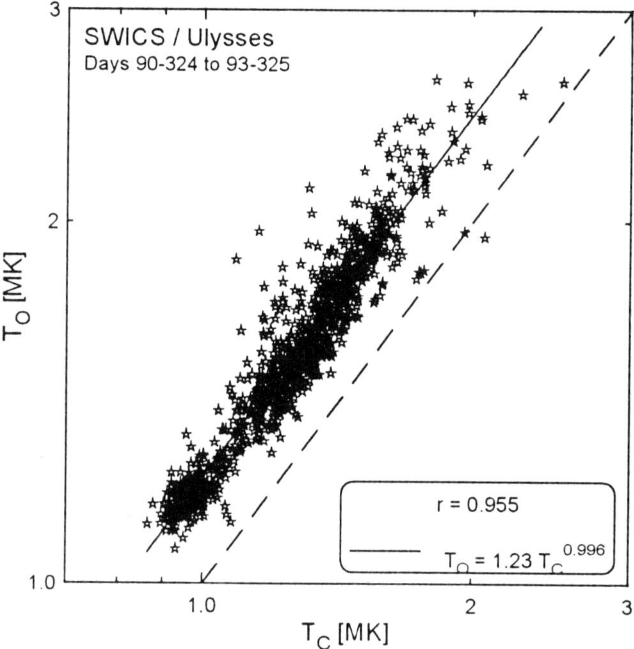

Figure 6.   Freezing-in temperatures calculated from the O^{7+}/O^{6+} and the C^{6+}/C^{5+} charge state ratios (labeled T_O and T_C, respectively), measured by SWICS/Ulysses. Each data point corresponds to a daily average. The same two populations as in Figs. 3 and 4 may be distinguished: coronal hole associated fast streams at uniform, low T, and interstream solar wind at varying, high T. Both follow a strong power-law correlation with T_O 23% higher than T_C.

indicator, such as the freezing-in temperature, is needed to identify the origin of high-speed solar wind. This is illustrated in Fig. 7. For example, the speed increase in March 1991 is accompanied by a high T_O; indeed, it was a transient event produced by a coronal mass ejection associated with a series of large flares, not by a coronal hole. Similarly, the speed increase on top of the fast stream in November 1992 (see Fig. 1 for more detail) is also identified easily as a transient event by its high T_O.

High-speed streams and interstream solar wind not only differ in the absolute freezing-in temperatures, but also in the temperatures obtained from the different charge state pairs of one element (such as Si or Fe). In the slow, interstream solar wind, these temperatures differ considerably from each other, indicating that freezing in occurs at a different altitude for each charge state. On the other hand, in a fast-stream the freezing-in temperatures obtained from different charge state pairs of an element are observed to agree much

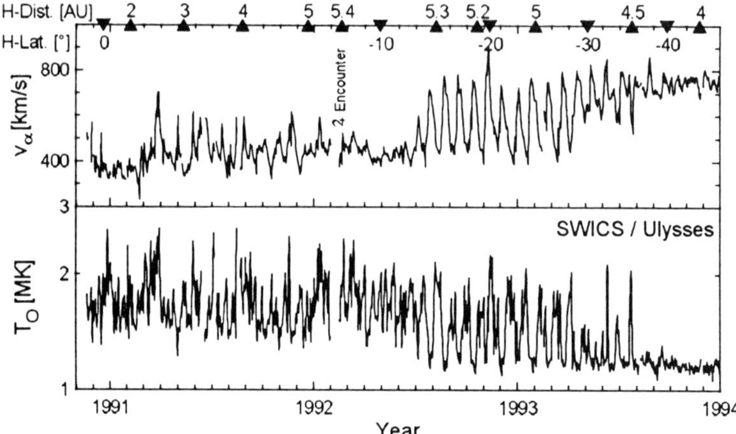

Figure 7.   Solar wind velocity of α particles as a function of time measured by SWICS/Ulysses between launch and late 1993 (upper panel). As Ulysses approaches high southern heliographic latitudes, a recurrent fast stream from an equatorial extension of the southern polar coronal hole (indicated by a low freezing-in temperature, lower panel) can be seen to emerge, and by late 1993, the spacecraft is completely immersed in it due to its high latitude.

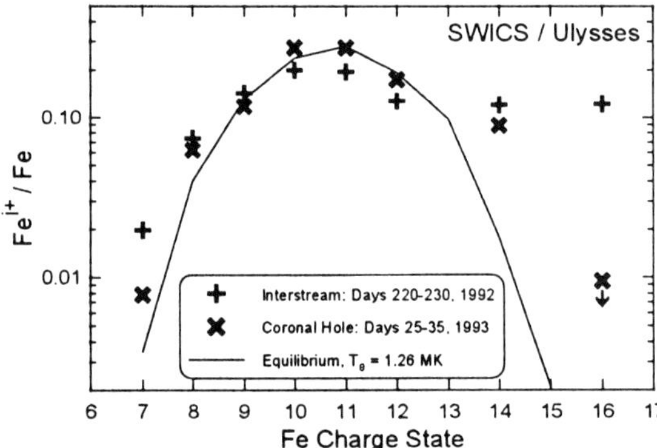

Figure 8.   Comparison of the charge states Fe^{i+} (normalized to total Fe) in the slow, interstream solar wind, and in a coronal hole associated fast stream. A uniform freezing-in temperature may only be found for the fast stream (indicated by the curve), but not for the slow solar wind. The coronal hole value for Fe^{16+} is an upper limit.

better with each other. In Fig. 8, we have determined average freezing-in temperatures from the charge states Fe^{7+} to Fe^{16+} in the slow solar wind and in a fast stream. In the latter case, a unique temperature may be found which fits nicely to the observed charge state ratios, but no such temperature exists for the slow solar wind, because both the low charge states and particularly the high charge states up to the Ne-like Fe^{16+} are abundant there (Ipavich et al. 1992). Such a behavior was predicted in the case of Si by Bürgi and Geiss (1986) from their dynamical calculation of charge states in an expanding solar atmosphere. In their model 4 (slow solar wind), they obtain widely differing Si temperatures from different charge state ratios, but these temperatures are almost uniform in their model 5 (coronal hole).

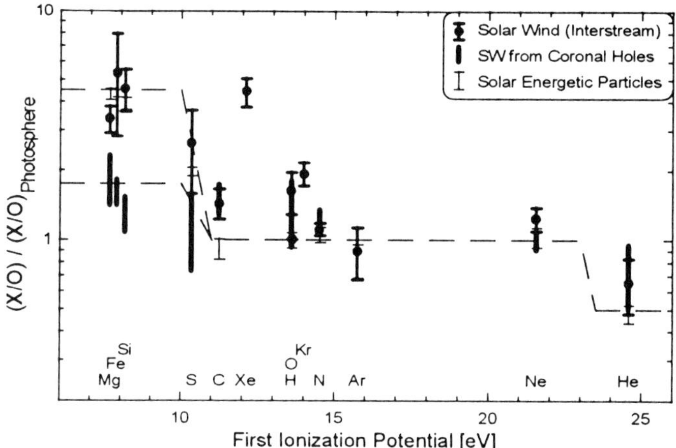

Figure 9. Solar wind (in ecliptic and from coronal holes) and solar energetic particle abundance ratios, relative to their photospheric values, as a function of the first ionization potential. A basic FIP fractionation pattern is present both in the in-ecliptic solar wind and in the SEPs (dashed line). Low-FIP elements are enriched, and He is depleted with respect to the high-FIP group, C-Ne. In the fast solar wind from coronal holes, the step between low- and high-FIP elements is significantly reduced. See text for a discussion of exceptions from the basic FIP pattern, such as Kr and Xe.

III. THE FIP FRACTIONATION EFFECT

A. FIP Fractionation Data

When we compare the normalized elemental abundances (relative to oxygen) of the solar wind and of the corona (as inferred from the solar energetic particle population, SEP) to the photospheric abundance ratios, we notice a significant fractionation between them. If this fractionation is plotted as a function of the first ionization potential, a regular pattern emerges (see Fig. 9). Elements with a low FIP ($\lesssim 10$ eV) lie on a plateau which is enhanced by a factor of ~ 4 to 5 with respect to elements with a high FIP ($\gtrsim 10$ eV). The intermediate elements with a FIP near 10 eV, C and S, are also enriched, but by smaller factors. The high-FIP elements again lie on a plateau up to, and including, Ne, but He, the highest FIP element, is underabundant by a factor of ~ 2 relative to the high-FIP elements. This basic FIP pattern, outlined by a dashed line in Fig. 9, is obeyed by the SEP and, with some exceptions

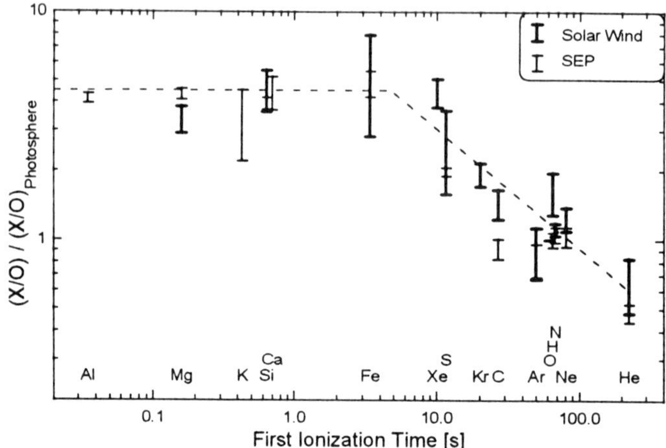

Figure 10. Abundance data from Fig. 9 but plotted vs the first ionization time (FIT). This results in a simpler pattern, with the low-FIT elements on a plateau and the high-FIT elements on a power-law with an exponent of about −0.5. Note that the FIP exceptions, Kr and Xe, now follow this pattern.

discussed below, by the in-ecliptic solar wind. It was first described for the SEP by Hovestadt et al. (1973); the SEP data reported here (see Table I) are from the recent compilation by Garrard and Stone (1993). Meyer (1981) and Veck and Parkinson (1981) first noted the same effect to hold in the in-ecliptic solar wind (see also the review by Meyer 1991).

With the abundances of more elements becoming available, and with the possibility to determine abundances as a function of the solar wind flow type, exceptions from the basic FIP pattern now become apparent. The most important one is a reduction of the fractionation factor between low- and high-FIP elements in the solar wind from coronal holes to a factor of ∼1.5 to 2 instead of ∼4 to 5, which is obeyed by all known low-FIP elements there (Mg, Fe, and Si) as discussed in Sec. II.B. It is interesting to note that carbon, which is slightly enriched relative to O in the interstream, seems to remain unchanged in fast streams, i.e., its slight enrichment is not reduced proportionally with that of the low-FIP elements. It may therefore be that C behaves as a (slightly enriched) high-FIP element in the interstream solar wind, but as a low-FIP element in fast streams, even though its enrichment factor does not change. This is not the case for S (Shafer et al. 1993), the other element near the FIP step. Its enrichment factor is ∼1.5 to 2 in the SEP and interstream solar wind, but drops to one in fast streams, joining the high-FIP elements, even though its FIP is smaller than that of C. From these emerging systematics, we would expect that Si would remain enriched also by a factor of 2 in fast streams while it is observed, in a preliminary measurement, to drop to the solar ratio (cf. Table I). However, the survey period where this factor has been obtained is too short to draw firm conclusions from this apparent discrepancy.

The heavy noble gases, Kr and Xe, do not follow the basic FIP picture either. Both are high-FIP elements, but both are enriched in the solar wind by factors of 2 (Kr) or even 4 (Xe). This may be understood to be a consequence of their high mass (see next section), and of their relatively short ionization time as compared to other high-FIP elements. The first ionization time (FIT) may therefore be the better ordering parameter than the FIP (cf. Geiss et al. 1994b). Ionization times are calculated as described in Geiss and Bochsler (1985); for elements not considered there, such as Kr and Xe, we use absorption cross sections from Hudson and Kieffer (1971), and assume 100% ionization efficiency to determine the FIT. In Fig. 10, we have plotted the abundance data from Fig. 9 vs FIT, where the new fractionation pattern is easily visible. Low-FIT ($\lesssim 10$ s) elements again lie on a plateau, but the fractionation factor of high-FIT elements now is roughly a power law in FIT with an exponent of about -0.5, and both Kr and Xe follow this pattern. The reason for the existence of a high-FIP plateau also becomes clear in Fig. 10. The high-FIP elements H, N, O, Ne, and Ar all have very similar first ionization times (Geiss and Bochsler 1985).

B. Theoretical FIP Fractionation Models

From the shape of the FIP fractionation curve (Fig. 9), we obtain two key ingredients for the construction of models: (1) any mechanism is to operate at a region in the solar atmosphere where it is (at least partially) neutral, or else it is not clear how the FIP (or FIT) could be a relevant parameter; (2) the step near 10 eV suggests that photoionization by Lyman-α photons (10.2 eV) may play an important role (particularly for C, which has low-lying metastable levels that can be ionized by Lyman-α, whereas the ground state cannot). The flux in the Lyman-α line is more than a factor of 2 higher than the integrated flux at shorter wavelengths. Photoionization should therefore be included in any attempt to model the FIP fractionation.

Over the past decade, several models have been proposed to explain the observed FIP fractionation (see also the review by Meyer 1993), most of which operate by ion-neutral separation in the chromosphere, predominantly by diffusion perpendicularly to a magnetic field. Geiss (1982) has shown that ion-neutral separation is much more effective than ion-ion separation, and gives quantitative estimates for the conditions under which it could take place: temperature $T \lesssim 10^4$ K and density $n \lesssim 10^{16}$ m^{-3}, i.e., in the upper chromosphere and lower transition region. The density estimate was obtained with gravity as the only driving force for the diffusion of neutrals; if other driving forces are considered, the limit on n would be higher.

Vauclair and Meyer (1985) considered a model in which ions and neutrals separate in the chromospheric canopy, where the magnetic field is horizontal (1000–2000 km above the photosphere). Neutral atoms may settle downward driven by gravity, but the diffusion of ions is impeded by the field, which reduces their diffusion velocity by a factor of

$$1/(1 + \omega^2 t_{\text{col}}^2) \qquad (2)$$

where ω is the gyrofrequency, and t_{col} is the typical time between collisions. The degree of ionization of the elements considered was not calculated self-consistently, but rather taken from the non-LTE model calculations of Vernazza et al. (1981) (cf. Fig. 2 of Geiss and Bochsler 1985). The authors find that a FIP-dependent fractionation may indeed be obtained in this model, but the resulting diffusion times to separate the species (hours to days) are much larger than the typical lifetime of structures in the solar atmosphere. This is not surprising, as the density in the model atmosphere of Vernazza et al. (1981) is higher throughout the considered region than the limiting density for gravity-driven ion-neutral separation of Geiss (1982), $n \lesssim 10^{16}$ m^{-3}. Recently, Vauclair (1994) has taken a new, dynamical approach along the same line, in which the magnetic field is rising from the photosphere through the chromosphere and corona, lifting an overabundance of low-FIP elements (the "skimmer" model). Preliminary results look quite promising if a field of 5 Gauss at a velocity of 5 km s^{-1} is considered (her figure 3). Ip and Axford (1991) have also considered a similar scenario with a magnetic flux tube rising through the chromosphere.

Quite complementary to Vauclair and Meyer (1985), Geiss and Bochsler (1985) explicitly investigated the excitation and ionization of nine elements under chromospheric conditions, but did not specify the mechanism of separation. A neutral gas mixture was exposed to the solar spectrum in an optically thin environment. Excitation and ionization by photons made the main contribution, but collisional ionization by electrons ($n_e = 10^{16}$ m^{-3}, $T_e = 10^4$ K) as well as radiative recombination were also taken into account. The resulting first ionization times (see Fig. 10) suggest that any separation of neutrals from ions has to occur on short time scales of the order of a minute, the typical ionization time of high-FIP elements under these conditions. This in turn suggests that the separation is to operate on small length scales, in order to keep the relative velocities within reasonable limits.

The first attempt to to construct a model considering both ionization and separation was made by von Steiger and Geiss (1989). They considered an initially neutral gas mixture of photospheric composition and uniform temperature, concentrated in narrow, vertical or nearly vertical magnetic structures (see Fig. 11). As the gas rises, it becomes excited and ionized by photons and electrons, and radiative recombination is also considered. The newly created ions remain bound to the magnetic field, i.e., trapped in the upward flow. The neutrals, on the other hand, are driven across the field lines by an initial density gradient, and get lost in a downward stream outside the structure under consideration. The resulting gas mixture at the top of the structure is so enriched in low-FIP elements, because they have been ionized more quickly, that less neutrals were lost from the structure. In Fig. 12, the resulting gas

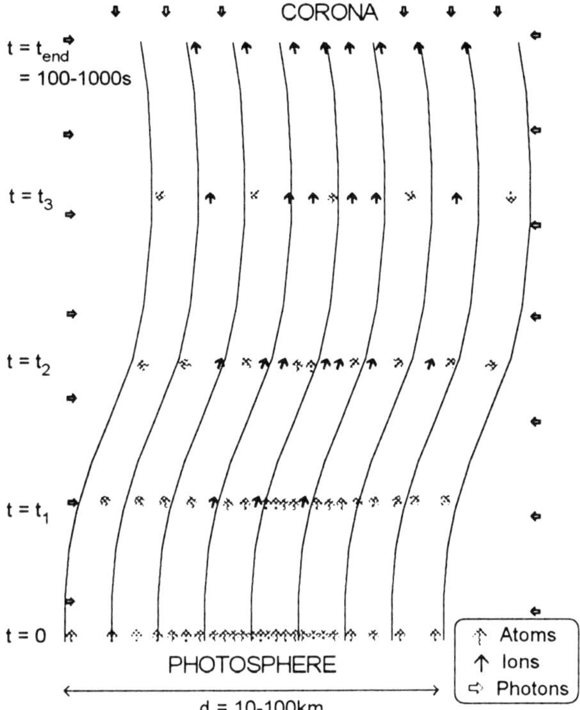

Figure 11. Sketch of the geometry used in the FIP fractionation model of von Steiger and Geiss (1989). An initially ($t = 0$) neutral gas (light arrows) is rising from the bottom in a more or less vertical magnetic structure. As it is irradiated by solar EUV (double arrows), the gas is gradually ionized (dark arrows) and thus trapped to the field, whereas the neutrals are driven across the field by an initial density gradient and get lost ($t = t_{1,2,3}$). The resulting gas mixture being fed to the corona is enriched in low-FIP elements ($T = t_{end}$).

mixture obtained in a typical model run is compared to the (in-ecliptic) solar wind composition. The agreement is reasonably good. Not only is the step between low-FIP and high-FIP elements reproduced, but the critical elements near this step, C and S, are obtained partially enriched, as is observed. Moreover, even the "FIP exceptions," Kr and Xe, result roughly with their observed enrichment from the model (Fig. 12, model O). The reduced step height observed in fast streams may also be accounted for by the model. It is sufficient to consider less narrow structures, of the order of 100 km, to obtain a step height of a factor of 2 only (Fig. 12, model R). This indicates that the chromospheric structure may be coarser beneath coronal holes than under closed field regions. However, the model has been criticized, less for the need of an (as yet) unobservable chromospheric structure and specific geometry

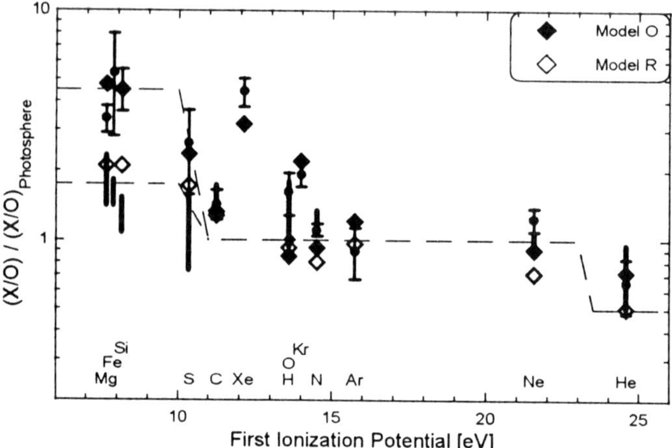

Figure 12.   Comparison of the results of a typical model run of von Steiger and Geiss (1989) (full symbols) to the solar wind composition from Fig. 9. Model parameters are $n = 10^{15} \mathrm{m}^{-3}$, $T = 6000$ K, $d = 20$ km (width of the structure), and $\theta = 10°$ (mean inclination of the structure). The agreement is quite good, even for the "FIP exceptions," Kr and Xe. Considering a wider model structure with $d = 120$ km (open symbols) results in a lesser step height, as it is observed in fast streams.

in which the mechanism operates, but for the fact that it fails to produce the correct fractionation pattern at densities higher than $\sim 10^{16}$ m^{-3} because of the reduced mobility of the neutrals leaving the structure. In the model atmosphere of Vernazza et al. (1981), densities are higher than this limit throughout the chromospheric temperature plateau, making it difficult to point at the exact location where this process may operate. However, chromospheric structures are generally not described by the homogeneous Vernazza model atmosphere.

Recently, Marsch et al. (1994) (see also von Steiger and Marsch 1994) have tried to overcome this difficulty by constructing a similar model, retaining the careful treatment of ionization and collisions by von Steiger and Geiss (1989), but considering a less specific model geometry. The fractionation is assumed to be by diffusion across a horizontal slab and along a (nearly) vertical magnetic field. For specific, but naturally chosen boundary conditions at the slab interfaces, the observed FIP fractionation pattern in the slow solar wind with a step height of ~ 5 is obtained at the upper slab interface, as well as the depletion of He relative to the high-FIPs. The crucial boundary condition is that the flux at the top of the slab is the same for each heavy ion species in natural units (density normalized to one at the slab bottom, and velocity expressed in units of the natural diffusion speed, $\sqrt{D\tau}$, where D is the diffusion constant in the ambient main gas, and τ is the ionization time). The main improvement is that the fractionation obtained in this model is almost independent of density and temperature for realistic chromospheric conditions. An interesting feature of the model is the prediction of an inversion layer inside the diffusive slab, in which the fractionation pattern is reversed with respect to the upper slab interface. On the other hand, the agreement for Kr and Xe is less good than it is in von Steiger and Geiss (1989). An adaption of the model to the reduced FIP step height observed in fast streams is possible by assuming that diffusive equilibrium is not quite reached, resulting in an incompletely fractionated mixture being output at the top of the slab. The similarity of the predictions of the two models is not very surprising, as both use the same mode of ionization, and differ only in the mode of ion-neutral separation.

A class of recently proposed models works by diffusion driven by electromagnetic forces, either (1) by dragging ions into a chromospheric gas parcel to be supplied to the corona [as in the models of Vauclair (1994), and of Ip and Axford (1991) mentioned above], or (2) by expelling neutrals, e.g., by pinching, out of such a parcel. S. Antiochos (as cited in Schmelz [1993]) has qualitatively investigated a mechanism of type (1). A spot of the chromosphere is heated by downward heat flux from the corona along a flux tube, leading to localized evaporation and lifting of chromospheric material. This, in turn, produces a horizontal thermoelectric field, dragging ions from the surroundings into the uplifting region. This leads to an accumulation of low-FIP elements in this region, because they are already ionized in the surroundings, whereas the high-FIPs still are predominantly neutral there. A mechanism of type (2) was investigated, again qualitatively, by Hénoux and Somov (1992). Considering a cylindrically symmetric magnetic field structure in a partially ionized atmosphere, the authors calculate the system of DC currents created by such a configuration. They show that, near the periphery of the flux tube, there is a strong diffusion of neutrals relative to the ions out of the structure, which may produce the observed FIP segregation. However, in order to justify their promising status (Meyer 1993), these models have yet to be worked out quantitatively and make predictions that can be compared to the observations. Finally, the work of Tagger et al. (1995) should be mentioned, who investigate an FIP fractionation by waves in a magnetic field in a partially ionized gas mixture, an idea put forward by Bochsler and Geiss (1989). Again, in order to assess the value of this idea, quantitative predictions are needed.

In summary, none of the nearly dozen models presented here is without its shortcomings, and much detailed work remains to be done in order to pinpoint the location and the mechanism(s) producing the observed FIP fractionation.

IV. OUTLOOK AND CONCLUSIONS

A. Future Solar Wind Missions

In the near future, two missions with comprehensive solar wind instruments are scheduled to be launched: Wind and SOHO. Both the SMS triple sensor on Wind (Gloeckler et al. 1995) and the CELIAS triple sensor on SOHO (Hovestadt et al. 1989) contain a novel type of high-resolution mass spectrometer. After passing a classical E/q deflection system, the ions pass a thin carbon foil and enter a time-of-flight section with a harmonic retarding potential (isochronous TOF). The time for half a harmonic oscillation is recorded and can be converted to the ion mass, because a known fraction of the ions leave the carbon foil singly charged (or neutral, but these are lost). The resolution of this sensor type is $m/\Delta m > 100$, which is one and a half orders of magnitude higher than that of SWICS. This will allow composition measurements of most elements up to Ni, and even isotope measurements of many of them.

An integrated three-dimensional space plasma instrument with high mass resolution has been described by Möbius et al. (1990). This sensor combines an E/q analyzer with two TOF sections, one isochronous for the ions and one conventional for the neutrals exiting the entrance carbon foil. Thereby, apart from the elemental and isotopic composition of the plasma, even its content in molecular ions may be determined, which makes this instrument type ideal for dusty planetary or cometary environments.

Finally, a mission to the solar corona with a perihelion of $\sim 4\ R_\odot$, carrying a comprehensive particles and waves payload (Feldman et al. 1989; Marsch et al. 1993), is considered essential for future improvement of our understanding of the solar wind. Many poorly understood phenomena, such as the solar wind acceleration, or the source region of the slow solar wind, could be investigated *in situ* by such a mission.

B. Conclusions

We have reviewed, after briefly discussing the principal experimental techniques, the current status of knowledge on the solar wind elemental composition. It has been shown that there is a systematic difference between the composition of the slow, interstream solar wind, and the fast streams emanating from coronal holes. While the former is enriched in low-FIP elements with respect to the photospheric composition, such an enrichment is much weaker in fast streams. The two solar wind types also differ substantially in their charge state composition, and it is shown that generally the frozen-in charge states, which act as coronal thermometers, are an ideal indicator for distinguishing different solar wind types when combined with the solar wind speed. Of the several models that have been proposed to explain the observed FIP fractionation, none can claim to explain the effect in a self-consistent way, indicating the cause for the effect, and evaluating its dynamics. Good progress has been made regarding the mode of ionization, and for the mode of ion-neutral separation several promising proposals have been made. Many models have yet to come up with results that can be compared to the observations, both in the solar wind and in the solar atmosphere.

We have not reviewed in this chapter the solar wind isotopic composition. This is partly due to the fact that the recent data are not superior to those obtained by the Apollo-SWC foil collection experiment (reviewed in Gloeckler and Geiss 1989), which were already confirmed by the ISEE-3 results for the He isotopes, or to the isotopic data on heavy noble gases derived from investigations of the lunar soils. Future missions with high mass resolution sensors, or a resumption of the foil collection technique, will advance our knowledge of the isotopic composition in the solar wind and the Sun. As it forms the baseline to which isotopic ratios found in other solar system objects are to be related, its study is considered one of the key questions for future solar wind missions (cf. Bochsler and Möbius 1993).

Acknowledgments. We thank A. B. Galvin, P. Bochsler, and R. Wimmer Schweingruber for many discussions, and W. C. Feldman and F. M. Ipavich for useful comments on the manuscript. This work was supported by the Swiss National Science Foundation.

REFERENCES

Anders, E., and Grevesse, N. 1989. Abundances of the elements: Meteoritic and solar. *Geochim. Cosmochim. Acta* 53:197–214.

Arnaud, M., and Rothenflug, R. 1985. An updated evaluation of recombination and ionization rates. *Astron. Astrophys. Suppl.* 60:425–457.

Bame, S. J., Asbridge, J. R., Feldman, W. C., and Kearney, P. D. 1974. Solar quiet corona: Temperature and temperature gradient. *Solar Phys.* 35:137–152.

Bame, S. J., Asbridge, J. R., Feldman, W. C., Montgomery, M. D., and Kearney, P. D. 1975. Solar wind heavy ion abundances. *Solar Phys.* 43:463–473.

Bame, S. J., Asbridge, J. R., Feldman, W. C., and Gosling, J. T. 1977. Evidence for a structure-free state at high solar wind speeds. *J. Geophys. Res.* 82:1487–1492.

Bame, S. J., Asbridge, J. R., Feldman, W. C., Fenimore, E. E., and Gosling, J. T. 1979. Solar wind heavy ions from flare-heated coronal plasma. *Solar Phys.* 62:179–201.

Bochsler, P. 1989. Velocity and abundance of silicon ions in the solar wind. *J. Geophys. Res.* 94:2365–2373.

Bochsler, P., and Geiss, J. 1989. Composition of the solar wind. In *Solar System Plasma Physics*, eds. J. H. Waite, J. L. Burch and R. L. Moore (Washington: American Geophysical Union), pp. 133–141.

Bochsler, P., and Möbius, E. 1993. Prospects for future solar wind missions. In *Scientific Requirements for Future Solar-Physics Space Missions*, eds. P. Maltby and B. Battrick, ESA SP-1157, pp. 43–51.

Bochsler, P., Geiss, J., and Kunz, S. 1986. Abundances of carbon, oxygen and neon in the solar wind during the period from August 1978 to June 1982. *Solar Phys.* 102:177–201.

Boesgaard, H. M., and Steigman, G. 1985. Big bang nucleosynthesis: Theories and observations. *Ann. Rev. Astron. Astrophys.* 23:319–378.

Borrini, G., Gosling, J. T., Bame, S. J., Feldman, W. C., and Wilcox, J. M. 1981. Solar wind helium and hydrogen structure near the heliospheric current sheet: A signal of coronal streamers at 1 AU. *J. Geophys. Res.* 86:4565–4573.

Borrini, G., Gosling, J. T., Bame, S. J., and Feldman, W. C. 1982. Helium abundance enhancements in the solar wind. *J. Geophys. Res.* 87:7370–7378.

Bürgi, A. 1987. Effects of non-Maxwellian electron velocity distribution functions
    and nonspherical geometry on minor ions in the solar wind. *J. Geophys. Res.*
    92:1057–1066.
Bürgi, A. 1992. Dynamics of alpha particles in coronal streamer type geometries. In
    Solar Wind Seven, eds. E. Marsch and R. Schwenn (Oxford: Pergamon Press),
    pp. 333–336.
Bürgi, A., and Geiss, J. 1986. Helium and minor ions in the corona and solar wind:
    Dynamics and charge states. *Solar Phys.* 103:347–383.
Cerutti, H. 1974. Die Bestimmung des Argons im Sonnenwind aus Messungen an den
    Apollo-SWC-Folien. Ph.D. Thesis, Univ. of Bern.
Coles, W. A., Rickett, B. J., Rumsey, V. H., Kaufmann, J. J., Turley, D. G., Ananthakr-
    ishnan, S., Armstrong, J. W., Harmon, J. K., Scott, S. L., and Sime, D. G. 1980.
    Solar cycle changes in the polar solar wind. *Nature* 286:239–241.
Coplan, M. A., Ogilvie, K. W., Bochsler, P., and Geiss, J. 1978. Ion composition
    experiment. *IEEE Trans. Geosci. Electron.* 16:185.
Feldman, W. C., Asbridge, J. R., Bame, S. J., and Gosling, J. T. 1977. Plasma and
    magnetic fields from the Sun. In *The Solar Output and Its Variation*, ed. O. R.
    White (Boulder: Colorado Assoc. Univ. Press), pp. 351–381.
Feldman, W. C., Anderson, J., et al. 1989. *Solar Probe—Scientific Rationale and
    Mission Concept*, JPL D-6797.
Galvin, A. B., Ipavich, F. M., Gloeckler, G., von Steiger, R., and Wilken, B. 1992.
    Silicon and oxygen charge state distributions and relative abundances in the solar
    wind measured by SWICS on Ulysses. In *Solar Wind Seven*, eds. E. Marsch and
    R. Schwenn (Oxford: Pergamon Press), pp. 337–340.
Galvin, A. B., Gloeckler, G., Ipavich, F. M., Shafer, C. M., Geiss, J., and Ogilvie, K. W.
    1993. Solar wind composition measurements by the Ulysses SWICS experiment
    during transient solar wind flows. *Adv. Space Res.* 13:(6)75–78.
Garrard, T. L., and Stone, E. C. 1993. New SEP-based solar abundances. *Proc. 23rd
    Intl. Cosmic Ray Conf.* 3:384–387.
Geiss, J. 1982. Processes affecting abundances in the solar wind. *Space Sci. Rev.*
    33:201.
Geiss, J. 1985. Diagnostics of corona by in-situ composition measurements at 1 AU. In
    Future Missions in Solar, Heliospheric and Space Plasma Physics, ESA SP-235,
    pp. 37–50.
Geiss, J., and Bochsler, P. 1985. Ion composition in the solar wind in relation to
    solar abundances. In *Rapports Isotopiques dans le Système Solaire*, (Toulouse:
    Cepadues-Editions), pp. 213–228.
Geiss, J., and Reeves, H. 1972. Cosmic and solar system abundances of deuterium
    and helium-3. *Astron. Astrophys.* 18:126–132.
Geiss, J., Hirt, P., and Leutwyler, H. 1970a. On acceleration and motion of ions in
    corona and solar wind. *Solar Phys.* 12:458.
Geiss, J., Eberhardt, P., Bühler, F., Meister, J., and Signer, P. 1970b. Apollo 11 and 12
    solar wind composition experiments: Fluxes of He and Ne isotopes. *J. Geophys.
    Res.* 75:5972–5979.
Geiss, J., Bühler, F., Cerutti, H., Eberhardt, P., and Filleux, Ch. 1972. Solar wind
    composition experiment. In *Apollo-16 Preliminary Science Report*, NASA-SP
    315, pp. 14/1–14/10.
Geiss, J., von Steiger, R., Gloeckler, G., and Galvin, A. B. 1994a. The neon abundance
    in the solar wind measured by SWICS on Ulysses. *Eos: Trans. AGU* 75:278
    (abstract).
Geiss, J., Gloeckler, G., and von Steiger, R. 1994b. Solar and heliospheric processes
    from solar wind measurements. *Phil. Trans. Roy. Soc. London A* 349:213–226.
Gloeckler, G. 1990. Ion composition measurement techniques for space plasmas. *Rev.*

Sci. Instrum. 61:3613–3620.

Gloeckler, G., and Geiss, J. 1989. The abundances of elements and isotopes in the solar wind. In *AIP Conf. Proc.*, vol. 183, ed. C. J. Waddington (New York: American Inst. of Physics), pp. 49–71.

Gloeckler, G., Ipavich, F. M., Hamilton, D. C., Wilken, B., Stüdemann, W., Kremser, G., and Hovestadt, D. 1986. Solar wind carbon, nitrogen and oxygen abundances measured in the Earth's magnetosheath with AMPTE/CCE. *Geophys. Res. Lett.* 13:793.

Gloeckler, G., Ipavich, F. M., Hamilton, D. C., Wilken, B., and Kremser, G. 1989. Heavy ion abundances in coronal hole solar wind flows. *Eos: Trans. AGU* 70:424 (abstract).

Gloeckler, G., Geiss, J., Balsiger, H., Bedini, P., Cain, J. C., Fischer, J., Fisk, L. A., Galvin, A. B., Gliem, F., Hamilton, D. C., Hollweg, J. V., Ipavich, F. M., Joos, R., Livi, S., Lundgren, R., Mall, U., McKenzie, J. F., Ogilvie, K. W., Ottens, F., Rieck, W., Tums, E. O., von Steiger, R., Weiss, W., and Wilken, B. 1992. The Solar Wind Ion Composition Spectrometer. *Astron. Astrophys. Suppl.* 92:267–289.

Gloeckler, G., Balsiger, H., Bochsler, P., Bürgi, A., Galvin, A. B., Geiss, J., Gliem, F., Hamilton, D. C., Holzer, T., Hovestadt, D., Ipavich, F. M., Kirsch, E., Lundgren, R., Ogilvie, K. W., and Wilken, B. 1995. The solar wind and suprathermal ion composition investigation on the Wind spacecraft. *Space Sci. Rev.* 71:79–124.

Grevesse, N., and Anders, E. 1991. Solar element abundances. In *Solar Interior and Atmosphere*, eds. A. N. Cox, W. C. Livingston and M. S. Matthews (Tucson: Univ. of Arizona Press), pp. 1227–1234.

Grevesse, N., Noels, A., and Sauval, A. J. 1992. Photospheric abundances. In *Coronal Streamers, Coronal Loops, and Coronal and Solar Wind Composition*, ed. C. Mattok, ESA SP-348, pp. 305–308.

Grünwaldt, H., and Rosenbauer, H. 1978. Study of He and H velocity differences as derived from HEOS-2 S-210 solar wind measurements. In *Pleins Feux sur la Physique Solaire*, (Paris: Editions CNRS), pp. 377–388.

Hannaford, P., Lowe, R. M., Grevesse, N., and Noels, A. 1992. Lifetimes of Fe II and the solar abundance of iron. *Astron. Astrophys.* 259:301–306.

Henoux, J.-C., and Somov, B. V. 1992. First ionization potential fractionation. In *Coronal Streamers, Coronal Loops, and Coronal and Solar Wind Composition*, ed. C. Mattok, ESA SP-348, pp. 325–330.

Hirshberg, J., Bame, S. J., and Robbins, D. E. 1972. Solar flares and solar wind helium enrichments: July 1965–July 1967. *Solar Phys.* 23:467.

Hovestadt, D., Vollmer, O., Gloeckler, G., and Fan, C. Y. 1973. Measurement of elemental abundance of very low energy solar cosmic rays. *Proc. 13th Intl. Cosmic Ray Conf.*, pp. 1498–1503.

Hovestadt, D., Gloeckler, G., Fan, C. Y., Fisk, L. A., Ipavich, F. M., Klecker, B., O'Gallagher, J. J., Scholer, M., Arbinger, H., Cain, J., Hofner, H., Kunneth, E., Laeverenz, P., and Tums, E. 1978. The nuclear and ionic charge distribution particle experiments on the ISEE-1 and ISEE-C spacecraft. *IEEE Trans. Geosci. Electron.* 16:166.

Hovestadt, D., Geiss, J., Gloeckler, G., Möbius, E., Bochsler, P., Gliem, F., Ipavich, F. M., Wilken, B., Axford, W. I., Balsiger, H., Bürgi, A., Copland, M., Dinse, H., Galvin, A. B., Gringauz, K. I., Grünwaldt, H., Hsieh, K. C., Klecker, B., Lee, M. A., Managadze, G. G., Marsch, E., Neugebauer, M., Rieck, W., Scholer, M., and Stüdemann, W. 1989. 'CELIAS'—charge, element and isotope analysis system for SOHO. In *The SOHO Mission*, ed. V. Domingo, ESA SP-1104, pp. 69–74.

Hudson, R. D., and Kieffer, L. J. 1971. Compilation of atomic ultraviolet photoabsorption cross sections for wavelengths between 3000 and 10 Ångstrøm. *Atomic*

Data 2:205–225.

Hundhausen, A. J. 1972. *Coronal Expansion and Solar Wind* (Berlin: Springer-Verlag).

Ip, W.-H., and Axford, W. I. 1991. On the first ionization potential effect in the solar corona. *Adv. Space Res.* 11:(1)247–250.

Ipavich, F. M., Galvin, A. B., Gloeckler, G., Hovestadt, D., Bame, S. J., Klecker, B., Scholer, M., Fisk, L. A., and Fan, C. Y. 1986. Solar wind Fe and CNO measurements in high-speed flows. *J. Geophys. Res.* 91:4133–4141.

Ipavich, F. M., Galvin, A. B., Geiss, J., Ogilvie, K. W., and Gliem, F. 1992. Solar wind iron and oxygen charge states and relative abundances measured by SWICS on Ulysses. In *Solar Wind Seven*, eds. E. Marsch and R. Schwenn (Oxford: Pergamon Press), pp. 369–374.

Kunz, S., Bochsler, P., and Geiss, J. 1983. Determination of solar wind elemental abundances from m/q observations during three periods in 1980. *Solar Phys.* 88:359–376.

Marsch, E., Mühlhäuser, K. H., Pilipp, W., Schwenn, R., and Rosenbauer, H. 1981. Initial results on solar wind alpha particle distributions as measured by Helios between 0.3 and 1 AU. In *Solar Wind Four*, ed. H. Rosenbauer, Report No. MPAE-W-100-81-31 (Katlenburg-Lindau: Max-Planck-Institut für Extraterrestrische Physik), pp. 443–449.

Marsch, E., Roux, A., et al. 1993. *Solar Corona Probe*, Proposal for an ESA M3 Mission (MPAE Lindau, CRPE Issy).

Marsch, E., von Steiger, R., and Bochsler, P. 1995. Element fractionation by diffusion in the solar chromosphere. *Astron. Astrophys.*, in press.

McKenzie, J. F., and Feldman, U. 1992. Variations in the relative elemental abundances of oxygen, neon, magnesium, and iron in high-temperature solar active-region and flare plasmas. *Astrophys. J.* 389:764–776.

Meyer, J.-P. 1981. A tentative ordering of all available solar energetic particle abundance observations. *Proc. 17th Intl. Cosmic Ray Conf.* 3:145–152.

Meyer, J.-P. 1991. Diagnostic methods for coronal abundances. *Adv. Space Res.* 11:(1)269–280.

Meyer, J.-P. 1993. Element fractionation at work in the solar atmosphere. In *Origin and Evolution of the Elements*, eds. N. Prantzos, E. Vangioni-Flam and M. Cassé (Cambridge: Cambridge Univ. Press), pp. 26–62.

Mitchell, D. G., and Roelof, E. C. 1980. Thermal iron ions in high speed solar wind streams: Detection by the IMP 7/8 energetic particle experiments. *Geophys. Res. Lett.* 7:661–664.

Möbius, E., Bochsler, P., Ghielmetti, A. G., and Hamilton, D. 1990. High mass resolution isochronous time-of-flight spectrograph for three-dimensional space plasma measurements. *Rev. Sci. Instr.* 61:3609–3612.

Neugebauer, M. 1981. Observations of solar wind helium. *Fund. Cosmic Phys.* 7:131–199.

Neugebauer, M., and Snyder, C. W. 1966. Mariner 2 observations of the solar wind. *J. Geophys. Res.* 71:4469.

Ogilvie, K. W., and Wilkerson, T. D. 1969. Helium abundance in the solar wind. *Solar Phys.* 8:435–449.

Owocki, S. P., and Scudder, J. D. 1983. The effect of a non-Maxwellian electron distribution on oxygen and iron ionization balances in the solar corona. *Astrophys. J.* 270:758–768.

Owocki, S. P., Holzer, T. E., and Hundhausen, A. J. 1983. The solar wind ionization state as a coronal temperature diagnostic. *Astrophys. J.* 275:354–366.

Reames, D. V. 1994. Coronal element abundances derived from solar energetic particles. *Adv. Space Res.* 14:(4)177–180.

Reames, D. V., Ramaty, R., and von Rosenvinge, T. T. 1988. Solar neon abundances from gamma-ray spectroscopy and ^{3}He-rich particle events. *Astrophys. J. Lett.* 332:87–91.

Rosenbauer, H., Schwenn, R., Marsch, E., Meyer, B., Miggenrieder, H., Montgomery, M. D., Mühlhäuser, K. H., Pilipp, W., Voges, W., and Zink, S. M. 1977. A survey on initial results of the Helios plasma experiment. *J. Geophys.* 42:561–580.

Schmelz, J. T. 1993. Elemental abundances of flaring solar plasma: Enhanced neon and sulfur. *Astrophys. J.* 408:373–381.

Schmid, J., Bochsler, P., and Geiss, J. 1988. Abundance of iron ions in the solar wind. *Astrophys. J.* 329:956–966.

Schwenn, R. 1990. Large-scale structure of the interplanetary medium. In *Physics of the Inner Heliosphere* vol. 1, eds. R. Schwenn and E. Marsch (Berlin: Springer-Verlag), pp. 99–181.

Shafer, C. M., Gloeckler, G., Galvin, A. B., Ipavich, F. M., Geiss, J., von Steiger, R., and Ogilvie, K. W. 1993. Sulfur abundances in the solar wind measured by SWICS on Ulysses. *Adv. Space Res.* 13:(6)79–82.

Shemi, A. 1991. The high Ne/C and Ne/O abundance ratios in the footprints of solar flares. *Mon. Not. Roy. Astron. Soc.* 251:221–226.

Smith, E. J., Neugebauer, M., Balogh, A., Bame, S. J., Erdös, G., Forsyth, R. J., Goldstein, B. E., Phillips, J. L., and Tsurutani, B. T. 1993. Disappearance of the heliospheric sector structure at Ulysses. *Geophys. Res. Lett.* 20:2327–2330.

Songaila, A., Cowie, L. L., Hogan, C. J., and Rugers, M. 1994. Deuterium abundance and background radiation temperature in high-redshift primordial clouds. *Nature* 368:599–603.

Tagger, M., Falgarone, E., and Shukurov, A. 1995. Ambipolar filamentation of turbulent magnetic fields. *Astron. Astrophys.* 299:940–946.

Vauclair, S. 1994. Element segregation in the solar chromosphere and the FIP bias: The "skimmer" model. *Astron. Astrophys.*, in press.

Vauclair, S., and Meyer, J.-P. 1985. Diffusion in the chromosphere, and the composition of the solar corona. *Proc. 19th Intl. Cosmic Ray Conf.* 4:233–236.

Veck, N. J., and Parkinson, J. H. 1981. Solar abundances from x-ray flare observations. *Mon. Not. Roy. Astron. Soc.* 197:41–55.

Vernazza, J. E., Avrett, E. H., and Loeser, R. 1981. Structure of the solar chromosphere III. *Astrophys. J. Suppl.* 45:635–725.

Villanueva, L., McNutt, R. L., Lazarus, A. J., and Steinberg, J. T. 1994. Voyager observations of O^{6+} and other minor ions in the solar wind. *J. Geophys. Res.* 99:2553–2565.

von Steiger, R., and Geiss, J. 1989. Supply of fractionated gases to the corona. *Astron. Astrophys.* 225:222–238.

von Steiger, R., and Geiss, J. 1993. Solar wind composition and expectations for high solar latitudes. *Adv. Space Res.* 13:(6)63–74.

von Steiger, R., and Marsch, E. 1994. Diffusive fractionation in the chromosphere. *Space Sci. Rev.* 70:341-346.

von Steiger, R., Christon, S. P., Gloeckler, G., and Ipavich, F. M. 1992a. Variable carbon and oxygen abundances in the solar wind as observed in Earth's magnetosheath by AMPTE/CCE. *Astrophys. J.* 389:791–799.

von Steiger, R., Geiss, J., Gloeckler, G., Balsiger, H., Galvin, A. B., Mall, U., and Wilken, B. 1992b. Magnesium, carbon, and oxygen abundances in different solar wind flow types, as measured by SWICS on Ulysses. In *Solar Wind Seven*, eds. E. Marsch and R. Schwenn (Oxford: Pergamon Press), pp. 399–404.

von Steiger, R., Wimmer Schweingruber, R. F., Geiss, J., and Gloeckler, G. 1995. Abundance variations in the solar wind. *Adv. Space Res.* 15(7):3–12.

Widing, K. G., and Feldman, U. 1992. Elemental abundances and their variations in the upper solar atmosphere. In *Solar Wind Seven*, eds. E. Marsch and R. Schwenn (Oxford: Pergamon Press), pp. 405–410.

Wieler, R., Baur, H., and Signer, P. 1993. A long-term change of the Ar/Kr/Xe fractionation in the solar corpuscular radiation. *Lunar Planet. Sci.* XXIV:1519–1520 (abstract).

Wimmer Schweingruber, R. F. 1994. Oxygen, Helium, and Hydrogen in the Solar Wind: SWICS/Ulysses Results. Ph.D. Thesis, Univ. of Bern.

Wimmer Schweingruber, R. F., von Steiger, R., Geiss, J., Balsiger, H., and Gloeckler, G. 1994. H, He, and O in the slow solar wind and in fast streams from coronal holes: SWICS/Ulysses results. *J. Geophys. Res.*, submitted.

Zirker, J. B., ed. 1977. *Coronal Holes and High Speed Streams* (Boulder: Colorado Associated Univ. Press).

KINETIC PHENOMENA IN THE SOLAR WIND

W. C. FELDMAN
Los Alamos National Laboratory

and

E. MARSCH
Max Planck Institut für Aeronomie

This review surveys shapes of solar wind ion and electron velocity distribution functions observed between 0.29 AU and 5.4 AU and their possible theoretical explanation. Because of the recent comprehensive review of results returned by the Helios probes between 0.29 and 1 AU (Marsch 1991*a*), attention is here concentrated on later results, in particular those for solar wind electrons returned by the ISEE 3 and Ulysses missions, and their assimilation with the earlier Helios results. We also outline our present state of theoretical understanding of aspects of these measurements in terms of: (1) distortions from single convecting Maxwellian distributions caused by expansion through the macroscopic gravitational, electrostatic, and magnetic fields that permeate interplanetary space; and (2) relaxation back to Maxwellians through Coulomb and wave-particle interactions. We conclude with a discussion of the connection of solar wind kinetic-scale observations with our present understanding of coronal heating and acceleration processes. Many of the *in-situ* observations of shapes of ion and electron velocity distributions can be understood if coronal heating and acceleration mechanisms are fundamentally time dependent and of small (supergranule size or smaller) spatial scale.

I. INTRODUCTION

A characteristic feature of the solar wind throughout most of interplanetary space is that binary Coulomb interactions are rare. Velocity distributions of ions and electrons are therefore allowed to diverge from their most probable configuration, a Maxwell-Boltzmann distribution. Impetus to diverge from a Maxwellian is provided by a variety of factors including the mix of non-thermal heating mechanisms that create and support a hot corona, the forced coronal expansion through inhomogeneous macroscopic fields that fill all of interplanetary space, and the local mixing of particles of different origin that are accelerated in distant source regions of the solar wind. All these effects give rise to strong multi-fluid interactions and/or plasma instabilities.

There are several reasons to study the shapes of nonequilibrium velocity distributions of solar wind particles and the resultant spontaneous relaxation processes that redistribute particle positions and velocities in phase space toward a more probable configuration . Parametric models of the solar wind expansion, if they are to match measured conditions at the coronal base and at 1 AU, clearly need extended heating and/or momentum addition beyond the sonic point (see, e.g., the chapter by Holzer et al.). Neither the nature nor the time and spatial scales of the relevant plasma processes are known. Regardless of their details, non-Maxwellian shapes of both ion and electron velocity distributions are expected and should sensitively reflect the action of both macroscopic fields and internal plasma processes, because the plasma is collisionless throughout much of the acceleration region and has to support a substantial energy flux.

Nonthermal distributions, then, form the inner boundary state for the subsequent expansion and thereby determine the nature and intensity of dissipation-scale waves in overlying layers. The resultant fluctuations, in turn, regulate the plasma transport from the Sun, and hence influence the asymptotic flow state of the solar wind. Specifically, the rate of energy supply to the wind by electron heat conduction will be a sensitive function of the spectrum of dissipation-scale turbulence in the corona. Through shaping of the velocity distributions, the turbulence will affect the dielectric properties and thus the kinetic normal modes of the plasma. Changes in these properties will, in turn, affect the ability of the solar wind to transmit outward propagating waves generated at or below the coronal base.

Another reason for interest in kinetic phenomena in the solar wind is that they regulate the mass, momentum and energy transport into and through plasma boundary layers and regimes that surround most planetary and cometary bodies in the heliosphere. Specific examples include ion pickup and assimilation, upstream shock waves and associated phenomena, magnetic reconnection, and plasma and energetic particle acceleration. Kinetic processes also mediate the interaction between, and subsequent radial evolution of, plasma streams evolving from spatially distinct regions of the corona.

On the other hand, knowledge of the local kinetic state and of its spatial evolution and relaxation in the wind plasma, can be used to infer the existence and nature of remote plasma processes that have local effects. Of course such inferences must necessarily be qualitative, because most often, considerable information is lost through entropy-increasing collisions or wave-particle interactions that occur in transit between the site of origin and the site of observation. All such interpretations, therefore, suffer from essential ambiguities. Nevertheless, self-consistent and comprehensive (and therefore credible) arguments can, at times, be constructed to support the action of hypothesized processes that occur nonlocally in the wind.

Because of space limitations, we will concentrate on kinetic effects that shape ion and electron velocity distributions. Although we aspire to the goal of developing a self-consistent, complete kinetic description for the radial evolution of particle velocity distributions of the solar wind, this goal is beyond our reach at this time. Main impediments to approaching this goal are: (1) the mechanisms that heat the corona and accelerate the solar wind are incompletely known; (2) the variety of wave modes possible in a multi-component plasma such as the solar wind is very large; (3) the onset of micro-instabilities and their eventual saturation depend sensitively on detailed shapes of particle velocity distribution functions; (4) the ranges of relevant distance and time scales in the solar wind are very large; (5) non-locally generated, background electromagnetic fields (which are often variable and turbulent) are present throughout the heliosphere, which scatter plasma particles and thereby change the shapes of particle velocity distributions; and (6) global interaction regions that are driven by the spatially nonuniform and time-dependent macroscopic solar wind expansion determine the asymptotic boundary state of the plasma, which shapes suprathermal electron velocity distributions throughout the inner heliosphere. Our present contribution to this discipline makes use of the many relatively recent comprehensive reviews of this subject. These reviews can guide the reader to the original literature regarding relaxation through plasma instabilities and their reaction back on the nonequilibrium particle populations that triggered them. In particular we refer to Hollweg (1975), Feldman (1979), Schwartz (1980), Shoub (1988), Marsch (1991*a*), Gurnett (1991), Roberts and Goldstein (1991), and Gary (1993). The evolution of plasma turbulence is reviewed in a separate chapter by Goldstein et al., and has also been dealt with in the review by Marsch (1991*b*).

In this chapter, observed electron and ion velocity distributions will be presented in Sec. II. The effectiveness of Coulomb collisions in limiting distortions and returning distributions to a Maxwellian is reviewed in Sec. III. In Sec. IV plasma instabilities and consequent wave-particle interactions are reviewed. We conclude with a discussion of the implications of observed interplanetary velocity distributions for the state of the corona and its heating mechanisms in Sec. V.

II. OBSERVED PARTICLE VELOCITY DISTRIBUTIONS

A. Introduction

In-situ space probes have explored the solar wind between 0.29 AU (Helios) and 56 AU (Voyager 2). However, limitations on plasma instrumentation have limited comprehensive surveys of shapes of particle velocity distributions to distances between 0.29 and 5.4 AU. In general, measured shapes are non-Maxwellian and vary systematically with distance from the Sun and with phase relative to the magnetic sector boundary and large-scale stream structure. They also develop distinctive shapes in association with transient disturbances driven by various forms of solar activity. These shapes and their inferred origins are reviewed next.

B. Electron Velocity Distributions

Electron velocity distribution functions, $F(v)$ of the solar wind can generally be decomposed into two components, a thermal core and a hot halo population

(Montgomery et al. 1968; Feldman et al. 1975; Pilipp et al. 1981,1987b), as illustrated in Fig. 1. Whereas the core population generally conforms closely to a bi-Maxwellian, the hot population can only be adequately described some of the time by a convecting bi-Maxwellian. Oftentimes, however, the hot component supports enhanced high-energy extensions that decreases asymptotically with increasing energy as a power law, $F(E) \propto E^{-p}$. Shapes of the core distributions vary much less than do those of the halo.

Ranges of the two shape parameters that characterize core distributions, average temperature, $T_c = (T_\parallel + 2T_\perp)/3$, and thermal anisotropy, $T_\parallel/T_\perp$, depend on both radial distance from the Sun and on position relative to the global sector structure of the solar wind (Feldman et al. 1977; Sittler and Scudder 1980; Phillips et al. 1989; Pilipp et al. 1990; Schwenn 1991; Phillips and Gosling 1990; Bame et al. 1992). Here parallel and perpendicular refer to directions relative to the local magnetic field B. In general, core temperatures decrease with increasing radial distance, are lowest in the high-speed flows, at magnetic sector boundaries, and in driver gas contained within interplanetary shock-wave disturbances. The anisotropy is highest when the density is lowest. Another parameter of interest is the bulk speed of the core electron population. It generally lags behind that of the protons at relative velocities (aligned with the local magnetic field direction) ranging between zero and about the local Alfvén speed.

In contrast, the shapes of the halo distributions not only require several different functional forms for an adequate representation, but also a wider range of parameter values for a given form. Three distinct classes of shapes are illustrated in Fig. 2 (Pilipp et al. 1987b). The distribution at the right is nearly isotropic and Maxwellian with a bulk speed close to that of the core population. These shapes are oftentimes observed in the slow-speed solar wind. The one in the middle is an intermediate variety being moderately anisotropic and resembling a pear in cross section (the thermal anisotropy of the electrons traveling away from the Sun is >1 and that of the returning electrons is <1). The bulk velocity of the entire halo population is greater than that of both the core electrons and the ions (note the relative offsets in Fig. 1). Its velocity relative to the core is aligned with the local magnetic field direction and has a magnitude such that the total electron velocity is equal to that of the ions, within experimental errors (Feldman et al. 1975).

The foregoing intermediate types of electron velocity distributions are most often observed. They can be generally modeled by a product of a modified Lorentzian energy function and an expansion to fourth order of Legendre polynomials (Dum et al. 1980; Feldman et al. 1982b). These distributions vary asymptotically in energy as a power law, with an index p generally in the range between 4 and 6. The coefficients in the Legendre expansions are dominated by the odd orders, needed to characterize the heat flux adequately, which is carried by electrons along the magnetic field but away from the Sun. However, all index orders are often significant as needed to describe the pear shapes mentioned above.

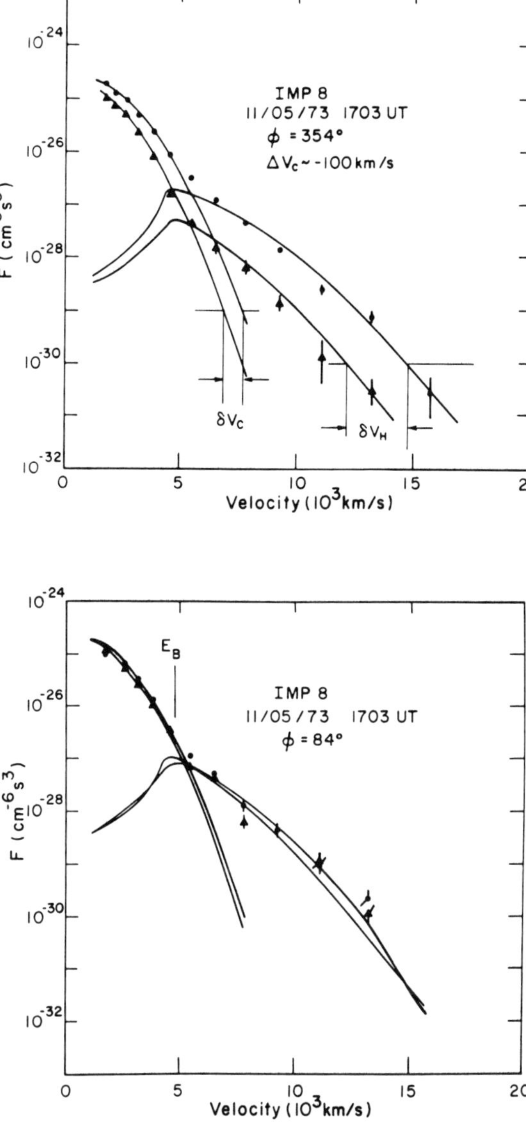

Figure 1.   Parallel to the magnetic field B (top) and perpendicular (bottom) cuts through a typical two-component electron velocity distribution consisting of a cold, Maxwellian core, and a relatively hot halo. The solid curves give the best-fitting Maxwellians to the separate components. The close overlap of the two folded perpendicular cuts at the bottom and the offset of the parallel cuts at the top reveal a bulk convection along B with the velocity of the halo larger than that of the core, both traveling away from the Sun. The breakpoint between the core and halo, denoted by E_B at the bottom, is interpreted to be the local value (relative to infinity) of the interplanetary electrostatic potential (figure from Feldman et al. 1975).

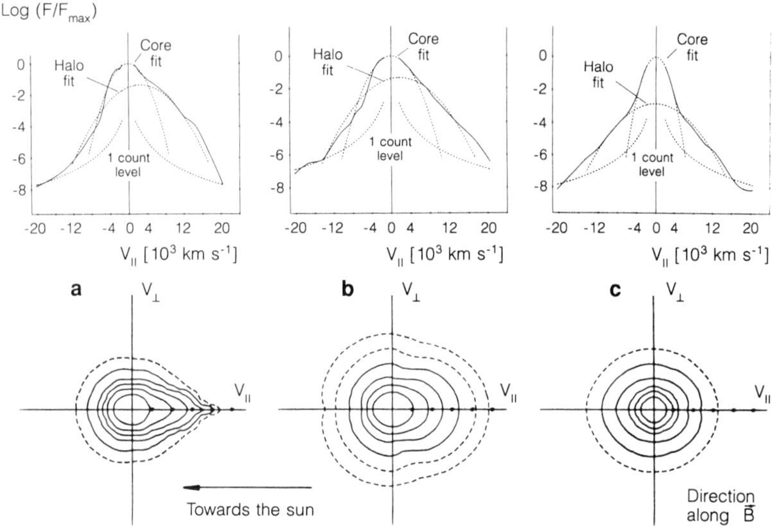

Figure 2.   Parallel to B cuts (top) through, and contours (bottom) of selected examples of three classes of solar wind electron velocity distributions, $F(v)$. Contours correspond to fractions of 10^{-1}, $10^{-2}\ldots10^{-6}$ of the maximum phase-space density. The right-hand distribution is typical of the slow-speed solar wind near the sector boundary, the middle disribution is typically observed at intermediate speeds, and the left-hand distribution is typical of the high-speed solar wind (figure from Pilipp et al. 1987b).

At high bulk speeds, the component of halo electrons that stream away from the Sun focus down to a narrow beam, termed the strahl (Rosenbauer et al. 1977), as shown by the left-hand distribution in Fig. 2. The contrast between distributions in the low- and high-speed solar wind is shown clearly in Fig. 3 by plotting several angular cuts through two different velocity distributions at constant energy. An extended survey of ISEE 3 data shows that angular widths of halo distributions in the slow-speed wind average about 90° FWHM and those in the high-speed wind average about 25° (Phillips et al. 1994). When the strahl is present, the flux of non-strahl halo electrons is usually very low (although nonzero) and nearly isotropic, as shown in Fig. 3. The ratio of the peak flux in the strahl to this background has been used to categorize the strahl strength and is typically between 10 and 100 at 1 AU and between 100 and 1000 at 0.3 AU (Pilipp et al. 1981). The shapes of electron strahl components parallel to the magnetic field in the high-speed solar wind can be adequately represented by a Maxwellian with temperatures close to, but less than 10^6K, as shown in Fig. 4. We note that this temperature is consistent with measurements of the corona within coronal holes (Habbal et al. 1993) and with models of the solar wind from coronal holes (see, e.g., Withbroe 1988) at an altitude between about 10 and 20 solar radii, ($R_\odot$). An estimate for this altitude is given by fits to measured angular widths of electron strahl

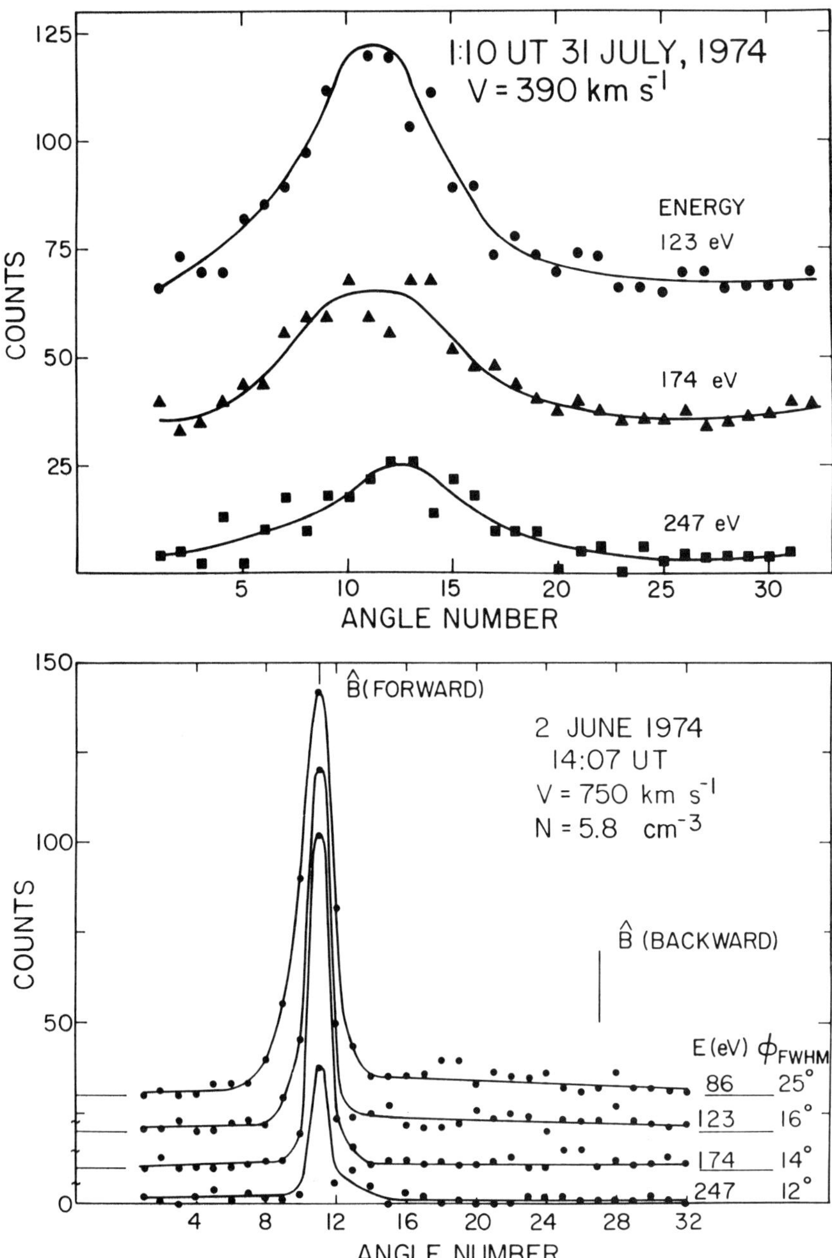

Figure 3.   Angular cuts at constant energy through electron velocity distributions typically observed in the low-speed (top) and high-speed (bottom) solar wind. The narrow peaks at the bottom define the strahl component. Their widths decrease monotonically with increasing energy, consistent with expansion of an isotropic velocity distribution in the neighborhood of 10 to 20 $R_\odot$ to 1 AU through radially decreasing magnetic and electric fields, including Coulomb collisions (figure from Feldman et al. 1978).

distributions (see, e.g., Fig. 3; Lemons and Feldman 1983), using exospheric theory (Jockers 1970; Lemaire and Scherer 1971; Schultz and Eviatar 1972; Scudder and Olbert 1979*a,b*; Shoub 1988).

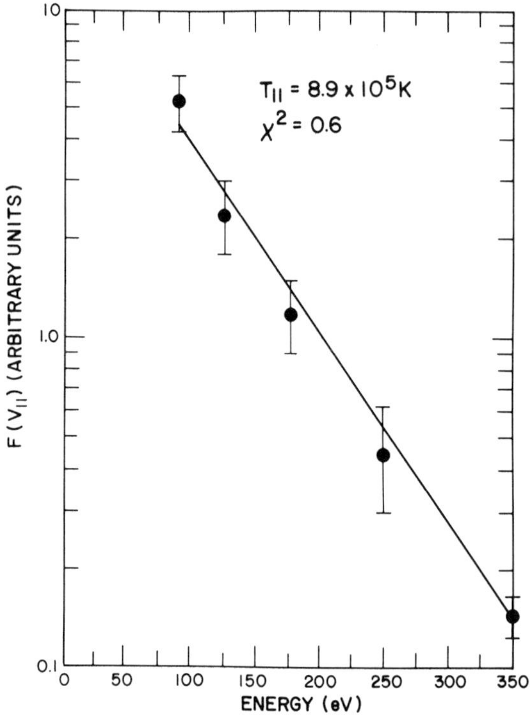

Figure 4.   Maxwellian fit to the parallel (integrated over $V_\perp$) electron velocity distribution measured at 1407 UT on 2 June 1974, which is a typical example of that measured in the high-speed solar wind. The fit is seen to be excellent, yielding a parallel temperature of 8.9×10^5 K. According to exospheric theory, this temperature should reflect that in the corona between 10 and 20 $R_\odot$ (figure from Feldman et al. 1978).

Another type of flow that supports a very narrow strahl is that typified by very low density. Although these flows have been observed only rarely, they are not anomalies (Gosling et al. 1982). One such event repeated during two successive solar rotations in July 1979, suggesting that it may be a quasi-stationary type of coronal expansion. An electron velocity distribution measured during this flow type on 4 July 1979 (0652:31 UT) is compared with one that is usually associated with a more normal type of solar wind measured earlier in the same day (0414:14 UT) in Fig. 5. The sharp rise in counts at the lowest energy channel of the spectrum at 0652 UT is due to photoelectrons, which dominates the plasma electrons in this spectrum because of the extremely low density of the solar wind at the time. Although both distributions in Fig. 5 have pronounced strahls, the core component at the left is nearly isotropic in azimuth, whereas the one on the right has a pronounced maximum along B and two distinct minima at $90°$ from B and $-B$ in azimuth. A closer look at distributions similar to the one at 0652 UT reveals that the pronounced maximum at backward angles from the strahl (along B) cuts off abruptly at about 100 eV, yet has a very similar angular shape to that of the strahl at lower energies, as shown in Figs. 6 and 7.

This effect can be simply explained in terms of reflection of outward traveling strahl electrons from the interplanetary electrostatic potential in an environment where both Coulomb collisions and wave-particle interactions are rare. This effect can happen because the density is low (therefore Coulomb collisions are rare) and the Alfvén speed is high (therefore the threshold of a class of instabilities that could lead to whistler-mode turbulence and enhanced electron scattering is high). Note that what we may call the "core" in the left-hand distribution in Fig. 5 may in reality be the reflected strahl in the right-hand distribution. The reflected strahl in the right-hand spectrum has a shape in phase space that can be fit locally with a Maxwellian having a temperature of about 10^6 K, as is shown in the lower left panel of Fig. 6. This unusually enhanced "core" temperature leads, in turn, to an unusually enhanced electrostatic potential difference between 1 AU and very large distances, in this example about 100 eV.

The halo appears as an isotropic background that underlies both the forward and reflected strahl and is a separate component. We note that this component can be fit by a power law in energy, $F(E) \propto E^{-3.0}$, shown in the upper right panel of Fig. 6. Similar fits to other spectra measured during this event yield power law exponents between 3.0 and 3.5, values very close to that which fits the quiet time interplanetary electron energy flux spectrum between 10 and 200 keV, $J(E) = E * F(E) \propto E^{-2.5}$ (Lin 1980). Neither the origin nor the relationship between both of these electron components is presently known. They may, for example, be generated: (1) in the zone of enhanced dissipation-scale turbulence that encircles the Sun, which extends to about 30 $R_\odot$; (2) as a result of multiple Coulomb (Scudder and Olbert 1979a,b; Shoub 1988) and, perhaps, wave-particle interactions within a large volume of the inner heliosphere; or (3) within corotating shocks forming beyond about 2.5 AU along the boundaries of corotating, stream-stream interaction regions and injected back into the inner heliosphere (Gosling et al. 1993).

Events such as the above low-density one, although rare, support the interpretation (Rosenbauer et al. 1977; Feldman et al. 1978; Pilipp et al. 1981; Lemons and Feldman 1983) that the strahl, which sometimes extends from subthermal to suprathermal energies, corresponds to the extension of the hot coronal electron population into interplanetary space. The core electron component in the solar wind is then an evolved state of that portion of the strahl that is bound to local interplanetary regions by the combined action of the interplanetary electrostatic potential and the magnetic field through conservation of energy and the electron magnetic moment (Jockers 1970;

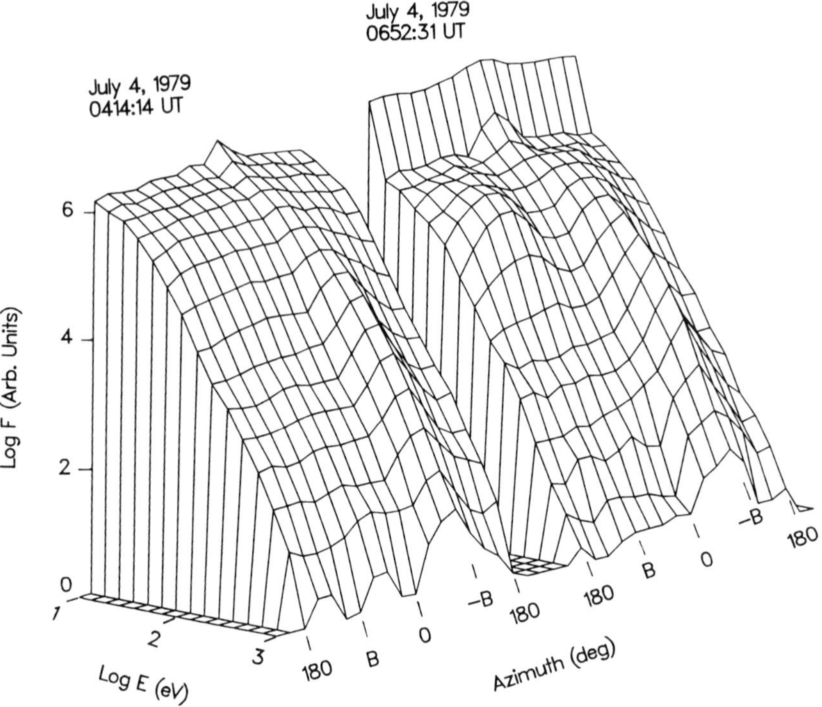

Figure 5.   Three-dimensional plots of two-dimensional (energy and azimuth) electron velocity distributions comparing one that is typical of the very low-density wind (at the right) and one that is typical of a more moderate density (at the left). Note that the right-most angular enhancement at high energy in the right-hand spectrum (corresponding to the heat-flux carrying electrons traveling along $-B$) is more sharply peaked than is that in the left-hand spectrum. Note also a distinct angular bump at moderate-to-low energies in the right-hand spectrum aligned with B, which does not exist in the left-hand spectrum. Both effects are consistent with the very low-density (and large Alfvén speed) conditions that were obtained on 4 July 1979 at 0652 UT, leading to nearly collision-free propagation from the Sun to the Earth. The sharp enhancement at the lowest energy at right is due to locally generated photoelectrons that dominate the plasma electrons here because of the low density (figure from Phillips et al. 1989).

Lemaire and Scherer 1971; Schulz and Eviatar 1972; Perkins 1973; Scudder and Olbert 1979*a,b*; Shoub 1988), as illustrated in Fig. 8. This portion of the distribution is filled by the combined action of Coulomb collisions and wave-particle interactions, needed to trap electrons locally (by increasing the electron magnetic moment, and hence the mirror force, thereby preventing electrons that come from the corona to return to the corona).

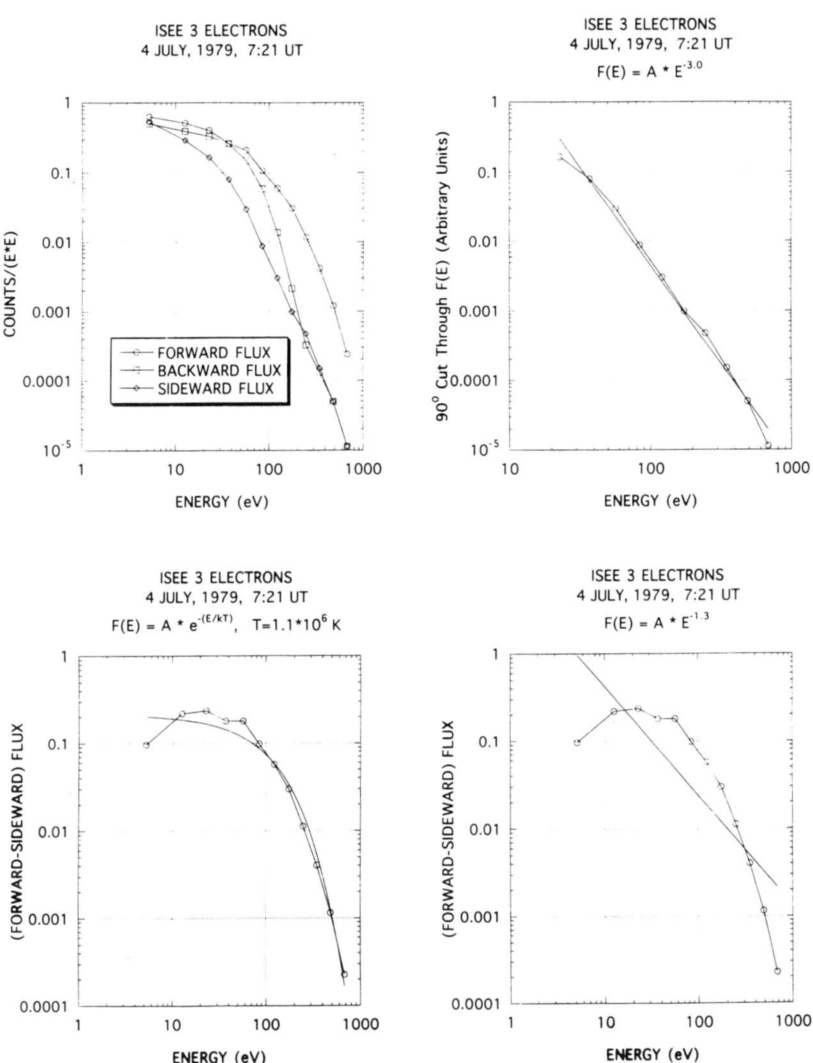

Figure 6.   Parallel and perpendicular cuts through an electron velocity distribution (proportional to counts E^{-2}) typical of the low-density solar wind. Forward and backward cuts along B are folded over one another at the upper left and superimposed on the average of the two cuts perpendicular to B. The high-energy portion of the perpendicular cut is shown at the upper right to be well charaterized by a power law with index $p = -3$. Fits to other spectra measured during this event have indices that range between -3 and -3.5. The fits to the (forward-sideward) cuts in the bottom two panels show that a Maxwellian fits the data above 100 eV much better than does a power law.

Individual core electrons then move stochastically from location to location in six-dimensional phase space (including both position and velocity) in such a way that their loss from a specific location due to Coulomb and wave-particle collisions is microscopically balanced by other collisions and by their injection from the strahl, as needed to maintain charge neutrality with the ions. We note though, that a Maxwellian provides a relatively good fit, and a power law function provides a very bad fit, to the measured spectrum of the strahl, as shown in the lower two panels of Fig. 6. A similar situation was obtained for strahl spectra measured in the high-speed solar wind, as shown in Fig. 4. These observations are significant in that the only times when it is possible to sample coronal electron distributions directly from a distance, provide no evidence to support the notion that the high-energy tails of coronal electron distributions are non-Maxwellian. In particular, they are not well represented by kappa distributions, which decrease asymptotically as $E^{-(\kappa+1)}$ (Scudder 1992a,b). Note that the potential difference between the corona and 1 AU is about 1 keV (the Pannekoek-Rosseland potential for a static corona; see, e.g., Lemaire and Scherer [1971, and references therein]), which is about 10 times the coronal electron thermal energy.

Interpretation of the spectra in Fig. 6 also suggests that a minimum, separate halo component exists at all times and has an origin distinct from that of the strahl. It underlies both the strahl and the core in velocity phase space. If correct, then the breakpoint between the core and the halo provides a useful but rough measure of the local interplanetary electrostatic potential (Feldman et al. 1975,1977; McComas et al. 1992). A more precise estimate of the potential is provided by Pilipp et al. (1990). They used the electron momentum equation (neglecting inertia) to determine potential differences by radial integration of the empirically determined electron temperature profiles for various types of solar wind. The resultant values ranged between 30 eV for fast wind, and 80 eV for slow wind, in the heliocentric distance range between 0.3 and 1 AU.

The halo component appears as a permanent feature of electron distributions, and as such is most clearly visible in cuts through electron velocity distributions perpendicular to the magnetic field (see, e.g., Pilipp et al. 1987b, their Fig. 4). It is understood to be composed of untrapped electrons, generated from the medium-energy strahl population by rare scattering events in the outer heliosphere. In contrast, the core electrons are trapped within the electrostatic potential well and thus are prone to suffer sufficient collisions to become fairly isotropic. The halo-to-core temperature ratio is about 7, a value predicted from Coulomb scattering by Scudder and Olbert (1979a,b), and recently found by McComas et al. (1992) to hold over a large region of the heliosphere from 1 to 4 AU. The ratio between the core/halo breakpoint energy (roughly equal to the electrostatic potential) and the core thermal energy is also found to be about 7 between 1 and 4 AU (McComas et al. 1992).

On occasion, halo distributions assume a bi-directional configuration in contrast to the shapes shown in Fig. 5 (Montgomery et al. 1974; Temnyi

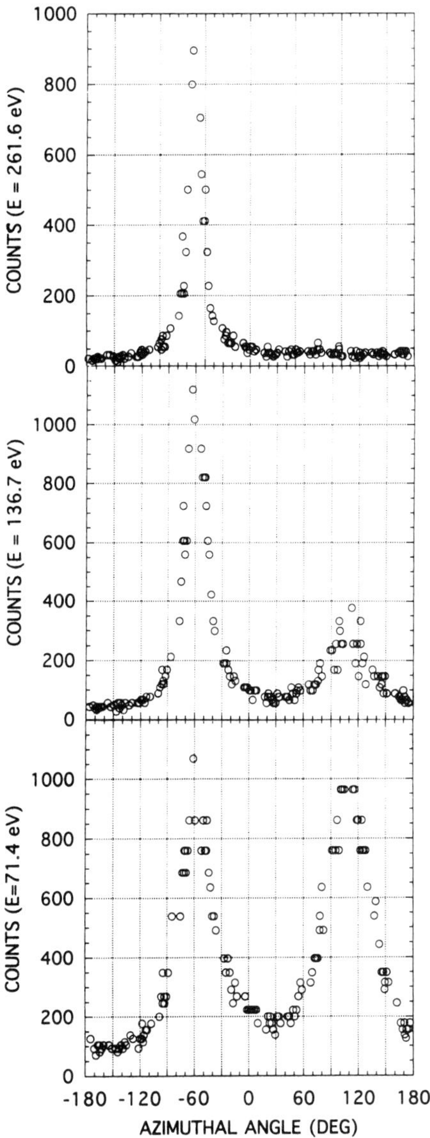

Figure 7. High angular resolution cuts at constant energy through low-density electron velocity distributions measured between 0711 and 0726 UT on 4 July 1979. We note that only one angular peak appears at the highest energy (on top) while a second peak grows with decreasing energy until both peaks are equal at 71 eV. The peaks are centered on B and $-B$. Their closely equal angular widths and heights at 71 eV are consistent with interpretation of the right-hand peak as resulting from the nearly collision-free reflection of outgoing electrons (the left-hand peak) from the interplanetary electric potential. Comparison between the plots in the bottom two panels fixes this potential between about 70 and 130 eV.

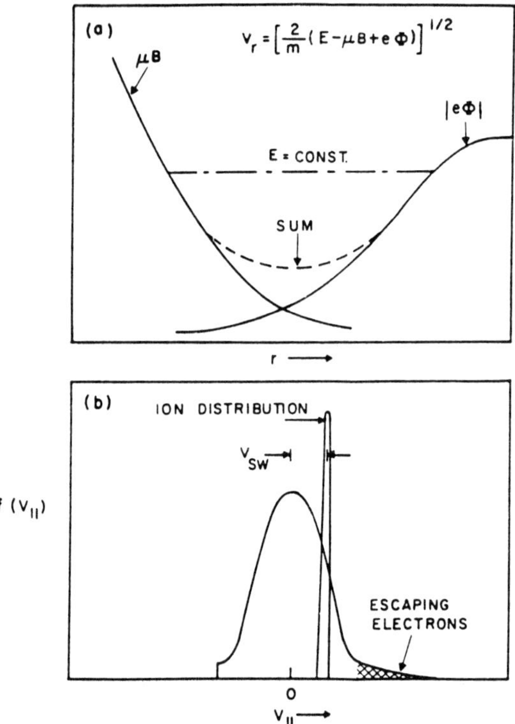

Figure 8. A schematic illustration of the two macroscopic fields that shape electron velocity distributions in exospheric theory from Perkins (1973). Electrons having energy less than the interplanetary electrostatic potential are bound to the local neighborhood by a magnetic mirror force closer to the Sun and an electric field at larger heliocentric distances. This configuration provides a natural explanation for observed two-component electron distributions consisting of bound (core) electrons and unbound (halo) electrons that are free to escape the heliosphere. In the absence of collisions (as assumed by exospheric theory), the bound electrons should have zero velocity relative to the Sun. Predicted ion and electron distributions are illustrated in the bottom panel. The fact that electrons are observed to flow away from the Sun must then reflect action of Coulomb and wave-particle collisions, which are not included in exospheric theory.

and Vaisberg 1979; Bame et al. 1981; Gosling et al. 1987; Pilipp et al. 1987*a,b,c*; Phillips et al. 1992*a,b*,1994). Many of these occurrences were observed within interplanetary shock wave disturbances, prompting their association with coronal mass ejections. Three-dimensional views of two two-dimensional electron distributions that illustrate the difference between uni-directional and bi-directional halo electron streaming is given in Fig. 9. In this case, both electron streams were symmetric. Most often though, the streams are asymmetric in intensity yet comparable in angular width. Only

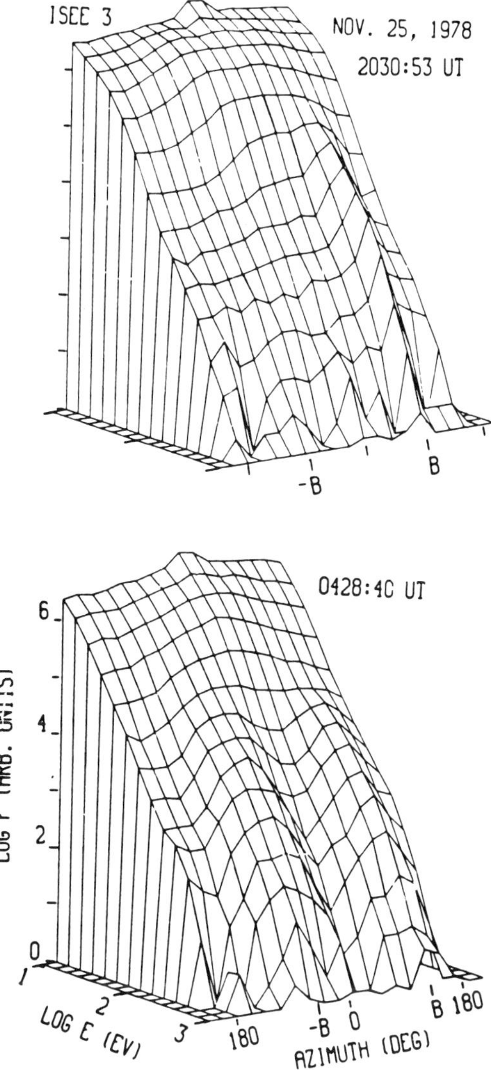

Figure 9.   Two three-dimensional views of two-dimensional (energy and azimuth) electron velocity distributions that contrast bi-directional flow conditions (at the bottom) with the more usual uni-directional flow conditions (at the top). Note the single angular enhancement at energies above about 100 eV denoting the heat-flux carrying electrons that travel along B away from the Sun in the topmost spectrum. This type of distribution has been interpreted to indicate an open field topology in which one end of the magnetic field is rooted in the Sun and the other end is connected to the local interstellar medium. In contrast, bi-directional flows along B and $-B$ must indicate either a loop magnetic topology, completely disconnected from the Sun, or a magnetic tongue (or more complicated flux rope) in which both ends of the local magnetic field loop back to the Sun where the magnetic mirror force can enforce symmetry (or near symmetry if collisions are important) along B and $-B$ (figure from Gosling et al. 1987).

rarely, a sharp angular enhancement is superimposed on one of the streams, which locally flows away from the Sun along the magnetic field, as shown in Fig. 10.

The bi-streaming electron halo configuration has been interpreted to indicate the passage past the spacecraft of either a closed magnetic loop or a generalized magnetic tongue because then both directions along the magnetic field lead back toward the Sun where the electron temperature and density are both higher. However, the compound distribution shown in Fig. 10 strongly suggests that magnetic connection to the Sun is maintained during this event. The large-scale magnetic topology would then have the structure of a tongue or general flux rope rather than as a detached loop, as shown in Fig. 11.

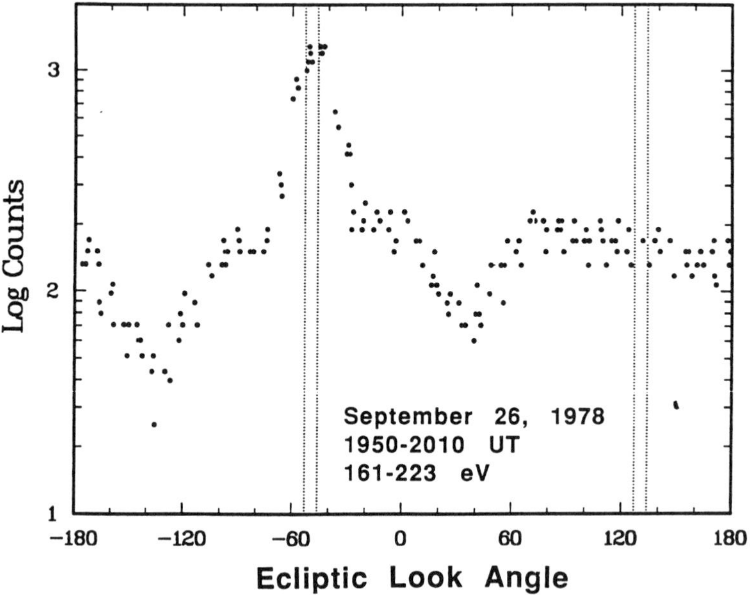

Figure 10. An angular cut at constant energy through a bi-streaming electron velocity distribution showing a narrow strahl-like enhancement superimposed on a broader, mesa-like configuration that is symmetric back to front along B. During the time required to accumulate the counts in this distribution, the magnetic field direction remained within the two vertical dotted lines. This composite configuration is interpreted to be a magnetic tongue or a flux rope, connected to the Sun at least at one, and perhaps both ends of the field line (figure from Phillips et al. 1992a,b).

An observation that prefers a flux rope interpretation of these types of events was observed using ISEE 3 on 7 July 1979, as shown in Fig. 12. Data showing arrival of a shock at 1850 UT on 6 July 1979 are: (1) sudden enhancements of whistler waves at 17.8 hz; (2) sudden increases of the total (B_T) and $z(B_z)$ components of the interplanetary magnetic field; and (3) a sudden enhancement in suprathermal electron fluxes between 2.0 and 7.4 keV,

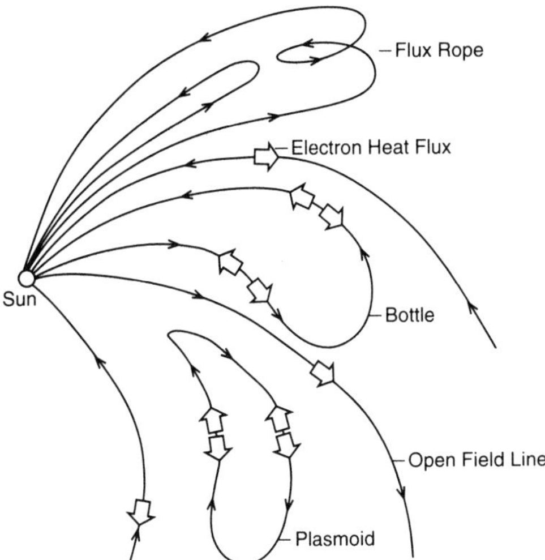

Figure 11.   Schematic illustrations of four generic types of interplanetary magnetic topology. At bottom is a simple open topology, connected at one end to the Sun and at the other end to the interstellar medium. Heat flux is expected to be uni-directional away from the Sun as shown by the arrow. Just above is a completely disconnected plasmoid. Heat flux is expected to be either absent or bi-directional. Next above is a magnetic bottle or tongue, connected at both ends to the Sun. The heat flux is expected to be bi-directional with, perhaps, a strahl-like enhancement superimposed on one of the two broad bi-directional streams if pitch-angle scattering near the Sun is sufficiently low. At the top is pictured the simplest flux-rope topology. It is also connected at both ends to the Sun but contains one or more loops where secondary reconnection has occurred, as expected if a sheared magnetic arcade erupts from the Sun. Electron pitch-angle distributions can be complicated and depend sensitively both on position of measurement and time dependence of injection at the footpoints from transient events (figure after Phillips et al. 1992*b*).

all shown in the upper left-hand panel of Fig. 12. Bi-streaming electron distributions starting at 0230 UT on 7 July as shown in the upper right- and lower left-hand panels were observed between about 0230 and 1130 UT on 7 July 1979. A solar radio burst occurred at 0430 UT on 7 July and Type III electrons at 7.4 keV energy arrived at ISEE 3 starting at about 0530 UT. Although not shown here, Type III electrons having energies below 1 keV became visible in the solar wind electron spectrum starting at about 0800 UT in the anti-sunward component of the bi-stream (center stream in the lower left-hand panel of Fig. 12). From this fact we infer that this lobe of the electron velocity distribution is the one connected most directly to the Sun.

Also during this event, the pitch-angle distribution of 4.2 keV electrons evolved from a narrowly focused beam (upper-left quadrant in the lower-right panel of Fig. 12) to one with a narrow waist and backward pointing ears. Such a loss cone configuration of backward-propagating electrons strongly suggests a flux rope interpretation (as pictured at the top in Fig. 11) in this

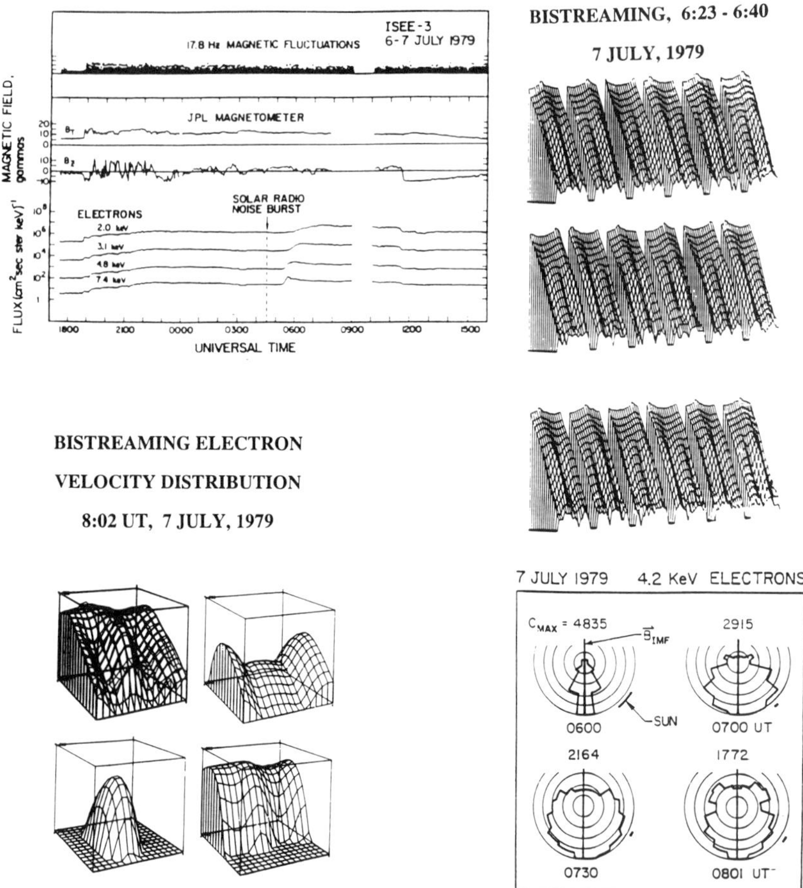

Figure 12. An instructive interplanetary bi-streaming electron event proving magnetic connection to the Sun and strongly suggesting a magnetic flux-rope topology. See the text for a detailed discussion. The data come from a combination of Anderson et al. (1981) and the Los Alamos ISEE 3 plasma electron experiment. The only part that needs special explanation is presented in the upper right- and lower left-hand panels. At the upper right is plotted $\log(F(E))$ as a function of energy and azimuth and at the lower left is plotted measured counts as a function of energy and azimuth. The upper-left quadrant of the lower-left panel gives an overlay of the measured counts and the best fit function consisting of three components; clockwise starting to the right are: (1) a diffuse halo population modeled in terms of the product of a generalized Lorentzian and an expansion in Legendre polynomials up to order 4; (2) a convecting bi-Maxwellian core population; and (3) a single-ended bi-Maxwellian strahl. The fit is seen to be good.

particular event because the large pitch-angle portions of the forward-going beam would be expected to mirror from that part of the central magnetic loop of the flux rope before it reaches its minimum distance to the Sun (near the asterisk, indicating magnetic reconnection in Fig. 11). The rest of the distribution would keep on going if it got past this point, thereby accounting for the hole in the return portion of the electron distribution.

C. Ion Velocity Distributions

Shapes of solar wind proton velocity distributions depend both on distance from the Sun and on phase with respect to the magnetic sector and bulk-flow stream structure. An overview of the full range of observations is provided by a representative sample of two-dimensional contour diagrams in Fig. 13. Magnetic sector and stream phase at a given heliocentric distance varies horizontally, and heliocentric distance for a given phase decreases from top to bottom in the figure. We start first with the left-hand column, which corresponds to flows near the heliospheric current sheet, or in the relatively dense, cold, and slow-speed solar wind. One notes a spectrum at the upper left consisting of tightly nested concentric circles, indicating an isotropic, cold plasma. These distributions heat with decreasing distance to the Sun, eventually developing a strahl-like tail at 0.29 AU. Some of the time the strahl portions of these distributions appear as resolved double (interpenetrating) proton streams at 1 AU, as shown in Fig. 14.

The spectra in the right-hand column in Fig. 13, measured within solar wind flows marked by low densities and high proton temperatures and bulk speed, show distinctive anisotropies. These spectra become hotter and ever more distorted from isotropic Maxwellians as the distance to the Sun decreases. The character of these distortions is such that the high-density core of the distributions have a $T_\perp > T_\parallel$ anisotropy and a resolved double-stream strahl population. These shapes can often be well represented by two relatively streaming bi-Maxwellians, as shown in Fig 15 (Bame et al. 1975; Feldman et al. 1976,1993; Leubner and Viñas 1986). Typical parameters at 1 AU are: (1) the density of the faster component is 20 to 30% of the total; (2) the relative velocities are aligned with the direction of the local magnetic field and are close to the local Alfvén speed; and (3) the thermal anisotropy of the primary component is such that $T_\perp = 2.5T_\parallel$.

An observation that underscores the control exercised by the local Alfvén speed occurred on 4 June 1991 within a 1.5-day period of low-speed flow on the trailing edge of a high-speed stream. Although double-proton streams were observed throughout this period, the magnetic field increased from 8 to 12 nT, while the density remained constant during an 8-hr interval. The proton velocity distribution changed immediately to one where the relative velocity of the secondary proton beam increased to 150 km s^{-1}, in step with the increase of the local Alfvén speed to 150 km s^{-1}, as shown in Fig. 16.

Although much less is known about the shapes of alpha-particle and heavier ion velocity distributions in the solar wind because they have a rela-

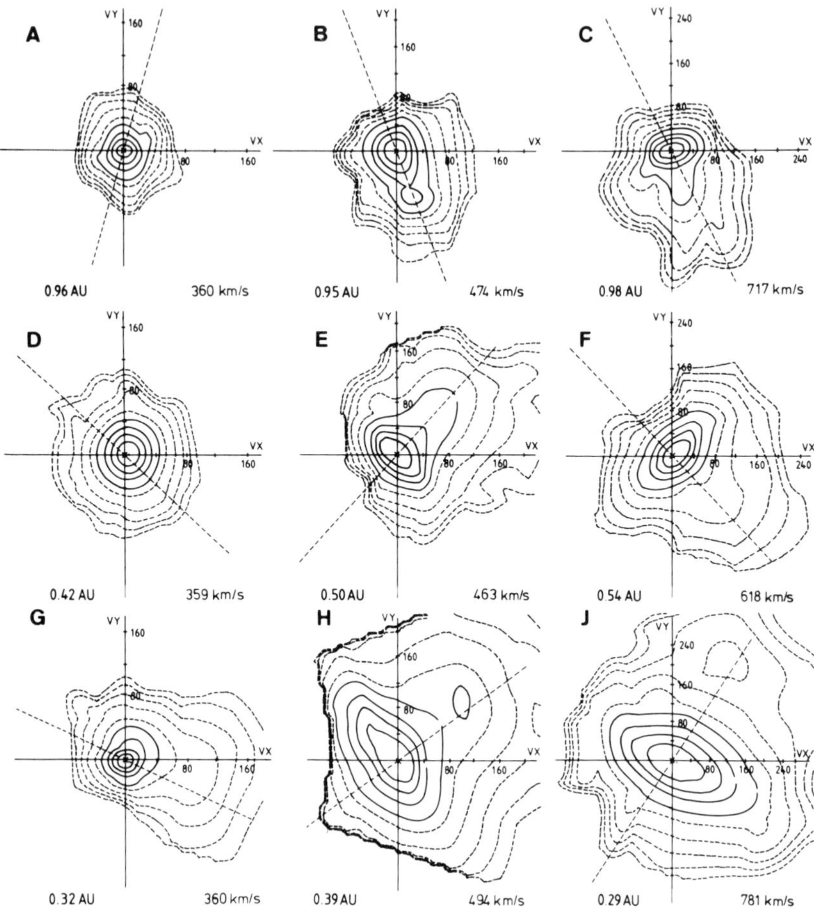

Figure 13.   A collage of contours of representative solar wind proton velocity distributions from the Helios plasma analyzer as a function of heliocentric distance (decreasing from top to bottom) and stream phase (from low speed at the left to high speed at the right). Contour levels correspond ot 0.8, 0.6, 0.4, 0.2, and 0.1, 0.03, 0.01, 0.003, and 0.001 of the maximum phase-space density. We note that non-Maxwellian shapes predominate, increasing in distortion from a single, isotropic cold distribution in the low-speed solar wind at 1 AU (upper left) to a two-component anisotropic ($T_\perp > T_\parallel$ core) warm distribution at 0.29 AU (lower right) (figure from Marsch et al. 1982a).

tively low abundance, they too are known to show a nonequilibrium behavior. Specifically, alpha-particle and heavier ion streams most often travel faster than the protons at relative velocities aligned with the magnetic field (Asbridge et al. 1976; Neugebauer and Feldman 1979; Schmidt et al. 1980,1987; Neugebauer et al. 1994; Feldman et al. 1993). Speed differences vary systematically with stream phase, ranging from near zero in the slow-speed solar wind to

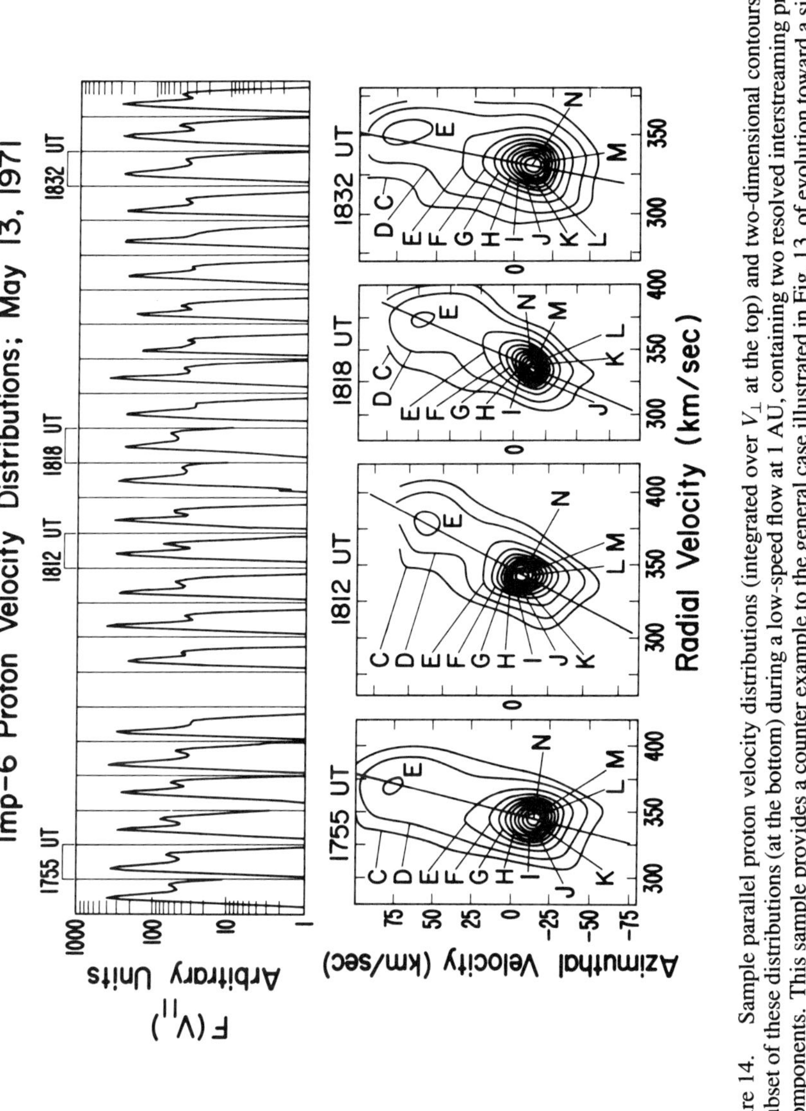

Figure 14. Sample parallel proton velocity distributions (integrated over $V_\perp$ at the top) and two-dimensional contours of a subset of these distributions (at the bottom) during a low-speed flow at 1 AU, containing two resolved interstreaming proton components. This sample provides a counter example to the general case illustrated in Fig. 13, of evolution toward a single, cold Maxwellian distribution with increasing distance in the low-speed solar wind (figure from Feldman et al. 1973*b*).

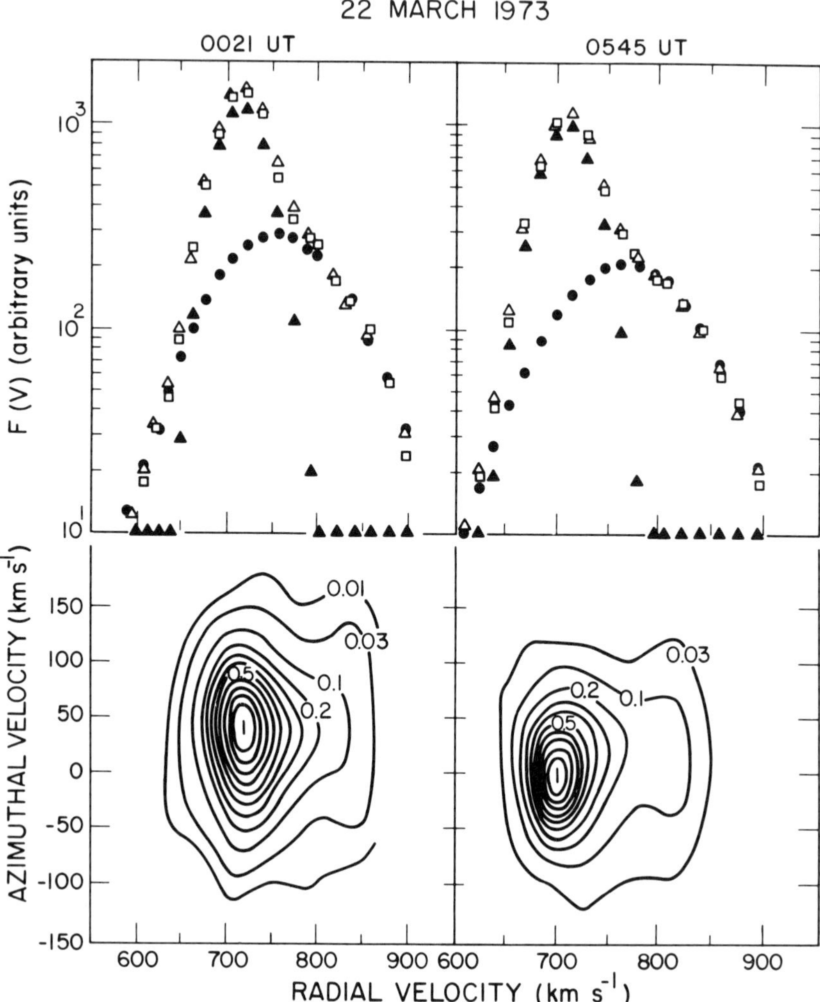

Figure 15.   Two examples of proton velocity distributions typically observed in the high-speed solar wind. Inspection of the radially projected measured distributions given by open-square symbols at the top, with the sum of two convecting bi-Maxwellians (open triangles), shown separately as a relatively cold core (filled triangles) and a faster hot beam (filled circles) reveals an excellent fit. The core component is strongly anisotropic having $T_\perp > T_\parallel$, as can be seen by the contour levels of the two-dimensional spectra in the bottom two panels (figure from Feldman et al. 1976,1993).

slightly more than the local Alfvén speed in the high-speed solar wind. The largest speed differences are recorded at the lowest heliocentric distances because the Alfvén speed is largest there (Marsch et al. 1982b; Marsch 1991a). During times of relative ion streaming, the alpha particles appear to be surfing

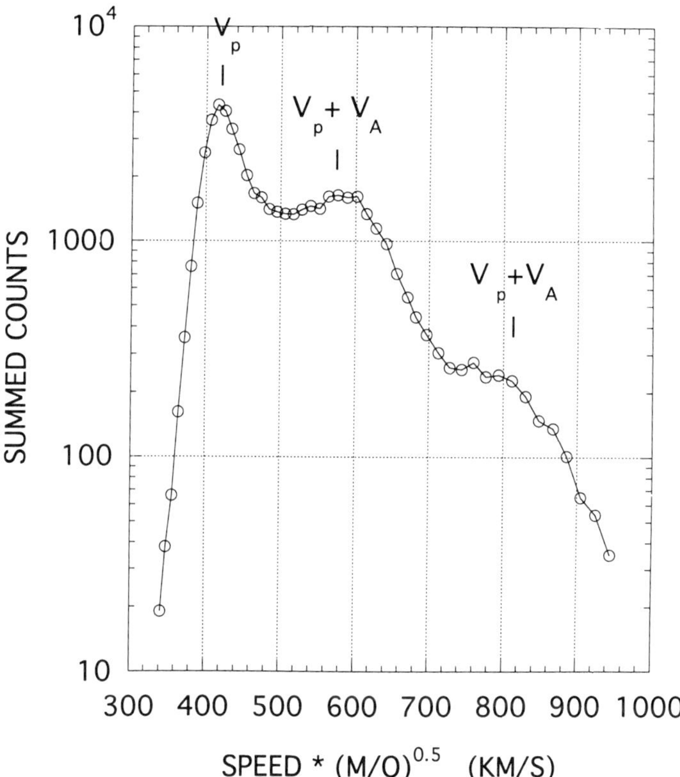

Figure 16.   A radially projected proton count-rate distribution (proportional to $E^2 F(V)$) measured using the Los Alamos IMP 8 plasma experiment at 1 AU, showing a resolved secondary beam traveling faster than the primary beam at the local Alfvén speed, here being 150 km s^{-1}. Although this beam may contain some contributions from alpha particles, its high intensity relative to that of the primary beam suggests that it is composed primarily of protons. The shoulder at the right labeled $V_p + V_A$ is located at the place where alpha particles traveling faster than the primary proton component by the local Alfvén speed would be expected. This distribution was measured within an extended time period of multi-stream flow for which the local Alfvén speed (and also the secondary-to-primary proton speed difference) was less than 100 km s^{-1}. Just before the spectrum shown above was measured, the magnetic field (and consequently the Alfvén speed) increased by 50%, accompanied by a simultaneous and immediate increase in relative beam streaming speed. Because no other changes in solar wind bulk flow parameters changed at this time, this example suggests that the relative streaming speed is limited locally by a plasma instability (figure from Hammond et al. 1995a).

on the local Alfvén wave field, which generally has a high amplitude in the high-speed solar wind (see, e.g., reviews by Schwenn 1991; Marsch 1991a).

Another indicator of nonequilibrium conditions is that temperatures of all heavy ions in the low-density and high-speed solar wind are proportional to their mass (Feldman et al. 1974,1976; Schmidt et al. 1980; Marsch et al. 1982b; Bochsler et al. 1985). At lower speeds and higher densities, the heavy ion temperatures all approach that of the protons. These observations are consistent with a picture in which ions are heated by the damping of interplanetary dissipation-scale turbulence (Jokipii and Davis 1969; Feldman et al. 1974; Neugebauer 1976; Tu 1988; Marsch 1991a), thereby equalizing heavy ion thermal speeds. Thermal energy is equipartitioned through Coulomb collisions, which is most effective at high densities and low temperatures (typically found in the low-speed solar wind). However, this cannot be the whole picture because it does not readily account for the fact that heavy ions generally travel faster than the protons at speeds comparable to the local Alfvén speed (see, e.g., Isenberg 1983,1984, and references therein).

A further complication that underscores the complexity underlying solar wind heavy ion velocity distributions is the observation of simultaneous double proton and double alpha-particle distributions at low bulk speeds (Asbridge et al. 1974; Feldman et al. 1993; Hammond et al. 1995a), as shown in Fig. 17. When present, the higher-energy alpha beam has the higher abundance. For example, the ratio of alpha abundances of the high to the low energy beams in Fig. 17 is 3.0. Such observations occur only at low bulk speeds. In contrast, distributions in the high-speed solar wind generally show two velocity-unresolved proton streams and only one alpha stream, traveling faster than the proton center of mass speed by the Alfvén speed, as shown in Fig. 18. A clear example showing a monotonic change from one type of configuration to the other was recorded in March 1974 on the trailing edge of a high-speed stream, as shown in Fig. 19 (Feldman et al. 1993; see, also, Feldman et al. 1973a). Inspection shows that here too, the secondary, high-energy alpha beam has the higher abundance. This situation also holds at times in the neighborhood of magnetic sector boundaries, and within coronal mass ejections, as shown in Fig. 20 (also see examples in Hammond et al. [1995b]). Note that in this last example, both proton and alpha-particle high-energy (secondary?) beams have the higher densities although the alpha abundance of the secondary beam is the larger. Interpretation of these observations involves discussion of coronal heating mechanisms, which is given in Sec. V.

III. COULOMB COLLISIONS AND TRANSPORT

A. Introduction

As we have seen in the previous section, the large-scale inhomogeneity of the interplanetary medium, its variability and complex magnetic field geometry,

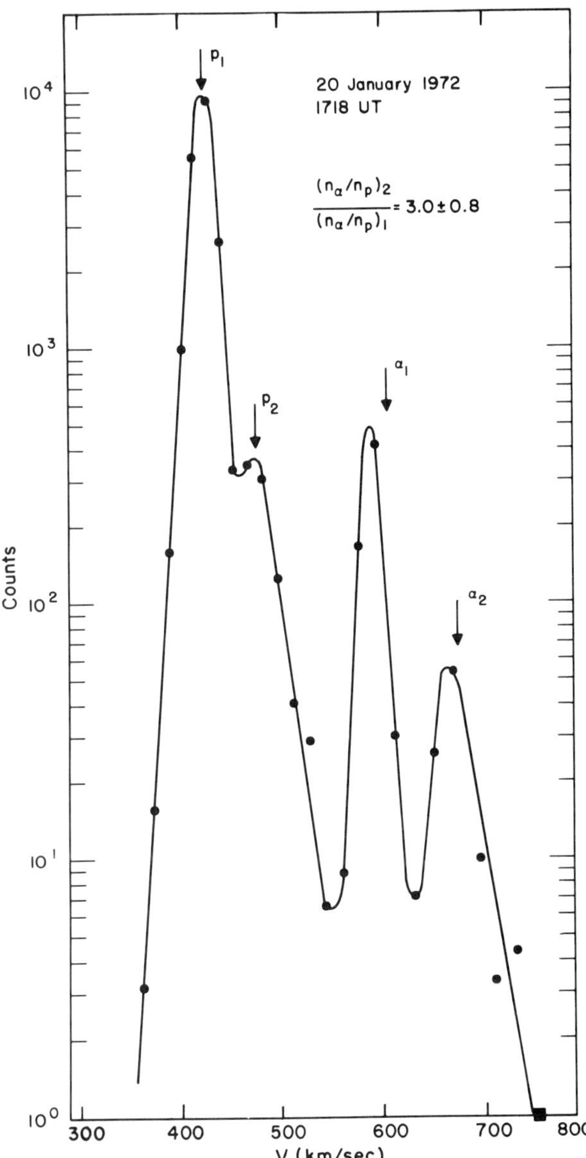

Figure 17.   A radially projected ion count-rate distribution showing simultaneous, resolved double-proton and alpha-particle streams. Although this is an exceptionally clear example, it illustrates a general result that, when present, the alpha abundance of the secondary beam is higher than that in the primary beam, in this case by a factor of 3 (figure from Asbridge et al. 1974).

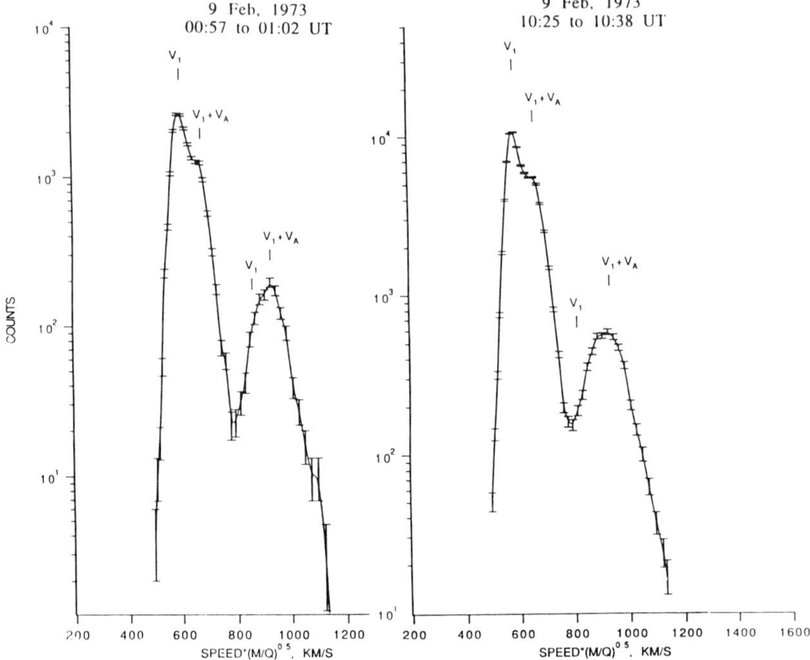

Figure 18.   The radially projected ion count-rate distributions typical of the high-speed solar wind that illustrate a general result. Whereas the proton distributions (left-hand peak) have pronounced high-energy shoulders indicating a secondary beam traveling at the Alfvén speed faster than the primary beam, the alpha-particle peaks (the right-hand peaks) appear as a single component traveling at the same speed ($V_1 + V_A$) as the secondary proton beams (figure from Feldman et al. 1993).

and the radial variations of plasma parameters leave their imprints on the particle distributions. Acting together in a collisionless expansion, these effects would lead to very strong deviations from Maxwellian shapes. The anticipated nonthermal features do, in fact, prevail in observed distributions, yet they are less distinctly developed than would be expected for a collisionless plasma. Consequently, dissipative processes must exist, which partly isotropize the distributions and redistribute internal energy between the various species and kinetic degrees of freedom, thus ensuring the observed fluid-like behavior of the solar wind.

Among such processes Coulomb collisions are the classical and best understood means of thermalizing particle distributions (see, e.g., Braginskii 1966). However, the collisional free path of thermal particles in the solar wind is usually found to be much larger (with the exception of the vicinity of the heliospheric current sheet) than the density scale height, and therefore classical transport theory is certainly not applicable (see, e.g., Kopp and Orral 1977; Olbert 1983). Yet this does not mean that collisions can be neglected entirely.

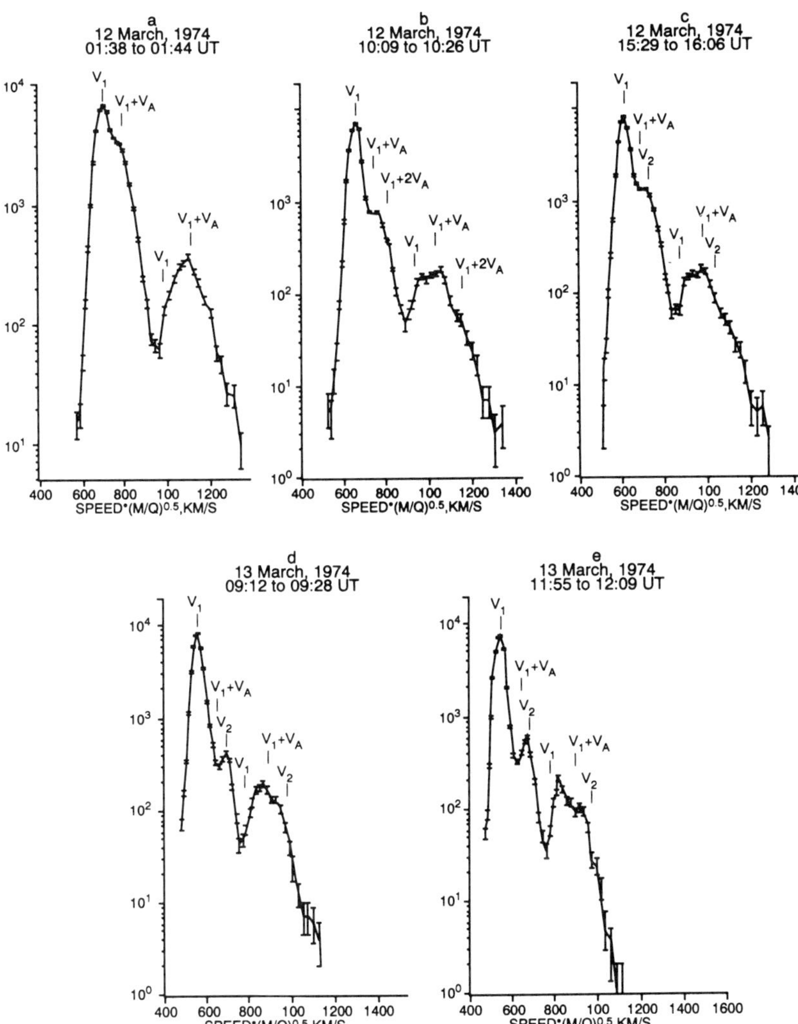

Figure 19.   A representative sample of radially projected count-rate distributions clearly showing the evolution from high-speed to low-speed flows along the trailing edge of a high-speed stream. This edge maps into the boundary layer of a coronal hole. Note the decreasing intensity of the secondary proton beam with decreasing solar wind speed (and increasing time) and the appearance at the lowest bulk speed at the lower right, of a secondary, high-energy alpha-particle beam. Here, as in Fig. 17, the secondary beam has the higher alpha-particle abundance (figure from Feldman et al. 1993).

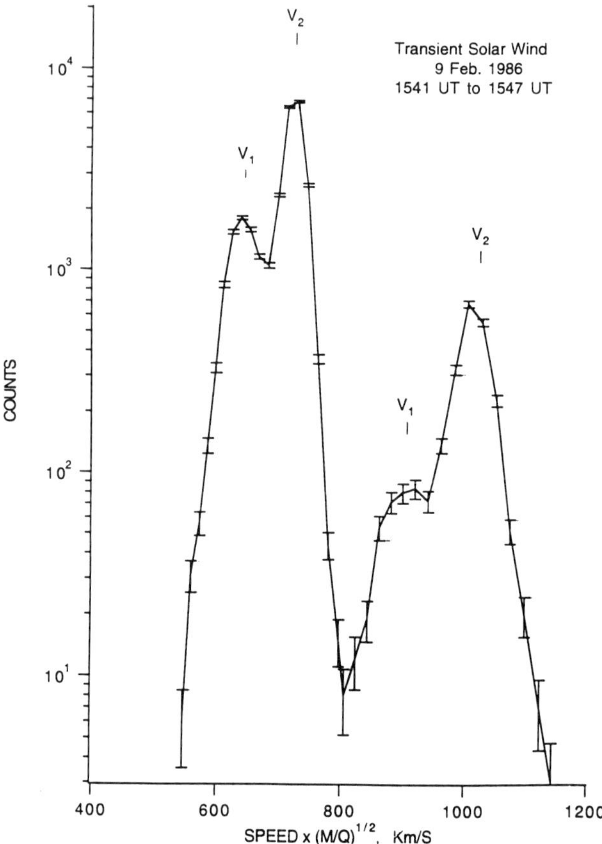

Figure 20.   A radially projected ion count-rate distribution measured within a magnetic cloud imbedded in a solar-driven shock-wave disturbance. This case is unusual in that both the higher-speed proton and alpha beams have the higher intensity. Here again, the higher-speed beam has the higher alpha abundance. Bi-streaming ion configurations are measured often within bi-streaming electron flows that are also associated with coronal mass ejections.

Even a few collisions per AU suffice to reduce the temperature anisotropies as compared to the extreme values obtained by assuming strict conservation of particle magnetic moments (Griffel and Davis 1969; Phillips and Gosling 1990). On the other hand, collisional relaxation times are strongly energy dependent, and therefore "runaway" is expected to occur above a certain threshold energy (see, e.g., Hinton 1983). This relaxation is manifest in the electron strahl population discussed previously. Solid observational evidence has also been provided for the role of Coulomb collisions in shaping ion velocity distributions (Marsch and Goldstein 1983). Generally, however, other dissipative processes like wave-particle interactions are more important and often dominate the transport (Dum 1983; Schwartz 1980). They are discussed separately in Sec. IV.

If wave-particle interactions are neglected, the basic, purely collisional evolution scenario of solar wind particles is as follows: Ions and electrons moving away from the corona along the diverging interplanetary spiral magnetic field B tend to conserve their total energy and magnetic moment. The effect of the decreasing magnetic field with increasing distance dominates, thereby forcing distribution functions to develop shapes that are elongated along B. Binary Coulomb collisions tend to return these distributions toward isotropy. Because the Coulomb cross section decreases strongly with increasing particle velocity, those particles having speeds higher than a few thermal speeds are not effectively scattered and will "run away" from the thermal population to form a separate, strongly focused beam or strahl population. These particles will then evolve under the sole influence of the conservative, macroscopic magnetic and electric fields that permeate all of interplanetary space. This evolution can be described by the guiding center limit of the Vlasov equation (see, e.g., Kulsrud [1983] for a general introduction), i.e., by a kinetic transport equation for the velocity distribution supplemented by an appropriate collision operator (Hinton 1983). Models ranging from simple relaxation time approximations to the full Fokker-Planck collision integral have been employed (Livi and Marsch 1986a,b,1987; Scudder and Olbert 1979a,b; Shoub 1987,1988). However, a fully self-consistent integration of the Boltzmann equation still remains to be performed.

B. Observations on the Role of Coulomb Collisions

Ample observational evidence exists for the regulative effects of Coulomb collisions on ion differential temperatures and speeds (Feldman et al. 1974; Neugebauer 1976,1981; Marsch et al. 1982b; Klein et al. 1985). Whereas Coulomb collisions seem to play a minor role in high-speed solar wind because of high-temperature and low-density conditions, it appears that in the slow wind, collisional friction couples the different ionic species more strongly to one another and enforces almost equal temperatures and speeds. Whenever the number of ion collisions N (defined as the ratio of the solar wind expansion time to the collision time) is larger than unity, the corresponding particle mean free path is considerably shorter than 1 AU at 1 AU. An overview of the variation of N with the solar wind stream structure can be found in Fig. 21, in which the Coulomb collisional domains are clearly delimited. How the particle distributions vary in their shapes and are affected by collisions is shown in Marsch and Goldstein (1983). A detailed analysis of measured ion spectra reveals that all kinds of forms ranging from isotropic near Maxwellians to extremely elongated (pancake or cigar-type) distributions are possible (see again Fig. 13), depending on collisionality. However, Coulomb collisions can neither reproduce the magnitude of many of the effects observed in the solar wind, nor their dependence on heliocentric distance; e.g. non-Maxwellian distortions seem to decrease with increasing heliocentric distance.

The collisional effects in slow solar wind on the He^{2+} to H^+ temperature

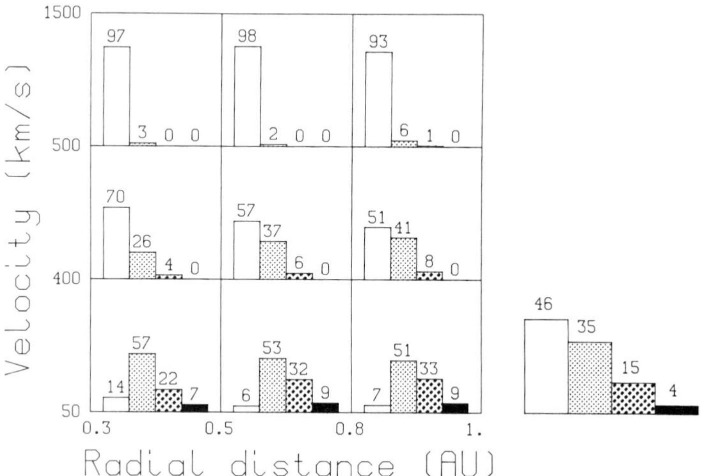

Figure 21. Histograms of the number of collisions N for three regimes of solar wind speeds and three distance intervals as indicated. The code is as follows: open boxes, $N < 0.1$; small dots, $0.1 < N < 1.0$; bold dots, $1 < N < 5$; and finally, black boxes, $N > 5$ (figure from Livi et al. 1986).

ratio T were first recognized (Feldman et al. 1974) and then investigated in much detail (Hernández and Marsch 1985; Hernández et al. 1987) on the basis of a simplified three-fluid model. Including energy-exchange and heat-flux-degradation, the ion temperature ratio T has been integrated as a function of heliocentric distance. The resulting dynamic equilibrium value of T in the case of a steady energy and momentum exchange between differentially streaming protons and alpha particles could be shown to be mainly a function of the relative helium to hydrogen abundance and the local number of collisions N. As the slow solar wind is inhomogenous and weakly collisional (N near 1), forces related to differential pressure gradients and the rest-frame electric field will always tend to produce a finite ion differential speed whenever T is larger than unity. Coulomb collisions cannot fully counteract this tendency. Observationally, T ranges between 1.5 and 2.5 and its equilibrium value of 1 is practically never reached.

C. Kinetic Effects of Coulomb Collisions

Kinetic models of velocity distributions have shown that under various collisional conditions, even a few collisions per AU suffice to prevent the formation of very extended tails and extreme temperature anisotropies (Livi and Marsch 1987). These studies also indicate that a few ion collisions are sufficient to produce an isotropic thermal core in the velocity distributions. However, at higher than normal energies, ion runaway still occurs and leads naturally to the formation of double beams, which cannot be slowed down by Coulomb friction. Some results of these numerical simulations are presented in Fig. 22, showing the evolution of a tail and, subsequently, of a second peak in the proton distribution while the particles travel from 0.05 to 1 AU.

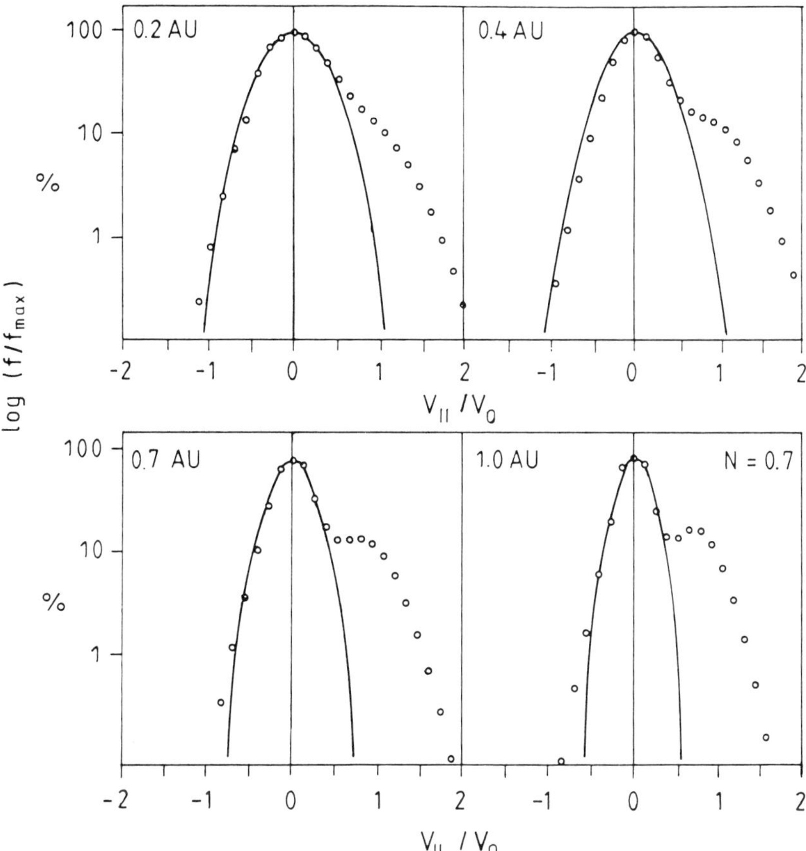

Figure 22.   One-dimensional cuts through model proton distribution functions calculated numerically from a kinetic model which takes magnetic mirror forces into account as well as Coulomb collisions represented by a relaxation time collisional operator in the Boltzmann equation. The parabolas have been inserted in the core to make the field-aligned skewness more evident. Note the occurrence of a heat-flux-carrying tail, which further evolves into a resolved second peak with increasing distance from the Sun. Near 1 AU the phase space density of the beam corresponds to about 20% of the maximum. This numerical run illustrates how a double-beam configuration can be generated in a weakly collisional medium by runaway in a magnetic mirror (figure from Livi and Marsch 1987).

The combined action of the mirror force associated with the interplanetary magnetic field and collisional scattering, can generate variously skewed proton velocity distributions. These distributions, in turn, can undergo considerable reshaping while expanding from the outer corona to 1 AU, if the number of collisions is sufficiently small. In Fig. 22, one sees that at 0.4 AU a marked heat-flux-carrying tail has developed, and at 1 AU even a resolved secondary proton peak in the one-dimensional cut through the distribution function has occurred at a significant fraction of the maximum phase space density. Modeling under different coronal or 1 AU constraints yields satisfying qualitative and even quantitative agreement between some of the observations (see Fig. 13 again) and simulations for the collisional domains of the wind. However, as mentioned previously, Coulomb collisions cannot reproduce the magnitude of the effects displayed in Figs. 17 and 20. Instead, these strongly non-Maxwellian shapes seem to reflect conditions in the inner corona that are intermittent and subject to abundance variations, rather than to interplanetary collision processes. The observation that non-Maxwellian distortions (specifically, the perpendicular proton temperature anisotropy), decrease with increasing heliocentric distance as shown in Fig. 13, clearly requires an explanation in terms of local wave-particle interactions.

A similar phenomenon is observed in the case of the electrons. Basically, the strahl owes its existence to the almost collisionless escape of suprathermal electrons from the corona (Jockers 1970; Lemaire and Scherer 1971). Exospheric theory predicts some of the primordial features of the electron distribution functions, remnants of which like the strahl are indeed observed. Yet, the quantitative agreement is quite often poor. Observed velocity distributions are less anisotropic and skewed than predicted for a collisionless expansion. Coulomb collision- mediated transport has been invoked in kinetic models of solar wind electrons to explain these observations (Feldman et al. 1979*a*; Scudder and Olbert 1979*a,b*). However, observations do not show all the features predicted by these theories (Feldman et al. 1982*a*; Lemons and Feldman 1983; Pilipp et al. 1987*b*). Numerical solutions of the Boltzmann equation with a Krook collision operator have also been shown capable of reproducing observed suprathermal and strahl populations (Olbert 1981). Shoub (1983) has outlined a more general kinetic framework for solar wind modeling, including the full Fokker-Planck collision operator, but detailed results do not exist.

A possibly serious shortcoming of existing models is the form of collision operators they used (Livi and Marsch 1986*b*). The full Landau collision integral, or equivalently the Fokker-Planck operator, yields very different collision frequencies than does the relaxation-time operator. A thorough parametric evaluation of velocity distributions, adopted from the measured ones shown in previous figures, reveals the high sensitivity of the effective collision rates to the detailed shape of the distributions (Marsch and Livi 1985*a,b*). Related calculations of the collisional reshaping and decay in time of ion distributions towards thermodynamic equilibrium, can be found in Livi and Marsch (1986*a,b*). Similar studies have also been carried out for more realistic electron model distributions by Marsch and Livi (1985*b*) with similar conclusions.

The validity of the Fokker-Planck equation itself, or of the Landau collision integral for dilute space plasmas, has been put into question by Shoub (1987,1988) through an evaluation of the underlying Boltzmann collision integral when expanded to higher order in the post-collision values of the particle velocities. The main result is that far from equilibrium, large-angle scattering plays a more important role in the collisional relaxation than assumed in the Fokker-Planck diffusion-like approach. The observation that a significant number of particles are distributed in a nonequilibrium fashion in velocity space, appears to favor the need for large-angle scattering events. This should have a special effect on the collisional properties of the electron strahl and ion beam, especially in the formation of the quiet-time diffuse electron halo population. Future modeling of these effects is highly desirable.

D. Towards a Collisional Transport Theory for Dilute Space Plasma

Observed collision times that are longer than convection expansion times in the solar wind force us to abandon assumption of local thermodynamic equilibrium (LTE) in modeling these distributions above the coronal base. Indeed, angle-averaged electron distribution functions in the solar transition zone where the density is relatively high, need not remain Maxwellian, but instead can form anisotropic high-velocity tails (Shoub 1983). These facts have suggested adoption of an isotropic kappa-distribution (a Maxwellian-like function at low energies that merges into a power law shape at high energies; see, e.g., Cuperman et al. [1980], and Scudder [1992a,b]) as a starting point for nonlocal formulations of the solar wind transport problem. Yet observed electron pitch-angle distributions at several 100 eV in interplanetary space can vary by orders of magnitude between the solar and antisolar direction. Furthermore, the strahl may become as narrow as 12° in the interior of a magnetic sector. Therefore, it is hard to conceive how an expansion about an isotropic, "locally" determined quasi-Maxwellian or kappa distribution should ever work (Dum 1983). Cuperman et al. (1980) have proposed to remedy the failure of classical transport theory by incorporating many higher moments of the velocity distributions. As demonstrated by Dum et al. (1980) with the help of measured particle velocity distributions, such Legendre-type expansions are usually at best semi-convergent and may require up to more than ten polynomials when the strahl is present. Even then, they often fail badly to provide acceptable fits to observed ion beams and electron strahls. Such extensions of classical theory are only reliable for very moderately nonthermal distributions and thus do not go beyond standard transport theory of multi-species anisotropic magnetized plasmas (Barakat and Schunk 1982; Hinton 1983).

Although considerable progress has been made toward a better understanding of nonclassical Coulomb transport in the inhomogeneous and dilute solar wind, a coherent theory or practical recipes for explicitly calculating transport coefficients are still lacking. Some general ideas on how to proceed towards a combined global-local concept for electron heat flow have been formulated under the assumption of test-particle Coulomb collision rates

(Scudder and Olbert 1979*a,b*). Unfortunately, the theory is incomplete as it stands; many of its specific predictions are not verified in the solar wind (Feldman et al. 1982*a*; Lemons and Feldman 1983). Better kinetic theories are direly needed that rely on the full Fokker-Planck collision operator in a realistic interplanetary field.

IV. WAVE-PARTICLE INTERACTIONS AND PLASMA INSTABILITIES

A. Introduction

As shown in Sec. II, both solar wind ion and electron velocity distributions develop strong nonequilibrium configurations in interplanetary space. Section III showed that Coulomb collisions can only rarely return such distributions to singly convecting, isotropic Maxwellians. Yet exospheric theory for the simplest form of coronal expansion—that from a spherically symmetric, spatially homogeneous, and time-stationary corona—predicts even more radical departures from equilibrium than are observed. Indeed, some of the observed effects, such as the relative velocity between heavy ions and the protons, have the wrong sense (see, e.g., Nakada 1970; Geiss et al. 1970; Joselyn and Holzer 1978). The system can be driven further from equilibrium in response to more complex states of the inner corona, such as expected to result from time-dependent coronal heating and acceleration mechanisms suggested by some remote-sensing observations of the solar transition zone (see, e.g., Brückner and Bartoe 1983; Dere et al. 1991). Clearly, other effects must be dominant at times.

In this section we consider effects of the damping of the remnants of the coronal wave field in interplanetary space and of the rearrangement of particles in phase space due to the spontaneous generation of waves in response to "free energy" in particle velocity distributions. Of the many distinct classes of possible free energy that have been identified in Sec. II, a large body of theoretical analysis has identified three as potential, dominant sources of instability (see, e.g., Gary [1993] for a comprehensive review): (1) the thermal anisotropies of proton core components in the high-speed solar wind; (2) the relative streaming between multiple ion beams of individual ion components and between different ion species; and (3) the relative streaming between halo and core electron components that provides the dominant mechanism of electron heat transport in the solar wind.

B. Wave Damping

Ion bulk convection accounts for the dominant form of energy transport in the solar wind beyond 0.3 AU. The remainder is split nearly equally between convected enthalpy, electron heat flux, and hydromagnetic waves. These modes of energy transport sum to less than 8% of the total (Feldman et al. 1977; Schwenn 1991). Of these various transport modes, wave propagation and damping has received much attention in the literature because the salient property of the solar convection zone is its strong spectrum of fluid motion. Comparison of the tenuous contents of the corona and solar wind with that of the plasma and magnetic field within the convection zone leaves no doubt that the entire character of the corona and solar wind must reflect both the form and magnitude of energy flux that leaks outward through the photosphere to the solar corona. The strong fluid motions that dominate the visible photosphere has consequently driven a long and deep pursuit of wave-heating models of the solar corona (see, e.g., Hollweg 1990; Narain and Ulmschneider 1990, and references therein).

Indeed, the solar wind supports a well-developed wave field whose evolution in interplanetary space reflects the approach to fully developed, hydromagnetic turbulence (see, e.g., the chapter by Goldstein et al. and Marsch [1991b] for comprehensive reviews). Most of the energy density of this turbulence is carried at the larger scales and in the fluctuation of magnetic components perpendicular to B. Its total energy density is highest far from the Sun and has been postulated to be generally high at the footpoints of these flows in the corona.

Application of a recent theory of the evolution of Alfvénic fluctuations in the high-speed solar wind (which provides an excellent fit to wave observations between 0.29 and 5 AU), to the heating of protons in this flow regime yielded excellent fits there as well (Tu 1988). In this model, energy cascading from the longer to shorter scales damps on the protons at length scales near the proton gyroradius, thereby increasing its magnetic moment. The theoretically predicted cascade energy flux from long to short wavelengths was shown by Tu (1988) to match closely that needed to account for the measured radial evolution of the proton magnetic moment (Schwartz and Marsch 1983), as shown in Fig. 23. A further prediction of the theory is that the rms fluctuations of the solar wind bulk velocity (providing a measure of the integrated energy density in Alfvén waves) should closely follow ion thermal speeds. This prediction is also verified by observations, as shown in Fig. 24.

These successes clearly support the following picture, valid on average in the high-speed solar wind. Alfvén waves originating below the Alfvén critical point continuously transfer energy from large scales, where most of the wave energy is contained, to the proton gyroradius, where it is, in turn, transferred to the protons through ion-cyclotron damping. This process proceeds most strongly in the high-speed solar wind, which is characterized by nonequilibrium ion velocity distributions having: (1) a $T_\perp > T_\parallel$ core thermal anisotropy; (2) a secondary, high-energy proton beam; and (3) a single alpha-particle beam traveling at the Alfvén speed faster than the protons. The net effect is the maintenance of a nonlinearly stable state in which the cyclotron-wave component of the large-scale MHD wave field is in steady state with the ion velocity distributions (Abraham-Shrauner and Feldman 1977). Unfortunately, the power spectrum of solar wind fluctuations in the dissipation range is not well known. In addition, the possible contribution to ion heating from compressive waves that suffer Landau and transit-time damping (Barnes 1979),

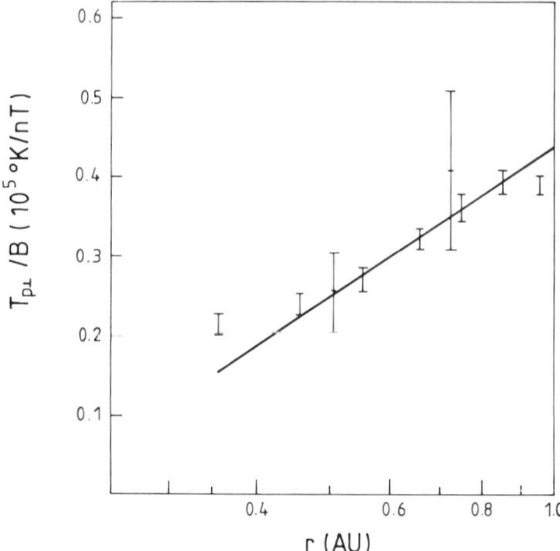

Figure 23.   The radial variation of proton magnetic moment m within high-speed flows measured using the Helios plasma analyzer (data with error bars). These data are overlayed with the increase in m predicted from a theory that fits the energy cascaded from large to small scales in measured solar wind Alfvénic turbulence, if all this energy goes into heating $T_\perp$ at the proton gyroradius (figure from Tu 1988).

needs to be evaluated in more theoretical detail and checked experimentally.

C. Plasma Instabilities

From time to time the power density of the solar wind wave field over discrete frequency intervals is considerably enhanced. One class of such events consists of traveling forward and reverse shocks associated both with solar-driven transients and with corotating interaction regions at heliocentric distances beyond about 2 AU. The net effect of these shock waves is to heat the ambient plasma and to accelerate a few particles to very high energies. This topic has been reviewed recently (see, e.g., Gurnett 1985,1991, and references therein) and so will not be considered further here.

Another class of wave enhancements consists of bursts of Langmuir waves observed in association with Type III electrons in interplanetary space (see, e.g., Lin et al. 1981; Gurnett et al. 1993). It is generally believed that these waves result from a bump-on-tail instability through a Landau resonance with electrons accelerated at the Sun having energies in the range between about 10 keV and 100 keV. We do not consider this instability further in this chapter because it is driven by particles within the positive slope portion of energetic electron flux distributions, which is far from the plasma regime of interest here. However, see the recent review by Gurnett (1991) for more details.

Of more direct interest here are isolated occurrences of enhanced left- and right-hand polarized ion cyclotron waves as shown in Fig. 25. In addition, bursts of ion acoustic-like waves are observed at times (Gurnett and Frank 1978; Gurnett 1991), as are bursts of correlated whistler and electron plasma

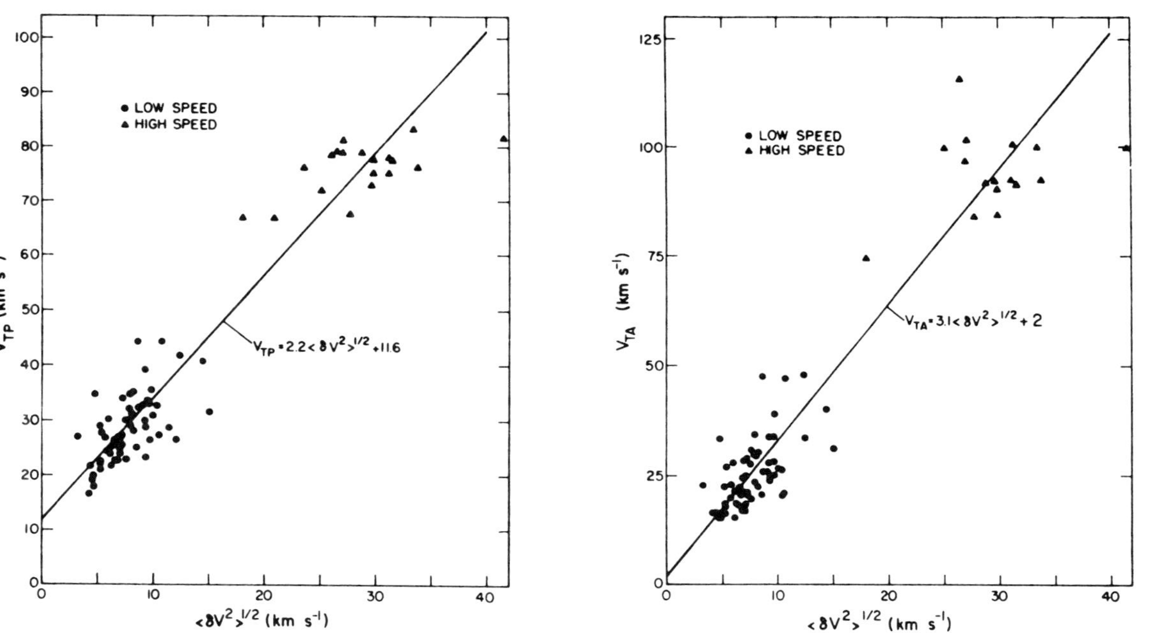

Figure 24.   The correlation between proton (left panel) and alpha-particle (right panel) thermal speeds and the rms value of solar wind bulk velocity variations in successive 3-hr intervals. These data were averaged over 61 low-speed and 19 high-speed flow conditions selected from measurements between March 1971 and July 1974 (figure from Feldman et al. 1976).

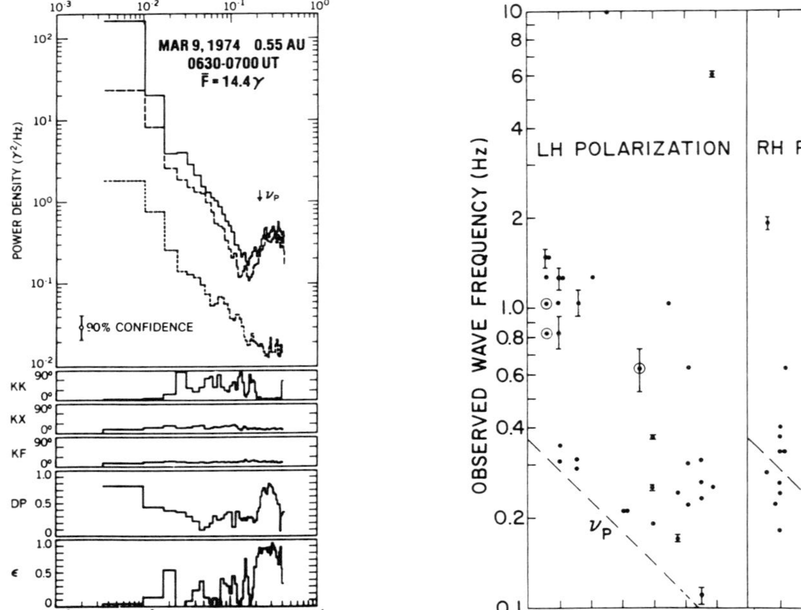

Figure 25.   At the left is a clear example of enhanced power density in transverse magnetic fluctuations near the proton gyrofrequency in the high-speed solar wind (from Behannon 1976). The two upper traces at the left give the power density of transverse (to B) fluctuations and the lower trace gives the power density of longitudinal fluctuations. Also shown at the lower left are the angle KK between the wave vector directions determined from the real and imaginary parts of the spectral density matrix, the angle KX between the unit wave vector and the radial, the angle KF between the unit wave vector and the magnetic field vector, and the fraction polarization of power density and its eccentricity. At right is a summary of examples of discreet wave activity such as that shown in the left-hand panel, arranged in accordance with observed wave frequency and heliocentric distance. Left-hand and right-hand polarizations are separated in the two halves of the panel, and the dashed lines mark the proton gyrofrequency.

oscillations (Kennel et al. 1980), and periods of enhanced whistler mode turbulence (Neubauer et al. 1977; Beinroth and Neubauer 1981; Coroniti et al. 1982). The time dependence of these events and/or their association with nonstationary aspects of the bulk flow state of the solar wind, leaves little doubt that they result from generation mechanisms operating locally. The near universal occurrence of non-Maxwellian velocity distributions in the solar wind as reviewed in Sec. II, and the burstiness of the events, leaves little doubt that these mechanisms are plasma instabilities driven by "free energy" carried by solar wind plasma particles.

The term "free energy" as used here is not synonymous with non-Maxwellian. For particle distributions to contain free energy, they must be sufficiently far from equilibrium that particles can spontaneously rearrange themselves in phase space leading to an approach toward equilibrium and the generation of waves. In other words, a particle distribution contains free energy if it leads to the generation of waves through a plasma instability. This process often involves very few particles (such as in resonant instabilities) and hence leads to relatively low wave intensities and only minor changes in the unstable distributions. An example is plateau formation in bump-on-tail particle distributions (see Gary [1993] for more details). Nevertheless, the net result is a partial return toward equilibrium and the generation of waves that can subsequently affect plasma transport through their interaction with other particles.

$T_\perp > T_\parallel$ *Proton Core Instability.* Much theoretical work has been done in studying instabilities driven by this configuration because it is universally observed in the high-speed solar wind (see, e.g., Gary [1993] for a review of the extensive literature). The end result of these studies is that observed configurations are always close to marginal stability, but often linearly unstable, as shown in Fig. 26. However, as mentioned in Sec. IV.B, this configuration is driven by the damping of Alfvénic turbulence pre-existing in the wind and so cannot be unstable to the generation of waves. Instead, it must always be close to a state of nonlinear marginal stability in which the particle distributions and waves coexist (Abraham-Shrauner and Feldman 1977).

Secondary Proton and Alpha-Particle Beams. Most of the ion velocity distributions shown in Figs. 13 through 20 exhibit secondary proton and/or alpha-particle beams. Quantitative modeling of these types of distributions (such as that shown in Fig. 26) reveal that during those times when secondary-beam velocities exceed that of the primary proton component by about the local Alfvén speed, then right-hand polarized magnetoacoustic waves are driven unstable. This process was, perhaps, responsible for the generation of the waves shown in the far right-hand panel of Fig. 25. Proton velocity distributions observed using Helios near perihlion (shown in the bottom row of Fig. 13) indicate that this situation must occur routinely beyond about 0.3 AU because the relative speeds of the secondary to primary proton beams are at, or just beyond, the Alfvén speed at 0.3 AU. Beams having speeds larger than the local Alfvén speed must develop at large distances because the

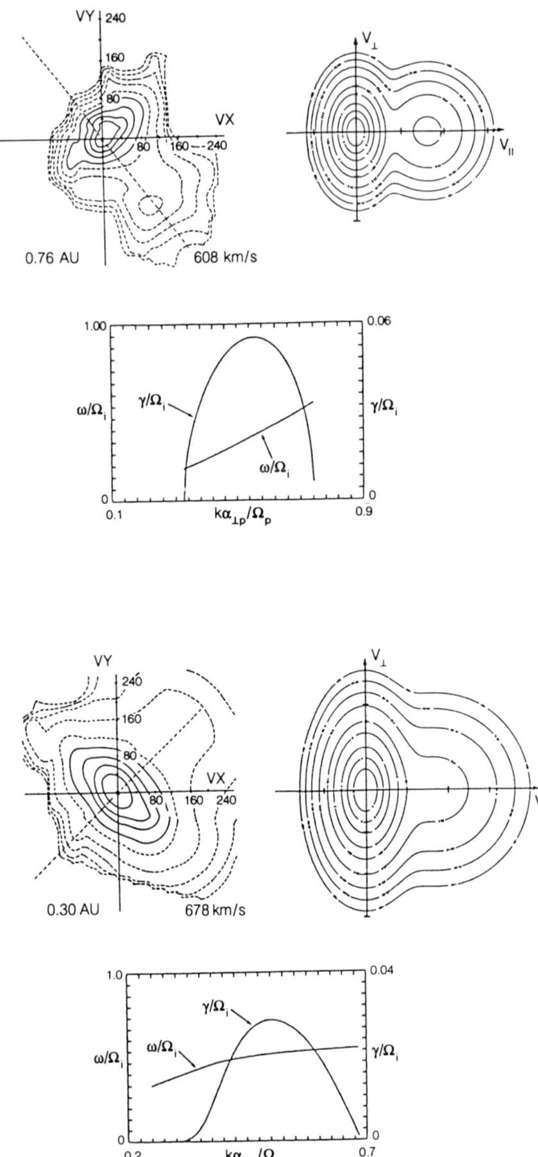

Figure 26.   Results of a calculation of instability thresholds for right-hand polarized magnetosonic waves driven by a secondary proton beam (above), and left-hand polarized ion-cyclotron waves driven by a $T_\perp > T_\parallel$ temperature anisotropy (below), determined using drifting bi-Maxwellian fits to measured proton velocity distributions observed using Helios (figure from Leubner and Viñas 1986).

Alfvén speed decreases monotonically with increasing heliocentric distance.

Electron Halo-Core Relative Streaming. A large body of evidence has been amassed to indicate that solar wind electron velocity distributions at 1 AU are often close to marginal stability, thereby limiting electron heat flux (see, e.g., Feldman et al. 1979b; Schwartz 1980; Marsch 1991a; Gary 1993, and references therein). However, there is currently no agreement as to the exact form of the free energy, and hence the type of instability that enforces this situation.

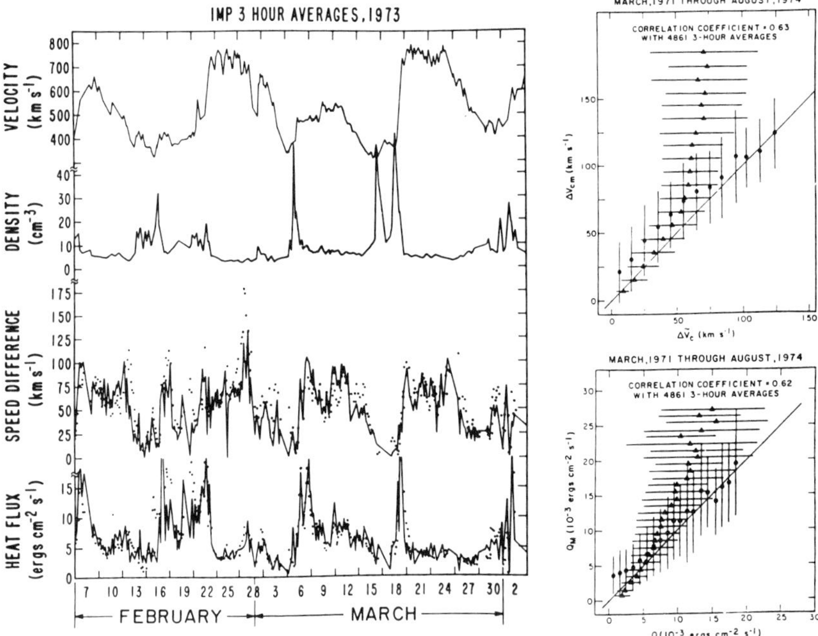

Figure 27.   A collection of IMP plasma data that support existence of a closure relation for solar wind heat flux at 1 AU. The top two traces at left give the bulk speed and proton density. Just below is an overlay of the measured (solid line) and model (dots) velocity difference between the electron core population and the proton bulk velocity, along B. At the bottom is a comparison between measured (solid line) and model (dots) electron heat flux. The two panels at the right show correlations between model and measured core-electron and proton velocity differences (top) and heat fluxes (bottom) for an extended period spanning March 1971 to August 1974. The triangles correspond to a sorting of the model values averaged over successive 3-hr intervals and the circles give the sorting of the corresponding 3-hr-averaged measured values (figure from Feldman et al. 1979b).

Local control over electron heat conduction is expressed most clearly by a relationship between some parameter characterizing the shapes of electron velocity distributions and the low-order bulk flow moments of the wind. A striking example is presented in Fig. 27. Shown in the panel at left, second curve from the bottom is the observed correlation between 3-hr averages of the measured speed difference between the core electrons and the solar wind bulk speed (ΔV_c, solid line) and a simple function of the bulk speed of the plasma as measured in a reference frame corotating with the Sun (dots). At the bottom is an overlay of a simple model for the heat flux (dots),

$$Q_m = g N_c k T_c \Delta V_{cm} \tag{1}$$

and the measured heat flux (solid line). Here $g = 10.7$ is a constant, N_c is the core electron density, k is Boltzmann's constant, T_c is the core electron temperature, and ΔV_{cm} is the foregoing model speed difference between the core and total electron populations. The correlations at the right show that these relations are generally valid over extended periods of time, here the 2.5-yr period between March 1971 and August 1974.

Regulation of solar wind heat flux by micro-instabilities was first predicted from exospheric theory by Schulz and Eviatar (1972) and Perkins (1973). Their analyses indicated that instabilities on the magnetosonic branch driven by the condition, $\Delta V_c > V_A$ (where V_A is the local Alfvén speed) would have the lowest threshold. However, because these waves do not interact strongly with the halo electrons (which carry the heat flux), regulation of the heat flux would need to be enforced indirectly through an increase of the interplanetary potential.

A more direct connection between instability driven waves and heat-flux regulation can occur if the right-hand polarized whistler heat flux instability is triggered (see Gary 1993, Sec. 8.3). This instability is driven by the relative core-halo drift velocity but quenched by a $T_\parallel > T_\perp$ anisotropy. Although a correlation between $\Delta V_c / V_A$ and the ratio between halo and total electron densities, N_h / N (predicted by theory whenever $T_\parallel / T_\perp < 1.2$ and the ratio of plasma to magnetic pressure, $b > 0.5$), is observed, these conditions do not generally obtain in the solar wind. More damaging is that the field-aligned whistlers predicted by theory have yet to be positively identified. Instead, obliquely propagating whistlers seem to be generally present in the wind (Neubauer et al. 1977; Beinroth and Neubauer 1981; Coroniti et al. 1982). Also, the possibility exists that lower-hybrid waves (Marsch and Chang 1982,1983) play a role in regulating the electron heat flux. They provide an effective means for coupling the thermal ions and electrons by Landau damping and thus for transferring energy and momentum between these species through wave-particle interactions, given of course that the wave amplitudes are sufficiently large.

Clearly more work is required to resolve the issue of heat flux regulation. Perhaps the background, obliquely propagating whistler wave field in the solar wind, acts on the ambient electrons to shape local electron velocity distributions in much the same way as the eventual damping of the energy cascade of Alfvénic turbulence in the high-speed solar wind acts to shape proton velocity distributions there. This "solution" of course begs the question as to the origin of the ambient whistler-mode turbulence. One possibility is that it is the remnant of the wave flux from the Sun that has not yet completely damped on ambient electrons. However, the observation by Kennel et al. (1980) of bursts of correlated Langmuir waves and whistler waves present an alternate and intriguing possibility. Based on the morphology of these events, the authors suggest that Langmuir-whistler correlations result not from direct wave-wave coupling, but from the sharing of a common source of free energy in ambient electron distributions. For example, a time-dependent burst of fast electrons caused by the sudden onset of a transient coronal and/or inner solar wind heating event, would produce a bump-on-tail distribution farther from the Sun through a time-of-flight effect, as shown in Fig. 28. Similar distributions observed during interplanetary Type III events, are known to generate Langmuir waves quasi-continuously all along affected magnetic flux tubes (Lin et al. 1981; Gurnett 1991). Another consequence of a transient pulse of electrons is that its leading edge must carry a wave of enhanced interplanetary potential with it. Such an increase is needed to ensure that the electron flux is equal to the ion flux everywhere in order to maintain charge neutrality all along the affected magnetic flux tube. It will, in turn, accelerate the core population sunward thereby increasing ΔV_c.

This chain of events should then set up a situation similar to that routinely observed upstream of the Earth's bow shock. Here incoming solar wind electrons having large pitch angles are reflected back into the upstream solar wind by the magnetic field ramp within the shock. This process creates velocity distributions in the electron foreshock having a pear-like shape at intermediate energies. A one-to-one association between these distributions and obliquely propagating whistler waves is observed (Feldman et al. 1982b,1983). Detailed models have shown these distributions to be unstable to the generation of observed (Hoppe et al. 1981,1982) obliquely propagating whistler waves (Sentman et al. 1983). We can only speculate that similar effects occur in the down-stream portion of the solar wind during the Kennel et al. (1980) correlated wave events. Here, the mirror force presented by the ambient interplanetary magnetic field replaces that of the magnetic ramp within the Earth's bow shock to reflect that portion of sunward-accelerated core electrons having large pitch angles.

V. IMPLICATIONS CONCERNING CORONAL HEATING MECHANISMS

A. Summary of Solar Wind Particle Velocity Distributions

Evolution of the shapes of particle velocity distributions of the solar wind is known only between 0.29 and 5.4 AU. Within this distance range both ion and electron distributions display a wide variety of non-Maxwellian shapes. We summarize first those of the electrons. Their overall character is set by the two dominant macroscopic fields, the ambipolar electrostatic potential and the interplanetary magnetic field. The variable topology of B, Coulomb

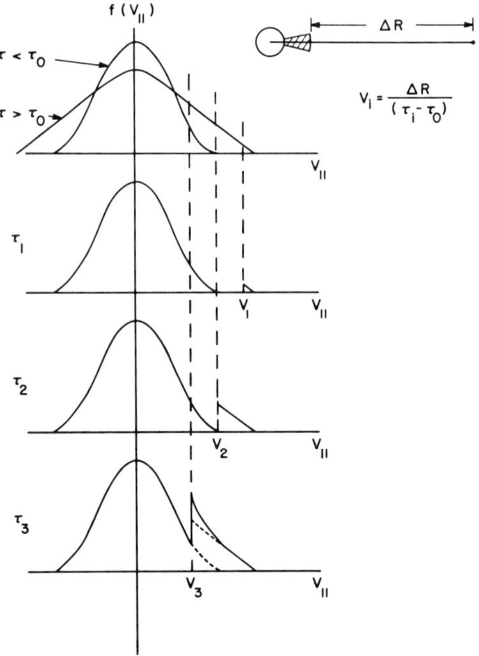

Figure 28.   A schematic illustration of how a transient heating of electrons at the coronal base can lead to bump-on-tail velocity distributions at an altitude of ΔR above the base due to a time-of-flight effect (figure from Feldman 1979).

collisions, and wave-particle interactions also make essential contributions to the observed shapes. These factors lead to:

1. Two readily discernible components; a thermal electrostatically bound core, and a suprathermal unbound halo;
2. A nearly isotropic Maxwellian core velocity distribution whose bulk velocity trails that of the protons by speeds up to about the local Alfvén speed;
3. An anisotropic halo population having $T_\parallel > T_\perp$, which generally flows away from the Sun and is skewed along the magnetic field (the so-called strahl), thereby carrying large parts of the solar wind heat flux;
4. Intermittent bi-directional halo flows, generally associated with a closed-loop, tongue or flux rope topology of the magnetic field;
5. Suprathermal strahl distributions that are strongly focused parallel to B and locally have an equivalent Maxwellian temperature close to 1×10^6 K. This situation prevails generally in high-speed solar wind and abnormally low-density bulk flows where both Coulomb and wave-particle collisions are rare;

6. Halo distributions that can be fit with a generalized Lorentzian in energy, times an expansion in Legendre polynomials in angle up to fourth order, during other, generally low-bulk-speed flow types;
7. A separate, nearly isotropic background component of halo electrons in the low-density wind, distributed with energy as a power law having an index close to that of the "quiet-time" solar electron spectrum observed between 10 keV and 200 keV (Lin 1980).

Solar wind ion velocity distributions can also be generally described in terms of two components whose character depends on stream phase and heliocentric distance. When two components are present, the resultant non-Maxwellian shapes and/or relative streaming velocities become less pronounced with increasing heliocentric distance. Their salient features include:

1. Single, nearly isotropic Maxwellians having nearly equal proton and alpha-particle bulk speeds in cold, dense solar wind flows usually observed near the heliospheric current sheet;
2. Double-proton and alpha-particle streams that are co-moving such that both secondary components travel at the local Alfvén speed relative to the primary components and have the higher alpha-particle abundance. This configuration often obtains near to, but not within, the current sheet, and within some bi-streaming electron flows and magnetic clouds;
3. Double-proton and single alpha-particle streams throughout high-speed flows characterized by (a) a proton primary component having a large $T_\perp > T_\parallel$ anisotropy; (b) a hotter, velocity-unresolved secondary proton component traveling at about the Alfvén speed faster than the primary proton component; and (c) a single alpha component traveling faster than the protons by about the Alfvén speed and having a temperature about 4 times larger than the protons.

B. Connection with Coronal Heating and Acceleration Mechanisms

Although much theoretical and experimental work has been devoted during the last 50 years to the identification of coronal heating and acceleration mechanisms, none have produced conclusive results. Early work concentrated on acoustic waves stemming from convective motions below the photosphere (Billings 1966, and references therein). However, severe observational limits to the outward energy flux set by analyses of OSO-8 observations (Athay and White 1978), and theoretical problems regarding refraction and shock formation of outward-propagating acoustic waves (see, e.g., Narain and Ulmschneider 1990, and references therein), lead to the conclusion that this mechanism can account for, at most, a small fraction of the energy required to maintain the corona. A different set of problems has been obtained for magnetosonic and Alfvén waves evolving upward through the photosphere from below. Indeed, the most recent theoretical studies conclude that the long-period waves that carry most of the energy in interplanetary space do not reach the middle corona from the photosphere because they are evanescent

(Moore et al. 1991), and that those generated at sufficient altitudes to reach the corona do not damp there, but continue propagating to higher altitudes (Parker 1991,1992). However, a careful analysis of solar wind Alfvén-wave spectra measured at distances beyond 0.29 AU and their extrapolation back to the Sun without dissipation in between, seems to indicate that they do not carry sufficient energy flux at the coronal base to accelerate the high-speed wind (Roberts 1989).

A host of heating mechanisms associated with various forms of solar activity, involving, e.g., current-sheet formation and magnetic reconnection, has been proposed in the literature (for a comprehensive recent account see the Heidelberg conference proceedings edited by Ulmschneider et al. [1991]). For a schematic picture of some of the dynamic plasma processes we show Fig. 29, which depicts low-altitude features such as polar plumes, macrospicules, bright-point flares, microflares, and explosive events as specific examples. Unfortunately, none of these events occur sufficiently often or with enough observed energy released per event to replenish all known coronal energy losses (Habbal 1992). In addition, mechanisms have been proposed that have no obvious observational underpinnings. Examples include the presence of a resonant cavity for Alfvén waves near the Sun (Moore et al. 1991), and enhanced heat flux carried by suprathermal electrons generated by some unknown mechanism in the chromosphere and/or transition region. This mechanism is assumed to produce a kappa-distribution (Scudder 1992a,b), one half of which fills the collisionless coronal exosphere and creates the temperature inversion by velocity filtration of the electrons in the combined gravitational and electrostatic potential.

None of the aforementionend activity-related heating mechanisms has been substantiated by data returned from remote measurement techniques or shown to supply the energy needed to support the corona. Their common feature is that they all involve small-scale phenomena. Fine, three-dimensional spatial resolution will therefore be needed to resolve the problem experimentally. Unfortunately, all available remote-sensing techniques suffer from a common deficiency; interpretation of data is limited by essential ambiguities. Although line-of-sight observations of the disk using high wavelength resolution spectrographs can have sufficient spatial resolution in two dimensions, they integrate along the line of sight and are biased to favor the lowest, densest layers of the solar atmosphere at a given temperature. Interpretations of emissions from plasma that is far from local thermodynamic equilibrium, such as have been proposed for the chromosphere-corona transition region, are therefore strongly model dependent (Esser 1992; Habbal et al. 1993). On the other hand, observations of the outer corona using coronagraph techniques necessarily integrate over a very large volume, thus preventing resolution of individual, small-scale filamentary structures, which are, for example, revealed in contrast-enhanced corona pictures taken during solar eclipses (Koutchmy and Livshits 1992).

Enhancements in the power of density fluctuations near the local proton

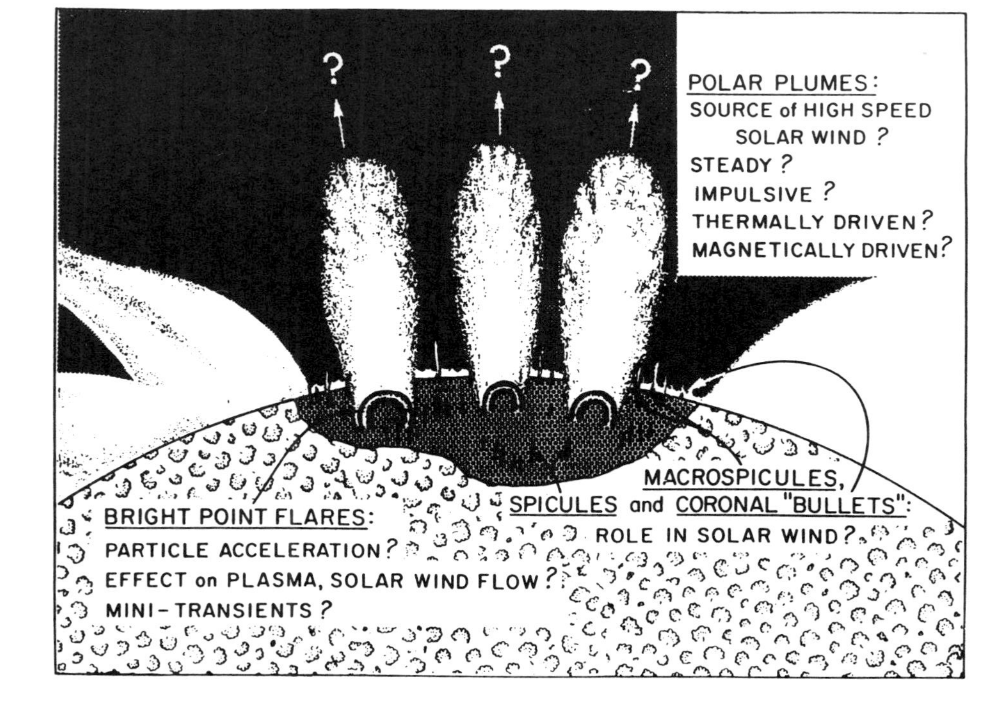

Figure 29.   A schematic drawing of a polar coronal hole (dark shaded area) showing many transient structures that may contribute importantly to coronal heating (figure from Withbroe et al. [1991]).

inertial length have been inferred from interplanetary scintillation data obtained between 7 and 20 $R_\odot$ (Coles et al. 1991a). Do these data reflect the presence of an ensemble of small-spatial-scale density filaments, such as can result from the parametric decay of large-amplitude Alfvén waves (Viñas and Goldstein 1992)? Or do they perhaps represent an ensemble of dense ion jets coursing through a less dense ambient medium (Feldman et al. 1993), or a jumble of low-speed diamagnetic plasmoids (Pneuman 1983; Mullen 1990), or an ensemble of large-amplitude compressive waves and fluctuations that gradually shock, damp, and heat the ambient plasma?

The problem of solar wind acceleration is even less resolved than that of coronal heating because it is affected fundamentally by the fact that the corona must be hot to begin with. Recent measurements of solar wind speed at radially aligned positions in the extended corona using interplanetary scintillation observations (Coles et al. 1991b) indicate that a driving mechanism by pure thermal-pressure-gradient forces is never adequate to explain the data. Only half their data can be fit by adding Alfvén waves to thermal pressure gradients, and the other half defy explanation using any of the standard wave-driven theories of solar wind acceleration. Does this failure point to the existence of a filamentary nature of the nascent solar wind, consisting of many spatially intermittent meso-scale flow tubes at low altitudes, which then intermingle and smooth out to create the "quiet" interplanetary wind? Although nonstandard additions to the conventional theories abound (see, e.g., Hollweg 1990; Withbroe et al. 1991, and references therein), they too have difficulties fitting existing interplanetary scintillation speed data. Less standard acceleration mechanisms are therefore needed, which also include complex and realistic flow geometries.

Another constraint on any acceleration mechanism is that it needs to result in a relatively constant mass flux, as observed (Feldman et al. 1977; Schwenn 1991). Perhaps both the temperature and density at the coronal base are relatively constant, as suggested by Withbroe (1989). This possibility is difficult to test using remote-sensing observations because of essential ambiguities in spectroscopic interpretations (Esser 1992). Another suggested explanation is that alpha particles buffer the solar wind mass flux (Leer et al. 1992; Bürgi 1992). This explanation requires that alpha particles collect in the low corona where their number density may amount to 30% or more of the total ion number density.

Interpretation of measured shapes of solar wind particle velocity distributions in terms of coronal heating and acceleration processes is also subject to ambiguities, but different than those encountered using remote-sensing photon measurements. In particular, many of the signatures of coronal heating and acceleration imprinted on solar wind particle distributions at the coronal base are erased by Coulomb and wave-particle interactions while the plasma transits from the Sun to interplanetary space, where the distributions are observed. Nevertheless, the exercise of direct inferences is worthwhile, because they result in a different set of constraints than that achieved by optical plasma diagnostics.

We start first with measured shapes of strahl electrons in high-speed streams and low-density flows. Their low angular widths argue for little scattering beyond the coronal base, indicating that their parallel temperatures provide a good proxy of the coronal electron temperature between 10 and 20 $R_\odot$. The resultant measured temperature at 1 AU, ranging between 0.85 and 1.15×10^6 K, is consistent with temperatures derived for the coronal base within coronal holes from a variety of remote sensing observations (Habbal et al. 1993) and models of the solar wind outflow (Withbroe 1988). A further result is that the shapes of reduced distributions parallel to B are closely Maxwellian. This result is rather surprising because the 100 to 300 eV electrons that display this shape at 1 AU all correspond to >1 keV electrons at the coronal base, a factor of more than 10 times the thermal energy there. Measurements of solar wind electrons within high-speed and low-density flows therefore do not support the existence of extended, power law tails in electron distributions below the coronal base, a characteristic that was hypothesized as a means of accounting for the temperature inversion that produces a million degree corona (Scudder 1992a,b).

Details of ion velocity distributions in the high-speed solar wind indicate that fundamentally time-dependent phenomena must play a significant role in the heating of coronal holes. Relevant observations are the universal presence of a secondary proton beam and a single alpha-particle beam, both traveling faster than the primary proton component by about the local Alfvén speed. This configuration evolves into sporadic extended intervals of coincident, double proton and alpha-particle streams at low bulk speeds. In nearly every case, the secondary beam has the higher helium abundance. These results together, can be understood naturally if time-dependent ion jets are produced episodically in the open corona (Feldman et al. 1976,1993) in accord with the observations of explosive events at low altitudes (Brückner and Bartoe 1983; Dere et al. 1991; Dere 1992). In this picture, the primary beam corresponds to the outflow of the low-density, million-plus-degree corona, which evolves into a low-speed and low-mass-flux solar wind. Coulomb friction in this ambient flow is only marginally able to couple the alpha particles to the protons, thereby yielding a low solar wind alpha-particle abundance. In other words, within coronal holes the density and temperature are both sufficiently low that the ambient expansion cannot by friction alone carry any alpha particles to large distances from the Sun. The alpha particles thus fall back and accumulate in the low corona. Episodic occurrences of dense ion jets through this medium could then couple the ambient plasma with high alpha-particle abundance at low-altitude to the jets, to form the secondary beams observed at large heliocentric distances.

A last constraint on processes active close to the Sun concerns the zone of dissipation-scale turbulence inferred from observations of interplanetary scintillations to exist between about 10 and 30 $R_\odot$ (Coles et al. 1991a). Electron strahl distributions during flows other than that in the high-speed solar wind are generally quite broad in angle and decrease asymptotically as a power law in energy. Such distributions are a natural consequence of a momentum diffusion process caused by interactions with waves having a power law spectrum in wave number (Borovsky and Eilek 1986). Unfortunately, nothing specific can be concluded about the coronal source region at this time because the very nature of the diffusion process does not allow a unique inversion. However, the wave field must be sufficiently strong to reduce the ambient conductivity significantly and hence strongly affect the bulk flow state of the solar wind.

Further progress on all of the foregoing issues will require direct *in-situ* exploration of the corona. This is needed to resolve or avoid the many essential ambiguities resulting from interpretations of remote-sensing data. To date, these have been provided by observations of the solar wind at large heliocentric distances, as well as of the corona through its photon emissions.

VI. CONCLUSIONS

The assimilation of all kinetic-scale measurements throughout interplanetary space between 0.29 AU and 5.4 AU and their theoretical interpretations are summarized schematically in Fig. 30. Electron velocity distributions of the solar wind are shaped by coronal heating and acceleration mechanisms, macroscopic global electric and magnetic fields, stream-stream bulk-flow interactions, and local Coulomb and wave-particle interactions. They generally contain two, and sometimes three components. Electrons in the thermal core population have energies less than the local electrostatic potential and develop a local configuration that reflects:

1. Direct injection from the Sun and a possible extended zone of strong plasma turbulence (depicted in Fig. 30 by the innermost coarse and fine stipled zones that encircle the Sun) and reflection from the electrostatic potential;
2. Injection back into the inner heliosphere from forward and reverse shocks that bound corotating interactions beyond about 2 AU (the outermost stipled zone in Fig. 30);
3. Coulomb and/or wave-particle scattering leading to trapping between magnetic and electrostatic mirror points;
4. Pitch-angle and energy diffusion in the general solar wind turbulence wave field (that permeates the heliosphere), thereby allowing a slow spatial migration of the trapped orbit zones of individual electrons, perhaps leading to a general sweeping of solar wind electrons to large heliocentric distances;
5. Cooling (heating) through Coulomb interactions with ambient ions in the low- (high-) speed solar wind;
6. Heating through Coulomb interactions with the electrostatically unbound electron population.

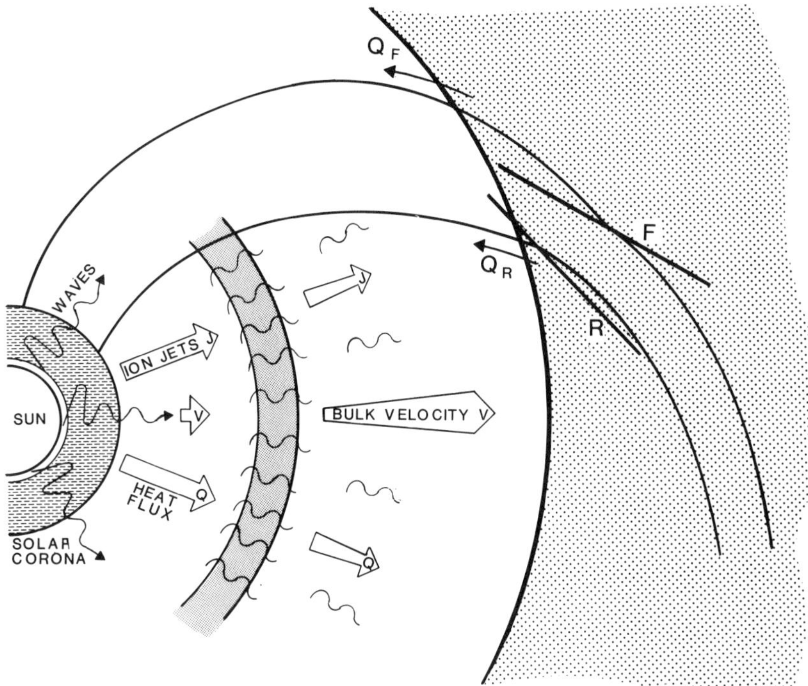

Figure 30.   A schematic drawing summarizing the many effects that help shape ion and electron velocity distributions within the inner heliosphere. A host of mechanisms presently suggested as supplying sufficient energy to support the solar corona result in the outward leakage of mass, momentum, and energy in the forms of: (1) both primary and secondary MHD waves; (2) ion jets; (3) a low-speed, low-mass flux ambient bulk expansion; and (4) an electron heat flux. Starting at some altitude above the coronal base where Coulomb collisions become unimportant, but below about 30 $R_\odot$, velocity distributions will develop sufficiently non-Maxwellian configurations that plasma instabilities will be triggered (denoted by the finely dotted annulus in the middle of the figure) thereby generating waves and limiting the distortions of these configurations from Maxwellians. This process has the effect of converting the energy in the forms of ion jets and electron heat flux, through enhanced thermal energy, into increased bulk convection. Another region outside of about 2.5 AU from the Sun also affect electron velocity distributions by injecting suprathermal electrons into the inner heliosphere from the forward and reverse shocks that bound corotating interactions, driven by collisions between solar wind streams having different bulk velocities.

As such, core electrons are generally cut off from both the local interstellar medium and the Sun and so form local heliospheric populations. However, a modest connection is maintained to both the Sun and the outer heliosphere through pitch-angle diffusion into the loss cone near the Sun and energy diffusion to higher kinetic energies, thereby extending the heliocentric range of core electrons. The rate of electron loss from all individual electron trapping zones within the inner heliosphere must exactly balance the rate of continuous injection from the diffusion-degraded strahl population if the wind is to be stationary in time. The difficulty in maintaining an exact balance (needed for charge neutrality) throughout interplanetary space (compounded by the probable existence of time-dependent, fine-scale coronal heating mechanisms) leads us to believe that a time-stationary description of the kinetic state of solar wind electrons will, perhaps, only be possible in a mean statistical sense. The very small and fast kinetic fluctuations observed in the solar wind need somehow to be averaged over. One can then consider electron propagation in smoothed, though still variable fields, which comprise the background medium. Such a description follows the same philosophy adopted in MHD turbulence theories, which consider statistically stationary ensembles of fluctuations and their correlation functions (see the chapter by Goldstein et al.), which are allowed to vary in space.

Perturbations in the electrostatic potential (and hence electron density) will develop throughout the region close to the Sun in order to respond to small changes in boundary conditions at the coronal base. These perturbations must, in turn, induce changes in the spectrum of solar wind dissipation-scale turbulence and hence its plasma transport properties. Perhaps these changes contribute to the inner-scale turbulence detected in interplanetary scintillation. Additional coarse-scale perturbations induced by bright-point flares, explosive events, macrospicules, etc., add to the problem by producing bump-on-tail distributions similar to those observed in Type III events. These distributions are known to be unstable to the generation of Langmuir, acoustic, and perhaps whistler waves. The net effect should be establishment of a zone of strong plasma turbulence that encircles the Sun, as depicted by the fine stipled region in Fig. 30.

Shapes of velocity distributions of solar wind ions generally indicate a nonlinear local steady state that coexists with the spectrum of ambient MHD waves throughout the inner heliosphere. However, the existence of multiple ion beams that become less pronounced with increasing heliocentric distance, strongly suggests their production by a mix of fundamentally time-dependent processes that are active at or below the coronal base. Known small-scale transient activity such as explosive events and bright-point flares no doubt contribute to this mix. The enhanced alpha-particle abundance of the secondary ion beams in the solar wind is then consistent with this picture, which predicts the gravitational settling of helium in the low corona within the low-flux ambient flow of the primary beam.

The sizable temperature anisotropies in the core part of fast proton distributions indicate ongoing interplanetary heating thereby requiring the existence of wave-particle interactions throughout the inner heliosphere. This heating may also be the agent that maintains solar wind proton temperatures above adiabatic values. The necessary energy no doubt comes from the large reservoir of low-frequency Alfvén fluctuations, which deliver part of their energy by way of a nonlinear cascade (Tu et al. 1984; Tu 1988; Marsch 1991a,b) to the particles through kinetic cyclotron wave damping. The same process may also occur in the corona thereby contributing to its heating. Thus, *in-situ* measurements of particle velocity distributions and kinetic wave spectra provide an indirect, albeit invaluable, means to diagnose coronal generation processes remotely, and hence lead to a more complete understanding of the solar corona and solar wind.

Acknowledgments. We wish to thank our colleagues, S. P. Gary, J. L. Phillips, J. T. Gosling, and D. J. McComas for many useful conversations. Work at Los Alamos was performed under the auspices of the U. S. Department of Energy.

REFERENCES

Abraham-Shrauner, B., and Feldman, W. C. 1977. Nonlinear Alfvén waves in high-speed solar wind streams. *J. Geophys. Res.* 82:618–624.

Anderson, K. A., McFadden, J. P., and Lin, R. P. 1981. Propagation of low energy solar electrons. *Geophys. Res. Lett.* 8:173–176.

Asbridge, J. R., Bame, S. J., Feldman, W. C., and Montgomery, M. D. 1976. Helium and hydrogen velocity differences in the solar wind. *J. Geophys. Res.* 81:2719–2727.

Athay, R. G., and White, O. R. 1978. Chromospheric and coronal heating by sound waves. *Astrophys. J.* 226:1135–1139.

Bame, S. J., Asbridge, J. R., Feldman, W. C., Gary, S. P., and Montgomery, M. D. 1975. Evidence for local ion heating in solar wind high speed streams. *Geophys. Res. Lett.* 2:373–375.

Bame, S. J., Asbridge, J. R., Feldman, W. C., Gosling, J. T., and Zwickl, R. D. 1981. Bi-directional streaming of solar wind electrons >80 eV: ISEE evidence for a closed-field structure within the driver gas of an interplanetary shock. *Geophys. Res. Lett.* 8:173–176.

Bame, S. J., Phillips, J. L., McComas, D. J., Gosling, J. T., and Goldstein, B. E. 1992. The Ulysses solar wind plasma investigation: Experiment description and initial in-ecliptic results. In *Solar Wind Seven*, eds. E. Marsch and R. Schwenn (Oxford: Pergamon Press), pp. 139–142.

Barakat, A. R., and Schunk, R. W. 1982. Transport equations for multicomponent anisotropic space plasmas: A review. *Plasma Phys.* 24:389–418.

Barnes, A. 1979. Hydromagnetic waves and turbulence in the solar wind. In *Solar System Plasma Physics*, eds. E. N. Parker, C. F. Kennel and L. J. Lanzerotti (Amsterdam: North-Holland), pp. 249–319.

Behannon, K. W. 1976. *Observations of the Interplanetary Magnetic Field Between 0.41 and 1 AU by the Mariner 10 Spacecraft*. GSFC Doc. 692-76-2 (Greenbelt, Md.: Goddard Space Flight Center).

Beinroth, H. J., and Neubauer, F. M. 1981. Properties of whistler-mode waves between 0.3 and 1 AU from Helios observations. *J. Geophys. Res.* 86:7755–7760.

Billings, D. E. 1966. *A Guide to the Solar Corona* (San Diego: Academic Press), pp. 1–14.

Bochsler, P., Geiss, J., and Joos, R. 1985. Kinetic temperatures of heavy ions in the solar wind. *J. Geophys. Res.* 90:10779–10789.

Borovsky, J. E., and Eilek, J. A. 1986. A study of the stochastic energization of charged particles with and without synchroton energy loss. *Astrophys. J.* 308:929–953.

Braginskii, S.I. 1966. Transport processes in plasma. In *Review of Plasma Physics*, vol. 1, ed. M. A. Leonotovich (New York: Consultants Bureau), pp. 205–311.

Brückner, G. E., and Bartoe, J. D. F. 1983. Observations of high energy jets in the corona above the quiet sun, the heating of the corona and the acceleration of the solar wind. *Astrophys. J.* 272:329–348.

Bürgi, A. 1992. Proton and alpha-particle fluxes in the solar wind: results of a three-fluid model. *J. Geophys. Res.* 97:3137–3150.

Coles, W. A., Lui, W., Harmon, J. K., and Martin, C. L. 1991a. The solar wind density spectrum near the sun: Results from Voyager radio measurements. *J. Geophys. Res.* 96:1745–1755.

Coles, W. A., Esser, R., Lovhaug, U.-P., and Markkanen, J. 1991b. Comparison of solar wind velocity measurements with a theoretical acceleration model. *J. Geophys. Res.* 96:13849–13859.

Coroniti, F. V., Kennel, C. F., Scarf, F. L., and Smith, E. J. 1982. Whistler mode turbulence in the disturbed solar wind. *J. Geophys. Res.* 87:6029–6044.

Cuperman, S., Weiss, I., and Dryer, M. 1980. Higher order fluid equations for multicomponent nonequilibrium stellar (plasma) atmospheres and star clusters. *Astrophys. J.* 239:345–359.

Dere, K. P. 1992. Explosive events and magnetic reconnection in the solar atmosphere. In *Solar Wind Seven*, eds. E. Marsch and R. Schwenn (Oxford: Pergamon Press), pp. 11–20.

Dere, K. P., Bartoe, J.-D., Brückner, G. E. 1991. Explosive events and magnetic reconnection in the solar atmosphere. *J. Geophys. Res.* 96:9399–9407.

Dum, C. T. 1983. Electrostatic waves and anomalous transport in the solar wind. In *Solar Wind Five*, ed. M. Neugebauer, NASA CP-2280, pp. 369–376.

Dum, C. T., Marsch, E., and Pilipp, W. G. 1980. Determination of wave growth from measured distribution functions and transport theory. *J. Plasma Phys.* 12:91–113.

Esser, R. 1992. Interpreting observations in the solar wind acceleration region. In *Solar Wind Seven*, E. Marsch and R. Schwenn (Oxford: Pergamon Press), pp. 31–36.

Feldman, W. C. 1979. Kinetic processes in the solar wind. In *Solar System Plasma Physics*, vol. I, eds. E. N. Parker, C. F. Kennel and L. J. Lanzerotti (Amsterdam: North-Holland), pp. 331–344.

Feldman, W. C., Asbridge, J. R., Bame, S. J., and Montgomery, M. D. 1973a. Double ion streams in the solar wind. *J. Geophys. Res.* 78:2017–2027.

Feldman, W. C., Asbridge, J. R., Bame, S. J., and Montgomery, M. D. 1973b. On the origin of solar wind proton thermal anisotropy. *J. Geophys. Res.* 78:6451–6468.

Feldman, W. C., Asbridge, J. R., and Bame, S. J. 1974. The solar wind He^{2+} to H^+ temperature ratio. *J. Geophys. Res.* 79:2319–2323.

Feldman, W. C., Asbridge, J. R., Bame, S. J., Montgomery, M. D., and Gary, S. P. 1975. Solar wind electrons. *J. Geophys. Res.* 80:4181–4196.

Feldman, W. C., Abraham-Shrauner, B., Asbridge, J. R., and Bame, S. J. 1976. The internal plasma state of the high speed solar wind at 1 AU. In *Physics of Solar Planetary Environments*, vol. 1, ed. D. J. Williams (Washington, D. C.: American Geophysical Union), pp. 413–427.

Feldman, W. C., Asbridge, J. R., Bame, S. J., and Gosling, J. T. 1977. Plasma and magnetic fields from the sun. In *The Solar Output and Its Variations*, ed. O. R. White (Boulder: Colorado Associated Univ. Press), pp. 351–382.

Feldman, W. C., Asbridge, J. R., Bame, S. J., Gosling, J. T., and Lemons, D. S. 1978.

Characteristic electron variations across simple high-speed solar wind streams. *J. Geophys. Res.* 83:5285–5295.

Feldman, W. C., Asbridge, J. R., Bame, S. J., Gosling, J. T., and Lemons, D. S. 1979*a*. The core electron temperature profile between 0.5 and 1 AU in the steady-state high speed solar wind. *J. Geophys. Res.* 84:4463–4467.

Feldman, W. C., Asbridge, J. R., Bame, S. J., Gosling, J. T., Lemons, D. S. 1979*b*. A possible closure relation for heat transport in the solar wind. *J. Geophys. Res.* 84:6621–6632.

Feldman, W. C., Asbridge, J. R., Bame, S. J., and Gosling, J. T. 1982*a*. Quantitative tests of a steady state theory of solar wind electrons. *J. Geophys. Res.* 87:7355–7362.

Feldman, W. C., Anderson, R. C., Asbridge, J. R., Bame, S. J., Gosling, J. T., and Zwickl, R. D. 1982*b*. Plasma electron signature of magnetic connection to the earth's bow shock: ISEE 3. *J. Geophys. Res.* 87:632–642.

Feldman, W. C., Anderson, R. C., Bame, S. J., Gary, S. P., Gosling, J. T., McComas, D. J., Thomsen, M. F., Paschmann, G., Hoppe, M. M. 1983. Electron velocity distributions near the earth's bow shock. *J. Geophys. Res.* 88:96–110.

Feldman, W. C., Gosling, J. T., McComas, D. J., and Phillips, J. L. 1993. Evidence for ion jets in the high-speed solar wind. *J. Geophys. Res.* 98:5593–5605.

Gary, S. P. 1993. *Theory of Space Plasma Microinstabilities* (Cambridge: Cambridge Univ. Press).

Geiss, J., Hirt, P., and Leutwyler, H. 1970. On acceleration and motion of ions in corona and solar wind. *Solar Phys.* 12:458–483.

Gosling, J. T., Asbridge, J. R., Bame, S. J., Feldman, W. C., Zwickl, R. D., Paschmann, G., Sckopke, N., and Russell, C. T. 1982. A sub-Alfvénic solar wind: interplanetary and magnetosheath observations. *J. Geophys. Res.* 87:239–245.

Gosling, J. T., Baker, D. N., Bame, S. J., Feldman, W. C., and Zwickl, R. D. 1987. Bidirectional solar wind electron heat flux events. *J. Geophys. Res.* 92:8519–8535.

Gosling, J. T., Bame, S. J., Feldman, W. C., McComas, D. J., Phillips, J. L., and Goldstein, B. E. 1993. Counterstreaming suprathermal electron events upstream of corotating shocks in the solar wind beyond 2 AU: Ulysses. *Geophys. Res. Lett.* 20:2335–2338.

Griffel, D. H., and Davis, L. 1969. The anisotropy of the solar wind. *Planet. Space Sci.* 17:1009–1020.

Gurnett, D. A. 1985. Plasma waves and instabilities. In *Collisionless Shocks in the Heliosphere: Reviews of Current Research*, eds. B. T. Tsurutani and R. G. Stone (Washington, D. C.: American Geophysical Union), pp. 207–224.

Gurnett, D. A. 1991. Waves and Instabilities. In *Physics of the Inner Heliosphere*, vol. II, eds. R. Schwenn and E. Marsch (Berlin: Springer-Verlag), pp. 135–157.

Gurnett, D. A., and Frank, L. A. 1978. Ion-acoustic waves in the solar wind. *J. Geophys. Res.* 83:58–74.

Gurnett, D. A., Hospodarsky, G. B., Kurth, W. S., Williams, D. J., and Bolton, S. J. 1993. Fine structure of Langmuir waves produced by a solar electron event. *J. Geophys. Res.* 98:5631–5637.

Habbal, S. R. 1992. Variable EUV emission in the quiet sun and coronal heating. In *Solar Wind Seven*, eds. E. Marsch and R. Schwenn (Oxford: Pergamon Press), pp. 41–48.

Habbal, S. R., Esser, R., and Arndt, M. B. 1993. How reliable are coronal hole temperatures deduced from observations? *Astrophys. J.* 413:435–444.

Hammond, M., Feldman, W. C., Bame, S. J., Phillips, J. L., Goldstein, B. E., and Balogh, A. 1995*a*. Double ion beams surrounding heliospheric current sheets. *J. Geophys. Res.* 100:7881–7889.

Hammond, M., Feldman, W. C., Phillips, J. L., and Balogh, A. 1995*b*. Observations of double ion beams associated with coronal mass ejections. *Adv. Space Res.*, in preparation.

Hernández, R., and Marsch, E. 1985. Collisional time scales for temperature and velocity exchange between drifting Maxwellians. *J. Geophys. Res.* 90:11062–11066.

Hernández, R., Livi, S., and Marsch, E. 1987. On the He^{2+} to H^+ temperature ratio in slow solar wind. *J. Geophys. Res.* 92:7723–7727.

Hinton, F. L. 1983. Collisional transport in plasma. In *Basic Plasma Physics I*, eds. A. A. Galeev and R. N. Sudan (Amsterdam: North-Holland), pp. 147–197.

Hollweg, J. V. 1975. Waves and instabilities in the solar wind. *Rev. Geophys. Space Phys.* 13:263–289.

Hollweg, J. V. 1990. Heating of the solar corona. *Computer Phys. Rept.* 12:205–232.

Hoppe, M.M., Russell, C. T., Frank, L. A., Eastman, T. E., and Greenstadt, E. W. 1981. Upstream hydromagnetic waves and their association with backstreaming ion populations: ISEE 1 and 2 observations. *J. Geophys. Res.* 86:4471–4492.

Hoppe, M. M., Russell, C. T., Eastman, T. E., and Frank, L. A. 1982. Characteristics of the ULF waves associated with upstream ion beams. *J. Geophys. Res.* 87:643–650.

Isenberg, P. A. 1983. Acceleration of heavy ions in the solar wind. In *Solar Wind Five*, ed. M. Neugebauer, NASA CP-2280, pp. 655–661.

Isenberg, P. A. 1984. The ion-cyclotron dispersion relation in a proton-alpha solar wind. *J. Geophys. Res.* 89:2133–2141.

Jockers, K. 1970. Solar wind models based on exospheric theory. *Astron. Astrophys.* 6:219–239.

Jokipii, J. R., and Davis, L., Jr. 1969. Long-wavelength turbulence and the heating of the solar wind. *Astrophys. J.* 156:1101–1106.

Joselyn, J., and Holzer, T. E. 1978. A steady three-fluid coronal expansion for nonspherical geometries. *J. Geophys. Res.* 83:1019–1026.

Kennel, C. F., Scarf, F. L., Coroniti, F. V., Fredericks, R. W., Gurnett, D. A., and Smith, E. J. 1980. Correlated whistler and electron plasma oscillation bursts detected on ISEE 3. *Geophys. Res. Lett.* 7:129–132.

Klein, L. W., Ogilvie, K. W., and Burlaga, L. F. 1985. Coulomb collisions in the solar wind. *J. Geophys. Res.* 90:7389–7395.

Kopp, R. A., and Orrall, F. Q. 1977. Models of coronal holes above the transition region. In *Coronal Holes and High Speed Streams*, ed. J. B. Zirker (Boulder: Colorado Associated Univ. Press), pp. 179–224.

Koutchmy, S., and Livshits, M. 1992. Coronal streamers. *Space Sci. Rev.* 61:393–417.

Kulsrud, R. M. 1983. MHD description of plasma. In *Basic Plasma Physics I*, eds. A. A. Galeev and R. N. Sudan (Amsterdam: North-Holland), pp. 115–146.

Leer, E., Holzer, T. E., and Shoub, E. C. 1992. Solar wind from a corona with a large helium abundance. *J. Geophys. Res.* 97:8183–8201.

Lin, R. P. 1980. Energetic particles in space. *Solar Phys.* 67:393–399.

Lin, R. P., Potter, D. W., Gurnett, D. A., and Scarf, F. L. 1981. Energetic electrons and plasma waves associated with a solar type III radio burst. *Astrophys. J.* 251:364–373.

Lemaire, J., and Scherer, M. 1971. Kinetic models of the solar wind. *J. Geophys. Res.* 76:7479–7490.

Lemons, D. S., and Feldman, W. C. 1983. Collisional modification to the exospheric theory of solar wind halo electron pitch angle distributions. *J. Geophys. Res.* 88:6881–6887.

Leubner, M. P., and Viñas A. F. 1986. Stability analysis of double peaked proton distribution functions in the solar wind. *J. Geophys. Res.* 91:13366–13372.

Livi, S., and Marsch, E. 1986*a*. On the collisional relaxation of solar wind velocity distributions. *Ann. Geophys.* 4A:333–340.

Livi, S., and Marsch, E. 1986*b*. Comparison of the Bhatnagar-Gross-Krook-Approximation with the exact Coulomb collision operator. *Phys. Rev. A* 34:533–540.

Livi, S., and Marsch, E. 1987. Generation of solar wind proton tails and double beams by Coulomb collisions. *J. Geophys. Res.* 92:7255–7261.

Livi, S., Marsch, E., and Rosenbauer, H. 1986. Coulomb collisional domains in the solar wind. *J. Geophys.* 91:8045–8050.

Marsch, E. 1991*a*. Kinetic physics of the solar wind plasma. In *Physics of the Inner Heliosphere 2: Particles, Waves and Turbulence*, eds. R. Schwenn and E. Marsch (Berlin: Springer-Verlag), pp. 45–133.

Marsch, E. 1991*b*. MHD turbulence in the solar wind. In *Physics of the Inner Heliosphere 2: Particles, Waves and Turbulence*, eds. R. Schwenn and E. Marsch (Berlin: Springer-Verlag), pp. 159–244.

Marsch, E., and Chang, T. 1982. Lower hybrid waves in the solar wind. *Geophys. Res. Lett.* 9:1155–1158.

Marsch, E., and Chang, T. 1983. Electromagnetic lower hybrid waves in the solar wind. *J. Geophys. Res.* 88:6869–6880.

Marsch E., and Goldstein, H. 1983. The effects of Coulomb collisions on solar wind ion velocity distributions. *J. Geophys. Res.* 88:9933–9940.

Marsch, E., and Livi, S. 1985*a*. Coulomb collision rates for the self-similar and kappa distributions. *Phys. Fluids.* 28:1379–1386.

Marsch, E., and Livi, S. 1985*b*. Coulomb self-collision frequencies for non-thermal velocity distributions in the solar wind. *Ann. Geophys.* 3:545–556.

Marsch, E., Mühlhäuser, K.-H., Schwenn, R., Rosenbauer, H., Pilipp, W., and Neubauer, F. M. 1982*a*. Solar wind protons: Three-dimensional velocity distributions and derived plasma parameters measured between 0.3 and 1 AU. *J. Geophys. Res.* 87:52–72.

Marsch, E., Mühlhäuser, K.-H., Rosenbauer, H., Schwenn, R., and Neubauer, F. M. 1982*b*. Solar wind helium ions: Observations of the Helios solar probes between 0.3 and 1 AU. *J. Geophys. Res.* 87:35–51.

McComas, D. J., Bame, S. J., Feldman, W. C., Gosling, J. T., and Phillips, J. L. 1992. Solar wind halo electrons from 1–4 AU. *Geophys. Res. Lett.* 19:1291–1294.

Montgomery, M. D., Bame, S. J., and Hundhausen, A. J. 1968. Solar wind electrons: Vela 4 measurements. *J. Geophys. Res.* 73:4999–5003.

Montgomery, M. D., Asbridge, J. R., Bame, S. J., and Feldman, W. C. 1974. Solar wind electron temperature depressions following some interplanetary shock waves: Evidence for magnetic merging. *J. Geophys. Res.* 79:3103–3110.

Moore, R. L., Musielak, Z. E., Suess, S. T., and An, C. H. 1991. Alfvén wave trapping, network microflaring, and heating in solar coronal holes. *Astrophys. J.* 378:347–359.

Mullen, D. 1990. Sources of the solar wind: what are the smallest-scale structures. *Astron. Astrophys.* 232:520–535.

Nakada, M. P. 1970. A study of the composition of the solar corona and solar wind. *Solar Phys.* 14:457–479.

Narain, W., and Ulsmschneider, P. 1990. Chromospheric and coronal heating mechanisms. *Space Sci. Rev.* 54:377–445.

Neubauer, F. M., Musmann, G., and Dehmel, G. 1977. Fast magnetic fluctuations in the solar wind: Helios 1. *J. Geophys. Res.* 82:3201–3212.

Neugebauer, M. 1976. The role of Coulomb collisions in limiting differential flow and temperature differences in the solar wind. *J. Geophys. Res.* 81:78–82.

Neugebauer, M. 1981. Observations of solar wind helium. *Fund. Cosmic Phys.* 7:131–199.

Neugebauer, M., and Feldman, W. C. 1979. Relation between superheating and superacceleration of helium in the solar wind. *Solar Phys.* 63:201–205.

Neugebauer, M., Goldstein, B. E., Bame, S. J., and Feldman, W. C. 1994. Ulysses near-ecliptic observations of differential flow between protons and alphas in the solar wind. *J. Geophys. Res.* 99:2503–2511.

Olbert, S. 1981. Inferences about the solar wind dynamics from observed distributions of electrons and ions. In *Proc. Intl. School on Plasma Astrophysics*, ESA SP-161, pp. 135–144.

Olbert, S. 1983. The role of thermal conduction in the acceleration of the solar wind. In *Solar Wind Five*, ed. M. Neugebauer, NASA CP-2280, pp. 149–162.

Parker, E. N. 1991. Heating solar coronal holes. *Astrophys. J.* 372:719–727.

Parker, E. N. 1992. Heating coronal holes and accelerating the solar wind. In *Solar Wind Seven*, eds. E. Marsch and R. Schwenn (Oxford: Pergamon Press), pp. 79–86.

Perkins, F. W. 1973. Heat conductivity, plasma instabilities, and radio star scintillations in the solar wind. *Astrophys. J.* 179:637–642.

Phillips, J. L., and Gosling, J. T. 1990. Radial evolution of solar wind thermal electron distributions due to expansion and collisions. *J. Geophys. Res.* 95:4217–4228.

Phillips, J. L., Gosling, J. T., McComas, D. J., Bame, S. J., and Gary, S. P. 1989. Anisotropic thermal electron distributions in the solar wind. *J. Geophys. Res.* 94:6563–6579.

Phillips, J. L., Gosling, J. T., McComas, D. J., Bame, S. J., and Feldman, W. C. 1992*a*. Magnetic topology of coronal mass ejections based on ISEE-3 observations of bidirectional electron fluxes at 1 AU. In *Proc. First SOLTIP Symposium*, vol. 2, eds. S. Fischer and M. Vandes (Prague: Astron. Inst. Czechoslovak Academy of Science Press), pp. 165–170.

Phillips. J. L., Gosling, J. T., McComas, D. J., Bame, S. J., and Feldman, W. C. 1992*b*. Quantitative analysis of bidirectional electron fluxes within coronal mass ejections at 1 AU. In *Solar Wind Seven*, eds. E. Marsch and R. Schwenn (Oxford: Pergamon Press), pp. 651–656.

Phillips, J. L., Feldman, W. C., Bame, S. J., Gosling, J. T., McComas, D. J., and Smith, E. J. 1994. Strahl-on-strahl electron distributions within coronal mass ejections at 1 AU: Evidence for attached magnetic topology. *Geophys. Res. Lett.*, in preparation.

Pilipp, W. G., Schwenn, R., Marsch, E., Mühlhäuser, K.-H., and Rosenbauer, H. 1981. Electron characteristics in the solar wind as deduced from Helios observations. In *Solar Wind Four*, ed. H. Rosenbauer, Report No. MPAE-W-100-81-31 (Katlenburg-Lindau: Max-Planck-Institut für Extraterrestrische Physik), pp. 241–249.

Pilipp, W. G., Miggenrieder, H., Mühlhäuser, K.-H., Rosenbauer, H., Schwenn, R., and Neubauer, F. M. 1987*a*. Variations of electron distribution functions in the solar wind. *J. Geophys. Res.* 92:1103–1118.

Pilipp, W. G., Miggenrieder, H., Montgomery, M. D., Mühlhäuser, K.-H., Rosenbauer, H., and Schwenn, R. 1987*b*. Characteristics of electron velocity distribution functions in the solar wind derived from the Helios plasma experiment. *J. Geophys. Res.* 92:1075–1092.

Pilipp, W. G., Miggenrieder, H., Montgomery, M. D., Mühlhäuser, K.-H., Rosenbauer, H., and Schwenn, R. 1987*c*. Unusual electron distribution functions in the solar wind derived from the Helios plasma experiment: Double-strahl distributions and distributions with an extremely anisotropic core. *J. Geophys. Res.* 92:1093–1101.

Pilipp, W. G., Miggenrieder, H., Mühlhäuser, K.-H., Rosenbauer, H., and Schwenn, R. 1990. Large scale variations of thermal electron parameters in the solar wind between 0.3 and 1 AU. *J. Geophys. Res.* 95:6305–6329.

Pneuman, G. W. 1983. Ejection of magnetic fields from the sun: Acceleration of a solar wind containing diamagnetic plasmoids. *Astrophys. J.* 265:468–482.

Roberts, D. A. 1989. Interplanetary observational constraints on Alfvén wave acceleration of the solar wind. *J. Geophys. Res.* 94:6899–6905.

Roberts, D. A., and Goldstein, M. 1991. *Turbulence and Waves in the Solar Wind.* U. S. Natl. Rept. to IUGG, 1987–1990 (Washington, D. C.: American Geophysical Union), pp. 932–943.

Rosenbauer, H., Schwenn, R., Marsch, E., Meyer, B., Miggenrider, H., Montgomery, M.D., Mühlhäuser, K.-H., Pilipp, W., Voges, W., and Zink, S. M. 1977. Helios plasma experiment. *J. Geophys.* 42:561–580.

Schmidt, J., Bochsler, P., and Geiss, J. 1987. Velocity of iron ions in the solar wind. *J. Geophys. Res.* 92:9901–9906.

Schmidt, W. K. H., Rosenbauer, H., Shelley, E. G., and Geiss, J. 1980. On temperature and speed of He^{2+} and O^{6+} ions in the solar wind. *Geophys. Res. Lett.* 7:697–700.

Schultz, M., and Eviatar, A. 1972. Electron-temperature asymmetry and the structure of the solar wind. *Cosmic Electrodyn.* 2:402–422.

Schwartz, S. J. 1980. Plasma instabilities in the solar wind: A theoretical review. *Rev. Geophys. Space Phys.* 18:313–336.

Schwartz, S. J., and Marsch, E. 1983. The radial evolution of a single solar wind plasma parcel. *J. Geophys. Res.* 88:9919–9932.

Schwenn, R. 1991. Large-scale structure of the interplanetary medium. In *Physics of the Inner Heliosphere 1: Large-Scale Phenomena*, eds. R. Schwenn and E. Marsch (Berlin: Springer-Verlag), pp. 99–181.

Scudder, J. D. 1992*a*. On the causes of temperature change in inhomogeneous low-density astrophysical plasmas. *Astrophys. J.* 398:299–318.

Scudder, J. D. 1992*b*. Why all stars should possess circumstellar temperature inversions. *Astrophys. J.* 398:319–349.

Scudder, J. D., and Olbert, S. 1979*a*. A theory of local and global processes which affect solar wind electrons, 1. The origin of typical 1 AU velocity distribution functions—Steady state theory. *J. Geophys. Res.* 84:2755–2772.

Scudder, J. D., and Olbert, S. 1979*b*. A theory of local and global processes which affect solar wind electrons, 2. Experimental support. *J. Geophys. Res.* 84:6603–6620.

Sentman, D.D., Thomsen, M. F., Gary, S. P., Feldman, W. C., and Hoppe, M. M. 1983. The oblique whistler instability in the earth's foreshock. *J. Geophys. Res.* 88:2048–2056.

Shoub, E. C. 1983. Invalidity of local thermodynamic equilibrium for electrons in the solar transition region. I. Fokker-Planck results. *Astrophys. J.* 266:339–369.

Shoub, E. C. 1987. Failure of the Fokker-Planck approximation to the Boltzmann integral for 1/r potentials. *Phys. Fluids* 30:1340–1352.

Shoub, E. C. 1988. Kinetic theory of solar wind acceleration. In *Proc. of the Sixth Intl. Solar Wind Conf.*, vol. 1, eds. V. J. Pizzo, T. E. Holzer and D. G. Sime, NCAR TN-306 (Boulder: Natl. Center for Atmos. Res.), pp. 59–85.

Sittler, E. C., Jr., and Scudder, J. D. 1980. An empirical polytrope law for solar wind thermal electrons between 0.45 and 4.76 AU: Voyager 2 and Mariner 10. *J. Geophys. Res.* 85:5131–5137.

Temnyi, V. V., and Vaisberg, O. L. 1979. A dumbbell distribution of epithermal electrons in the solar wind based on observations on the Prognoz 7 satellite. *Cosmic Res.* 17:476–481.

Tu, C.-Y. 1988. The damping of interplanetary Alfvénic fluctuations and the heating of the solar wind. *J. Geophys. Res.* 93:7–20.

Tu, C.-Y., Pu, Z.-Y., and Wei, F.-S. 1984. The power spectrum of interplanetary

Alfvénic fluctuations: Derivation of the governing equations and its solution. *J. Geophys. Res.* 89:9695–9702.

Ulmschneider, P., Priest, E. R., and Rosner, R., eds. 1991. *Mechanisms of Chromospheric and Coronal Heating* (Berlin: Springer-Verlag).

Viñas, A. F., and Goldstein, M.L. 1992. Parametric instabilities of large amplitude Alfvén waves with obliquely propagating sidebands. In *Solar Wind Seven*, eds. E. Marsch and R. Schwenn (Oxford: Pergamon Press), pp. 577–581.

Withbroe, G. L. 1988. The temperature structure, mass and energy flow in the corona and inner solar wind. *Astrophys. J.* 325:442–467.

Withbroe, G. L. 1989. The solar wind mass flux. *Astrophys. J. Lett.* 337:49–52.

Withbroe, G. L., Feldman, W. C., and Ahluwalia, H. S. 1991. The solar wind and its origins. In *Solar Interior and Atmosphere*, eds. A. N. Cox, W. C. Livingston and M. S. Matthews (Tucson: Univ. of Arizona Press), pp. 1087–1106.

ANGULAR MOMENTUM EVOLUTION IN LATE-TYPE STARS

P. CHARBONNEAU
High Altitude Observatory

C. J. SCHRIJVER
Astronomical Institute, Utrecht

K. B. MacGREGOR
High Altitude Observatory

Evidence is compelling that angular momentum loss associated with magnetized winds emanating from the coronae of late-type stars is the primary agent determining the observed distribution and evolution of their rotation rates on and near the main sequence. The rotational history of a star involves many different aspects, ranging from structural evolution, through dynamo action and internal angular momentum transport, to angular momentum loss in a magnetized stellar wind. We first present the ideas behind each of these links in the chain, outlining the insights gained thus far, and with emphasis on physical principles. Turning then to angular momentum evolution, a simple formulation is used to illustrate robust features of rotational evolution models, first on the pre-main and main sequence, and then on the post-main sequence. This is followed by a critical review of specific, detailed models. The chapter closes with a few brief comments on some aspects of angular momentum evolution not adressed in the main body of the text.

I. INTRODUCTION: THE OBSERVATIONAL PICTURE

Angular momentum evolution refers to the variation with time of stellar rotation rates. Rotation is commonly neglected in the standard picture of stellar evolution; given the success of such standard models at reproducing the observed properties of most stars, in particular their distribution in the Hertzsprung-Russell (HR) diagram, this may appear to be, *a posteriori*, a justifiable assumption. The Sun is perhaps the most obvious example of the incompleteness of the standard model, however. While most of its global properties (radius, luminosity, etc.) can be adequately reproduced within standard models, the Sun exhibits a wide range of phenomena which have no place in standard stellar structure theory. The presence of a $\sim 10^6$ K corona overlying the solar atmosphere and the wind-like outflow it powers, as well as the various manifestations of magnetic activity, are some of the more notorious examples of this very unsatisfactory state of affairs. The energy required to heat the corona and power the solar wind is by no means excessive, amounting to a fraction $\sim 10^{-5}$ of the solar radiative luminosity. However, it has emerged that the diversion of even such a minute fraction of the Sun's energy output is ultimately difficult, if not impossible, to accomplish in the absence of rotation. Rotation is required to break the large-scale symmetry that would characterize, in its absence, the interaction of magnetic fields and bulk fluid motions. Without rotation, generation of large-scale magnetic fields via dynamo action is impossible, and without magnetic fields there would presumably be no solar and stellar activity, and maybe not even a corona.

Solar rotation, as revealed by the motion of sunspots across the solar disk, has been known at least since the early seventeenth century (see, e.g., Galilei [1613] *Letters on Sunspots*). However, it is only in recent years that photometric and spectroscopic techniques (and hardware) have reached sufficient sensitivity to allow the detection of rotational modulation due to surface inhomogeneities (or "starspots") in late-type stars other than the Sun. The determination of the Sun's rotation by direct measurement of the Doppler shift at the solar limbs was first performed by Vogel (1871), while the first, and somewhat serendipitous, stellar measurement was carried out by Schlesinger (1909), in the bright component of the eclipsing binary δ Librae at occultation. Most modern determinations of stellar rotation rely on Doppler broadening of strong spectral lines (see Gray 1992, ch. 17), as apparently first suggested by Abney (1877), but successfully carried out only much later (Elvey 1929). For a single star, this yields only a lower limit to the true rotation rate, as the angle i between the line of sight and the star's rotation axis is generally unknown. However, a large sample of measured projected rotational velocities ($v \sin i$) clearly contains valuable information, in a statistical sense, concerning the true distribution of rotation rates.

In one of the earliest such data sets for single stars, assembled by Struve and collaborators in the early 1930s, the existence of systematic differences between average rotation rates for late-type and early-type stars was almost immediately recognized (Elvey 1930; Struve 1930; see Struve [1945] for an early review). The sharpness of the transition from rapid to slow rotation in proceeding from early-F to late-F spectral types was also noted (Westgate 1934). In his classic work on stellar rotation, Kraft (1967*a*,*b*; 1970, and references therein; see also Wilson 1966) illustrated this perhaps better than anyone before him. Figure 1 herein is a remake of Fig. 1 in Kraft (1967*b*), showing the distribution of rotation rates in a HR diagram, for a sample of main-sequence and post-main-sequence field stars. As one runs down the main sequence, there is a precipitous drop in $v \sin i$ at spectral type F5, possibly extending to higher luminosity classes (although at somewhat later spectral types). Stars on the cool side of this so called rotational dividing line rotate rather slowly, while on the hot side rapid rotation is the rule. Kraft went on to show that under the assumption of internal solid-body rotation, in the interval $1.5 \lesssim M/M_\odot \lesssim 20$ the observed rotation rates are consistent with a

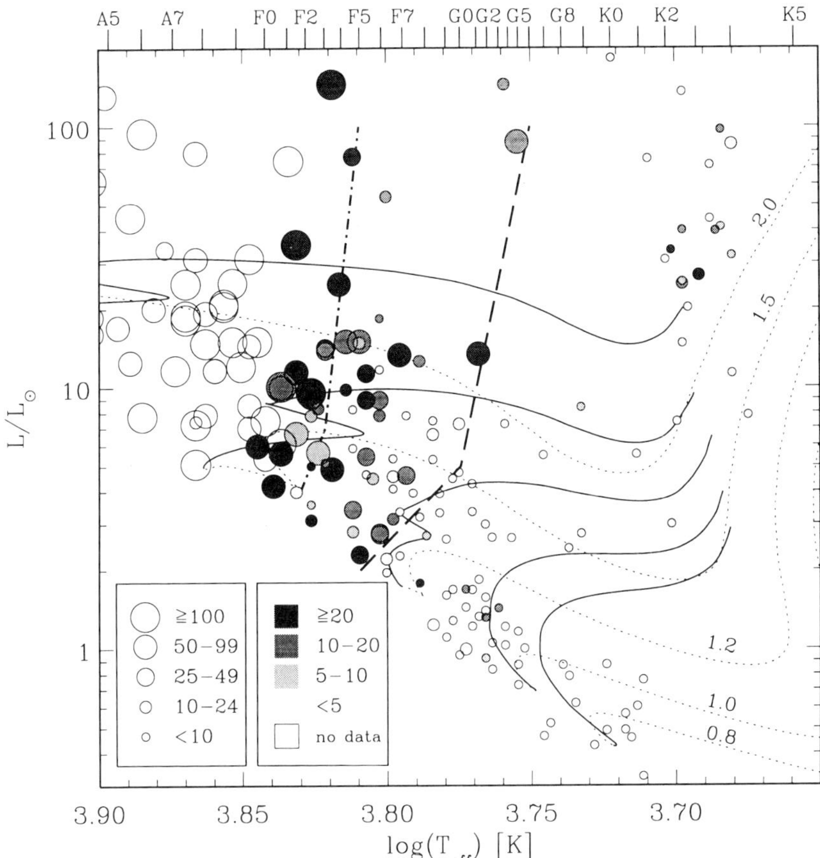

Figure 1.   Distribution of rotation rates in the HR diagram. The size of the symbols increase with $v \sin i$, and their degree of shading with chromospheric emission [as measured by the normalized emission index $R(\mathrm{CIV}) \equiv \mathrm{flux}/\sigma T^4$ in units of 10^{-7}]. Spectral types are for luminosity class V, taken from Gray (1992). Evolutionary tracks courtesy of D. Vandenberg (1993, personal communication), are labeled according to $M/M_\odot$, with pre-main-sequence portions plotted as dotted lines. Data are from the compilation of Simon and Drake (1989) and Soderblom (1985), complemented by data from the *Bright Star Catalogue* (Hoffleit 1982). The dashed line is the rotational dividing line, and the dash-dotted line the so called granulation boundary, marking the onset of envelope convection in moving from left to right in the HR diagram (see Gray 1992, ch. 18). Note that all stars in the upper and center right of the diagram are post-main-sequence objects, *not* pre-main-sequence stars.

power law dependence for the total angular momentum content J on stellar mass M of the form $J \propto M^n$, with $n = 1.57$, since then commonly referred to as Kraft's relation. Figure 1 also hints at a relationship between rotation and chromospheric activity, measured here by the emission index $R(\mathrm{CIV})$; at any location in the cool half of the HR diagram, stronger emission occurs preferentially in the more rapid rotators.

A number of hypotheses can be considered to explain the sharp decrease of rotation rates around spectral type F5. Late-type stars could divert a considerable fraction of their initial angular momentum content into planetary orbital angular momentum, or hide it from view in very rapidly rotating cores. Alternatively, late-type stars could simply emerge from the star formation process with intrinsically less angular momentum per unit mass than early-type stars. The preferred alternative, already then and still now, is that late-type stars gradually lose angular momentum via a mechanism that operates much less efficiently, or not at all, in early-type stars.

The idea that a magnetized wind can extract considerable angular momentum from a rotating star is usually attributed to Schatzman (1959,1962). In contrast to the stationary wind-like outflows *Ansatz* being developed at the time by Parker (1963; see also his chapter), Schatzman considered angular momentum loss associated with episodic magnetized outflow events occurring on the *pre*-main sequence as the primary explanation for the observed difference between the observed rotational state of late- and early-type main-sequence stars. He nevertheless coined the physical mechanism responsible for the enhanced angular momentum loss rate associated with magnetized—as opposed to unmagnetized—outflows; in a rotating reference frame, Coriolis forces in the expanding gas generate magnetic stresses in any pre-existing magnetic field emanating from the star, the associated magnetic torque density in general largely exceeding the specific angular momentum advected by the gas. The process is naturally restricted to late-type stars upon noting, as Schatzman did, that early-type stars lack a deep convective envelope and so presumably cannot sustain a solar-like dynamo capable of driving magnetically coupled outflows.

Motivated in part by the success of Parker's solar wind model, Weber and Davis (1967) and Mestel (1968) constructed magnetohydrodynamical (MHD) models for stationary, magnetized winds, and showed that the process envisioned by Schatzman can operate quite efficiently in steady, continuous outflows from stars on the main sequence. Observationally, Skumanich (1972) suggested that the decline with time t of both rotation and magnetic activity follows a $t^{-1/2}$ relation, while Durney (1972) showed that such a time dependence is compatible with predictions from simple MHD wind and dynamo theories. Further observational work carried out since then has amply confirmed the general decline of rotation rate with age, but has also revealed that the situation is much more complex than initially suggested by Skumanich's $t^{-1/2}$ relation (for recent and thorough reviews see Stauffer [1991,1994] and Soderblom [1991]). Figure 2 shows observed $v \sin i$ values for late-type stars in three young clusters: α Persei (age $\simeq 5 \times 10^7$ yr), the Pleiades ($\simeq 7 \times 10^7$ yr), and the Hyades ($\simeq 6 \times 10^8$ yr). The spin-down rate is not only seen to vary with effective temperature (or mass), but a large spread in rotation exists at any T_{eff} up to at least the age of the Pleiades. Additionally, in both α Persei and the Pleiades there also exists an excess of slow rotators as compared to what one would expect from, say, the combination of randomly oriented rotation axes and a Maxwellian distribution of true rotational velocities. This is particularly troublesome in the case of late-K and M dwarfs, because at the ages of α Persei and the Pleiades these stars have not yet reached the main sequence, and so should presumably still be spinning *up*. The presence of a small population of very rapid rotators among the Pleiades K dwarfs is also difficult to reproduce, from the modeling standpoint (see more on this subject in Sec. V below). Nevertheless, by the age of the Hyades all K, G and late-F stars have spun down, and the sharp drop in rotations rates from early to mid-F spectral types is clearly delineated.

It will become apparent in what follows that angular momentum evolution is inseparably linked to aspects of stellar structure and evolution, such as coronal heating and magnetic activity, that are themselves poorly understood and currently the subject of intense observational and theoretical research efforts. In order to keep the size of this chapter within reasonable bounds, we have been somewhat selective in the material to be covered, as well as in the angle of attack, so to speak, of our discussion. First, whenever possible the emphasis is placed on physical effects that can be quantified and modeled in a deterministic way, rather than on broader phenomenological/parametric constructs (which, incidentally, is not meant to imply that the former provide an inherently superior representation of reality than the latter). For example, the discussion of wind theory is placed exclusively in the context of steady-state, single fluid magnetohydrodynamics, as opposed to, e.g., stochastic or empirical modeling. Second, in what follows we are concerned almost exclusively with the rotational evolution of single stars. By this we not only exclude binarity, but also any interaction with, e.g., circumstellar accretion disks (although we comment briefly on the latter in Sec. VIII). Moreover, the greater part of the discussion focuses on late-type stars, and on the Sun in particular. Third, we focus on rotational evolution on and near the main sequence, i.e., from the bottom of the Hayashi tracks to the base of the giant branch. Fourth, rather than providing an exhaustive coverage of the technical literature, entry points are provided in the form of references to a limited selection of classical and/or key papers.

The remainder of the chapter is organized as follows. Section II begins by examining the circumstances under which evolutionary variations of stellar moments of inertia can, in themselves, alter a star's surface rotation rate, which leads rather naturally into the problem of internal angular momentum redistribution. Section III discusses dynamo theory. This being a particularly vast topic, the section focuses on aspects of dynamo theory that are most relevant to the issue of angular momentum evolution, introducing relevant observations of magnetic activity in the Sun and stars along the way. Section IV is primarily a detailed discussion of the Weber-Davis MHD wind solution. This provides the central ingredient necessary for rotational evolution modeling, namely an estimate of angular momentum loss rates in the Sun and stars as a function of rotation rate, surface magnetic field strength, etc. With

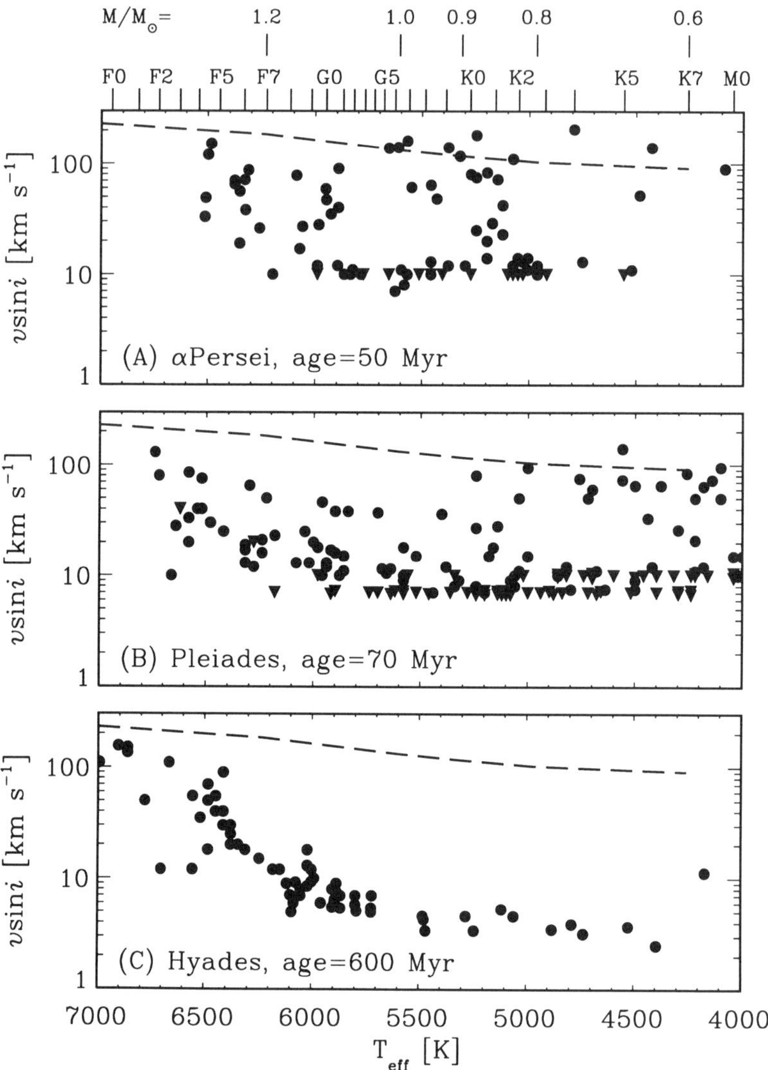

Figure 2.   Projected rotational velocities vs effective temperature in three young clusters. Filled circles are measured values and triangles are upper limits. Data is courtesy of D. Soderblom, and a few updates follow Prosser (1994). The dashed line is the run of rotation rate with mass predicted by Kraft's relation extrapolated to lower masses, assuming solid-body rotation and using a revised power law index $n = 2.02$ (Kawaler 1987).

all major building blocks of the problem in place, the following two sections make use of a simplified model to tackle the problem of angular momentum evolution *per se*, first on the pre-main sequence and main sequence (Sec. V), and then on the subgiant and giant branches (Sec. VI). Section VII is a critical review of more elaborate models to be found in the recent literature.

II. STRUCTURAL EVOLUTION AND ANGULAR MOMENTUM REDISTRIBUTION

A. Structural Evolution

The moment of inertia I of a spherically symmetric mass distribution is

$$I(t) = \frac{8\pi}{3} \int_0^{R(t)} \rho(r,t)\, r^4 \, \mathrm{d}r \tag{1}$$

a time-dependent quantity in the present context because of variations in the star's radius and internal mass distribution, both of these being brought about by nuclear burning in the core. The total angular momentum content J of the star is

$$J(t) = 4\pi \int_0^{\pi/2} \int_0^{R(t)} \Omega(r,\theta,t)\rho(r,t)\, r^4 \sin^3\theta \, \mathrm{d}r\, \mathrm{d}\theta \tag{2}$$

for an arbitrary axisymmetric rotation law $\Omega(r,\theta,t)$. In the absence of angular momentum loss and in the limiting case of solid-body rotation, $J = \Omega I$. Conservation of angular momentum then requires that

$$\frac{\Omega(t)}{\Omega_0} = \frac{I_0}{I(t)} \tag{3}$$

where Ω_0 and I_0 are evaluated at some reference time t_0. Figure 3 shows evolutionary tracks and the corresponding time evolution of I for stars of various masses. These variations are expected, by themselves, to lead to strong spin up on the pre-main sequence and strong spin down on the subgiant and giant branches. Note that despite a $\sim 15\%$ increase in radius, I remains essentially constant as solar-type stars evolve from the ZAMS to the end of the main sequence; this is because the conversion of hydrogen to helium increases the density in the core at the expense of the envelope, almost exactly offsetting the increase of I that would otherwise result from the slow expansion of the envelope.

Stars with $M/M_\odot \lesssim 1.3$ remain redward of the rotational dividing line throughout their main- and post-main-sequence evolution up to the giant branch. In contrast, more massive stars spend their main-sequence lifetime blueward of the dividing line, crossing it more or less horizontally after leaving the main sequence, the crossing time being a rather sensitive function of mass (cf., Fig. 3). This is also the time interval where the largest increase of moment of inertia occurs as a consequence of rapid changes in the internal

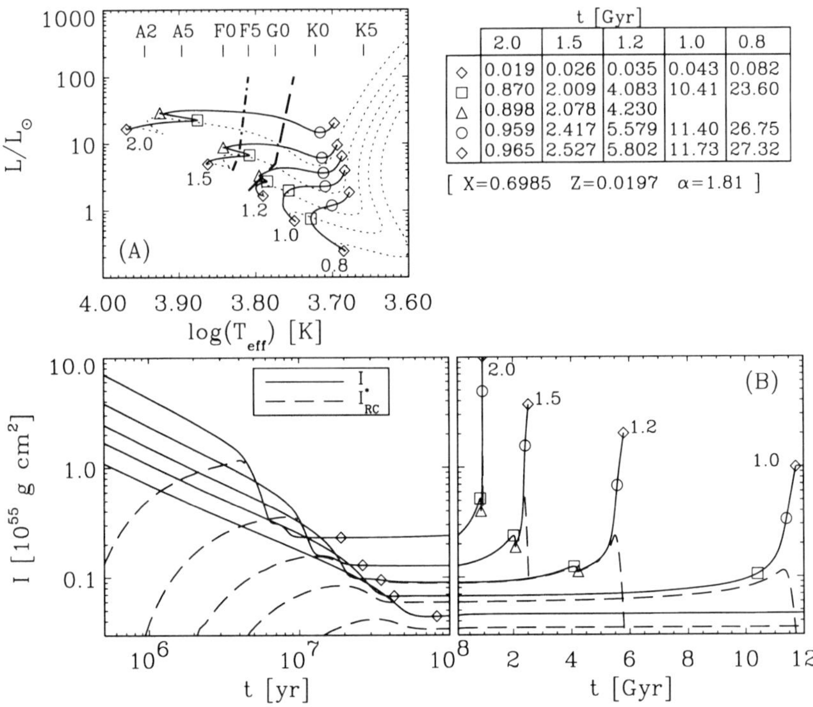

Figure 3.   Evolutionary tracks (plot A) and evolutionary variations of the total (solid lines) and core (dashed lines) moments of inertia for late-type stars of various masses, from the pre-main sequence to the base of the giant branch (plot B). The diamonds ($\diamond$) indicate the ZAMS, and the dashed line is the rotational dividing line (cf., Fig. 1). Ages at selected points along the tracks are listed in the Table.

mass distribution, even though the bulk of the increase in radius occurs later, on the giant branch itself. In this evolutionary phase, the relative ages of the stars can be estimated fairly easily because age is (almost) uniquely related to the position in the HR diagram for single (sub)giants between the granulation boundary and the upturn onto the giant branch. The T_{eff} dependence of the rotation rate along evolutionary tracks consequently allows the study of the rotational evolution (assuming, of course, that the initial properties of the sample do not change systematically with position in the HR diagram).

Figures 4 and 5, modified after Schrijver and Pols (1993), display observed $v \sin i$ values together with the run of the average velocities as a function of spectral color (the latter is corrected for the expected average projection factor $4/\pi$), for luminosity class (LC) III ($2 \lesssim M_*/M_\odot \lesssim 5$) and LC IV ($1.2 \lesssim M_*/M_\odot \lesssim 2$) stars, respectively. The figures show that the mean rotational velocities decrease monotonically as stars age, with the exception of the coolest giants (a phenomenon not yet properly studied). The dashed lines correspond to the surface rotational evolution predicted by Eq. (3) for $M_* =$

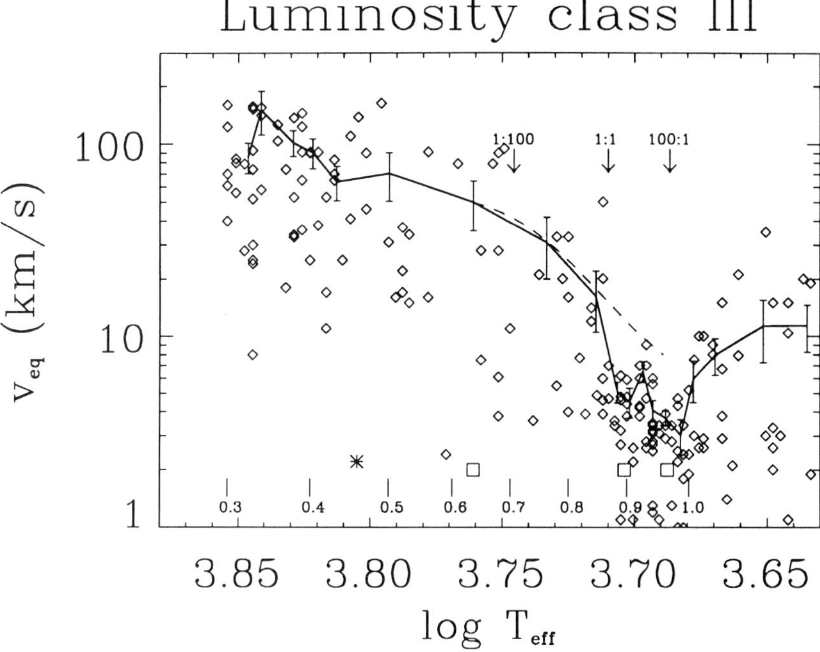

Figure 4. Observed rotation velocity v_{eq} vs T_{eff} for 2.5 $M_\odot$ LC III stars (modified after Schrijver and Pols 1993). Bars near the abscissa mark $B - V$ color. Squares above the color markers show the age in units of 2.5×10^6 yr of evolving 2.5 $M_\odot$ stars starting when the convection zone is four density scale heights deep. The asterisk locates the granulation boundary (Gray and Nagel 1989). The solid line represents the average velocity $\frac{4}{\pi}\bar{v}_{eq}$ (with error bars for the standard deviation) obtained by ordering the data by decreasing T_{eff} and binning them into adjacent groups of ten stars. The T_{eff} where the ratio of the moment of inertia of the envelope and of the interior equal 1:100, 1:1, and 100:1 are marked on the upper right-hand side of the panel. The computed run of $v_{eq}(T_{eff})$ for rigid rotation without magnetic brake, as described by Eq. (3), is indicated by a dashed line.

2.5 $M_\odot$ (Fig. 4) and for 1.5 $M_\odot$ (see Fig. 5). This mass is close to the lower end of the mass range for giants, because the more massive stars evolve more rapidly, biasing the sample towards lower masses. Subsequent core–helium burning phases, not addressed in these studies, may add stars of lower masses to the cool side of the diagram, complicating the interpretation). That these lines lie above the running mean indicates that the increase in moment of inertia is too small to account for the observed drop in rotation rates (see also Rutten and Pylyser 1988). This suggests that angular momentum loss does take place as these stars evolve toward the giant branch (see, e.g., Gray and Nagar 1985). Further support for this conclusion comes from a study of the distribution of rotation rates at different T_{eff}. Gray (1989,1991) notes that if the initial distribution of rotation rates, $H_0(v_{eq})$, and of the orientations of rota-

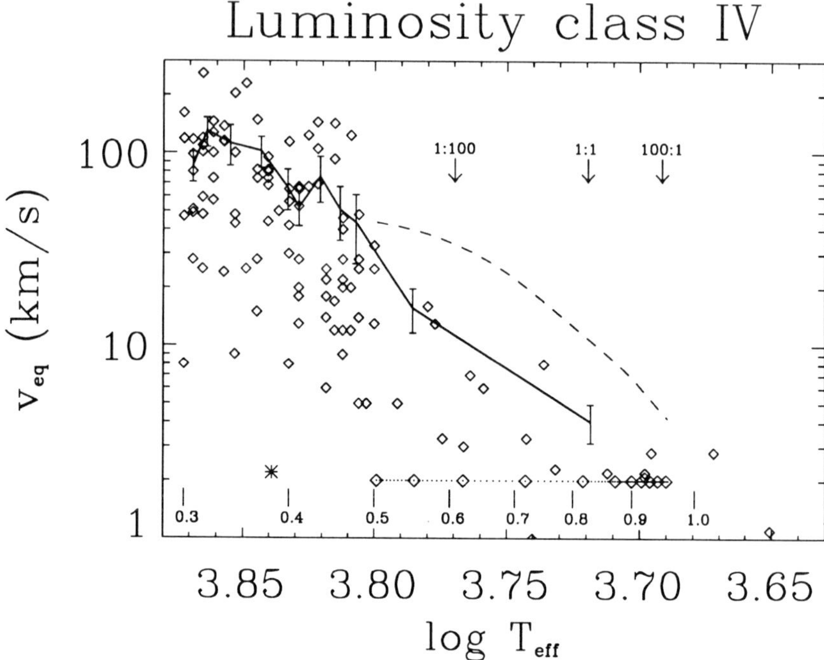

Figure 5.　As Fig. 4, but for luminosity class IV stars of 1.5 $M_\odot$. The stellar age is indicated above the $B - V$ markers by diamonds placed 25×10^6 yr apart and dots placed 2.5×10^6 yr apart.

tion axes do not change with time, the rotational history of stars can be derived from the comparison of distributions $H(v_{eq}, T_{eff})$ at different effective temperatures, because of the monotonically redward evolution during the phases relevant to this problem. If only the increase in the moments of inertia were to play a role, $H(v_{eq}, T_{eff})$ would be separable into $F(v_{eq}) \cdot g(T_{eff})$, which is contradicted by the observations. Hence we can conclude that angular momentum loss is taking place.

For LC III giants, $\bar{v}_{eq}$ decreases by a factor of about 10 between $B - V = 0.5$ and $B - V = 0.9$. It is unclear at present how drastic this change actually is. Gray (1991), for instance, suggests that the transition happens around G2III ($B - V \approx 0.77$ or $T_{eff} \approx 5280$ K according to Gray's scales), within only a few subdivisions in spectral type. The rotational data compiled by Schrijver and Pols (1993) suggest a smoother decrease in rotational velocities, although this is based on a running-mean average velocity of data that moreover include some results from relatively old catalogs. Rutten and Pylyser (1988) point out that in the much larger sample of Ca II H+K fluxes no clear discontinuity is seen (recently supported by cluster data; cf., Beasley and Cram 1993) so that if the rotational discontinuity exists, one must explain why it is not apparent in the chromospheric indicator.

For the 1.5 $M_\odot$ stars the decrease in $\bar{v}_{\mathrm{eq}}$ with decreasing T_{eff} is clearly rather pronounced, reaching a factor of about 10 by the time the stars reach $B - V \approx 0.6$. Redward of that point, the run of $\bar{v}_{\mathrm{eq}}(T_{\mathrm{eff}})$ for the LC IV stars is probably strongly affected by the existence of a detection limit at about $v \sin i \approx 1\text{--}2$ km s^{-1} so that only the fastest rotators have a detectable rotation rate: subgiants redward of $B - V \approx 0.6$ presumably rotate much slower on average than suggested by the data in Fig. 5. Modeling rotational evolution in this part of the HR diagram (cf., Sec. VI below) is a delicate affair, complicated by (1) the very short evolutionary time scales, and (2) the possible contamination by stars on (and/or evolving toward) the horizontal branch (see, e.g., Maeder and Meynet 1989).

B. Internal Angular Momentum Redistribution

In general, Eq. (3) is inadequate even in the absence of angular momentum loss, as it holds only under the assumption of internal solid-body rotation. For solar-type stars at least, the vigorous, large-scale fluid motions in stellar convective envelopes are expected to redistribute angular momentum rather efficiently (Kippenhahn and Weigert 1990, Sec. 42.3). Using typical length scales ℓ and velocities u_{ML} from the mixing length theory of convection, the time scale for dissipative redistribution of angular momentum throughout the solar convection zone is $\tau \sim (\Delta R)^2/(u_{\mathrm{ML}}\ell) \simeq 1$ yr, where ΔR is the thickness of the convective envelope. This turns out to be much shorter than typical angular momentum loss time scales (cf., Sec. IV), or any other relevant internal redistribution time scale (this subject is treated further below). Consequently, in terms of its response to an externally applied torque, the solar convective envelope may be assumed to behave as a solid body, with an average rotation rate equal to the (observed) surface rate. Doppler measurement and observations of sunspot motions across the solar disk do reveal that there exists a latitudinal gradient in the Sun's surface angular velocity. Helioseismological observations further demonstrate that this gradient is maintained all the way to the base of the convective envelope, with essentially no angular velocity gradient existing in the radial direction. This is most likely due to *internal* forcing arising from the complex way in which rotation, and possibly magnetic fields, interact with turbulent flows in convection zones. The response to *external* forcing is the relevant aspect in the present context.

What goes on in the radiative core is understood to a much lesser extent. Although there are a number of physical mechanisms capable of affecting angular momentum transport in a radiative zone, it is unlikely that any can be as efficient as fully developed turbulent convection, as in the overlying envelope. The internal redistribution of angular momentum is described by the ϕ-component of the momentum equation; in an axisymmetric system $(\partial/\partial\phi = 0)$,

$$\frac{\partial}{\partial t}(\rho\tilde{\omega}^2\Omega) = -\rho\mathbf{U}_p \cdot \nabla(\tilde{\omega}^2\Omega) + \nabla \cdot [\rho\varpi^2 v_{\mathrm{eff}}\nabla\Omega]$$

$$+\frac{1}{4\pi}\mathbf{B}_p \cdot \nabla(B_\phi \varpi) \tag{4}$$

where $\varpi = r \sin\theta$, $\mathbf{U}_p$ ($= u_r \hat{\mathbf{e}}_r + u_\theta \hat{\mathbf{e}}_\theta$) is a large-scale meridional velocity field, $\mathbf{B}_p$ and B_ϕ are poloidal and toroidal magnetic fields, and ν_{eff} an effective viscosity, i.e., including eventual contribution from turbulent transport. The three terms on the right-hand side correspond respectively to: (1) advection by large-scale bulk fluid motions; (2) "turbulent viscosity" or enhanced viscous-like transport associated with turbulent motions; (3) torques transmitted by a large-scale internal magnetic field. Even for the Sun, the transport efficiency of these mechanisms cannot yet be modeled from first principles, because we have very little knowledge of the (magneto)hydrodynamical state of the solar interior. Dimensional analysis of Eq. (4) yields the following three time scales for angular momentum redistribution by large-scale internal flows τ_U, viscous-like transport τ_ν, and large-scale magnetic fields τ_B throughout a volume of linear dimensions $\sim \ell$:

$$\tau_U = \ell/|\mathbf{U}_p| \tag{5a}$$

$$\tau_\nu = \ell^2/\nu_{\mathrm{eff}} \tag{5b}$$

$$\tau_B = \ell/u_A \tag{5c}$$

where $u_A = |\mathbf{B}_p|/\sqrt{4\pi\rho}$ is the Alfvén speed based on the internal poloidal field strength. Keeping in mind that Eqs. (5) are at best order-of-magnitude estimates, it is possible to calculate the flow speed, effective viscosity, and poloidal magnetic field strength required to bring about angular momentum transport on a time scale shorter than, say, 10^9 yr; for $\ell = \mathrm{R}_\odot$, this exercise yields $|\mathbf{U}_p| \sim 10^{-6}$ cm s^{-1}, $\nu_{\mathrm{eff}} \sim 10^7$ cm^{2}s^{-1}, and $|\mathbf{B}_p| \sim 10^{-5}$ G (for $\rho \sim 1$ g cm^{-3}).

None of these numbers is unreasonably high, so that no class of processes can be dismissed *a priori* with any confidence. The lower bound for $|\mathbf{U}_p|$ is probably the more restrictive. For example, the speed of large-scale flow associated with the thermally driven Eddington-Sweet-Vogt circulation (see, e.g., Tassoul 1978, ch. 8; Tassoul and Tassoul 1982, and references therein) is of order $L_*\Omega^2 R_*/G^2 M_*^3 \sim 10^{-9}$ cm s^{-1} ($L_* =$ luminosity) for the present-day Sun, indicating that this transport process is negligible in the solar interior (see also Tassoul and Tassoul 1984). The flow speed associated with the mechanically driven Ekman circulation is more difficult to estimate; flow speeds $\lesssim 10^{-5}$ cm s^{-1} have been suggested for the secondary Ekman flow, with larger flow velocities in the Ekman boundary layer proper, but numerous uncertainties remain (see also Sec. VII.A below). The lower bound for ν_{eff} corresponds to viscous transport enhanced by a factor of about 10^5 over that associated with simple microscopic viscosity, but still 3 to 5 orders of magnitude smaller than the effective viscosity in the convective envelope ($\nu_{\mathrm{eff}} \sim 10^{11}$—$10^{12}$ cm^{2}s^{-1}, according to mixing length theory). While this may appear a large factor, it is actually not inconceivable that small-scale turbulence fed by rotational instabilities could lead to enhanced viscous-like transport at that level. The lower bound for $|\mathbf{B}_p|$ is of the order of the interstellar magnetic field. This is almost absurdly low, because conservation of magnetic flux during stellar formation is expected to intensify any pre-existing field. The situation is further complicated by the possibilities of: (1) *destroying* a significant fraction of the primordial flux during the fully convective phase of pre-main-sequence evolution (see, however, Cattaneo and Vainshtein 1991) and/or via other magnetic instabilities operating through any/all evolutionary phases (see, e.g., Mestel and Weiss 1987; Balbus and Hawley 1994); (2) *producing* additional magnetic flux via dynamo action occurring on the pre-main sequence (Schüssler 1975; Parker 1981).

If angular momentum redistribution proceeds very efficiently throughout the radiative interior and at the core-envelope interface, then a star behaves as a solid body, with moment of inertia I_*. If it does not, then any externally applied torque will only spin the envelope down, with the radiative interior retaining its initial angular velocity. In what follows, the former situation is referred to as strong core-envelope coupling, and the latter as core-envelope decoupling. Because $I_{CE} \ll I_*$ for $M_* \gtrsim M_\odot$ on the main sequence, these two extreme cases lead to markedly different surface rotational evolution, a point to which we shall return in due time below. Presently, we turn instead to other "building blocks" of rotational evolution modeling.

III. MAGNETIC FIELDS IN COOL STARS

A. Dynamo Theory

All rotating stars with a convective envelope immediately underneath the photosphere (the so-called cool stars) appear to have nonradiatively heated outer atmospheres with radiative properties that resemble those observed in the complex topology of the solar outer atmosphere. In analogy to the Sun, this activity is assumed to be largely magnetic in origin; this assumption is supported by direct measurements of magnetic fields on a subset of these stars (see, e.g., Saar 1991).

The mean magnetic field strength in stellar coronae is related to the stellar rotation as a result of the so called dynamo operating in stellar interiors. Dynamo theory attempts to explain the properties of the global stellar magnetic fields. In itself, the existence of magnetic fields in stars is not surprising: a weak seed field in the interstellar gas would result in a strong field in the stellar interior owing to the compression of the material in the star formation process. Such an initial field may survive for a long time in the radiative interior. Consider, for instance, a star with a nonconvective interior of radius R. The time scale for a large-scale magnetic field to diffuse out of the star as a result of merely resistive dissipation is $\tau_\eta \approx R^2/\eta$, with η the magnetic diffusivity (inversely proportional to the electrical conductivity). For plasma at temperatures in excess of 10^6 K, η is rather low while the typical system size is so large that τ_η exceeds 10^{10} yr, i.e., longer than the main-sequence life time of the Sun. Consequently, dynamo theory is not essential to explain the presence of a magnetic field, but to explain why the observed time scales for changes in the photospheric magnetic fields of the Sun and other solar-like stars is many orders of magnitude shorter than the resistive dissipation time, namely of the order of only a decade. In addition to that, there are the observed migration of bands of maximum magnetic activity from mid-latitudes towards the equator, and the inversion of the direction of the magnetic axes of photospheric bipolar regions that occurs after each sunspot cycle that need explaining.

Magnetic fields obey the magnetohydrodynamic induction equation that, for constant η, reads:

$$\frac{\partial \mathbf{B}}{\partial t} = \nabla \times (\mathbf{u} \times \mathbf{B}) + \eta \nabla^2 \mathbf{B} \tag{6}$$

with $\mathbf{B}$ the magnetic field, and $\mathbf{u}$ the fluid velocity. The first term on the right-hand side describes the tendency of magnetic field lines to be convected with the fluid (a tendency that decreases in relative importance with increasing magnetic diffusivity) and the second term describes the diffusive dissipation of the magnetic field (increasing with increasing magnetic diffusivity).

If one takes ℓ as a characteristic length scale and u as a characteristic amplitude in the fluid velocity, the ratio of the first to the second term on the right-hand side of Eq. (6) yields the dimensionless magnetic Reynolds number:

$$\Re_{\mathrm{m}} \equiv \frac{u\ell}{\eta}. \tag{7}$$

In typical astrophysical plasmas the values of $\Re_{\mathrm{m}}$ are very much larger than unity, so that the magnetic field lines are frozen into the fluid; that is to say, the field lines are moved around by flow components perpendicular to the field and vice versa. The term in Eq. (6) associated with resistive diffusion is often negligible, but other forms of diffusion (see below) are often important.

Many kinematic dynamo models use the "mean field approach" that was originally suggested by Parker (1955). In this approach both the magnetic field and the velocity field are split into two parts, $\mathbf{B} = [\mathbf{B}] + \mathbf{B}_1$ and $\mathbf{u} = \mathbf{u}_0 + \mathbf{u}_1$. The separation into a "mean field" and a fluctuating field relies on the assumption that fluctuations occur on widely different length scales, while the mean $[\cdot]$ is taken over an intermediate scale. This separation leads to a dynamo equation that can be written as:

$$\frac{\partial [\mathbf{B}]}{\partial t} = \nabla \times (\mathbf{u}_0 \times [\mathbf{B}]) + \nabla \times \alpha [\mathbf{B}] + (\beta + \eta)\nabla^2 [\mathbf{B}]. \tag{8}$$

The first term on the right-hand side describes the effects of large-scale flows that reflect that the Sun does not rotate as a solid body: the angular velocity Ω depends on the position in the solar interior, hence the term "differential rotation." The origin of the differential rotation lies in the interplay of rotation and convection, although the precise physics is as yet not well understood (see, e.g., Rüdiger 1989).

The next term in Eq. (8), describing helicity, also has its origin in the interaction of the rotation and convection: rising (descending) gas in the solar convection zone expands (is compressed) as it reaches regions of lower (higher) density. Owing to Coriolis forces, the expansion (compression) results in a preferred rotation of the rising and descending material. This preferred rotation is called helicity. Magnetic field lines dragged along by the fluid motion will be subjected to the same sense of rotation, thereby changing direction. Steenbeck and Krause (1969) estimate $\alpha = -\frac{1}{3}\langle \mathbf{u}_1 \cdot \nabla \times \mathbf{u}_1 \rangle \tau_c$ to be of the order of $\Omega \ell^2 / L$, where ℓ is the size scale of the convective eddies, and L the macroscale of the system (see also Durney and Robinson 1982).

The last term in Eq. (8) describes the effects of ordinary Ohmic dissipation (η) and turbulent diffusion (β): turbulent convective motions will disperse the field lines in a kind of random walk. Steenbeck and Krause (1969) estimate $\beta = \frac{1}{3}\langle u_1^2 \rangle \tau_c$ to be of the order of $u_{\mathrm{ML}}\ell$, where u_{ML} is the typical convective velocity; note that generally $\beta \gg \eta$, so that η is usually omitted in the dynamo equation.

In analyzing Eq. (8) one may either work with kinematic models in which the magnetic field is assumed not to react back upon the fluid motions, or one may make $\mathbf{u}_0$, α, and β dependent on $\mathbf{B}$ although the quantitative knowledge of such a dependence is limited. Kinematic dynamos are linear and hence yield no information on the strength of the magnetic field. In reality, the field reacts back on the fluid motion due to the Lorentz force: the fluid motion is hampered, so that both the helicity and the differential rotation are quenched when the field strength increases. The solutions of the nonlinear dynamo equation have complex properties, that include long-term modulations, grand minima, periodic solutions, but also quasi-periodic or even chaotic solutions (see papers in Krause et al. 1993). The main weakness of most of the models published to date is that the choice of the nonlinearity is rather arbitrary, because the physics of the interactions is not well known. Hence, the nonlinear dynamo models have yet a long way to go; at present we are merely exploring the possible types of behavior that nonlinear systems can exhibit.

Let us turn back to Eq. (8) to study some of its properties. If one assumes $[\mathbf{B}] \propto \exp(\Lambda t)$, one can separate the temporal and spatial parts of Eq. (8), then yielding an eigenvalue problem, that allows the analysis of the stability of dynamo modes:

$$\Lambda[\mathbf{B}] = \nabla \times (\mathbf{u}_0 \times [\mathbf{B}] + \alpha\,[\mathbf{B}] - \beta\nabla \times [\mathbf{B}]). \tag{9}$$

This equation can be made dimensionless by multiplying it with the time scale for turbulent diffusion $\tau_\beta \approx L^2/\beta_0$ (where the index 0 indicates that some typical value is chosen). Two dimensionless dynamo numbers can then be formed:

$$C_\alpha \equiv \frac{\alpha_0 L}{\beta_0} \quad \text{and} \quad C_\Omega \equiv \frac{\omega_0 L^2}{\beta_0} \tag{10}$$

where L is the macroscale of the region where the dynamo operates, and ω_0 is a characteristic value of the differential rotation (following Durney and Latour [1978], see below).

The two numbers in Eq. (10) measure the relative importance of helicity to turbulent diffusion and of advection to turbulent diffusion, respectively. They can be combined into a single dynamo number:

$$D \equiv C_\alpha \cdot C_\Omega. \tag{11}$$

Equation (9) has a discrete spectrum of complex eigenvalues Λ_i. For a certain critical dynamo number D_{cr}, there is one marginally stable solution of Eq. (9) (i.e., with growth rate $\Re e(\Lambda_0) \equiv 0$), called the fundamental mode. All higher modes are not excited because they have a negative growth rate. If $D > D_{\mathrm{cr}}$, then $\Re e(\Lambda_0) > 0$ and at least one mode will grow. If, on the other hand, $D < D_{\mathrm{cr}}$, then all modes have negative growth rates and $\langle \mathbf{B} \rangle$ eventually goes to zero.

Durney and Latour (1978) argue that the Reynolds stresses that create the velocity shear are of order $\rho(\ell\Omega)^2$ (with ρ the density), so that their divergence $\rho(\ell\Omega)^2/L$ creates zonal macroscopic flows of order $\ell^2\Omega/L$, hence ω should be of order $(\ell/L)^2\Omega$. Then both C_α and C_Ω are comparable to $\ell\Omega/u_{\mathrm{ML}}$, so that:

$$D = \left(\frac{\ell\Omega}{u_{\mathrm{ML}}}\right)^2 \propto \left(\frac{\tau_c}{P}\right)^2 = R_o^{-2} \tag{12a}$$

where R_o is the so called Rossby number and P the rotation period of the star. The value of R_o for the Sun, for instance, is approximately 1.7 (for a mixing length of about 2 pressure scale heights, the convective turnover time scale one pressure scale height above the bottom of the convective envelope is of the order of fifteen days; see, e.g., Gilman [1980]), or $D \approx 0.35$, i.e. of order unity.

Estimates leading to Eq. (12a) are based on dimensional arguments, and different authors may arrive at different estimates. Rosner and Weiss (1992), for example, estimate the dynamo number to be:

$$D' = R_o^{-2}\left(\frac{L}{\ell}\right)^2. \tag{12b}$$

This expression suggests a strong direct dependence of the dynamo number on the relative size of the convective cells. For ℓ equal to two pressure scale heights as measured one pressure scale height above the bottom of the convective zone and for a Rossby number of 1.7 (see below Eq. 12a), $D' \approx 6$.

The Rossby number, the ratio of rotation period to the turnover time τ_c of convective eddies measures the importance of the Coriolis forces on stellar convection. In the empirical determination of relationships between rotation rate and magnetic activity, the Rossby number has often been used as one of the parameters. The Rossby number, or rather the dynamo number, does not contain information on the field strength; however, the value of the dynamo number as given in Eq. (12) is a measure for the growth rate of the field if initially there were only a seed field. The growth rate will be largest in the most rapidly rotating stars, given Eq. (12), until at some point the field will react back on the flow, limiting the growth. The Rossby number cannot be expected, therefore, *a priori*, to be a useful indicator of field strength in stellar studies (see following sections). Moreover, when stars of different τ_c are compared, many other parameters will be different as well (see below), which may or may not affect the dynamo number directly (cf., Eqs. 12a and 12b).

B. Measuring Magnetic Fields on Stellar Surfaces

With nonkinematic dynamo models still being developed, one has to use observational data to derive the dependence of activity on rotation. Classical methods used in solar physics to measure the magnetic field by studying line profiles in circularly polarized light are of little use in the case of stars owing to the complex magnetic geometry in their photospheres: in analogy to the Sun, stellar magnetic fields are expected to emerge from the interior in a large number of bipolar regions of various sizes distributed in active latitude bands. As a result, the right- and left-hand circularly polarized signals of the opposite line-of-sight components of the bipolar regions will nearly cancel in the disk-integrated light of a star. Linear polarization does not cancel, but is hard to interpret quantitatively.

The method used in recent years to measure magnetic fields on cool stars basically relies on the comparison of high-resolution, high-quality line profiles of a magnetically sensitive line in unpolarized light with some reference line (see, for instance, the review by Hartmann [1987]). The reference line can either be a suitably chosen magnetically insensitive line in the spectrum of the same star or the same line in a similar star of very low magnetic activity. With this method, the magnetic broadening of lines in spectra of late-type stars can be used to measure stellar magnetic fields. The calibration of the magnetic line broadening to yield estimates of the magnetic flux density in stellar photospheres relies on proper analysis of the radiative transfer in these lines, and assumptions about the topology of the field in the stellar photosphere and the emission from the magnetic features (which in spots, for instance, differs substantially from the emission from the quiet photosphere). The method can only be applied to stars in a limited range of rotation rates; if a star rotates too slowly, the photospheric magnetic flux appears to be too small to yield a measurable signal, while if the star rotates rapidly, the rotational Doppler broadening masks the magnetic signal. Stars to which this method can be applied successfully range in rotation periods from typically several days up to about 15 days (note that there are exceptions to this range; if a rapidly rotating star is viewed nearly pole-on, the Doppler broadening is weak, so that field measurements can be performed succesfully down to very short rotation periods).

Despite these problems, the method has proved relatively successful in recent years, allowing the determination of the intrinsic photospheric field strength B_0 and the fraction f of the stellar surface covered by magnetic fields. The intrinsic field strength B_0 appears to be determined by the pressure balance between the gas pressure in the photosphere surrounding the magnetic feature and the magnetic pressure within the magnetic feature:

$$B_{eq} = \sqrt{8\pi\, P_{gas}(\tau_{5000} = 1)}. \qquad (13)$$

This equality is the result of the so called convective collapse. When magnetic field surfaces in the photosphere, the gas in the interior of the tube radiates its energy away, so that it cools and begins to flow down. The consequent evacuation of the field region leads to a concentration of the field. This concentration stops when the internal magnetic and the external gas pressure are about equal; B_0 is therefore largely determined by the photospheric density and temperature.

The surface-averaged filling factor, on the other hand, does clearly depend on the rotation rate of the star. For a limited set of dwarf stars it was found that the mean photospheric flux density $\langle|\phi|\rangle$ and the angular velocity Ω are related through

$$\langle|\phi|\rangle \propto \Omega \qquad (14)$$

(see, e.g., Saar 1991). The amount of data is too small to allow significant conclusions concerning the role of the convective turnover time in the relationship between rotation rate and mean magnetic flux density.

C. Radiative Losses and Stellar Magnetic Fields

Because of the limited range to which the spectroscopic method to determine photospheric magnetic flux densities can be applied, other radiative diagnostics of the effects of magnetic fields in stellar outer atmospheres are commonly used. The magnetic field that is generated in cool stars through dynamo action enables subphotospheric mechanical energy to reach the plasma trapped in the field high above the photosphere. The dissipation of that energy heats the plasma to temperatures ranging from the minimum observed just above the photosphere to values in excess of 10^6 K in the corona. In the present context it is important to note that the observed stellar coronal temperatures represent temperatures of the closed loop structures in which the plasma density is much higher than in the open field regions (equivalent to solar coronal holes) from which the wind predominantly originates. The emission from the loops therefore dominates in the observed spectra, so these temperatures should be interpreted with care in studies of the angular momentum loss. Things may not be all bad, however, the temperatures in solar coronal holes appear to be somewhat lower, but of the same order of magnitude, as those of closed loops.

The mechanism(s) responsible for the heating of stellar outer atmospheres remain unknown, despite many efforts to identify them. The only mechanism that has been excluded as the sole source of atmospheric heating is the dissipation of purely acoustic waves, because on the one hand this mechanism cannot explain the highly structured topology of the solar chromosphere, transition region, and corona, while on the other hand computations suggest that the strong density gradient inhibits sufficient acoustic power to reach the corona to explain its radiative losses. Nevertheless, acoustic heating has been suggested to contribute measurably to the total radiative output of many stellar outer atmospheres. The so called basal flux inferred from stellar data corresponds to the emission level in extremely slow rotators where the dynamo is presumed to be very weak; it does not show signs of cyclic variation or of rotational modulation. On the Sun the basal emission level is observed over the centers of supergranulation cells where there are no signs of intrinsically strong magnetic fields. Recent theoretical work reproduces the approximate level of the basal chromospheric emission, and explains why there is little dependence on surface gravity or on metal abundances. An alternative, or perhaps supplementary, explanation of the basal heating may be that it reflects the presence of intrinsically weak magnetic fields that are generated by the convective turbulence.

The prime heating mechanisms, however, must be related to the magnetic field in some way or another. Two distinct groups of mechanisms may be distinguished. On the one hand, the heating may be associated with the dissipation of waves (one or more types of magnetohydrodynamic waves), either because of shock formation as they travel outward into the more and more tenuous stellar atmosphere, because of some kind of resonance, or because of small-scale phase mixing. On the other hand, the heating may be the result of the dissipation of small-scale field discontinuities introduced into the atmospheric field by the motions of the photospheric foot points caused by the displacements associated with convective turbulence in and below the photosphere. At present there is no consensus on the prime heating mechanism, and it is likely that different mechanisms may be operating at different temperature regimes in the stellar outer atmosphere and perhaps even in different magnetic topologies, such as coronal holes and closed loop structures. What is beyond any doubt, is that the field is involved, either actively or passively, in determining the high degree of structure that is observed in the solar atmosphere, and that the entire atmosphere is highly dynamic down to the smallest observable scales (for the Sun this is presently one to a few arc seconds, depending on the wavelengths that are most suited for the observations).

The causal link between the magnetic field and the atmospheric radiative losses suggests that the latter can be used to measure the magnetic flux in cases where other more direct spectral methods are doomed to fail. The detailed study of radiative losses from the outer atmospheres of the Sun and other magnetically active cool stars produced a puzzling result: scatter diagrams comparing radiative surface flux densities originating from different temperature regimes in stellar atmospheres reveal power laws with a power law index that differs increasingly from unity with increasing difference in formation temperature for the pair of spectral diagnostics being studied. The nonproportionality holds important clues on the structure of stellar outer atmospheres, but complicates a direct determination of the relationship between photospheric magnetic fluxes and atmospheric radiative fluxes, because many different radiative diagnostics are used to monitor stellar activity and these must all be brought to a common scale. Extensive studies suggest that the use of surface flux densities F yields simple power law relationships that are, surprisingly, independent of stellar effective temperature and gravity. This is true, provided that the basal flux is subtracted from the observed stellar fluxes (see Rutten et al. [1991] for a study based on a large set of stellar data, and Schrijver [1992] for solar data).

Power laws are not only found when comparing hemisphere-averaged stellar emissions, but also for the solar atmosphere when it is observed with moderate spatial resolution. The flux-flux relationships for spatially resolved solar data and for hemisphere-averaged stellar data appear to be virtually identical, despite the surface averaging over nonlinear, locally valid relationships that is involved in transforming the first into the second. Schrijver and Harvey (1989) studied the transformation of flux-flux relationships derived from moderate-resolution solar observations into relationships between disk-integrated fluxes at different phases in the solar cycle. This transformation involves the convolution of the locally valid relationships between radiative and magnetic flux densities with the time-dependent histograms of magnetic flux densities. To illustrate these transformation properties, let there be a relationship

$$\Delta F_{\rm i} = a_{\rm i} \cdot |\phi|^{\alpha_{\rm i}} \tag{15}$$

between the photospheric magnetic flux density $|\phi|$ and some radiative flux density $\Delta F_{\rm i}$ in excess of the basal flux, both measured on a scale of, say, a few arcseconds, and let the mean magnetic flux densities have a time-dependent surface distribution of $H_{\rm t}(|\phi|)d|\phi|$ for a comparable spatial resolution. The surface-averaged mean magnetic flux density is then

$$\langle |\phi| \rangle = \int |\phi| \cdot H_{\rm t}(|\phi|)\, d|\phi| \tag{16}$$

and the surface-averaged radiative flux density equals

$$\langle \Delta F_{\rm i} \rangle = \int a_{\rm i} \cdot |\phi|^{\alpha_{\rm i}} H_{\rm t}(|\phi|)\, d|\phi|. \tag{17}$$

Schrijver and Harvey (1989) point out that for the Sun throughout the solar cycle the distribution function $H_{\rm t}(|\phi|)d|\phi|$ is such that to a good approximation

$$\langle \Delta F_{\rm i} \rangle = a'_{\rm i} \cdot \langle |\phi| \rangle^{\alpha_{\rm i}} \tag{18}$$

with $a_i' \approx a_i$ for values of α_i in the observed range between about 0.5 and 1 (see below). The cause for the virtually identical power law indices for the disk-integrated and spatially resolved flux-flux relationships lies in the special character of the distribution of magnetic flux densities over the solar surface (and presumably also over stellar surfaces): as the Sun becomes more active, the fraction of the surface that is densely covered by magnetic flux tubes increases at the expense of the coverage by less active portions, with associated changes over the entire range of observed flux densities. Surprisingly, the changes in the distribution function are such that the power law index is unaffected, with only a weak change in the constant of proportionality. Note that the transformation properties also hold for relationships between pairs of radiative flux densities.

The comparability of expressions (15) and (18) opens up the possibility of a direct calibration between radiative and magnetic flux densities for cool stars using spatially resolved solar observations, thus circumventing the need to rely on stellar magnetic field measurements. At present only one relationship between magnetic and radiative flux densities has been determined by direct observation; Schrijver et al. (1989), following Skumanich et al. (1975), derived the relationship between chromospheric Ca II H+K flux densities, $\Delta F_{\mathrm{Ca\,II}}$, in excess of the nonmagnetic basal component and magnetic flux densities $|\phi|$ using KPNO/MacMath data with a resolution of 2.4″. For a plage and surrounding network (excluding spot umbrae and penumbrae) they find a mean relationship $\Delta F_{\mathrm{Ca\,II}}(|\phi|) \propto |\phi|^{0.60\pm0.10}$, exhibiting a substantial scatter that is probably caused by temporal variations and details in the field geometry.

Combining this directly determined dependence of radiative on magnetic flux densities with the set of relationships between radiative flux densities determined for both the spatially resolved Sun and for samples of cool stars, other slopes α_i of power laws relating radiative and magnetic flux densities, $\Delta F_i \propto |\phi|^{\alpha_i}$ can be estimated. The slopes α_i range from about $\alpha_{\mathrm{chrom}} \approx 0.6$ for the relationship between radiative chromospheric and mean photospheric magnetic flux densities, up to $\alpha_{\mathrm{cor}} \approx 1$ for that between radiative coronal soft X-ray and mean photospheric magnetic flux densities (cf., Schrijver 1991).

D. The Rotation-Activity Relationship

Empirical studies have demonstrated that rotation rate and atmospheric activity (and therefore, as discussed in Sec. III.C, the mean photospheric magnetic flux density) are indeed related for stars with convective envelopes; when comparing any radiative measure of stellar magnetic activity ΔF_i as discussed in Sec. III.C, with, for instance, angular velocity Ω for stars of similar effective temperature and gravity, one finds that ΔF_i increases monotonically with increasing Ω for stars with rotation periods P_{rot} exceeding about one day. The precise form of these relationships is unclear, partly because of the relatively small samples and the substantial intrinsic stellar variability, but in the comparison of the radiative measure of activity with rotation period for stars of a given spectral type and luminosity class, the relationships generally cannot be approximated by power laws over their entire range (see, e.g., Noyes et al. 1984; Basri 1987; Rutten and Schrijver 1987). This behavior implies that the simple dependence of $\langle|\phi|\rangle \propto \Omega$, that is often assumed, does not apply to all stars. If one limits the range from a few days up to 20 days, however, the relationship between coronal radiative losses and rotation periods, for instance, can be approximated by $\langle F_X\rangle \propto \Omega^{0.8\pm0.2}$ (see, e.g., Schrijver and Zwaan 1991). Combining this approximation with the relationship $\langle F_X\rangle \propto \langle\Delta F_{Ca\,II}\rangle^{1.5\pm0.2}$, found for stars, and $\Delta F_{Ca\,II} \propto |\phi|^{0.6\pm0.1}$ found (as described above) for solar spatially resolved data, one finds $\langle|\phi|\rangle \propto \Omega^{0.9\pm0.3}$, consistent with a linear dependence.

The linear approximation is not valid for extremely rapid rotators, or for very slow rotators. At the slow end, the activity seems to drop off more rapidly, in a way that is sometimes approximated by an exponential fit $\langle|\phi|\rangle \propto \exp(-\gamma P)$, while current observational studies suggest that some dynamo process operates *whenever* a star rotates, which would exclude the theoretically proposed need for some minimal rate of rotation to excite the dynamo (Schrijver 1993a). For the fastest rotators there are clear indications for a saturation in the atmospheric activity in stars with rotation periods below a few days (see, e.g., Vilhu 1984), although it is unclear whether this saturation reflects a saturation in the dynamo action, a saturation in the heating mechanism, an increased flux cancellation rate in the photosphere, or some combination of these. Radiative losses in the short-period W UMa binaries do even appear to decrease with increasing rotation rate for rotation periods below about one day (see, e.g., Vilhu and Rucinsky 1983; Rutten and Schrijver 1987).

The simple approximation of Eq. (14) that $\langle|\phi|\rangle \propto \Omega$ is also inappropriate when comparing stars of different masses or stars in different stages of their evolution: magnetic activity also depends on the properties of the convective envelope. This dependence is, for instance, reflected by the gradual onset of activity redward of the effective temperature where envelope convection first occurs (the dividing line for envelope convection intersects the main sequence at about 1.3 $M_\odot$) extending over an interval of (at least) some 0.2 in $B - V$. This gradual onset has been suggested to be caused by the change in the convective turnover time, τ_c, computed for the bottom of the convective envelope. While this is a possible interpretation (based on a remark by Durney and Latour [1978] that the Rossby number, P_{rot}/τ_c, is roughly inversely proportional to the square root of the dynamo number [cf., Sec. III.A]) one should not over-interpret the observational constraints. The support for the role of τ_c in the dependence of activity on rotation stands or falls with the invariance of the shape of the rotation-activity relationship under variations in spectral type. Rutten and Schrijver (1987) have argued that the relationship $A(P_{rot}, T_{eff}, \ldots)$ between rotation period P_{rot} and activity A in, e.g., Ca II H+K, is not simply a separable function of rotation rate and effective temperature, i.e., $A(P_{rot}, T_{eff}) \neq A_0(P_{rot}) \cdot A_1(T_{eff})$, although most studies that support the dependence of activity on τ_c implicitly assume such a separability to hold (see discussion in Schrijver 1993a,b).

We are thus lead to conclude that: (1) the relationship between rotation and activity in single stars involves at least the rotation rate and a basic stellar parameter such as the effective temperature (possibly indirectly related to τ_c); (2) the functional dependence on stellar effective temperature does not appear to support a simple dependence on the dimensionless Rossby number; (3) the approximation that $\langle|\phi|\rangle$ is proportional to Ω is valid in only a restricted range of rotation periods. It is important to note that the latter approximation relates the mean photospheric magnetic flux density with the rotation rate, while models for angular momentum loss through a stellar wind require the related, but not directly observable, mean coronal magnetic flux density (see Sec. IV.D). These last two comments should be kept in mind when modeling the rotational histories of stars (see Secs. V and VI).

IV. ANGULAR MOMENTUM LOSS FROM ROTATING, MAGNETIZED WINDS

The focal point of this section is the MHD wind model initially developed by Weber and Davis (1967), as adapted by Belcher and MacGregor (1976). This model involves a number of rather restrictive assumptions, but does yield consistent solutions of the (reduced) system of MHD equations, and as such can be used to compute deterministically the mass and angular momentum loss rates associated with a rotating, magnetized wind.

A. The Weber-Davis Solution

Given (1) a high temperature corona extending outward from some fiducial radius r_0 ($> R_*$) and corotating with the stellar surface at angular velocity Ω; (2) a poloidal magnetic flux distribution at r_0; (3) a density distribution at r_0; and (4) a spherically symmetric gravitational field $\mathbf{g} = -GM_*/r^2\hat{\mathbf{e}}_r$, steady-state, axisymmetric solutions are sought in the equatorial plane, assuming that neither the flow nor the magnetic field have components out of that plane. In other words, $\mathbf{u} = [u_r(r), 0, u_\phi(r)]$ and $\mathbf{B} = [B_r(r), 0, B_\phi(r)]$ in spherical polar coordinates (r, θ, ϕ). The plasma is assumed to be a perfect gas of fully ionized hydrogen, and the energy equation is replaced by a polytropic relation of the form $p/p_0 = (\rho/\rho_0)^\alpha$. The polytropic index α is in the range $1 < \alpha \le 3/2$, and considered given. Note that this implies a very specific spatial dependence for heating throughout the flow. The quantities ρ_0, T_0, Ω and B_{r0} are also taken as given, a 0 subscript referring to quantities evaluated at r_0. Under these assumptions, the r and ϕ-components of the equations of motion reduce to

$$\rho\left(u_r\frac{du_r}{dr} - \frac{u_\phi^2}{r}\right) = -\frac{GM_*\rho}{r^2} - \frac{dp}{dr} - \frac{B_\phi}{4\pi r}\frac{d}{dr}(rB_\phi) \qquad (19a)$$

$$\rho\left(u_r\frac{\mathrm{d}u_\phi}{\mathrm{d}r} + \frac{u_r u_\phi}{r}\right) = \frac{B_r}{r}\frac{\mathrm{d}}{\mathrm{d}r}(rB_\phi) \tag{19b}$$

while the continuity equation and Maxwell's equation $\nabla \cdot \mathbf{B} = 0$ immediately yield conservations statements for the mass and magnetic fluxes. Using these results, the ϕ-component of the induction equation can be manipulated into

$$ru_\phi - \frac{rB_\phi B_r}{4\pi\rho u_r} = L \tag{20}$$

expressing conservation of the total angular momentum per unit mass L carried away by the wind (see Weber and Davis 1967; Belcher and MacGregor 1976). The first term on the left-hand side is the specific angular momentum of the moving fluid element, while the second is the torque density associated with the magnetic field. Introducing expressions for the Alfvén velocities associated with each component of the magnetic field $A_r = B_r/(4\pi\rho)^{1/2}$, $A_\theta = B_\theta/(4\pi\rho)^{1/2}$, and $A_\phi = B_\phi/(4\pi\rho)^{1/2}$, further algebraic manipulations allow Eq. (20) to be cast into the form

$$u_\phi = \Omega r\frac{(u_r^2 L/\Omega r^2) - A_r^2}{u_r^2 - A_r^2}. \tag{21}$$

Regularity at the Alfvén point (the location $r = r_A$ where $u_r = A_r$) requires the numerator on the right-hand side of this expression also to vanish there, implying

$$L = \Omega r_A^2. \tag{22}$$

This indicates that the total angular momentum per unit mass transported by the wind (i.e., including the torque density associated with the magnetic field) is equal to the specific angular momentum that would be carried away by an unmagnetized wind flowing strictly radially ($u_\phi = 0$), and corotating out to a radius r_A.

Using these results, the r-component of the momentum equation can be cast into either of the following forms:

$$\frac{\mathrm{d}}{\mathrm{d}r}\left[\frac{1}{2}(u_r^2 + u_\phi^2) - \frac{GM_\odot}{r} + \frac{c_s^2}{\alpha - 1} - \frac{(\Omega r)A_r A_\phi}{u_r}\right] = 0 \tag{23a}$$

$$\frac{\mathrm{d}u_r}{\mathrm{d}r} = \left(\frac{u_r}{r}\right)\frac{(u_r^2 - A_r^2)(2c_s^2 + u_\phi^2 - GM_\odot/r) + 2u_r u_\phi A_r A_\phi}{(u_r^2 - A_r^2)(u_r^2 - c_s^2) - u_r^2 A_\phi^2}. \tag{23b}$$

Equation (23a) is the Bernoulli equation for the flow, i.e., a conservation statement for the total energy per unit mass E. It resembles the corresponding expression obtained for a nonrotating, unmagnetized wind, but the solution topology of the equation of motion is much more complex; the denominator of Eq. (23b) vanishes when u_r becomes equal to the phase speed of either the fast

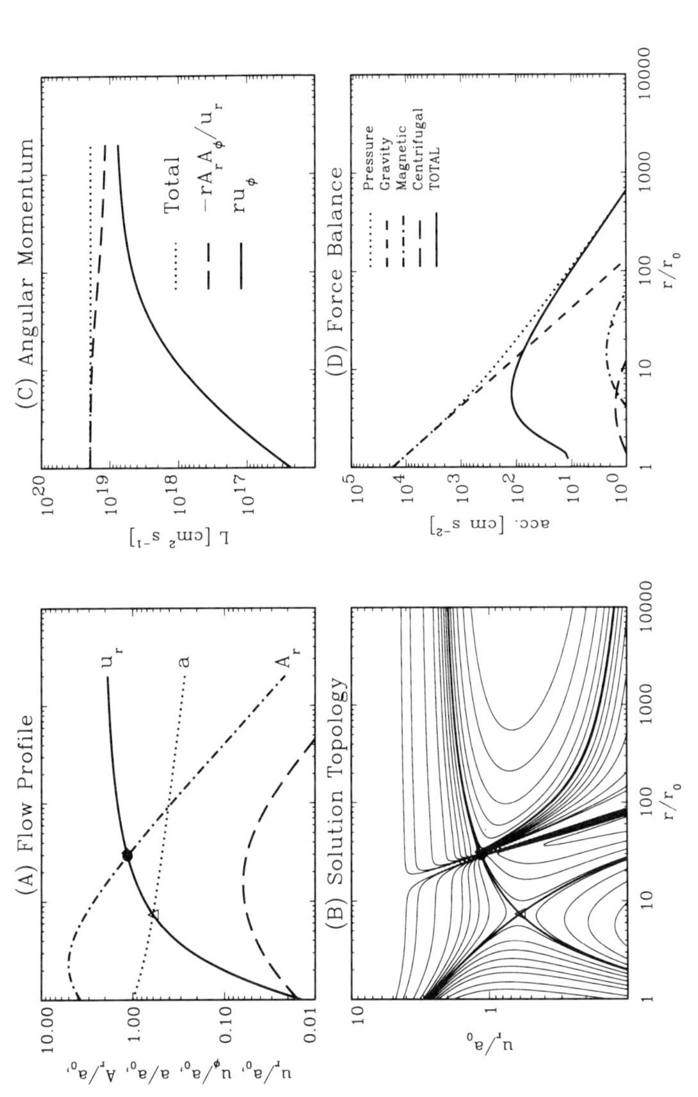

Figure 6. A rotating, magnetized wind solution for the present-day Sun. Solution input parameters are $\alpha = 1.13$, $r_0 = 1.25\,R_\odot$, $T_0 = 1.5 \times 10^6$ K, $\rho_0 = 5 \times 10^7$ protons per cubic cm. Plot A shows the variations with radius of u_r (solid line), u_ϕ (dashed line), the sound speed c_s (dotted line), and the radial Alfvén speed A_r (dash-dotted line). The triangle, solid dot and diamond indicate the location of the slow, Alfvén, and fast point, respectively (the Alfvén and fast point nearly coincide here). Plot B illustrates the solution topology, while plot C shows the contributions to the total angular momentum loss associated with the wind, and plot D the force balance in the wind.

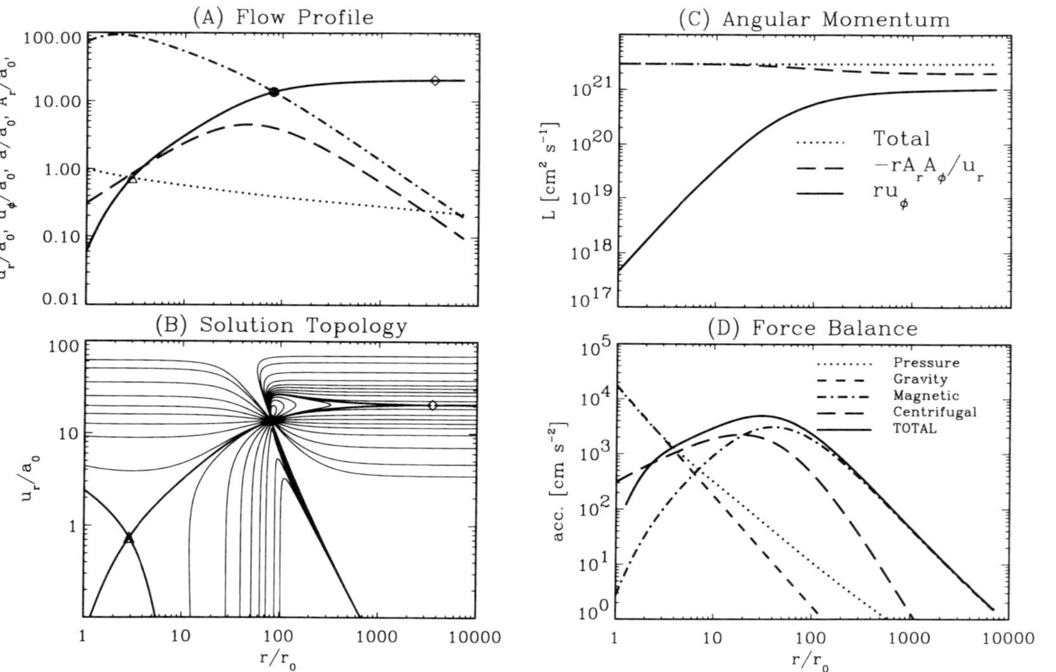

Figure 7.   Similar to Fig. 6, but for a rotation rate and surface field strength 20 times larger. Comparing to Fig. 6, note how the flow speed now largely exceeds the sound speed at large distance (plot A). This can be traced to magnetic and centrifugal contributions to the net acceleration, that now dominate beyond the slow point (plot D). Note also how the Alfvén and fast points are now clearly distinct, and how the former is located at much greater radial distance than for the solution shown in Fig. 6.

or slow magnetosonic wave mode, which occurs at distinct radial positions r_f and r_s. There are thus three critical points in the flow, as opposed to one for the unmagnetized thermally driven wind solution. Solutions can be obtained by requiring that the numerator and denominator of Eq. (23b) simultaneously vanish at the fast and slow point, and that the Bernoulli constant E evaluated at those points be equal to its value at r_0. This yields six coupled transcendental equations, which are solved numerically for the quantities u_{r0}, $u_{\phi 0}$, r_f, u_{rf}, r_s, and u_{rs}. For a given set of input parameters (r_0, M_*, α, T_0, Ω, and B_{r0}), these six quantities fully define the rotating magnetized wind solution. Figures 6 and 7 depict two such solutions for $M_* = M_\odot$. The solution on Fig. 6 has $B_{r0} = 2\,\mathrm{G}$, and $P_{\mathrm{rot}} = 2\pi/\Omega = 24.3$ days, i.e., parameters corresponding to the present-day Sun. The solution shown on Fig. 7, on the other hand, has $B_{r0} = 40\,\mathrm{G}$ and $P_{\mathrm{rot}} = 1.21$ days. Such values are more characteristic of the young Sun, as it may have appeared at, say, the age of the α Per cluster.

The Alfvén point (r_A, solid dot on Figs. 6 and 7) moves outward as the rotation rate and surface field strength are increased. In view of Eq. (22), an enhanced angular momentum loss rate ensues. In general, the angular momentum loss rate is a rapidly increasing function of rotation rate and surface field strength. Note, however, that the mass loss rate increases only by a factor of $\sim$3 between Figs. 6 and 7. (Compare the radial flow velocities at r_0.) This was to be expected, because momentum deposition by magnetic and centrifugal forces occurs mostly beyond the sonic point (cf., Figs. 6D and 7D); the effect on the flow is thus an increase in the asymptotic flow speed rather than the mass flux (see the chapter by Holzer et al.).

It must be emphasized that these two solutions differ markedly in their dynamics. The former is mostly thermally driven, while in the latter magnetocentrifugal driving dominates at large distances. In these two regimes, the dependence of quantities such as r_A on Ω, B_{r0}, etc., differs markedly (see Belcher and MacGregor 1976; MacGregor and Charbonneau 1994).

B. Mass and Angular Momentum Loss Rates

The solution shown in Fig. 6 is characterized by mass and angular momentum loss rates $dM/dt = 2.9 \times 10^{-14}\,\mathrm{M_\odot\,yr^{-1}}$ and $dJ/dt \simeq 2 \times 10^{31}$ dyne cm. Both compare favorably to *in-situ* observations at the Earth's orbit (Pizzo et al. 1983). The time scale for mass and angular momentum loss are then

$$\tau_M = \mathrm{M_\odot} \left(\frac{dM}{dt}\right)^{-1} \sim 10^{13}\,\mathrm{yr} \tag{24a}$$

$$\tau_J = J_\odot \left(\frac{dJ}{dt}\right)^{-1} \sim 10^{9}\,\mathrm{yr} \tag{24b}$$

where it was further assumed, in computing τ_J, that the present-day Sun is rotating as a solid body ($J_\odot = \Omega I_\odot$), with a moment of inertia $I_\odot \simeq 7 \times 10^{53}$ g cm^2. These are to be compared to the Sun's main-sequence lifetime $\tau_\odot \simeq$

10^{10} yr. While τ_M is far too large for the Sun's structure to be affected by mass loss over its evolutionary lifetime, τ_J is of the order of the solar age, indicating that significant angular momentum loss is taking place. For the solution shown in Fig. 7, one obtains $\tau_J \simeq 5 \times 10^7$ yr, i.e., very short compared to the main-sequence lifetime, indicating that angular momentum loss can drive rapid changes in rotation rates after stars arrive on the main sequence.

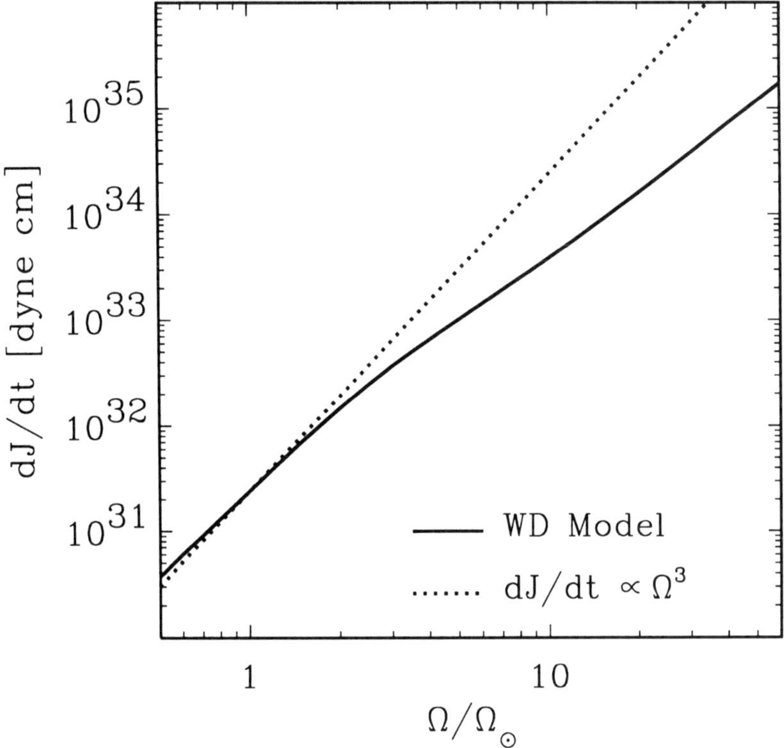

Figure 8.   Rate of angular momentum loss predicted by the Weber-Davis solution with $B_{r0} \propto \Omega$ (solid line). Going from high to low rotation rate corresponds essentially to a time sequence, from the ZAMS onward. The dotted line is $dJ/dt \propto \Omega^3$ calibrated to the present-day Sun, which yields the Skumanich relation for a time-independent moment of inertia.

C. The Skumanich $t^{-1/2}$ Relation

Having established that the Sun has lost angular momentum over the course of its main-sequence lifetime, it remains to derive the functional form of $\Omega(t)$ for a solar-type star. For this purpose, the key result obtained above is Eq. (22). Because the Weber-Davis solution is valid only in the equatorial plane, an equivalent expression valid over the whole sphere must first be constructed. This turns out to be easy, if a single (drastic) additional assumption is made, namely that the magnetic field has no meridional component at any latitude (i.e., that the poloidal part of the magnetic field is monopolar). Under these conditions, and provided that the mass flux is independent of latitude, the equatorial Weber-Davis solution can be "mapped" directly onto a cone of apex angle θ, with r_A independent of θ. The *total* specific angular momentum carried away by the magnetically coupled wind is then that of a thin, rigidly rotating spherical shell of radius r_A, namely $L_{\rm sph} = (2/3)\Omega r_A^2$. Therefore,

$$\frac{dJ}{dt} = 4\pi \rho_A r_A^2 u_{rA} \left(\frac{2}{3}\Omega r_A^2 \right).$$ (25)

Now, $u_{rA} = A_{rA}$ at the Alfvén radius. Conservation of mass and magnetic flux implies $r_0^2 B_{r0} = r_A^2 B_{rA}$ and $B_{rA}^2 = 4\pi \rho_A A_{rA}^2$, so that

$$\frac{dJ}{dt} = -\frac{2}{3} B_{r0}^2 r_0^4 \Omega A_{rA}^{-1}.$$ (26)

As discussed in Sec. III, there are theoretical and observational reasons to believe that a relationship of the form $B_{r0} \propto \Omega$ should hold up to rotation rates of at least a few times the present solar rate. Because $u_{rA} = A_{rA} \simeq c_s$ for mostly thermally driven winds (cf., Fig. 6) and further assuming that the coronal temperature is independent of surface magnetic field strength and/or rotation rate, Eq. (26) takes the general form

$$\frac{d\Omega}{dt} \propto -\Omega^3$$ (27)

for a time-independent moment of inertia, which integrates immediately to

$$\frac{1}{\Omega^2(t)} - \frac{1}{\Omega^2(t_0)} \propto t - t_0.$$ (28)

In the asymptotic limit $t \gg t_0$, $\Omega(t) \ll \Omega(t_0)$ and

$$\Omega(t) \propto t^{-1/2}.$$ (29)

This is usually referred to as Skumanich's relation, and was first inferred observationally (Skumanich 1972; see also Durney 1972). While such a $t^{-1/2}$ dependence seems to fit reasonably well the observed rotational evolution of solar-type stars older than $\sim 10^9$ yr, the situation at earlier times is much more complex, as discussed in Sec. I. This should not come as a surprise, as the preceding derivation is based on a very restrictive set of assumptions, *all* of which must be met if the Skumanich relation, whether in the form of Eq. (27) or Eq. (29), is to yield an accurate representation of rotational evolution. Even if these assumptions were valid, the Skumanich relation cannot be used indiscriminately. Figure 8 illustrates the variations of dJ/dt with rotation rate predicted by the Weber-Davis model, with $B_{r0} \propto \Omega$. Clearly, Eq. (27)

provides a reasonably accurate representation of the angular momentum loss rate only at low rotation rates ($\Omega/\Omega_\odot \lesssim 5$, say). This can be traced directly to the growing importance of magnetic and centrifugal driving in the overall wind dynamics (cf., Fig. 7D) as Ω (and thus B_{r0}) are increased, i.e., the wind is no longer mostly thermally driven. Clearly, even within the simplified framework of the Weber-Davis model, a parametric representation of the form $dJ/dt \propto \Omega^N$ with constant N is only applicable over a restricted range of rotation rates.

D. The Spin Down of ZAMS Late-Type Stars

The swift spin down of rapidly rotating solar-type stars between the ages of α Persei and the Pleiades implies a spin down time scale of the order of the age difference between these two clusters, i.e., $\sim$20 Myr (cf., Fig. 2; Stauffer and Hartmann 1987). This represents a powerful constraint; is such a short spin down time scale compatible with the angular momentum loss rates predicted by the Weber-Davis model? Calculating the spin-down time scale τ_J (cf., Eq. 24b) requires an estimate of the effective moment of inertia of the object on which the torque is applied. Two extreme cases were considered in Sec. II: (1) strong core-envelope coupling, whereby the star behaves as a solid body, or (2) core-envelope decoupling, with only the convective envelope being spun down. Denoting the corresponding spin-down time scales by $\tau_{J,*}$ and $\tau_{J,CE}$ respectively, application of the Weber-Davis model to ZAMS stars yields $\tau_{J,*} \sim 10^8$ to 10^9 yr, and $\tau_{J,CE} \sim 10^7$ to 10^8 yr (for $0.8 \lesssim M_*/M_\odot \lesssim 1.2$ and assuming in all cases a coronal temperature of 1.5×10^6 K, a surface magnetic field strength of 50 G and a rotation period of one day); at least for solar-type stars, a spin-down time scale of a few 10^7 yr is compatible with observations, while a few 10^8 yr is clearly too high (cf., Fig. 2). The observational argument for core-envelope decoupling, initially put forth by Stauffer and Hartman (1987) on the basis of moment of inertia considerations, thus remains valid.

E. Beyond the Weber-Davis Model

The assumption of a monopolar magnetic field is perhaps one of the more obvious shortcomings of the Weber-Davis (WD) solution. Even for a monopolar magnetic field, the WD solutions actually cannot be consistently extended out of the equatorial plane, as there then exists an unbalanced magnetic pressure component in the θ-direction (see Suess and Nerney 1973; Nerney and Suess 1975). This is expected to lead to a bending of field lines towards the symmetry axis, except possibly in the limit of high rotation, where centrifugal effects can cause, at low latitudes, a bending of field lines toward the equatorial plane. Furthermore, in the lower quiet solar minimum corona the large-scale magnetic structure is much closer to dipolar than monopolar. However, even a purely dipolar magnetic flux distribution makes the axisymmetric problem inherently two-dimensional, with the consequence that the WD formalism developed above cannot be applied. An alternative is of course to obtain fully

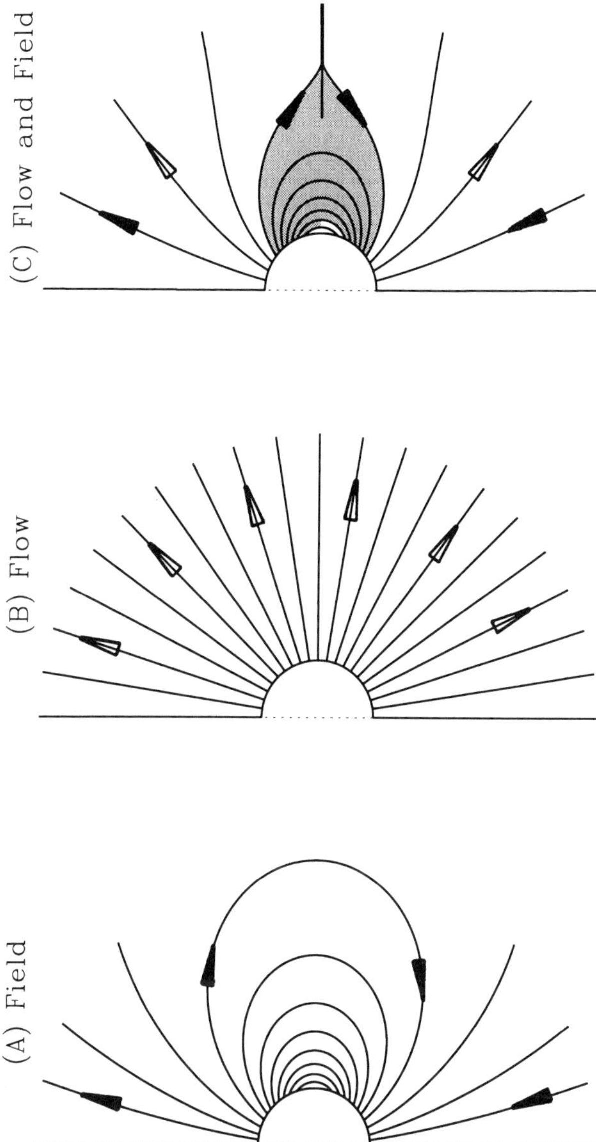

Figure 9. Phenomenological MHD wind models for a dipolar field geometry. Diagram A shows poloidal field lines in the absence of the flow, diagram B the flow stream lines in the absence of a poloidal field. Solid (open) arrows indicate the direction of the magnetic field (flow). Diagram C illustrates a possible "compromise" configuration, consisting of a wind zone, where plasma does flow along field lines, and a dead zone (shaded area), where $\mathbf{u} = 0$.

numerical steady-state solutions. For axisymmetric configurations, this is certainly within the reach of current generations of computers and numerical techniques. While two- and three-dimensional MHD wind models abound, very few truly consistent solutions have actually appeared in the literature (nevertheless, see Sakurai 1985; Washimi and Sakurai 1993).

The main difficulty arising when considering complex field geometries is of a topological nature; in the perfectly conducting limit, the plasma must either flow along magnetic field lines, or advect the magnetic field. Which of the two possibilities materializes is a function of the relative energy densities associated with the magnetic field and the flow (or, alternately, of the so called plasma β). If the former dominates over the latter, the plasma is channeled along the magnetic field lines. In the opposite limit, the magnetic field is advected by the flow. Consider the case of a dipolar magnetic field, as shown in Fig. 9; in the absence of the flow, the magnetic field is as shown in diagram A. Without the magnetic field, a spherically symmetric flow would ensue, as shown in diagram B. These two configurations are clearly incompatible in the equatorial regions; either the (closed) field lines open up, or the flow is completely choked.

For a pure dipole, the field strength decreases with distance along the equator as $B_\theta \propto 1/r^3$. The flow speed for a thermally driven wind, on the other hand, is expected to increase monotonically with distance. One may envision that close to r_0, where the poloidal field is strongest and $u_r \ll c_s$, the flow is effectively channeled and choked by the magnetic field, while beyond a certain critical distance the energy density of the flow becomes greater than the magnetic energy density, and the magnetic field lines open up, allowing the flow of plasma. This is illustrated schematically in Fig. 9C. The configuration is characterized by a "dead zone" where magnetic field lines are closed and $\mathbf{u} = 0$, and a "wind zone" where plasma flow does occur, dragging along field lines more or less radially. The presence of the dead zone reduces the total mass loss, and thus, via Eq. (25), the net angular momentum loss. Approximate wind solutions constructed within this framework (see Mestel 1968; Mestel and Spruit 1987; Cameron et al. 1991) suggest that in the limit of high rotation rate and surface field strength, the effective angular momentum loss may deviate significantly from that predicted by WD-type solutions. In terms of magnetic field geometry effects, the WD solution likely provides an upper limit to the true angular momentum loss rate, for a given coronal temperature, gravity, rotation rate and surface field strength. A comprehensive set of numerical solutions for the axisymmetric MHD wind problem is sorely needed, in order to quantify these effects systematically.

V. ROTATIONAL EVOLUTION: PRE-MAIN SEQUENCE AND MAIN SEQUENCE

A. The Initial Condition Problem

Having assembled all of the required building blocks, the task now at hand is to construct rotational evolution models. This begins with the choice of a suitable initial condition. Using simple virial theorem arguments, it is easy to show that for star formation to occur at all, protostars must manage to shed most of their initial magnetic flux (probably via ambipolar diffusion; see, e.g., Mouschovias 1991), as well as their initial angular momentum (likely via disk formation and/or a form of magnetic braking akin to that presented in Sec. IV; see Mouschovias 1991; Bodenheimer 1991). Theory and models have, however, not yet reached the point where quantitative estimates of the initial angular momentum content $J_0(M)$ of stars can be confidently made from first principles.

One possibility for estimating J_0 for solar-type stars is simply to extrapolate Kraft's relation to lower masses. Another option is to use the observed rotation rate distribution for T Tauri stars directly. There are a number of practical difficulties associated with this approach, however. Mass and radius for T Tauri stars are typically not known with high accuracy, introducing considerable uncertainties in the conversion of observed rotation rates to angular momentum content. This problem is compounded by our complete lack of knowledge regarding the internal rotational state of T Tauri stars. Even if this situation were to be dramatically improved upon, using the observed T Tauri velocity distributions for an initial condition still amounts to making an implicit assumption, namely that the same distribution is characteristic of all clusters at the time of their formation. This difficulty carries over to the comparison of rotation rates between clusters; attempting to evolve, say, Fig. 2A into Fig. 2B is only meaningful if the *present* rotation rate distribution of G stars in α Per is characteristic of that of the Pleiades 20 Myr ago. With these caveat in mind, we now (finally) proceed with the modeling of rotational evolution.

B. Basic Properties of Rotational Evolution Models

Consider first a simple two-component model, whereby both the convective envelope and radiative core are assumed to rotate as solid bodies with angular velocities Ω_{CE} and Ω_{RC}, and exchange angular momentum on a (given) time scale τ_C. Further assuming that the wind-induced torque extracts angular momentum only from the convective envelope, the time evolutions of Ω_{CE} and Ω_{RC} are governed by the following coupled ordinary differential equations:

$$\frac{dJ_{CE}}{dt} = \frac{\Delta J}{\tau_C} - \frac{J_{CE}}{\tau_J} + \frac{dJ_d}{dt} \tag{30a}$$

$$\frac{dJ_{RC}}{dt} = -\frac{\Delta J}{\tau_C} - \frac{dJ_d}{dt} \tag{30b}$$

with $J_{\mathrm{CE}} = I_{\mathrm{CE}}\Omega_{\mathrm{CE}}$, $J_{\mathrm{RC}} = I_{\mathrm{RC}}\Omega_{\mathrm{RC}}$, $\Delta J = I_{\mathrm{CE}}I_{\mathrm{RC}}(\Omega_{\mathrm{RC}} - \Omega_{\mathrm{CE}})/(I_{\mathrm{RC}} + I_{\mathrm{CE}})$, and e.g. $\tau_J = 3I_{\mathrm{CE}}/(2r_A^2\dot{\mathrm{M}})$ from the WD wind model (see MacGregor and Brenner 1991; MacGregor and Charbonneau 1994). The last term on the right-hand sides of Eqs. (30) corresponds to (instantaneous) angular momentum transfer associated with a receding or deepening convective envelope, an important effect on the pre-main sequence as well as on the subgiant and giant branches. Once $I_{\mathrm{CE}}(t)$ and $I_{\mathrm{RC}}(t)$ are known along the evolutionary tracks, the quantity dJ_{d}/dt can be constructed, and Eqs. (30) integrated to yield rotational evolution solutions.

Figure 10 shows one such solution. The evolution of the various relevant time scales is shown in plot A, and the rotation rates of the core and envelope in plot B. This solution was computed assuming $B_{r0} \propto \Omega_{\mathrm{CE}}$ and a coupling time $\tau_C = 20$ Myr. At early ($t \lesssim 10$ Myr) times, the structural evolutionary time scale (τ_E) is shorter than either τ_J or τ_C. This, combined with the large moment of inertia of the envelope (cf., Fig. 3) means that neither angular momentum loss due to the wind nor exchange between the core and envelope can significantly affect the spin up associated with the rapidly decreasing I_{CE}. Once $\tau_J \lesssim \tau_E$, angular momentum loss begins to dominate the right-hand side of Eq. (30a), and surface spin down ensues. The envelope remains decoupled from the underlying core for a time interval of order τ_C, after which angular momentum resupply from below starts balancing the loss due to the wind. From a few 10^8 yr onward, $\tau_J \gtrsim \tau_C$, and the rotation rates of the core and envelope gradually converge, with, in this case, little differential rotation persisting beyond $\sim 10^9$ yr.

Figure 11 illustrates the effects of varying model parameters. Unless otherwise noted, all solutions are for a 1 $M_\odot$ star, assuming $B_{r0} \propto \Omega_{\mathrm{CE}}$ and a coupling time $\tau_C = 20$ Myr (solid lines on all panels, corresponding to the solution shown in Fig. 9). In Fig. 11, plot A illustrates the dependence on the assumed coupling time τ_C; plot B illustrates the consequences of adopting various dynamo relationships; and plot C of various initial rotation rates. Plot D shows the rotational evolution of stars of various masses. In most cases, the surface rotation reaches a maximum either on or shortly prior to the ZAMS. The "production" of rapid rotators at the ages of α Persei and the Pleiades is therefore a problematic affair. Increasing the initial rotation rate is no solution; the turning point $\tau_J = \tau_E$ then occurs sooner, with a correspondingly larger angular momentum loss occurring prior to the arrival on the ZAMS. Saturating the dynamo (i.e., using $B_{r0} \propto \Omega_{\mathrm{CE}}$ up to a certain rotation rate, and keeping B_{r0} constant thereafter) can produce rapid rotators for the first few 10^7 yr on the main sequence, but tends to yield rotation rates too high by the age of Hyades. The difficulty in making rapid rotators is, to a large extent, a consequence of the angular momentum loss rate being a rapidly increasing function of rotation rate; faster rotators simply spin down faster. Using the same two-component model, Keppens et al. (1995) have investigated how the evolution of rotation rate *distributions* for solar-type stars is influenced by the adopted coupling time, dynamo relationship, etc.

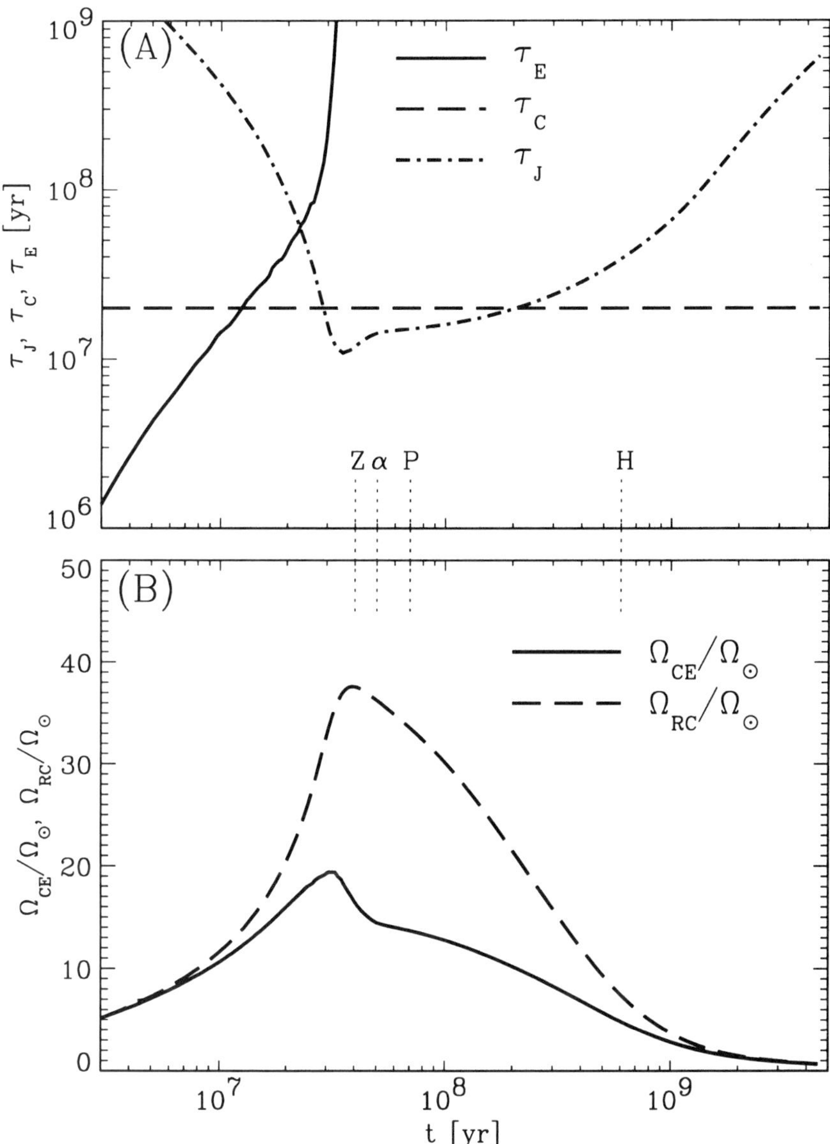

Figure 10.   Rotational evolution of a 1 $M_\odot$ stars computed using the two-shell formulation described in the text with $\tau_C = 20$ Myr. Plot A shows variations with time of relevant time scales, and plot B the corresponding rotational evolution of the core and envelope (see text). The dotted tickmarks indicate the ages corresponding to the 1 $M_\odot$ ZAMS (Z) α Persei (α), Pleiades (P) and Hyades (H).

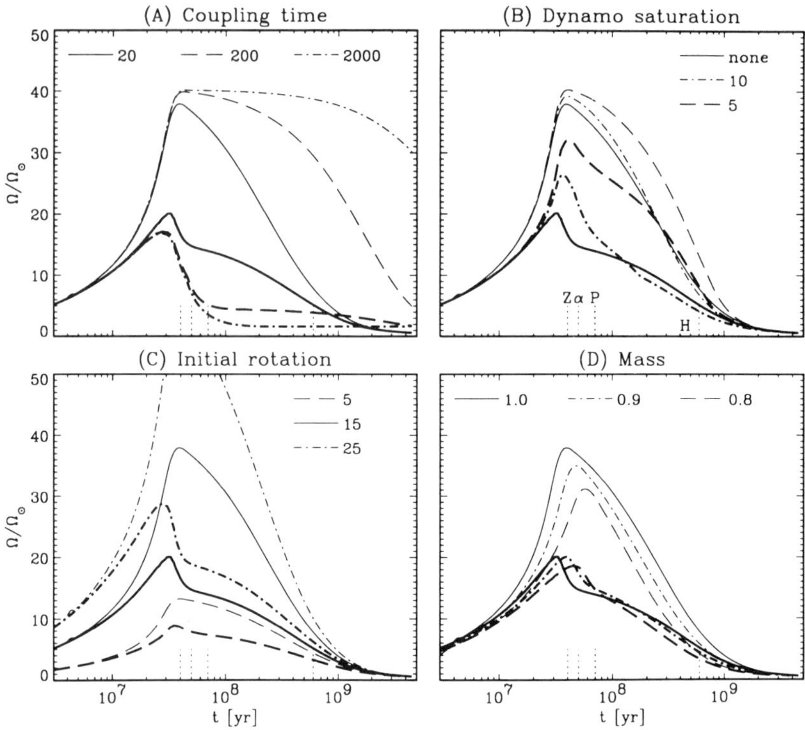

Figure 11.   Rotational evolution of solar-type stars, for various parameter choices and dynamo prescriptions. The thick and thin lines correspond to $\Omega_{\rm CE}(t)$ and $\Omega_{\rm RC}(t)$, respectively. Curves in plot A are labeled according to τ_C, in Myr, and in plot B in terms of the critical rotation rate (in units of $\Omega_\odot$) above which the dynamo saturates. Curves in plot C are labeled according to equatorial rotational velocity (in km s^{-1}) at an age of $\sim$2 Myr, and in plot D according to mass, in solar units.

Using the observed T Tauri rotation rate distribution as initial condition, they find that a coupling time $\tau_C = 10$ Myr combined with dynamo saturation at 20 times the present solar rotation rates reproduces reasonably well the observed evolution of the rapid rotator population in young clusters.

The computations of Keppens et al. also illustrate that the making of a slow α Persei rotator ($\Omega_{\rm CE}/\Omega_\odot \lesssim 5$, say) is an even harder task. Assuming a low initial rotation rate obviously helps to some extent, but the associated reduction of pre-main-sequence angular momentum loss makes it correspondingly more difficult to offset the spin up due to the decreasing moment of inertia; note, in Fig. 11C, how decreasing the initial rotation rates causes the surface rotation to peak at somewhat later times on the main sequence. Slow rotators on the early main sequence can be produced by drastically increasing τ_C (cf., Fig. 11A). Such models, however, retain rapidly rotating cores all the way to the solar age. This is ruled out by helioseismology unless, of course, the weak solar internal differential rotation is the exception rather than the rule.

Another important feature of these solutions is the convergence of rotation rates occurring beyond ~ 1 Gyr. By the solar age, "memory" of the assumed initial condition and dynamo relationship is effectively lost, in terms of both surface rotation rate and internal differential rotation.

Despite the simplistic treatment of internal angular momentum redistribution implicit in the formulation of the two-component model, the various aspects discussed above turn out to be fairly robust properties of rotational evolution models in general, and also carry over to models of different masses (cf., Fig. 11D). The interested reader is referred to MacGregor and Charbonneau (1994) and Keppens et al. (1995) for further discussion.

VI. ROTATIONAL EVOLUTION: THE SUBGIANT AND GIANT BRANCHES

By the time cool stars evolve off the main sequence they have spun down substantially, leaving them with little magnetic activity, and thus with only a weak magnetic brake. Mature stars more massive than about 1.3 $M_\odot$, on the other hand, do not support a solar-like dynamo until they evolve off the main sequence and their effective temperature drops enough to cross the line in the HR diagram redward of which a convective envelope develops (see Figs. 1 and 3). They consequently retain (most of) their main-sequence angular momentum until that phase.

A. A Model for the Rotational History of Evolving Stars

The rotational history of stars that are evolving on the (sub)giant branch can be modeled in a way that is similar to the description given in Sec. V. The rotational history of a rigidly rotating star conserving its angular momentum is described by Eq. (3). Figures 4 and 5 demonstrate that the decrease in the observed rotational velocities is stronger than expected from the change in the moment of inertia in both models. For LC III giants of 2.5 $M_\odot$ $\bar{v}_{eq}(\log(T_{eff}) \approx 3.69)$ is about a factor of 2 lower than expected in the absence of magnetic braking, assuming that the sample is dominated by "first crossers." Unless the rotation rate would increase very strongly with depth, these stars must have lost some 50% of their angular momentum since the onset of convection, because $I_{CE} \gg I_{RC}$. This angular momentum must have been lost in the $\approx 5 \times 10^6$ yr that have passed since the onset of envelope convection.

Continuing initial studies by Endal and Sofia (1981), by Gray and Endal (1982), and by Rutten and Pylyser (1988), Schrijver and Pols (1993) model the angular momentum loss from the convective envelope by the magnetized wind through

$$\frac{dJ_{CE}}{dt} = \alpha \, g(T_{eff}) \left(\frac{R_*}{R_\odot}\right)^n \Omega_{CE}^3 \tag{31}$$

which is consistent with the Skumanich relation $v_{eq}(t) \propto t^{-1/2}$ (cf., Eq. 29) for stars with a constant moment of inertia. For the particular case of Eq. (26) $n = 4$, but the value of n can be varied in the simulations to take into account, for instance, dead zone variations and other effects discussed in Sec. IV.D. The constant of proportionality for Eq. (31) that is consistent with the Skumanich relation $v(t) \propto t^{-1/2}$ is $\alpha = 2.6\,10^{-21} I_0 R_0^2$ (for t in years and other quantities in cgs units). In this expression for α, I_0 and R_0 are the moment of inertia and the radius of the stars for which the transformation from the Skumanich relationship for velocity into the relationship relating angular momentum and torque (such as Eq. [31]) is performed. Schrijver and Pols (1993) (in which the expression for α is erroneously printed, but properly used in the numerical code), evaluated I_0 and R_0 for a G0 ZAMS star.

The description by Eq. (31) also incorporates a function $g(T_{eff})$ that allows for a dependence of the braking efficiency on the phase of the stellar evolution through, for instance, a dependence of the dynamo strength on properties of the convective envelope (as discussed in Sec. III.D).

B. General Evolutionary Considerations

Consider the time scales for magnetic braking implied by Eq. (31) for $n = 4$:

$$\tau_J = \frac{J}{|\dot{J}|} = \frac{I_j R_\odot^4}{\alpha g(T_{eff}) R_*^2 v_{eq}^2} \tag{32}$$

where $I_j = I_*$ if the star rotates rigidly or $I_j = I_{CE}$ if the convective envelope is slowed down separately from the radiative core (note that the factor $R_\odot^4$ was omitted in the expression printed in Schrijver and Pols [1993]). Figure 12 shows that if the brake reaches full strength $g(T_{eff}) = 1$ when the envelope is only four density scale heights deep (a somewhat arbitrary point where Schrijver and Pols (1993) started their computations), τ_J is of the order of a few years for subgiants if angular momentum is drained only from the envelope. For giants τ_J is about 10^3 yr. This time scale is orders of magnitude less than the evolutionary time scale because of the low moment of inertia of such a shallow envelope as compared to that of the entire star. Note that the braking time scales are short compared to the numbers found for young solar-like cluster stars because of (1) the low moment of inertia of the shallow convective envelope, and (2) the strong radius dependence of the magnetic brake.

If the time scale τ_C for the angular-momentum coupling between the interior and the envelope were large compared to the braking time scale, τ_J, derived from Eq. (31) for the model in which envelope and interior are not coupled rigidly, the envelope would be slowed rapidly until the shear between interior and envelope would be large enough for the coupling to compensate for the magnetic braking. The shear depends on the ratio of τ_J and τ_C, as can be seen by making an approximation to Eq. (30a). Assume that the evolutionary time scale is long enough that the term dJ_d/dt may be ignored,

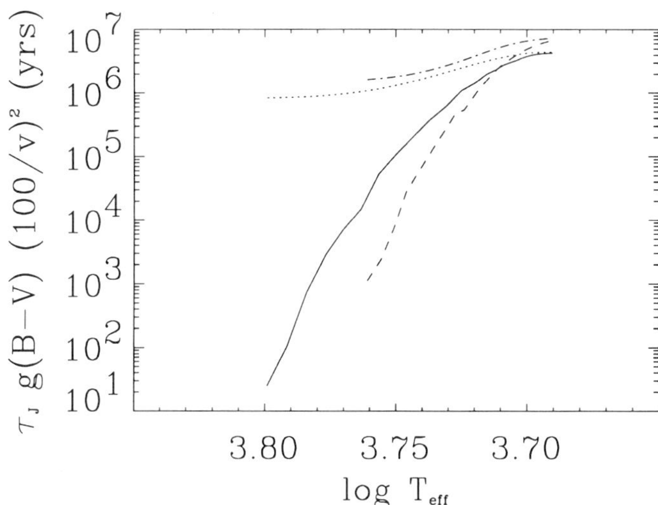

Figure 12.   Time scale for magnetic braking τ_J scaled by $g(T_{\mathrm{eff}})$. $(100/v_{\mathrm{eq}})^2$ (i.e., for a parameterization as in Eq. (32) with $n = 4$, relative to a mean value of $v_{\mathrm{eq}} = 100$ and $g(T_{\mathrm{eff}}) = 1$, for a rigidly rotating convective envelope (LC IV, 1.5 $M_\odot$: solid, LC III, 2.5 $M_\odot$: dashed) and for a rigidly rotating star (LC IV: dotted, LC III: dashed-dotted).

that $I_{\mathrm{CE}} \ll I_{\mathrm{RC}}$, and that $\Omega_{\mathrm{CE}} \ll \Omega_{\mathrm{RC}}$. In that case the angular momentum loss from the interior can be ignored, allowing for a (quasi)stationary situation in which the torques acting on the envelope associated with the magnetic brake and the interior-envelope coupling cancel. This cancellation occurs when

$$\frac{\Omega_{\mathrm{CE}}}{\Omega_{\mathrm{RC}}} = \frac{I_{\mathrm{CE}} R_\odot^n}{\alpha g(T_{\mathrm{eff}}) R_*^n \Omega_{\mathrm{CE}}^2 \tau_{\mathrm{C}}} = \frac{\tau_J}{\tau_{\mathrm{C}}} \tag{33}$$

a posteriori consistent with the assumptions.

A low value of τ_J/τ_{C} would therefore lead to a rapid spin down of the envelope immediately redward of the onset of convection. This is ruled out by the observations summarized in Figs. 4 and 5 for subgiants and giants, respectively, although a rapid spin down may follow later for the giants if the decrease in rotation velocities is as pronounced as advocated by Gray (1991) (cf., Sec. II.A). Hence, either the coupling time scale τ_{C} must not exceed τ_J as computed for $g(T_{\mathrm{eff}}) = 1$, or the function $g(T_{\mathrm{eff}})$ describing the braking efficiency must be strongly reduced, or both. Even with a τ_{C} so small that the star would rotate (nearly) rigidly, a magnetic brake at full strength as given by Eq. (31) with $n = 4$, would slow the subgiants down far too rapidly. This follows from a comparison of the time scales in Fig. 12 and the evolutionary time scales plotted in Figs. 4 and 5. The relative values of the time scales suggests that the braking efficiency should be substantially decreased (by factors up 10^6, depending on τ_{C}) upon onset of convection as compared to the strength in stars with deeper convective envelopes.

The strong decrease in rotational velocities observed for subgiants occurs within a period of roughly 5×10^7 yr, in a phase of the stellar evolution in which the moment of inertia of the envelope I_{CE} is less than 1% of the stellar value I_*. As the stars evolve, the convective envelope deepens, and dredges up angular momentum from the radiative interior. It is therefore possible that the envelope spins up again. Such a hypothetical resurgence of rotation rate is not observed, and so must have been prevented by not only slowing down the envelope, but by also removing much of the interior's angular momentum before I_{CE} begins to dominate I_*. Removing angular momentum from the interior requires an interior-envelope interaction time τ_C that is at most comparable to the time scale for angular momentum loss from the convective envelope (shown in Fig. 12).

C. Results of Simulations

The models by Schrijver and Pols (1993) demonstrate that a magnetic brake with $n = 4$ in Eq. (31) is too weak to drain sufficient angular momentum from rapidly evolving, rigidly rotating giants, even if $g(T_{\mathrm{eff}}) = 1$ from the moment envelope convection sets in. On the other hand, as argued above, $g(T_{\mathrm{eff}})$ needs to be substantially reduced for subgiants just after the onset of convection. If Eq. (31) is to hold for giants and subgiants alike with a comparably steep $g(T_{\mathrm{eff}})$, agreement with observed rotational velocities is found for $n \approx 6$ and a rapidly increasing $g(T_{\mathrm{eff}})$ over an interval $\Delta(B - V)$ of at least 0.2, consistent with the results for main-sequence stars discussed in Sec. III.D. Schrijver and Pols (1993) point out, however, that tradeoffs of n and $g(T_{\mathrm{eff}})$ can be made, that make the description ambiguous.

As mentioned above, the time required by a 2.5 $M_\odot$ giant to cross from the onset of convection to the giant branch is some 5×10^6 yr, so in order to lose sufficient angular momentum from the interior in the two-component model while the surface rotates fast, the coupling time scale τ_C must be shorter than that. In order to prevent a dip in the rotational history, a value close to $n \approx 6$, and a much steeper increase in $g(T_{\mathrm{eff}})$ is required for the two-component model than for the model of rigid rotation, while τ_C should be at most a few Myr. The results do not uniquely determine the steepness of $g(T_{\mathrm{eff}})$ and the value of τ_C; a steeper $g(T_{\mathrm{eff}})$ reduces the braking of the envelope at low values of I_{CE}/I_{RC}, but a lower value of τ_C would yield the same result. If, however, τ_C were drastically reduced, then this would result in rigid rotation; this case has been already discussed above.

If the envelope could decelerate more rapidly than the interior, the rotation rate could in principle increase during some time in the evolution as angular momentum is dredged up from the interior. Avoiding this phenomenon, as required by the data in Figs. 4 and 5, constrains $g(T_{\mathrm{eff}})$ and n (the apparent need to avoid a dredge–up of angular momentum leaves other problems unaddressed, such as why horizontal branch stars rotate relatively rapidly, despite their advanced evolutionary phase [see, e.g., Peterson 1985a,b]). Schrijver and Pols (1993) argue that models can be found that are just consistent with the observational data in which the angular velocity of the interior of sub-giants can be up to four times as large as that of the envelope in some phase of the evolution. If there is a "rotation discontinuity" for giants, as suggested by Gray (1991) (see Sec. II.A and Fig. 4 herein), a strong resurgence can be prevented even if the envelope and the interior are only weakly coupled. The position of such a discontinuity at about log $T_{\mathrm{eff}} \approx 3.723$ then requires: (1) a very weak brake (or dynamo) up to that point; (2) a rapid onset of the magnetic brake (within about 0.01 in log T_{eff}); (3) a very strong brake beyond that (implying a strong radius dependence with n at least equal to about 7); and (4) a core-envelope coupling time not much shorter than a few times 10^5 yr. The value of $\Omega_{\mathrm{RC}}/\Omega_{\mathrm{CE}}$ in these simulations peaks when the envelope extends over a substantial fraction of the stellar radius. Asteroseismology combined with measurement of rotation periods should allow the detection of such shears.

The main conclusions from these simulations are that (1) it is likely that the magnetic braking is strongly reduced at the onset of convection in at least subgiants, and (2) that angular momentum coupling between the envelope and at least some substantial part of the interior must occur at least for subgiants, and for giants also if there is no pronounced rotation discontinuity. Possible causes for a relatively rapid exchange of angular momentum between convective envelope and radiative interior are discussed in the next section.

VII. ROTATIONAL EVOLUTION: DETAILED MODELS

Within the two-component model of rotational evolution used in the preceding two sections, the choice of coupling time incorporates into a single number, the core-envelope coupling time τ_C, all aspects of what was referred to in Sec. II as internal angular momentum redistribution. Results thus obtained for pre-main-sequence, main-sequence and post-main-sequence rotational evolution indicate that the choice of τ_C is often critical in determining the overall behavior of the models. It was also argued in Sec. II.B that many classes of mechanisms can potentially provide that coupling, on time scales shorter than evolutionary. In this section we review recent rotational evolution models that incorporate more elaborate formulations of internal angular momentum distribution based on specific physical mechanisms.

A. Internal Redistribution by Hydrodynamical Mechanisms

One of the most confounding effects of (moderate) rotation in otherwise stably stratified radiative interiors is its potential for destabilization. The number of possible rotationally fed instabilities is indeed enormous, and grows even larger in the presence of angular velocity gradients. For the purpose of the present discussion one may first establish a distinction between *dynamical* and *secular* instabilities (see, e.g., Endal and Sofia 1978). The former are violent instabilities, with very short growth rates. Many of these are discussed extensively in Tassoul (1978, ch. 7). It may be safely assumed that in the quasi-static evolutionary phases considered here, whenever stars become unstable with respect to such instabilities they readjust their structure (in particular their internal rotation profile) almost instantly so as to restore stability. Secular instabilities are of a milder nature, and can manifest themselves as slow, large-scale bulk fluid motion (see, e.g., the Eddington-Sweet-Vogt meridional circulation), and/or small-scale turbulence.

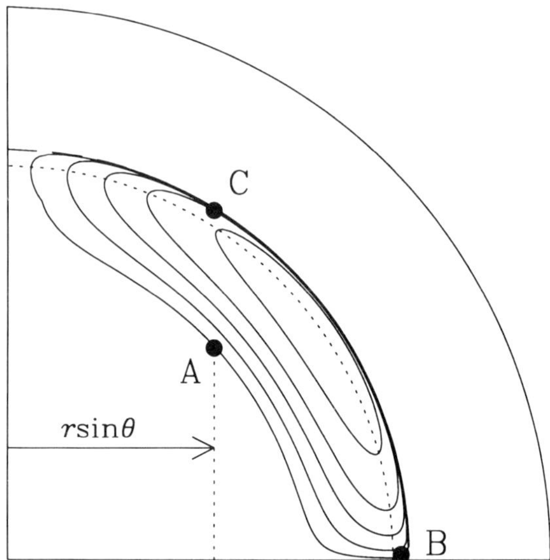

Figure 13.   Schematic representation of Ekman circulation in a meridional quadrant. The core-envelope interface is indicated by a dashed line, and the extent of the Ekman pumping layer (its thickness exaggerated for illustrative purposes) by a dotted line. The strong density stratification is expected to deflect the Ekman circulation a few scale heights below the pumping layer; here this depth has been (arbitrarily) set at 0.5 $R_\odot$. The rotation axis is vertical and coincides with the left quadrant boundary.

One of the earliest angular momentum redistribution mechanisms to be studied in detail is the large-scale circulation associated with shear layers (see, e.g., Howard et al. 1967). This flow is driven by unbalanced viscous stresses in regions of strong velocity gradients (for a general discussion see Pedlosky [1987, Sec. 4.3]). The idea is illustrated schematically in Fig. 13, in a situation where a shear $\partial\Omega/\partial r < 0$ exists within a boundary layer (or "Ekman layer" in the present context) located below the core-envelope interface. Under such circumstances, the flow in the boundary layer is θ-directed and poleward. Mass conservation requires that the underlying fluid be sucked into the layer in the equatorial regions, and ejected downward in the polar region, so that a large-scale flow must exist in the interior. An advected fluid parcel, moving outward from A to B on Fig. 13, spins down so as to conserve angular momentum because $\tilde\omega = r\sin\theta$ is increasing. However, it does not spin up again upon moving back towards the rotation axis (even though $\tilde{\omega}$ is now decreasing from B to C), because it does so within the boundary layer, where viscous stresses keep its rotation rate essentially constant. In this manner, angular momentum is continuously transferred from the core to the envelope; first from the interior to the Ekman layer by the flow, and then from the Ekman layer to the convective envelope by viscous stresses. While no truly satisfactory theory applicable to solar/stellar radiative interiors is available as yet, the aforementioned papers should give the interested reader a good flavor of the complexities and difficulties involved.

In a related context, Spiegel and Zahn (1992) have constructed quasi-steady state models of hydrodynamical shear layers that allow a smooth transition at the core-envelope interface between the states of latitudinal differential rotation in the solar envelope and of solid-body rotation in the upper part of the solar core, as inferred from helioseismology. The large-scale currents they obtain also arise from imbalances in the viscous stresses, but in this case these have a more complex latitudinal dependence than for the simple Ekman layer illustrated schematically in Fig. 13. The shear layer is assumed to be in a highly anisotropic turbulent state, with horizontal turbulent momentum transport largely exceeding vertical transport. The *ad hoc* nature of this assumption notwithstanding, it remains to be seen whether the application of such models to rotational evolution *per se* can satisfy all observational constraints.

Modeling turbulent mass and momentum transport associated with a given instability requires, to begin with, an understanding of the conditions under which that instability begins to manifest itself. For example, a plane parallel, stratified shear flow $\mathbf{u} = (u(z), 0, 0)$, $\rho(z)$ with $\mathbf{g} = -g\hat{\mathbf{e}}_z$ is found to become unstable whenever the Richardson number

$$R_i = -g\frac{(\partial \ln \rho/\partial z)}{(\partial u/\partial z)^2} \tag{34}$$

drops below a certain critical value R_{crit}, equal to $1/4$ in the absence of other stabilizing and/or destabilizing mechanisms. In practice, gradients in mean molecular weight are expected to stabilize the flow, while radiative diffusion is expected to destabilize it (see, e.g., Kippenhahn and Weigert 1990, Sec. 43.2). For conditions characterizing stellar interiors, this makes any determination of R_{crit} from first principles highly uncertain.

Once a given flow configuration is deemed unstable, one faces the additional difficulty of following the nonlinear evolution of the instability and/or constructing appropriate expressions for the transport coefficients associated with the turbulent motions. Perhaps the simplest approach is to retain the functional form of the viscous stress tensor, while replacing the microscopic viscosity ν by an "effective viscosity" ν_{eff} with, typically, $\nu_{\mathrm{eff}}/\nu \gg 1$. Assuming *a priori* that the surface rotation rate obeys the Skumanich relation (cf., Eq. 28), Tassoul and Tassoul (1989) find that even moderately enhanced viscous transport, i.e., $\nu_{\text{eff}}/\nu = 500$ to 1000, can produce by the solar age a level of internal differential rotation compatible with helioseismological data.

A number of more elaborate prescriptions have been proposed for stability criteria and associated transport coefficients in the present context (see, e.g., Zahn 1974; Knobloch and Spruit 1982). These most often rely on dimensional analysis and order-of-magnitude arguments, and so any numerical values thus computed cannot be used directly for quantitative modeling. A more pragmatic approach consists in treating quantities such as R_{crit} as adjustable parameters, to be calibrated by confronting the resulting rotational evolution models with observations. Because many adjustable parameters are usually needed to define a given turbulent transport model, additional observables are required for this strategy to be practical. One such set of secondary observables is the abundances of the light elements lithium, beryllium and boron. These elements are destroyed by proton capture at relatively low temperatures, and so can be used as probes of the extent of turbulent mixing in stellar radiative interiors. The link with angular momentum transport is constructed by *assuming* that mass and angular momentum transport are proportionally related. The quality of the available data (particularly for Li) is such that this procedure can yield constraints on the rotational evolution models themselves.

This approach has become the trademark of at least one school of rotational evolution modeling. Pinsonneault et al. (1989,1990,1991; see also Endal and Sofia 1981; Chaboyer et al. 1995a,b) have computed an extensive set of parametric models taking into consideration a large number of turbulent transport processes fed by rotational instabilities. Using a parametric representation for angular momentum loss due to Kawaler (1988), they follow simultaneously the rotational state of the surface and interior as well as the abundances of Li and Be. A number of model parameters can be constrained by requiring that the present-day Sun rotates at the observed surface rate and depletes Li and Be in the observed proportions. To date these models are markedly more successful at predicting the time-evolution of light element abundances than at reproducing the observed rotation rates for the first few 10^8 yr on the main sequence, or the internal solar rotation profile inferred by helioseismology. Whether this is due to the adopted angular momentum redistribution *Ansatz* or to the specific angular momentum loss parametrization remains unclear at this juncture. Yet this state of affairs is somewhat troubling; it is difficult to imagine how mass transport could be modeled with greater accuracy than angular momentum transport, because, in the framework of these computations, the former is a parallel manifestation of the latter.

B. Internal Redistribution by Magnetic Fields

A number of authors have considered how the internal solar rotation might be affected by the presence of a large-scale magnetic field within the radiative portion of the Sun's interior. Consider an axisymmetric system, i.e., one in which the rotation and magnetic axes coincide. An externally imposed shear

$\nabla\Omega$ acting on a pre-existing poloidal magnetic field $\mathbf{B}_p$ induces a toroidal component, as described by the ϕ-component of the induction equation (with $\eta \to 0$):

$$\frac{\partial B_\phi}{\partial t} = \varpi\,\mathbf{B}_p \cdot \nabla\Omega. \tag{35}$$

For time-independent shear and poloidal field, Eq. (35) yields a growth of B_ϕ that is linear in time, which gives rise to a growing azimuthally directed Lorentz force component

$$[(\nabla \times \mathbf{B}) \times \mathbf{B}]_\phi \propto \varpi^{-1}\mathbf{B}_p \cdot \nabla(\varpi\,B_\phi). \tag{36}$$

It is easy to verify that this Lorentz force acts in such a way as to oppose the driving shear. In the absence of dissipation and external forcing, the only steady states allowed by Eq. (35) must satisfy

$$\mathbf{B}_p \cdot \nabla\Omega = 0. \tag{37}$$

Equation (37) constrains Ω to be constant *along* individual poloidal magnetic field lines, but does not rule out angular velocity differences *across* field lines. A configuration satisfying Eq. (37) is termed isorotational. The time scale τ_B (cf., Eq. 5c) obtained previously is expected to characterize the evolution towards an isorotational state, but evidently this can be quite distinct from a state of strict solid-body rotation, i.e., Ω constant along *and* across magnetic field lines. In the context of the solar spin down, the rotational state of the interior is thus likely to depend rather sensitively on the relative time scales for: (1) angular momentum redistribution along field lines; (2) coupling across field lines; and (3) external forcing by the wind-mediated torque. Hence order-of-magnitude arguments for the existence of a state of strict solid-body rotation in solar and stellar interiors based exclusively on estimates of the time scale τ_B should be considered cautiously (see also Charbonneau and MacGregor 1992).

Charbonneau and MacGregor (1993) have computed a large set of rotational evolution models based on the Weber-Davis estimate for angular momentum loss and internal angular momentum redistribution by large-scale axisymmetric magnetic fields. Their numerical computations take into account both the generation of a toroidal magnetic field via shearing of the pre-existing poloidal component, as well as the back reaction of the associated azimuthal Lorentz force on the driving shear. After a short ($\sim 10^6$ yr for $|\mathbf{B}_p| = 1$ G) period of internal field buildup and torsional Alfvén wave damping via phase mixing, the interior settles in a quasi-steady state of near dynamical balance, whereby the total magnetic and viscous stress along the spherical surface formed by the core-envelope shell interface nearly equilibrates the externally applied torque due to the magnetized wind:

$$r_{\mathrm{CE}}^3 \int\int B_r B_\phi \sin\theta \, \mathrm{d}\theta \mathrm{d}\phi \sim \frac{\mathrm{d}J}{\mathrm{d}t}. \tag{38}$$

This state of dynamical balance has important consequences for the rotational evolution of the system. Because the toroidal field tends to grow until Eq. (38) is approximately satisfied independently of the magnitude of B_r, the surface rotational evolution curves show little (if any) dependence on the internal poloidal field strength. This suggests that the present solar rotation rate is a poor measuring stick with which to estimate the strength of any large-scale magnetic field in the solar radiative interior. The rotational evolution sequences computed to date are limited to the time interval from the ZAMS to the solar age; as such, they cannot be used to attempt detailed modeling of the observed rotation rates in young clusters because the assumed initial condition on the ZAMS (internal solid-body rotation and zero toroidal field) is most likely unrealistic. But they do indicate that there exist classes of poloidal field configurations that allow both rapid spin down near the ZAMS *and* weak internal differential rotation by the solar age.

In the preceding computations, the growth of the toroidal magnetic field is limited dynamically by the associated Lorentz force, as described by Eq. (38). There are other mechanisms that can limit the growth of magnetic fields, however. Flux removal by magnetic buoyancy has been widely discussed in the context of solar and stellar dynamo models, but this process operates most efficiently in the upper parts of the radiative core and only for magnetic fields of near equipartition strength. Enhanced dissipation associated with global, dynamical instabilities of large-scale magnetic fields (see Pitts and Tayler 1985, and references therein; Balbus and Hawley 1994) may also limit the growth of B_ϕ long before Eq. (38) is actually realized, but such effects are difficult to quantify. Mestel and Weiss (1987) have argued that once large-scale instabilities set in, the effective magnetic diffusivity increases until the system settles in a quasi-steady state characterized by a magnetic Reynolds number of order 1. On this basis, they suggest that the buildup of the toroidal field from shearing of the poloidal component should proceed until

$$B_\phi^2 \simeq \left(\frac{\varpi}{D}\right)\left(\frac{\varpi\,\Delta\Omega}{u_A}\right)|\mathbf{B}_p|^2 \qquad (39)$$

where $\Delta\Omega$ is the angular velocity difference between the core and envelope, D the length scale over which this shear is acting, and u_A the Alfvén speed based on the poloidal field (compare this expression to Eq. 38). Using the two-component formulation described in Sec. V.B, Li and Cameron (1993) have constructed a set of rotational evolution models incorporating a two-parameter prescription for the coupling time τ_C based on Eq. (39). Only their "weak coupling" model subset is apparently capable of providing a reasonable fit to the observed $v \sin i$ distribution of young clusters. Its application to the solar case, however, leads to a rather awkward conclusion: in order for the present internal solar differential rotation to lie within the bounds set by helioseismology, the rotation rate of the young Sun cannot have exceeded its present value by more than 25%. While this cannot be strictly ruled out, it appears very difficult to reconcile with available observational data (cf., Sec. I).

Not surprisingly, no serious attempts have been made at modeling internal angular momentum redistribution in the presence of turbulence *and* magnetic fields. Neither the effects of magnetic fields on the onset and growth of turbulence, nor the effect of turbulence on the growth and decay of magnetic fields, can be reliably estimated at this juncture, even in an order-of-magnitude sense. This appears to be an area where any progress is likely to arise from insight gained from the study of large-scale numerical simulations.

Whether hydro- or magnetohydrodynamical effects are invoked, the behavior of the coupling mechanism is often such that the net angular momentum transport from the core to the envelope tends to equilibrate the externally applied torque. To a large extent, internal angular momentum redistribution (and any associated mass transport) is effectively regulated by the wind-mediated torque. It is thus appropriate to bring this section to a close by stressing once more that an accurate determination of the angular momentum loss rate due to magnetized winds, for realistic magnetic field geometries and coronal conditions, remains a truly central aspect of angular momentum evolution modeling.

VIII. CONCLUSIONS

Rotational evolution has come a long way in recent years; yet from the modeling standpoint the situation is still far from satisfactory. One important weakness of current models is the relatively simplistic nature of the MHD wind models currently available to compute angular momentum loss rates. This is a well-defined and tractable problem that merits immediate attention. Improvements in dynamo models are also much needed, as this currently represents, from the theoretical standpoint at least, one of the weakest links in the causal chain of events associated with angular momentum evolution. It is, however, unlikely that all of the failures of current rotational evolution models lie along those lines.

The models presented in Sec. V highlighted the difficulty of producing either very rapid ($\Omega/\Omega_\odot \gtrsim 50$) or very slow ($\Omega/\Omega_\odot \lesssim 5$) rotators. The fact that even the very young α Persei cluster contains a large population of slow rotators suggests looking toward the pre-main sequence for an explanation. In recent years a number of authors have proposed scenarios whereby coupling (magnetic or otherwise) between a pre-main-sequence star and a circumstellar accretion disk regulates the star's rotation rate and prevents, to some extent, pre-main-sequence spin up associated with the decreasing stellar moment of inertia (Königl 1990; Bouvier 1993; Cameron and Campbell 1993). Observations do suggest that T Tauri stars showing evidence for the presence of accretion disks do not spin up significantly as they approach the main sequence (see, e.g., Edwards et al. 1993). At this writing, truly quantitative models of star-disk coupling are yet to be developed, but this appears to be a promising avenue toward the solution of the slow rotator problem (see Keppens et al. 1995). Nevertheless, the formation of planetary systems should also be kept in mind as a viable alternative for disposing of a large fraction of a star's initial angular momentum.

Solutions to the difficulties encountered in Sec. V may lie, at least in part, with internal angular momentum redistribution. This, interestingly, can lead far beyond the immediate task of modeling the observed internal evolution of surface rotation rates. By confronting results of detailed models to observations, rotational evolution modeling has the potential to open a new window into stellar interiors, in the sense of establishing probable values for quantities such as internal magnetic fields, or bounds on the efficiency of turbulent momentum and mass transport. While this approach is still in its infancy and currently does not have the probing power of, e.g., helioseismology, the work reviewed in Sec. VII suggests that valuable information concerning the (magneto)hydrodynamical state of the solar interior can indeed be obtained in this manner.

Similar comments apply to dynamo theory; admittedly, the problems occupying solar dynamo modelers are in most instances rather far removed from rotational evolution considerations. Nevertheless, the various manifestations of solar-like magnetic activity in stars other than the Sun are also in dire need of a quantitative explanation. This task is not only interesting in and of itself, but offers the means of obtaining information concerning the large-scale, open components of stellar magnetic fields that are largely invisible in spectroscopic chromospheric and coronal activity tracers. For example, the apparent need to invoke some form of dynamo saturation in order to produce rapidly rotating Pleiades, if unambiguously demonstrated, holds clues to dynamo efficiency and/or relative coverage of magnetically closed vs open regions, at high rotation rates.

The difficulties involved in modeling the rotational evolution of single stars carry over to the components of binary systems, with some additional (and fascinating) complexities. Evolutionary effects take on a new dimension in close binaries systems, with the possibility of mass exchange between the components. The global structure of a star's wind can be significantly affected by the presence of a nearby magnetized companion, with important consequences for the angular momentum loss rate of the system as a whole (see, e.g., Li et al. 1994*a*). Perhaps the more far-reaching difference, as compared to single stars, is the conversion of orbital angular momentum to spin angular momentum via tidal friction and/or tidally induced internal flows. This means that as the orbital separation decreases, the components of close binaries can spin up even if subjected to strong wind-mediated angular momentum loss, leading to enhanced activity levels. Some of the better studied classes of late-type binary systems (such as the RS CVn binaries) are becoming popular targets for quantitative modeling of magnetic activity and rotational evolution (see, e.g., Li et al. 1994*a,b*)

Many of the recent advances in rotational evolution modeling detailed in the preceding sections have been made possible, to a significant extent, by the availability of increasingly powerful computers and numerical techniques. This is an ongoing trend that will continue to open new territory for quantitative exploration. Consider for example the angular momentum loss models used to date for rotational evolution modeling; essentially all such models are based on steady-state wind solutions/approximations, yet evidence is accumulating for the ubiquitous existence of transient, episodic mass loss events in young, rapidly rotating stars. A number of authors (cf., Cameron and Robinson 1989; Jeffries 1993, and references therein) have argued that the observed spectral variability in Hα can be interpreted in terms of extended clouds of cool gas being magnetocentrifugally maintained in a state of approximate corotation out to a few stellar radii above the photosphere. As the (cool) matter comprising these clouds ultimately leaks out of the confining magnetic field, it carries away considerable specific angular momentum; the very simple model of Cameron and Robinson (1989) yields braking time scales in the range of 10^7 to 10^8 yr. The Sun once again provides a useful paradigm, in a somewhat different yet related context. Coronal mass ejections (CMEs; see the chapter by Hundhausen) contribute a few 10^{14} to a few 10^{16} g, once a day on average over the solar cycle, to the Sun's mass output. This adds up to only $\sim 10^{-15} M_\odot$ yr^{-1}, only a few percent of the mass loss associated with the quasi-steady component of the solar wind. It was noted in Sec. IV that for solar-like coronal conditions, the mass loss predicted by the WD model is hardly enhanced as the rotation rate and surface magnetic field strength are increased. It was also noted in Sec. III that the various spectroscopic diagnostics of magnetic activity suggest a linear increase of activity with rotation rate, at least up to $\Omega/\Omega_\odot \sim 10$. If the yield and/or frequency of CMEs increases correspondingly, one may envision that there may have been a time in the Sun's past when the contribution from transient events was comparable to that from quasi-steady wind-like outflows, making the models presented in Sec. IV of limited applicability. This, interestingly, takes us back to Schatzman's original suggestion, albeit in a much sharper focus.

Acknowledgments. The wind solutions shown in Figs. 6 and 7, as well as the rotational evolutions solutions depicted in Fig. 10 and 11, were computed by R. Keppens as part of ongoing work on rotational evolution modeling. We are indebted to D. Vandenberg for providing evolutionary tracks, and D. Soderblom for supplying in electronic form the data shown in Fig. 2. We wish to thank P. Hoyng, R. Keppens, M. Pinsonneault, D. Soderblom, J. Stauffer and K. Zwaan for constructive comments on the first draft of this chapter. CJS acknowledges a fellowship by the Royal Netherlands Academy of Arts and Sciences. The National Center for Atmospheric Research is sponsored by the National Science Foundation.

REFERENCES

Abney, W. de W. 1877. Effect of a star's rotation on its spectrum. *Mon. Not. Roy. Astron. Soc.* 37:278–279.

Balbus, S. A., and Hawley, J. F. 1994. The stability of differentially rotating weakly magnetized stellar radiative zones. *Mon. Not. Roy. Astron. Soc.* 266:769–774.

Basri, G. 1987. Stellar activity in synchronized binaries. II. A correlation analysis with single stars. *Astrophys. J.* 316:377–388.

Beasley, A. J., and Cram, L. E. 1993. Magnetic acceleration of winds from solar type stars. *Astrophys. J.* 417:157–169.

Belcher, J. W., and MacGregor, K. B. 1976. Chromospheric activity in galactic open clusters. *Astrophys. J.* 210:498–507.

Bodenheimer, P. 1991. Angular momentum effects in star formations. In *Angular Momentum Evolution of Young Stars*, eds. S. Catalano and J. R. Stauffer (Dordrecht: Kluwer), pp. 1–20.

Bouvier, J. 1993. The rotational evolution of low-mass pre-main squence stars. In *Eighth Cambridge Workshop on Cool Stars, Stellar Systems, and the Sun*, ed. J.-P. Caillault (San Francisco: Astron. Soc. of the Pacific), pp. 151–162.

Cameron, A. C., and Campbell, C. G. 1993. Rotational evolution of magnetic T Tauri stars with accretion disks. *Astron. Astrophys.* 274:309–318.

Cameron, A. C., and Robinson, R. D. 1989. Fast Hα variations on a rapidly rotating cool main sequence star. I. Circumstellar clouds. *Mon. Not. Roy. Astron. Soc.* 236:57–87.

Cameron, A. C., Li, J., and Mestel, L. 1991. Theory of magnetic braking of late-type stars. In *Angular Momentum Evolution of Young Stars*, eds. S. Catalano and J. R. Stauffer (Dordrecht: Kluwer), pp. 297–314.

Cattaneo, F., and Vainshtein, S. I. 1991. Suppression of turbulent transport by a wak magnetic field. *Astrophys. J. Lett.* 376:21–24.

Chaboyer, B., Pinsonneault, M. H., and Demarque, P. 1995a. Stellar models with microscopic diffusion and rotational mixing. I. Application to the Sun. *Astrophys. J.* 441:865–875.

Chaboyer, B., Pinsonneault, M. H., and Demarque, P. 1995b. Stellar models with microscopic diffusion and rotational mixing. II. Application to open clusters. *Astrophys. J.* 441:876–885.

Charbonneau P., and MacGregor K. B. 1992. Angular momentum transport in magnetized stellar radiative zones. I. Numerical solutions to the core spin-up model problem. *Astrophys. J.* 387:639–661.

Charbonneau P., and MacGregor K. B. 1993. Angular momentum transport in magnetized stellar radiative zones. II. The solar spin-down. *Astrophys. J.* 417:762–780.

Durney, B. R. 1972. Comments. In *Solar Wind*, eds. C. P. Sonnet, P. J. Coleman and L. M. Wilcox, NASA SP-308, pp. 282–286.

Durney, B. R., and Latour, J. 1978. On the angular momentum loss of late-type stars. *Geophys. Astrophys. Fluid Dyn.* 9:241–255.

Durney, B. R., and Robinson, R. D. 1982. On an estimate of the dynamo-generate magnetic field in late-type stars. *Astrophys. J.* 253:290–297.

Edwards, S., Strom, S. E., Hartigan, P., Strom, K. M., Herbst, W., Attridge, J., Merril, K. M., Probst, R., and Gatley, I. 1993. Angular momentum regulation in low-mass young stars surrounded by accretion disks. *Astron. J.* 106:372–382.

Elvey, C. T. 1929. The contours of helium lines in stellar spectra. *Astrophys. J.* 70:141–159.

Elvey, C. T. 1930. The rotation of stars and the contours of Mg$^+$4481. *Astrophys. J.* 71:221–230.

Endal, A. S., and Sofia, S. 1978. The evolution of rotationg stars. II. Calculations with time-dependent redistribution of angular momentum for 7 and 10 $M_\odot$ stars. *Astrophys. J.* 220:279–290.

Endal, A. S., and Sofia, S. 1981. The evolution of rotating stars. I. Method and exploratory calculations of a 7M star. *Astrophys. J.* 210:184–198.

Galilei, G. 1613. History and demonstrations concerning sunspots and their phenomena. In *Discoveries ans Opinions of Galileo*, trans. ed. S. Drake (New York: Doubleday, 1957), p. 87.

Gilman P. A.. 1980. Differential rotation in stars with convective zones. In *Stellar Turbulence*, eds. D. F. Gray and J. L. Linsky (Berlin: Springer), pp. 19–38.

Gray, D. F. 1989. The rotational break for G girants. *Astrophys. J.* 347:1021–1029.

Gray, D. F. 1991. Rotation of evolved stars. In *Angular Momentum Evolution of Young Stars*, eds. S. Catalano and J. R. Stauffer (Dordrecht: Kluwer), pp. 183–199.

Gray, D. F. 1992. *The Observation and Analysis of Stellar Photospheres*, 2nd ed. (Cambridge: Cambridge Univ. Press).

Gray, D. F., and Endal, A. S. 1982. The angular momentum history of the Hyades K giants. *Astrophys. J.* 254:162–167.

Gray, D. F., and Nagar, P. 1985. The rotational discontinuity shown by luminosity Class IV stars. *Astrophys. J.* 298:756–760.

Gray, D. F., and Nagel, T. 1989. The granulation boundary in the H-R diagram. *Astrophys. J.* 341:421–426.

Hartmann, L. 1987. Stellar magnetic fields: Optical observations and analysis. In *Cool Stars, Stellar Systems, and the Sun*, eds. J. L. Linsky and R. E. Stencel (Berlin: Springer) pp. 1–9.

Hoffleit, D. 1982. *The Bright Star Catalogue*, 4th rev. ed. (New Haven, Conn.: Yale Univ. Observatory).

Howard, L. N., Moore, D. W., and Spiegel, E. A. 1967. *Nature* 214:1297.

Jeffries, R. D. 1993. Prominence activity on the rapidly rotating field star HC 197890. *Mon. Not. Roy. Astron. Soc.* 262:369–376.

Kawaler, S. D. 1987. Angular momentum in stars: The Kraft curve revisited. *Publ. Astron. Soc. Pacific* 99:1322–1328.

Kawaler, S. D. 1988. Angular momentum loss in low-mass stars. *Astrophys. J.* 333:236–247.

Keppens, R., MacGregor, K. B., and Charbonneau, P. 1995. On the evolution of rotational velocity distributions for solar-type stars. *Astron. Astrophys.* 294:469–487.

Kippenhahn, R., and Weigert, A. 1990. *Stellar Structure and Evolution* (Berlin: Springer-Verlag).

Königl, A. 1990. Disk accretion onto magnetic T Tauri stars. *Astrophys. J. Lett.* 370:39–43.

Knobloch, E., and Spruit, H. C. 1982. Stability of differential rotation in stars. *Astron. Astrophys.* 113:261–268.

Kraft, R. P. 1967*a*. Studies of stellar rotation. IV. A comparison of rotational velocities in the Alpha Persei cluster and the Pleiades. *Astrophys. J.* 148:129–139.

Kraft, R. P. 1967*b*. Studies of stellar rotation. V. The dependence of rotation on age among solar-type stars. *Astrophys. J.* 150:551–569.

Kraft, R. P. 1970. Stellar rotation. In *Spectroscopic Astrophysics*, ed. G. H. Herbig (Berkeley: Univ. of California Press), p. 385.

Krause, F., Rädler K.-H., and Rüdiger G., eds. 1993. *The Cosmic Dynamo* (Dordrecht: Kluwer).

Li, J., and Cameron, A. C. 1993. Rotational evolution of solar-type stars with core-envelope decoupling. *Mon. Not. Roy. Astron. Soc.* 261:766–782.

Li, J., Wu, K., and Wickramasinghe, D. T. 1994*a*. Reduced magnetic braking in synchronously rotating magnetic cataclysmic variables. *Mon. Not. Roy. Astron. Soc.* 268:61–68.

Li, J., Wu, K., and Wickramasinghe, D. T. 1994*b*. On the rate of orbital angular momentum loss of synchronously rotating magnetic cataclysmic variables. *Mon. Not. Roy. Astron. Soc.* 270:769–773.

MacGregor, K. B., and Brenner, M. 1991. Rotational evolution of solar-type stars. I. Main-sequence evolution. *Astrophys. J.* 376:204–213.

MacGregor, K. B., and Charbonneau, P. 1994. Angular momentum evolution of late-type stars: A theoretical perspective. In *Eighth Cambridge Workshop on Cool Stars, Stellar Systems, and the Sun*, ed. J.-P. Caillault (San Francisco: Astron. Soc. of the Pacific), pp. 174–186.

Maeder, A., and Meynet G. 1989. Grids of evolutional models from 0.85 to 120 $M_\odot$. Observational tests and the mass limits. *Astron. Astrophys.* 210:155–173.

Mestel, L. 1968. Magnetic braking by a stellar wind. *Mon. Not. Roy. Astron. Soc.* 138:359–391.

Mestel, L., and Spruit, H. C. 1987. On magnetic braking of late-type stars. *Mon. Not. Roy. Astron. Soc.* 226:57–66.

Mestel, L., and Weiss, N. O. 1987. Magnetic fields and non-uniform rotation in stellar radiative zones. *Mon. Not. Roy. Astron. Soc.* 226:123–135.

Mouschovias, T. Ch. 1991. Single-stage fragmentation and a modern theory of star formation. In *The Physics of Star Formation and Early Stellar Evolution*, eds. C. J. Lada and N. D. Kylafis (Dordrecht: Kluwer), pp. 449–468.

Nerney, S. F., and Suess, S. T. 1975. Restricted three-dimensional stellar wind modeling. I. Polytropic case. *Astrophys. J.* 196:837–847.

Noyes, R. W., Hartmann, L., Baliunas, S. L., Duncan, D. K., and Vaughan, A. H. 1984. Rotation, convection and magnetic activity in lower main-sequence stars. *Astrophys. J.* 279:763–777.

Parker, E. N. 1955. Hydrodynamic dynamo models. *Astrophys. J.* 122:293–314.

Parker, E. N. 1963. *Interplanetary Dynamical Processes* (New York: J. Wiley).

Parker, E. N. 1981. Residual fields from extinct dynamos. *Geophys. Astrophys. Fluid Dyn.* 18:175–195.

Pedlosky, J. 1987. *Geophysical Fluid Dynamics*, 2nd ed. (Berlin: Springer-Verlag).

Peterson, R. C. 1985*a*. The rotation of horizontal-branch stars. III. Members of the globular cluster M4[1]. *Astrophys. J.* 289:320–325.

Peterson, R. C. 1985*b*. The rotation of horizontal-branch stars. IV. Members of the globular cluster NGC 288[1]. *Astrophys. J. Lett.* 294:35–37.

Pinsonneault, M. H., Kawaler, S. D., Sofia, S., and Demarque, P. 1989. Evolutionary models of the rotating Sun. *Astrophys. J.* 338:424–452.

Pinsonneault, M. H., Kawaler, S. D., and Demarque, P. 1990. Rotation of low-mass stars: A new probe of stellar evolution. *Astrophys. J. Suppl.* 74:501–550.

Pinsonneault, M. H., Deliyannis, C. P., and Demarque, P. 1991. Evolutional models of halo stars with rotation. I. Evidence for differential rotation with depth in stars. *Astrophys. J.* 367:239–252.

Pitts, E., and Tayler, R. J. 1985. The adiabatic stability of stars containing magnetic fields. VI. The influences of rotation. *Mon. Not. Roy. Astron. Soc.* 216:139–154.

Pizzo, V., Schwenn, R., Marsch, E., Rosenbauer, H., Mülhäuser, K.-H., and Neubauer, F. M. 1983. Determination of the solar wind angular momentum flux from the HELIOS data—an observational test of the Weber and Davis theory. *Astrophys. J.* 271:335–354.

Prosser, C. F. 1994. Photometry and spectroscopy in the open cluster α Persei. II. *Astron. J.* 107:1422–1432.

Rosner, R., and Weiss, N. O. 1992. The origin of the solar cycle. In *The Solar Cycle*, ed. K.L. Harvey (San Francisco: Astron. Soc. of the Pacific), pp. 511–531.

Rüdiger, G. 1989. *Differential Rotation and Stellar Convection* (New York: Gordon and Breach).

Rutten, R. G. M., and Pylyser, E. 1988. Magnetic structure in cool stars. XV. The evolution of rotation rates and chromospheric activity of giants. *Astron. Astrophys.* 191:227–236.

Rutten, R. G. M., and Schrijver, C. J. 1987. Magnetic structure in cool stars. XIII. Appropriate units for the rotation-activity relation. *Astron. Astrophys.* 177:155–162.

Rutten, R. G. M., Schrijver, C. J., Lemmens, A. F. P., and Zwaan, C. 1991. Magnetic structure in cool stars. XVII. Minimum radiative losses from the outer atmosphere. *Astron. Astrophys.* 252:203–219.

Saar, S. H. 1991. Recent advances in the observation and analysis of stellar magnetic fields. In *The Sun and Cool Stars: Activity, Magnetism and Dynamos*, eds. I. Tuominen, D. Moss and G. Rüdiger (Berlin: Springer-Verlag), p. 389.

Sakurai, T. 1985. Magnetic stellar winds: A 2-D generalization of the Weber-Davis model. *Mon. Not. Roy. Astron. Soc.* 152:121–129.

Schatzman, E. 1959. Sur la perte de masse et les processus de freinage de la rotation. In *The Hertzsprung-Russell Diagram*, ed. J. L. Greestein, p. 129.

Schatzman, E. 1962. A theory of the role or magnetic activity during star formation. *Ann. Astrophys.* 25:18–29.

Schlesinger, F. 1909. The Algol variable σ Librae. *Publ. Allegheny Obs.* 1:123–134.

Schrijver, C. J. 1991. Rotational Velocities of low mass stars in young clusters. In *Mechanisms of Chromospheric and Coronal Heating*, eds. P. Ulmschneider, E. R. Priest and R. Rosner (Berlin: Springer-Verlag), pp. 257–272.

Schrijver, C. J. 1992. The basal and strong-field component of the solar atmosphere. *Astron. Astrophys.* 258:507–520.

Schrijver, C. J. 1993a. Observational constraints on dynamos in cool stars. In *Inside the Stars*, eds. W. W. Weiss and A. Baglin (San Francisco: Astron. Soc. of the Pacific), pp. 591–604.

Schrijver, C. J. 1993b. Magnetic activity in dwarf stars with shallow convective envelopes. *Astron. Astrophys.* 269:446–456.

Schrijver, C. J., and Harvey, K. L. 1989. The distribution of solar magnetic fluxes and the nonlinearity of stellar flux-flux relations. *Astrophys. J.* 343:481–488.

Schrijver, C. J., and Pols, O. R. 1993. Rotation, magnetic braking and dynamos in cool giants and subgiants. *Astron. Astrophys.* 278:51–67.

Schrijver, C. J., and Zwaan, C. 1991. Activity in tidally interacting binaries. *Astron. Astrophys.* 251:183–198.

Schrijver, C. J., Coté, J., Zwaan, C., and Saar, S. H. 1989. Reoations between the photospheric magnetic field and the emission from the outer atmospheres of cool stars. I. The solar CaII K line core emission. *Astrophys. J.* 337:964–976.

Schüssler, M. 1975. Axisymmetric χ^2-dynamos in the Hayash-phase. *Astron. Astrophys.* 38:263–270.

Simon, T., and Drake, S. A. 1989. The evolution of chrmospheric activity of cool giant and subgiant stars. *Astrophys. J.* 346:303–329.

Skumanich, A. 1972. Time scales for CaII emission decay, rotational braking, and lithium depletion. *Astrophys. J.* 171:565–567.

Skumanich, A., Smythe, C., and Frazier, E. N. 1975. On the statistical description of inhomogeneities in the quiet solar atmosphere. I. Linear regression analysis and absolute calibration of multichannel observations of the Ca+ emission network. *Astrophys. J.* 200:747–764.

Soderblom, D. R. 1985. A survey of chromospheric emission and rotation among solar-type stars in the solar neighborhood. *Astron. J.* 90:2103–2115.

Soderblom, D. R. 1991. Main sequence angular momentum loss in low-mass stars. In

Angular Momentum Evolution of Young Stars, eds. S. Catalano and J. R. Stauffer (Dordrecht: Kluwer), pp. 151–165.

Spiegel, E. E., and Zahn, J.-P. 1992. The solar tachocline. *Astron. Astrophys.* 265:106–114.

Stauffer, J. R. 1991. Rotational velocities of low-mass stars in young clusters. In *Angular Momentum Evolution of Young Stars*, eds. S. Catalano and J. R. Stauffer (Dordrecht: Kluwer), pp. 117–134.

Stauffer, J. R. 1994. The angular momentum evolution of young main sequence stars. In *Eighth Cambridge Workshop on Cool Stars, Stellar Systems, and the Sun*, ed. J.-P. Caillault (San Francisco: Astron. Soc. of the Pacific), pp. 163–173.

Stauffer, J. R., and Hartmann, L. W. 1987. The distribution of rotational velocities of low mass stars in the Pleiades. *Astrophys. J.* 318:337–355.

Steenbeck, M., and Krause, F. 1969. Zur Dynamotheorie stellarer und planetarer Magnetfelder. I. Berechnung sonnenahnlicher wechselfeldgeneratoren. *Astron. Nachr.* 291:49.

Struve, O. 1930. On the axial rotation of stars. *Astrophys. J.* 72:1–17.

Struve, O. 1945. The cosmogonical significance of stellar rotation. *Popular Astron.* 53:201–218.

Suess, S. T., and Nerney, S. F. 1973. Meridional flow and the validity of the two-dimensional approximation in stellar-wind modeling. *Astrophys. J.* 184:17–25.

Tassoul, J.-L. 1978. *Theory of Rotating Stars* (Princeton: Princeton Univ. Press).

Tassoul, J.-L., and Tassoul, M. 1982. Meridional circulation in rotating stars. I. A boundary layer analysis of mean steady motions in early-type stars. *Astrophys. J. Suppl.* 49:317–350.

Tassoul, M., and Tassoul, J.-L. 1984. Meridional circulation in rotationg stars. VIII. The solar spin-down problem. *Astrophys. J.* 286:350–358.

Tassoul, J.-L., and Tassoul, M. 1989. The internal rotation of the Sun. *Astron. Astrophys.* 213:397–401.

Vilhu, O. 1984. The nature of magnetic activity in lower main sequence stars. *Astron. Astrophys.* 133:117–136.

Vilhu, O., and Rucinsky S. M. 1983. Period-activity relations in close binaries. *Astron. Astrophys.* 127:5–14.

Vogel, H. 1871. Resultate spectralanalytischer Beobachtugnen, Angestellt auf der Sternwarte zu Bothkamp. *Astron. Nachr.* 78:241–252.

Washimi, H., and Sakurai, T. 1993. A simulation study of the solar wind including the solar rotation effect. *Solar Phys.* 143:173–186.

Weber, E. J., and Davis, L., Jr. 1967. The angular momentum of the solar wind. *Astrophys. J.* 148:217–227.

Westgate, C. 1934. A statistical study of rotational broadening in 112 stars of class F. *Astrophys. J.* 79:357–360.

Wilson, O. C. 1966. Stellar convection zones, chromospheres, and rotation. *Astrophys. J.* 144:695–708.

Zahn, J.-P. 1974. Rotational instabilities and stellar evolution. In *Stellar Instabilities and Evolution*, eds. P. Ledoux, A. Noels and A. W. Rodgers (Dordrect: D. Reidel), pp. 185–195.

PART V
Interactions With Surrounding Media

THE INTERSTELLAR CLOUD SURROUNDING THE SOLAR SYSTEM

P. C. FRISCH
The University of Chicago

Ultraviolet spectral data of nearby stars indicate that the cloud surrounding the solar system has an average neutral density $<n(\mathrm{H\,I})>\sim 0.1$ cm^{-3}, temperature ~ 6800 K, and turbulence $\xi \sim 1.7$ km s^{-1}. Comparisons between the anomalous cosmic ray data and ultraviolet data suggest that the electron density is in the range $n(\mathrm{e}^-)\sim 0.22$ to 0.44 cm^{-3}. This cloud is flowing past the Sun from a position centered in the Norma-Lupus region. The cloud properties are consistent with interstellar gas which originated as material evaporated from the surfaces of embedded clouds in the Scorpius-Centaurus Association, and which was then displaced towards the Sun by a supernova event about 400,000 years ago. The Sun and surrounding cloud velocities are nearly perpendicular in space, and this cloud is sweeping past the Sun. The morphology of this cloud can be reconstructed by assuming that the cloud moves in a direction parallel to the surface normal. With this assumption, the Sun entered the surrounding cloud 2000 to 8000 years ago, and is now about 0.05 to 0.16 pc from the cloud surface. Prior to its recent entry into the surrounding cloud complex, the Sun was embedded in a region of space with average density lower than 0.0002 cm^{-3}. If a denser cloud velocity component seen towards α Cen A,B is real, it will encounter the solar system within 50,000 yr. The nearby magnetic field seen upwind has a spatial orientation that is parallel to the cloud surface. The nearby star Sirius is viewed through the wake of the solar system, but this direction also samples the hypothetical cloud interface. Comparisons of anomalous cosmic ray and interstellar absorption line data suggest that trace elements in the surrounding cloud are in ionization equilibrium. Data towards nearby white dwarfs indicate partial helium ionization, $<N(\mathrm{H\,I})/N(\mathrm{He\,I})>\sim 13.7$, which is consistent with pickup ion data within the solar system if less than 40% hydrogen ionization occurs in the heliopause region. However, the white dwarfs may sample additional ionized gas beyond the cloud surrounding the solar system.

I. INTRODUCTION

It has long been known that the interstellar cloud surrounding the solar system determines the boundary conditions of the heliosphere (Axford 1972; Holzer 1972; Fahr 1971; Baranov et al. 1971). The shock configuration of the heliopause region is determined by the charge density, magnetic field, temperature and relative velocity of the surrounding interstellar cloud. This cloud feeds neutral gas into the solar system, where resonantly scattered solar emission produces H I Lyman α and He I 584 Å backscattered radiation features (see, e.g., Thomas and Krassa 1971; Bertaux and Blamont 1971; Adams and Frisch 1977; Ajello et al. 1987,1994). Ions and atoms of interstellar origin have been measured in the inner solar system, the solar wind and the terrestrial magnetosphere as pickup ions and the anomalous cosmic ray component (see, e.g., Möbius et al. 1985,1988; Adams et al. 1991; Cummings and Stone 1990; Gloeckler et al. 1993; Witte et al. 1993; Cummings et al. 1993).

For the past 5 Myr, the Sun has been immersed in a region of space with very low average interstellar densities, $n < 0.0002$ cm^{-3}. This conclusion is derived from the extremely low interstellar column densities seen towards β CMa data (Gry et al. 1995), combined with the position of this star in the anti-apex direction. Between 2000 and 8000 years ago, the Sun appears to have entered the surrounding interstellar cloud, which is warm and low density, $T \sim 6800$ K, $<n(\mathrm{H\,I} + \mathrm{H\,II}) \sim 0.1$ to 0.5 cm^{-3}. The low density region in the third galactic quadrant ($\ell = 180°{-}270°$), from which the Sun is emerging, corresponds to an interarm region of the Galaxy. The space trajectory of the Sun is shown in Fig. 1.

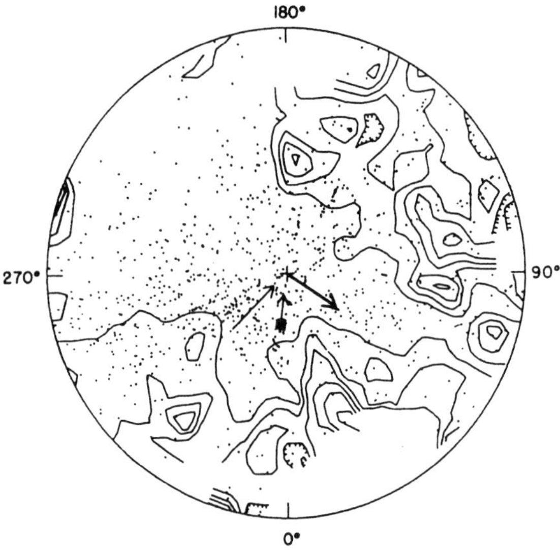

Figure 1.   The solar trajectory is shown plotted on the distribution of nearby interstellar matter (Frisch and York 1986; Lucke 1978). Plot radius is 500 pc, with the galactic center at the bottom of the figure and $\ell = 90°$ to the right. The minimum contour depicted corresponds to $N(\mathrm{H\,I}) \sim 5 \times 10^{20}$ cm^{-2}, and the contour scale is exponential corresponding to the magnitude system (see Lucke 1978). The solar trajectory through space is shown as a bold arrow, the LSR upwind velocity direction as a light arrow, and the upwind direction in the solar rest frame as a feathered arrow.

As will be seen below, the space trajectory of the Sun is nearly perpendicular to the flow of interstellar material past the Sun, which appears to be driven by the asymmetric expansion of superbubble shells associated with Scorpius-Centaurus Association star formation epochs (Frisch 1995). Therefore, during the next 3 to 4 Myr changes in the boundary conditions of the solar system are expected to be due to interstellar clouds expanding from this association. The local flow velocity is about 20 km s^{-1} in the local standard of rest (LSR); during the next 4 Myr the nearest 80 pc of interstellar gas in the upwind direction is expected to sweep past the Sun. The upwind region contains both molecular and ionized clouds, and there is intriguing evidence that a wisp of cool denser material is found in front of one of our nearest stellar neighbors, α Cen A,B, at a distance of 1.3 pc (Landsman et al. 1984,1986; J. L. Linsky, personal communication). The position of this star is very close to the LSR upwind direction, and if confirmed this denser wisp of material will pass over the Sun within 50,000 yr.

Our understanding of the interstellar cloud surrounding the solar system (nicknamed the "local fluff") is sketchy. Accurate information about the density, temperature, ionization, velocity, turbulence and morphology of the surrounding cloud can only be determined through high spectral resolution observations of the ultraviolet spectra (900 Å–3000 Å) of nearby stars. The only instrument currently capable of providing these data is the Goddard High Resolution Spectrometer on the Hubble Space Telescope (HST). However, telescope time has not been available for the needed observations, so that our understanding of the cloud surrounding the solar system is data limited.

High spectral resolution measurements of interstellar absorption lines seen in nearby stars actually measure sight line averaged properties, or, where several clouds are present and distinguished by velocity, averaging is over the region corresponding to the velocity interval. If the cloud density and/or relative ionization are not uniform, recombination states (e.g., Ca II, Na I, Mg I), with abundances which increase with density squared, will preferentially sample denser regions. In addition, particularly for low total column densities such as found locally, the ratio of ions to neutrals will vary throughout the cloud due to the attenuation of ionizing radiation within the cloud. The nearest stars for which we have interstellar line data, α Cen A,B, are 1.3 pc distant, or over 2700 times more distant than the heliopause. Even these stars show two foreground interstellar clouds (McClintock et al. 1978). Thus, the sight line averaged properties to nearby stars do not necessarily describe the cloud surrounding the solar system, and must be used in combination with *in-situ* data for an accurate picture.

Previous useful reviews or papers with summary information on nearby interstellar matter ($d < 100$ pc) are McClintock et al. (1978), Vidal-Madjar et al. (1978), Frisch and York (1983,1986,1990), Paresce (1984), Frisch (1990,1994), Bochkarev (1987), Murthy et al. (1987), Lallement (1993), and Linsky et al. (1993*b*). Also, there are many reviews in several symposia dedicated to the characteristics of nearby interstellar gas, IAU Colloquium

No. 81 (Kondo et al. 1984) and COSPAR XXVI held in Toulouse, France, in 1986 (see, e.g., Cox and Snowden 1986). Results based on the diffuse soft X-ray and extreme ultraviolet (EUV) background radiation provide a valuable perspective on nearby interstellar matter, but are not discussed here; the reviews by Cox and Reynolds (1987), McCammon and Sanders (1990), and Frisch (1995) can be consulted. Much of the material contained in this chapter is discussed in more detail in Frisch (1995).

In this review, the "basic" solar motion through the LSR is used to convert between heliocentric and LSR velocity frames. The basic solar vector corresponds to a solar motion of 15.4 km s^{-1} towards galactic coordinates $\ell = 51°$, $b = +23°$. The "basic" velocity frame of reference is defined so that the average motion of nearby K giant stars is zero, bypassing uncertainties inherent when using younger star populations where the velocity distribution may not have relaxed to an equilibrium state (Mihalas and Binney 1981).

II. VELOCITY FIELD OF NEARBY CLOUDS

I begin the discussion by establishing that the velocity vector of the cloud surrounding the solar system is nonzero in the LSR, in contrast to $\sim$100 pc distant cool interstellar clouds which tend to be at rest in the LSR. Using available data on interstellar Ca II and Na I absorption line profiles in nearby stars (Lallement et al. 1986; Vallerga et al. 1993; Lallement and Bertin 1992; Bertin et al. 1993; Welty et al. 1994), it can easily be shown that the local fluff is part of a flow of nearby interstellar gas through the local standard of rest at the location of the Sun and that the "local flow" velocity rest frame provides a better fit to the Doppler velocities of nearby weak (0.4–15 mA) interstellar Ca II and Na I absorption components than does the LSR velocity frame. Figures 2a and 2b show a single absorption component plotted per star, for 26 nearby stars [a] within 50 pc of the Sun. The abscissa for each point is the velocity the component has in the LSR velocity frame; the ordinate for each point is the velocity the component has in the local flow velocity frame [b]. Although 45 absorption components are found in this 26 star sample, only one velocity component per star is plotted in each figure. In Fig. 2a, the plotted component is selected to be the component which has the smallest absolute velocity in the local flow velocity frame. In Fig. 2b, the plotted component is selected to have the smallest absolute value in the LSR velocity frame. In other words, Fig. 2a shows the subset of components most likely to be formed in the local flow, and Fig. 2b shows the subset most likely to be

---

[a] The 26 stars are 51 Oph, α PsA, β Oph, α Oph, δ Her, ζ Aql, α Aql, γ Aqr, η Aqr, τ Her, ζ Peg, α Peg, η UMa, α Lac, α Cep, 2 And, α And, δ Cas, 3 Eri, δ Vel, HR4023, β Crt, α Hyi, ι Cen, ϵ Gru and α Gru.

[b] Note that if each component were at rest in the LSR frame, the points would be aligned along a vertical line which would intersect the LSR axis at 0 km s^{-1}. If each component were at rest in the local flow frame, the points would be aligned along a horizontal line which would intersect the ordinate at 0 km s^{-1}.

LSR gas. Obviously the components plotted in Fig. 2a have less scatter in the local flow velocity frame than the components in Fig. 2b do in the LSR frame. The obvious conclusion is that local flow gas is more likely to be seen in front of these 26 stars than gas at rest in the LSR. This conclusion may be reached despite the fact that absorption due to circumstellar disk, mass loss or photospheric phenomena is known to contribute to the observed absorption profiles for many of these stars.

This statement can be made more quantitative. The rms variation from the local flow velocity vector of the 26 components plotted in Fig. 2a is 5.5 km s^{-1}. However, half of this rms comes from two outlying components (τ Her and 2 And). When these outlying components are omitted from the rms calculation, the rms variation of the remaining 24 components is 2.4 km s^{-1}. This variation can be compared with typical measurement errors of ± 1.5 to ± 3 km s^{-1}. For Fig. 2b, the rms variation of the 26 components in the LSR velocity frame is 8.7 km s^{-1}. When the three most discrepant points are omitted from the calculation (ι Cen, γ Oph and τ Her), the rms deviation for the remaining 23 components from the LSR velocity is 6.5 km s^{-1}.

The local flow velocity vector used to make Figs. 2a and 2b is the heliocentric vector $\mathbf{V} = -26.8 \pm 1.3$ km s^{-1}, arriving from the direction of $\ell,b = 6.2°,+11.7°$. This vector corresponds to an LSR upwind vector of $\mathbf{V},\ell,b = -19.0$ km s^{-1}, 334.5°, $-1.8°$ in the LSR. This velocity frame is the best fit vector to interstellar Ca II K absorption lines in 12 nearby stars [c] located in both upwind and downwind directions (Frisch 1995). Many efforts have been made to derive the flow vector of interstellar matter in the solar vicinity (Crutcher 1982; Frisch and York 1986; Lallement et al. 1986; Bzowski 1988; Lallement and Bertin 1992). Nonuniform motions over 20 to 100 scales, and confusing stellar or circumstellar contributions have prevented unambiguous results. The best effort to isolate the cloud surrounding the solar system is that of Lallement and Bertin (1992), where a small localized star sample was selected for fitting, omitting absorption features with possible confusion from circumstellar gas. However, until the surrounding cloud is observed in absorption in all directions, the results remain ambiguous. When the velocities of the local cloud material in the Galactic center hemisphere (V_g) and longitude interval $\ell = 60°$ to 170° (V_{ag}) are determined independently, best fit heliocentric vectors are $V_g = \mathbf{V},\ell,b = -29.4$ km s^{-1}, 5°, +21°, and $V_{ag} = \mathbf{V}, \ell,b = -25.7 \pm 0.5$ km s^{-1}, 6.1°$\pm 13°$, +16.4°$\pm 13°$ [d] (Lallement and Bertin 1992). These vectors differ mainly in velocity, suggesting there may be a velocity gradient of about 3 km s^{-1} per 10 pc between the upwind and downwind directions, which is consistent with a flow driven from behind. Both components appear to be seen towards α Oph. The direction and velocity of the anti-center vector appear to agree best with the velocity vector determined

---

[c] These 12 stars are α Oph, δ Her, γ Aqr, η Aqr, ζ Peg, α Peg, η UMa, α Lac, α Cep, α And, α CVn, and δ Cas.

[d] The stars used to drive the vector V_{ag} were: δ Cas, η Aur, α Cep, γ Aqr, α Lac, α Peg.

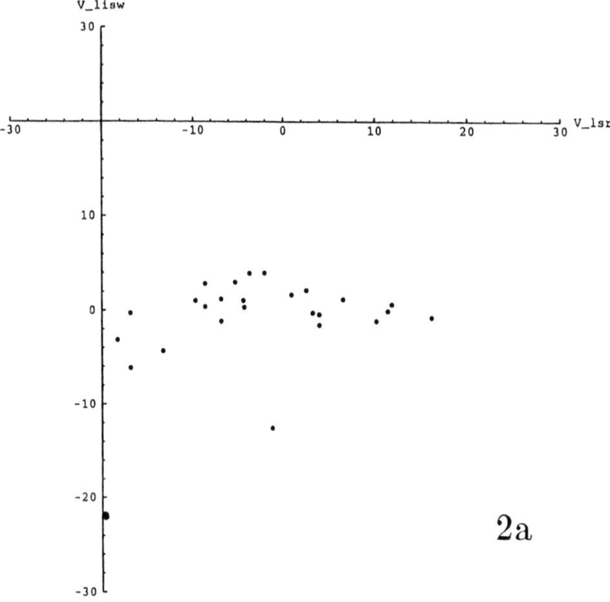

from the arrival angle of interstellar He I in the solar system near 5 AU (Witte et al. 1993; Lallement and Bertin 1992), but additional data are needed. Interestingly, the LSR upwind direction is within $\sim$20° of one of our nearest stars α Cen A,B ($\ell = 316°$, $b = -1°$).

The large scatter of components in the local flow velocity frame for negative LSR velocity points in Fig. 2a reflects the fact that in the upwind direction many nearby stars have multiple absorption components, consistent with the greater column densities found in this direction.

III. AVERAGE H I DENSITY

An average density $<n(\mathrm{H\,I})>\sim$0.09 to 0.15 cm^{-3} is found from measurements of interstellar H I Lyman α absorption features in the spectra of the nearest ($d<4$ pc) stars, using both GHRS and Copernicus data. This average is based on α Cen A,B (1.3 pc), Sirius (2.7 pc), ϵ Eri (3.3 pc), ϵ Ind (3.4 pc), and Procyon (3.5 pc) data (summarized in Linsky et al. 1993a,b; Frisch 1994,1995). Stars in the Galactic center hemisphere with distances $D<30$ pc show $<n(\mathrm{H\,I})>\sim$0.1 cm^{-3}, while in the opposite hemisphere the average density falls dramatically for star distances greater than 5 pc. Plots of the distribution of average sight line densities versus star distance for nearby stars are found in Frisch and York (1983), Paresce (1984), and Frisch and York (1990). These show that in the galactic center hemisphere of the sky, an average neutral hydrogen density $<n(\mathrm{H\,I})>\sim$0.1 cm^{-3} is found extending to

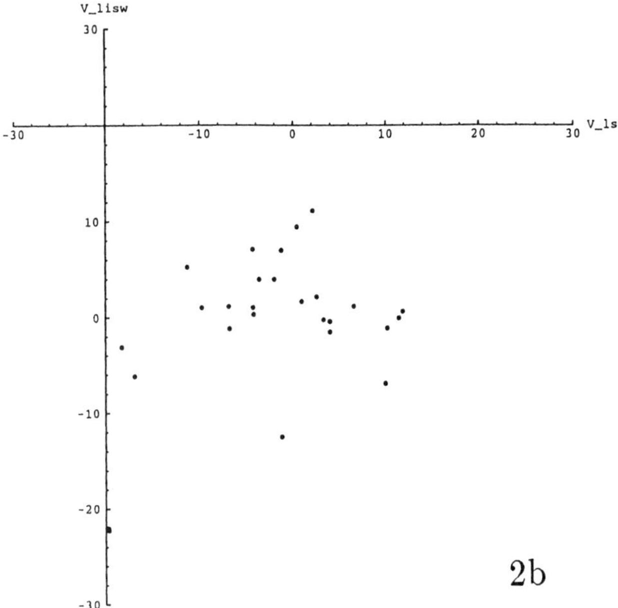

Figure 2.   The velocites of weak Ca II K or Na I D absorption features in 26 nearby
(d <50 pc) stars are plotted. Only one component per star is plotted in each figure,
although several components may be visible. The ordinate is the velocity of the
component transformed into the heliocentric "local flow" velocity frame (V =
-26.8 ± 1.3 km s^{-1}, arriving from the heliocentric direction, ℓ, b = 6.2°, +11.7°).
The abscissa gives the cloud velocity transformed into LSR (using "basic" solar
motion). In Fig. 2a, the plotted component is selected to be the cloud component
with the smallest absolute velocity in the "local flow" velocity frame. In Fig. 2b,
the component is selected to have the smallest value in the LSR velocity frame.
Measurement errors are typically ±1.5 to 3.0 km s^{-1}. The two outlying components
are seen in the stars 2 And (x, y = 1.3, -12.5 km s^{-1}) and τ Her (x, y = -19.9, -22.1
km s^{-1}). The "local flow" velocity frame clearly provides a better fit to the velocities
of nearby interstellar gas than does the LSR frame. The data used to make these
plots are from Bertin et al. (1993), Vallerga et al. (1993), Lallement et al. (1986),
Lallement and Bertin (1992), and Welty et al. (1994). Absorption components seen
in the stars 51 Oph, α PsA, γ Oph, α Oph, δ Her, ζ Aql, α Aql, γ Aqr, ϵ Aqr, τ Her,
ζ Peg, α Peg, η UMa, α Lac, α Cep, 2 And, α And, δ Cas, τ3 Eri, δ Vel, HR 4023,
β Crt, α Hyi, ι Cen, ϵ Gru, and α Gru are plotted.

distances of over 30 pc, despite absolute space density variations of factors of
10^3. The denser regions may be formed by the compression of gas with initial
density <n>~0.1 cm^{-3}, such as expected if a superbubble shell expanded
into a uniform n~0.1 cm^{-3} medium (see below).

Multiple interstellar velocity components seen towards the nearest stars
show that interstellar material within 3 to 5 pc of the Sun is not homogeneous
or uniform. Thus, the density of the cloud surrounding the solar system may
deviate from the average value. Towards α Cen A (ℓ = 316°, b = -1°, d =

1.3 pc) Landsman et al. (1984,1986) found preliminary evidence that D I is formed in gas which is cooler than, and blue shifted by, 8 ± 2 km s^{-1} from the H I gas. Towards Sirius ($\ell = 227°$, $b = -9°$, $d = 2.7$ pc), two absorption features at heliocentric velocities 18.7 ± 1.5 km s^{-1} and 13.0 ± 1.5 km s^{-1} are seen in the ultraviolet Mg II and Fe II lines (Lallement et al. 1994). The Sirius data are consistent with recent GHRS observations of ϵ CMa which is $13°$ from the Sirius sight line (Gry et al. 1995). Towards α Aql ($\ell = 48°$, $b = -9°$, $d = 5$ pc) three absorption components have been fit to the optical Ca II line, although some may be circumstellar (Lallement et al. 1986). Additional data are required to determine the magnitude of possible local deviations from the average neutral density of 0.1 cm^{-3}.

IV. TEMPERATURE AND TURBULENCE

The temperature of the local fluff is found by assuming that the widths of interstellar absorption lines are composed of a mass independent turbulent contribution ($\xi^2 = 2v_{turb}^2$) and a mass dependent thermal contribution ($v_{therm}^2 = kT/Am_p$) so that $b^2 = v_{therm}^2 + \xi^2$. The variable b is the Doppler broadening constant in a Voigt profile. For temperatures derived in this way to be accurate, the velocity component structure of the absorption profile must be resolved. Using GHRS observations of H I, Mg II and Fe II absorption lines, temperature and turbulent parameters found for the nearest gas towards Capella are $T = 7000\pm200$ K, $\xi = 1.66\pm0.03$ km s^{-1}, towards Procyon $T = 6700\pm200$ K, $\xi = 2.30\pm0.03$ km s^{-1} (Linsky et al. 1993a,b), and for the local fluff component towards Sirius $T = 7600\pm3000$ K, $\xi = 1.4$ (+0.6, -1.4) km s^{-1} (Lallement et al. 1994). These temperatures are within the range of the temperature determined for the interstellar He I and H I atoms which have penetrated the solar system (see, e.g., Ajello et al. 1987; Witte et al. 1993; Lallement 1990). The higher local fluff temperature inferred for the surrounding cloud from earlier Copernicus observations of β CMa (Gry et al. 1985), which is $\sim5°$ from the Sirius sight line, is now seen to be due to spectrally unresolved interstellar velocity components.

For derived cloud temperatures and turbulence to be accurate, the H I, Mg II and Fe II lines must sample the same region of space. Because Mg II and Fe II lines are formed in both warm ionized and neutral gas (Mg I and Fe I have ionization potentials of 7.6 eV and 7.9 eV respectively), while the Lyman α H I line is heavily saturated and samples all neutral gas, in principle these lines may not equivalently sample the matter distribution of individual sight lines. In fact, because significant ionized gas is present and sampled by Mg II but not H I, unless neutrals and ions are fully mixed, Mg II and H I will not trace the same matter distribution.

The turbulence decay in the material surrounding the solar system has been discussed by Ferrara (1994), where he assumed the turbulence was caused by a supernova event which injected 10^{51} ergs about 2×10^5 to 10^6 years ago into an interstellar medium with density $n\sim4 \times 10^{-3}$ cm^{-3}. A time dependent Kolmogorov turbulent energy spectrum was assumed, with viscosity attributed to particle interactions (atoms and ions). For turbulence $\xi \sim 1.66 \pm 0.03$ km s^{-1} (Linsky et al. 1993a), Ferrara constrained the outer scale of the turbulence to the range $0.9 \times 10^{18} \leq \ell_o < 6 \times 10^{18}$ cm. For a supernova $> 10^6$ years ago, such as is consistent with the ionization equilibrium inferred for the cloud surrounding the solar system (Frisch 1994), an outer scale $\ell_o \sim 6 \times 10^{18}$ cm ~ 2 pc is found. This value is also consistent with both the turbulent spectrum in diffuse ionized galactic ISM (Spangler 1991), and the multicomponent cloud structure seen in stars 1 to 5 pc from the Sun. Ferrara assumed incompressible, homogeneous, isotropic turbulence, neglecting magnetic fields. The inner scale is comparable to the proton gyroradius, which is ~ 500 km for the local fluff properties and $B \sim 1.4\,\mu$G.

V. CLOUD IONIZATION

Because H II is not directly observable, ascertaining the electron density in the cloud surrounding the solar system requires a model for the mechanism which maintains the ionization. The model must identify the ionization mechanism (e.g., photoionization versus collisional ionization), and whether the gas is in equilibrium (e.g., the ionization rate equals the recombination rate). Within the limits imposed by sparse data and an assumed cloud model, independent theoretical and observational studies suggest that the surrounding cloud is significantly ionized, that helium is more ionized than hydrogen, and that the trace elements in the cloud are in ionization equilibrium.

In a neutral interstellar cloud, the ionization radiation field hardens towards the cloud interior because the effective cross section per hydrogen atom is higher for softer photons with $\lambda \sim 912$ Å due to absorption in the ISM (see, e.g., Cruddace et al. 1974). For a flat incident spectrum, the ratio of the hydrogen ionizing flux to the helium ionizing flux will decrease as $\exp[-N(\text{H I})\ 5 \times 10^{-18}]$ towards cloud interior. However, the extreme ultraviolet photon field from white dwarfs and hot stars is hardened significantly above 400 Å during propagation through interstellar medium, resulting in relatively more helium than hydrogen ionizing photons incident on the cloud surface. Because of the low column densities of interstellar gas in the third galactic quadrant ($\ell = 180°-270°$; see Fig. 1), and at high galactic latitudes, EUV sources located in these regions will experience the least attenuation.

A model of the local fluff ionization by Slavin (1989), including an evaporative interface between the cloud and postulated (but puzzling) surrounding hot plasma, illustrates the main considerations in modeling the local ionization. The putative interface produces helium ionizing photons; and the resulting recombination of ionized helium would provide a hydrogen ionizing source internal to the cloud which may dominate EUV sources. External EUV sources such as the B2 II star ϵ CMa and nearby white dwarfs (see, e.g., Cassinelli et al. 1995; Reynolds 1990; Cheng and Bruhweiler 1990) will also contribute. Including all of these sources except for ϵ CMa, Slavin predicts a fractional hydrogen ionization of 14% in the cloud core, and 42% at the cloud surface for magnetic field tangential to the cloud surface such as observed. The fractional helium ionization is, respectively, 29% at the core and 36% at the surface. If the radiation field of ϵ CMa field were included, these ionizations would increase. Observed Si III, Si IV and C IV column densities towards ϵ CMa (Gry et al. 1995) are consistent with predicted interface values (Slavin 1989). The observed EUVE upper 3σ upper limit on He II 256 Å emission due to the conductive interface matches the value predicted by Slavin for magnetic field with an angle 45° to the radial (Jelinsky et al. 1995). However, the nearby magnetic field has a direction approximately parallel to the local cloud surface (Frisch 1994), so conduction would be inhibited. The morphology of the cloud surrounding the solar system indicates that the Sun has recently entered the local cloud and may be near the cloud surface, or possible interface, and that the star α CMa lies in a sight line tangential to the cloud interface (Fig. 3). The Slavin local cloud model yields a predicted ratio $N(\text{H I})/N(\text{He I})\sim12.1$ to 15.4 at cloud core, and $N(\text{H I})/N(\text{He I})\sim9.1$ to 18.3 at the cloud surface, which is within the observed range seen towards nearby white dwarfs (see below).

If the surrounding cloud is treated as a hypothetical spherical cloud 3 pc in radius and $n\sim0.1$ to 0.4 cm^{-3} density, immersed in a plasma ($T = 10^6$ K, $n = 0.005$ cm^{-3}), it will evaporate in less than 1.5 Myr if not inhibited by a magnetic field. This result is problematical. However, the stellar polarization data indicate that the nearby magnetic field is parallel to the cloud surface (Frisch 1995; Fig. 3) such that cloud evaporation may be inhibited.

Two classes of observations of stars within 20 to 70 pc of the Sun support high local ionization and preferential helium ionization. The first examples are Copernicus observations of N II I in η UMa ($\ell = 101°$, $b = +65°$, $d = 45$ pc) and α Leo ($\ell = 226°$, $b = +49°$, $d = 21$ pc), which indicate substantial fractions of ionized gas nearby (29% and 49%, respectively; Frisch and York 1990). The second group of observations are HUT and EUVE observations of nearby white dwarfs located at distances of 40 to 70 pc (These data are summarized in Frisch [1995].) The relative hydrogen to helium ionization has been inferred from observations of the five white dwarfs (HZ43, GD246, G191-B2B, GD71, and GD153). The interstellar medium towards these white dwarfs, located 40 to 70 pc from the Sun, has $<N(\text{H I})/N(\text{He I})>\sim13.7$, which when compared to the nominal cosmic ratio ~10, suggests preferential helium ionization (see, e.g., Vennes et al. 1993a,b; Finley et al. 1993; Finley and Kimble 1993; Paerels and Heise 1989; Kimble et al. 1993a,b). This ratio implies more helium than hydrogen ionization, but gives no information on hydrogen ionization. However, for each star, the cosmic ratio $N(\text{H I})/N(\text{He I})\sim10$ is included within the error bars. This average value is comparable to the 25% helium ionization, and the upper limit on hydrogen ionization of 27%, found towards GD246 [e] (Vennes et al. 1993a). Some of the white dwarfs included

---

[e] The white dwarf GD246 is located at $\ell = 87°$, $b = -45°$ with a distance of 65 pc.

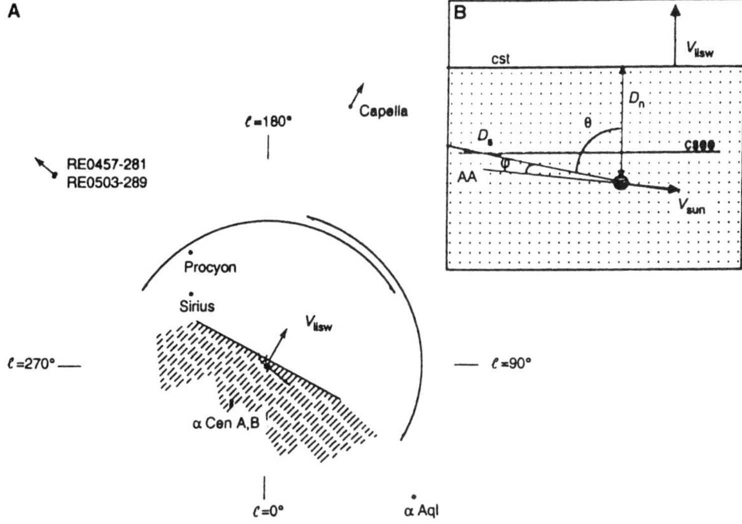

Figure 3.　(A) The morphology of the cloud surrounding the solar system is shown as inferred from a few simple assumptions. It is assumed that the cloud velocity vector is normal to the cloud surface and that the distance to the local fluff cloud surface in the anti-apex direction is 0.3 to 0.5 pc, as is found from data for Sirius A local fluff component and the white dwarfs RE0457-281 and RE0503-289 (Barstow et al. 1993). With these assumptions, the solar system encountered the surrounding cloud 2000 to 8000 years ago. Because the solar and cloud trajectories are nearly perpendicular in space, this cloud system is sweeping past the Sun. The interstellar magnetic field which polarizes the light of nearby stars in the upwind direction is parallel to this cloud surface to within 5° to 15°. An artists view of the solar wake is included. Several nearby stars are shown plotted at their respective positions (the Capella and white dwarf positions are not shown to scale). The solid arc indicates the angular region containing the stars Lallement and Bertin (1992) used in order to derive the "anti-galactic" velocity vector. The dotted arc illustrates the "low Mg II column density" region of Genova et al. (1990), at $b = 0°$, which is centered approximately on the downwind direction (in the LSR). This figure is a starting point for discussion, and not a final map, because cloud multiplicity is not illustrated (e.g., towards α Cen, α Aql and Sirius A). If the cloud motion is not parallel to the surface normal as is assumed, then this cloud morphology is not correct. (B) An enlargement of the relationship of the solar velocity (bold vector labeled V_{sun}) and the cloud surface today (line labeled "cst") is shown. For additional details see Frisch (1994).

in this average may sample additional ionized gas not connected with the local material.

The electron density can also be estimated by evaluating the relative abundances of dominate and subordinate ionization states in the surrounding cloud using an assumed ionization equilibrium model (Frisch 1986; Frisch et al. 1990,1994; Lallement et al. 1994). GHRS data give the ratio $N(\text{Mg II})/N(\text{Mg I}) = 220$ (+70, −40) in the local cloud towards Sirius

($\ell = 227°$, $b = -9°$, $d = 2.7$ pc), which yields $n(e^-) = 0.08$ to 0.55 cm^{-3} for a range of assumed cloud temperatures (Lallement et al. 1994). Gry et al. (1995) measured a larger ratio $N(\text{Mg II})/N(\text{Mg I}) = 414$ for the local fluff component towards ϵ CMa than seen towards Sirius, and found $n(e^-) = 0.12$ cm^{-3} for $T = 7200$ K. More GHRS spectral data are needed to reduce the uncertainties. Frisch (1994) combined observations of $N(\text{C I})/N(\text{O I})$ determined for the source population of the anomalous cosmic ray component (using the fact that oxygen and hydrogen ionization are linked by charge exchange) with the $N(\text{Mg II})/N(\text{Mg I})$ Sirius data, and found that the two disparate sets of observations constrained the electron density in the surrounding interstellar cloud to $n(e^-) = 0.22$ to 0.44 cm^{-3} for $T = 6700$ to 8000 K and an assumed $n(\text{H I}){\sim}0.1$ cm^{-3}, yielding a fractional hydrogen ionization of 69 to 81%. The general consistency between the electron densities derived from these disparate data suggests that the underlying assumption of ionization equilibrium for the trace elements is accurate. However, uncertainties in the acceleration of the pickup ion population to anomalous cosmic ray energies, and the propagation and lifetimes of the anomalous cosmic rays in the solar system, remain. Independent absorption line data of additional ions are needed before the conclusion of ionization equilibrium is proven. Lallement et al. (1994) hypothesized that the high ionization seen towards Sirius may be the result of the sight line sampling the cloud interface. This speculation is consistent with the cloud morphology shown in Fig. 3 (which I derived from other data) where the Sirius sight line is seen to be nearly parallel to the cloud surface. In any case, because the Sun appears to be within about 0.1 pc of the cloud surface, it may be embedded in an interface region.

VI. MAGNETIC FIELD

There are no direct data on the magnetic field strength in the cloud surrounding the solar system. The nearest region for which data on the interstellar magnetic field exist is a 100° by 100° region of space centered near the upwind direction $\ell = 333°$, $b = 0°$. In this region, weakly polarized stars ($P{\sim}0.02\%$) within 30 pc of the Sun have been observed, with the nearest interstellar polarized stars ζ Ind (3.4 pc) and 36 Oph (5.3 pc) (Tinbergen 1982). The polarization is produced by magnetically aligned interstellar dust grains in the upwind cloud. The polarization is patchy and difficult to detect (Leroy 1993); although some of the polarization detections are less then 3σ, the correlation between the angles of polarization of different stars, which give a local magnetic field directed towards $\ell{\sim}70°$, provides additional evidence the polarizations are real. There is no significant evidence of correlation of polarization strength with star distance, so the magnetic field must extend to the solar location (Frisch 1990). The magnetic field sampled by Tinbergen's data is parallel to the surface of the local cloud (Fig. 3) to within 5 to 15°, so that cloud evaporation may be inhibited. There are two possibilities for the field origin: (1) the field could be embedded in an expanding shell structure from the Scorpius-Centaurus association, in which case the Sun would be at the edge of this structure (as other data indicates); or (2) the field is the generic Galactic magnetic field associated with the solar location, with the absence of downwind polarization due to the absence of polarizing dust grains.

The interstellar magnetic field strength in the solar vicinity has been derived by comparing rotation and dispersion measures of over 200 pulsars within several kpc (Rand and Kulkarni 1989; Rand and Lyne 1994). The nearby global ordered field has a circular geometry, with local field strength $B = 1.4\pm0.15\ \mu$G directed towards $\ell = 88°$, and is formed in the warm mainly ionized low density component of the ISM. There is also a nonuniform, random, component with strength ~3 to $5\ \mu$G, which is probably associated with the enhanced magnetic strengths found in HI 21-cm Zeeman splitting in superbubble shells (Heiles 1986). In the North Polar Spur region, $\ell = 0°$ to $60°$ and $b = 0°$ to $60°$, a field strength of $\sim5\ \mu$G is found. This sight line is a tangential sight line through the local superbubble that drives the flow of interstellar matter near the Sun.

Additional indirect arguments supporting weak field strengths, $B \sim 1.4\ \mu$G, for the ambient cloud are the existence of the local flow and Zeeman splitting limits. The observed flow of interstellar matter past the Sun implies that a weak magnetic field ($B\sim1.4\ \mu$G) is more likely then a strong field ($B\sim5\ \mu$G). The ratio of gas to magnetic pressure is $\beta\sim4$ to 7 (Table I), so the surrounding cloud would drag a captured galactic field through space. If the magnetic field strength were larger ($B>3.5\ \mu$G), then $\beta<1$, and the magnetic field should confine the surrounding cloud and prevent the observed flow. HI Zeeman splitting limits on 21-cm features sampling diffuse gas in the solar neighborhood indicate an upper limit on the nearby field of $2\ \mu$G (Verschuur 1993). Finally, a magnetic field strength of $B\sim1\ \mu$G field appears adequate to align the polarizing dust grains (Mathis 1986 1990; Duley 1978).

Other arguments about field strength have been made. If turbulence in the local cloud is determined by the magnitude of the Alfvén velocity, a field strength of $B\sim0.2$ to $0.6\ \mu$G would be implied for $\xi\sim1$ to 2 km s^{-1} and $n(e^-)\sim0.22$ to 0.44 cm^{-3}. In contrast, if equipartition of energy exists, then $B\sim3.8\ \mu$G for $n_{\rm tot}\sim0.4$ cm^{-3}, $T = 7000$ K. Boulares and Cox (1990) present arguments that the magnetic pressure of a field with strength $\sim5\ \mu$G is needed locally to support high latitude interstellar matter.

This magnetic field direction shown by the Tinbergen data ($\ell\sim70°$) is close to the magnetic field direction $\ell\sim80°$ inferred from the "focusing" direction of optical polarization vectors of hundreds of nearby ($d<200$ pc) stars (Mathewson and Ford 1970; Heiles 1986). It is also near the global nearby field direction, $\ell = 88°$, inferred from observations of 200 pulsars within several kps (Rand and Kulkarni 1989; Rand and Lyne 1994). The magnetic field direction points towards the large dust complex called the "Great Rift," which is associated with a sight line sampling the local Orion Spiral Arm.

The Tinbergen optical polarization data suggests the Sun is embedded in the edge of the upwind cloud producing the polarization. The absence of polarization in nearby stars in the anti-center hemisphere may be due to the fact we are about 0.1 pc from the surface of the local cloud in this direction (Fig. 3). Because the alignment mechanism strength is proportional to $B^2/nT^{3/2}$ (see, e.g., Spitzer 1978), another possibility is a higher temperature or disrupted grain population near the cloud edge. This is also consistent with the distribution of nearby interstellar matter shown by interstellar Mg II absorption lines towards nearby stars (Genova et al. 1990), which give a local cloud column density minimum aligned with the downwind LSR flow vector direction. The upwind cloud system has been termed the "squall line," and it shows embedded features with density contrasts on the order of $\sim 10^3$ (Frisch 1995). Because interstellar magnetic field strengths do not appear correlated with gas density for $n = 0.1$ to 100 cm^{-3} (Heiles 1986), there is no *a priori* reason to assume that upwind field strengths differs from local values.

VII. CLOUD MORPHOLOGY

The morphology of the cloud surrounding the solar system has been inferred from the total local fluff column density towards Sirius and the assumption that the local fluff velocity vector is perpendicular to the cloud surface. In the LSR, the motion of the Sun and local flow are nearly perpendicular. For $N(\text{H I} + \text{H II}) = 5.1 \times 10^{17}$ cm^{-2} towards Sirius and the assumption that the cloud motion is perpendicular to the surface, the cloud morphology can be reconstructed as shown in Fig. 3. If the underlying assumptions are correct, some useful conclusions appear. The relative Sun-cloud motion and these assumptions imply that the Sun has entered the local fluff 2000 to 8000 years ago. Although the model in Fig. 3 will be updated as more and better data become available, the general conclusion that the solar system has entered this cloud during historical epochs should remain valid (Frisch 1994,1995).

The high ionization seen in the Sirius direction can be explained by the fact that this sight line samples a tangential path through the interface region of the local fluff where relatively high ionizations are predicted ($N(\text{H II})/N(\text{H I} + \text{H II})\sim 42\%$ to 54%; Slavin 1989), although the ionization process is poorly understood.

Figure 3 is a simplified picture because nearby cloud multiplicity is not included. The blue-shifted cloud towards Sirius is not illustrated; also, the hatched region material is neither smoothly flowing nor homogeneous, because two absorption components separated by -8 ± 2 km s^{-1} have been identified towards α Cen A at 1.3 pc (Landsman et al. 1984,1986) and three optical Ca II K line absorption components have been identified in α Aql at 5 pc (Lallement et al. 1986). A key issue is the filling factor of the additional velocity components; if these features are due to wisps of gas, they may provide minimal absorption for external EUV and soft X-ray photons.

Also, included in Fig. 3 is a hypothetical solar wake. The Sun traveling through space at the velocity ~ 15.4 pc Myr^{-1} (~ 3.2 AU yr^{-1}), and with

TABLE I
The Cloud Surrounding the Solar System*

Quantity	Value
Average neutral space density	0.09–0.15 cm^{-3}
Average electron density	0.22–0.44 cm^{-3}
Fractional ionization for	
homogeneous cloud	59%–83%
Temperature	6800±200 K
Sound velocity	9.7 km s^{-2}
Electron plasma frequency	4.2–6.0 kHz
Magnetic field strength	(1.4±0.5 μG)
Alfvén velocity	(4.6–7.4 km s^{-1})
Fast mode velocity	(10.7–12.2 km s^{-1})
Solar system Mach number for	
relative Sun-cloud velocity 26 km s^{-1}	(2.1–2.4)
Electron gyroradius	(1400 km)
Proton gyroradius	(500 km)
Proton mean free path	(0.14 AU)
Plasma β parameter	(4–7)
Upwind LSR velocity vectors	
12-star average	$V = -19.0$ km s^{-1},
	$\ell = 334.5°, b = -1.8°$
7 stars in Galactic	$V = -21.0$ km s^{-1},
center hemisphere	$\ell = 334.5°, b = +11.8°$
6 stars in anti-center hemisphere	$V = -18.0$ km s^{-1},
	$\ell = 331.9°, b = +4.6°$

* The quantities in parantheses are based on estimates of the magnetic field strength. The average electron density is based on comparisons between the Sirius sight line and anomalous cosmic ray data.

respect to the surrounding cloud at $\sim$26 km s^{-1} ($\sim$5.4 AU yr^{-1}), will leave a small wake consisting of cooling and recombining solar wind plasma (average density $\sim$0.003 cm^{-3}), gravitationally focused, accelerated and heated interstellar hydrogen neutrals that avoided ionization in the solar system (density $\sim$0.1 cm^{-3}), as well as a population of slowly recombining interstellar atoms that were ionized in the solar system. Because the H I Lyman α line becomes optically thick at $N(H I) \sim 10^{13}$ cm^{-2}, for local fluff Doppler constant values, including turbulence, $b \sim 12$ km s^{-1}, an H I population like the local fluff atoms will become optically thick over a path length of $\sim$10 AU for a nominal density $n \sim 0.1$ cm^{-3}. If such a population were hotter, it would contribute H I Lyman α absorption with an anomalous velocity dispersion towards background stars such as Sirius (which is about 15° from the anti-apex of the solar motion, and offset about 4° downwind).

Whether such a population would distort the Lyman α absorption line analysis towards background stars in the anti-apex direction would depend on the magnitude of the heating of the hydrogen population that passed through the solar system without becoming ionized. For instance, heating of H I to $T \sim 40,000$ K (or giving Doppler value $b \sim 25$ km s^{-1}) would create a fast H I population with Lyman α absorption in the wings of the interstellar Lyman α line. A crude estimate of the opening of the solar wake can be made, $\Theta \sim 2 \tan^{-1}(v_{\mathrm{lisw}}/v_{\mathrm{sound}}) \sim 42°$, where $v_{\mathrm{lisw}} \sim 26$ km s^{-1} and $v_{\mathrm{sound}} \sim 10$ km s^{-1} (because wave propagation perpendicular to the heliocentric upwind direction is also parallel to the local magnetic field lines). The local fluff motion, of course, will sweep the wake into the downwind direction.

VIII. CLOUD ORIGIN

The cloud surrounding the solar system is in a low density interstellar region of space often referred to as the "local bubble." It has been suggested that the local bubble was created by the explosive event that formed the Geminga pulsar (Garmire et al. 1992; Bignami et al. 1993; Gehrels and Chen 1993). However, I have estimated that when the uncertainties on the proper motion, velocity and distance are considered, Geminga is 4 to 25 times more likely to originate in the Orion Association (perhaps the young Trapezium region or the supernova event that formed the Orion's Cloak supernova remnant) than it is to originate in a supernova event causing the local bubble (Frisch 1993a,1995). Smith et al. (1993) concur with this view on the Geminga origin, suggesting the λ Ori group as the formation site. Improvements in the determination of both the space trajectory and distance of Geminga can be expected to clarify the relation between Geminga and the local bubble. Because the Orion Association seems a much more likely origin region for Geminga, the possibility that it has a causal relation to the local bubble is not considered further here.

The concept of the local bubble arose because both diffuse 0.1 to 0.5 keV thermal X-ray emission (see, e.g., Sanders et al. 1977; Davelaar et al. 1980) and optical and ultraviolet absorption line data (Frisch 1981) gave evidence of supernova shock front activity around the Sun. The apparent conflict between the symmetric bubble geometry derived from early soft X-ray results, which centered the Sun in a compact supernova remnant, and the asymmetric geometry derived from optical and ultraviolet absorption line data which placed the supernova remnant on the Galactic hemisphere side of the Sun, has always been puzzling. However, this question will not be resolved until the origin of the soft X-ray background is fully understood. Another confusion in the term is that the local bubble has also been used to refer to the low-density interarm region seen in the third galactic quadrant.

The LSR upwind direction of the cloud surrounding the solar system is located at $\ell = 330°\pm17°$, $b = 4°\pm17°$[f]. This direction is centered in the Norma constellation, and the error circle includes Lupus; the upwind direction coincides with the patch of polarization in nearby stars discovered by Tinbergen (1982). The LSR downwind direction, $\ell = 150°$, $b = -4°$, is aligned with the minimum local cloud column density region discovered by Genova et al. (1990), supporting the view the Sun is close to the cloud surface in this direction. The upwind direction coincides with the region of the Scorpius-Centaurus Association (SCA), and epochs of star formation in the SCA appear to drive the flow of interstellar matter near the Sun. The SCA subgroups are Upper Centaurus Lupus (at $\ell = 320°\pm13°$, $b = 10°\pm13°$), Lower Centaurus Crux (at $\ell = 297°\pm15°$, $b = 15°\pm15°$), and Upper Scorpius (at $\ell = 347°\pm12°$, $b = 21°\pm12°$), with nuclear ages of 14 to 15, 11 to 12 and 5 to 6 Myr, respectively (de Geus 1991; Jones 1971). De Geus (1991) modeled the Loop I 21-cm filaments as a superbubble shell due to successive epochs of star formation in SCA. However, this model did not include the fact that stellar wind bubble and supernova remnant expansion in these subgroups would have yielded an asymmetrically shaped superbubble (Frisch 1981).

Gas expanding in a direction parallel to galactic rotation would have collided with the dense material associated with the Aquila Rift molecular cloud [g] and the local spiral arm within distances of less than 100 pc, stopping the expansion. De Geus noted that the expansion of the expanding shell formed by the first epoch of star formation (which created the Upper Centaurus Lupus subgroup) may have initiated the most recent epoch which formed the Scorpius subgroup. The intense radio continuum source known as the North Polar Spur may have resulted from the collision of the hot X-ray emitting plasma expanding from the recent ($\sim$250,000 yr ago) Loop I supernova event with the dense Aquila Rift molecular cloud, located in the Gould's Belt plane at a distance of 100 to 150 pc from the Sun.

Expansion in a direction perpendicular to the direction of galactic rotation, corresponding to expansion into the low density interarm region surrounding the Sun, would have maintained the low interarm densities and yielded much larger shell radii.

A self-consistent quantitative description of this scenario is presented in Frisch (1995). In this scenario, the first epoch of star formation yielded a shell feature which has swept down the interarm region past the Sun. Cooled gas associated with this shell is seen downwind in the general Orion-Auriga-Perseus constellations about 40 to 50 pc from the Sun; this feature bounds the local bubble on the downwind side, and is observed shadowing EUV sources by the ROSAT Wide Field Camera (Warwick et al. 1993). This gas has a current shell radius and velocity of 205 pc and 8.6 km s^{-1}, and manifests the

---

[f] The uncertainties on the upwind position do not include errors in the solar velocity vector, which are unknown. When the Hipparcos astrometric data become available, it will be possible to reduce the uncertainties in the solar motion vector.

[g] The Aquila Rift molecular cloud in turn may be left over from the parent giant molecular cloud out of which the Scorpius-Centaurus association formed.

warp of Gould's Belt and the local arm. The second epoch of star formation yielded expanding shell features which swept quickly through the evacuated cavity formed by the first star formation epoch. In the 5-Myr interval between the second and third star formation epochs, interstellar material was evaporated from the surface of molecular clouds embedded in the SCA end of the elongated cavity, in a classic cloud evaporation scenario (see, e.g., McKee and Cowie 1977; White and Long 1991). The final epoch of star formation and subsequent explosive events would have displaced this evaporated material into additional shell-like structures, one of which has expanded to the location of the Sun. This picture is quantitatively consistent with the cloud surrounding the solar system (average density 0.1 cm^{-3}) corresponding to the pre-shell gas, and the squall line gas (velocity vector in the LSR $V_g = V, \ell, b = -21.0$ km s^{-1}, 334.5°, +11.8°) corresponding to the expanding shell, if the age of the most recent supernova explosion is $\sim$400,000 yr, and the shell radius of 130 to 160 pc (see Frisch 1995). (The youngest supernova remnant associated with Loop I, and which produces the soft X-ray hot spots, has not yet expanded to the solar vicinity.) Superbubble shell models (see, e.g., MacLow and McCray 1988) give a predicted radius of the shell affecting the solar environment of 160 pc and predicted velocity of 22.4 km s^{-1}, values in agreement with the squall line gas and SCA subgroup distances of 100 to 180 pc.

The enhanced Ca II/Na I ratios and Fe II abundances seen in interstellar material close to the Sun (see, e.g., Frisch 1981; Vallerga et al. 1993; Bertin et al. 1993) are consistent with expected properties of gas consisting partly of destroyed mantle material from dust grains evaporated from cloud surfaces in supernova remnants (White and Long 1991).

The possibility that the cloud surrounding the solar system is a part of a cooling supernova remnant or superbubble shell permits the possibility that the cloud is not in ionization equilibrium. Although both anomalous cosmic ray and interstellar magnesium data give similar values for the electron density, which suggests that the underlying assumption of ionization equilibrium is correct at least for the trace elements, additional data are needed.

IX. COMPARISONS WITH IN-SITU DATA

Inferred cloud properties based on observations of interstellar matter within the solar system can be compared with results based on absorption line studies of nearby stars as a means of resolving inherent uncertainties due to using sight line averaged properties. For the first comparison, recent results from Ulysses He I arrival angle data at 1 to 5 AU (Witte et al. 1993) and earlier He I 584 backscatter data at 1 AU (see, e.g., Ajello et al. 1987, and references therein) give a downwind direction of $\lambda = 72\pm2.4°$, $\beta = -2.5\pm2.7°$ in ecliptic coordinates, interstellar cloud temperature $T = 6700\pm1500$ K, and velocity $v = 26\pm1.0$ km s^{-1}. In heliocentric coordinates (the solar rest frame), this position corresponds to an upwind direction in galactic coordinates of $(\ell, b) = $

$(359.8°, +16.2°)$. In the local standard of rest this upwind vector becomes $(v, \ell, b) = (-19.5$ km s^{-1}, 325.5°, 3.6°$)$. The direction and velocity of the "anti-galactic" vector (heliocentric, $V_{ag} = V, \ell, b = -25.7\pm0.5$ km s^{-1}, 5.9°$\pm\pm$3°, $+16.7°\pm13°$) agrees better with the velocity vector determined from the arrival angle of interstellar He I in the solar system near 5 AU (Witte et al. 1993), suggesting this is the cloud surrounding the solar system (Lallement and Bertin 1992; Lallement et al. 1994). The overall vector derived from 12 nearby stars (heliocentric, $V, \ell, b = -26.8\pm1.3$ km s^{-1}, 6.2° 1$\pm$3°, $+11.7°\pm13°$) is also consistent with the He I arrival angle data. Comparisons of the interstellar velocity derived from Lyman α backscattered data are tricky, however, because substantial deceleration of hydrogen at the heliopause occurs, and it may be combined with re-acceleration within the solar system (Baranov and Malama 1993).

The second comparison can be made between the "pickup ion" population, consisting of neutral H, He, Ne, N and O interstellar atoms that have become ionized by charge exchange with the solar wind (see, e.g., Möbius et al. 1988; Gloeckler et al. 1993). Ulysses observations of interstellar pickup ions in the outer heliosphere give the ratio H/He = 8.4$\pm$2.0 (Gloeckler et al. 1993; G. Gloeckler 1995, personal communication). However, hydrogen may be up to 50% ionized by resonant charge exchange with interstellar protons in the heliopause region "filter" (Baranov and Malama 1993), giving a range for the interstellar ratio H/He = 6.4 to 21. Because some hydrogen filtering at the heliopause is believed to occur, the pickup ion data suggest that preferential helium ionization is present in the surrounding cloud. The pickup ion data are consistent with the average white dwarf value $<N(\mathrm{H\,I})/N(\mathrm{He\,I})>\sim13.7$ previously discussed.

The anomalous cosmic ray component consisting of He, C, N, O, Ne, and Ar are believed to originate as neutral interstellar atoms, which are ionized (becoming pickup ions) and accelerated to cosmic ray energies in the solar system (see, e.g., Garcia-Munoz et al. 1973; Cummings and Stone 1990; Mewaldt et al. 1993a,b). When the anomalous cosmic ray C/O ratio is compared with the Mg I/Mg II ratio towards Sirius, consistent results are found for $n(\mathrm{e}^-) = 0.22$ to 0.44 cm^{-3} if the trace elements in the cloud are in ionization equilibrium (Frisch 1994).

In-situ observations of micron-sized dust grains with an interstellar origin have been detected in the outer solar system by the Ulysses spacecraft (Grün et al. 1992,1993), and are consistent with the high local ionization inferred from absorption line data. These grains are in retrograde orbits with velocities larger than the solar system escape velocity and also consistent with the velocity of the interstellar cloud surrounding the solar system. The Ulysses dust data yield a mean particle mass flux of 6×10^{-13} g cm^{-2} s^{-1}, corresponding to a gas-to-dust mass ratio of $\sim$70 for an ambient neutral density of 0.1 cm^{-3}. This is over 30% lower than the global interstellar gas-to-dust values of 100 to 150 (Devereux and Young 1990), and is therefore consistent with the high inferred local cloud ionization.

X. CONCLUSION

The cloud surrounding the solar system establishes the galactic environment of the Sun, constrains the nature of the solar system interactions with the ambient gas, and influences the characteristics of the solar wind. The characteristics of this cloud are summarized in Table I. The surrounding cloud is part of the leading edge of a cloud complex driven by expanding gas from the Scorpius-Centaurus Association which is sweeping past the Sun. For the past 5 Myr, the Sun has been immersed in a region of space with very low average interstellar densities, $n < 0.0002$ cm^{-3}. Within the last 10,000 yr, the Sun entered the warm, low density surrounding interstellar cloud ($T \sim 7000$ K, $<n(\mathrm{H\,I} + \mathrm{H\,II})> \sim 0.1$ to 0.5 cm^{-3}), and due to the flow of interstellar material near the Sun, the Sun can be expected to encounter much denser regions in the next 5 Myr. The galactic environment of the Sun is changing. Data towards one of our nearest neighbors, α Cen A,B, indicates that within 50,000 yr the Sun may encounter a still denser cloud. Many outstanding questions remain about the cloud surrounding the solar system, such as the space density, excitation, ionization state, morphology, and abundances. Unfortunately, although data from the GHRS spectrograph on the Hubble Space Telescope would allow us to answer these questions, observing time has been unavailable. Our knowledge of this cloud, and the environmental future of the Sun, is data limited.

REFERENCES

Adams, J. H., Jr., Garcia-Munoz, M., Grigorov, N. L., Klecker, Bl., Kondratyeva, M. A., Mason, G. M., McGuire, R. E., Mewaldt, R. A., Panasyuk, M. I., and Tretyakova, Ch. A. 1991. The charge state of the anomalous component of cosmic rays. *Astrophys. J. Lett.* 375:45–49.

Adams, T. F., and Frisch, P. C. 1977. High-resolution observations of the Lyman alpha sky background. *Astrophys. J.* 212:300–308.

Ajello, J. M., Stewart, A. I., Thomas, G. E., and Graps, A. 1987. Solar cycle study of interplanetary La variations: Pioneer Venus Orbiter sky background results. *Astrophys. J.* 317:964–1022.

Ajello, J. M., Pryor, W. R., Barth, C. A., Hord, C. W., Stewart, A. I. F., Simmons, K. E., and Hall, D. T. 1994. Galileo ultraviolet spectrometer: Multiple scattering effects at solar maximum. *Astron. Astrophys.* 317:964–1022.

Axford, W. I. 1972. The interaction of the solar wind with the interstellar medium. In *Solar Wind II*, ed. C. P. Sonett, NASA SP-308, pp. 609–660.

Baranov, V. B., Krasnobaev, K. V., and Kulikovsky, A. G. 1971. A model of the interaction of the solar wind with the interstellar medium. *Soviet Phys. Doklady* 15(9):791–793.

Baranov, V. B., and Malama, Y. G. 1993. Model of the solar wind interaction with the local interstellar medium. *J. Geophys. Res.* 98:15157–15163.

Barstow, M. A., Wesemael, F., Holberg, J. B., Werner, K., Buckley, D. A. H., Stobie, R. S., Fontaine, G., Rosen, S. R., Demers, S., Lamontagne, R., Irwin, M. J., Bergeron, P., Kepler, S. O., and Vennes, S. 1993. Two new hot white dwarfs in a region of exceptionally low H I density. *Adv. Space Res.* 13:281–285.

Bertaux, J. L., and Blamont, J. E. 1971. Evidence for a source of an extraterrestrial hydrogen Lyman-α emission: The interstellar cloud. *Astron. Astrophys.* 11:200–217.

Bertin, P., Lallement, R., Ferlet, R., and Vidal-Madjar, A. 1993. Na I/Ca II ratio in the local interstellar medium. *Astron. Astrophys.* 278:549–560.

Bignami, G. F., Caraveo, P. A., and Mereghetti, S. 1993. The proper motion of Geminga's optical counterpart. *Nature* 361:704–706.

Bochkarev, N. G. 1987. Local interstellar medium. *Astrophys. Space Sci.* 138:229–302.

Boulares, A., and Cox, D. P. 1990. Galactic hydrostatic equilibrium with magnetic tension and cosmic ray diffusion *Astrophys. J.* 365:544–555.

Bzowski, M. 1988. Local Interstellar wind velocity from doppler shifts of interstellar matter lines. *Acta Astron.* 38:443–453.

Cassinelli, J. P., Cohen, D. H., MacFarlane, J. J., Drew, J. E., Lynas-Gray, A. E., Hoare, M. G., Vallerga, J. V., Welsh, B. Y., Veder, P. W., Hubeny, I., and Lanz, T. 1995. EUVE spectroscopy of epsilon CMa from 70 Å to 730 Å. *Astrophys. J.* 438:932–941.

Cheng, K.-P., and Bruhweiler, F. C. 1990. Ionization processes in the local interstellar medium effects of the hot coronal substrate *Astrophys. J.* 364:573–580.

Cox, D. P., and Reynolds, R. J. 1987. The local interstellar medium. *Ann. Rev. Astron. Astrophys.* 25:303–344.

Cox, D. P., and Snowden, S. L. 1986. Perspective on the local interstellar medium. *Adv. Space Res.* 6:97–108.

Cruddace, R., Paresce, F., Bowyer, S., and Lampton, M. 1974. Opacity of ISM to soft X-rays and EUV radiation. *Astrophys. J.* 187:497–504.

Crutcher, R. M. 1982. The local interstellar medium. *Astrophys. J.* 254:82–87.

Cummings, A. C., and Stone, E. C. 1990. Elemental composition of the very local interstellar medium as deduced from observations of anomalous cosmic rays. *Proc. 21st Intl. Cosmic Ray Conf.* 6:190–192.

Cummings, J. R., Cummings, A. C., Mewaldt, R. A., Selesnic, R. S., Stone, E. C., and von Rosenvinge, T. T. 1993. New evidence for anomalous cosmic rays trapped in the magnetosphere. *Geophys. Res. Lett.* 20:2003–2006.

Davelaar, J., Bleeker, J. A. M., and Deerenberg, A. J. M. 1980. X-ray characteristics of Loop I and the local interstellar medium. *Astron. Astrophys.* 92:231–237.

De Geus, E. J. 1991. Interaction of stars and interstellar matter in Scorpio Centaurus. *Astron. Astrophys.* 262:258–270.

Devereux, N. A., and Young, J. S. 1990. The gas/dust ratio in spiral galaxies. *Astrophys. J.* 359:42–55.

Duley, W. W. 1978. Magnetic alignment of interstellar grains. *Astrophys. J. Lett.* 219:129–132.

Fahr, H. J. 1971. The interplanetary hydrogen cone and its solar cycle variation. *Astron. Astrophys.* 14:263–280.

Ferrara, A. 1994. On turbulence decay in the local bubble and its implications for the galactic ISM. In *Unsolved Problems of the Milky Way*, ed. L. Blitz (Dordrecht: Kluwer).

Finley, D. S., and Kimble, R. A. 1993. EUV spectra of hot DA white dwarfs. *Bull. Amer. Astron. Soc.* 25:878 (abstract).

Finley, D. S., Jelinsky, P., Dupuis, J., and Koester, D. 1993. Initial white dwarf results from the Extreme Ultraviolet Explorer. *Astrophys. J.* 417:259–264.

Frisch, P. C. 1981. The nearby interstellar medium. *Nature* 293:377–379.

Frisch, P. C. 1986. The physical properties of the "local fluff." *Adv. Space Res.* 6:345–352.

Frisch, P. C. 1990. Characteristics of the local interstellar medium. In *Physics of the Outer Heliosphere*, eds. S. Grzedzielski and E. Page (Oxford: Pergamon Press), pp. 19–21.

Frisch, P. C. 1993. Whence Geminga medium. *Nature* 364:395–396.

Frisch, P. C. 1994. Morphology and ionization of the interstellar cloud surrounding the solar system. *Science* 265:1423–1427.

Frisch, P. C. 1995. Characteristics of nearby interstellar matter. *Space Sci. Rev.* 72:499–591.

Frisch, P. C., and York, D. G. 1983. Synthesis maps of ultraviolet observations of neutral interstellar gas. *Astrophys. J. Lett.* 271:59–63.

Frisch, P. C., and York, D. G. 1986. Interstellar clouds near the Sun. In *The Galaxy and the Solar System*, eds. R. Smoluchowski, J. N. Bahcall and M. S. Matthews (Tucson: Univ. of Arizona Press), pp. 83–100.

Frisch, P. C., and York, D. G. 1990. The distribution of nearby H I and H II gas. In *Extreme Ultraviolet Astronomy*, eds. R. F. Malina and S. Bowyer (Oxford: Pergamon Press), pp. 322–332.

Frisch, P. C., Welty, D. E., York, D. G., and Fowler, J. R. 1990. Ionization in nearby interstellar gas. *Astrophys. J.* 357:514–523.

Garcia-Munoz, M., Mason, G. M., and Simpson, J. A. 1973. A new test for solar modulation theory the 1972 May–July low-energy galactic cosmic-ray proton and helium spectra. *Astrophys. J. Lett.* 182:81–84.

Garmire, G. P., Nousek, J. A., Apparao, K. M. V., Burrows, D. N., Fink, R. L., and Kraft, R. P. 1992. The soft X-ray diffuse background observed with the HEAO 1 low-energy detectors. *Astrophys. J.* 399:694–703.

Gehrels, N., and Chen, W. 1993, The Geminga supernova as a possible cause of the local interstellar bubble. *Nature* 361:706–707.

Genova, R., Molaro, P., Vladilo, G., and Beckman, J. E. 1990. Mg II observed in the local interstellar medium the local cloud. *Astrophys. J.* 355:150–158.

Gloeckler, G., Geiss, J., Balsiger, H., Fisk, L. A., Galvin, A. B., Ipavich, F. M., Ogilvie, K. W., von Steiger, R., and Wilken, B. 1993. Detection of interstellar pick-up hydrogen in the solar system. *Science* 261:70–75.

Grün, E., Zook, H. A., Baguhl, M., Fechtig, H., Hanner, M. S., Kissel, J., Lindblad, B. A., Linkert, D., Linkert, G., Mann, I. B., McDonnell, J. A. M., Morfill, G. E., Polanskey, C., Riemann, R., Schwehm, G., and Siddique, N. 1992. Ulysses dust measurements near Jupiter. *Science* 257:1550–1552.

Grün, E., Zook, H. A., Baguhl, M., Balogh, A., Bame, S. J., Fechtig, H., Forsyth, R., Hanner, M. S., Horanyi, M., Kissel, J., Lindblad, B. A., Linkert, D., Linkert, G., Mann, I., McDonnell, J. A. M., Morfill, G. E., Phillips, J. L., Polanskey, C., Schwehm, G., Siddique, N., Staubach, P., Svestka, J., and Taylor, A. 1993. Discovery of Jovian dust streams and interstellar grains by the Ulysses spacecraft. *Nature* 362:428–430.

Gry, C., York, D. G., and Vidal-Madjar, A. 1985. The exceptionally vacant line of sight to Beta Canis Majoris. *Astrophys. J.* 296:593–599.

Gry, C., Lemonon, L., Vidal-Madjar, A., Lemoine, M., and Ferlet, R. 1995. The interstellar void in the direction to epsilon CMa local clouds and hot gas. *Astron. Astrophys.* 302:497–508.

Heiles, C. 1986. Interstellar magnetic fields. In *Interstellar Processes*, eds. D. J. Hollenbach and H. A. Thronson, Jr. (Dordrecht: D. Reidel), pp. 171–194.

Holzer, T. E. 1972. Interaction of the solar wind with the neutral component of the interstellar gas. *J. Geophys. Res.* 77:5407–5420.

Jelinsky, P., Vallerga, J. V., and Edelstein, J. 1995. First spectral observations of the diffuse background with the Extreme Ultraviolet Explorer. *Astrophys. J.* 442:653–661.

Jones, D. H. P. 1971. Kinematics of the Scorpio-Centaurus association and Gould's Belt. *Mon. Not. Roy. Astron. Soc.* 152:231–259.

Kimble, R. A., Davidsen, A. F., Blair, W. P., Bowers, C. W., Dixon, W. V., Durrance, S. T., Feldman, P. D., Ferguson, H. C., Henry, R. C., Kriss, G. A., Druk, J. W., Long, K. S., Moos, H. W., and Vancura, O. 1993*a*. Extreme ultraviolet observations of G191-B2B and the local interstellar medium with the Hopkins Ultraviolet Telescope. *Astrophys. J.* 404:663–670.

Kimble, R. A., Davidsen, A. F., Long, K. S., and Feldman, P. D. 1993*b* Extreme ultraviolet observations of HZ43 and the local H/He ratio with the Hopkins Ultraviolet Telescope. *Astrophys. J. Lett.* 408:41–44.

Kondo, Y., Bruhweiler, F. C., and Savage, B. D. 1984. *Local Interstellar Medium*, NASA CP-2345.

Lallement, R. 1990. Scattering of solar UV on local neutral gases. In *Physics of the Outer Heliosphere*, eds. S. Grzedzielski and E. Page (Oxford: Pergamon Press), pp. 49–60.

Lallement, R. 1993, Measurements of the interstellar gas. *Adv. Space Res.* 13:113–120.

Lallement, R., and Bertin, P. 1992. Northern-hemisphere observations of nearby interstellar gas possible detection of the local cloud. *Astron. Astrophys.* 266:479–484.

Lallement, R., Vidal-Madjar, A., and Ferlet, R. 1986. Multi-component velocity structure of the local interstellar medium. *Astron. Astrophys.* 168:225–236.

Lallement, R., Bertin, P., Ferlet, R., Vidal-Madjar, A., and Bertaux, J. L. 1994. Interstellar clouds toward Sirius and local cloud ionization. *Astron. Astrophys.* 286:898–908

Landsman, W. B., Henry, R. C., Moos, H. W., and Linsky, J. L. 1984. Observation of interstellar hydrogen and deuterium toward Alpha Centauri A *Astrophys. J.* 285:801–815.

Landsman, W. B., Murthy, J., Henry, R. C., Moos, H. W., Linsky, J. L., and Russell, J. L. 1986. IUE observations of interstellar hydrogen and deuterium toward Alpha Cen B. *Astrophys. J.* 303:791–855.

Leroy, J. L. 1993. A polarimetric investigation on interstellar dust within 50 pc from the Sun. *Astron. Astrophys.* 274:203–213.

Linsky, J. L., Diplas, A., Ayres, T., Wood, B., and Brown, A. 1993*a*. A reanalysis of the interstellar medium along the Capella line of sight. *Bull. Amer. Astron. Soc.* 25:1464 (abstract).

Linsky, J. L., Brown, A., Gayley, K., Diplas, A., Savage, B. D., Ayres, T. R., Landsman, W., Shore, S. N., and Heap, S. R. 1993*b*. GHRS observations of the local interstellar medium and the deuterium/hydrogen ratio along the line of sight toward Capella. *Astrophys. J.* 402:694–709.

Lucke, P. B. 1978. Distribution of color excesses and interstellar reddening material in the solar neighborhood. *Astron. Astrophys.* 64:367–377.

MacLow, M.-M., and McCray, R. 1988. Superbubbles in disk galaxies. *Astrophys. J.* 324:776–785.

Mathewson, D. S., and Ford, V. L. 1970. Polarization observations of 1800 stars. *Mon. Not. Roy. Astron. Soc.* 74:139–182.

Mathis, J. S. 1986. The alignment of interstellar grains. *Astrophys. J.* 308:281–294.

Mathis, J. S. 1990. Interstellar dust and extinction. *Ann. Rev. Astron. Astrophys.* 28:37–70.

McCammon, D., and Sanders, W. T. 1990. The soft X-ray background and its Origins.

Ann. Rev. Astron. Astrophys. 28:657–688.

McClintock, W., Henry, R. C., Linsky, J. L., and Moos, H. W. 1978. Ultraviolet observations of cool stars. VII. Local interstellar hydrogen and deuterium Lyman alpha. *Astrophys. J.* 225:465–481.

McKee, C., and Cowie, L. L. 1977. The evaporation of spherical clouds in a hot gas. II. Effects of radiation. *Astrophys. J.* 215:213–225.

Mewaldt, R. A., Cummings, A. C., Cummings, J. R., Stone, E. C., and von Rosenvinge, T. T. 1993*a* The return of the anomalous cosmic ray component to 1 AU in 1992. *Geophys. Res. Lett.* 20:2263–2269.

Mewaldt, R. A., Cummings, A. C., Cummings, J. R., Stone, E. C., and von Rosenvinge, T. T. 1993*b* The return of the anomalous cosmic ray component to 1 AU in 1992. *Proc. 23rd Intl. Cosmic Ray Conf.* 3:401–403.

Mihalas, D., and Binney, J. 1981. *Galactic Astronomy: Structure and Kinematics* (San Francisco: W. H. Freeman), pp. 398–399.

Möbius, E., Hovestadt, D., Klecker, B., Scholer, M., Gloeckler, G., and Ipavich, F. M. 1985. Direct Observation of He II pick-up ions of interstellar origin in the solar wind. *Nature* 318:426–429.

Möbius, E., Klecker, B., Hovestadt, D., Scholer, M. 1988. Interaction of Interstellar pick-up ions with the solar wind. *Astron. Space Sci.* 144:487–496.

Murthy, J., Henry, R. C., Moos, H. W., Landsman, W. B., Linsky, J. L., Vidal-Madjar, and Gry, C. 1987. IUE observations of hydrogen and deuterium in the local interstellar medium. *Astrophys. J.* 315:675–684.

Paerels, F. B. S., and Heise, J. 1989. A soft X-ray survey of hot white dwarfs with EXOSAT. *Astrophys. J.* 339:1000–1012.

Paresce, F. 1984. On the distribution of interstellar matter around the Sun. *Astron. Astrophys.* 89:1022–1037.

Rand, R. J., and Kulkarni, S. R. 1989. Local galactic magnetic field. *Astrophys. J.* 343:760–772.

Rand, R. J., and Lyne, A. G. 1994. New rotation measures of distant pulsars in the inner galaxy and magnetic field reversals. *Mon. Not. Roy. Astron. Soc.*, 268:497–505.

Reynolds, R. J. 1990. Lower limit on the ionization fraction of the very local interstellar medium. In *Physics of the Outer Heliosphere*, eds. S. Grzedzielski and E. Page (Oxford: Pergamon Press), pp. 101–104.

Sanders, W. T., Kraushaar, W. Al., Nousek, J. A., and Fried, P. M. 1977. Soft diffuse x-rays in the southern galactic hemisphere. *Astrophys. J. Lett.* 217:87–91.

Slavin, J. D. 1989. Consequences of a conductive boundary on the local cloud. I. No dust. *Astrophys. J.* 346:718–727.

Smith, V. V., Cunha, K., Plez, B. 1993. Is Geminga a runaway member of the Orion association? *Astron. Astrophys.* 281:L41–L44.

Spangler, S. R. 1991. The dissipation of magnetohydrodynamic turbulence responsible for interstellar scintillation and the heating of the interstellar medium. *Astrophys. J.* 376:540–555.

Spitzer, L. 1978. *Physical Processes in the Interstellar Medium* (New York: Wiley and Sons).

Thomas, G. E., and Krassa, R. F. 1971. OGO 5 measurements of the Lyman alpha sky background. *Astron. Astrophys.* 11:218–233.

Tinbergen, J. 1982. Interstellar polarization in the immediate solar neighborhood. *Astron. Astrophys.* 105:53–64.

Vallerga, J. V., Vedder, P. W., Craig, N., and Welsh, B. Y. 1993. High-resolution Ca II observations of the local interstellar medium. *Astrophys. J.* 411:729–749.

Vennes, S., Dupuis, J., Rumph, T., Drake, J., and Bowyer, S. 1993*a*. The first detection of ionized helium in the local ISM EUVE and IUE spectroscopy of the hot DA white dwarf GD 246. *Astrophys. J. Lett.* 410:119–122.

Vennes, S., Dupuis, J., Bowyer, S., Fontaine, G., Wiercigroch, A., Jelinsky, P., We-semael, F., and Malina, R. 1993*b*. Low density of neutral hydrogen and helium in the local ISM EUVE photometry of the Lyman continuum of the hot white dwarfs MCT 0501-2858, MCT 0455-2812, HZ 43, and GD 153. *Astrophys. J. Lett.* 421:35–38.

Verschuur, G. L. 1993. Zeeman Effect Observations of HI Emission Profiles: 3 Low Field Limits in the Diffuse Interstellar Medium. Preprint.

Vidal-Madjar, A., Laurent, C., Burston, P., and Audouze, J. 1978. Is the solar system entering a nearby interstellar cloud? *Astrophys. J.* 223:589–600.

Warwick, R. S., Barber, C. R., Hodgkin, S. T., and Pye, J. P. 1993. The EUV source population and the local bubble. *Mon. Not. Roy. Astron. Soc.*, 262:289–300.

Welty, D. E., Hobbs, L. M., and Kulkarni, V. P. 1994. High resolution survey of interstellar Na I D1 lines. *Astrophys. J.* 436:152–175.

White, R. L., and Long, K. S. 1991. Supernova remnant evolution in an interstellar medium with evaporating clouds. *Astrophys. J.* 373:543–555.

Witte, M., Rosenbauer, H., Banaszkiewicz, M., and Fahr, H. 1993. Ulysses neutral gas experiment determination of the velocity and temperature of the interstellar neutral helium. *Adv. Space Res.* 13:(6)121–130.

THE TERMINATION SHOCK AND THE HELIOSHEATH

S. T. SUESS
NASA Marshall Space Flight Center

and

STEVEN NERNEY
Ohio University at Lancaster

The solar system is moving through the local interstellar medium at approximately 25 km s^{-1}. An important component of the local interstellar medium is a 10^4 K plasma that, because of the motion of the solar system, is effectively an interstellar wind blowing on the heliosphere. The heliosphere provides resistance to this flow due to the presence of the solar wind so that the interaction between the two is a type of stagnation point flow. There are many other interesting ingredients to the system, including cosmic rays, a neutral interstellar gas, the interplanetary magnetic field, and the interstellar magnetic field. It is possible to construct analytic and numerical models that, to first order, show the character of the interaction. The main features include a shock transition as the supersonic solar wind slows and is turned to flow down a heliotail, a contact surface dividing this subsonic solar wind flow from the adjacent interstellar medium, and the possibility of a bow shock in the external interstellar flow. The solar wind is strongly time dependent on time scales from less than a solar rotation to a solar cycle, and has large latitudinal gradients, both of which result in distortions of the termination shock away from steady, spherical symmetry. The external flow also imposes large asymmetries in the shape of the termination shock.

I. INTRODUCTION

The heliosphere is that volume of space dominated by the solar wind. The heliopause is the contact discontinuity dividing interstellar plasma from the solar wind; the solar wind flows down the heliotail eventually to become parallel to the local interstellar wind. There is possibly a bow shock in the interstellar medium (if the interstellar wind is supersonic and superAlfvénic), and the heliosheath is the volume between the termination shock and the heliopause. The termination shock arises because the solar wind must accommodate itself to the pressure, both gas and dynamic, of the interstellar gas. This produces subsonic flow (in the Sun's frame of reference) in the heliosheath. To illustrate this terminology, it is helpful to use the schematic diagram of the expected configuration shown in Fig. 1 (Suess 1990). The termination shock (labeled "terminal shock" in this figure) is shown as a wavy line because the shock is probably time-dependent, asymmetric, and thickened by dissipative processes. This figure is based on physical arguments and model calculations influenced by early ideas of what the interaction should be like. Presently, these ideas are still weakly constrained by observations; therefore our preconceived notions may not be accurate.

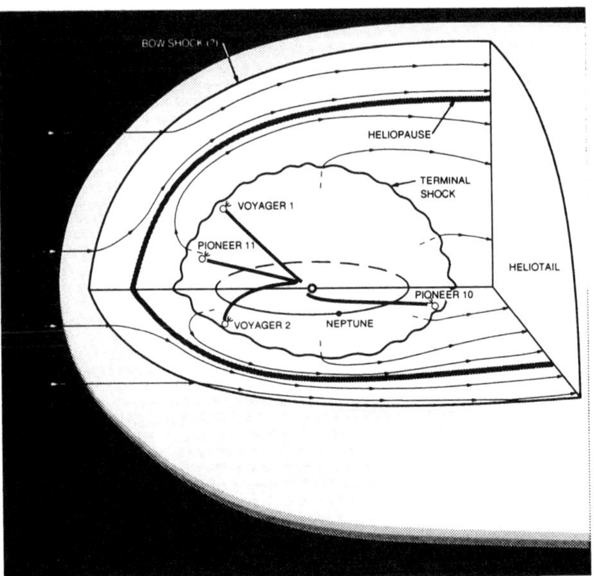

Figure 1. Schematic diagram of the major interfaces in the heliosphere. The drawing is roughly to the scale of what is believed to be the true size of the heliosphere, and the orbit of Neptune is shown to give dimensions to that scale. Pioneer 11 and Voyagers 1 and 2 are traveling roughly upstream, with Voyagers 1 and 2 moving in opposite directions out of the ecliptic plane. Pioneer 10 is moving downstream (Suess 1990).

The configuration in Fig. 1 exists because of the motion of the heliosphere through the local interstellar medium. Evidence for this motion comes from the detection of solar Lyman α and helium 304 Å radiation that has been backscattered off of interstellar neutral atoms entering the solar system (Quemerais et al. 1993; Lallement et al. 1990). There is supporting evidence from *in situ* detection of ions derived from interstellar atoms by the Ulysses and AMPTE missions (Gloeckler et al. 1993; Möbius et al. 1985) and from direct interstellar neutral helium detection (Witte et al. 1992). The backscatter phenomenon originates with interstellar neutral atoms entering the solar system essentially unimpeded by the plasma and magnetic fields. They are acted upon only by photon pressure and gravity, so their orbits are easily calculated. They are eventually ionized through charge exchange or photoionization. Before being ionized, they resonantly scatter solar Lyman α radiation, which can be detected by Earth-orbiting and interplanetary spacecraft. The collision time between interstellar neutral atoms and ions is $\sim 10^3$ yr and the mean free path is 10^3 to 10^4 AU, both of which are short enough so the neutral atoms are moving together with the interstellar plasma in the vicinity of the heliosphere. Therefore, the map of the scattered radiation carries information on the velocity vector of the interstellar neutral atoms, which then tells us from which direction and how fast local interstellar plasma is coming into the heliosphere (measurements of stellar spectra are used to estimate the density and temperature of the plasma and the strength and direction of the interstellar magnetic field).

The conclusion is that there is an effective interstellar wind moving approximately in the plane of the ecliptic at a speed of about 25 km s^{-1}. The plasma density and temperature, poorly known, are about 0.1 cm^{-3} and 10^4K, respectively (Frisch 1994; chapter by Frisch; Bertin et al. 1993; Lallement et al. 1990). The Lyman α backscatter maximum is in the vicinity of 5 AU, depending on the phase of the solar cycle, so its parallax is easily measured from 1 AU and the above numbers for the velocity vector can be accepted with some confidence.

The properties of the solar wind at 1 AU are well known; the density and proton temperature are of the order 7 cm^{-3} and 5×10^4 K, respectively. The flow is essentially radial and the speed ranges from 350 to more than 1000 km s^{-1} due to the presence of high-speed coronal hole flows and transient solar activity. There is also a magnetic field which is drawn into a spiral due to the rotation of the Sun (Parker 1963). Because it is a magnetized, ionized gas, the solar wind presents an obstacle to the plasma in the local interstellar medium, resulting in a kind of stagnation point flow. Because the solar wind is supersonic, it is possible to deflect it easily as it approaches the stagnation point only if it is first slowed to subsonic speed, therefore leading to a termination shock that surrounds the solar system. The position and shape of this shock is determined by the total pressures in the solar wind and the local interstellar medium.

The scale size of this heliospheric stagnation point flow is suggested by the location of the orbit of Neptune in Fig. 1. Recently, direct evidence has been found that this suggested scale is correct, in the form of a detection of low-frequency radio waves in the outer solar system by Voyagers 1 and 2. This has been used to infer that the shock lies at 80 to 120 AU and the heliopause at 100 to 150 AU in the upstream direction (Gurnett et al. 1993). The trajectories of Pioneers 10 and 11 and Voyagers 1 and 2 are roughly shown in Fig. 1 to indicate how they are located and moving with respect to the direction of arrival of the interstellar wind. The Voyager spacecraft will almost certainly survive not only to encounter the termination shock, but also the heliopause (Pesses et al. 1993). They are moving generally in the upstream direction, but Voyager 1 and 2 are moving north and south at heliographic angles that will eventually exceed 30°N (V1) and S (V2). Their heliocentric distances in mid-1994 are approximately 59 AU, 55 AU, and 42 AU for Pioneer 10 and Voyagers 1 and 2, respectively.

Efforts to model the heliosphere began in the 1950s, when the nature of the interaction was first suggested (Davis 1955), and have continued sporadically up to the present. Now, with the anticipated encounter of the termination shock and heliopause, several specific problems have been identified that will be necessary to solve in order to interpret the resulting data.

Section II is a summary of the dominant hydrodynamic processes, with a calculation of the termination shock standoff distance in the case of the most simplifying assumptions. It also includes a description of some numerical models, time-dependent motion of the termination shock, and termination shock asymmetry. Section III describes what is known about the magnetic field in the heliosheath. Section IV contains a brief summary. Recent comprehensive reviews of these topics have been published by Suess (1990), Holzer (1989), Baranov (1990), and in a conference proceedings on the same topic (Grzedzielski and Page 1990).

II. A HYDRODYNAMIC DESCRIPTION

The interaction between the heliosphere and the interstellar medium is most easily thought of as being a hydrodynamic problem modified by magnetic fields. The magnetic field can dominate the dynamics locally, and modify the shapes of the heliopause and termination shock, but probably does not alter the overall topology from that illustrated in Fig. 1. Therefore, the flows will be described from a hydrodynamic point of view here and magnetic fields will be addressed in Sec. III.

A. Termination Shock Standoff Distance

Without the solar wind, interstellar plasma and magnetic fields would fill the solar system inward to the point where pressure equilibrium was reached with the static solar atmosphere. That such a situation is untenable was realized when it was found that the temperature of the corona is over 10^6 K (Parker 1963). This temperature, for a spherically symmetric, thermally conducting corona, leads to a finite gas pressure infinitely far from the Sun that is many orders of magnitude larger than any possible interstellar pressure. No equilibrium exists and the corona expands to form the solar wind which, in the absence of an interstellar medium, leads to zero temperature and pressure infinitely far from the Sun. The interstellar pressure is somewhere between zero and that predicted for a static corona. Therefore, there is a contact discontinuity between the two media, the heliopause, plus the shock transition that slows the solar wind as it runs into the interstellar medium. In the simplest case, the expanding solar wind would blow a bubble of continuously increasing size in the interstellar plasma and the problem is spherically symmetric. Motion of the heliosphere through the local interstellar medium gives a preferred direction to the configuration and the solar wind turns to flow downstream, out the heliotail.

TABLE I
Properties of the Local Interstellar Medium[a]

Neutral Component	
Flow speed	23 ± 3 km s^{-1} [b]
Flow direction	75.4° ecliptic longitude
	$-7.5°$ ecliptic latitude
Hydrogen density	0.10 ± 0.03 cm^{-3}
Helium density	0.010 ± 0.005 cm^{-3}
Hydrogen temperature	$(9\pm2)\times10^3$ K
Helium temperature	$(9\pm2)\times10^3$ K

Ionized Component	
Electron density	<0.3 cm^{-3} (estimated to be 0.1 cm^{-3})
Flow speed	Assumed the same as the neutral component
Flow direction	Assumed the same as the neutral component
Ion temperature	Assumed the same as the neutral component

Magnetic Field	
Magnitude	$0.3-0.5\ nT$ [c]
Direction	Unknown

Cosmic Rays	
Total pressure	$(1.3\pm0.2)\times10^{-12}$ dyn cm^{-2} [d]

[a] Sources are as follows: for the neutral component, Chassefiere et al. (1986) and Holzer (1989); for the ionized component and the magnetic field, Frisch (1994 and chapter by Frisch), Frisch et al. (1987), Cox and Reynolds (1987), and Bochkarev (1987); and for cosmic rays, Ip and Axford (1985).
[b] The large uncertainty is due to the difference between investigators.
[c] The total uncertainty may be as large as an order of magnitude.
[d] The pressure from cosmic rays having <300 meV/nucleon is $\sim(3\pm2)\times10^{-13}$ dyn cm^{-2}.

An estimate of the distance to the termination shock begins with the assumption of a balance between the total pressures in the solar wind and the local interstellar medium at the stagnation point (anything else would imply a moving stagnation point). Flow in the heliosheath is subsonic so, to a first approximation, it can be assumed that this flow is incompressible and the pressure at the stagnation point is the same as the pressure at the "sub-stagnation point," the point on the termination shock directly towards the Sun from the stagnation point. This is the pressure of the shocked solar wind, which equals the total pressure of the solar wind just inside of the shock. The dynamic pressure, $\rho_{sw}v_{sw}^2$, dominates the total pressure so long as the solar wind is supersonic and super-Alfvénic, which it is at 1 AU and, presumably, inside the termination shock at the sub-stagnation point. Therefore solar wind properties at 1 AU can be used to estimate the pressure at the stagnation point. At 1 AU, the average speed and density of the solar wind (v_{sw}, ρ_{sw}) are ~400 km s^{-1} and 7 cm^{-3}. As the wind flows outward, its velocity remains nearly constant while its density decreases as the inverse square of the distance. Thus, at the termination shock the solar wind pressure is approximately

$$\left(\rho_{sw} v_{sw}^2\right) \big|_{1\ \mathrm{AU}} \times \left(\frac{1}{R_S^2}\right) \tag{1}$$

where R_S is the upstream distance to the shock in AU. If the pressure in the interstellar medium is given by p_I, then the pressure balance can be approximated by

$$\frac{1}{R_S^2} \left(\rho_{sw} v_{sw}^2\right) \big|_{1\ \mathrm{AU}} = p_I. \tag{2}$$

Coefficients in this equation can be used to allow for the amplification of the interstellar magnetic field as it drapes over the heliosphere and of the compressibility in the heliosheath (Axford 1972; Suess 1990). It is believed that the main contributions to p_I come from the thermal pressure, dynamic pressure, and magnetic pressure. Potential additional contributions come from galactic cosmic rays and interstellar neutral gas. Estimates of the numerical values and some indications of the uncertainties for the various contributions to p_I are given in Table I. More recent and comprehensive information can be found in Frisch (1994; chapter by Frisch; and Bertin et al. 1993). Substituting numbers from Table I into Eq. (2) gives a broad range for R_S (Axford and Suess 1994). As an example, if the cosmic ray pressure is neglected, and we use $B_I = 0.3$ nT, $n_I = 0.1$ cm^{-3}, a mass per particle of 2×10^{-24} g to account for the presence of some helium, $T_I = 7000$ K, $v_I = 25$ km s^{-1}, and the above numbers for the solar wind at 1 AU, then $R_S = 100$ AU. With these numbers, the relative importances of the dynamic, magnetic and thermal pressures in the interstellar medium are in the ratios ~6.5:1.6:1. Thus, the magnetic field plays an important role in the pressure balance.

B. A Potential Flow Model

A simple global model for flow in the heliosheath and outside the heliopause is found by assuming the heliosheath and external flow are subsonic. Inspection of Table I shows that this is a poor approximation for the interstellar medium, but a rationale is offered by noting that the external flow is probably sub-Alfvénic and hence behaves somewhat like subsonic flow. This may not be a strong rationale, because the calculation proceeds by ignoring the interstellar magnetic field. Nevertheless, such a calculation can be completely analytic and, hence, lead to insight into physical aspects of the problem, even if incorrect in detail.

Figure 2 shows, more quantitatively, the definitions made in Fig. 1. However, unlike that figure, the streamlines shown here have been computed

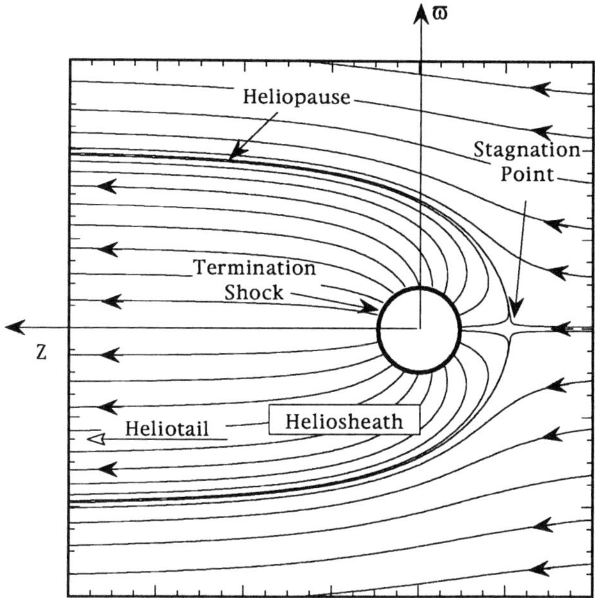

Figure 2.   Streamline plot for the irrotational flow solution, for $\epsilon = 0.125$, an appropriate choice for the heliosphere. The heliopause, heliosheath, heliotail, termination shock, and stagnation point are defined. There is no external heliosheath or bow shock because the flow is incompressible beyond the termination shock. The flow is symmetric about the heliotail so the streamlines are the same in any plane containing the axis of the heliotail (Suess and Nerney 1990,1991).

from the analytic potential flow model. Besides incompressibility, this model incorporates other simplifying assumptions. First, the shock is assumed spherical with axisymmetric flow about the heliotail. The flow is further assumed to be nonrotational in the heliosheath and the interstellar medium and time variations are ignored. The problem is thereby reduced to one of simple stationary, potential, stagnation point flow. The boundary conditions are that the flow is aligned with the interstellar wind infinitely far down the heliotail and that the radial flow speed on the downstream side of the shock is determined by the shock jump condition. The solar wind is taken to be spherically symmetric inside the termination shock and magnetic fields are ignored everywhere.

One more boundary condition remains; i.e., what will be the nature of the flow along the heliopause on the heliotail. Two choices are a free-slip boundary (no viscosity) and a non-slip boundary. The choice depends on whether one believes the flows in the heliotail and interstellar medium are coupled through some sort of viscous-like interaction at the heliopause. For the time being, we will assume that this is a non-slip boundary—the alternative will be discussed at the end of this section.

This potential flow problem is easily solved in terms of a scalar velocity potential Φ, such that (Parker 1963; Suess and Nerney 1990,1991)

$$\nabla^2 \Phi = 0 \tag{3}$$

$$\mathbf{v} = -\nabla\Phi. \tag{4}$$

The solution is axisymmetric about the heliotail, so the velocity is logically written in cylindrical, heliotail coordinates $(\tilde{\omega}, \phi, z)$, where $\tilde{\omega}$ is the cylindrical distance from the z-axis, which in turn is along the symmetry axis of the heliotail (see Fig. 2). The angle ϕ is measured counterclockwise, from the equatorial plane, about the z-axis. The components of the flow field in these coordinates are:

$$v_{\tilde{\omega}} = \frac{v_s R_S^2 \tilde{\omega}}{r^3}\left[1 - \frac{3R_S\epsilon^{2/3}z}{2r^2}\right] \tag{5a}$$

$$v_z = v_s\left[\frac{R_S^2 z}{r^3} - \epsilon^{2/3}\left(R_S^3\frac{(z^2 - \tilde{\omega}^2/2)}{r^5} - 1\right)\right] \tag{5b}$$

$$v_\phi = 0 \tag{5c}$$

$$\psi = -v_s\left[\frac{R_S^2 z}{r} + \frac{1}{2}\epsilon^{2/3}\tilde{\omega}^2\left(\frac{R_S^3}{r^3} - 1\right)\right] \tag{5d}$$

where

$$r = (\tilde{\omega}^2 + z^2)^{1/2} \tag{6a}$$

$$\tilde{\omega} = (x^2 + y^2)^{1/2} \tag{6b}$$

and

$$\epsilon = \left(\frac{v_{i\infty}}{v_s}\right)^{3/2}. \tag{7}$$

v_s is the solar wind speed on the downstream side of the termination shock and $v_{i\infty}$ is the velocity of the far-field interstellar wind. Equation (5d) is an equation for a streamline constant ψ and is derived from Eqs. (5a) and (5b). The streamlines shown in Fig. 2 are computed from these equations for $\epsilon = 0.125$. Because of this assumed symmetry, Fig. 2 applies equally well to any plane that includes the heliotail axis.

$\epsilon = 0.125$ results from using $v_s = 100$ km s^{-1} and $v_{i\infty} = 25$ km s^{-1} in Eq. (7). These numbers are appropriate for the heliosphere (see Table I) if the termination shock is a strong shock so that the solar wind speed is reduced by a factor of 4. This solution therefore represents, under the assumptions listed above, a realistic picture of the relative positions of the termination shock and heliopause, the shape of the heliopause, and the relative cross-sectional area of the heliotail. It is important to note that for this value of ϵ, the ratio of the distance to the stagnation point to the termination shock is 2. Therefore, the heliopause is twice as far from the Sun as the termination shock, in the upstream direction. Evidence will be given below that this is probably an upper bound. The effect of changing ϵ is illustrated in Fig. 3. The top half of this figure is plotted for $\epsilon = 0.5$ and the bottom half is for $\epsilon = 0.05$. In the top diagram, the streamlines are noticeably nonradial on the termination shock and the upstream heliosheath is thinner than the shock radius. In the bottom diagram, the streamlines are approximately radial on the termination shock and the heliosheath is thicker than the shock radius.

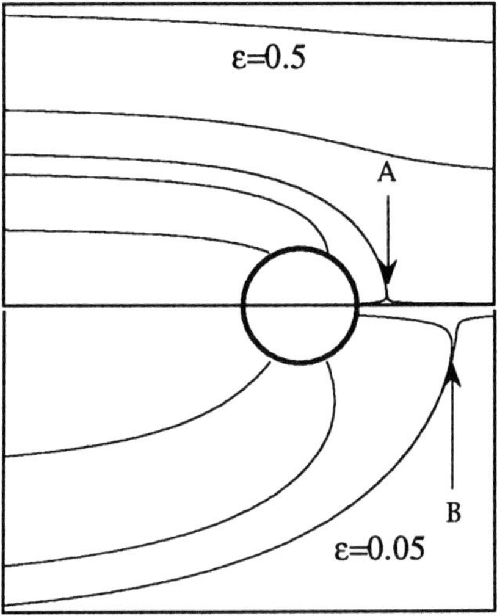

Figure 3.   Two sets of streamlines for different values of the parameter ϵ (Eqs. 7 and 9) in the irrotational flow solution. The top half of the figure shows the solution for $\epsilon = 0.5$, and the bottom half for $\epsilon = 0.05$, demonstrating graphically the point that small ϵ implies a thick upstream heliosheath.

ϵ must be small for the irrotational flow approximation to be globally valid, because ϵ describes the degree to which the flow is nonradial on R_S. That is (Suess and Nerney 1990)

$$\left[\frac{v_\theta}{v_r}\right]_{R_S} = -\frac{3}{2}\epsilon^{2/3}\sin\theta. \tag{8}$$

The potential flow model admits specifying a scalar function on the termination shock and we have assumed spherically symmetric, radial solar wind flow inside R_S. Equation (8) implies the presence of non-radial flow at the shock and hence, because v_θ is discontinuous at R_S, there is also finite vorticity at the shock. Therefore, this is a measure of the degree to which the irrotational flow approximation is accurate. It has been shown recently in numerical simulations (Steinolfson et al. 1994) that the change in flow direction is accommodated within a thin boundary layer just beyond R_S. Thus it is only in this thin layer that the approximation is inaccurate.

The above calculation was done assuming a non-slip boundary condition on the flanks of the heliotail, so that $v_z(z \to \infty) = v_{i\infty}$. An alternative is to assume a free-slip boundary in which there is a balance in dynamic pressure across the heliopause in the distant tail (Parker 1963). Mathematically, this results in a different definition for ϵ, which we will call ϵ_P, such that

$$\epsilon_P = \left[\frac{\rho_i^{1/2} v_{i\infty}}{\rho_s^{1/2} v_s} \right]^{3/2} \tag{9}$$

where ρ_i is the undisturbed density in the interstellar medium and ρ_s is the density in the heliosheath. From the numbers already given for the solar wind, and assuming $R_S = 100$ AU, one finds from Table I that $\epsilon_P > \epsilon$. The modeling consequence is that the upstream heliosheath is thinner for free-slip conditions on the boundary of the heliotail. Referring to Fig. 3, a larger ϵ also means the cross-section of the heliotail is smaller. The physical reason for this is easy to see; there is no longer the constraint that the flow down the heliotail must slow to $v_{i\infty}$ from v_s just downstream of the termination shock. Therefore, the flow down the heliotail is more rapid for the free-slip boundary condition. Both solutions have the same mass flux through the termination shock, as well as a constant velocity across the heliotail as $z \to \infty$, so the resulting cross-section of the heliotail in the free-slip case is smaller— corresponding to ϵ_P being larger than ϵ. The implication is that the non-slip boundary condition produces the largest possible heliotail cross-section or, equivalently, the thickest heliosheath in the upstream direction. Equating the mass flux across the termination shock to that flowing down the heliotail, and solving Eq. (5d) for streamlines in the limit $z \to \infty$, gives the transverse radius of the heliotail

$$R_T = 2R_S \epsilon^{-1/3}. \tag{10}$$

This gives $R_T = 4\,R_S$ for $v_{i\infty} = 25$ km s^{-1} and $v_s = 100$ km s^{-1}.

C. R_S in the Potential Flow Model

The absolute value of R_S in Eqs. (5) is as yet undetermined in the potential flow model. To find this number requires doing the equivalent calculation as was done in Sec. II.A, which will also entail evaluating how incompressible the flow is in the heliosheath. There are four steps to the process of finding R_S. The first is to utilize the Rankine-Hugoniot shock jump conditions. The second is to evaluate Bernoulli's equation along a streamline to calculate the pressure difference between the stagnation point and the sub-stagnation point on the termination shock. The third is to require pressure equilibrium across the heliopause. The fourth is to specify the boundary condition on the flanks of the heliotail (or some equivalent condition).

The shock jump conditions are easily found and will not be repeated here (see, e.g., Parker 1963, Eqs. 9.3). In the present context, the Bernoulli equation for steady, nondissipative flow is (Batchelor 1967, Eq. 3.5.4)

$$\frac{1}{2} v_s^2 + \frac{p_s}{\rho_s} \frac{\gamma}{(\gamma - 1)} = \frac{1}{2} v_{i\infty}^2 + \frac{p_\infty}{\rho_\infty} \frac{\gamma}{(\gamma - 1)} \tag{11}$$

where ρ_s and p_s are the solar wind density and pressure on the downstream side of the termination shock; ρ_∞ and p_∞ are the same in the distant heliotail, and the fourth of the above conditions has been invoked to set $v_\infty = v_{i\infty}$ (i.e., the non-slip condition has been invoked on the flanks of the distant heliotail). Pressure equilibrium across the heliopause requires

$$p_\infty = p_{i\infty} \qquad (12)$$

where the non-slip condition has again been invoked to eliminate dynamical effects. The free-slip condition would have resulted in additional terms in Eq. (11) that depend on the difference in dynamic pressure on either side of the heliopause.

Taking a constant solar wind speed and density varying as the inverse square of the radius inside the termination shock allows these equations to be solved for both the amount the density varies in the heliosheath and the dimensional value of R_S. Generally, for acceptable values of ϵ, flow in the heliosheath is incompressible under the non-slip condition (Suess and Nerney 1990,1991; Parker 1963). In addition, the value of R_S is reasonably well approximated by Eq. (2), or the various extensions that have been published to that calculation to include effects such as cosmic ray pressure (Suess 1990; Holzer 1989; Axford 1972).

An important result of the pressure balance calculation is that heliotail plasma is much less dense than interstellar plasma. This is because the temperature of the local interstellar medium (LISM) is $O[10^4]$ degrees while heliosheath plasma is $O[10^6]$ degrees, which is the temperature solar wind plasma is heated to when it passes through the termination shock. The heliopause, being a contact discontinuity, is an isobaric surface. For this surface to be in equilibrium, the above ratio of temperatures in the heliotail and the LISM implies an inverse of this ratio in the densities. Therefore, the density is two orders of magnitude smaller in the heliotail than in the LISM. This is also true for free-slip boundary conditions—the heliotail plasma is still a low-density, hot plasma. The difference is that the plasma moves down the tail much faster, so the diameter of the tail is smaller.

D. Supersonic Flow in the Interstellar Medium

If the interstellar flow is assumed to be supersonic, then it is impossible to do the potential flow calculation rigorously because the external flow cannot be assumed incompressible. In this case, it is necessary to undertake a numerical solution. Such a solution might start with supersonic solar and interstellar winds, determine the locations of the discontinuities (bow shock, termination shock, heliopause), and find the nature of the asymptotic solar wind flow down the heliotail. The equations solved vary in their complexity, depending on the ambition of the project. The simplest calculations start with time-dependent, polytropic, ideal hydrodynamics. The first calculations of this sort were done by Shima et al. (1986) and Matsuda et al. (1989). A recent solution has added the effect of interstellar neutral atoms (Baranov and Malama 1993). Numerical solutions are easier to do if the interstellar flow is supersonic because information cannot propagate upstream from the heliosphere—a condition which, if not satisfied, often affects the stability of the computation.

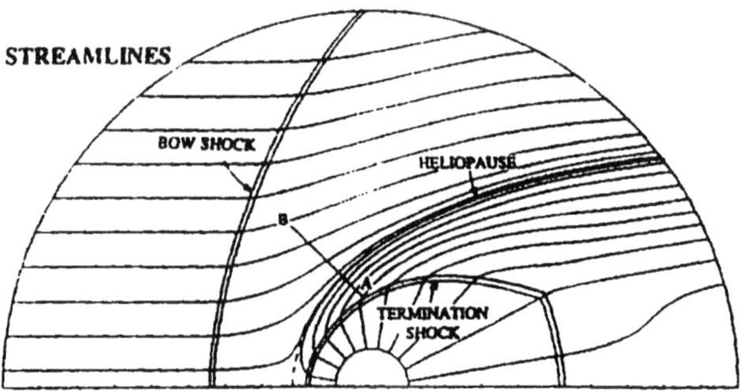

Figure 4.   Velocity streamlines in the relaxed dynamic equilibrium of the numerical simulation of interstellar wind, solar wind interaction by Steinolfson et al. (1994). The interstellar flow is from left to right. The shock locations were determined from the corresponding contour plots of the thermodynamic variables.

An example of the result from a numerical model by Steinolfson et al. (1994) is shown in Fig. 4. The reader is referred to that paper for detailed information on the calculation; but, to summarize, it was done as an initial-boundary value problem in which a nonequilibrium initial state was selected and allowed to relax in time. Thermal conduction was ignored and the size of the computational box (the semicircle in Fig. 4) was 600 AU. The approximate locations of the bow and termination shocks (as determined from the contour plots of the thermodynamic variables) have been superimposed on the computer-generated plot of the velocity streamlines. The physical quantities for this simulation were, for the solar wind, $v_r = 300$ km s^{-1}, $v_\theta = 0$, $n_e = 3$ cm^{-3}, and temperature $T = 10^5$ K at 1 AU, and, for the interstellar medium $v_{i\infty} = 24.88$ km s^{-1}, $n_e = 0.2$ cm^{-3}, and $T = 10^4$ K. The choice for the interstellar density reflects both recent measurements indicating a higher value than in Table I and the high value of the interstellar dynamic pressure that is needed to confine the shock to be inside 100 AU (actually at 45 AU in this example).

Several things are to be noted in this solution. Most obvious is the shape of the termination shock. In the downstream direction the shock is elongated and flat across the downstream side, and a similar feature, called a Mach disk, appears in the model of Baranov and Malama (1993). This shape was first predicted by Wallis and Dryer (1976). Next, the heliosheath on the upstream side is even thinner than predicted by the potential flow model for the same interstellar dynamic pressure. The reason for this is the striking change in the location of the shock with respect to the Sun. The

Sun is now upstream of the geometrical center of the shock. The shock and the heliopause do not reach their maximum distance from the z-axis (Fig. 2) until well downstream of the Sun, which was not the case in the potential flow model when a spherical, Sun-centered shock was assumed. Finally, the heliosheath in the upstream direction is further thinned due to the high flow speed down the tail compared to the potential flow model. This high flow speed comes about because the heliopause turns out to be a free-slip boundary instead of more nearly a non-slip boundary in this example—a consequence of small numerical dissipation. There is less shear at the heliopause in the model of Baranov and Malama (1993), but presently it cannot be said which model is most realistic physically. In either case, these models reinforce our conclusion in Sec. II.B that the potential flow model with non-slip boundaries on the heliotail produces the thickest plausible upstream heliosheath.

Steinolfson (1994) has also published a numerical simulation for external flow having a Mach number M_I of 0.8. Although this is nearly transonic, the shock is almost spherical, with its center displaced only slightly downstream from the Sun. This new result largely justifies the potential flow analysis when $M_I < 1.0$.

E. Asymmetry of the Termination Shock

The results from Sec. II.D on the shape of the heliospheric termination shock and the heliopause are easily understood in terms of well-known aerodynamic principles. Furthermore, these principles allow extrapolation to infer how asymmetries might appear for subsonic interstellar flow. We also consider that asymmetries in the solar wind itself may impose asymmetries on the termination shock.

First, consider the flow in the heliosheath (e.g., Fig. 2 or Fig. 4). The flow slows as it approaches the stagnation point and then accelerates after it turns the corner and goes down the heliotail. As the flow accelerates down the heliotail, then: (1) if the heliopause in the tail is a non-slip boundary, the flow speed remains small and the flow field is smooth; and (2) if the boundary is free-slip, then the flow speed can become large and a standing wave may replace the termination shock in the downstream direction (Wallis and Dryer 1976). Dissipative effects, or numerical effects simulating dissipation can smooth the expansion wave and Mach disk so that the overall flow field remains smooth and the shock more rounded than illustrated in Fig. 4 (Baranov and Malama 1993). This acceleration and possible expansion wave in the flanks and tail of the heliosheath may lead to large density variations, invalidating the incompressible flow approximation used in the potential flow calculation. Such is the case for highly supersonic external flow.

The flow in the interstellar medium, which is deflected around the heliopause, around the nose, and past the flanks, gains speed as it regains its original flow direction, and eventually reaches a transverse pressure equilibrium far down the heliotail. However, the details of this flow can be quite different, depending on whether the flow is supersonic or not.

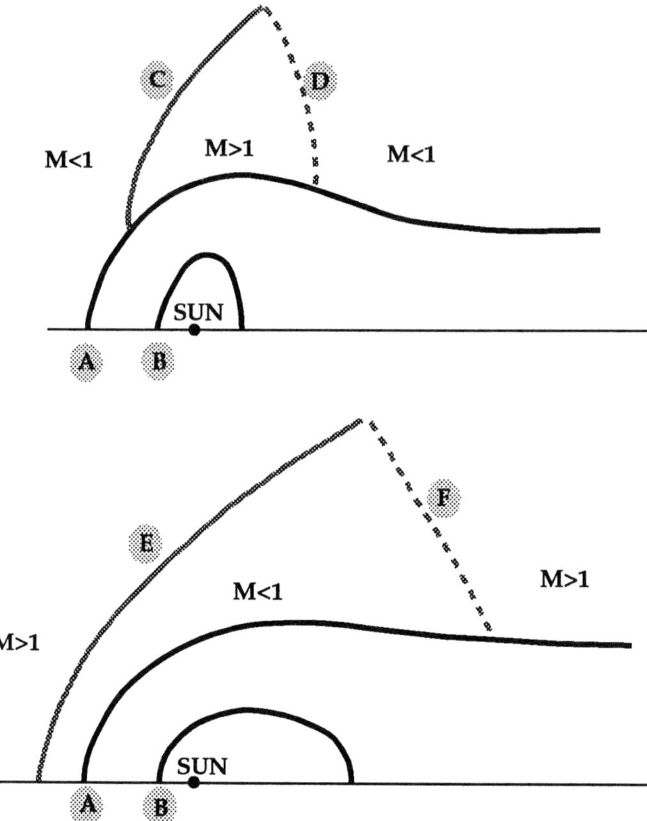

Figure 5.   Schematic diagrams of the heliosphere and the local interstellar medium when the external flow is subsonic (a) and supersonic (b). The interfaces A and B are the heliopause and the termination shock, respectively. In (a), the intially subsonic interstellar flow accelerates and becomes supersonic along C, and then makes a transition back to subsonic flow in a shock at D. In (b), the intially supersonic interstellar flow transitions to subsonic flow by a shock at E (figure after Suess et al. 1987).

Figure 5 summarizes the differences (Suess et al. 1987). In the top panel of Fig. 5, the external Mach number is assumed to be less than one, so there is no bow shock ahead of the heliopause at A. As the flow accelerates around A, it is nevertheless possible that it becomes supersonic as suggested by the dashed interface C. If so, then the flow must undergo a transition to subsonic flow as it exits down the flanks of the heliotail, near interface D. As the external flow accelerates around the flanks, especially if it becomes supersonic, the effective pressure is reduced, which pulls the flanks outwards. This causes both the flanks of the heliopause and the flanks of the termination shock to be extended in the transverse direction. The effect is enhanced for a thinner heliosheath. The schematic diagrams in Fig. 5 were drawn before it was known how thin the heliosheath might be, and now it appears that the effect may be quite large. The consequence is that the termination shock, in the absence of any internal effects, might be quite disk-shaped and nearly centered on the Sun if the external flow is effectively subsonic.

Conversely, the bottom panel of Fig. 5 shows the situation for an external Mach number that is larger than unity. Here, the external flow passes first through a bow shock at interface E before it is deflected around the heliosphere and then accelerated on the flanks to again become supersonic in the vicinity of interface F. The distinct difference in this high Mach number example is that the downstream distance to the termination shock can become effectively infinite as the external flow becomes hypersonic. The reason for this is that the pressure on the downstream side of the termination shock is reduced as the Mach number increases. When the interstellar velocity is subsonic, the back pressure is essentially identical to the interstellar pressure. Conversely, this pressure information is unable to immediately propagate to the back side of the termination shock for supersonic interstellar flow (Suess et al. 1987; Steinolfson et al. 1994). Therefore, for supersonic interstellar flow, the termination shock is likely to be bullet shaped, elongated in the upstream-downstream direction, with the Sun located nearer to the upstream side.

The solar wind itself may also contribute to asymmetry of the termination shock. The solar wind speed has long been known to be latitude dependent, implying the solar wind dynamic pressure may also depend on latitude. For the time being (see Sec. II.G below), consider only a steady-state picture in which the solar wind dynamic pressure depends on heliographic latitude. If it increases with latitude, the termination shock will be elongated over the poles of the Sun. Conversely, if it decreases with latitude, the termination shock will be more disk-shaped with the disk axis aligned with the solar rotation axis (Suess 1990).

Because the interstellar wind is approaching essentially from the heliographic equator, the asymmetries combine to produce a rather complicated three-dimensional object. It is impossible to say what the net result will be without knowing more about the strength of the interstellar magnetic field and the latitude dependence of the solar wind dynamic pressure. However, spacecraft presently in the outer heliosphere (Voyagers 1/2, Pioneers 10/11) and at high heliographic latitudes (Ulysses) are providing some of the answers and a more complete picture will soon evolve.

F. Time-Dependent Solar Wind

The heliospheric termination shock is expected to move in response to variations in upstream solar wind conditions. That the solar wind conditions do vary in time and space is without question: it is difficult to identify any time period in the solar wind when it can be said to be perfectly steady (see, e.g., Hundhausen 1972). These fluctuations take many forms, but the largest are due to large solar eruptions, coronal mass ejections (CMEs) and to corotating high-speed streams which originate in coronal holes. Taken together, CMEs and coronal holes can produce O[1] fluctuations in the solar wind over time scales from minutes to years.

An example serves to motivate this section. In Sec. II.A, Eq. (2) was used to estimate the termination shock distance; the example mentioned produced $R_S = 130$ AU. In the solar wind, there are high speed streams in which the dynamic pressure can double (Belcher et al. 1993; Suess 1993a). These streams are due to coronal holes at the Sun, and hence recur approximately every 27 days over their lifetime, which can be several rotations to more than a year. A doubling of the solar wind dynamic pressure, in Eq. (2), leads to a factor of $\sqrt{2}$ increase in R_S. In the example of Sec. II.A, this means R_S would change from 130 AU to 184 AU, if the shock were in equilibrium with the solar wind. However, because the streams are recurring every 27 days, the equilibrium value of R_S changes from 130 AU to 184 AU and back every 27 days. The average speed at which the termination shock would have to move is almost 7000 km s^{-1}, or 2% of the speed of light and more than ten times the speed of typical high-speed solar wind. This is far greater than any physically realistic speed attainable by the shock and therefore it will generally not be near an equilibrium position.

Nevertheless, the termination shock will move in response to changing solar wind conditions. Its speed will be determined by the magnitude of the change, the solar wind ambient density, temperature, and flow speed, the shock strength, and possibly the thickness of the heliosheath. If a change in the solar wind lasts long enough, the shock may approach a new equilibrium position and begin slowing down. Otherwise, solar wind fluctuations will be continually pumping the shock inwards and outwards. Because these solar wind fluctuations are not necessarily well correlated in space and extend only to limited solid angles, the shock motion will also be highly variable with position; while the shock is moving outward in some locations, it will be moving inward at other locations.

First, consider a simple kinematic calculation to illustrate how shock motion might appear. Assume that under an increase in solar wind dynamic pressure, the shock will move outward at the solar wind speed, slowing as it comes within 5 AU of its new equilibrium position. Assume, conversely, that under a decrease in solar wind dynamic pressure, the shock will move inward at approximately the Alfvén speed, again slowing as it comes within 5 AU of its new equilibrium position. Figure 6 illustrates what would happen under forcing from simulated solar wind conditions (Suess 1993a).

Figure 6 shows a simulation lasting for one year. The top two panels show the simulated solar wind, which contains a high-speed stream recurring every 27 days and lasting 4 days. Superimposed on this systematic variation is a set of quasi-random shorter-period fluctuations of decreasing amplitude with increasing frequency. The net result is a data set which has the "appearance" of typical solar wind during the declining phase of the solar sunspot cycle. The bottom panel of Fig. 6 shows the resulting position of the termination shock as a function of time. The shock radius changes quasi-periodically in response to

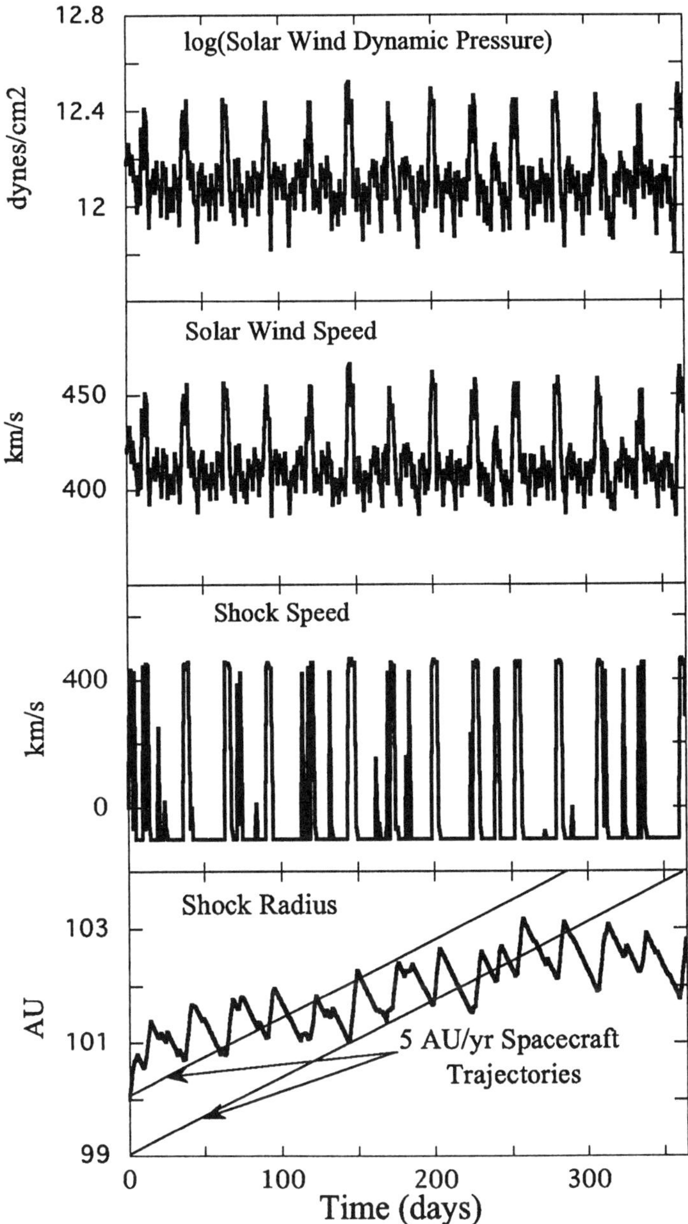

Figure 6.   A kinematic simulation of termination shock motion in response to periodic 27 day changes in solar wind dynamic pressure, with further shorter-period fluctuations superimposed to construct a synthetic data set that has the character of the observed solar wind. The simulation assumes that: (1) the maximum outward and inward shock speeds are the solar wind speed and 100 km s^{-1}, respectively; (2) the shock begins moving instantaneously with its maximum speed; (3) it takes 1 AU (4 days) for the shock to decelerate to zero speed as it approaches its new equilibrium position. Superimposed on this figure are trajectories showing the radial motion of a hypothetical spacecraft moving at 5 AU yr^{-1}, beginning at two different times (Suess 1993*a*).

the high speed streams and the higher frequency forcing. In addition, there is a tendency for the shock to move outward because its location at the beginning of the simulation was not the equilibrium position for the average conditions in the solar wind shown at the top of Fig. 6.

One of the most important results to be gained from Fig. 6 comes from superimposing trajectories of hypothetical spacecraft. A typical spacecraft radial speed in the outer solar system is 5 AU yr^{-1} and two such trajectories are shown. What is seen is that once the shock is encountered, it is likely to be encountered several more times. It is difficult to anticipate accurately the first shock encounter, but subsequent encounters can be predicted from the knowledge of the first encounter, together with the solar wind properties sampled subsequent to that encounter.

This simple kinematic example only invokes plausible assumptions about the movement of the termination shock in response to the solar wind. What it says is that shock motion can be very important. Recently, analytic and numerical dynamic models have been developed for the shock motion that supplement this kinematic model and allow a critical appraisal of the assumptions. Barnes (1993) derives an analytic expression for the initial motion of the shock after encounter with a contact discontinuity in the solar wind, while Grzedzielski and Lazarus (1993) do the same for the response to a train of waves. Whang and Burlaga (1993) calculate the history of termination shock motion using a numerical model and characteristic theory for MHD shocks along with measured solar wind parameters.

Turning first to Barnes' calculation, he begins by writing the shock jump conditions and considering an initial termination shock S_o, choosing a reference frame in which this shock is at rest. The unshocked gas is characterized by ρ_u, v_u, p_u, and the shocked gas by ρ_s, v_s, p_s. The interaction between a contact discontinuity C_o and the shock, can then be described using the schematic diagram shown in Fig. 7 (Barnes 1993). There are two cases: either a density increase or a density decrease at the contact discontinuity, driving the termination shock either outward or inward. Fig. 7a is for a density increase ($\rho_1 > \rho_u$) and Fig. 7b is for a density decrease ($\rho_1 < \rho_u$). In the former case, there are two outwardly propagating shocks with a contact discontinuity between them. The left shock S_1 is strong and is identified as the termination shock. In the later case, there is only one shock, strong and inwardly propagating (the termination shock) and an outwardly propagating rarefaction wave R. This interaction can be solved analytically for the initial motion of the termination shock. What Barnes (1993) finds is that the inward and outward speed of the shock, which depends on the magnitude of the change in upstream dynamic pressure, is typically of the order of 100 km s^{-1}. This is comparable to the inward speed assumed by Suess (1993a), but considerably less than the outward speed he assumed. One consequence, if this reflects a typical shock speed, as well as initial speed, is that a spacecraft would encounter the shock fewer times than suggested by Fig. 6.

Whang and Burlaga's (1993) numerical simulation addresses the last

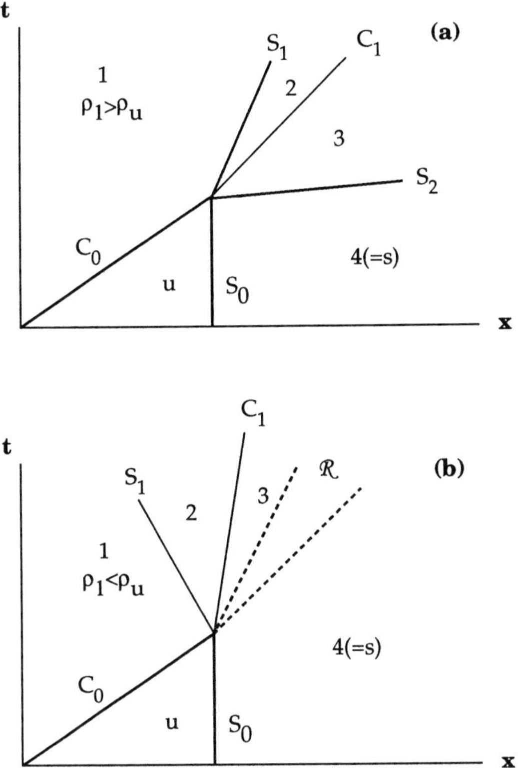

Figure 7.   Schematic diagram in the $x-t$ (space-time) plane of the interaction of an upstream contact discontinuity C_o with the termination shock S_o when $\rho_1 > \rho_u$. After the interaction there are two outwardly propagating shock waves, S_1 and S_2 with a contact discontinuity C_1 between them. (b) The same, but for $\rho_1 < \rho_u$. After the interaction there is a single inwardly propagating shock wave, to its right an outwardly moving contact discontinuity, and to the right of that, instead of a second shock, is a simple-wave rarefaction R propagating into the downstream gas (figure after Barnes 1993).

question raised above, namely what the shock motion might be like after the initial interaction with a discontinuity. They find shock speeds comparable to those derived by Barnes and proceed to integrate the equations over a long period of time using solar wind data collected at Voyagers 1 and 2 from 1978 to 1988. Considering the longer period evolution of shock distance, in keeping with the suggestions made earlier in Sec. II.E on shock asymmetry, Whang and Burlaga find that the equatorial termination shock moves between 88 and 102 AU over the 11-yr period, with the distance being anticorrelated with sunspot number. During the declining phase of the solar cycle, the shock speed is often greater than or comparable to Voyager 1 and 2 speeds so that multiple encounters with the shock would be possible at those times.

G. Summary

It is clear from Secs. II.B, II.D, and II.E that the subsonic and supersonic external flow limits produce quite different results, even in the heliosheath if the external flow is sufficiently supersonic. Conversely, the subsonic external flow case is extremely attractive to the modeler because it is largely analytic (and can probably be generalized to considerably more complex geometries than assumed in Sec. II.B) and because it serves as the basis for quantitative modeling of the magnetic field in the heliosheath in Sec. III.B. However, solvability is not an adequate rationale for avoiding the issue of relevance. It is $\mathbf{B}_{LISM}$ which may result in the subsonic external flow model being relevant to the heliosphere.

The reason for this has to do with the reasons for the pressure reduction on the downstream side of the termination shock, which was described in Sec. II.E, in connection with Fig. 5. In the presence of $\mathbf{B}_{LISM}$, it is possible that while the external flow is supersonic, it may be sub-Alfvénic. From Table I, it can be seen that this is the case for the suggested range of plausible $\mathbf{B}_{LISM}$. The consequence of this is that, with respect to Fig. 5, the pressure on the downstream side of the termination shock would be similar to that for subsonic external flow and the shock shape would be more like that in Fig. 5a, than in 5b. This, then, is the rationale for continuing investigation into both subsonic and supersonic external flow models. Until more is known about the local interstellar medium and the shape of the termination shock, it will be important to continue both classes of models.

There is yet another interesting consequence, however, of a nonzero $\mathbf{B}_{LISM}$. This is that it introduces an additional asymmetry into the problem (Holzer 1989). Unless $\mathbf{B}_{LISM}$ is aligned with the interstellar wind, the field will exert an anisotropic pressure on the heliopause. This will cause, for example, the cross-section of the heliotail to be more elliptical than circular. A preliminary analytic study of the effect of $\mathbf{B}_{LISM}$ on the shape of the heliopause was described by Parker (1963) in the limit of no interstellar wind. However, recent approaches invoking what is known as the Newtonian approximation are able to reduce the problem of the shape of the heliopause to the solution of ordinary differential equations (Fahr et al. 1988,1993). In this approach, the heliopause is considered to be a free boundary whose shape is approximated by solving for a pressure balance between the *undisturbed plasma flows* that would exist both in the absence of the boundary and in the absence of the opposing flow. This hypothesis therefore ignores all the dynamics of the interaction in computing the constant pressure surface between the two totally independent flow fields. The reported results seem to support the validity of the approximation. Accepting the principle, the calculation is able to take into account magnetic fields, as well as thermal and dynamic pressures. One interesting result is that for flow-aligned $\mathbf{B}_{LISM}$, the distance to the stagnation point is actually slightly increased due to the magnetic curvature force in the vicinity of the flow just outside the stagnation point.

III. THE MAGNETIC FIELD IN THE HELIOSHEATH

The magnetic field downstream of the termination shock has been a subject of some interest, but has only recently been amenable to quantitative modeling (Nerney et al. 1991,1993; Suess and Nerney 1993). There are at least two motives behind interest in the magnetic field. The first is that it may be dynamically important (Axford 1972). The second is that it may modify cosmic ray propagation through the heliosphere (Suess and Nerney 1993).

A. Spherical Flow

Inside the termination shock, the interplanetary magnetic field, on average, exhibits a very clear behavior first predicted by Parker (1963). Consider the field to be radial at the Sun and carried, passively, by a solar wind of constant speed, $\mathbf{v}_{sw} = \hat{e}_r v_{sw}$, with a solar angular rotation speed of Ω. Then, this field is found to be

$$B_r = B_o(\theta, \phi) \left(\frac{r_o}{r}\right)^2 \tag{13a}$$

$$B_\phi = -B_o(\theta, \phi) \left(\frac{r_o}{r}\right)^2 \frac{\Omega}{v_{sw}} (r - r_o) \sin\theta \tag{13b}$$

using only the equations for the conservation of magnetic flux and steady-state induction for an infinitely conducting medium:

$$\nabla \cdot \mathbf{B} = 0 \tag{14a}$$

$$\nabla \times (\mathbf{v} \times \mathbf{B}) = 0. \tag{14b}$$

This solution predicts an Archimedian spiral in which the fieldlines are spirals on cones whose half-angle is the polar angle and which spiral angle decreases from the equator ($\theta = 90°$) to the pole. Observations have confirmed that this is an excellent description of the average field from 35 $R_\odot$ to 50 AU. The immediate question is: what happens at and beyond the termination shock.

The first question is relatively simply answered. For a strong shock in an ideal gas, the transverse component of the field is amplified by a factor of 4 and the radial component is unaffected. Beyond the shock, the situation changes dramatically, even in the case of spherically symmetric solar wind flow. To see this, consider the solution to Eq. (14a) and the continuity equation, $\nabla \cdot (\rho \mathbf{v}) = \rho \nabla \cdot \mathbf{v} = 0$, for incompressible, radial flow. One finds that

$$v_{sw} \propto \frac{1}{r^2} \tag{15a}$$

$$\rho = \text{constant} \tag{15b}$$

$$B_r \propto \frac{1}{r^2} \tag{15c}$$

$$B_\phi \propto r \tag{15d}$$

in which B_ϕ is to be compared to what was found inside the shock, in Eq. (13b), i.e., that $B_\phi(r < R_S) \propto (1/r)$. Figure 8 illustrates how the field described by Eqs. (14) and (15) appears, showing an equatorial fieldline and a group of fieldlines lying above $12°$ polar angle. A termination shock has been arbitrarily placed at 50 AU, beyond which the magnetic field is given by Eqs. (15). What is apparent is that the high latitude fieldlines are straighter and, hence, shorter. The four-fold compression of the transverse field at the shock is also clearly apparent. What is not so visible in this plot is the growing azimuthal field beyond the shock, which grows without bound, leading to a magnetic field catastrophe. This problem was realized early, and the solar wind equations, with an added term representing the azimuthal field beyond the termination shock, were solved to show that the asymptotic state is quite different from that represented above (Axford 1972).

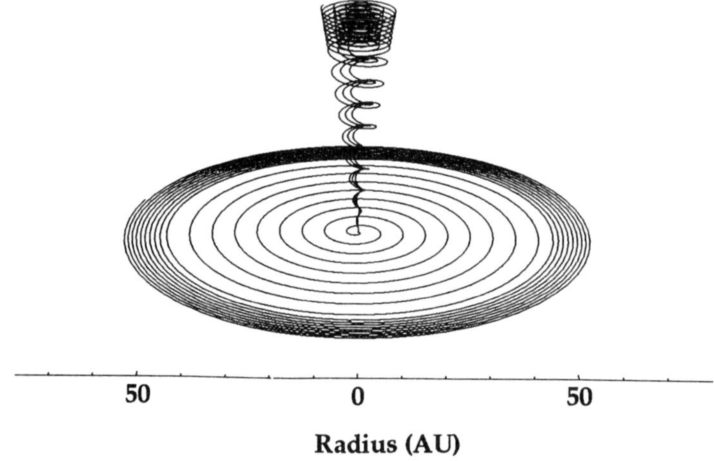

50   **0**   **50**

Radius (AU)

Figure 8.   Magnetic fieldlines upstream and downstream of a spherical termination shock, here arbitrarily placed at 50 AU, in spherically symmetric solar wind flow. One fieldline is in the heliographic equatorial plane and a group of fieldlines is above heliographic latitude $78°$. The four-fold compression of the radial field at the shock is apparent, along with the near-polar fieldlines being shorter (figure after Nerney et al. 1991). The equatorial fieldline illustrates the Archimedian spiral inside the shock that was predicted by Parker (1963).

The solution to this one-dimensional problem is completely analytic, and can be found in Nerney et al. (1991). Figure 9 shows the resulting effect of the magnetic field. The left side of this figure shows the properties of the average solar wind flow speed, density, and azimuthal magnetic field inside the termination shock. On the right side, as dashed lines, is shown the solution in Eqs. (15) leading to the magnetic field catastrophe. But, the amplification eventually leads to a magnetic field which is strong enough to begin accelerating the solar wind flow and, correspondingly, causing the density to fall off. The final state is analogous to the supersonic solar wind; the flow speed reaches a constant, the density is again falling off as r^{-2}, and the azimuthal magnetic field is again falling off as r^{-1}. Extending this model to three dimensions would allow the magnetic field build-up to be further relieved by flow in the meridional direction. Recently, numerical modeling of these MHD effects has been initiated by Washimi and Nozawa (1992), but detailed results for a heliosphere-like geometry have yet to be reported.

Quantitative solutions to the spherically symmetric MHD problem are shown in Fig. 10. Panel 10a shows the nondimensional density, pressure, total pressure, magnetic field, Alfvén Mach number, and mass flux, plotted versus distance in terms of termination shock radii. Panel 10b shows the dependence of the total pressure on plasma $\beta_v \equiv 8\pi \rho_{\mathrm{sw}} v_{\mathrm{sw}}^2 / B_{\mathrm{sw}}^2$, evaluated on the upstream side of the termination shock. Panel 10c again shows the total pressure, here at three different polar angles, and it is this panel which illustrates the important new feature introduced by this solution. The pressure beyond the shock at the pole is essentially constant, like the hydrodynamic solution in Sec. II, while the pressure at the equator now falls off with distance. This means that the pressure at the heliopause can be significantly less than at the termination shock due to MHD effects if the heliosheath thickness is more than ~ 6 shock radii. Consequently, the pressure at the termination shock can be much larger than p_I, modifying the evaluation in Eq. (2) such that the termination shock might lie far inside 100 AU for the parameters given in the example in Sec. II.A. This discovery elevated interest in this problem as there were tentative reported detections of the termination shock, using radio wave, galactic cosmic ray, and backscattered solar Lyman α observations, placing it inside the distance suggested by Eq. (2) (Gurnett et al. 1993; Czechowski and Grzedzielksi 1990; Gangopadhyay et al. 1989).

B. A Kinematic Magnetic Field Model

Given the flow field described in Sec. II.B, it is possible to do a kinematic magnetic field calculation in the heliosheath much like that done in Eqs. (13) and (14). The calculation will be outlined here, with details available in Nerney et al. (1991,1993) and Suess and Nerney (1993). It is essentially a Lagrangian calculation because the field is passively advected by the plasma. The magnetic field at each point in the heliosheath depends only on the past history of the plasma at that point.

The solution proceeds by reducing the problem to evaluating two streamline constants and one ordinary differential equation along a streamline. To begin, the components of the magnetic field are written in streamline coordinates as

$$\mathbf{B} = B_s \hat{e}_s + B_t \hat{e}_t + B_\phi \hat{e}_\phi \tag{16a}$$

$$\hat{e}_s \equiv \mathbf{v}/v \tag{16b}$$

$$\hat{e}_t \equiv \hat{e}_\phi \times \mathbf{v}/v \tag{16c}$$

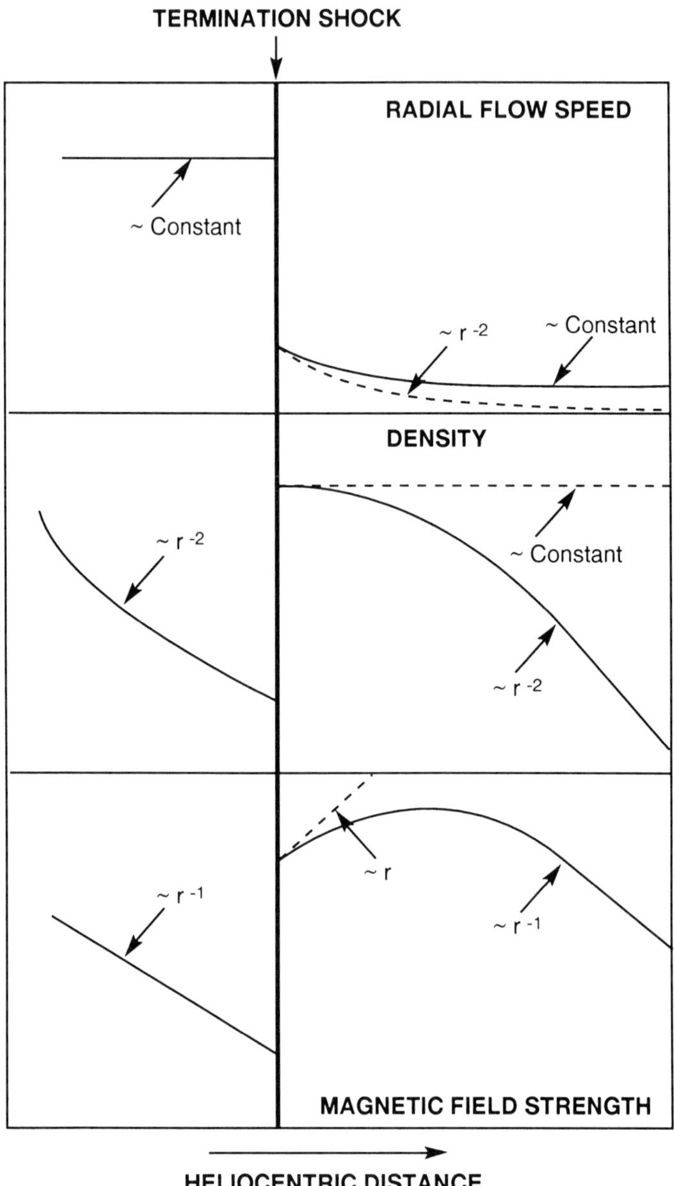

Figure 9.   Schematic diagram of solar wind density, magnetic field, and radial velocity upstream and downstream of the termination shock, with dashed lines showing what the density, velocity, and field would be like in the absence of MHD effects in the heliosheath, in spherically symmetric flow with the field strength approaching infinity. Eventually, MHD effects become important, producing the profiles shown as solid lines on the right-hand portion of this figure (Suess 1990).

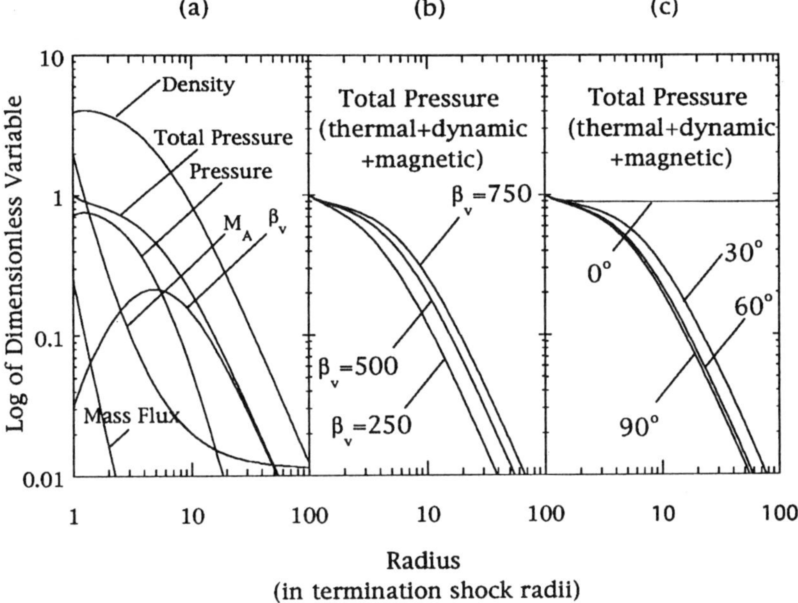

Figure 10.   The solution to MHD flow downstream of the termination shock in a spherically symmetric solar wind (Axford 1972; Nerney et al. 1991). In panel (a), the dynamic plasma β (see text for definition) is 500 and all variables are dimensionless. The radius has been scaled by an arbitrary value for the radius of the termination shock. In panel (b), the total pressure p_T for $\beta = 250, 500, 750$; as the field strength increases, the total pressure decreases. In panel (c), the total pressure at the polar angles $\theta = 0, 30, 60, 90$ degrees. The 0-degree case is the same as the zero-field solution, and the solution at 90 degrees is the same as shown in panels (a) and (b).

where $v = |\mathbf{v}|$. The new unit vectors required to define streamline coordinates, $\hat{e}_s$ and $\hat{e}_t$, are illustrated in Fig. 13 (Nerney et al. 1993). The scalar functions B_s, B_t, and B_ϕ are then found to be given by:

$$\frac{B_\phi}{\tilde{\omega}} = f(\psi, \phi) \tag{17a}$$

$$B_t v \tilde{\omega} = F(\psi, \phi) \tag{17b}$$

$$\mathbf{v} \cdot \nabla \left(\frac{B_s}{v} \right) = \frac{2 B_t}{\tilde{\omega} v^2} \nabla \psi \cdot \nabla v \tag{17c}$$

where ψ is the streamline constant in Eq. (5d) and where f and F are functions only of their arguments, and are fully determined by the boundary conditions on the magnetic field. The streamline constants in Eqs. (17a) and (17b) can be derived, for example, using the method of Jacobians. Equations (17a) and (17b) came from the induction equation, while (17c) came from the solenoidal condition for the magnetic field.

Recently, a more direct derivation of Eq. (17a) has been found and is reproduced here. Starting with the induction Eq. (14b), and using both the solenoidal condition Eq. (14a) and $\nabla \cdot \mathbf{v} = 0$, the equation reduces to

$$v_{\tilde{\omega}} \frac{\partial}{\partial \tilde{\omega}} B_\phi + v_z \frac{\partial}{\partial z} B_\phi = \frac{B_\phi v_{\tilde{\omega}}}{\tilde{\omega}}. \tag{18}$$

Substitute $b = \ln B_\phi$, divide by $v_{\tilde{\omega}}$, and use $dz/d\tilde{\omega} = v_z/v_{\tilde{\omega}}$ along streamlines, to give

$$\frac{\partial b}{\partial \tilde{\omega}} + \frac{dz}{d\tilde{\omega}} \frac{\partial b}{\partial z} = \frac{1}{\tilde{\omega}}. \tag{19}$$

The left hand side of Eq. (19) is the total derivative of b so that $b \propto \ln \tilde{\omega}$, or (using the definition of b),

$$\ln B_\phi \propto \ln \tilde{\omega} \tag{20}$$

and Eq. (17a) follows.

Equation (17c) is integrated numerically along streamlines beginning with the magnetic field boundary condition at the termination shock. It can be seen from this equation that B_s will be large both when B_t is large and when $\nabla \psi$ is in the same direction as ∇v. This occurs near the stagnation point, where the gradient in the flow speed points toward the instantaneous center of curvature of the streamline.

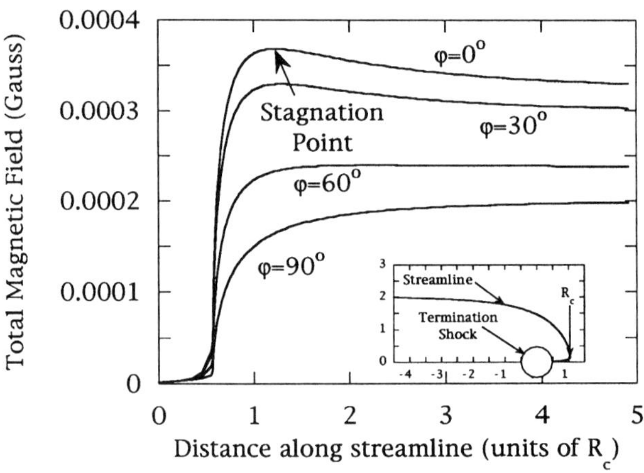

Figure 11.   Magnetic field along streamlines that approach the heliopause for various angles about the heliotail (the inset shows the streamline upon which the magnetic field is calculated). The field for $\phi = 0°$ is dominated by the slowing of the flow near the stagnation point while the field at $\phi = 90°$ increases linearly with $\tilde{\omega}$ as per Eq. (17a) (Nerney et al. 1993).

Figures 11 and 12 show the kinematic magnetic field solution, under the assumption that the termination shock lies at 100 AU and the field at the inside of the shock is that described by an Archimedian spiral. Figure 11 is a plot of the total magnetic field strength along streamlines lying very close to the heliopause. Because the streamlines are identical for any ϕ-angle about the heliotail, only one streamline needs to be shown, which is done in the inset panel. The angle $\phi = 0$ lies in the solar equatorial plane, and field strengths are shown at angles $\phi = 0, 30, 60,$ and 90 degrees. Two effects appear in this plot. The first is the amplification of the field as the flow approaches the stagnation point. This is easy to understand as it is analogous to the magnetic field catastrophe described in Sec. III.A. It happens because the component of the field perpendicular to the flow is amplified as the flow slows when it approaches the stagnation point. It is a singular solution in the sense that the field strength is therefore infinite at the stagnation point. The effect is most pronounced for streamlines lying in the equatorial plane and decreases in the plane containing the solar rotation axis, the "meridional plane."

The other phenomenon seen in Fig. 11 is an amplification of the field along the heliopause at high latitudes. This occurs in a very thin layer as seen by examining the field strength along several streamlines, each of which lies slightly further from the heliopause. This is done in Fig. 12, where the top panel shows seven numbered streamlines that start near the sub-stagnation point, but, with increasing numerical label, are successively further from the stagnation point. The second panel shows the speed along these streamlines, clearly demonstrating the slowing of the flow as it approaches the stagnation point, with the slowing being most pronounced along the streamline closest to the stagnation point. Finally, in the bottom panel, the field strength along these fieldlines is shown. This shows that on the flanks of the heliopause, the field strength becomes infinite at high latitudes, but the layer over which this occurs is extremely thin, with a thickness that decreases to zero as the equator is approached.

The physical significance of the strong field near the stagnation point and the strong field on the high latitude heliopause in the heliotail are quite different. In the latter case, the main effect will be to thicken the heliopause boundary layer, but the total volume and total flux contained in this layer are relatively small. All the flux contained in the layer originates in the small region in the vicinity of the sub-stagnation point. A geometrical description of the reason for the existence of the strong polar heliopause field is given in Nerney et al. (1993), where the physical consequences are further discussed. Conversely, the strong field in the vicinity of the stagnation point probably has important dynamical consequences in that region. As stated above, this amplification is analogous to the amplification occurring in the magnetic field catastrophe described in Sec. III.A. It is what remains of that effect in the presence of the turning of the flow down the heliotail. Examination of the value of the plasma β in the region of the stagnation point leads to the conclusion that MHD effects are probably important in a cone of approximately $30°$

S. T. SUESS AND S. NERNEY

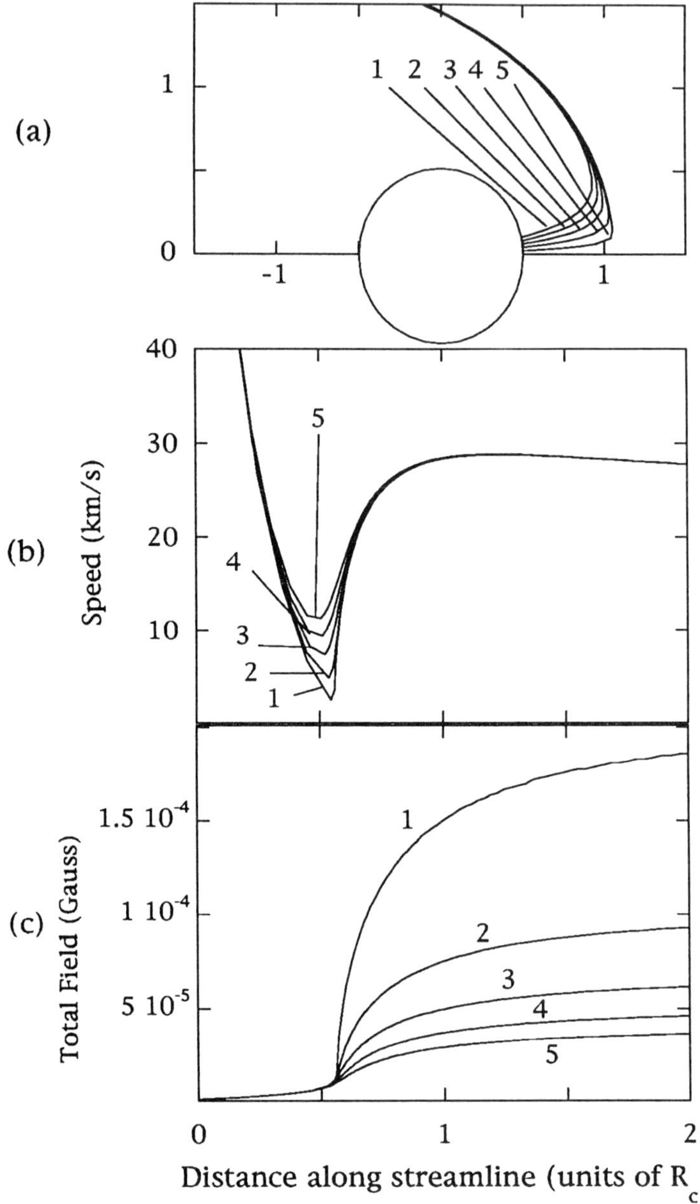

Figure 12. (a) Streamlines used for the velocity and field plots in (b) and (c), in the $\phi = 90°$ plane, all originating near the sub-stagnation point. (b) Total velocity along the streamlines shown in panel (a). The streamline distance is measured in units of R_c, the distance to the stagnation point in the limit $\epsilon \rightarrow 0$ (Suess and Nerney 1990). (c) Total field strength along the streamlines shown in (a). The field decreases quickly away from the heliopause, indicating that the magnetic pressure ridge found there is extremely thin (Nerney et al. 1993).

half-angle, centered around the stagnation point. Elsewhere, MHD effects are probably of little direct importance (Nerney et al. 1991).

This difference between the stagnation point amplification and the polar heliopause amplification is a result of competing effects that dominate at different streamline lengths and originate because the magnetic field on the termination shock is a function of azimuth about the heliotail. The competing effects are due to the consequent varying geometry between the flow and magnetic vectors on the termination shock as a function of ϕ. In heliotail coordinates, $\mathbf{B}$ on the termination shock, for vanishingly small values of $\tilde{\omega}$, is in the $\hat{e}_{\tilde{\omega}}$ direction in the $\phi = 0$ plane and in the negative $\hat{e}_\phi$ direction in the $\phi = 90°$ plane while $\vec{v}$ lies in the $\tilde{\omega} - z$ plane for all ϕ angles. The amplification of B_ϕ is described by Eq. (17a) and shows mathematically why the magnetic field is enhanced on the heliopause. If $B_\phi{\neq}0$ on the termination shock, a ridge is produced at the heliopause because for those streamlines approaching the heliopause, $\tilde{\omega}_o{\approx}0$, where $\tilde{\omega}_o$ is the starting value on the termination shock. Therefore, $B_\phi \rightarrow \infty$ at the heliopause for all $\phi{\neq}0, 180°$. The B_t component is selectively amplified via the slowing of the flow near the stagnation point (Eq. 17b), which does not have a similar singularity on the flanks of the heliopause. However, this further enhances B_s because it depends on B_t through Eq. (17c). Because all streamlines from the region near the sub-stagnation point map to the heliopause, these streamlines end up at large values of $\tilde{\omega}$, relative to the very small values of $\tilde{\omega}$ near the sub-stagnation point. Equation (17b) then requires that B_t must approach zero on the heliopause while (17c) shows that B_s/v is a streamline constant once the heliopause streamlines reach large values of $\tilde{\omega}$. These effects are most noticeable in the $\phi = 0°$ plane where B_t is largest on the termination shock.

C. Field Reversals, the Solar Cycle, and Reconnection on the Heliopause

Figures 13 and 2 show that all streamlines from the vicinity of the sub-stagnation point pass close to the heliopause, which means that the character of the magnetic field polarity near the sub-stagnation point defines the polarity pattern on the heliopause. Figure 14 illustrates this point for a time in the solar cycle when the heliospheric current sheet (HCS) is approximately a great circle tilted at about 45° to the solar equatorial plane (the HCS, a surface dividing oppositely directed magnetic polarities in the interplanetary medium and, except near solar maximum, approximately the surface defined by a tilted dipole at the Sun as deformed in the solar wind by solar rotation [Suess 1993b]). Streamlines approaching the heliopause all begin equatorward of the maximum latitude of the HCS and therefore the magnetic field on these streamlines reverses twice every 27 day solar rotation. These reversals are then painted, as stripes, onto the inside of the heliopause and the shape of the stripes is quasi-circular because of the proximity of their source to the sub-stagnation point (the shape of the stripes, of course, will change with depth into the heliosheath). Figure 13 shows these circular stripes on a much expanded scale, with the polarity reversals indicated where the field vectors are meeting nose-to-tail. The changing structure of the reversal with angle ϕ about the heliotail, reflecting the changing relationship between the flow field and the magnetic field near the sub-stagnation point for different ϕ angles, is also clearly shown. What is not obvious from Fig. 13 is the scale of these polarity stripes. Assuming a flow speed down the tail of 25 km s^{-1}, which would be the case for the non-slip boundary condition, these stripes are each only 0.2 AU across. This is to be compared with the transverse dimension of the heliotail, which is 200 to 400 AU. Finally, except for a very brief period around solar minimum that lasts less than a year, the field at the sub-stagnation point will always undergo reversals every solar rotation.

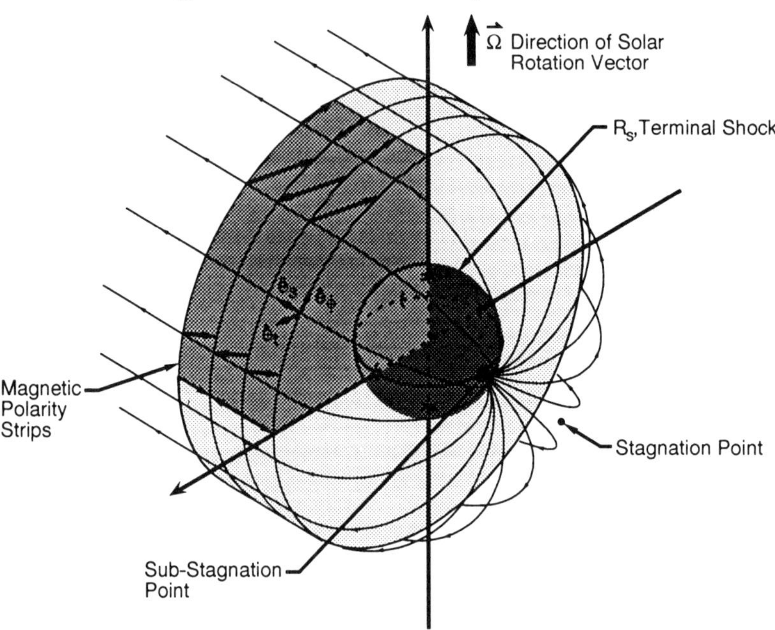

Figure 13.   A schematic diagram showing that streamlines outlining the heliopause arise near the sub-stagnation point on the termination shock. The unit vectors for streamline and magnetic field computation are shown, as are the magnetic polarity stripes that are painted onto the heliopause by the heliospheric current sheet crossing the sub-stagnation point. The orientation of the magnetic field is shown for four ϕ angles (Nerney et al. 1993).

This raises the question of reconnection and how it might proceed on the heliopause. The simplest assumption is that the interstellar magnetic field $\mathbf{B}_{\mathrm{LISM}}$ has an orientation which is steady on a solar cycle time scale and whose length scale is at least comparable to the transverse dimension of the heliosphere. Thus, alternate polarity stripes are pressed against a quasi-

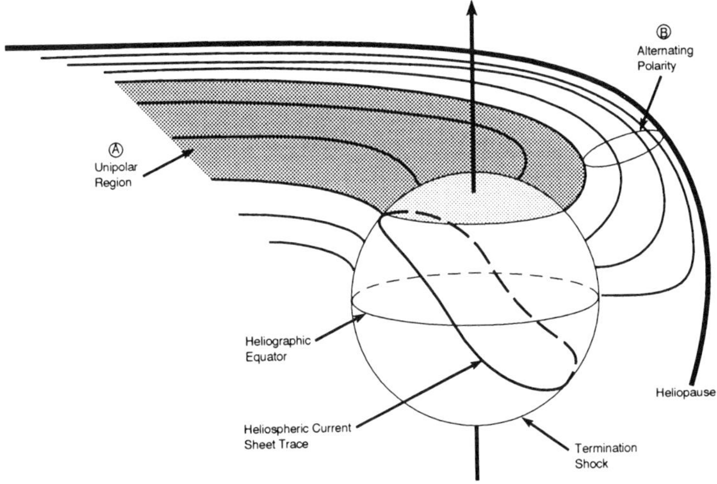

Figure 14.   Schematic diagram of streamlines in the heliosheath and their relationship to the interplanetary magnetic field on the termination shock. The magnetic field is unipolar at latitudes above the tilt of the heliospheric current sheet and hence the field on these streamlines is unipolar. At lower latitudes, streamlines carry an alternating field, with reversals occurring twice or more per solar rotation. The trace of the heliospheric current sheet at the termination shock is indicated here as it would appear in an ideal case, but the existence of reversals every solar rotation is independent of this idealization (Suess and Nerney 1993).

steady $\mathbf{B}_{\mathrm{LISM}}$ at the stagnation point. Reconnection occurs on the Earth's magnetopause preferentially on the day side when the interplanetary magnetic field is southward and when the plasma β or Mach number is low. By analogy for the heliopause, the equivalent preferential field polarity will be achieved in one or the other or the alternating polarity stripes. It has already been shown that the Mach number and the plasma β are small near the stagnation point, so the location where reconnection between $\mathbf{B}_{\mathrm{LISM}}$ and the heliosheath magnetic field occurs is on small quasi-circular regions centered on the stagnation point. The scale of the reconnection will be the width scale of the stripes, a fraction of an AU (Suess and Nerney 1993). There are solar cycle variations in the stripe structure but the general pattern holds throughout the solar cycle. Therefore, by the time heliosheath plasma is carried into the heliotail, the heliopause will be threaded by continuous fieldlines on scale sizes less than 0.2 AU, closely coupling the flows across this boundary and making the non-slip boundary condition invoked in Sec. II.B a compelling choice. It also means that the heliopause will be porous to galactic cosmic rays along fieldlines which eventually end up near the heliographic equator. Conversely, from Fig. 14 it is apparent that heliosheath magnetic fieldlines which originate on polar portions of the termination shock never closely approach the heliopause and it is only through cross-field diffusion that galactic cosmic rays can reach these lines.

The solar cycle imprint on the picture of field polarity illustrated in Fig. 14 is now understood to be different than that of sectors in the interplanetary magnetic field. The phase of the small-scale reversals will alternate every 11 years, but this is probably a physically insignificant effect. What does change is the cross-sectional area of the unipolar magnetic region shaded in Fig. 14 (Nerney et al. 1995). This will grow and shrink over the solar cycle. At 25 km s^{-1}, the interstellar wind speed, the length down the tail of these unipolar regions is only about 50 AU, smaller than the cross-sectional area of the heliotail. If there is any significance at all to the solar cycle imprint, it will probably be at the interface between these lenses, which might allow the freer access of galactic cosmic rays to high heliographic latitude magnetic fieldlines.

IV. SUMMARY

A general hydrodynamic view has been described of the interaction between the local interstellar medium and the solar wind. This has been supplemented by a kinematic description of the magnetic field in the heliosheath. This chapter has not been presented as a review. Rather, it is tutorial in nature, providing a basis for understanding the physics of the interaction and the analytic and numerical approaches that have been used to model that interaction. The large, and growing, body of data that is being accumulated on the outer heliosphere is not addressed here but will nevertheless provide the ultimate test of the ideas that have been presented.

Acknowledgments. This research was supported by the NASA Office of Cosmic and Heliospheric Physics and by the Ulysses Project. Portions of this chapter were written by S. F. Nerney while he was a NRC/NAS Resident Research Associate working at the NASA Marshall Space Flight Center.

REFERENCES

Axford, W. I. 1972. The interaction of the solar wind with the interstellar medium. In *Solar Wind*, eds. C. P. Sonett, P. J. Coleman, Jr., and J. M. Wilcox, NASA SP-308, pp. 609–658.

Axford, W. I., and Suess, S. T. 1994. Spacecraft to explore the outer heliosphere. *EOS: Trans. AGU* 75:587–594.

Baranov, V. B. 1990. Gas dynamics of the solar wind interaction with the interstellar medium. *Space Sci. Rev.* 52:89–120.

Baranov, V. B., and Malama, Yu. G. 1993. Model of the solar wind interaction with the local interstellar medium: Numerical solution of self-consistent problem, *J. Geophys. Res.* 98:15157–15164.

Barnes, A. 1993. Motion of the heliospheric termination shock: A gasdynamic model. *J. Geophys. Res.* 98:15137–15146.

Batchelor, G. K. 1967. *An Introduction to Fluid Dynamics* (New York: Cambridge Univ. Press).

Belcher, J. W., Lazarus, A. J., McNutt, R. L. , Jr., and Gordon, G. S., Jr. 1993. Solar wind conditions in the outer heliosphere and the distance to the termination shock. *J. Geophys. Res.* 98:15177–15184.

Bertin, P., Lallement, R., Ferlet, R., and Vidal-Madjar, A. 1993. Detection of the local interstellar cloud from high-resolution spectroscopy of nearby stars: Inferences on the heliospheric interface. *J. Geophys. Res.* 98:15193–15198.

Bochkarev, N. 1987. Local interstellar medium. *Astrophys. Space Sci.* 138:229–302.

Chassefiere, E., Bertaux, J. L., Lallement, R., and Kurt, V. G. 1986. Atomic hydrogen and Helium densities of the interstellar material in front of the nearby star α Ophiuchi. *Astron. Astrophys.* 160:229–242.

Cox, D. P., and Reynolds, R. J. 1987. The local interstellar medium. *Ann. Rev. Astron. Astrophys.* 25:303–344.

Czechowski, A., and Grzedzielski, S. 1990. Frequency drift of 3-kHz interplanetary radio emissions: evidence of Fermi accelerated trapped radiation in a small heliosphere? *Nature* 344:640–641.

Davis, L. E., Jr. 1955. Interplanetary magnetic fields and cosmic rays. *Phys. Rev.* 100:1440–1444.

Fahr, H. J., Grzedzielski, S., and Ratkiewicz, R. 1988. Magnetohydrodynamic modeling of the 3-dimensional heliopause using the Newtonian approximation. *Annales Geophysicae* 6(4):337.

Fahr, H. J., Fichtner, J., and Scherer, K. 1993. Determination of the heliospheric shock and the supersonic solar wind geometry by means of the interstellar wind parameters. . *Astron. Astrophys.* 277:249–264.

Frisch, P. C. 1994. Morphology and ionization of the interstellar cloud surrounding the solar system. *Science* 265:1423–1427.

Frisch, P. C., York, D. G., and Fowler, J. R. 1987. The local interstellar medium, VII, The local interstellar wind and interstellar material in front of the nearby star α Ophiuchi. *Astrophys. J.* 320:842–849.

Gangopadhyay, P., Ogawa, H. S., and Judge, D. L. 1989. Evidence of a nearby solar wind shock as obtained from distant Pioneer 10 ultraviolet glow data. *Astrophys. J.* 336:1012–1021.

Gloeckler, G., Geiss, J., Balsiger, H., Fisk, L. A., Galvin, A. B., Ipavich, F. M., Ogilvie, K. W., von Steiger, R., and Wilken, B. 1993. Detection of interstellar pick-up hydrogen in the solar system. *Science* 261:70.

Grzedzielski, S., and Lazarus, A. J. 1993. 2- to 3- kHz continuum emissions as possible indications of global heliospheric "breathing." *J. Geophys. Res.* 98:5551–5558.

Grzedzielski, S., and Page, D. E., eds. 1990. *Physics of the Outer Heliosphere*, (New York: Pergamon Press).

Gurnett, D. A., Kurth, W. S., Allendorf, S. C., and Poynter, R. L. 1993. Radio emission from the heliopause triggered by an interplanetary shock. *Science* 262:199-203.

Holzer, T. E. 1989. Interaction between the solar wind and the interstellar medium. *Ann. Rev. Astron. Astrophys.* 27:199–234.

Hundhausen, A. J. 1972. *Coronal Expansion and Solar Wind* (Berlin: Springer-Verlag).

Ip, W.-H., and Axford, W. I. 1985. Estimates of galactic cosmic ray spectra at low energies. *Astron. Astrophys.* 149:7–10.

Lallement, R., Ferlet, R., Vidal-Madjar, A., and Gry, C. 1990. Velocity structure of the local interstellar medium. In *Physics of the Outer Heliosphere*, eds. S. Grzedzielski and D. E. Page (New York: Pergamon Press), pp. 37–42.

Matsuda, T., Fujimoto, Y., Shima, E., Sawada, K., and Inaguchi, T. 1989. Numerical simulations of interaction between stellar wind and interstellar medium. *Prog. Theor. Phys.* 81:810.

Möbius, E., Hovestadt, D., Klecker, B., Scholer, M., Gloeckler, G., and Ipavich, F. M. 1985. Direct observation of He^+ pick-up ions of interstellar origin in the solar wind. *Nature* 318:426.

Nerney, S. F., Suess, S. T., and Schmahl, E. J. 1991. Flow downstream of the heliospheric termination shock: Magnetic field kinematics. *Astron. Astrophys.* 250:556–564.

Nerney, S. F., Suess, S. T., and Schmahl, E. J. 1993. Flow downstream of the heliospheric terminal shock: The magnetic field on the heliopause. *J. Geophys. Res.* 98:15169–15176.

Nerney, S. F., Suess, S. T., and Schmahl, E. J. 1995. Flow downstream of the heliospheric terminal shock: Magnetic field line topology and solar cycle imprint. *J. Geophys. Res.* 100:3463–3471.

Parker, E. N. 1963. *Interplanetary Dynamical Processes* (New York: Wiley).

Pesses, M. E., Jones, W. V., and Forman, M. 1993. Voyager and Pioneer missions to the boundaries of the heliosphere. *J. Geophys. Res.* 98:15123–15128.

Quemerais, E., Lallement, R., and Bertaux, J.-L. 1993. Lyman α observations as a possible means for the detection of the heliospheric interface. *J. Geophys. Res.* 98:15199–15220.

Shima, E., Takuyo, M., and Inaguchi, T. 1986. Interaction between a stellar wind and an accretion flow. *Mon. Not. Roy. Astron. Soc.* 221:687.

Steinolfson, R. S. 1994. Termination shock response to large-scale solar wind fluctuations. *J. Geophys. Res.* 99:13307–13314.

Steinolfson, R. S., Pizzo, V. J., and Holzer, T. E. 1994. Gasdynamic models of the solar wind/interstellar medium interaction. *Geophys. Res. Lett.* 21:245–248.

Suess, S. T. 1990. The heliopause. *Rev. Geophys.* 28:97–115.

Suess, S. T. 1993a. Temporal variations in the termination shock distance. *J. Geophys. Res.* 98:15147–15156.

Suess, S. T. 1993b. The relationship between coronal and interplanetary magnetic fields. *Adv. Space Res.* 13:31–42.

Suess, S. T., and Nerney, S. F. 1990. Flow downstream of the heliospheric termination shock, 1. Irrotational flow. *J. Geophys. Res.* 95:6403–6412.

Suess, S. T., and Nerney, S. F. 1991. Correction to "Flow Downstream of the Heliospheric Termination Shock, 1. Irrotational Flow." *J. Geophys. Res.* 96:1883.

Suess, S. T., and Nerney, S. F. 1993. The polar heliospheric magnetic field. *Geophys. Res. Lett.* 20:329–332.

Suess, S. T., Hathaway, D. H., and Dessler, A. J. 1987. Asymmetry of the heliosphere. *Geophys. Res. Lett.* 14:977–980.

Wallis, M. K., and Dryer, M. 1976. Sun and comets as sources in an external flow. *Astrophys. J.* 205:895–899.

Washimi, H., and Nozawa, S. 1992. A simulation study of the outer heliosphere including the solar rotation effect. In *Solar Wind Seven*, eds. E. Marsch and R. Schwenn (Oxford: Pergamon Press), pp. 315–318.

Whang, Y. C., and Burlaga, L. F. 1993. Termination shock: Solar cycle variations of location and speed. *J. Geophys. Res.* 98:15221-15230.

Witte, M., Rosenbauer, H., Banaszkiewicz, M., and Fahr, H. 1992. The Ulysses neutral gas experiment: determination of the velocity and temperature of the interstellar neutral helium. COSPAR/World Space Congress, Washington D. C., 28 Aug.–5 Sept., Abstract book, p. 425.

OUTER HELIOSPHERIC RADIO EMISSIONS

W. S. KURTH and D. A. GURNETT
The University of Iowa

In this chapter we review the observations and various interpretations of the 2 to 3 kHz heliospheric radio emissions that have been detected by the plasma wave receivers on the two Voyager spacecraft over the last decade or so. A new event which began in mid-1992 has offered unprecedented information on the details of the emission. We concentrate on interpretations that involve emission at the plasma frequency or its harmonic near the heliopause. The emissions appear to be triggered by a system of interplanetary shocks associated with a global merged interaction region that had its origins in a series of spectacular solar events in the spring and early summer of 1991. Using a model for the propagation of these disturbances throughout the heliosphere and the observed delay from the activity on the Sun to the onset of the radio emission, estimates of the distance to the source region are derived ranging from 110 to 160 AU. We demonstrate the rich nature of the radio emission spectrum and the type of information it may provide about the heliopause and the triggering interplanetary shocks.

I. INTRODUCTION

From our perspective on Earth, it is natural to think of the Sun's atmosphere as extending to the upper reaches of the corona, primarily because with proper visual inspection the corona appears analogous to the Earth's upper atmosphere and ionosphere. However, another way to view the solar atmosphere is to trace it to its farthest reaches, to the limit of its primary influence on the surrounding medium. In this definition, the Sun's atmosphere extends to the heliopause which is the boundary of the region of space dominated by magnetic fields and plasmas originating in the Sun. Certainly in the context of "cosmic winds" it is more appropriate to take this more expansive point of view than the former.

Low-frequency 1.8 to 3.6 kHz radio emissions observed in the outer heliosphere are the subject of this chapter. We will make a strong case that these radio emissions are the result of the interaction of the Sun with the interstellar medium, similar in some respects to the more familiar solar type II and III radio bursts which are the result of the interaction of certain types of solar activity, coronal mass ejections, high-speed plasma streams, and other types of solar activity with the solar wind within 2 AU. By analyzing these high-frequency radio emissions (from about 30 kHz to above 400 MHz) a great deal has been inferred about the physics of the inner heliosphere; for example, the radial variation of the solar wind plasma density and the trajectory of disturbances propagating outward from the Sun.

The low-frequency radio emissions discussed in this chapter are a relatively new phenomenon, first reported in 1984 (Kurth et al. 1984a). Because they are restricted to the frequency range between about 1.8 and 3.6 kHz, they cannot normally propagate in the solar wind inside of about Saturn's orbit ($\sim$10 AU). This limit arises due to the cutoff of electromagnetic radio waves propagating in the so called free-space modes at the electron plasma frequency $f_p = 9000(n_e)^{1/2}$ where n_e is in cm^{-3} and f_p is in Hz. Because the Voyager spacecraft are the first to carry very low frequency receivers significantly beyond the orbit of Earth, they have provided our first (and only) opportunity to explore the very low-frequency end of the solar spectrum. Current evidence strongly suggests that these radio emissions are the result of the Sun's interaction with the surrounding interstellar medium and herald that most-distant extreme of the Sun's upper atmosphere, the heliopause.

The first observations of the low-frequency heliospheric radio emissions were of a major emission event which began in the latter part of 1983 and which extended into 1984 (Kurth et al. 1984a). This event was detected while Voyager 1 was at a heliocentric radial distance of about 17 AU and Voyager 2 was at about 13 AU. After the recognition of this event, it was also noted that a number of weaker events had been observed as early as 1982 by Voyager 1 as close as 15 AU to the Sun (Kurth et al. 1984b). As we will discuss at length, the events of the early 1980s were still strongly influenced by the relatively high density of the solar wind surrounding the spacecraft and some of the important aspects of the emissions remained largely masked for nearly a decade.

The early reports of the radio emissions discussed a number of different possible sources but most of the sources had substantial problems in explaining the observations. While speculative, the favored source had to do with the complex interaction of the solar wind with the interstellar medium and focused on the region near the termination shock where the supersonic solar wind slows to subsonic speed in response to the slowly moving surrounding interstellar medium. However, literally years passed with no major re-occurrences of the radio emission while the Voyagers moved outward through the heliosphere; it became more and more difficult to consider a source in the very distant heliosphere.

During the period of 1989–1991 a small number of very weak events were detected (Kurth and Gurnett 1991) that fueled continued interest in the phenomena and, further, suggested a consistent morphology for the events consisting of a relatively stable low-frequency component centered near 2 kHz and a more transient higher frequency component which appeared as one or more upwardly drifting tones extending to the 3 to 3.6 kHz range. Finally, beginning in 1992, nearly a decade (and perhaps more importantly, a solar cycle) later, another major event appeared, stronger than even the 1983–1984 event (Gurnett et al. 1993). This event, observed from the vantage points of 39 and 51 AU by Voyagers 2 and 1, respectively, was apparently unaffected by any local effects, hence, provided an unadulterated view of the entire spectrum and eliminated most possibilities for a source in the inner heliosphere. Simultaneous observations by Ulysses of the low-frequency Jovian radio spectrum were also available during this recent event (M. L. Kaiser, personal communication) and the Ulysses observations are used below to argue against a Jovian source (Gurnett and Kurth 1994*a*).

Given the strong new event, the lack of local effects, arguments that the source was not located in the inner heliosphere (<10–15 AU), and a concerted effort by the Voyager Project to obtain more high-resolution observations than were available during the 1983–84 event, Gurnett et al. (1993) offered a new model for the source of the radio waves and a new proposal for the triggers of the events. This model involved a system of interplanetary shocks associated with unusual levels of solar activity in the declining phase of the solar cycle that propagate outward through the solar wind and excite radio emissions in the cold plasma just beyond the heliopause. Two of the strongest Forbush decreases ever monitored are thought to be indicators of strong, global regions of turbulent magnetic fields and plasmas which interact with the interstellar medium to act as triggers for the two major radio emission events observed to date. Subsequently Czechowski et al. (1995), Zank et al. (1994), and Whang and Burlaga (1994) have proposed somewhat different models involving the interaction of solar wind disturbances (shocks) with plasmas between the termination shock and the heliopause to generate the radio emissions.

In this chapter we will review the salient features of the observations which bear on the interpretation of the source of the radio emissions and the clues they provide on various possible sources. Then, we review models which have been proposed to explain the frequency-time structure of the emissions. We will provide a summary of the basic generation mechanism which has been discussed at length for the radio waves. We then discuss the different types of triggering events that have been proposed and demonstrate why the events associated with two major Forbush decreases would seem to be the most likely. The identification of plausible triggering events naturally leads to an estimate of the distance to the source region through time-of-flight considerations. We will summarize the results of these analyses to date. Because these radio emissions are the first direct observations of an interaction associated with the outer most regions of the heliosphere, and because a number of questions are raised about certain aspects of the models, it follows that additional work is suggested and we discuss avenues of continued research which should help to clarify the situation.

II. OBSERVATIONS

The observations of heliospheric radio emissions are limited to the two Voyager spacecraft over the time interval beginning in about 1982 and continuing to the present. During this time, the spacecraft have traversed some 40 AU and have moved in significantly different directions. Figure 1 shows the trajectories of the two spacecraft relative to the outer planets in two orthogonal views. The left panel shows the trajectories projected into the ecliptic plane. The right panel shows the trajectories rotated into a common meridional plane. Note the marked departure of Voyager 1 to northern ecliptic latitudes after its encounter with Saturn in 1980 and the deflection of Voyager 2 into the southern hemisphere subsequent to its Neptune encounter in 1989. Voyager 1 is moving about 3.5 AU yr^{-1} while Voyager 2 is traveling at about 3.0 AU yr^{-1}. The vector V_S represents the velocity of the Sun with respect to the local interstellar medium (Ajello et al. 1987).

The observations of the heliospheric radio emissions are best summarized by the frequency-time spectrograms in Color Plates 13 and 14. The Voyager 1 observations are provided in the top panels of Plates 13 and 14 and those from Voyager 2 are in the bottom panel of each. In both plates, the wave observations are presented as intensity as a function of time (abscissa) and frequency (ordinate) over the range of 1 to 4 kHz. Intensity is encoded according to the accompanying color bar wherein red represents the most intense waves and blue the least intense. Relative amplitudes are provided by the decibel scale accompanying the color bar, but some caution should be exercised in the use of this scale. The wideband receiver which provides these observations utilizes an automatic gain control (AGC) circuit, hence, absolute amplitudes are not directly available from these data (see Scarf and Gurnett 1977). In particular, because the input signal strength could vary enough, particularly from interference below about 1 kHz (not shown), to affect the gain of the AGC amplifier, the amplitudes from two different times cannot be accurately compared. However, we believe that the presentations in Plates 13 and 14 do not suffer from significant shifts in gain and the qualitative comparisons we discuss herein should be faithfully represented in the plates. The relative amplitudes across frequency can be accurately compared for a single time period, but because only a narrow range of dynamic range (10 dB in Plate 13 and 13 dB in Plate 14) has been mapped to the color bar, signals stronger than the maximum of this range will all appear as red.

While the data in Plates 13 and 14 appear to be continuous, the actual measurements are only available at times indicated with tics along the top of the spectrogram. Specifically, prior to about mid-1987 measurements are only available about once per month on each spacecraft and the spectra are linearly interpolated from one measurement to the next. Later, measurements are made once per week or even more often. Each measurement consists of a spectrum averaged over about 15 s. For the 2 to 3 kHz emissions, the improved temporal resolution (up to 1 spectrum per 2 hr in one case) give us confidence that the interpolation across month-long intervals prior to mid-1987 does not introduce misleading features in the spectrograms. The narrowband line at 2.4 kHz is an interference band from the spacecraft power supply; a notch filter is centered on this frequency to minimize the intensity of this interference and may impose a gap in an otherwise continuous spectrum

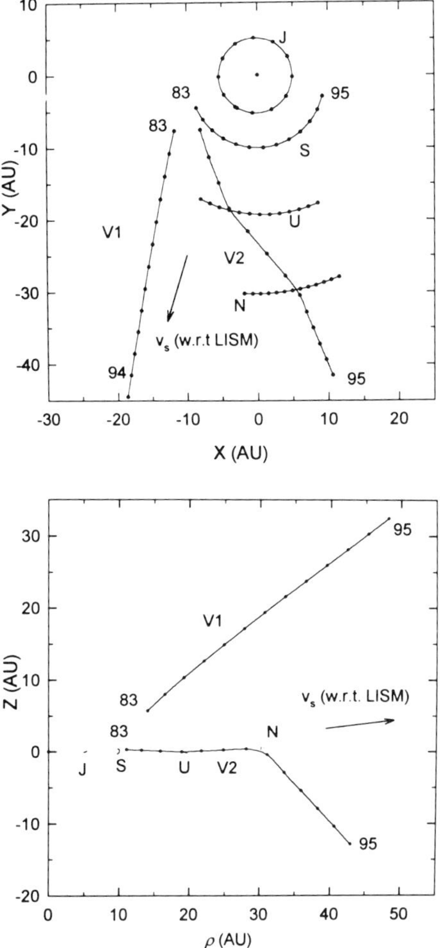

Figure 1. A summary of the geometry of the observations presented in Color Plate 13 including the positions of the outer planets and the two Voyager spacecraft from 1983–1995.

for real signals at this frequency.

Plate 13 covers the period from 1983 through 1994 and, subject to the caveats above, shows the relative strength of the various radio emission events over the 12-yr-long interval. The original event begins on about day 242 of 1983 and lasts for 6 to 9 months, although there is some evidence of the 2-kHz component extending into 1987. The most intense period of activity commences on about day 188 of 1992 and extends in the 3-kHz range until late 1994; the low-frequency component is still visible in the most recent observations which extend through the end of 1994. This event appears to comprise secondary events, such as the renewed activity beginning on about day 195 of 1993, seen most prominently near 3 kHz; at low frequencies this renewed activity overlaps the decaying phase of the 1992 event. The re-intensification in 1993 qualifies as the third most intense period of activity observed to date. Still another secondary event appears in early 1994 and extends to the middle of the year.

Four more events can be seen at much weaker intensities in the interval from 1989 through 1991 and one isolated event in late 1985. These weaker events, especially those observed in 1989–1991 are only marginally detectable in the Voyager 1 spectrogram in Plate 13; there is only a hint of the 1991 event in the Voyager 2 observations.

The lowest frequency extent of any of the observed events is about 1.8 kHz. The low-frequency component is much more prominent and is strikingly steady in the 1992 event and we will argue below that the lack of similar low-frequency emission in the 1983 event is due to significant local plasma effects associated with the relatively high local plasma frequencies at the locations of the spacecraft (between about 13 and 17 AU). The maximum frequency extent of any of the emissions is about 3.6 kHz. As described by Kurth and Gurnett (1991) the two components of these several radio emissions can be described as a relatively continuous low-frequency component and transient high-frequency components consisting primarily of narrowband emissions which rise in frequency with time. Kurth et al. (1987) reported that the 3-kHz emission beginning in 1983 increased in frequency at a rate of about 1 kHz yr^{-1}. Kurth and Gurnett (1991) reported frequency drift rates ranging from 1.4 to 1.8 kHz hr^{-1} for the events in 1989 and 1990 and the maximum drift rate in the 1992–1993 event is about 3.0 kHz yr^{-1} (Gurnett et al. 1993).

The radio emission events depicted in Plates 13 and 14 are remarkable for a number of reasons. These represent the longest-lived radio emission events ever detected from a solar system source excluding quasi-permanent phenomena such as Jovian decimetric emissions and continuum radiation trapped in planetary magnetospheres. The fact that the heliospheric events show discrete, coherent frequency-time structure over time scales of several months to over a year indicate a dynamical process operating over very long distance scales for any reasonable range of velocities expected in the interplanetary medium whether related to bulk plasma motion or characteristic speeds such as the Alfvén or sound speeds. An alternative way of viewing the very long dynamical structures is relative to a process which evolves in a monotonic manner over the time scale of many solar rotations. This would seem to be incompatible with the normal stream structure observed in the interplanetary medium out to the most distant spacecraft some 60 AU from the Sun.

Another point to notice from Plates 13 and 14 is the strong similarity between the observations made at the two spacecraft. Over the interval of observation the spacecraft were separated by distances ranging from about 10 to 54 AU. This is conclusive evidence that the phenomena must be a radio emission propagating from a distant source and not an effect associated with the plasma environment near the spacecraft. This, coupled with the fact that the emissions, especially near the end of the intervals presented, are at frequencies well above the highest characteristic frequency of the plasma $\sim f_p$ near a few hundred Hz confirm the identification of these emissions as freely propagating radio emissions.

There are some subtle differences between the spectra observed by the two spacecraft. Some of these may be due to the slight differences in sensitivity between the two receivers and some may be actual differences in intensity possibly due to differing relative distances from the source. Kurth et al. (1994) have performed preliminary analyses of these differences from the assuming they are due to differing distances between the source and the two spacecraft. While such analyses suggest that these differences might be used to locate the source (in conjunction with the direction-finding studies discussed below), the cross-calibration between the two instruments is not known well enough to be confident in the results at this time.

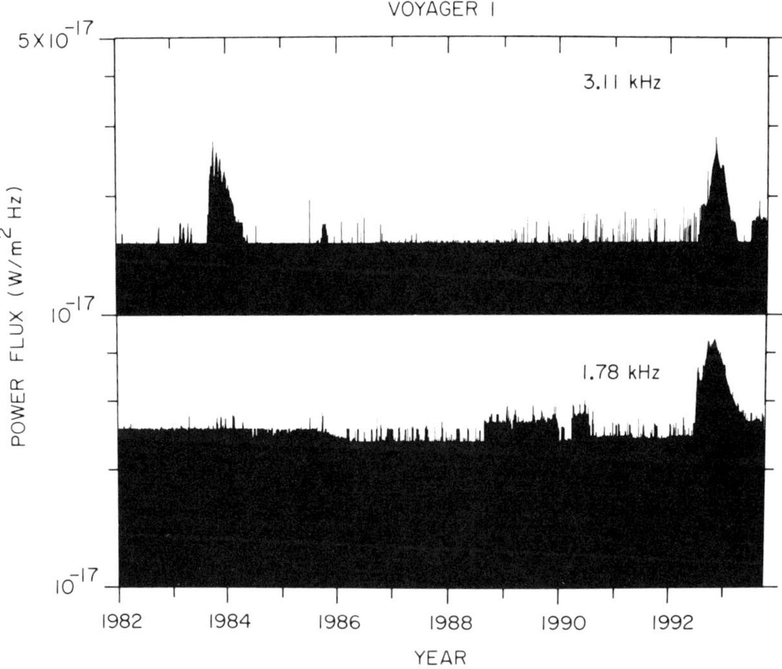

Figure 2.   The time history of intensities in the Voyager 1 1.78- and 3.11-kHz spectrum analyzer channels from 1983 through late 1993.

Figure 2 provides calibrated intensity as a function of time for the time period from 1982 through late 1993 for discrete spectrum analyzer channels at 1.78 and 3.11 kHz as observed by Voyager 1. Unfortunately, because of a failure in the Voyager 2 spacecraft data system a few months after launch, similar observations are not available from Voyager 2. Nevertheless, the intensity information in Fig. 2 is useful to obtain absolute amplitude information and provides a more continuous set of observations in that the intensities in these channels are sampled every 4 to 96 seconds, depending on the telemetry mode of the spacecraft, at least when the spacecraft is being tracked which ranges from nearly continuously to as little as about 12 hours per day over this time interval. Because of the smooth variations in the intensity of these radio emissions, we linearly interpolate over the tracking gaps in this figure. The 1983-1984 event and the 1992–1994 event are most prominent in this presentation, but the late 1985 event and the most recent (1993–1994) activity are also easily discerned. The weaker events between 1989 and 1991 are not visible in these data; the effective integration of the signal from the wideband receiver utilized for Plates 13 and 14 make the wideband observations a few dB more sensitive than the spectrum analyzer channels.

Figure 2 shows the primary difference in the two major events as a near absence of any response in the 1.78-kHz channel for the earlier event. Figure 3 provides a comparison of the 3.11-kHz observations of the 1983 and 1992 events on common time and amplitude scales. In these temporally expanded views, another important difference in the two events is apparent. As reported by Kurth et al. (1984a), there are quasi-periodic decreases in the amplitude of the earlier event with a period of about 25 days. Kurth et al. speculated that these decreases were due to the periodic increase in the local solar wind plasma density often associated with the stream structure in the outer heliosphere. This speculation was later confirmed by R. L. McNutt (personal communication) and also by L. F. Burlaga (personal communication) by noting that these diminutions occurred during times of high solar wind plasma densities at Voyager 1. Now, with the benefit of the 1992 observations, it is also apparent that the low-frequency spectrum of the 1983 event was significantly affected by the relatively high solar wind plasma densities inside of 20 AU. This explanation would predict little or no such effect (periodic decreases in intensity and an attenuated low-frequency component) at the larger radial distances where the more recent observations were made. Assuming an r^{-2} variation in the local plasma density, the latter observations would be made with average plasma densities a factor of about 6 lower than the earlier ones; the associated plasma frequencies would be 2.5 times lower. Hence, one aspect of the earlier observations which was local in nature has been shown to be unimportant for observations in the more distant heliosphere.

Because the peak intensity is only a few times the background (instrumental) noise level, accurate estimates of the wave intensities must be obtained by subtracting this background level from the observed intensities. Hence, the apparent peak of the 1992 event at 1.78 kHz of 4×10^{-17} W m^{-2}Hz^{-1} actually corresponds to about 1.8×10^{-17} W m^{-2}Hz^{-1}. By assuming an isotropic source at a distance of at least the relative distance between the two spacecraft (because they both observe approximately the same intensity according to

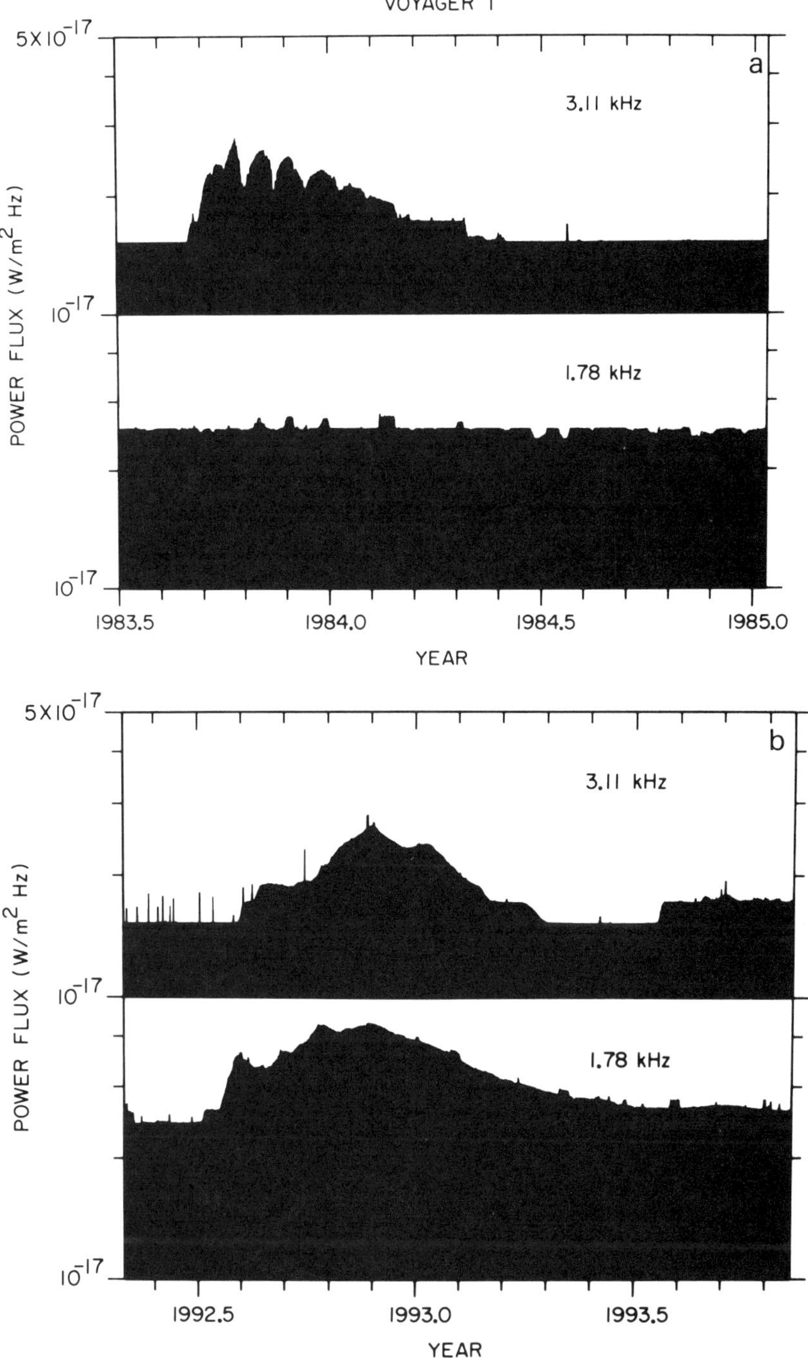

Figure 3. Expanded time scale plots showing the differences in the appearance of the (a) 1983 and (b) 1992 events.

Plate 14), Gurnett et al. (1993) estimate a total power of about 10^{13} W for this event at its peak, making it perhaps the most intense radio emission in the solar system. From observations obtained by Helios, the most intense solar type III radio emissions are just under 10^{13} W and the integrated average Jovian radio emission spectrum is 4×10^{11} W (Alexander et al. 1981).

Plate 14 provides an expanded view of the observations since early 1992, including the major event of 1992 and the renewed activity of 1993 and 1994 as viewed by Voyager 1 (top panel) and Voyager 2 (bottom panel) in the same format as in Plate 13. Here, the highly complex structure of the 1992–1994 event becomes quite clear. Below about 2.5 kHz the spectrum is dominated by the so called 2-kHz component that is generally more intense than the higher-frequency component. The 2-kHz component exhibits less spectral and temporal variation than the higher-frequency component, as was described as a general characteristic of the emissions by Kurth and Gurnett (1991). Even so, it is possible to see the onset of at least two narrowband structures near the leading edge of the low-frequency component and there is some evidence that these structures increase at least slightly in frequency over time. Toward the end of the event, near day 100 of 1993, there appear to be two distinct, constant-frequency emissions centered near 2 kHz and separated by 300 Hz or so. Hence, the "2-kHz continuum" component is not entirely continuous and exhibits some of the same, if less pronounced, drifting structure as the higher-frequency transient component.

The transient components at higher frequency earned this adjective because of their narrowband, increasing frequency natures. It is possible to identify clearly at least three of these features in the frequency range above 2.5 kHz during 1992, but there is evidence of one or more bands whose identity is more questionable due to their proximity to other bands. The first transient rising band forms the leading edge of the higher-frequency component of the 1992 event, beginning shortly after the 2-kHz component. The first two bands appear to drift to about 3.6 kHz as an asymptotic limit. Other bands do not rise to as high a frequency before they drop below detectability. The renewed activity in 1993 and 1994 consists of a pair of transient rising tones near 3 kHz in both cases, but with the maximum frequency attained in 1993 at about 3.4 kHz and only slightly above 3 kHz in 1994. The bands drifted at about 1 kHz y^{-1} in 1993 and those in 1994 drifted at about 1.5 kHz y^{-1}. In both pairs of bands, the band separation is approximately 200 Hz. Upon close inspection, the upper band of the 1994 re-intensification shows some evidence of fine structure (band splitting) with a separation of about 30 Hz. Both the 1993 and 1994 re-intensifications are accompanied by renewed or intensified emission in the 2 kHz band, as well. The renewed 2-kHz activity is superimposed on the decaying phase of the 1992 event and is, therefore, somewhat less noticeable. The new 1993 and 1994 activity is not as intense as the 1992 event.

As we will discuss in detail later, the narrowband emissions provide important information about the source of the waves. Because the emission frequency is almost certainly associated with the electron plasma frequency in the source region, the narrowband nature of the individual bands suggests that there is very little variation in density across the emitting region and that there is likely more than one emitting region. Because of natural variations in plasma densities at a variety of scale sizes in all known solar system plasmas, the narrow bands strongly suggest the source regions for these narrowband emissions are fairly localized and might be thought of as radio "hot spots" where conditions are optimal for wave generation.

III. SOURCE

Kurth et al. (1984a) considered several different possible sources for the 1983–1984 radio emission. Jupiter was thought to be the most reasonable of the possible planetary sources because it was known to be the source of rather intense radio emissions in the same spectral range as the heliospheric emissions (Scarf et al. 1979). However, because the two Voyagers observed virtually the same intensity despite their separation of several AU and, therefore, differing distances to the outer planets, radiation from a compact source such as a planetary magnetosphere was considered unlikely. Further arguments against the Jovian source included the fact that no 10-hour variations in the intensity of the heliospheric emissions were observed while such variations are characteristic of the Jovian continuum radiation and the spectral shape of the heliospheric emission was significantly different from the Jovian spectrum. These arguments were questioned, however, after it was suggested that the rising tones in the radio emission spectrum might be explained by a Fermi scattering effect caused by repeated reflections off of density irregularities moving outward from the Sun at an average of 400 km s^{-1} (Czechowski and Grzedzielski 1990,1992). Because this model for the frequency drifts required no ad hoc assumptions about the source of the radio emissions, it gained reasonable respect in the literature and was used to argue that the heliospheric cavity was relatively small, with a scale size of about 100 AU. Given the possibility that the radio emissions are trapped in a cavity, Kurth (1993) reconsidered the original arguments used to rule Jupiter out as a source and found that cavity effects might explain each of the observed characteristics which had been used to rule out this source. Kaiser et al. (1992) also thought that Jupiter should be reconsidered as the source of these emissions. A more complete discussion of the reconsideration of a Jovian source is provided in Kurth (1993).

Kurth (1993) also outlined a method by which a Jovian source might be verified or disproved via the use of Ulysses and Voyager observations during the time when Ulysses could still detect radio emission from Jupiter, specifically the continuum radiation below about 10 kHz. The proposed method basically relied upon the detection of an anomalously strong outburst of Jovian radio activity in the few-kHz range concurrent with the detection of a heliospheric radio event. Such a set of observations would provide a strong

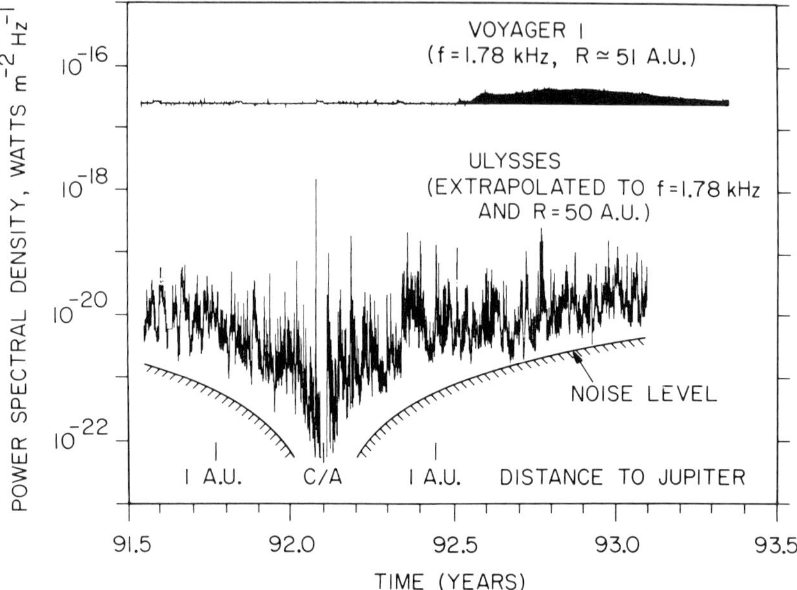

Figure 4. A comparison of radio emission observations from Ulysses and Voyager showing that there is a very large difference between the Voyager and Ulysses radio emission intensities. Also, note that there is no outstanding feature in the Jovian observations which correlates with the onset of the heliospheric emissions (figure from Gurnett and Kurth 1994*a*).

Jovian connection. Fortunately, the strong 1992 event provided an ideal case to form the basis for such a study. The results of the joint Ulysses-Voyager observations are given by Gurnett and Kurth (1994*a*). Figure 4 summarizes the results of this study. In this illustration, the Ulysses observations of Jovian continuum radiation at 10 kHz (M. L. Kaiser, personal communication; see also Stone et al. 1992) are used as an indication of activity at lower frequencies, because the solar wind plasma frequency at Ulysses for the several months around the Jupiter encounter ranged from less than about 2 to more than 5 kHz; it would be difficult to remove local solar wind density effects from the observations at 2 to 3 kHz. However, both Gurnett et al. (1980) and Kaiser (personal communication) report that Jovian continuum radiation falls off as $\sim f^{-4}$, hence, the 10-kHz intensities observed easily by Ulysses could be extrapolated to 1.78 kHz with reasonable assurance that no major Jovian radio events would be missed. Further, the Ulysses intensities were extrapolated to a distance of 50 AU assuming that they propagate freely from Jupiter as from a compact source. The lower plot in Fig. 4 shows the extrapolated Ulysses data. The effective noise level of these observations is shown as a hatched line below the observations. It might be argued that the r^{-2} extrapolation is incorrect based on an inspection of the trend in the Ulysses data to see weaker emissions closer to Jupiter. It could be that Jupiter is actually an extended source and the emission drops off somewhat more weakly than r^{-2}. However, this possibility does not generally affect the conclusions of the study.

There is no obvious intensification of the Jovian emissions which correlates with the onset of the heliospheric emission which is plotted on the same time and intensity scale. Furthermore, the intensities at Voyager with the background properly subtracted are some 300 times greater than the Jovian emissions extrapolated to 1.78 kHz. The only way to reconcile the differences in intensity would be to assume that the heliosphere forms a high-Q cavity, with Q of order 300 (Gurnett and Kurth 1994), in order to allow the energy density of the Jovian emissions to build up to the level observed by Voyager.

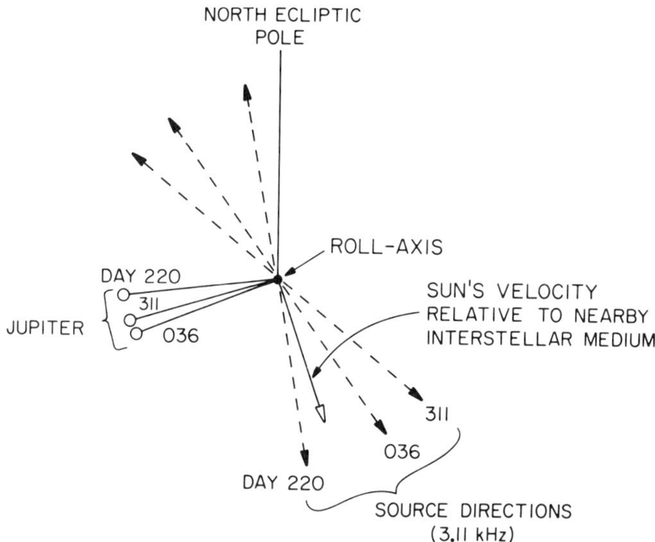

Figure 5. The results of direction-finding analysis performed for data obtained while Voyager 1 was executing roll maneuvers in order to calibrate the magnetometer. In general, the results are consistent with a source in the general direction of the arriving interstellar wind (figure from Gurnett et al. 1993).

Direction-finding observations obtained during the 1992–1994 event also argue against a Jovian source. Figure 5 shows the results of three such measurements reported by Gurnett et al. (1993), but we have added the position of Jupiter. In the plane of the sky as observed from Voyager 1, Jupiter does not seem to be consistent with the observed directions. From Fig. 1 it is clear that Jupiter is very close to the direction to the Earth, hence, the rotating dipole technique would not provide a reasonable direction to a possible source so close to the rotation axis. It is just this point, however, which may be the best evidence to exclude Jupiter. Along with measuring a roll phase angle which identifies the plane including the source of the radio emissions, the spin modulated signal also provides one other piece of information. The peak-to-peak amplitude of the modulation pattern is also measured. In some cases, this modulation is as much as 50% and in all cases measured so far, is a minimum of 10%. The strongest modulation would be for a point source located in a plane exactly perpendicular to the rotation axis of the spacecraft with no scattering. Scattering, an extended source, and a location closer to the axis of rotation are all effects which tend to decrease the modulation of the signal as the dipole antenna rotates. If we attribute all of the decrease in modulation to only the angle of the source from the rotation axis, then this would define a minimum angle with respect to the rotation axis of the spacecraft at which the source would have to be. If one includes the effect of an extended source (for which we have no estimate) or scattering (which Cairns [1995] estimates can account for all of the observed decrease in modulation), then the source must lie even farther from the axis of rotation than the limiting angle and closer to the plane perpendicular to that axis to account for the observed modulation depth. Because in 1992 the Voyager 1 was greater than 47 AU from the Sun and because the maximum perpendicular separation of the Earth (toward which the spacecraft rotation axis points) and Jupiter is about 6 AU, then the maximum angular separation of Earth and Jupiter is about $7°$. The measured depth of modulation, if attributed only to the angle from the rotation axis, implies angular separations between the Earth and the source ranging from about $15°$ to $45°$ for events analyzed. Therefore, the modulation depth precludes Jupiter as a source.

Because an explanation for the drifting structures in the heliospheric radio emissions had been given by Czechowski and Grzedzielski (1990,1992) as the result of repeated reflections within a heliospheric cavity, it was necessary to determine the probability that the radio emissions were, indeed, trapped. Gurnett et al. (1993) showed that the 3.11-kHz component of the wave spectrum shows significant spin modulation due to the anisotropic distribution of wave vectors arriving at the Voyager 1 spacecraft. Such anisotropies would not be expected to exist in a high-Q cavity unless very stringent symmetries were present which would organize multiple reflections in some way. One known example of a natural radio cavity is the Earth's magnetosphere in which nonthermal continuum is trapped. While there is definite spin modulation observed for these terrestrial waves due to the fact they are propagating down the magnetotail and eventually leaving the system, the Q is very small, estimated to be only about 3 (Gurnett 1975). Hence, only a small number of reflections keep the continuum radiation trapped in any fixed volume. The spin modulation observation, then, suggests that at least the 3-kHz component of the heliospheric radio emission is not trapped. The 1.78-kHz channel does not show any evidence of spin modulation. While this result is consistent with trapping, there is also the possibility that scattering (Cairns 1995) or interference in this channel due to the operation of the spacecraft gyros, which are always operating when the spacecraft is rotated, explains the lack of spin modulation.

Inspection of Plate 14 shows a number of subtle (few dB) variations in intensity between the two spacecraft. For example, the leading edge of the

3-kHz component at the start of the activity is more prominent at Voyager 2 than Voyager 1. The upper-most band of the 3-kHz pair in the 1993 re-intensification is more prominent at Voyager 1. While the cross-calibration of the two instruments is known only to a few dB at present, that the relative intensities vary in such a manner can only be explained by true differences in the spectra. Such differences are not expected for trapped emissions, which would be virtually the same amplitude throughout the low-density volume of the heliosphere. We also point out that Czechowski and Grzedzielski (1990) determined that a 1 kHz yr^{-1} drift in the frequency of the radio emissions would imply a relatively small heliosphere, with a scale size of 100 AU or so. For some of the more recent events the drift rates are higher, up to about 3 kHz yr^{-1} for the 1992 event. Because it is the reflections off of the outward moving density irregularities and the number of these reflections over an interval of time which determine the frequency drift rate, it seems clear that the Czechowski and Grzedzielski (1990) model runs into difficulty with the more recent higher drift rates of up to 3 kHz yr^{-1}. This would almost certainly imply even shorter distance scales in the heliospheric cavity and the higher drift rates become increasingly difficult to support. However, Zank et al. (1994) and Czechowski et al. (1995) have suggested that transitory density structures downstream of the termination shock could serve to confine the waves in a volume of scale 70 AU, hence, drift rates up to the observed values could be supported, even though the heliopause is much more distant. In this case, though, it would seem important for the transitory structures to have reasonably long lifetimes of a few months.

Farrell (1993) has suggested another mechanism for the drifting features which also involves multiple reflections between the inner heliosphere and the heliopause. In this theory, however, the reflections are required to be resonant with monochromatic oscillations of the heliopause such that the wave travel time is an integer ratio of the boundary oscillation period for fixed frequency bands. The bands could drift in response to a gradually changing cavity size or to changing heliopause oscillation period. While this work offers a tantalizing new mechanism for the drifting features, it would seem that the resonance conditions are just too severe to expect them to hold over the 100+ AU of heliocentric distance scales and over the several-month periods which represent the lifetime for many of the observed bands. The irregularities observed in the interplanetary medium out to the most distant spacecraft would seem to be sufficient to disrupt any such resonances long before a discrete band could form.

We conclude that the waves are not trapped at the highest frequencies (>2.5 kHz). This still leaves the possibility that the lower-frequency waves might be trapped. However, if the source strength of the radio emission is roughly constant in frequency, then the observations in Plate 14 argue that the Q of the cavity is very small, because the intensity of the 2-kHz component is not more than 5 or 6 dB greater than the 3-kHz component.

The anisotropic distribution of wave electric fields allows a one-dimensional determination of the source direction following the technique of Kurth et al. (1975) and Baumback et al. (1976). Gurnett et al. (1993) reported this analysis for the Voyager spin-modulated observations at 3.11 kHz early in the 1992–1993 event and determined that the emissions were arriving from a direction which is consistent with the direction of arrival of the interstellar wind as determined by Ajello et al. (1987). Figure 5 shows the directions of arrival for three different determinations made possible by Voyager 1 roll maneuvers in late 1992 and early 1993 from Gurnett et al. (1993). Because the determinations of direction from the rotating dipole antenna technique only provide information on the plane of the source, we can show the results by showing the intersection of this plane with the plane of the sky. The reasonable agreement between the source directions shown in Fig. 5 with the direction of arrival of the interstellar wind is significant and contributes strongly to the Gurnett et al. (1993) source in the pile up region at the nose of the heliosphere. The variation in directions between the various determinations are much larger than the ~1 degree uncertainties in the directions. By comparing the times when these directions were determined with the portion of the spectrum which was within the 3.11-kHz channel's passband at the time, it is clear that each direction determination is of a different transient narrowband emission. It follows that these bands are limited in size and are in somewhat different directions; this aspect contributes to the concept of radio "hot spots" introduced above. Further, additional direction-finding results show a systematic trend for the source to rotate from a direction consistent with the direction from which the interstellar wind comes to a direction some 90° away. This trend has been interpreted by Gurnett and Kurth (1994b) as a movement of the triggering shock's interaction region progressively moving from the nose toward the flanks of the heliosphere. Cairns (1995) point out that scattering might account for apparent changes in direction of arrival, but one would expect random variations from scattering, not systematic ones. The only direction-finding results from the 1983 event were reported by Kurth (1988); this result was consistent with radio emissions propagating radially either inward toward the Sun or outward from the Sun (a 180° ambiguity is an attribute of the rotating dipole direction-finding technique); however, an asymmetric modulation pattern prohibited any detailed analysis of this result.

Kurth et al. (1984a) also considered a distributed source in the inner heliosphere (in the heliocentric distance range of 10–15 AU) where the solar wind plasma frequency drops through the 2 to 3 kHz range. This source was eliminated because of a lack of *in-situ* observations in this region of significant Langmuir wave activity which might constitute a source for the radio waves (see the section on generation mechanism, Sec. IV below) and because it seemed unreasonable to expect the highly variable stream structure to generate a stable (on time scales of several solar rotations), narrowband radio emission.

Millisecond pulsars were also considered by Kurth et al. (1984a) as potential sources, but were ruled out on energetic arguments even though

Lipunov (1983) had estimated that a power flux similar to that observed might be expected. The increasing frequency of the emission and its complex spectrum, though, did not seem plausible for a fast pulsar source.

In considering other possible sources external to the solar system, Kurth et al. (1984a) hypothesized that average stellar systems might be the source of very low-frequency radio emissions at large radial distances from the parent star where the characteristic frequencies of the stellar plasmas would be in the few kHz range. However, if one considers such a source as being commonplace, it is only natural to look at the closest star, and in this case, the Sun is the obvious candidate. The Sun is also the obvious choice simply from the point of view of received intensity; for another star to dominate the solar radio spectrum, the emission intensity of the other star would have to be exceptionally high. The low frequency of the radio emissions and the known density profile of the solar wind immediately suggested that such radio emissions might be the result of solar wind interactions with the surrounding interstellar medium. Therefore, they proposed that a possible source might be associated with this interaction region and concentrated on the termination shock, the innermost portion of the interaction region.

The termination shock source was favored for several years. Several difficulties existed with this interpretation, dealing primarily with the details of the emission spectrum. Generating the emission at the observed frequency for reasonable termination shock locations was perhaps the most severe problem, because for source locations beyond the Pioneer and Voyager locations, it was difficult to understand how the source plasma frequency could be high enough to explain the emission frequencies. One tack was taken by Kurth et al. (1984a) in assuming the source had to be on the downstream side of the termination shock and the characteristic frequency would then increase by a factor of 2 if the shock were strong. Cairns and Gurnett (1992) pointed out that the normal generation regions for bow shock-associated radio emissions and some interplanetary shocks were not on the downstream, high-density side, but the upstream, low-density side of the shock. They suggested density variations of order 4 to 10 above ambient could boost the source frequency on the upstream side of the termination shock high enough to explain the observed frequencies. In updating this discussion, the Voyager 1 and Pioneer 10 are now at or beyond 60 AU from the Sun and the usual r^{-2} model for the solar wind plasma frequency places the local plasma frequency at about 300 Hz. To consider generation at the plasma frequency in the supersonic solar wind requires extraordinary conditions and special models. Partially in response to this problem, Zank et al. (1994) have studied the propagation of density enhancements through the termination shock and the possibility of generating radio emissions in the observed frequency range.

Another problem, not necessarily limited to the termination shock source, is the observed spectrum of emissions including two general bands near 2 and 3 kHz. The generation mechanism to be discussed below can result in emission at both the plasma frequency and its harmonic. The two observed bands often suggest an $f_p - 2f_p$ pair, but the ratio of the observed frequencies is generally considerably less than 2:1.

Previous to the 1992 event, there was also no satisfactory (or at least unique) explanation for the differences in the low-frequency, quasi-steady component and the higher-frequency transient (drifting) component. Different source regions for the two components were suggested, but this did not seem compatible with nearly simultaneous onsets for the two components.

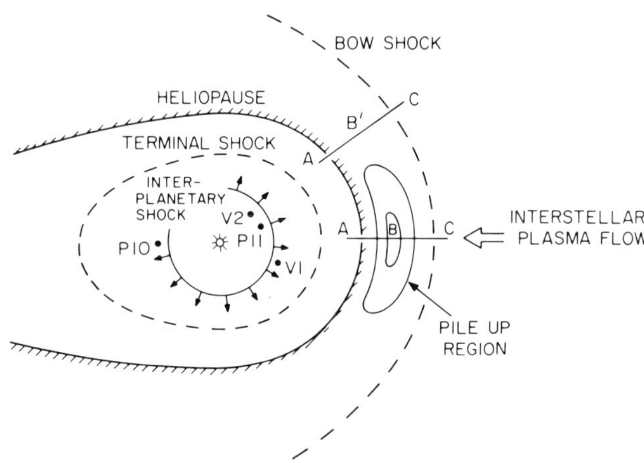

Figure 6. A model heliosphere showing the propagation of an interplanetary shock towards the heliospheric boundary (figure reprinted with kind permission from Elsevier Science Ltd.; from Gurnett et al. 1993).

Fahr et al. (1986) suggested that the heliopause itself might be the source of radio emissions. While they suggested a number of plasma wave modes involving ions as possible wave turbulence which could result in the emission of radio waves, these modes are almost certainly too low in frequency to be of importance here. However, Fahr et al. also suggest that there could be an "electrostatically turbulent heliopause layer" in which the characteristic frequency could be about 1.3 kHz (assuming an electron density of about 0.02 cm^{-3}). They suggested that radio emissions at such a frequency could propagate inwards to within about 16 AU of the Sun. They further suggested that the local magnetic field conditions at the heliopause might only be favorable for the generation of the electrostatic plasma waves at certain times, possibly associated with the phase of the solar cycle.

Subsequent to the detection and analysis of the 1992 event shown in Plate 14, Gurnett et al. (1993) interpreted the radio emissions as the result of an interplanetary shock moving through the plasma just beyond the heliopause generating electrostatic waves at the local plasma frequency. The electrostatic waves, as described below, then couple into electromagnetic waves at either the local plasma frequency or its harmonic, or both. This interpretation explains the rising frequency bands in the transient, higher-frequency component of the radio emission as the motion of the excitation source moving through a region of plasma with increasing density, possibly the interstellar plasma which is piled up at the nose of the heliosphere. The steadier, low-frequency component is interpreted by Gurnett et al. as a similar process as above but occurring near the flanks of the heliopause where there is little or no plasma pile up. This model is illustrated in Figs. 6 and 7.

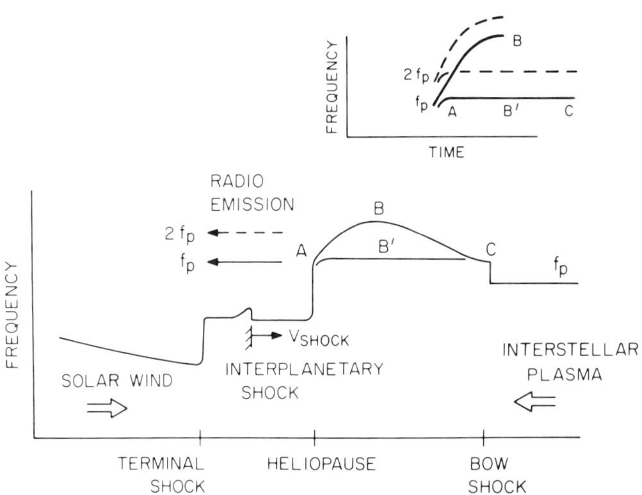

Figure 7. A schematic of the type of observations the model of Gurnett et al. (1993) predicts assuming a general structure for the outer heliosphere including a plasma pile up region near the nose (figure reprinted with kind permission from Elsevier Science Ltd.; Gurnett et al. 1993).

Figure 6 shows a model of the heliosphere including a termination shock and heliopause. The model is motivated by the work of Steinolfson et al. (1994) and includes a region upstream of the heliopause where interstellar plasma piles up as it is slowed and deflected around the heliopause. Shown in the interior of the supersonic solar wind is a more-or-less spherical shock wave moving outward from the Sun past the various outer heliospheric spacecraft and towards the interstellar medium. Two different cross sections are shown through the post-heliopause interstellar medium. First, section A-B-C crosses through the pile up region near the nose of the heliosphere. Second, section A-B'-C crosses through the post-heliopause interstellar medium closer to the flank of the heliosphere.

Figure 7 shows the profile of the plasma frequency (proportional to the square root of the plasma density) in the outer heliosphere through the very local interstellar medium downstream of the heliospheric bow shock. Between the heliopause and the bow shock, two plasma frequency profiles are shown, corresponding to cross sections A-B-C and A-B'-C in Fig. 6. Also indicated in

Fig. 7 is an interplanetary shock moving outward through the solar wind. The inset of Fig. 7 shows a schematic frequency-time spectrogram in the form of Plate 14 showing the form of narrowband emissions which might be expected for radio waves being generated along the two cross sections at either f_p or $2f_p$. Together, Figs. 6 and 7 form the basis for the Gurnett et al. (1993) model for the generation of the heliospheric radio emissions.

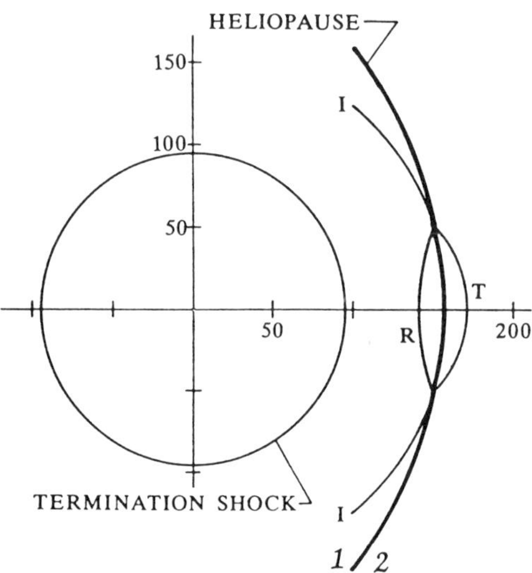

Figure 8. A schematic from Whang and Burlaga (1994) highlighting the salient features of their model. As an incident interplanetary shock (I) encounters the heliopause, both a reflected (R) and a transmitted (T) shock are formed. The transmitted shock is the triggering agent for the 3-kHz component of the radio emission spectrum and the reflected shock is responsible for the 2-kHz component in the Whang and Burlaga theory.

Whang and Burlaga (1994) have addressed the existence of different bands of emissions near 2 and 3 kHz in a different way than Gurnett et al. (1993). Whang and Burlaga point out that a shock which impinges on the heliopause, which they model as a tangential discontinuity with a strong density jump, would result in both a transmitted shock which continues into the interstellar medium and a reflected shock which propagates sunward in the post-termination shock solar wind as shown in Fig. 8. They calculate the plasma frequency in the vicinities of the transmitted and reflected shocks as a function of the incident shock speed and the position of the termination shock (which depends on the magnitude of the interstellar magnetic field). In this way, Whang and Burlaga show that the emissions near 2 kHz can be generated at the reflected shock while emissions near 3 kHz can be generated near the transmitted shock for a fixed location of the termination shock. They provide solutions for radiation generated at either f_p or at $2f_p$. This model suffers from at least two difficulties. First, is the reflected shock strong enough to generate radio emissions? The second, related problem is concerned with how the reflected shock could produce radio waves in the shocked solar wind downstream of the termination shock, but the original shock would not when it was in the same region prior to encountering the heliopause. We assume that the transmitted shock moves into a much cooler plasma in the interstellar medium, hence, we would expect Landau damping to be significantly reduced and radio emission to be more likely (Gurnett et al. 1993).

Zank et al. (1994) have studied the possibility of generating the heliospheric radio emissions in the region between the termination shock and the heliopause, where the solar wind is subsonic. This region, under stable conditions, would not normally be favored as a source regions because the nominal plasma density is only four times that of the supersonic solar wind just inside the termination shock assuming a strong gas-dynamic shock. However, this group has explored the modification of the downstream solar wind resulting from the propagation of shocks and density enhancements as one would expect from global merged interaction regions (GMIRs) and other outer heliospheric structures through the termination shock. They argue that the downstream solar wind would have significant density variations, including enhancements which could bring the perturbed density more in line with those required to support f_p or $2f_p$ radiation in the 2 to 3 kHz range. Simply stated, the propagation of a density enhancement through the termination shock will result in a propagating region of significantly elevated densities. Subsequently, should a solar wind shock follow such a disturbance, it could generate radio emissions as it propagates through the density enhancement, causing emissions with the appropriate frequency and frequency drift to explain the observed signals. Figure 9 illustrates the salient features of the Zank et al. model. It should be noted that Zank et al. calculate an expected distance to the termination shock in the range of 60 to 70 AU. Furthermore, they rely on emission at $2f_p$ in order to support frequencies as high as those observed. Given the current distances of the Voyager 1 and Pioneer 10 spacecraft (between 60 and 65 AU) with no clear indication of the termination shock in the near vicinity of the spacecraft, one would have to question this model simply on the basis of the quoted distance to the shock. The situation is even more questionable if the emission is at f_p; some arguments favoring emission at f_p are given in the discussion below.

Czechowski et al. (1995) have considered the interaction of the neutral interstellar hydrogen with the heliosphere and suggest that there could be a thin, cold, dense layer of plasma just inside of the heliopause resulting from the charge exchange interaction of the neutrals with the solar wind protons. They demonstrate that a shock propagating through such a layer with reasonable propagation speeds could generate radio emissions with the observed drift rates. Figure 10 shows the frequency drifts which could be obtained for reasonable conditions near the heliopause. In a sense, this model

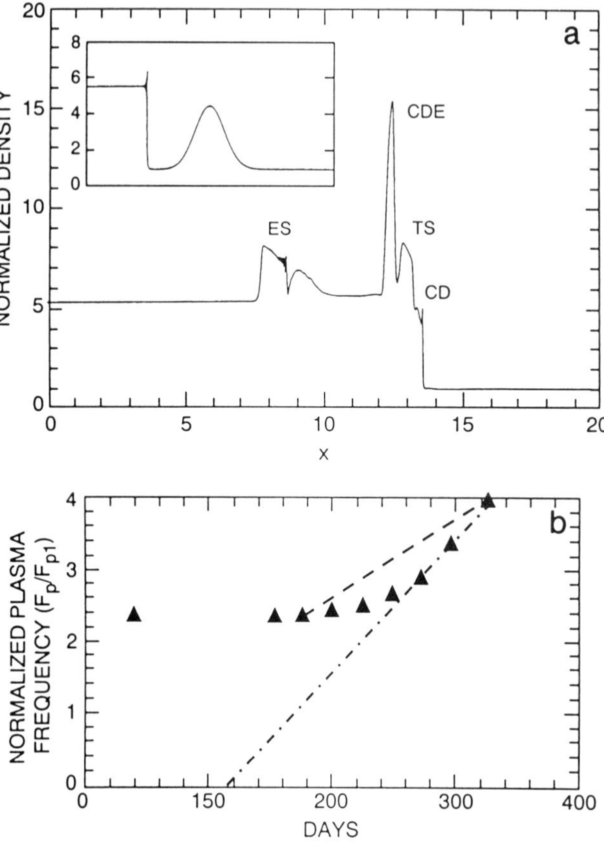

Figure 9.　(a) A representation from Zank et al. (1994) depicting how an interplanetary shock propagating in the post-termination shock region can produce emissions drifting upward in frequency by exciting emissions near f_p as it moves through an increasing density structure set up by a previous density enhancement in the solar wind. A shock (*ES*) is emitted by the initial interaction of the density enhancement and the termination shock. The compressed density enhancement is denoted by *CDE*. *TS* is the transmitted interplanetary shock. *CD* is a new contact discontinuity. The inset shows the original density pulse prior to its interaction with the termination shock. (b) The frequency-time evolution of radio emissions generated by the transmitted shock as it crosses the *CDE* using two methods for calculating frequency drift.

does not deviate greatly from the of Gurnett et al. (1993). What is new here is the suggestion that a cold plasma with a suitable density gradient may exist just inside the heliopause and that the source region could lie sunward of the heliopause.

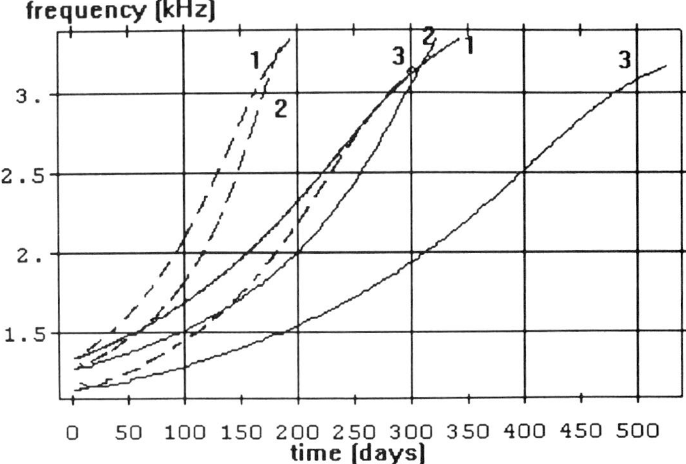

Figure 10.   An illustration from Czechowski et al. (1995) showing rising frequency tones which might be emitted from the propagation of a shock through a proposed cool, dense layer just inside the heliopause. The cold plasma is the result of charge exchange processes with interstellar neutrals. The dashed (solid) lines represent shocks with an initial speed of 500 km s^{-1} (300 km s^{-1}) moving through cold plasma layers computed for three different heliospheric models.

IV. GENERATION MECHANISM

Given a source in the outer heliosphere, the only generation mechanism which has been addressed with any detail is a mechanism which is similar to that which is thought to be responsible for radio emissions from planetary bow shocks, interplanetary shocks, and solar type II radio bursts. The basic model relies on electrons being scattered or accelerated by the shock forming a bump-on-tail distribution which is unstable to Langmuir waves (also known as electron plasma oscillations). The Langmuir waves are local oscillations in the plasma and do not propagate appreciable distances. However, a number of theories have been developed that involve nonlinear interactions between the Langmuir waves and low-frequency waves such as ion-acoustic waves which result in the emission of an electromagnetic wave near the electron plasma frequency, hence, a radio wave which can freely propagate away from the source region.   Another possible scenario is that two Langmuir waves nonlinearly interact to form an electromagnetic wave at about twice the electron plasma frequency. This basic picture was invoked by Kurth et al. (1984a) and Kurth (1988). The theory was seriously developed through a series of papers (Macek et al. 1991a,b; Cairns and Gurnett 1992; Cairns et al. 1992; Macek 1994) which attempted to determine the feasibility of generating the observed brightness of the radio waves within a reasonably sized generation region and with Langmuir waves of reasonable intensity. With the assumption of a source extending over a significant region of the termination shock with a thickness of no more than about an AU, the required amplitude of Langmuir waves is reasonably consistent with the expected amplitude obtained by extrapolating observed Langmuir waves upstream of planetary bow shocks to ~100 AU. Further, Kurth and Gurnett (1993) argued that such a layer of Langmuir waves is a reasonable expectation for the region just upstream from the termination shock.

The strong suggestion of radio "hot spots" based on multiple, narrowband emissions may invalidate somewhat the assumption of a source which extends over a significant region of the termination shock, but no quantitative analysis of the impact of a smaller source exists. Also, we assume in this review that the theoretical work mentioned above which was applied to the region near the termination shock is more or less transferable to the region near the heliopause. Of course, detailed treatment of such a transfer is required in order to justify this.

The Gurnett et al. (1993), Zank et al. (1994), Whang and Burlaga (1994), and Czechowski et al. (1995) models all rely on some mechanism which is likely similar to that described here, except that it is an interplanetary shock which serves as the source of the energetic electrons which produce the Langmuir waves just ahead of the shock. Gurnett et al. (1993) suggest that when the shock enters the much colder plasma beyond the heliopause (Czechowski et al. [1995] would argue the cold plasma layer should exist inside the heliopause) Landau damping is reduced, hence, generation of Langmuir waves is more likely. This line of reasoning would not be consistent with generation in the relatively hot subsonic solar wind as required by Zank et al. (1994) or at the reflected shock in the Whang and Burlaga (1994) model.

V. TRIGGERS

The 1983-1984 and 1992-1994 radio events were clearly uncommon events. Most radio emissions in the solar system exhibit several orders of magnitude variations in intensity over time, so it is reasonable to assume the heliospheric radio emissions do, also. It follows, though, that the limited Voyager sensitivity actually acts as filter for only those most intense events, whether or not weaker emissions are present. The question which arises, then, is what is the mechanism causing the emission to turn on rather suddenly, or at least reach the intensities which are observed only once in a decade. With this line of reasoning, a number of efforts were undertaken to find a signature in the solar wind which might explain the onset of the heliospheric radio emissions. Lanzerotti et al. (1985) reported a decrease in solar wind-accelerated protons which implied a quiet solar wind condition and reasoned that a decrease in solar wind structure between the Voyagers and the source might make it easier for low-frequency waves to propagate inwards toward smaller heliocentric distances. A similar comment was made by J. A. Van Allen (personal communication). But this line of reasoning did not explain the duration of the 1983 event nor could it explain the brief reappearances of the emission in

1989–1991. Other attempts to find concurrent features in the mid-heliospheric plasma which would correlate with the radio emission were unsuccessful.

In 1988, McNutt (1988) searched the entire Voyager plasma data set from the beginning of the mission in order to identify a signature of some solar wind event which might correlate with the radio emissions, with no constraint on temporal concurrence. The result was the identification of two very fast high-speed streams in 1979 and in early 1983. McNutt suggested that these high-speed streams might be reasonable triggers for the radio emissions when they reached the source region, which was presumed to be the termination shock. The delay from the earliest Voyager 2 detection of the 1983 stream at 11 AU to the onset of the radio emission was about 250 days. The maximum stream speed reported by McNutt was 780 km s^{-1}. Hence, the maximum distance from the Sun to the source region using simple time-of-flight considerations is about 123 AU. McNutt determined time-of-flight using the peak of the radio emission and reported a maximum distance of about 140 AU. McNutt also considered that the shock might slow to a minimum of about 400 km s^{-1} by the time it reached the source, hence estimated a minimum distance of about 70 AU. The 1979 stream would have reached a similar heliocentric distance while the Voyagers were well inside 10 AU; hence they would not have been in a sufficiently low-density plasma to observe whether there was any corresponding radio event. The maximum distance reported by McNutt is in the range of the possible distances reported by Gurnett and Kurth (1995), however, by assuming the source was the termination shock, McNutt was forced to hypothesize transient density enhancements in the shocked solar wind and second harmonic emission to explain even the low-frequency component of the emissions.

McNutt's attempt at finding a trigger was only partially successful. First, with only a single event to study, it was impossible to verify the hypothesis of a high-speed stream trigger. Second, when the 1985 event appeared, there was no obvious high-speed stream as the 1983 feature with which to associate the new event. He continued to search for streams and predicted the onset of the radio emission on the basis of new streams. Notably, upon detecting fast streams associated with the 1991 solar activity, McNutt and colleagues predicted the onset of radio emissions in late 1991 or early 1992 (McNutt et al. 1991).

Grzedzielski and Lazarus (1993) refined McNutt's trigger model by noting that the 1983 (and subsequent radio emission events) followed an extended interval of increasing solar wind pressure. They further reasoned that the trigger might be the result of a number of smaller streams which coalesce at or near the source region, thus making a stronger effect near the source than they would individually in the middle heliosphere where Voyager detects them. They plotted the trajectories of a number of streams and found a number of combinations which merged near 90 AU at about the time of onset for three of the events.

After the detection of the 1992 event, Gurnett et al. (1993) appealed to the notion that the very intense events of 1983 and 1992 must have highly unusual triggering events associated with them. They noted that the most intense Forbush decrease ever recorded with the Deep River neutron monitor occurred on day 164 of 1991, 1.1 yr prior to the onset of the 1992 event. Analyses of this event by Van Allen and Fillius (1992) and Webber and Lockwood (1993), showed the association of this Forbush decrease with very strong cosmic ray intensity decreases and/or interplanetary shocks at the Pioneer 10 and 11 and Voyager spacecraft as well as with the very intense solar activity in the March–June 1991 time frame. Using the velocity range estimated by Van Allen and Fillius and Webber and Lockwood of between 600 and 800 km s^{-1} and using an estimate of the speed of the shock in the outer heliosphere associated with this activity based on modeling performed by R. S. Steinolfson (personal communication), Gurnett et al. established the distance to the source using time-of-flight considerations which ranged between 116 and 177 AU. Based on the Gurnett et al. model of the source region lying just beyond the heliopause, this locates the heliopause within the same distance range.

While the association of the most intense Forbush decrease on record with the most intense heliospheric radio emission meets the criterion of having a highly unusual event for a trigger, a single correlation such as this remains open to conjecture just as was the case with McNutt's (1988) high-speed stream trigger hypothesis. However, Gurnett et al. (1993) also reported a very similar association between a very intense Forbush decrease with the 1982-1983 event. In this case, a Forbush decrease beginning on day 195 of 1982, the second most intense on record at the Deep River monitor, came 1.1 yr prior to the onset of the radio emission. It should be noted that this Forbush decrease occurs nearly 6 months prior to the high-speed stream reported by McNutt (1985). Hence, even taking into account the time-of-flight to the 11 AU position of Voyager 2, the solar activity responsible for the 1982 Forbush decrease cannot be associated with McNutt's high-speed stream. Estimates for the velocity of the solar wind disturbance associated with the Forbush decrease (Van Allen and Randall 1985; Webber et al. 1986) were 810 to 850 km s^{-1}. Figure 11 shows the Deep River neutron monitor data including these two Forbush decreases at Earth in relation to the onset of the two major radio emission events as seen in the Voyager 1 3.11-kHz channel (Gurnett and Kurth 1996). That a similar case can be made for both of these events with similarly significant solar wind disturbances significantly strengthens the hypothesis that the Forbush decreases mark solar wind disturbances which can be identified as triggers for the radio emissions.

Gurnett and Van Allen (1993) have looked at the other, weaker radio emissions and the neutron monitor data in an attempt to find similar triggers for the remaining events. By doing a superposed epoch analysis, they have identified possible features of lesser magnitude than the previously discussed Forbush decreases which lead some of the other radio emission events by an interval of time somewhat larger than the 1.1 yr found for the two major event.

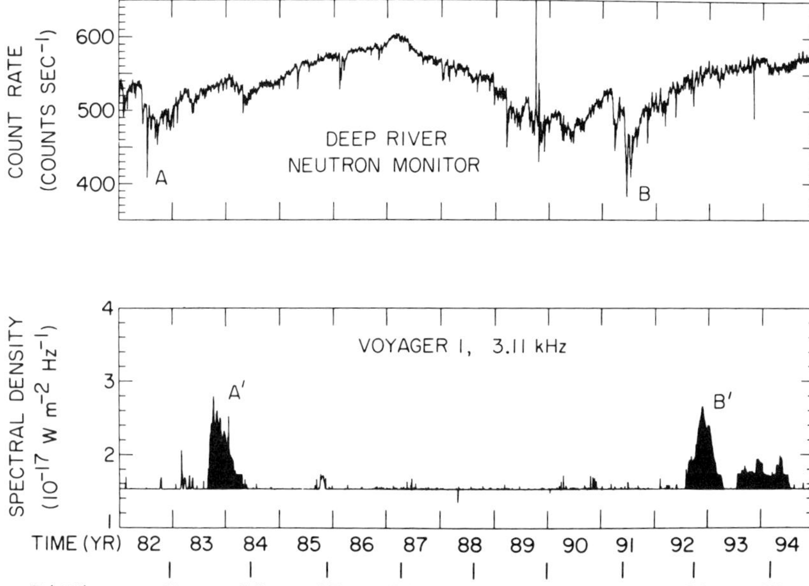

Figure 11.   Deep River neutron monitor data (upper panel) and Voyager 1 3.11-kHz
   intensities as functions of time showing the delay between the two deepest Forbush
   decreases on record and the onset of the two most intense heliospheric radio events
   (figure from Gurnett and Kurth 1996).

Because no velocity information is currently available for these disturbances,
it is not possible to determine whether the distance to the source is similar to
the two major events or not.

   Using the two major events, Gurnett et al. (1993) proposed that an interplanetary shock associated with the Forbush decrease propagated out through
the heliosphere and triggered the emission when the shock arrived in the
cold, post-heliopause plasma.   The rising frequency bands in the dynamic
spectra shown most clearly in Plate 14 are, therefore, due to the propagation
of the triggering shock through a density gradient, probably associated with
plasma piled up near the nose of the heliopause due to the interaction with the
oncoming interstellar wind based on the model of Steinolfson et al. (1994).

   Because the shock speed downstream of the termination shock is probably
somewhat smaller than it is in the upstream region and because the distance
between the termination shock and the heliopause is not known, Gurnett et al.
(1993) estimated the distance to the heliopause by using Eq. (1).

$$R_H = \frac{V_1 T \alpha}{1 - (1 - \alpha)\delta} \tag{1}$$

Here, V_1 is the shock speed in the supersonic solar wind, $\alpha = V_1/V_2$ where
V_2 is the speed in the downstream region, T is the total time required for the

shock to propagate to the source, and δ is the ratio of the distances to the termination shock R_T and the heliopause R_H. Using values of α of 0.7 ± 0.1 and δ of 0.75 ± 0.05 based on a gas-dynamic model of the heliosphere by R. S. Steinolfson (personal communication) and a range of 600 to 800 km s^{-1} for V_1, a range of heliopause distances of 116 to 177 AU is obtained for the 1992–1993 event. Using this range for R_H and the same range for δ as above, the distance to the termination shock can be between 81 and 142 AU.

Gurnett and Kurth (1995) recalculated source distances ranging from 116 to 196 AU using the time-of-flight technique described above, but considered all published values of shock speeds for the shocks associated with the two Forbush decreases. However, they point out that the shocks do not likely propagate at constant velocities in the solar wind as implied by Eq. (1), but slow down with time, hence, the calculated time-of-flight distances could be high by on the order of 20%. They suggest that a more reasonable range of distances to the source, then, is 110 to 160 AU. Using a value of $R_T/R_H = 0.73$ (for the ratio of the termination shock distance to the distance of the heliopause) as a typical value from various numerical simulations, corresponding termination shock distances range from 80 to 115 AU. Other models (see, for example, the chapter by Suess and Nerney) have different values for R_T/R_H, in particular, smaller than 0.73 which would have the effect of decreasing the range of termination shock distances.

The termination shock distance range of 80 to 115 AU is considerably larger than that deduced by Cummings et al. (1994) using the observations of the intensity of anomalous helium and carbon cosmic rays during the last solar minimum (1986–1989). Of course, it is reasonable to expect the heliopause and termination shock to "breathe" in and out in response to changes in internal (or external) pressure. For example, Whang and Burlaga (1993) suggest the termination shock could fluctuate between about 88 and 102 AU throughout a solar cycle, for a 16% variation. Because the Gurnett and Kurth (1995) distance is determined just past solar maximum in mid-1992, there is no reason to expect agreement with Cummings et al. whose results were determined from data obtained near solar minimum. Whang and Burlaga (1993) suggest that the heliosphere should be of maximum size around solar minimum because the solar wind has relatively large momentum during the declining phase of the solar cycle. The shock moves inward during the rising phase and is near minimum just past solar maximum. As 1992 is in the declining phase of the solar cycle, this line of reasoning would imply that the heliosphere should be near minimum size when Gurnett and Kurth determined the distance to the heliopause, but the 1986–1989 period is one in which the heliosphere is moving inward from its maximum distance. Hence, the differences in termination shock distance determined by these two experimental techniques would seem to be even more disparate than at first glance. It should be noted that by modeling the spectrum of anomalous cosmic rays observed in 1994, A. C. Cummings and R. G. Stone (personal communication) have derived a new distance to the termination shock of about 98 AU. This value is certainly

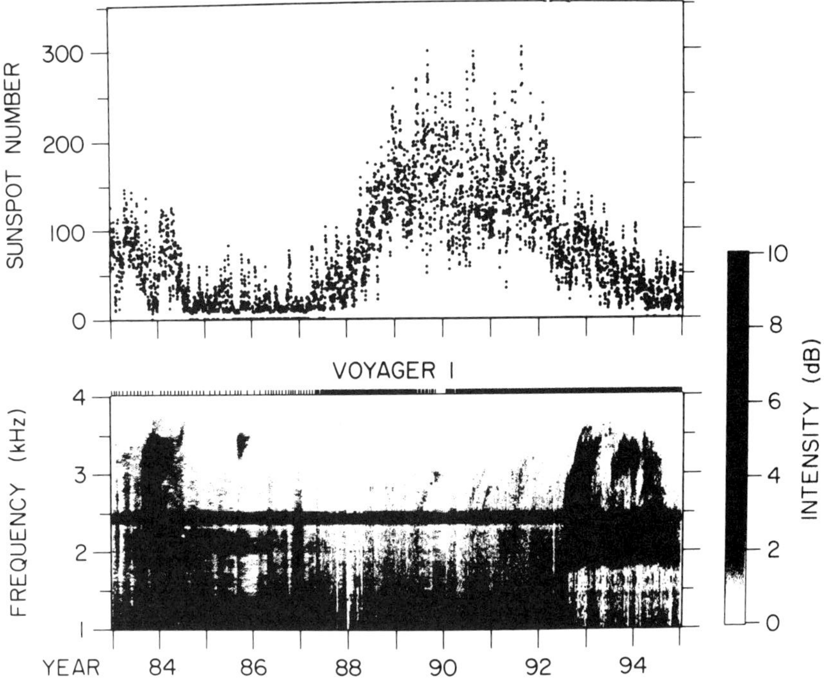

Figure 12.   A comparison of solar activity as indicated by daily sunspot numbers and the heliospheric radio emission activity for the period 1983–1995. Both major radio events occur in the declining phase of the 11-yr solar cycle; the weak events in 1989–1991 occur near sunspot maximum.

comparable to that determined by Gurnett and Kurth.

The apparent causal relationship between the solar activity responsible for the two large Forbush decreases and the onsets of the two major radio events suggests a correlation between solar activity and the generation of the radio emissions. Further, the decade-long interval between radio emission onsets is similar enough to the $\sim$11-yr solar activity cycle period to suggest a correlation with the solar cycle. In Fig. 12, the upper panel shows daily sunspot numbers obtained from the National Geophysical Data Center and the lower panel is a gray-scale version of the Voyager 1 spectrogram given in Plate 13. While there is barely more than one 11-yr solar cycle shown, some evidence of a pattern is apparent, although the correlation is not straightforward. There is definitely a correspondence between the two major radio events and the declining phase of the solar cycle. This amounts to a correlation with the peak of sunspot activity with a lag of some 2 yr. Given the 1.1-yr time-of-flight delay demonstrated in Fig. 11, one might attribute some of the observed lag to this. However, there is no simple explanation of the remaining 1-yr lag. The weak events between 1989 and 1991 occur during the broad peak in sunspot activity.

VI. DISCUSSION

The new interpretations of Gurnett et al. (1993), Zank et al. (1994), Czechowski et al. (1995) and Whang and Burlaga (1994) provide a number of solutions for previous uncertainties in the explanation of the heliospheric radio emissions. However, these theories rely on models of the heliosphere and its interaction with the interstellar medium for some critical aspects of the interpretations. For example, Gurnett et al. suggest that the increasing frequency bands are related to emission at the plasma frequency or its harmonic in the source region, which implies a density gradient through which the triggering shock propagates. Using the Steinolfson et al. (1994) model for the solar wind interaction with the interstellar medium (as well as others such as Baranov and Malama's [1993]), the current interpretation identifies this density gradient with the region of interstellar plasma pile up near the nose of the heliosphere. The 2-kHz component, which does not show frequency drifts as large as those at higher frequencies, then, would apparently come from the flanks of the post-heliopause interstellar wind as shown in Figs. 6 and 7. The radio emissions, then, seem to verify the existence of a pile up region; however, there is no incontrovertible evidence which ties the radio emission directly to the post-heliopause region. The Czechowski et al. (1995) model including a cold dense layer due to interstellar hydrogen charge exchanging with solar wind protons is an example of another possibility for generating the density gradient.

In the Whang and Burlaga (1994) model, the drifts of 3-kHz transient emissions are also explained by propagation of the transmitted triggering shock through the post-heliopause pile up region. In this model, however, the reflected shock propagates back into the subsonic solar wind and the lower plasma frequency there is responsible for the lower frequency component of the radio emission.

Zank et al. (1994) use the propagation of triggering shocks through elevated density regions in the subsonic solar wind caused by the earlier passage of enhanced plasma structures to generate transient radio emissions in the required frequency range. In this case, however, use of the two major Forbush decreases as triggers probably does not work; the earlier density enhancements associated with the GMIRs led by the shock causing the Forbush decrease would be required to build up the subsonic solar wind density and shocks embedded in the GMIRs would then be necessary to act as triggers. This would seem to lose the compelling temporal correlation between the Forbush decreases and the onset of the two major emissions of which the Gurnett et al. (1993) model takes advantage. Further, placing the source region just beyond the termination shock effectively moves this boundary out to the range of 110 to 160 AU based on the time-of-flight considerations of Gurnett and Kurth (1995); the heliopause would then be another 30% more distant.

An open issue is the question of how the trigger acts to "turn on" the radio emissions. Even if one presumes the general model of emission at the plasma frequency or its harmonic as described above, something must change radically at the source region to increase the efficiency of radio emission generation, otherwise one would expect radio emissions to be generated by the shock all along its trajectory, including that portion through the supersonic and subsonic solar wind. Gurnett et al. (1993) suggest that the abrupt increase in radio emission efficiency may arise just beyond the heliopause when the triggering shock enters a region of greatly reduced plasma temperature. Such a plasma would be less prone to Landau damping of the Langmuir waves than the hotter subsonic solar wind. Should this concept be valid, then it provides another tie point to the post-heliopause region (or the cold layer just inside the heliopause) as the source region.

The distances to the source region calculated from the trigger/time-of-flight technique are dependent on models for the propagation speed of the trigger along its trajectory, and models of the thickness and plasma characteristics of the subsonic solar wind. Steinolfson and Gurnett (1995) provide just such a study by simulating the propagation of a shock through the heliosphere and varying the external pressure in order to achieve the observed 1.1-yr time-of-flight. Even given the observed speed of the shock in the inner heliosphere as provided by multiple spacecraft observations, the speed beyond the most distant spacecraft is model dependent, particularly relative to the path length through the shocked solar wind, i.e., the distance from the termination shock to the heliopause, and the speed with which it propagates through this totally unobserved region. While the bulk speed of the solar wind would decrease at the termination shock, the Alfvén speed would increase. These two effects tend to oppose each other in terms of the shock speed, but it is almost certainly not the case that there is no change in speed. Gurnett et al. (1993), using results from R. S. Steinolfson (personal communication), estimate that the termination shock is 73% of the distance to the heliopause and that the downstream shock speed is about 60% of the observed speed (but see Steinolfson and Gurnett [1995] for updated values). While Steinolfson and Gurnett (1995) report relatively small uncertainties for these values, other models for the heliosphere have significantly different values, at least for the relative distance to the termination shock (see the chapter by Suess and Nerney). In summary, then, the model dependencies of the Gurnett et al. interpretation leave substantial uncertainty as to the distance to the heliopause and the termination shock.

Gurnett et al. (1993) have identified two major Forbush decreases observed at Earth as indicators of unusually disturbed solar wind conditions which would provide the unique trigger seemingly implied by the once-per-decade occurrence of strong heliospheric radio emissions. They show in detail that the 1991 Forbush decrease is associated with effects including a shock and other major disruptions such as sharp decreases in the galactic cosmic ray flux which were observed to propagate out to the distances of the Voyager and Pioneer spacecraft. They argued that such evidence was consistent with a near global shock and associated solar wind turbulence which helps to qualify the disturbance as the trigger. The discovery of a similar event associated with the 1983-1984 event certainly reinforces this line of reasoning. However, the narrowband emissions seen in Plate 14 and the direction-finding results in Fig. 5 suggest that there are local radio "hot spots" in the source region. The narrow bands imply that there is little variation in the source electron plasma frequency or electron density over the emitting region. Provided even a general model for the post-heliopause density structure suggested in Fig. 6 and described in Steinolfson et al. (1994), it is clear that the "hot spots" cannot subtend a very large region, either in radial distance or in azimuth and meet the requirements of a narrowband emission. Hence, while it may be that a global disturbance is at the heart of the triggering mechanism, it is also the case that much of the radio emission is actually being emitted in relatively small regions. In fact, Fahr et al. (1986) have already suggested that the magnetic field configuration at the heliopause may strongly influence the region's ability to generate radio waves. We would also point out that variations in the conditions along the triggering shock, including magnetic field orientation, may also be important in the generation of radio "hot spots."

The Gurnett et al. (1993) model suggests that single shocks acted as triggers for the two major radio emission events shown in Fig. 11. However, close inspection of Plate 14 reveals the existence of a very complex set of narrowband emissions which must be explained in terms of such a triggering shock. It is not possible to deconvolve these narrowband emissions fully into a clear picture of how such a shock would result in the observed emission, but it is illustrative to consider some limiting cases of hypothetical radio spectra and ask how they might be generated under the general ideas of the Gurnett et al. model. Figure 13 consists of a set of schematic frequency-time spectrograms in the format of Plate 14 and a simple model of the heliospheric interaction with the interstellar medium based on Fig. 6. In the top two panels, the spectrograms are hypothetical sets of narrowband emissions. On the right side of each panel, "trajectories" of at least a portion of a shock front through radio hot spots are shown. In the top panel, we provide the hypothetical case of nearly simultaneous onset of three bands at different frequencies, each of which rises in frequency to varying degrees, and asymptotically arrive at three different frequencies. These bands are labeled A, B, and C and correspond to the three "trajectories" in the heliospheric model in the top panel. This hypothetical situation closely resembles the Gurnett et al. model for three "hot spots" cutting through three different regions of the post-heliopause interstellar medium. Band A, at the highest frequencies must propagate most closely through the peak of the pile up region. Band C, at the lowest frequencies and showing the smallest frequency drift, must propagate through the flank region. Band B is an intermediate case. Spherical symmetry in both the triggering shock and the distance to the heliopause results in the simultaneous onset times.

In the middle panel, we show a different set of hypothetical radio emission bands. In this case, the three bands have the same beginning and ending

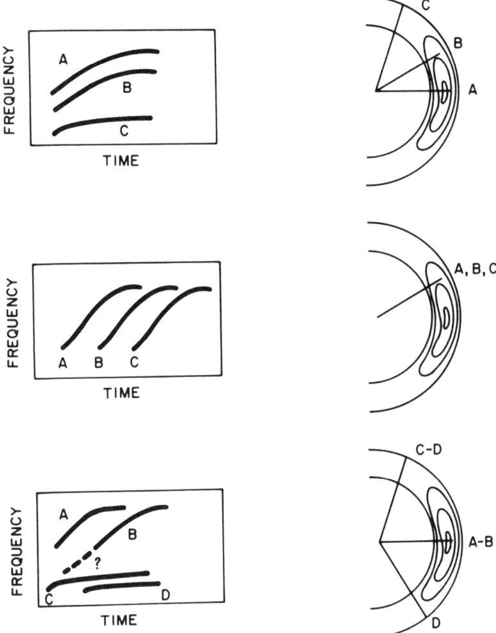

Figure 13.   Sets of hypothetical dynamic spectra and what they might mean in terms of the source and triggering of the radio emissions. In the bottom panel, a schematic of some of the bands based on the data in Plate 14. Using these constructs, we can conclude that it is likely that the 1992–1993 radio emission event was triggered by a complex global merged interaction region with several imbedded shocks.

frequency but have different onset times. The fact that they all have the same ending frequency can only mean that they move through the same region of the pile up region, hence, the "trajectory" of the hot spot must be the same for all three bands. Because the onset times for the three bands is different, the conclusion is that there are three different triggering shocks which move through the same region.

In the bottom panel of Fig. 13, we have tried to represent schematically a number of the features seen in the actual spectrograms in Plate 14. While these are idealized to a minor extent, the reader should agree that we have selected features which are obvious upon close inspection of the Plate. Bands A and B represent the first two high-frequency drifting bands. Bands C and D represent two bands with different start times in the 2-kHz portion of the emission spectrum. The dashed line connecting band B to lower frequencies is included to indicate that it is difficult to know whether or not this band may have started at lower frequencies or not. For the purposes of this discussion, we will assume that the band starts at the higher frequency. Bands A and B exhibit the same type of characteristics as those in the hypothetical case in the middle panel of Fig. 13, having different start times but the same asymptotic ending frequency. We interpret these two bands as evidence of multiple shocks acting as triggers in approximately the same region of the post-heliopause interstellar medium. Bands A and C seem to fit the pattern in the top panel of the figure, that is, having similar onset times but different drift rates and different asymptotic frequencies. Therefore, we interpret these two pairs of bands as evidence that a single shock could excite hot spots in different regions, very similar to the original Gurnett et al. model. The fact that band C actually begins a bit earlier than band A suggests an asymmetry in either the shock front or the heliopause. Bands C and D show different starting times and possibly slightly different asymptotic frequencies. Therefore, these two bands could be due to multiple shocks exciting slightly different regions or a single shock exciting different hot spots, but with definite asymmetries in either the shock front or the heliopause distance. Finally, bands A and C might also be construed as fundamental and harmonic radiation from the same shock moving through a single emitting region and the difference in frequency is related to the $f_p - 2f_p$ emitting frequencies. The primary problem with this interpretation is that the ratio of the frequencies of these two bands is not 2:1 but closer to 1.3:1 and varies as a function of time.

The foregoing discussion of the hypothetical and observed dynamic spectra in Fig. 13 serves to help the reader understand how the spectrum of the radio emissions might be used to determine information from the source region or the triggering shock fronts themselves. Furthermore, the discussion leads to the distinct possibility that the radio emissions observed in the 1992–1994 event are the result of multiple shocks. In spite of the fact that a single, very intense Forbush decrease has been identified with this event, the fact that the solar activity during the March–June 1991 time frame was extremely complex including multiple coronal mass ejections and interplanetary shocks, it is reasonable to consider the existence of two or more triggering shocks associated with this general interval of disturbance. In fact, McDonald and Burlaga (see their chapter) and McDonald et al. (1994) provide evidence of a global merged interaction region (GMIR) on the basis of Pioneer and Voyager measurements at several tens of AU. This GMIR includes a number of imbedded shocks and is a very complex system. McDonald and Burlaga have argued that some of the observed structure in this GMIR may, indeed, explain some of the details of the extended 1992–1993 radio emission event. While we agree with this conclusion in general, it is important in interpreting temporal variations in the 3.11-kHz channel intensities (such as provided in Fig. 3b) to recall that the narrowband emissions drift through the passband of the channel (which has a bandwidth of about 300 Hz). Hence, at least some of the temporal variations exhibited in Fig. 3b are due to the appearance of different narrowband emissions at 3.11 kHz at different times. A simple comparison of the data in Fig. 3b to the GMIR structure is presented by McDonald et al. (1994). While we agree that there is reason to expect some features of the GMIR to correlate with some features in the radio emission intensity vs time profile, the frequency drift effect mentioned here must be considered when interpreting intensity variations in the 3.11-kHz channel.

If it is true that radio emissions such as the 1992-1994 event are due to multiple triggers embedded in a GMIR, and if it is true that the hot spots are rather localized regions, then it becomes even more difficult to compute accurately the time-of-flight of the triggering shocks in order to arrive at distance estimates for the source regions. Instead of using the average propagation of the entire GMIR as would be derived from times-of-flight to the several different spacecraft in different directions, it would be more important to know the propagation of an individual shock moving toward the hot spot. Belcher et al. (1993) have calculated the propagation speed of a shock associated with the spring 1991 solar activity as about 550 km s^{-1}, somewhat less than the previously published estimates for the cosmic ray intensity decrease associated with this time period. Perhaps it is important to recall a simulation of a series of solar wind disturbances and how they merge into a GMIR as performed by Akasofu and Hakamada (1983). The resulting structure, an example of which is given in Fig. 14, includes several different interplanetary shocks and a GMIR which exhibits strong asymmetries.

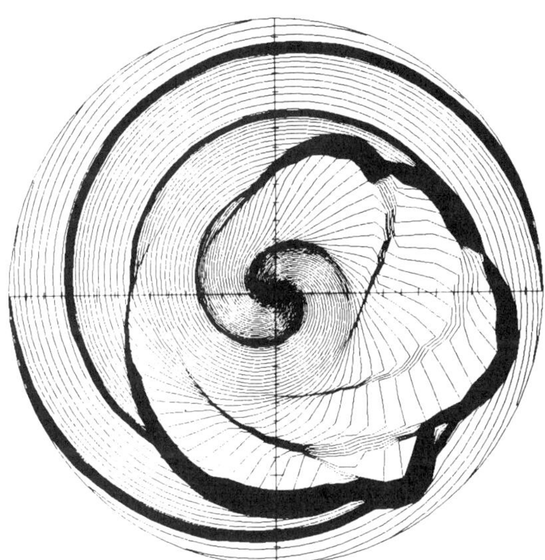

Figure 14.   A contour plot taken from simulation studies by Akasofu and Hakamada (1983) showing the nature of asymmetries which can appear in the solar wind due to a series of solar events taking place a various times during the solar rotation. While this simulation only extends to 15 AU, the result emphasizes the difficulty in trying to determine the time a triggering event might reach any given distance in the outer heliosphere. Given that the heliospheric radio emissions tend to suggest that there are relatively small emitting regions or "hot spots," it is clear that great care must be taken in using observation times of a specific shock at spacecraft located at various azimuthal positions in the solar wind (not to mention various latitudes) to determine average shock speeds.

Yet another question to be answered has to do with whether the heliospheric emissions are generated at the local plasma frequency in the source or its harmonic, or both. This question has as its premise that the basic generation mechanism described above is correct. However, the efficiency of generation at f_p relative to $2f_p$ is dependent upon details of the mechanism and the detailed conditions in the source. Some of the considerations bearing on which emission is more efficient are discussed in Cairns and Gurnett (1992) and Cairns et al. (1992). The importance of answering this question bears directly on how one interprets the spectrograms in Plate 14 relative to the plasma densities in the source region. Given that the general model of Gurnett et al. is correct, this uncertainty leads to an uncertainty in the source (interstellar) plasma frequency of a factor of 2 or in the plasma density of a factor of 4. There is some evidence in the data consistent with radiation at f_p. First, there is no obvious down-turning in frequency of the emission bands. If the radiation were generated at $2f_p$ then it would be possible for the second harmonic waves to continue to propagate inward towards the Sun, even after the triggering shock crossed the peak in the pile up and moved into a negative gradient region. At f_p, any waves generated as the shock propagated to lower densities than the peak would not be able to propagate back towards the Sun. We do not consider this to be conclusive evidence of radiation at f_p, however. Second, Lallement et al. (1993) report local interstellar plasma densities in the range of 0.06 to 0.1 cm^{-3} which corresponds to plasma frequencies from 2.2 to 2.8 kHz. Generation at $2f_p$ would result in emission frequencies well above the observed range. Third, Whang and Burlaga (1994) conclude that f_p emission is consistent with their model of emission from shocks transmitted and reflected at the heliopause. Using a termination shock distance of 60 AU (implying an interstellar $|B|$ of 0.8 nT in their model), they can explain emission bands near 3 kHz from the transmitted shock and near 2 kHz from the reflected shock.

The final question we raise here has to do with the nearly constant lower-frequency limit in the 1992–1994 activity at about 1.8 kHz seen in Plate 14. Such a constant limit often implies a propagation cutoff. That is, a high-density region between the observer and the source with a plasma frequency of 1.8 kHz would create a cutoff in the spectrum at 1.8 kHz provided the source generated waves to frequencies down to 1.8 kHz or lower. Gurnett et al. (1993) suggest this cutoff is due to the interstellar plasma frequency being 1.8 kHz; this implies an interstellar plasma density of 0.04 cm^{-3} and is close to recent estimates of the very local interstellar plasma density of 0.06 to 0.1 cm^{-3} (Lallement et al. 1993). Of course, this would be the density in the flanks of the post-heliopause interstellar wind where there is no significant pile up. Belcher et al. (1993) suggest that the cutoff could be due to the subsonic solar wind density. While the average density in the subsonic region should be considerably less than this, normal density irregularities in the solar wind would be preserved as they propagate through the termination shock; hence, Belcher et al. argue that a significantly higher cutoff frequency could be supported than that derived from the normal densities at this distance from the Sun. In order for these density irregularities to be so effective at creating a radio cutoff, however, the density irregularities would almost certainly be required to be extended completely around the Sun, otherwise lower frequencies could leak into the inner heliosphere through gaps or holes in such a density structure.

One other interpretation of the lower-frequency limit of the emission is that it is symptomatic of the conditions in the source region. That is, the 1.8 kHz cutoff could be a minimum in the emission spectrum at the source. For waves being generated in the flanks of the post-heliosphere interstellar medium where there is no pile up, the local plasma frequency profile could be basically flat and the lower-frequency limit of the emission could simply represent this fact.

VII. CONCLUSIONS

Based on the critical information provided by the 1992–1994 heliospheric radio emission activity, considerable progress has been made in understanding the source of the emissions. It seems unlikely that the highest frequency portion of the radio emission spectrum (>2.5 kHz) is trapped. This implies that the interstellar plasma density is less than about 0.08 cm^{-3}, entirely consistent with the values obtained by Lallement et al. (1993). If the lower portion of the spectrum is trapped, it is unlikely that the Q of the cavity is more than 2 or 3. It also seems quite certain that Jupiter cannot be the source of the radio emission, especially if the radiation is not durably trapped. The elimination of the Jovian source removes the most viable of the originally suggested solar system sources except for the interaction of the solar wind with the interstellar medium. Further, the great distances implied by the time-of-flight considerations and the high densities in the source region strongly suggest the source of radio emissions is in the vicinity of the heliopause. The very good correlation between large Forbush decreases and the onsets of the two strongest radio emission episodes observed over a 12-yr period makes a compelling case that the most intense radio events are triggered by solar activity. We suspect that in the case of the 1992 event it is a complex set of shocks associated with a global merged interaction region which is responsible. The source appears to be in the range of 110 to 160 AU, suggesting that the heliopause is at similar distances if one assumes the model of Gurnett et al. (1993) is generally correct.

Further studies of the radio emissions, of the interplanetary shocks that trigger them, and the improvement of heliospheric models should decrease the uncertainties in the current estimates of the distance to the heliopause and termination shock. These studies should also allow us to determine the density profile in the source region, implying remote sensing of the plasma pile up region near the nose of the heliosphere. The fact that we observe multiple radio "hot spots" suggests that the radio emissions may eventually allow a more global view of the heliopause.

Acknowledgment. Research at the University of Iowa was supported by NASA through a contract with the Jet Propulsion Laboratory.

REFERENCES

Ajello, J. M., Stewart, A. I., Thomas, G. E., and Graps, A. 1987. Solar cycle study of interplanetary Lyman-alpha variations: Pioneer Venus Orbiter sky background results. *Astrophys. J.* 317:964–986.

Akasofu, S.-I., and Hakamada, K. 1983. Solar wind disturbances in the outer heliosphere, caused by six successive solar flares from the same active region. *Geophys. Res. Lett.* 10:577–579.

Alexander, J. K., Carr, T. D, Thieman, J. R., Schauble, J. J., and Riddle, A. C. 1981. Synoptic observations of Jupiter's radio emissions: Average statistical properties observed by Voyager. *J. Geophys. Res.* 86:8529–8545.

Baranov, V. B., and Malama, Y. G. 1993. Model of the solar wind interaction with the local interstellar medium: Numerical solution of self-consistent problem. *J. Geophys. Res.* 98:15157–15163.

Baumback, M. M., Kurth, W. S., and Gurnett, D. A. 1976. Direction-finding measurements of type III radio bursts out of the ecliptic plane. *Solar Phys.* 48:361–380.

Belcher, J. W., Lazarus, A. J., Gordon, J. S., Jr., Szabo, A., and McNutt, R. L., Jr. 1993. Voyager 2 observations of the solar wind plasma responsible for the 1992–1993 heliospheric radio event. Cosmic Winds and the Heliosphere Conference, Oct. 18–22, Tucson, Ariz., Abstract book, p. 9.

Cairns, I. H. 1995. Radio wave scattering in the outer heliosphere: Preliminary calculations. *Geophys. Res. Lett.*, in press.

Cairns, I. H., and Gurnett, D. A. 1992. The outer heliospheric radio emissions: (1) Constraints on emission processes and the source region. *J. Geophys. Res.* 97:6235–6244.

Cairns, I. H., Kurth, W. S., and Gurnett, D. A. 1992. The outer heliospheric radio emissions: (2) A foreshock source model. *J. Geophys. Res.* 97:6245–6259.

Cummings, A. C., Stone, E. C., and Webber, W. R. 1994. Distance to the solar wind termination shock and the source flux of anomalous cosmic rays during 1986–1988. *J. Geophys. Res.* 99:11547–11552.

Czechowski, A., and Grzedzielski, S. 1990. Frequency drift of 3-kHz interplanetary radio emissions: Evidence of Fermi accelerated trapped radiation in a small heliosphere? *Nature* 344:640–641.

Czechowski, A., and Grzedzielski, S. 1992. Evolution of trapped radiation in a 3-dimensional heliosphere: A computer simulation. In *Solar Wind Seven*, eds. E. Marsch and R. Schwenn (Oxford: Pergamon Press), pp. 443–446.

Czechowski, A., Grzedzielski, S, and Macek, W. M. 1995. Frequency drift of the VLF emissions generated by transient shocks near the heliopause. *Adv. Space Res.* 16(9):297–301.

Fahr, H. J., Neutsch, W., Grzedzielski, S., Macek, W., and Ratkiewicz-Landowska, R. 1986. Plasma transport across the heliopause. *Space Sci. Rev.* 43:329–381.

Farrell, W. M. 1993. The heliospheric cavity radio emission: Generation of discrete tones by Fermi acceleration via oscillating boundary. *Geophys. Res. Lett.*

20:2011–2014.

Grzedzielski, S., and Lazarus, A. J. 1993. 2- to 3-kHz continuum emissions as possible indications of global heliospheric "breathing." *J. Geophys. Res.* 98:5551–5558.

Gurnett, D. A. 1975. The Earth as a radio source: The nonthermal continuum. *J. Geophys. Res.* 80:2751–2763.

Gurnett, D. A., and Kurth, W. S. 1994*a*. Evidence that Jupiter is not the source of the 2–3 kHz heliospheric radio emission. *Geophys. Res. Lett.* 21:1571–1574.

Gurnett, D. A., and Kurth, W. S. 1994*b*. Radio emissions at 2–3 kHz and their relationship to heliospheric structure. Second Pioneer–Voyager Symp. on Energetic Particles and Fields in the Outer Heliosphere, May 31–June 3, Durham, N. H., Abstract book.

Gurnett, D. A., and Kurth, W. S. 1995. Heliospheric 2–3 kHz radio emissions and their relationship to large Forbush decreases. *Adv. Space Sci.* 16(9):279–290.

Gurnett, D. A., and Kurth, W. S. 1996. Observations and analyses of heliospheric 2–3 kHz radio emissions. In *Proceedings of Solar Wind 8*, eds. D. Winterhalter, J. Gosling, S. Habbal, W. Kurth and M. Neugebauer (New York: American Inst. of Physics), in press.

Gurnett, D. A., Kurth, W. S., and Scarf, F. L. 1980. The structure of the Jovian magnetotail from plasma wave observations. *Geophys. Res. Lett.* 7:53–56.

Gurnett, D. A., Kurth, W. S., Allendorf, S. C., and Poynter, R. L. 1993. Radio emission from the heliopause triggered by an interplanetary shock. *Science* 262:199–203.

Gurnett, D. A., and Van Allen, J. A. 1993. Heliospheric radio emission events and their relationship to large Forbush decreases. Cosmic Winds and the Heliosphere Conference, Oct. 18–22, Tucson, Ariz., Abstract book, p. 17.

Kaiser, M. L., Desch, M. D., and Farrell, W. M. 1992. Jupiter as the likely source of the heliospheric "3 kHz" noise. Magnetospheres of the Outer Planets Symposium, June 22–26, Los Angeles, Calif., Abstract book, p. R1.6.

Kurth, W. S. 1988. The low frequency interplanetary radio emission: Evidence of the solar wind-interstellar wind interaction? In *Proceedings of the Sixth International Solar Wind Conference*, vol. II, eds. V. J. Pizzo, T. Holzer and D. G. Sime (Boulder: High Altitude Observatory of the National Center for Atmospheric Res.), pp. 667–679.

Kurth, W. S. 1993. The low-frequency interplanetary radiation. *Adv. Space. Res.* 13(6):209–215.

Kurth, W. S., and Gurnett, D. A. 1991. New observations of the low frequency interplanetary radio emissions. *Geophys. Res. Lett.* 18:1801–1804.

Kurth, W. S., and Gurnett, D. A. 1993. Plasma waves as indicators of the termination shock. *J. Geophys. Res.* 98:15129–15136.

Kurth, W. S., Gurnett, D. A., and Allendorf, S. C. 1994. Dual spacecraft determination of the source of heliospheric radio emissions. Second Pioneer–Voyager Symp. on Energetic Particles and Fields in the Outer Heliosphere, May 31–June 3, Durham, N. H., Abstract book, p. 15.

Kurth, W. S., Baumback, M. M., and Gurnett, D. A. 1975. Direction-finding measurements of auroral kilometric radiation. *J. Geophys. Res.* 80:2764–2770.

Kurth, W. S., Gurnett, D. A., Scarf, F. L., and Poynter, R. L. 1984*a*. Detection of a radio emission at 3 kHz in the outer heliosphere. *Nature* 312:27–31.

Kurth, W. S., Gurnett, D. A., Scarf, F. L., and Poynter, R. L. 1984*b*. A new radio emission at 3 kHz in the outer heliosphere. In *Proceedings, Course and Workshop on Plasma Astrophysics*, ESA SP-207, pp. 285–288.

Kurth, W. S., Gurnett, D. A., Scarf, F. L., Poynter, R. L. 1987. Long-period dynamic spectrograms of low-frequency interplanetary radio emissions. *Geophys. Res. Lett.* 14:49–52.

Lallement, R., Bertaux, J.-L., and Clark, J. T. 1993. Deceleration of interstellar hydrogen at the heliospheric interface. *Science* 260:1095–1098.

Lanzerotti, L. J., Maclennan, C. G., and Gold, R. E. 1985. Interplanetary conditions during 3-kHz radio-wave detections in the outer heliosphere. *Nature* 316:243–244.

Lipunov, V. M. 1983. Detection of magnetomultipole radiation from neutron stars. *Astron. Astrophys.* 127:L1–L2.

Macek, W. M. 1994. Mechanism of low-frequency radio emissions in the heliosphere. *Geophys. Res. Lett.* 21:249–252.

Macek, W. M., Cairns, I. H., Kurth, W. S., and Gurnett, D. A. 1991*a*. Plasma wave generation near the inner heliospheric shock. *Geophys. Res. Lett.* 18:357–360.

Macek, W. M., Cairns, I. H., Kurth, W. S., and Gurnett, D. A. 1991*b*. Low-frequency radio emissions in the outer heliosphere: Constraints on emission processes. *J. Geophys. Res.* 96:3801–3806.

McDonald, F. B., Barnes, A., Burlaga, L. F., Gazis, P., Mihalov, J., and Selesnick, R. S. 1994. Effects of the intense solar activity of March/June 1991 observed in the outer heliosphere. *J. Geophys. Res.* 99:14705–14715.

NcNutt, R. L. 1988. A solar-wind "trigger" for the outer heliosphere radio emissions and the distance to the termination shock. *Geophys. Res. Lett.* 15:1307–1310.

McNutt, R. L., Grzedzielski, S., Lazarus, A. J., and Belcher, J. W. 1991. A possible trigger for the 3 kHz radio emission in the outer heliosphere from the solar events of March, 1991. *Eos: Trans. AGU* 72:388 (abstract).

Scarf, F. L., and Gurnett, D. A. 1977. A plasma wave investigation for the Voyager mission. *Space Sci. Rev.* 21:289–308.

Scarf, F. L., Gurnett, D. A., and Kurth, W. S. 1979. Jupiter plasma wave observations: An initial Voyager 1 overview. *Science* 204:991–995.

Steinolfson, R. S., and Gurnett, D. A. 1995. Distances to the termination shock and heliopause from a simulation analysis of the 1992–93 heliospheric radio emission event. *Geophys. Res. Lett.* 22:651–654.

Steinolfson, R. S., Pizzo, V. J., and Holzer, T. 1994. Gasdynamic models of the solar wind/interstellar medium interaction. *Geophys. Res. Lett.* 21:245–248.

Stone, R. G., Pedersen, B. M., Harvey, C. C., Canu, P., Cornilleau-Wehrlin, N., Desch, M. D., de Villedary, C., Fainberg, J., Farrell, W. M., Goetz, K., Hess, R. A., Hoang, S., Kaiser, M. L., Kellogg, P. J., Lecacheux, A., Lin, N., MacDowall, R. J., Manning, R., Meetre, C. A., Meyer-Vernet, N., Moncuquet, M., Osherovich, V., Reiner, M. J., Tekle, A., Thiessen, J., and Zarka, P. 1992. Ulysses radio and plasma wave observations in the Jupiter environment. *Science* 257:1524–1531.

Van Allen, J. A., and Fillius, R. W. 1992. Propagation of a large Forbush decrease in cosmic-ray intensity past the Earth, Pioneer 11 at 34 AU, and Pioneer 10 at 53 AU. *Geophys. Res. Lett.* 19:1423–1430.

Van Allen, J. A., and Randall, B. A. 1985. Interplanetary cosmic ray intensity: 1972-1984 and out to 32 AU. *J. Geophys. Res.* 90:1399–1412.

Webber, W. R., and Lockwood, J. A. 1993. Giant transient decreases of cosmic rays in the outer heliosphere in September 1991. *J. Geophys. Res.* 98:7821–7825.

Webber, W. R., Lockwood, J. A., and Jokipii, J. R. 1986. Characteristics of large Forbush-type decreases in the cosmic radiation 2. Observations at different heliocentric radial distances. *J. Geophys. Res.* 91:4103–4110.

Whang, Y. C., and Burlaga, L. F. 1993. Termination shock: Solar cycle variations of location and speed. *J. Geophys. Res.* 98:15221–15230.

Whang, Y. C., and Burlaga, L. F. 1994. Interaction of global merged interaction region shock with the heliopause and its relation to the 2- and 3-kHz radio emissions. *J. Geophys. Res.* 99:21457–21465.

Zank, G. P., Cairns, I. H., Donohue, D. J., and Matthaeus, W. H. 1994. Radio emissions and the heliospheric termination shock. *J. Geophys. Res.* 99:14729–14735.

TRANSPORT AND ACCELERATION OF ENERGETIC PARTICLES IN WINDS

J. R. JOKIPII
University of Arizona

There are three principal types of interaction of a collisionless wind from a condensed object, such as a star or galaxy, with energetic particles. First, the wind interacts strongly with the low-energy part of externally produced cosmic rays which come from outside of the system; most of these are reflected after a few scattering mean free paths, effectively shielding the interior of the wind from most of the incident particles (this phenomenon is called *modulation* of the cosmic ray flux). Second, the central object itself generates varying amounts of energetic particles, some continuously and some in short bursts. Third, the wind itself can, and in most cases will, accelerate charged particles to high energies, principally by the mechanism of shock acceleration, at the various shocks which are found in the winds. Acceleration by waves may also be important for lower-energy particles. The energetic particles generated at the central object or in the wind propagate through the wind to the external medium, and can be sufficiently numerous and energetic to modify significantly the dynamics of the wind. These phenomena are illustrated in this chapter, primarily with reference to observed energetic particles in the heliosphere; however, the principles are generally applicable and similar phenomena should occur in other winds.

I. INTRODUCTION

Figure 1 illustrates the cosmic ray flux observed at quiet times in the solar system. The spectrum extends quite smoothly from an energy of less than 10^5 eV, over some 15 decades, to more than 10^{20} eV. Over most of the high-energy part of the spectrum, from energies of the order of 10^{11} or 10^{12} eV up to the maximum energies, the solar wind has little impact on the particles. In this chapter, the terms *energetic particle* and *cosmic rays* are used interchangeably for all particles with energies higher than thermal energies.

In the low-energy part of the spectrum, below ≈ 1 GeV, one sees significant time variations and spectral features which reflect the various interactions of the solar wind with the cosmic ray flux. During quiet periods (low solar activity) the observed spectrum is believed to consist primarily of galactic particles and particles accelerated in the solar wind, mostly at various heliospheric shocks. At intervals, during periods of high solar activity, the low-energy part of this spectrum is overshadowed by very intense, short-lived bursts of solar energetic particles, that are accelerated in the lower solar atmosphere. The energy of these solar particles seldom exceeds 1 GeV. The

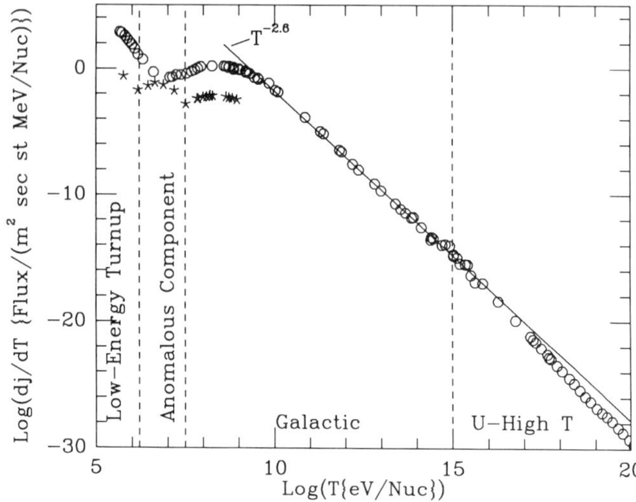

Figure 1.   The cosmic ray intensity as a function of energy. The $\circ$ correspond to protons for $T \lesssim 10^{15}$ eV. The composition at higher energies is not certain, but contains considerable heavier nuclei. The $*$ correspond to oxygen. The oxygen peak at $T < 10^7$ eV/nuc is due to anomalous oxygen. Data compiled from Meyer (1969), Linsley (1980) and Gloeckler (1979). The turnup at lower energies and the ultra-high-energy particles are indicated.

particles observed below ≈ 1 MeV may be in part a continuous solar component. The energy spectra of energetic particles in other winds will probably show similar effects, but at different energies, depending on the parameters of the wind.

The spectral features and time variations discussed above are determined by physical processes occurring in the solar and heliospheric plasma. Figure 2 illustrates the anticipated overall structure of the solar wind: a similar structure is likely to apply to many other winds. We must remember, however, that the outermost part of this illustrated structure is based on theoretical inference only, and has not yet been confirmed by observations.

Consider first the effects of the wind on the essentially time-independent flux of galactic cosmic rays entering the heliosphere from outside. The solar wind affects significantly all the energetic particles that we see in the inner heliosphere below energies of the order of 10^{11} eV. We may expect observable effects at any given heliocentric radius if the particle gyroradius in the local magnetic field is of the order of the heliocentric radius or smaller, which occurs at some 10^{12} eV in the inner heliosphere. Illustrated in Fig. 3 is an example of the effects of the solar wind on the observed time dependence of the intensity of $\approx$ several GeV cosmic rays at Earth (as measured by a groundbased neutron monitor) over the past four decades. The intensity of the cosmic rays is "modulated" by the solar wind in antiphase with the sunspot

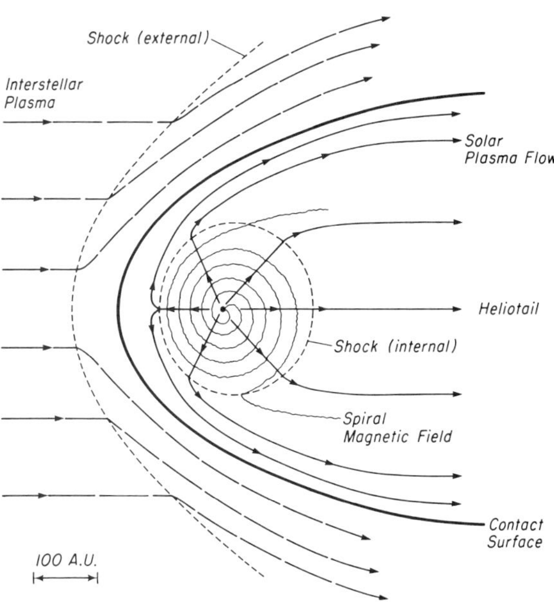

Figure 2.   Schematic illustration of a plausible structure for the heliosphere, based on a fluid-mechanical extrapolation of observations in the inner part of the heliosphere.

number. Readily apparent are pronounced minima in the intensity occurring roughly each 11 yr. These periods of cosmic ray minimum correspond to sunspot maxima, when the Sun is most effective at excluding these cosmic rays from the inner solar system. Conversely, the cosmic ray maxima occur approximately midway between the minima, near sunspot minimum. Notice also the alternating sharp and flat shapes of successive cosmic ray maxima. This solar "modulation" of the galactic cosmic ray flux by the wind provides a valuable probe of physical processes occurring in regions of the heliosphere that have not been observed directly. Also, understanding the effects of the heliosphere on cosmic rays is a necessary pre-requisite to reconstructing the undisturbed galactic flux that is necessary to studying the physics of cosmic rays in the Galaxy.

In addition to the *externally* produced particles, an observer inside the heliosphere will in general see energetic particles from a variety of sources *inside* the heliosphere. Figure 4 illustrates the currently known sources such as (a) the Sun, which emits solar energetic particles (or "solar cosmic rays" on a variety of time scales and energies); (b) propagating shocks which produce energetic storm particle events and corotating ion events; (c) bow shocks upstream of planets and other objects such as comets that produce upstream events, and magnetospheres, which can leak copious amounts of energetic particles into the heliosphere; and (d) the solar wind termination shock, which is now believed to be the source of the anomalous cosmic rays.

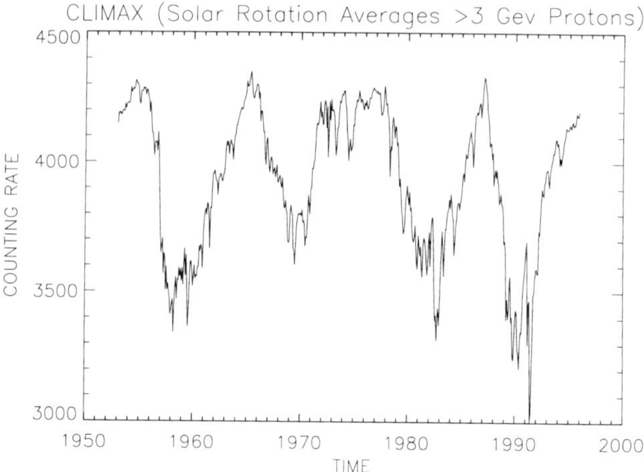

Figure 3.   The counting rate of the Climax neutron monitor (which responds primarily to cosmic ray protons with energies of several GeV) over the past 40 yr (K. R. Pyle, personal communication).

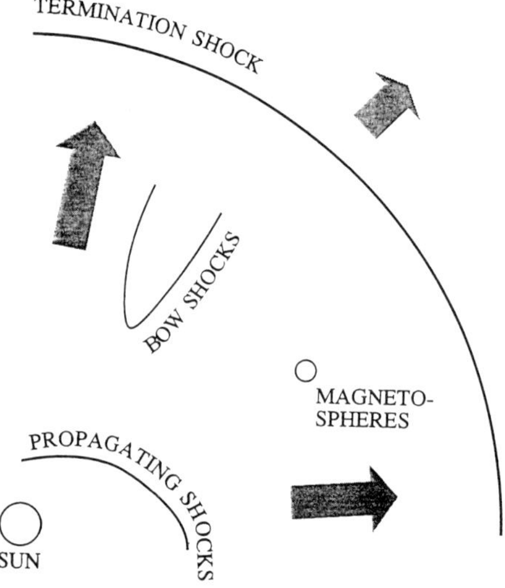

Figure 4.   Diagram showing known sources of energetic particles in the solar wind. Sun: solar energetic particles (solar cosmic rays). Propagating shocks: energetic storm particles, corotating ions. Bow shocks: up-stream events. Magnetosphere: Jovian electrons. Termination shock: anomalous cosmic rays.

The Sun sporadically emits energetic particles in association with coronal mass ejections and solar flares. The passage of a shock wave in interplanetary space (generated by solar activity) also has characteristic effects on energetic particles. It will, in general, depress the intensity of high-energy (>100

MeV) particles (the Forbush decrease) and accelerate lower-energy ones. In addition, magnetospheres will release energetic particles into the solar wind. In fact, all cosmic ray electrons observed in the inner heliosphere that have energies below ≈ 30 MeV are now believed to be electrons emitted from Jupiter's magnetosphere which have filled the inner heliosphere.

The various populations of energetic particles discussed above are an important part of the heliosphere, and presumably have their analogs in other winds. Because of their high mobility, energetic particles provide valuable probes of the plasmas and magnetic fields in distant regions of the heliosphere that are not accessible to direct observation. Furthermore, in some cases, the particles can be sufficiently numerous to modify the dynamics and flow patterns of the ambient plasma.

II. TRANSPORT AND ACCELERATION OF COSMIC RAYS IN COLLISIONLESS PLASMAS

In order to understand the various aspects of energetic particles in the heliosphere we must understand their transport. Almost 30 yr ago a very general transport equation was developed (initially by Parker 1965; see also Axford 1965; Gleeson and Axford 1967; Jokipii and Parker 1969). This equation describes the transport of these particles in the *diffusive* approximation (where scattering by magnetic irregularities is sufficiently rapid to maintain near isotropy in the angular distribution) and appears to account very well for most of the transport phenomena that have been observed.

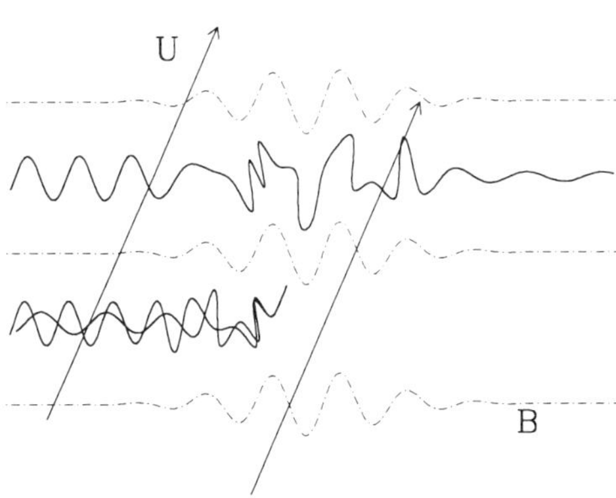

Figure 5.   Computed trajectories of two particles (solid lines) in a magnetic field (illustrated by the dot-dashed line) which varies only in the direction along the field. The two particles were started with the same energy, but with slightly different pitch angles.

The two solid curves in Fig. 5 illustrates the interaction of two energetic particles with an irregular magnetic field. The charged particles do not collide with the very rarefied ambient plasma, but instead gyrate around the magnetic field and interact with the magnetic field. As illustrated the magnetic irregularities cause the orbit of each particle to shift and to change its pitch angle with respect to the magnetic field by an amount that is sensitive to the particle parameters. To obtain the transport equation, the particle motions are analyzed for a statistical ensemble of magnetic fields and particles. The scattering by the magnetic field isotropizes the particles in pitch angle relative to the fluid frame, and in most cases their resulting transport can be described in terms of a diffusion tensor κ_{ij}. In principle, κ_{ij} can be determined from the observed power spectrum of magnetic irregularities (see, e.g., Jokipii 1971). Until now the accuracy of this determination has not been adequate and the theory has been used for guidance only. For the most part, κ_{ij} is treated as a phenomenological parameter, chosen to fit the data using the theoretical values obtained from the magnetic power spectrum.

In addition to the scattering and resulting diffusion, the entire magnetic field structure and the particles gyrating around it are convected by the fluid at the fluid velocity $\mathbf{U}(\mathbf{r}, t)$. The convection of the magnetic field also gives rise to a $\mathbf{U} \times \mathbf{B}$ electric field which can cause energy change. Hence, changes in the density of the plasma caused by the divergence of $\mathbf{U}$ cause the cosmic ray energies to change to adiabatic compression or expansion, and this then changes their intensity. Finally, the particle orbits respond to *large-scale* variations in the magnetic field. These are distinct from the stochastic variations causing the scattering, and which result in *coherent* gradient and curvature drifts of the particles, effectively competing with diffusion caused by the small angle scattering. These four major effects were combined first by Parker in (1965) to obtain the transport equation for the phase space distribution $f(\mathbf{r}, p, t)$, where $\mathbf{r}$ is position, t is time and p is the particle momentum,

$$
\begin{aligned}
\frac{\partial f}{\partial t} = {} & \frac{\partial}{\partial x_i}\left[\kappa_{ij}^{(s)}\frac{\partial f}{\partial x_j}\right] && \text{(diffusion)} \\
& - U_i \frac{\partial f}{\partial x_i} && \text{(convection)} \\
& - V_{d,i}\frac{\partial f}{\partial x_i} && \text{(guiding center drift)} \\
& + \frac{1}{3}\frac{\partial U_i}{\partial x_i}\left[\frac{\partial f}{\partial \ln p}\right] && \text{(energy change)} \\
& + Q(x_i, t, p). && \text{(source)}
\end{aligned}
\tag{1}
$$

Here, the diffusion coefficient and drift velocity may be written in terms of the particle momentum, charge q, speed w, the speed of light c and the perpendicular and parallel diffusion coefficients, $\kappa_\perp$ and $\kappa_\parallel$, as

$$\kappa_{ij}^{(s)} = \kappa_\perp \delta_{ij} - \frac{(\kappa_\perp - \kappa_\parallel) B_i B_j}{B^2}$$
$$V_{d,i} = \frac{pcw}{3q} \epsilon_{ijk} \frac{\partial}{\partial x_j} \left(\frac{B_k}{B^2} \right).$$

(2)

It should be mentioned that in reality, the solar wind and cosmic rays are a coupled plasma, so that in addition to the ambient plasma and magnetic field affecting the particles as in Eq. (1), the pressure of the cosmic rays will in general affect the flow. The cosmic rays are very energetic compared to the thermal energies of the solar wind, but because of their high energies and in spite of their low density, they can exert a finite pressure. This pressure provides a coupling between the fluid and the cosmic rays and in general must be included in the fluid equations, although in most parts of the solar system it is apparently negligible. In addition, one should really think in terms of a single distribution function for any given particle species, in which the cosmic rays are the high-energy tail of the distribution, although in most cases the distinction between thermal gas and cosmic rays is clear. Also, over most of the heliosphere, the energy density of the cosmic rays is low enough that the effects of the cosmic rays on the plasma can be neglected (the "test particle" approximation). We will concentrate on this approximation in this chapter. Inclusion of the effects of the cosmic rays is considered in the chapter by Lee.

III. ILLUSTRATIVE ANALYTIC SOLUTIONS TO THE TRANSPORT EQUATION

Equation (1) is sufficiently complex that to obtain solutions in most cases requires that numerical methods be applied. However, analytic solutions may be obtained for some simple one-dimensional cases, and these illuminate the general nature of energetic particle phenomena encountered in the heliosphere. For this purpose, consider first a simple one-dimensional, uniform flow of the thermal plasma and magnetic field in the x-direction, and take the diffusion coefficient κ_{xx} to be constant. If the x-axis is an angle θ to the average magnetic field, we have from Eq. (2), $\kappa_{xx} = \kappa_\parallel \cos^2(\theta) + \kappa_\perp \sin^2(\theta)$. Equation (1) in this case reduces to the simple form $\partial f/\partial t = \kappa_{xx} \partial^2 f/\partial x^2$.

An impulsive injection of particles at some point x_0 at time t_0 then results in the standard diffusion solution $f = C/\sqrt{(t-t_0)} \exp[-(x - x_0 - Ut)^2/(\kappa_{xx}(t-t_0))]$. In this one-dimensional picture with uniform velocity the particles do not change their energy and remain at their initial momentum. This solution is similar to the impulsive solar particle events and suggests that diffusion is a major component of the transport of these solar particles. Of course, this simple model only provides a partial explanation of the basic physics of the phenomenon.

The solution for time-*independent* injection at x_0 can be written immediately for points upstream as $f = f_0 \exp(U(x - x_0)/\kappa_{xx}$, where f_0 is the constant value of f at $x = x_0$. Downstream, the solution is independent of x. This solution exhibits the characteristic behavior where the energetic particles are swept toward positive values of x but diffusive upstream. In a steady state, one requires a balance between the convected flux and the diffusive flux, which leads directly to the simple solutions shown. Parker's original explanation of the modulation of galactic cosmic rays by the solar wind (Parker 1958) can be related simply to this solution, in which the interstellar value is assumed to be reached at some finite $x = 0$ and the intensity upstream is then decreased by the flow of the solar wind. The slow variation of the cosmic ray intensity with time through the sunspot cycle can then be related to variations in either the diffusion coefficient or the flow velocity. We now know that this simple model does not account for all of the phenomena, but it does illustrate the basic effects of diffusion and convection. The energy change in the expanding solar plasma and the large-scale drift motions, neglected in this simple example, are of comparable importance to the convection.

Finally, we illustrate the nature of the phenomena associated with a shock wave. As shown in a series of important papers by Axford et al. (1978), Krymsky (1977), Bell (1978) and Blandford and Ostriker (1978), shocks can accelerate particles very efficiently in cosmic plasmas. For more recent reviews of shock acceleration see Drury (1983) and Jones and Ellison (1991). This "diffusive" shock acceleration may be derived from the Parker transport equation. Introduce a stationary shock wave at the position $x = 0$. The fluid has velocity U_1 upstream of the shock and velocity $U_2 = U_1/r_s$ downstream. The shock ratio r_s must be less than 4 for a gas-dynamical shock. To proceed, we solve the transport equation (as in the previous paragraph) in the upstream and downstream regions separately, and then apply a *jump* condition to relate the two solutions. This condition is obtained simply by integrating Eq. (1) from a short distance upstream of the shock to a short distance downstream, and may be written

$$\left[\kappa_{xx} \frac{\partial f}{\partial x} + \frac{U_x}{3} \frac{\partial f}{\partial \ln p} - \frac{pcw}{3q} \frac{B_z}{B^2} \frac{\partial f}{\partial y} \right]_1^2 = Q_* \tag{3}$$

where Q_* represents that part of the source that is singular at the shock, and κ_{xx}, the coefficient of diffusion normal to the shock face.

Illustrated in Fig. 6 is a typical distribution of particles around a shock. Here particles are injected near the shock starting at some time t_0. One finds that the distribution function at any given *spatial* location asymptotically approaches a power law in momentum p^{-q} with a spectral index that is related simply to the shock ratio r_s through $q = (r_s + 1)/(r_s - 1)$; it is 4 for a strong shock. Moreover, at any finite time, there is a high-momentum cutoff that moves to higher energies with a time scale at any energy given by

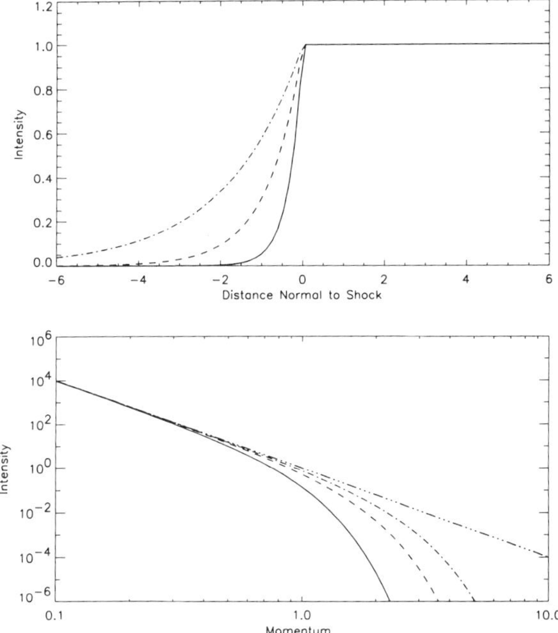

Figure 6. Illustration of the nature of the solution of the transport equation at a plane, time-independent shock. Particles are injected at a low momentum and beginning at some time t_0, at the shock. The three plots in the upper figure show the time-asymptotic spatial dependence of the intensity at three energies, showing the exponential decrease upstream of the shock and the constant value downstream. The rate of decrease upstream is smaller for higher energies because of their more rapid diffusion. The lower figure shows the momentum spectrum at a given point at three successive times after t_0, and the time-asymptotic power law.

(Forman and Morfill 1979).

$$\tau_a = \frac{3}{(U_1 - U_2)} \int_{p_o}^{p} \left\{ \frac{\kappa_1}{U_1} + \frac{\kappa_2}{U_2} \right\} \frac{\mathrm{d}p}{p}$$

or

$$\frac{1}{p} \frac{\mathrm{d}p}{\mathrm{d}\tau_a} = \frac{U_{sh}^2 (r - 1)}{3r[\kappa_1 + r\kappa_2]}.$$

(4)

The spatial dependence of the time-asymptotic shock solution is illustrated in Fig. 7 in which we see the exponential decrease with characteristic scale κ_{xx}/U upstream of the shock and a constant solution down stream of the shock, just as in the case of a simple time-independent source discussed earlier. At any energy above the injection energy, the shock acts like a source of energetic particles at that energy.

Figure 7.   Perspective drawings of an idealized heliospheric current sheet, drawn out
from the Sun by the solar wind and twisted by solar rotation. The drawings are all
about 30 AU across, and show the increase in the waviness as time progresses from
near sunspot minimum (upper left) toward sunspot maximum.

The physical origin of this acceleration at shocks is in general quite subtle. Most discussion has related to the case where the shock is quasi-parallel, in which case the energy change is caused by the compression (converging motion of the scatterers) at the shock, commonly called first-order Fermi acceleration. However, if the shock is not quasi-parallel, the average magnetic field changes direction and magnitude at the shock, and the drift along the shock face in the direction of the $\mathbf{U} \times \mathbf{B}$ electric field becomes important. This drift effect dominates for quasi-perpendicular shocks and Fermi acceleration is not the proper term. In fact, it may be shown (Jokipii 1990) that in general, the energy change in Eq. (1) has contributions from both drift and compression/expansion.

In any application to a wind, these simple one-dimensional solutions must be modified to take account of the actual flow geometry, magnetic field and boundary conditions. But they serve to display the essential nature of cosmic ray behavior in the heliosphere. The effects of adiabatic cooling due to the spherical expansion and the particle drift motions are now known, in the case of the heliosphere, to be very important. Unfortunately, in all the cases we know of, the drift motions are not simple, and analytic techniques are insufficient. We must turn to numerical simulations.

IV. GLOBAL MODELS OF COSMIC RAYS IN THE HELIOSPHERE

A. The Model

We next consider the application of the transport equation to more realistic model systems. Modern computers make it possible to develop quite realistic three-dimensional, time-dependent solutions, although resolution can be limited because the distribution depends on the three space variables and momentum, as a function of time. Hence two-dimensional and time-independent solutions also have a role to play. More importantly, because knowledge of the structure of the outer heliosphere is still quite limited, the outer part of the heliosphere beyond the termination shock is generally treated very crudely in these models. In fact, only a few recent papers have even included the region beyond the shock in any meaningful way. Because of this, models can only give information about those aspects of cosmic ray phenomena that are not very sensitive to the detailed nature of the interface with the interstellar medium.

The general structure of the computational models that have been employed by our group at Arizona in studying global models of cosmic ray transport and acceleration in the heliosphere generally are similar to that in Fig. 2, but with spherical boundaries. We have used similar models in considering transport in supernova blast waves and in the galactic wind. The two outer spheres in the model correspond to the termination shock and the contact surface separating the solar wind plasma from the interstellar medium. These surfaces are taken to be simple spheres because of our lack of knowledge of the actual structure. Between the spheres, beyond the solar-wind termination shock, we simply assume that there is a shock transition at some specified radius R_{sh}, generally taken to be 70 to 90 AU. Beyond this point we assume that the radial velocity drops suddenly by a factor of 4 (strong shock) and then falls off as $1/r^2$, out to the contact surface. This constant surface is taken to be the outer spherical boundary of the system, where it is assumed that the cosmic ray distribution takes on the uniform, constant galactic value. This value is generally taken to be a power law in total particle energy per nucleon with exponent -2.65. This spectrum is not well known except at higher energies where modulation is unimportant.

Our knowledge of the heliosphere inside the termination shock is more reliable. Because of the supersonic nature of the solar wind, it is reasonable to extrapolate our knowledge of the inner heliosphere out to the assumed shock boundary. The large-scale magnetic field (see, e.g., Smith and Thomas 1986; see also the chapter by Smith) is well ordered in the years around sunspot minimum, and consists of a Parker Archimedean spiral, with the field directed in one direction in the northern heliospheric hemisphere and in the opposite direction in the southern. The two fields are separated by a thin current sheet. The shape of this current sheet is illustrated in Fig. 7. At sunspot minimum, it is nearly equatorial, and its waviness increases with time until it is quite large (maximum excursion in latitude greater than 60 deg) near sunspot maximum, where it is likely that the entire structure becomes much more complex than indicated. The magnetic field direction in each hemisphere changes sign near each sunspot maximum, with the northern hemisphere field around the 1986 sunspot minimum pointed inward toward the Sun. In the present discussion it will be assumed that the simple wavy current-sheet magnetic structure is given by this model, recognizing that it may not be valid in the few years around sunspot maximum. The most recent Arizona models allow for considerable deformation of the simple current sheet due to flow interactions in the solar wind. Jokipii and Kóta (1989) have also suggested that the as yet unobserved *polar* magnetic field may differ from the classical model even at sunspot minimum.

Observations (see, e.g., Newkirk and Fisk 1985) show that the solar wind velocity increases with heliomagnetic latitude λ_{mag}, from 350 to 400 km s^{-1} near the current sheet, to some 700 to 800 km s^{-1} for $\lambda_{mag} \gtrsim 20°$. Recent observations on the Ulysses spacecraft as it moves to high latitudes (see, e.g., Bame et al. 1993) support this overall picture.

Both two- and three-dimensional simulations of cosmic ray transport in this model heliosphere have been developed. The virtue of the two-dimensional codes is the ability to consider more detailed effects because of the smaller demands of array size and computation time. Such codes are at present being used both here at Arizona and by the group associated with Potchefstroom University in South Africa. In the two-dimensional codes the current sheet must in general be flat. The group in Potchefstroom has modified the flat current sheet near the equator in attempt to simulate the effect of a wavy sheet in a two-dimensional model (see, e.g., Potgieter and Moraal 1988).

Three-dimensional codes are quite large and complex, and at present are available only at Arizona, where two versions have been developed by Kóta. Within the termination shock the magnetic field and flow structure can be made quite realistic. In the models, solar rotation results in regions of fast solar wind overtaking previously emitted slower solar wind, which then form corotating shocks (corotating interaction regions or CIRs); this whole dynamical system is allowed to interact dynamically with and modify the magnetic field and current sheet. The major simplification in the dynamics of the model is that only a radial flow velocity is allowed (this is a good approximation as observed deviations from radial flow are small). In the models, it was assumed that the heliospheric current sheet at the Sun is a plane inclined at an angle α to the equator. The magnetic field at the Sun is taken to be uniform in magnitude and to change sign at the current sheet. This magnetic field structure is then carried out by the wind and evolves with radius. The wind is radial near the Sun and one may write it as a function of latitude as

$$V_w = V_{\text{pole}} - (V_{\text{pole}} - V_{\text{eq}})/[1 + (\sin(\lambda_{\text{mag}})/sin(\lambda_{1/2}))^2] \qquad (5)$$

where the V_{pole}, V_{eq} and $\lambda_{1/2}$ specify the increase of the wind with heliographic latitude and are chosen to give the values inferred over the poles. Because transverse fluid stresses are small, the transverse velocities were neglected and the radial momentum equation was solved with a frozen-in magnetic field, including the pressure and magnetic forces. The fluid is followed using the equations of motion until it settles into a corotating pattern and the resulting velocity and magnetic field structure is used in solving the cosmic ray transport equation. This simplified model wind is found to be quite similar to the observed solar wind (Kóta 1992). A fully time-dependent three-dimensional transport model, including propagating shocks, is not yet available.

The Forbush decrease, a depression in the galactic cosmic ray intensity caused by a propagating shock has been studied using the two-dimensional model of Lockwood et al. (1986). The solution illustrates the passage of a region of depressed diffusion coefficient, corresponding to a corotating interaction region propagating past an observer. In this code the velocity of the fluid was not varied so there is no diffusive shock acceleration ahead of the disturbance.

B. Modulation of Galactic Cosmic Rays

The particle motions in the heliosphere are composed of the diffusive random walk and convection as discussed in the simple solutions discussed previously. But in addition, because of the large-scale structure of the magnetic field imposed by the solar wind, the drift motions of the particles due to the large-scale variation of the average magnetic field must also be included. The nature of drift motions is illustrated in Fig. 8 for a simple, constant velocity, azimuthally symmetric wind. The typical magnitude of the drift velocity in the inner heliosphere at intermediate latitudes is $\approx 10^8$ cm s^{-1}. Apparent in the figure is the characteristic transport in latitude and the rapid drift along the current sheet. The arrows on the curves illustrate the direction of the motions during the periods when the northern heliosphere magnetic field is outward from the Sun and the southern field is inward, as was the case in 1975 and is the case now. In this case, the motion of positively charged particles such as protons are inward from the polar regions of the heliosphere toward the equator then ejected from the heliosphere along the current sheet. Because the divergence of the drift velocity is zero, the streamlines entering from poles are exactly compensated for by the outward motion along the current sheet. During the last sunspot minimum, on the other hand, particles came inward along the current sheet and were ejected out at the poles. These coherent drifts are superimposed upon the diffusion, and the diffusion coefficient does *not* change sign with the sign of the overall magnetic field.

First consider the results from our standard models using the Archimedean spiral magnetic field. This field falls off as $1/r$ in the regions of the equatorial

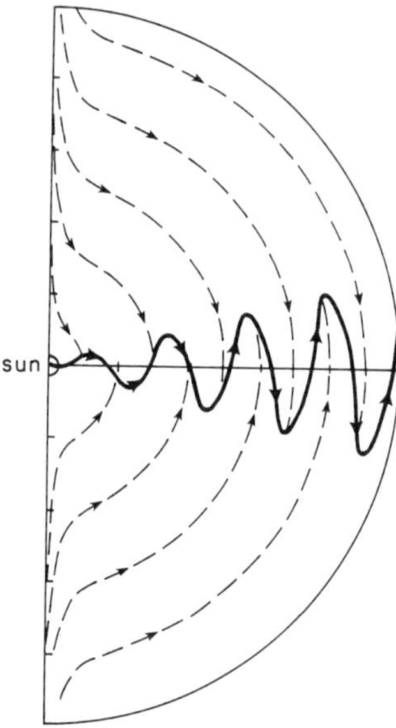

Figure 8.   Drift trajectories of ≈ 1 GeV protons in an Archimedean magnetic field
with a wavy current sheet, for a time when the northern heliospheric magnetic field
was pointing inward toward the Sun (1975). During the next sunspot minimum
(1986) the trajectories are reversed. The general magnitude of the velocity at 1 AU
as is $\approx 10^8$ cm s^{-1}.

plane beyond about 1 AU radius, because the spiral is tightly wrapped, and
the radial field is small. But because there is no wrapping up of the field in
the polar regions, the polar Archimedean spiral field falls off as $1/r^2$. So,
over the poles, in the outer heliosphere (at distances of the order of 40 or
50 AU) the magnetic field intensity is extremely weak ($\sim 10^{-8}$ Gauss). The
diffusion coefficient must reflect the magnitude of the ambient magnetic field
(we set the diffusion mean free path as a factor of order 10 times the local
gyroradius). Hence, because of the small magnetic field magnitude over the
poles, we have extremely rapid diffusion of the particles into the inner solar
system. Also, and for the same reason, the gradient and curvature drifts are
very rapid over the poles. Hence, when the northern magnetic field is outward
from the Sun, as in 1975, the drifts and the polar diffusion act together to bring
positively charged particles in very rapidly from the polar regions. On the
other hand, in 1965 or in 1986, when the field has the opposite direction, the
drift motions are fighting the diffusion in the polar regions. The effects of the
drift are therefore somewhat suppressed in the latter case. The particles both
diffuse very rapidly in from the poles and are sucked in very rapidly from the
equator by the drift motions, to be subsequently ejected at some intermediate
latitude.   There is actually a minimum in the intensity at an intermediate
latitude because particles coming in by either route lose energy continuously.

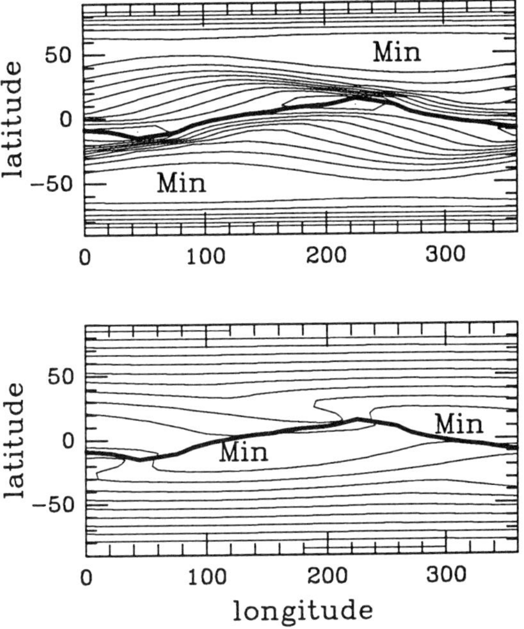

Figure 9.   Contour plots of the intensity of $\approx$1 GeV protons as a function of latitude and longitude at a heliocentric radius of 16 AU. In each figure the maximum occurs at the poles and the minimum is indicated. The top figure is for a period when the northern solar magnetic field was inward toward the Sun, as around the 1986 sunspot minimum. The bottom figure is for the opposite sign of the magnetic field, which occurred around the 1975 minimum.

These effects are illustrated clearly in Fig. 9, which shows contours of $\approx$1 GeV protons in latitude and longitude at a heliocentric radius of 16 AU, for the two opposite signs of the classical Archimedean spiral magnetic field. Inclusion of a modified polar magnetic field (Jokipii and Kóta 1989) results in a quantitative change in the model, but the overall picture is similar to that shown here.

A significant number of observed cosmic ray effects are in very good agreement with this model. These include the variation of the intensity with time during the 22-yr solar magnetic cycle, the correlation of the intensity with the geomagnetic *aa* index, the correlation of the intensity with the tilt of the interplanetary current sheet, etc. (for a discussion, see McKibben [1988]). These effects are natural predictions of the model with no necessity to carefully tune the parameters. The 22-yr variations in particular are a unique consequence of the drift motions, that are necessary in any quantitative model.

Perhaps the most unique basic prediction of the theory, including the coherent drift motions, is that because the sense of the particle drift changes from one sunspot cycle to the next, one expects changes in the cosmic ray intensity and its spatial distribution from cycle to cycle as well. In particular, in the last (1986) sunspot minimum, protons drift inward along the current sheet and then are scattered off of it to drift outward towards the pole. As they drift away from the current sheet they lose energy (cool) because of the divergence in the flow velocity of the wind. Another, equivalent, way of saying this is that they drift against the interplanetary $\mathbf{V} \times \mathbf{B}$ electric field. Because of the cooling, the intensity will *decrease* for some distance away from the current sheet. Conversely, in 1975 during the previous sunspot cycle, and now, the particles drift in from the *polar* regions, striking the current sheet, and then were ejected outward along the sheet. As they drift in from the polar regions they are again drifting against the electric field, and the intensity will decrease as they get closer to the current sheet. This produces a *positive* latitudinal gradient away from the vicinity of the current sheet.

A measurement which gives strong support to the general predictions of the model, was carried out on the Voyager spacecraft near the 1986 sunspot minimum (Cummings et al. 1987). At this time, the intensity of the cosmic rays at the current sheet was found to be higher than the intensity of the cosmic rays at some distance from the current sheet as predicted by the model. This is in contrast to observations carried out earlier at the previous sunspot cycle by the Pioneer spacecraft in which the intensity of the cosmic rays was increasing away from the current sheet. This change in sign of the latitudinal gradient in the vicinity of the current sheet was a striking prediction of the models, and its verification must be regarded as providing strong support for the general picture.

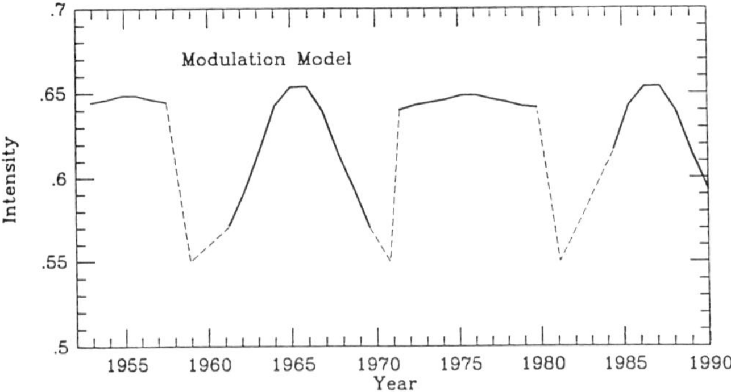

Figure 10.   Illustration of the prediction of the modulation model using nominal parameters. The dashed lines indicate periods around sunspot maximum where the idealized magnetic field and solar wind configuration assumed breaks down. This should be compared with the observations shown in Fig. 3. The correspondence between the predicted flat-topped and more sharply peaked maxima is apparent.

In addition to the change in sign of the latitudinal gradients the model also suggests that successive intensity maxima in the cosmic rays at successive sunspot minimum would have different shapes. As is illustrated in Fig. 10, in 1965 the observed intensity of the cosmic rays had a very sharply peaked maximum while in 1975 the maximum was much more broad and again in 1985 the peak was again sharp. We are not yet at the next cosmic ray maximum, but the most-recent data suggest that the next sunspot minimum will also be broad. This observed behavior of successive intensity maxima is also predicted by the model discussed above. This may be seen as follows; disturbances occur in the solar wind and magnetic field at low latitudes as sunspot activity increases from minimum, and will affect particles coming in along the equatorial regions more than if they are coming in from the poles. Hence, when the magnetic field is such that the particles are drifting inward along the current sheet (1965,1986), the intensity will respond more strongly to the increasing solar activity. In 1975 or 1997, on the other hand, when the particles are coming in over the poles, they do not come near the current sheet until reaching the inner heliosphere, and hence will not respond as readily to the initial increase in sunspot activity.

It should be noted also in Fig. 10 that the model shows dashed lines at periods around sunspot maximum, when the cosmic ray intensity is at its minimum. This is because we do not believe that the well-ordered current structure and flow model used in our computations is applicable around sunspot maximum. During these periods, we expect that the intensity changes will be caused primarily by dynamic structures in the heliosphere—probably by the merged interaction regions discussed in the chapter by McDonald and Burlaga.

As a final illustration of comparison of the model, we mention the correlation of the inclination of the current sheet with the cosmic ray intensity observed during the last sunspot cycle (Smith and Thomas 1986; Lockwood and Webber 1992). It is clear that there is a significantly smaller correlation with the current sheet inclination (determined from corona observations) in the 1975 sunspot minimum and a much stronger correlation in the 1985 sunspot minimum. These are to be expected in the model because of the direction of the drift motions illustrated in Fig. 8. During the 1975 sunspot minimum the positively charged particles were drifting in over the poles and were not sensitive to the tilt of the current sheet, whereas in 1985 they were drifting in along the current sheet and were quite sensitive to its tilt.

These observations of cosmic ray ions lend significant support to the over all picture of modulation which has been developed over the past two decades. However, recent observations of cosmic ray electrons suggest that there are some gaps in our understanding. Observations reported by Tuska et al. (1991) are difficult to reconcile with the predictions of the standard model. The intensity of the nucleonic component of the cosmic ray flux, as indicated by a neutron monitor returned in 1986 to approximately the same intensity as at the previous sunspot minimum in 1976, whereas the electron intensity rose to a value some 40% higher than the value in 1976. To lowest order, one would expect that the time variation of the electron intensity at these energies would be similar to that of ions, except that the drift motions differ in sign because of the opposite charge of the electrons. This is clearly not observed in the data. A difficulty with presently available observations is that most of them refer to only the sum of electrons plus positrons. The ratio of positrons to electrons is not measured and may vary with time. A resolution of the difficulties in interpreting the electron measurements awaits observations of both positrons and electrons over a period of time.

In summary, a comprehensive model of the effect of the solar wind on galactic cosmic rays has been developed. It gives a picture of modulation of nuclei in which large-scale drift effects combine with shocks and interaction regions to produce the observed variations. The behavior of electrons remains a problem.

C. Heliospheric Shocks and Cosmic Rays

Although shocks are fully contained in the most recent modulation models discussed above, they do not play a major role in the modulation of galactic cosmic rays. They do, however, play a major role in lower-energy particles with energies on the order of a few hundred MeV/nucleon and below. Shocks play two roles: (1) the enhanced velocity and turbulence behind the shock act to sweep away particles and decrease their intensity as in Forbush decreases (discussed above), and (2) the shocks accelerate particles as we have seen above in Sec. III. Many different kinds of shocks are found in the heliosphere. There are propagating shock waves associated with coronal mass ejections, solar flares, or merged interaction regions. The interaction of fast and slow streams in the solar wind combined with the solar rotation produces corotating shocks and there are bow shocks around many objects. Finally in the interaction of the solar wind with the interstellar medium arises the strongest and largest shock by far, the termination shock of the solar wind.

Particle acceleration to energies significantly higher than the ambient thermal energies is observed at all collisionless shocks in the heliosphere. However, except for the termination shock, the resulting particle energies are quite low compared to galactic or anomalous cosmic rays (less than or approximately equal to 10 MeV). However, the termination shock is now believed to accelerate particles to more than 1 GeV.

We may understand why the termination shock is so much more effective in accelerating particles by considering Eq. (4) expressing the rate of acceleration of energetic particles in diffusive shock acceleration. The rate scales with the square of the flow speed of the upstream gas relative to the shock and inversely with the diffusion coefficient. Clearly, then, in the same medium (and hence the same diffusion coefficient) a faster shock has a great advantage. Also, for a given acceleration rate, the maximum energy attained will scale with the time available. A fast, long-lived shock is clearly best at producing high-energy particles. Now, the speed of propagating shocks in the solar wind is of the order of 100 to 200 km s^{-1} relative to the solar wind. The speed of the solar wind relative to the termination shock, on the other hand, is of the order of 800 km s^{-1} in the polar regions. Moreover, the termination shock is stationary and has been much the same for a very long time. More detailed considerations, based on Eq. (4) and typical solar wind parameters lead to the conclusion that propagating shocks can accelerate particles to some tens of MeV per charge, whereas the termination shock can readily accelerate particles to several hundreds of MeV per charge or even higher.

D. Anomalous Cosmic Rays

The anomalous component observed in the quiet time cosmic ray flux has been the subject of considerable discussion and speculation since its discovery in 1972. A recent review of this phenomenon is in Jokipii (1990). In the anomalous component the fluxes of helium, nitrogen, oxygen, neon, and more recently protons and low levels of carbon are observed to be enhanced in a region of the energy spectrum ranging from a kinetic energy of 20 MeV to perhaps 300 MeV. Anomalous oxygen is seen in Fig. 11 as the bump in the spectrum at an energy of about 200 MeV (similar bumps are seen for the other species). The observed radial intensity gradient of these particles is positive out to the maximum distance reached by current spacecraft, indicating that this component is not of solar origin, and probably originates in the outer solar system. It now seems quite certain that the anomalous cosmic rays are accelerated at the termination shock at the solar wind. They are initially neutral interstellar atoms that have streamed into the heliosphere as a consequence of its motion. They have become ionized and picked up by the solar wind (suggested by Fisk et al. 1974) and then carried with its magnetic field out to the termination shock, and there accelerated to the observed energies (suggested by Pesses et al. 1981).

The spectra of anomalous cosmic ray oxygen, helium and protons are shown as the data points in Fig. 11. It is clear that the spectra have similar shapes, suggesting that the particles are accelerated to approximately the same characteristic energy. Because they all have a charge of +1 (the ionization process removes only one electron), this suggests that the acceleration mechanism is related to their charge.

The termination shock of the solar wind is quasi-perpendicular (shock normal nearly perpendicular to the magnetic field). Most discussion of shock acceleration in the literature has concentrated on quasi-parallel shocks, where the major energy gain comes from compression along the field lines. However, as was pointed out by Jokipii (1986), diffusive shock acceleration can occur for any angle between the shock normal and the magnetic field and is in general due to both compression and drift. The results presented above for acceleration at a shock (Eq. 4 and related discussion) are valid for all cases as long as diffusion is a good approximation. The acceleration process at quasi-perpendicular shocks is more closely related to drift in the $\mathbf{V} \times \mathbf{B}$ electric field than to compressive Fermi acceleration. In fact, it is readily demonstrated

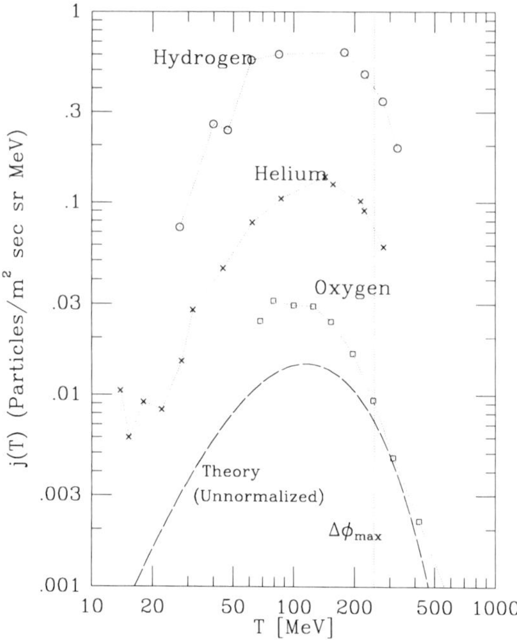

Figure 11.   Observed and computed energy spectra of anomalous oxygen, hydrogen
and helium observed near 21 AU in 1985.   The anomalous hydrogen is from
Cummings and Stone (1988), and the helium and oxygen were computed from
digitized values provided by A. C. Cummings and correspond to the values in
Cummings and Stone (1987). The vertical line indicates the electrostatic potential
difference between the equator and pole of the heliosphere. The dashed line is the
result of a two-dimensional numerical simulation for singly charged oxygen.

that if the scattering frequency is significantly less than the gyrofrequency,
the energy ΔT gained by a particle having electric charge $q = Ze$, at a quasi-
perpendicular shock, is approximately Z times the electrostatic potential $\Delta\phi$
gained in drifting along the shock face. In general, for a shock having an
arbitrary angle to the magnetic field, the energy gain is a combination of
both the first-order Fermi acceleration due to compression and the drift in the
electric field.

Hence, it is of interest to consider the electrostatic potential difference
between the heliospheric equator and pole, which is readily shown to be
independent of the wind velocity and heliocentric radius. We may write

$$\Delta\phi = \frac{eB_r r^2 \Omega_\odot}{c}$$

$$\approx 240 \text{ MV} \tag{6}$$

where the numerical value results from using a radial magnetic field of 3.5
γ at a radius of 1 AU and a solar rotational angular velocity $\Omega_\odot = 2.9 \times$

(corresponding to a magnetic field magnitude of 5 γ at 1 AU). Singly-charged particles accelerated at the termination shock would then have a spectrum which exhibits a decrease above an energy between 200 and 300 MeV. Clearly, the spectra in Fig. 11 are quite similar to each other in shape and exhibit a decrease beyond about 250 to 300 MeV, giving support to the idea that they are accelerated at the termination shock.

Jokipii (1986) presented the results of two-dimensional numerical simulations which verify the above general considerations. The model used was the same as that used in the modulation studies discussed above, except that low-energy ($\approx$100 keV) particles were injected at the shock and the resulting acceleration followed in time for a period of several years to attain a quasi-stationary solution. The theoretical curve in Fig. 11 illustrates a typical energy spectrum for singly charged oxygen. These simulations can readily reproduce many of the features of the observed anomalous cosmic rays, and it now seems likely this indeed is the explanation of this part of the cosmic ray spectrum.

Acceleration at the termination shock of the solar wind is so effective that it is likely that other species of particles accelerated at the shock will become apparent as our understanding of the termination shock improves. One particularly intriguing possibility is that the $\approx$10 MeV electrons which are copiously emitted from the magnetosphere of Jupiter and fill the inner heliosphere, may be carried out to the termination shock and there accelerated to energies of the order of 1 GeV. It is possible that the anomalies in the observations of galactic cosmic ray electrons at 5 GeV, mentioned earlier, may be at least in part due to this acceleration of Jovian electrons at the termination shock (Jokipii and Kóta 1991).

It has been suggested that acceleration at the termination shock may also be so intense that the energy density of cosmic rays at the shock may be comparable to that of the solar wind. In this case, the shock itself will be considerably modified by the cosmic rays and may in fact consist of a smooth cosmic ray shock with a gas sub-shock. Observations and modeling of this phenomena are still at an early stage (see the chapter by Lee).

Consider finally some general properties of acceleration at wind termination shocks. We may estimate the characteristic energy at such shocks by considering the electrostatic potential drop or potential energy change between the equator of the wind and the polar regions of the wind as we did above for the solar wind. This energy may be written in terms of the wind speed, shock radius and azimuthal magnetic field at the shock as

$$T_{\text{sh}} \approx q \frac{V_w B_\phi}{c} R_{\text{sh}}. \tag{7}$$

This energy is the order of 10^{18} to 10^{19} eV per charge for the galactic wind (see Jokipii and Morfill [1987] for a discussion). It can be as large as 10^{15} or 10^{16} eV for OB stars and 10^{11} to 10^{12} eV for M dwarfs. Because it now seems that the anomalous cosmic rays are indeed accelerated by the solar wind's termination shock, acceleration at other termination shocks should be expected and may play a significant role in our general picture of cosmic rays.

V. SUMMARY AND CONCLUSIONS

By using specific examples observed in the solar wind and heliosphere, this chapter has illustrated the three principal types of interaction of a collisionless wind from a condensed object such as a star or galaxy with energetic particles: (1) the interaction with the low-energy part of externally produced cosmic rays which come from outside of the system which reduces the intensity in the wind; (2) the transport of energetic particles produced by the central object; and (3) the acceleration of important cosmic-ray components by the wind itself, primarily by the means of diffusive shock acceleration.

The particle behavior observed in the heliosphere can be reasonably well understood using the general diffusive transport equation first written down in 1965 (Parker 1965). It is clear that the large-scale transport of cosmic rays depends in important ways on both the large-scale magnetic structure and transient large shocks and interaction regions. The picture of acceleration of anomalous cosmic rays at the solar wind termination shock appears to account well for the observed features of this component, and suggests that acceleration at termination shocks of other winds may be important.

Acknowledgments. This work was supported, in part by the National Science Foundation and by the National Aeronautics and Space Administration. I am grateful to K. R. Pyle for providing me with the latest Climax neutron monitor data in digital form.

REFERENCES

Axford, W. I. 1965. The modulation of galactic cosmic rays in the interplanetary medium. *Planet. Space Sci.* 13:115–130.

Axford, W. I., Leer, E., and Skadron, G. 1977. The acceleration of cosmic rays by shock waves. *Proc. 15th Intl. Cosmic Ray Conf.* 11:132–137.

Bame, S. J., Goldstein, B. E., Gosling, J. T. , Harvey, J. W., McComas, D. J., Neugebauer, M., and Phillips, J. L. 1993. Ulysses observations of a recurrent high speed solar wind stream and the heliomagnetic streamer belt. *Geophys. Res. Lett.* 20:2323–2326.

Bell, A. R. 1978. The acceleration of cosmic rays in shock fronts—I. *Mon. Not. Roy. Astron. Soc.* 182:147–156.

Blandford, R. D., and Ostriker, J. P. 1978. Particle acceleration by astrophysical shocks. *Astrophy. J. Lett.* 221:29–32.

Cummings, A. C., and Stone, E. C. 1987. *Proc. 20th Intl. Cosmic Ray Conf.* 3:421–424.

Cummings, A. C., and Stone, E. C. 1988. Evidence for anomalous cosmic-ray hydrogen. *Astrophys. J. Lett.* 334:77–80.

Drury, L. O'C. 1983. An introduction to the theory of diffusive shock acceleration of energetic particles in tenuous plasmas. *Rept. Prog. Phys.* 46:973–1027.

Fisk, L. A., Kozlovsky, B., and Ramaty, R. 1974. An interpretation of the observed oxygen and nitrogen enhancements in low-energy cosmic rays. *Astrophys. J. Lett.* 190:35–37.

Forman, M. A., and Morfill, G. 1979. *Proc. 16th Intl. Cosmic Ray Conf.* 5:328–332.

Gleeson, L. J., and Axford, W. I. 1967. Cosmic rays in the interplanetary medium. *Astrophy. J. Lett.* 149:115–118.

Gloeckler, G. 1979. Compositions of energetic particle populations in interplanetary space. *Rev. Geophys. Space Phys.* 17:569–582.

Jokipii, J. R. 1971. Propagation of cosmic rays in the solar wind. *Rev. Geophys. Space Phys.* 9:27–87.

Jokipii, J. R. 1986. Particle acceleration at a termination shock 1. Application to the solar wind and anomalous component. *J. Geophys. Res.* 91:2929–2932.

Jokipii, J. R. 1990. The anomalous components of cosmic rays. In *Physics of the Outer Heliosphere*, eds. S. Grzedzielski and D. E. Page (New York: Pergamon Press), pp. 169–178.

Jokipii, J. R., and Morfill, G. E. 1987. Ultra-high energy cosmic rays in a galactic wind and its termination shock. *Astrophys. J.* 312:170–177.

Jokipii, J. R., and Kóta, J. 1989. The polar heliospheric magnetic field. *Geophys. Res. Lett.* 16:1–4.

Jokipii, J. R., and Kóta, J. 1991. *Proc. 22nd Intl. Cosmic Ray Conf.* 3:569–573.

Jokipii, J. R., and Parker, E. N. Cosmic-ray life and the stochastic nature of the galactic magnetic field. 1969. *Astrophys. J.* 155:799–806.

Jones, F. C., and Ellison, D. C. 1991. The plasma physics of shock acceleration. *Space Sci. Rev.* 58:259–346.

Kóta, J. 1992. A numerical model of the large-scale solar wind in the outer heliosphere. In *Solar Wind 7*, eds. E. Marsch and R. Schwenn (Oxford: Pergamon Press), pp. 205–208.

Krymsky, G. F. 1977. A regular mechanism for the acceleration of charged particles on the front of a shock wave.

Lockwood, J. A., Webber, W. R., and Jokipii, J. R. 1986. *J. Geophys. Res.* 91:2851–2857.

McKibben, R. B. 1988. Cosmic ray modulation. In *Proc. of the Sixth Intl. Solar Wind Conf.*, vol. 2, eds. V. J. Pizzo, T. E. Holzer and D. E. Sime, NCAR TN-306 (Boulder: Natl. Center for Atmos. Res.), pp. 615–633.

Meyer, P. 1969. Cosmic rays in the galaxy. *Ann. Rev. Astron. Astrophys.* 7:1–38.

Newkirk, G., and Fisk, L. A. 1985. *J. Geophys. Res.* 90:3391–3414.

Parker, E. N. 1958. Cosmic ray modulation by the solar wind. *Phys. Rev.* 110:1445–1557.

Parker, E. N. 1965. The passage of energetic charge particles through interplanetary space. *Planet. Space Sci.* 13:9–49.

Pesses, M. E., Jokipii, J. R., and Eichler, D. 1981. Cosmic ray drift, shock wave acceleration, and the anomalous component of cosmic rays. *Astrophys. J. Lett.* 246:85–88.

Potgieter, M. S., and Moraal, H. 1988. Acceleration of cosmic rays in the solar wind termination shock. I. A steady state technique in a spherically symmetrical model. *Astrophys. J.* 330:445–455.

Smith, E. J., and Thomas, B. T. 1986. Latitudinal extent of the heliospheric current sheet and modulation of galactic cosmic rays. *J. Geophys. Res.* 91:2933–2942.

Tuska, E., Evenson, P., and Meyer, P. 1991. Solar modulation of cosmic electrons: Evidence for dynamic regulation. *Astrophys. J. Lett.* 373:27–30.

EFFECTS OF COSMIC RAYS AND INTERSTELLAR GAS ON THE DYNAMICS OF A WIND

MARTIN A. LEE
University of New Hampshire

The modulation of galactic cosmic rays and other energetic particles by a stellar wind produces a gradient in their pressure, which in turn influences the wind dynamics. The basic equations describing this interaction are presented and discussed in the "hydrodynamic" approximation, in which the diffusive transport of a population of energetic particles is described by an effective diffusion tensor independent of particle energy. Previous and current work on this system is reviewed. A solution of the equations is presented for the case in which both the galactic cosmic ray pressure, and the pressure of the anomalous cosmic ray component accelerated at the wind termination shock, are small compared with the wind ram pressure. Analytical expressions for the deceleration of the wind and the modification of the shock and its location are derived in this case. These effects are estimated to have a relative magnitude of several percent in the solar wind. The interstellar neutral gas which penetrates a stellar cavity may be ionized, predominantly by photoionization and charge exchange with the wind, and may also have a significant dynamical effect on the wind. The basic equations describing this interaction are also presented and discussed. A solution is presented under the assumption that the mass-loading and momentum-loading of the wind by the interstellar pickup ions is small compared with the wind ram pressure. Analytical expressions for the deceleration of the wind, the contribution of the pickup ions to the wind pressure, and the modification of the termination shock location are derived. Again, with the exception of pickup ion pressure which is large compared with solar wind thermal pressure, the effects are estimated to be several percent in relative magnitude in the solar wind.

I. INTRODUCTION

The solar, or a stellar, wind is illustrated schematically in Fig. 1. The wind flows away from the central star with approximate spherical symmetry as shown by the inner vectors, and with a supersonic speed $u(r)$ at radial distances r beyond the region near the star where the wind is accelerated. At these distances the gravitational field of the star is negligible and the wind streams freely with $u \cong u_0$ and mass density $\rho(r) \cong \rho_0(r_0/r)^2$, where subscript o refers to a reference radial distance, usually $r_0 = 1$ AU in the case of the solar wind. The ram pressure of the wind, $P_{ram} = \rho u^2$, therefore decreases with increasing r proportional to r^{-2}.

Eventually the ram pressure of the wind can no longer expel the magnetic field and ionized gas of the very local interstellar medium (VLISM). Because

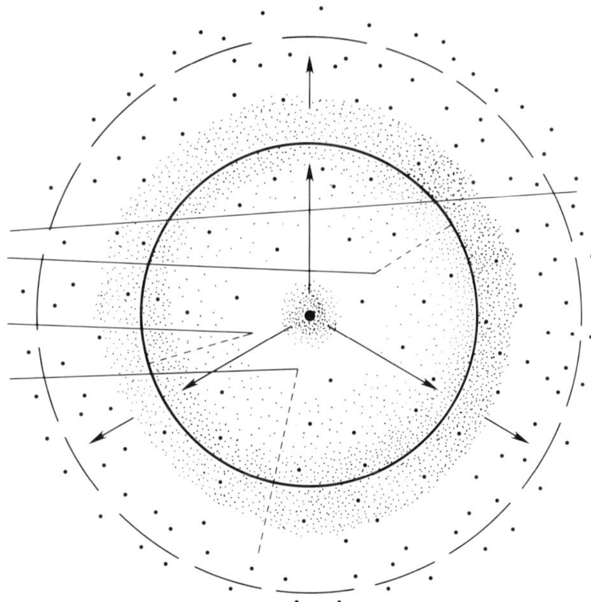

Figure 1.   Schematic diagram of the Sun (or a star), the solar wind (denoted by
vectors), the wind termination shock (solid circle), and the heliopause (dashed
circle). Also shown are particle populations which dynamically influence the wind:
galactic cosmic rays modulated by the heliospheric cavity (large dots), anomalous
cosmic rays accelerated at the shock (light dots straddling the shock), solar cosmic
rays (light dots about the Sun), interstellar neutral gas and interstellar pickup ions
(whose trajectories are denoted by straight solid and dashed lines, respectively).

the wind is supersonic (technically super-fast-magnetosonic), the presence
of the interstellar gas and field cannot be communicated to the wind via
hydromagnetic signals, and deceleration of the wind at a fast magnetosonic
shock occurs. This termination shock of the wind, illustrated by the solid
circle in Fig. 1 and described in the chapter by Suess and Nerney, occurs
approximately where the wind ram pressure balances the pressure P_∞ of the
VLISM. Beyond the shock the wind is submagnetosonic and flows towards
the heliopause (in the case of the Sun), the boundary between solar and
interstellar plasma indicated by a dashed circle in Fig. 1. Of course in steady
state, the "pause" cannot be enclosed. It is generally thought to exhibit a
tail in the direction of the flow of the VLISM past the star, or indeed two
oppositely directed tails along the interstellar magnetic field. Eventually
stellar and interstellar gas should mix along the flanks of the tail, leaving
a distinct and generally nonspherical heliosphere (in the case of the Sun).
Figure 1 does not show an outer shock beyond the "pause" in the interstellar
medium, which exists if the VLISM flow is supermagnetosonic or if the stellar
wind is young and still pushing back the VLISM to create the stellar wind

cavity. The structure of the heliosphere of the Sun has been discussed in a number of recent reviews (Lee 1988; Holzer 1989; Suess 1990). A very early description of the heliosphere by Parker (1963) is still invaluable.

This view of the structure of a stellar wind, however, is modified substantially by the presence of interstellar constituents which *can* communicate their presence to the stellar wind. Galactic cosmic rays (denoted in Fig. 1 by heavy dots) are sufficiently mobile that they can penetrate the heliosphere and influence the stellar wind, in spite of an unfavorable wrapped magnetic field morphology due to stellar rotation. Their incomplete penetration is known in the heliosphere as the solar modulation of galactic cosmic rays, which is described in the chapter by Jokipii.

The other constituent which can penetrate the stellar wind is the neutral component of the interstellar gas. As shown by the straight lines in Fig. 1 for the Sun, neutral atoms flow through the heliosphere with a bulk velocity of 20 to 25 km s^{-1} toward approximately 74.5° ecliptic longitude and −7.5° ecliptic latitude (Holzer 1989). The lines are not parallel because the inferred neutral atom temperature yields thermal speeds comparable to the flow speed (see, e.g., Chassefière et al. 1986). The trajectories are nearly straight until the atoms are affected by gravitation and radiation pressure near the Sun. As the atoms approach the Sun they are more likely to be ionized by solar ultraviolet photons, electron impact, or charge exchange with solar wind ions. The newborn ions are picked up by the solar wind and advected to the termination shock and beyond (as shown by light dashed lines in Fig. 1). The pickup ions have a substantial effect on the solar wind because they mass-load and momentum-load the wind, and dominate its internal pressure in the outer heliosphere (Isenberg 1986).

A fraction of the interstellar pickup ions are injected at the termination shock into the process of diffusive shock acceleration. The accelerated ions reach energies of ∼200 MeV per charge, characteristic of the potential difference between the pole and the solar ecliptic in the outer heliosphere, and diffuse back into the solar wind as the anomalous cosmic ray component (Garcia-Munoz et al. 1973; Hovestadt et al. 1973; McDonald et al. 1974; Fisk et al. 1974; Pesses et al. 1981). The anomalous component (indicated by light dots straddling the termination shock in Fig. 1) may influence dynamically the solar wind and modify the structure of the solar wind termination shock.

Finally, there is another population of energetic particles which can influence the solar wind. Solar energetic particles, or solar cosmic rays, produced in the corona at flares, sites of magnetic reconnection, or shocks, may contribute to solar wind acceleration. This cosmic ray component is denoted by light dots near the Sun in Fig. 1.

Although in this chapter we emphasize the Sun, certainly interstellar gas, galactic cosmic rays, energetic particles accelerated at a wind termination shock, and stellar energetic particles can all influence wind dynamics at any star in a similar fashion. The specific interaction depends, of course, on the interstellar environment of the star, the star type, and the characteristics of the wind. For example, O and B stars ionize the hydrogen surrounding the star to form large H II regions partially expelled from the star by the stellar wind. In this environment atoms cannot penetrate the wind and therefore have no dynamical effect on it.

The purpose of this chapter is to present and discuss the basic equations describing the influence of cosmic rays and interstellar gas on wind dynamics, and to illustrate the interaction by presenting solutions to the equations within simplifying assumptions. For example, although cosmic rays and the interstellar pickup ions exhibit complex temporal and spatial dependence, we shall limit our treatment to stationary configurations with spherical symmetry. Finally we shall attempt to include references to both key past work and current research.

II. COSMIC RAYS: BASIC EQUATIONS

In fact, the influence of cosmic rays on the dynamics of a wind is very simple. Because the cosmic ray mass density is negligible compared with that of the wind, the influence is completely described by including the cosmic ray pressure gradient $-\nabla P_c$ as a force in the momentum equation describing the stellar wind plasma. The single-fluid polytrope equations describing the wind are then

$$\frac{\partial \rho}{\partial t} + \nabla \cdot (\rho \mathbf{u}) = 0 \tag{1}$$

$$\rho \frac{\partial \mathbf{u}}{\partial t} + \rho (\mathbf{u} \cdot \nabla)\mathbf{u} + \rho \frac{GM}{r^2}\hat{e}_r + \nabla P_g = -\nabla P_c \tag{2}$$

$$\frac{\partial P_g}{\partial t} + \mathbf{u} \cdot \nabla P_g + \gamma_g (\nabla \cdot \mathbf{u}) P_g = 0 \tag{3}$$

where ρ, $\mathbf{u}$ and P_g are the wind mass density, velocity, and pressure, respectively, M is the mass of the star, and G is the gravitational constant. The constant γ_g is the polytrope index of the gas ($P_g \propto \rho^{\gamma_g}$); to reproduce the essential effects of heat conduction near the star $1 < \gamma_g < 3/2$, but γ_g should relax to its adiabatic value, $\gamma_g = 5/3$, at larger radial distances. The polytrope equation of state for the wind clearly only approximates the effects of heating by waves, heat conduction, or other processes, which require an energy equation for a more accurate description. Equation (2) also neglects acceleration of the wind by a wave pressure gradient. Because the emphasis of this chapter is the influence of energetic particles, we restrict ourselves to polytrope winds. In Sec. VI we outline the inclusion of Alfvén waves.

Equations (1) through (3) must be supplemented with an equation for P_c. If the cosmic rays are nearly isotropic and have speeds v which satisfy $v^2 \gg u^2$, then their transport is described by the equation

$$\frac{\partial f}{\partial t} + (\mathbf{u} + \mathbf{u}_D) \cdot \nabla f - \nabla \cdot \kappa \cdot \nabla f - \frac{1}{3}(\nabla \cdot \mathbf{u})p\frac{\partial f}{\partial p} = Q \tag{4}$$

where $f(\mathbf{x}, p, t)$ is the cosmic ray omnidirectional distribution function [$n = \int f d^3\mathbf{p} = 4\pi \int p^2 \mathrm{d}p f$, where $n(\mathbf{x}, t)$ is cosmic ray number density], p is particle momentum, and $Q(\mathbf{x}, p, t)$ is the source term. The tensor $\kappa(\mathbf{x}, p, t)$ is the symmetric spatial diffusion tensor, and $\mathbf{u}_D$ is the drift velocity, given in terms of the particle charge q and the average magnetic field $\mathbf{B}$ by $\mathbf{u}_D = pvc(3q)^{-1}\nabla \times (B^{-2}\mathbf{B})$. Generally Q vanishes everywhere except at the star and at shocks where energetic particles can be accelerated out of the thermal wind plasma. Equation (4) neglects several higher-order effects including the phase speeds of the scattering waves, cosmic ray viscosity, and stochastic acceleration, all of which are generally negligible. Equation (4) was first written down by Parker (1965), and clarified or generalized by Dolginov and Toptygin (1967), Gleeson and Axford (1967), and Jokipii and Levy (1977).

The cosmic ray pressure is given by

$$P_c(\mathbf{x}, t) = \frac{1}{3} \int d^3\mathbf{p}\, pvf(\mathbf{x}, p, t). \tag{5}$$

In general, Eq. (4) must be solved for $f(\mathbf{x},p,t)$ so that the pressure may be constructed via Eq. (5). Equivalently the pressure moment of Eq. (4) yields

$$\partial P_c/\partial t + \mathbf{u} \cdot \nabla P_c + (1/3) \int d^3\mathbf{p}\, pv(\mathbf{u}_D \cdot \nabla f - \nabla \cdot \kappa \cdot \nabla f)$$

$$+(\nabla \cdot \mathbf{u})(1/3) \int d^3\mathbf{p}\, pvf[(1/3)p^{-3}v^{-1}\mathrm{d}/\mathrm{d}p(p^4 v)] = 1/3 \int d^3\mathbf{p}\, pvQ. \tag{6}$$

Substantial simplification occurs if the drift and diffusion terms are replaced by $\bar{\mathbf{u}}_D \cdot \nabla P_c$ and $\nabla \cdot \bar{\kappa} \cdot \nabla P_c$, and the quantity in square brackets by a constant γ_c. With those replacements Eq. (6) becomes

$$\partial P_c/\partial t + (\mathbf{u} + \bar{\mathbf{u}}_D) \cdot \nabla P_c - \nabla \cdot \bar{\kappa} \cdot \nabla P_c + \gamma_c(\nabla \cdot \mathbf{u})P_c = \bar{Q} \tag{7}$$

where $\bar{Q} = (1/3) \int d^3\mathbf{p}\, pvQ$. The effective drift velocity $\bar{\mathbf{u}}_D$ and diffusion tensor $\bar{\kappa}$ are equal to $\mathbf{u}_D$ and κ, respectively, if $\mathbf{u}_D$ and κ are independent of p, which is generally not a good approximation. For nonrelativistic and relativistic particle distributions γ_c equals 5/3 and 4/3, respectively. In spite of its limitations, if γ_c is chosen to satisfy $4/3<\gamma_c<5/3$ and $\bar{\kappa}$ and $\bar{\mathbf{u}}_D$ are chosen to be representative of the bulk of the energetic particles, then Eq. (7) should provide an adequate description of the cosmic ray pressure. Virtually all work on the dynamical influence of cosmic rays on a wind has adopted this "hydrodynamical" description of the cosmic ray transport. Furthermore, to be consistent with Eqs. (1) through (3), which neglect the ambient magnetic field, we neglect the drift transport in Eq. (7). The drift term has never been considered in the context of the dynamical influence of cosmic rays on a wind, because it requires a model which depends on latitude and/or longitude as well as on r.

Equations (1), (2), (3), and (7) describe the coupled behavior of wind and cosmic rays. Indeed the coupled system is quadratically nonlinear and not so simple. The cosmic rays decelerate the wind, and the wind in turn determines the "modulated" cosmic ray distribution.

Assuming a stationary configuration with spherical symmetry, Eqs. (1), (2), (3), and (7) become

$$r^{-2}d/dr\,(r^2\rho u) = 0 \tag{8}$$

$$\rho u\,du/dr + d/dr(P_g + P_c) + \rho GMr^{-2} = 0 \tag{9}$$

$$u\,dP_g/dr + \gamma_g P_g r^{-2}d/dr\,(r^2 u) = 0 \tag{10}$$

$$u\,dP_c/dr + \gamma_c P_c r^{-2}d/dr\,(r^2 u) - r^{-2}d/dr\,(r^2\kappa\,dP_c/dr) = 0 \tag{11}$$

where we neglect the "bar" on the radial diffusion coefficient κ. Equation (8) yields conservation of the total mass flux with $r^2\rho u = (4\pi)^{-1}F$, where the total mass flux $F = -dM/dt$, the rate of mass loss of the star. Equation (11) may be rewritten as

$$\frac{1}{r^2}\frac{d}{dr}\left[r^2\left(\frac{\gamma_c}{\gamma_c-1}uP_c - \frac{1}{\gamma_c-1}\kappa\frac{dP_c}{dr}\right)\right] = u\frac{dP_c}{dr}. \tag{12}$$

Clearly the quantity in parentheses is cosmic ray energy flux, with cosmic ray energy density E_c given by $E_c = (\gamma_c - 1)^{-1}P_c$. The energy flux consists of advected enthalpy and the diffusion of energy density. Because a wind expels galactic cosmic rays, $dP_c/dr > 0$ and $u\,dP_c/dr > 0$, so that the wind does positive work on galactic cosmic rays. For solar cosmic rays, on the other hand, $dP_c/dr < 0$ so that they do work on the wind. Equations (8) through (11) may be combined to yield

$$\frac{1}{r^2}\frac{d}{dr}\left[r^2 u\left(\frac{1}{2}\rho u^2 - \rho\frac{GM}{r} + \frac{\gamma_g P_g}{\gamma_g - 1} + \frac{\gamma_c P_c}{\gamma_c - 1}\right) - \frac{r^2\kappa}{\gamma_c - 1}\frac{dP_c}{dr}\right] = 0 \tag{13}$$

which establishes the conservation of total energy flux consisting of the flux of bulk kinetic energy density, gravitational potential energy, and gas enthalpy, and the cosmic ray energy flux.

Historically, Axford (1965) and Axford and Newman (1965) were the first to consider the deceleration of a wind due to the cosmic ray pressure gradient. Their work, published prior to or contemporary with the original formulation of the cosmic ray transport equation, employed a transport equation which includes only convection with the wind and diffusion, and omits the term describing the energy exchange of cosmic rays with the wind. Sousk and Lenchek (1969) correctly included energy exchange with the wind, but also used a cosmic ray transport equation including only convection and diffusion. Babayan and Dorman (1977*a,b*,1990) also used an incomplete transport equation. Lee and Axford (1988) presented the solution of Eqs. (8) through

(11) neglecting gravity and setting $P_g = 0$; we shall return to their work in Sec. V.

III. COSMIC RAYS: SOLUTION TOPOLOGIES AND SINGULARITIES

The first correct, and comprehensive, investigation of the system of Eqs. (8) through (11) was conducted by Ko and colleagues in a series of three papers (Ko and Webb 1987, 1988; Ko et al. 1988). The first paper described the singularities of the equations and outlined the topology of the solutions, the second developed a perturbation solution based on small cosmic ray pressure, and the third presented an iterative numerical solution. Figure 2 shows solution curves (panel a) for a particular choice of parameters with κ constant, in terms of the dimensionless variables λ ($\propto r^{-1}$), Ψ ($\propto u^2$) and $\bar{P}_c$ ($\propto P_c$). Panels b, c and d show the projections of the solution curves into the three independent planes. The singular points of the system of equations all lie on the dotted line shown in panel a. The projection of this critical line in the (λ, Ψ)-plane gives the gas sonic curve on which $u^2 = \gamma_g P_g \rho^{-1}$. Depending on parameters (γ_g, γ_c, κ, entropy and energy flux), a particular point along the critical line may be a node, a focus, or a saddle point. The solution curves shown in Fig. 2 may clearly approach (or intersect) the critical line at a saddle point (at $\lambda \simeq 1$, about which nearby solutions are "repelled") and a focus (at $\lambda \simeq 0.25$, at which solutions spiral into the critical point).

We note in Eq. (11) that the limit, $\kappa \to \infty$, in general requires that $dP_c/dr \to 0$. From Eq. (9) we then note that cosmic rays decouple from the wind in this limit. The solution curves in Fig. 2 assume a large value of κ so that when they are projected into the (λ, Ψ)-plane in panel b they approximate the solution curves in the absence of cosmic rays, at least for $\lambda \gtrsim 0.5$. The saddle-point critical point is clearly that obtained for the one-fluid polytropic stellar wind (see the chapters by Axford and McKenzie, and by Holzer et al.). The physically acceptable wind solution passes through that critical point from the lower right of panel b to the upper left. However, even for large κ, we see that the wind solution decelerates at large r (small λ) due to the cosmic ray pressure gradient into a focus critical point. A physically acceptable solution requires that a shock be inserted before the focus critical point to connect the wind solution to one of the family of curves shown near the origin of panel b, with an appropriate change in the gas entropy. This shock is the stellar wind termination shock described in Sec. I, modified by cosmic ray pressure.

IV. COSMIC RAYS: A PERTURBATION SOLUTION

In order to illustrate the dynamical influence of cosmic rays on a stellar wind, we consider a simpler system than that described by Eqs. (8) through (11) with solutions illustrated in Fig. 2. The wind ram pressure near the star is

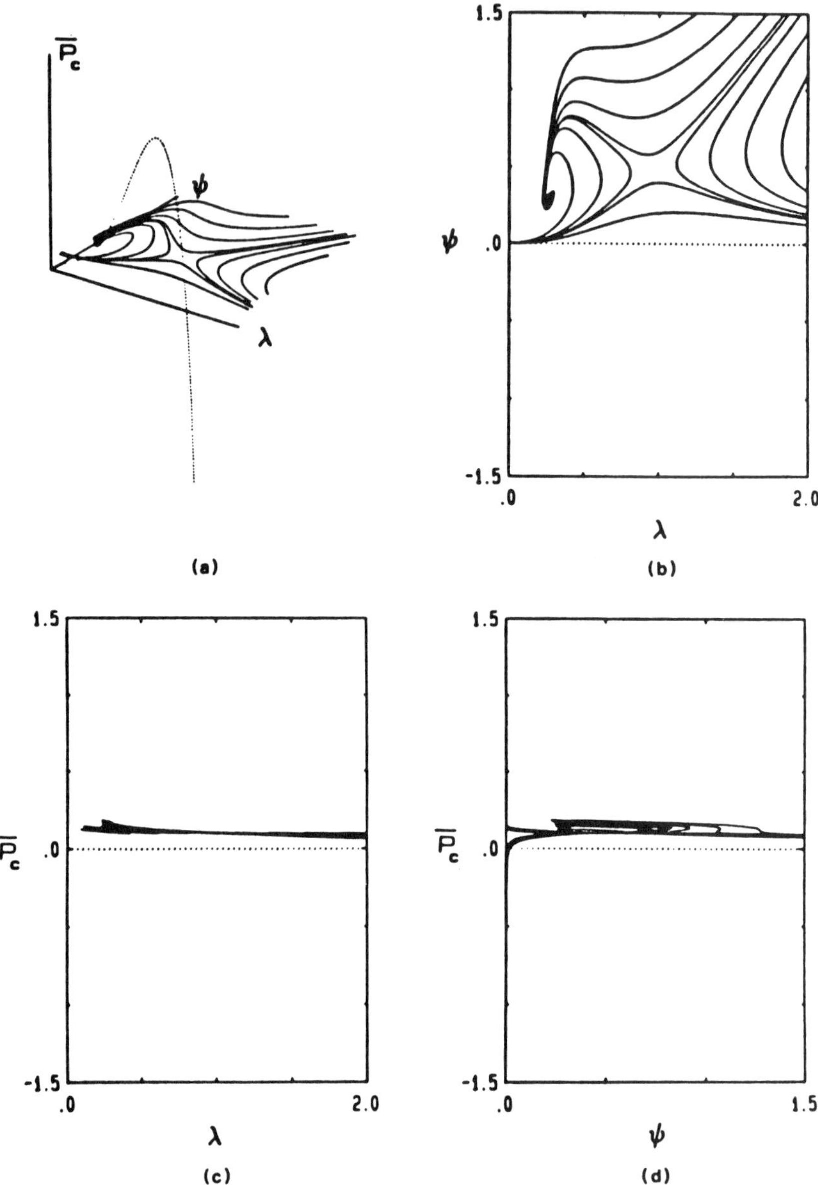

Figure 2.   Solution curves of the system of Eqs. (8) through (11). Panel a shows the solutions in the space of dimensionless variables $\lambda(\propto r^{-1})$, $\Psi(\propto u^2)$, and $\bar{P}_c(\propto P_c)$; panels b, c and d show their projections into the three independent planes. The dotted line in panel a is the critical line of the system (figure from Ko and Webb 1987).

generally much greater than the cosmic ray pressure. At the Sun the wind ram pressure at Earth is $\sim 2 \times 10^{-8}$ dyne cm^{-2}, whereas the cosmic ray pressure is about 1×10^{-12} dyne cm^{-2} (Holzer 1989). Therefore, in the region of wind acceleration the cosmic rays have no influence, and we may consider equations which in the absence of cosmic rays describe the "free-streaming" wind. Thus we neglect gravity, and the gas pressure P_g upstream of the termination shock. Equations (8) through (11) then become

$$r^2 \rho u = (4\pi)^{-1} F \tag{14}$$

$$(4\pi)^{-1} F r^{-2} \mathrm{d}u/\mathrm{d}r + \mathrm{d}/\mathrm{d}r (P_g + P_c) = 0 \tag{15}$$

$$u \mathrm{d}P_g/\mathrm{d}r + \gamma_g P_g r^{-2} \mathrm{d}/\mathrm{d}r (r^2 u) = 0 \tag{16}$$

$$u \mathrm{d}P_c/\mathrm{d}r + \gamma_c P_c r^{-2} \mathrm{d}/\mathrm{d}r (r^2 u) - r^{-2} \mathrm{d}/\mathrm{d}r (r^2 \kappa \mathrm{d}P_c/\mathrm{d}r) = 0. \tag{17}$$

Furthermore, we consider a perturbation expansion in $\epsilon \cong P_c(\rho u^2)^{-1} \ll 1$. Upstream of the shock to zeroth order in ϵ, we obtain solutions describing the free-streaming wind as

$$u = u_\mathrm{o} \tag{18}$$

$$\rho = \rho_\mathrm{o}(r_\mathrm{o}/r)^2 \tag{19}$$

where subscript o refers to a reference radial distance. If the termination shock is at $r = r_{so}$ (to zeroth order) then the Rankine-Hugoniot jump conditions at a strong shock yield

$$u_2 = u_\mathrm{o}(\gamma_g - 1)(\gamma_g + 1)^{-1} \tag{20}$$

$$\rho_2 = \rho_\mathrm{o}(r_\mathrm{o}/r_{so})^2(\gamma_g + 1)(\gamma_g - 1)^{-1} \tag{21}$$

$$P_{g2} = \rho_\mathrm{o}(r_\mathrm{o}/r_{so})^2 u_\mathrm{o}^2 2(\gamma_g + 1)^{-1} \tag{22}$$

where subscript 2 indicates values adjacent to, and downstream of, the shock. The zeroth-order energy flux integral of Eqs. (14) through (16) yields Bernoulli's Law as

$$1/2\, u^2 + \gamma_g(\gamma_g - 1)^{-1} P_g \rho^{-1} = 1/2\, u_\mathrm{o}^2. \tag{23}$$

Eliminating u and ρ in favor of P_g by using Eq. (14), and Eq. (16) in the form $P_g = P_{g2}(\rho/\rho_2)^{\gamma_g}$, and requiring that $P_g \to P_\infty$ as $r \to \infty$, we obtain for the shock distance

$$(r_{so}/r_\mathrm{o})^2 = \rho_\mathrm{o} u_\mathrm{o}^2 P_\infty^{-1} 2(\gamma_g + 1)^{-1}[(\gamma_g + 1)^2(4\gamma_g)^{-1}]^{\gamma_g/(\gamma_g - 1)} \tag{24}$$

a result originally derived by Parker (1963). The factor following P_∞^{-1} on the right-hand side of Eq. (24) equals ~ 0.88 for $\gamma_g = 5/3$. Thus the shock is located approximately where the stellar wind ram pressure equals the pressure of the interstellar medium P_∞. The above factor accounts for the fact that

P_{g2} is somewhat smaller than P_∞, and there exists a small ram pressure just downstream of the shock.

Now to obtain P_c at first order in ϵ, we must solve Eq. (17) using the wind speed derived at zeroth order. We shall consider two classes of "cosmic rays": galactic cosmic rays P_c and cosmic rays accelerated at the termination shock P_a, which in the heliosphere are known as the anomalous cosmic ray component (Garcia-Munoz et al. 1973; Hovestadt et al. 1973; McDonald et al. 1974; Fisk et al. 1974; Jokipii 1986). We denote the galactic cosmic rays as GCR and the anomalous cosmic rays as ACR. For both classes of particles, we make the common assumption (see, e.g., Le Roux and Potgieter 1991) that the diffusion coefficient in the wind κ_1 is proportional to r: $\kappa_{1c} = \kappa_{oc}(r/r_{so})$ and $\kappa_{1a} = \kappa_{oa}(r/r_{so})$.

Because the ACR are produced at the shock and, at least in the heliosphere, are nonrelativistic, we choose $\gamma_a = \gamma_g$ and we choose their diffusion coefficient downstream of the shock to vanish. Subject to the boundary conditions that P_a is bounded as $r \to 0, \infty$ and continuous at $r = r_{so}$, the solution of Eq. (17) for the ACR is then

$$r < r_{so} : P_a = P_{ao}(r/r_{so})^{\alpha_a} \tag{25}$$

$$r > r_{so} : P_a = P_{ao}(\rho/\rho_2)^{\gamma_g} \tag{26}$$

where

$$\alpha_a = -1 + (1/2)r_{so}u_o\kappa_{oa}^{-1} + [(1 - 1/2r_{so}u_o\kappa_{oa}^{-1})^2 + 2\gamma_g r_{so}u_o\kappa_{oa}^{-1}]^{1/2} \tag{27}$$

and $P_{ao} = P_a(r_{so})$. The particular value of P_{ao} is obtained by integrating Eq. (17) across the shock, but including a source term on the right-hand side to account for cosmic ray injection at the shock. A technical difficulty arises in calculating P_{ao}: if $\kappa_{oa} \ll u_o r_{so}$ then the shock appears planar, and the momentum spectral index of the cosmic ray distribution function approaches 4 in the strong shock limit. However, such a "hard" power law is inconsistent with the nonrelativistic choice $\gamma_a = \gamma_g$. As a result P_{ao} may become negative. In reality the shock is not a strong shock, by virtue of the energetic particles themselves and the interstellar pickup hydrogen, which we discuss in detail in Sec. VIII, and drift limits the ACR momentum spectrum at high momentum. Also allowing the downstream diffusion coefficient to be finite would affect the calculation of P_{ao}. Accordingly, we simply take Eqs. (25) and (26) to be the correct solution without specifying the relation between P_{ao} and the cosmic ray injection rate at the shock.

For GCR the diffusion coefficient downstream of the shock must be chosen to allow their access into the stellar wind. We take $\kappa = \kappa_2$, a constant, in this region. From Eq. (23) we may infer that $\rho_\infty/\rho_2 \cong 1.1$, where $\rho_\infty = \rho(r \to \infty)$, so that the flow is nearly incompressible downstream of the termination shock (this is not necessarily true in a hydromagnetic model). Accordingly, we set $u = u_2(r_{so}/r)^2$ in Eq. (17). Then, setting $P_{c\infty} = P_c(r \to$

∞), requiring that both P_c and the cosmic ray energy flux are continuous at the shock (we therefore assume there is no additional injection of GCR at the shock), and requiring that $P_c(r \to 0)$ be bounded, we obtain the solution

$$r < r_{so} : P_c = P_{co}(r/r_{so})^{\alpha_c} \tag{28}$$

$$r > r_{so} : P_c = P_{c\infty} - (P_{c\infty} - P_{co})[1 - \exp(-u_2 r_{so}^2 \kappa_2^{-1} r^{-1})]$$

$$\cdot [1 - \exp(-u_2 r_{so}\kappa_2^{-1})]^{-1} \tag{29}$$

where α_c is given by Eq. (27) for α_a, but with κ_{oa} replaced by κ_{oc} and γ_g replaced by γ_c, and where

$$P_{co} = u_2 P_{c\infty}\{\gamma_c(u_o - u_2) + u_2 - \alpha_c \kappa_{oc} r_{so}^{-1}$$

$$+ \exp(u_2 r_{so}\kappa_2^{-1})[\alpha_c \kappa_{oc} r_{so}^{-1} - \gamma_c(u_o - u_2)]\}^{-1} \tag{30}$$

is the GCR pressure at the shock. The term containing $\exp(-u_2 r_{so}^2 \kappa_2^{-1} r^{-1})$ on the right-hand side of Eq. (29) is a solution of the "convection-diffusion" equation of cosmic ray transport describing a balance between outward convection and inward diffusion. This equation was used by Axford and Newman (1965), Sousk and Lenchek (1969), and Babayan and Dorman (1977a,b,1990). The constant term on the right-hand side of Eq. (29) does not satisfy the convection-diffusion equation but is a solution of Eq. (17) which is required to describe the transport of cosmic rays in the inner heliosphere and those which are re-accelerated at the termination shock and convected back into the interstellar medium.

The deceleration of the solar wind is obtained by substituting Eqs. (25) and (28) into Eq. (15) with $P_g = 0$ and $P_c \to P_c + P_a$. There obtains, correct to first order in ϵ

$$u = u_o - 4\pi r^2 F^{-1}[\alpha_c(\alpha_c + 2)^{-1} P_c + \alpha_a(\alpha_a + 2)^{-1} P_a]. \tag{31}$$

The cosmic rays will also affect the location of the stellar wind termination shock. With $\delta u_1 = u_o - u(r_{so})$, from Eq. (31), the stellar wind mass density just upstream of the shock at $r = r_s$ is $\rho_1 = (4\pi)^{-1} F r_s^{-2}(u_o - \delta u_1)^{-1}$.

Deriving the Rankine-Hugoniot jump conditions, correct to first order, requires some care. Because the cosmic ray pressure is continuous at the gas subshock, we still have the usual relations in terms of the shock compression ratio X

$$\rho_2 = X\rho_1 \tag{32}$$

$$u_2 = X^{-1}u_1 \tag{33}$$

$$P_{g2} = \rho_1 u_1^2(1 - X^{-1}). \tag{34}$$

The compression ratio is determined by requiring that the energy flux

$$F[1/2u^2 + \gamma_g(\gamma_g - 1)^{-1}(P_a + P_g)\rho^{-1} + \gamma_c(\gamma_c - 1)^{-1} P_c\rho^{-1}]$$

$$-4\pi r^2[\kappa_a(\gamma_g - 1)^{-1}\mathrm{d}P_a/\mathrm{d}r + \kappa_c(\gamma_c - 1)^{-1}\mathrm{d}P_c/\mathrm{d}r] \qquad (35)$$

is conserved across the gas subshock. The GCR energy flux involving P_c is conserved by construction. Evaluating Eq. (35) on both sides of the shock using Eqs. (25), (26), (32), (33) and (34), we obtain the compression ratio as

$$X = (\gamma_g + 1)(\gamma_g - 1)^{-1} - (\gamma_g + 1)^2(\gamma_g - 1)^{-1}4\pi r_{so}^2(u_o F)^{-1}$$

$$\{2\gamma_g(\gamma_g^2 - 1)^{-1}P_{ao} - \kappa_{oa}\alpha_a(\gamma_g - 1)^{-1}u_o^{-1}r_{so}^{-1}P_{ao}\}. \qquad (36)$$

Substituting X into Eqs. (32) through (34) yields ρ_2, u_2 and P_{g2} correct to first order in ϵ.

Determination of r_s^2 to first order is similar to the method used at zeroth order. Equation (23) may be generalized by writing the conservation of energy flux downstream of the shock as

$$\mathrm{d}/\mathrm{d}r[1/2u^2 + \gamma_g(\gamma_g - 1)^{-1}\rho^{-1}(P_g + P_a)] = -\rho^{-1}\mathrm{d}P_c/\mathrm{d}r \qquad (37)$$

where we have utilized the fact that the diffusion coefficient of the ACR is negligible downstream. As we have already neglected the compressibility of the downstream flow in calculating P_c, we may set $\rho = \rho_\infty$ on the right-hand side of Eq. (37) and obtain the integral

$$1/2u^2 + \gamma_g(\gamma_g - 1)^{-1}(P_g + P_a)\rho^{-1} + P_c\rho_\infty^{-1} =$$

$$1/2u_2^2 + \gamma_g(\gamma_g - 1)^{-1}(P_{g2} + P_{ao})\rho_2^{-1} + P_{co}\rho_\infty^{-1}. \qquad (38)$$

From Eqs. (16), (22) and (24), we find $\rho_\infty = \rho_2[(\gamma_g + 1)^2(4\gamma_g)^{-1}]^{1/(\gamma_g-1)}$. Eliminating u and ρ in favor of $P_g + P_a$ by using Eq. (14) and $P_g + P_a = (P_{g2} + P_{ao})(\rho/\rho_2)^{\gamma_g}$, and requiring $P_g + P_a \to P_\infty$ and $P_c \to P_{c\infty}$ as $r \to \infty$, we obtain an equation for r_s^2 correct to first order in P_c/P_∞, P_a/P_∞:

$$(r_s/r_{so})^2 = 1 + 4\pi r_{so}^2(F u_o)^{-1}\{-\alpha_a(\alpha_a + 2)^{-1}P_{ao} + 1/2P_{ao}$$

$$+1/2\kappa_{oa}\alpha_a u_o^{-1}r_{so}^{-1}P_{ao} - \alpha_c(\alpha_c + 2)^{-1}P_{co}$$

$$-2\gamma_g(\gamma_g + 1)^{-1}[(\gamma_g + 1)^2(4\gamma_g)^{-1}]^{-1/(\gamma_g-1)}(P_{c\infty} - P_{co})\}. \qquad (39)$$

In the heliosphere the ACR have typical energies of 100 MeV/charge and correspondingly small values of the diffusion coefficient, which satisfies $r_{so}u_o \gg \kappa_{oa}$ so that $\alpha_a \gg 1$. In any case, the ACR moves the shock outward. This can be understood by noting that the ACR gains its energy from the wind. Thus, it cannot change the location of the complete shock transition, which depends only on F and P_∞. However, because the transition of the ACR to its pressure at the shock is distributed over a foreshock of lengthscale κ_{oa}/u_o, the gas subshock is pushed out a distance of ϵ times that order. As $\kappa_{oa} \to 0$ the behavior of the ACR is identical to that of the thermal gas; in this limit the ACR does not affect shock location.

The GCR, on the other hand, move the shock inwards; they not only decelerate the stellar wind, reducing the ram pressure, but also their modulation provides an additional confining pressure. In the limit $\kappa_{oc} \to \infty$ and $\kappa_2 \to \infty$, in which cosmic rays readily penetrate the inner heliosphere, their gradients vanish, they have no dynamical influence, and they do not affect the location of the shock. Interestingly they are still expelled from the immediate vicinity of the star due to our assumption that $\kappa_c \propto r$, but there the wind ram pressure completely dominates that of the GCR. In the opposite limit in which $\kappa_2 \to 0$, P_{co} is negligible, and the only GCR term which survives in curly brackets in Eq. (39) is $-2\gamma_g(\gamma_g + 1)^{-1}[(\gamma_g + 1)^2(4\gamma_g)^{-1}]^{-1/(\gamma_g-1)} P_{c\infty}$. The effect of this term is to modify the shock location according to Eq. (24) by replacing P_∞ by $P_\infty + P_{c\infty}$, which implies that the cosmic rays provide only an additional confining pressure.

We illustrate the perturbation solution by choosing parameters representative of the heliosphere. We take $\gamma_g = 5/3$, $u_0 = 400$ km s^{-1} and $\rho_0 m_p^{-1} = 8$ cm^{-3} ($r_0 = 1$ AU), where m_p is the proton mass. We take $r_{so} = 80$ AU which requires $P_\infty = 2.94 \times 10^{-12}$ dyne cm^{-2}. If this pressure were provided by the interstellar magnetic field $\mathbf{B}_I$ and the heliosphere were approximately spherical, then $|\mathbf{B}_I| = 5.7\,\mu$G, a very reasonable value. For the GCR we take $\gamma_c = 1.5$, $r_{so}u_0\kappa_{oc}^{-1} = 1$ (so that modulation within the termination shock is of order 1 as observed), $\kappa_2 = \kappa_{oc}$, and $P_{c\infty} = 1/3 P_\infty$. The last choice yields $P_{c\infty} = 0.98 \times 10^{-12}$ dyne cm^{-2}, consistent with inferred cosmic ray pressures in interstellar space. For the ACR we take $r_{so}u_0\kappa_{oa}^{-1} = 10$, consistent with the inferred energies of ACR hydrogen, and $P_{ao} = 0.67 \times 10^{-12}$ dyne cm^{-2}, which yields the reasonable result that the ACR pressure is one fifth the ram pressure of the solar wind at the termination shock. With these values we obtain $\alpha_c = 1.3$, $\alpha_a = 11$, and $P_{c\infty} = 1.2\,P_{co}$ (so that modulation in the heliosheath is small). With these parameters the gas subshock compression ratio X, given by Eq. (36), is reduced from 4 to 3.5 due to cosmic ray pressure. Equation (39) then yields

$$(r_s/r_{so})^2 = 1 + 0.041 - 0.096 - 0.056 \tag{40}$$

where the three small terms correspond to terms dependent on P_{ao}, P_{co} and $P_{c\infty} - P_{co}$, respectively. Clearly, as discussed above, the ACR move the shock out, and GCR modulation both upstream and downstream of the termination shock moves the shock in. All effects, in this example, are at the few percent level, with $r_s = 75.4$ AU.

Figure 3 shows the perturbation solution for this choice of parameters. Shown is the zeroth order flow speed $u_0(r)$ with a shock at $r = r_{so} = 80$ AU, the pressure of galactic cosmic rays P_c, the pressure of the anomalous cosmic ray component P_a, and the flow speed $u(r)$, correct to first order in P_c/P_∞, P_a/P_∞. P_c increases with lengthscale κ_c/u_0 in the wind and with lengthscale κ_2/u_2 beyond the shock to match the interstellar GCR pressure, $P_{c\infty}$. $P_c(r \to 0)$ vanishes because $\kappa_{1c}(r \to 0) \to 0$. Similarly, P_a increases

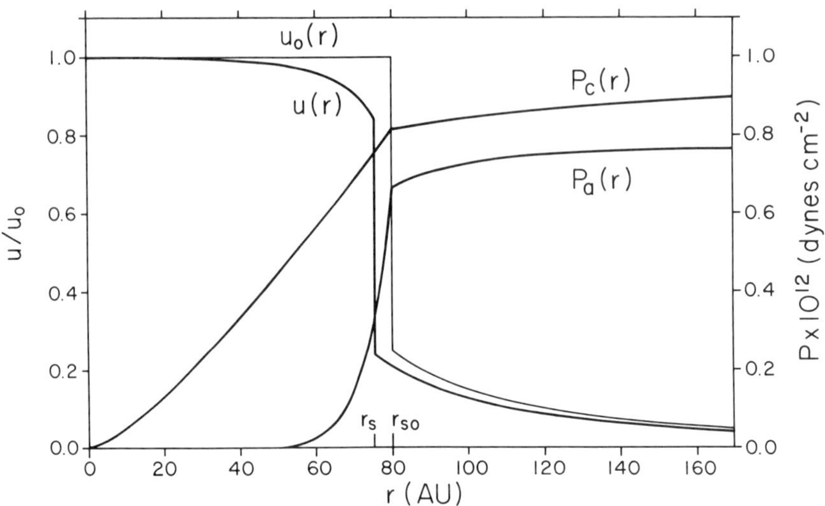

Figure 3.   Perturbation solution of Eqs. (14) through (17) showing the zeroth-order wind speed $u_0(r)$ (normalized to u_0, left scale) with a strong termination shock at $r = r_{so}$, the (first-order) galactic cosmic ray pressure $P_c(r)$ and anomalous cosmic ray pressure $P_a(r)$ (right scale), and the wind speed $u(r)$ correct to first order with a weaker termination shock at $r = r_s$. Parameters (see text) were chosen to be representative of the heliosphere.

with lengthscale κ_a/u_0 in the wind. P_a continues to increase beyond the shock because the diffusion coefficient is assumed to vanish, so that P_a is proportional to the gas pressure P_g. A finite diffusion coefficient would result eventually in a decrease in P_a as the particles diffuse away from the shock. The flow decelerates first on the lengthscale κ_c/u_0 due to the GCR pressure gradient, then on the lengthscale κ_a/u_0 due to the ACR pressure gradient, undergoes a gas subshock at $r_s < r_{so}$, and decelerates as a nearly incompressible flow, $u \propto r^{-2}$, at large distances. A very similar figure is presented by Ko et al. (1988, their Figure 3) based on numerical calculations and a similar choice of parameters.

V. COSMIC RAYS: THE GAS SUBSHOCK

The perturbation solution derived in the last section and presented in Fig. 3 exhibits a gas subshock. This cannot be avoided in a perturbation solution because the cosmic ray perturbation cannot eliminate the shock which appears in the zeroth-order solution. In general, however, Ko and Webb (1987) show that a second critical point of the system described by Eqs. (8) through (11) could be a saddle point, which would provide for a shockless transition to the nearly incompressible flow at large distances. They did not explore the conditions under which a shock is, or is not, present.

A system which has no shock was explored by Lee and Axford (1988). They considered the system of Eqs. (14) through (17) with vanishing gas pressure P_g:

$$\rho u r^2 = (4\pi)^{-1} F \tag{41}$$

$$\rho u \, du/dr = -dP_c/dr \tag{42}$$

$$u \, dP_c/dr + \gamma_c P_c r^{-2} d/dr(r^2 u) - r^{-2} d/dr(r^2 \kappa \, dP_c/dr) = 0. \tag{43}$$

Under the assumption $\kappa = (\gamma_c - 1)ku_o r$, they derived solutions depicted in Fig. 4 by the radial profiles $u(r)/u_o$ was a function of r/r_o, where in this figure $r_o \equiv r_{so}$. The profiles are shown for $\gamma_c = 5/3$ and several values of the constant k. For $k = 0$ the cosmic rays are effectively equivalent to a thermal gas and the profile exhibits a shock at $r = r_o$. For finite k the profiles exhibit a shockless transition from $u = u_o$ to nearly incompressible flow ($u \propto r^{-2}$) at large r. This model, of course, has only a cosmic ray pressure at large distance. It cannot simply be matched to an interstellar pressure which has cosmic ray and gas components, because the resulting cosmic ray gradient could not be maintained. Thus, it would seem that the wind gas pressure and the interstellar gas pressure would have to be carefully chosen to yield a shockless transition. This careful balance certainly does not occur in the heliosphere, because a gas subshock is required to accelerate the interstellar pickup ions into the anomalous component.

VI. COSMIC RAYS: THEORETICAL EXTENSIONS

The work of Ko and colleagues (Ko and Webb 1987,1988; Ko et al. 1988) was generalized by Ko (1991) to include the Alfvén waves responsible for scattering the cosmic rays. Ko generalized Eqs. (8) through (11) to

$$r^{-2} d/dr(r^2 \rho u) = 0 \tag{44}$$

$$\rho u \, du/dr = -\rho G M r^{-2} - d/dr(P_g + P_c + P_w) \tag{45}$$

$$r^{-2} d/dr[r^2 u(1/2\rho u^2 - \rho G M r^{-1} + \gamma_g(\gamma_g - 1)^{-1} P_g)] + u \, d/dr(P_c + P_w) = 0 \tag{46}$$

$$(u + V_A) dP_c/dr + \gamma_c P_c r^{-2} d/dr[r^2(u + V_A)] - r^{-2} d/dr(r^2 \kappa \, dP_c/dr) = 0 \tag{47}$$

$$r^{-2} d/dr[r^2(3u + 2V_A)P_w] = u \, dP_w/dr - V_A \, dP_c/dr \tag{48}$$

where the gas energy flux conservation Eq. (46) replaces the gas equation of state Eq. (10). The wave pressure P_w is given by $\langle |\delta \mathbf{B}|^2 \rangle (8\pi)^{-1}$, where $\delta \mathbf{B}$ is the magnetic field fluctuation. This model assumes a radial "monopole" ambient magnetic field $\mathbf{B}_o$ ($B_o \propto r^{-2}$) which does not exert a Lorentz force on the gas. The Alfvén speed is defined by $V_A^2 = B_o^2 (4\pi\rho)^{-1}$. The Alfvén waves all propagate either outward from the star ($V_A > 0$) or inward ($V_A < 0$). The wave pressure gradient provides an additional force in the gas momentum

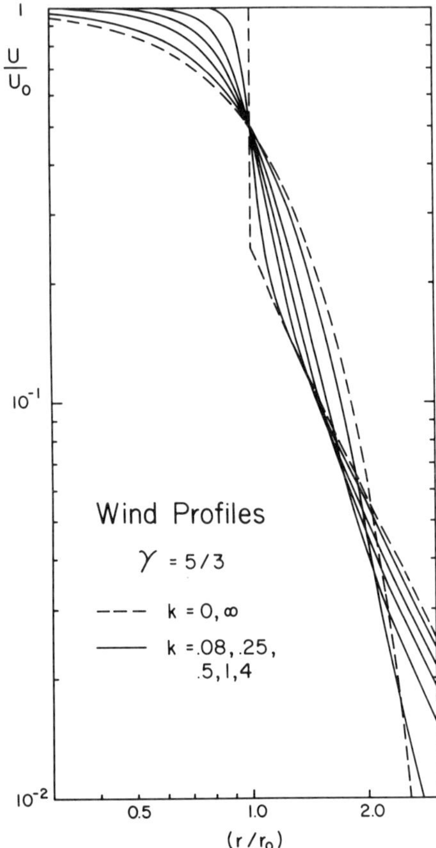

Figure 4.   Wind speed profiles $u(r)/u_0$ of the system of Eqs. (41) through (43) for $\gamma_c = 5/3$ and various values of $\kappa (\propto r)$ as a function of r/r_{so}. Note that the transition at $r \cong r_{so}$ is shockless unless $\kappa \to 0$, in which limit the cosmic rays are equivalent to a thermal gas (figure from Lee and Axford 1988).

Eq. (45), and an additional energy transfer rate between gas and waves in Eq. (46). In the cosmic ray equation of state (47), the gas flow speed u is replaced by $u + V_A$ because the cosmic rays are scattered elastically in the wave frame. Equation (48) is an equation of state for the waves: the divergence of the wave energy flux $(3u + 2V_A)P_w$, is determined by the energy transfer rate to the gas $-u\,dP_w/dr$, and the rate at which work is done on the waves by the cosmic rays $-V_A\,dP_c/dr$, which includes the Alfvén streaming instability (Lerche 1967). The wave energy flux $(3u + 2V_A)P_w$, consists of advection of the wave kinetic energy density $u(1/2\rho|\delta\mathbf{u}|^2)$ $(= uP_w$ for Alfvén waves), and the wave Poynting flux $2(u + V_A)P_w$. The system of Eqs. (44) through (48) has the "wave-action" integral

$$(\gamma_c - 1)^{-1} P_c + 8\pi r^2 F^{-1} P_w (u + V_A)^2 V_A^{-1} = \text{const} \qquad (49)$$

which facilitates the solution.

Ko (1991) uses the numerical iterative algorithm of Ko et al. (1988) to solve Eqs. (44) through (48) under the assumption that all waves propagate inward ($V_A < 0$) and κ is constant upstream and downstream of the termination shock. The waves reduce the advection of cosmic rays out of the inner heliosphere, which leads to a higher cosmic ray pressure there. Also they are amplified to substantial levels at the shock (see, e.g., McKenzie and Westphal 1969), which leads downstream of the shock to reduced diffusion for a radial field, for which $\kappa \propto P_w^{-1}$ (Jokipii 1971). However, it should be noted that a radial field implies no stellar rotation. Stellar rotation produces an azimuthal magnetic field B_ϕ; the Lorentz force must then be included, the Alfvén speed in the above equations must be reduced to its radial component ($V_{Ar}^2 = B_r^2/4\pi\rho$), and cosmic ray diffusion may be dominated by diffusion perpendicular to the magnetic field which satisfies $\kappa \propto P_w$. Also, the coupling between forward and backward propagating waves due to turbulent processes is generally substantial, and reduces the effect of the waves. Nevertheless, the work of Ko (1991) illustrates the fundamental role that waves play in coupling the cosmic rays to a stellar wind.

The remarks above illustrate the importance of the ambient magnetic field $\mathbf{B}_0$ in determining the wind flow, wave propagation, and the cosmic ray diffusion tensor. Zirakashvili et al. (1991) include the pressure of the ambient magnetic field in the gas momentum Eq. (45), but neglect magnetic tension, terms of order V_A/u, and deviations from a spherical flow geometry. The field is determined by being frozen into the gas. They determine the radial spatial diffusion coefficient, including diffusive transport both parallel and perpendicular to the field, by solving wave kinetic Eq. (48). Finally, reverting to a planar geometry in the outer heliosphere, they argue that a shockless transition to a state dominated by cosmic ray pressure may occur. However, as we discussed in Sec. V, we suspect that a correct matching to conditions in interstellar space will generally forbid the shockless transition.

Recent work by Le Roux and Ptuskin (1995) revisits the calculation of Lee and Axford (1988), but explicitly includes the dependence of the diffusion coefficient on particle momentum. They first solve Eq. (4) for the cosmic ray omnidirectional distribution function in a given flow $u(r)$. They then compute the cosmic ray pressure via Eq. (5) and then recompute the flow $u(r)$ using Eq. (9). By iterating this procedure the final solution is obtained. It is clear that such a procedure is essential for a quantitatively correct solution because κ depends sensitively on P. In the calculation presented in Sec. IV, for example, the diffusion coefficients appropriate to galactic cosmic rays and the anomalous component in the heliosphere are quite different. If these two species were lumped together and described by a single effective diffusion coefficient, much of the structure exhibited by the solutions derived in that section would be lost.

An important extension of the theory has been made by Fahr et al. (1992) and Fichtner et al. (1993), who considered the nonspherically symmetric stationary system based on Eqs. (1), (2), (3), and (7) ($\bar{\mathbf{u}}_D = 0$). They computed the asymmetric distribution of anomalous cosmic rays in the heliosphere arising from the asymmetric distribution of their seed ions, the interstellar pickup ions, and the expected asymmetry of the solar wind termination shock due to the flow of the interstellar gas relative to the Sun. They then calculated the dynamical influence of this distribution of the anomalous component on the solar wind flow. However, a major deficiency in this model is the assumption of an isotropic spatial diffusion tensor: $\kappa_{ij} = \kappa\delta_{ij}$. In fact, the ratio of the tensor components parallel and perpendicular to the field may be as large as $\kappa_{\parallel}/\kappa_{\perp} \sim B_o{}^4|\delta\mathbf{B}|^{-4}$, where $\delta\mathbf{B}$ refers to field fluctuations cyclotron resonant with the energetic particles. Particularly for the lower-energy anomalous component, $|\delta\mathbf{B}| \ll B_o$, so that particles diffuse much more readily along $\mathbf{B}_o$ than across it. Because the magnetic field is nearly azimuthal in the outer heliosphere, they chose $\kappa \sim \kappa_{\perp}$ to account for the radial gradients inferred for the anomalous component. Their assumption of isotropy then completely underestimates diffusive transport along $\mathbf{B}$ which should eliminate the azimuthal dependence of the anomalous component intensity they derive.

A final question arises as to whether the solution of the system of Eqs. (14) through (17), for example, for the stationary cosmic ray modified stellar wind is stable. As might be expected from the instability of Alfvén waves discussed above, the answer is "no." Drury (1984) showed that a sound wave propagating parallel to the cosmic ray pressure gradient is unstable if $|\nabla P_c|/P_c > \gamma_c\kappa^{-1}(\gamma_g P_g/\rho)^{1/2}$, assuming that κ does not depend on ρ. The threshold for the instability is determined by damping due to the friction between the sound waves and the cosmic rays. The instability is due to the fact that the cosmic ray pressure gradient contributes to the acceleration of a fluid element by an amount which is proportional to the fluctuation velocity of the element, so that the wave is amplified. Numerical calculations by Drury and Falle (1986) showed that the instability can grow to significant levels upstream of strong fast shocks and contribute to small-scale structure and eventually heating of the thermal gas. Subsequently, Zank et al. (1990) included the magnetic field $\mathbf{B}$ and the drift velocity $\bar{\mathbf{u}}_d$ and investigated the general stability of waves in the presence of a cosmic ray pressure gradient. They found a variety of instabilities due to cosmic ray "squeezing" of a fluid element, due to "buoyancy" set up by the fluid stratification in the cosmic ray gradient, and due to cosmic ray drift. These fluid instabilities may even have to be augmented or modified by kinetic instabilities due to the cosmic ray gradient but depending on the structure of the cosmic ray distribution function.

VII. COSMIC RAYS: RELATED WORK

It should be noted that the system of equations described here to treat the coupling of a stellar wind and cosmic rays also has other applications. Indeed, the cosmic ray fluid Eq. (7) was derived by Axford et al. (1977) and Drury and Völk (1981) to use in combination with the gas fluid equations to investigate the general structure of planar cosmic ray modified shocks. Specific modified shock profiles were derived by Axford et al. (1982). Alfvén waves were included by McKenzie and Völk (1982). Duffy (1992) computed numerical solutions of shock structure for the case $\kappa = \kappa(p)$ in which the transport Eq. (4) for the omnidirectional distribution function must be solved in conjunction with the gas fluid equations. The cosmic ray modified wind with termination shock is essentially the spherical analog of the planar cosmic ray modified shock. Donohue and Zank (1993) and Donohue et al. (1994) have investigated the time-dependent behavior of a cosmic ray modified shock in response to advected structures or other shocks, with specific application to a stellar wind termination shock. Chalov and Fahr (1994) have investigated the stationary structure of the termination shock modified by the pressure of the anomalous cosmic ray component.

The same basic system of equations also describes a cosmic ray-driven galactic wind, but with possibly generalized geometry and with the cosmic ray flux specified in the Galaxy as well as in intergalactic space. Early work on cosmic ray-driven galactic winds was done by Johnson and Axford (1971) and Ipavich (1975). More recent work by Zank (1989), Breitschwerdt et al. (1991) and Fichtner et al. (1991) is reviewed in the chapter by Morfill and Jokipii.

An application similar to that of the cosmic ray-driven galactic wind was proposed by Jokipii and Morfill (1993). They suggested that solar energetic particles may well be continuously present in the solar corona as a result of small-scale flare activity. These particles diffuse away from the Sun resulting in a negative radial pressure gradient which provides additional acceleration of the solar wind. Because they may have a large diffusion coefficient they can accelerate the wind beyond the critical point, which yields faster and less dense wind as required to reproduce the observed characteristics of fast wind. Jokipii and Morfill (1993) argue that the required particle intensities are not inconsistent with observed upper limits on X-ray and γ-ray emission from the corona during quiet times.

VIII. INTERSTELLAR GAS AND PICKUP IONS

As described in Sec. I, the ionization of interstellar gas that penetrates the stellar wind may have a significant dynamical influence on the wind. In the description which follows of the mass-loading, momentum-loading and heating of the wind by interstellar pickup ions, we neglect the dynamical influence of the energetic particles described previously. To the extent that the dynamical effects of energetic particles and pickup ions are both small and accessible with perturbation analysis, their influences are additive to order ϵ. We also focus on conditions in the heliosphere, recognizing that the assumptions we make may not be valid at all stars with winds. Neutral helium penetrates closest to the Sun with a characteristic ionization distance of ~ 0.5 AU. Because at this heliocentric radial distance the influence of solar gravity is negligible, we use the "free streaming" wind Eqs. (14) through (16) as a starting point. We furthermore neglect the bulk velocity $\mathbf{V_o}$ of passage of the neutral gas into the heliosphere as well as its thermal velocity compared with the wind velocity $u\hat{e}_r$, while keeping in mind that the distribution of the neutral gas depends crucially on $\mathbf{V_o}$.

We consider only hydrogen and helium; the other species have a much smaller influence. Helium is ionized almost exclusively by photoionization at a rate $\nu_{He}{}^o (r_o/r)^2$, where $\nu_{He}{}^o$ is the ionization rate at reference distance r_o. The rate $\nu_{He}{}^o$ depends on the flux of ionizing solar ultraviolet radiation. Although $\nu_{He}{}^o$ is highly variable in both time and heliographic latitude and longitude, we take it to be constant in this stationary, spherically symmetric treatment. Similarly hydrogen may be photoionized at a rate $\nu_H{}^o (r_o/r)^2$. However, hydrogen may also be ionized by charge exchange with a solar wind proton at a rate $\sigma u \rho m_p{}^{-1}$, where m_p is the proton mass and we neglect the contribution of solar wind helium to ρ. The quantity σ is the charge exchange cross section, which is approximately independent of u for characteristic wind speeds. Finally, we note that the neutral gas densities depend on both r and heliographic latitude and longitude. However, beyond $\sim$10 AU, where the dynamical influence of pickup ions on the wind is largest, the neutral gas densities are approximately equal to their constant "interstellar" values $n_{H,\infty}$ and $n_{He,\infty}$. Actually these values may be somewhat modified from their interstellar values by ionization in the heliosheath or just beyond the heliopause where there may be an enhanced density of ionized gas as it flows around the heliosphere. In what follows we neglect these effects and simply take the neutral gas densities to be $n_{H,\infty}$ and $n_{He,\infty}$.

In the absence of the neutral gas and setting $P_c = 0$, we write Eqs. (14) through (16) in the form

$$r^{-2}\mathrm{d}/\mathrm{d}r\,(r^2 \rho u) = 0 \tag{50}$$

$$r^{-2}\mathrm{d}/\mathrm{d}r\,(r^2 \rho u^2) = -\mathrm{d}P_g/\mathrm{d}r \tag{51}$$

$$r^{-2}\mathrm{d}/\mathrm{d}r\,[r^2 u(1/2\rho u^2 + \gamma_g(\gamma_g - 1)^{-1} P_g)] = 0 \tag{52}$$

which gives explicit evolutionary equations for mass flux, momentum flux, and energy flux. Only the photoionization of neutral gas contributes to the mass flux, because charge exchange leaves the number of protons unchanged. On the other hand, because the photoionized pickup ions are initially approximately at rest, they do not contribute to the momentum and energy flux. In the charge-exchange process, however, the newly created neutral at the solar wind speed u is a loss of momentum flux and energy flux to the wind. Including these sources and sinks on the right-hand sides of Eqs. (50) through (52) we obtain

$$r^{-2}\mathrm{d}/\mathrm{d}r\,(r^2 \rho u) = m_p(r_o/r)^2(\nu_H{}^o n_{H,\infty} + 4\nu_{He}{}^o n_{He,\infty}) \tag{53}$$

$$r^{-2}\mathrm{d}/\mathrm{d}r\,(r^2 \rho u^2) = -\mathrm{d}P_g/\mathrm{d}r - \sigma n_{H,\infty}\rho u^2 \tag{54}$$

$$r^{-2}d/dr[r^2 u(1/2\rho u^2 + \gamma_g(\gamma_g - 1)^{-1} P_g)] = -\sigma n_{H,\infty} 1/2\rho u^3. \qquad (55)$$

A more general form of these equations was derived by Holzer (1972) including the bulk velocity $\mathbf{V}_o$ and thermal velocity of the neutrals, and the ambient magnetic field, but assuming a spherically symmetric flow.

We solve Eqs. (53) through (55) in the solar wind by employing a perturbation expansion in the same spirit as that employed in Sec. IV. The gas pressure P_g and the terms involving ion pickup are taken to be first order in $\epsilon \sim P_g(\rho u^2)^{-1}$. At zeroth order in ϵ we obtain the standard solution (18) and (19) for the "'free streaming" solar wind

$$u = u_o \qquad (56)$$

$$\rho = \rho_o(r_o/r)^2. \qquad (57)$$

At first order we obtain for P_g

$$P_g = 1/2(\gamma_g - 1)(2\gamma_g - 1)^{-1} r_o^2 \rho_o u_o r^{-1}\beta + Cr^{-2\gamma_g} \qquad (58)$$

$$\beta \equiv \sigma n_{H,\infty} u_o + m_p \rho_o^{-1}(\nu_H{}^o n_{H,\infty} + 4\nu_{He}{}^o n_{He,\infty}) \qquad (59)$$

where C is a constant. The term $Cr^{-2\gamma_g}$ is the pressure of the solar wind gas, which decreases rapidly with increasing r in the outer heliosphere. We neglect its contribution in what follows. The first term on the right-hand side of Eq. (58) is the pressure of the pickup hydrogen and helium. Of course this term is not valid at small r when $n(r)$ differs from n_∞, within ~ 1 AU for helium and within ~ 5 AU for hydrogen. Substituting Eq. (58) into Eq. (54) then yields the deceleration of the solar wind as

$$du/dr = [1/2(\gamma_g - 1)(2\gamma_g - 1)^{-1} - 1]\beta. \qquad (60)$$

Actually the first term in square brackets describes the acceleration due to the pickup ion pressure gradient, whereas the -1 describes the deceleration due to mass loading and momentum loading. Integrating Eq. (60) we obtain correct to first order in ϵ

$$u = u_o - 1/2\beta(3\gamma_g - 1)(2\gamma_g - 1)^{-1} r \qquad (61)$$

indicating a net deceleration of the wind. The sound speed C_s of the wind is given by

$$C_s{}^2 = \gamma_g P_g/\rho = 1/2u_o\gamma_g(\gamma_g - 1)(2\gamma_g - 1)^{-1}\beta r \qquad (62)$$

which increases with heliocentric radial distance r as a result of "heating" associated with ion pickup. The heating is due to the accumulation of pickup ions rather than an actual transfer of thermal energy to the solar wind gas.

It is interesting to relate P_g in Eq. (58) to the number density $N(r)$ of pickup ions in the wind. Considering only hydrogen we have

$$N(r) = \int_0^r dr' u_o^{-1}(\nu_H{}^\circ + \sigma u_o \rho_o m_p{}^{-1})(r_o/r')^2 n_H(r')(r'/r)^2 \qquad (63)$$

which in the outer heliosphere is approximately

$$N(r) = u_o^{-1} r_o{}^2 r^{-1} n_{H,\infty}(\nu_H{}^\circ + \sigma u_o \rho_o m_p{}^{-1}). \qquad (64)$$

From Eq. (58) for hydrogen we obtain the energy density of pickup hydrogen as

$$E = (\gamma - 1)^{-1} P = 1/2(2\gamma - 1)^{-1} r_o{}^2 u_o r^{-1} m_p n_{H,\infty}(\nu_H{}^\circ + \sigma u_o \rho_o m_p{}^{-1}) \qquad (65)$$

where we now omit the subscript g on E, P and γ. Dividing Eq. (65) by Eq. (64) we obtain the energy per pickup ion as

$$EN^{-1} = 1/2 m_p u_o{}^2 (2\gamma - 1)^{-1}. \qquad (66)$$

The factor $(2\gamma - 1)^{-1}$ equals $3/7$ for $\gamma = 5/3$. This result also follows from kinetic theory which describes ion pickup at speed u_o in the solar wind frame and the subsequent adiabatic deceleration of the ions in the divergent solar wind (Vasyliunas and Siscoe 1976).

The dynamical influence of the interstellar gas on the solar wind is significant. The effects could be comparable or larger for winds at other stars. Appropriate average values for the relevant parameters are $\sigma = 2 \times 10^{-15}$ cm^2 (Axford 1972), $\nu_H{}^\circ = 0.9 \times 10^{-7}$ s^{-1} and $\nu_{He}{}^\circ = 0.8 \times 10^{-7}$ s^{-1} ($r_o = 1$ AU) for low to moderate solar activity (Möbius 1993), $n_{H,\infty} = 0.06$ cm^{-3} and $n_{He,\infty} = 0.006$ cm^{-3} (see, e.g., Chassefière et al. 1986), $\rho_o m_p{}^{-1} = 8$ cm^{-3} (see, e.g., Lazarus and McNutt 1990), and $u_o = 4 \times 10^7$ cm s^{-1}. With $u = u_o - \delta u$, we find at $r = 50$ AU, $\delta u = 37$ km s^{-1}, and $C_s = 64$ km s^{-1}. The deceleration of the solar wind is admittedly small compared with typical variations of 50 km s^{-1} for the average solar wind throughout the solar activity cycle (Lazarus and McNutt 1990). Nevertheless, a careful comparison of observations by IMP8 at 1 AU and Voyager 2 at $\sim$40 AU has revealed a decrease in the solar wind speed at Voyager 2 by about 30 km s^{-1}, of the same order as that prediction (Richardson et al. 1995).

However, the increase in the sound speed is substantial. Neglecting the pressure of the solar wind gas in the outer heliosphere, and noting that a typical Alfvén speed there is 25 km s^{-1}, the pickup ions reduce the fast magnetosonic Mach number of the wind $[\equiv u_o(C_s{}^2 + V_A{}^2)^{-1/2}]$ near 50 AU from 16 to 6. This reduction of the Mach number weakens the wind termination shock (which we took to be strong in Sec. IV) and should have a substantial effect on the shape of planetary bow shocks and the propagation of weak interplanetary traveling shocks. Burlaga et al. (1994) have recently confirmed the expected large pressure of pickup hydrogen at Voyager 2 near ~35 AU. Noting that pickup hydrogen should be approximately proportional to solar wind proton density due to its primary ionization by charge exchange, they infer the pickup ion pressure required for pressure balance in apparently stable structures and find that these pressures are in reasonable agreement with the prediction.

The behavior and composition of pickup ions in the solar wind is a major focus of the SWICS investigation on Ulysses; the results to date are described in the chapter by Gloeckler and Geiss.

IX. INTERSTELLAR GAS: TERMINATION SHOCK LOCATION

The mass loading and momentum loading of the solar wind by the interstellar gas will also modify the location of the termination shock. We now provide an approximate calculation of that modification in the spirit of the perturbation expansion employed in Sec. IV. The zeroth order solution and shock location are derived and presented in Eq. (18) through (24). At first order in ϵ the gas pressure and wind speed upstream of the shock are given by Eqs. (58) (with $C = 0$) and (61) with $r = r_s$. At first order the gas density upstream of the shock follows from Eq. (53) as

$$\rho_1 = \rho_0 r_0{}^2 r_s{}^{-2}(1+\delta u_1 u_0{}^{-1})+m_p r_0{}^2 (r_s u_0)^{-1}(\nu_H{}^\circ n_{H,\infty}+4\nu_{He}{}^\circ n_{He,\infty}) \quad (67)$$

where $\delta u_1 = u_0 - u(r_s)$. The hydrodynamic Rankine-Hugoniot jump conditions then give

$$u_2 = (\gamma - 1)(\gamma + 1)^{-1}u_1[1 + 2\gamma(\gamma - 1)^{-1}P_1 u_0{}^{-2}\rho_1{}^{-1}] \quad (68)$$

$$\rho_2 = (\gamma + 1)(\gamma - 1)^{-1}\rho_1[1 - 2\gamma(\gamma - 1)^{-1}P_1 u_0{}^{-2}\rho_1{}^{-1}] \quad (69)$$

$$P_2 = 2(\gamma + 1)^{-1}\rho_1 u_1{}^2 - (\gamma - 1)(\gamma + 1)^{-1}P_1 \quad (70)$$

where subscript 1(2) refers to quantities adjacent to the shock upstream (downstream).

The basic Eqs. (53) through (55) must be modified downstream of the shock in the subsonic flow. Because the density increases at the shock by a factor of somewhat less than 4 and remains close to that value, photoionization, which was smaller than charge-exchange ionization for the dominant species hydrogen anyway, is completely negligible. Furthermore, the charge-exchange ionization rate depends on the relative speed of solar wind protons and neutral gas, which is dominated by the proton thermal speed (rather than u as was the case in the wind). Similarly, the energy loss rate is dominated by the thermal energy. Downstream of the shock Eqs. (53) through (55) then become

$$r^2 \rho u = r_s{}^2 \rho_2 u_2 \quad (71)$$

$$\rho u \, du/dr = -dP/dr - \sigma n_{H,\infty}\alpha(P/\rho)^{1/2}\rho u \quad (72)$$

880   M. A. LEE

$$\rho u \, d/dr[1/2u^2 + \gamma(\gamma-1)^{-1}P\rho^{-1}] = -\sigma n_{\mathrm{H},\infty}\alpha(P/\rho)^{1/2}(\gamma-1)^{-1}P \quad (73)$$

where $\alpha^2 = 128(9\pi)^{-1}$ according to Holzer (1972).

At this point the spherically symmetric model leads to untenable results. If we divide Eq. (73) by ρu and integrate from r_s to $r > r_s$ we obtain

$$1/2u^2 + \gamma(\gamma-1)^{-1}P\rho^{-1} = 1/2u_1^2 + \gamma(\gamma-1)^{-1}P_1\rho_1^{-1} - \chi_1(r) \quad (74)$$

$$\chi_1(r) \equiv \sigma n_{\mathrm{H},\infty}\alpha(\gamma-1)^{-1}(r_s^2\rho_2 u_2)^{-1}\int_{r_s}^{r} dr'(r')^2 P(P/\rho)^{1/2}. \quad (75)$$

Because P and ρ are nearly constant downstream of the shock, χ_1 increases without limit as r increases. This occurs in the spherically symmetric model because there is unlimited time available for charge exchange particularly because $u \propto r^{-2}$, which yields the factor $(r')^2$ in the integral of χ_1. In fact, the interstellar gas and/or field sweep the shocked wind away and create a heliosheath of finite extent r_b. Thus, at $r = r_b$ we set $u = 0$, $P = P_\infty$, and $\rho = \rho_\infty$ to obtain

$$\gamma(\gamma-1)^{-1}P_\infty\rho_\infty^{-1} = 1/2u_1^2 + \gamma(\gamma-1)^{-1}P_1\rho_1^{-1} - \chi_1(r_b). \quad (76)$$

Combining Eqs. (72) and (73) to find ρ as a function of P we obtain

$$\rho_\infty = \rho_2(P_\infty/P_2)^{1/\gamma}[1 - (\gamma-1)\gamma^{-1}\chi_2(r_b)] \quad (77)$$

$$\chi_2(r_b) \equiv \sigma n_{\mathrm{H},\infty}\alpha P_2^{-1}\rho_2^{\gamma}\int_{r_s}^{r_b} dr(P/\rho)^{1/2}\rho^{1-\gamma}[u - P(\gamma-1)^{-1}(\rho u)^{-1}] \quad (78)$$

where we take $\chi_2(r_b)$ [and also $\chi_1(r_b)$] to be first order in ϵ.

Now the procedure is analogous to that of Sec. IV. The derived expressions for ρ_1, u_1, P_1, ρ_2, u_2 and P_2, along with Eq. (77), are substituted into Eq. (76). The only remaining unknown variable is r_s. To zeroth order we obtain expression (24): $r_s = r_{so}$.

At first order we require the expression $\chi_1 + P_\infty\rho_2^{-1}(P_\infty/P_2)^{-1/\gamma}\chi_2$. In this expression the second term in square brackets in χ_2 nearly cancels χ_1. Thus the dominant term arises from the first term (u) in the square brackets. Setting P and ρ in the integral of the resulting expression by the zeroth-order parts of P_2 and ρ_2 (as P and ρ are nearly constant), taking in this term $P_\infty \cong P_2$, and at this point letting $r_b \to \infty$, we obtain

$$\chi_1 + P_\infty\rho_2^{-1}(P_\infty/P_2)^{-1/\gamma}\chi_2$$

$$\cong \sigma n_{\mathrm{H},\infty}\alpha u_o^2 r_{so}(\gamma-1)(\gamma+1)^{-2}[2(\gamma-1)]^{1/2}. \quad (79)$$

With Eq. (79) we then obtain from Eq. (76) correct to first order

$$r_s = r_{so} + r_{so}\{-8^{-1}(5\gamma - 1)(2\gamma - 1)^{-1}\sigma n_{H,\infty}r_{so}+$$

$$(3/8)(\gamma - 1)(2\gamma - 1)^{-1}r_{so}m_p(u_o\rho_o)^{-1}(\nu_H{}^o n_{H,\infty} + 4\nu_{He}{}^o n_{He,\infty})$$

$$-\gamma(\gamma + 1)^{-2}[2(\gamma - 1)]^{1/2}\alpha\sigma n_{H,\infty}r_{so}\}. \tag{80}$$

In order, the terms in curly brackets correspond to charge exchange in the wind, photoionization in the wind, and charge exchange in the heliosheath. Charge exchange in the wind clearly leads to a loss of wind ram pressure which causes the shock to move inwards. At first sight perhaps paradoxically, photoionization moves the shock outwards. The mass loading decreases the wind speed but increases the mass density; the latter effect dominates and increases the wind ram pressure. The fact that charge exchange in the sheath causes the shock to move in appears to result from a subtle combination of effects. The loss of energy flux in the sheath apparent in Eqs. (74) and (75) implies that the flux is larger at the shock under the action of charge exchange. In order to maintain continuity of energy flux at the shock, the shock moves inwards. However, the charge exchange "softens" the gas equation of state so that ρ_∞ is larger at large distance and the resulting energy flux is smaller, which requires the shock to move outwards. The first of these effects dominates the second.

To illustrate the magnitude of these effects, we use the characteristic parameter values quoted in the last section and take r_{so} = 80 AU. Equation (80) then becomes

$$r_s = r_{so}[1 - 0.056 + 0.003 - 0.082] = 69 \text{ AU} \tag{81}$$

with the three corrections corresponding in order to the three terms in Eq. (80). Clearly photoionization is inconsequential. However, charge exchange contributes more than a 10% reduction in the predicted location of the shock.

Two recent numerical calculations have addressed the interaction of the interstellar hydrogen gas with the solar wind and its effect on the location of the termination shock. Whang et al. (1995) find that the interaction causes the shock to move out by $\sim$30%, whereas Pauls et al. (1995) find that it causes the shock to move in by $\sim$30%, qualitatively in agreement with our results. Clearly this discrepancy needs to be resolved.

X. CONCLUSIONS

We have presented the basic equations which describe the interaction between a stellar wind and galactic cosmic rays, energetic particles accelerated at the stellar wind termination shock, and neutral interstellar gas which penetrates the stellar cavity. In the basic presentation we have restricted ourselves to spherical symmetry, neglected the stellar magnetic field, and treated the energetic particles as a fluid with a single effective diffusion coefficient. These restrictions do not obscure the nature of the basic interaction. The energetic particles decelerate the wind, weaken the wind termination shock, and shift its location. The wind in turn "modulates" the galactic cosmic rays. The ionized interstellar gas decelerates and heats the wind and also weakens the termination shock and shifts its location.

We have first briefly reviewed work on this subject by others. We have then presented two new calculations of the dynamical influence of energetic particles and interstellar gas on a wind assuming, as is the case at the Sun, that these effects are small and only become important for the "free streaming" wind beyond the reach of stellar gravity. We developed expansions in the energetic particle pressure and the loading by the ionized interstellar neutral gas in order to generate analytical expressions for wind deceleration, wind heating, and the modification of the termination shock. We also have reviewed more recent work in this area including the explicit inclusion of the waves (and their instability) responsible for the coupling of the energetic particles to the wind, initial attempts to include the effects of the ambient magnetic field, and the suggestion that energetic solar cosmic rays may contribute to the acceleration of the solar wind. We have pointed out that essentially the same theory or set of equations describes the structure of cosmic ray modified shocks and the acceleration of cosmic ray-driven galactic winds.

We have evaluated these dynamical effects on the solar wind and in the heliosphere by choosing the parameters appropriately. They are by and large not large effects, generally resulting in changes of several percent, for example, in the decrease of the wind speed or the location of the solar wind termination shock. One large effect is the increase in the internal pressure of the solar wind due to the pickup of ionized interstellar hydrogen. Beyond a heliocentric radial distance of ~ 10 to 20 AU pickup ions dominate the pressure and substantially reduce the Mach number of the solar wind.

However, the dynamical effects of energetic particles and interstellar gas on the winds of other stars may be much larger. This requires in the case of cosmic rays that their interstellar pressure be comparable with that of the interstellar ionized gas and magnetic field, *and* that much of the modulation occurs in the outer wind. This would appear to be possible. In the case of the interstellar gas this requires simply that $\beta r_{so}/u_o$ be comparable with unity, which requires, if charge exchange dominates, that the charge exchange lifetime of a wind proton is comparable with its lifetime in the wind. However, V_o must also satisfy $V_o n_{H,\infty} \gg u_o \rho_1/m_p$ in order that the neutrals indeed penetrate deeply into the wind.

Much work remains to be done in this area of stellar wind research. Cosmic ray drift, the stellar magnetic field, Alfvén wave intensities in realistic field configurations, and a diffusion coefficient appropriate to the field and wave intensities, all need to be included. It must also be recognized that the anomalous cosmic ray component and the interstellar gas are connected; some fraction of the interstellar pickup ions, which depends on the termination shock structure, is accelerated at the shock to form the anomalous component. Finally, even in this hydrodynamical spherically symmetric model, a clear prediction of the wind profile including the termination subshock would be instructive for a variety of diffusion coefficients, neutral interstellar gas densities, and possible ratios of cosmic ray to interstellar ionized gas pressures.

Acknowledgments. The author is indebted to G. M. Webb for bringing to his attention recent work in this area and for several discussions concerning the existence and location of the termination shock. He also wishes to express his gratitude to R. Cowsik for a very enjoyable visit to the Indian Institute of Astrophysics in Bangalore, where a portion of this work was accomplished. He appreciates the careful preparation of the typescript by K. Theall. This work was supported, in part, by grants from the NASA Space Physics Theory Program and NASA Supporting Research and Technology, and a grant from the National Science Foundation.

REFERENCES

Axford, W. I. 1965. The modulation of galactic cosmic rays in the interplanetary medium. *Planet. Space Sci.* 13:115–130.

Axford, W. I. 1972, The interaction of the solar wind with the interstellar medium. In *Solar Wind*, eds. C. P. Sonett, P. J. Coleman, Jr., and J. M. Wilcox, NASA SP-308, pp. 609–660.

Axford, W. I., and Newman, R. C. 1965. The effect of cosmic ray friction on the solar wind. *Proc. Intl. Conf. Cosmic Rays IX* 1:173–175.

Axford, W. I., Leer, E., and Skadron, G. 1977. The acceleration of cosmic rays by shock waves. *Proc. Intl. Conf. Cosmic Rays XV* 11:132–137.

Axford, W. I., Leer, E., and McKenzie, J. F. 1982. The structure of cosmic ray shocks. *Astron. Astrophys.* 111:317–325.

Babayan, V. K., and Dorman, L. I. 1977*a*. The nonlinear theory of cosmic ray modulation by solar wind. I. Spherically-symmetrical model. *Proc. Intl. Conf. Cosmic Rays XV* 3:107–112.

Babayan, V. K., and Dorman, L. I. 1977*b*. The nonlinear theory of cosmic ray modulation by solar wind. II. The focusing effect in the asymmetric model. *Proc. Intl. Conf. Cosmic Rays XV* 3:113–118.

Babayan, V. K., and Dorman, L. I. 1990. Nonlinear effects of cosmic ray interaction with solar wind in the outer heliosphere. In *Physics of the Outer Heliosphere*, eds. S. Grzedzielski and D. E. Page (New York: Pergamon Press), pp. 191–193.

Breitschwerdt, D., McKenzie, J. F., and Völk, H. J. 1991. Galactic winds I. Cosmic ray and wave-driven winds from the Galaxy. *Astron. Astrophys.* 245:79–98.

Burlaga, L. F., Ness, N. F., Belcher, J. W., Szabo, A., Isenberg, P. A., and Lee, M. A.

1994. Pickup protons and pressure balanced structures: Voyager 2 observations in MIRs near 35 AU. *J. Geophys. Res.* 99:21511–21524.

Chalov, S. V., and Fahr, H.-J. 1994. A two-fluid model of the solar wind termination shock modified by shock-generated cosmic rays including energy losses. *Astron. Astrophys.* 288:973–980.

Chassefière, E., Bertaux, J. L., Lallement, R., and Kurt, V. G. 1986. Atomic hydrogen and helium densities of the interstellar medium measured in the vicinity of the Sun. *Astron. Astrophys.* 160:229–242.

Dolginov, A. Z., and Toptygin, I. 1967. Multiple scattering of particles in a magnetic field with random inhomogeneities. *Soviet Phys.-JETP* 24:1195–1202.

Donohue, D. J., and Zank, G. P. 1993. Steady-state and dynamical structure of a cosmic-ray-modified termination shock. *J. Geophys. Res.* 98:19005–19025.

Donohue, D. J., Zank, G. P., and Webb, G. M. 1994. Time-dependent evolution of cosmic-ray-modified shock structure: Transition to steady state. *Astrophys. J.* 424:263–274.

Drury, L. O'C. 1984. Reaction effects in diffusive shock acceleration. *Adv. Space Res.* 4(2–3):185–191.

Drury, L. O'C., and Falle, S. A. E. G. 1986. On the stability of shocks modified by particle acceleration. *Mon. Not. Roy. Astron. Soc.* 223:353–376.

Drury, L. O'C., and Völk, H. J. 1981. Hydromagnetic shock structure in the presence of cosmic rays. *Astrophys. J.* 248:344–351.

Duffy, P. 1992. The self-consistent acceleration of cosmic rays in modified shocks with Bohm-type diffusion. *Astron. Astrophys.* 262:281–294.

Fahr, H.-J., Fichtner, H., and Grzedzielski, S. 1992. The influence of the anomalous cosmic-ray component on the dynamics of the solar wind. *Solar Phys.* 137:355–383.

Fichtner, H., Fahr, H.-J., Neutsch, W., Schlickeiser, R., Crusius-Wätzel, A., and Lesch, H. 1991. Cosmic-ray-driven galactic wind. *Nuovo Cimento* 106:909–925.

Fichtner, H., Fahr, H.-J., and Sreenivasan, S. R. 1993. The influence of a non-spherical solar wind termination shock on the pressure distribution of the anomalous component of cosmic rays in the heliosphere. *Proc. Intl. Conf. Cosmic Rays XXIII* 3:423–426.

Fisk, L. A., Kozlovsky, B., and Ramaty, R. 1974. An interpretation of the observed oxygen and nitrogen enhancements in low-energy cosmic rays. *Astrophys. J. Lett.* 190:35–37.

Garcia-Munoz, M., Mason, G.M., and Simpson, J. A. 1973. A new test for solar modulation theory: The 1972 May-July low-energy galactic cosmic-ray proton and helium spectra. *Astrophys. J. Lett.* 182:81–84.

Gleeson, L. J., and Axford, W. I. 1967. Cosmic rays in the interplanetary medium. *Astrophys. J. Lett.* 149:115–118.

Holzer, T. E. 1972. Interaction of the solar wind with the neutral component of the interstellar gas. *J. Geophys. Res.* 77:5407–5431.

Holzer, T. E. 1989. Interaction between the solar wind and the interstellar medium. *Ann. Rev. Astron. Astrophys.* 27:199–234.

Hovestadt, D., Vollmer, O., Gloeckler, G., and Fan, C. Y. 1973. Differential energy spectra of low-energy (<8.5 MeV per nucleon) heavy cosmic rays during solar quiet times. *Phys. Rev. Lett.* 31:650–653.

Ipavich, F. M. 1975. Galactic winds driven by cosmic rays. *Astrophys. J.* 196:107–120.

Isenberg, P. A. 1986. Interaction of the solar wind with interstellar neutral hydrogen: Three-fluid model. *J. Geophys. Res.* 91:9965–9972.

Johnson, H. E., and Axford, W. I. 1971. Galactic winds. *Astrophys. J.* 165:381–390.

Jokipii, J. R. 1971. Propagation of cosmic rays in the solar wind. *Rev. Geophys. Space Phys.* 9:27–87.

Jokipii, J. R. 1986. Particle acceleration at a termination shock 1. Application to the solar wind and the anomalous component. *J. Geophys. Res.* 91:2929–2932.

Jokipii, J. R., and Levy, E. H. 1977. Effects of particle drifts on the solar modulation of galactic cosmic rays. *Astrophys. J. Lett.* 213:85–88.

Jokipii, J. R., and Morfill, G. M. 1993. Cosmic rays and the acceleration of the solar wind. *Proc. Intl. Conf. Cosmic Rays XXIII* 3:179–182.

Ko, C.-M. 1991. Cosmic-ray-modified stellar winds IV. The effect of self-generated Alfvén waves. *Astron. Astrophys.* 251:713–722.

Ko, C.-M., and Webb, G. M. 1987. Cosmic-ray-modified stellar winds I. Solution topologies and singularities. *Astrophys. J.* 323:657–671.

Ko, C.-M., and Webb, G. M. 1988. Cosmic-ray-modified stellar winds II. A perturbation approach. *Astrophys. J.* 325:296–313.

Ko, C.-M., Jokipii, J. R., and Webb, G. M. 1988. Cosmic-ray-modified stellar winds III. A numerical iterative approach. *Astrophys. J.* 326:761–768.

Lazarus, A. J., and McNutt, R. L., Jr. 1990. Plasma observations in the distant heliosphere: a view from Voyager. In *Physics of the Outer Heliosphere*, eds. S. Grzedzielski and D. E. Page (New York: Pergamon Press), pp. 229–234.

Lee, M. A. 1988. The solar wind termination shock and the heliosphere beyond. In *Solar Wind Six*, eds. V. J. Pizzo, T. E. Holzer and D. G. Sime, NCAR TN-306 (Boulder: National Center for Atmospheric Res.), pp. 635–650.

Lee, M. A., and Axford, W. I. 1988. Model structure of a cosmic-ray mediated stellar or solar wind. *Astron. Astrophys.* 194:297–303.

Lerche, I. 1967. Unstable magnetosonic waves in a relativistic plasma. *Astrophys. J.* 147:689–696.

Le Roux, J. A., and Potgieter, M. S. 1991. The simulation of Forbush decreases with time-dependent cosmic-ray modulation models of varying complexity. *Astron. Astrophys.* 243:531–545.

Le Roux, J. A., and Ptuskin, V. S. 1995. Galactic cosmic-ray mediation of a spherical solar wind flow 1. Steady state cold gas hydrodynamical approximation. *Astrophys. J.* 438:427–433.

McDonald, F. B., Teegarden, B. J., Trainor, J. H., and Webber, W. R. 1974. The anomalous abundance of cosmic-ray nitrogen and oxygen nuclei at low energies. *Astrophys. J. Lett.* 187:105–108.

McKenzie, J. F., and Westphal, K. O. 1969. Transmission of Alfvén waves through the Earth's bow shock. *Planet. Space Sci.* 17:1029–1037.

McKenzie, J. F., and Völk, H. J. 1982. Non-linear theory of cosmic ray shocks including self-generated Alfvén waves. *Astron. Astrophys.* 116:191–200.

Möbius, E. 1993. Gases of non-solar origin in the solar system. In *Landolt-Börnstein Tables of Physics*, ed. H. H. Voigt (Berlin: Springer-Verlag), pp. 186–191.

Parker, E. N. 1963. *Interplanetary Dynamical Processes* (New York: Wiley-Interscience), pp. 113–128.

Parker, E. N. 1965. The passage of energetic charged particles through interplanetary space. *Planet. Space Sci.* 13:9–49.

Pauls, H. L., Zank, G. P., and Williams, L. L. 1995. Interaction of the solar wind with the local interstellar medium. *J. Geophys. Res.* 100:21595–21604.

Pesses, M. E., Jokipii, J. R., and Eichler, D. 1981. Cosmic ray drift, shock wave acceleration, and the anomalous component of cosmic rays. *Astrophys. J. Lett.* 246:85–88.

Richardson, J. D., Paularena, K. I., Lazarus, A. J., and Belcher, J. W. 1995. Evidence for a solar wind slowdown in the outer heliosphere? *Geophys. Res. Lett.* 22:1469–1472.

Sousk, S. F., and Lenchek, A. M. 1969. The effect of galactic cosmic rays upon the dynamics of the solar wind. *Astrophys. J.* 158:781–795.

Suess, S. T. 1990. The heliopause. *Rev. Geophys.* 28:97–115.

Vasyliunas, V. M., and Siscoe, G. L. 1976. On the flux and the energy spectrum of interstellar ions in the solar system. *J. Geophys. Res.* 81:1247–1252.

Whang, Y. C., Burlaga, L. F., and Ness, N. F. 1995. Locations of the termination shock and the heliopause. *J. Geophys. Res.* 100:17015–17023.

Zank, G. P. 1989. Solution topologies for cosmic ray modified galactic winds I. Spherical symmetry. *Astron. Astrophys.* 225:37–47.

Zank, G. P., Axford, W. I., and McKenzie, J. F. 1990. Instabilities in energetic particle modified shocks. *Astron. Astrophys.* 233:275–284.

Zirakashvili, V. N., Dorman, L. I., and Ptuskin, V. S. 1991. Cosmic ray nonlinear modulation in the outer heliosphere. *Proc. Intl. Conf. Cosmic Rays XXII* 3:585–588.

SUPERNOVA BLAST WAVES AND PRE-SUPERNOVA WINDS: THEIR COSMIC RAY CONTRIBUTION

PETER L. BIERMANN
Max Planck Institut für Radioastronomie

Shocks in stellar winds can accelerate particles; energetic particles are observed through nonthermal radio emission in novae, OB stars and Wolf Rayet stars. Supernova explosions into predecessor stellar winds can lead to particle acceleration, which we suggest can explain most of the observed cosmic rays of the nuclei of helium and heavier elements, from GeV in particle energies up to near 3×10^9 GeV, as well as electrons above about 30 GeV. To make our case, we go through the following steps: (1) Using a postulate for an underlying principle that leads to transport coefficients in a turbulent plasma, we derive the properties of energetic particles accelerated in spherical shocks in a stellar wind. (2) We suggest that a dynamo working in the inner convection zone of an upper main sequence star can lead to high magnetic field strengths, which may become directly observable in massive white dwarfs, massive red giant stars and Wolf Rayet stars. (3) Such magnetic fields may put additional momentum into stellar winds from the pressure gradient of the toroidal field, with reduced angular momentum loss. (4) We use the statistics of Wolf Rayet stars and radio supernovae to derive a lower limit for the magnetic field strengths. This limit gives support to the wind-driving argument as well as the derivation of the maximum particle energy that can be reached. (5) From a comparison of the radio-luminosities of various stars, radio supernovae, as well as supernova remnants, there appears to be a critical Alfvénic Mach number for electron injection. With this concept we propose an explanation for the observed proton/electron ratio in galactic cosmic rays at GeV energies. (6) We check the model prediction quantitatively in the cosmic ray spectrum and chemical composition against air shower data from (a) Akeno, (b) a world data set, (c) Fly's Eye, and (d) against further cosmic ray data available from other experiments. (7) Finally, we summarize various important caveats, and outline important next steps as well as checks regarding the implications of these concepts for stars and stellar evolution.

I. INTRODUCTION

Most supernovae are explosions of massive stars (see, e.g., Wheeler 1989), often stars that have a stellar wind prior to the explosion. Thus the physics of these stellar winds becomes important for a discussion of what happens when the star explodes and a shockwave travels down such a stellar wind. Optical and X-ray data have been interpreted as due to shock structures and shock heating (see, e.g., Owocki 1992). Observationally, there is also evidence for shock waves in stellar winds prior to the explosion, such as suggested by nonthermal radio emission from OB and Wolf Rayet stars, as well by the nonthermal radio emission from the nova GK Per (Seaquist et al. 1989). An interpretation of this nonthermal radio emission is an important test for any theory of particle acceleration in shocks in stellar winds.

In this chapter we review recent work on the acceleration process in shock waves in stellar winds, and argue that a large part of the observed cosmic rays can be attributed to supernova shocks in stellar winds.

A. Stellar Winds

Stellar winds are observed in many cases, in low-mass stars such as the Sun, as well as high-mass stars such as OB stars. For the latter the wind driving can be explained as an effect of radiation pressure (Lucy and Solomon 1970; Castor et al. 1975; Pauldrach et al. 1986; Owocki 1990). For Wolf Rayet stars the momentum in the wind is generally believed to be too large to be explainable due to radiation driving in the limit of single scattering, and so, as one possible solution, multiple scattering models have been devised. Magnetic fields have been argued to contribute to the driving through the fast magnetic-rotator concept (Cassinelli 1982,1991); this idea, however, has been criticized on the basis that the corresponding large angular momentum loss would lead to a severe self-limitation of the process (Nerney and Suess 1987) and that therefore the process could not be general. There is a modified magnetic-rotator model, for which the Alfvénic surface is close to the stellar surface, and hence the angular momentum loss is strongly reduced (Biermann and Cassinelli 1993). In this case, the lifetime is not limited by angular momentum loss, but an initial driving of the wind is required, and the pressure gradient of the tangential magnetic field is argued to provide an amplification of the momentum of the wind.

B. Cosmic Rays

After the discovery of cosmic rays by Hess (1912) and Kohlhörster (1913), Baade and Zwicky (1934) proposed that supernova explosions produce cosmic rays. Alfvén (1939) argued early for a local origin in our Galaxy, which is confirmed by the age determinations of the cosmic rays (Garcia-Munoz et al. 1977). Fermi (1949,1954) proposed the basic concept of acceleration still being used. Shklovskii (1953) and Ginzburg (1953) made a convincing case for particle acceleration in supernova remnants. Ginzburg (1953) emphasized the interesting role of novae; and indeed, the nova GK Per is a test case for the evolution of shocks in winds and their particle acceleration. Hayakawa (1956) proposed that stellar evolution gives rise to an enhancement of heavy elements and pointed out the importance of spallation in the interstellar medium; and again, the enrichment found in the cosmic ray contribution from supernova explosions into winds is indeed believed today to be due to just this enrichment. And finally, Cocconi (1956) argued convincingly that the most energetic cosmic ray particles are from an extragalactic origin; the GRO observations (Sreekumar et al. 1993) of a neighboring galaxy, the Small Magellanic Cloud, provided the last and a very strong argument, that indeed the cosmic rays in the lower energy range are not universal, and thus have to be galactic. An early seminal form of some of the ideas expressed in the following can be found in Peters (1959,1961). A brief historical review is given by Ginzburg (1993). The form of Fermi's process used today was discovered nearly simultaneously by Axford et al. (1977), Bell (1978a, b), Blandford and Ostriker (1978), and Kryskii (1977). In this picture the main particle energy gain is from repeated scattering off magnetic irregularities on the two sides of a shock front. Because those two sides can be considered as a continuously contracting system relative to each other, particles that remain in the system gain energy. There are many important reviews and books on cosmic ray physics. We just mention the classical book by Hayakawa (1969), the recent book by Berezinsky et al. (1990), and the reviews by Drury (1983), Blandford and Eichler (1987), and Jones and Ellison (1991).

Today there are only some well-accepted arguments about the origins of cosmic rays: (a) the cosmic rays below about 10^4 GeV are believed to be predominantly due to the explosion of stars into the normal interstellar medium (Lagage and Cesarsky 1983); (b) the cosmic rays from near 10^4 GeV up to the knee, at 5×10^6 GeV, are likely predominantly due to explosions of massive stars into their former stellar wind (Völk and Biermann 1988); (c) while this latter claim is not undisputed, there has been certainly no agreement yet on the origin of the cosmic rays of higher particle energy.

Direct observation of cosmic rays, either from goundbased instrumentation, from satellites or from balloons has been the driving input for nearly all considerations in cosmic ray work. These data demonstrate that

1. The overall spectrum of cosmic rays is a power law up to an energy, commonly referred to as the knee, near 5×10^6 GeV, to continue with a steeper power law to the ankle, near 3×10^9 GeV, with a slight turn up beyond and an apparent cutoff near 10^{11} GeV;
2. The spectra are about $E^{-2.74}$ for hydrogen at moderate energy, and slightly flatter for helium and heavier elements, while the electron spectrum is consistent with the hydrogen spectrum at low energy and then changes over to about $E^{-3.3}$;
3. The chemical composition at low energy is crudely similar to that of the interstellar medium, with hydrogen and helium underabundant relative to silicon. All such properties require explanation.

We will assume in the following that the correction from the observed cosmic ray spectrum to the source spectrum is a change in spectral slope by exactly $1/3$ for relativistic particles (see Biermann 1993a,1994a,b,1995c), so that we are looking at source spectra of approximately $E^{-2.4}$ below the knee, and of approximately $E^{-2.8}$ above the knee. This can be argued on the basis of a Kolmogorov spectrum of the irregularities of the interstellar magnetic field, from plasma simulations, from analogies with *in-situ* measurements of the solar wind, and other observations of the interstellar medium (see, e.g., Biermann 1993*a*). It is clear, however, that a Kolmogorov spectrum of interstellar turbulence is insufficient to explain in a straightforward manner the abundance ratios in cosmic rays (see, e.g., Garcia-Munoz et al. 1987; Biermann 1995*c*). Such properties of cosmic rays may require a deeper understanding of the interstellar medium than we currently have. However, we note that the secondary to primary ratio of spallation products in cosmic rays does not simply yield a spectrum of interstellar turbulence in a medium, which varies on time scales equivalent to those of cosmic ray transport, and which has known inhomogeneities of density contrasts of many powers of ten.

C. Shocks in Winds

There are various kinds of evidence for the presence of shock structures in stellar winds, from *in-situ* observations in the solar wind, to optical line profiles to X-ray emission from massive stars (Lucy 1982; Owocki and Rybicki 1985; Owocki et al. 1988; Owocki 1990,1992,1994; MacFarlane and Cassinelli 1989; Usov and Melrose 1992; Feldmeier 1993). Furthermore, there are shocks to be expected in the colliding wind zones between the binary star components of two massive stars (Eichler and Usov 1993); we do not consider such cases here. In this chapter, we wish to concentrate on particle acceleration and the ensuing nonthermal radio emission from shock zones in the winds of single stars.

Nonthermal radio emission is usually attributed to the synchrotron emission of energetic electrons gyrating in magnetic fields. OB and Wolf Rayet stars are both stars with strong powerful winds. They are usually thermal radio emitters from free-free emission, and in many cases, they are also nonthermal radio emitters. Similarly, nonthermal radio emission is detected from some novae, in particular, from the nova GK Per. Thus, any theory to explain particle acceleration in spherical shocks has to account for the properties of such nonthermal radio emission; if, as a result, one finds an explanation for a component of cosmic rays, it provides support for the coherence of a theory to account for the origin of cosmic rays.

For this nonthermal radio emission, we can determine its luminosity, spectrum, time dependence, spatial distribution, and, what turns out to be especially useful, its polarization. In supernova remnants and in the nova GK Per we can even determine the spatial arrangement of polarization. This allows us to derive the topology of the magnetic field. For young supernova remnants this spatial arrangement is always predominantly radial, while the compression of an arbitrarily arranged interstellar magnetic field would naturally lead to a tangential configuration. Also, for the wind of the nova GK Per, we expect an unperturbed tangential magnetic field, and a strengthening of such a field in a shock wave; the observations also in this case show a radial field (Seaquist et al. 1989). The observed topology of the magnetic field is thus at right angles to that expected on very simple grounds. This implies a very general reason, and we suggest that this is best understood as the consequence of a rapid convective radial motion in the shocked region of the plasma. If this notion is the correct interpretation of these data, then the diffusive transport of energetic particles in the shocked region is possibly dominated by this rapid convective motion, and its properties then influence the spectrum of the energetic particles in the shock.

This leads to the basic thesis underlying all the arguments made here. We propose that *a principle of the smallest dominant scale*, either in real space or in velocity space (Biermann 1993*a*), allows us to determine the relevant transport coefficients, which describe the overall transport of particles in the shocked region, both parallel and transverse to the shock direction, and to derive drift energy gains.

D. Explosions: Interstellar Medium Versus Winds

One might ask what the essential difference actually is between explosions into a stellar wind and an explosion into the interstellar medium. There are three important differences. (1) a stellar wind has a density gradient, asymptotically of density $\rho(r) \sim r^{-2}$ with radius r, leading to persistent high shock velocities; (2) the possibility that stellar winds have much stronger magnetic fields than the interstellar medium; and (3) the winds of massive stars are enriched in heavy elements in the last stages of their evolution, thus strongly biasing any energetic particle population.

As a consequence, the first phase of an explosion, when the ejected mass M_{ej} is in free expansion until an approximately equal amount of mass is snowplowed together, has a quite different characteristic radius R_e. For an interstellar medium of ion density n_i this radius is

$$R_e(\text{ISM}) = 1.97 \text{ pc} \left(\frac{M_{ej}}{M_\odot} \right)^{1/3} n_i^{-1/3}. \tag{1}$$

Usually, this radius is exceeded even for low densities of the interstellar medium, so that the phase of expansion with constant energy is relevant for any discussion, the Sedov phase. This means that the shock speed U_1 is steadily decreasing with radius, as $U_1 \sim r^{-3/2}$.

In contrast, the corresponding radius for the expansion into a wind is given by

$$R_e(\text{wind}) = 306 \text{ pc} \frac{M_{ej}}{M_\odot} \frac{V_{W,-2}}{\dot{M}_{-5}}. \tag{2}$$

Here $V_{W,-2}$ is the wind velocity in units of $10^{-2}c$, and $\dot{M}_{-5}$ is the wind mass loss in units of 10^{-5} $M_\odot$ per year. This radius is, obviously, reduced for the explosive expansion into a red giant wind, with a wind speed of only 30 km s^{-1}, by a factor of 100 to only a few pc, everything else being equal. However, even then this radius will usually not be reached or at least not be surpassed by a considerable factor.

As a result, the shock velocity of an explosion into the interstellar medium is typically slowing down steadily, while in striking contrast the explosion into a wind can be expected to be in an approximately free expansion phase and so at nearly constant high velocity.

Another question, one might ask, is whether an explosion into a wind does not naturally lead to an explosion into the interstellar medium, when the edge of the wind bubble is reached. From evolutionary calculations of massive stars, it is clear that there is considerable mass loss before the final explosion, and all this mass forms, together with snowplowed interstellar medium material, a thick shell around the wind zone. Hence for those stars, which have a large amount of mass loss through a wind, the supernova shock has some difficulty penetrating this shell around the wind bubble without considerable energy loss. One difference between wind supernovae and interstellar medium supernovae is then that in the first case the explosion dissipates a significant fraction of its energy in the wind bubble shell, while in the second case the explosion can break through and form a remnant in adiabatic expansion. Wheeler (1989) shows that the dividing line is likely to be near a zero-age main-sequence mass of about 15 solar masses; we discuss this point below (Sec. VI) in the context of the energy requirements of the galactic cosmic rays attributed to the two source populations.

E. Outline

This review is based on earlier work, summarizes it and expands upon it (Biermann 1993*a*; Biermann and Cassinelli 1993; Biermann and Strom 1993; Stanev et al. 1993; Rachen and Biermann 1993; Rachen et al. 1993; Nath and Biermann 1993,1994*a, b*; Biermann et al. 1995) with earlier reviews in Biermann (1993*b*,1994*a,b*,1995*c*) with, however, an entirely different emphasis.

Much of the new material described here is based on discussions the author had with other participants at the meetings in Raleigh, N. C. (September 1993, organized by D. Ellison and S. Reynolds), in Tucson, Ariz. (October 1993, organized by C. P. Sonett, M. S. Giampapa and J.R. Jokipii), in Budapest, Hungary (March 1994, organized by Zs. Nemeth and E. Sormorjai), in Vulcano, Italy (May 1994, organized by F. Giovannelli and G. Mannocchi), in Nandaihe, China (August 1994, organized by Z. G. Deng, X.-Y. Xia and G. Börner), in St. Petersburg, Russia (September 1994, organized by D. A. Varshalovich, A. Bykov and others), and in Stockholm, Sweden (September 1994, organized by L. Bergström, P. Carlson, P. O. Hulth, and H. Snellman). Not all reports written for these meetings are mentioned.

In the following we describe in Sec. II the basic concept for particle acceleration in spherical shocks; in Sec. III we derive the properties of the energetic particle population; in Sec. IV we discuss the implications for stellar winds; in Sec. V we discuss the consequences for energetic electrons; in Sec. VI the consequences for energetic nuclei; and Sec. VII gives a summary.

II. THE SHOCK REGION

A. A Basic Instability

The normal asymptotic configuration for an embedded magnetic field in a stellar wind has been derived by Parker (1958), and gives a magnetic field, which is tangential, decreases with radius r as $1/r$, and with colatitude θ as $\sin\theta$.

Consider a spherical shock wave in such a wind. We will use the approximation throughout, that the shock wave is spherically symmetric, and that any asymmetry is introduced by the orientation and latitude dependence of the magnetic field only. This is a strong simplification, but is necessary to keep the issue clear which we discuss here, the physics of shock waves which are at least locally spherical, and which run through a stellar wind.

Then a spherical shock wave, centered on the star in its symmetry, compresses the magnetic field, this being tangential, by the full compression factor of 4 for a strong shock in an adiabatic gas of index 5/3.

Particles can be injected into an acceleration process, and give a large proportion of the overall pressure and energy density (see, e.g., Ellison et al. 1990; Jones and Ellison 1991). Then we have the case of a cosmic ray modified shock wave, such as treated by Zank et al. (1990). In the unperturbed state we have here a configuration, where the magnetic field is perpendicular to the shock direction, i.e., parallel to the shock surface. Zank et al. demonstrated that for this case the configuration is neutrally stable, with an increasing instability as soon as the shock becomes more oblique relative to the underlying magnetic field, leading to maximum instability for the case of a parallel shock configuration. In the oblique shock configuration as many as three instabilities may operate (Zank et al. 1990, see their Table 1). In the case which we consider here, this means that the slightest perturbation of the shock surface relative to the underlying magnetic field is leading to an instability. Such an instability then increases the deformation of the surface, which in turn strengthens the instability. Ultimately, the shock surface is maximally deformed, and changes its shape continuously, because the configuration is stuck in an unstable mode. Thus, strong turbulence is expected in the shock region, including downstream (Ratkiewicz et al. 1994).

This means then, for instance, that at the tip of an outward bulge of the shock surface the bulge can move outward with respect to the time averaged shock frame, until a column density is encountered equal to the column density behind the shock; then the bulge, driven by the instability, has to slow down in its local motion. At the sides of the bulge, energetic particles readily can leak into the upstream region, going through the shock just once in a locally highly oblique configuration (i.e., magnetic field vs shock normal), before the shock overruns them again. This is equivalent to saying, that the local acceleration efficiency is strongly reduced, because here we envisage an energetic particle to spend a fair amount of time on either side of the shock, gaining an appreciable amount of energy from drifts, before going back through the shock.

Such a picture then leads to a concept, where the shock surface is jumping around an average location, a sphere in our case. The upstream length scale of this jumping is a length corresponding to the same column density as the average downstream region, which corresponds to all the matter snowplowed in the expansion of the spherical shock. It is important to note that the length scales associated with the instability are hydrodynamic and therefore we believe it is justified to use hydrodynamic scales below; we also adopt below the gross simplification of the maximum density jump of 4 valid for a normal gas with adiabatic coefficient of $5/3$, here ignoring any cosmic ray modification (see, e.g., Duffy et al. 1994). In this concept the acceleration is a combination of (a) the drifts the particles experience in the upstream or downstream regions, because the averaged magnetic fields are, of course, still perpendicular to the shock direction, and (b) the energy gain from the Lorentz transformation each time a particle goes through the shock. At the same time, particles also lose energy from adiabatic expansion, because in the expansion of a spherical shock the local length scales always increase. We caution the reader, that we use results here from cosmic ray transport theory in an environment for which this theory was not made, i.e., where the magnetized ionized plasma is dominantly turbulent.

B. Radio Polarization

How can we test such a concept with observations? The radio observations of young supernova remnants and the nova GK Per can be a guide here. Radio polarization observations of supernova remnants clearly indicate what the typical local structure of these shocked plasmas is. Obviously, the normal expansion of a supernova remnant is not into a stellar wind, but into the interstellar medium. However, the typical magnetic field is highly oblique on average in a random field, and so the essential issue remains the same as in the case of a shock into a wind. The observational evidence (Milne 1971; Downs and Thompson 1972; Reynolds and Gilmore 1986; Milne 1987; Dickel et al. 1988) has been summarized by Dickel et al. (1991) in the statement that all shell type supernova remnants less than 1000 yr old show dominant radial structure in their magnetic fields near their boundaries. There are several possible ways to explain this; we concentrate here on the idea, that this polarization pattern is due to rapid convective motion which induces locally strong shear.

These examples are for supernova explosions into the interstellar medium; there is also an observation demonstrating the same effect for an explosion into a wind; Seaquist et al. (1989) find for the spatially resolved shell of the nova GK Per a radially oriented magnetic field in the shell, while the overall dependence of the magnetic field on radial distance r is deduced to be $1/r$, just as expected for the tightly wound up magnetic field in a wind. The interpretation given is that the shell is the higher density material behind a shock wave caused by the nova explosion in 1902 and now traveling through a wind. Seaquist et al. (1989) note the similarity to young supernova remnants.

The important conclusion for us here is that there appear to be strong radial differential motions in perpendicular shocks which provide the possibility that particles get *convected* parallel to the shock direction. The instability described earlier (Zank et al. 1990) may be the physical reason for this rapid convective motion, we would like to suggest. We assume this to be a diffusive process, and note that others have also pointed out that this may be a key to shock acceleration (see, e.g., Falle 1990).

It is important to emphasize that older supernova remnants do not show a clear pattern as described above. For instance, the data of the radio knot motions in Cas A (Tuffs 1986) clearly show a rather chaotic behavior; it appears that the radio knots move erratically with a speed of the order of the shock speed itself. This is actually important, it turns out; we will use this erratic motion below to limit the drift effect along the shock sphere.

C. Tycho vs Cygnus Loop

One important consequence of the concept introduced here is that for spherical shocks, that do not accelerate a new particle population to high energy, the overall shell thickness should be just that due to the snowplow; for a spherical shock in a homogeneous medium this is $r/12$ in the case of a strong shock. Obviously, if cooling becomes important, then the thickness is even less. For a shock that does accelerate a new energetic particle population in a shock in the interstellar medium, the instability described above leads to an estimate of the shell thickness of $5r/16$, again for a strong shock, and referred to the outer edge (see below).

The Cygnus Loop as well as Tycho appear to show indeed such properties. The Cygnus Loop does not require a new particle population for its radio emission, the squeezed interstellar medium is well sufficient, and the shell is thin (J. Raymond 1993, personal communication; Green 1990). At the same time, the supernova remnant Tycho (Dickel et al. 1991) does require a newly accelerated particle population to explain the radio observations and its shell thickness over most of its circumference is close to prediction, i.e., fairly thick. Also here we note that such an agreement does not *prove* our concept to be correct, but it does show consistency. We will discuss this point at more length below.

III. THE SPECTRUM

A. Derivation

The important conclusion for us here is that both observations and theoretical arguments suggest the existence of strong radial differential motions in those perpendicular shocks which are mediated by cosmic rays; this in turn suggests that particles get *convected* parallel to the shock direction. We emphasize that convective motion at a given scale entails that particle diffusion is independent of energy. We assume this convective turbulence with associated particle transport to be a diffusive process, for which we have to derive a natural velocity and a natural length scale, which can be combined to yield a diffusion coefficient. A classical prescription is the method of Prandtl (1925). In his argument an analogy to kinetic gas theory is used to derive a diffusion coefficient from a natural scale and a natural velocity of the system. Despite many weaknesses of this generalization Prandtl's theory has held up remarkably well in many areas of physics far beyond the original intent. In a leap of faith we will use a similar prescription here.

In order to generalize, we introduce the notion of the *smallest dominant scale*. This can be a scale in length or in velocity, may refer to an anisotropic transport, and thus may be different in orthogonal directions. This *principle* does not say that nature lets convective transport compete on the basis of different scales, because then the longest scale would be the fastest transport, and would win; it rather implies that nature chooses for the effective convective transport the smallest dominant scales. The idea, that the smallest dominant scale may, e.g., be related to the distance to a wall in a fluid current, has already been used for a long time (see section III§5 in Prandtl 1949). We use the assumption that the smallest dominant scale is the relevant scale several times in the course of this chapter in order to derive diffusion coefficients and other scalings. In some cases, when the smallest dominant scale is energy dependent, there may be a switch from one physical scale to another, say, to an energy-independent scale. Such a switch defines critical particle energies.

Consider the structure of a layer shocked by a supernova explosion into a stellar wind in the case that the adiabatic index of the gas is 5/3 and the shock is strong. Then there is an inherent length scale in the system, namely the thickness of the shocked layer, in the spherical case for a shock velocity much larger than the wind speed and in the strong shock limit $r/4$. This is the thickness of the matter snowplowed all the way from the star to the current location of the shock front, in the simplified picture of constant density throughout the shell. There is also a natural velocity scale, namely the velocity difference of the flow with respect to the two sides of the shock. Both are the smallest dominant scale, in velocity and in length.

Our basic conjecture, **argument 1**, *based on observational evidence as well as theoretical arguments*, is then that the convective random walk of energetic particles perpendicular to the unperturbed magnetic field can be described by a diffusive process with a downstream diffusion coefficient $\kappa_{rr,2}$ which is given by the thickness of the shocked layer and the velocity difference across the shock, and is independent of energy:

$$\kappa_{rr,2} = \frac{1}{3}\frac{U_2}{U_1}r(U_1 - U_2). \tag{3}$$

This is in apparent contradiction to the normal argument that (for relativistic particles) $\kappa_{rr} = c\,\lambda(E)/3 > c\,r_g/3$, where $\lambda(E)$ is the mean free path for resonant scattering of a particle of energy E and Larmor radius r_g.

We do not invoke resonant scattering, but fast convective turbulence as the dominant process. We do not ignore resonant scattering, but use Jokipii's (1987) argument on permissible scattering coefficients in oblique geometries (see below) to demonstrate that we are within bounds here.

The upstream diffusion coefficient can be derived in a similar way, but with a larger scale. We take here the second critical step, **argument 2**, namely that the upstream length scale is just U_1/U_2 times larger, and so is r. This is the relevant scale for the same column density on both sides of the average shock location, and can be argued on the basis of what limits the instability (see above). This, also, is the same ratio as the mass density and the ratio of the gyroradii of the same particle energy. Because the magnetic field is lower by a factor of U_1/U_2 upstream, that means that the upstream gyroradius of the maximum energy particle that could be contained in the shocked layer, is also r. Hence the natural scale is just r, and so the upstream diffusion coefficient is

$$\kappa_{rr,1} = \frac{1}{3} r (U_1 - U_2).$$

(4)

It immediately follows that

$$\frac{\kappa_{rr,1}}{r U_1} = \frac{\kappa_{rr,2}}{r U_2} = \frac{1}{3} \left(1 - \frac{U_2}{U_1} \right).$$

(5)

For these diffusion coefficients, it also follows that the residence times (Drury 1983) on both sides of the shock are equal and are

$$\frac{4\kappa_{rr,1}}{U_1 c} = \frac{4\kappa_{rr,2}}{U_2 c} = \frac{4}{3} \frac{r}{c} \left(1 - \frac{U_2}{U_1} \right).$$

(6)

Here it must be said that the residence time is normally derived from a diffusion argument which is obviously not valid here, because it requires that the scattering length is much smaller than the dominant scale, while we basically identify those two scales. We have not proven, but surmise that the concept of the residence time could be rederived from a time-averaging of the probability that a given particle is still on the same side of the jumping and wobbling shock surface after some time t. It is an assumption, based on dimensional arguments, that the result would be the same.

Adiabatic losses then cannot limit the energy reached by any particle as they run directly with the acceleration time, both being independent of energy, and so the limiting size of the shocked layer limits the energy that can be reached to that where the gyroradius just equals the thickness of the shocked layer, provided the particles can reach this energy. We assume here that the average of the magnetic field $\langle B \rangle$ is not changed very much by all this convective motion. This then leads to a maximum energy of

$$E_{\max} = \frac{U_2}{U_1} Z e r B_2 = Z e r B_1$$

(7)

where Ze is the particle charge and $B_{1,2}$ is the magnetic field strength on the two sides of the shock. This means, that the energy reached corresponds to the maximum gyroradius the system will allow on both sides of the shock. It also says that we push the diffusive picture right up its limit where on the downstream side the diffusive scale becomes equal to the mean free path and the gyroradius of the most energetic particles.

Jokipii (1987) has derived a general condition for possible values of the diffusion coefficient. Its value has to be larger than the gyroradius multiplied by the shock speed. This condition is fulfilled here, for the maximum energy particles only by a factor of $1 - U_2/U_1 < 1$, because here the shock speed and the radial scale of the system give both the largest gyroradius as well as the diffusion coefficient. This also counters part of the criticism of Völk in Ellison et al. (1994)—the other part of his criticism is dealt with in Biermann (1993a, Sec. II).

Observations can now give information of possible values for the diffusion coefficient as well; Smith et al. (1994) estimate the cosmic ray diffusion coefficient at shocks in the Large Magellenic Cloud and find severe upper limits. However, first of all, they use the assumption that the underlying unperturbed magnetic field is parallel to the shock normal; second, they measure an instantaneous situation, whereas I have argued above that strong and fast convective motion dominates the shock region. The diffusion coefficient derived above is an effective temporal and geometric average over this fast convection. Hence there is no contradiction.

There is an important consequence of this picture for the diffusion laterally. From the residence time scale and the velocity difference across the shock, we find a distance which can be traversed in this time of

$$\frac{4}{3}\frac{r}{c}\left(1 - \frac{U_2}{U_1}\right)(U_1 - U_2). \tag{8}$$

As the convective turbulence in the radial direction also induces motion in the other two directions, with maximum velocity differences of again $U_1 - U_2$, this distance is also the the typical lateral length scale. We noted above that the observed motions of radio knots in the supernova remnant Cas A support such an argument (Tuffs 1986). From this scale and again the residence time we can construct an upper limit to the diffusion coefficient in lateral directions of

$$\kappa_{\theta\theta,\max} = \frac{4}{9}\left(1 - \frac{U_2}{U_1}\right)^3\left(\frac{U_1}{c}\right)^2 rc \tag{9}$$

which is for strong shocks equal to

$$\kappa_{\theta\theta,\max} = \frac{1}{3}\left(\frac{3}{4}\frac{U_1}{c}\right)^2 rc. \tag{10}$$

Again in the spirit of the idea, that the smallest dominant scale determines the effective transport, this then will begin to dominate as soon as the θ-diffusion coefficient reaches this maximum at a critical energy. As long as the θ-diffusion coefficient is smaller, it will dominate particle transport in θ and the upper limit derived here is irrelevant. When the θ-diffusion coefficient reaches and passes this maximum, then the particle in its drift will no longer see an increased curvature due to the convective turbulence because of averaging and the part (here we have to account for both losses and gains by drifts) of drift acceleration due to increased curvature is eliminated. This then reduces the energy gain, and the spectrum becomes steeper from that energy on. The critical particle energy thus implied will be identified below with the particle energy at the knee of the observed cosmic ray spectrum.

1. Drifts. Consider particles that are either upstream of the shock, or downstream; as long as the gyrocenter is upstream we will consider the particle to be there, and similarly downstream. In general, the energy gain of the particles will be governed primarily by their adiabatic motion in the electric and magnetic fields. The expression for the energy gain is then (Northrop 1963, Eq. 1.79), for an isotropic angular distribution

$$\frac{dE}{dt} = Ze\mathbf{V_d}\frac{\mathbf{U} \times \mathbf{B}}{c_i} + \frac{pw}{c}\frac{\partial \ln B}{\partial t} \tag{11}$$

where the first term arises from the drifts and the second from the induced electric field. This equation is valid in any coordinate frame. We explicitly work in the shock frame, and separate the two terms above and consider the drift term first. The second term is accounted for further below.

The θ-drift velocity in a normal stellar wind is given above. The θ-drift can be understood as arising from the asymmetric component of the diffusion tensor, the θr-component. The natural scales there are the gyroradius and the speed of light, and so we note that for (Forman et al. 1974)

$$\kappa_{\theta r} = \frac{1}{3}r_g c \tag{12}$$

the exact limiting form derived from ensemble averaging, we obtain the drift velocity by taking the proper covariant divergence (Jokipii et al. 1977); this is not simply (spherical coordinates) the r-derivative of $\kappa_{\theta r}$. The general drift velocity is given by (see, e.g., Jokipii 1987)

$$V_{d,\theta} = c\frac{E}{3Ze}curl_\theta\frac{\mathbf{B}}{B^2}. \tag{13}$$

The θ-drift velocity is thus

$$V_{d,\theta} = \frac{2}{3}cr_g/r \tag{14}$$

where r_g is now taken to be positive. This drift velocity is just that due to the gradient as well as the curvature, and in fact both effects contribute here equally.

It must be remembered that there is a lot of convective turbulence which increases the curvature. The characteristic scale of the turbulence is $r/4$ for strong shocks, and thus the curvature is $4/r$ maximum. Here we have to consider the balance of drift energy gains and losses. We call the motion of a convective element that moves in the upstream direction upflow, and for a downstream direction of motion, we call it downflow. Then the direction of an upflow convective element is not in general just radial, and therefore we must take the average of an assumed omnidirectional distribution of upflow directions, weighting it with the probability of that particular direction, and obtain 2/3 of the maximum curvature (same averaging as done in Eq. 2.47 of Drury [1983] for the averaging of the momentum gain). It is plausible that for upflow directions we have to take the maximum curvature, because the velocity difference to the surroundings is maximum. For downflow, however, the velocity difference to the surroundings is only 1/4 of that upflow, and so we take for the downflow convection a curvature of 1/4 of the maximum, multiplied also by 2/3. The difference of gains and losses is then the net gain, thus giving a factor of $(2/3)(1 - 1/4) = 1/2$ of the maximum curvature (**argument 3**). This is equivalent to taking half the maximum as average for the net energy gain due to drifts. We obtain then for the curvature a factor of $2/r$ which is twice the curvature without any turbulence; this increases the curvature term by a factor of 2 thus changing its contribution from 1/3 to 2/3 in the numerical factor in the expression above. Hence the total drift velocity, combining now again the curvature (2/3) and gradient (1/3) terms, is

$$V_{d,\theta} = \frac{1}{3}\left(1 + \frac{U_1}{2U_2}\right) c r_g / r \tag{15}$$

now written for arbitrary shock strength. It is easily verified that the factor in front is unity for strong shocks where $U_1/U_2 = 4$. The energy gain associated with such a drift is given by the product of the drift velocity, the residence time, and the electric field. Upstream this energy gain is given by

$$\Delta E_1 = \frac{4}{3} E \frac{U_1}{c} f_d \left(1 - \frac{U_2}{U_1}\right) \tag{16}$$

where

$$f_d = \frac{1}{3}\left(1 + \frac{U_1}{2U_2}\right). \tag{17}$$

Thus, $f_d = 1$ for strong shocks. The corresponding expression downstream is

$$\Delta E_2 = \frac{4}{3} E \frac{U_2}{c} f_d \left(1 - \frac{U_2}{U_1}\right) \tag{18}$$

giving a total energy gain of

$$\Delta E/E = \frac{4}{3} \frac{U_1}{c} f_d \left(1 + \frac{U_2}{U_1}\right)\left(1 - \frac{U_2}{U_1}\right). \tag{19}$$

The drift energy gain averages over the magnetic field strength during the gyromotion. We emphasize that this energy gain is independent of this average magnetic field, so that even variations of the magnetic field strength due to convective motions do not change this energy gain.

It can easily be shown, that this treatment can be extended to subrelativistic energies, and provides power laws in momentum as a result, just as in the standard case (see below; Drury 1983).

It is of interest to note here, that the net distance traveled (i.e., drifted) by the particle, e.g., upstream, is given by

$$l_{\perp 1} = \frac{4}{3} \frac{E}{Z e B_1} f_d \left(1 - \frac{U_2}{U_1} \right) \tag{20}$$

which is the gyroradius itself for $U_2/U_1 = 1/4$, corresponding to a strong shock. This then says that we are at a gyroradius limit for the drift distance. Just as in isotropic turbulence the gyroradius is a lower limit to the mean free path for particle scattering parallel to the magnetic field in a turbulent plasma, suggesting that it may be useful to think of the plasma also as maximally turbulent perpendicular both to the flow and to the magnetic field. We emphasize that during this drift the particle makes many gyromotions. It is also important to note that the magnetic field structure in the shocked region—as discussed above on the basis of observations—will contain local regions of opposite magnetic field and so the drift itself will be erratic and be the sum of many single element drift movements. What we have derived is the average net energy gain due to drifts, with the drift distance corresponding to the average magnetic field strength.

2. The Energy Gain of Particles. Shock acceleration in its standard form just uses the Lorentz transformation for an energetic particle at a velocity v much larger than the shock velocity to compute the energy gain as the particle goes between scattering in a weakly turbulent magnetic field from downstream to upstream and back. In practice we will consider the case when v is close to c.

We assume the shock to be subrelativistic and so the phase space distribution of the particles to be nearly isotropic. Then downstream (see Drury [1983] for the exact derivation) the particles have a finite chance to escape. A detailed discussion yields a power law for the distribution $p^2 f(p)$, where $f(p)$ is the particle distribution function of momentum p in phase space, with the power law index -4 for strong shocks for a gas of adiabatic index $5/3$ (i.e., $p^2 f(p) \sim p^{-2}$).

However, this assumes that there are no other energy losses or energy gains during a cycle of a particle going back and forth between upstream and downstream. In a configuration, where there is a component of the magnetic field perpendicular to the shock normal, there can be energy gains by drifts parallel to the electric field seen by a particle moving with the shock system, and also losses due to adiabatic expansion in the expansion of a curved shock. Because drift acceleration is a rate, the resulting particle energy gain is proportional to the time spent on either side of the shock. Furthermore, there can be spectral changes due to the fact, that particles were injected at a different rate in the past, when some particle under consideration now was injected; again, this is likely to happen in a spherical shock.

Let us consider then one full cycle of a particle remaining near the shock and cycling back and forth from upstream to downstream and back. The energy gain just due to the Lorentz transformations in one cycle can then be written as

$$\frac{\Delta E}{E}\bigg|_{LT} = \frac{4}{3}\frac{U_1}{c}\left(1 - \frac{U_2}{U_1}\right). \tag{21}$$

Adding the energy gain due to drifts we obtain

$$\frac{\Delta E}{E} = \frac{4}{3}\frac{U_1}{c}\left(1 - \frac{U_2}{U_1}\right) x \tag{22}$$

where

$$x = 1 + \frac{1}{3}\left(1 + \frac{U_1}{2U_2}\right)\left(1 + \frac{U_2}{U_1}\right) \tag{23}$$

which is 9/4 for a strong shock, when $U_1/U_2 = 4$.

It is easy to show that the additional energy gain flattens the particle spectrum by

$$\frac{3U_2}{U_1 - U_2}\left(1 - \frac{1}{x}\right). \tag{24}$$

3. Expansion and Injection History. Consider how long it takes a particle to reach a certain energy:

$$\frac{dt}{dE} = \left\{8\frac{\kappa_{rr,1}}{U_1 c}\right\} \Big/ \left\{\frac{4}{3}\frac{U_1}{c}\left(1 - \frac{U_2}{U_1}\right) x E\right\}. \tag{25}$$

Here we have used that $\kappa_{rr,1}/U_1 = \kappa_{rr,2}/U_2$. Because we have

$$r = U_1 t \tag{26}$$

this leads to

$$\frac{dt}{t} = \frac{dE}{E}\frac{3U_1}{U_1 - U_2}\frac{2}{x}\frac{\kappa_{rr,1}}{r U_1} \tag{27}$$

and so to a dependence of

$$t(E) = t_o\left(\frac{E}{E_o}\right)^{\beta} \tag{28}$$

with

$$\beta = \frac{3U_1}{U_1 - U_2}\frac{2}{x}\frac{\kappa_{rr,1}}{r U_1} \tag{29}$$

which is a constant independent of r and t.

Particles that were injected some time ago were injected at a different rate, say, proportional to r^b. This then leads to a correction factor for the abundance of

$$\left(\frac{E}{E_o}\right)^{-b\beta}. \tag{30}$$

However, in a d-dimensional space, particles have r^d more space available to them than when they were injected, and so we have another correction factor which is

$$\left(\frac{E}{E_o}\right)^{-d\beta}. \tag{31}$$

The combined effect is a spectral change by

$$-\frac{3U_1}{U_1 - U_2}\frac{2}{x}(d+b)\frac{\kappa_{rr,1}}{rU_1}. \tag{32}$$

Thus we have a density correction factor, which depends on the particle energy, and so changes the spectrum.

This expression can be compared with a limiting expansion derived by Drury (1983, Eq. 3.58), who also allowed for a velocity field; Drury (1983) generalized earlier work on spherical shocks by Krymskii and Petukhov (1980) and Prishchep and Ptuskin (1981). Drury's expression agrees with the more generally derived expression given here for $x = 1$. The comparison with Drury's work clarifies that for $\kappa \sim r$ the inherent time dependence drops out except, obviously, for the highest energy particles, discussed further below; the same comparison shows that the statistics of the process are properly taken into account in our simplified treatment. We note that $\kappa_{rr,1}/rU_1$ is $1/4$ in the wind case, and $1/12$ in the ISM case, both for strong shocks, and thus still small compared to unity. *A fortiori*, the comparison shows that in this limit of small $\kappa_{rr,1}/rU_1$, the derivation by Drury, using the properly derived classical cosmic ray transport theory, and our heuristic derivation, both agree. This agreement does not provide proof of our microscopic picture, but it is an important check.

If the expansion is linear, as it is the case here, then the r^d-term also describes the adiabatic losses in their effect on the spectrum, due to the general expansion of the shock layer and thus accounts for the second term in Eq. (11) above. Hence the total spectral difference, as compared with the planeparallel case, is given by

$$\frac{3U_1}{U_1 - U_2}\left\{\frac{U_2}{U_1}\left(\frac{1}{x}-1\right)+\frac{2}{x}(b+d)\frac{\kappa_{rr,1}}{rU_1}\right\}. \tag{33}$$

Here we use the following sign convention. For this expression positive, the spectral index of the particle distribution is steeper than without this correction; this then takes the minus sign in Eqs. (30–32) properly into account.

Expression (33) constitutes the basic result of this section. For a wind we have $b = -2$ and $d = +3$, and so $b + d = 1$. For a Sedov-type expansion into a homogeneous medium the additional differential adiabatic losses steepen the spectrum further (Biermann and Strom 1993). The total spectral change is then for $U_1/U_2 = 4$ given by $1/3$, so that the spectrum obtained is

$$\text{Spectrum (source)} \; = \; E^{-7/3}. \tag{34}$$

This is what we wanted to derive. Generalizing to transrelativistic energies, we note that the spectrum is a power law in relativistic momentum p, as can easily be seen from the detailed derivation.

Generalizing now for arbitrary wind speed and arbitrary shock strength, we obtain a reduced thickness of the shocked layer. This reduction of the thickness of the layer for a finite wind velocity is due to the fact that the material that is snowplowed together is not all gas between zero radius and the current radius r, but between zero radius and $r(1 - V_W/(V_W + U_1))$, because the gas keeps moving while the shock moves out towards r. For the sequence of $V_W/U_1 = 0$, 1.0, and $\gg 1$ we thus obtain particle spectral index differences, in addition to the index of $7/3$, of $0.0, 0.136, 0.303$, corresponding to synchrotron emission spectral index of an electron population with the same spectrum, of $-0.667, -0.735, -0.818$ (using the convention that flux density $S_\nu \sim \nu^\alpha$). The work of Owocki et al. (1988) suggests that typical shocks in winds have a velocity in the wind frame similar to the wind velocity in the observers frame itself, which implies that, in the simplified picture here, only spectral indices for the synchrotron emission between -0.667 and -0.735 are relevant, with an extreme range of spectral indices up to -0.818 for strong shocks. Obviously, for weaker shocks with $U_1/U_2 < 4$ the spectrum can be steeper, e.g., for $U_1/U_2 = 3.5$ we obtain an optically thin spectral index for the synchrotron emission of -0.734 for $V_W \ll U_1$ and -0.815 for $V_W/U_1 = 1$. This then is the prediction for the spectral index of the nonthermal radio emission from massive stars like OB stars and Wolf Rayet stars, as well as for other shocks in winds like those around novae.

We can estimate the uncertainty only with difficulty, because the argument is a limit, the limit of a strong shock for instance, and also in all other steps of the argument we have used the conceptual limit of the scales involved. However, we can estimate one uncertainty, which arises from the finite wind speed of Wolf Rayet stars, or similar stars with strong winds which explode as supernovae. These wind speeds can go up to several thousand km s^{-1}, while the supernova shock is variously estimated to 10^4 km s^{-1} to twice that much. As a limiting argument, we use the fact that the ratio of the wind speed to the supernova shock speed is <0.2; this gives a steepening of the derived spectral index of the particle distribution by 0.04. This uncertainty also may correspond to curvature of the spectrum, because there is a time-evolution as the shock progresses out through the stellar wind. As more energy of the shock is dissipated and more mass of the stellar wind is snowplowed, the shock slows down; then those particles already accelerated keep their flatter spectrum (see Drury 1983, Eq. 2.44), while those particles freshly injected and accelerated will have a steeper spectrum. Thus, in the range $V_W/U_1 = 0...0.2$, we obtain a spectral index in the range $7/3...7/3+0.04$. We note that the evolved stars which are most common have a slow wind. The distribution of stars through this adopted range of wind velocities is likely to be biased towards small numbers. Therefore we ascribe to the spectral index derived here an uncertainty of -0.02 ± 0.02, which describes both the uncertainty in an assumed power law, and the possible curvature. After correcting for leakage from the Galaxy, the spectrum is

$$\text{Spectrum (Earth)} = E^{-8/3-0.02\pm0.02} \tag{35}$$

which is very close to that observed near Earth at particle energies below the knee. We plan to discuss the error estimates in a separate contribution.

It is of interest to note, that an injection spectrum of $E^{-7/3}$ can also lead via pp-collisions in a synchrotron dominated regime to a spectrum of pair-secondaries of $E^{-10/3}$, which translates in the synchrotron spectrum to a $-7/6$ power law flux density spectrum and a $-13/6$ power law photon number spectrum, very close to that observed by GRO for the Crab pulsar (Schönfelder 1992, unpublished). If the curvature of a shock is not exactly spherical, obviously, some differences to these particular numerical values will occur. Such a speculative interpretation would place the origin of the pulses in periodically excited shocks traveling down a perpendicular magnetic field configuration in a pulsar wind as considered here.

Such an injection spectrum of $-7/3$ of relativistic particles in strong and fast shocks propagating through a stellar wind leads to an unambiguous radio synchrotron emission spectrum of $\nu^{-2/3}$ in flux density S_ν (compare, e.g., the nonthermal radioemission of OB stars, WR stars, novae, especially GK Per [Reynolds and Chevalier 1983,1984] radio supernovae, and supernova remnants). The observed variety of spectral indices of the radio emission of supernova remnants is discussed in some detail in Biermann and Strom (1993).

B. The Maximum Energy of Particles

The maximum energy particles can reach is given above, and depends linearly on the magnetic field. Thus, we require estimates for the magnetic field in the stellar winds of Wolf Rayet stars and other massive stars, that explode as supernovae, like red and blue supergiants. Comparing at first the corresponding estimates that Völk and Biermann (1988) used, we note that the energies implied here are larger by approximately c/U_1 for the same given magnetic field strength, because their expression for the maximum energy that particles could reach contains an additional factor of approximately $U_1/c \ll 1$ as compared with our Eq. (7). This modified limit of Völk and Biermann to the particle energy, which can be reached, corresponds well to the limit valid for supernova explosions into the interstellar medium (see Biermann and Strom 1993).

Cassinelli (1982) and Maheswaran and Cassinelli (1988,1992) have argued that Wolf Rayet stars have very much larger magnetic fields than that which Völk and Biermann used, in order to drive their winds. The magnetic fields given by Cassinelli and coworkers are of order a few thousand Gauss on the surface of the star. We introduce the conjecture here, discussed in more detail in Biermann and Cassinelli (1993), and below, that the Alfvén radius of the stellar wind is close to the stellar surface itself. Then it follows that the product Br has approximately the same value on the surface as in the wind, and is of order $3 \times 10^{14} B_{0.5}$ cm Gauss, where $B_{0.5}$ is the strength of the magnetic field at 10^{14} cm radial distance in units of 3 Gauss. From this number we infer a maximum energy of particles of

$$E_{\max}(\text{protons}) \; = \; 9 \times 10^7 B_{0.5} \, \text{GeV} \tag{36}$$

and

$$E_{\max}(\text{iron}) \; = \; 3 \times 10^9 \, B_{0.5} \, \text{GeV}. \tag{37}$$

This suggests that the highest energy particles from the acceleration process discussed here are mostly iron or other heavy nuclei. The chemical composition is expected to change abruptly to mostly protons again when the extragalactic component takes over (Rachen and Biermann 1993) somewhere near 3×10^9 GeV. The dependence of the maximum particle energy on the magnetic properties of the stellar winds implies a smearing due to the distribution of magnetic field strengths from all the different stars which contribute, and hence leads to a steepening of the spectrum, possibly well before the cut-off. If this mechanism provides the largest particles energies, then obviously other contributions are not excluded, by pulsars, neutron star binaries, or even from a hypothetical termination of the galactic wind.

C. The Knee in the Cosmic Ray Spectrum

We wish to discuss here the bend in the spectrum of cosmic rays at the knee, near 5×10^6 GeV.

Let us consider the structure of the wind through which the supernova shock is running. The maximum energy a particle can reach is proportional to $\sin^2\theta$, because the space available for the gyromotion from a particular latitude is limited in the direction of the pole by the axis of symmetry. Hence, the maximum energy attainable is lowest near the poles. Then, consider the pole region itself, where the radial dependence of the magnetic field is $1/r^2$, and the magnetic field is mostly radial. We can make two arguments here. Either we put the upstream diffusive scale $4\kappa_{rr,1}/(cU_1)$ equal to r/c in the strong shock limit, or we can put acceleration time and flow time equal to each other. Both arguments lead to the same result. We use here the Bohm limit in the diffusion coefficient $\kappa_{rr,1} = \frac{1}{3}cE/ZeB(r)$, because we have a shock configuration near the pole, where the direction of propagation of the shock is *parallel* to the magnetic field—often referred to as a parallel shock configuration. This then leads to a maximum energy for the particles of

$$E = \frac{3}{4} Z e B(r) r \frac{U_1}{c} \tag{38}$$

which is proportional to $1/r$ near the pole, where the magnetic field is parallel to the direction of shock propagation; the corresponding gyroradius is then given by $\frac{3}{4}\frac{U_1}{c}r$. Putting this equal to the gyroradius of particles that are accelerated further out at some colatitude θ, where the magnetic field is nearly perpendicular to the direction of shock propagation, gives the limit where the latitude-dependent acceleration breaks down. This then gives the critical angle as

$$\sin \theta_{\text{crit}} = \frac{3}{4} \frac{U_1}{c}. \tag{39}$$

The angular range of $\theta < \theta_{\text{crit}}$ we refer to as the *polar cap* below. We note that this angle is usually much larger than the angle where $B_r = B_\phi$ at a given radial distance r. If, as observed in the solar wind, the equatorial sheet of the magnetic field structure oscillates around the geometric symmetry plane, then this critical angle may either be effectively larger or the argument may fail altogether. The energy at the location of the critical angle as defined above is then given by

$$E_{\text{knee}} = Z e B(r) r \left(\frac{3}{4} \frac{U_1}{c} \right)^2. \tag{40}$$

We identify this energy with the knee feature in the cosmic ray spectrum, because all latitudes outside the polar cap contribute the same spectrum up to this energy; from this energy to higher particle energies a smaller part of the hemisphere contributes and also, the energy gain is reduced, as argued below; this reduction of the energy gain in each cycle of acceleration happens at the same critical energy. This is valid in the region where the magnetic field is nearly perpendicular to the shock, and thus this knee energy is independent of radius.

All this immediately implies that the chemical composition at the knee changes so that the gyroradius of the particles at the spectral break is the same, implying that the different nuclei break off in order of their charge Z, considered as particles of a certain energy (and not as energy per nucleon). In a spectrum in energy per particle, this introduces a considerable smearing.

In the polar cap the acceleration is a continuous mix between the regime where the diffusion coefficient is determined by the thickness of the shell, and the regime where it is dominated by turbulence parallel to the magnetic field; this latter regime is rather small in angular extent. Thus, $\kappa_{rr,1}/rU_1$ might be quite a bit smaller than 1/4. Hence the polar cap will have a spectrum which is determined by a range of

$$0 < \frac{\kappa_{rr,1}}{rU_1} < \frac{1}{3} \left(1 - \frac{U_2}{U_1} \right) \tag{41}$$

as well as by a rather reduced role for the extra energy gain due to drifts. Thus the spectrum is harder in the polar cap region, because we are close to the standard parallel shock configuration, for which the particle spectrum is well approximated by E^{-2}. On the other hand, the polar cap is small relative to 4π with about $(U_2/U_1)^2$ and only a spectrum much flatter than $E^{-7/3}$ (like, for example, E^{-2}) will make it possible for the polar cap to contribute appreciably near the knee energy, because then the spectral flux near the knee is increased relative to 1 GeV by $(E_{\mathrm{knee}}/m_p c^2)^{1/3}$ which approximately compensates for its small area. The uncertainty in the spectrum of the polar cap component particles is not determinable at present; should the spectrum be significantly steeper than E^{-2}, then this component almost certainly is altogether insignificant. Therefore we do not ascribe any uncertainty to this spectral index; it must remain an assumption, derived from the limiting argument. During an episode with drift towards the poles, a larger part of the sphere can contribute for larger energy particles, and so there is an additional tendency to flatten the spectrum of the polar cap contribution. The combination of the polar cap with the rest of the stellar hemisphere might lead to a situation where up to, say, $600Z$ TeV the entire hemisphere excluding the polar cap dominates, while from $600Z$ TeV up to the knee the polar cap begins to contribute appreciably. Near the knee energy the polar caps might thus contribute equally to the rest of the 4π steradians. Because of spatial limitations most of the hemisphere has to dominate again above the knee, although with a fraction of the hemisphere that decreases with particle energy. This introduces a weak progressive steepening of the spectrum with energy, which we discuss elsewhere. The superposition of such spectra for different chemical elements, including the polar cap contribution, has been tested successfully (see Stanev et al. 1993), and is briefly summarized in Sec. VI. The results of these checks suggest that the polar cap may be a reason for the flattening of the cosmic ray spectrum as one approaches the knee feature.

Note that we are using a limiting argument to derive the spectrum below the knee, and again use a limiting argument (see below) for the spectrum above the knee. Close to the knee, such an argument breaks down on either side, and so a softening of the knee feature is to be expected. On top of such a softened knee feature the polar cap gives an additional component.

The expression for the particle energy at the knee also implies by the observed relative sharpness of the break of the spectrum that the actual values of the combination $B(r)rU_1^2$ must be very nearly the same for all supernovae that contribute appreciably in this energy range; however, it is difficult to put a numerical value to this argument (see Stanev et al. [1993] for the systematics that go into any parameter estimate from fitting the air shower data). Please note that $B(r)r$ is evaluated in the Parker regime, and so is related to the surface magnetic field by $B(r)r = B_s r_s^2 \Omega_s / V_W$, where the values with index s refer to the surface of the star and V_W is the wind velocity. Thus, in our picture, the expression

$$B_s r_s^2 \frac{\Omega_s}{V_W} U_1^2 \tag{42}$$

is approximately a universal constant for all stars that explode as supernovae after a Wolf Rayet phase. It may also hold for all massive stars of lower mass that explode as supernovae.

This then implies that we may have identified a functional relationship for the mechanical energy of exploding stars connecting the magnetic field, the angular momentum, and the ejection energy. Such a relationship could be fortuitous, because all massive stars become very similar to each other near the end of their evolution. But it could also be an indication for an underlying physical cause. Related ideas have been expressed and discussed by Kardashev (1971), Bisnovatyi-Kogan (1971), LeBlanc and Wilson (1970), Ostriker and Gunn (1971), Amnuel et al. (1972), Bisnovatyi-Kogan et al. (1976), and Kundt (1976), with Bisnovatyi-Kogan (1970) the closest to the argument below. All this leads to the following interesting suggestion (see Biermann 1993a).

The source of the mechanical energy observable in supernova explosions may then be the gravitational energy of a core accretion disk at a scale determined by angular momentum and mediated by the magnetic field. We leave a discussion of how this argument may lead to a surface magnetic field strength to another time.

The distribution in the knee particle energy from the different stellar properties and explosive energies clearly leads to a rounding of the effective average knee shape, which softens the power laws on the both sides of the knee (see Biermann 1995a).

D. The Latitude Distribution of the Particles

Consider the derivation of the spectrum beyond the knee. Because the maximum energy particles can attain is a strong function of colatitude, the spectrum beyond the knee requires a discussion of the latitude distribution, which we have first to derive. The latitude distribution is established by the drift of particles which builds up a gradient which in turn leads to diffusion down the gradient. Hence it is clear that drifts towards the equator lead to higher particle densities near the equator, and drifts towards the poles lead to higher particle densities there. Thus the equilibrium latitude distribution is given by the balancing of the θ-diffusion and the θ-drift.

The diffusion tensor component $\kappa_{\theta\theta}$ can be derived similar to our heuristic derivation of the radial diffusion term κ_{rr}, again by using the smallest dominant scales. The characteristic velocity of particles in θ is given by the erratic part of the drifting, corresponding to spatial elements of different magnetic field direction. This is on average the value of the drift velocity $|V_{d,\theta}|$, possibly modified by the locally increased values of the magnetic field strength. The characteristic distance is the distance to the symmetry axis $r \sin\theta$; this is the smallest dominant scale as soon as the thickness of the shocked layer is larger than the distance to the symmetry axis, i.e., $\sin\theta < U_2/U_1$. Thus we can write in this approximation, **argument 4**,

$$\kappa_{\theta\theta,1} = \frac{1}{3} \mid V_{d,\theta} \mid r(1 - \mu^2)^{1/2}. \tag{43}$$

Here μ is the cosine of the colatitude on the sphere we consider for the shock in the wind. Interestingly, this can also be written in the form

$$\frac{1}{3}r_g c(1 - \mu^2)^{1/2} \tag{44}$$

where r_g is taken as positive; we also note that $c(1 - \mu^2)^{1/2}$ is the maximum drift speed at a given latitude, valid for the local maximum particle energy. This suggests that the latitude diffusion might be usefully thought of as diffusion with a length scale of the gyroradius, and the particle speed, to within the angle dependent factor, which just cancels out the latitude dependence of the magnetic field strength in the denominator of the gyroradius.

We assume then for the colatitude dependence a power law $(1 - \mu^2)^{-a}$ and first match the latitude dependence of the diffusion term and the drift, and then use the numerical coefficients to determine the exponent in this law. The diffusion term and the drift term have the same colatitude dependence because the double derivative and the internal factor of $(1 - \mu^2)$ lead to a $(1 - \mu^2)^{-a-1}$ for the diffusive term, while the drift term is just the simple derivative giving the same expression. For $(1 - \mu^2) \ll 1$ the condition then is $\frac{2}{3}a^2 = \pm a$. The diffusive term is always positive, while the θ-drift term is negative for ZB_s negative. This means for positive particles and a magnetic field directed inward, the θ-drift is towards the pole. In that case, the exponent a is either zero or $a = 3/2$. Because the drift itself clearly produces a gradient, the case with $a = 0$ is of no interest here. It follows that for positive particles and an inwardly directed magnetic field, the latitude distribution is strongly biased towards the poles, emphasizing in its integral the lower energies, and thus making the overall spectrum steeper beyond the knee energy. Such a configuration may influence the polar cap structure in terms of magnetic fields and energetic particles. The radial drift in this case is directed outward, which means that particles drift ahead of the shock by a small amount only to be caught up again by the diffusive region ahead of the shock. For the magnetic field directed outward and for positive particles, the radial drift is inwards, taking particles out of the system at a slightly increased rate and thus steepening the overall spectrum by a small amount.

When the magnetic field is directed outward and the particles have a positive charge, the drift is towards the equator with then a positive gradient with $(1 - \mu^2)^{3/2}$, again in the limit $(1 - \mu^2) \ll 1$.

We note that this exponent $3/2$ is reduced in the case that the erratic part of the drift is increased over the steady net drift component.

One important consequence exists for the resulting radio emission. When the density of energetic particles is largest near the equator, where the magnetic field is strongest, then we have the case of maximum radio emission. In the other extreme case, that the density of energetic particles is maximal near the poles, where the magnetic field is minimal, then the radio emission is minimal, i.e., usually unobservable. All this is relevant, of course, only in the simplified picture of a homogeneous structure of the stellar wind. If there were a rapid cycling of polarities, e.g., much faster than on the Sun, then the period may interfere with this picture; the consequences have not yet been worked out.

E. The Spectrum Beyond the Knee

This means that from E_{knee} the energy gain of all particles in the entire colatitude range, that is affected by diffusion, has to be considered together. In our model for the diffusion in θ we have used the drift velocity and the distance to the symmetry axis as natural scales in velocity and in length. When the θ-drift reaches the maximum, derived earlier, (see Eq. 10), then the latitude drift changes character. This happens at a critical energy, which is reached at

$$\kappa_{\theta\theta,1} = \kappa_{\theta\theta,\mathrm{max}} \tag{45}$$

which translates into (see Eq. 40)

$$E_{\mathrm{crit}} \left(\frac{3}{4} \frac{U_1}{c} \right)^2 E_{\mathrm{max}} = E_{\mathrm{knee}}. \tag{46}$$

We emphasize that two different basically geometric arguments lead to the same critical energy E_{knee} because we have derived the same critical energy from arguments about the polar cap. However, it is from this argument here, that the spectrum beyond the knee follows.

Above this critical particle energy then the particles average in their path over the small curvature scales, and perceive dominantly the full curvature of the system. Because at full curvature $1/r$ there is no direction ambiguity; the same argument applied as above indicates that we have a net balance of $1 - 1/4$ for the curvature term, reducing it from its value (at $U_1/U_2 = 4$) of 2 to 3/4, i.e., in Eq. (23); the numerical value of the first bracket goes from 3 to 1.75, thus reducing the value of x from 2.25 to 1.729. This then leads to an overall spectrum of $E^{-2.735}$, before taking leakage into account. We emphasize that this result is based on putting all the parameters which go into it at their most simple or most extreme limit. Again, using the correction implied by a finite wind speed of $V_W/U_1 = 0.2$ gives a modified spectrum of $E^{-2.866}$. Therefore we associate an error range of 0.13 with the spectral index derived above. We note in addition that in our model where the supernova explosions into the interstellar medium and the supernova explosions into stellar winds both contribute to moderate particle energies, these differences in the sources may imply differences in the propagation, which can only be quantified after a thorough investigation of the cosmic ray propagation in the

Galaxy. Thus, finally, the spectrum is

$$E^{-3.07-0.07\pm0.07} \tag{47}$$

with leakage accounted for. This is what we wanted to derive. As noted above the use of limiting arguments to derive the spectrum on either side of the knee implies that the knee itself may be quite soft, and thus curvature is to be expected (see Biermann 1994c).

F. Assumptions and Systematic Uncertainties

The assumptions adopted are inspired by Prandtls mixing length approach; all use the key proposition that the *smallest dominant scale*, either in geometric length, or in velocity space, gives the natural transport coefficient. In this sense the assumptions are *derived from a basic principle* which we postulate.

Our basic **argument 1**, *based on observational evidence as well as theoretical reasoning*, is that for a cosmic ray mediated shock the *convective random walk* of energetic particles perpendicular to the *unperturbed* magnetic field can be described by a diffusive process with a downstream diffusion coefficient $\kappa_{rr,2}$ which is given by the thickness of the shocked layer and the velocity difference across the shock, and is independent of energy.

The upstream diffusion coefficient can be derived in a similar way, but with a larger scale based on the *same column density as in the downstream layer*. This leads to the second critical **argument 2**, namely that the upstream length scale is just U_1/U_2 times larger.

It must be remembered that the large intensity of convective turbulence increases the curvature. The characteristic scale of the turbulence is $r/4$ for strong shocks, again, as an example, in the case of the wind-supernova, and thus the curvature is $4/r$ maximum. Half the maximum of the curvature allows for the net balance of gains and losses for the energy gain due to drifts (**argument 3**), and so we obtain then for the curvature $2/r$ which is twice the curvature without any turbulence; this increases the curvature term for the spectral range below the knee.

The diffusion tensor component $\kappa_{\theta\theta}$ can be derived similar to our heuristic derivation of the radial diffusion term κ_{rr}, again by using the smallest dominant scales. The characteristic velocity of particles in θ is given by the erratic part of the drifting, corresponding to spatial elements of different magnetic field direction. This is on average the value of the drift velocity $|\,V_{d,\theta}\,|$, possibly modified by the locally increased values of the magnetic field strength. The characteristic length is the distance to the symmetry axis $r\sin\theta$ (**argument 4**); this is the smallest dominant scale as soon as the thickness of the shocked layer is larger than the distance to the symmetry axis, i.e., $\sin\theta < U_2/U_1$.

Rapid convection also gives a competing diffusion in the θ-direction, independent of particle energy; this will begin to dominate as soon as the energy dependent θ-diffusion coefficient reaches this maximum at a critical energy. As long as the θ-diffusion coefficient is smaller, it will dominate particle transport in θ and the upper limit derived here is irrelevant. When the θ-diffusion coefficient reaches and passes this maximum given by the fast convection, then the particle in its drift will no longer see an increased curvature due to the convective turbulence due to averaging. The part of drift acceleration due to increased curvature is eliminated. Again, a detailed consideration of gains and losses of the drift energy gains leads to the spectrum of particles beyond the knee. The critical energy derived in this way is the same as that derived from a phase-space argument near the poles.

All these arguments are inspired by Prandtl's mixing length approach; all use the key proposition that the **smallest dominant scale**, either in geometric length, or in velocity space, gives the diffusive transport discussed. We assume this to be true even for the anisotropic transport parallel and perpendicular to the shock.

In addition, we (1) use the simplified notion of a purely spherical shock; (2) ignore the modifications of the shock introduced by the cosmic rays themselves, except in the conceptual derivation of the initial argument, where the cosmic rays are critical for the instability; (3) use a test particle approach; (4) have not checked with a full three-dimensional calculation on a supercomputer system that our conceptualization of the local physics actually yields what we argue; and (5) also have not checked with such a calculation that our concept of the *smallest dominant scale* is as general as we use it. We note, however, that this concept is already contained in early descriptions of Prandtl's mixing length scale argument.

We should emphasize very strongly that these uncertainties mean that the spectral indices derived for the power law section of the various components of the cosmic rays correspond to a limiting argument. If things were really not as simple—and they are likely to be much more complicated— then the spectrum derived and any comparison with data must be taken with considerable caution.

G. Summary of the Predictions

The proposal is that three sites of origin account for the cosmic rays observed: (1) supernova (SN) explosions into the interstellar medium (ISM), ISM-SN; (2) supernova explosions into the stellar wind of the predecessor star, wind-SN; and (3) radio galaxy hot spots. Here the cosmic rays attributed to supernova-shocks in stellar winds, wind-SN, produce an important contribution at all energies up to 3×10^9 GeV.

Particle energies go up to $100Z$ TeV for ISM-SN, and to $100Z$ PeV with a bend at $600Z$ TeV for wind-SN. Radio galaxy hot spots go up to nearly or slightly beyond 100 EeV at the source. These numerical values are estimates with uncertainties of surely larger than a factor of 2, because they derive from an estimated strength of the magnetic field, and estimated values of the effective shock velocity (see above).

The spectra are predicted to be $E^{-2.75\pm0.04}$ for ISM-SN (Biermann and Strom 1993), and for wind-SN $E^{-2.67-0.02\pm0.02}$ and $E^{-3.07-0.07\pm0.07}$ below and above the knee, respectively, and $E^{-2.0}$ at injection for radio galaxy hot spots. The polar cap of the wind-SN contributes an $E^{-2.33}$ component (allowing for leakage from the Galaxy), which, however, contributes significantly only near and below the knee, if at all. The uncertainty in the radio galaxy spectra is discussed elsewhere. These spectra are for nuclei and are corrected for leakage from the Galaxy. Electron spectra are discussed below.

The chemical abundances are near normal for the injection from ISM-SN, and are strongly enriched for the contributions from wind-SN. Helium and heavier elements are dominantly from wind-SN already at GeV particle energies. At the knee the spectrum bends downwards at a given rigidity, and the heavier elements bend downwards at higher energy per particle. Thus beyond the knee the heavy elements dominate all the way to the switchover to the extragalactic component, which is, once again, mostly hydrogen and helium, corresponding to what is expected to contribute from the interstellar medium of a radio galaxy, as well as from any intergalactic contribution mixed in (Biermann 1993b). This continuous mix in the chemical composition at the knee renders the overall knee feature in a spectrum in energy per particle unavoidably quite smooth, a tendency which can only partially be offset by the possible polar cap contribution, because that component also is strongest at a given rigidity.

We note that further uncertainties of the spectrum derive from (a) the time evolution of any acceleration process as the shock races outward; (b) the match between ISM-SN and wind-SN; (c) the mixing of different stellar sources with possibly different magnetic properties; and (d) the differences in propagation in any model which uses different source populations. These uncertainties translate into a distribution of power law indices of the spectra, to curvature of the spectra, to a smearing of the knee feature, and to a smoothing of the cutoffs. Obviously, this is in addition to the underlying uncertainty associated with the concept of the *smallest dominant scale* itself.

There is one important prediction for stars. The particle energies of cosmic rays in the wind component go up to $100Z$ PeV, which *necessarily* implies that those massive stars which have strong winds into which they explode as supernovae, have appreciable magnetic fields at that stage. The magnetic field is required to be of order 3 Gauss at a distance of 10^{14} cm from the hydrostatic surface of the star. If the magnetic field topology is of a Parker type, i.e., is mostly tangential and depends on radius r as r^{-1}, right down to near the surface of the star, as argued in Biermann and Cassinelli (1993) and below, then the surface magnetic fields are still quite moderate, e.g., of order a few thousand Gauss for Wolf Rayet stars. We explore some of the consequences below.

IV. WINDS OF MASSIVE STARS

In this section we discuss the winds of massive stars, and the consequences of the proposition that the magnetic fields may not be totally negligible, i.e., may be of the order of 3 Gauss at 10^{14} cm radial distance in the stellar wind.

A. Dynamo

First we propose that high magnetic fields can be generated inside massive stars so that later, when these interiors get exposed to become the surface of Wolf Rayet stars these high magnetic fields can help drive the wind. The dynamo mechanism is believed to act in the turbulent zone in the interior of rotating massive stars and, given a seed field such as may be produced by a battery mechanism in a rotating star, increases the magnetic field strength up to a maximum—which can be estimated in different ways. One possible limit is the dynamic pressure of the convective motions. The analogy with the interstellar medium suggests the limit where the Coriolis force equals the magnetic stresses (Ruzmaikin et al. 1988; Gilbert and Childress 1990; Gilbert 1991). The strength of the magnetic field can then be estimated from the condition, that the magnetic torque is limited by the Coriolis forces over the size of the convective region. This condition can be written as

$$B = (\Omega \rho_c R_{cc} v_t f_1)^{1/2} \tag{48}$$

where Ω is the rotation rate of the star, ρ_c is the average density of the convection zone, R_{cc} is the radius of the convective core, v_t is the characteristic turbulent velocity, and f_1 is a correction factor of order unity in order to allow for (a) structural variations ignored here, and also for (b) the fact that the dominant length scale is likely to be smaller than the radius of the convective region. The turbulent velocity is estimated from the condition that turbulent convection transports all the luminosity L:

$$L = 4\pi R_{cc}^2 \rho_c v_t^3 f_2 \tag{49}$$

where f_2 is also a correction factor of order unity to allow for structural variations (we use the average density and the radius of the entire convective region).

Models of N. Langer (1992, personal communication) provide the input together with data from the textbook of Cox and Giuli (1968):

$$R = 9.0 \, \text{R}_\odot \left(\frac{M_\star}{40 \, \text{M}_\odot} \right)^{0.512} \tag{50}$$

$$M_{cc}/M_\star = 0.628 \left(\frac{M_\star}{40 \, \text{M}_\odot} \right)^{0.466} \tag{51}$$

$$R_{cc}/R = 0.384 \left(\frac{M_\star}{40 \, \text{M}_\odot} \right)^{0.377} \tag{52}$$

where R and $M_\star$ are the stellar radius and mass, respectively. This leads to a magnetic field of

$$B = 1.9 10^6 \alpha_{rot}^{1/4} f_1^{1/2} f_2^{-1/6} \left(\frac{M_\star}{40 \, \text{M}_\odot} \right)^{-0.058} \, \text{Gauss} \tag{53}$$

where $\alpha_{\rm rot}$ is the fraction of critical rotation at the surface, assumed to be solid body rotation. The same argument for Wolf Rayet stars (using data of Langer [1989], with helium mass fraction $Y = 1$) gives

$$B = 2.3 10^7 \alpha_{\rm rot}^{1/4} f_1^{1/2} f_2^{-1/6} \left(\frac{M_W}{5\,{\rm M}_\odot}\right)^{0.035} \text{Gauss.} \tag{54}$$

Hence this readily produces magnetic fields, dependent on the rotation rate of the star, of up to 2×10^6 Gauss for O stars, and about 10 times more for Wolf Rayet stars. In both cases, the induced magnetic field is nearly independent of stellar mass. Critical rotation here means that the convective core has to rotate at an *angular* velocity which would correspond at the surface to critical rotation there; but in fact, as we will see below, it is not required that the actual surface rotates this fast.

In conclusion, we find that we can rather easily generate magnetic fields at levels deep inside the star beyond the local virial limit at the surface of the star, which gives of the order of a few times 10^4 Gauss nearly independent of the rotation rate (Maheswaran and Cassinelli 1988,1992). Rotationally induced circulations may be able to carry these magnetic fields to the surface of the stars on a time scale much shorter than the main sequence life time. In this transport, the magnetic field is weakened by flux conservation and so surface fields in the range 10^3 to 10^4 Gauss are quite plausible, and are below the local virial theorem limit. Such values are all we require in the following. We will use these weaker magnetic field strengths, in the following, so that the uncertainty of what really limits the dynamo mechanism, can be checked more from observation than theoretical faith. These estimates are valid for massive stars and extend clearly to below the mass range where we have Wolf Rayet stars as an important final phase of evolution. We thus expect many of the arguments in the following to hold generally for all massive single stars with extended winds, whether slow or fast, whether red or blue supergiant preceding the supernova explosion.

There is a further consequence for the generation of magnetic fields in white dwarfs. The most massive stars that do not become supernovae, but white dwarfs, are sufficiently massive to contain also convective cores which get exposed when the white dwarf is formed. Thus there ought to be a correlation between massive white dwarfs and the detection of strong magnetic fields; this is consistent with the observational data (J. Liebert 1992, personal communication; Schmidt et al. 1992).

An interesting additional observation is the high incidence of kilo-Gauss fields on Am stars (Lanz and Mathys 1993); three of four Am stars searched have such strong fields, and there is evidence that the field is fairly disordered. The important point is that such stars have an outer radiative zone, and a core convection zone just like the more massive stars. Ultimately we may find a connection in the build-up of core magnetic fields in all upper main sequence stars with an inner convection zone.

It is not clear at present whether the magnetic fields on the surface of OB stars are fossil remnants from a pre-main-sequence phase, or are fields that have been carried through the radiative zone by circulations. Calculations for this process similar to those of Charbonneau and MacGregor (1992) remain to be done and are required to demonstrate that this latter process is really possible. Such a transport mechanism must remain an assumption at this stage.

B. Wind-Driving

In the standard version of the fast magnetic rotator theory (Hartmann and MacGregor 1982; Lamers and Cassinelli 1994, ch. 8) the magnetic field is calculated self-consistently and turns out to be radial close to the stellar surface and then starts bending at the critical point where the radial Alfvén velocity, the local corotation velocity and the wind velocity all coincide. This produces a long lever arm for the loss of angular momentum, and is the essence of the criticism of Nerney and Suess (1987) against the model. The spindown of fast magnetic rotator Wolf-Rayet stars is a topic addressed by Poe et al. (1989), where it is assumed that the field is radial close to the stellar surface. However, the magnetically driven winds do not require the magnetic field to be radial near the surface. In a convection zone near the surface, it may be plausible to assume that the magnetic field is nearly isotropic in its turbulent character below the region where the wind gets started, and radial in the wind zone near to the star, leaving the radial magnetic field lines dominant for as long as the flow velocity is below the radial Alfvén speed. We note that the surface magnetic field structure on the Sun is mostly radial at the level where we can observe it, and we can only infer its structure below the surface from detailed modeling.

However, in a star, where the outer layers are radiative, it is not at all clear that the magnetic field is initially radial. Even the slightest differential rotation will tend to make a magnetic field, which originates in the central convective region, to be tangential. Even the acceleration into a wind does not completely overturn this tendency; we simply do not know. In fact, the recent model calculations of Wolf Rayet star winds by Kato and Iben (1992) suggest that the critical point of the wind where the wind becomes supersonic may be already inside the photosphere, and so it is reasonable to suppose that the other critical point where the wind speed exceeds the local radial Alfvén velocity may also be inside the star, if there is such a point at all—the radial flow velocity may be faster than the radial Alfvén velocity throughout. Similarly, the arguments by Lucy and Abbott (1993) which discuss the effect of multiple scattering in radiation driving, also lead to an effective initial acceleration of the wind, which may lead to a fast wind which is already near the photosphere.

In the following we propose to discuss such a wind, generalizing from Parker (1958) and Weber and Davis (1967). As in Weber and Davis we limit ourselves here to the equatorial region. We use magneto-hydrodynamics and Maxwells equations and thus have for the angular momentum transport L_J

per unit mass:

$$L_J = r v_\phi - \left(\frac{B_r}{4\pi\rho v_r} \right) r B_\phi \; = \; \text{const.} \tag{55}$$

Here the components of the magnetic field are B_ϕ and B_r, and the components of the wind velocity are v_ϕ and v_r, while ρ is the density and the index a refers to a reference radius, which may or may not correspond to the stellar surface.

Introducing for the magnetic flux

$$F_B = r^2 B_r = r_a^2 B_{\text{ra}} \; = \; \text{const} \tag{56}$$

with flux freezing

$$r(v_r B_\phi - v_\phi B_r) = -\Omega F_B \; = \; \text{const} \tag{57}$$

and mass flux

$$\dot{M} = 4\pi\rho r^2 v_r = 4\pi\rho_a r_a^2 v_{\text{ra}} \; = \; \text{const} \tag{58}$$

the tangential velocity can be written as

$$v_\phi = \frac{L_J}{r} \left(1 - \frac{F_B^2}{\dot{M}} \frac{\Omega}{L_J v_r} \right) \Big/ \left(1 - \frac{F_B^2}{\dot{M}} \frac{1}{r^2 v_r} \right). \tag{59}$$

The angular momentum loss can be written as

$$L_J = \epsilon \Omega r_a^2 \tag{60}$$

where ϵ defines the location from where the angular momentum loss is actually occurring. If there is no Alfvén critical point outside the star, then obviously

$$\epsilon < 1. \tag{61}$$

We note that the term $F_B^2/(\dot{M} v_r r^2)$ appearing in the expression for the tangential velocity can be rewritten with the surface radial Alfvén Mach number

$$M_{\text{Ara}} = v_{\text{ra}}/v_{\text{Ara}} \tag{62}$$

where

$$v_{\text{Ar}} = B_r/(4\pi\rho)^{1/2} \tag{63}$$

as

$$\frac{F_B^2}{\dot{M}} \frac{1}{v_r r^2} = \frac{1}{M_{\text{Ara}}^2} \frac{v_{\text{ra}}}{v_r} \frac{r_a^2}{r^2} = \frac{1}{M_{\text{Ar}}^2}. \tag{64}$$

Similarly we have the relationship

$$\frac{F_B^2}{\dot{M}} \frac{\Omega}{L_J v_r} = \frac{1}{M_{\text{Ar}}^2} \frac{r^2}{r_a^2} \frac{1}{\epsilon} = \frac{1}{M_{\text{Ara}}^2} \frac{v_{\text{ra}}}{v_r} \frac{1}{\epsilon}. \tag{65}$$

Thus the tangential velocity can be written as

$$v_\phi = \frac{L_J}{r}\left(1 - \frac{v_{\mathrm{ra}}}{M_{\mathrm{Ara}}^2 \epsilon v_r}\right)\Big/\left(1 - \frac{1}{M_{\mathrm{Ar}}^2}\right). \tag{66}$$

We can reasonably assume that the radial velocity is steadily increasing with radius. The tangential velocity should neither be negative nor exceed the rotational velocity of the star itself, and so we derive the conditions

$$M_{\mathrm{Ara}} > 1 \tag{67}$$

and

$$M_{\mathrm{Ara}}^{-2} < \epsilon < 1 \tag{68}$$

for the conditions envisaged here, that there is no Alfvén critical point outside the star. These conditions translate into

$$\frac{1}{M_{\mathrm{Ara}}^2}\frac{v_{\mathrm{ra}}}{v_r}\frac{1}{\epsilon} < 1 \tag{69}$$

and

$$\frac{1}{M_{\mathrm{Ar}}^2} < 1. \tag{70}$$

It follows, e.g., that $M_{\mathrm{Ar}} \sim r$ asymptotically.

We define U_ϵ and U_M and obtain

$$0 < U_\epsilon = 1 - \frac{\epsilon r_a^2}{r^2} < 1 \tag{71}$$

and

$$0 < U_M = 1 - \frac{1}{M_{\mathrm{Ar}}^2} < 1. \tag{72}$$

The tangential magnetic field can then be written as

$$B_\phi = -\frac{F_B \Omega}{r v_r}\left(1 - \frac{\epsilon r_a^2}{r^2}\right)\Big/\left(1 - \frac{1}{M_{\mathrm{Ar}}^2}\right) \tag{73}$$

and is thus also without change of sign outside the star. It is easy to verify that no magnetic flux is transported to infinity (see Parker 1958).

The radial momentum equation is

$$\left(v_r \frac{\mathrm{d}}{\mathrm{d}r} v_r\right)\left(1 - \zeta \frac{c_s^2}{v_r^2}\right) =$$

$$2\zeta \frac{c_s^2}{r} - \frac{GM_\star}{r^2} + \frac{F_{\mathrm{rad}}\sigma_T N}{m_p c} + \tag{74}$$

$$\frac{v_\phi^2}{r} - \frac{1}{8\pi\rho r^2}\frac{\mathrm{d}}{\mathrm{d}r}(r B_\phi)^2.$$

where ζ indicates the nonadiabacity of the flow:

$$\zeta = \left(\frac{dP}{d\rho}\right)_{\text{along } r} \Big/ \left(\frac{dP}{d\rho}\right)_{\text{adiabatic}} . \tag{75}$$

We will adopt $\zeta = 1$ for simplicity in the following.

In this equation the radiation force (flux F_{rad}) on both lines and continuum is given with the correction factor N over the Thompson cross section, also including the effect due to the chemical element composition being different from pure hydrogen; this factor N depends on the physical state of the gas. The adiabatic speed of sound is c_s. We here proceed to evaluate the last term with the expressions already derived and discuss it in the context of the momentum equation.

The gradient term of $(r B_\phi)^2$ in the momentum equation can then be rewritten as (on the right-hand side of the momentum equation)

$$-\frac{1}{8\pi\rho r^2}\frac{d}{dr}(r B_\phi)^2 =$$

$$\left(v_r\frac{d}{dr}v_r\right)\frac{1}{M_{\text{Ara}}^2}\left(\frac{r_a\Omega}{v_{\text{ra}}}\right)^2\left(\frac{v_{\text{ra}}}{v_r}\right)^3\frac{U_\epsilon^2}{U_M^3}$$

$$-2\frac{1}{M_{\text{Ara}}^2}r_a\Omega^2\left(\frac{r_a}{r}\right)^3\frac{v_{\text{ra}}}{v_r}\frac{U_\epsilon}{U_M^3}\left(\epsilon-\frac{1}{M_{\text{Ara}}^2}\frac{v_{\text{ra}}}{v_r}\right). \tag{76}$$

A second term on the right-hand side of the momentum equation is the centrifugal force, which can be written as

$$\frac{v_\phi^2}{r} = r_a\Omega^2\left(\frac{r_a}{r}\right)^3\frac{1}{U_M^2}\left(\epsilon-\frac{1}{M_{\text{Ara}}^2}\frac{v_{\text{ra}}}{v_r}\right)^2. \tag{77}$$

The last term in brackets never goes through zero outside the star because of the conditions we have set above for ϵ and M_{Ara}. Comparing now the centrifugal term with the second term from the gradient of the tangential magnetic field, we note that for dv_r/dr approaching 0 these two terms differ by the factor

$$-\frac{1}{2}\frac{U_M}{U_\epsilon}\left(\epsilon M_{\text{Ara}}^2\frac{v_r}{v_{\text{ra}}}-1\right) \tag{78}$$

where $\epsilon M_{\text{Ara}}^2 > 1$. If v_r/v_{ra} exceeds a value of 3 at large radius r, then the centrifugal term, which provides some acceleration, dominates over the other term at large r. Here we have ignored all the terms of order unity that approach unity with both the radius r and the radial velocity v_r becoming large. Even close to the star the centrifugal force may dominate.

The first term in the gradient of the tangential magnetic field is more interesting, however. This term becomes a summand to the various terms multiplying $(v_r\frac{d}{dr}v_r)$ and has there, on the left-hand side, a minus sign and is to be compared with unity in the case that we are already at supersonic speed. This may be attained through radiation driving with multiple scattering.

We define an Alfvén Mach number with respect to the total magnetic field with

$$M_A^2 = \frac{v_r^2 4\pi\rho}{B_\phi^2 + B_r^2}.$$

(79)

Generally we have obviously

$$M_A < M_{\mathrm{Ar}}.$$

(80)

With this generalized Alfvén Mach number we can rewrite the entire factor to $(v_r \frac{\mathrm{d}}{\mathrm{d}r} v_r)$ in a simple form

$$1 - \frac{\zeta}{M_s^2} - \left(\frac{M_{\mathrm{Ar}}^2}{M_A^2} - 1\right)/(M_{\mathrm{Ar}}^2 - 1)$$

(81)

where M_s is the sonic Mach number for the radial flow velocity. Critical points appear whenever this expression goes through zero or a singularity. This expression is easily seen to be equivalent to Eq. (6) of Hartmann and MacGregor (1982); they, however, find a magnetic field which is initially radial and neglected radiative forces. This also shows that we have the critical point of Weber and Davis (1967) again for $\frac{\zeta}{M_s^2} \ll 1$, when and if $M_{\mathrm{Ar}} = 1$, but we have in our case another critical point at $M_A = 1$, which is the fast magnetosonic point (see Hartmann and MacGregor 1982). In order to make a realistic judgment on this question, we would have to combine the model of Kato and Iben (1992) or a detailed calculation of multiple scattering (Lucy and Abbott 1993) to drive winds initially with a realistic treatment of the magnetic field inside the star, in order to derive the range of possible properties of the magnetic field near the surface of the star.

However, there are a few general conclusions one can already draw. If the magnetic field inside the star begins as a dominantly tangential field, then it is by no means clear whether there is any point at which we have $M_{\mathrm{Ar}} < 1$, either inside or outside the star going outward; the unwinding of the magnetic field is intimately coupled to the initially weak radial flow, and so $M_{\mathrm{Ar}} > 1$ may hold throughout. In that case, the term derived from the radial gradient of the tangential magnetic field goes through unity only when the generalized Alfvén Mach number goes through unity. If, as we argued, $M_{\mathrm{Ar}} > 1$ possibly throughout, then, again going outward with radius, we start with a low generalized Alfvén Mach numbers M_A, either supersonic or subsonic flow, the entire expression is negative, matching the negative gravitational force on the right-hand side. Far out, both sonic and generalized Alfvén Mach numbers are large, and so the expression is positive. It follows that the expression must go through zero. The condition $M_A = 1$ here is the fast magnetosonic point, and the radial velocity there corresponds to the

Michel velocity. Or, in other words, the radial velocity is equal to the total Alfvén velocity.

We know from observations that the winds in OB and Wolf Rayet stars are strongly supersonic; using our model, we deduce that the radial velocity is weakly super-Alfvénic with respect to the tangential (dominant) magnetic field component, and so we expect far outside the star a configuration where $M_s >> 1$, $M_{\mathrm{Ar}} >> 1$ and $M_A \gtrsim 1$, but this latter condition may not necessarily be satisfied by much. This means that the expression approaches unity and somewhat smaller than unity. This entails the condition that the right hand side has to be positive, which is readily interpreted as possibly arising from line radiation, because that term is the only one which has the same radial dependence as the gravitational force. Comparing then the effect of line driving (Lucy and Solomon 1970; Castor et al. 1975) with and without such a magnetic field, the net effect is an amplification in the sense that for $M_{\mathrm{Ar}}^2 \gg 1$, the velocity gradient is asymptotically increased by

$$1 \Big/ \left(1 - \frac{1}{M_A^2} \right). \tag{82}$$

For M_A close to unity, this factor can be arbitrarily large, and thus illustrates the amplification possible from the pressure gradient of a tangential magnetic field. Obviously, such a large amplification is present only for a small radial region, and so the final wind velocity can still be not much more than the tangential Alfvén velocity (see below). Thus we argue that there is a magnetic field configuration where the field is mostly tangential inside the star, and remains mostly tangential outside the star, where the initial acceleration of the wind occurs by the opacity mechanism discussed by Kato and Iben (1992) or by multiple scattering (Lucy and Abbott 1993) in a line-driving modus; outside the star we have a line-driven wind, but the amplification by the effect of the pressure gradient of the tangential magnetic field actually produces the large momentum in the wind.

To see the properties of the equations better, we simplify by dropping all right-hand terms in the differential equation except for the gravitational and the radiative force, and introduce characteristic length and velocity scales

$$r_\star = \left(\frac{GM_\star}{v_{\mathrm{ra}}^2} \right) \left(\frac{v_{\mathrm{ra}}}{v_{\mathrm{Ara}}} \right)^{4/3} \left(\frac{v_{\mathrm{ra}}}{r_a \Omega} \right)^{4/3} \tag{83}$$

and

$$v_\star = v_{\mathrm{ra}} \left(\frac{v_{\mathrm{Ara}}}{v_{\mathrm{ra}}} \right)^{2/3} \left(\frac{r_a \Omega}{v_{\mathrm{ra}}} \right)^{2/3} \tag{84}$$

which is easily recognized as the Michel velocity (Hartmann and MacGregor 1982). This equation corrects a misprint in the corresponding expression in Stanev et al. (1993), in that the second exponent is 2/3 instead of 1/3. These normalizations lead to the same overall dimensionless form of the momentum equation as in Owocki (1990), because we have to multiply the entire Eq. (73) with

$$\frac{r^2}{r_\star v_\star^2} = \frac{r^2}{GM_\star}.$$

(85)

The Eddington luminosity is given by

$$L_{\rm edd} = \frac{4\pi GM_\star m_p c}{\sigma_T}$$

(86)

where σ_T is the Thompson cross section. In the approximation that $U_\epsilon = U_M \simeq 1$ and $r\Omega \gg v_r$, the Michel velocity corresponds to the overall Alfvén velocity.

Next we write down the whole differential wind equation in dimensionless form:

$$-\frac{1}{2}\frac{dy^2}{d(1/x)}\left[-\frac{\zeta}{M_s^2} - \left(\frac{M_{\rm Ar}^2}{M_A^2} - 1\right)\Big/(M_{\rm Ar}^2 - 1)\right] =$$

$$+ 2\zeta c_{\rm sa}^2 r_a \left(\frac{r_a}{r}\right)^{1/3}\left(\frac{v_{\rm ra}}{v_r}\right)^{2/3}\frac{1}{GM_\star}$$

$$+ \frac{LN}{L_{\rm edd}} - 1$$

$$+ \Omega^2 r_a^3 \left(\frac{r_a}{r}\right)\frac{1}{U_M^2}\left(\epsilon - \frac{1}{M_{\rm Ara}^2}\frac{v_{\rm ra}}{v_r}\right)$$

$$\left[-2\frac{U_\epsilon}{U_M}\frac{v_{\rm ra}}{v_r}\frac{1}{M_{\rm Ara}^2} + \epsilon - \frac{1}{M_{\rm Ara}^2}\frac{v_{\rm ra}}{v_r}\right]\frac{1}{GM_\star}.$$

(87)

We note that in a full treatment of this equation the temperature enters in the speed of sound, here approximated by an adiabatic law with gas constant 5/3, the state of ionization enters into the factor N, which can be highly space dependent (see, e.g., Lucy and Abbott 1993), and that hence we can discuss here only a very simplified form of this equation. We approximate this expression in several steps in order to write down only two limiting forms.

In the approximation that the sonic and the radial Alfvénic Mach number are both large on the surface, the left-hand side simplifies to

$$-\frac{1}{2}\frac{dy^2}{d(1/x)}\left(1 - \frac{1}{M_A^2}\right)$$

(88)

showing only the fast magnetosonic point as a remaining singularity. At that point, the right-hand side must go from negative to positive, which could happen in several ways. First of all, the pressure term is clearly unimportant, because we already neglected the subsonic regime. The last term, which derives from the centrifugal force and the second part of the magnetic field pressure gradient, is asymptotically positive; on the surface this term may be negative. Writing

$$\epsilon = \frac{E}{M_{\text{Ara}}^2} < 1 \tag{89}$$

with the condition

$$M_{\text{Ara}}^2 > E > 1 \tag{90}$$

we find the condition on the surface

$$E < \frac{3M_{\text{Ara}}^2 - 1}{M_{\text{Ara}}^2 + 1} \tag{91}$$

for the last term in Eq. (87) to be negative on the surface. It thus depends on a detailed consideration of the inner structure of the star and its rotation law to determine whether this term can be sufficiently large and negative on the surface to compensate the radiative driving term, and so to locate the fast magnetosonic point.

Another possibility is that the radial dependence of the radiative driving term, i.e., the radial variation of the factor N determines the location of the fast magnetosonic point. In that case we may be allowed to neglect the centrifugal and the remaining term of the magnetic pressure gradient.

The differential wind equation then reads finally in dimensionless length x and velocity y

$$\frac{1}{2}\left(1 - \frac{1}{y^3}\right)\frac{\mathrm{d}y^2}{\mathrm{d}(1/x)} = -f_{\text{rad}} \tag{92}$$

with

$$f_{\text{rad}} = \frac{NL}{L_{\text{edd}}} - 1. \tag{93}$$

We note that the radial dependence of the magnetic field strength leads asymptotically to

$$\left(\frac{M_{\text{Ar}}^2}{M_A^2} - 1\right) / \left(M_{\text{Ar}}^2 - 1\right) = \left(y^3 \frac{U_M^3}{U_\epsilon^2}\right)^{-1} \simeq y^{-3} \tag{94}$$

in the limit, that $U_M^3/U_\epsilon^2 \to 1$.

Clearly we require then that the radiative force dominates over gravity, by however little, to obtain the correct sign of the right-hand side, thus $f_{\text{rad}} > 0$. We assume here for illustration that N is constant with r.

The analytic solution is

$$\frac{1}{2}(y^2 - y_i^2) + \frac{1}{y} - \frac{1}{y_i} = -f_{\text{rad}}\left(\frac{1}{x} - \frac{1}{x_i}\right) \tag{95}$$

where the index i refers to an initial state. For $y \gg 1 > y_i$ and $x \gg x_i$, the limiting solution is

$$y = \sqrt{2\left(\frac{1}{y_i} + f_{\text{rad}}\frac{1}{x_i}\right)}. \tag{96}$$

We note right away that both effects contribute to a faster wind. Thus, the analytic solution yields two extreme possibilities, either

$$v_{r\infty} = v_e \left(\frac{NL}{L_{\text{edd}}} - 1 \right)^{1/2} \tag{97}$$

or

$$v_{r\infty} = \sqrt{2} \left(\frac{v_{\text{Ara}}}{v_{\text{ra}}} \right)^{1/3} \left(\frac{r_a \Omega}{v_{\text{ra}}} \right)^{1/3} v_\star \tag{98}$$

where we have used the definition

$$v_e = \left(\frac{2GM_\star}{r_a} \right)^{1/2} \tag{99}$$

for the surface escape velocity from the star. Obviously, we require that

$$2^{3/2} \left(\frac{v_{\text{Ara}}}{v_{\text{ra}}} \right) \left(\frac{r_a \Omega}{v_{\text{ra}}} \right) > 1 \tag{100}$$

in order to have a final velocity above the Michel velocity. We note that

$$\frac{\Omega r}{v_r} = - \frac{B_\phi}{B_r} \frac{U_M}{U_\epsilon} \tag{101}$$

for any r and also on the surface, and thus the condition is likely to be fulfilled for a magnetic field configuration which is highly tangential at the stellar surface. By the same condition, the initial surface velocity is below the Michel velocity (except for the factor 2 which is of order unity). This is necessary for the flow to have a transition through the fast magnetosonic point. This sharpens the approximations introduced above. We note that in the magnetic driving case, the final velocity far outside can also be written as

$$v_{r\infty} = v_e \sqrt{2} \frac{v_{\text{Ara}}}{v_{\text{ra}}} \frac{r_a \Omega}{v_e} \tag{102}$$

where both factors to $v_e \sqrt{2}$ are less than unity in our approximation, and so the case of pure magnetic driving in this approximation may yield lower velocities than radiation driving at large distances from the star in this very simple first approximation. The transition between the two extreme cases is given by

$$2 \left(\frac{v_{\text{Ara}}}{v_{\text{ra}}} \frac{r_a \Omega}{v_e} \right)^2 \simeq \frac{NL}{L_{\text{edd}}} - 1. \tag{103}$$

We have therefore in this approximation two possible extreme solutions. First, for a small magnetic field, the Alfvén velocity drops out and the line driving is the regulating agency. Second, for a large magnetic field, the terminal wind velocity is not far from the tangential Alfvén speed, and the line driving effect is important but not dominant in that it provides the source of momentum to be amplified. This latter picture is the concept we propose to explain the momentum in the winds of Wolf Rayet stars. It is different from the solutions of Hartmann and MacGregor (1982) in the sense, that in their wind solutions the main acceleration of the wind is between the slow magnetosonic point and the Alfvén point, both of which are not outside the star in the case which we discuss. Our solutions are similar, however, in that both in their case as here, the terminal velocity of the wind is not far from the the total Alfvén velocity. In this picture the angular momentum loss of the star refers to a characteristic level inside the star and so the criticism of Nerney and Suess (1987) is countered; the angular momentum loss of the star through mass loss is reduced.

Introducing then the radiative driving in the form of Castor et al. (1975), and using the nomenclature of Owocki (1990), we have for the radiation driving the following form

$$\frac{NL}{L_{\text{edd}}} = C_{\text{CAK}} \left(-y \frac{dy}{d(1/x)} \right)^{\alpha} + \Gamma \qquad (104)$$

where C_{CAK} is related to the mass loss rate (Owocki 1990, their Eq. 14) and $\Gamma = L/L_{\text{edd}}$ describes the effect of the continuum radiation, and thus in a second approximation

$$\frac{1}{2} \left(1 - \frac{1}{y^3} \right) \frac{dy^2}{d(1/x)} = -C_{\text{CAK}} \left(-y \frac{dy}{d(1/x)} \right)^{\alpha} + 1 - \Gamma. \qquad (105)$$

This, then, replaces Eq. (13) of Owocki (1990). Here $C_{\text{CAK}}(-y \frac{dy}{d(1/x)})^{\alpha}$ describes the driving by radiative forces in the approximation by Castor et al. (1975), and describes the spatial dependence of N above. This shows that we have a critical point where $y = 1$ and where $C_{\text{CAK}}(-y \frac{dy}{d(1/x)})^{\alpha} = 1 - \Gamma$. We must ask here, whether we have a critical point at all, where $M_A = 1$. Consider the case of low magnetic field; then the initial surface flow velocity may well be super-Alfvénic, and so $y_a > 1$, and if in fact $y_a >> 1$, then the magnetic field can be neglected. However, if the initial velocity is sub-Alfvénic and the final radiation-driven wind is super-Alfvénic, then magnetic driving certainly cannot be neglected. These conditions can be written as

$$\frac{v_{\text{ra}}}{v_{\star}} = \left(\frac{v_{\text{ra}}}{v_{\text{Ara}}} \right)^{2/3} \left(\frac{v_{\text{ra}}}{r_a \Omega} \right)^{2/3} < 1 \qquad (106)$$

and

$$\frac{v_{\text{CAK}}}{v_{\star}} = \frac{v_e}{v_{\text{ra}}} \left(\frac{v_{\text{ra}}}{v_{\text{Ara}}} \right)^{2/3} \left(\frac{v_{\text{ra}}}{r_a \Omega} \right)^{2/3} \sqrt{\frac{\alpha}{1-\alpha}} > 1. \qquad (107)$$

If these two conditions are fulfilled, then we certainly have a critical point, where $y = 1$. Because $\Omega r_a / v_{\text{ra}}$ is proportional to the ratio of the tangential over radial component of the magnetic field on the reference level of the star (see Eq. 101), a sufficiently dominant tangential field will always ensure that the first condition is fulfilled. Strong wind driving corresponds to α close to unity, and so in many cases the second condition will also be fulfilled. In the case that the first condition is fulfilled, but not the second, the effects of the magnetic field also cannot be neglected, but a full solution is required to discuss this at the required depth. A full solution of Eq. (105) remains to be done.

Using the observed wind velocities gives an estimate for the Alfvén velocity, and thus implies (assuming our wind-driving theory to be correct) that the magnetic field is quite high, of the order of 3 Gauss at the fiducial radius of 10^{14} cm. Or, to turn the argument around, the magnetic field strengths implied by our cosmic ray arguments provide a possible explanation for the origin of the momentum of Wolf Rayet star winds. On the surface of the star the magnetic field strength implied is of order a few thousand Gauss, quite easily within the limits implied by the surface virial theorem. Also the surface of the star may not rotate as fast, but the inside could still rotate at an angular velocity corresponding to near critical at the surface. We emphasize that here our proposed wind-driving theory (if true) provides an independent argument on the magnetic field strength on the surface of Wolf Rayet stars. If it can be shown to be the correct physical interpretation, then these magnetic field strengths follow. On the other hand, if observations show such magnetic field strengths to be too high, then both the wind theory proposed here, as well as some of the cosmic ray arguments may fail.

C. Radio Emission from Stars

In this section we derive the basic expressions for the luminosity, spectrum, and time dependence of the nonthermal radio emission from single spherical shocks in the winds of single massive stars. In subsequent sections we then use these expressions to discuss radio supernovae, young radio supernova remnants in starburst galaxies, Wolf Rayet stars, and OB stars. The radio emission of the nova GK Per (Seaquist et al. 1989) remains to be discussed in the detail required to match it with what we know about cataclysmic variables, close binary star systems with an accretion disk around a white dwarf (Biermann et al. 1995).

Electrons suffer massive losses due to synchrotron radiation and so they can achieve only energies up to that point where the acceleration and loss times become equal; this is a strong function of latitude because the magnetic field varies rather strongly with latitude. The shock transfers a fraction η of the bulk flow energy into relativistic electrons. The emissivity of an electron population with the spectrum

$$N(\gamma)d\gamma = C\gamma^{-p}d\gamma \tag{108}$$

is given by (Rybicki and Lightman 1979) in cgs units for $p = 7/3$ by

$$\epsilon_\nu = 3.90 \, 10^{-18} C B^{5/3} \nu^{-2/3} \mathrm{erg\,s^{-1}cm^{-3}Hz^{-1}}. \tag{109}$$

For a wind speed equal to the shock velocity and the limit of a strong shock in a gas of adiabatic index $5/3$ leads to an optically thin synchrotron spectrum of -0.735, and thus $p = 2.47$. The corresponding expressions for $p = 2.47$ are given in Biermann and Cassinelli (1993). Similarly the absorption coefficient for synchrotron self-absorption is for $p = 7/3$

$$\kappa_{\nu,\text{syn}} = 1.23 \times 10^{12} C B^{13/6} \nu^{-19/6} \text{cm}^{-1}. \tag{110}$$

In all these expressions we have averaged the aspect angle. We will use here, for illustration, the limit of large shock speeds for the discussion of the radio supernovae, where the radio spectral index is $-2/3$. The free-free opacity is given by

$$\kappa_{\nu,\text{ff}} = 0.21 T_e^{-1.35} \nu^{-2.1} n_e^2 \text{cm}^{-1} \tag{111}$$

in an approximation originally derived by Altenhoff et al. (1960) and widely disseminated by Mezger and Henderson (1967). Here T_e and n_e are the electron temperature and density, and the approximation has been used that the ion and electron density are equal, and that the effective charge of the ions is $Z = 1$. We note that for cosmic abundances we have the following approximations (Schmutzler 1987) assuming full ionization: $\rho = 1.3621 m_H n$, $n_e = 1.181n$, and $n_i = 1.086n$, where n is the total hydrogen density (particles of mass m_H per cc), and n_i the ion density. With these approximations for full ionization the free-free opacity would increase a factor of 2 over the approximation by Altenhoff et al. (1960), which, however, is compensated by the incomplete ionization (see, again, the calculations by Schmutzler [1987]) in the photon ionized regions (here, winds) near massive stars. In the following we will use the approximation by Altenhoff et al. replacing $n_e = n$.

The synchrotron luminosity in the optically thin limit is then given by an integration over the emitting volume, which we take in the context of our simplified picture to be a spherical shell of thickness $r/4$. The radio emission only arises from that part of the shell where the shock velocity is larger than the local Alfvén velocity and where the synchrotron loss time is longer than the local acceleration time. These conditions can lead to a restricted range in latitude for the emission (see, e.g., Nath and Biermann 1994b; Biermann et al. 1995). The luminosity is then given by the integral over the latitude dependent emissivity. Here we must normalize the cosmic ray electron density to the shock energy density, using our induced latitude dependence. We will concentrate on the case, when the energetic particle density is stronger near the equator, and discuss the opposite case briefly at the end. Using a lower bound for the electron spectrum much below the rest mass energy and an upper bound much above that value we have for the constant C, then the expression

$$C(\mu) = C_o(1 - \mu^2)^{3/2} \tag{112}$$

with $\mu_\star \geq \mu \geq 0$ where $\mu_\star$ refers to that latitude where the latitude dependent acceleration breaks down:

$$(1 - \mu_\star^2)^{1/2} = \frac{3}{4}\frac{U_1}{c} \ll 1 \tag{113}$$

from our argument about the knee energy (see Biermann 1993a, Sec. 8) for protons and other nuclei. At the equator the energy density of the electrons can be written as

$$C_o = (p - 2)\eta \rho U_1^2 / (m_e c^2) \tag{114}$$

where we use just the relativistic part of the distribution function and incorporate the uncertainty on the existence and strength of any subrelativistic part of the electron distribution function in the factor η. The integration over latitudes leads to an integral correction factor of $\frac{1}{2}B(\frac{p+11}{4}, \frac{1}{2})$, where $B(z_1, z_2)$ is the β-function, and p the power law index of the electron distribution function. For the power law index used here, $p = 7/3$, this integral is very close to 0.50 in value. We also have

$$\rho = \frac{\dot{M}}{4\pi r^2 V_W}. \tag{115}$$

Assuming that all latitudes contribute up to $\mu_\star \lesssim 1$, the final expression for the luminosity is then

$$L_\nu(nth) = 8.1 \times 10^{24} \mathrm{erg\ s^{-1} Hz^{-1}}$$
$$\eta_{-1} \frac{\dot{M}_{-5}}{V_{W,-2}} U_{1,-2}^2 B_{0.5}^{5/3} r_{14}^{-2/3} \nu_{9.7}^{-2/3}. \tag{116}$$

Here the mass loss is in units of 10^{-5} $M_\odot$ yr^{-1}, the unperturbed magnetic field strength at the reference radius of 10^{14} cm in units of of 3 Gauss, the wind and the shock velocity in units of 0.01 c, and as reference frequency we use 5 GHz.

Here we note that this emission might become optically thick both to free-free absorption in the lower temperature region outside the shock, or to synchrotron self-absorption inside the shocked region. In the approximation that we use for the equatorial region in the slab model (i.e., direct central axis radial integration), we obtain for the critical radius $r_{1,\mathrm{ff}}$, where the free-free absorption has optical thickness unity:

$$r_{1,\mathrm{ff}} = 1.35 \times 10^{14} \left(\frac{\dot{M}_{-5}}{V_{W,-2}}\right)^{2/3} \nu_{9.7}^{-0.70} \mathrm{cm}. \tag{117}$$

We have used here the analytical approximation for the free-free opacity introduced above and have assumed a temperature of $T_e = 2 \times 10^4$ K.

Similarly, for synchrotron self-absorption the critical radius $r_{1,\mathrm{syn}}$ is given for $p = 7/3$:

$$r_{1,\mathrm{syn}} = 3.99 \times 10^{14} \mathrm{cm}$$

$$\eta_{-1}^{0.316} \left(\frac{\dot{M}_{-5}}{V_{W,-2}} \right)^{0.316} U_{1,-2}^{0.632} B_{0.5}^{0.684} v_{9.7}^{-1.} \tag{118}$$

The maximum synchrotron luminosity is given by the dominant absorption process; which one is stronger is given by comparing the relevant radii; in the numerical example synchrotron self-absorption happens to be stronger. Free-free absorption is dominant for $p = 7/3$ if

$$\frac{\dot{M}_{-5}}{V_{W,-2}} > 22.0 \eta_{-1}^{0.901} U_{1,-2}^{1.802} B_{0.5}^{1.951} v_{9.7}^{-0.856}. \tag{119}$$

We note that the parameter η might well be quite low.

In the following we give the maximum luminosities calculated by using the proper optical depth for a central axis approximation in a slab geometry for the radiative transfer at the equator; this gives an additional factor of about 2/3 (see below). However, for the total emission we do integrate properly over all latitudes, while for the absorption we approximate by using the equatorial values. This is a fair approximation, because both emission and absorption decrease towards the poles, but the absorption decreases even faster. The exact numerical value in our approximation is used here.

In the case that free-free absorption dominates, the maximum luminosity is then given for $p = 7/3$ by

$$L_\nu(nth) = 3.8 \times 10^{24} \mathrm{erg\ s}^{-1} \mathrm{Hz}^{-1}$$

$$\eta_{-1} \left(\frac{\dot{M}_{-5}}{V_{W,-2}} \right)^{0.556} U_{1,-2}^2 B_{0.5}^{1.667} v_{9.7}^{-0.200}. \tag{120}$$

In the case that synchrotron self-absorption dominates, these maximum luminosities are given by

$$L_\nu(nth) = 2.2 \times 10^{24} \mathrm{erg\ s}^{-1} \mathrm{Hz}^{-1}$$

$$\eta_{-1}^{0.789} \left(\frac{\dot{M}_{-5}}{V_{W,-2}} \right)^{0.789} U_{1,-2}^{1.579} B_{0.5}^{1.211}. \tag{121}$$

Here we do not consider mixed cases.

Obviously, the ratio of mass loss rate and wind velocity can be different by many orders of magnitude among predecessor supernova stars, and their numerical values must be argued on the basis of observations.

It is useful also to calculate the thermal radio emission, using the same standard parameters, adjust our parameters to a realistic range, and then compare the nonthermal luminosities. The thermal emission can be properly integrated, allowing for the sphericity of the wind structure (Biermann et al. 1990) to give:

$$L_\nu(th) = 4\pi^2 B_\nu(T) r_{1,\mathrm{ff}}^2 \Gamma\left(\frac{1}{3}\right). \tag{122}$$

With our standard parameters this luminosity in a steady wind,

$$L_\nu(th) = 3.0 \times 10^{17} \left(\frac{\dot{M}_{-5}}{V_{W,-2}}\right)^{4/3} \nu_{9.7}^{+0.60} \mathrm{erg\ s^{-1} Hz^{-1}} \tag{123}$$

is weakly dependent on electron temperature.

In the following we make a number of consistency checks on the basic notions used above. First, we note that shocks can only exist when the shock velocity is larger than the Alfvén velocity. Because in our wind driving theory developed above the wind velocity is just slightly larger than the Alfvén velocity itself, this implies that the shock velocity in the frame of the flow must be at least the same velocity as the wind, and so (within our analytical approximations) we have the condition that always

$$U_1 \geq V_W \tag{124}$$

where the limit of the equality corresponds to a spectral index for the synchrotron emission of -0.735 and in the limit of large shock velocity to $-2/3$. Note that the Alfvén velocity is colatitude θ dependent and is proportional to $\sin\theta$. Thus, even for lower shock speeds, there is a latitude range where a shock can be formed, but then the luminosity is very much reduced.

Second, we must check that the synchrotron loss time is larger than the acceleration time, because otherwise we would not have any electrons at the appropriate relativistic energies. This condition can be rewritten as

$$\nu_{9.7} B_{0.5}^3 \lesssim U_{1,-2}^2 r_{14}. \tag{125}$$

This is the condition at the equator, and the condition becomes weaker at higher latitudes. From this condition it is obvious that for reasonable ranges of the parameters, acceleration can succeed to the required electron energies. It also shows that at higher frequencies or smaller radii the condition would fail. At smaller radii the emission is usually optically thick (see above), and higher frequencies are as yet difficult to observe at the required sensitivity. The implied high-frequency cutoff in the observed synchrotron spectra would, however, be an important clue.

Third, we must check whether the implication that the shock speeds are typically similar to the wind speeds, is supported by data on Wolf Rayet stars, and their theoretical understanding. The wind calculations with shocks suggest that the typical shock velocities are indeed of order of the wind speed itself or somewhat higher (Owocki et al. 1988), and so we expect a range in radio spectral indices of -0.667 to -0.735 or steeper if the shocks are not strong, i.e., if $U_1/U_2 < 4$. We note that the actual value of the nonthermal luminosity is changed only moderately in this range of spectral indices.

Fourth, we need to discuss optical thickness effects in more detail. A shock travels from the region inside of where free-free absorption dominates through this region to the outside. The emission then is first weak and has a steep spectrum due to the strong frequency dependence of the free-free absorption and the exponential cutoff induced; then it approaches a peak in emission with a spectral index approaching -0.67 to near about -0.735 or steeper as discussed above, and finally it becomes weaker with the optically thin spectrum. At the location where our ray encounters the shock, the temperature of the gas increases drastically, and the differential optical depth for free-free absorption goes to zero. Hence we have the simple case that we have emission inside the shock and absorption outside. We limit ourselves to the central axis as a first approximation, and also neglect the spatial variation of the emission itself (see above). This is equivalent to pure screen-like absorption, but with a screen which extends from the shock to the outside and so is variable. The radio luminosity is given by (using free-free absorption)

$$L_\nu(nth) = e^{-\tau_\nu} L_\nu(nth, \text{no abs}). \tag{126}$$

We have the spectral index $-\alpha_{\text{thin}}$ in the optically thin regime, and the time dependence of the spectral index given by

$$\alpha(\nu) = -\alpha_{\text{thin}}(1 - (t_\nu^\star/t)^3). \tag{127}$$

The spectral index is positive and very steep at first and then goes through zero to become slowly negative approaching asymptotically the optically thin spectral index. The nonthermal luminosity considered as a function of time sharply rises at first, then peaks at optical depth $\alpha_{\text{thin}}/3$, obviously strongly dependent on frequency, and finally drops off as $1/t^{\alpha_{\text{thin}}}$:

$$\frac{d\ln L_\nu(nth)}{d\ln t} = -\alpha_{\text{thin}}\left(1 - \frac{10}{7}(t_\nu^\star/t)^3\right). \tag{128}$$

Here we see that the time dependence of the flux density at a given frequency and the time dependence of the spectral index are closely related. The time of luminosity maximum depends on frequency as

$$t_{\text{max } L} \sim \nu^{-0.7} \tag{129}$$

from the frequency dependence of the optical depth

$$\tau_\nu \sim t^{-3}\nu^{-2.1}. \tag{130}$$

These relationships can be used to check on the importance of free-free absorption in our approximations.

Synchrotron self-absorption is due to internal absorption inside the shell, so again in the central axis approximation we have

$$L_\nu(nth) = \frac{1 - e^{-\tau_\nu}}{\tau_\nu} L_\nu(nth, \text{ no abs}). \tag{131}$$

This results in the time dependence for the spectral index of

$$\alpha(\nu) = \frac{5}{2} + \frac{5 + 2\alpha_{\text{thin}}}{2}\tau_\nu - \frac{5 + 2\alpha_{\text{thin}}}{2}\frac{\tau_\nu}{1 - e^{-\tau_\nu}} \tag{132}$$

with

$$\tau_\nu \sim t^{-(5+2\alpha)/2}\nu^{-(5+2\alpha)/2}. \tag{133}$$

The time dependence of the double logarithmic derivative of luminosity on time is exactly the same

$$\frac{d \ln L_\nu(nth)}{d \ln t} = \alpha(\nu). \tag{134}$$

It follows that the time of maximum depends on frequency as

$$t_{\max L} \sim \nu^{-1}. \tag{135}$$

This is a characteristic feature for synchrotron self-absorption in the approximation used (and well known from radio quasars) and differs from the case considered above, for free-free absorption. Thus, the true maximum of the luminosity is given by an optical depth less than unity, which gives a correction factor to the luminosities introduced above (Eqs. 120 and 121) of about 2/3; we have corrected the luminosities introduced there for this factor.

Finally, we should comment on the sign of the magnetic field. We have used here throughout the assumption that the magnetic field is oriented such that the drifts are towards the equator for the particles considered. Clearly, because we consider stars with a magnetic field driven by turbulent convection in a rotating system, we can expect that there are sign reversals of the magnetic field just as on the Sun. For the other sign of the magnetic field and the other drift direction, there is by many powers of ten less nonthermal emission so it is to be expected that at any given time we should detect at most half of the stars in nonthermal radio emission; occasionally we might even catch a shock traveling through the region in the wind where the sign is reversing itself, because the wind mirrors the time history of the star in terms of magnetic field. If the shock travels through a layer bounded by magnetic field reversals both inside and outside, it then becomes important to ask what the time scale of particle acceleration is relative to the time scale of traversing this layer; particle acceleration at high energies may be severely limited, if such layers are thin. In the following we will first discuss the radio observations of Wolf Rayet stars, then OB stars, and finally supernovae.

The theoretical luminosities derived can be compared with the observations of Wolf Rayet stars, which yield (Abbott et al. 1986) nonthermal luminosities up to about 5×10^{19} erg s^{-1} Hz^{-1} at 5 GHz and thermal luminosities up to 7×10^{18} erg s^{-1} Hz^{-1}. Because the most important parameter, that enters here is the density of the wind, or in terms of wind parameters, the ratio of the mass loss to the wind speed $\dot{M}/V_W$, we induce that the most extreme stars have a higher value for $\dot{M}/V_W$ by 10.6. This translates into a higher nonthermal emission as well, by a factor of 3.3 to give 6.0×10^{24} erg s^{-1} Hz$^{-1}\eta_{-1}U_{1,-2}^{2}B_{0.5}^{1.735}$, using the case when free-free absorption dominates (for synchrotron self-absorption the corresponding luminosity is very nearly the same). This is much higher than the observed nonthermal luminosities and suggests that there is a limiting factor. We implicitly assume here, that some of the massive stars that exploded as the observed supernovae, either were Wolf Rayet stars before their explosion, or that their properties were not significantly different. Because we derived the wind density from the timing of the lightcurve above, and the shock velocity both from the models of Owocki et al. (1988) and the argument that the shock velocity be larger than the Alfvén velocity (which in turn we argued is not very much lower than the wind velocity) the only other important parameter is the magnetic field strength, and that we assume to be of similar magnitude. Because Wolf Rayet stars do not distinguish themselves from stars that somewhat later in life explode as supernovae, we can use all the same parameters, except for the shock speed. Then the only parameter left is the highly uncertain efficiency of electron injection η.

We can thus ask whether the efficiency η might depend on shock speed. The shock speeds in supernovae are of order $0.03\ c$ or greater, while those in Wolf Rayet star winds are of order $0.01\ c$ or less. Somewhere between these speeds there appears to be a critical shock speed, at which the character of electron injection changes. Here we do not wish to discuss injection, but merely note that the data suggest efficiencies of order $\eta \simeq 10^{-6\pm1}$ for the shock speeds thought to be normal in Wolf Rayet star winds, and of order $\eta \simeq 0.1$ for supernova explosions. The data thus suggest that the injection of electrons into the diffusive shock acceleration appears to exhibit a step function property at a critical shock velocity. Comparing normal supernova remnants, the nova GK Per, and the sources discussed above, a natural choice is a critical Alfvénic Mach number, which we can only estimate to be in the range of 4 to 40. If true, this might be important also for electron injection in quasars, which also exhibit a dichotomy into radio-loud and radio-weak objects. In Biermann and Strom (1993), and below, we suggest that this concept leads to an estimate for the electron/proton ratio of the lower energy cosmic rays consistent with observations.

Now we can go back and ask again, whether free-free absorption or synchrotron self-absorption dominates in the various cases considered. Clearly, when η, the efficiency for electron injection, is very small, then free-free absorption always dominates, and so the maximum luminosities during a radio variability episode of a Wolf Rayet (or OB) star should be frequency dependent. Also, for slow winds in supernova predecessor stars, again free-free absorption will dominate normally (e.g., in the two examples used above). On the other hand, for fast winds of the supernova predecessor stars (as in the case of a Wolf Rayet star exploding), synchrotron self-absorption is likely to be stronger than free-free absorption and the maximum luminosities at different radio frequencies should be independent of frequency. This is a testable prediction, as it relates radio and optical properties of a young supernova.

D. OB Stars

The theoretical luminosities derived can be compared with the observations of OB stars, which yield (Bieging et al. 1989) nonthermal luminosities up to about 7×10^{19} erg s^{-1} Hz^{-1} at 5 GHz, and thermal luminosities up to 2.5×10^{19} erg s^{-1} Hz^{-1}. Because the most important parameter, that enters here is the density of the wind, or in terms of wind parameters, the ratio of the mass loss to the wind speed $\dot{M}/V_W$, we induce that the most extreme stars have a higher value for $\dot{M}/V_W$ by 27.6. This translates into a higher nonthermal emission as well, by a factor of 5.4 to give 9.8×10^{24} erg s^{-1} Hz$^{-1} \eta_{-1} U_{1,-2}^2 B_{0.5}^{1.735}$. This is much more than the observed nonthermal luminosities. Because the shock properties are likely to be similar to Wolf Rayet stars, this is again consistent with the idea that the injection efficiency of electrons might be quite low, in conjunction with magnetic field values not too far from what we argued to be valid for Wolf Rayet stars.

We can also compare the detection statistics and observed spectral indices. Because, in a variability episode, a shock comes from below through the region of optical thickness near unity, the nonthermal emission increases rapidly to its maximum, when its spectral index is

$$\alpha(\nu, \text{max}) = -0.3\alpha_{\text{thin}} \tag{136}$$

which is approximately -0.2. Thereafter, as the luminosity decreases more slowly, the spectral index gradually approaches the optically thin index of -0.67 or slightly steeper. Therefore we expect the spectral index distribution of detected sources to be a broad distribution from near -0.2 to near -0.7; this is what has been found (Bieging et al. 1989). Because of the two possible signs of the magnetic field orientation, and the associated drift energy gains for particles, we also expect that at any given time at most half of all sources are detectable with nonthermal emission, even at extreme sensitivity. This is consistent with the observations. We predict a similar behaviour for Wolf Rayet stars.

Now we must ask how the magnetic fields can penetrate the radiative region from below. This question cannot presently be tackled by simulations on a very large computer, but it is likely that the circulations induced by rotation transport magnetic fields to the surface, where even a rather slight differential rotation draws out the magnetic field into a mostly tangential configuration. In this case, clearly the origin of the momentum of the wind can be readily accounted for from line driving (Lucy and Solomon 1970; Castor et al. 1975), so we suspect that the magnetic driving adds only little (which is the "other" case discussed above near the end of the wind section, Sec. IV.B).

We can also make a comparison with the only existing theory to explain the nonthermal radio emission of single massive stars with winds (White 1985). White's theory is based on a concept involving a large number of shocks and does not explain the data as already demonstrated by Bieging et al. (1989) in terms of (i) radio spectral index, (ii) time variability nor (iii) the statistics of detection. As a consequence any estimate of the strengths of magnetic fields based on White's theory is in doubt (see Bieging et al. 1989). The theory of diffusive particle acceleration by an ensemble of shock waves has been properly derived by Schneider (1993), and remains to be compared with the data of stars. Our theory is based on using single shocks and readily provides spectral indices in the entire range that is observed; it explains the time variability and easily accounts for a fair fraction of undetected sources.

Shocks in stellar winds as a consequence of supernova explosions give rise to γ-ray emission from hadronic interactions (Berezinsky and Ptuskin 1989); this latter conclusion was reached also for normal shocks in stellar winds by Chen and White (1991b) and White and Chen (1992). For supernova explosions, it is important as a check, that the resulting predicted γ-ray production scales with the square of the wind-density, and thus with $(\dot{M}/V_W)^2$ (Berezinskii et al. 1990, Eq. 7.57 in VII§4).

The binary system WR140 has been detected by GRO (Hermsen et al. seminar at the GRO symposium in Maryland, 1993; OSSE and BATSE instruments), confirming a detailed prediction by Eichler and Usov (1993), which is based on a model of colliding wind shocks. This latter agreement of observation and prediction is a consistency check on the concept, that freshly accelerated protons produce high-energy photons from hadronic interaction, as opposed to purely leptonic processes.

E. Radio Emission from Radio Supernovae

The adopted value for the magnetic field, however, was derived from the notion that the subsequent supernova shocks accelerate particles to extremely high energies. This argument can be checked with the observations of those supernovae of which the radio emission in the wind was observed, five sources discussed by Weiler and colleagues in a number of papers (Weiler et al. 1986,1989,1990,1991; Panagia et al. 1986) and the supernova 1987A (Turtle et al. 1987; Jauncey et al. 1988; Staveley-Smith et al. 1992) and related radio sources in starburst galaxies. However, we restrict ourselves to those supernovae for which we can reasonably assume that the predecessor star was indeed a star with a strong wind, and this we will do using statistical arguments in the subsequent section.

In fact, our model can be paraphrased as a numerical version of Chevalier's (1982) model with the parameters fixed. In the screen approximation valid for free-free absorption (see above for details), the time evolution of the nonthermal emission in Chevalier's model can be written as

$$L_\nu(nth) = K_1 \nu^\alpha t^\beta e^{-K_2 t^\delta} \tag{137}$$

with α, β and δ to be fitted to the data. For fast strong shocks our model predicts that $\alpha = \beta = -2/3$ and $\delta = -3$, and for slower shocks that still $\alpha = \beta$ but larger in number, for $U_1/V_W = 1$, $\alpha = \beta = -0.735$, for instance. This is consistent with the detailed fits for three supernovae (Weiler et al. 1986), which appear to arise from stars with stellar winds. Using the numbers from the fit from Chevalier's model, which requires $\beta = -3 + \alpha - \delta$, the averages are $\langle\alpha\rangle = -0.71 \pm 0.20$, $\langle\beta\rangle = -0.70 \pm 0.04$, $\langle\alpha, \beta\rangle = -0.70 \pm 0.13$, $\langle\alpha/\beta\rangle = 1.01 \pm 0.24$, and $\langle\delta\rangle = -3.01 \pm 0.17$ from these somewhat sparse data.

For supernova 1986J the data clearly cover a sufficiently large time to test this numerical model in more detail; Weiler et al. (1989) find that this simple model in its screen approximation does not provide a good fit to the early epochs. We suspect similarly to Weiler et al., that this lack of a good fit can be traced to the simplification that we consider the external absorption as a simple screen, and disregard the lateral structure. Further possible reasons for a failure to adhere strictly to the simplified model are the following: (a) the pre-shock wind may not be smooth in its radial behavior, there might have been a weaker shock running through earlier which nevertheless can disturb the radial density profile (see the calculations by MacFarlane and Cassinelli [1989]); (b) the mixing between free-free absorption (outside the shock region) and synchrotron self-absorption (inside the shock region); (c) the structure of acceleration in its latitude dependence is considered here only for the acceleration of particles (Biermann 1993a), but not for absorption and radiative transfer. Given a very detailed multi-frequency data set it would be interesting to model the data fully.

In fact, we can use the numerical values for the radii for maximum luminosity derived above to obtain the pre-shock wind density, which is proportional to $\dot{M}/V_W$. With the data given by Weiler et al. (1986), this yields for the supernova 1979C

$$\frac{\dot{M}_{-5}}{V_{W,-2}}(1979\mathrm{C}) = 3.010^3 U_{1,-1.5}^{3/2} \tag{138}$$

and for the supernova 1980K

$$\frac{\dot{M}_{-5}}{V_{W,-2}}(1980\mathrm{K}) = 3.410^2 U_{1,-1.5}^{3/2}. \tag{139}$$

We note here that the nomenclature for these supernovae has changed between Weiler et al. (1986) and Weiler et al. (1989); we use here the more recent version.

Clearly, these numbers tell us that the predecessor stars had a slow wind, and thus were probably red supergiants (see Weiler et al. 1986). This then implies for shock speeds of 0.03 c, $\eta = 0.1$, and free-free absorption being dominant, that the nonthermal luminosities at 5 GHz expected vs observed are

$$L_{\max}(1979C) = 3.0 \times 10^{27} \text{ vs } 2 \times 10^{27} \text{erg s}^{-1} \text{Hz}^{-1} \qquad (140)$$

and

$$L_{\max}(1980K) = 9.0 \times 10^{26} \text{ vs } 1 \times 10^{26} \text{erg s}^{-1} \text{Hz}^{-1} \qquad (141)$$

both for the assumed strength of the magnetic field. Because the luminosity is proportional to the magnetic field strength to the power 5/3, and directly proportional to the electron efficiency parameter η, we derive thus a lower limit to the magnetic field, given an upper limit on η. Because $\eta = 0.1$ is unlikely to be surpassed by much, using the ratio of the luminosities expected/observed of the two cases above yields an estimated lower limit to the magnetic field strength at our reference radius of

$$B > 1.5\eta_{-1}^{-3/5}U_{1,-1.5}^{-6/5} \text{ Gauss.} \qquad (142)$$

This implies a strong lower limit to the magnetic field from using $\eta = 1$ of 0.4 Gauss. The uncertainty in these estimates is clearly at least a factor of 2.

The argument is often made, that hydrodynamic instabilities could increase the magnetic field strength in the post-shock flow (Reynolds and Chevalier 1981); in such a case the estimate given here would only signify how much the magnetic field has increased behind the shock. However, the work by Galloway and Proctor (1992) suggests that dynamo time scales are not fast enough to do this. As emphasized in the summary, a direct observational check on the magnetic field strength in stellar winds is very desirable; it may be possible with data from a pulsar in a binary system with a massive early-type star.

For 1987A, the initial radio luminosity was indeed of order 10^{25} erg s^{-1} Hz^{-1} so, applying our model with a shock speed of order 0.03 c is consistent with our expectation of $3.4 \times 10^{25} \eta_{-1}$ erg s^{-1}Hz^{-1}. Similarly, for the new radio supernova 1993J in the galaxy M81 the observed maximum radio luminosity of 2.2×10^{26} erg s^{-1}Hz^{-1} at 99.4 GHz (Phillips and Kulkarni 1993) is quite compatible with a reasonable shock speed, allowing for the fact that the predecessor star was a red giant (thus $V_{W,-2} \ll 1$); both optically thin time evolution and optically thin spectrum are also consistent with the arguments presented here (Panagia et al. 1993).

This demonstrates that strong magnetic fields also exist in the winds of massive stars in the red part of the HR diagram, where the winds are slow, as argued earlier; note that the predecessor to supernova 1987A was a blue supergiant with a fast wind.

The only unknown parameter in all these predictions is the efficiency of electron acceleration η and the strength of the magnetic field; with η close to 0.1, clearly close to the maximum reasonable number, leads then to the requirement that the magnetic field strength is near to what we assumed, 3 Gauss at 10^{14} cm, but even higher, if η is very much less than unity. This again confirms independently that indeed the magnetic field must be as high as argued by Cassinelli (1982,1991) to drive the winds, as is required to accelerate cosmic rays particles to energies near 3×10^9 GeV.

For supernova predecessor stars with fast winds like Wolf Rayet and OB stars, it is of interest to ask whether the expected luminosity violates the Compton limit, famous from the study of radio quasars. At the Compton limit the first order inverse Compton X-ray luminosity becomes equal to the synchrotron luminosity, and it has been found from observations that compact radio quasars are close to this limit and indeed have strong X-ray emission. This question can be formulated as a limit to the brightness temperature of the radio source (using here synchrotron self-absorption) which then gives the limit

$$\eta_{-1} \left(\frac{\dot{M}_{-5}}{V_{W,-2}} \right) U^2_{1,-1.5} B_{0.5}^{-1} \nu_{9.7}^{1.27} < 1.5 \times 10^7. \tag{143}$$

For free-free absorption the limit is

$$\eta_{-1} \left(\frac{\dot{M}_{-5}}{V_{W,-2}} \right)^{-0.777} U^2_{1,-1.5} B_{0.5}^{5/3} \nu_{9.7}^{-0.6} < 0.18. \tag{144}$$

The first of these conditions is almost certainly always fulfilled, while the second one may be so tight as to suggest that inverse Compton X rays might be observable. This has indeed been checked with modeling successfully the observed X-ray spectra beyond photon energies of 2 keV by Chen and White (1991a) for Orion OB stars. This suggests that the inverse Compton X-ray luminosity ought to be less, usually considerably less, than the synchrotron luminosity. Most of the X-ray emission from hot stars is thought to arise from those same shocks in free-free X-ray emission which we consider for particle acceleration; a modeling of this was done by White and Long (1986) and MacFarlane and Cassinelli (1989).

F. The Statistics of Wolf Rayet Stars and Supernovae

There are a variety of ways to estimate the relative frequency of Wolf Rayet star supernova explosions relative to supernova explosions of lower-mass stars (Hidayat 1991; Leitherer 1991; Massey and Armandroff 1991; Shara et al. 1991). In our Galaxy there are between 300 and 1000 Wolf Rayet stars, which have an average lifetime of about 10^5 yr. This gives an estimated occurrence of Wolf Rayet supernovae of about one every 100 to 300 yr. As the total rate of supernovae in our Galaxy is estimated at about one every 30 yr, this means that roughly one in 3 to one in 10 supernovae should represent the explosion of a Wolf Rayet star.

The numbers of stars on the main sequence between 8 solar masses and about 25 solar masses, and between 25 solar masses and the upper end of the main sequence also should correspond to the ratio of Wolf Rayet stars and the rest of those stars which explode as supernovae. Using for the simple estimate the Salpeter mass function gives an estimated ratio of about 1 in 5 supernova events which originate from a Wolf Rayet star. Approximately 1 in 4 supernova events come from a star with a strong wind (Wheeler 1989), where the change to stars with only a weak wind is estimated to be near a main-sequence mass of 15 $M_\odot$.

The model for the origin of the high energy population of relativistic cosmic rays proposed in Biermann (1993) and tested successfully in Stanev et al. (1993) suggests from the energetics a ratio of about 1 in 3, assuming the amount of energy pumped into cosmic rays per supernova to be the same for all kinds. However, here we lump all supernova explosions into stellar winds together, both with fast and slow winds. Hence in the sample of radio supernovae of type II (6 sources) presented by Weiler et al. (1986), there ought to be between zero and 2 events based on Wolf Rayet star explosions; in the sample of radio sources likely to be very young supernova remnants (28 sources with radio luminosities at 5 GHz) in the starburst galaxy M82 (Kronberg et al. 1985) there should be roughly between 3 and 10 sources which originate from a Wolf Rayet star explosion. On the other hand, all of these objects are possibly explosions of stars into former stellar winds (there is evidence that the initial mass function in starburst galaxies is biased in favor of massive stars), so the proportion of the stars among them that are due to Wolf Rayet star explosions, could be even higher than estimated here. The radio luminosities are very similar for the sources in M82 and the radio supernovae of Weiler et al. (1986); the average luminosity calculated in a variety of ways is always in the range of 3×10^{25} erg s^{-1}Hz^{-1} and 10^{26} erg s^{-1}Hz^{-1}, fitting our expectation (see above) for $U_1 \simeq 0.03\ c$ rather well. We note, that for the same wind density, which is proportional to $\dot{M}/V_W$, the range of expected radio maximum luminosities is similar for supernova explosions into slow and fast winds; on the other hand, if the mass loss rates are similar for slow and fast winds, then the densities can be much higher in slow winds, and the expected nonthermal radio luminosities are much higher for slow wind stellar explosions. Thus, observationally, we may only detect radio supernovae of stars exploding into slow winds. There is one observational signature, which may be difficult to measure accurately. For a given mass loss rate, slow winds are usually dominated by free-free absorption, which leads to a frequency-dependent maximum luminosity (see Eq. 119), while fast winds are likely to be dominated by synchrotron self-absorption, which leads to a frequency independent maximum radio luminosity. In Weiler et al. (1986) the data for the supernovae 1979C and 1980K illustrate this possible difference, where the best fit to 1979C shows a frequency dependent maximum luminosity, while for 1980K the maxima at 1.4 and 5 GHz appear to be the same.

For our arguments on cosmic rays it is not relevant whether the predecessor stars were Wolf Rayet stars or other massive stars with extended winds permeated by strong magnetic fields. We expect from the similarity in the internal structure of the stars on the upper main sequence, that their global properties such as the generated magnetic field strength should not be drastically different. Here we assume that we can use the implied properties for Wolf Rayet stars just as for stars of slightly lower mass which explode into slow winds.

V. ELECTRONS

A. The Injection of Relativistic Electrons

Above, we argued on the basis of a comparison of Wolf Rayet stars and radio supernovae, that there appears to be a critical Alfvénic Mach number for the injection of electrons. This critical Mach number cannot be pinpointed to one specific number at this time, but appears to be in the range of values of 4 to 40.

Levinson (1992,1994) has argued on theoretical grounds, that indeed there is such a critical Alfvénic Mach number for electron injection; his prediction contains a free parameter, so that only consistency with our argument can be verified.

Feldman (1993, personal communication) noted that data from solar wind shocks also show that there is critical Alfvénic Mach number for electron injection (see Edmiston and Kennel [1984] and Kennel et al. [1985] for a possible physical argument).

What we lack now is a generalized theory to give this result, a definite numerical value, as well as an argument as to what determines or eliminates electron injection below this critical Alfvénic Mach number. It is possible to interpret the radio data of stars with this concept, because it leads to a latitude restriction and thereby to a decrease in the radio emission; a quantitative check suggests that the radio emission is confined to a fairly small polar region, while energetic protons can interact over a large part of the hemisphere (Nath and Biermann 1994b; Biermann et al. 1995).

B. The Maximum Energy of Relativistic Electrons

For ISM-SN, the maximum energy for accelerated electrons ranges between 30 and 100 GeV, using a low-density interstellar medium, and synchrotron losses vs net acceleration (acceleration minus adiabatic losses) as the limiting factor. It follows that the observed high-energy electron tail cannot be attributed to normal supernova explosions into the interstellar medium.

For wind-SN, the maximum energy is given by the consideration already used to derive the maximum emission frequency for nonthermal radio emission (above, Eq. 125). The corresponding maximum electron energy is approximately given by

$$E_{e,\max} = 0.3 r_{pc} B_{0.5}^{-2} U_{1,-2} \text{TeV} \tag{145}$$

where r_{pc} is the radius where the stellar wind changes character, due to the shell formed by interaction with the interstellar medium. This radius may be of order a few parsec, so the expected maximum energy, for a few parsec and a shock speed of order 10^4 km s^{-1} is about 3 TeV. This is consistent with the observations (summarized by Wiebel 1992) that detect energetic electrons up to a few TeV particle energies. We leave the problem of how to get such high energy electrons through the Galaxy to us and the implied consequences for another occasion. We conclude for here, that for electrons above about 30 to 100 GeV, wind-supernovae are required to explain particle energies and spectrum.

C. The Spectrum of Relativistic Electrons

At particle energies of order GeV and slightly higher, the radio emission of external galaxies is the most reliable indicator of this spectrum. Golla (1989) has discussed the best data available, accounting for all the possible contribution from thermal radio emission and finds that all excellent data (a sample of seven galaxies with a well-determined nonthermal radio spectrum with an error less than 0.1 in the spectral index) are compatible with a single spectral index of a power law energy distribution for the electrons, in this energy range, of 2.76±0.12. This is very nearly the same spectral index as found for hydrogen in cosmic rays.

Wiebel (1992) compiled all the data available in the literature, and finds above about 30 GeV a spectrum of $E^{-3.26\pm0.06}$, which is to be compared with $E^{-3.33-0.02\pm0.02}$ as the expected spectrum from wind shocks of, steeper than unity, than the injected spectrum due to synchrotron losses.

The particle energy of the switch between the two source sites, and the maximal electron energy observed, as well as the positron fraction, remain to be discussed in detail.

D. The Proton/Electron Ratio in Cosmic Rays

During the expansion, the energy of any individual relativistic electron decreases by adiabatic expansion as the ratio of the radii from the time when electron injection ceases to the time of when proton injection ceases. After this point, the energy densities of both particles decrease together. The simple dilution of the electron population is paralled by the steadily decreasing injection rate of protons, because both the volume and the ram pressure of the shock go down as radius to the -3 power during the adiabatic phase (see Shklovsky 1968, Eq. 7.27); should the injection of protons cease at a later stage, then this differential dilution must also be reckoned. The energy density ratio of the electron population relative to that of the protons is given by ($p = 2.420$)

$$(R_{\mathrm{crit},p}/R_{\mathrm{crit},e})^{p-1}. \tag{146}$$

This intermediate switch from electron injection with steady acceleration to a simple adiabatic loss regime determines the net scaling of the power of the electron population to that of the proton population in cosmic rays. The observations suggest that from 1 GeV the energetic electron density is only about 1% (see, e.g., Wiebel 1992) of the density of the protons. From this observed ratio the energy density ratio integrated over the entire relativistic part of the particle spectrum, protons relative to electrons, for the spectral index of -2.42, is given by 4.3. In the standard leaky box model, the ratio of energy densities of protons and electrons is not influenced by propagation effects. This suggests an expansion of a factor of order 3 in radius between the time when electron injection ceases and the time when proton injection ceases; here we assume that electrons and protons originally have comparable energy densities of their relativistic particle populations. We may have thus identified the origin of the observed electron/proton ratio in cosmic rays.

E. The Shell Thickness of Supernova Remnants in the Interstellar Medium

One clear prediction of the concept of fast convective turbulence in the shock region is the fairly large thickness of the shell; this shell is the thickness of all the matter snowplowed together downstream, and in addition the length scale with the same column density upstream. Thus, what we refer to as *upstream* in our model is fully contained in the emission shell observed; the average shock location is deep inside the emission shell. The outer edge of the observed emission then, in this picture, is the location of the presently locally protuding shock, seen from the side. Therefore, the shell thickness in this model is both upstream and downstream in the language used earlier and, when referred to the outer radius, has the value

$$\Delta r/r = \frac{1 + U_1/U_2}{4U_1/U_2} \qquad (147)$$

for supernova explosions into the interstellar medium.

This gives $\Delta r/r = 5/16$ for a strong shock in a medium of constant density. It is important to note, that in the concept discussed here, this large thickness of the shell is traced to an instability of a cosmic ray mediated shock front, i.e., a shock strongly influenced in its structure by cosmic ray protons and other nuclei. As a consequence, a shock front, which does not inject new particles into the system, but just squeezes the existing energetic particle population, will not show this effect, and should therefore be much thinner, in the simple limit of a strong shock $\Delta r/r = 1/12$. This is indeed what is consistent with the Cygnus Loop (J. Raymond 1993, personal communication); however, one problem with the Cygnus Loop is the open possibility that it is a warped sheet which may appear broader than it really is.

The data for some sources can be read off published graphs (Pye et al. 1981; Dickel et al. 1982,1988,1991; Seward et al. 1983), and are close to this value for parts of the sources Kepler, SN1006, Tycho, and RCW 103. Apparently, the new radio emission of supernova 1987A also appears to fit this prediction (Staveley-Smith et al. 1993). A more detailed analysis of such data is clearly required; it is obvious, that many supernova remnants are not nicely circularly symmetric, nor show a clear shell structure. After all, we know the interstellar medium to be extremely inhomogeneous. The data are also not always in agreement between radio and X rays, which after all, trace nonthermal particles and thermal hot gas; when cooling becomes important, then the simple shell argument may not be sufficiently accurate, because it assumes constant density throughout the shell. The analogous argument for winds remains to be done, and may be necessary to interpret the radio data for the nova GK Per.

VI. AIR SHOWER DATA, OTHER CHECKS AND CONSEQUENCES

We have discussed and reviewed the tests with airshower data elsewhere, and it suffices to summarize here the predictions and tests.

We predict for protons a spectrum of $E^{-2.75\pm0.04}$ (Biermann and Strom 1993); the Akeno data fit gives $E^{-2.75}$ (Stanev et al. 1993). Radio data of normal galaxies give $E^{-2.76\pm0.12}$ (Golla 1989).

We predict for helium and heavier elements $E^{-2.67-0.02\pm0.02}$ below the knee; the Akeno data also here give a spectrum very close to the prediction, of $E^{-2.66}$.

We predict for the nuclei beyond the knee $E^{-3.07-0.07\pm0.07}$. The Akeno data give $E^{-3.07}$. The world data set of all good high-energy data also gives this spectrum, as well as the cutoff at the predicted particle energy (Rachen et al. 1993). The Fly's Eye data (Bird et al. 1993; Gaisser et al. 1993) demonstrate that the chemical composition switches rapidly from a heavy composition to a light composition near 3×10^{18} eV, as predicted in 1990 (Biermann 1993b), and in a brief form earlier by Biermann and Strittmatter (1987).

We predict the particle energies at the bend, or knee, and the energies of the various cutoffs; the test with the Akeno data gives fitted values for these numbers close to prediction. From the fit the numerical values are rather strongly constrained, to within 20%, because we fit both vertical and slanted showers simultaneously in their shower-size distribution.

It is obvious, that we have difficulty estimating the systematic error resulting from the quite general use of limiting arguments, such as always strong shocks, always maximal curvature in the elements of the fast convection, always ignoring possible enhancements of the magnetic field strength (important for the drifts). The predictions and fits, which appear to agree generally quite well, illustrate the possible refinements required within the context of the approximations made.

Further checks are possible with (i) the data analysis of Seo et al. (1991), who give a spectrum of $E^{-2.74\pm0.02}$ for hydrogen, and $E^{-2.68\pm0.03}$ for helium, very close to our predictions; (ii) the analysis of Freudenreich et al. (1990) who have argued for some time, that the chemical composition near the knee becomes heavily enriched; (iii) the newest Fly's Eye data (Bird et al. 1994) which give a spectrum of $E^{-3.07\pm0.01}$ beyond the knee in the energy range 2×10^{17} eV to 8×10^{19} eV for the mono-ocular data, and $E^{-3.18\pm0.02}$ in the energy range 2×10^{17} eV to 4×10^{19} eV for stereo data, while the classical data from Haverah Park (Cunningham et al. 1980; also see, e.g., Sun et al. 1993) give a spectrum of $E^{-3.09\pm0.02}$ below 10^{19} eV; and (iv) the JACEE data, presented at Calgary (Asakimori et al. 1993), which also give spectra for hydrogen and helium at GeV particle energies, of $E^{-2.77\pm0.06}$ and $E^{-2.67\pm0.08}$, respectively.

Recently, we have discussed all the low-energy data for these spectra (Biermann et al. 1995) and have shown, that also the overall normalization between ISM-SN and wind-SN is close to that expected on the basis of which stars have strong winds on the main sequence and which do not; this gives an overall ratio of energy contained in the two populations of 3 to 1. There we also attribute the underabundance of hydrogen and helium to (a) the two different source sites, and (b) the enrichment in the winds of evolved massive stars (see Silberberg et al. 1990).

In other recent work, we have shown that the low-energy cosmic rays may re-ionize the intergalactic medium after leaving normal galaxies in galactic winds (Nath and Biermann 1993). We have used the cosmic ray ionization rate implied by molecular cloud data to estimate the lower energy cutoff of galactic cosmic rays to between 30 and 60 MeV kinetic energy (Nath and Biermann 1994a).

Furthermore, we have used the concept of particle acceleration in shocks running through stellar winds and then hitting the surrounding molecular shells to propose an explanation for the strong γ-ray lines observed by COMPTEL in the Orion star forming region (Bloemen et al. 1994; Nath and Biermann (1994b); for competing models see Bykov and Bloemen (1994) and Ramaty et al. 1994). We consider this an important test of the entire picture, because we obtain a satisfactory match at once for the nonthermal radio emission of early type stars, the strong γ-ray line emission, and the very low γ-ray continuum emission. This successful match has required a consideration of the latitude-dependent particle acceleration of a shock running through a stellar wind. We conclude, that the model has passed a large number of tests quite successfully, allowing first quantitative checks to be made. However, many questions remain to be answered, especially regarding the transport of cosmic rays, and the secondary to primary ratio.

VII. CAVEATS

With our basic postulate of the *the smallest dominant scale* we have made a giant leap of faith in treating convection and turbulence in an ionized magnetic medium. This is the most glaring step in our argument, which may take a long time to verify. Similarly, in our derivation we used approximations from cosmic ray transport theory far beyond its proven range of validity. The formal agreement with the derivation of Drury (1983) gives reason to hope that we may not be too far from a proper description.

The agreement claimed between prediction and measurement of cosmic ray spectra depends critically on the correction for galactic transport, here derived from a Kolmogorov spectrum. The secondary to primary ratio in cosmic ray nuclei as a function of energy suggests clearly otherwise. We have not demonstrated that the structure of the interstellar medium and its temporal behaviour really allows the secondary to primary ratio to be understood in the context of our theory. The data give conflicting evidence as to the chemical composition of the cosmic rays across the knee, whether it is dominantly light or increasingly heavy. We only show that the Akeno air shower data are consistent with the second possibility. However, it can be expected that MACRO-EASTOP data will shed considerable light on this issue.

For intergalactic transport of cosmic rays we have used an approximation of nearly straight line paths; a connection with the bubble structure of the galaxy distribution may exist. A comparison with the existing data base of high-energy events is still outstanding.

For massive stars we have suggested a theory for the winds based on somewhat stronger magnetic fields than hitherto used, and have emphasized the possibility of a more tangential magnetic field geometry near the surface of the star. However, we have not actually calculated the structure of such a wind, nor have we been able to compare it with observations to the detail desirable. While the apparent agreement of the predictions with data gives hope that it is worthy of pursuing the development of the theory, much work remains to be done.

VIII. SUMMARY

Here we concentrate on the consequences for stars; the implications for cosmic rays have been described elsewhere (Biermann 1993*b*,1994*a,b*,,1995*d*); the preceding section gives the more important implications and further questions.

1. Novae, OB, and Wolf Rayet stars show evidence for particle acceleration in winds; the theory proposed can account for what is known of the spectra, luminosities, and temporal behaviour. Better and more complete data are needed.
2. White dwarfs provide a check on the notion, that a magnetic dynamo operates inside the inner convection zone in massive stars; this dynamo may provide the fairly strong magnetic field, which we argue is present in the stellar winds of massive stars.
3. Wolf Rayet stars may partially derive the momentum in their wind from the pressure gradient of the tangential magnetic field; their pressure gradient acts as an amplifier on a wind, which has some initial acceleration most likely from line driving, possibly in a multiple scattering mode.
4. The comparison of the various stellar radio sources leads to the concept of a critical Alfvénic Mach number for electron injection; this concept may be the basis for understanding of the proton/electron ratio in the

observed galactic cosmic rays.

5. Supernova shocks in the stellar winds may provide the sources of galactic cosmic rays in (a) helium and heavier elements, (b) electrons beyond about 30 GeV, (c) all the way across the knee to about 3×10^9 GeV, where the chemical composition is mostly heavy. We have been able to quantitatively test the theory using air shower data and other recent cosmic ray data.

Various alternative models exist:

1. A postulated galactic wind model (Jokipii and Morfill 1987) may accelerate particles at a galactic wind termination shock; this is argued to contribute particles over the entire range of particle energies, from low energies to the end of the cosmic ray spectrum. At low energies, the distance which particles can reach upstream from the shock, when they stream back to our Galaxy, is limited by κ / V_W, where V_W refers to the galactic wind velocity and κ is the diffusion coefficient for energetic particles at the relevant energies. The distance is so small for any reasonable value for the diffusion coefficient, that low-energy particles cannot reach the Galaxy. This requires that we have two different source populations, which merge near the knee, which in turn implies considerable fine tuning of the source parameters. At the high particle energies, it is difficult to see that the particles can be contained in the Galaxy (Berezinskii et al. 1990, IV§3).

2. The multiple shocks in the environment of OB superbubbles and young supernova remnants (Bykov and Toptygin 1990,1992; Polcaro et al. 1991,1993; Bykov and Fleishman 1992; Ip and Axford 1992) may also contribute.

3. The cosmic background of active galactic nuclei may contribute through the production of energetic neutrons which convert back to protons (Protheroe and Szabo 1992).

For all such models, a clear prediction of the spectrum and its chemical abundance distribution, as well as a detailed check with the air shower size distribution, both for slanted and oblique showers, is desirable.

We emphasize that our proposal, as far as the galactic cosmic rays are concerned, rests on a plausible but nevertheless speculative assumption about the nature of the transport of energetic particles in perpendicular shock waves, namely that there are large fast convective motions across the average location of the shock interface; this notion is, however, supported by radio polarization observations of young supernova remnants, as well as theoretical arguments, as briefly described above. The model has *predictive power*. We have given predictions and checks above. However, a large amount of work remains to be done.

There are many obvious tasks to be done next; as we have given a number of important steps and checks for the cosmic ray aspect elsewhere, we concentrate here on the ramifications concerning stars and stellar evolution:

1. Direct observational tests for the strength of the magnetic fields in OB stars, Wolf Rayet stars, red supergiant winds, and radio novae are critically important. The model presented above unequivocally depends on the magnetic field strengths proposed. One possibility to determine the strength of the magnetic field in winds appears to be the time dependent depolarization of the radio emission from a pulsar orbiting an upper main-sequence star such as B1259-63 (Thorsett [1994] reporting on radio observations made by R. N. Manchester et al.).

2. The radio spectral evolution of novae, single OB and WR stars as well as radio supernovae, including the latitude distribution and radiative transfer, should be calculated to make the predictions more accurate, and thus more testable.

3. The dynamo producing the magnetic field in the inner convective zone in upper main-sequence stars needs to be modeled as well as the transport of magnetic flux through the radiative zone outwards, in order to see whether the magnetic fields and their geometry proposed can really be generated and transported.

4. We need a detailed model of a stellar wind, which starts with, possibly, line driving in the multiple scattering mode, and then receives additional outward momentum from the gradient of the tangential magnetic field.

5. Finally, we need a proper stellar evolution calculation taking into account maximal rotation and maximum magnetic fields, which may play an important role in the final stages of stellar evolution.

Acknowledgments. First of all, I wish to thank my graduate student H. Seemann for checking all equations in this chapter , and going over it with a finely toothed comb. Second, I have to thank my collaborators in the ongoing work partially reported here: J. P. Cassinelli, H. Falcke, T. K. Gaisser, J. R. Jokipii, H. Meyer, B. B. Nath, J. Rachen, E.-S. Seo, T. Stanev, R. G. Strom, and B. Wiebel. Further evolution of these ideas was prompted by intense discussions and exchanges again with my collaborators as well as G. Auriemma, V. S. Berezinsky, P. Bhattacharjee, J. H. Bieging, A. Bykov, P. Charbonneau, J. Cronin, L. Dedenko, J. R. Dickel, R. Diehl, V. Dogiel, J. Eilek, P. Evenson, M. Giller, F. Giovannelli, C. Jarlskog, J. Kirk, G. Mann, J. F. McKenzie, M. Nagano, S. P. Owocki, M. I. Pravdin, V. S. Ptuskin, R. Ratkiewicz, J. Raymond, E. R. Seaquist, F. D. Seward, V. V. Usov, G. Webb, J. Wdowczyk, and G. Zank. Helpful comments on various parts of this manuscript were received from V. S. Berezinsky, P. Charbonneau, J. R. Dickel, R. N. Manchester, J. F. McKenzie, S. P. Owocki, E. R. Seaquist, P. Song, J. Raymond, F. D. Seward, and G. P. Zank as well as two unknown referees. I noted the contribution from the interaction at various meetings above; I am grateful for the organizers for inviting me to their conferences, always in beautiful settings, such as the island Vulcano. Important help in finding some of the older references was received from J. Pfleiderer and the librarian of the MPI for Fluid Mechanics. High Energy Physics with the author is supported by a NATO travel grant.

REFERENCES

Abbott, D. C., Bieging, J. H., Churchwell, E., Torres, A. V. 1986 Radio emission from galactic Wolf Rayet stars and the structure of Wolf Rayet star winds. *Astrophys. J.* 303:239–261.

Alfvén, H. 1939. On the motion of cosmic rays in interstellar space. *Phys. Rev.* 55(5):425–429.

Altenhoff, W., Mezger, P. G., Wendker, H., Westerhout, G. 1960 Die Durchmusterung der Milchstraße und die Quellendurchmusterung bei 2.7 GHz. *Publ. Obs. Bonn* 59:48–86.

Amnuel, P. R., Guseinov, O. Kh., Kasumov, F. K. 1973. Ejection of supernova envelopes by magnetic pumping. *Soviet Astron. A. J.* 16:932–937.

Asakimori, K., Burnett, T. H., Cherry, M. L., Christl, M. J., Dake, S., Derrickson, J. H., Fountain, W. F., Fuki, M., Gregory, J. C., Hayashi, T., Holynski, R., Iwai, J., Iyono, A., Jones, W. V., Jurak, A., Miyamura, O., Moon, K. H., Oda, H., Ogata, T., Parnell, T. A., Roberts, R. E., Strausz, S. C., Takahashi, Y., Tominaga, T., Watts, J. W., Wefel, J. P., Wilczynska, B., Wilczynski, H., Wilkes, R. J., Wolter, W., and Wosiek, B. 1993 Cosmic ray composition and spectra: (1). Protons. *Proc. 23rd Intl. Cosmic Ray Conf.* 2:21–24, 25–29.

Axford, W. I., Leer, E., Skadron, G. 1977. The acceleration of cosmic rays by shock waves. In *Proceedings of 15th Intl. Cosmic Ray Conf.* 11:132–137.

Baade, W., and Zwicky, F. 1934. Cosmic rays from supernovae. In *Proc. Natl. Academy Sci.* 20(5):259–263.

Bell, A. R. 1978.*a*. The acceleration of cosmic rays in shock fronts—I. *Mon. Not. Roy. Astron. Soc.* 182:147–156.

Bell, A. R. 1978*b*. The acceleration of cosmic rays in shock fronts—II. *Mon. Not. Roy. Astron. Soc.* 182:443–455.

Berezinsky, V. S., and Ptuskin, V. S. 1989. Radiation from young SN shells produced by cosmic rays accelerated in shock waves. *Astron. Astrophys.* 215:399–408.

Berezinskii, V. S., Bulanov, S. V., Dogiel, V. A., Ginzburg, V. L., Ptuskin, V. S. 1990. *Astrophysics of Cosmic Rays* (Amsterdam: North-Holland).

Bieging, J. H., Abbott, D. C., and Churchwell, E. B. 1989. A survey of radio emission from galactic OB stars. *Astrophys. J.* 340:518–536.

Biermann, P. L. 1993*a*. Cosmic rays I. The cosmic ray spectrum between 10^4 GeV and 3×10^9 GeV. *Astron. Astrophys.* 271:649–661.

Biermann, P. L. 1993*b*. Highly luminous radiogalaxies as sources of cosmic rays. In *Currents in Astrophysics and Cosmology*, eds. G. G. Fazio and R. Silberberg (Cambridge: Cambridge Univ. Press), pp. 12–19.

Biermann, P. L. 1994*a*. AGN and galactic sites of cosmic ray origin. In *High Energy Astrophysics*, ed. J. Matthews (Singapore: World Scientific), pp. 217–286.

Biermann, P. L. 1994*b*. Cosmic rays: Origin and acceleration—What we can learn from radioastronomy. In *Invited and Rapporteur Lectures of the 23rd International Cosmic Ray Conference*, eds. R. B. Hicks and D. Leahy (Singapore: World Scientific), pp. 45–83.

Biermann, P. L. 1994*c*. Production and acceleration of cosmic rays. In *Frontier Objects in Astrophysics and Particle Physics*, eds. F. Giovannelli (Nuovo Cimento), in press.

Biermann, P. L. 1995*a*. On the knee of the cosmic ray spectrum. In *Frontier Objects in Astrophysics and Particle Physics*, eds. F. Giovanell and G. Mannocchi (Bologna:

Italian Physical Soc.), pp. 593–597.

Biermann, P. L. 1995*b*. Production and acceleration of cosmic rays. In *Frontier Objects in Astrophysics and Particle Physics*, eds. F. Giovanell and G. Mannocchi (Italian Physical Soc.), pp. 469–481.

Biermann, P. L. 1995*c*. The origin of cosmic rays. *Space Sci. Rev.* 74:385–396.

Biermann, P. L. 1995*d*. The origin of galactic cosmic rays. *Nucl. Phys. B* 43:221–228.

Biermann, P. L., and Cassinelli, J. P. 1993. Cosmic rays II. Evidence for a magnetic rotator Wolf-Rayet star origin. *Astron. Astrophys.* 277:691–706.

Biermann, P. L., and Strittmatter, P. A. 1987. Synchrotron emission from shock waves in active galactic nuclei. *Astrophys. J.* 322:643–649.

Biermann, P. L., and Strom, R. G. 1993. Cosmic rays III. The cosmic ray spectrum between 1 GeV and 10^4 GeV and the radio emission from supernova remnants. *Astron. Astrophys.* 275:659–669.

Biermann, P. L., Chini, R., Greybe-Götz, A., Haslam, G., Kreysa, E., Mezger, P. G. 1990 Supernova 1987A at 1.3 mm. *Astron. Astrophys.* 227:L21–L24.

Biermann, P. L., Gaisser, T. K., and Stanev, T. 1994. The origin of galactic cosmic rays. *Phys. Rev. D*, in press.

Biermann, P. L., Strom, R. G., and Falcke, H. 1995. Cosmic rays V: The nonthermal radioemission of the old nova GK Per—a signature of hadronic interactions? *Astron. Astrophys.*, in press.

Bird, D. J., Corbató, S. C., Dai, H. Y., Dawson, B. R., Elbert, J. W., Gaisser, T. K., Green, K. D., Huang, M. A., Kieda, D. B., Ko, S., Larsen, C. G., Loh, E. C., Luo, M., Salamon, M. H., Smith, J. D., Sokolsky, P., Sommers, P., Stanev, T., Tang, J. K. K., Thomas, S. B., and Tilav, S. 1993. Evidence for correlated changes in the spectrum and composition of cosmic rays at extremely high energies. *Phys. Rev. Lett.* 71:3401–3404.

Bird, D. J., Corbató, S. C., Dai, H. Y., Dawson, B. R., Elbert, J. W., Emerson, B. L., Green, K. D., Huang, M. A., Kieda, D. B., Luo, M., Ko, S., Larsen, C. G., Loh, E. C., Salamon, M. H., Smith, J. D., Sokolsky, P., Sommers, P., Tang, J. K. K., and Thomas, S. B. 1994. The cosmic ray energy spectrum observed by the Fly's Eye. *Astrophys. J.* 424:491–502.

Bisnovatyi-Kogan, G. S. 1971. The explosion of a rotating star as a supernova mechanism. *Soviet Astron. A. J.* 14:652–655.

Bisnovatyi-Kogan, G. S., Popov, Yu.P., and Samochin, A. A. 1976. The magneto-hydrodynamic rotational model of supernova explosion. *Astrophys. Space Sci.* 41:287–320.

Blandford, R. D., and Ostriker, J. P. 1978. Particle acceleration by astrophysical shocks. *Astrophys. J. Lett.* 221:29–32.

Blandford, R., and Eichler, D. 1987. Particle acceleration at astrophysical shocks: A theory of cosmic ray origin. *Phys. Rept.* 154:1–75.

Bloemen, H., Wijnands, R., Bennett, K., Diehl, R., Hermsen, W., Lichti, G., Morris, D., Ryan, J., Schönfelder, V., Strong, A. W., Swanenburg, B. N., de Vries, C., and Winkler. 1994. COMPTEL observations of the Orion complex: Evidence for cosmic ray induced gamma-ray lines. *Astron. Astrophys.* 281:L5–L8.

Bykov, A. M., and Fleishman, G. D. 1992. On non-thermal particle generation in superbubbles. *Mon. Not. Roy. Astron. Soc.* 255:269–275.

Bykov, A. M., and Bloemen, H. 1994. Gamma-ray spectroscopy of the interstellar medium in the Orion complex. *Astron. Astrophys.* 283:L1–L4.

Bykov, A. M., and Toptygin, I. N. 1990. Theory of charged-particle acceleration by a collection of shock waves in a turbulent medium. *Soviet Phys. JETP* 71:702–708.

Bykov, A. M., and Toptygin, I. N. 1992. Diffusion of charged particles in a large-scale stochastic magnetic field. *Soviet Phys. JETP* 74:462–468.

Cassinelli, J. P. 1982. Theories for the winds from Wolf Rayet stars. In *Wolf-Rayet stars: Observations, Physics, Evolution*, eds. C. W. H. de Loore and A. J. Willis (Dordrecht: D. Reidel), pp. 173–183.

Cassinelli, J. P. 1991. Wolf-Rayet stellar wind theory. In *Wolf-Rayet Stars and Interrelations with Other Massive Stars in Galaxies*, eds. K. A. van der Hucht and B. Hidayat (Dordrecht: Kluwer), pp. 289–307.

Castor, J. I., Abbott, D. C., and Klein, R. I. 1975. Radiation-driven winds in Of stars. *Astrophys. J.* 195:157–174.

Charbonneau, P., and MacGregor, K. B. 1992. Angular momentum transport in magnetized stellar radiative zones. I. Numerical solutions to the core spin-up model problem. *Astrophys. J.* 387:639–661.

Chen, W., and White, R. L. 1991*a*. Inverse-Compton gamma-ray emission from chaotic, early-type stellar winds and its detectability by Gamma Ray Observatory. *Astrophys. J. Lett.* 381:63–66.

Chen, W., and White, R. L. 1991*b*. Nonthermal X-ray emission from winds of OB supergiants. *Astrophys. J.* 366:512–528.

Chevalier, R. A. 1982. The radio and X-ray emission from Type II supernovae. *Astrophys. J.* 259:302–310.

Cocconi, G. 1956. Intergalactic space and cosmic rays. *Nuovo Cimento* 3:1433–1442.

Cox, J. P., and Giuli, R. T. 1968. *Principles of Stellar Structure* (New York: Gordon and Breach).

Cunningham, G., Lloyd-Evans, J., Pollock, A. M. T., Reid, R. J. O., and Watson, A. A. 1980. The energy spectrum and arrival direction distribution of cosmic rays with energies above 10^{19} eV. *Astrophys. J. Lett.* 236:L71–L75.

Dickel, J. R., Murray, S. S., Morris, J., and Wells, D. C. 1982 A multiwavelength comparison of Cassiopeia A and Tycho's supernova remnant. *Astrophys. J.* 257:145–150 [erratum 264:746 (1983)].

Dickel, J. R., Sault, R., Arendt, R. G., Matsui, Y., and Korista, K. T. 1988. The evolution of the radio emission from Kepler's supernova remnant. *Astrophys. J.* 330:254–263.

Dickel, J. R., van Breugel, W. J. M., and Strom, R. G. 1991. Radio structure of the remnant of Tychos's supernova (SN 1572). *Astron. J.* 101:2151–2159.

Downs, G. S., and Thompson, A. R. 1972. The distribution of linear polarization in Cassiopeia A at wavelengths of 9.8 and 11.1 cm. *Astron. J.* 77:120–133.

Drury, L. O'C. 1983. An introduction to the theory of diffusive shock acceleration of energetic particles in tenuous plasmas. *Rept. Prog. Phys.* 46:973–1027.

Duffy, P., Drury, L. O'C., and Völk, H. 1994. Cosmic ray hydrodynamics at shock fronts. *Astron. and Astrophys.* 291:613–621.

Edmiston, J. P., and Kennel, C. F. 1984. A parametric survey of the first critical Mach number for a fast MHD shock. *J. Plasma Phys* 32:429.

Eichler, D., and Usov, V. 1993. Particle acceleration and nonthermal radio emission in binaries of early-type stars. *Astrophys. J.* 402:271–279.

Ellison, D. C., Möbius, E., and Paschmann, G. 1990. Particle injection and aceleration at earth's bow shock: Comparison of upstream and downstream events. *Astrophys. J.* 352:376–394.

Ellison, D. C., Reynolds, S. P., Borkowski, K., Chevalier, R., Cox, D. P., Dickel, J. R., Pisarski, R., Raymond, J., Spangler, S. R., Völk, H. J., and Wefel, J. P. 1994. Supernova remnants and the physics of strong shock waves. *Publ. Astron. Soc. Pacific* 106:780–797.

Falle, S. A. E. G. 1990. Diffusive shock acceleration of relativistic particles. In *Neutron Stars and their Birth Events*, ed. W. Kundt (Dordrecht: Kluwer), pp. 303–318.

Feldmeier, A. 1993. Zeitabhängige Struktur und Energietransfer der Winde heißer, massereicher Sterne. Ph.D. Thesis Univ. of Munich.

Fermi, E. 1949. On the origin of the cosmic radiation. *Phys. Rev.* 75(8):1169–1174.

Fermi, E. 1954. Galactic magnetic fields and the origin of cosmic radiation. *Astrophys. J.* 119:1–6.

Forman, M. A., Jokipii, J. R., and Owens, A. J. 1974. Cosmic-ray streaming perpendicular to the mean magnetic field. *Astrophys. J.* 192:535–540.

Freudenreich, H. T., Mincer, A. I., Berley, D., Goodman, J. A., Tonwar, S., Wrotniak, A., and Yodh, G. B. 1990. Study of hadrons at the cores of extensive air showers and the elemental composition of cosmic rays at 10^{15} eV. *Phys. Rev. D* 41:2732–2750.

Gaisser, T. K., Stanev, T., Tilav, S., Corbato, S. C., Dai, H. Y., Dawson, B. R., Elbert, J. W., Emerson, B., Kieda, D. B., Luo, M., Ko, S., Larsen, C., Loh, E. C., Salamon, M. H., Smith, J. D., Sokolsky, P., Sommers, P., Tang, J., Thomas, S. B., and Bird, D. J. 1993. Cosmic ray composition around 10^{18} eV. *Phys. Rev. D* 47:1919–1932.

Galloway, D. J., and Proctor, M. R. E. 1992. Numerical calculations of fast dynamos in smooth velocity fields with realistic diffusion. *Nature* 356:691–693.

Garcia-Munoz, M., Mason, G. M., and Simpson, J. A. 1977 The age of galactic cosmic rays derived from the abundance of ^{10}Be. *Astrophys. J.* 217:859–877.

Garcia-Munoz, M., Simpson, J. A., Guzik, T. G., Wefel, J. P., and Margolis, S.H. 1987. Cosmic-ray propagation in the Galaxy and in the heliosphere: The path-length distribution at low energy. *Astrophys. J. Suppl.* 64:269–304.

Gilbert, A. D. 1991. Fast dynamo action in a steady chaotic flow. *Nature* 350:483–485.

Gilbert, A. D., and Childress, S. 1990. Evidence for fast dynamo action in a chaotic web. *Phys. Rev. Lett.* 65:2133–2136.

Ginzburg, V. L. 1953. Supernovae and novae as sources of cosmic and radio radiation. *Dokl. Akad. Nauk SSSR* 92:1133–1136 (NSF-Transl. 230).

Ginzburg, V. L. 1993. The origin of cosmic rays (forty years later). *Physics-Uspekhi* 36:587–591.

Golla, G. 1989. Beobachtungen von M31 bei 327 MHz mit dem Apertursyntheseteleskop Westerbork/NL. M.Sc. Thesis, Univ. of Bonn.

Green, D. A. 1990. The Cygnus loop at 408 MHz: Spectral variations and a better overall view. *Astron. J.* 100:1927–1942.

Hartmann, L., and MacGregor, K. B. 1982. Protostellar mass and angular momentum loss. *Astrophys. J.* 259:180–192.

Hayakawa, S. 1956. Supernova origin of cosmic rays. *Progr. Theoret. Phys.* 15:111–121.

Hayakawa, S. 1969. *Cosmic Ray Physics* (New York: Wiley-Interscience).

Hess, V. F. 1912 "Über Beobachtungen der durchdringenden Strahlung bei sieben Freiballonfahrten. *Phys. Z.* 13:1084–1091.

Hidayat, B. 1991. Galactic Wolf-Rayet distribution and IMF. In *Wolf-Rayet Stars and Interrelations with Other Massive Stars in Galaxies*, eds. K. A van der Hucht and B. Hidayat (Dordrecht: Kluwer), pp. 587–589.

Ip, W.-H., and Axford, W. I. 1992. Particle acceleration up to 10^{20} eV. In *Particle Acceleration in Cosmic Plasmas*, eds. G. P. Zank, and T. K. Gaisser (New York: American Inst. of Physics), pp. 400–405.

Jauncey, D. L., Kemball, A., Bartel, N., Whitney, A. R., Rogers, A. E., Shapiro, I. I., Preston, R. A., Clark, T. A., Harvey, B. R., Jones, D. L., Nicolson, G. D., Nothnagel, A., Phillips, R. B., Reynolds, J. E., and Webber, J. C. 1988. Supernova 1987A: Radiosphere resolved with VLBI five days after the neutrino burst. *Nature* 334:412–415.

Jokipii, J. R. 1987. Rate of energy gain and maximum energy in diffusive shock acceleration. *Astrophys. J.* 313:842–846.

Jokipii, J. R., and Morfill, G. 1987. Ultra-high energy cosmic rays in a galactic wind and its termination shock. *Astrophys. J.* 312:170–177.

Jokipii, J. R., Levy, E. H., and Hubbard, W. B. 1977. Effects of particle drift on cosmic ray transport I. General properties, application to solar modulation. *Astrophys. J.* 213:861–868.

Jones, F. C., and Ellison, D. C. 1991. The plasma physics of shock acceleration. *Space Sci. Rev.* 58:259–346.

Kardashev, N. S. 1970 Pulsars and nonthermal radiosources. *Soviet Astron. A. J.* 14:375–384.

Kato, M., and Iben, I., Jr. 1992. Self-consistent models of Wolf-Rayet stars as helium stars with optically thick winds. *Astrophys. J.* 394:305–312.

Kennel, C. F., Edmiston, J. P., and Hada, T. 1985. A quarter century of collisonless shock research. In *Collisionless Shocks in the Heliosphere: Reviews of Current Research*, eds. B. T. Tsurutani and R. G. Stone (Washington, D. C.: American Geophysical Union), p. 1.

Kohlhörster, W. 1913. Messungen der durchdringenden Strahlung im Freiballon in größeren Höhen. *Phys. Z.* 14:1153–1156.

Kronberg, P. P., Biermann, P. L., and Schwab, F. R. 1985. The nucleus of M82 at radio and X-ray bands: Discovery of a new radio population of supernova candidates. *Astrophys. J.* 291:693–707.

Krymskii, G. F. 1977. A regular mechanism for the acceleration of charged particles on the front of a shock wave. *Dokl. Akad. Nauk SSSR* 234:1306–1308.

Krymskii, G. F., and Petukhov, S. I. 1980. Particle acceleration by the "regular" mechanism at a spherical shock front. *Soviet Astron. A. J. Lett.* 6:124—126.

Kundt, W. 1976. Supernova explosions driven by magnetic springs. *Nature* 261:673–674.

Lagage, P. O., and Cesarsky, C. J. 1983. Cosmic ray acceleration in the presence of self-excited waves. *Astron. Astrophys.* 118:223–228.

Lamers, H. J. G. L. M., and Cassinelli, J. P. 1994. *Introduction to Stellar Winds*, in press.

Langer, N. 1989. Standard models of Wolf-Rayet stars. *Astron. Astrophys.* 210:93–113.

Lanz, T., and Mathys, G. 1993. A search for magnetic fields in Am stars. *Astron. Astrophys.* 280:486–492.

LeBlanc, J. M., and Wilson, J. R. 1970. A numerical example of the collapse of a rotating magnetized star. *Astrophys. J.* 161:541–551.

Leitherer, C. 1991. Observational aspects of stellar evolution in the upper HRD. In Wolf-Rayet Stars and Interrelations with Other Massive Stars in Galaxies, eds. K. A. van der Hucht and B. Hidayat (Dordrecht: Kluwer), pp. 465–478.

Levinson, A. 1992. Electron injection in collisionless shocks. *Astrophys. J.* 401:73–80.

Levinson, A. 1994. Electron injection and acceleration at nonlinear shocks: Results of numerical simulations. *Astrophys. J.* 426:327–333.

Lucy, L. B. 1982. The formation of resonance lines in locally nonmonotonic winds. *Astrophys. J.* 255:278–285.

Lucy, L. B., and Abbott, D. C. 1993. Multiline transfer and the dynamics of Wolf-Rayet winds. *Astrophys. J.* 405:738–746.

Lucy, L. B., and Solomon, P. M. 1970. Mass loss by hot stars. *Astrophys. J.* 159:879–893.

MacFarlane, J. J., and Cassinelli, J. P. 1989. Shock-generated X-ray emission in radiatively drivein winds: A model for Tau Scorpii. *Astrophys. J.* 347:1090–1099.

Maheswaran, M., and Cassinelli, J.P. 1988 On the surface magnetic fields of rapidly rotating stars with winds. *Astrophys. J.* 335:931–939.

Maheswaran, M., and Cassinelli, J. P. 1992. Constraints on the surface magnetic fields of hot stars with winds. *Astrophys. J.* 386:695–702.

Massey, P., and Armandroff, T. E. 1991 Wolf-Rayet stars in nearby galaxies: Facts and fancies. In *Wolf-Rayet Stars and Interrelations with Other Massive Stars in Galaxies*, eds. K. A. van der Hucht and B. Hidayat (Dordrecht: Kluwer), pp. 575–586

Mezger, P. G., and Henderson, A. P. 1967. Galactic HII regions. I. Observations of their continuum radiation at the frequency 5 GHz. *Astrophys. J.* 147:471–489.

Milne, D. K. 1971. The supernova of 1006 AD. *Australian J. Phys.* 24:757–767.

Milne, D. K. 1987. An atlas of supernova remnant magnetic fields. *Australian J. Phys.* 40:771–787.

Nath, B. B., and Biermann, P. L. 1993. Did cosmic rays reionize the intergalactic medium? *Mon. Not. Roy. Astron. Soc.* 265:241–249.

Nath, B. B., and Biermann, P. L. 1994*a*. Cosmic ray ionization of the interstellar medium. *Mon. Not. Roy. Astron. Soc.* 267:447–451.

Nath, B. B., and Biermann, P. L. 1994*b* Gamma line radiation from stellar winds in Orion complex: A test for cosmic ray origin theory. *Mon. Not. Roy. Astron. Soc.* 270:L33–L36.

Nerney, S., and Suess, S. T. 1987. Radiatively driven winds from magnetic, fast-rotating stars: Wolf Rayet stars? *Astrophys. J.* 321:355–369.

Northrop, T. G. 1963. *The Adiabatic Motion of Charged Particles* (New York: Inter-science.

Ostriker, J. P., and Gunn, J. E. 1971. Do pulsars make supernovae? *Astrophys. J. Lett.* 164:95–104.

Owocki, S. P. 1990. Winds from hot stars. *Rev. Mod. Astron.* 3:98–123.

Owocki, S. P. 1992. Instabilities in hot-star winds: Basic physics and recent developments. In *The Atmospheres of Early-Type Stars*, eds. U. Heber and C. S. Jeffrey (Berlin: Springer-Verlag), pp. 393–408.

Owocki, S. P. 1994. In *Instabilities and Variability in Hot-Star Winds*, eds. A. Moffat, S. Owocki, A. Fullerton and N. St-Louis (Dordrecht: Kluwer), in press).

Owocki, S. P., and Rybicki, G. B. 1985. Instabilities in line-driven stellar winds. II. Effects of scattering. *Astrophys. J.* 299:265–276.

Owocki, S. P., Castor, J. I., and Rybicki, G. B. 1988. Time dependent models of radiatively driven stellar winds. I. Nonlinear evolution of instabilities for a pure absorption model. *Astrophys. J.* 335:914–930.

Panagia, N., Sramek, R. A., and Weiler, K. W. 1986 Subluminous, radio emitting Type I supernovae. *Astrophys. J. Lett.* 300:55–58.

Panagia, N., Van Dyk, S. D., and Weiler, K. W. 1993. Supernova 1993J in NGC3031. *IAU Circ. No.* 5762.

Parker, E. N. 1958. Dynamics of the interplanetary gas and magnetic fields. *Astrophys. J.* 128:664–676.

Pauldrach, A., Puls, J., and Kudritzki, R. P. 1986. Radiation-driven winds of hot luminous stars. Improvements of the theory and first results. *Astron. Astrophys.* 164:86–100.

Peters, B. 1959. Origin of cosmic radiation. *Nuovo Cimento Suppl.* XIV:436–456.

Peters, B. 1961. Primary cosmic radiation and extensive airshowers. *Nuovo Cimento* XXII:800–819.

Phillips, J. A., and Kulkarni, S. R. 1993. *IAU Circ. No. 5884.*

Poe, C. H., Friend, D. B., and Cassinelli, J. P. 1989. A rotating, magnetic, radiation-driven wind model for Wolf-Rayet stars. *Astrophys. J.* 337:888–902.

Polcaro, V. F., Giovannelli, F., Manchanda, R. K., Norci, L., and Rossi, C. 1991. Evidence of an interacting stellar wind and possible high energy emission in the open cluster Berkeley 87. *Astron. Astrophys.* 252:590–596.

Polcaro, V. F., Brinkmann, W., Giovannelli, F., Manchanda, R. K., Norci, L., Persi, P., and Rossi, C. 1993. High energy gamma ray emission from open clusters. *Astron. Astrophys. Suppl.* 97:139.

Prandtl, L. 1925. Bericht zu Untersuchungen zur ausgebildeten Turbulenz. *Zeitschrift angew. Math. und Mech.* 5:136–139.

Prandtl, L. 1949. *Strömungslehre*, 5th ed. (Vieweg: Braunschweig).

Prishchep, V. L., and Ptuskin, V. S. 1981. Fast-particle acceleration at a spherical shock front. *Soviet Astron. A. J.* 25:446–450.

Protheroe, R. J., and Szabo, A. P. 1992. High energy cosmic rays from active galactic nuclei. *Phys. Rev. Lett.* 69:2885–2888.

Pye, J. P., Pounds, K. A., Rolf, D. P., Seward, F. D., Smith, A., and Willingale. 1981. An X-ray map of SN 1006 from the Einstein Observatory. *Mon. Not. Roy. Astron. Soc.* 194:569–582.

Rachen, J. P., and Biermann, P. L. 1993. Extragalactic ultra-high energy cosmic rays I. Contributions from hot spots in FR-II radio galaxies. *Astron. Astrophys.* 272:161–175.

Rachen, J. P., Stanev, T., and Biermann, P. L. 1993. Extragalactic ultra-high energy cosmic rays II. Comparison with experimental data. *Astron. Astrophys.* 273:377–382.

Ramaty, R., Kozlovsky, B. and Lingenfelter, R. E. 1994. Gamma-ray lines from the Orion complex. *Astrophys. J. Lett.*, in press.

Ratkiewicz, R., Axford, W. I., and McKenzie, J. F. 1994. Similarity solutions for synchrotron emission from a supernova blast wave. *Astron. Astrophys.* 291:935–942.

Reynolds, S. P., and Chevalier, R. A. 1981. Nonthermal radiation from supernova remnants in the adiabatic stage of evolution. *Astrophys. J.* 245:912–919.

Reynolds, S. P., and Chevalier, R. A. 1983. Nonthermal radioemission from the old nova GK Per. *Bull. Amer. Astron. Soc.* 15:965.

Reynolds, S. P., and Chevalier, R. A. 1984. A new type of extended nonthermal radioemitter: Detection of the old nova GK Persei. *Astrophys. J. Lett.* 281:33–35.

Reynolds, S. P., and Gilmore, D. M. 1986. Radio observations of the remnant of the supernova of A.D. 1006 I. Total intensity observations. *Astron. J.* 92:1138–1144.

Rybicki, G. B., and Lightman, A. P. 1979. *Radiative Processes in Astrophysics* (New York: Wiley Interscience).

Ruzmaikin, A., Sokoloff, D., and Shukurov, A. 1988. Magnetism of spiral galaxies. *Nature* 336:341–347.

Schmidt, G. D., Bergeron, P., Liebert, J., Saffer, R. A. 1992. Two ultramassive white dwarfs found among candidates for magnetic fields. *Astrophys. J.* 394:603–608.

Schmutzler, T. 1987. Zum thermischen Zustand dünner Plasmen unter Einfluß beliebiger Photonenspektren. Ph.D. Thesis, Univ. of Bonn

Schneider, P. 1993. Diffusive particle acceleration by an ensemble of shock waves. *Astron. Astrophys.* 278:315–327.

Seaquist, E. R., Bode, M. F., Frail, D. A., Roberts, J. A., Evans, A., and Albinson, J. S. 1989. A detailed study of the remnant of Nova GK Persei and its environs. *Astrophys. J.* 344:805–825.

Seo, E.-S., Ormes, J. F., Steitmatter, R. E., Stochaj, S. J., Jones, W. V., Stephens, S. A., and Bowen, T. 1991. Measurement of cosmic-ray proton and Helium spectra during the 1987 solar minimum. *Astrophys. J.* 378:763–772.

Seward, F., Gorenstein, P., and Tucker, W. 1983. The mass of Tycho's supernova remnant as determined from a high resolution X-ray map. *Astrophys. J.* 266:287–297.

Shara, M. M., Potter, M., Moffat, A. F. J., and Smith, L. F. 1991. A deep survey for faint galactic Wolf-Rayet stars. In *Wolf-Rayet Stars and Interrelations with Other Massive Stars in Galaxies*, eds. K. A. van der Hucht and and B. Hidayat (Dordrecht: Kluwer), pp. 591–594.

Shklovskii, I. S. 1953. On the origin of cosmic rays. *Dokl. Akad. Nauk, SSSR* 91(3):475–478 (Library of Congress Transl. RT-1495).

Shklovsky, I. S. 1968. *Supernovae* (London: Wiley-Interscience).

Silberberg, R., Tsao, C. H., Shapiro, M. M., and Biermann, P. L. 1990. Source composition of cosmic rays at energies 1–1000 TeV per nucleon. *Astrophys. J.* 363:265–269.

Smith, R. C., Raymond, J. C., and Laming, J. M. 1994. High resolution spectroscopy of Balmer-dominated shocks in the Large Magellanic Cloud. *Astrophys. J.* 420:286–293.

Sreekumar, P., Bertsch, D. L., Dingus, B. L., Fichtel, C. E., Hartman, R. C., Hunter, S. D., Kanbach, G., Kniffen, D. A., Lin, Y. C., Mattox, J. R., Mayer-Hasselwander, H. A., Michelson, P. F., von Montigny, C., Nolan, P. L., Pinkau, K., Schneid, E. J., Stone, R. G., and Thompson, D. J. 1993. Constraints on the cosmic rays in the Small Magellanic Cloud. *Phys. Rev. Lett.* 70:127–129.

Stanev, T., Biermann, P. L., and Gaisser, T. K. 1993. Cosmic rays IV. The spectrum and chemical composition above 10^4 GeV. *Astron. Astrophys.* 274:902–908.

Staveley-Smith, L., Manchester, R. N., Kesteven, M. J., Campbell-Wilson, D., Crawford, D. F., Turtle, A. J., Reynolds, J. E., Tzioumis, A. K., Killeen, N. E. B., and Jauncey, D. L. 1992. Birth of a radio supernova remnant in 1987A. *Nature* 355:147–149.

Staveley-Smith, L., Briggs, D. S., Rowe, A. C. H., Manchester, R. N., Reynolds, J. E., Tzioumis, A. K., and Kesteven, M. J. 1993. Structure of the radio remnant of supernova 1987A. *Nature* 366:136–138.

Sun, L., Wang, S., and Watson, A. A. 1993. The galactic latitude and energy dependence of cosmic ray primary mass above 10^{19} eV. *Proc. 23rd Intl. Cosmic Ray Conf.*, pp. 43–46.

Thorsett, S. E. 1994. Millisecond pulsars—the times are a-changing. *Nature* 367:684–685.

Tuffs, R. J. 1986. Secular changes within Cassiopeia A at 5 GHz—I. *Mon. Not. Roy. Astron. Soc.* 219:13–38.

Turtle, A. J., Campbell-Wilson, D., Bunton, J. D., Jauncey, D. L., Kesteven, M. J., Manchester, R. N., Norris, R. P., Storey, M. C., and Reynolds, J. E. 1987. A prompt radioburst from supernova 1987A in the Large Magellanic Cloud. *Nature* 327:38–40.

Usov, V. V., and Melrose, D. B. 1992. X-ray emission from single magnetic early-type stars. *Astrophys. J.* 395:575–581.

Völk, H. J., and Biermann, P. L. 1988. Maximum energy of cosmic ray particles accelerated by supernova remnant shocks in stellar wind cavities. *Astrophys. J. Lett.* 333:65–68.

Weber, E. J., and Davis, L., Jr. 1967. The angular momentum of the solar wind. *Astrophys. J.* 148:217–227.

Weiler, K. W., Sramek, R. A., Panagia, N., van der Hulst, J. M., and Salvati, M. 1986. Radio supernovae. *Astrophys. J.* 301:790–812.

Weiler, K. W., Panagia, N., Sramek, R. A., van der Hulst, J. M., Roberts, M. S., and Nguyen, L. 1989. Radio emission from supernovae. I. One to twelve year old supernovae. *Astrophys. J.* 336:421–428.

Weiler, K. W., Panagia, N., and Sramek, R. A. 1990. Radio emission from supernovae. II. SN 1986J: A different kind of Type II. *Astrophys. J.* 364:611–625.

Weiler, K. W., van Dyk, S. D., Panagia, N., Sramek, R. A., and Discenna, J. L. 1991. The 10 year radio light curves for SN 1979C. *Astrophys. J.* 380:161–166.

Wheeler, J. C. 1989. Introduction to supernovae. In *Supernovae*, eds. J. C. Wheeler, T. Piran and S. Weinberg (Singapore: World Scientific), pp. 1–93.

White, R. L. 1985. Synchrotron emission from chaotic stellar winds. *Astrophys. J.* 289:698–708.

White, R. L., and Long, K. S. 1986. X-ray emission from Wolf Rayet stars. *Astrophys. J.* 310:832–837.

White, R. L., and Chen, W. 1992. π^0-decay gamma-ray emission from winds of massive stars. *Astrophys. J. Lett.* 387:81–84.

Wiebel, B. 1992. Chemical composition in high energy cosmic rays. HEGRA-Note, Univ. of Wuppertal.

Zank, G. P., Axford, W. I., and McKenzie, J. F. 1990. Instabilities in energetic particle modified shocks. *Astron. Astrophys.* 233:275–284.

SOLAR WIND INTERACTION WITH COMETS: LESSONS FOR MODELING THE HELIOSPHERE

TAMAS I. GOMBOSI, DARREN L. De ZEEUW, TIMUR Y. LINDE and
KENNETH G. POWELL
University of Michigan

Both cometary and stellar wind flows can be visualized as nearly point-sources of fluid
within a streaming flow. The interaction of the solar wind with the very local interstellar
medium and the interaction of comets with the solar wind show a number of important
similarities. The solar wind-cometary interaction is well studied both experimentally
and theoretically, while the interaction of the heliosphere with the interstellar medium
is the object of an emerging field. This chapter explores the possible analogies between
the comet-solar wind and heliosphere-interstellar medium interactions and discusses
new multi-scale methods for modeling these complex phenomena.

I. INTRODUCTION

Comets represent an astrophysical "terrella" experiment which can help us
to understand the interaction of expanding stellar winds with the surrounding
interstellar medium. In spite of the obvious differences, the analogy is quite
striking. Both cometary and stellar wind flows can be visualized as nearly
point-sources of fluid within a streaming flow. In the simplest approximation
a comet can be considered to be a source of partially ionized gas expanding
into the solar wind plasma, while stellar winds are expanding plasma flows
within the partially ionized interstellar gas.

Compared to stellar winds, cometary environments are both experimentally and theoretically well studied. Three widely different comets have been
visited by seven spacecraft so far. The *in-situ* measurements nicely com-
plement the large body of groundbased and near Earth-orbit observations of
cometary phenomena. The main physical and chemical processes controlling
the solar wind interaction with cometary atmospheres are reasonably well
understood. On the other hand, our understanding of the interaction between
the heliosphere and the interstellar medium hardly goes beyond the "cartoon"
level. Only very recently have we seen the emergence of two-dimensional
models (though even these calculations oversimplify the gas dynamics and
either neglect the magnetic field altogether or use a potential flow model).
For details we refer to Baranov and Malama (1993), Nerney et al. (1993)
and Steinolfson et al. (1994). Several chapters of this book also deal with

the interaction of the heliosphere with the surrounding interstellar medium, therefore here we only briefly summarize the main elements of the interaction of the heliosphere with the surrounding medium.

This chapter is not a review of the cometary plasma environment or of the solar wind interaction with the interstellar medium. There are several detailed reviews (observations and theory) in the literature and we refer the reader to these (see, e.g., Holzer 1989; Axford 1990; Baranov 1990; Neugebauer 1990; Suess 1990; Gombosi 1991). This chapter discusses some new developments in multi-scale modeling of the heliosphere and of cometary environments and draws parallels between cometary atmospheres and the heliosphere. In Sec. II we briefly discuss the main features of the interaction between the heliosphere and the interstellar medium. Section III presents a brief summary of the main regions in cometary environments, while Sec. IV discusses new multi-scale models of the heliosphere and of the interaction of the solar wind with comets. Section V presents a three-dimensional gas-dynamic model of the heliosphere. The chapter closes with a brief summary.

II. THE HELIOSPHERE

The heliosphere is the region of space dominated by the presence of solar wind plasma. In a very general sense the heliosphere contains three large-scale boundaries.

The Sun is a point source of supersonic magnetized plasma which radially expands into a low pressure external medium. This point source moves with a velocity of ≈ 26 km s^{-1} with respect to the very local interstellar medium. It is generally assumed that the partially ionized interstellar gas is fairly cold (even though the thermodynamic properties of this region are only poorly known), and consequently this relative motion is supersonic (see, e.g., Frisch 1994; chapter by Frisch). An immediate consequence of the supersonic motion of the heliosphere in the interstellar plasma is the formation of an upstream bow shock, which decelerates and deflects the interstellar charged particles.

Flow lines of the interstellar plasma do not penetrate into the region dominated by the solar wind flow, but flow around a "contact surface," also called the "heliopause." This surface is a direct analogy to the cometary "contact surface," to be discussed in Sec. III. At the nose of the heliopause the interstellar gas is nearly stagnating and a density pile-up is expected. The heliopause is considered to be the outer boundary of the heliosphere.

The third large-scale discontinuity is expected to form inside the heliopause. On the upwind side of the heliosphere the initially radially expanding supersonic solar wind must somehow be diverted to the downstream direction, otherwise it would intersect the heliopause. This diversion can only take place in subsonic flow, therefore the supersonic expansion of the solar wind must be terminated by an inner shock or termination shock. On the downwind side, a termination shock is also formed due to the expansion of the supersonic solar wind into a low, but finite pressure external medium.

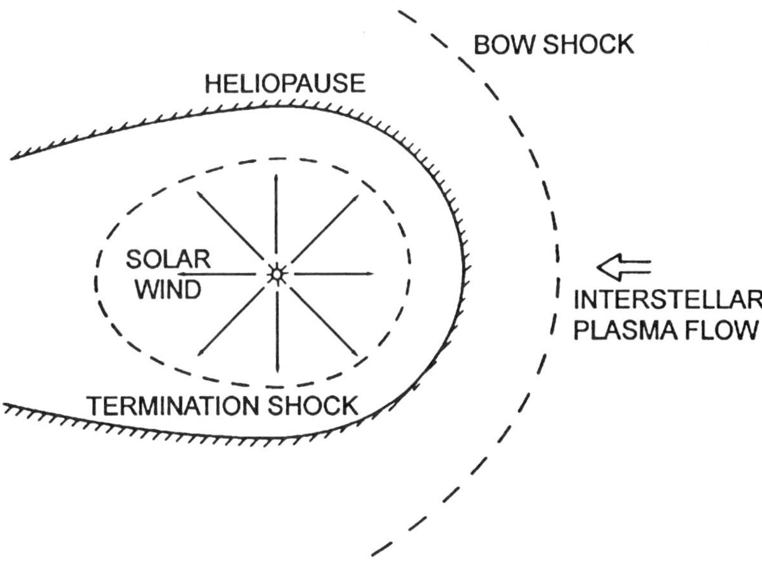

Figure 1.   Schematic representation of the interaction between the heliosphere and the interstellar medium.

A schematic representation of this scenario is shown in Fig. 1. The reader is referred to the other chapters in this book for more detailed accounts and also to the reviews by Holzer (1989), Axford (1990), Baranov (1990) and Suess (1990).

III. BOUNDARIES IN COMETARY PLASMA ENVIRONMENTS

It was recognized about a quarter century ago (Biermann et al. 1967) that the expanding cometary exosphere represents an extensive, "soft" obstacle for the supersonic and super-Alfvénic solar wind flow. The resulting interaction is very different from that with other solar system bodies with gravitationally bound dense atmospheres and/or significant intrinsic magnetic fields. Neutral atoms and molecules of cometary origin become ionized (because of photoionization, charge transfer or electron impact ionization) with characteristic ionization scale lengths of 10^5 to 10^7 km. The ionization process introduces a new, practically stationary particle into the high-speed magnetized flow of the solar wind. Spacecraft instrumentation at comets Giacobini-Zinner, Grigg-Skjellerup and Halley detected large-amplitude, low-frequency magnetic field fluctuations. These fluctuations grow from the relatively low solar wind turbulence level (at large cometocentric distances) to very large amplitudes in the vicinity of the bow shock. The enhanced fluctuation level is due to instabilities associated with the solar wind interaction with ionized cometary material.

In their pioneering work Biermann et al. (1967) assumed that the plasma flow rapidly accommodates the new ions, i.e., the entire plasma population can be characterized by a single temperature and flow velocity. Biermann et al.

(1967) had also predicted that the deceleration of the solar wind flow by mass loading leads to the formation of a weak shock and the flow is impulsively decelerated to subsonic velocities. Later Galeev et al. (1985) recognized that implanted cometary ions carry most of the hydrodynamic pressure and that charge exchange cooling of the implanted plasma population can play a very important role in the dynamics of the contaminated solar wind flow. Their model predicted a weak and highly structured shock with a viscous subshock, a continuously decelerated and cooled plasma flow behind the shock and finally a stagnation region. The *in-situ* measurements later confirmed the main features predicted by this model.

Downstream of the bow shock the plasma population is a varying mixture of shocked plasma (solar wind contaminated with upstream pick-up particles) and cometary plasma ionized in the subsonic region. This region is called the cometosheath. The cometosheath is characterized by a rapidly increasing rate of ion pick-up (as the plasma moves towards the comet), resulting in continuous deceleration (and eventual stagnation) of the plasma flow, accompanied by increasing plasma density and magnetic field magnitude. The inner, nearly stagnating region of the cometosheath is primarily photochemically controlled and the plasma density varies as r^{-1}.

A few hundred comet radii from the nuclei of active comets, a "contact surface" or diamagnetic cavity boundary is formed (this is the surface separating the magnetized cometosheath plasma from the magnetic field-free inner cometary ionosphere). The "diamagnetic cavity boundary" is formed at a location where the inward-pointing (towards the comet) total magnetic force (the sum of the magnetic pressure gradient and magnetic curvature forces) is balanced by the outward-pointing ion-neutral drag force. An inner shock was predicted inside the contact surface to decelerate the supersonic outflow of the cometary ions and divert them toward the tail (Wallis and Dryer 1976). However, recombination between the "inner shock" and the diamagnetic cavity boundary greatly diminishes the distance between these two boundaries, so much so, that it is very difficult to distinguish them (Cravens 1989; Goldstein et al. 1989).

For a more detailed description of our present understanding of the solar wind interaction with active comets the reader is referred to detailed reviews by Neugebauer (1990) and Gombosi (1991).

IV. MULTI-SCALE MODELING OF THE SOLAR WIND-COME-TARY INTERACTION

In the last decade several large-scale, multi-dimensional models have been developed to describe the solar wind interaction with an expanding cometary atmosphere (see, e.g., Schmidt and Wegmann 1982; Fedder et al. 1986; Schmidt-Voigt 1987; Schmidt et al. 1988; Huebner et al. 1991; Wegmann et al. 1993; Lindgren and Cravens 1993). These models significantly advanced our understanding of the global nature of the interaction. However, these large-scale MHD models suffer from one or more of three general limitations typical of the present generation of global MHD calculations: (i) the modeled volume is not large enough to include the entire upstream mass loading region, the flaring shock and the long antisunward plasma tail; (ii) the resolution is not fine enough to resolve many important plasma regions (such as the diamagnetic cavity, the shock structure, or the cometopause); and (iii) it is difficult to choose appropriate boundary conditions.

Recently Gombosi et al. (1994) presented the first results of a new generation of global MHD models which solve the governing equations of solar wind mass loading on an adaptively refined unstructured grid. The model successfully addresses all three major weaknesses of earlier calculations:

1. It models a 1/3 AU×1/6 AU axisymmetric volume, which is large enough to encompass all physically important regions;
2. It uses very high resolution in high gradient regions, such as the shock, the stagnation region, and near the nucleus and is capable of resolving the shock and the inner cometary ionosphere;
3. It uses finite volume MUSCL-type schemes which are robust and accurate, minimize numerical dissipation and dispersion, faithfully represent shocks and other discontinuities, and, most importantly, provide a framework in which boundary conditions can be applied in a systematic way consistent with the local physics.

A. Model Equations

The transport model used by Gombosi et al. (1994) solves the ideal MHD equations with mass loading. These equations are the following:

$$\frac{\partial \rho}{\partial t} + \nabla \cdot (\rho\,\mathbf{u}) = \dot{\rho}$$

$$\frac{\partial (\rho\,\mathbf{u})}{\partial t} + \nabla \cdot \left(\rho\,\mathbf{u}\mathbf{u} + p\,I + \frac{B^2}{2\mu_0}I - \frac{\mathbf{B}\,\mathbf{B}}{\mu_0} \right) = \dot{\mathbf{M}}$$

$$\frac{\partial}{\partial t}\left(\varepsilon + \frac{B^2}{2\mu_0} \right) + \nabla \cdot \left((\varepsilon + p)\,\mathbf{u} + \frac{(\mathbf{B}\cdot\mathbf{B})\,\mathbf{u} - \mathbf{B}\,(\mathbf{B}\cdot\mathbf{u})}{\mu_0} \right) = \dot{\varepsilon}$$

$$\frac{\partial \mathbf{B}}{\partial t} + \nabla \cdot (\mathbf{u}\,\mathbf{B} - \mathbf{B}\,\mathbf{u}) = 0.$$

Here $\varepsilon = \frac{1}{2}\rho\,u^2 + \frac{3}{2}p$ is the internal energy density of the plasma, $\mathbf{u}$ and $\mathbf{u}_n$ are the plasma and cometary neutral gas velocities, ρ and p are the mass density and pressure of the mass loaded plasma, $\mathbf{B}$ is the magnetic field vector. Other quantities in these equations are the 3×3 unit matrix I, the permeability of free space μ_0, and the mass momentum and energy addition rates $\dot{\rho}$, $\dot{\mathbf{M}}$ and $\dot{\varepsilon}$.

B. Numerical Solution on an Adaptive Grid

The numerical procedure used in solving the above MHD equations is built from two pieces that are extremely well suited to the cometary mass-loading problem. The first is a data structure that allows for adaptive refinement of the mesh in regions of interest, and the second is a second-order MUSCL-type scheme based on an a new approximate Riemann solver for ideal MHD. These components are described in detail by Gombosi et al. (1994).

Adaptive refinement and coarsening of a mesh is a very attractive way to make optimal use of computational resources. It becomes particularly attractive for problems in which there are disparate spatial scales. For problems like the cometary mass-loading problem, in which the ionization length scale and the radius of the comet differ by several orders of magnitude, an adaptive mesh is a virtual necessity. The primary difficulty in implementing an adaptive scheme is related to the way in which the solution data are stored; the data structure must be much more flexible than the simple array-type storage used in the majority of scientific computations.

The basic data structure used is a hierarchical cell-based tree structure—"parent" cells are refined by division into "children" cells. Each cell has a pointer to its parent cell (if one exists) and to its children cells (if they exist). The cells farthest down the hierarchy, that is, the ones with no children, are the cells on which the calculation takes place.

This tree structure contains all the connectivity information necessary to carry out the flow calculations; no other connectivity information is stored. While it involves a more sophisticated implementation than a simple array-type data structure, it is an efficient and natural way to implement adaptive refinement and coarsening. The mesh is generated in such a way that important geometric and flow features are resolved. The geometry is resolved by recursively dividing cells in the vicinity of the comet until a specified cell-size is obtained. A larger cell-size is specified for the remainder of the mesh. Flow features are resolved by obtaining a solution on this original mesh, and then automatically refining cells in which the flow gradients are appreciable, and coarsening cells in which the flow gradients are negligible.

Previous upwind-based schemes for MHD have been based on the one-dimensional Riemann problem obtained by noting that, for one-dimensional problems, the $\nabla \cdot \mathbf{B} = 0$ condition reduces to the constraint that $B_x = $ constant. Applying schemes based on this one-dimensional Riemann problem to MHD problems in more than one dimension leads to a serious problem, however. Because there is no evolution equation for the component of the magnetic field normal to a cell face, a separate procedure for updating this portion of the magnetic field must be implemented, and must be implemented in such a way as to meet the $\nabla \cdot \mathbf{B} = 0$ constraint. Typically, this is done by the use of a projection scheme.

A new approach has been recently developed and applied by Gombosi et al. (1994). It is based on the observation that adding source terms proportional to $\nabla \cdot \mathbf{B}$ to the three-dimensional MHD equations can modify the resulting one-dimensional Riemann problem in such a way as to give a consistent update for the component of the magnetic field normal to a cell face. With these modified equations, all of the components of the magnetic field may be updated in a consistent manner, such that $\nabla \cdot \mathbf{B} = 0$ is satisfied to within truncation error, once it is imposed as an initial condition to the problem. The resulting scheme is stable to these truncation-level errors in the divergence of $\mathbf{B}$, and does not require a projection scheme.

C. Model Results

MHD equations were solved for a Halley class comet. Normalization was done with the help of the ionization scale length λ, the upstream mass density ρ_∞, and the upstream acoustic speed $a_\infty = (\gamma p_\infty / \rho_\infty)^{1/2}$, with $\gamma = 5/3$. In the calculations the cometary gas was assumed to consist of water group ions with a gas production rate of $10^{30} \mathrm{s}^{-1}$, the ionization scale length was $\lambda = 10^6$ km, while the upstream solar wind density and temperature were $10 \mathrm{\ cm}^{-3}$ and 10^5 K, respectively. The radial outflow velocity of cometary particles was assumed to be $u_n = 1$ km s^{-1}. At the boundaries of the simulation box (tens of millions of km from the nucleus in all directions), free-streaming solar wind conditions were applied, while the distributed mass loading source (described above) was applied inside the simulation domain.

We carried out several two-dimensional and three-dimensional MHD calculations modeling the solar wind interaction with an expanding cometary atmosphere. In the three-dimensional case the solar wind magnetic field was assumed to be pointing outward along the nominal Parker spiral (45° with respect to the radial direction). The plasma flow velocity was along the Sun-comet axis. In the two-dimensional axisymmetric model it is assumed that far upstream the solar wind velocity and magnetic field vectors are parallel to the Sun-comet axis and that the entire solution is rotationally symmetric around this axis. We used the numerical method described in the previous section to obtain a solution to the set of MHD equations. Plate 15 in the Color Section shows the three-dimensional cometary interaction region of the final grid structure under steady-state conditions, which already incorporates a large amount of information about the interaction itself. All distances are measured in units of ionization scale length λ. (A simple conversion to a Halley type case can be obtained by assuming that distance is measured in units of 10^6 km.) The grid is shown as it is seen from north of the equatorial plane: the X axis points from the Sun to the nucleus, the Y axis is in the equatorial plane, and the Z axis points north. The nucleus is centered around the origin of the coordinate system.

The nucleus size is many orders of magnitude smaller than the ionization scale length; this fact is one of the major obstacles in traditional cometary MHD calculations. In our particular case, however, the adaptive grid methodology makes it possible to resolve the immediate vicinity of the nucleus, the shock structure, and the global scale, in the same calculation. It should be noted that the size of the entire simulation box is $60 \times 40 \times 40$ ionization scale lengths (for Halley conditions this would correspond to a simulation volume of $0.4 \mathrm{\ AU} \times 0.26 \mathrm{\ AU} \times 0.26 \mathrm{\ AU}$), while the insert shows the grid structure near the shock and the inner tail.

Color Plate 16 shows the variation of Mach number in the cometary plasma environment. The plate shows only a small fraction of the simulation region near the bow shock ($9\times8\times8$ million km). Again, we are looking down to the equatorial plane from the north with the Sun on the left. Several cuts perpendicular to the Sun-nucleus axis are also shown; these intersect the Sun-nucleus axis at distances $X = -3 \times 10^6$, $X = -1.5 \times 10^6$, $X = 0$, $X = +1.5 \times 10^6$, $X = +3 \times 10^6$, $X = +4.5 \times 10^6$, and $X = +6 \times 10^6$ km. Inspection of Plate 16 reveals that a shock wave is formed upstream of the comet (marked by a solid white line). It is interesting to mention that all comet Halley shock crossings are surprisingly close to our calculated mass loading shock. Behind the shock, the continuously mass loaded plasma flow continues to decelerate gradually, and eventually the flow stagnates in the near nucleus region. Finally, the motional electric field gradually accelerates the plasma in the tail to solar wind velocity. Plate 16 also shows that the plasma starts decelerating due to mass loading ahead of the shock. In effect, the shock occurs at the location where the total amount of mass loading of a given flow line exceeds a small critical value.

One of the most interesting questions about the interaction of the solar wind with comets is the formation of the plasma tail. With the help of our new numerical model we investigated the effect of the interplanetary magnetic field orientation on the magnetic field in the tail. Plates 17 and 18 show the magnetic field magnitude in the case of a 45° Parker spiral angle (Plate 17) and in the case of a nearly radial interplanetary magnetic field (Plate 18). The two plates show the same volume as Plate 16, but the color code represents the normalized magnetic field magnitude (normalized to the free streaming solar wind value). Magnetic field lines originating in the equatorial plane are shown as solid white lines.

It is interesting to compare Plates 17 and 18. In the parallel case (Plate 17) there is some magnetic field pile-up behind the shock, but the cometary tail is practically field free, and field line draping is minimal (one sees significant draping only in the innermost coma, near the contact surface). The situation is dramatically different in the case of the 45° Parker spiral (Plate 18). Inspection of Plate 18 reveals two very interesting features: (i) the field lines not only become draped in the equatorial plane, but they also "slip" above (and below) the plane, because this is the path of "least resistance" to pass through the dense mass loading region; and (ii) a well-pronounced magnetic tail is formed with magnetic field magnitudes up to four times the ambient solar wind value. The magnetic tail is slightly tilted towards the Parker spiral due to magnetic curvature effects.

This result suggests an intriguing possibility: cometary plasma tails "disappear" when the interplanetary magnetic field turns radial, and "reform" when the magnetic field lines return to the Parker spiral configuration. This effect may offer a new explanation to the poorly understood "tail disconnection events," when the plasma tails of comet detaches and eventually disappears (see, e.g., Brandt 1982). In the past, tail disconnection events were interpreted as consequences of passages of sector boundaries and other major interplanetary discontinuities. Our model results, however, imply that more "benign" effects, such as near radial interplanetary magnetic field lines may trigger such events. In the near future we intend to carry out more detailed modeling efforts to investigate this possibility further.

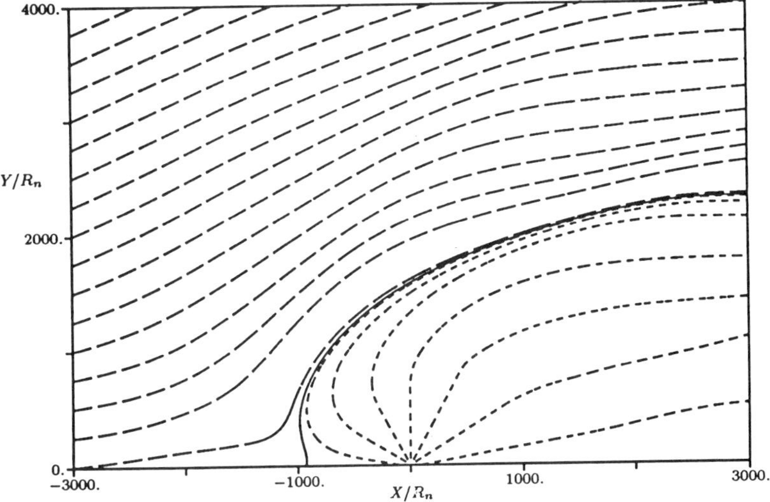

Figure 2.   Near-nucleus plasma flow lines. Dotted lines represent ionospheric plasma flow lines, while solid lines represent contaminated solar wind flow lines. The heavy line separating the two sets of flow lines is the cometary contact surface.

Figure 2 shows plasma streamlines in the inner coma for the parallel magnetic field line case. Long dashed lines represent mass-loaded solar wind streamlines, while the short dashed lines show flow lines originating at the nucleus. The figure clearly shows the formation of a very sharp contact surface between the cometary ionosphere and the mass loaded solar wind plasma. The location of the contact surface at the subsolar point is about 930 nucleus radii. The equivalent radius of comet Halley was about 3 km (the corresponding radius of a sphere with surface area equal to the active area of comet Halley), so this result indicates a subsolar contact surface distance of about 2800 km. This seems to be quite comparable to the distance observed by Giotto (Neubauer et al. 1986).

Figure 3 shows the normalized velocity vectors in the inner coma. Figure 3 also shows diversion of the contaminated solar wind around the contact surface. The most interesting feature in this figure is the near nucleus flow structure inside the contact surface. The nucleus acts as a point source of ionospheric plasma which radially expands with supersonic velocities to about 600

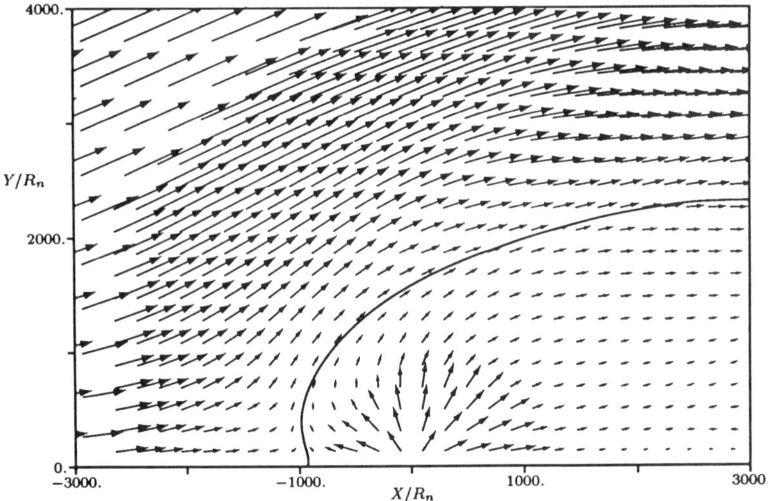

Figure 3.　Flow vector field near the nucleus. The arrows represent the flow velocity vectors; their direction and length characterize the direction and magnitude of the plasma velocity. The solid line represents the contact surface. One can see the formation of the inner shock.

comet radii on the day side and 1100 radii on the night side, where an inner shock is formed. Such a shock has been predicted for the day side and near the terminator, because it is needed to divert the ionospheric plasma flow towards the tail (see, e.g., Wallis and Dryer 1976; Mendis et al. 1985; Wegmann et al. 1987). However, as a result of recombination of the shocked ionospheric plasma, the inner shock moved very close to the diamagnetic cavity boundary at comet Halley (Cravens 1989; Goldstein et al. 1989). The present model does not include recombination; therefore the inner shock is clearly separated from the diamagnetic cavity boundary.

V.　THREE-DIMENSIONAL GAS-DYNAMIC MODEL OF THE HE-LIOSPHERE

As a first step in multi-scale modeling of the solar wind interaction with the local interstellar medium, we performed a three-dimensional hydrodynamic calculation (the magnetic field was neglected). The spherically symmetric simulation domain extended from 1 AU to 500 AU. The total number of grid points was about 35,000. In the future we will make simulations with greater refinement. The Sun was assumed moving with a velocity of 26 km s^{-1} with respect to the interstellar medium. The adopted values for the interstellar ion density and temperature were 0.1 proton cm^{-3} and 1 eV, respectively. The solar wind parameters at 1 AU were $n = 8$ protons cm^{-3}, $u = 400$ km s^{-1}, and $T = 10^5$ K. With these parameters the apex motion was supersonic (the ion-acoustic Mach number was approximately 1.5) and a shock wave was formed in the interstellar plasma.

Color Plate 19 shows the interstellar and solar wind flow lines. Inspection of this plate reveals that the two plasma flows are separated by a contact discontinuity, the heliopause. The color coding represents the plasma density. One can see the formation of the heliospheric shock wave far upstream and the diversion of the interstellar streamlines behind the shock. The innermost dark blue region represents the domain of supersonic solar wind; this region is terminated by an asymmetric "termination shock."

Color Plate 20 shows a blowup of the heliospheric interface surfaces. The solid red lines mark the termination shock, the heliopause, and the upstream interstellar shock from the Sun outward. It can be seen that in this particular simulation the termination shock is formed at 60 AU in the upwind and at 120 AU in the downwind directions, the nose of the heliopause is at 95 AU, and the interstellar shock is formed at 200 AU. These values are not inconsistent with present observations.

It is interesting to underline again the similarity between the cometary and heliospheric simulation. In spite of the fact the basic physical processes are different, the global structures of the interaction regions are quite similar. However, multi-scale heliosphere models are only in a fairly early stage of development. In the next few years one can expect multi-scale three-dimensional MHD models which use more realistic inner boundary conditions near the Sun and extend the simulation volume well into the interstellar medium. It is critical that these models be able to account for the effects of high-speed solar wind originating from the polar regions and from coronal holes, the effects of the coronal magnetic fields, as well as coronal mass ejections. We will do our best to contribute to this challenging effort.

VI. SUMMARY

Results of two new three-dimensional models describing the effects of cometary mass loading on the magnetized solar wind, and the interaction of the solar wind with the interstellar medium have been presented.

The results of the comet simulation show remarkable similarities to the interaction of the heliosphere with the local interstellar medium. In the near future we are planning to use our new adaptive MHD methodology to model various aspects of the three-dimensional structure of the heliosphere.

Acknowledgments. This work was supported by grants from the NASA Planetary Atmospheres Program, the NSF Solar System Astronomy Program, and by an NSF-NASA-AFOSR interagency grant.

REFERENCES

Axford, W. I. 1990. Introductory lecture—the heliosphere. In *Physics of the Outer Heliosphere*, eds. S. Grzedzielski and D. E. Page (Oxford: Pergamon Press), pp. 7–15.

Baranov, V. B. 1990. Gas dynamics of the solar wind interaction with the interstellar medium. *Space Sci. Rev.* 52:89–120.

Baranov, V. B., and Malama, Yu. G. 1993. The model of the solar wind interaction with the local interstellar medium. *J. Geophys. Res.* 98:15157–15163.

Biermann, L., Brosowski, B., and Schmidt, H. U. 1967. The interaction of the solar wind with a comet. *Solar Phys.* 1:254–284.

Brandt, J. C. 1982. Observations and dynamics of plasma tails. In *Comets*, ed. L. L. Wilkening (Tucson: Univ. of Arizona Press), pp. 519–537.

Cravens, T. E. 1989. A magnetohydrodynamical model of the inner coma of comet Halley. *J. Geophys. Res.* 94:15025–15040.

Fedder, J. A., Lyon, J. G., and Giuliani, J. 1986. Numerical simulations of comets: Predictions for comet Giacobini-Zinner. *Eos: Trans. AGU* 67:17–18.

Frisch, P. C. 1994. Morphology and ionization of the interstellar cloud surrounding the solar system. *Science* 265:1423–1427.

Galeev, A. A., Cravens, T. E., and Gombosi, T. I. 1985. Solar wind stagnation near comets. *Astrophys. J.* 289:807–819.

Goldstein, B. E., Altwegg, K., Balsiger, H., Fuselier, S. A., Ip, W.-H., Meier, A., Neugebauer, M., Rosenbauer, H., and Schwenn, R. 1989. Observations of a shock and a reconnection layer at the contact surface of comet Halley. *J. Geophys. Res.* 94:17251–17257.

Gombosi, T. I. 1991. The plasma environment of comets. *Rev. Geophys.* 29:976–984.

Gombosi, T. I., Powell, K. G., and DeZeeuw, D. L. 1994. Axisymmetric modeling of cometary mass loading on an adaptively refined grid: MHD results. *J. Geophys. Res.* 99:21525–21539.

Holzer, T. E. 1989. Interaction between the solar wind and the interstellar medium. *Ann. Rev. Astron. Astrophys.* 27:199–234.

Huebner, W. F., Boice, D. C., Schmidt, H. U., and Wegmann, R. 1991. Structure of the coma: Chemistry and solar wind interaction. In *Comets in the Post-Halley Era*, eds. R. L. Newburn, M. Neugebauer and J. Rahe (Dordrecht: Kluwer), pp. 907–936.

Lindgren, C. J., and Cravens, T. E. 1993. A magnetohydrodynamic model of a comet. In *Plasma Environments of Non-Magnetic Planets*, ed. T. I. Gombosi (Oxford: Pergamon Press), pp. 7–15.

Mendis, D. A., Houpis, H. L. F., and Marconi, M. L. 1985. The physics of comets. *Fund. Cosmic Phys.* 10:1–380.

Neubauer, F. M., Glassmeier, K.-H., Pohl, M., Raeder, J., Acuna, M. H., Burlaga, L. F., Ness, N. F., Musmann, G., Mariani, F., Wallis, M. K., Ungstrup, E., and Schmidt, H.-U. 1986. First results from the Giotto magnetometer experiment at comet Halley. *Nature* 321:352–355.

Neugebauer, M. 1990. Spacecraft observations of the interaction of active comets with the solar wind *Rev. Geophys.* 28:231–252.

Nerney, S., Suess, S. T., and Schmahl, E. J. 1993. Flow downstream of the heliospheric terminal shock: The magnetic field on the heliopause. *J. Geophys. Res.* 98:15169–15176.

Schmidt, H. U., and Wegmann, R. 1982. Plasma flow and magnetic fields in comets. In *Comets*, ed. L. L. Wilkening (Tucson: Univ. of Arizona Press), pp. 538–560.

Schmidt, H. U., Wegmann, R., Huebner, W. F., and Boice, D. C. 1988. Cometary gas and plasma flow with detailed chemistry. *Comp. Phys. Comm.* 49:17–59.

Schmidt-Voigt, M. 1987. Time dependent MHD models for the cometary magnetosphere. In *Proc. Symposium on the Diversity and Similarity of Comets*, eds. E. J. Rolfe and B. Battrick, ESA SP-278, pp. 127–131.

Steinolfson, R. S., Pizzo, V. J., and Holzer, T. 1994. Gasdynamic models of the solar wind/interstellar medium interaction. *Geophys. Res. Lett.* 21:245–248.

Suess, S. T. 1990. The heliopause. *Rev. Geophys.* 28:97–115.

Wallis, M. K., and Dryer, M. 1976. Sun and comets as sources in an external flow. *Astrophys. J.* 205:895–899.

Wegmann, R., Schmidt, H. U., Huebner, W. F., and Boice, D. C. 1987. Cometary MHD and chemistry. *Astron. Astrophys.* 187:339–350.

Wegmann, R., Rauer, H., and Schmidt, H. U. 1993. The effect of discontinuities in the solar wind on the plasma tails of comets. In *Plasma Environments of Non-Magnetic Planets*, ed. T. I. Gombosi (Oxford: Pergamon Press), pp. 81–84.

GLOSSARY

GLOSSARY*

aa index	a geomagnetic index from Greenwich and Melbourne; nearly antipodal in latitude, corrected for several small differences, and then averaged. The purpose is to eliminate some local and/or regional effects.
acoustic waves	pressure waves in a compressible medium. Convective turbulence in stars acts as a source of acoustic waves with a broad spectrum of frequencies. As these waves propagate upwards, they steepen into shocks and deposit energy, possibly giving rise to a chromosphere and corona.
AGB	asymptotic giant branch; locus of stars in the HR diagram which have consumed all of the available hydrogen and helium in the core and are powered by hydrogen and helium burning in shells.
AL index	a short time resolution magnetic variation index intended to monitor the auroral electrojet from auroral latitude stations.
Alfvén velocity	the characteristic velocity at which noncompressive disturbances propagate along a magnetic field, $v_A = B/(4\pi\rho)^{0.5}$.
Alfvén wave	a type of magnetohydrodynamic wave in which the wave perturbation is transverse to the direction of the magnetic field lines.
alpha particle	the nucleus of a ^{4}He atom.

---

* We have used some definitions from *Glossary of Astronomy and Astrophysics* by J. Hopkins (by permission of The University of Chicago Press, copyright 1980 by the University of Chicago) and from *Astrophysical Quantities* by C. W. Allen (London: Athlone Press, 1973). for further information regarding the Hapke function. We also acknowledge definitions and helpful comments from various chapter authors.

ambipolar diffusion	the process whereby ions and neutrals drift relative to each other in the presence of a variable magnetic field.
AMPTE	Active Magnetospheric Particle Tracers Explorer; an Earth-orbiting spacecraft in operation from August, 1984 until early 1989.
Archimedes spiral	a classical spiral defined mathematically; it corresponds to the lines of force of the solar wind magnetic field stretching from the rotating Sun into space.
astrosphere	the cavity which a star carves out for itself from the local interstellar medium through the action of its wind; analog of heliosphere for the Sun.
AU	astronomical unit; the mean distance of the Earth from the Sun, equal to 1.496×10^{13} cm.
Bartels rotation number	a system for numbering 27-day solar rotations; introduced by Julius Bartels in order to study recurrent geomagnetic activity.
Bernoulli equation	the integral form of the fluid momentum equation, evaluated along a streamline. It is a statement of the work–energy theorem for fluid flow.
Bernoulli's law	for steady ideal (inviscid) flow, $1/2u^2 + \phi + h$ is constant along a fluid streamline. Here u is the fluid flow speed, ϕ is the gravitational potential per unit mass, and h is the enthalpy per unit mass. For a polytrope equation-of-state for the fluid $p \propto \rho^\gamma$, where p and ρ are fluid pressure and mass density, respectively, and γ is the polytrope index, we obtain $h = \gamma(\gamma-1)^{-1}p\rho^{-1}$. For an incompressible fluid, $h = p\rho^{-1}$.
BGK Boltzmann equation	In the 1950s, Bhatnagar, Gross and Krook introduced a modified version of the Boltzmann equation (now known as the BGK Boltzmann equation). The BGK Boltzmann equation has a modified form for the scattering term, in which the distribution function relaxes towards isotropy on a time scale characteristic of the scattering process. *See also* Boltzmann's equation.
blackbody	a perfect absorber of radiation.

Boltzmann's constant (k)
the constant of proportionality between energy per degree of freedom and temperature: $k = 1.38 \times 10^{-16}$ erg per degree Kelvin.

Boltzmann's equation
a fundamental transport equation in the kinetic theory of gases and plasmas. *See also* BGK Boltzmann equation.

bow shock
a shock wave formed upstream of an obstacle in supersonic flow, where the flow abruptly slows and where density increases proportionately.

calorisphere
a region of a stellar atmosphere where the temperature remains above the radiative equilibrium temperature at all times.

Carrington rotation
a numbering of successive 27-day solar rotations indroduced in the last century by solar astronomer R. C. Carrington.

cataclysmic variable
a short-period semi-detached mass-exchanging binary composed of a white dwarf and, typically, a late-type (G, K, or M) main-sequence secondary.

CCD
charged-coupled device.

Chapman layer
a theoretical model of an ionosphere's structure. The density of charged particles is calculated by equating the rate of photoionization and the rate of recombination.

charge transfer collisions
charge exchange; a collision between an ion and a neutral atom which results in the ion recombining and the atom being ionized, i.e., $x^+ + y \rightarrow x^+ y^+$.

chromosphere
a region in a cool star's atmosphere where the gas temperature rises above the photospheric values to between 6000 and 8000 K. The temperature rise indicates deposition of mechanical energy (either acoustic or magnetic), and energy balance is achieved mainly by radiative losses in the continua and lines of H, H^- and metals.

CIR   corotating interaction region; a region in interplane-
                    tary space bounded by a forward–reverse shock pair,
                    created by a high-speed solar wind stream ramming
                    into previously ejected slow solar wind due to the solar
                    rotation.

CME   coronal mass ejection; a large body of magnetically
                    confined coronal material being released from the co-
                    rona by magnetic reconnection at its base, often in
                    conjunction with solar flares.

CMIR   corotating merged interaction region.

COBE   Cosmic Background Explorer; a space mission to mea-
                    sure the temperature and the anisotropy of the primor-
                    dial cosmic background radiation.

collisionopause   the region near a comet which is collisionally thick,
                    i.e., where electron-neutral collisions affect the plasma
                    significantly.

convection elec-   the electric field measured in the rest frame of a plasma
   tric field   fluid, generally the field induced by the fluid's motion,
                    $E = v \times B$.

corona   that portion of the solar atmosphere with temperature
                    above about 25,000 K. Coronal emission is generally
                    seen in the EUV and X-ray portion of the spectrum.

coronal hole   a large-scale, low-density, low-temperature region in
                    the solar corona where the EUV and X-ray coronal
                    emission is abnormally low or absent; a coronal region
                    apparently associated with diverging magnetic fields.
                    It is a source of high-speed solar wind streams.

coronal streamer   a magnetically confined, loop-like coronal structure
                    straddling the magnetic neutral line on the solar surface,
                    thus situated mostly near the solar equator, where the
                    slow solar wind originates.

Coulomb drag   electrostatic interactions resulting in net transfer of
                    momentum between charged particles, normally in a
                    plasma.

cross-tail potential	the electric potential induced across the magnetotail.
dissociative recombination	recombination of a molecular ion which results in its chemical dissociation, e.g., $xy^+ + e^- \rightarrow x + y$.
dMe stars	red dwarfs with hydrogen lines seen in emission in the visible spectrum with strong chromospheres and coronae, where the L_x/L_{bol} ratio may be as large as 0.01 (larger than for any other class of single star). These stars exhibit frequent flares, indicators of strong magnetic activity.
dM stars	red dwarfs with hydrogen lines seen in absorption in the visible spectrum, usually regarded as stars with weak chromospheres and low magnetic activity.
dwarf novae	nonmagnetic cataclysmic variables which experience outbursts in the infrared through X-ray flux. The outbursts have recurrence times of tens to hundreds of days, and increases of between 3 and 5 magnitudes in the visible bands.
Eddington factor	the ratio between the second and zeroeth moment of the radiation field (defined through the specific intensity). In the case of an isotropic radiation field, the Eddington factor is $1/3$. For a number of numerical methods the Eddington factor is used to close the system of the radiative moment equations.
Ekman layer	an idealized model of the planetary boundary layer in which a constant interior geostrophic wind is modified near the ground by a turbulent shear stress per unit mass given by the eddy viscosity $k\,\partial u/\partial z$, with constant eddy coefficient k. The resulting variation of the wind with height is described by the Ekman spiral, in which the wind speed increases with height while the wind direction, angled towards the low pressure near the surface, becomes aligned with the geostrophic wind aloft.
electrons, type III	a transient, field-aligned stream of energetic electrons having energies in the range generally between 10 and 100 keV.

Elässer variables	$z^{\pm} = 3D\delta v \pm \delta b$, where z^+ and z^- refer to waves propagating "outward" and "inward" with respect to B_0, and δv and δb are the fluctuating velocity field and magnetic field (in Alfvén) speed units, respectively.
EUV	extreme ultraviolet radiation.
eV	electron volt; a unit of energy for ionizing radiation, with 1 eV = 1.62×10^{-12} erg. Nuclei in the cosmic rays typically have energies from about a mega-electron volt (MeV, 10^6 eV) to many giga-electon volts (GeV, sometimes BeV, 10^9 eV per nucleon.
F corona	the dust scattered component of the coronal white light intensity. *See also* K corona.
FIP effect	first ionization potential; an effect operating in the solar chromosphere, enhancing the abundance of elements with a low first ionization potential in the corona and solar wind relative to the high-FIP elements.
Forbush decrease	the abrupt (12-hr) large (10%) decrease in cosmic ray intensity a day or two after an explosive outburst on the Sun, recovering over a period of days and weeks. The decrease is caused by a blast wave in the solar wind which sweeps back the cosmic rays in its outward passage.
freezing-in	the decoupling of ion charge states from electron collision equilibrium near the coronal temperature maximum, due to dereasing electron density. Charge states remain unaltered ("frozen-in") in the solar wind and thus may be used as coronal thermometers. Also, the magnetic field which, if frozen into, a large-scale plasma moves with the flow of the plasma.
FWHM	full width at half maximum (minimum), e.g., the full width of a spectral line at half-maximum intensity.
Gauss	unit of magnetic field strength; 10^4 Gauss = 1 Tesla.
GCR	galactic cosmic rays; highly energetic (usually about 0.1 to 10 GeV) particles (mainly protons) incident on the solar system.
geomagnetic field	the Earth's magnetic field.

Gyr

gigayear; 10^9 yr.

gyrotropic

symmetric about the magnetic field direction.

heliopause

the surface representing the outer boundary of the heliosphere, separating the solar wind from interstellar plasma.

heliosheath

a region inside the heliosphere between the solar wind termination shock and the heliopause in which the shocked solar wind is deflected to flow into the tail.

heliosphere

the volume of space in the vicinity of the Sun where plasma processes are controlled by the solar wind.

heliospheric current sheet

a thin sheet of current that forms at the interface between oppositely directed magnetic fields in the solar wind from the north and south solar hemispheres.

heliotail

the elongated tail of the heliosphere extending downwind of the approaching interstellar gas formed by the solar wind escaping from the heliosphere.

Helmet streamers

large, long-lived emission features seen in the K corona. They extend above large quiescent prominences or older, extended active regions. They are most often seen in high latitudes during mid-solar cycle. They have broad bases, often extending to a width of nearly one solar radius, and rise to a pointed top one or two solar radii above the limb.

HR diagram

Hertzsprung-Russell diagram; a plot of luminosity vs spectral type; more loosely, a plot of magnitude vs color.

Hubble constant

the expansion rate of the universe, $H_o = \dot{R}/R$, where R is the scale parameter, i.e., the distance to any remote galaxy. Today's value is approximately $H_o = 50\,\mathrm{km\,s^{-1}}$ $\mathrm{M_o} c^{-1}$.

HST

Hubble Space Telescope.

HUT

Hopkins Ultraviolet Telescope.

hybrid stars

stars with evidence for cool winds and warm material in their atmospheres.

ICI | ion composition instrument flown on ISEE-3; combining an electrostatic analyzer and a Wien filter to obtain mass per charge spectra of solar wind ions.

IMF | interplanetary magnetic field; the Sun's magnetic field, advanced outwards by the solar wind. Near the Sun, the field is largely radial. Near Pluto, it is azimuthal, roughly perpendicular to the Sun–Pluto line.

information gradient | the variation of our knowledge of properties of stellar winds across the HR diagram.

intrinsic magnetosphere | the planet–solar wind interaction originating from a planet's internal magnetic field.

ionosphere | a layer of ionized particles in a planet's upper atmosphere, generally produced by photoionization.

IPS | interplanetary scintillation; modification of the wavefront from a compact radio source (such as a quasar) by electron density irregularities in the solar wind. As the irregularities move outward in the solar wind, their diffraction patterns sweep across the Earth-based antenna system, providing an estimate of the speed of the solar wind.

IRAS | Infrared Astronomical Satellite.

ISEE-3 | International Sun-Earth Explorer; spacecraft launched in 1978 to study particles and magnetic fields predominantly from the Sun and the magnetosphere. In 1983, ISEE-3 was moved from its near-Earth orbit to interplanetary space in order to study particles and fields associated with comet Giacobini-Zinner. In this latter role, ISEE-3 was renamed ICE, International Cometary Explorer.

IUE | International Ultraviolet Explorer satellite.

kappa distribution | a class of functions of a single variable, such as particle energy, that approach a constant value at low energies and a power law having index, $k + 1$, at asymptotic energies.

K corona	the electron scattered component of the coronal white light intensity from which the electron density is derived. *See also* F corona.
Kelvin-Helmholtz instability	the contraction of a star contemplated as a consequence of a star's radiating its thermal energy in the absence of thermonuclear sources; also, the tendency of waves to grow on a shear boundary between two regions in relative motion to the boundary.
KeV	energy in units of thousands of electron volts.
kinetic effects	effects related to the energy distribution of particles rather than the average properties of the fluid. At smaller scales, kinetic effects become significant. In order to describe them accurately, the Boltzmann equation, which includes the details of the particles' energy distribution, must be used.
Kolmogoroff's hypothesis	the supposition that medium-scale lengths in a turbulent cascade are only affected by energy input at the large, energy-containing scales. This leads to a self-similar spectrum of fluctuations.
kpc	kiloparsec = 3.086×10^{21} cm.
Landau damping	the conversion of wave energy to particle heating, which is concentrated on those particles that have velocities equal to the phase velocity of the wave.
Landau resonance	the characteristic of Landau damping that confines the energy transfer to a narrow range of particle energies centered on particle velocities equal to the wave phase velocity.
Langmuir waves	electrostatic waves characterized by oscillatory relative motions between electrons and ions.
Larmor radius	The radius of gyration of a charged particle in a magnetic field.
LC	luminosity class.
Lighthill expression	flux of acoustic power emitted by quadrupole stresses in turbulent flow: $F_a \sim v^2$ where v = rms turbulent speed.

LIR   local interaction region.

LISM   local interstellar medium.

LMIR   local merged interaction region.

long-period vari-
ables   variable stars which occur in two groups: (i) massive supergiants $M \geq 9$ $M_\odot$ ascending the red giant branch; (ii) AGB stars. In some stars, the variability is due to pulsation, which may be regular (as in the Miras), semi-regular, or irregular, while in other stars (e.g., R CrB), the variability is due to eruptions.

LTE   local thermodynamic equilibrium; the assumption that all the distribution functions characterizing the material and its interaction with the radiation field (but not that of the radiation itself) at a point in the star are given by thermodynamic equilibrium relations at local values of the temperature and density.

Lucy approxima-
tion   in the case of an extended stellar atmosphere and spherical geometry, the radiation field can be described by a semi-analytical method to close the moment equations of the radiation field. This allows the calculation of the temperature structure in radiative equilibrium. Due to the Lucy approximation, the specific intensities are split into two contributions. One part covers the radiation field within the solid angle under which the star is seen, whereas the other part comprises the rest of the radiation field that results from scattering and/or emission outside the stellar photosphere.

luminosity   the integral of the surface flux F over the stellar surface, $4\pi R^2 \sigma T^4$, where R is the radius of the star, T is its effective temperature, and σ is the Stefan-Boltzmann constant.

$M_\odot$ solar mass $= 1.989 \times 10^{33}$ g.

magnetic dynamo   motions in a conducting fluid which generates or sustains a magnetic field. To be a dynamo, the fluid involved must be both conducting and convecting.

magnetic pressure
: a measure of the magnetic forces acting on a palsma. The force is equal to the gradient of B^2/μ_0. Because the pressure force is also the gradient of the pressure, B^2/μ_0 may be regarded as a type of pressure inherent to the magnetic fields.

MHD
: magnetohydrodynamics; a description of a plasma as a single, conducting fluid. To be accurate, the time and length scales of the system must be much larger than the Larmor radii of the plasma particles.

magnetopause
: the surface which separates the magnetosphere from the solar wind. At the magnetopause, the magnetic pressure of the planet's internal magnetic field deflects the flow of the solar wind around the magnetosphere. A magnetopause only forms if the planet has a sufficiently strong internal magnetic field.

magnetosheath
: the region of solar wind flow between the bow shock and the planet's magnetopause or ionopause. Here the plasma and magnetic fields are of solar origin, but have been heated and deflected.

magnetosphere
: the region influenced by the planet's magnetic field. The magnetosphere may contain substantial densities of plasma if there are strong sources (from satellite or the planet's atmospheres) within the magnetosphere.

magnetotail
: an extension of the magnetosphere on a planet's night side. In ordinary fluid mechanics, this would be called a wake. A magnetotail also implies elongated magnetic field lines stretching out behind the planet. Magnetotails can extend hundreds of planetary radii.

main sequence
: the principal sequence of stars on the HR diagram, containing more than 90% of the stars we observe, that runs diagonally from upper left (high temperature, high luminosity) to lower right (low temperature, low luminosity). Most of the lifetime of a star is spent on the main sequence.

maser
: microwave amplification by stimulated emission of radiation in an environment where local conditions favor the creation of a population inversion among the energy levels of a particular transition.

mass loading | the drag and deceleration of the solar wind as a result of pick-up ionization.

mass loss | the process of ejection of matter from a star, either by thermal pressure gradients from a hot corona, or by magnetic driving, or by radiation pressure on either grains (cool stars) or strong spectral lines (hot stars).

mass spectrometer | an analytical instrument in which the sample is converted into a beam of ions which can be separated from each other on the basis of their mass-to-charge ratio, generally by a magnetic or electrostatic field, so that the relative proportions of entities of different mass, commonly isotopes, can be determined.

Maunder minimum | the period roughly between 1645 and 1715, identified with an apparent lack of sunspots and aurorae; named after E. W. Maunder (1851–1928).

Maxwellian | unit of magnetic flux equal to 1 Gauss per cm^2.

M dwarfs | low-mass stars at the bottom of the main sequence, cooler than the Sun, with masses ≤ 0.5 M$_\odot$, burning hydrogen in the core through $p - p$ cycle, and with convective envelopes which may (at the lowest masses) extend throughout the entire star.

MEM | maximum entropy method.

MHD shocks | magnetohydrodynamic shocks; shock waves mediated by the magnetohydrodynamic properties of a plasma rather than by molecular collisions alone, as for a shock in a neutral gas. MHD shocks form when disturbances move through the gas faster than the characteristic wave speed. In many cases (for MHD shocks), the appropriate wave velocity is the Alfvén velocity, which is the speed of one of the most commonly found wave modes in magnetized and ionized gases.

Michel velocity | An electromagnetic wind has a Poynting flux, the radial component of which can be approximated as $S_r \approx \rho V_M^3$. The velocity V_M, called the Michel velocity, is an important scale in the expansion of magnetic winds, and was invented by Michel (1969. *Astrophy. J.* 158:727).

MIR

merged interaction region; solar wind structure formed in the outer heliosphere when two interaction regions (caused either by corotating streams or by successive coronal mass ejections) combine as one overtakes the other.

Mira variables

radially pulsating cool giant stars of spectral class M or C or S with regular pulsation periods in the range of 2 months to 1 year or more, and amplitudes of 2 to 5 magnitudes or more in visible light. These stars lie near the top of the AGB, named for the prototype σ Ceti (Mira) which varies from 2 to 20 magnitudes on a period of 331 days, and was first noted to be variable by Fabricius in 1596.

μ G

microgauss = 10^{-6} Gauss; a unit of magnetic field strength.

MUSCL-type schemes

monotone upstream-centered schemes for conservation laws. This numerical method for solving hyperbolic differential equations was developed by Bram van Leer to achieve second-order spacial accuracy with Godunov schemes. In this extension of the Godunov approach, the state variables at a cell interface are obtained from an extrapolation between the neighboring cells.

Myr

million years, 10^6 yr.

noble gases

the elements He, Ne, Ar, Kr, Xe, and Rn, which are almost always monoatomic and rarely undergo chemical reactions. They are also called rare gases or inert gases.

NSSDC

National Space Science Data Center; a NASA organization located at Goddard Space Flight Center (Greenbelt, Md.) that is responsible for archiving planetary mission data and making them available upon request to qualified researchers.

PANDORA radiative transfer code
a sophisticated computer code which solves the coupled equations of radiative transfer and statistical equilibrium in a self-consistent manner. It is able to account for radiative transfer problems related to effects of non-local thermodynamic equilibrium. PANDORA is based on the so-called equivalent two-level atom approach and is well-suited to calculate synthetic spectra of stellar atmospheres and winds. The computer code PANDORA is very large as it contains over 3500 subroutines. Created by E. H. Avrett and R. Loeser (Harvard Smithsonian Center for Astrophysics) in 1966, the code is constantly updated to comply with needs of the scientific community.

Parker wind
a steady, spherically symmetric stellar wind with velocity increasing monotonically with radius and becoming supersonic beyond a critical point.

pc
parsec; a parallax second. The distance at which one astronomical unit subtends an angle of 1 second of arc. 1 pc = 206,265 AU = 2.086×10^{13} lightyears.

P Cygni profile
spectral line shape caused by the presence of a massive wind escaping from a star. It consists of a redward emission and a blueward absorption.

P Cygni star
a type of star named after the 5th magnitude peculiar variable Ble star P Cygni (HD 193272), about 1200 pc distant, whose spectrum shows strong emission lines, like those of Be and Wolf-Rayet stars, with blueshifted absorption components and redshifted emission components which are presumed to arise in an expanding shell of low-density matter.

pick-up ions
neutral particles which have become ionized while moving with respect to a magnetized plasma. These newly created ions are accelerated by the convection magnetic field up to the rest frame of the background plasma, which is correspondingly decelerated (mass-loaded).

pitch angle | the angle between a particle's velocity and the magnetic field. Particles with a high pitch angle can be trapped by certain magnetic field geometries. Asymmetries in pitch angle distribution result in particle-wave instabilities.

Planck function | the energy distribution of blackbody radiation under conditions of thermal equilibrium at a temperature T: $B_\nu = 2h\nu^3/c^2[\exp(h\nu/kT)-1]^{-1}$, where h is Planck's constant and ν is the frequency.

plasmoid | a quasi-stable toroidal aggregate of magnetized plasma.

population I and II | a rough classification system of stars in galaxies. Population I stars are young stars, formed primarily in the disk of spiral galaxies. Populaion II stars are stars generally formed in the spheroid of spirals or ellipticals, primarily at the time the galaxy formed (hence, these stars are old and generally have low metallicity).

Prandtl mixing length | a parameter in the theory of heat transport by turbulent convection, taken to be the distance over which a heated "bubble" moves radially before dissipating its heat; usually set equal to a pressure scale height or some fraction thereof.

pre-main-sequence stars | young stars still undergoing gravitational contraction prior to their arrival on the main sequence.

proton | the hydrogen nucleus.

proton Larmor radius | the proton Larmor radius is defined as $\langle v_e \rangle / \Omega_p$, where $\langle v_e \rangle$ is the proton thermal speed and Ω_p is the proton gyrofrequency (eB/m_pc).

quasi-perpendicular shock | a shock in a magnetized plasma, where the local normal to the shock surface is approximately perpendicular to the direction of the upstream magnetic field.

Raleigh-Taylor instability | a type of boundary instability between two static fluids (e.g., cold dense gas above hot rarified gas) with different thermodynamic properties.

Rankine-Hugoniot shock jump conditions | the relation governing conservation of mass, energy and momentum across a shock wave.

re-acceleration   the process whereby an already energetic particle gains further energy when it encounters a suitable transition between one hydromagnetic flow and another.

red dwarf   low-mass star near the lower end of the main sequence, typically of spectral class M, with a mass ≤ 0.5 M$_\odot$, radius ≤ 0.5 R$_\odot$, $T_{\mathrm{eff}} \lesssim 4000$ K, and number densities in excess of any other class of star in the Galaxy.

red giant branch   locus of stars in the HR diagram in which hydrogen is no longer available as fuel in the core, but hydrogen burning occurs in a shell.

regolith   a fragment layer found on the surface of many planetary or subplanetary objects. It is created from the local competent lithologies by meteoroid impact and subsequently comminuted and turned over by such impacts.

Reimers' relation   the parameterized mass loss formula of the form $\dot{M} \propto LR/M$ where L is stellar luminosity, R is stellar radius, and M is stellar mass. This is equivalent to assuming $(\dot{M} v_{\mathrm{escape}}^2)/L =$ constant, i.e., that a constant fraction of the stellar luminosity is going into providing wind material with the energy it needs to escape.

remanent magnetization   magnetic fields retained within an object, as a result of previous exposure to a magnetic field while the object was above its Curie temperature.

Reynolds number   a dimensionless number ($R_e = \frac{Lv}{\nu}$), where L is a typical dimension of the system, v is a measure of the velocities that prevail, and ν is the kinematic viscosity that governs the conditions for the occurrence of turbulence in fluids.

ring/beam distribution   a distribution of particle velocities which is like a ring distribution, except that the particles have a significant velocity along the magnetic field lines. All particles have the same parallel velocity (as in a beam).

ring distribution   a distribution of particle velocities in which all particles have the same speed perpendicular to the magnetic field and negligible speed along the magnetic field. The perpendicular velocity is gyrotropic, so that the same number of particles are moving in any given direction. Ring distributions are strongly unstable, both to thermalization by collisions and to wave-particle instabilities.

ROSAT   Röntgen Satellite and X-ray Observatory.

Rossby number (R_o) the product of the rotation period and the convective overturn time R/o is inversely proportional to the square root of the dynamo number n_D.

SCR   solar cosmic ray; energetic (usually $\gtrsim 1$ MeV but seldom > 100 MeV) particles, mainly protons and electrons, accelerated as a result of very energetic processes on the Sun; also called solar energetic particles or solar-flare-associated particles.

Sedov phase   the second phase in the expansion of a supernova remnant, when the energy is adiabatically conserved. This is the phase when the partial differential equations describing the expansion can be scaled with the expansion law and can thus be reduced to ordinary differential equations: Writing the radius as r, and its rate of change, the velocity as v thus gives in this phase: $v^2 r^3 =$ constant, which leads to $v =$ constant $r^{-3/2}$, and then to $r =$ constant $t^{-2/5}$ where t is the time.

SNR   supernova remnant; the debris of an exploded star and the material that it has swept up.

SOHO   Solar Heliosphere Observatory.

solar flare   an explosive outburst of material on the solar photosphere that is observationally related to local magnetic fields on the Sun.

solar wind   the expansion of the solar corona to form supersonic plasma streaming away from the Sun.

SPARTAN 201 — an autonomous, detached Space Shuttle payload that observes the solar corona at far ultraviolet and visible wavelength. The two instruments on the SPARTAN 201 spacecraft are the Ultraviolet Spectrometer (UVCSi/SPARTAN) and the White Light Coronagraph (WLC).

spherical shell distribution — a distribution of particle velocities which is spherically symmetric, and in which all the particles have the same speed.

sputtering yield — the average number of atoms sputtered off a surface as a result of a single, impacting particle.

stand-off distance — the distance between an obstacle and the sub-solar magnetopause or ionopause.

Strahl — a portion of the velocity distribution of suprathermal electrons that is characterized by a narrowly focused streaming along the magnetic field.

substorm — a disturbance within a planet's magnetosphere, triggered by changes in the solar wind, that results in strong, induced electric fields, particle precipitation and large-scale changes in magnetic field geometry.

suprathermal — an energy distribution of particles that has a greater number of high-energy particles than would a thermal (Maxwell-Boltzmann) distribution.

SWICS — a solar wind mass spectrometer flown aboard Ulysses to determine the composition of the solar wind as well as of interstellar pick-up ions. SWICS is using time-of-flight and energy detectors to resolve ions with identical mass-per-charge ratio.

terminal velocity — the velocity of wind outflow at great distance from the star.

termination shock — the discontinuity where the solar wind encounters the interstellar medium in a supersonic encounter.

Tesla — a unit of magnetic flux density, 1 Tesla $= 10^4$ Gauss.

thermalization	the process by which collisions alter the energy distribution of a gas, driving it towards a Boltzmann or thermal distribution.
TIR	transient interaction region.
Ulysses	formally ISPM; a spacecraft on a highly inclined orbit around the Sun, carrying a comprehensive particles and fields payload.
Virial theorem	a relation between the averaged values of total kinetic and potential energies of a dynamical system, and certain averaged values of its variation with time and at its surface.
$v \sin i$	the rotational velocity of a star projected onto the line of sight; v = equatorial rotational velocity of a star and i = inclination of its rotational access with respect to the observer's line of sight.
wave-particle interactions	any coupling between electrostatic or electromagnetic waves and individual particles. Such interactions can generate waves if the initial particles are in an untable distribution, or create a suprathermal distribution if the initial wave amplitude was sufficient.
white dwarf	an old star where the material is supported against gravity by degenerate electron pressure.
WKB theory	Wentzel-Kramers-Brillouin theory; linearized small-wave length approximation for wave propagation in an inhomogeneous medium.
Wolf-Rayet stars	very hot stars (spectral class O) in an extremely active phase, apparently ejecting shells of matter at high speed.

X-ray binary
the X-ray source in a binary star system. Generally assumed to be a compact star (neutron star, black hole, or white dwarf) in orbit with a noncompact star in which the X rays are emitted as the result of accretion of gas from the noncompact star onto the compact star. Massive X-ray binary (MXB): X-ray binary in which the noncompact star is an early type (O or B) star. Low-mass X-ray binary (LMXB): X-ray binary in which the noncompact star is a late-type (F or later) star.

Yohkoh mission
Japanese spacecraft observing the Sun with a high-quality X-ray telescope.

ZAMS
zero-age main-sequence stars.

ζ Aurigae systems
eclipsing binaries containing a large cool star and a small hot star. The cool star is typically of spectral class G, K, or M, and of luminostiy class II or I. As a small star goes into eclipse, spectral lines due to the wind and chromosphere of the cool star are seen in absorption against the background of the hot star continuum.

CARRINGTON ROTATION 1775 (WSO)

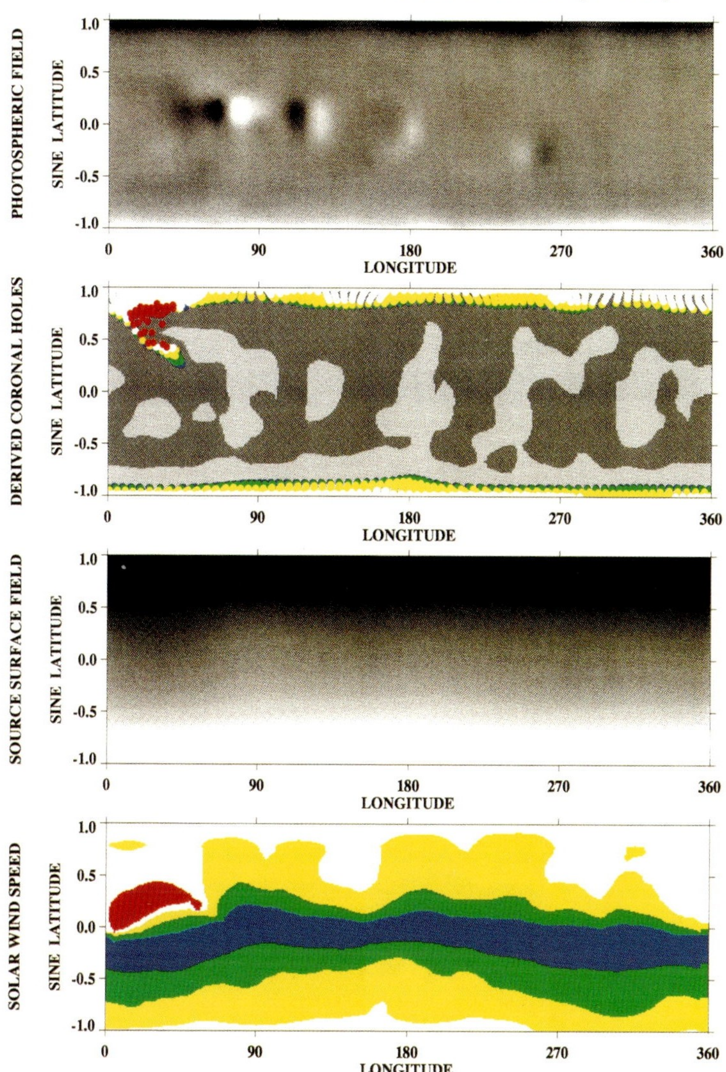

Plate 1. Sunspot minimum. Global maps of the radial component of photospheric field (top), the photospheric footpoints of open field lines (second from top), the source-surface magnetic field (third from top), and the source-surface expansion factor (fourth), all derived from observations at the Wilcox Solar Observatory using a current-free coronal model. Positive magnetic polarities are indicated by light shading and negative polarities by dark shading. Open field lines are assigned colors corresponding to the amount of expansion their associated flux tubes experience in going from photosphere to source surface; expansion factors (3, 4.8, 8, 18) mark the boundaries between the colors (red, white, yellow, green, blue). (See the chapter by Sheeley et al.)

CARRINGTON ROTATION 1741 (WSO)

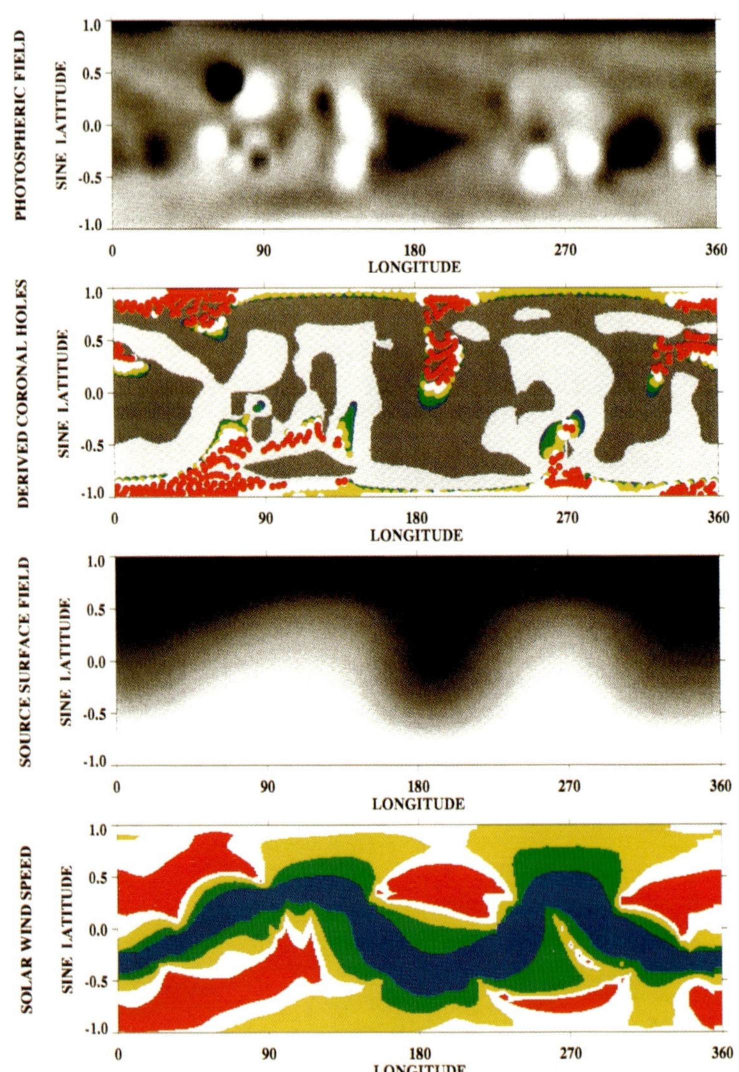

Plate 2.   Declining phase. The format is the same as for Plate 1. (See the chapter by Sheeley et al.)

CARRINGTON ROTATION 1813 (WSO)

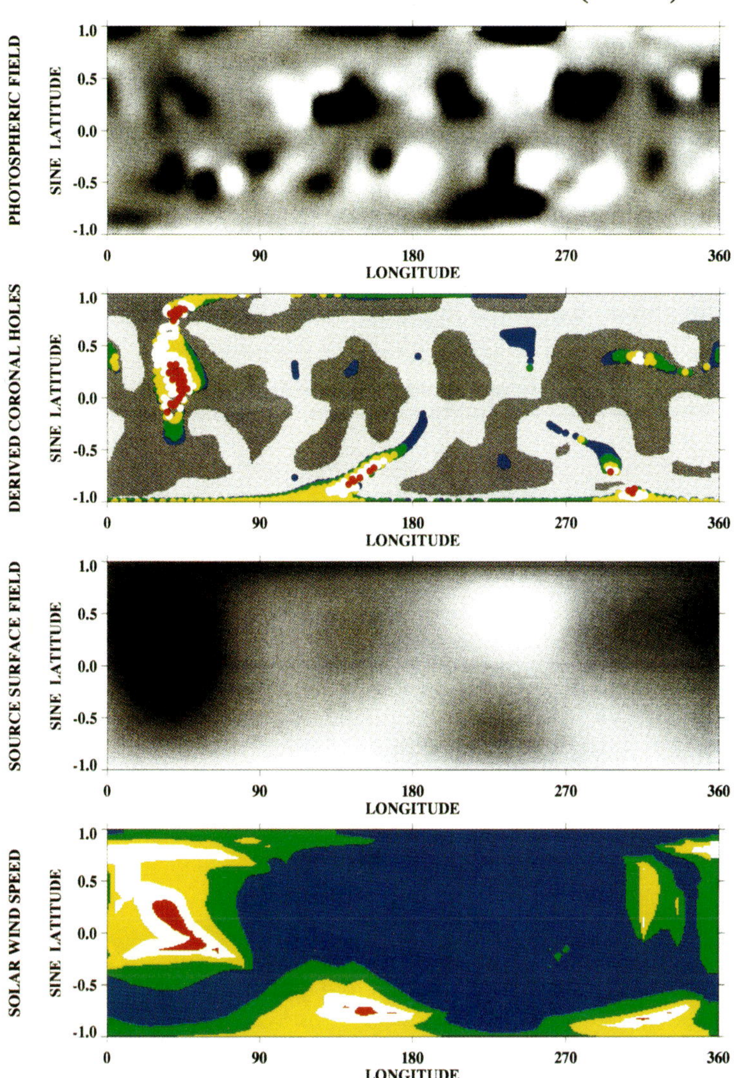

Plate 3.   Sunspot maximum. The format is the same as for Plate 2. (See the chapter by Sheeley et al.)

CARRINGTON ROTATION 1847 (WSO)

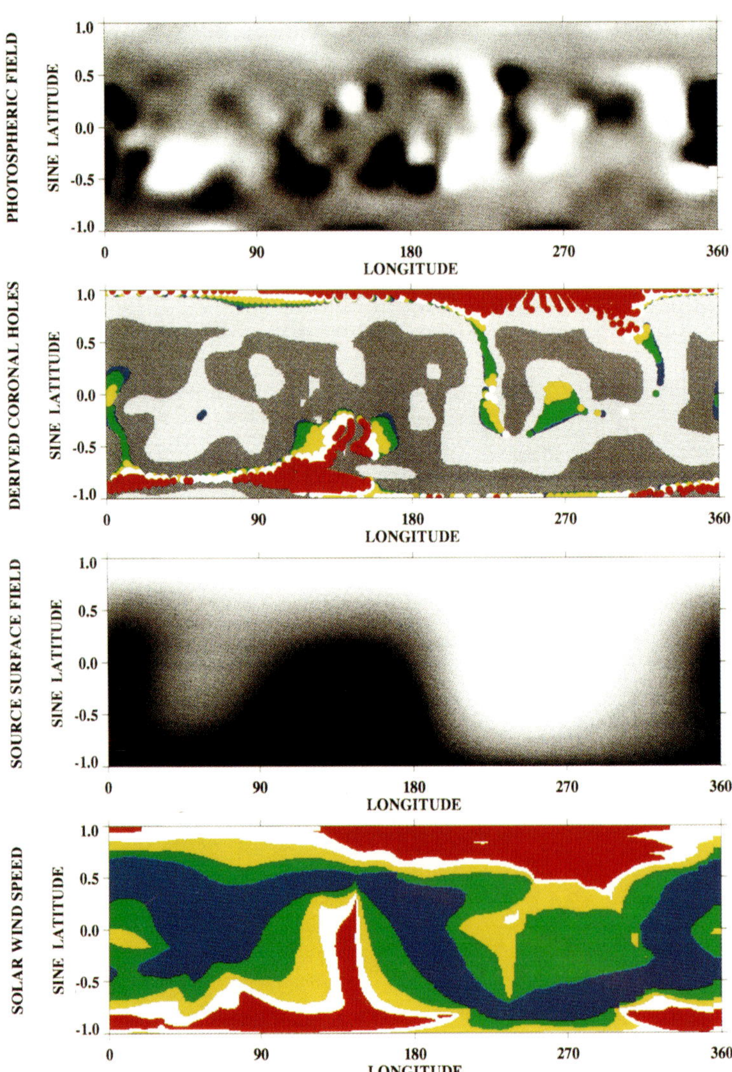

Plate 4.   After polar-field reversal. The format is the same as for Plate 1. (See the chapter by Sheeley et al.)

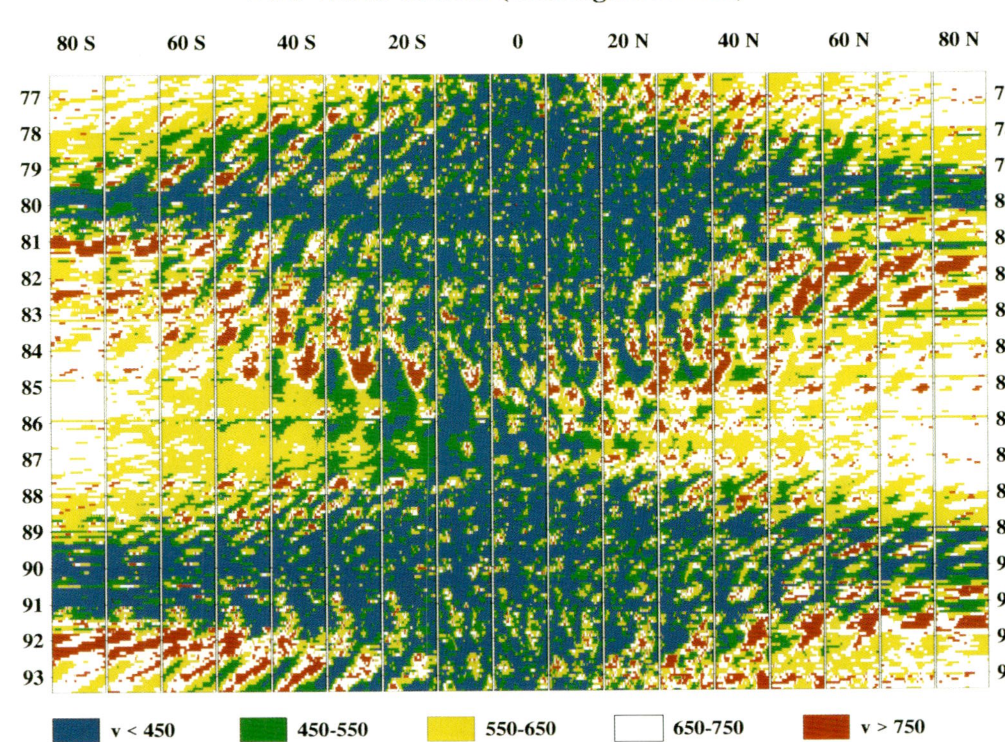

Plate 5.   Solar-cycle summary. A latitude vs time display of solar wind speed derived from maps of source-surface expansion factor (like those of Plate 1) using an in-ecliptic calibration. (See the chapter by Sheeley et al.)

ECLIPTIC WIND SPEED & IMF POLARITY

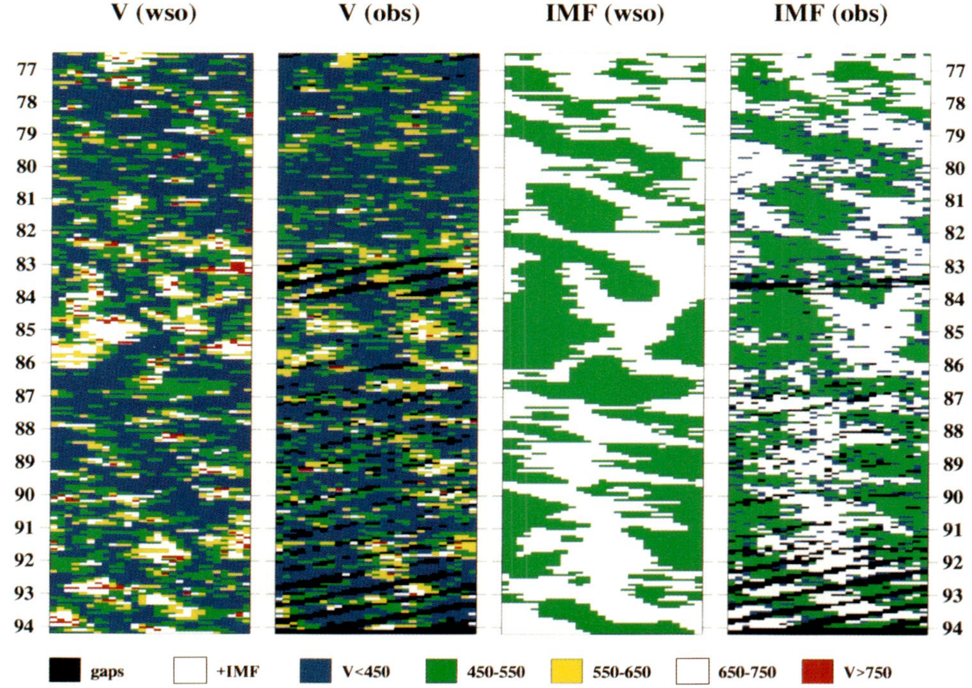

Plate 6.   In-ecliptic comparisons of derived and observed wind speed (left two columns) and magnetic polarity (third and fourth from left), illustrating the extent to which the current-free extrapolation reproduces the observed polarities and wind speeds. The color coding is the same as in Plate 5, except that values of mixed polarity or indeterminate polarity have been assigned a shade of blue in the fourth column. (See the chapter by Sheeley et al.)

ULYSSES-LATITUDE CARRINGTON PLOTS

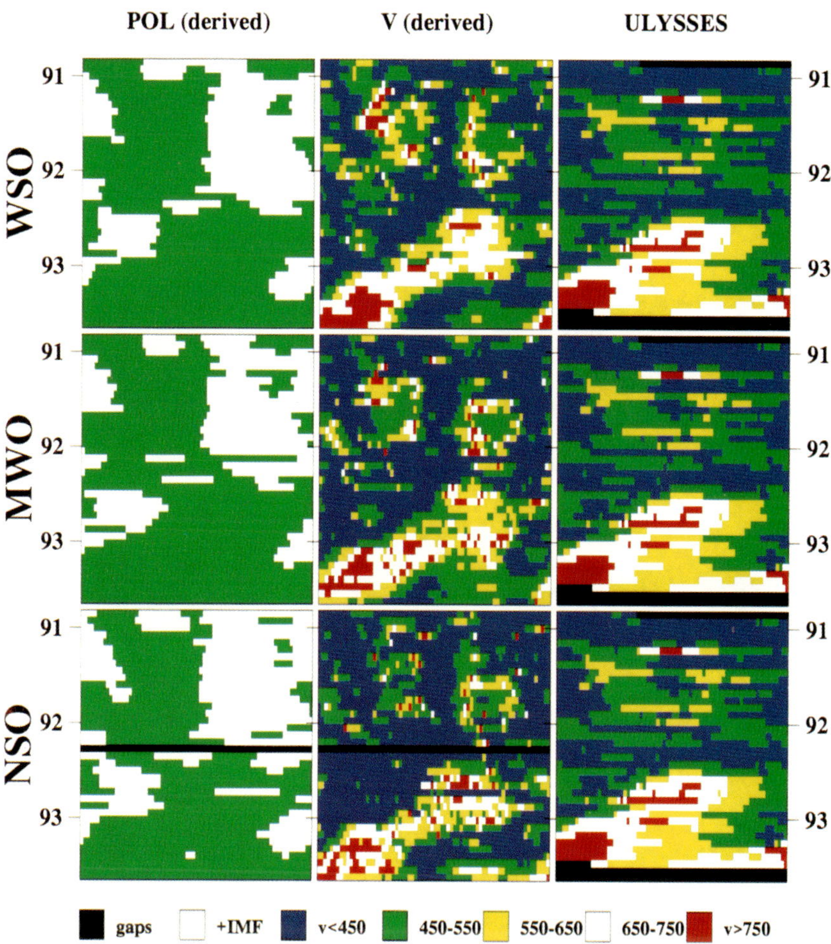

Plate 7.   Stackplots of magnetic polarity (left), source-surface expansion factor (middle), and Ulysses wind speed (right), all displayed in Carrington strips at the latitude of the Ulysses spacecraft during rotations 1835–1878. The upper, middle, and lower panels refer to magnetic observations obtained at WSO, MWO, and NSO, respectively, and the color-coding is the same as used in Plates 1 through 5. (See the chapter by Sheeley et al.)

CARRINGTON ROTATION 1862 (WSO)

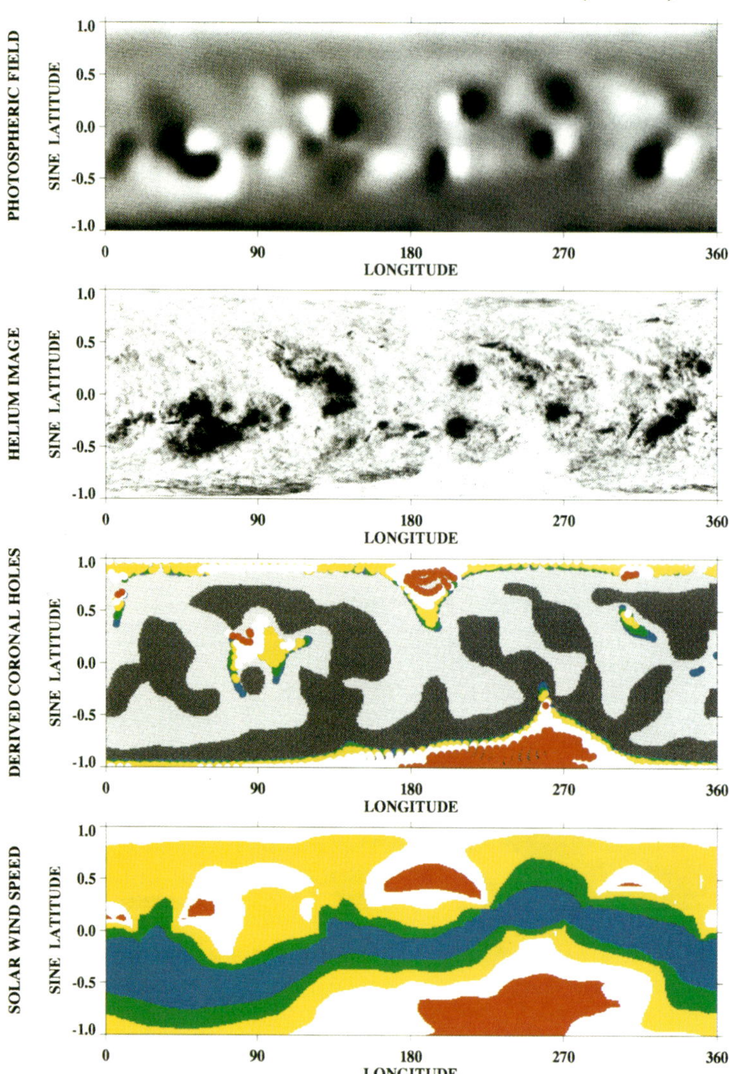

Plate 8. Solar observations during Carrington rotation 1862, in the same format as Plates 1 through 4, except that source-surface field has been replaced by helium intensity to show observed coronal holes (white areas). (See the chapter by Sheeley et al.)

CARRINGTON ROTATION 1869 (WSO)

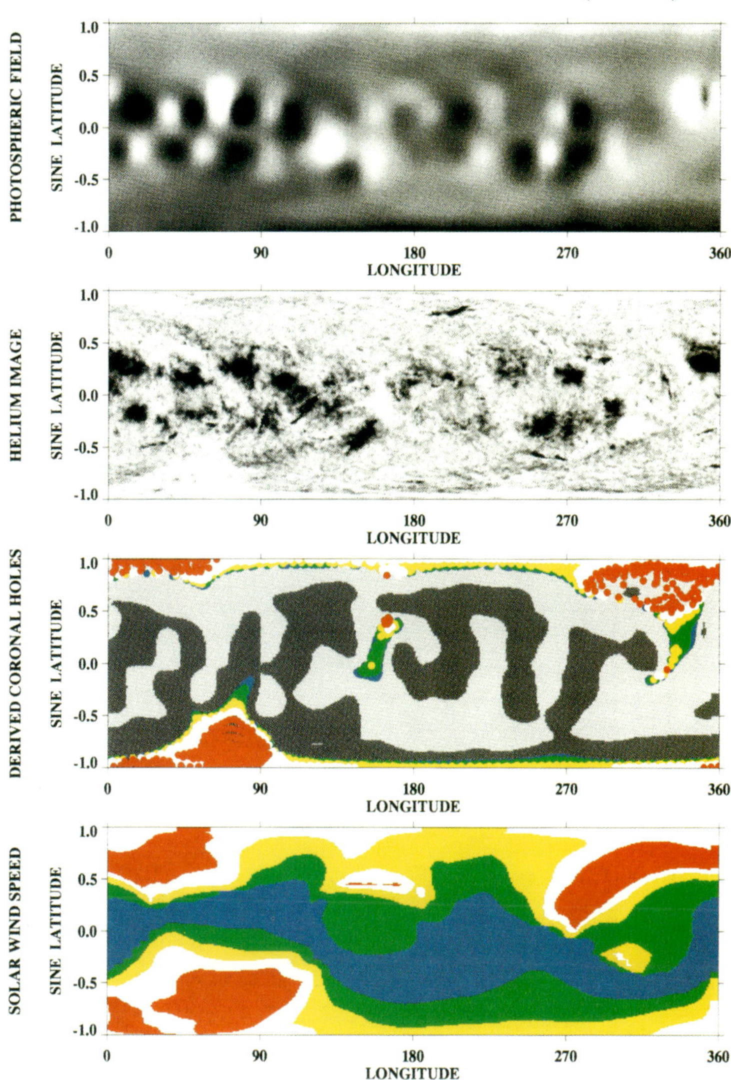

Plate 9.   Solar observations during Carrington rotation 1869 when Ulysses observed its fastest speeds. Same format as for Plate 8. (See the chapter by Sheeley et al.)

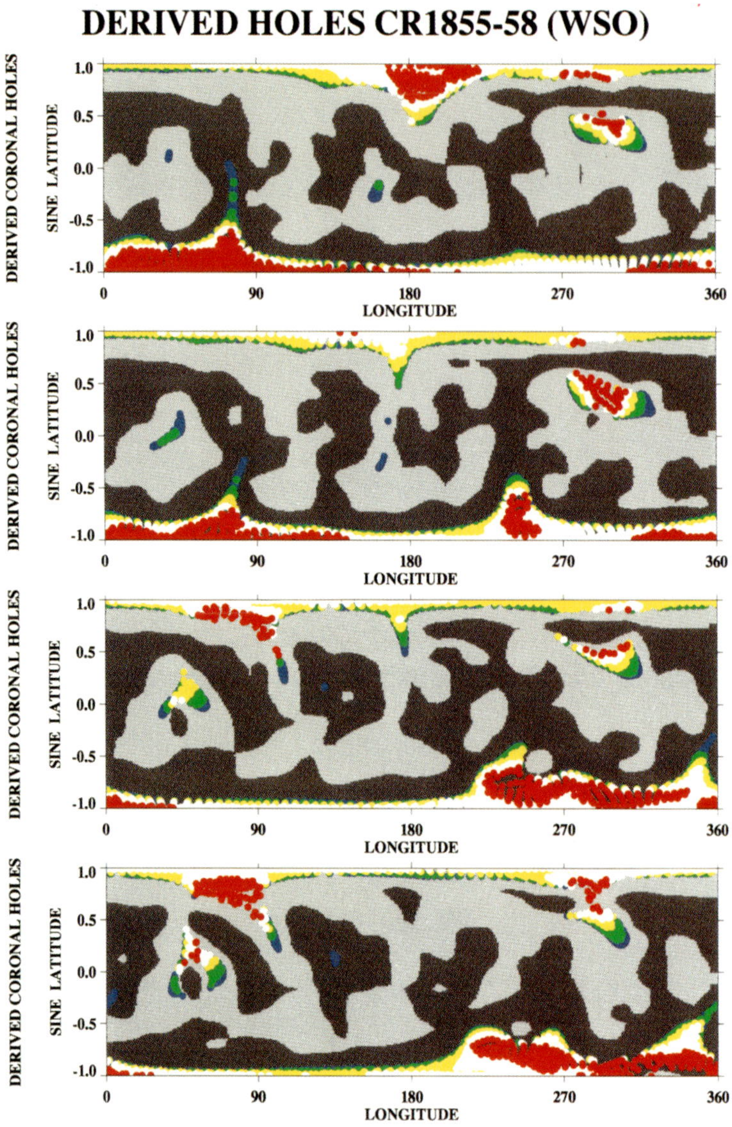

DERIVED HOLES CR1855-58 (WSO)

Plate 10. Derived coronal holes during Carrington rotations 1855–1858, showing the transition that gave rise to Ulysses high-speed stream in mid-1992. (See the chapter by Sheeley et al.)

DERIVED SPEED CR1855-58 (WSO)

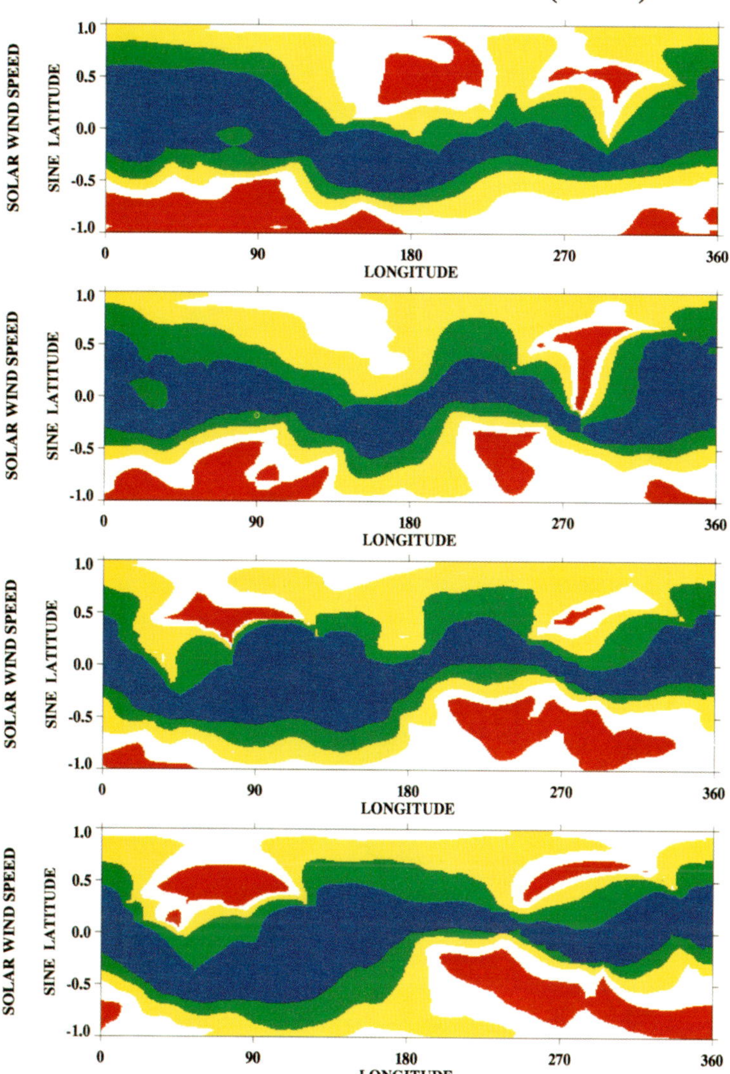

Plate 11. Derived source-surface expansion factor during Carrington rotations 1855–1858, showing the transition that gave rise to Ulysses high-speed stream in mid-1992. (See the chapter by Sheeley et al.)

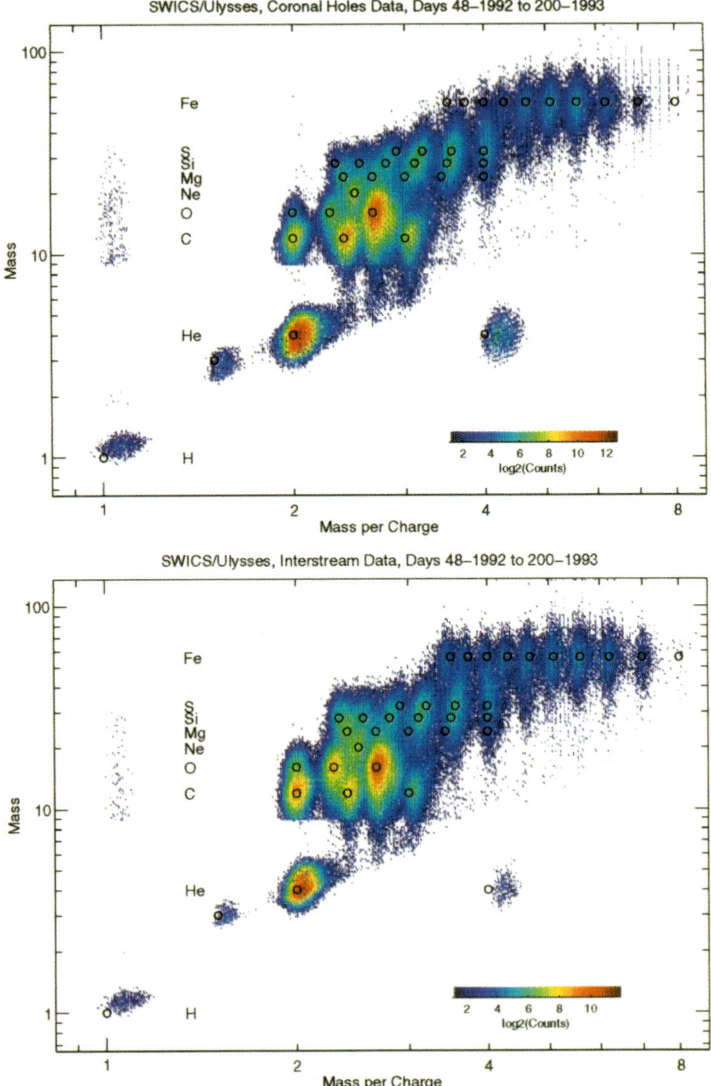

Plate 12.   SWICS/Ulysses pulse height matrices, accumulated in high-speed solar wind ($v_\alpha > 600$ km s^{-1}, upper panel) and in low-speed solar wind ($v_\alpha < 600$ km s^{-1}, lower panel) between Feb. 1992 and July 1993. Significant differences in charge states and even composition are evident. High charge states (such as C^{6+}, O^{8+}, or Fe^{16+}) are much less abundant in high-speed streams, indicating a lower coronal temperature at their source region. The reduced Mg/O ratio in these streams is also clearly visible, particularly from the Mg^{10+} to Ne^{8+} ratio, because the Ne/O ratio is the same in both panels. (The horizontal edge just below mass 10 is due to an instrument priority scheme favoring the selection of uplse height data with $m > 9$, which of course can be accounted for in the data analysis.) (See the chapter by von Steiger et al.)

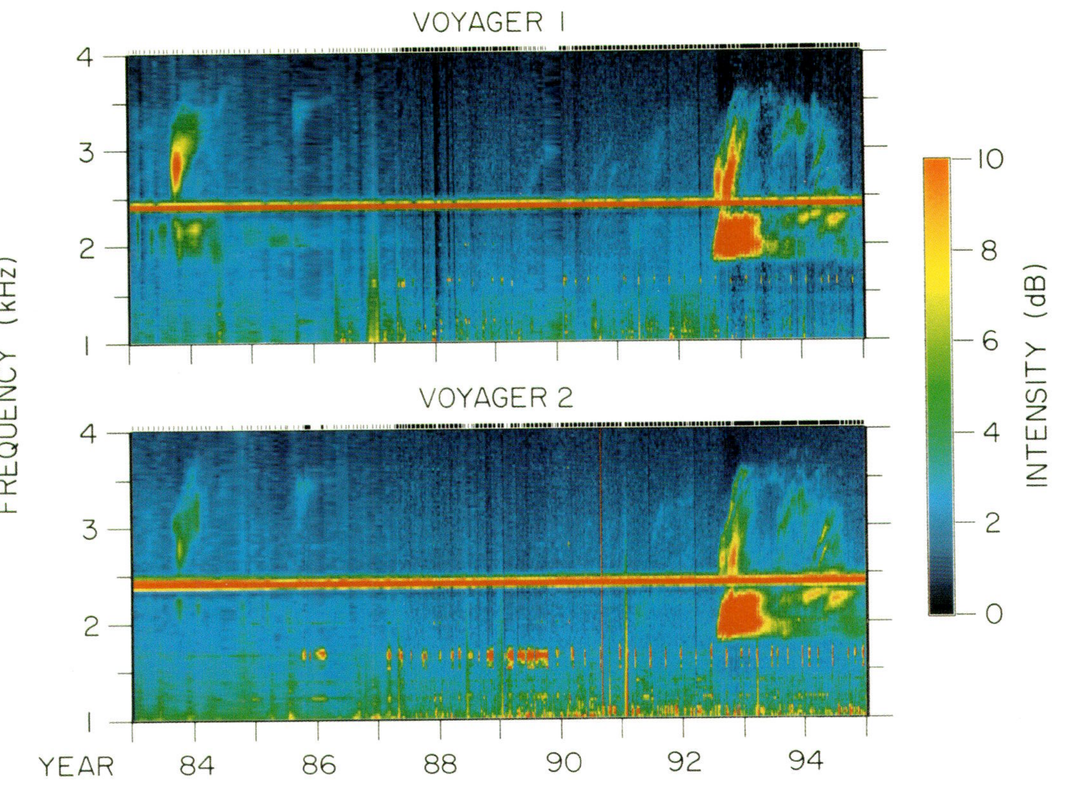

Plate 13.   Dynamic spectrograms from Voyager 1 (top) and Voyager 2 (bottom) during the time period from 1983 through 1994 showing the several heliospheric radio emissions observed over the time interval. (See the chapter by Kurth and Gurnett.)

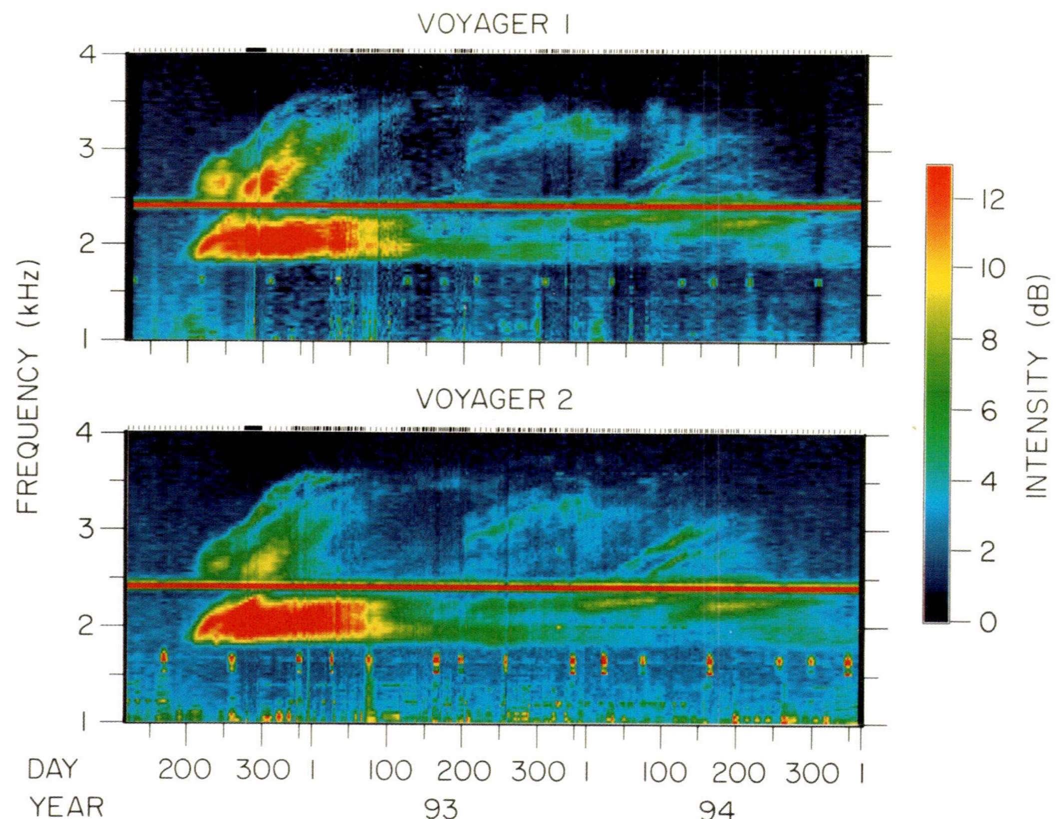

Plate 14.   Spectrograms similar in format to those in Plate 13, but expanded in time to show the detailed evolution of the 1992 heliospheric radio emission event. (See the chapter by Kurth and Gurnett.)

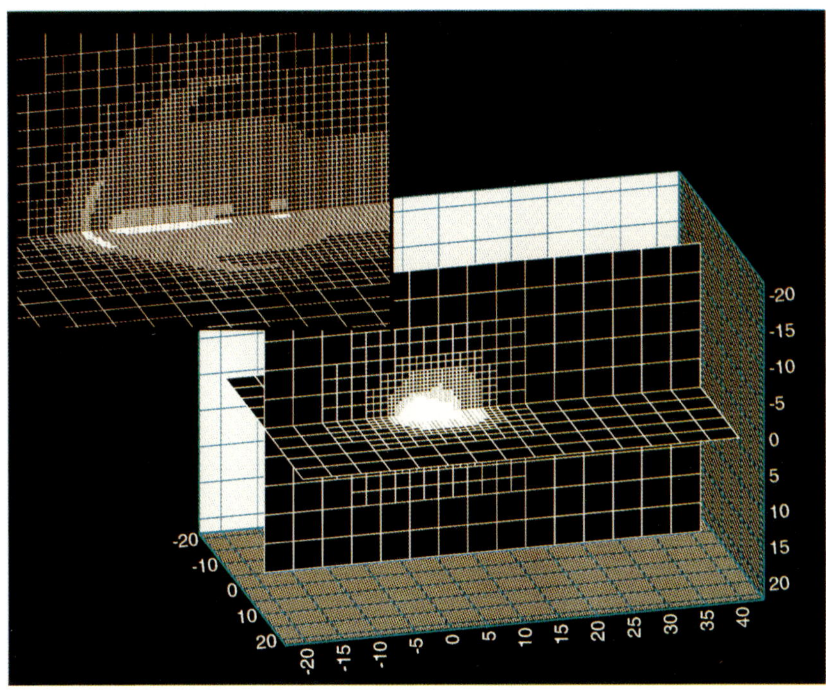

Plate 15.   Steady-state solution adapted grid for the three-dimensional cometary MHD model. The grid shows a large amount of information about the solution. All distances are measured in units of ionization scale length λ (In our case $\lambda \approx 10^6$ km). (See the chapter by Gombosi et al.)

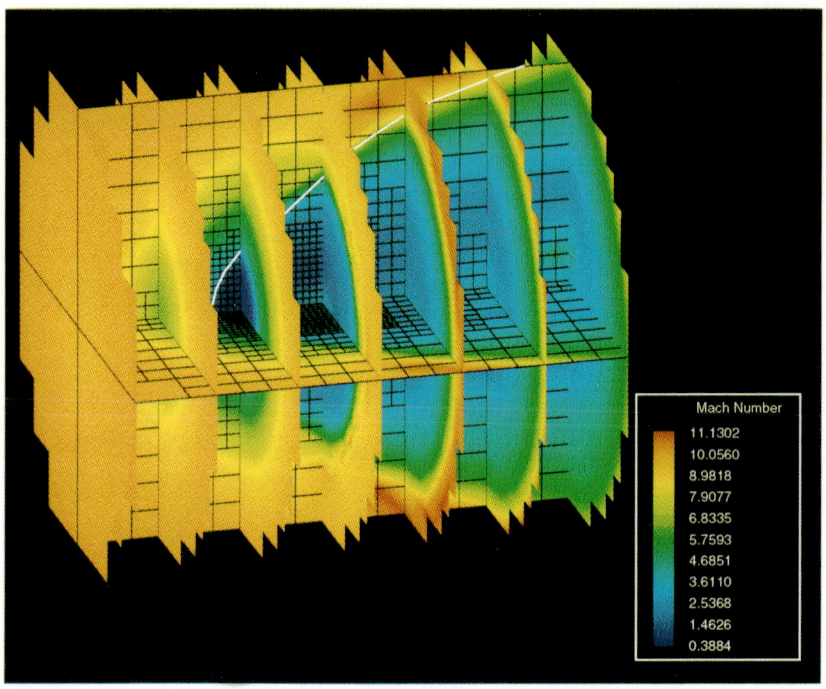

Plate 16.   The distribution of Mach number in the cometary interaction region, looking down to the equatorial plane from the north with the Sun on the left. Several cuts perpendicular to the Sun-nucleus axis are shown. These cuts are separated by a distance of 1.5λ. The white solid line represents the location of the cometary shock. (See the chapter by Gombosi et al.)

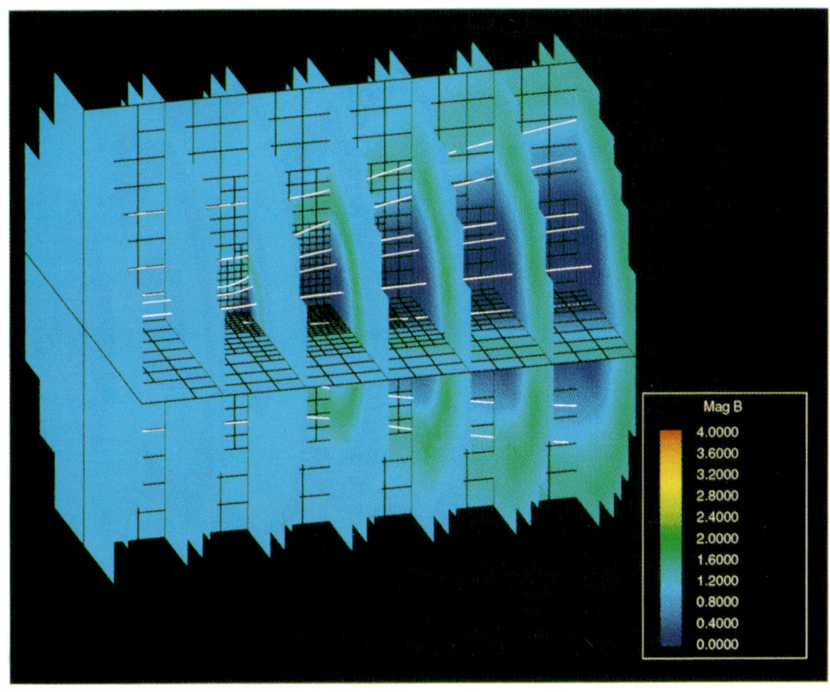

Plate 17.   The distribution of magnetic field magnitude for parallel solar wind geometry in the cometary interaction region. Solid white lines represent magnetic field lines. The viewing geometry is the same as for Plate 16. (See the chapter by Gombosi et al.)

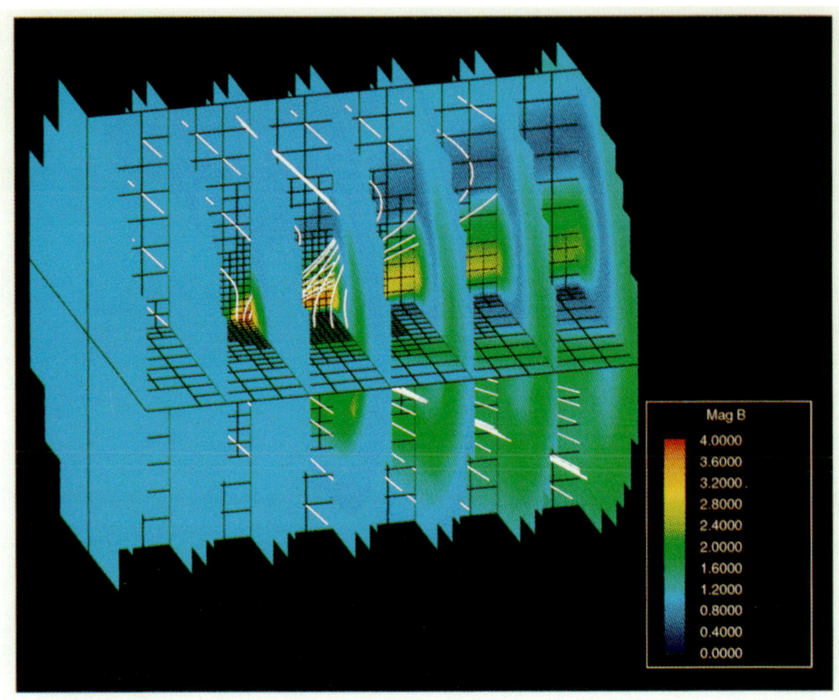

Plate 18.   The distribution of magnetic field magnitude for a 45° Parker spiral geometry in the cometary interaction region. Solid white lines represent magnetic field lines. The viewing geometry is the same as for Plate 16. (See the chapter by Gombosi et al.)

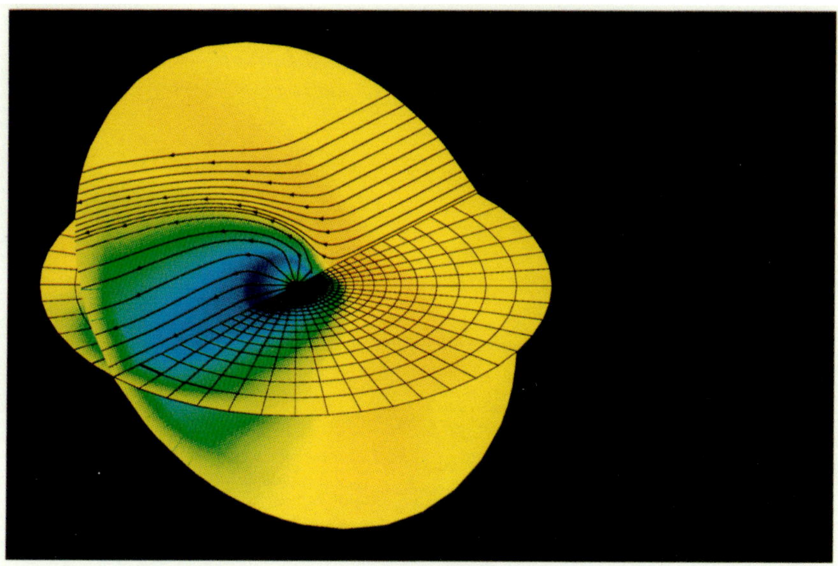

Plate 19. Interstellar and solar wind stream lines for the three-dimensional heliosphere calculation. The upstream bow shock and the heliopause are evident. The
termination shock is hidden in the dark blue region. (See the chapter by Gombosi
et al.)

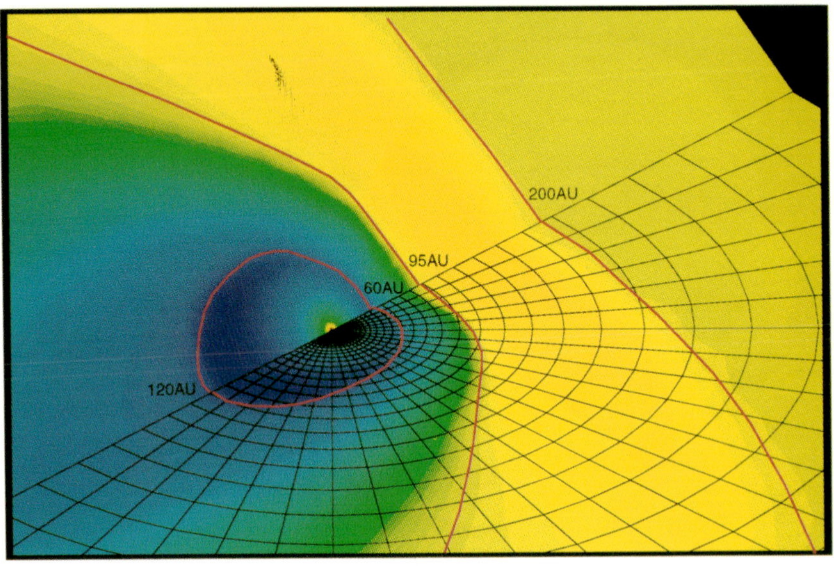

Plate 20.   Blowup of the heliospheric discontinuities showing the asymmetry of the termination shock. (See the chapter by Gombosi et al.)

ACKNOWLEDGMENTS

ACKNOWLEDGMENTS

The editors acknowledge the assistance of NASA Grant NAGW-1931 in funding a portion of this project. The editors also wish to thank J. E. Frecker, who volunteered as one of the proofreaders of this book and M. Schuchardt and K. Swarthout for their invaluable assistance with graphics and photograph reductions. They also would also like to acknowledge G. E. Morfill and S. Koutchmy for their assistance in obtaining the cover materials. The following authors wish to acknowledge specific funds involved in supporting the preparation of their chapters.

Bally, J.: NASA Grants NAGW-3192 *and* NAGW-4590

Balogh, A.: JPL Contract 959377

Biermann, P. L.: NATO Travel Grant CRG 9100072

Cuntz, M.: German Research Foundation Grant Cu 19-2/1, NASA Grant NAGW-2904 *and* Austrian Fonds zur Förderung der wissenschaftlichen Forschung P8758-PHY and P09694-TEC

Dorfi, E. A.: Austrian Fonds zur Förderung der wissenschaftlichen Forschung P8758-PHY and P09694-TEC

Esser, R.: NASA Grants NAGW-249 *and* NAGW-3513

Gombosi, T. I.: NASA Planetary Atmospheres Program Grant NAGW-1366, NSF Solar System Astronomy Program Grant AST-9313712 *and* NSF-NASA-AFOSR Interagency Grant NSF-ATM-9318181

Habbal, S. R.: NASA Grants NAGW-249 and NAGW-3513 *and* Air Force Grant AFOSR-91-0244

Hansteen, V. H.: Norwegian Research Council Contracts 107781/43 and 100839/432

Holzer, T. E.: NASA Grant NASW-4574

Jokipii, J. R.: NSF Grant ATM-8618260 *and* NASA Grant NSG-7101

Lee, M. A.: NASA Space Physics Theory Program Grant NAG5-1479, NASA Supporting Research and Technology Grant NAGW-2579 *and* NSF Grant ATM-9215279

Leer, E.: Norwegian Research Council Contracts 107781/43 and 100839/432

Mauche, C. W.: U. S. Dept. of Energy Contract No. W-7405-Eng-48 to the Lawrence Livermore National Laboratory *and* NASA Grant NAGW-528 to the Smithsonian Astrophysical Observatory

Mullan, D. J.: NASA Grant NAGW-2456

Parker, E. N.: NASA Grant NAGW-2122

Raymond, J. C.: NASA Grant NAGW-528

Riley, P.: JPL Contract 959377

Stahler, S. W.: NSF Grant AST-90-14479

INDEX

INDEX